Algebra for College Students

SEVENTH EDITION

R. David Gustafson

Rock Valley College

Peter D. Frisk

Rock Valley College

D0068944

BROOKS/COLE
CENGAGE Learning

Australia • Brazil • Japan • Korea • Mexico • Singapore • Spain • United Kingdom • United States

BROOKS/COLE
CENGAGE Learning™

Algebra for College Students
R. David Gustafson and Peter D. Frisk

Mathematics Editor: Jennifer Huber

Publisher: Robert Pirtle

Assistant Editor: Rebecca Subity

Editorial Assistant: Sarah Woicicki

Technology Project Manager: Christopher Delgado

Marketing Assistant: Jessica Perry

Advertising Project Manager: Bryan Vann

Project Manager, Editorial Production: Hal Humphrey

Print/Media Buyer: Kristine Waller

Permissions Editor: Sarah Harkrader

Production Service: Ellen Brownstein, Chapter Two

Text Designer: Tani Hasegawa

Photo Researcher: Sue C. Howard

Copy Editor: Ellen Brownstein

Illustrator: Lori Heckelman

Cover Designer: Harold Burch

Cover Image: Lester Lefkowitz/CORBIS

Compositor: Graphic World, Inc.

For product information and technology assistance, contact us at
Cengage Learning Customer & Sales Support, 1-800-354-9706

For permission to use material from this text or product, submit all requests online at cengage.com/permissions
Further permissions questions can be emailed to
permissionrequest@cengage.com

Library of Congress Control Number: 2003111207

ISBN-13: 978-0-534-46387-8

ISBN-10: 0-534-46387-8

Brooks/Cole
10 Davis Drive
Belmont, CA 94002-3098
USA

Cengage Learning is a leading provider of customized learning solutions with office locations around the globe, including Singapore, the United Kingdom, Australia, Mexico, Brazil, and Japan. Locate your local office at:
international.cengage.com/region

Cengage Learning products are represented in Canada by Nelson-Education, Ltd.

For your course and learning solutions, visit **academic.cengage.com**

Purchase any of our products at your local college store or at our preferred online store **www.ichapters.com**

Printed in Canada
3 4 5 6 7 11 10

To our students,
especially

Daniel, Tyler, Spencer,
Skyler, and Garrett

Books in the Gustafson/Frisk Series

Contents

Chapter 9

More Functions and Operations on Functions   545

Chapter 10

Exponential and Logarithmic Functions   608

Chapter 11

Solving Polynomial Equations   672

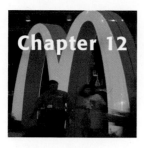

Chapter 12

Conic Sections and Quadratic Systems   *709*

Chapter 13

Natural-Number Functions and Probability   *767*

Preface

To the Instructor

Algebra for College Students, Seventh Edition, includes the topics usually associated with intermediate and college algebra. This text can be used to teach this material in a one-semester four- or five-hour course, or it can be used in a two-semester sequence.

Our goal has been to write a book that

1. is enjoyable to read,
2. is easy to understand,
3. is relevant, and
4. develops the necessary skills for success in future courses or on the job.

The Seventh Edition retains the philosophy of the highly successful previous editions. The revisions include several improvements in line with the NCTM standards, the AMATYC Crossroads, and the current trends in mathematics reform. For example, emphasis continues to be placed on graphing and problem solving. In fact, most sections contain applications problems.

General Changes in the Seventh Edition

In the Seventh Edition, the following general improvements have been made.

- *We have made the text more inviting to students* by incorporating a new design that is more open and easier to read. All calculator art has been rendered by an artist to be more visually appealing and to allow the labeling of graphs. Each section now includes a few sentences of introduction. We continue to use color not just as a design feature, but to highlight terms that instructors would point to in a classroom discussion.
- *We have made the chapter openers more effective* by redesigning them and including a photograph for visual interest. The relevant applications problems have been placed in the exercise sets.

- *We have given students real-life applications of the material in the chapters* by including an InfoTrac® College Edition project in each chapter opener.
- *We have increased the emphasis on problem solving through realistic applications* by increasing the number of applications problems. The text now includes an index of applications.
- *We have fine tuned the presentation of many topics* for a better flow of ideas and for clarity.

Specific Changes in the Seventh Edition

Chapter One remains a review of basic algebra. It covers the real number system, arithmetic, the properties of real numbers, exponents, and basic equation solving. The first section has been rewritten to make it more streamlined and easier for students to understand. Although solution sets are still written in interval notation, there is equal emphasis on writing solution sets using set notation.

Chapter Two covers graphs, equations of lines, and functions. In Exercise Set 2.1, problems have been added where the graph passes through the origin.

Chapter Three covers systems of equations. This chapter contains only minor changes.

Chapter Four covers inequalities. In Section 4.1, the material on solving linear and compound inequalities has been rewritten to make is easier to understand. There is an expanded discussion of interval notation, and a table is included comparing interval notation and set notation. The practice problems in Exercise Set 4.1 have been extensively revised.

Chapter Five covers polynomials and polynomial functions. This chapter contains only minor changes.

Chapter Six covers rational expressions. This chapter contains only minor changes.

Chapter Seven covers radicals, rational exponents, and complex numbers. Additional problems have been added to Exercise Sets 7.1 and 7.2. The topic of complex numbers has been moved to this chapter as Section 7.7. In this section, the material on powers of i has been moved later in the section. A new project has been added.

Chapter Eight covers quadratic functions and nonlinear inequalities. This chapter has been revised extensively.

   The work on completing the square and the quadratic formula is now covered in two sections. Section 8.1 reviews solving quadratics by factoring and covers completing the square. The material has been expanded to cover quadratic equations with imaginary and complex roots. A graphical explanation of completing the square is now included. The exercise set has been greatly revised and includes new applications.

   Section 8.2 covers solving quadratic equations using the quadratic formula. The material has been expanded to cover quadratic equations with complex roots. Again, the exercise set has been greatly revised.

   To improve the transition from the work on the quadratic formula to the work on the discriminant, the material on the discriminant and equations that can be written in quadratic form has been moved to Section 8.3.

   A better introduction to the construction of sign graphs has been included in Section 8.5.

Chapter Nine covers functions and operations on functions. The chapter contains only minor changes.

Chapter Ten covers exponential and logarithmic functions. This chapter contains only minor changes.

Chapter Eleven covers solving polynomial equations. It contains only minor changes.

Chapter Twelve covers conic sections and quadratic systems. More work with graphing calculators has been included. In the work on parabolas, p is now considered to be an undirected distance. This reduces the number of formulas that students will need to memorize.

Chapter Thirteen covers the binomial theorem, sequences, and permutations and combinations. This chapter contains only minor changes.

Calculators

The use of calculators is assumed throughout the text. We believe that students should learn calculator skills in the mathematics classroom. They will then be prepared to use calculators in science and business classes and for nonacademic purposes. The directions within each exercise set indicate which exercises require the use of a calculator.

Since many algebra students now have graphing calculators, keystrokes are given for both scientific and graphing calculators.

Ancillaries for the Instructor

Annotated Instructor's Edition (0-534-46388-6)
This special version of the complete student text contains a Resource Integration Guide and all answers are printed next to their respective exercises.

Test Bank (0-534-46392-4)
The *Test Bank* includes 8 tests per chapter, as well as 3 final exams. The tests contain a combination of multiple-choice, free-response, true/false, and fill-in-the-blank questions.

Complete Solutions Manual (0-534-46391-6)
The *Complete Solutions Manual* provides worked-out solutions to all of the problems in the text.

Text-Specific Videotapes (0-534-46393-2)
The text-specific videotape set, available at no charge to qualified adopters of the text, features 10- to 20-minute problem-solving lessons that cover each section of every chapter.

BCA/iLrn 3.5 Instructor Version (0-534-46394-0)
BCA/iLrn Instructor Version is made up of two components: *BCA/iLrn Testing* and *BCA/iLrn Tutorial. BCA/iLrn Testing* is a revolutionary, Internet-ready, text-specific testing suite that allows you to customize exams and track student progress in an accessible, browser-based format. *BCA/iLrn* offers full algorithmic generation of problems and free-response mathematics. *BCA/iLrn Tutorial* is a text-specific, interactive tutorial that is delivered via the Web (at http://bca.brookscole.com) and is offered in

both student and instructor versions. The tracking program built into the instructor version enables you to monitor student progress carefully. The complete integration of the testing, tutorial, and course management components simplifies your routine tasks. Results flow automatically to your gradebook and you can easily communicate with individuals, sections, and entire courses.

BCA/iLrn has been updated with a diagnostic tool to assess and place students in any course and provide a personalized study plan. Other updates include:

- The new rich text form (RTF) conversion tool that allows you to open tests in most word processors for further formatting and customization,
- Several new problem types that provide greater variety and more diverse challenges in your tests, and
- A redesigned *Tutorial* interface that effectively engages students and helps them learn math concepts faster.

TLE Labs (0-534-46403-3)
Think of *TLE Labs* as electronic labs that upgrade any *BCA/iLrn Tutorial. TLE Labs* introduce and explore key concepts. *BCA/iLrn Tutorials* reinforce those concepts with unlimited practice. Exploring these core concepts interactively at their own pace prepares students for the work of the traditional course, allows them to perform better in the course overall, and prepares them for the next level of mathematics.

- *TLE Labs* are now available with all © 2005 developmental mathematics titles.
- *TLE Labs* include 15 lessons adapted from the full version of *The Learning Equation*. The lessons correlate to the key concepts of this book and the course in general.
- In addition to the printed textbook, students receive everything they need to succeed in the course: an online version of the text, access to the *TLE* lessons, text-specific interactive tutorials, and access to *vMentor* (Web-based, real-time, live tutors), all in the same single, unified environment.

Ask your Cengage•Brooks/Cole representative about *TLE*.

WebTutor™ ToolBox on WebCT and *Blackboard*:
0-534-21620-X (WebCT), 0-534-21620-1 (Blackboard)
Preloaded with content and available free via PIN code when packaged with this text, *WebTutor ToolBox for WebCT* and *Blackboard* pairs the content of this text's rich Book Companion Web Site with all the sophisticated course management functionality of a WebCT and Blackboard product. You can assign materials (including online quizzes) and have the results flow automatically to your gradebook. *ToolBox* is ready to use as soon as you log on. You also can customize its preloaded content by uploading images and other resources, adding Web links, or creating your own practice materials. Students have access only to student resources on the Web site. Instructors can enter a PIN code for access to password-protected instructor resources.

MyCourse 2.1
Ask us about our new free online course builder. Whether you want only the easy-to-use tools to build it or the content to furnish it, Brooks/Cole offers you a simple solution for a custom course Web site that allows you to assign, track, and report on student progress; load your syllabus; and more. Highlights of Version 2.1 include new features that let you import your class roster and grade book, see assignments as students will see them using the Student Preview Mode, create individual performance

grade books, and copy content from an existing course to a new course. Contact your Cengage•Brooks/Cole representative for details or visit www.academic.cengage.com.

Ancillaries for the Student

Student Solutions Manual (0-534-46390-8)
The *Student Solutions Manual* provides worked out solutions to the odd-numbered problems in the text.

Brooks/Cole Mathematics Web site, http://mathematics.brookscole.com
When you adopt a Cengage•Brooks/Cole mathematics text, you and your students will have access to a variety of teaching and learning resources. This Web site features everything from book-specific resources to newsgroups. It's a great way to make teaching and learning an interactive and intriguing experience.

InfoTrac® College Edition
Incorporate an array of current applications or the latest research into your lectures with four months of free access to InfoTrac College Edition. In addition to robust tutorial services, your students also receive anytime, anywhere access to InfoTrac College Edition. This online library offers the full text of articles from almost 5,000 scholarly and popular publications, updated daily and going back as far as 22 years. Both adopters and their students receive unlimited access for four months.

BCA/iLrn Tutorial Student Version (0-534-46397-5)
Free access to this text-specific, interactive, Web-based tutorial system is included with the text. *BCA/iLrn Tutorial Student Version* is browser-based, making it an intuitive mathematical guide, even for students with little technological proficiency. *BCA/iLrn Tutorial* allows students to work with real math notation in real time, providing instant analysis and feedback. The entire textbook is available in PDF format through *BCA/iLrn Tutorial,* as are section-specific video tutorials, unlimited practice problems, and additional student resources, such as a glossary, Web links, and more. *BCA/iLrn Tutorial* is also offered on a dual-platform CD-ROM. And, when students get stuck on a particular problem or concept, they need only log on to *vMentor,* accessed through *BCA/iLrn Tutorial,* where they can talk (using their own computer microphones) to *vMentor* tutors who will skillfully guide them through the problem using the interactive whiteboard for illustration.

A personalized study plan also can be generated by the course-specific diagnostic built into *BCA/iLrn Tutorial.* With a personalized study plan, students can focus their time where they need it the most, creating a positive learning environment and paving a pathway to success in their mathematics course.

vMentor
Packaged free with every text and accessed seamlessly through *BCA/iLrn Tutorial,* *vMentor* provides tutorial help that can substantially improve student performance, increase test scores, and enhance technical aptitude. Students will have access, via the Web, to highly qualified tutors. When students get stuck on a particular problem or concept, they need only log on to *vMentor,* where they can talk (using their own computer microphones) to *vMentor* tutors who will skillfully guide them through the problem using the interactive whiteboard for illustration. Brooks/Cole also offers *vClass,* an online virtual classroom environment that is customizable and easy to use. *vClass* keeps students engaged with full two-way audio, instant messaging, and an interactive

whiteboard—all in one intuitive, graphical interface. For information about obtaining a *vClass* site license, please contact your Cengage•Brooks/Cole representative.

Interactive Video Skillbuilder CD-ROM (0-534-46389-4)

The *Interactive Video Skillbuilder* CD-ROM contains more than eight hours of video instruction. The problems worked during each video lesson are shown next to the viewing screen so that students can try working them before watching the solution. To help students evaluate their progress, each section contains a 10-question Web quiz, the results of which can be emailed to the instructor, and each chapter contains a chapter test, with the answer to each problem on each test. Also included is *MathCue* tutorial and testing software, which is keyed to the text and includes these components:

- *MathCue Skill Builder* presents problems to solve, evaluates answers, and tutors students by displaying complete solutions with step-by-step explanations.

- *MathCue Quiz* allows students to generate large numbers of quiz problems keyed to problem types from each section of the book.

- *MathCue Chapter Test* also provides large numbers of problems keyed to problem types from each chapter.

- *MathCue Solution Finder* allows students to enter their own basic problems and receive step-by-step help as if they were working with a tutor.

- Score reports for any *MathCue* session can be printed and handed in for credit or extra-credit, or sent by email via *MathCue's* secure email score system.

To the Student

Congratulations! You now own a state-of-the-art textbook that has been written especially for you. We have tried to write a book that you can read and understand. The text includes carefully written narrative and an extensive number of worked examples with Self Checks.

To get the most out of this course, you must read and study the textbook properly. We recommend that you work the examples on paper first, and then work the Self Checks. Only after you thoroughly understand the concepts taught in the examples should you attempt to work the exercises. A *Student Solutions Manual* is available, which contains the worked out solutions to the odd-numbered exercises.

Since the material presented in *Algebra for College Students, Seventh Edition,* will be of value to you in later years, we suggest that you keep this text. It will be a good source of reference in the future and will keep at your fingertips the material that you have learned here.

We wish you well.

Hints on Studying Algebra

The phrase "Practice makes perfect" is not quite true. It is "*Perfect* practice that makes perfect." For this reason, it is important that you learn how to study algebra to get the most out of this course.

Although we all learn differently, here are some hints on studying algebra that most students find useful.

Plan a Strategy for Success

To get where you want to be, you need a goal and a plan. Your goal should be to pass this course with a grade of A or B. To earn one of these grades, you must have a plan to achieve it. A good plan involves several points:

- Getting ready for class,
- Attending class,
- Doing homework,
- Arranging for special help when you need it, and
- Having a strategy for taking tests.

Getting Ready for Class

To get the most out of every class period, you will need to prepare for class. One of the best things you can do is to preview the material in the text that your instructor will be discussing in class. Perhaps you will not understand all of what you read, but you will be better able to understand your instructor when he or she discusses the material in class.

Do your work every day. If you get behind, you will become frustrated and discouraged. Make a promise that you will always prepare for class, and then keep that promise.

Attending Class

The classroom experience is your opportunity to learn from your instructor. Make the most of it by attending every class. Sit near the front of the room where you can easily see and hear. Remember that it is your responsibility to follow the discussion, even though that takes concentration and hard work.

Pay attention to your instructor, and jot down the important things that he or she says. However, do not spend so much time taking notes that you fail to concentrate on what your instructor is explaining. Listening and understanding the big picture is much better than just copying solutions to problems.

Don't be afraid to ask questions when your instructor asks for them. Asking questions will make you an active participant in the class. This will help you pay attention and keep you alert and involved.

Doing Homework

It requires practice to excel at tennis, master a musical instrument, or learn a foreign language. In the same way, it requires practice to learn mathematics. Since practice in mathematics is homework, homework is your opportunity to practice your skills and experiment with ideas.

It is important for you to pick a definite time to study and do homework. Set a formal schedule and stick to it. Try to study in a place that is comfortable and quiet. If you can, do some homework shortly after class, or at least before you forget what was discussed in class. This quick follow-up will help you remember the skills and concepts your instructor taught that day.

Each formal study session should include three parts:

1. Begin every study session with a review period. Look over previous chapters and see if you can do a few problems from previous sections. Keeping old skills alive will greatly reduce the amount of time you will need to prepare for tests.

2. After reviewing, read the assigned material. Resist the temptation of diving into the problems without reading and understanding the examples. Instead, work the examples and Self Checks with pencil and paper. Only after you completely understand the underlying principles behind them should you try to work the exercises.

Once you begin to work the exercises, check your answers with the printed answers in the back of the text. If one of your answers differs from the printed answer, see if the two can be reconciled. Sometimes, answers have more than one form. If you decide that your answer is incorrect, compare your work to the example in the text that most closely resembles the exercise, and try to find your mistake. If you cannot find an error, consult the *Student Solutions Manual.* If nothing works, mark the problem and ask about it in your next class meeting.

3. After completing the written assignment, preview the next section. This preview will be helpful when you hear that material discussed during the next class period.

You probably already know the general rule of thumb for college homework: two hours of practice for every hour you spend in class. If mathematics is hard for you, plan on spending even more time on homework.

To make doing homework more enjoyable, study with one or more friends. The interaction will clarify ideas and help you remember them. If you must study alone, a good study technique is to explain the material to yourself out loud.

Arranging for Special Help   Take advantage of any special help that is available from your instructor. Often, your instructor can clear up difficulties in a short period of time.

Find out whether your college has a free tutoring program. Peer tutors can often be of great help.

Taking Tests   Students often get nervous before taking a test because they are afraid that they will do poorly. To build confidence in your ability to take tests, rework many of the problems in the exercise sets, work the exercises in the Chapter Summaries, and take the Chapter Tests. Check all answers with the answers printed at the back of the text.

Then guess what the instructor will ask, build your own tests, and work them. Once you know your instructor, you will be surprised at how good you can get at picking test questions. With this preparation, you will have some idea of what will be on the test, and you will have more confidence in your ability to do well.

When you take a test, work slowly and deliberately. Scan the test and work the easy problems first. Tackle the hardest problems last.

Acknowledgments

We are grateful to the following people, who reviewed the manuscript at various stages of its development. They all had valuable suggestions that have been incorporated into the text.

Barbara C. Armenta
Pima Community College, East Campus

Jack Brillhart
Tri-State University

David Byrd
Enterprise State Junior College

Joseph F. Conrad
Solano Community College

Ann Corbeil
Massasoit Community College

Harold Farmer
Wallace Community College-Hanceville

Mark Foster
Santa Monica College

Dorothy K. Holtgrefe
Seminole Community College

Mike Judy
Fullerton College

Daniel F. Mussa
Southern Illinois University

Kamilia S. Nemri
Spokane Community College

Victoria Paaske
University of Wisconsin–Waukesha

Joseph Phillips
Sacramento City College

Dennis Ragan
Kishwaukee College

Ari Ruselink
Tri-State University

Steve Schonefeld
Tri-State University

Alan Snowden
Mississippi Delta Community College

George J. Witt
Glendale Community College

We are grateful to the staff at Brooks/Cole, especially our editor Jennifer Huber. We also thank Bob Pirtle, Rachael Sturgeon, Leah Thomson, Hal Humphrey, Vernon Boes, and Rebecca Subity.

We are indebted to Ellen Brownstein, our production service, Diane Koenig, who read the entire manuscript and worked every problem, and Mike Welden, who prepared the *Student Solutions Manual.* Finally, we thank Lori Heckelman for her fine artwork and Graphic World for their excellent typesetting.

R. David Gustafson
Peter D. Frisk

Math Tutorial Quick Start Guide

BCA/iLrn Can Help You Succeed in Math

BCA/iLrn is an online program that facilitates math learning by providing resources and practice to help you succeed in your math course. Your instructor chose to use *BCA/iLrn* because it provides online opportunities for learning (Explanations found by clicking **Read Book**), practice (Exercises), and evaluating (Quizzes). It also gives you a way to keep track of your own progress and manage your assignments.

The mathematical notation in *BCA/iLrn* is the same as that you see in your textbooks, in class, and when using other math tools like a graphing calculator. *BCA/iLrn* can also help you run calculations, plot graphs, enter expressions, and grasp difficult concepts. You will encounter various problem types as you work through *BCA/iLrn,* all of which are designed to strengthen your skills and engage you in learning in different ways.

Logging in to BCA/iLrn

Registering with the PIN Code on the BCA/iLrn Card

Situation: Your instructor has not given you a PIN code for an online course, but you have a textbook with a BCA/iLrn product PIN code.

Initial Login

1. Go to **http://iLrn.com.**
2. In the menu at the left, click on **Student Tutorial.**
3. Make sure that the name of your school appears in the "School" field. If your school name does not appear, follow steps a–d below. If your school is listed, go to step 4.

   a. Click on **Find Your School.**

   b. In the "State" field, select your state from the drop-down menu.

    c. In the "Name of school" field, type the first few letters of your school's name; then click on **Search.** The school list will appear at the right.

    d. Click on your school. The "First Time Users" screen will open.

4. In the "PIN Code" field, type the BCA/iLrn PIN code supplied on your BCA/iLrn card.

5. In the "ISBN" field, type the ISBN of your book (from the textbook's back cover), for example, 0-534-46332-0.

6. Click on **Register.**

7. Enter the appropriate information. Fields marked with a red asterisk must be filled in.

8. Click on **Register** and **Begin BCA/iLrn.** The "My Assignments" page will open.

> You will be asked to select a user name and password. Save your user name and password in a safe place. You will need them to log in the next time you use BCA/iLrn. Only your user name and password will allow you to reenter BCA/iLrn.

Subsequent Login

1. Go to **http://iLrn.com.**

2. Click on **Login.**

3. Make sure the name of your school appears in the "School" field. If not, then follow steps 3a–d under "Initial Login" to identify your school.

4. Type your user name and password (see boxed information above); then click on **Login.** The "My Assignments" page will open.

Navigating through BCA/iLrn

To navigate between chapters and sections, use the drop-down menu below the top navigation bar. This will give you access to the study activities available for each section.
    The view of a tutorial in BCA/iLrn looks like this.

Math Toolbar

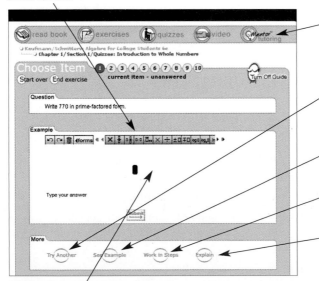

vMentor: Live online tutoring is only a click away. Tutors can take screen shots of your book and lead you through a problem with voice-over and visual aids.

Try Another: Click here to have BCA/iLrn create a new question or a new set of problems.

See Examples: Preworked examples provide you with additional help.

Work in Steps: BCA/iLrn can guide you through a problem step-by-step.

Explain: Additional explanation from your book can help you with a problem.

Type your answer here.

Online Tutoring with *vMentor*

Access to BCA/iLrn also means access to online tutors and support through vMentor, which provides live homework help and tutorials. To access vMentor while you are working in the Exercises or "Tutorial" areas in BCA/iLrn, click on the **vMentor Tutoring** button at the top right of the navigation bar above the problem or exercise.

Next, click on the **vMentor** button; you will be taken to a Web page that lists the steps for entering a vMentor classroom. If you are a first-time user of vMentor, you might need to download Java software before entering the class for the first class. You can either take an Orientation Session or log in to a vClass from the links at the bottom of the opening screen.

All vMentor Tutoring is done through a vClass, an Internet-based virtual classroom that features two-way audio, a shared whiteboard, chat, messaging, and experienced tutors.

You can access vMentor Sunday through Thursday, as follows:

5 p.m. to 9 p.m. Pacific Time

6 p.m. to 10 p.m. Mountain Time

7 p.m. to 11 p.m. Central Time

8 p.m. to midnight Eastern Time

If you need additional help using vMentor, you can access the Participant Guide at this Website: **http://www.elluminate.com/support/guide.pdf.**

Index of Applications

Applications that are Examples are shown with boldface page numbers.
Applications that are Exercises are shown with lightface page numbers.

1

A Review of Basic Algebra

InfoTrac Project

Do a keyword search on "cholesterol," and find the article "Cooking oil to fight fat and cholesterol." Write a summary of the article.

A manufacturer wants to make 12.6 ounces of a mixture of flaxseed oil, olive oil, and coconut oil. They will use 2.2 ounces of coconut oil costing $1.56 an ounce. How much flaxseed oil at $0.46 an ounce and olive oil at $0.78 an ounce must be added for the final mixture to cost $11?

Complete this project after studying Section 1.7.

©Lois Ellen Frank/CORBIS

Mathematics in Electronics

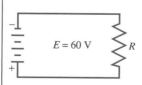

$E = 60$ V

R

The illustration to the left is a schematic diagram of a resistor connected to a voltage source of 60 volts. As a result, the resistor dissipates power in the form of heat. The power P lost when a voltage V is placed across a resistance R is given by the formula

$$P = \frac{E^2}{R}$$

Solve the formula for R. If P is 4.8 watts and E is 60 volts, find R.

**Exercise Set 1.5
Problem 116**

*The language of mathematics is algebra. The word **algebra** comes from the title of a book written by the Arabian mathematician Al-Khowarazmi around A.D. 800. Its title **Ihm al-jabr wa'l muqabalah** means restoration and reduction, a process then used to solve equations.*

In algebra, we will work with expressions that contain numbers and letters. The letters that stand for numbers are called **variables.** Some examples of algebraic expressions are

$$x + 45, \qquad \frac{8x}{3} - \frac{y}{2}, \qquad \frac{1}{2}bh, \qquad \frac{a}{b} + \frac{c}{d}, \qquad \text{and} \qquad 3x + 2y - 12$$

It is the use of variables that distinguishes algebra from arithmetic.

1.1 The Real Number System

- **Sets of Numbers** ❚ **Inequality Symbols**
- **Graphs of Real Numbers** ❚ **Absolute Value of a Number**

Getting Ready **1.** Name some different kinds of numbers. **2.** Is there a largest number?

We begin the study of algebra by reviewing various subsets of the real numbers. We will also show how to graph these sets on a number line.

Sets of Numbers

A **set** is a collection of objects. To denote a set, we often enclose a list of its **elements** with braces. For example,

$\{1, 2, 3, 4\}$ denotes the set with elements 1, 2, 3, and 4.

$\left\{\frac{1}{2}, \frac{2}{3}, \frac{3}{4}\right\}$ denotes the set with elements $\frac{1}{2}$, $\frac{2}{3}$, and $\frac{3}{4}$.

Three sets of numbers commonly used in algebra are as follows:

Natural Numbers   The set of **natural numbers** includes the numbers we use for counting:

$$\{1, 2, 3, 4, 5, 6, 7, 8, 9, \ldots\}$$

Whole Numbers   The set of **whole numbers** includes the natural numbers together with 0:

$$\{0, 1, 2, 3, 4, 5, 6, 7, 8, 9, \ldots\}$$

Integers   The set of **integers** includes the natural numbers, 0, and the negatives of the natural numbers:

$$\{\ldots, -5, -4, -3, -2, -1, 0, 1, 2, 3, 4, 5, \ldots\}$$

Each group of three dots, called an **ellipsis,** indicates that the numbers continue forever.

In Figure 1-1, we graph each of these sets from -6 to 6 on a number line.

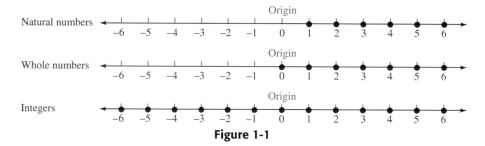

Figure 1-1

Since every natural number is also a whole number, we say that the set of natural numbers is a **subset** of the set of whole numbers. Note that the set of natural numbers and the set of whole numbers are also subsets of the integers.

To each number x in Figure 1-1, there corresponds a point on the number line called its **graph,** and to each point there corresponds a number x called its **coordinate.** Numbers to the left of 0 are **negative numbers,** and numbers to the right of 0 are **positive numbers.**

 Comment   0 is neither positive nor negative.

There are two important subsets of the natural numbers.

Prime Numbers   The **prime numbers** are the natural numbers greater than 1 that are divisible only by 1 and themselves.

$$\{2, 3, 5, 7, 11, 13, 17, 19, 23, \ldots\}$$

Composite Numbers   The **composite numbers** are the natural numbers greater than 1 that are not prime.

$$\{4, 6, 8, 9, 10, 12, 14, 15, 16, 18, 20, 21, \ldots\}$$

Figure 1-2 shows the graphs of the primes and composites that are less than or equal to 14.

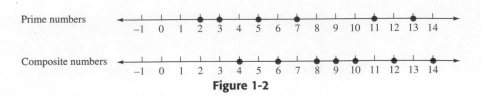

Figure 1-2

Comment   1 is the only natural number that is neither prime nor composite.

There are two important subsets of the integers.

Even Integers	The **even integers** are the integers that are exactly divisible by 2.

$$\{\ldots, -8, -6, -4, -2, 0, 2, 4, 6, 8, \ldots\}$$

Odd Integers	The **odd integers** are the integers that are not exactly divisible by 2.

$$\{\ldots, -9, -7, -5, -3, -1, 1, 3, 5, 7, 9, \ldots\}$$

Figure 1-3 shows the graphs of the even and odd integers from -6 to 6.

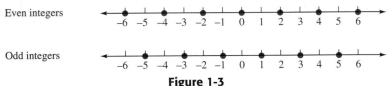

Figure 1-3

So far, we have written sets by listing their elements within braces. This method is called the **roster method.** A set can also be written in **set builder notation.** In this method, we write a rule that describes what elements are in a set. For example, the set of even integers can be written in set builder notation as follows:

$$\{x\,|\,x \text{ is an integer that can be divided exactly by 2.}\}$$

The set of all x such that   a rule that describes membership in the set.

To find coordinates of more points on the number line, we need the **rational numbers.**

Rational Numbers	The set of **rational numbers** is

$$\left\{x\,\middle|\,x \text{ can be written in the form } \frac{a}{b}\ (b \neq 0), \text{ where } a \text{ and } b \text{ are integers.}\right\}$$

Each of the following is an example of a rational number.

$$\frac{8}{5},\quad \frac{2}{3},\quad -\frac{44}{23},\quad \frac{-315}{476},\quad 0 = \frac{0}{7},\quad \text{and}\quad 17 = \frac{17}{1}$$

Because each number has an integer numerator and a nonzero integer denominator.

Comment   Note that $\frac{0}{5} = 0$, because $5 \cdot 0 = 0$. However, $\frac{5}{0}$ is undefined, because there is no number that, when multiplied by 0, gives 5.

The fraction $\frac{0}{0}$ is indeterminate, because all numbers, when multiplied by 0, give 0. Remember that the denominator of a fraction cannot be 0.

EXAMPLE 1   Explain why each number is a rational number: **a.** -7,   **b.** 0.125,   and **c.** $-0.666.$ . . .

Solution   **a.** The integer -7 is a rational number, because it can be written as $\frac{-7}{1}$, where -7 and 1 are integers and the denominator is not 0. Note that all integers are rational numbers.

b. The decimal 0.125 is a rational number, because it can be written as $\frac{1}{8}$, where 1 and 8 are integers and the denominator is not 0.

c. The decimal $-0.666.$ . . is a rational number, because it can be written as $\frac{-2}{3}$, where -2 and 3 are integers and the denominator is not 0.

Self Check   Explain why each number is a rational number: **a.** 4   and   **b.** 0.5.   ∎

The next example illustrates that every rational number can be written as a decimal that either terminates or repeats a block of digits.

EXAMPLE 2   Change each fraction to decimal form and tell whether the decimal terminates or repeats: **a.** $\frac{3}{4}$ and   **b.** $\frac{421}{990}$.

Solution   **a.** To change $\frac{3}{4}$ to a decimal, we divide 3 by 4 to obtain 0.75. This is a terminating decimal.

b. To change $\frac{421}{990}$ to a decimal, we divide 421 by 990 to obtain 0.4252525. . . . This is a repeating decimal, because the block of digits 25 repeats forever. This decimal can be written as $0.4\overline{25}$, where the overbar indicates the repeating block of digits.

Self Check   Change each fraction to decimal form and tell whether the decimal terminates or repeats: **a.** $\frac{5}{11}$ and   **b.** $\frac{2}{5}$.   ∎

The rational numbers provide coordinates for many points on the number line that lie between the integers (see Figure 1-4). Note that the integers are a subset of the rational numbers.

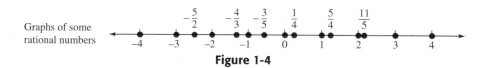

Graphs of some rational numbers

Figure 1-4

Points on the number line whose coordinates are nonterminating, nonrepeating decimals have coordinates that are called **irrational numbers.**

Irrational Numbers   The set of **irrational numbers** is

$\{x \mid x$ is a nonterminating, nonrepeating decimal$\}$

Some examples of irrational numbers are

$$0.313313331\ldots \qquad \sqrt{2} = 1.414213562\ldots \qquad \pi = 3.141592653\ldots$$

If we unite the set of rational numbers (the terminating or repeating decimals) and the set of irrational numbers (the nonterminating, nonrepeating decimals), we obtain the set of **real numbers.**

Real Numbers

The set of **real numbers** is

$$\{x \mid x \text{ is a terminating, a repeating, or a nonterminating, nonrepeating decimal.}\}$$

The number line in Figure 1-5 shows several points on the number line and their real-number coordinates. The points whose coordinates are real numbers fill up the number line.

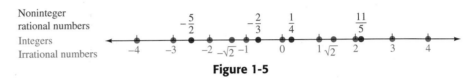

Figure 1-5

Figure 1-6 shows how several of the previous sets of numbers are related.

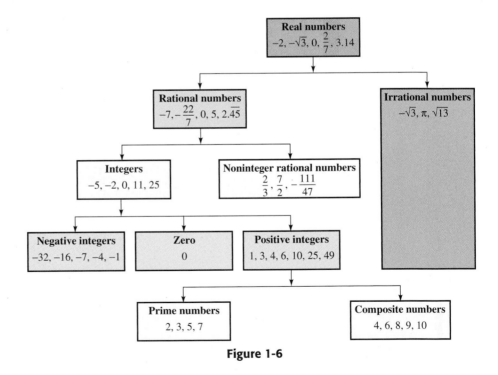

Figure 1-6

Inequality Symbols

To show that two quantities are not equal, we can use an **inequality symbol.**

Symbol	Read as	Examples
$\neq$	"is not equal to"	$6 \neq 9$ and $0.33 \neq \frac{3}{5}$
$<$	"is less than"	$22 < 40$ and $0.27 < 3.1$
$>$	"is greater than"	$19 > 5$ and $\frac{1}{2} > 0.3$
$\leq$	"is less than or equal to"	$1.8 \leq 3.5$ and $35 \leq 35$
$\geq$	"is greater than or equal to"	$25.2 \geq 23.7$ and $29 \geq 29$

We can always write an inequality with the inequality symbol pointing in the opposite direction. For example,

$17 < 25$ is equivalent to $25 > 17$

$5.3 \geq -2.9$ is equivalent to $-2.9 \leq 5.3$

On the number line, the coordinates of points get larger as we move from left to right, as shown in Figure 1-7. Thus, if a and b are the coordinates of two points, the one to the right is the greater. This suggests the following general principle:

If $a > b$, point a lies to the right of point b on a number line.

If $a < c$, point a lies to the left of point c on a number line.

Figure 1-7

Graphs of Real Numbers

Graphs of sets of real numbers are often portions of a number line called **intervals.** For example, the graph shown in Figure 1-8(a) is the interval that includes all real numbers x that are greater than -5. Since these numbers make the inequality $x > -5$ true, we say that they satisfy the inequality. The parenthesis on the graph indicates that -5 is not included in the interval.

To express this interval in **interval notation,** we write $(-5, \infty)$. Once again, the parentheses indicate that the endpoints are not included.

The interval shown in Figure 1-8(b) is the graph of the inequality $x \leq 7$. It contains all real numbers that are less than or equal to 7. The bracket at 7 indicates that 7 is included in the interval. To express this interval in interval notation, we write $(-\infty, 7]$. The bracket indicates that 7 is in the interval.

$$-5 \qquad\qquad 7$$
$$(-5, \infty) \qquad\qquad (-\infty, 7]$$
$$\{x \mid x > -5\} \qquad\qquad \{x \mid x \leq 7\}$$

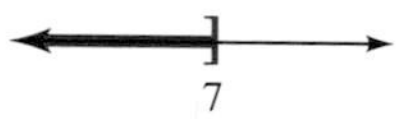

(a)   (b)

Figure 1-8

Comment The symbol ∞ (infinity) is not a real number. It is used to indicate that the graph in Figure 1-8(a) extends infinitely far to the right, and the graph in Figure 1-8(b) extends infinitely far to the left.

The graphs in Figure 1-8 can also be drawn using open and filled circles. An open circle indicates that an endpoint is not included, and a filled circle indicates that an endpoint is included.

EXAMPLE 3 Graph each set on the number line and then write it in interval notation:
a. $\{x \mid x < 9\}$ and   **b.** $\{x \mid x \geq 6\}$.

Solution **a.** $\{x \mid x < 9\}$ includes all real numbers that are less than 9. The graph is shown in Figure 1-9(a). This is the interval $(-\infty, 9)$.

b. $\{x \mid x \geq 6\}$ includes all real numbers that are greater than or equal to 6. The graph is shown in Figure 1-9(b). This is the interval $[6, \infty)$.

Figure 1-9

Self Check Graph $\{x \mid x \geq 5\}$ and write it in interval notation.

Two inequalities can often be written as a single expression. For example,

$2 < x < 15$

is a combination of the inequalities $2 < x$ and $x < 15$. It is read as "2 is less than x, and x is less than 15." It means that x is between 2 and 15. Its graph is shown in Figure 1-10.

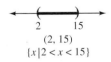

Figure 1-10

EXAMPLE 4 Graph each set on the number line and then write it in interval notation:
a. $\{x \mid -5 \leq x \leq 6\}$ **b.** $\{x \mid 2 < x \leq 8\}$ and   **c.** $\{x \mid -4 \leq x < 3\}$.

Solution **a.** The set $\{x \mid -5 \leq x \leq 6\}$ includes all real numbers from -5 to 6, as shown in Figure 1-11(a). This is the interval $[-5, 6]$.

b. The set $\{x\,|\,2 < x \le 8\}$ includes all real numbers between 2 and 8, including 8, as shown in Figure 1-11(b). This is the interval (2, 8].

c. The set $\{x\,|\,-4 \le x < 3\}$ includes all real numbers between −4 and 3, including −4, as shown in Figure 1-11(c). This is the interval [−4, 3).

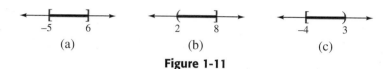

(a)   (b)   (c)

Figure 1-11

Self Check   Graph $\{x\,|\,-6 < x \le 10\}$ and then write it in interval notation.   ∎

The graph of the set

$\{x\,|\,x < -2 \text{ or } x \ge 3\}$ Read as "the set of all real numbers x, such that x is less than −2 or greater than or equal to 3."

is shown in Figure 1-12. This graph is called the **union** of two intervals and can be written in interval notation as

$(-\infty, -2) \cup [3, \infty)$ Read the symbol ∪ as "union."

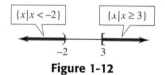

Figure 1-12

EXAMPLE 5   Graph the set on the number line and then write it in interval notation: $\{x\,|\,x \le -4 \text{ or } x > 5\}$.

Solution   The set $\{x\,|\,x \le -4 \text{ or } x > 5\}$ includes all real numbers less than or equal to −4 together with all real numbers greater than 5, as shown in Figure 1-13. This is the interval $(-\infty, -4] \cup (5, \infty)$.

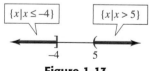

Figure 1-13

Self Check   Graph $\{x\,|\,x < -1 \text{ or } x > 5\}$ and then write it in interval notation.   ∎

The following table shows three ways to describe an interval.

Set Notation	Graph	Interval Notation
$\{x \mid x > a\}$		(a, ∞)
$\{x \mid x < a\}$		$(-\infty, a)$
$\{x \mid x \geq a\}$		$[a, \infty)$
$\{x \mid x \leq a\}$		$(-\infty, a]$
$\{x \mid a < x < b\}$		(a, b)
$\{x \mid a \leq x < b\}$		$[a, b)$
$\{x \mid a < x \leq b\}$		$(a, b]$
$\{x \mid a \leq x \leq b\}$		$[a, b]$
$\{x \mid x < a \text{ or } x > b\}$		$(-\infty, a) \cup (b, \infty)$

Absolute Value of a Number

The **absolute value** of a real number a, denoted as $|a|$, is the distance on a number line between 0 and the point with coordinate a. For example, the points shown in Figure 1-14 with coordinates of 3 and -3 both lie 3 units from 0. Thus, $|3| = |-3| = 3$.

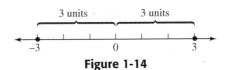

Figure 1-14

In general, for any real number a, $|a| = |-a|$.

The absolute value of a number can be defined more formally.

Absolute Value   For any real number x, $\begin{cases} \text{If } x \geq 0, \text{ then } |x| = x. \\ \text{If } x < 0, \text{ then } |x| = -x. \end{cases}$

If x is positive or 0, then x is its own absolute value. However, if x is negative, then $-x$ (which is a positive number) is the absolute value of x. Thus, $|x| \geq 0$ for all real numbers x.

EXAMPLE 6   Find each absolute value:

a. $|3| = 3$ **b.** $|-4| = 4$

c. $|0| = 0$ **d.** $-|-8| = -(8) = -8$ Note that $|-8| = 8$.

Self Check   Find each absolute value: **a.** $|-9|$ and   **b.** $-|-12|$.

Self Check Answers

1. **a.** $4 = \dfrac{4}{1}$,   **b.** $0.5 = \dfrac{1}{2}$ 2. **a.** $0.\overline{45}$, repeating decimal,   **b.** 0.4, terminating decimal

3. $[5, \infty)$ 4. $(-6, 10]$ 5. $(-\infty, -1) \cup (5, \infty)$

6. **a.** 9,   **b.** -12

Orals   1. Define a natural number.   2. Define a whole number.
3. Define an integer.   4. Define a rational number.
5. List the first four prime numbers.   6. List the first five positive even integers.
7. Find $|-6|$.   8. Find $|10|$.

1.1   EXERCISES

REVIEW   *To simplify a fraction, factor the numerator and the denominator and divide out common factors. For example, $\frac{12}{18} = \frac{6 \cdot 2}{6 \cdot 3} = \frac{\cancel{6} \cdot 2}{\cancel{6} \cdot 3} = \frac{2}{3}$. Simplify each fraction.*

1. $\dfrac{6}{8}$ $\dfrac{3 \cdot 2}{4 \cdot 2} = {}^{3}/_{4}$ 2. $\dfrac{15}{20}$ $\dfrac{3 \cdot 5}{4 \cdot 5} = {}^{3}/_{4}$

3. $\dfrac{32}{40}$ $\dfrac{8 \cdot 4}{8 \cdot 5} = {}^{4}/_{5}$ 4. $\dfrac{56}{72}$ $\dfrac{7 \cdot 8}{9 \cdot 8} = {}^{7}/_{9}$

To multiply fractions, multiply the numerators and multiply the denominators. To divide fractions, invert the divisor and multiply. Always simplify the result if possible.

5. $\dfrac{1}{4} \cdot \dfrac{3}{5}$ $= \dfrac{3}{20}$ 6. $\dfrac{3}{5} \cdot \dfrac{20}{27}$ $\dfrac{60}{135}$

7. $\dfrac{2}{3} \div \dfrac{3}{7}$ $\dfrac{14}{9}$ 8. $\dfrac{3}{5} \div \dfrac{9}{15}$

To add (or subtract) fractions, write each fraction with a common denominator and add (or subtract) the numerators and keep the same denominator. Always simplify the result if possible.

9. $\dfrac{5}{9} + \dfrac{4}{9}$ $\dfrac{9}{9}$ R $\boxed{1}$ 10. $\dfrac{16}{7} - \dfrac{2}{7}$

11. $\dfrac{2}{3} + \dfrac{4}{5}$ $\dfrac{6}{8}$ ${}^{3}/_{4}$ 12. $\dfrac{7}{9} - \dfrac{2}{5}$

VOCABULARY AND CONCEPTS   *Fill in the blanks.*

13. A ____ is a collection of objects.

14. The numbers $1, 2, 3, 4, 5, 6, \ldots$ form the set of _Real_ numbers.

15. An ____ integer can be divided exactly by 2.

16. An ____ integer cannot be divided exactly by 2.

17. A prime number is a _____ number that is larger than __ and can only be divided exactly by ____ and 1.

18. A _____ number is a natural number greater than __ that is not _____.

19. __ is neither positive nor negative.

20. The denominator of a fraction can never be __.

21. A repeating decimal represents a _____ number.

22. A nonrepeating, nonterminating decimal represents an _____ number.

23. The symbol __ means "is less than."

24. The symbol __ means "is greater than or equal to."

25. The symbol __ means "is approximately equal to."

26. If x is negative, $|x| = $ ____.

List the elements in the set $\left\{ -3, 0, \frac{2}{3}, 1, \sqrt{3}, 2, 9 \right\}$ that satisfy the given condition.

27. natural number   28. whole number

29. integer   30. rational number

31. irrational number   32. real number

33. even natural number   34. odd integer

35. prime number   36. composite number

37. odd composite number   **38.** even prime number

PRACTICE   *Graph each set on the number line.*

39. The set of prime numbers less than 8

0 1 2 3 4 5 6 7 8

40. The set of integers between -7 and 0

-7 -6 -5 -4 -3 -2 -1 0

41. The set of odd integers between 10 and 18

10 11 12 13 14 15 16 17 18

42. The set of composite numbers less than 10

0 1 2 3 4 5 6 7 8 9 10

Change each fraction into a decimal and classify the result as a terminating or a repeating decimal.

43. $\dfrac{7}{8}$ **44.** $\dfrac{7}{3}$

45. $-\dfrac{11}{15}$ **46.** $-\dfrac{19}{16}$

Insert an $<$ or an $>$ symbol to make a true statement.

47. 5   9   **48.** 9   0

49. -5 -10 **50.** -3 10

51. -7 7   **52.** 0 -5

53. 6 -6 **54.** -6 -2

Write each statement with the inequality symbol pointing in the opposite direction.

55. $19 > 12$ **56.** $-3 \geq -5$

57. $-6 \leq -5$ **58.** $-10 < 13$

59. $5 \geq -3$ **60.** $0 \leq 12$

61. $-10 < 0$ **62.** $-4 > -8$

Graph each set on the number line.

63. $\{x | x > 3\}$ **64.** $\{x | x < 0\}$

65. $\{x | x \leq 7\}$ **66.** $\{x | x \geq -2\}$

67. $[-5, \infty)$ **68.** $(-\infty, 9]$

69. $\{x | 2 < x < 5\}$ **70.** $[0, 5)$

71. $[-6, 9]$ **72.** $\{x | -1 < x \leq 3\}$

73. $\{x | x < -3 \text{ or } x > 3\}$ **74.** $(-\infty, -4] \cup (2, \infty)$

75. $(-\infty, -6] \cup [5, \infty)$ **76.** $\{x | x < -2 \text{ or } x \geq 3\}$

Write each expression without using absolute value symbols. Simplify the result when possible.

77. $|20|$ **78.** $|-20|$

79. $-|-6|$ **80.** $-|8|$

81. $|-5| + |-2|$ **82.** $|12| + |-4|$

83. $|-5| \cdot |4|$ **84.** $|-6| \cdot |-3|$

85. Find x if $|x| = 3$.   **86.** Find x if $|x| = 7$.

87. What numbers x are equal to their own absolute values?

88. What numbers x when added to their own absolute values give a sum of 0?

APPLICATIONS

89. On the thermometer in the illustration, graph each of the following temperature readings: $12°, 8°, 0°, -6°$.

90. On the number line in the illustration, the origin is the point B.C./A.D.
  a. What happened in the year 1441?
  b. What happened in the year -500?

MAYA CIVILIZATION

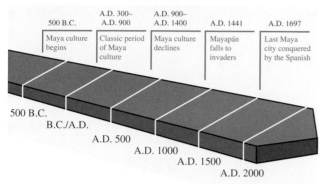

Based on data from *People in Time and Place, Western Hemisphere* (Silver Burdett & Ginn Inc., 1991), p. 129

WRITING

91. Explain why the integers are a subset of the rational numbers.

92. Explain why every integer is a rational number, but not every rational number is an integer.

93. Explain why the set of primes together with the set of composites is not the set of natural numbers.

94. Is the absolute value of a number always positive? Explain.

SOMETHING TO THINK ABOUT

95. How many integers have an absolute value that is less than 50?

96. How many odd integers have an absolute value between 20 and 40?

97. The **trichotomy property** of real numbers states that

If a and b are two real numbers, then

$$a < b \qquad \text{or} \qquad a = b \qquad \text{or} \qquad a > b$$

Explain why this is true.

98. Which of the following statements are always true?
  a. $|a + b| = |a| + |b|$
  b. $|a \cdot b| = |a| \cdot |b|$
  c. $|a + b| \le |a| + |b|$

1.2 Arithmetic and Properties of Real Numbers

- **Adding Real Numbers** ▪ **Subtracting Real Numbers**
- **Multiplying Real Numbers** ▪ **Dividing Real Numbers**
- **Order of Operations** ▪ **Measures of Central Tendency**
- **Evaluating Algebraic Expressions** ▪ **Properties of Real Numbers**

Getting Ready *Perform each operation.*

1. $5 + 4$ **2.** $4 + 5$ **3.** $3 \cdot 4$ **4.** $4 \cdot 3$
5. $12 - 7$ **6.** $15 \div 3$ **7.** $21 \div 7$ **8.** $25 - 19$

In this section, we will show how to add, subtract, multiply, and divide real numbers. We will then discuss several properties of real numbers.

Adding Real Numbers

When two numbers are added, we call the result their **sum.** To find the sum of $+2$ and $+3$, we can use a number line and represent the numbers with arrows, as shown in Figure 1-15(a). Since the endpoint of the second arrow is at $+5$, we have $+2 + (+3) = +5$.

To add -2 and -3, we can draw arrows as shown in Figure 1-15(b). Since the endpoint of the second arrow is at -5, we have $(-2) + (-3) = -5$.

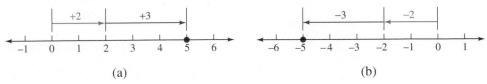

Figure 1-15

To add -6 and $+2$, we can draw arrows as shown in Figure 1-16(a). Since the endpoint of the second arrow is at -4, we have $(-6) + (+2) = -4$.

To add $+7$ and -4, we can draw arrows as shown in Figure 1-16(b). Since the endpoint of the final arrow is at $+3$, we have $(+7) + (-4) = +3$.

Figure 1-16

These examples suggest the following rules.

Adding Real Numbers

With like signs: Add the absolute values of the numbers and keep the common sign.

With unlike signs: Subtract the absolute values of the numbers (the smaller from the larger) and keep the sign of the number with the larger absolute value.

EXAMPLE 1 Add the numbers:

 a. $+4 + (+6) = +10$ Add the absolute values and use the common sign: $4 + 6 = +10$.
 b. $-5 + (-3) = -8$ Add the absolute values and use the common sign: $-(5 + 3) = -8$.
 c. $+9 + (-5) = +4$ Subtract the absolute values and use a $+$ sign: $+(9 - 5) = +4$.
 d. $-12 + (+5) = -7$ Subtract the absolute values and use a $-$ sign: $-(12 - 5) = -7$.

Self Check   Add: **a.** $-7 + (-2)$,   **b.** $-7 + 2$,   **c.** $7 + 2$,   and   **d.** $7 + (-2)$.   ∎

Subtracting Real Numbers

When one number is subtracted from another number, we call the result their **difference.** To find a difference, we can change the subtraction into an equivalent addition. For example, the subtraction $7 - 4$ is equivalent to the addition $7 + (-4)$, because they have the same result:

$$7 - 4 = 3 \qquad \text{and} \qquad 7 + (-4) = 3$$

This suggests that to subtract two numbers, we change the sign of the number being subtracted and add.

Subtracting Real Numbers

If a and b are real numbers, then $a - b = a + (-b)$.

EXAMPLE 2   Subtract:

    a. $12 - \mathbf{4} = 12 + (\mathbf{-4})$ Change the sign of 4 and add.

              $= 8$

    b. $-13 - \mathbf{5} = -13 + (\mathbf{-5})$ Change the sign of 5 and add.

              $= -18$

    c. $-14 - (\mathbf{-6}) = -14 + (\mathbf{+6})$ Change the sign of -6 and add.

              $= -8$

Self Check   Subtract: **a.** $-15 - 4$,   **b.** $8 - 5$,   and   **c.** $-12 - (-7)$.   ■

Multiplying Real Numbers

When two numbers are multiplied, we call the result their **product.** We can find the product of 5 and 4 by using 4 in an addition five times:

    $5(4) = 4 + 4 + 4 + 4 + 4 = 20$

We can find the product of 5 and -4 by using -4 in an addition five times:

    $5(-4) = (-4) + (-4) + (-4) + (-4) + (-4) = -20$

Since multiplication by a negative number can be defined as repeated subtraction, we can find the product of -5 and 4 by using 4 in a subtraction five times:

    $-5(4) = -4 - 4 - 4 - 4 - 4$

              $= -4 + (-4) + (-4) + (-4) + (-4)$ Change the sign of each 4 and add.

              $= -20$

We can find the product of -5 and -4 by using -4 in a subtraction five times:

    $-5(-4) = -(-4) - (-4) - (-4) - (-4) - (-4)$

              $= 4 + 4 + 4 + 4 + 4$ Change the sign of each -4 and add.

              $= 20$

The products $5(4)$ and $-5(-4)$ both equal $+20$, and the products $5(-4)$ and $-5(4)$ both equal -20. These results suggest the first two of the following rules.

Multiplying Real Numbers

With like signs: Multiply their absolute values. The product is positive.

With unlike signs: Multiply their absolute values. The product is negative.

Multiplication by 0: If x is any real number, then $x \cdot 0 = 0 \cdot x = 0$.

EXAMPLE 3   Multiply:

a. $4(-7) = -28$ Multiply the absolute values: $4 \cdot 7 = 28$. Since the signs are unlike, the product is negative.

b. $-5(-6) = +30$ Multiply the absolute values: $5 \cdot 6 = 30$. Since the signs are alike, the product is positive.

c. $-7(6) = -42$ Multiply the absolute values: $7 \cdot 6 = 42$. Since the signs are unlike, the product is negative.

d. $8(6) = +48$ Multiply the absolute values: $8 \cdot 6 = 48$. Since the signs are alike, the product is positive.

Self Check   Multiply: **a.** $(-6)(5)$,   **b.** $(-4)(-8)$,   **c.** $(17)(-2)$,   and   **d.** $(12)(6)$.   ▮

Dividing Real Numbers

When two numbers are divided, we call the result their **quotient.** In the division $\frac{x}{y} = q \ (y \neq 0)$, the quotient q is a number such that $y \cdot q = x$. We can use this relationship to find rules for dividing real numbers. We consider four divisions:

$$\frac{+10}{+2} = +5, \text{ because } +2(+5) = +10 \qquad \frac{-10}{-2} = +5, \text{ because } -2(+5) = -10$$

$$\frac{-10}{+2} = -5, \text{ because } +2(-5) = -10 \qquad \frac{+10}{-2} = -5, \text{ because } -2(-5) = +10$$

These results suggest the first two rules for dividing real numbers.

Dividing Real Numbers   *With like signs:* Divide their absolute values. The quotient is positive.

With unlike signs: Divide their absolute values. The quotient is negative.

Division by 0: Division by 0 is undefined.

❗ Comment   If $x \neq 0$, then $\frac{0}{x} = 0$. However, $\frac{x}{0}$ is undefined for any value of x.

EXAMPLE 4   Divide:

a. $\dfrac{36}{18} = +2$ Divide the absolute values: $\frac{36}{18} = 2$. Since the signs are alike, the quotient is positive.

b. $\dfrac{-44}{11} = -4$ Divide the absolute values: $\frac{44}{11} = 4$. Since the signs are unlike, the quotient is negative.

c. $\dfrac{27}{-9} = -3$ Divide the absolute values: $\frac{27}{9} = 3$. Since the signs are unlike, the quotient is negative.

d. $\dfrac{-64}{-8} = +8$ Divide the absolute values: $\frac{64}{8} = 8$. Since the signs are alike, the quotient is positive.

Self Check   Divide: **a.** $\dfrac{55}{-5}$,   **b.** $\dfrac{-72}{-6}$,   **c.** $\dfrac{-100}{10}$,   **d.** $\dfrac{50}{25}$.   ▮

Order of Operations

Suppose you are asked to contact a friend if you see a rug for sale while traveling in Turkey. After locating a nice one, you send the following message to your friend.

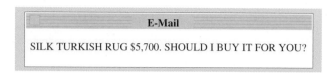

The next day, you receive this response.

The first statement in your friend's message says to buy the rug at any price. The second says not to buy it, because it is too expensive. The placement of the exclamation point makes these statements read differently, resulting in different interpretations.

When reading mathematical statements, the same kind of confusion is possible. To illustrate, we consider the expression $5 + 3 \cdot 7$, which contains the operations of addition and multiplication. We can calculate this expression in two different ways. However, we will get different results.

Method 1: add first

$\mathbf{5 + 3} \cdot 7 = \mathbf{8} \cdot 7$ Add 5 and 3.

$= 56$ Multiply 8 and 7.

Method 2: multiply first

$5 + \mathbf{3 \cdot 7} = 5 + \mathbf{21}$ Multiply 3 and 7.

$= 26$ Add 5 and 21.

⎯⎯⎯⎯ Different results ⎯⎯⎯⎯

To eliminate the possibility of getting different answers, we will agree to do multiplications before additions. So Method 1 above is incorrect. The correct calculation of $5 + 3 \cdot 7$ is

$5 + \mathbf{3 \cdot 7} = 5 + \mathbf{21}$ Do the multiplication first.

$= 26$ Then do the addition.

To indicate that additions should be done before multiplications, we must use **grouping symbols** such as parentheses ( ), brackets [ ], or braces { }. In the expression $(5 + 3)7$, the parentheses indicate that the addition is to be done first:

$\mathbf{(5 + 3)}7 = \mathbf{8} \cdot 7$

$= 56$

To guarantee that calculations will have one correct result, we will always do calculations in the following order.

Rules for the Order of Operations for Expressions without Exponents

Use the following steps to do all calculations within each pair of grouping symbols, working from the innermost pair to the outermost pair.

1. Do all multiplications and divisions, working from left to right.

2. Do all additions and subtractions, working from left to right.

When all grouping symbols have been removed, repeat the rules above to finish the calculation.

In a fraction, simplify the numerator and the denominator separately. Then simplify the fraction, whenever possible.

EXAMPLE 5 Evaluate each expression:

a. $4 + 2 \cdot 3 = 4 + 6$ Do the multiplication first.

$= 10$ Then do the addition.

b. $2(3 + 4) = 2 \cdot 7$ Because of the parentheses, do the addition first.

$= 14$ Then do the multiplication.

c. $5(3 - 6) \div 3 + 1 = 5(-3) \div 3 + 1$ Do the subtraction within parentheses.

$= -15 \div 3 + 1$ Then do the multiplication: $5(-3) = -15$.

$= -5 + 1$ Then do the division: $-15 \div 3 = -5$.

$= -4$ Finally, do the addition.

d. $5[3 - 2(6 \div 3 + 1)] = 5[3 - 2(2 + 1)]$ Do the division within parentheses: $6 \div 3 = 2$.

$= 5[3 - 2(3)]$ Do the addition: $2 + 1 = 3$.

$= 5(3 - 6)$ Do the multiplication: $2(3) = 6$.

$= 5(-3)$ Do the subtraction: $3 - 6 = -3$.

$= -15$ Do the multiplication.

e. $\dfrac{4 + 8(3 - 4)}{6 - 2(2)} = \dfrac{-4}{2}$ Simplify the numerator and denominator separately.

$= -2$

Self Check Evaluate: **a.** $5 + 3 \cdot 4$, **b.** $(5 + 3) \cdot 4$, **c.** $3(5 - 7) \div 6 + 3$, and **d.** $\dfrac{5 - 2(4 - 6)}{9 - 2 \cdot 3}$. ∎

Measures of Central Tendency

Three types of averages are commonly used in newspapers and magazines: the **mean,** the **median,** and the **mode.**

Mean

The **mean** of several values is the sum of those values divided by the number of values.

$$\text{Mean} = \frac{\text{sum of the values}}{\text{number of values}}$$

EXAMPLE 6   **Football**   Figure 1-17 shows the gains and losses made by a running back on seven plays. Find the mean number of yards per carry.

Solution   To find the mean number of yards per carry, we add the numbers and divide by 7.

$$\frac{-8 + (+2) + (-6) + (+6) + (+4) + (-7) + (-5)}{7} = \frac{-14}{7} = -2$$

The running back averaged -2 yards (or lost 2 yards) per carry.

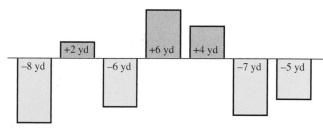

Figure 1-17

Median   The **median** of several values is the middle value. To find the median,
1. Arrange the values in increasing order.
2. If there is an odd number of values, choose the middle value.
3. If there is an even number of values, find the mean of the middle two values.

Mode   The **mode** of several values is the value that occurs most often.

EXAMPLE 7   Ten workers in a small business have monthly salaries of

$2,500, $1,750, $2,415, $3,240, $2,790,
$3,240, $2,650, $2,415, $2,415, $2,650

Find   **a.** the median   and   **b.** the mode of the distribution.

Solution   **a.** To find the median, we first arrange the salaries in increasing order:

$1,750, $2,415, $2,415, $2,415, $2,500 ,
$2,650 , $2,650, $2,790, $3,240, $3,240

Because there is an even number of salaries, the median will be the mean of the middle two scores, $2,500 and $2,650.

$$\text{Median} = \frac{\$2,500 + \$2,650}{2} = \$2,575$$

b. Since the salary $2,415 occurs most often, it is the mode.

If two different numbers in a distribution tie for occurring most often, the distribution is called *bimodal*.

Although the mean is probably the most common measure of average, the median and the mode are frequently used. For example, workers' salaries are often compared to the median (average) salary. To say that the modal (average) shoe size is 9 means that a shoe size of 9 occurs more often than any other size.

Evaluating Algebraic Expressions

Variables and numbers can be combined with the operations of arithmetic to produce **algebraic expressions.** To evaluate algebraic expressions, we substitute numbers for the variables and simplify.

EXAMPLE 8   If $a = 2$, $b = -3$, and $c = -5$, evaluate **a.** $a + bc$ and **b.** $\dfrac{ab + 3c}{b(c - a)}$.

Solution   We substitute 2 for a, -3 for b, and -5 for c and simplify.

a. $a + bc = 2 + (-3)(-5)$
$\qquad\qquad = 2 + (15)$
$\qquad\qquad = 17$

b. $\dfrac{ab + 3c}{b(c - a)} = \dfrac{2(-3) + 3(-5)}{-3(-5 - 2)}$

$\qquad\qquad\quad = \dfrac{-6 + (-15)}{-3(-7)}$

$\qquad\qquad\quad = \dfrac{-21}{21}$

$\qquad\qquad\quad = -1$

Self Check   If $a = 2$, $b = -5$, and $c = 3$, evaluate **a.** $b - ac$ and **b.** $\dfrac{ab - 2c}{ac + 2b}$.   ∎

Table 1-1 shows the formulas for the perimeters of several geometric figures. The distance around a circle is called a **circumference.**

Figure	Name	Perimeter	Figure	Name	Perimeter/circumference
	Square	$P = 4s$		Trapezoid	$P = a + b + c + d$
	Rectangle	$P = 2l + 2w$		Circle	$C = \pi D$ (π is approximately 3.1416)
	Triangle	$P = a + b + c$			

Table 1-1

EXAMPLE 9   Find the perimeter of the rectangle shown in Figure 1-18.

Solution   We substitute 2.75 for l and 1.25 for w into the formula $P = 2l + 2w$ and simplify.

$$P = 2l + 2w$$
$$P = 2(\mathbf{2.75}) + 2(\mathbf{1.25})$$
$$= 5.50 + 2.50$$
$$= 8.00$$

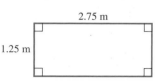

Figure 1-18

The perimeter is 8 meters.

Self Check   Find the perimeter of a rectangle with a length of 8 meters and a width of 5 meters.

Properties of Real Numbers

When we work with real numbers, we will use the following properties.

Properties of Real Numbers

> If a, b, and c are real numbers, the following properties apply.
>
> **The associative properties for addition and multiplication**
>
> $$(a + b) + c = a + (b + c) \qquad (ab)c = a(bc)$$
>
> **The commutative properties for addition and multiplication**
>
> $$a + b = b + a \qquad ab = ba$$
>
> **The distributive property of multiplication over addition**
>
> $$a(b + c) = ab + ac$$

The associative properties enable us to group the numbers in a sum or a product in any way that we wish and still get the same result. For example,

$$(\mathbf{2 + 3}) + 4 = \mathbf{5} + 4 \qquad\qquad 2 + (\mathbf{3 + 4}) = 2 + \mathbf{7}$$
$$= 9 \qquad\qquad\qquad = 9$$
$$(\mathbf{2 \cdot 3}) \cdot 4 = \mathbf{6} \cdot 4 \qquad\qquad 2 \cdot (\mathbf{3 \cdot 4}) = 2 \cdot \mathbf{12}$$
$$= 24 \qquad\qquad\qquad = 24$$

 Comment   Subtraction and division are not associative, because different groupings give different results. For example,

$$(\mathbf{8 - 4}) - 2 = \mathbf{4} - 2 = 2 \quad \text{but} \quad 8 - (\mathbf{4 - 2}) = 8 - \mathbf{2} = 6$$
$$(\mathbf{8 \div 4}) \div 2 = \mathbf{2} \div 2 = 1 \quad \text{but} \quad 8 \div (\mathbf{4 \div 2}) = 8 \div \mathbf{2} = 4$$

The commutative properties enable us to add or multiply two numbers in either order and obtain the same result. For example,

$$2 + 3 = 5 \qquad \text{and} \qquad 3 + 2 = 5$$
$$7 \cdot 9 = 63 \qquad \text{and} \qquad 9 \cdot 7 = 63$$

Comment   Subtraction and division are not commutative, because doing these operations in different orders will give different results. For example,

$$8 - 4 = 4 \qquad \text{but} \qquad 4 - 8 = -4$$

$$8 \div 4 = 2 \qquad \text{but} \qquad 4 \div 8 = \frac{1}{2}$$

The distributive property enables us to evaluate many expressions involving a multiplication over an addition. We can add first inside the parentheses and then multiply, or multiply over the addition first and then add.

$$2(\mathbf{3 + 7}) = 2 \cdot \mathbf{10} \qquad\qquad\qquad 2(3 + 7) = \mathbf{2 \cdot 3 + 2 \cdot 7}$$
$$= 20 \qquad\qquad\qquad\qquad\quad = 6 + 14$$
$$\qquad\qquad\qquad\qquad\qquad\qquad = 20$$

We can interpret the distributive property geometrically. Since the area of the largest rectangle in Figure 1-19 is the product of its width a and its length $b + c$, its area is $a(b + c)$. The areas of the two smaller rectangles are ab and ac. Since the area of the largest rectangle is equal to the sum of the areas of the smaller rectangles, we have $a(b + c) = ab + ac$.

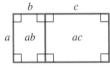

Figure 1-19

A more general form of the distributive property is called the **extended distributive property.**

$$a(b + c + d + e + \cdots) = ab + ac + ad + ae + \cdots$$

EXAMPLE 10   Use the distributive property to write each expression without parentheses:

a. $2(x + 3)$ and   **b.** $2(x + y - 7)$.

Solution   **a.** $2(x + 3) = 2x + 2 \cdot 3$ **b.** $2(x + y - 7) = 2x + 2y - 2 \cdot 7$
$\qquad\qquad\quad = 2x + 6 \qquad\qquad\qquad\qquad\qquad = 2x + 2y - 14$

Self Check   Remove parentheses: $-5(a - 2b + 3c)$.   ∎

The real numbers 0 and 1 have important special properties.

Properties of 0 and 1

Additive identity: The sum of 0 and any number is the number itself.

$$0 + a = a + 0 = a$$

Multiplication property of 0: The product of any number and 0 is 0.

$$a \cdot 0 = 0 \cdot a = 0$$

Multiplicative identity: The product of 1 and any number is the number itself.

$$1 \cdot a = a \cdot 1 = a$$

For example,

$$7 + 0 = 7, \qquad 7(0) = 0, \qquad 1(5) = 5, \qquad \text{and} \qquad (-7)1 = -7$$

If the sum of two numbers is 0, the numbers are called **additive inverses, negatives,** or **opposites** of each other. For example, 6 and -6 are negatives, because $6 + (-6) = 0$.

The Additive Inverse Property

For every real number a, there is a real number $-a$ such that

$$a + (-a) = -a + a = 0$$

The symbol $-(-6)$ means "the negative of negative 6." Because the sum of two numbers that are negatives is 0, we have

$$-6 + [-(-6)] = 0 \qquad \text{and} \qquad -6 + 6 = 0$$

Because -6 has only one additive inverse, it follows that $-(-6) = 6$. In general, the following rule applies.

The Double Negative Rule

If a represents any real number, then $-(-a) = a$.

If the product of two numbers is 1, the numbers are called **multiplicative inverses** or **reciprocals** of each other.

The Multiplicative Inverse Property

For every nonzero real number a, there exists a real number $\frac{1}{a}$ such that

$$a \cdot \frac{1}{a} = \frac{1}{a} \cdot a = 1 \quad (a \neq 0)$$

Some examples of reciprocals are

- 5 and $\frac{1}{5}$ are reciprocals, because $5\left(\frac{1}{5}\right) = 1$.

- $\frac{3}{2}$ and $\frac{2}{3}$ are reciprocals, because $\frac{3}{2}\left(\frac{2}{3}\right) = 1$.

- -0.25 and -4 are reciprocals, because $-0.25(-4) = 1$.

The reciprocal of 0 does not exist, because $\frac{1}{0}$ is undefined.

Self Check Answers

1. a. -9, **b.** -5, **c.** 9, **d.** 5   **2. a.** -19, **b.** 3, **c.** -5 **3. a.** -30, **b.** 32, **c.** -34, **d.** 72
4. a. -11, **b.** 12, **c.** -10, **d.** 2   **5. a.** 17, **b.** 32, **c.** 2, **d.** 3   **8. a.** -11, **b.** 4   **9.** 26 m
10. $-5a + 10b - 15c$

Orals *Perform each operation.*

1. $+2 + (-4)$ **2.** $-5 - 2$ **3.** $7(-4)$

4. $(-7)(-4)$ **5.** $3 + (-3)(2)$ **6.** $\dfrac{-4 + (-2)}{-3 + 5}$

1.2 EXERCISES

REVIEW *Graph each set on the number line.*

1. $\{x|x > 4\}$

2. $(-\infty, -5]$

3. $(2, 10]$

4. $\{x|-4 \le x \le 4\}$

5. A man bought 32 gallons of gasoline at $1.29 per gallon and 3 quarts of oil at $1.35 per quart. The sales tax was included in the price of the gasoline, but 5% sales tax was added to the cost of the oil. Find the total cost.

6. On an adjusted income of $57,760, a woman must pay taxes according to the schedule shown in the table. Compute the tax bill.

Over	But not over	Tax	Of the amount over
$0	$10,000	. . . . . 10%	$0
10,000	37,450	1,000.00 + 15%	10,000
37,450	96,700	5,117.50 + 27%	37,450

VOCABULARY AND CONCEPTS *Fill in the blanks.*

7. To add two numbers with like signs, we add their _____ values and keep the _____ sign.

8. To add two numbers with unlike signs, we _____ their absolute values and keep the sign of the number with the larger absolute value.

9. To subtract one number from another, we _____ the sign of the number that is being subtracted and ____.

10. The product of two real numbers with like signs is _____.

11. The quotient of two real numbers with unlike signs is _____.

12. The denominator of a fraction can never be __.

13. The three measures of central tendency are the _____, _____, and _____.

14. Write the formula for the circumference of a circle. _____

15. Write the associative property of multiplication. _____

16. Write the commutative property of addition. _____

17. Write the distributive property of multiplication over addition. _____

18. What is the additive identity? __

19. What is the multiplicative identity? __

20. $-(-a) = $ __

PRACTICE *Perform the operations.*

21. $-3 + (-5)$

22. $2 + (+8)$

23. $-7 + 2$

24. $3 + (-5)$

25. $-3 - 4$

26. $-11 - (-17)$

27. $-33 - (-33)$

28. $14 - (-13)$

29. $-2(6)$

30. $3(-5)$

31. $-3(-7)$

32. $-2(-5)$

33. $\dfrac{-8}{4}$

34. $\dfrac{25}{-5}$

35. $\dfrac{-16}{-4}$

36. $\dfrac{-5}{-25}$

37. $\dfrac{1}{2} + \left(-\dfrac{1}{3}\right)$

38. $-\dfrac{3}{4} + \left(-\dfrac{1}{5}\right)$

39. $\dfrac{1}{2} - \left(-\dfrac{3}{5}\right)$

40. $\dfrac{1}{26} - \dfrac{11}{13}$

41. $\dfrac{1}{3} - \dfrac{1}{2}$

42. $\dfrac{7}{8} - \left(-\dfrac{3}{4}\right)$

43. $\left(-\dfrac{3}{5}\right)\left(\dfrac{10}{7}\right)$

44. $\left(-\dfrac{6}{7}\right)\left(-\dfrac{5}{12}\right)$

45. $\dfrac{3}{4} \div \left(-\dfrac{3}{8}\right)$

46. $-\dfrac{3}{5} \div \dfrac{7}{10}$

47. $-\dfrac{16}{5} \div \left(-\dfrac{10}{3}\right)$

48. $-\dfrac{5}{24} \div \dfrac{10}{3}$

49. $3 + 4 \cdot 5$

50. $5 \cdot 3 - 6 \cdot 4$

51. $3 - 2 - 1$

52. $5 - 3 - 1$

53. $3 - (2 - 1)$

54. $5 - (3 - 1)$

55. $2 - 3 \cdot 5$

56. $6 + 4 \cdot 7$

57. $8 \div 4 \div 2$

58. $100 \div 10 \div 5$

59. $8 \div (4 \div 2)$

60. $100 \div (10 \div 5)$

61. $2 + 6 \div 3 - 5$

62. $6 - 8 \div 4 - 2$

63. $(2 + 6) \div (3 - 5)$

64. $(6 - 8) \div (4 - 2)$

65. $\dfrac{3(8 + 4)}{2 \cdot 3 - 9}$ **66.** $\dfrac{5(4 - 1)}{3 \cdot 2 + 5 \cdot 3}$

67. $\dfrac{100(2 - 4)}{1,000 \div 10 \div 10}$ **68.** $\dfrac{8(3) - 4(6)}{5(3) + 3(-7)}$

Use the distribution: 7, 5, 9, 10, 8, 6, 6, 7, 9, 12, 9. *You may use a calculator.*

69. Find the mean.   **70.** Find the median.

71. Find the mode.

Use the distribution: 8, 12, 23, 12, 10, 16, 26, 12, 14, 8, 16, 23.

72. Find the median.   **73.** Find the mode.

74. Find the mean.

Let $a = 3$, $b = -2$, $c = -1$, and $d = 2$ and evaluate each expression.

75. $ab + cd$ **76.** $ad + bc$

77. $a(b + c)$ **78.** $d(b + a)$

79. $\dfrac{ad + c}{cd + b}$ **80.** $\dfrac{ab + d}{bd + a}$

81. $\dfrac{ac - bd}{cd - ad}$ **82.** $\dfrac{bc - ad}{bd + ac}$

Sorting records is a common task in data processing. A selection sort requires C comparisons to sort N records, where C and N are related by the formula $C = \frac{N(N - 1)}{2}$.

83. How many comparisons are needed to sort 200 records?

84. How many comparisons are needed to sort 10,000 records?

85. Perimeter of a triangle Find the perimeter of a triangle with sides that are 23.5, 37.2, and 39.7 feet long.

86. Perimeter of a trapezoid Find the perimeter of a trapezoid with sides that are 43.27, 47.37, 50.21, and 52.93 centimeters long.

Tell which property of real numbers justifies each statement.

87. $3 + 7 = 7 + 3$

88. $2 \cdot (9 \cdot 13) = (2 \cdot 9) \cdot 13$

89. $3(2 + 5) = 3 \cdot 2 + 3 \cdot 5$

90. $4 \cdot 3 = 3 \cdot 4$

91. $81 + 0 = 81$

92. $3(9 + 2) = 3 \cdot 9 + 3 \cdot 2$

93. $5 \cdot \dfrac{1}{5} = 1$

94. $3 + (9 + 0) = (9 + 0) + 3$

95. $a + (7 + 8) = (a + 7) + 8$

96. $1 \cdot 3 = 3$

97. $(2 \cdot 3) \cdot 4 = 4 \cdot (2 \cdot 3)$

98. $8 + (-8) = 0$

Use a calculator to verify each statement. Identify the property of real numbers that is being illustrated.

99. $(37.9 + 25.2) + 14.3 = 37.9 + (25.2 + 14.3)$

100. $7.1(3.9 + 8.8) = 7.1 \cdot 3.9 + 7.1 \cdot 8.8$

101. $2.73(4.534 + 57.12) = 2.73 \cdot 4.534 + 2.73 \cdot 57.12$

102. $(6.789 + 345.1) + 27.347 = (345.1 + 6.789) + 27.347$

APPLICATIONS

103. Earning money One day Scott earned $22.25 tutoring mathematics and $39.75 tutoring physics. How much did he earn that day?

104. Losing weight During an illness, Wendy lost 13.5 pounds. She then dieted and lost another 11.5 pounds. What integer represents her change in weight?

105. Changing temperatures The temperature rose 17° in 1 hour and then dropped 13° in the next hour. Find the overall change in temperature.

106. Displaying the flag Before the American flag is displayed at half-mast, it should first be raised to the top of the flagpole. How far has the flag in the illustration traveled?

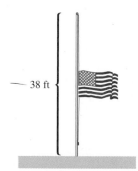

— 38 ft

107. **Changing temperatures** If the temperature has been dropping 4° each hour, how much warmer was it 3 hours ago?

108. **Playing slot machines** In Las Vegas, Harry lost $30 per hour playing the slot machines. How much did he lose after gambling for 15 hours?

109. **Filling a pool** The flow of water from a pipe is filling a pool at the rate of 23 gallons per minute. How much less water was in the pool 5 hours ago?

110. **Draining a pool** If a drain is emptying a pool at the rate of 12 gallons per minute, how much more water was in the pool 2 hours ago?

Use a calculator to help solve the following problems.

111. **Military science** An army retreated 2,300 meters. After regrouping, it moved forward 1,750 meters. The next day, it gained another 1,875 meters. What integer represents the army's net gain (or loss)?

112. **Grooming horses** John earned $8 an hour for grooming horses. After working for 8 hours, he had $94. How much did he have before he started work?

113. **Managing a checkbook** Sally started with $437.37 in a checking account. One month, she had deposits of $125.18, $137.26, and $145.56. That same month, she had withdrawals of $117.11, $183.49, and $122.89. Find her ending balance.

114. **Stock averages** The illustration shows the daily advances and declines of the Dow Jones average for one week. What integer represents the total gain or loss for the week?

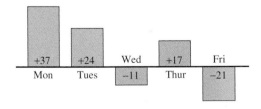

115. **Selling clothes** If a clerk had the sales shown in the table for one week, find the mean of daily sales.

Monday	. . . .$1,525
Tuesday	. . . .$ 785
Wednesday	. . . .$1,628
Thursday	. . . .$1,214
Friday	. . . .$ 917
Saturday	. . . .$1,197

116. **Size of viruses** The table gives the approximate lengths (in centimicrons) of the viruses that cause five common diseases. Find the mean length of the viruses.

Polio	2.5
Influenza	105.1
Pharyngitis	74.9
Chicken pox	137.4
Yellow fever	52.6

117. **Calculating grades** A student has scores of 75, 82, 87, 80, and 76 on five exams. Find his average (mean) score.

118. **Averaging weights** The offensive line of a football team has two guards, two tackles, and a center. If the guards weigh 298 and 287 pounds, the tackles 310 and 302 pounds, and the center 303 pounds, find the average (mean) weight of the offensive line.

119. **Analyzing ads** The businessman who ran the following ad earns $100,000 and employs four students who earn $10,000 each. Is the ad honest?

> **HIRING**
> Hard-working, intelligent students
> Good pay: average wage of
> $28,000

120. **Averaging grades** A student has grades of 78%, 85%, 88%, and 96%. There is one test left, and the student needs to average 90% to earn an A. Does he have a chance?

121. **Perimeter of a square** Find the perimeter of the square shown in the illustration.

7.5 cm

122. Circumference of a circle To the nearest hundredth, find the circumference of the circle shown in the illustration.

25 m

WRITING

123. The symmetric property of equality states that if $a = b$, then $b = a$. Explain why this property is often confused with the commutative properties. Why do you think this is so?

124. Explain why the mean of two numbers is halfway between the two numbers.

SOMETHING TO THINK ABOUT

125. Pick five numbers and find their mean. Add 7 to each of the numbers to get five new numbers and find their mean. What do you discover? Is this property always true?

126. Take the original five numbers in Exercise 125 and multiply each one by 7 to get five new numbers and find their mean. What do you discover? Is this property always true?

127. Give three applications in which the median would be the most appropriate average to use.

128. Give three applications in which the mode would be the most appropriate average to use.

1.3   Exponents

■ **Exponents** ■ **Properties of Exponents** ■ **Zero Exponents**
■ **Negative Exponents** ■ **Order of Operations**
■ **Evaluating Formulas**

Getting Ready   *Find each product:*

1. $2 \cdot 2$ **2.** $3 \cdot 3 \cdot 3$

3. $(-4)(-4)(-4)$ **4.** $(-3)(-3)(-3)(-3)$

5. $\dfrac{1}{3} \cdot \dfrac{1}{3} \cdot \dfrac{1}{3}$ **6.** $-\left(\dfrac{2}{5} \cdot \dfrac{2}{5} \cdot \dfrac{2}{5} \cdot \dfrac{2}{5}\right)$

In this section, we will review exponents, a shortcut way of indicating repeated multiplication.

Exponents

Exponents indicate repeated multiplication. For example,

$y^2 = y \cdot y$ Read y^2 as "y to the second power" or "y squared."

$z^3 = z \cdot z \cdot z$ Read z^3 as "z to the third power" or "z cubed."

$x^4 = x \cdot x \cdot x \cdot x$ Read x^4 as "x to the fourth power."

These examples suggest the following definition.

Natural-Number Exponents

If n is a natural number, then

$$x^n = \overbrace{x \cdot x \cdot x \cdot \cdots \cdot x}^{n \text{ factors of } x}$$

The exponential expression x^n is called a **power of x,** and we read it as "x to the nth power." In this expression, x is called the **base,** and n is called the **exponent.**

$$\text{Base} \rightarrow x^n \leftarrow \text{Exponent}$$

 Comment   A natural-number exponent tells how many times the base of an exponential expression is to be used as a factor in a product.

 EXAMPLE 1   Write each number without exponents:

a. $2^5 = 2 \cdot 2 \cdot 2 \cdot 2 \cdot 2$
$= 32$

b. $(-2)^5 = (-2)(-2)(-2)(-2)(-2)$
$= -32$

c. $-4^4 = -(4^4)$
$= -(4 \cdot 4 \cdot 4 \cdot 4)$
$= -256$

d. $(-4)^4 = (-4)(-4)(-4)(-4)$
$= 256$

e. $\left(\dfrac{1}{2}a\right)^3 = \left(\dfrac{1}{2}a\right)\left(\dfrac{1}{2}a\right)\left(\dfrac{1}{2}a\right)$
$= \dfrac{1}{8}a^3$

f. $\left(-\dfrac{1}{5}b\right)^2 = \left(-\dfrac{1}{5}b\right)\left(-\dfrac{1}{5}b\right)$
$= \dfrac{1}{25}b^2$

Self Check   Write each number without exponents: **a.** 3^4,   **b.** $(-5)^3$,   and   **c.** $\left(-\frac{3}{4}a\right)^2$.   ∎

Comment   Note the difference between $-x^n$ and $(-x)^n$.

$$-x^n = -\overbrace{(x \cdot x \cdot x \cdot \ \cdots \ \cdot x)}^{n \text{ factors of } x} \qquad\qquad (-x)^n = \overbrace{(-x)(-x)(-x) \cdot \ \cdots \ \cdot (-x)}^{n \text{ factors of } -x}$$

Also, note the difference between ax^n and $(ax)^n$

$$ax^n = a \cdot \overbrace{x \cdot x \cdot x \cdot \ \cdots \ \cdot x}^{n \text{ factors of } x} \qquad\qquad (ax)^n = \overbrace{(ax)(ax)(ax) \cdot \ \cdots \ \cdot (ax)}^{n \text{ factors of } ax}$$

Properties of Exponents

Since x^5 means that x is to be used as a factor five times, and since x^3 means that x is to be used as a factor three times, $x^5 \cdot x^3$ means that x will be used as a factor eight times.

$$x^5 x^3 = \overbrace{x \cdot x \cdot x \cdot x \cdot x}^{5 \text{ factors of } x} \cdot \overbrace{x \cdot x \cdot x}^{3 \text{ factors of } x} = \overbrace{x \cdot x \cdot x \cdot x \cdot x \cdot x \cdot x \cdot x}^{8 \text{ factors of } x}$$

In general,

$$x^m x^n = \overbrace{x \cdot x \cdot x \cdot \ \cdots \ \cdot x}^{m \text{ factors of } x} \overbrace{x \cdot x \cdot x \cdot \ \cdots \ \cdot x}^{n \text{ factors of } x} = \overbrace{x \cdot x \cdot x \cdot x \cdot \ \cdots \ \cdot x}^{m + n \text{ factors of } x}$$

Thus, *to multiply exponential expressions with the same base, we keep the same base and add the exponents.*

The Product Rule of Exponents

If m and n are natural numbers, then

$$x^m x^n = x^{m+n}$$

 Comment The product rule of exponents applies only to exponential expressions with the same base. The expression $x^5 y^3$, for example, cannot be simplified, because the bases of the exponential expressions are different.

EXAMPLE 2 Simplify each expression:

a. $x^{11} x^5 = x^{11+5}$
$\qquad\qquad = x^{16}$

b. $a^5 a^4 a^3 = (a^5 a^4) a^3$
$\qquad\qquad = a^9 a^3$
$\qquad\qquad = a^{12}$

c. $a^2 b^3 a^3 b^2 = a^2 a^3 b^3 b^2$
$\qquad\qquad = a^5 b^5$

d. $-8x^4 \left(\dfrac{1}{4} x^3\right) = \left(-8 \cdot \dfrac{1}{4}\right)(x^4 x^3)$
$\qquad\qquad\qquad = -2x^7$

Self Check Simplify each expression: **a.** $a^3 a^5$, **b.** $a^2 b^3 a^3 b^4$, and **c.** $-8a^4\left(-\frac{1}{2} a^2 b\right)$. ∎

To find another property of exponents, we simplify $(x^4)^3$, which means x^4 cubed or $x^4 \cdot x^4 \cdot x^4$.

$$(x^4)^3 = x^4 \cdot x^4 \cdot x^4 = \overbrace{x \cdot x \cdot x \cdot x}^{x^4} \cdot \overbrace{x \cdot x \cdot x \cdot x}^{x^4} \cdot \overbrace{x \cdot x \cdot x \cdot x}^{x^4} = x^{12}$$

In general, we have

$$(x^m)^n = \overbrace{x^m \cdot x^m \cdot x^m \cdot \cdots \cdot x^m}^{n \text{ factors of } x^m} = \overbrace{x \cdot x \cdot x \cdot x \cdot x \cdot \cdots \cdot x}^{mn \text{ factors of } x} = x^{mn}$$

Thus, *to raise an exponential expression to a power, we keep the same base and multiply the exponents.*

To find a third property of exponents, we square $3x$ and get

$$(3x)^2 = (3x)(3x) = 3 \cdot 3 \cdot x \cdot x = 3^2 x^2 = 9x^2$$

In general, we have

$$(xy)^n = \overbrace{(xy)(xy)(xy) \cdot \cdots \cdot (xy)}^{n \text{ factors of } xy} = \overbrace{xxx \cdot \cdots \cdot x}^{n \text{ factors of } x} \cdot \overbrace{yyy \cdot \cdots \cdot y}^{n \text{ factors of } y} = x^n y^n$$

To find a fourth property of exponents, we cube $\frac{x}{3}$ to get

$$\left(\frac{x}{3}\right)^3 = \frac{x}{3} \cdot \frac{x}{3} \cdot \frac{x}{3} = \frac{x \cdot x \cdot x}{3 \cdot 3 \cdot 3} = \frac{x^3}{3^3} = \frac{x^3}{27}$$

In general, we have

$$\left(\frac{x}{y}\right)^n = \overbrace{\left(\frac{x}{y}\right)\left(\frac{x}{y}\right)\left(\frac{x}{y}\right) \cdot \cdots \cdot \left(\frac{x}{y}\right)}^{n \text{ factors of } x/y} \quad (y \neq 0)$$

$$= \frac{\overbrace{xxx \cdot \, \cdots \, \cdot x}^{n \text{ factors of } x}}{\underbrace{yyy \cdot \, \cdots \, \cdot y}_{n \text{ factors of } y}} \qquad \text{Multiply the numerators and multiply the denominators.}$$

$$= \frac{x^n}{y^n}$$

The previous results are called the **power rules of exponents.**

The Power Rules of Exponents

If m and n are natural numbers, then

$$(x^m)^n = x^{mn} \qquad (xy)^n = x^n y^n \qquad \left(\frac{x}{y}\right)^n = \frac{x^n}{y^n} \quad (y \neq 0)$$

EXAMPLE 3   Simplify each expression:

a. $(3^2)^3 = 3^{2 \cdot 3}$ **b.** $(x^{11})^5 = x^{11 \cdot 5}$
$\qquad\qquad = 3^6$ $\qquad\qquad\quad = x^{55}$
$\qquad\qquad = 729$

c. $(x^2 x^3)^6 = (x^5)^6$ **d.** $(x^2)^4(x^3)^2 = x^8 x^6$
$\qquad\qquad\; = x^{30}$ $\qquad\qquad\qquad = x^{14}$

Self Check   Simplify each expression: **a.** $(a^5)^8$,   **b.** $(a^4 a^3)^3$,   and   **c.** $(a^3)^3(a^2)^3$.   ▮

EXAMPLE 4   Simplify each expression. Assume that no denominators are zero.

a. $(x^2 y)^3 = (x^2)^3 y^3$ **b.** $(x^3 y^4)^4 = (x^3)^4 (y^4)^4$
$\qquad\qquad\; = x^6 y^3$ $\qquad\qquad\quad\; = x^{12} y^{16}$

c. $\left(\dfrac{x}{y^2}\right)^4 = \dfrac{x^4}{(y^2)^4}$ **d.** $\left(\dfrac{x^3}{y^4}\right)^2 = \dfrac{(x^3)^2}{(y^4)^2}$
$\qquad\qquad = \dfrac{x^4}{y^8}$ $\qquad\qquad\quad = \dfrac{x^6}{y^8}$

Self Check   Simplify each expression: **a.** $(a^4 b^5)^2$ and   **b.** $\left(\dfrac{a^5}{b^7}\right)^3$ $(b \neq 0)$.   ▮

Zero Exponents

Since the rules for exponents hold for exponents of 0, we have

$$x^0 x^n = x^{0+n} = x^n = 1x^n$$

Because $\mathbf{x^0 x^n = 1x^n}$, it follows that $x^0 = 1$ $(x \neq 0)$.

Zero Exponents   If $x \neq 0$, then $x^0 = 1$.

❗   **Comment** 0^0 is undefined.

Because of the previous definition, any nonzero base raised to the 0th power is 1. For example, if no variables are zero, then

$$5^0 = 1, \qquad (-7)^0 = 1, \qquad (3ax^3)^0 = 1, \qquad \left(\frac{1}{2}x^5y^7z^9\right)^0 = 1$$

Negative Exponents

Since the rules for exponents are true for negative integer exponents, we have

$$x^{-n}x^n = x^{-n+n} = x^0 = 1 \quad (x \neq 0)$$

Because $x^{-n} \cdot x^n = 1$ and $\frac{1}{x^n} \cdot x^n = 1$, we define x^{-n} to be the reciprocal of x^n.

Negative Exponents

If n is an integer and $x \neq 0$, then

$$x^{-n} = \frac{1}{x^n} \qquad \text{and} \qquad \frac{1}{x^{-n}} = x^n$$

 Comment By the definition of negative exponents, the base cannot be 0. Thus, an expression such as 0^{-5} is undefined.

Because of this definition, we can write expressions containing negative exponents as expressions without negative exponents. For example,

$$5^{-2} = \frac{1}{5^2} = \frac{1}{25} \qquad\qquad 10^{-3} = \frac{1}{10^3} = \frac{1}{1,000}$$

and if $x \neq 0$, we have

$$(2x)^{-3} = \frac{1}{(2x)^3} = \frac{1}{8x^3} \qquad 3x^{-1} = 3 \cdot \frac{1}{x} = \frac{3}{x}$$

EXAMPLE 5 Write each expression without negative exponents:

a. $x^{-5}x^3 = x^{-5+3}$ $\qquad\qquad\qquad\qquad$ **b.** $(x^{-3})^{-2} = x^{(-3)(-2)}$

$\qquad\qquad = x^{-2}$ $\qquad\qquad\qquad\qquad\qquad\qquad = x^6$

$\qquad\qquad = \dfrac{1}{x^2}$

Self Check Write each expression without negative exponents: **a.** $a^{-7}a^3$ and **b.** $(a^{-5})^{-3}$. ∎

To develop a rule for dividing exponential expressions, we proceed as follows:

$$\frac{x^m}{x^n} = x^m\left(\frac{1}{x^n}\right) = x^m x^{-n} = x^{m+(-n)} = x^{m-n}$$

Thus, *to divide exponential expressions with the same nonzero base, we keep the same base and subtract the exponent in the denominator from the exponent in the numerator.*

The Quotient Rule   If m and n are integers, then

$$\frac{x^m}{x^n} = x^{m-n} \quad (x \neq 0)$$

EXAMPLE 6   Simplify each expression:

a. $\dfrac{a^5}{a^3} = a^{5-3}$

$= a^2$

b. $\dfrac{x^{-5}}{x^{11}} = x^{-5-11}$

$= x^{-16}$

$= \dfrac{1}{x^{16}}$

c. $\dfrac{x^4 x^3}{x^{-5}} = \dfrac{x^7}{x^{-5}}$

$= x^{7-(-5)}$

$= x^{12}$

d. $\dfrac{(x^2)^3}{(x^3)^2} = \dfrac{x^6}{x^6}$

$= x^{6-6}$

$= x^0$

$= 1$

e. $\dfrac{x^2 y^3}{xy^4} = x^{2-1} y^{3-4}$

$= xy^{-1}$

$= \dfrac{x}{y}$

f. $\left(\dfrac{a^{-2}b^3}{a^2 a^3 b^4}\right)^3 = \left(\dfrac{a^{-2}b^3}{a^5 b^4}\right)^3$

$= (a^{-2-5} b^{3-4})^3$

$= (a^{-7} b^{-1})^3$

$= \left(\dfrac{1}{a^7 b}\right)^3$

$= \dfrac{1}{a^{21} b^3}$

Self Check   Simplify each expression: **a.** $\dfrac{(a^{-2})^3}{(a^2)^{-3}}$ and   **b.** $\left(\dfrac{a^{-2}b^5}{b^8}\right)^{-3}$.   ∎

To illustrate one more property of exponents, we consider the simplification of $\left(\frac{2}{3}\right)^{-4}$.

$$\left(\frac{2}{3}\right)^{-4} = \frac{1}{\left(\frac{2}{3}\right)^4} = \frac{1}{\frac{2^4}{3^4}} = 1 \div \frac{2^4}{3^4} = 1 \cdot \frac{3^4}{2^4} = \frac{3^4}{2^4} = \left(\frac{3}{2}\right)^4$$

The example suggests that to raise a fraction to a negative power, we can invert the fractional base and then raise it to a positive power.

Fractions to Negative Powers   If n is an integer, then

$$\left(\frac{x}{y}\right)^{-n} = \left(\frac{y}{x}\right)^n \quad (x \neq 0, y \neq 0)$$

EXAMPLE 7   Write each expression without using parentheses. Write answers without negative exponents.

a. $\left(\dfrac{3}{5}\right)^{-4} = \left(\dfrac{5}{3}\right)^{4}$

$= \dfrac{625}{81}$

b. $\left(\dfrac{y^2}{x^3}\right)^{-3} = \left(\dfrac{x^3}{y^2}\right)^{3}$

$= \dfrac{x^9}{y^6}$

c. $\left(\dfrac{2x^2}{3y^{-3}}\right)^{-4} = \left(\dfrac{3y^{-3}}{2x^2}\right)^{4}$

$= \dfrac{81y^{-12}}{16x^8}$

$= \dfrac{81}{16x^8} \cdot y^{-12}$

$= \dfrac{81}{16x^8} \cdot \dfrac{1}{y^{12}}$

$= \dfrac{81}{16x^8 y^{12}}$

d. $\left(\dfrac{a^{-2}b^3}{a^2a^3b^4}\right)^{-3} = \left(\dfrac{a^2a^3b^4}{a^{-2}b^3}\right)^{3}$

$= \left(\dfrac{a^5b^4}{a^{-2}b^3}\right)^{3}$

$= (a^{5-(-2)}b^{4-3})^3$

$= (a^7b)^3$

$= a^{21}b^3$

Self Check Write $\left(\dfrac{3a^3}{2b^{-2}}\right)^{-5}$ without using parentheses.   ∎

We summarize the rules of exponents as follows.

Properties of Exponents

If there are no divisions by 0, then for all integers m and n,

$$x^m x^n = x^{m+n} \qquad (x^m)^n = x^{mn} \qquad (xy)^n = x^n y^n \qquad \left(\dfrac{x}{y}\right)^n = \dfrac{x^n}{y^n}$$

$$x^0 = 1 \quad (x \neq 0) \qquad x^{-n} = \dfrac{1}{x^n} \qquad \dfrac{x^m}{x^n} = x^{m-n} \qquad \left(\dfrac{x}{y}\right)^{-n} = \left(\dfrac{y}{x}\right)^n$$

The same rules apply to exponents that are variables.

EXAMPLE 8 Simplify each expression. Assume that $a \neq 0$ and $x \neq 0$.

a. $\dfrac{a^n a}{a^2} = a^{n+1-2}$

$= a^{n-1}$

b. $\dfrac{x^3 x^2}{x^n} = x^{3+2-n}$

$= x^{5-n}$

c. $\left(\dfrac{x^n}{x^2}\right)^2 = \dfrac{x^{2n}}{x^4}$

$= x^{2n-4}$

d. $\dfrac{a^n a^{-3}}{a^{-1}} = a^{n+(-3)-(-1)}$

$= a^{n-3+1}$

$= a^{n-2}$

Self Check Simplify each expression (assume that $t \neq 0$): **a.** $\dfrac{t^n t^2}{t^3}$ and **b.** $\left(\dfrac{2t^n}{3t^3}\right)^3$.   ∎

FINDING POWERS

To find powers of numbers with many calculators, we use the y^x key. For example, to find 5.37^4, we enter these numbers and press these keys:

5.37 y^x 4 $=$ Some calculators have an x^y key.

The display will read 831.5668016.

To use a graphing calculator, we enter these numbers and press these keys:

5.37 $\wedge$ 4 ENTER

The display will read $5.37\char`\^4$
$$831.5668016$$

If neither of these methods works, consult your owner's manual.

Order of Operations

When simplifying expressions containing exponents, we find powers before performing additions and multiplications.

EXAMPLE 9   If $x = 2$ and $y = -3$, find the value of $3x + 2y^3$.

Solution
$$
\begin{aligned}
3x + 2y^3 &= 3(\mathbf{2}) + 2(\mathbf{-3})^3 && \text{Substitute 2 for } x \text{ and } -3 \text{ for } y.\\
&= 3(2) + 2(-27) && \text{First find the power: } (-3)^3 = -27.\\
&= 6 - 54 && \text{Then do the multiplications.}\\
&= -48 && \text{Then do the subtraction.}
\end{aligned}
$$

Self Check   Evaluate $-2a^2 - 3a$ if $a = -4$.

Evaluating Formulas

Table 1-2 shows the formulas used to compute the areas and volumes of many geometric figures.

EXAMPLE 10   Find the volume of the sphere shown in Figure 1-20.

Solution   The formula for the volume of a sphere is $V = \frac{4}{3}\pi r^3$. Since a radius is half as long as a diameter, the radius of the sphere is half of 20 centimeters, or 10 centimeters.

$$V = \frac{4}{3}\pi r^3$$

$$V = \frac{4}{3}\pi(\mathbf{10})^3 \qquad \text{Substitute 10 for } r.$$

$$\approx 4188.790205 \qquad \text{Use a calculator.}$$

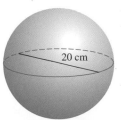

20 cm

Figure 1-20

To two decimal places, the volume is 4,188.79 cm^3.

Self Check   Find the volume of a pyramid with a square base, 20 meters on each side, and a height of 21 meters.

Figure	Name	Area	Figure	Name	Volume
	Square	$A = s^2$		Cube	$V = s^3$
	Rectangle	$A = lw$		Rectangular solid	$V = lwh$
	Circle	$A = \pi r^2$		Sphere	$V = \dfrac{4}{3}\pi r^3$
	Triangle	$A = \dfrac{1}{2}bh$		Cylinder	$V = Bh^*$
	Trapezoid	$A = \dfrac{1}{2}h(b_1 + b_2)$		Cone	$V = \dfrac{1}{3}Bh^*$
				Pyramid	$V = \dfrac{1}{3}Bh^*$

*B represents the area of the base.

Table 1-2

1. a. 81,   **b.** -125,   **c.** $\dfrac{9}{16}a^2$ **2. a.** a^8,   **b.** a^5b^7,   **c.** $4a^6b$ **3. a.** a^{40},   **b.** a^{21},   **c.** a^{15}

4. a. a^8b^{10},   **b.** $\dfrac{a^{15}}{b^{21}}$ **5. a.** $\dfrac{1}{a^4}$,   **b.** a^{15} **6. a.** 1,   **b.** a^6b^9 **7.** $\dfrac{32}{243a^{15}b^{10}}$ **8. a.** t^{n-1},

b. $\dfrac{8t^{3n-9}}{27}$ **9.** -20 **10.** 2,800 m^3

Orals   *Simplify each expression:*

1. 4^2 **2.** 3^3 **3.** x^2x^3 **4.** y^3y^4

5. 17^0 **6.** $(x^2)^3$ **7.** $(a^2b)^3$ **8.** $\left(\dfrac{b}{a^2}\right)^2$

9. 5^{-2} **10.** $(x^{-2})^{-1}$ **11.** $\dfrac{x^5}{x^2}$ **12.** $\dfrac{x^2}{x^5}$

1.3   EXERCISES 📼

REVIEW   *If $a = 4$, $b = -2$, and $c = 5$, find each value.*

1. $a + b + c$

2. $a - 2b - c$

3. $\dfrac{ab + 2c}{a + b}$

4. $\dfrac{ac - bc}{6ab + b}$

VOCABULARY AND CONCEPTS   *Fill in the blanks.*

5. In the exponential expression x^n, x is called the _____, and n is called the _____.

6. A natural-number exponent tells how many times the base is used as a _____.

7. $x^m x^n = $ _____

8. $(x^m)^n = $ _____

9. $(xy)^n = $ _____

10. $\left(\dfrac{x}{y}\right)^n = $ ___ $(y \neq 0)$

11. If $a \neq 0$, then $a^0 = $ __.

12. If $a \neq 0$, then $a^{-1} = $ __.

13. If $x \neq 0$, then $\dfrac{x^m}{x^n} = $ _____.

14. $\left(\dfrac{4}{5}\right)^{-3} = \left(\dfrac{\ }{\ }\right)^3$.

Write the formula to find each quantity.

15. Area of a square: _____

16. Area of a rectangle: _____

17. Area of a triangle: _____

18. Area of a trapezoid: _____

19. Area of a circle: _____

20. Volume of a cube: _____

21. Volume of a rectangular solid: _____

22. Volume of a sphere: _____

23. Volume of a cylinder: _____

24. Volume of a cone: _____

25. Volume of a pyramid: _____

26. In Exercises 23–25, B represents the area of the _____ of a solid.

PRACTICE   *Identify the base and the exponent.*

27. 5^3

28. -7^2

29. $-x^5$

30. $(-t)^4$

31. $2b^6$

32. $(3xy)^5$

33. $(-mn^2)^3$

34. $(-p^2q)^2$

Simplify each expression. Assume that no denominators are zero.

35. 3^2 **36.** 3^4

37. -3^2 **38.** -3^4

39. $(-3)^2$ **40.** $(-3)^3$

41. 5^{-2} **42.** 5^{-4}

43. -5^{-2} **44.** -5^{-4}

45. $(-5)^{-2}$ **46.** $(-5)^{-4}$

47. 8^0 **48.** -9^0

49. $(-8)^0$ **50.** $(-9)^0$

51. $(-2x)^5$ **52.** $(-3a)^3$

53. $(-2x)^6$ **54.** $(-3y)^5$

55. x^2x^3 **56.** y^3y^4

57. k^0k^7 **58.** x^8x^{11}

59. $x^2x^3x^5$ **60.** $y^3y^7y^2$

61. p^9pp^0 **62.** z^7z^0z

63. aba^3b^4 **64.** $x^2y^3x^3y^2$

65. $(-x)^2y^4x^3$ **66.** $-x^2y^7y^3x^{-2}$

67. $(x^4)^7$ **68.** $(y^7)^5$

69. $(b^{-8})^9$

70. $(z^{12})^2$

71. $(x^3y^2)^4$

72. $(x^2y^5)^2$

73. $(r^{-3}s)^3$

74. $(m^5n^2)^{-3}$

75. $(a^2a^3)^4$

76. $(bb^2b^3)^4$

77. $(-d^2)^3(d^{-3})^3$

78. $(c^3)^2(c^4)^{-2}$

79. $(3x^3y^4)^3$

80. $\left(\dfrac{1}{2}a^2b^5\right)^4$

81. $\left(-\dfrac{1}{3}mn^2\right)^6$

82. $(-3p^2q^3)^5$

83. $\left(\dfrac{a^3}{b^2}\right)^5$

84. $\left(\dfrac{a^2}{b^3}\right)^4$

85. $\left(\dfrac{a^{-3}}{b^{-2}}\right)^{-2}$

86. $\left(\dfrac{k^{-3}}{k^{-4}}\right)^{-1}$

87. $\dfrac{a^8}{a^3}$

88. $\dfrac{c^7}{c^2}$

89. $\dfrac{c^{12}c^5}{c^{10}}$

90. $\dfrac{a^{33}}{a^2a^3}$

91. $\dfrac{m^9m^{-2}}{(m^2)^3}$

92. $\dfrac{a^{10}a^{-3}}{a^5a^{-2}}$

93. $\dfrac{1}{a^{-4}}$

94. $\dfrac{3}{b^{-5}}$

95. $\dfrac{3m^5m^{-7}}{m^2m^{-5}}$

96. $\dfrac{(2a^{-2})^3}{a^3a^{-4}}$

97. $\left(\dfrac{4a^{-2}b}{3ab^{-3}}\right)^3$

98. $\left(\dfrac{2ab^{-3}}{3a^{-2}b^2}\right)^2$

99. $\left(\dfrac{3a^{-2}b^2}{17a^2b^3}\right)^0$

100. $\dfrac{a^0+b^0}{2(a+b)^0}$

101. $\left(\dfrac{-2a^4b}{a^{-3}b^2}\right)^{-3}$

102. $\left(\dfrac{-3x^4y^2}{-9x^5y^{-2}}\right)^{-2}$

103. $\left(\dfrac{2a^3b^2}{3a^{-3}b^2}\right)^{-3}$

104. $\left(\dfrac{3x^5y^2}{6x^5y^{-2}}\right)^{-4}$

105. $\dfrac{(3x^2)^{-2}}{x^3x^{-4}x^0}$

106. $\dfrac{y^{-3}y^{-4}y^0}{(2y^{-2})^3}$

107. $\dfrac{a^na^3}{a^4}$

108. $\dfrac{b^9b^7}{b^n}$

109. $\left(\dfrac{b^n}{b^3}\right)^3$

110. $\left(\dfrac{a^2}{a^n}\right)^4$

111. $\dfrac{a^{-n}a^2}{a^3}$

112. $\dfrac{a^na^{-2}}{a^4}$

113. $\dfrac{a^{-n}a^{-2}}{a^{-4}}$

114. $\dfrac{a^n}{a^{-3}a^5}$

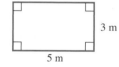 *Use a calculator to find each value.*

115. 1.23^6

116. 0.0537^4

117. -6.25^3

118. $(-25.1)^5$

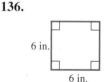 *Use a calculator to verify that each statement is true.*

119. $(3.68)^0 = 1$

120. $(2.1)^4(2.1)^3 = (2.1)^7$

121. $(7.2)^2(2.7)^2 = [(7.2)(2.7)]^2$

122. $(3.7)^2 + (4.8)^2 \neq (3.7 + 4.8)^2$

123. $(3.2)^2(3.2)^{-2} = 1$

124. $[(5.9)^3]^2 = (5.9)^6$

125. $(7.23)^{-3} = \dfrac{1}{(7.23)^3}$

126. $\left(\dfrac{5.4}{2.7}\right)^{-4} = \left(\dfrac{2.7}{5.4}\right)^4$

Evaluate each expression when $x = -2$ and $y = 3$.

127. x^2y^3

128. x^3y^2

129. $\dfrac{x^{-3}}{y^3}$

130. $\dfrac{x^2}{y^{-3}}$

131. $(xy^2)^{-2}$

132. $-y^3x^{-2}$

133. $(-yx^{-1})^3$

134. $(-y)^3x^{-2}$

Find the area of each figure. Round all answers to the nearest unit.

135.

3 m, 5 m

136.

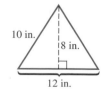

6 in., 6 in.

137.

6 cm

138.

10 in., 8 in., 12 in.

139.

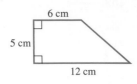

140.

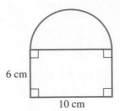

141.

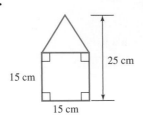

142.

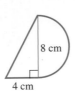

Find the volume of each figure. Round all answers to the nearest unit.

143.

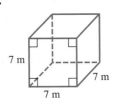

144.

145.

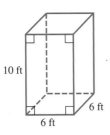

146.

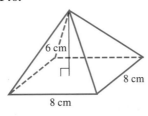

147.

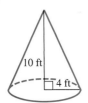

148.

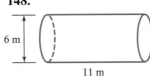

149.

150.

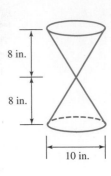

APPLICATIONS *Use a calculator to find each value.*

151. The formula $A = P(1 + i)^n$ gives the amount A in an account when P dollars is the amount originally deposited (the principal), i is the annual interest rate, and n is the number of years. If $5,000 is deposited in an account paying 11% compounded annually, how much will be in the account in 50 years?

152. The formula $P = A(1 + i)^{-n}$ gives the principal P that must be deposited at an annual rate i to grow to A dollars in n years. How much must be invested at 9% annual interest to have $1 million in 50 years?

WRITING

153. Explain why a positive number raised to a negative power is positive.

154. Explain the rules that determine the order in which operations are performed.

155. In the definition of x^{-1}, x cannot be 0. Why not?

156. Explain why $(xyz)^2 = x^2y^2z^2$.

SOMETHING TO THINK ABOUT

157. Find the sum: $2^{-1} + 3^{-1} - 4^{-1}$.

158. Simplify: $(3^{-1} + 4^{-1})^{-2}$.

159. Construct an example using numbers to show that $x^m + x^n \neq x^{m+n}$.

160. Construct an example using numbers to show that $x^m + y^m \neq (x + y)^m$.

1.4 Scientific Notation

- **Scientific Notation**
- **Using Scientific Notation to Simplify Computations**
- **Significant Digits** ▪ **Problem Solving**

Getting Ready *Evaluate each expression:*

1. 10^1 **2.** 10^2 **3.** 10^3 **4.** 10^4

5. 10^{-2} **6.** 10^{-4} **7.** $4(10^3)$ **8.** $7(10^{-4})$

Very large and very small numbers occur often in science. For example, the speed of light is approximately 29,980,000,000 centimeters per second, and the mass of a hydrogen atom is approximately 0.00000000000000000000001673 gram.

Because these numbers contain a large number of zeros, they are hard to read and remember. In this section, we will discuss a notation that will enable us to express these numbers in a more compact form.

Scientific Notation

With exponents, we can write very large and very small numbers in a form called **scientific notation.**

Scientific Notation   A number is written in **scientific notation** when it is written in the form $N \times 10^n$, where $1 \le |N| < 10$ and n is an integer.

EXAMPLE 1   Change 29,980,000,000 to scientific notation.

Solution   The number 2.998 is between 1 and 10. To get 29,980,000,000, the decimal point in 2.998 must be moved ten places to the right. We can do this by multiplying 2.998 by 10^{10}.

$$29{,}980{,}000{,}000 = 2.998 \times 10^{10}$$

Self Check   Change 150,000,000 to scientific notation.   ▪

EXAMPLE 2   Write 0.00000000000000000000001673 in scientific notation.

Solution   The number 1.673 is between 1 and 10. To get 0.00000000000000000000001673, the decimal point in 1.673 must be moved twenty-four places to the left. We can do this by multiplying 1.673 by 10^{-24}.

$$0.00000000000000000000001673 = 1.673 \times 10^{-24}$$

Self Check   Change 0.000025 to scientific notation.   ▪

EXAMPLE 3   Change -0.0013 to scientific notation.

Solution   The absolute value of -1.3 is between 1 and 10. To get -0.0013, we move the decimal point in -1.3 three places to the left by multiplying by 10^{-3}.

$$-0.0013 = -1.3 \times 10^{-3}$$

Self Check   Change $-45{,}700$ to scientific notation.   ∎

We can change a number written in scientific notation to **standard notation.** For example, to write 9.3×10^7 in standard notation, we multiply 9.3×10^7.

$$9.3 \times 10^7 = 9.3 \times 10{,}000{,}000 = 93{,}000{,}000$$

EXAMPLE 4   Change   **a.** 3.7×10^5 and   **b.** -1.1×10^{-3} to standard notation.

Solution   **a.** Since multiplication by 10^5 moves the decimal point 5 places to the right,

$$3.7 \times 10^5 = 370{,}000$$

b. Since multiplication by 10^{-3} moves the decimal point 3 places to the left,

$$-1.1 \times 10^{-3} = -0.0011$$

Self Check   Change   **a.** -9.6×10^4 and   **b.** 5.62×10^{-3} to standard notation.   ∎

Each of the following numbers is written in both scientific and standard notation. In each case, the exponent gives the number of places that the decimal point moves, and the sign of the exponent indicates the direction that it moves:

$5.32 \times 10^4 = 5\,3\,2\,0\,0.$
4 places to the right

$6.45 \times 10^7 = 6\,4\,5\,0\,0\,0\,0\,0.$
7 places to the right

$2.37 \times 10^{-4} = 0.\,0\,0\,0\,2\,3\,7$
4 places to the left

$9.234 \times 10^{-2} = 0.\,0\,9\,2\,3\,4$
2 places to the left

$4.89 \times 10^0 = 4.\,8\,9$

No movement of the decimal point

Comment   Numbers such as 47.2×10^3 and 0.063×10^{-2} appear to be written in scientific notation, because they are the product of a number and a power of 10. However, they are not in scientific notation, because 47.2 and 0.063 are not between 1 and 10.

EXAMPLE 5   Change   **a.** 47.2×10^3 and   **b.** 0.063×10^{-2} to scientific notation.

Solution   Since the first factors are not between 1 and 10, neither number is in scientific notation. However, we can change them to scientific notation as follows:

a. $\mathbf{47.2 \times 10^3} = (\mathbf{4.72 \times 10^1}) \times 10^3$ Write 47.2 in scientific notation.
$= 4.72 \times (10^1 \times 10^3)$
$= 4.72 \times 10^4$

The ancient Egyptians developed two systems of writing. In hieroglyphics, each symbol was a picture of an object. Because hieroglyphic writing was usually inscribed in stone, many examples still survive today. For daily life, Egyptians used hieratic writing. Similar to hieroglyphics, hieratic writing was done with ink on papyrus sheets.

One papyrus that survives, the Rhind Papyrus, was discovered in 1858 by a British archaeologist, Henry Rhind. Also known as the Ahmes Papyrus after its ancient author, it begins with a description of its contents: *Directions for Obtaining the Knowledge of All Dark Things.*

The Ahmes Papyrus and another, the Moscow Papyrus, together contain 110 mathematical problems and their solutions. Many of these were probably for education, because they represented situations that scribes, priests, and other government and temple administration workers were expected to be able to solve.

The Ahmes Papyrus
© Copyright The British Museum

b. $0.063 \times 10^{-2} = (6.3 \times 10^{-2}) \times 10^{-2}$ Write 0.063 in scientific notation.
$$= 6.3 \times (10^{-2} \times 10^{-2})$$
$$= 6.3 \times 10^{-4}$$

Self Check   Change   **a.** 27.3×10^2 and   **b.** 0.0025×10^{-3} to scientific notation. ∎

Using Scientific Notation to Simplify Computations

Scientific notation is useful when simplifying expressions containing very large or very small numbers.

EXAMPLE 6   Use scientific notation to simplify $\dfrac{(0.00000064)(24{,}000{,}000{,}000)}{(400{,}000{,}000)(0.0000000012)}$.

Solution   After changing each number into scientific notation, we can do the arithmetic on the numbers and the exponential expressions separately.

$$\frac{(0.00000064)(24{,}000{,}000{,}000)}{(400{,}000{,}000)(0.0000000012)} = \frac{(6.4 \times 10^{-7})(2.4 \times 10^{10})}{(4 \times 10^8)(1.2 \times 10^{-9})}$$
$$= \frac{(6.4)(2.4)}{(4)(1.2)} \times \frac{10^{-7}10^{10}}{10^8 10^{-9}}$$
$$= 3.2 \times 10^4$$

In standard notation, the result is 32,000.

Self Check   Simplify: $\dfrac{(320)(25{,}000)}{0.00004}$. ∎

USING SCIENTIFIC NOTATION

Scientific and graphing calculators often give answers in scientific notation. For example, if we use a calculator to find 301.2^8, the display will read

> **6.77391496 ¹⁹**   On a scientific calculator.

> **301.2 ^8**
> **6.7739149b1E19**   On a graphing calculator.

In either case, the answer is given in scientific notation and is to be interpreted as

$$6.77391496 \times 10^{19}$$

Numbers can also be entered into a calculator in scientific notation. For example, to enter 24,000,000,000 (which is 2.4×10^{10} in scientific notation), we enter these numbers and press these keys:

2.4 EXP 10 ⎫
               ⎬ Whichever of these keys is on your calculator.
2.4 EE 10 ⎭

To use a calculator to simplify

$$\frac{(24,000,000,000)(0.00000006495)}{0.00000004824}$$

we must enter each number in scientific notation, because each number has too many digits to be entered directly. In scientific notation, the three numbers are

$$2.4 \times 10^{10} \qquad 6.495 \times 10^{-8} \qquad 4.824 \times 10^{-8}$$

To use a scientific calculator to simplify the fraction, we enter these numbers and press these keys:

2.4 EXP 10 × 6.495 EXP 8 +/− ÷ 4.824 EXP 8 +/− =

The display will read **3.231343284 ¹⁰**. In standard notation, the answer is 32,313,432,840.

The steps are similar on a graphing calculator.

Significant Digits

If we measure the length of a rectangle and report the length to be 45 centimeters, we have rounded to the nearest centimeter. If we measure more carefully and find the length to be 45.2 centimeters, we have rounded to the nearest tenth of a centimeter. We say that the second measurement is more accurate than the first, because 45.2 has three *significant digits* but 45 has only two.

It is not always easy to know how many significant digits a number has. For example, 270 might be accurate to two or three significant digits. If 270 is rounded to the nearest ten, the number has two significant digits. If 270 is rounded to the nearest unit, it has three significant digits. This ambiguity does not occur when a number is written in scientific notation.

Finding Significant Digits

If a number M is written in scientific notation as $N \times 10^n$, where $1 \leq |N| < 10$ and n is an integer, the number of significant digits in M is the same as the number of digits in N.

In a problem where measurements are multiplied or divided, the final result should be rounded so that the answer has the same number of significant digits as the least accurate measurement.

Problem Solving

EXAMPLE 7   Earth is approximately 93,000,000 miles from the sun, and Jupiter is approximately 484,000,000 miles from the sun. Assuming the alignment shown in Figure 1-21, how long would it take a spaceship traveling at 7,500 mph to fly from Earth to Jupiter?

Solution   When the planets are aligned as shown in the figure, the distance between Earth and Jupiter is $(484,000,000 - 93,000,000)$ miles or $391,000,000$ miles. To find the length of time in hours for the trip, we divide the distance by the rate.

$$\frac{391,000,000 \text{ mi}}{7,500 \frac{\text{mi}}{\text{hr}}} = \frac{3.91 \times 10^8 \text{ mi}}{7.5 \times 10^3 \frac{\text{mi}}{\text{hr}}}$$

There are three significant digits in the numerator and two in the denominator.

$$\approx 0.5213333 \times 10^5 \text{mi} \cdot \frac{\text{hr}}{\text{mi}}$$

$$\approx 52,133.33 \text{ hr}$$

Since there are 24×365 hours in a year, we can change this result from hours to years by dividing $52,133.33$ by (24×365).

$$\frac{52,133.33 \text{ hr}}{(24 \times 365) \frac{\text{hr}}{\text{yr}}} \approx 5.95129376 \text{ hr} \cdot \frac{\text{yr}}{\text{hr}} \approx 5.95129376 \text{ yr}$$

Rounding to two significant digits, the trip will take about 6.0 years.

Figure 1-21

Self Check   How long would it take if the spaceship could travel at 12,000 mph?

Orals   *Give each numeral in scientific notation.*

1. 352   **2.** 5,130

3. 0.002   **4.** 0.00025

Give each numeral in standard notation.

5. 3.5×10^2 **6.** 4.3×10^3

7. 2.7×10^{-1} **8.** 8.5×10^{-2}

1.4   EXERCISES

REVIEW   *Write each fraction as a terminating or a repeating decimal.*

1. $\dfrac{3}{4}$ **2.** $\dfrac{4}{5}$

3. $\dfrac{13}{9}$ **4.** $\dfrac{14}{11}$

5. A man raises 3 to the second power, 4 to the third power, and 2 to the fourth power and then finds their sum. What number does he obtain?

6. If $a = -2$, $b = -3$, and $c = 4$, evaluate

$$\frac{5ab - 4ac - 2}{3bc + abc}$$

VOCABULARY AND CONCEPTS   *Fill in the blanks.*

7. A number is written in scientific notation when it is written in the form $N \times$ ____, where $1 \le |N| < 10$ and n is an integer.

8. To change 6.31×10^4 to standard notation, we move the decimal point in 6.31 ____ places to the right.

9. To change 6.31×10^{-4} to standard notation, we move the decimal point four places to the ____.

10. The number 6.7×10^3 ___ ($>$ or $<$) the number $6,700,000 \times 10^{-4}$.

PRACTICE   *Write each numeral in scientific notation.*

11. 3,900   **12.** 1,700

13. 0.0078   **14.** 0.068

15. −45,000   **16.** −547,000

17. −0.00021   **18.** −0.00078

19. 17,600,000   **20.** 89,800,000

21. 0.0000096   **22.** 0.000046

23. 323×10^5 **24.** 689×10^9

25. $6,000 \times 10^{-7}$ **26.** 765×10^{-5}

27. 0.0527×10^5 **28.** 0.0298×10^3

29. 0.0317×10^{-2} **30.** 0.0012×10^{-3}

Write each numeral in standard notation.

31. 2.7×10^2 **32.** 7.2×10^3

33. 3.23×10^{-3} **34.** 6.48×10^{-2}

35. 7.96×10^5 **36.** 9.67×10^6

37. 3.7×10^{-4} **38.** 4.12×10^{-5}

39. 5.23×10^0 **40.** 8.67×10^0

41. 23.65×10^6 **42.** 75.6×10^{-5}

Write each numeral in scientific notation and perform the operations. Give all answers in scientific notation.

43. $\dfrac{(4,000)(30,000)}{0.0006}$ **44.** $\dfrac{(0.0006)(0.00007)}{21,000}$

45. $\dfrac{(640,000)(2,700,000)}{120,000}$

46. $\dfrac{(0.0000013)(0.000090)}{0.00039}$

Write each numeral in scientific notation and perform the operations. Give all answers in standard notation.

47. $\dfrac{(0.006)(0.008)}{0.0012}$

48. $\dfrac{(600)(80,000)}{120,000}$

49. $\dfrac{(220,000)(0.000009)}{0.00033}$

50. $\dfrac{(0.00024)(96,000,000)}{640,000,000}$

51. $\dfrac{(320,000)^2(0.0009)}{12,000^2}$

52. $\dfrac{(0.000012)^2(49,000)^2}{0.021}$

Use a scientific calculator to evaluate each expression. Round each answer to the appropriate number of significant digits.

53. $23,437^3$

54. 0.00034^4

55. $(63,480)(893,322)$

56. $(0.0000413)(0.0000049)^2$

57. $\dfrac{(69.4)^8(73.1)^2}{(0.0043)^3}$

58. $\dfrac{(0.0031)^4(0.0012)^5}{(0.0456)^{-7}}$

APPLICATIONS *Use scientific notation to find each answer. Round all answers to the proper number of significant digits.*

59. Wavelengths Transmitters, vacuum tubes, and lights emit energy that can be modeled as a wave. List the wavelengths shown in the table in order, from shortest to longest.

Type	Use	Wavelength (m)
visible light	lighting	9.3×10^{-6}
infrared	photography	3.7×10^{-5}
x-ray	medical	2.3×10^{-11}
radio wave	communication	3.0×10^{2}
gamma ray	treating cancer	8.9×10^{-14}

60. Distance to the sun The sun is about 93 million miles from Earth. Find this distance in feet. (1 mi = 5,280 ft.)

61. Speed of sound The speed of sound in air is 3.31×10^4 centimeters per second. Find the speed of sound in centimeters per hour.

62. Volume of a tank Find the volume of the tank shown in the illustration.

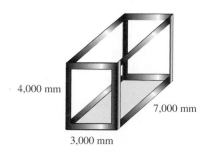

4,000 mm

7,000 mm

3,000 mm

63. Mass of protons If the mass of 1 Proton is 0.00000000000000000000000167248 gram, find the mass of 1 million protons.

64. Speed of light The speed of light in a vacuum is about 30,000,000,000 centimeters per second. Find the speed of light in mph. (*Hint:* 160,000 cm ≈ 1 mile.)

65. Distance to the moon The moon is about 235,000 miles from Earth. Find this distance in inches.

66. Distance to the sun The sun is about 149,700,000 kilometers from Earth. Find this distance in miles. (*Hint:* 1 km ≈ 0.6214 mile.)

67. Solar flares Solar flares often produce immense loops of glowing gas ejected from the sun's surface. The flare in the illustration extends about 95,000 kilometers into space. Express this distance in miles. (*Hint:* 1 km ≈ 0.6214 mile.)

© Stocktrek/CORBIS

68. Distance to the moon The moon is about 378,196 kilometers from Earth. Express this distance in inches. (*Hint:* 1 km ≈ 0.6214 mile.)

69. Angstroms per inch One **angstrom** is 0.0000001 millimeter, and one inch is 25.4 millimeters. Find the number of angstroms in one inch.

70. Range of comet One **astronomical unit** (AU) is the distance from Earth to the sun—about 9.3×10^7 miles. Halley's comet ranges from 0.6 to 18 AU from the sun. Express this range in miles.

71. Flight to Pluto The planet Pluto is approximately 3,574,000,000 miles from Earth. If a spaceship can travel 18,000 mph, how long will it take to reach Pluto?

72. Light year Light travels about 300,000,000 meters per second. A **light year** is the distance that light can travel in one year. How many meters are in one light year?

73. Distance to Alpha Centauri Light travels about 186,000 miles per second. A **parsec** is 3.26 light years. The star Alpha Centauri is 1.3 parsecs from Earth. Express this distance in miles.

74. Life of a comet The mass of the comet shown in the illustration is about 10^{16} grams. When the comet is close to the sun, matter evaporates at the rate of 10^7 grams per second. Calculate the life of the comet if it appears every 50 years and spends ten days close to the sun.

© Stocktrek/CORBIS

WRITING

75. Explain how to change a number from standard notation to scientific notation.

76. Explain how to change a number from scientific notation to standard notation.

SOMETHING TO THINK ABOUT

77. Find the highest power of 2 that can be evaluated with a scientific calculator.

78. Find the highest power of 7 that can be evaluated with a scientific calculator.

1.5 Solving Equations

▐ **Equations** ▐ **Properties of Equality** ▐ **Solving Linear Equations**
▐ **Combining Like Terms** ▐ **Identities and Contradictions** ▐ **Formulas**

Getting Ready *Fill in the blanks.*

1. $+ 3 = 5$ **2.** $8 - \blacksquare = 4$ **3.** $\dfrac{12}{\blacksquare} = 4$ **4.** $\blacksquare \cdot 5 = 30$

In this section, we show how to solve equations, one of the most important concepts in algebra. Then, we will apply these equation-solving techniques to solve formulas for various variables.

Equations

An **equation** is a statement indicating that two quantities are equal. The equation $2 + 4 = 6$ is true, and the equation $2 + 4 = 7$ is false. If an equation has a variable (say, x) it can be either true or false, depending on the value of x. For example, if $x = 1$, the equation $7x - 3 = 4$ is true.

$$7(\mathbf{1}) - 3 = 4 \quad \text{Substitute 1 for } x.$$
$$7 - 3 = 4$$
$$4 = 4$$

However, the equation is false for all other values of x. Since 1 makes the equation true, we say that 1 *satisfies* the equation.

The set of numbers that satisfies an equation is called its **solution set.** The elements of the solution set are called **solutions** or **roots** of the equation. Finding the solution set of an equation is called *solving the equation.*

EXAMPLE 1 Determine whether 3 is a solution of $2x + 4 = 10$.

Solution We substitute 3 for x and see whether it satisfies the equation.

$$2x + 4 = 10$$
$$2(\mathbf{3}) + 4 \stackrel{?}{=} 10 \quad \text{Substitute 3 for } x.$$
$$6 + 4 \stackrel{?}{=} 10 \quad \text{First do the multiplication on the left-hand side.}$$
$$10 = 10 \quad \text{Then do the addition.}$$

Since $10 = 10$, the number 3 satisfies the equation. It is a solution.

Self Check Is -5 a solution of $2x - 3 = -13$?   ■

Properties of Equality

To solve an equation we replace the equation with simpler ones, all having the same solution set. Such equations are called **equivalent equations.**

Equivalent Equations | Equations with the same solution set are called **equivalent equations.**

We continue to replace each resulting equation with an equivalent one until we have isolated the variable on one side of an equation. To isolate the variable, we can use the following properties:

Properties of Equality

If a, b, and c are real numbers and $a = b$, then

$$a + c = b + c \quad \text{and} \quad a - c = b - c$$

and if $c \neq 0$, then

$$ac = bc \quad \text{and} \quad \frac{a}{c} = \frac{b}{c}$$

In words, we can say: *If any quantity is added to (or subtracted from) both sides of an equation, a new equation is formed that is equivalent to the original equation.*

If both sides of an equation are multiplied (or divided) by the same nonzero quantity, a new equation is formed that is equivalent to the original equation.

Solving Linear Equations

The most basic equations that we will solve are **linear equations.**

Linear Equations

A linear equation in one variable (say, x) is any equation that can be written in the form.

$ax + c = 0$ (a and c are real numbers and $a \neq 0$)

EXAMPLE 2   Solve: $2x + 8 = 0$.

Solution   To solve the equation, we will isolate x on the left-hand side.

$$2x + 8 = 0$$
$$2x + 8 - \mathbf{8} = 0 - \mathbf{8} \quad \text{To eliminate 8 from the left-hand side, subtract 8 from both sides.}$$
$$2x = -8 \quad \text{Simplify.}$$
$$\frac{2x}{\mathbf{2}} = \frac{-8}{\mathbf{2}} \quad \text{To eliminate 2 from the left-hand side, divide both sides by 2.}$$
$$x = -4 \quad \text{Simplify.}$$

Check: We substitute -4 for x to verify that it satisfies the original equation.

$$2x + 8 = 0$$
$$2(-4) + 8 \stackrel{?}{=} 0 \quad \text{Substitute } -4 \text{ for } x.$$
$$-8 + 8 \stackrel{?}{=} 0$$
$$0 = 0$$

Since -4 satisfies the original equation, it is the solution. The solution set is $\{-4\}$.

Self Check   Solve $-3a + 15 = 0$ and give the solution set.   ∎

EXAMPLE 3   Solve: $3(x - 2) = 20$.

Solution   We isolate x on the left-hand side.

$$3(x - 2) = 20$$
$$3x - 6 = 20 \quad \text{Use the distributive property to remove parentheses.}$$
$$3x - 6 + \mathbf{6} = 20 + \mathbf{6} \quad \text{To eliminate } -6 \text{ from the left-hand side, add 6 to both sides.}$$
$$3x = 26 \quad \text{Simplify.}$$
$$\frac{3x}{\mathbf{3}} = \frac{26}{\mathbf{3}} \quad \text{To eliminate 3 from the left-hand side, divide both sides by 3.}$$
$$x = \frac{26}{3} \quad \text{Simplify.}$$

Check: $3(x - 2) = 20$

$$3\left(\frac{26}{3} - 2\right) \overset{?}{=} 20 \qquad \text{Substitute } \tfrac{26}{3} \text{ for } x.$$

$$3\left(\frac{26}{3} - \frac{6}{3}\right) \overset{?}{=} 20 \qquad \text{Get a common denominator: } 2 = \tfrac{6}{3}.$$

$$3\left(\frac{20}{3}\right) \overset{?}{=} 20 \qquad \text{Combine the fractions: } \tfrac{26}{3} - \tfrac{6}{3} = \tfrac{20}{3}.$$

$$20 = 20 \qquad \text{Simplify.}$$

Since $\frac{26}{3}$ satisfies the equation, it is the solution. The solution set is $\left\{\frac{26}{3}\right\}$.

Self Check   Solve $-2(a + 3) = 18$ and give the solution set.   ∎

Combining Like Terms

To solve many equations, we will need to combine like terms. An **algebraic term** is either a number or the product of numbers (called **constants**) and variables. Some examples of terms are $3x$, $-7y$, y^2, and 8. The **numerical coefficients** of these terms are 3, -7, 1, and 8 (8 can be written as $8x^0$), respectively.

In algebraic expressions, terms are separated by $+$ and $-$ signs. For example, the expression $3x^2 + 2x - 4$ has three terms, and the expression $3x + 7y$ has two terms.

Terms with the same variables with the same exponents are called **like terms** or **similar terms:**

$5x$ and $6x$ are like terms.

$27x^2y^3$ and $-326x^2y^3$ are like terms.

$4x$ and $-17y$ are unlike terms.   Because they have different variables.

$15x^2y$ and $6xy^2$ are unlike terms.   Because the variables have different exponents.

By using the distributive law, we can combine like terms. For example,

$$\mathbf{5}x + \mathbf{6}x = (\mathbf{5 + 6})x = 11x \qquad \text{and} \qquad \mathbf{32}y - \mathbf{16}y = (\mathbf{32 - 16})y = 16y$$

This suggests that to combine like terms, *we add or subtract their numerical coefficients and keep the same variables with the same exponents.*

EXAMPLE 4   Solve: $3(2x - 1) = 2x + 9$.

Solution

$$3(2x - 1) = 2x + 9$$

$6x - 3 = 2x + 9$ Use the distributive property to remove parentheses.

$6x - 3 + 3 = 2x + 9 + 3$ To eliminate -3 from the left-hand side, add 3 to both sides.

$6x = 2x + 12$ Combine like terms.

$6x - 2x = 2x - 2x + 12$ To eliminate $2x$ from the right-hand side, subtract $2x$ from both sides.

$4x = 12$ Combine like terms.

$x = 3$ To eliminate 4 from the left-hand side, divide both sides by 4.

Check: $3(2x - 1) = 2x + 9$

$$3(2 \cdot 3 - 1) \stackrel{?}{=} 2 \cdot 3 + 9 \qquad \text{Substitute 3 for } x.$$

$$3(5) \stackrel{?}{=} 6 + 9$$

$$15 = 15$$

Since 3 satisfies the equation, it is the solution. The solution set is {3}.

Self Check   Solve $-4(3a + 4) = 2a - 4$ and give the solution set.   ∎

To solve more complicated linear equations, we will follow these steps.

Solving Linear Equations

1. If the equation contains fractions, multiply both sides of the equation by a number that will eliminate the denominators.
2. Use the distributive property to remove all sets of parentheses and combine like terms.
3. Use the addition and subtraction properties to get all variables on one side of the equation and all numbers on the other side. Combine like terms, if necessary.
4. Use the multiplication and division properties to make the coefficient of the variable equal to 1.
5. Check the result by replacing the variable with the possible solution and verifying that the number satisfies the equation.

EXAMPLE 5   Solve: $\dfrac{5}{3}(x - 3) = \dfrac{3}{2}(x - 2) + 2$.

Solution   **Step 1:**   Since 6 is the smallest number that can be divided by both 2 and 3, we multiply both sides of the equation by 6 to eliminate the fractions:

$$\frac{5}{3}(x - 3) = \frac{3}{2}(x - 2) + 2$$

$$6\left[\frac{5}{3}(x - 3)\right] = 6\left[\frac{3}{2}(x - 2) + 2\right] \qquad \begin{array}{l}\text{To eliminate the fractions, multiply both} \\ \text{sides by 6.}\end{array}$$

$$6 \cdot \frac{5}{3}(x - 3) = 6 \cdot \frac{3}{2}(x - 2) + 6 \cdot 2 \qquad \begin{array}{l}\text{Use the distributive property on the right-} \\ \text{hand side.}\end{array}$$

$$10(x - 3) = 9(x - 2) + 12 \qquad \text{Simplify.}$$

Step 2:   We use the distributive property to remove parentheses and then combine like terms.

$$10x - 30 = 9x - 18 + 12$$

$$10x - 30 = 9x - 6$$

Step 3:   We use the addition and subtraction properties by adding 30 to both sides and subtracting $9x$ from both sides.

$$10x - 30 - 9x + 30 = 9x - 6 - 9x + 30$$

$$x = 24 \qquad \text{Combine like terms.}$$

Since the coefficient of x in the above equation is 1, Step 4 is unnecessary.

Step 5: We check by substituting 24 for x in the original equation and simplifying:

$$\frac{5}{3}(x - 3) = \frac{3}{2}(x - 2) + 2$$

$$\frac{5}{3}(\mathbf{24} - 3) \stackrel{?}{=} \frac{3}{2}(\mathbf{24} - 2) + 2$$

$$\frac{5}{3}(21) \stackrel{?}{=} \frac{3}{2}(22) + 2$$

$$5(7) \stackrel{?}{=} 3(11) + 2$$

$$35 = 35$$

Since 24 satisfies the equation, it is the solution. The solution set is $\{24\}$. ∎

EXAMPLE 6 Solve: $\dfrac{x + 2}{5} - 4x = \dfrac{8}{5} - \dfrac{x + 9}{2}$.

Solution

$$\frac{x + 2}{5} - 4x = \frac{8}{5} - \frac{x + 9}{2}$$

$$\mathbf{10}\left(\frac{x + 2}{5} - 4x\right) = \mathbf{10}\left(\frac{8}{5} - \frac{x + 9}{2}\right) \quad \text{To eliminate the fractions, multiply both sides by 10.}$$

$$2(x + 2) - 40x = 2(8) - 5(x + 9) \quad \text{Remove parentheses.}$$

$$2x + 4 - 40x = 16 - 5x - 45 \quad \text{Remove parentheses.}$$

$$-38x + 4 = -5x - 29 \quad \text{Combine like terms.}$$

$$-33x = -33 \quad \text{Add } 5x \text{ and } -4 \text{ to both sides.}$$

$$\frac{-33x}{-33} = \frac{-33}{-33} \quad \text{Divide both sides by } -33.$$

$$x = 1 \quad \text{Simplify.}$$

Check: $\dfrac{x + 2}{5} - 4x = \dfrac{8}{5} - \dfrac{x + 9}{2}$

$$\frac{\mathbf{1} + 2}{5} - 4(\mathbf{1}) \stackrel{?}{=} \frac{8}{5} - \frac{\mathbf{1} + 9}{2} \quad \text{Substitute 1 for } x.$$

$$\frac{3}{5} - 4 \stackrel{?}{=} \frac{8}{5} - 5$$

$$\frac{3}{5} - \frac{20}{5} \stackrel{?}{=} \frac{8}{5} - \frac{25}{5}$$

$$-\frac{17}{5} = -\frac{17}{5}$$

Since 1 satisfies the equation, it is the solution. The solution set is $\{1\}$.

Self Check Solve $\dfrac{a + 5}{5} + 2a = \dfrac{a}{2} - \dfrac{a + 14}{5}$ and give the solution set. ∎

Identities and Contradictions

The equations discussed so far are called **conditional equations.** For these equations, some numbers x satisfy the equation and others do not. An **identity** is an equation that is satisfied by every number x for which both sides of the equation are defined.

EXAMPLE 7   Solve: $2(x - 1) + 4 = 4(1 + x) - (2x + 2)$.

Solution

$$2(x - 1) + 4 = 4(1 + x) - (2x + 2)$$
$$2x - 2 + 4 = 4 + 4x - 2x - 2 \quad \text{Use the distributive property to remove parentheses.}$$
$$2x + 2 = 2x + 2 \quad \text{Combine like terms.}$$

The result $2x + 2 = 2x + 2$ is true for every value of x. Since every number x satisfies the equation, it is an identity.

Self Check   Solve: $3(a + 4) + 5 = 2(a - 1) + a + 19$.   ∎

A **contradiction** is an equation that has no solution.

EXAMPLE 8   Solve: $\dfrac{x - 1}{3} + 4x = \dfrac{3}{2} + \dfrac{13x - 2}{3}$.

Solution

$$\frac{x - 1}{3} + 4x = \frac{3}{2} + \frac{13x - 2}{3}$$
$$6\left(\frac{x - 1}{3} + 4x\right) = 6\left(\frac{3}{2} + \frac{13x - 2}{3}\right) \quad \text{To eliminate the fractions, multiply both sides by 6.}$$
$$2(x - 1) + 6(4x) = 3(3) + 2(13x - 2) \quad \text{Use the distributive property to remove parentheses.}$$
$$2x - 2 + 24x = 9 + 26x - 4 \quad \text{Remove parentheses.}$$
$$26x - 2 = 26x + 5 \quad \text{Combine like terms.}$$
$$-2 = 5 \quad \text{Subtract } 26x \text{ from both sides.}$$

Since $-2 = 5$ is false, no number x can satisfy the equation. The solution set is the **empty set,** denoted as $\varnothing$.

Self Check   Solve: $\dfrac{x + 5}{5} = \dfrac{1}{5} + \dfrac{x}{5}$.   ∎

Formulas

Suppose we want to find the heights of several triangles whose areas and bases are known. It would be tedious to substitute values of A and b into the formula $A = \frac{1}{2}bh$ and then repeatedly solve the formula for h. It is easier to solve for h first and then substitute values for A and b and compute h directly.

To solve a formula for a variable means to isolate that variable on one side of the equation and isolate all other quantities on the other side.

EXAMPLE 9   Solve $A = \frac{1}{2}bh$ for h.

Solution

$$A = \frac{1}{2}bh$$

$2A = bh$ To eliminate the fraction, multiply both sides by 2.

$\dfrac{2A}{b} = h$ To isolate h, divide both sides by b.

$h = \dfrac{2A}{b}$ Write h on the left-hand side.

Self Check   Solve $A = \frac{1}{2}bh$ for b.   ∎

EXAMPLE 10   For simple interest, the formula $A = p + prt$ gives the amount of money in an account at the end of a specific time. A represents the amount, p the principal, r the rate of interest, and t the time. We can solve the formula for t as follows:

Solution

$$A = p + prt$$

$A - p = prt$ To isolate the term involving t, subtract p from both sides.

$\dfrac{A - p}{pr} = t$ To isolate t, divide both sides by pr.

$t = \dfrac{A - p}{pr}$ Write t on the left-hand side.

Self Check   Solve $A = p + prt$ for r.   ∎

EXAMPLE 11   The formula $F = \frac{9}{5}C + 32$ converts degrees Celsius to degrees Fahrenheit. Solve the formula for C.

Solution

$$F = \frac{9}{5}C + 32$$

$F - 32 = \dfrac{9}{5}C$ To isolate the term involving C, subtract 32 from both sides.

$\dfrac{5}{9}(F - 32) = \dfrac{5}{9}\left(\dfrac{9}{5}C\right)$ To isolate C, multiply both sides by $\frac{5}{9}$.

$\dfrac{5}{9}(F - 32) = C$ $\dfrac{5}{9} \cdot \dfrac{9}{5} = 1$.

$C = \dfrac{5}{9}(F - 32)$

To convert degrees Fahrenheit to degrees Celsius, we can use the formula $C = \frac{5}{9}(F - 32)$.

Self Check   Solve $S = \dfrac{180(n - 2)}{5}$ for n.   ∎

1. yes   **2.** {5}   **3.** {−12}   **4.** $\left\{-\dfrac{6}{7}\right\}$ **6.** {−2}   **7.** all real numbers   **8.** ∅   **9.** $b = \dfrac{2A}{h}$

10. $r = \dfrac{A - p}{pt}$ **11.** $n = \dfrac{5S}{180} + 2$ or $n = \dfrac{5S + 360}{180}$

Orals   *Combine like terms.*

1. $5x + 4x$ 　　　　　　　　　　　　　**2.** $7s^2 - 5s^2$

Tell whether each number is a solution of $2x + 5 = 13$.

3. 3 　　　　　**4.** 4 　　　　　**5.** 5 　　　　　**6.** 6

Solve each equation.

7. $3x - 2 = 7$ 　　**8.** $\dfrac{1}{2}x - 1 = 5$ 　　**9.** $\dfrac{x - 2}{3} = 1$ 　　**10.** $\dfrac{x + 3}{2} = 3$

1.5   EXERCISES

REVIEW   *Simplify each expression.*

1. $(-4)^3$ 　　　　　　**2.** -3^3

3. $\left(\dfrac{x + y}{x - y}\right)^0$ 　　　**4.** $(x^2x^3)^4$

5. $\left(\dfrac{x^2x^5}{x^3}\right)^2$ 　　　**6.** $\left(\dfrac{x^4y^3}{x^5y}\right)^3$

7. $(2x)^{-3}$ 　　　　**8.** $\left(\dfrac{x^2}{y^5}\right)^{-4}$

VOCABULARY AND CONCEPTS   *Fill in the blanks.*

9. An _____ is a statement that two quantities are equal.

10. If a number is substituted for a variable in an equation and the equation is true, we say that the number _____ the equation.

11. If two equations have the same solution set, they are called _____ equations.

12. If a, b, and c are real numbers and $a = b$, then $a + c = b + $ __ and $a - c = b - $ __.

13. If a, b, and c are real numbers and $a = b$, then $a \cdot c = b \cdot$ __ and $\dfrac{a}{c} = $ __ $(c \neq 0)$.

14. A number or the product of numbers and variables is called an algebraic _____.

15. _____ terms are terms with the same variables and with the same exponents.

16. To combine like terms, add their _____ and keep the same _____ and exponents.

17. An _____ is an equation that is true for all values of its variable.

18. A contradiction is true for __ values of its variable.

PRACTICE   *Tell whether 5 is a solution of each equation.*

19. $3x + 2 = 17$ 　　　　**20.** $7x - 2 = 33$

21. $\dfrac{3}{5}x - 5 = -2$ 　　**22.** $\dfrac{2}{5}x + 12 = 8$

Solve each equation.

23. $x + 6 = 8$ 　　　　**24.** $y - 7 = 3$
25. $a - 5 = 20$ 　　　**26.** $b + 4 = 18$
27. $2u = 6$ 　　　　　**28.** $3v = 12$
29. $\dfrac{x}{4} = 7$ 　　　　　**30.** $\dfrac{x}{6} = 8$
31. $3x + 1 = 3$ 　　　**32.** $8x - 2 = 13$
33. $2x + 1 = 13$ 　　　**34.** $2x - 4 = 16$
35. $3(x - 4) = -36$ 　**36.** $4(x + 6) = 84$
37. $3(r - 4) = -4$ 　　**38.** $4(s - 5) = -3$

Tell whether the terms are like terms. If they are, combine them.

39. $2x, 6x$ 　　　　　**40.** $-3x, 5y$
41. $-5xy, -7yz$ 　　　**42.** $-3t^2, 12t^2$

43. $3x^2, -5x^2$ **44.** $5y^2, 7xy$

45. $xy, 3xt$ **46.** $-4x, -5x$

Solve each equation. If the equation is an identity or a contradiction, so indicate.

47. $3a - 22 = -2a - 7$ **48.** $a + 18 = 6a - 3$

49. $2(2x + 1) = 15 + 3x$ **50.** $-2(x + 5) = 30 - x$

51. $3(y - 4) - 6 = y$ **52.** $2x + (2x - 3) = 5$

53. $5(5 - a) = 37 - 2a$ **54.** $4a + 17 = 7(a + 2)$

55. $4(y + 1) = -2(4 - y)$ **56.** $5(r + 4) = -2(r - 3)$

57. $2(a - 5) - (3a + 1) = 0$

58. $8(3a - 5) - 4(2a + 3) = 12$

59. $3(y - 5) + 10 = 2(y + 4)$

60. $2(5x + 2) = 3(3x - 2)$

61. $9(x + 2) = -6(4 - x) + 18$

62. $3(x + 2) - 2 = -(5 + x) + x$

63. $-4p - 2(3p + 5) = -6p + 2(p + 2)$

64. $2q - 3(q - 5) = 5(q + 2) - 7$

65. $4 + 4(n + 2) = 3n - 2(n - 5)$

66. $4x - 2(3x + 2) = 2(x + 3)$

67. $\frac{1}{2}x - 4 = -1 + 2x$ **68.** $2x + 3 = \frac{2}{3}x - 1$

69. $\frac{x}{2} - \frac{x}{3} = 4$ **70.** $\frac{x}{2} + \frac{x}{3} = 10$

71. $\frac{x}{6} + 1 = \frac{x}{3}$ **72.** $\frac{3}{2}(y + 4) = \frac{20 - y}{2}$

73. $5 - \frac{x + 2}{3} = 7 - x$

74. $3x - \frac{2(x + 3)}{3} = 16 - \frac{x + 2}{2}$

75. $\frac{4x - 2}{2} = \frac{3x + 6}{3}$ **76.** $\frac{t + 4}{2} = \frac{2t - 3}{3}$

77. $\frac{a + 1}{3} + \frac{a - 1}{5} = \frac{2}{15}$

78. $\frac{2z + 3}{3} + \frac{3z - 4}{6} = \frac{z - 2}{2}$

79. $\frac{5a}{2} - 12 = \frac{a}{3} + 1$ **80.** $\frac{5a}{6} - \frac{5}{2} = -\frac{1}{2} - \frac{a}{6}$

81. $4(2 - 3t) + 6t = -6t + 8$

82. $2x - 6 = -2x + 4(x - 2)$

83. $\frac{a + 1}{4} + \frac{2a - 3}{4} = \frac{a}{2} - 2$

84. $\frac{y - 8}{5} + 2 = \frac{2}{5} - \frac{y}{3}$

85. $3(x - 4) + 6 = -2(x + 4) + 5x$

86. $2(x - 3) = \frac{3}{2}(x - 4) + \frac{x}{2}$

87. $y(y + 2) + 1 = y^2 + 2y + 1$

88. $x(x - 3) = x^2 - 2x + 1 - (5 + x)$

Solve each formula for the indicated variable.

89. $A = lw$ for w **90.** $p = 4s$ for s

91. $V = \frac{1}{3}Bh$ for B **92.** $b = \frac{2A}{h}$ for A

93. $I = prt$ for t **94.** $I = prt$ for r

95. $p = 2l + 2w$ for w **96.** $p = 2l + 2w$ for l

97. $A = \frac{1}{2}h(B + b)$ for B **98.** $A = \frac{1}{2}h(B + b)$ for b

99. $y = mx + b$ for x **100.** $y = mx + b$ for m

101. $l = a + (n - 1)d$ for n

102. $l = a + (n - 1)d$ for d

103. $S = \frac{a - lr}{1 - r}$ for l

104. $C = \frac{5}{9}(F - 32)$ for F

105. $S = \frac{n(a + l)}{2}$ for l **106.** $S = \frac{n(a + l)}{2}$ for n

APPLICATIONS

107. Force of gravity The masses of the two objects in the illustration on the next page are m and M. The force of gravitation F between the masses is

$$F = \frac{GmM}{d^2}$$

where G is a constant and d is the distance between them. Solve for m.

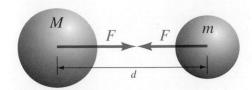

108. Thermodynamics The Gibbs free-energy formula is $G = U - TS + pV$. Solve for S.

109. Converting temperatures Solve the formula $F = \frac{9}{5}C + 32$ for C and find the Celsius temperatures that correspond to Fahrenheit temperatures of $32°$, $70°$, and $212°$.

110. Doubling money A man intends to invest \$1,000 at simple interest. Solve the formula $A = p + prt$ for t and find how long it will take to double his money at the rates of 5%, 7%, and 10%.

111. Cost of electricity The cost of electricity in a certain city is given by the formula $C = 0.07n + 6.50$, where C is the cost and n is the number of kilowatt hours used. Solve for n and find the number of kwh used for costs of \$49.97, \$76.50, and \$125.

112. Cost of water A monthly water bill in a certain city is calculated by using the formula $n = \frac{5,000C - 17,500}{6}$, where n is the number of gallons used and C is the monthly cost. Solve for C and compute the bill for quantities used of 500, 1,200, and 2,500 gallons.

113. Ohm's law The formula $E = IR$, called **Ohm's Law,** is used in electronics. Solve for R and then calculate the resistance R if the voltage E is 56 volts and the current I is 7 amperes. (Resistance has units of **ohms.**)

114. Earning interest An amount P, invested at a simple interest rate r, will grow to an amount A in t years, according to the formula $A = P(1 + rt)$. Solve for P. Suppose a man invested some money at 5.5%. If after 5 years, he had \$6,693.75 on deposit, what amount did he originally invest?

115. Angles of a polygon A regular polygon has n equal sides and n equal angles. The measure a of an interior angle is given by $a = 180°\left(1 - \frac{2}{n}\right)$. Solve for n. Find the number of sides of the regular polygon in the illustration if an interior angle is $135°$.

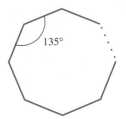

116. Power loss The illustration is the schematic diagram of a resistor connected to a voltage source of 60 volts. As a result, the resistor dissipates power in the form of heat. The power P lost when a voltage V is placed across a resistance R is given by the formula

$$P = \frac{E^2}{R}$$

Solve for R. If P is 4.8 watts and E is 60 volts, find R.

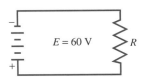

WRITING

117. Explain the difference between a conditional equation, an identity, and a contradiction.

118. Explain how you would solve an equation.

SOMETHING TO THINK ABOUT *Find the mistake.*

119. $3(x - 2) + 4 = 14$

$\quad 3x - 2 + 4 = 14$

$\quad\quad 3x + 2 = 14$

$\quad\quad\quad 3x = 12$

$\quad\quad\quad\quad x = 4$

120. $\quad\quad A = p + prt$

$\quad\quad A - p = prt$

$\quad A - p - pt = t$

$\quad\quad\quad t = A - p - pt$

1.6   Using Equations to Solve Problems

▌ **Recreation Problems**   ▌ **Business Problems**
▌ **Geometry Problems**   ▌ **Lever Problems**

Getting Ready   *Let x* = 18.

1. What number is 2 more than x?

2. What number is 4 times x?

3. What number is 5 more than twice x?

4. What number is 2 less than one-half of x?

In this section, we will apply our equation-solving skills to solve problems. When we translate the words of a problem into mathematics, we are creating a *mathematical model* of the problem. To create these models, we can use Table 1-3 to translate certain words into mathematical operations.

Addition (+)	Subtraction (−)	Multiplication (·)	Division (÷)
added to	subtracted from	multiplied by	divided by
plus	difference	product	quotient
the sum of	less than	times	ratio
more than	less	of	half
increased by	decreased by	twice	

Table 1-3

We can then change English phrases into algebraic expressions, as in Table 1-4.

English phrase	Algebraic expression
2 added to some number	$x + 2$
the difference between two numbers	$x - y$
5 times some number	$5x$
the product of 925 and some number	$925x$
5% of some number	$0.05x$
the sum of twice a number and 10	$2x + 10$
the quotient (or ratio) of two numbers	$\dfrac{x}{y}$
half of a number	$\dfrac{x}{2}$

Table 1-4

Once we know how to change phrases into algebraic expressions, we can solve many problems. The following list of steps provides a strategy for solving problems.

Problem Solving

1. *Analyze the problem* by reading it carefully to understand the given facts. What information is given? What vocabulary is given? What are you asked to find? Often a diagram will help you visualize the facts of the problem.

2. *Form an equation* by picking a variable to represent the quantity to be found. Then express all other quantities mentioned as expressions involving that variable. Finally, write an equation expressing a quantity in two different ways.

3. *Solve the equation.*

4. *State the conclusion.*

5. *Check the result* in the words of the problem.

Recreation Problems

EXAMPLE 1   **Cutting a rope**   A mountain climber wants to cut a rope 213 feet long into three pieces. If each piece is to be 2 feet longer than the previous one, where should he make the cuts?

Analyze the problem   If x represents the length of the shortest piece, the climber wants the lengths of the three pieces to be

$$x, \qquad x + 2, \qquad \text{and} \qquad x + 4$$

feet long. He knows that the sum of these three lengths can be expressed in two ways: as $x + (x + 2) + (x + 4)$ and as 213.

Form an equation   Let x represent the length of the first piece of rope. Then $x + 2$ represents the length of the second piece, and $x + 4$ represents the length of the third piece. (See Figure 1-22.)

213 ft

x ft $(x + 2)$ ft $(x + 4)$ ft

Figure 1-22

From Figure 1-22, we see that the sum of the individual pieces must equal the total length of the rope.

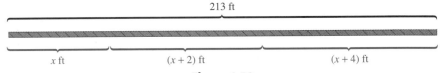

The length of the first piece	plus	the length of the second piece	plus	the length of the third piece	equals	the total length of the rope.
x	$+$	$x + 2$	$+$	$x + 4$	$=$	213

Solve the equation   We now solve the equation.

$$x + x + 2 + x + 4 = 213$$
$$3x + 6 = 213 \qquad \text{Combine like terms.}$$
$$3x = 207 \qquad \text{Subtract 6 from both sides.}$$
$$x = 69 \qquad \text{Divide both sides by 3.}$$
$$x + 2 = 71$$
$$x + 4 = 73$$

State the conclusion   He should make cuts 69 feet from one end and 73 feet from the other end, to get lengths of 69 feet, 71 feet, and 73 feet.

Check the result   Each length is 2 feet longer than the previous length, and the sum of the lengths is 213 feet.   ∎

Business Problems

When the regular price of merchandise is reduced, the amount of reduction is called the **markdown** (or discount).

$$\boxed{\text{Sale price}} \quad = \quad \boxed{\substack{\text{regular} \\ \text{price}}} \quad - \quad \boxed{\text{markdown}}$$

Usually, the markdown is expressed as a percent of the regular price.

$$\boxed{\text{Markdown}} \quad = \quad \boxed{\substack{\text{percent of} \\ \text{markdown}}} \quad \cdot \quad \boxed{\substack{\text{regular} \\ \text{price}}}$$

EXAMPLE 2   **Finding the percent of markdown**   A home theater system is on sale for $777. If the list price was $925, find the percent of markdown.

Analyze the problem   In this case, $777 is the sale price, $925 is the regular price, and the markdown is the product of $925 and the percent of markdown.

Form an equation   We can let r represent the percent of markdown, expressed as a decimal. We then substitute $777 for the sale price and $925 for the regular price in the formula

$$\boxed{\text{Sale price}} \quad \text{equals} \quad \boxed{\substack{\text{regular} \\ \text{price}}} \quad \text{minus} \quad \boxed{\text{markdown}}$$
$$777 \qquad\quad = \qquad\quad 925 \qquad\quad - \qquad\quad r \cdot 925$$

Solve the equation   We now solve the equation.

$$777 = 925 - r \cdot 925$$
$$777 = 925 - 925r$$
$$-148 = -925r \qquad \text{Subtract 925 from both sides.}$$
$$0.16 = r \qquad \text{Divide both sides by } -925.$$

State the conclusion   The percent of markdown is 16%.

Check the result   Since the markdown is 16% of $925, or $148, the sale price is $925 − $148, or $777.   ∎

EXAMPLE 3   **Portfolio analysis**   A college foundation owns stock in IBC (selling at $54 per share), GS (selling at $65 per share), and ATB (selling at $105 per share). The foundation owns equal shares of GS and IBC, but five times as many shares of ATB.

   If this portfolio is worth $450,800, how many shares of each type does the foundation own?

Analyze the problem   The value of the IBC stock plus the value of the GS stock plus the value of the ATB stock must equal $450,800.

- If x represents the number of shares of IBC, then $54x$ is the value of that stock.
- Since the foundation has equal numbers of shares of GS and of IBC, x also represents the number of shares of GS. The value of this stock is $65x$.
- Since the foundation owns five times as many shares of ATB, it owns $5x$ shares of ATB. The value of this stock is $105(5x)$.

We set the sum of these values equal to $450,800.

Form an equation   We let x represent the number of shares of IBC. Then x also represents the number of shares of GS, and $5x$ represents the number of shares of ATB.

The value of IBC stock	plus	the value of GS stock	plus	the value of ATB stock	equals	the total value of the portfolio.
$54x$	$+$	$65x$	$+$	$105(5x)$	$=$	$450,800$

Solve the equation   We now solve the equation.

$$54x + 65x + 105(5x) = 450,800$$
$$54x + 65x + 525x = 450,800 \qquad 105(5x) = 525x.$$
$$644x = 450,800 \qquad \text{Combine like terms.}$$
$$x = 700 \qquad \text{Divide both sides by 644.}$$

State the conclusion   The foundation owns 700 shares of IBC, 700 shares of GS, and 5(700), or 3,500, shares of ABT.

Check the result   The value of 700 shares of IBC at $54 per share is $37,800. The value of 700 shares of GS at $65 per share is $45,500. The value of 3,500 shares of ATB at $105 per share is $367,500. The sum of these values is $450,800.   ∎

Geometry Problems

Figure 1-23 illustrates several geometric figures. A **right angle** is an angle whose measure is 90°. A **straight angle** is an angle whose measure is 180°. An **acute angle** is an angle whose measure is greater than 0° but less than 90°.

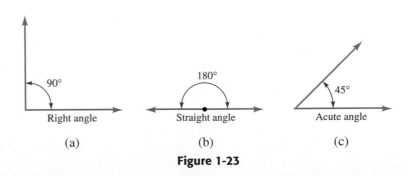

(a)   (b)   (c)

Figure 1-23

If the sum of two angles equals 90°, the angles are called **complementary,** and each angle is called the **complement** of the other. If the sum of two angles equals 180°, the angles are called **supplementary,** and each angle is called the **supplement** of the other.

A **right triangle** is a triangle with one right angle. In Figure 1-24(a), ∠C (read as "angle C") is a right angle. An **isosceles triangle** is a triangle with two sides of equal measure that meet to form the **vertex angle.** The angles opposite the equal sides, called the **base angles,** are also equal. An **equilateral triangle** is a triangle with three equal sides and three equal angles.

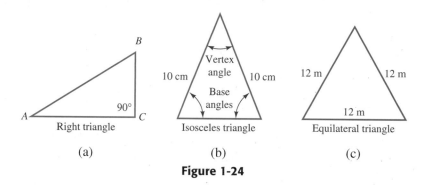

(a)   (b)   (c)

Figure 1-24

EXAMPLE 4   **Angles in a triangle**   If the vertex angle of the isosceles triangle shown in Figure 1-24(b) measures 64°, find the measure of each base angle.

Analyze the problem   We are given that the vertex angle measures 64°. If we let $x°$ represent the measure of one base angle, the measure of the other base angle is also $x°$. Thus, the sum of the angles in the triangle is $x° + x° + 64°$. Because the sum of the measures of the angles of any triangle is 180°, we know that $x° + x° + 64°$ is equal to 180°.

Form an equation   We can form the equation

The measure of one base angle	plus	the measure of the other base angle	plus	the measure of the vertex angle	equals	180°.
x	$+$	x	$+$	64	$=$	180

Solve the equation   We now solve the equation.

$$x + x + 64 = 180$$
$$2x + 64 = 180 \qquad \text{Combine like terms.}$$
$$2x = 116 \qquad \text{Subtract 64 from both sides.}$$
$$x = 58 \qquad \text{Divide both sides by 2.}$$

State the conclusion   The measure of each base angle is 58°.

Check the result   The sum of the measures of each base angle and the vertex angle is 180°:

$$58° + 58° + 64° = 180°$$

EXAMPLE 5   **Designing a dog run**   A man has 28 meters of fencing to make a rectangular kennel. If the kennel is to be 6 meters longer than it is wide, find its dimensions.

Analyze the problem   The perimeter P of a rectangle is the distance around it. If w is chosen to represent the width of the kennel, then $w + 6$ represents its length. (See Figure 1-25.) The perimeter can be expressed either as $2w + 2(w + 6)$ or as 28.

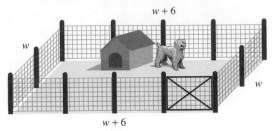

Figure 1-25

Form an equation   We let w represent the width of the kennel. Then $w + 6$ represents its length.

Two widths	plus	two lengths	equals	the perimeter.
$2 \cdot w$	$+$	$2 \cdot (w + 6)$	$=$	28

Solve the equation   We now solve the equation.

$$2 \cdot w + 2 \cdot (w + 6) = 28$$

$$2w + 2w + 12 = 28 \qquad \text{Use the distributive property to remove parentheses.}$$

$$4w + 12 = 28 \qquad \text{Combine like terms.}$$

$$4w = 16 \qquad \text{Subtract 12 from both sides.}$$

$$w = 4 \qquad \text{Divide both sides by 4.}$$

$$w + 6 = 10$$

State the conclusion   The dimensions of the kennel are 4 meters by 10 meters.

Check the result   If a kennel has a width of 4 meters and a length of 10 meters, its length is 6 meters longer than its width, and the perimeter is $2(4)$ meters $+ 2(10)$ meters $= 28$ meters. ∎

Lever Problems

EXAMPLE 6   **Engineering**   Design engineers must position two hydraulic cylinders as in Figure 1-26 to balance a 9,500-pound force at point A. The first cylinder at the end of the lever exerts a 3,500-pound force. Where should the design engineers position the second cylinder, which is capable of exerting a 5,500-pound force?

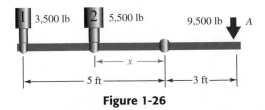

Figure 1-26

Analyze the problem   The lever will be in balance when the force of the first cylinder multiplied by its distance from the pivot (also called the **fulcrum**) added to the second cylinder's force multiplied by its distance from the fulcrum, is equal to the product of the 9,500-pound force and its distance from the fulcrum.

Form an equation   We let x represent the distance from the larger cylinder to the fulcrum.

Force of cylinder 1, times its distance	plus	force of cylinder 2, times its distance	equals	force to be balanced, times its distance.
$3,500 \cdot 5$	$+$	$5,500x$	$=$	$9,500 \cdot 3$

Solve the equation   We can now solve the equation.

$$3,500 \cdot 5 + 5,500x = 9,500 \cdot 3$$
$$17,500 + 5,500x = 28,500$$
$$5,500x = 11,000 \qquad \text{Subtract 17,500 from both sides.}$$
$$x = 2 \qquad \text{Divide both sides by 5,500.}$$

State the conclusion   The design must specify that the second cylinder be positioned 2 feet from the fulcrum.

Check the result $3,500 \cdot 5 + 5,500 \cdot 2 = 17,500 + 11,000 = 28,500$
$9,500 \cdot 3 = 28,500$ ∎

Orals   *Find each value.*

1. 20% of 500

2. $33\frac{1}{3}\%$ of 600

If a stock costs $54, find the cost of

3. 5 shares

4. x shares

Find the area of the rectangle with the given dimensions.

5. 6 meters long, 4 meters wide

6. l meters long, $l - 5$ meters wide

1.6   EXERCISES

REVIEW   *Simplify each expression.*

1. $\left(\dfrac{3x^{-3}}{4x^2}\right)^{-4}$

2. $\left(\dfrac{r^{-3}s^2}{r^2r^3s^{-4}}\right)^{-5}$

3. $\dfrac{a^m a^3}{a^2}$

4. $\left(\dfrac{b^n}{b^3}\right)^3$

VOCABULARY AND CONCEPTS   *Fill in the blanks.*

5. The expression _____ represents the phrase "4 more than 5 times x."

6. The expression _____ represents the phrase "6 percent of x."

7. The expression _____ represents the phrase "the value of x shares priced at $40 per share."

8. A right angle is an angle whose measure is ____.

9. A straight angle measures ____.

10. An _____ angle measures greater than 0° but less than 90°.

11. If the sum of the measures of two angles equals 90°, the angles are called _____ angles.

12. If the sum of the measures of two angles equals _____, the angles are called supplementary angles.

13. If a triangle has a right angle, it is called a _____ triangle.

14. If a triangle has two sides with equal measure, it is called an _____ triangle.

15. The equal sides of an isosceles triangle form the _____ angle.

16. An equilateral triangle has three _____ sides and three equal angles.

APPLICATIONS

17. **Cutting a rope** A 60-foot rope is cut into four pieces, with each successive piece twice as long as the previous one. Find the length of the longest piece.

18. **Cutting a cable** A 186-foot cable is to be cut into four pieces. Find the length of each piece if each successive piece is 3 feet longer than the previous one.

19. **Cutting a board** The carpenter in the illustration saws a board into two pieces. He wants one piece to be 1 foot longer than twice the length of the shorter piece. Find the length of each piece.

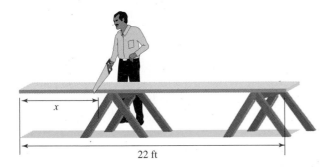

20. **Cutting a beam** A 30-foot steel beam is to be cut into two pieces. The longer piece is to be 2 feet more than three times as long as the shorter piece. Find the length of each piece.

21. **Buying a TV and a VCR** See the following ad. If the TV costs $55 more than the VCR, how much does the TV cost?

BUY *BOTH* FOR
$655

22. **Buying golf clubs** The cost of a set of golf clubs is $590. If the irons cost $40 more than the woods, find the cost of the irons.

23. **Buying a washer and dryer** Find the percent of markdown of the sale in the following ad.

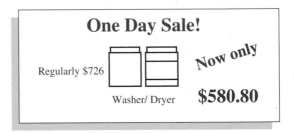

One Day Sale!

Regularly $726

Washer/ Dryer

Now only

$580.80

24. **Buying furniture** A bedroom set regularly sells for $983. If it is on sale for $737.25, what is the percent of markdown?

25. **Buying books** A bookstore buys a used calculus book for $12 and sells it for $40. Find the percent of markup.

26. **Selling toys** The owner of a gift shop buys stuffed animals for $18 and sells them for $30. Find the percent of markup.

27. **Value of an IRA** In an Individual Retirement Account (IRA) valued at $53,900, a student has 500 shares of stock, some in Big Bank Corporation and some in Safe Savings and Loan. If Big Bank sells for $115 per share and Safe Savings sells for $97 per share, how many shares of each does the student own?

28. **Assets of a pension fund** A pension fund owns 12,000 shares in a stock mutual fund and a mutual bond fund. Currently, the stock fund sells for $12 per share, and the bond fund sells for $15 per share. How many shares of each does the pension fund own if the value of the securities is $165,000?

29. Selling calculators   Last month, a bookstore ran the following ad and sold 85 calculators, generating $3,875 in sales. How many of each type of calculator did the bookstore sell?

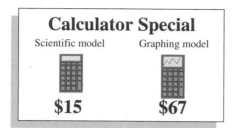

Calculator Special

Scientific model   Graphing model

$15   **$67**

30. Selling grass seed   A seed company sells two grades of grass seed. A 100-pound bag of a mixture of rye and bluegrass sells for $245, and a 100-pound bag of bluegrass sells for $347. How many bags of each are sold in a week when the receipts for 19 bags are $5,369?

31. Buying roses   A man with $21.25 stops after work to order some roses for his wife's birthday. If each rose costs $1.25 and there is a delivery charge of $5, how many roses can he buy?

32. Renting a truck   To move to Wisconsin, a man can rent a truck for $29.95 per day plus 19¢ per mile. If he keeps the truck for one day, how many miles can he drive for a cost of $77.45?

33. Car rental   While waiting for his car to be repaired, a man rents a car for $12 per day plus 10¢ per mile. If he keeps the car for 2 days, how many miles can he drive for a total cost of $30? How many miles can he drive for a total cost of $36?

34. Computing salaries   A student earns $17 per day for delivering overnight packages. She is paid $5 per day plus 60¢ for each package delivered. How many more deliveries must she make each day to increase her daily earnings to $23?

35. Finding dimensions   The rectangular garden shown in the illustration is twice as long as it is wide. Find its dimensions.

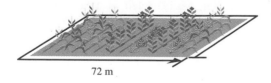

72 m

36. Finding dimensions   The width of a rectangular swimming pool is one-third its length. If its perimeter is 96 meters, find the dimensions of the pool.

37. Fencing a pasture   A farmer has 624 feet of fencing to enclose the pasture shown in the illustration. Because a river runs along one side, fencing will be needed on only three sides. Find the dimensions of the pasture if its length is double its width.

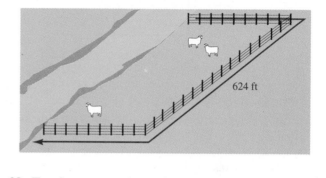

624 ft

38. Fencing a pen   A man has 150 feet of fencing to build the pen shown in the illustration. If one end is a square, find the outside dimensions.

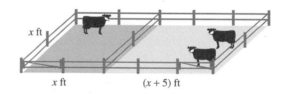

x ft

x ft $(x + 5)$ ft

39. Enclosing a swimming pool   A woman wants to enclose the swimming pool shown in the illustration and have a walkway of uniform width all the way around. How wide will the walkway be if the woman uses 180 feet of fencing?

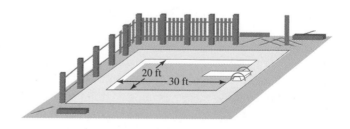

20 ft
30 ft

40. Framing a picture   An artist wants to frame the following picture with a frame 2 inches wide. How wide will the framed picture be if the artist uses 70 inches of framing material?

x in.

x + 5 in.

41. Supplementary angles   If one of two supplementary angles is 35° larger than the other, find the measure of the smaller angle.

42. Supplementary angles   Refer to the illustration and find x.

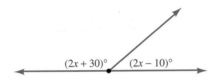

$(2x + 30)°$ $(2x - 10)°$

43. Complementary angles   If one of two complementary angles is 22° greater than the other, find the measure of the larger angle.

44. Complementary angles in a right triangle
Explain why the acute angles in the following right triangle are complementary. Then find the measure of angle A.

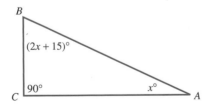

B

$(2x + 15)°$

90° $x°$

C   A

45. Supplementary angles and parallel lines   In the illustration, lines r and s are cut by a third line l to form ∠1 and ∠2. When lines r and s are parallel, ∠1 and ∠2 are supplementary. If m(∠1) = $(x + 50)°$ and m(∠2) = $(2x - 20)°$ and lines r and s are parallel, find x. Read m(∠1) as "the measure of ∠1."

46. In the illustration, find m(∠3).   (*Hint:* See Exercise 45.)

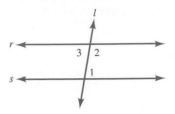

l

r   3 / 2

s   1

47. Angles of an equilateral triangle   Find the measure of each angle in an equilateral triangle.

48. Vertical angles   When two lines intersect as in the illustration, four angles are formed. Angles that are side-by-side, such as ∠1 and ∠2, are called **adjacent angles.** Angles that are nonadjacent, such as ∠1 and ∠3 or ∠2 and ∠4, are called **vertical angles.** From geometry, we know that if two lines intersect, vertical angles have the same measure. If m(∠1) = $(3x + 10)°$ and m(∠3) = $(5x - 10)°$, find x. Read m(∠1) as "the measure of ∠1."

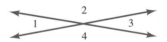

2
1   3
4

49. If m(∠2) = $(6x + 20)°$ in the previous illustration and m(∠4) = $(8x - 20)°$, find m(∠1). (See Exercise 48.)

50. Angles of a quadrilateral   The sum of the angles of any four-sided figure (called a *quadrilateral*) is 360°. The following quadrilateral has two equal base angles. Find x.

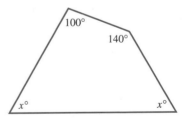

100°
140°

$x°$ $x°$

51. Height of a triangle   If the height of a triangle with a base of 8 inches is tripled, its area is increased by 96 square inches. Find the height of the triangle.

52. Engineering design The width, *w*, of the flange in the following engineering drawing has not yet been determined. Find *w* so that the area of the 7-in.-by-12-in. rectangular portion is exactly one-half of the total area.

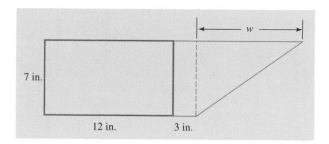

7 in.

12 in.   3 in.

53. Balancing a seesaw A seesaw is 20 feet long, and the fulcrum is in the center. If an 80-pound boy sits at one end, how far will the boy's 160-pound father have to sit from the fulcrum to balance the seesaw?

54. Establishing equilibrium Two forces—110 pounds and 88 pounds—are applied to opposite ends of an 18-foot lever. How far from the greater force must the fulcrum be placed so that the lever is balanced?

55. Moving a stone A woman uses a 10-foot bar to lift a 210-pound stone. If she places another rock 3 feet from the stone to act as the fulcrum, how much force must she exert to move the stone?

56. Lifting a car A 350-pound football player brags that he can lift a 2,500-pound car. If he uses a 12-foot bar with the fulcrum placed 3 feet from the car, will he be able to lift the car?

57. Balancing a lever Forces are applied to a lever as indicated in the illustration. Find *x*, the distance of the smallest force from the fulcrum.

58. Balancing a seesaw Jim and Bob sit at opposite ends of an 18-foot seesaw, with the fulcrum at its center. Jim weighs 160 pounds, and Bob weighs 200 pounds. When Kim sits 4 feet in front of Jim, the seesaw balances. How much does Kim weigh?

59. Temperature scales The Celsius and Fahrenheit temperature scales are related by the equation $C = \frac{5}{9}(F - 32)$. At what temperature will a Fahrenheit and a Celsius thermometer give the same reading?

60. Installing solar heating One solar panel in the illustration is to be 3 feet wider than the other, but to be equally efficient, they must have the same area. Find the width of each.

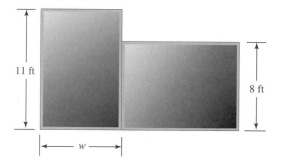

11 ft

8 ft

w

WRITING

61. Explain the steps for solving an applied problem.

62. Explain how to check the solution of an applied problem.

SOMETHING TO THINK ABOUT

63. Find the distance *x* required to balance the lever in the illustration.

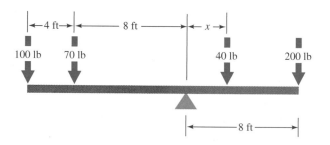

4 ft   8 ft   *x*

100 lb   70 lb   40 lb   200 lb

8 ft

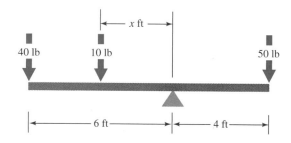

x ft

40 lb   10 lb   50 lb

6 ft   4 ft

64. Interpret the answer to Exercise 63.

1.7   More Applications of Equations

■ **Investment Problems**   ■ **Uniform Motion Problems**
■ **Mixture Problems**

Getting Ready   **1.** Find 8% of $500.

2. Write an expression for 8% of x.

3. If $9,000 of $15,000 is invested at 7%, how much is left to invest at another rate?

4. If x of $15,000 is invested at 7%, write an expression for how much is left to be invested at another rate.

5. If a car travels 50 mph for 4 hours, how far will it go?

6. If coffee sells for $4 per pound, how much will 5 pounds cost?

In this section, we will continue our investigation of problem solving by considering investment, motion, and mixture problems.

Investment Problems

EXAMPLE 1   **Financial planning**   A professor has $15,000 to invest for one year, some at 8% and the rest at 7%. If she will earn $1,110 from these investments, how much did she invest at each rate?

Analyze the problem   Simple interest is computed by the formula $i = prt$, where i is the interest earned, p is the principal, r is the annual interest rate, and t is the length of time the principal is invested. In this problem, $t = 1$ year. Thus, if x is invested at 8% for one year, the interest earned is $i = \$x(8\%)(1) = \$0.08x$. If the remaining $\$(15,000 - x)$ is invested at 7%, the amount earned on that investment is $\$0.07(15,000 - x)$. See Figure 1-27. The sum of these amounts should equal the total earned interest of $1,110.

	i	$=$	p	·	r	·	t
8% investment	$0.08x$		x		0.08		1
7% investment	$0.07(15,000 - x)$		$15,000 - x$		0.07		1

Figure 1-27

Form an equation   We can let x represent the number of dollars invested at 8%. Then $15,000 - x$ represents the number of dollars invested at 7%.

The interest earned at 8%	plus	the interest earned at 7%	equals	the total interest.
$0.08x$	$+$	$0.07(15,000 - x)$	$=$	$1,110$

Solve the equation   We now solve the equation.

$$0.08x + 0.07(15{,}000 - x) = 1{,}110$$

$8x + 7(15{,}000 - x) = 111{,}000$ To eliminate the decimals, multiply both sides by 100.

$8x + 105{,}000 - 7x = 111{,}000$ Use the distributive property to remove parentheses.

$x + 105{,}000 = 111{,}000$ Combine like terms.

$x = 6{,}000$ Subtract 105,000 from both sides.

$15{,}000 - x = 9{,}000$

State the conclusion   She invested $6,000 at 8% and $9,000 at 7%.

Check the result   The interest on $6,000 is 0.08($6,000) = $480. The interest earned on $9,000 is 0.07($9,000) = $630. The total interest is $1,110. ∎

Uniform Motion Problems

EXAMPLE 2   **Travel time**   A car leaves Rockford traveling toward Wausau at the rate of 55 mph. At the same time, another car leaves Wausau traveling toward Rockford at the rate of 50 mph. How long will it take them to meet if the cities are 157.5 miles apart?

Analyze the problem   In this case, the cars are traveling toward each other as shown in Figure 1-28(a). Uniform motion problems are based on the formula $d = rt$, where d is distance, r is rate, and t is time. We can organize the given information in the chart shown in Figure 1-28(b).

We know that one car is traveling at 55 mph and that the other is going 50 mph. We also know that they travel for the same amount of time—say, t hours. Thus, the distance that the faster car travels is $55t$ miles, and the distance that the slower car travels is $50t$ miles. The sum of these distances equals 157.5 miles, the distance between the cities.

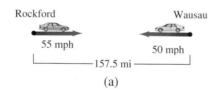

	Rate	·	Time	=	Distance
Faster car	55		t		$55t$
Slower car	50		t		$50t$

(a)   (b)

Figure 1-28

Form an equation   We can let t represent the time that each car travels. Then $55t$ represents the distance traveled by the faster car, and $50t$ represents the distance traveled by the slower car.

The distance the faster car goes	plus	the distance the slower car goes	equals	the distance between cities.
$55t$	$+$	$50t$	$=$	157.5

Solve the equation   We now solve the equation.

$55t + 50t = 157.5$

$105t = 157.5$ Combine like terms.

$t = 1.5$ Divide both sides by 105.

State the conclusion   The two cars will meet in $1\frac{1}{2}$ hours.

Check the result   The faster car travels $1.5(55) = 82.5$ miles. The slower car travels $1.5(50) = 75$ miles. The total distance traveled is 157.5 miles.   ∎

Mixture Problems

EXAMPLE 3   **Mixing nuts**   The owner of a candy store notices that 20 pounds of gourmet cashews are getting stale. They did not sell because of their high price of \$12 per pound. The store owner decides to mix peanuts with the cashews to lower the price per pound. If peanuts sell for \$3 per pound, how many pounds of peanuts must be mixed with the cashews to make a mixture that could be sold for \$6 per pound?

Analyze the problem   This problem is based on the formula $V = pn$, where V represents the value, p represents the price per pound, and n represents the number of pounds.

We can let x represent the number of pounds of peanuts to be used and enter the known information in the chart shown in Figure 1-29. The value of the cashews plus the value of the peanuts will be equal to the value of the mixture.

	Price	·	Number of pounds	=	Value
Cashews	12		20		240
Peanuts	3		x		$3x$
Mixture	6		$20 + x$		$6(20 + x)$

Figure 1-29

Form an equation   We can let x represent the number of pounds of peanuts to be used. Then $20 + x$ represents the number of pounds in the mixture.

The value of the cashews	plus	the value of the peanuts	equals	the value of the mixture.
240	+	$3x$	=	$6(20 + x)$

Solve the equation   We now solve the equation.

$$240 + 3x = 6(20 + x)$$
$$240 + 3x = 120 + 6x \qquad \text{Use the distributive property to remove parentheses.}$$
$$120 = 3x \qquad \text{Subtract } 3x \text{ and } 120 \text{ from both sides.}$$
$$40 = x \qquad \text{Divide both sides by 3.}$$

State the conclusion   The store owner should mix 40 pounds of peanuts with the 20 pounds of cashews.

Check the result   The cashews are valued at $\$12(20) = \$240$, and the peanuts are valued at $\$3(40) = \$120$, so the mixture is valued at $\$6(60) = \$360$.

The value of the cashews plus the value of the peanuts equals the value of the mixture.   ∎

EXAMPLE 4   **Milk production**   A container is partially filled with 12 liters of whole milk containing 4% butterfat. How much 1% milk must be added to get a mixture that is 2% butterfat?

Analyze the problem   Since the first container shown in Figure 1-30(a) contains 12 liters of 4% milk, it contains 0.04(12) liters of butterfat. To this container we will add the contents of the second container, which holds 0.01l liters of butterfat.

The sum of these two amounts of butterfat (0.04(12) + 0.01l) will equal the amount of butterfat in the third container, which is 0.02(12 + l) liters of butterfat. This information is presented in table form in Figure 1-30(b).

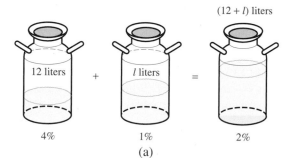

(12 + l) liters

12 liters   + l liters   =

4%   1%   2%

(a)

	Amount of butterfat	=	Amount of milk	·	Percent of butterfat
Whole milk	0.04(12)		12		0.04
1% milk	0.01(l)		l		0.01
2% milk	0.02(12 + l)		12 + l		0.02

(b)

Figure 1-30

Form an equation   We can let l represent the number of liters of 1% milk to be added. Then

The amount of butterfat in 12 liters of 4% milk	plus	the amount of butterfat in l liters of 1% milk	equals	the amount of butterfat in (12 + l) liters of mixture.
0.04(12)	+	0.01l	=	0.02(12 + l)

Solve the equation   We now solve the equation.

$$0.04(12) + 0.01l = 0.02(12 + l)$$
$$4(12) + 1l = 2(12 + l) \qquad \text{Multiply both sides by 100.}$$
$$48 + l = 24 + 2l \qquad \text{Use the distributive property to remove parentheses.}$$
$$24 = l \qquad \text{Subtract 24 and } l \text{ from both sides.}$$

State the conclusion   Thus, 24 liters of 1% milk should be added to get a mixture that is 2% butterfat.

Check the result   12 liters of 4% milk contains 0.48 liters of butterfat. 24 liters of 1% milk contains 0.24 liters of butterfat. This gives a total of 36 liters of a mixture that contains 0.72 liters of butterfat. This is a 2% solution.   ∎

Orals   *Assume that all investments are for one year.*

1. How much interest will $1,500 earn if invested at 6%?

2. Express the amount of interest $x will earn if invested at 5%.

3. If $x of $30,000 is invested at 5%, how would you express the amount left to be invested at 6%?

4. If Brazil nuts are worth $x per pound, express how much 20 pounds will be worth.

5. If whole milk is 4% butterfat, how much butterfat is in 2 gallons?

6. If whole milk is 4% butterfat, express how much butterfat is in x gallons.

1.7   EXERCISES

REVIEW   *Solve each equation.*

1. $9x - 3 = 6x$

2. $7a + 2 = 12 - 4(a - 3)$

3. $\dfrac{8(y - 5)}{3} = 2(y - 4)$ **4.** $\dfrac{t - 1}{3} = \dfrac{t + 2}{6} + 2$

VOCABULARY AND CONCEPTS   *Fill in the blanks.*

5. The formula for simple interest i is $i = prt$, where p is the _____, r is the ____ and t is the ____.

6. Uniform motion problems are based on the formula _____, where d is the _____, r is the ____ of speed, and t is the ____.

7. Dry mixture problems are based on the formula $V = pn$, where V is the ____, p is the ____ per unit, and n is the _____ of units.

8. The liquid mixture problem in Example 4 is based on the idea that the amount of _____ in the first container plus the amount of _____ in the second container is equal to the amount of _____ in the mixture.

APPLICATIONS

9. Investing money   Lured by the following ad, a woman invested $12,000, some in a money market account and the rest in a 5-year CD. How much was invested in each account if the income from both investments is $1,060 per year?

First Republic Savings and Loan	
Account	**Rate**
NOW	5.5%
Savings	7.5%
Money market	8.0%
Checking	4.0%
5-year CD	9.0%

10. Investing money   A man invested $14,000, some at 7% and some at 10% annual interest. The annual income from these investments was $1,280. How much did he invest at each rate?

11. Supplemental income   A teacher wants to earn $1,500 per year in supplemental income from a cash gift of $16,000. She puts $6,000 in a credit union that pays 7% annual interest. What rate must she earn on the remainder to achieve her goal?

12. Inheriting money   Paul split an inheritance between two investments, one paying 7% annual interest and the other 10%. He invested twice as much in the 10% investment as he did in the 7% investment. If his combined annual income from the two investments was $4,050, how much did he inherit?

13. Investing money   Kyoko has some money to invest. If she could invest $3,000 more, she can qualify for an 11% investment. Otherwise, she can invest the money at 7.5% annual interest. If the 11% investment yields twice as much annual income as the 7.5% investment, how much does she have on hand to invest?

14. Supplemental income   A bus driver wants to earn $3,500 per year in supplemental income from an inheritance of $40,000. If the driver invests $10,000 in a mutual fund paying 8%, what rate must he earn on the remainder to achieve his goal?

15. Concert receipts   For a jazz concert, student tickets are $2 each, and adult tickets are $4 each. If 200 tickets were sold and the total receipts are $750, how many student tickets were sold?

16. School plays   At a school play, 140 tickets are sold, with total receipts of $290. If adult tickets cost $2.50 each and student tickets cost $1.50 each, how many adult tickets were sold?

17. Computing time   One car leaves Chicago headed for Cleveland, a distance of 343 miles. At the same time, a second car leaves Cleveland headed toward Chicago. If the first car averages 50 mph and the second car averages 48 mph, how long will it take the cars to meet?

18. **Storm research**   During a storm, two teams of scientists leave a university at the same time to search for tornados. The first team travels east at 20 mph, and the second travels west at 25 mph, as shown in the illustration. If their radios have a range of 90 miles, how long will it be before they lose radio contact?

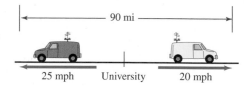

19. **Cycling**   A cyclist leaves Las Vegas riding at the rate of 18 mph. See the illustration. One hour later, a car leaves Las Vegas going 45 mph in the same direction. How long will it take the car to overtake the cyclist?

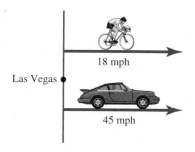

20. **Computing distance**   At 2 P.M., two cars leave Eagle River, WI, one headed north and one headed south. If the car headed north averages 50 mph and the car headed south averages 60 mph, when will the cars be 165 miles apart?

21. **Running a race**   Two runners leave the starting gate, one running 12 mph and the other 10 mph. If they maintain the pace, how long will it take for them to be one-quarter of a mile apart?

22. **Trucking**   Two truck drivers leave a warehouse traveling in opposite directions, one driving at 50 mph and one driving at 56 mph. If the slower driver has a 2-hour head start, how long has she been on the road when the two drivers are 683 miles apart?

23. **Riding a jet ski**   A jet ski can go 12 mph in still water. If a rider goes upstream for 3 hours against a current of 4 mph, how long will it take the rider to return?   (*Hint:* Upstream speed is $(12 - 4)$ mph; how far can the rider go in 3 hours?)

24. **Taking a walk**   Sarah walked north at the rate of 3 mph and returned at the rate of 4 mph. How many miles did she walk if the round-trip took 3.5 hours?

25. **Computing travel time**   Jamal traveled a distance of 400 miles in 8 hours. Part of the time, his rate of speed was 45 mph. The rest of the time, his rate of speed was 55 mph. How long did Jamal travel at each rate?

26. **Riding a motorboat**   The motorboat in the illustration can go 18 mph in still water. If a trip downstream takes 4 hours and the return trip takes 5 hours, find the speed of the current.

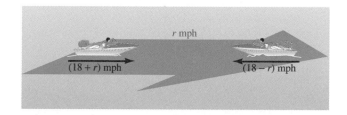

27. **Mixing candies**   The owner of a store wants to make a 30-pound mixture of two candies to sell for $1 per pound. If one candy sells for 95¢ per pound and the other for $1.10 per pound, how many pounds of each should be used?

28. **Computing selling price**   A mixture of candy is made to sell for 89¢ per pound. If 32 pounds of candy, selling for 80¢ per pound, are used along with 12 pounds of a more expensive candy, find the price per pound of the better candy.

29. **Diluting solutions**   In the illustration, how much water should be added to 20 ounces of a 15% solution of alcohol to dilute it to a 10% solution?

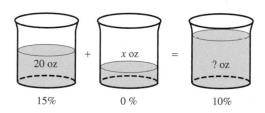

30. **Increasing concentration**   How much water must be boiled away to increase the concentration of 300 gallons of a salt solution from 2% to 3%?

31. **Making whole milk**   Cream is approximately 22% butterfat. How many gallons of cream must be mixed with milk testing at 2% butterfat to get 20 gallons of milk containing 4% butterfat?

32. **Mixing solutions**   How much acid must be added to 60 grams of a solution that is 65% acid to obtain a new solution that is 75% acid?

33. **Raising grades**   A student scores 70% on a test that contains 30 questions. To improve his score, the instructor agrees to let him work 15 additional questions. How many must he get right to raise his grade to 80%?

34. **Raising grades**   On a second exam, the student in Exercise 33 earns a score of 60% on a 20-question test. This time, the instructor allows him to work 20 extra problems to improve his score. How many must he get right to raise his grade to 70%?

35. **Computing grades**   Before the final, Estela had earned a total of 375 points on four tests. To receive an A in the course, she must have 90% of a possible total of 450 points. Find the lowest number of points that she can earn on the final exam and still receive an A.

36. **Computing grades**   A student has earned a total of 435 points on five tests. To receive a B in the course, he must have 80% of a possible total of 600 points. Find the lowest number of points that the student can make on the final exam and still receive a B.

37. **Managing a bookstore**   A bookstore sells a history book for $65. If the bookstore makes a profit of 30% on each book, what does the bookstore pay the publisher for each book?   (*Hint:* The retail price = the wholesale price + the markup.)

38. **Managing a bookstore**   A bookstore sells a textbook for $39.20. If the bookstore makes a profit of 40% on each sale, what does the bookstore pay the publisher for each book?   (*Hint:* The retail price = the wholesale price + the markup.)

39. **Making furniture**   A woodworker wants to put two partitions crosswise in a drawer that is 28 inches deep. (See the illustration.) He wants to place the partitions so that the spaces created increase by 3 inches from front to back. If the thickness of each partition is $\frac{1}{2}$ inch, how far from the front end should he place the first partition?

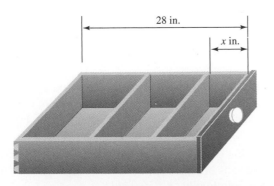

40. **Building shelves**   A carpenter wants to put four shelves on an 8-foot wall so that the five spaces created decrease by 6 inches as we move up the wall. (See the illustration.) If the thickness of each shelf is $\frac{3}{4}$ inch, how far will the bottom shelf be from the floor?

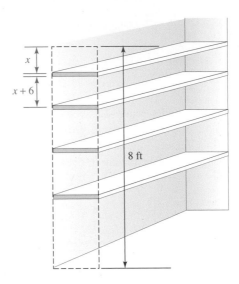

WRITING

41. What do you find most difficult in solving application problems, and why?

42. Which type of application problem do you find easiest, and why?

SOMETHING TO THINK ABOUT

43. Discuss the difficulties in solving this problem:

    A man drives 100 miles at 30 mph. How fast should he drive on the return trip to average 60 mph for the entire trip?

44. What difficulties do you encounter when solving this problem?

    Adult tickets cost $4, and student tickets cost $2. Sales of 71 tickets bring in $245. How many of each were sold?

Projects

Project 1

You are a member of the Parchdale City Planning Council. The biggest debate in years is taking place over whether or not the town should build an emergency reservoir for use in especially dry periods. The proposed site has room enough for a conical reservoir of diameter 141 feet and depth 85.3 feet.

As long as Parchdale's main water supply is working, the reservoir will be kept full. When a water emergency occurs, however, the reservoir will lose water to evaporation as well as supplying the town with water. The company that has designed the reservoir has given the town the following information.

If D equals the number of consecutive days the reservoir of volume V is used as Parchdale's water supply, then the total amount of water lost to evaporation after D days will be

$$0.1V \cdot \left(\frac{D - 0.7}{D}\right)^2$$

So the volume of water left in the reservoir after it has been used for D days is

Volume = V − (usage per day) · D − (evaporation)

Parchdale uses about 57,500 cubic feet of water per day under emergency conditions. The majority of the council members feel that building the reservoir is a good idea if it will supply the town with water for a week. Otherwise, they will vote against building the reservoir.

a. Calculate the volume of water the reservoir will hold. Remember to use significant digits correctly. Express the answer in scientific notation.

b. Show that the reservoir will supply Parchdale with water for a full week.

c. How much water per day could Parchdale use if the proposed reservoir had to supply the city with water for ten days?

Project 2

Theresa manages the local outlet of Hook n' Slice, a nationwide sporting goods retailer. She has hired you to help solve some problems. Be sure to provide explanations of how you arrive at your solutions.

a. The company recently directed that all shipments of golf clubs be sold at a retail price of three times what they cost the company. A shipment of golf clubs arrived yesterday, and they have not yet been labeled. This morning, the main office of Hook n' Slice called to say that the clubs should be sold at 35% off their retail prices. Rather than label each club with its retail price, calculate the sale price, and then relabel the club, Theresa wonders how to go directly from the cost of the club to the sale price. That is, if C is the cost to the company of a certain club, what is the sale price for that club?
  1. Develop a formula and answer this question for her.
  2. Now try out your formula on a club that cost the company $26. Compare your answer with what you find by following the company procedure of first computing the retail price and then reducing that by 35%.

b. All golf bags have been sale-priced at 20% off for the past few weeks. Theresa has all of them in a large rack, marked with a sign that reads "20% off retail price." The company has now directed that this sale price be reduced by 35%. Rather than relabel every bag with its new sale price, Theresa would like to simply change the sign to say "_____ % off retail price."
  1. Determine what number Theresa should put in the blank.
  2. Check your answer by computing the final sale price for a golf bag with a $100 retail price in two ways:
     • by using the percentage you told Theresa should go on the sign, and
     • by doing the two separate price reductions.
  3. Will the sign read the same if the retail price is first reduced by 35%, and then the sale price reduced by 20%? Explain.

Chapter Summary

CONCEPTS	REVIEW EXERCISES
	1.1 The Real Number System

A *set* is a collection of objects.

Natural numbers:
  $\{1, 2, 3, 4, 5, \ldots\}$

Whole numbers:
  $\{0, 1, 2, 3, 4, 5, \ldots\}$

Integers:
  $\{\ldots, -3, -2, -1, 0, 1, 2,$
  $3, \ldots\}$

Even integers:
  $\{\ldots, -6, -4, -2, 0, 2, 4,$
  $6, \ldots\}$

Odd integers:
  $\{\ldots, -5, -3, -1, 1, 3,$
  $5, \ldots\}$

Prime numbers:
  $\{2, 3, 5, 7, 11, 13, \ldots\}$

Composite numbers:
  $\{4, 6, 8, 9, 10, 12, \ldots\}$

Rational numbers: numbers that can be written as $\frac{a}{b}$ ($b \neq 0$), where a and b are integers

Irrational numbers: numbers that can be written as nonterminating, nonrepeating decimals

Real numbers: numbers that can be written as decimals

REVIEW EXERCISES

List the numbers in $\left\{-4, -\frac{2}{3}, 0, 1, 2, \pi, 4\right\}$ that satisfy the given condition.

1. whole number   **2.** natural number

3. rational number

4. integer

5. irrational number   **6.** real number

7. negative number   **8.** positive number

9. prime number   **10.** composite number

11. even integer   **12.** odd integer

13. Graph the set of prime numbers between 20 and 30.

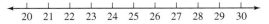
20 21 22 23 24 25 26 27 28 29 30

14. Graph the set of composite numbers between 5 and 13.

5   6   7   8   9   10 11 12 13

Graph each set on the number line.

15. $\left\{x \mid x \geq -4\right\}$

16. $\left\{x \mid -2 < x \leq 6\right\}$

17. $(-2, 3)$ **18.** $[2, 6]$

19. $\{x \mid x > 2\}$ **20.** $(-\infty, -1)$

21. $(-\infty, 0] \cup (2, \infty)$

Absolute value:

$$\begin{cases} \text{If } x \geq 0, \text{ then } |x| = x. \\ \text{If } x < 0, \text{ then } |x| = -x. \end{cases}$$

Write each expression without absolute value symbols.

22. $|0|$ **23.** $|-1|$

24. $|8|$ **25.** $-|8|$

1.2   Arithmetic and Properties of Real Numbers

Adding and subtracting real numbers:

With like signs:
Add their absolute values and keep the same sign.

With unlike signs:
Subtract their absolute values and keep the sign of the number with the greater absolute value.

$$x - y \text{ is equivalent to } x + (-y).$$

Multiplying and dividing real numbers:

With like signs: Multiply (or divide) their absolute values. The sign is positive.

With unlike signs: Multiply (or divide) their absolute values. The sign is negative.

Order of operations without exponents:

Do all calculations within grouping symbols, working from the innermost pair to the outermost pair.

1. Do multiplications and divisions, working from left to right.
2. Do additions and subtractions, working from left to right.

In a fraction, simplify the numerator and denominator separately. Then simplify the fraction.

Perform the operations and simplify when possible.

26. $3 + (+5)$ **27.** $-6 + (-3)$

28. $-15 + (-13)$ **29.** $25 + 32$

30. $-2 + 5$ **31.** $3 + (-12)$

32. $8 + (-3)$ **33.** $7 + (-9)$

34. $-25 + 12$ **35.** $-30 + 35$

36. $-3 - 10$ **37.** $-8 - (-3)$

38. $27 - (-12)$ **39.** $38 - (-15)$

40. $(+5)(+7)$ **41.** $(-6)(-7)$

42. $\dfrac{-16}{-4}$ **43.** $\dfrac{-25}{-5}$

44. $4(-3)$ **45.** $-3(8)$

46. $\dfrac{-8}{2}$ **47.** $\dfrac{8}{-4}$

Simplify each expression.

48. $-4(3 - 6)$ **49.** $3[8 - (-1)]$

50. $-[4 - 2(6 - 4)]$ **51.** $3[-5 + 3(2 - 7)]$

52. $\dfrac{3 - 8}{10 - 5}$ **53.** $\dfrac{-32 - 8}{6 - 16}$

Consider the numbers 14, 12, 13, 14, 15, 20, 15, 17, 19, 15.

54. ⊞ Find the mean.   **55.** Find the median.

56. Find the mode.

57. Can the mean, median, and mode of a group of numbers be the same?

Evaluate when $a = 5$, $b = -2$, $c = -3$, and $d = 2$.

58. $\dfrac{3a - 2b}{cd}$

59. $\dfrac{3b + 2d}{ac}$

60. $\dfrac{ab + cd}{c(b - d)}$

61. $\dfrac{ac - bd}{a(d + c)}$

Associative properties:

$$(a + b) + c = a + (b + c)$$
$$(ab)c = a(bc)$$

Commutative properties:

$$a + b = b + a$$
$$ab = ba$$

Distributive property:

$$a(b + c) = ab + ac$$
$$a + 0 = 0 + a = a$$
$$a \cdot 1 = 1 \cdot a = a$$
$$a + (-a) = 0$$

Double negative rule:

$$-(-a) = a$$

If $a \neq 0$, then

$$a\left(\frac{1}{a}\right) = \frac{1}{a} \cdot a = 1$$

Tell which property justifies each statement.

62. $3(4 + 2) = 3 \cdot 4 + 3 \cdot 2$

63. $3 + (x + 7) = (x + 7) + 3$

64. $3 + (x + 7) = (3 + x) + 7$

65. $3 + 0 = 3$

66. $3 + (-3) = 0$

67. $5(3) = 3(5)$

68. $3(xy) = (3x)y$

69. $3x \cdot 1 = 3x$

70. $a\left(\dfrac{1}{a}\right) = 1 \quad a \neq 0$

71. $-(-x) = x$

1.3   Exponents

Properties of exponents:
If there are no divisions by 0,

$$x^n = \overbrace{x \cdot x \cdot x \cdot \,\cdots\, \cdot x}^{n \text{ factors of } x}$$

$$x^m x^n = x^{m+n} \qquad (x^m)^n = x^{mn}$$

$$(xy)^n = x^n y^n \qquad \left(\frac{x}{y}\right)^n = \frac{x^n}{y^n}$$

$$x^0 = 1 \qquad\qquad x^{-n} = \frac{1}{x^n}$$

$$\frac{x^m}{x^n} = x^{m-n} \qquad \left(\frac{x}{y}\right)^{-n} = \left(\frac{y}{x}\right)^n$$

Simplify each expression and write all answers without negative exponents.

72. 3^6

73. -2^6

74. $(-4)^3$

75. -5^{-4}

76. $(3x^4)(-2x^2)$

77. $(-x^5)(3x^3)$

78. $x^{-4}x^3$

79. $x^{-10}x^{12}$

80. $(3x^2)^3$

81. $(4x^4)^4$

82. $(-2x^2)^5$

83. $-(-3x^3)^5$

84. $(x^2)^{-5}$

85. $(x^{-4})^{-5}$

86. $(3x^{-3})^{-2}$

87. $(2x^{-4})^4$

88. $\dfrac{x^6}{x^4}$ **89.** $\dfrac{x^{12}}{x^7}$

90. $\dfrac{a^7}{a^{12}}$ **91.** $\dfrac{a^4}{a^7}$

92. $\dfrac{y^{-3}}{y^4}$ **93.** $\dfrac{y^5}{y^{-4}}$

94. $\dfrac{x^{-5}}{x^{-4}}$ **95.** $\dfrac{x^{-6}}{x^{-9}}$

Simplify each expression and write all answers without negative exponents.

96. $(3x^2y^3)^2$ **97.** $(-3a^3b^2)^{-4}$

98. $\left(\dfrac{3x^2}{4y^3}\right)^{-3}$ **99.** $\left(\dfrac{4y^{-2}}{5y^{-3}}\right)^3$

1.4 Scientific Notation

Scientific notation:
$N \times 10^n$, where $1 \le |N| < 10$ and n is an integer.

Write each numeral in scientific notation.

100. 19,300,000,000

101. 0.0000000273

Write each numeral in standard notation.

102. 7.2×10^7

103. 8.3×10^{-9}

104. Each person in the United States uses approximately 1,640 gallons of water per day. If the population of the United States is 270 million people, how many gallons of water are used each day?

1.5 Solving Equations

If a and b are real numbers and $a = b$, then
$a + c = b + c$
$a - c = b - c$
$ac = bc$ $(c \ne 0)$
$\dfrac{a}{c} = \dfrac{b}{c}$ $(c \ne 0)$

Solve each equation.

105. $5x + 12 = 37$ **106.** $-3x - 7 = 20$

107. $4(y - 1) = 28$ **108.** $3(x + 7) = 42$

109. $13(x - 9) - 2 = 7x - 5$ **110.** $\dfrac{8(x - 5)}{3} = 2(x - 4)$

111. $\dfrac{3y}{4} - 13 = -\dfrac{y}{3}$ **112.** $\dfrac{2y}{5} + 5 = \dfrac{14y}{10}$

Solve for the indicated quantity.

113. $V = \dfrac{4}{3}\pi r^3$ for r^3

114. $V = \dfrac{1}{3}\pi r^2 h$ for h

115. $v = \dfrac{1}{6}ab(x + y)$ for x

116. $V = \pi h^2\left(r - \dfrac{h}{3}\right)$ for r

1.6   Using Equations to Solve Problems

To solve problems, use the following strategy:

1. Analyze the problem.

2. Form an equation.

3. Solve the equation.

4. State the conclusion.

5. Check the result.

117. Carpentry   A carpenter wants to cut a 20-foot rafter so that one piece is three times as long as the other. Where should he cut the board?

118. Geometry   A rectangle is 4 meters longer than it is wide. If the perimeter of the rectangle is 28 meters, find its area.

119. Balancing a seesaw   Sue weighs 48 pounds, and her father weighs 180 pounds. If Sue sits on one end of a 20-foot long seesaw with the fulcrum in the middle, how far from the fulcrum should her father sit to balance the seesaw?

1.7   More Applications of Equations

120. Investment problem   Sally has $25,000 to invest. She invests some money at 10% interest and the rest at 9%. If her total annual income from these two investments is $2,430, how much does she invest at each rate?

121. Mixing solutions   How much water must be added to 20 liters of a 12% alcohol solution to dilute it to an 8% solution?

122. **Motion problem**   A car and a motorcycle both leave from the same point and travel in the same direction. (See the illustration.) The car travels at an average rate of 55 mph and the motorcycle at an average rate of 40 mph. How long will it take before the vehicles are 5 miles apart?

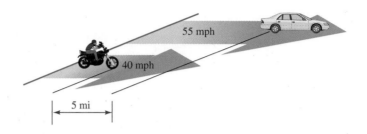

Chapter Test

Let $A = \left\{-2, 0, 1, \frac{6}{5}, 2, \sqrt{7}, 5\right\}$.

1. What numbers in A are natural numbers?

2. What numbers in A are irrational numbers?

Graph each set on the number line.

3. The set of odd integers between -4 and 6

    $\xleftarrow{\quad}\overset{\;\;-4\;\;-3\;\;-2\;\;-1\;\;\;0\;\;\;\;1\;\;\;\;2\;\;\;\;3\;\;\;\;4\;\;\;\;5\;\;\;\;6}{\rule{0pt}{0pt}}\xrightarrow{\quad}$

4. The set of prime numbers less than 12

    $\xleftarrow{\quad}\overset{\;\;0\;\;\;1\;\;\;2\;\;\;3\;\;\;4\;\;\;5\;\;\;6\;\;\;7\;\;\;8\;\;\;9\;\;\;10\;\;11\;\;12}{\rule{0pt}{0pt}}\xrightarrow{\quad}$

Graph each set on the number line.

5. $\{x \mid x > 4\}$

6. $[-3, \infty)$

7. $\{x \mid -2 \leq x < 4\}$

8. $(-\infty, -1] \cup [2, \infty)$

Write each expression without using absolute value symbols.

9. $-|8|$

10. $|-5|$

Perform the operations.

11. $7 + (-5)$

12. $-5(-4)$

13. $\dfrac{12}{-3}$

14. $-4 - \dfrac{-15}{3}$

Consider the numbers $-2, 0, 2, -2, 3, -1, -1, 1, 1, 2$.

15. Find the mean.

16. Find the median.

Let $a = 2$, $b = -3$, and $c = 4$ and evaluate each expression.

17. ab

18. $a + bc$

19. $ab - bc$

20. $\dfrac{-3b + a}{ac - b}$

Tell which property of real numbers justifies each statement.

21. $3 + 5 = 5 + 3$

22. $a(b + c) = ab + ac$

Simplify. Write all answers without using negative exponents. Assume that no denominators are zero.

23. $x^3 x^5$

24. $(2x^2 y^3)^3$

25. $(m^{-4})^2$

26. $\left(\dfrac{m^2 n^3}{m^4 n^{-2}}\right)^{-2}$

Write each numeral in scientific notation.

27. 4,700,000

28. 0.00000023

Write each numeral in standard notation.

29. 6.53×10^5

30. 24.5×10^{-3}

Solve each equation.

31. $9(x + 4) + 4 = 4(x - 5)$

32. $\dfrac{y - 1}{5} + 2 = \dfrac{2y - 3}{3}$

33. Solve $P = L + \dfrac{s}{f} i$ for i.

34. A rectangle has a perimeter of 26 centimeters and is 5 centimeters longer than it is wide. Find its area.

35. Jamie invests part of $10,000 at 9% annual interest and the rest at 8%. If his annual income from these investments is $860, how much does he invest at 8%?

36. How many liters of water are needed to dilute 20 liters of a 5% salt solution to a 1% solution?

2 Graphs, Equations of Lines, and Functions

✎ *InfoTrac Project*

Do a keyword search on "linear relationships" and find the article "Storage characteristics and nutritive value changes in Bermudagrass hay as affected by moisture content and density of rectangular bales." According to the article, when the moisture level was 325 g, the visible mold was 3.73. At a moisture level of 178 g, the visible mold was 1.13. Using this information, develop an equation showing the relationship between moisture level and visible mold. Round the slope to the nearest thousandth. Use your equation to predict the mold level for a moisture level of 248 g. How does your prediction compare with the actual measurement of mold reported in the article for a moisture level of 248 g? According to the article, what effect did bale density have on nutritive value?

Complete this project after studying Section 2.3.

©Owen Franken/CORBIS

Mathematics in Economics

An electronics firm manufacturers DVD players, receiving $120 for each unit it produces. If *x* represents the number of units produced, the income received is determined by the *revenue function*

$$R(x) = 120x$$

If the manufacturer has fixed costs of $12,000 per month and variable costs of $57.50 for each unit it produces, the *cost function* is

Exercise Set 2.4 Problem 83

$$C(x) = \textit{variable costs} + \textit{fixed costs}$$
$$= 57.50x + 12{,}000$$

How many units must the manufacturer sell for revenue to equal cost?

It is often said that "A picture is worth a thousand words." In this chapter, we will show how numerical relationships can be described by using mathematical pictures called **graphs.**

| 2.1 | Graphing Linear Equations |

- **The Rectangular Coordinate System**   ▮ **Graphing Linear Equations**
- **Horizontal and Vertical Lines**   ▮ **Applications**
- **The Midpoint Formula**

Getting Ready   *Graph each set of numbers on the number line.*

1. $\{-2, 1, 3\}$

$$\begin{array}{ccccccccc} & | & | & | & | & | & | & | & | \\ -3 & -2 & -1 & 0 & 1 & 2 & 3 & 4 \end{array}$$

2. $\{x \mid x > -2\}$

3. $\{x \mid x \le 3\}$

4. $(-3, 2)$

Many cities are laid out on a rectangular grid. For example, on the east side of Rockford, all streets run north and south and all avenues run east and west. (See Figure 2-1.) If we agree to list the street numbers first, every location can be identified by using an ordered pair of numbers. If Jose lives on the corner of Third Street and Sixth Avenue, his location is given by the ordered pair (3, 6).

This is the street. ────┐
                          ↓
          (3, 6)
                 ↑
                 └──── This is the avenue.

If Lisa has a location of (6, 3), we know that she lives on the corner of Sixth Street and Third Avenue. From the figure, we can see that

- Bob Anderson's location is (4, 1).
- Rosa Vang's location is (7, 5).
- The location of the store is (8, 2).

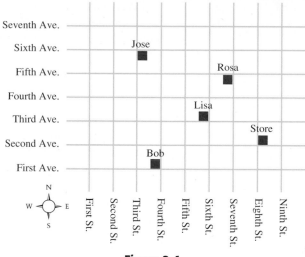

Figure 2-1

The idea of associating an ordered pair of numbers with points on a grid is attributed to the 17th-century French mathematician René Descartes. The grid is often called a **rectangular coordinate system,** or **Cartesian coordinate system,** after its inventor.

The Rectangular Coordinate System

A rectangular coordinate system (see Figure 2-2) is formed by two intersecting perpendicular number lines.

- The horizontal number line is usually called the *x*-axis.
- The vertical number line is usually called the *y*-axis.

The positive direction on the *x*-axis is the right, and the positive direction on the *y*-axis is upward. If no scale is indicated on the axes, we assume that the axes are scaled in units of 1.

The point where the axes cross is called the **origin.** This is the 0 point on each axis. The two axes form a **coordinate plane** and divide it into four regions called **quadrants,** which are numbered as in Figure 2-2.

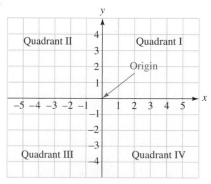

Figure 2-2

Every point in a coordinate plane can be identified by a pair of real numbers x and y, written as (x, y). The first number in the pair is the **x-coordinate,** and the second number is the **y-coordinate.** The numbers are called the **coordinates of the point.** Some examples of ordered pairs are $(-4, 6)$, $(2, 3)$, and $(6, -4)$.

$$(-4, 6)$$

In an ordered pair, the ⎯⎯↑ ⌐⎯ The y-coordinate
x-coordinate is listed first.   is listed second.

The process of locating a point in the coordinate plane is called **graphing** or **plotting** the point. In Figure 2-3(a), we show how to graph the point Q with coordinates of $(-4, 6)$. Since the **x-coordinate** is negative, we start at the origin and move 4 units to the left along the x-axis. Since the **y-coordinate** is positive, we then move up 6 units to locate point Q. Point Q is the **graph** of the ordered pair $(-4, 6)$ and lies in quadrant II.

To plot the point $P(2, 3)$, we start at the origin, move 2 units to the right along the x-axis, and then move up 3 units to locate point P. Point P lies in quadrant I. To plot point $R(6, -4)$, we start at the origin and move 6 units to the right and then 4 units down. Point R lies in quadrant IV.

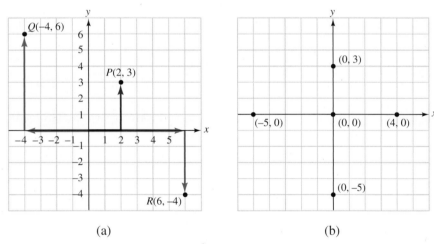

(a)   (b)

Figure 2-3

💬 **Comment**   Note that point Q with coordinates of $(-4, 6)$ is not the same as point R with coordinates $(6, -4)$. Since the order of the coordinates of a point is important, we call the pairs **ordered pairs.**

In Figure 2-3(b), we see that the points $(-5, 0)$, $(0, 0)$, and $(4, 0)$ all lie on the x-axis. In fact, every point with a y-coordinate of 0 will lie on the x-axis. We also see that the points $(0, -5)$, $(0, 0)$, and $(0, 3)$ all lie on the y-axis. In fact, every point with an x-coordinate of 0 will lie on the y-axis. The coordinates of the origin are $(0, 0)$.

EXAMPLE 1   **Tracking hurricanes**   The map shown in Figure 2-4 shows the path of a hurricane. The hurricane is currently located at 85° longitude and 25° latitude. If we agree to list longitude first, the location of the hurricane is given by the ordered pair $(85, 25)$. Use the map to answer the following questions.

a. What are the coordinates of New Orleans?

b. What are the estimated coordinates of Houston?

c. What are the estimated coordinates of the landfall of the hurricane?

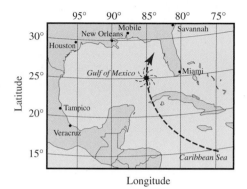

Figure 2-4

Solution   **a.** Since New Orleans is at 90° longitude and 30° latitude, its coordinates are (90, 30).

         b. Since Houston has a longitude that is a little more than 95° and a latitude that is a little less than 30°, its estimated coordinates are (96, 29).

         c. The hurricane appears to be headed for the point with coordinates of (83, 29).

Graphing Linear Equations

The equation $y = -\frac{1}{2}x + 4$ contains the two variables x and y. The solutions of this equation form a set of ordered pairs of real numbers. For example, the ordered pair $(-4, 6)$ is a solution, because the equation is satisfied when $x = -4$ and $y = 6$.

$$y = -\frac{1}{2}x + 4$$

$$6 = -\frac{1}{2}(-4) + 4 \quad \text{Substitute } -4 \text{ for } x \text{ and } 6 \text{ for } y.$$

$$6 = 2 + 4$$

$$6 = 6$$

This pair and others that satisfy the equation are listed in the table shown in Figure 2-5.

$$y = -\tfrac{1}{2}x + 4$$

x	y
-4	6
-2	5
0	4
2	3
4	2

Figure 2-5

The **graph of the equation** $y = -\frac{1}{2}x + 4$ is the graph of all points (x, y) on the rectangular coordinate system whose coordinates satisfy the equation.

EXAMPLE 2   Graph: $y = -\frac{1}{2}x + 4$.

Solution   To graph the equation, we plot the ordered pairs listed in the table shown in Figure 2-5. These points appear to lie on a line, as shown in Figure 2-6. In fact, if we were to plot many more pairs that satisfied the equation, it would become obvious that the resulting points will all lie on the line. This line is the graph of the equation.

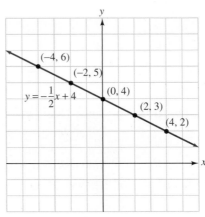

Figure 2-6

When we say that the graph of an equation is a line, we imply two things:

1. Every point with coordinates that satisfy the equation will lie on the line.
2. Any point on the line will have coordinates that satisfy the equation.

Self Check   Graph: $y = 2x - 3$.   ∎

When the graph of an equation is a line, we call the equation a **linear equation.** Linear equations are often written in the form $Ax + By = C$, where A, B, and C stand for specific numbers (called **constants**) and x and y are variables. This form is called the **general form of the equation of a line.**

EXAMPLE 3   Graph: $3x + 2y = 12$.

Solution   We can pick values for either x or y, substitute them into the equation, and solve for the other variable. For example, if $x = 2$,

$$3x + 2y = 12$$
$$3(2) + 2y = 12 \quad \text{Substitute 2 for } x.$$
$$6 + 2y = 12 \quad \text{Simplify.}$$
$$2y = 6 \quad \text{Subtract 6 from both sides.}$$
$$y = 3 \quad \text{Divide both sides by 2.}$$

The ordered pair (2, 3) satisfies the equation. If $y = 6$,

$$3x + 2y = 12$$
$$3x + 2(6) = 12 \qquad \text{Substitute 6 for } y.$$
$$3x + 12 = 12 \qquad \text{Simplify.}$$
$$3x = 0 \qquad \text{Subtract 12 from both sides.}$$
$$x = 0 \qquad \text{Divide both sides by 3.}$$

A second ordered pair that satisfies the equation is (0, 6).

These pairs and others that satisfy the equation lie on a line, as shown in Figure 2-7. The graph of the equation is the line shown in the figure.

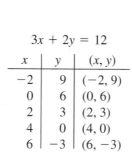

$$3x + 2y = 12$$

x	y	(x, y)
-2	9	$(-2, 9)$
0	6	$(0, 6)$
2	3	$(2, 3)$
4	0	$(4, 0)$
6	-3	$(6, -3)$

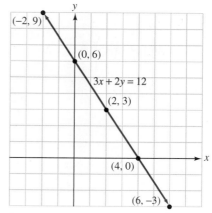

Figure 2-7

Self Check   Graph: $3x - 2y = 6$.

In Example 3, the graph intersects the y-axis at the point with coordinates (0, 6), called the **y-intercept,** and intersects the x-axis at the point with coordinates (4, 0), called the **x-intercept.** In general, we have the following definitions.

Intercepts of a Line

> The **y-intercept** of a line is the point $(0, b)$ where the line intersects the y-axis. To find b, substitute 0 for x in the equation of the line and solve for y.
>
> The **x-intercept** of a line is the point $(a, 0)$ where the line intersects the x-axis. To find a, substitute 0 for y in the equation of the line and solve for x.

EXAMPLE 4   Use the x- and y-intercepts to graph $2x + 5y = 10$.

Solution   To find the y-intercept, we substitute 0 for x and solve for y:

$$2x + 5y = 10$$
$$2(0) + 5y = 10 \qquad \text{Substitute 0 for } x.$$
$$5y = 10 \qquad \text{Simplify.}$$
$$y = 2 \qquad \text{Divide both sides by 5.}$$

The y-intercept is the point $(0, 2)$. To find the x-intercept, we substitute 0 for y and solve for x:

$$2x + 5y = 10$$
$$2x + 5(\mathbf{0}) = 10 \qquad \text{Substitute 0 for } y.$$
$$2x = 10 \qquad \text{Simplify.}$$
$$x = 5 \qquad \text{Divide both sides by 2.}$$

The x-intercept is the point $(5, 0)$.

Although two points are enough to draw a line, we plot a third point as a check. To find the coordinates of a third point, we can substitute any convenient number (such as -5) for x and solve for y:

$$2\mathbf{x} + 5y = 10$$
$$2(\mathbf{-5}) + 5y = 10 \qquad \text{Substitute } -5 \text{ for } x.$$
$$-10 + 5y = 10 \qquad \text{Simplify.}$$
$$5y = 20 \qquad \text{Add 10 to both sides.}$$
$$y = 4 \qquad \text{Divide both sides by 5.}$$

The line will also pass through the point $(-5, 4)$.

A table of ordered pairs and the graph are shown in Figure 2-8.

$$2x + 5y = 10$$

x	y	(x, y)
-5	4	$(-5, 4)$
0	2	$(0, 2)$
5	0	$(5, 0)$

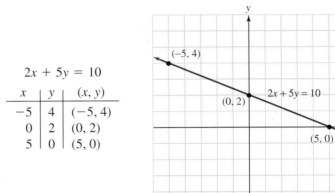

Figure 2-8

Self Check   Find the x- and y-intercepts and graph $y = 3x + 4$.   ∎

The method of graphing a line used in Example 4 is called the **intercept method.**

Horizontal and Vertical Lines

EXAMPLE 5   Graph: **a.** $y = 3$ and   **b.** $x = -2$.

Solution   **a.** Since the equation $y = 3$ does not contain x, the numbers chosen for x have no effect on y. The value of y is always 3.

After plotting the pairs (x, y) shown in Figure 2-9, we see that the graph is a horizontal line with a y-intercept of $(0, 3)$. The line has no x-intercept.

b. Since the equation $x = -2$ does not contain y, y can be any number.

After plotting the pairs (x, y) shown in Figure 2-9, we see that the graph is a vertical line with an x-intercept of $(-2, 0)$. The line has no y-intercept.

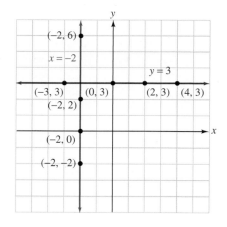

$y = 3$

x	y	(x, y)
-3	3	$(-3, 3)$
0	3	$(0, 3)$
2	3	$(2, 3)$
4	3	$(4, 3)$

$x = -2$

x	y	(x, y)
-2	-2	$(-2, -2)$
-2	0	$(-2, 0)$
-2	2	$(-2, 2)$
-2	6	$(-2, 6)$

Figure 2-9

Self Check   Graph $x = 4$ and $y = -3$ on one set of coordinate axes.   ∎

The results of Example 5 suggest the following facts.

Horizontal and Vertical Lines

If a and b are real numbers, then

The graph of $x = a$ is a vertical line with x-intercept at $(a, 0)$. If $a = 0$, the line is the y-axis.

The graph of $y = b$ is a horizontal line with y-intercept at $(0, b)$. If $b = 0$, the line is the x-axis.

Applications

EXAMPLE 6   The following table gives the number of miles (y) that a bus can go on x gallons of gas. Plot the ordered pairs and estimate how far the bus can go on 9 gallons.

x	2	3	4	5	6
y	12	18	24	30	36

Solution   Since the distances driven are rather large numbers, we plot the points on a coordinate system where the unit distance on the x-axis is larger than the unit distance on the y-axis. After plotting each ordered pair as in Figure 2-10, we see that the points lie on a line.

To estimate how far the bus can go on 9 gallons, we find 9 on the x-axis, move up to the graph, and then move to the left to locate a y-value of 54. Thus, the bus can go approximately 54 miles on 9 gallons of gas.

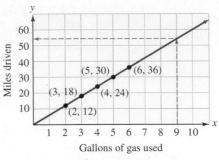

Figure 2-10

Self Check   How far can the bus go on 10 gallons?

EXAMPLE 7   **Depreciation**   A copy machine purchased for $6,750 is expected to depreciate according to the formula $y = -950x + 6,750$, where y is the value of the copier after x years. When will the copier have no value?

Solution   The copier will have no value when its value (y) is 0. To find x when $y = 0$, we substitute 0 for y and solve for x.

$$y = -950x + 6,750$$
$$0 = -950x + 6,750$$
$$-6,750 = -950x \qquad \text{Subtract 6,750 from both sides.}$$
$$7.105263158 = x \qquad \text{Divide both sides by } -950.$$

The copier will be worthless in about 7.1 years.

Self Check   When will the copier be worth $2,000?

The Midpoint Formula

To distinguish between the coordinates of two points on a line, we often use subscript notation. Point $P(x_1, y_1)$ is read as "point P with coordinates of x sub 1 and y sub 1." Point $Q(x_2, y_2)$ is read as "point Q with coordinates of x sub 2 and y sub 2."

If point M in Figure 2-11 lies midway between points $P(x_1, y_1)$ and $Q(x_2, y_2)$, point M is called the **midpoint** of segment PQ. To find the coordinates of M, we find the mean of the x-coordinates and the mean of the y-coordinates of P and Q.

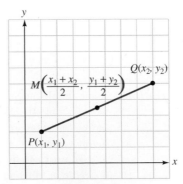

Figure 2-11

The Midpoint Formula

The **midpoint** of the line segment with endpoints at $P(x_1, y_1)$ and $Q(x_2, y_2)$ is the point M with coordinates of

$$\left(\frac{x_1 + x_2}{2}, \frac{y_1 + y_2}{2} \right)$$

EXAMPLE 8 Find the midpoint of the line segment joining $P(-2, 3)$ and $Q(7, -5)$.

Solution To find the midpoint, we find the mean of the x-coordinates and the mean of the y-coordinates to get

$$\frac{x_1 + x_2}{2} = \frac{-2 + 7}{2} \qquad \text{and} \qquad \frac{y_1 + y_2}{2} = \frac{3 + (-5)}{2}$$
$$= \frac{5}{2} \qquad\qquad\qquad = -1$$

The midpoint of segment PQ is the point $M\left(\frac{5}{2}, -1\right)$.

Self Check Find the midpoint of the segment joining $P(5, -4)$ and $Q(-3, 5)$. ∎

Accent on Technology

GRAPHING LINES

Courtesy Texas Instruments Incorporated

Figure 2-12

We have graphed linear equations by finding ordered pairs, plotting points, and drawing a line through those points. Graphing is much easier if we use a graphing calculator.

Graphing calculators have a window to display graphs (see Figure 2-12). To see the proper picture of a graph, we must decide on the minimum and maximum values for the x- and y-coordinates. A window with standard settings of

$$\text{Xmin} = -10 \qquad \text{Xmax} = 10 \qquad \text{Ymin} = -10 \qquad \text{Ymax} = 10$$

will produce a graph where the value of x is in the interval $[-10, 10]$ and y is in the interval $[-10, 10]$.

To graph $3x + 2y = 12$, we must first solve the equation for y.

$$3x + 2y = 12$$
$$2y = -3x + 12 \qquad \text{Subtract } 3x \text{ from both sides.}$$
$$y = -\frac{3}{2}x + 6 \qquad \text{Divide both sides by 2.}$$

To graph the equation, we enter the right-hand side of the equation after a symbol like $\backslash Y_1 =$ or $f(x) =$. After entering the right-hand side, the display should look like

$$\backslash Y_1 = -(3/2)X + 6 \qquad \text{or} \qquad f(x) = -(3/2)X + 6$$

We then press the GRAPH key to get the graph shown in Figure 2-13(a). To show more detail, we can draw the graph in a different window. A window with settings of $[-1, 5]$ for x and $[-2, 7]$ for y will give the graph shown in Figure 2-13(b).

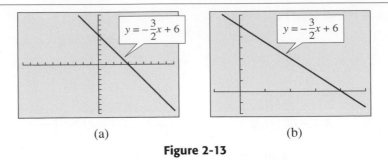

(a)   (b)

Figure 2-13

We can use the trace command to find the coordinates of any point on a graph. For example, to find the x-intercept of the graph of $y = -\frac{3}{2}x + 6$, we graph the equation as in Figure 2-13(a) and press the TRACE key to get Figure 2-14(a). We then use the $>$ and $<$ keys to move the cursor along the line toward the x-intercept until we arrive at the point with coordinates shown in Figure 2-14(b). To get better results, we can zoom in to get a magnified picture, trace again, and move the cursor to the point with the coordinates shown in Figure 2-14(c). Since the y-coordinate is nearly 0, this point is nearly the x-intercept. We can achieve better results with more zooms.

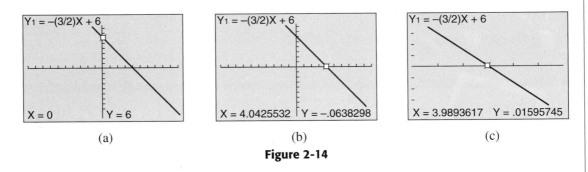

(a)   (b)   (c)

Figure 2-14

We can get the result more directly by using the zero command found in the CALC menu on a TI-83 Plus graphing calculator. First, we graph the equation as in Figure 2-13(a). Then we press 2nd CALC 2 to obtain Figure 2-15(a). We then enter a left-bound guess such as 2, press ENTER, enter a right-bound guess such as 5, press ENTER, and press ENTER again to obtain Figure 2-15(b). From the figure, we see that the x-intercept is the point $(4, 0)$.

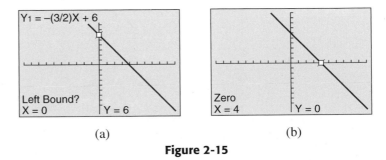

(a)   (b)

Figure 2-15

Self Check Answers

2.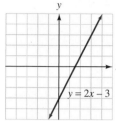

3.

4.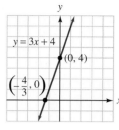

5.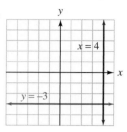

6. 60 miles   **7.** 5 years   **8.** $\left(1, \frac{1}{2}\right)$

Orals   *Find the x- and y-intercepts of each line.*

1. $x + y = 3$ **2.** $3x + y = 6$

3. $x + 4y = 8$ **4.** $3x - 4y = 12$

Tell whether the graphs of the equations are horizontal or vertical.

5. $x = -6$ **6.** $y = 8$

Find the midpoint of a line segment with endpoints at

7. $(2, 4), (6, 8)$ **8.** $(-4, 6), (4, -8)$

2.1   EXERCISES

REVIEW

1. Evaluate: $-3 - 3(-5)$.
2. Evaluate: $(-5)^2 + (-5)$.
3. Simplify: $\dfrac{-3 + 5(2)}{9 + 5}$.
4. Simplify: $|-1 - 9|$.
5. Solve: $-4x + 7 = -21$.
6. Solve $P = 2l + 2w$ for w.

VOCABULARY AND CONCEPTS   *Fill in the blanks.*

7. The pair of numbers $(6, -2)$ is called an _____.
8. In the ordered pair $(-2, -9)$, -9 is called the ___ coordinate.
9. The point with coordinates $(0, 0)$ is the _____.
10. The x- and y-axes divide the coordinate plane into four regions called _____.
11. Ordered pairs of numbers can be graphed on a _____ system.
12. The process of locating the position of a point on a coordinate plane is called _____ the point.

13. The point where a graph intersects the y-axis is called _____.
14. The point where a graph intersects the _____ is called the x-intercept.
15. The graph of any equation of the form $x = a$ is a _____ line.
16. The graph of any equation of the form $y = b$ is a _____ line.
17. The symbol x_1 is read as "x _____."
18. The midpoint of a line segment joining (a, b) and (c, d) is given by the formula _____.

PRACTICE   *Plot each point on the rectangular coordinate system.*

19. $A(4, 3)$
20. $B(-2, 1)$
21. $C(3, -2)$
22. $D(-2, -3)$
23. $E(0, 5)$
24. $F(-4, 0)$
25. $G(2, 0)$
26. $H(0, 3)$

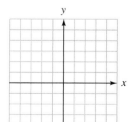

Give the coordinates of each point.

27. *A*

28. *B*

29. *C*

30. *D*

31. *E*

32. *F*

33. *G*

34. *H*

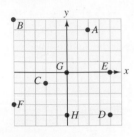

Complete each table.

35. $y = -x + 4$

x	y
-1	
0	
2	

36. $y = x - 2$

x	y
-2	
0	
4	

37. $y = 2x - 3$

x	y
-1	
0	
3	

38. $y = -\dfrac{1}{2}x + \dfrac{5}{2}$

x	y
-3	
-1	
3	

Graph each equation. See Exercises 35–38.

39. $y = -x + 4$

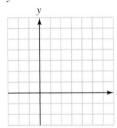

40. $y = x - 2$

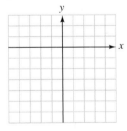

41. $y = 2x - 3$

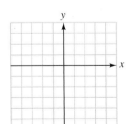

42. $y = -\dfrac{1}{2}x + \dfrac{5}{2}$

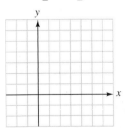

Graph each equation.

43. $3x + 4y = 12$

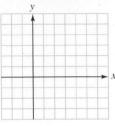

44. $4x - 3y = 12$

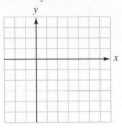

45. $y = -3x + 2$

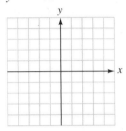

46. $y = 2x + 3$

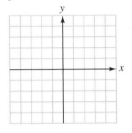

47. $y = \dfrac{3}{2}x$

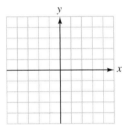

48. $y = -\dfrac{2}{3}x$

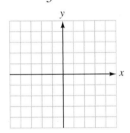

49. $3y = 6x - 9$

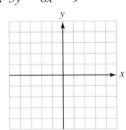

50. $2x = 4y - 10$

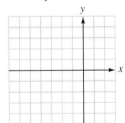

51. $3x + 4y - 8 = 0$

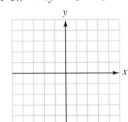

52. $-2y - 3x + 9 = 0$

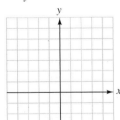

53. $x = 3$

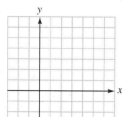

54. $y = -4$

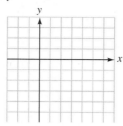

55. $-3y + 2 = 5$

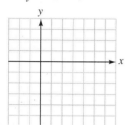

56. $-2x + 3 = 11$

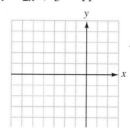

Find the midpoint of segment PQ.

57. $P(0, 0)$, $Q(6, 8)$

58. $P(10, 12)$, $Q(0, 0)$

59. $P(6, 8)$, $Q(12, 16)$

60. $P(10, 4)$, $Q(2, -2)$

61. $P(2, 4)$, $Q(5, 8)$

62. $P(5, 9)$, $Q(8, 13)$

63. $P(-2, -8)$, $Q(3, 4)$

64. $P(-5, -2)$, $Q(7, 3)$

65. $Q(-3, 5)$, $P(-5, -5)$

66. $Q(2, -3)$, $P(4, -8)$

67. Finding the endpoint of a segment If $M(-2, 3)$ is the midpoint of segment PQ and the coordinates of P are $(-8, 5)$, find the coordinates of Q.

68. Finding the endpoint of a segment If $M(6, -5)$ is the midpoint of segment PQ and the coordinates of Q are $(-5, -8)$, find the coordinates of P.

Use a graphing calculator to graph each equation, and then find the x-coordinate of the x-intercept to the nearest hundredth.

69. $y = 3.7x - 4.5$

70. $y = \dfrac{3}{5}x + \dfrac{5}{4}$

71. $1.5x - 3y = 7$

72. $0.3x + y = 7.5$

APPLICATIONS

73. Hourly wages The table gives the amount y (in dollars) that a student can earn for working x hours. Plot the ordered pairs and estimate how much the student will earn for working 8 hours.

x	2	4	5	6
y	12	24	30	36

74. Distance traveled The table shows how far y (in miles) a biker can go in x hours. Plot the ordered pairs and estimate how far the biker can go in 8 hours.

x	2	4	5	6
y	30	60	75	90

75. Value of a car The table shows the value y (in dollars) of a car that is x years old. Plot the ordered pairs and estimate the value of the car when it is 4 years old.

x	0	1	3
y	15,000	12,000	6,000

76. Earning interest The table shows the amount y (in dollars) in a bank account drawing simple interest left on deposit for x years. Plot the ordered pairs and estimate the value of the account in 6 years.

x	0	1	4
y	1,000	1,050	1,200

77. House appreciation A house purchased for $125,000 is expected to appreciate according to the formula $y = 7,500x + 125,000$, where y is the value of the house after x years. Find the value of the house 5 years later.

78. Car depreciation A car purchased for $17,000 is expected to depreciate according to the formula $y = -1,360x + 17,000$, where y is the value of the car after x years. When will the car be worthless?

79. Demand equation   The number of TV sets that consumers buy depends on price. The higher the price, the fewer people will buy. The equation that relates price to the number of TVs sold at that price is called a **demand equation.** If the demand equation for a 13-inch TV is $p = -\frac{1}{10}q + 170$, where p is the price and q is the number of TVs sold at that price, how many TVs will be sold at a price of $150?

80. Supply equation   The number of TV sets that manufacturers produce depends on price. The higher the price, the more TVs manufacturers will produce. The equation that relates price to the number produced at that price is called a **supply equation.** If the supply equation for a 13-inch TV is $p = \frac{1}{10}q + 130$, where p is the price and q is the number produced for sale at that price, how many TVs will be produced if the price is $150?

81. TV coverage   In the illustration, a TV camera is located at point $(-3, 0)$. The camera is to follow the launch of a space shuttle. As the shuttle rises vertically, the camera can tilt back to a line of sight given by $y = 2x + 6$. Estimate how many miles the shuttle will remain in the camera's view.

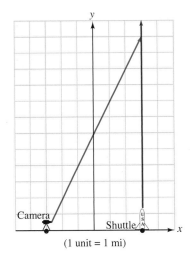

(1 unit = 1 mi)

82. Buying tickets   Tickets to the circus cost $5 each plus a $2 service fee for each block of tickets.
  a. Write a linear equation that gives the cost y for a student buying x tickets.
  b. Complete the table in the illustration and graph the equation.
  c. Use the graph to find the cost of buying 4 tickets.

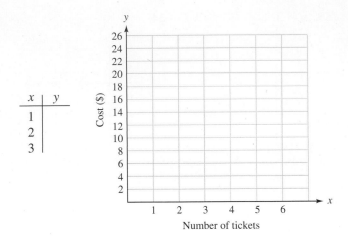

x	y
1	
2	
3	

Number of tickets

83. Telephone costs   In one community, the monthly cost of local telephone service is $5 per month, plus 25¢ per call.
  a. Write a linear equation that gives the cost y for a person making x calls.
  b. Complete the table in the illustration and graph the equation.
  c. Use the graph to estimate the cost of service in a month when 20 calls were made.

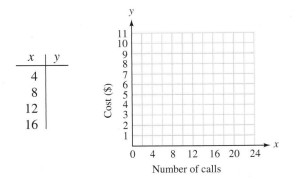

x	y
4	
8	
12	
16	

Number of calls

84. Crime prevention   The number n of incidents of family violence requiring police response appears to be related to d, the money spent on crisis intervention, by the equation $n = 430 - 0.005d$. What expenditure would reduce the number of incidents to 350?

WRITING

85. Explain how to graph a line using the intercept method.

86. Explain how to determine the quadrant in which the point $P(a, b)$ lies.

SOMETHING TO THINK ABOUT

87. If the line $y = ax + b$ passes only through quadrants I and II, what can be known about the constants a and b?

88. What are the coordinates of the three points that divide the line segment joining $P(a, b)$ and $Q(c, d)$ into four equal parts?

2.2	**Slope of a Nonvertical Line**

▌ **Slope of a Line**   ▌ **Interpretation of Slope**
▌ **Horizontal and Vertical Lines**   ▌ **Slopes of Parallel Lines**
▌ **Slopes of Perpendicular Lines**

Getting Ready   *Simplify each expression:*

1. $\dfrac{6 - 3}{8 - 5}$ **2.** $\dfrac{10 - 4}{2 - 8}$ **3.** $\dfrac{25 - 12}{9 - (-5)}$ **4.** $\dfrac{-9 - (-6)}{-4 - 10}$

We have seen that two points can be used to graph a line. Later, we will show that we can graph a line if we know the coordinates of only one point and the slant of the line. A measure of this slant is called the *slope* of the line.

Slope of a Line

A service offered by an online research company costs $2 per month plus $3 for each hour of connect time. The table in Figure 2-16(a) gives the cost y for certain numbers of hours x of connect time. If we construct a graph from this data, we get the line shown in Figure 2-16(b).

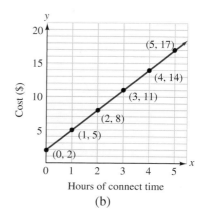

Hours of connect time					
x	0	1	2	3	4	5
y	2	5	8	11	14	17
Cost					

(a)

(b)

Figure 2-16

From the graph, we can see that if x changes from 0 to 1, y changes from 2 to 5. As x changes from 1 to 2, y changes from 5 to 8, and so on. The ratio of the change in y divided by the change in x is the constant 3.

$$\frac{\text{Change in } y}{\text{Change in } x} = \frac{5-2}{1-0} = \frac{8-5}{2-1} = \frac{11-8}{3-2} = \frac{14-11}{4-3} = \frac{17-14}{5-4} = \frac{3}{1} = 3$$

The ratio of the change in y divided by the change in x between any two points on any line is always a constant. This constant rate of change is called the **slope** of the line.

Slope of a Nonvertical Line

The slope of the nonvertical line passing through points (x_1, y_1) and (x_2, y_2) is

$$m = \frac{\text{change in } y}{\text{change in } x} = \frac{y_2 - y_1}{x_2 - x_1} \quad (x_2 \neq x_1)$$

EXAMPLE 1   Find the slope of the line shown in Figure 2-17.

Solution   We can let $(x_1, y_1) = (-2, 4)$ and $(x_2, y_2) = (3, -4)$. Then

$$m = \frac{\text{change in } y}{\text{change in } x}$$

$$= \frac{y_2 - y_1}{x_2 - x_1}$$

$$= \frac{-4 - 4}{3 - (-2)}$$

Substitute -4 for y_2, 4 for y_1, 3 for x_2, and -2 for x_1.

$$= \frac{-8}{5}$$

$$= -\frac{8}{5}$$

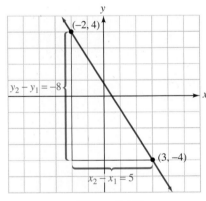

Figure 2-17

The slope of the line is $-\frac{8}{5}$. We would obtain the same result if we let $(x_1, y_1) = (3, -4)$ and $(x_2, y_2) = (-2, 4)$.

Self Check   Find the slope of the line joining the points $(-3, 6)$ and $(4, -8)$.

   Comment   When calculating slope, always subtract the y-values and the x-values in the same order.

$$m = \frac{y_2 - y_1}{x_2 - x_1} \qquad \text{or} \qquad m = \frac{y_1 - y_2}{x_1 - x_2}$$

However,

$$m \neq \frac{y_2 - y_1}{x_1 - x_2} \qquad \text{and} \qquad m \neq \frac{y_1 - y_2}{x_2 - x_1}$$

The change in y (often denoted as Δy) is the **rise** of the line between two points. The change in x (often denoted as Δx) is the **run.** Using this terminology, we can define slope to be the ratio of the rise to the run:

$$m = \frac{\Delta y}{\Delta x} = \frac{\text{rise}}{\text{run}} \quad (\Delta x \neq 0)$$

EXAMPLE 2   Find the slope of the line determined by $3x - 4y = 12$.

Solution   We first find the coordinates of two points on the line.

- If $x = 0$, then $y = -3$. The point $(0, -3)$ is on the line.
- If $y = 0$, then $x = 4$. The point $(4, 0)$ is on the line.

We then refer to Figure 2-18 and find the slope of the line between $(0, -3)$ and $(4, 0)$ by substituting 0 for y_2, -3 for y_1, 4 for x_2, and 0 for x_1 in the formula for slope.

$$m = \frac{\Delta y}{\Delta x}$$

$$= \frac{\mathbf{y_2 - y_1}}{\mathbf{x_2 - x_1}}$$

$$= \frac{\mathbf{0 - (-3)}}{\mathbf{4 - 0}}$$

$$= \frac{3}{4}$$

The slope of the line is $\frac{3}{4}$.

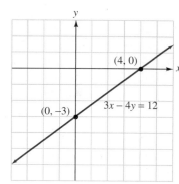

Figure 2-18

Self Check   Find the slope of the line determined by $2x + 5y = 12$.   ∎

Interpretation of Slope

Many applied problems involve equations of lines and their slopes.

EXAMPLE 3   **Cost of carpet**   A store sells a carpet for $25 per square yard, plus a $20 delivery charge. The total cost c of n square yards is given by the following formula.

c equals	cost per square yard	times	the number of square yards	plus	the delivery charge.
c =	25	·	n	+	20

Graph this equation and interpret the slope of the line.

Solution   We can graph the equation on a coordinate system with a vertical c-axis and a horizontal n-axis. Figure 2-19 shows a table of ordered pairs and the graph.

$$c = 25n + 20$$

n	c	(n, c)
10	270	(10, 270)
20	520	(20, 520)
30	770	(30, 770)
40	1,020	(40, 1,020)
50	1,270	(50, 1,270)

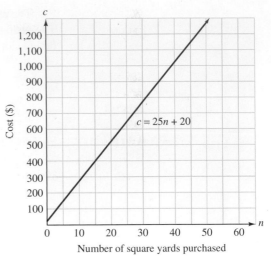

Figure 2-19

If we pick the points (30, 770) and (50, 1,270) to find the slope, we have

$$m = \frac{\Delta c}{\Delta n}$$

$$= \frac{c_2 - c_1}{n_2 - n_1}$$

$$= \frac{1{,}270 - 770}{50 - 30} \qquad \text{Substitute 1,270 for } c_2, \text{ 770 for } c_1, \text{ 50 for } n_2, \text{ and 30 for } n_1.$$

$$= \frac{500}{20}$$

$$= 25$$

The slope of 25 is the cost of the carpet in dollars per square yard.

Self Check   Interpret the y-intercept of the graph in Figure 2-19. ∎

EXAMPLE 4   **Rate of descent**   It takes a skier 25 minutes to complete the course shown in Figure 2-20. Find his average rate of descent in feet per minute.

Solution   To find the average rate of descent, we must find the ratio of the change in altitude to the change in time. To find this ratio, we calculate the slope of the line passing through the points (0, 12,000) and (25, 8,500).

$$\text{Average rate of descent} = \frac{12{,}000 - 8{,}500}{0 - 25}$$

$$= \frac{3{,}500}{-25}$$

$$= -140$$

The average rate of descent is -140 ft/min.

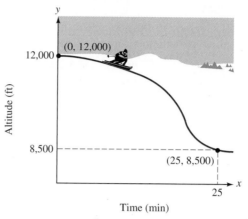

Figure 2-20

Self Check   Find the average rate of descent if the skier completes the course in 20 minutes.   ▌

Horizontal and Vertical Lines

If $P(x_1, y_1)$ and $Q(x_2, y_2)$ are points on the horizontal line shown in Figure 2-21(a), then $y_1 = y_2$, and the numerator of the fraction

$$\frac{y_2 - y_1}{x_2 - x_1} \qquad \text{On a horizontal line, } x_2 \neq x_1.$$

is 0. Thus, the value of the fraction is 0, and the slope of the horizontal line is 0.

If $P(x_1, y_1)$ and $Q(x_2, y_2)$ are two points on the vertical line shown in Figure 2-21(b), then $x_1 = x_2$, and the denominator of the fraction

$$\frac{y_2 - y_1}{x_2 - x_1} \qquad \text{On a vertical line, } y_2 \neq y_1.$$

is 0. Since the denominator cannot be 0, a vertical line has no defined slope.

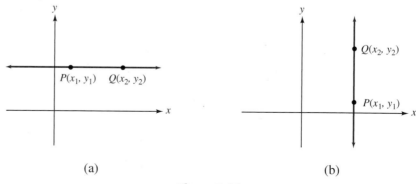

(a)   (b)

Figure 2-21

**Slopes of Horizontal and
Vertical Lines**

All horizontal lines (lines with equations of the form $y = b$) have a slope of 0.

All vertical lines (lines with equations of the form $x = a$) have no defined slope.

If a line rises as we follow it from left to right, as in Figure 2-22(a), its slope is positive. If a line drops as we follow it from left to right, as in Figure 2-22(b), its slope is negative. If a line is horizontal, as in Figure 2-22(c), its slope is 0. If a line is vertical, as in Figure 2-22(d), its slope is undefined.

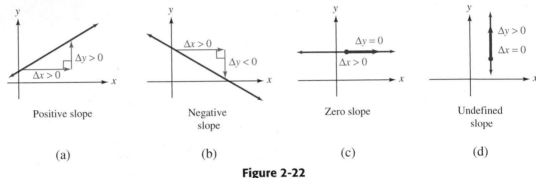

Positive slope	Negative slope	Zero slope	Undefined slope
(a)	(b)	(c)	(d)

Figure 2-22

Slopes of Parallel Lines

To see a relationship between parallel lines and their slopes, we refer to the parallel lines l_1 and l_2 shown in Figure 2-23, with slopes of m_1 and m_2, respectively. Because right triangles ABC and DEF are similar, it follows that

$$m_1 = \frac{\Delta y \text{ of } l_1}{\Delta x \text{ of } l_1}$$
$$= \frac{\Delta y \text{ of } l_2}{\Delta x \text{ of } l_2}$$
$$= m_2$$

Figure 2-23

Thus, if two nonvertical lines are parallel, they have the same slope. It is also true that when two lines have the same slope, they are parallel.

Slopes of Parallel Lines

Nonvertical parallel lines have the same slope, and lines having the same slope are parallel.

Since vertical lines are parallel, lines with no defined slope are parallel.

EXAMPLE 5   The lines in Figure 2-24 are parallel. Find y.

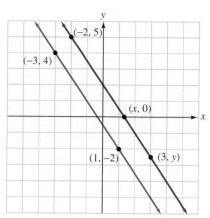

Figure 2-24

Solution   Since the lines are parallel, they have equal slopes. To find y, we find the slope of each line, set them equal, and solve the resulting equation.

Slope of blue line   *Slope of red line*

$$\frac{-2-4}{1-(-3)} = \frac{y-5}{3-(-2)}$$

$$\frac{-6}{4} = \frac{y-5}{5}$$

$-30 = 4(y-5)$ Multiply both sides by 20.

$-30 = 4y - 20$ Use the distributive property.

$-10 = 4y$ Add 20 to both sides.

$-\dfrac{5}{2} = y$ Divide both sides by 4 and simplify.

Thus, $y = -\frac{5}{2}$.

Self Check   In Figure 2-24, find x.   ∎

Slopes of Perpendicular Lines

Two real numbers a and b are called **negative reciprocals** if $ab = -1$. For example,

$$-\frac{4}{3} \quad \text{and} \quad \frac{3}{4}$$

are negative reciprocals, because $-\frac{4}{3}\left(\frac{3}{4}\right) = -1$.

The following theorem relates perpendicular lines and their slopes.

Slopes of Perpendicular Lines

> If two nonvertical lines are perpendicular, their slopes are negative reciprocals.
>
> If the slopes of two lines are negative reciprocals, the lines are perpendicular.

Because a horizontal line is perpendicular to a vertical line, a line with a slope of 0 is perpendicular to a line with no defined slope.

EXAMPLE 6 Are the lines shown in Figure 2-25 perpendicular?

Solution We find the slopes of the lines and see whether they are negative reciprocals.

$$\text{Slope of } OP = \frac{\Delta y}{\Delta x} \qquad\qquad \text{Slope of } PQ = \frac{\Delta y}{\Delta x}$$

$$= \frac{y_2 - y_1}{x_2 - x_1} \qquad\qquad = \frac{y_2 - y_1}{x_2 - x_1}$$

$$= \frac{-4 - 0}{3 - 0} \qquad\qquad = \frac{4 - (-4)}{9 - 3}$$

$$= -\frac{4}{3} \qquad\qquad = \frac{8}{6}$$

$$\qquad\qquad\qquad\qquad\qquad = \frac{4}{3}$$

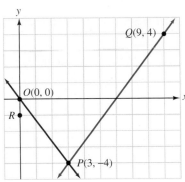

Figure 2-25

Since their slopes are not negative reciprocals, the lines are not perpendicular.

Self Check In Figure 2-25, is PR perpendicular to PQ?

Self Check Answers

1. -2 **2.** $-\frac{2}{5}$ **3.** The y-coordinate of the y-intercept is the delivery charge.   **4.** -175 ft/min   **5.** $\frac{4}{3}$ **6.** no

Orals *Find the slope of the line passing through*

1. $(0, 0)$, $(1, 3)$ **2.** $(0, 0)$, $(3, 6)$
3. Are lines with slopes of -2 and $-\frac{8}{4}$ parallel?
4. Find the negative reciprocal of -0.2.
5. Are lines with slopes of -2 and $\frac{1}{2}$ perpendicular?

2.2   EXERCISES

REVIEW *Simplify each expression. Write all answers without negative exponents.*

1. $(x^3y^2)^3$

2. $\left(\dfrac{x^5}{x^3}\right)^3$

3. $(x^{-3}y^2)^{-4}$

4. $\left(\dfrac{x^{-6}}{y^3}\right)^{-4}$

5. $\left(\dfrac{3x^2y^3}{8}\right)^0$

6. $\left(\dfrac{x^3x^{-7}y^{-6}}{x^4y^{-3}y^{-2}}\right)^{-2}$

VOCABULARY AND CONCEPTS   *Fill in the blanks.*

7. Slope is defined as the change in __ divided by the change in __.

8. A slope is a rate of _____.

9. The formula to compute slope is $m = $ _____.

10. The change in y (denoted as Δy) is the ____ of the line between two points.

11. The change in x (denoted as Δx) is the ____ of the line between two points.

12. The slope of a _____ line is 0.

13. The slope of a _____ line is undefined.

14. If a line rises as x increases, its slope is _____.

15. _____ lines have the same slope.

16. The slopes of _____ lines are negative _____.

PRACTICE   *Find the slope of the line that passes through the given points, if possible.*

17.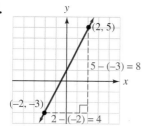

18.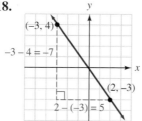

19. $(0, 0), (3, 9)$ 20. $(9, 6), (0, 0)$
21. $(-1, 8), (6, 1)$ 22. $(-5, -8), (3, 8)$
23. $(3, -1), (-6, 2)$ 24. $(0, -8), (-5, 0)$
25. $(7, 5), (-9, 5)$ 26. $(2, -8), (3, -8)$
27. $(-7, -5), (-7, -2)$
28. $(3, -5), (3, 14)$
29. $(2.5, 3.7), (3.7, 2.5)$ 30. $(1.7, -2.3), (2.3, -1.7)$

Find the slope of the line determined by each equation.

31. $3x + 2y = 12$ 32. $2x - y = 6$
33. $3x = 4y - 2$ 34. $x = y$
35. $y = \dfrac{x - 4}{2}$ 36. $x = \dfrac{3 - y}{4}$
37. $4y = 3(y + 2)$ 38. $x + y = \dfrac{2 - 3y}{3}$

Tell whether the slope of the line in each graph is positive, negative, 0, or undefined.

39.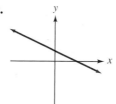

40.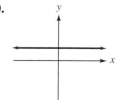

41.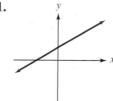

42.

43.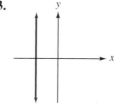

44.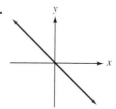

Tell whether the lines with the given slopes are parallel, perpendicular, or neither.

45. $m_1 = 3, m_2 = -\dfrac{1}{3}$

46. $m_1 = \dfrac{1}{4}, m_2 = 4$

47. $m_1 = 4, m_2 = 0.25$

48. $m_1 = -5, m_2 = \dfrac{1}{-0.2}$

49. $m_1 = \dfrac{5.5}{2.7}, m_2 = \left(\dfrac{2.7}{5.5}\right)^{-1}$

50. $m_1 = \dfrac{3.2}{-9.1}, m_2 = \dfrac{-9.1}{3.2}$

Tell whether the line PQ is parallel or perpendicular (or neither) to a line with a slope of -2.

51. $P(3, 4), Q(4, 2)$
52. $P(6, 4), Q(8, 5)$
53. $P(-2, 1), Q(6, 5)$
54. $P(3, 4), Q(-3, -5)$

55. $P(5, 4), Q(6, 6)$

56. $P(-2, 3), Q(4, -9)$

Find the slopes of lines PQ and PR and tell whether the points P, Q, and R lie on the same line. *(Hint: Two lines with the same slope and a point in common must be the same line.)*

57. $P(-2, 4), Q(4, 8), R(8, 12)$

58. $P(6, 10), Q(0, 6), R(3, 8)$

59. $P(-4, 10), Q(-6, 0), R(-1, 5)$

60. $P(-10, -13), Q(-8, -10), R(-12, -16)$

61. $P(-2, 4), Q(0, 8), R(2, 12)$

62. $P(8, -4), Q(0, -12), R(8, -20)$

63. Find the equation of the x-axis and its slope.

64. Find the equation of the y-axis and its slope, if any.

65. Show that points with coordinates of $(-3, 4), (4, 1)$, and $(-1, -1)$ are the vertices of a right triangle.

66. Show that a triangle with vertices at $(0, 0), (12, 0)$, and $(13, 12)$ is not a right triangle.

67. A square has vertices at points $(a, 0), (0, a), (-a, 0)$, and $(0, -a)$, where $a \neq 0$. Show that its adjacent sides are perpendicular.

68. If a and b are not both 0, show that the points $(2b, a), (b, b)$, and $(a, 0)$ are the vertices of a right triangle.

69. Show that the points $(0, 0), (0, a), (b, c)$, and $(b, a + c)$ are the vertices of a parallelogram. *(Hint: Opposite sides of a parallelogram are parallel.)*

70. If $b \neq 0$, show that the points $(0, 0), (0, b)$, $(8, b + 2)$, and $(12, 3)$ are the vertices of a trapezoid. *(Hint: A **trapezoid** is a four-sided figure with exactly two sides parallel.)*

APPLICATIONS

71. Grade of a road Find the slope of the road. *(Hint: 1 mi = 5,280 ft.)*

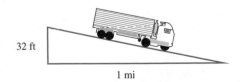

72. Slope of a roof Find the slope of the roof.

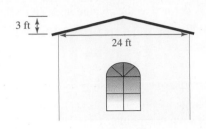

73. Physical fitness Find the slope of the treadmill for each setting listed in the table.

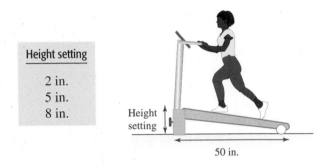

Height setting
2 in.
5 in.
8 in.

74. Wheelchair ramps The illustration shows two designs for a ramp to make a platform wheelchair accessible.
   a. Find the slope of the ramp shown in design 1.
   b. Find the slope of each part of the ramp shown in design 2.
   c. Give one advantage and one disadvantage of each design.

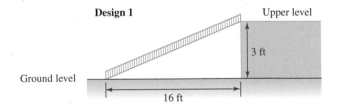

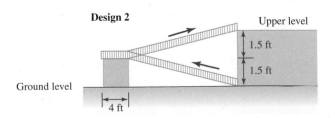

75. Slope of a ladder A ladder reaches 18 feet up the side of a building with its base 5 feet from the building. Find the slope of the ladder.

76. Rate of growth   When a college started an aviation program, the administration agreed to predict enrollments using a straight-line method. If the enrollment during the first year was 8, and the enrollment during the fifth year was 20, find the rate of growth per year (the slope of the line). (See the illustration.)

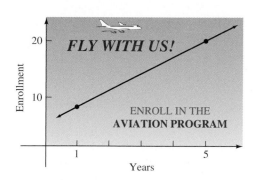

77. Rate of growth   A small business predicts sales according to a straight-line method. If sales were $85,000 in the first year and $125,000 in the third year, find the rate of growth in sales per year (the slope of the line).

78. Rate of decrease   The price of computer technology has been dropping for the past ten years. If a desktop PC cost $5,700 ten years ago, and the same computing power cost $1,499 two years ago, find the rate of decrease per year. (Assume a straight-line model.)

WRITING

79. Explain why a vertical line has no defined slope.

80. Explain how to determine from their slopes whether two lines are parallel, perpendicular, or neither.

SOMETHING TO THINK ABOUT

81. Find the slope of the line $Ax + By = C$. Follow the procedure of Example 2.

82. Follow Example 2 to find the slope of the line $y = mx + b$.

83. The points $(3, a)$, $(5, 7)$, and $(7, 10)$ lie on a line. Find a.

84. The line passing through points $(1, 3)$ and $(-2, 7)$ is perpendicular to the line passing through points $(4, b)$ and $(8, -1)$. Find b.

2.3   Writing Equations of Lines

▪ **Point-Slope Form of the Equation of a Line**
▪ **Slope-Intercept Form of the Equation of a Line**
▪ **Using Slope as an Aid in Graphing**
▪ **General Form of the Equation of a Line**
▪ **Straight-Line Depreciation**   ▪ **Curve Fitting**

Getting Ready   *Solve each equation.*

1. $3 = \dfrac{x - 2}{4}$

2. $-2 = 3(x + 1)$

3. Solve $y - 2 = 3(x - 2)$ for y.

4. Solve $Ax + By + 3 = 0$ for x.

We now apply our knowledge of slope to write the equation of a line passing through two fixed points. We will also use slope as an aid in graphing lines.

Point-Slope Form of the Equation of a Line

Suppose that the line shown in Figure 2-26 has a slope of m and passes through the point (x_1, y_1). If (x, y) is a second point on the line, we have

$$m = \frac{y - y_1}{x - x_1}$$

or if we multiply both sides by $x - x_1$, we have

(1) $y - y_1 = m(x - x_1)$

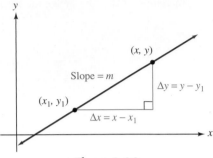

Figure 2-26

Because Equation 1 displays the coordinates of the point (x_1, y_1) on the line and the slope m of the line, it is called the **point-slope form** of the equation of a line.

Point-Slope Form

The equation of the line passing through $P(x_1, y_1)$ and with slope m is

$$y - y_1 = m(x - x_1)$$

EXAMPLE 1   Write the equation of the line with a slope of $-\frac{2}{3}$ and passing through $(-4, 5)$.

Solution   We substitute $-\frac{2}{3}$ for m, -4 for x_1, and 5 for y_1 into the point-slope form and simplify.

$y - \mathbf{y_1} = m(x - x_1)$

$y - \mathbf{5} = -\dfrac{2}{3}[x - (-4)]$ Substitute $-\frac{2}{3}$ for m, -4 for x_1, and 5 for y_1.

$y - \mathbf{5} = -\dfrac{2}{3}(x + 4)$ $-(-4) = 4.$

$y - \mathbf{5} = -\dfrac{2}{3}x - \dfrac{8}{3}$ Use the distributive property to remove parentheses.

$y = -\dfrac{2}{3}x + \dfrac{7}{3}$ Add 5 to both sides and simplify.

The equation of the line is $y = -\dfrac{2}{3}x + \dfrac{7}{3}$.

Self Check   Write the equation of the line with slope of $\frac{5}{4}$ and passing through $(0, 5)$.   ∎

EXAMPLE 2   Write the equation of the line passing through $(-5, 4)$ and $(8, -6)$.

Solution   First we find the slope of the line.

$$m = \frac{y_2 - y_1}{x_2 - x_1}$$

$$= \frac{-6 - 4}{8 - (-5)} \qquad \text{Substitute } -6 \text{ for } y_2, 4 \text{ for } y_1, 8 \text{ for } x_2, \text{ and } -5 \text{ for } x_1.$$

$$= -\frac{10}{13}$$

Because the line passes through both points, we can choose either one and substitute its coordinates into the point-slope form. If we choose $(-5, 4)$, we substitute -5 for x_1, 4 for y_1, and $-\frac{10}{13}$ for m and proceed as follows.

$$y - y_1 = m(x - x_1)$$

$$y - 4 = -\frac{10}{13}[x - (-5)] \qquad \text{Substitute } -\frac{10}{13} \text{ for } m, -5 \text{ for } x_1, \text{ and } 4 \text{ for } y_1.$$

$$y - 4 = -\frac{10}{13}(x + 5) \qquad -(-5) = 5.$$

$$y - 4 = -\frac{10}{13}x - \frac{50}{13} \qquad \text{Remove parentheses.}$$

$$y = -\frac{10}{13}x + \frac{2}{13} \qquad \text{Add 4 to both sides and simplify.}$$

The equation of the line is $y = -\dfrac{10}{13}x + \dfrac{2}{13}$.

Self Check   Write the equation of the line passing through $(-2, 5)$ and $(4, -3)$. ∎

Slope-Intercept Form of the Equation of a Line

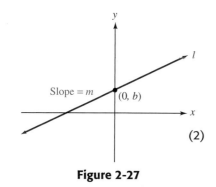

Figure 2-27

Since the y-intercept of the line shown in Figure 2-27 is the point $(0, b)$, we can write the equation of the line by substituting 0 for x_1 and b for y_1 in the point-slope form and simplifying.

$$y - y_1 = m(x - x_1) \qquad \text{The point-slope form of the equation of a line.}$$

$$y - b = m(x - 0) \qquad \text{Substitute } b \text{ for } y_1 \text{ and } 0 \text{ for } x_1.$$

$$y - b = mx$$

$$y = mx + b$$

Because Equation 2 displays the slope m and the y-coordinate b of the y-intercept, it is called the **slope-intercept form** of the equation of a line.

Slope-Intercept Form

The equation of the line with slope m and y-intercept $(0, b)$ is

$$y = mx + b$$

EXAMPLE 3   Use the slope-intercept form to write the equation of the line with slope 4 that passes through the point $(5, 9)$.

Solution   Since we are given that $m = 4$ and that the ordered pair $(5, 9)$ satisfies the equation, we can substitute 5 for x, 9 for y, and 4 for m in the equation $y = mx + b$ and solve for b.

$$y = mx + b$$
$$9 = 4(5) + b \qquad \text{Substitute 9 for } y, \text{ 4 for } m, \text{ and 5 for } x.$$
$$9 = 20 + b \qquad \text{Simplify.}$$
$$-11 = b \qquad \text{Subtract 20 from both sides.}$$

Because $m = 4$ and $b = -11$, the equation is $y = 4x - 11$.

Self Check   Write the equation of the line with slope -2 that passes through the point $(-2, 8)$.

■

Using Slope as an Aid in Graphing

It is easy to graph a linear equation when it is written in slope-intercept form. For example, to graph $y = \frac{4}{3}x - 2$, we note that $b = -2$ and that the y-intercept is $(0, b) = (0, -2)$. (See Figure 2-28.)

Because the slope of the line is $\frac{\Delta y}{\Delta x} = \frac{4}{3}$, we can locate another point on the line by starting at the point $(0, -2)$ and counting 3 units to the right and 4 units up. The line joining the two points is the graph of the equation.

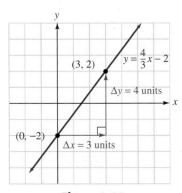

Figure 2-28

EXAMPLE 4   Find the slope and the y-intercept of the line with the equation $2(x - 3) = -3(y + 5)$. Then graph the line.

Solution   We write the equation in the form $y = mx + b$ to find the slope m and the y-intercept $(0, b)$.

$$2(x - 3) = -3(y + 5)$$
$$2x - 6 = -3y - 15 \qquad \text{Use the distributive property to remove parentheses.}$$
$$2x + 3y - 6 = -15 \qquad \text{Add } 3y \text{ to both sides.}$$
$$3y - 6 = -2x - 15 \qquad \text{Subtract } 2x \text{ from both sides.}$$
$$3y = -2x - 9 \qquad \text{Add 6 to both sides.}$$
$$y = -\frac{2}{3}x - 3 \qquad \text{Divide both sides by 3.}$$

The slope is $-\frac{2}{3}$, and the y-intercept is $(0, -3)$. To draw the graph, we plot the y-intercept $(0, -3)$ and then locate a second point on the line by moving 3 units to the right and 2 units down. We draw a line through the two points to obtain the graph shown in Figure 2-29.

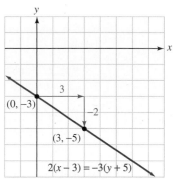

Figure 2-29

Self Check   Find the slope and the y-intercept of the line with the equation $2(y - 1) = 3x + 2$ and graph the line. ∎

EXAMPLE 5   Show that the lines represented by $4x + 8y = 10$ and $2x = 12 - 4y$ are parallel.

Solution   In the previous section, we saw that the lines will be parallel if their slopes are equal. So we solve each equation for y to see whether the lines are distinct and whether their slopes are equal.

$$4x + 8y = 10 \qquad\qquad 2x = 12 - 4y$$
$$8y = -4x + 10 \qquad\qquad 4y = -2x + 12$$
$$y = -\frac{1}{2}x + \frac{5}{4} \qquad\qquad y = -\frac{1}{2}x + 3$$

Since the values of b in these equations are different, the lines are distinct. Since the slope of each line is $-\frac{1}{2}$, they are parallel.

Self Check   Are lines represented by $3x - 2y = 4$ and $6x = 4(y + 1)$ parallel? ∎

EXAMPLE 6   Show that the lines represented by $4x + 8y = 10$ and $4x - 2y = 21$ are perpendicular.

Solution   Since two lines will be perpendicular if their slopes are negative reciprocals, we solve each equation for y to determine their slopes.

$$4x + 8y = 10 \qquad\qquad 4x - 2y = 21$$
$$8y = -4x + 10 \qquad\qquad -2y = -4x + 21$$
$$y = -\frac{1}{2}x + \frac{5}{4} \qquad\qquad y = 2x - \frac{21}{2}$$

Since the slopes are $-\frac{1}{2}$ and 2 (which are negative reciprocals), the lines are perpendicular.

Self Check   Are lines represented by $3x + 2y = 6$ and $2x - 3y = 6$ perpendicular? ∎

EXAMPLE 7   Write the equation of the line passing through $(-2, 5)$ and parallel to the line $y = 8x - 3$.

Solution   Since the equation is solved for y, the slope of the line given by $y = 8x - 3$ is the coefficient of x, which is 8. Since the desired equation is to have a graph that is parallel to the graph of $y = 8x - 3$, its slope must also be 8.
   We substitute -2 for x_1, 5 for y_1, and 8 for m in the point-slope form and simplify.

$$y - y_1 = m(x - x_1)$$
$$y - 5 = 8[x - (-2)] \quad \text{Substitute 5 for } y_1, \text{ 8 for } m, \text{ and } -2 \text{ for } x_1.$$
$$y - 5 = 8(x + 2) \qquad -(-2) = 2.$$
$$y - 5 = 8x + 16 \qquad \text{Use the distributive property to remove parentheses.}$$
$$y = 8x + 21 \qquad \text{Add 5 to both sides.}$$

The equation is $y = 8x + 21$.

Self Check   Write the equation of the line that is parallel to the line $y = 8x - 3$ and passes through the origin. ▮

EXAMPLE 8   Write the equation of the line passing through $(-2, 5)$ and perpendicular to the line $y = 8x - 3$.

Solution   The slope of the given line is 8. Thus, the slope of the desired line must be $-\frac{1}{8}$, which is the negative reciprocal of 8.
   We substitute -2 for x_1, 5 for y_1, and $-\frac{1}{8}$ for m into the point-slope form and simplify:

$$y - y_1 = m(x - x_1)$$
$$y - 5 = -\frac{1}{8}[x - (-2)] \quad \text{Substitute 5 for } y_1, -\frac{1}{8} \text{ for } m, \text{ and } -2 \text{ for } x_1.$$
$$y - 5 = -\frac{1}{8}(x + 2) \qquad -(-2) = 2.$$
$$y - 5 = -\frac{1}{8}x - \frac{1}{4} \qquad \text{Remove parentheses.}$$
$$y = -\frac{1}{8}x - \frac{1}{4} + 5 \qquad \text{Add 5 to both sides.}$$
$$y = -\frac{1}{8}x + \frac{19}{4} \qquad \text{Combine terms: } -\frac{1}{4} + \frac{20}{4} = \frac{19}{4}.$$

The equation is $y = -\frac{1}{8}x + \frac{19}{4}$.

Self Check   Write the equation of the line that is perpendicular to the line $y = 8x - 3$ and passes through $(2, 4)$. ▮

General Form of the Equation of a Line

Recall that any linear equation that is written in the form $Ax + By = C$, where A, B, and C are constants, is said to be written in **general form**.

Comment   When writing equations in general form, we usually clear the equation of fractions and make A positive. We will also make A, B, and C as small as possible. For example, the equation $6x + 12y = 24$ can be changed to $x + 2y = 4$ by dividing both sides by 6.

Finding the Slope and y-Intercept from the General Form

If A, B, and C are real numbers and $B \neq 0$, the graph of the equation

$$Ax + By = C$$

is a nonvertical line with slope of $-\dfrac{A}{B}$ and a y-intercept of $\left(0, \dfrac{C}{B}\right)$.

You will be asked to justify the previous results in the exercises. You will also be asked to show that if $B = 0$, the equation $Ax + By = C$ represents a vertical line with x-intercept of $\left(\frac{C}{A}, 0\right)$.

EXAMPLE 9   Show that the lines represented by $4x + 3y = 7$ and $3x - 4y = 12$ are perpendicular.

Solution   To show that the lines are perpendicular, we will show that their slopes are negative reciprocals. The first equation, $4x + 3y = 7$, is written in general form, with $A = 4$, $B = 3$, and $C = 7$. By the previous result, the slope of the line is

$$m_1 = -\frac{A}{B} = -\frac{4}{3}$$

The second equation, $3x - 4y = 12$, is also written in general form, with $A = 3$, $B = -4$, and $C = 12$. The slope of this line is

$$m_2 = -\frac{A}{B} = -\frac{3}{-4} = \frac{3}{4}$$

Since the slopes are negative reciprocals, the lines are perpendicular.

Self Check   Are the lines $4x + 3y = 7$ and $y = -\frac{4}{3}x + 2$ parallel?

We summarize the various forms for the equation of a line in Table 2-1.

General form of a linear equation	$Ax + By = C$ A and B cannot both be 0.
Slope-intercept form of a linear equation	$y = mx + b$ The slope is m, and the y-intercept is $(0, b)$.
Point-slope form of a linear equation	$y - y_1 = m(x - x_1)$ The slope is m, and the line passes through (x_1, y_1).
A horizontal line	$y = b$ The slope is 0, and the y-intercept is $(0, b)$.
A vertical line	$x = a$ There is no defined slope, and the x-intercept is $(a, 0)$.

Table 2-1

Straight-Line Depreciation

For tax purposes, many businesses use **straight-line depreciation** to find the declining value of aging equipment.

EXAMPLE 10   **Value of a lathe**   The owner of a machine shop buys a lathe for $1,970 and expects it to last for ten years. It can then be sold as scrap for an estimated **salvage value** of $270. If y represents the value of the lathe after x years of use, and y and x are related by the equation of a line,

  a. Find the equation of the line.

  b. Find the value of the lathe after $2\frac{1}{2}$ years.

  c. Find the economic meaning of the y-intercept of the line.

  d. Find the economic meaning of the slope of the line.

Solution   **a.** To find the equation of the line, we find its slope and use point-slope form to find its equation.

When the lathe is new, its age x is 0, and its value y is $1,970. When the lathe is 10 years old, $x = 10$ and its value is $y = \$270$. Since the line passes through the points (0, 1,970) and (10, 270), as shown in Figure 2-30, the slope of the line is

$$
\begin{aligned}
m &= \frac{y_2 - y_1}{x_2 - x_1} \\
&= \frac{270 - 1{,}970}{10 - 0} \\
&= \frac{-1{,}700}{10} \\
&= -170
\end{aligned}
$$

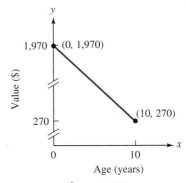

Figure 2-30

To find the equation of the line, we substitute -170 for m, 0 for x_1, and 1,970 for y_1 into the point-slope form and simplify.

$$
\begin{aligned}
y - y_1 &= m(x - x_1) \\
y - \mathbf{1{,}970} &= -\mathbf{170}(x - \mathbf{0}) \\
y &= -170x + 1{,}970
\end{aligned}
$$

(3)

The current value y of the lathe is related to its age x by the equation $y = -170x + 1{,}970$.

  b. To find the value of the lathe after $2\frac{1}{2}$ years, we substitute 2.5 for x in Equation 3 and solve for y.

$$
\begin{aligned}
y &= -170x + 1{,}970 \\
&= -170(\mathbf{2.5}) + 1{,}970 \\
&= -425 + 1{,}970 \\
&= 1{,}545
\end{aligned}
$$

In $2\frac{1}{2}$ years, the lathe will be worth $1,545.

  c. The y-intercept of the graph is $(0, b)$, where b is the value of y when $x = 0$.

$$y = -170x + 1,970$$
$$y = -170(\mathbf{0}) + 1,970$$
$$y = 1,970$$

Thus, b is the value of a 0-year-old lathe, which is the lathe's original cost, $1,970.

d. Each year, the value of the lathe decreases by $170, because the slope of the line is -170. The slope of the line is the **annual depreciation rate.** ∎

Curve Fitting

In statistics, the process of using one variable to predict another is called **regression.** For example, if we know a man's height, we can make a good prediction about his weight, because taller men usually weigh more than shorter men.

Figure 2-31 shows the result of sampling ten men at random and finding their heights and weights. The graph of the ordered pairs (h, w) is called a **scattergram.**

Man	Height (h) in inches	Weight (w) in pounds
1	66	140
2	68	150
3	68	165
4	70	180
5	70	165
6	71	175
7	72	200
8	74	190
9	75	210
10	75	215

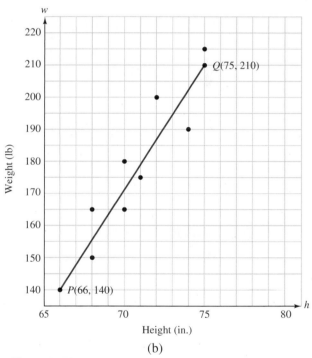

(a) (b)

Figure 2-31

To write a **prediction equation** (sometimes called a **regression equation**), we must find the equation of the line that comes closer to all of the points in the scattergram than any other possible line. There are exact methods to find this equation, but we can only approximate it here.

To write an approximation of the regression equation, we place a straightedge on the scattergram shown in Figure 2-31 and draw the line joining two points that seems to best fit all the points. In the figure, line PQ is drawn, where point P has coordinates of $(66, 140)$ and point Q has coordinates of $(75, 210)$.

Our approximation of the regression equation will be the equation of the line passing through points P and Q. To find the equation of this line, we first find its slope.

$$m = \frac{y_2 - y_1}{x_2 - x_1}$$

$$= \frac{210 - 140}{75 - 66}$$

$$= \frac{70}{9}$$

We can then use point-slope form to find its equation.

$$y - y_1 = m(x - x_1)$$

$$y - 140 = \frac{70}{9}(x - 66) \qquad \text{Choose } (66, 140) \text{ for } (x_1, y_1).$$

$$y = \frac{70}{9}x - \frac{4{,}620}{9} + 140 \qquad \text{Remove parentheses and add 140 to both sides.}$$

(4) $$y = \frac{70}{9}x - \frac{1{,}120}{3} \qquad\qquad -\frac{4{,}620}{9} + 140 = -\frac{1{,}120}{3}.$$

Our approximation of the regression equation is $y = \frac{70}{9}x - \frac{1{,}120}{3}$.

To predict the weight of a man who is 73 inches tall, for example, we substitute 73 for x in Equation 4 and simplify.

$$y = \frac{70}{9}x - \frac{1{,}120}{3}$$

$$y = \frac{70}{9}(73) - \frac{1{,}120}{3}$$

$$y \approx 194.4$$

We would predict that a 73-inch-tall man chosen at random will weigh about 194 pounds.

Self Check Answers

1. $y = \frac{5}{4}x + 5$ **2.** $y = -\frac{4}{3}x + \frac{7}{3}$ **3.** $y = -2x + 4$ **4.** $m = \frac{3}{2}, (0, 2)$

5. yes   **6.** yes   **7.** $y = 8x$ **8.** $y = -\frac{1}{8}x + \frac{17}{4}$ **9.** yes

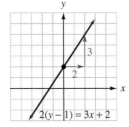
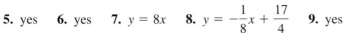

Orals   *Write the point-slope form of the equation of a line with $m = 2$, passing through the given point.*

1. $(2, 3)$ **2.** $(-3, 8)$

Write the equation of a line with $m = -3$ and the given y-intercept.

3. $(0, 5)$ **4.** $(0, -7)$

Tell whether the lines are parallel, perpendicular, or neither.

5. $y = 3x - 4, y = 3x + 5$ **6.** $y = -3x + 7, x = 3y - 1$

2.3   EXERCISES

REVIEW EXERCISES   *Solve each equation.*

1. $3(x + 2) + x = 5x$

2. $12b + 6(3 - b) = b + 3$

3. $\dfrac{5(2 - x)}{3} - 1 = x + 5$

4. $\dfrac{r - 1}{3} = \dfrac{r + 2}{6} + 2$

5. **Mixing alloys**   In 60 ounces of alloy for watch cases, there are 20 ounces of gold. How much copper must be added to the alloy so that a watch case weighing 4 ounces, made from the new alloy, will contain exactly 1 ounce of gold?

6. **Mixing coffee**   To make a mixture of 80 pounds of coffee worth $272, a grocer mixes coffee worth $3.25 a pound with coffee worth $3.85 a pound. How many pounds of the cheaper coffee should the grocer use?

VOCABULARY AND CONCEPTS   *Fill in the blanks.*

7. The point-slope form of the equation of a line is _____.

8. The slope-intercept form of the equation of a line is _____.

9. The general form of the equation of a line is _____.

10. Two lines are parallel when they have the _____ slope.

11. Two lines are _____ when their slopes are negative reciprocals.

12. The process that recognizes that equipment loses value with age is called _____.

PRACTICE   *Use point-slope form to write the equation of the line with the given properties. Write each equation in general form.*

13. $m = 5$, passing through $P(0, 7)$

14. $m = -8$, passing through $P(0, -2)$

15. $m = -3$, passing through $P(2, 0)$

16. $m = 4$, passing through $P(-5, 0)$

Use point-slope form to write the equation of each line. Write the equation in general form.

17.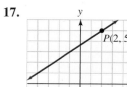

18.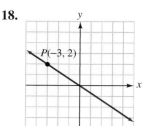

Use point-slope form to write the equation of the line passing through the two given points. Write each equation in slope-intercept form.

19. $P(0, 0)$, $Q(4, 4)$

20. $P(-5, -5)$, $Q(0, 0)$

21. $P(3, 4)$, $Q(0, -3)$

22. $P(4, 0)$, $Q(6, -8)$

Use point-slope form to write the equation of each line. Write each answer in slope-intercept form.

23.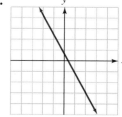

24.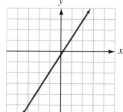

Use slope-intercept form to write the equation of the line with the given properties. Write each equation in slope-intercept form.

25. $m = 3$, $b = 17$

26. $m = -2$, $b = 11$

27. $m = -7$, passing through $P(7, 5)$

28. $m = 3$, passing through $P(-2, -5)$

29. $m = 0$, passing through $P(2, -4)$

30. $m = -7$, passing through the origin

31. Passing through $P(6, 8)$ and $Q(2, 10)$

32. Passing through $P(-4, 5)$ and $Q(2, -6)$

Write each equation in slope-intercept form to find the slope and the y-intercept. **Then use the slope and y-intercept to draw the line.**

33. $y + 1 = x$

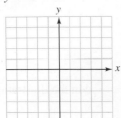

34. $x + y = 2$

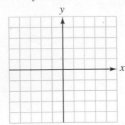

35. $x = \dfrac{3}{2}y - 3$

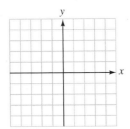

36. $x = -\dfrac{4}{5}y + 2$

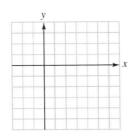

37. $3(y - 4) = -2(x - 3)$

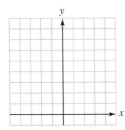

38. $-4(2x + 3) = 3(3y + 8)$

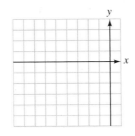

Find the slope and the y-intercept of the line determined by the given equation.

39. $3x - 2y = 8$

40. $-2x + 4y = 12$

41. $-2(x + 3y) = 5$

42. $5(2x - 3y) = 4$

43. $x = \dfrac{2y - 4}{7}$

44. $3x + 4 = -\dfrac{2(y - 3)}{5}$

Tell whether the graphs of each pair of equations are parallel, perpendicular, or neither.

45. $y = 3x + 4,\ y = 3x - 7$

46. $y = 4x - 13,\ y = \dfrac{1}{4}x + 13$

47. $x + y = 2,\ y = x + 5$

48. $x = y + 2,\ y = x + 3$

49. $y = 3x + 7,\ 2y = 6x - 9$

50. $2x + 3y = 9,\ 3x - 2y = 5$

51. $x = 3y + 4,\ y = -3x + 7$

52. $3x + 6y = 1,\ y = \dfrac{1}{2}x$

53. $y = 3,\ x = 4$

54. $y = -3,\ y = -7$

55. $x = \dfrac{y - 2}{3},\ 3(y - 3) + x = 0$

56. $2y = 8,\ 3(2 + x) = 2(x + 2)$

Write the equation of the line that passes through the given point and is parallel to the given line. Write the answer in slope-intercept form.

57. $(0, 0),\ y = 4x - 7$

58. $(0, 0),\ x = -3y - 12$

59. $(2, 5),\ 4x - y = 7$

60. $(-6, 3),\ y + 3x = -12$

61. $(4, -2),\ x = \dfrac{5}{4}y - 2$

62. $(1, -5),\ x = -\dfrac{3}{4}y + 5$

Write the equation of the line that passes through the given point and is perpendicular to the given line. Write the answer in slope-intercept form.

63. $(0, 0),\ y = 4x - 7$

64. $(0, 0),\ x = -3y - 12$

65. $(2, 5),\ 4x - y = 7$

66. $(-6, 3),\ y + 3x = -12$

67. $(4, -2),\ x = \dfrac{5}{4}y - 2$

68. $(1, -5),\ x = -\dfrac{3}{4}y + 5$

Use the method of Example 9 to find whether the graphs determined by each pair of equations are parallel, perpendicular, or neither.

69. $4x + 5y = 20, 5x - 4y = 20$

70. $9x - 12y = 17, 3x - 4y = 17$

71. $2x + 3y = 12, 6x + 9y = 32$

72. $5x + 6y = 30, 6x + 5y = 24$

73. Find the equation of the line perpendicular to the line $y = 3$ and passing through the midpoint of the segment joining $(2, 4)$ and $(-6, 10)$.

74. Find the equation of the line parallel to the line $y = -8$ and passing through the midpoint of the segment joining $(-4, 2)$ and $(-2, 8)$.

75. Find the equation of the line parallel to the line $x = 3$ and passing through the midpoint of the segment joining $(2, -4)$ and $(8, 12)$.

76. Find the equation of the line perpendicular to the line $x = 3$ and passing through the midpoint of the segment joining $(-2, 2)$ and $(4, -8)$.

77. Solve $Ax + By = C$ for y and thereby show that the slope of its graph is $-\frac{A}{B}$ and its y-intercept is $\left(0, \frac{C}{B}\right)$.

78. Show that the x-intercept of the graph of $Ax + By = C$ is $\left(\frac{C}{A}, 0\right)$.

APPLICATIONS *Assume straight-line depreciation or straight-line appreciation.*

79. Depreciation equations A truck was purchased for $19,984. Its salvage value at the end of 8 years is expected to be $1,600. Find the depreciation equation.

80. Depreciation equations A business purchased the computer shown. It will be depreciated over a 5-year period, when it will probably be worth $200. Find the depreciation equation.

$2,350

81. Appreciation equations A famous oil painting was purchased for $250,000 and is expected to double in value in 5 years. Find the appreciation equation.

82. Appreciation equations A house purchased for $142,000 is expected to double in value in 8 years. Find its appreciation equation.

83. Depreciation equations Find the depreciation equation for the TV in the want ad in the illustration.

For Sale: 3-year-old 65-inch TV, $1,750 new. Asking $800. Call 715-5588. Ask for Joe.

84. Depreciating a lawn mower A lawn mower cost $450 when new and is expected to last 10 years. What will it be worth in $6\frac{1}{2}$ years?

85. Salvage value A copy machine that cost $1,750 when new will be depreciated at the rate of $180 per year. If the useful life of the copier is 7 years, find its salvage value.

86. Annual rate of depreciation A machine that cost $47,600 when new will have a salvage value of $500 after its useful life of 15 years. Find its annual rate of depreciation.

87. Real estate A vacation home is expected to appreciate about $4,000 a year. If the home will be worth $122,000 in 2 years, what will it be worth in 10 years?

88. Car repair A garage charges a fixed amount, plus an hourly rate, to service a car. Use the information in the table to find the hourly rate.

A-1 Car Repair	
Typical charges	
2 hours	$143
5 hours	$320

89. Printer charges A printer charges a fixed setup cost, plus $15 for every 100 copies. If 300 copies cost $75, how much will 1,000 copies cost?

90. Predicting burglaries A police department knows that city growth and the number of burglaries are related by a linear equation. City records show that 575 burglaries were reported in a year when the local population was 77,000, and 675 were reported in a year when the population was 87,000. How many burglaries can be expected when the population reaches 110,000?

WRITING

91. Explain how to find the equation of a line passing through two given points.

92. In straight-line depreciation, explain why the slope of the line is called the *rate of depreciation.*

SOMETHING TO THINK ABOUT *Investigate the properties of the slope and the y-intercept by experimenting with the following problems.*

93. Graph $y = mx + 2$ for several positive values of m. What do you notice?

94. Graph $y = mx + 2$ for several negative values of m. What do you notice?

95. Graph $y = 2x + b$ for several increasing positive values of b. What do you notice?

96. Graph $y = 2x + b$ for several decreasing negative values of b. What do you notice?

97. How will the graph of $y = \frac{1}{2}x + 5$ compare to the graph of $y = \frac{1}{2}x - 5$?

98. How will the graph of $y = \frac{1}{2}x - 5$ compare to the graph of $y = \frac{1}{2}x$?

99. If the graph of $y = ax + b$ passes through quadrants I, II, and IV, what can be known about the constants a and b?

100. The graph of $Ax + By = C$ passes only though quadrants I and IV. What is known about the constants A, B, and C?

2.4 Introduction to Functions

- **Functions** - **Function Notation**
- **Finding Domains and Ranges of Functions** - **The Vertical Line Test**
- **Linear Functions**

Getting Ready *If $y = \frac{3}{2}x - 2$, find the value of y for each value of x.*

1. $x = 2$ **2.** $x = 6$ **3.** $x = -12$ **4.** $x = -\dfrac{1}{2}$

In this section, we will introduce functions, one of the most important ideas in mathematics. The farther you go in mathematics, the more you will study functions.

Functions

If x and y are real numbers, an equation in x and y determines a correspondence between the values of x and y. To see how, we consider the equation $y = \frac{1}{2}x + 3$. To find the value of y (called an **output value**) that corresponds to $x = 4$ (called an **input value**), we substitute 4 for x and simplify.

$$y = \frac{1}{2}x + 3$$

$$y = \frac{1}{2}(\mathbf{4}) + 3 \qquad \text{Substitute the input value of 4 for } x.$$

$$= 2 + 3$$

$$= 5 .$$

The ordered pair (4, 5) satisfies the equation and shows that a y-value of 5 corresponds to an x-value of 4. This ordered pair and others that satisfy the equation appear in the table shown in Figure 2-32. The graph of the equation also appears in the figure.

$y = \frac{1}{2}x + 3$

x	y	(x, y)
-2	2	$(-2, 2)$
0	3	$(0, 3)$
2	4	$(2, 4)$
4	5	$(4, 5)$
6	6	$(6, 6)$

↑ ↑
Inputs Outputs

A y-value of 2 corresponds to an x-value of -2.
A y-value of 3 corresponds to an x-value of 0.
A y-value of 4 corresponds to an x-value of 2.
A y-value of 5 corresponds to an x-value of 4.
A y-value of 6 corresponds to an x-value of 6.

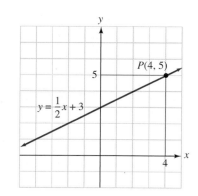

Figure 2-32

To see how the table in Figure 2-32 determines the correspondence, we simply find an input in the x-column and then read across to find the corresponding output in the y-column. For example, if we select 2 as an input value, we get 4 as an output value. Thus, a y-value of 4 corresponds to an x-value of 2.

To see how the graph in Figure 2-32 determines the correspondence, we draw a vertical and a horizontal line through any point (say, point P) on the graph, as shown in the figure. Because these lines intersect the x-axis at 4 and the y-axis at 5, the point $P(4, 5)$ associates 5 on the y-axis with 4 on the x-axis. This shows that a y-value of 5 corresponds to an x-value of 4.

In this example, the set of all inputs x is the set of real numbers. The set of all outputs y is also the set of real numbers.

When a correspondence is set up by an equation, a table, or a graph, in which only one y-value corresponds to each x-value, we call the correspondence a **function.** Since the value of y usually depends on the number x, we call y the **dependent variable** and x the **independent variable.**

Functions

A **function** is a correspondence between a set of input values x (called the **domain of the function**) and a set of output values y (called the **range of the function**), where exactly one y-value in the range corresponds to each number x in the domain.

EXAMPLE 1 Does $y = 2x - 3$ define y to be a function of x? If so, find its domain and range, and illustrate the function with a table and graph.

Solution   For a function to exist, every value of x must determine one value of y. To find y in the equation $y = 2x - 3$, we multiply x by 2 and then subtract 3. Since this arithmetic gives one result, each choice of x determines one value of y. Thus, the equation does define y to be a function of x.

Since the input x can be any real number, the domain of the function is the set of real numbers, denoted by the interval $(-\infty, \infty)$. Since the output y can be any real number, the range is also the set of real numbers, denoted as $(-\infty, \infty)$.

A table of values and the graph appear in Figure 2-33.

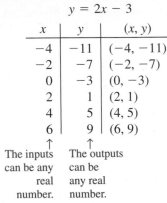

$$y = 2x - 3$$

x	y	(x, y)
-4	-11	$(-4, -11)$
-2	-7	$(-2, -7)$
0	-3	$(0, -3)$
2	1	$(2, 1)$
4	5	$(4, 5)$
6	9	$(6, 9)$

The inputs can be any real number. The outputs can be any real number.

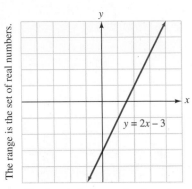

The range is the set of real numbers.

The domain is the set of real numbers.

Figure 2-33

Self Check   Does $y = -2x + 3$ define y to be a function of x?

EXAMPLE 2   Does $y^2 = x$ define y to be a function of x?

Solution   For a function to exist, each value of x must determine one value of y. If we let $x = 16$, for example, y could be either 4 or -4, because $4^2 = 16$ and $(-4)^2 = 16$. Since more than one value of y is determined when $x = 16$, the equation does not represent a function.

Self Check   Does $|y| = x$ define y to be a function of x?

Function Notation

We use a special notation to denote functions.

Function Notation   The notation $y = f(x)$ denotes that the variable y is a function of x.

The notation $y = f(x)$ is read as "y equals f of x." Note that y and $f(x)$ are two notations for the same quantity. Thus, the equations $y = 4x + 3$ and $f(x) = 4x + 3$ are equivalent.

Comment   The notation $f(x)$ does not mean "f times x."

The notation $y = f(x)$ provides a way of denoting the value of y (the dependent variable) that corresponds to some number x (the independent variable). For example, if $y = f(x)$, the value of y that is determined by $x = 3$ is denoted by $f(3)$.

EXAMPLE 3   Let $f(x) = 4x + 3$. Find   **a.** $f(3)$,   **b.** $f(-1)$,   **c.** $f(0)$,   and   **d.** $f(r)$.

Solution   **a.** We replace x with 3:

$$f(x) = 4x + 3$$
$$f(3) = 4(3) + 3$$
$$= 12 + 3$$
$$= 15$$

b. We replace x with -1:

$$f(x) = 4x + 3$$
$$f(-1) = 4(-1) + 3$$
$$= -4 + 3$$
$$= -1$$

c. We replace x with 0:

$$f(x) = 4x + 3$$
$$f(0) = 4(0) + 3$$
$$= 3$$

d. We replace x with r:

$$f(x) = 4x + 3$$
$$f(r) = 4r + 3$$

Self Check   If $f(x) = -2x - 1$, find   **a.** $f(2)$ and   **b.** $f(-3)$.   ∎

To see why function notation is helpful, consider the following sentences:

1. In the equation $y = 4x + 3$, find the value of y when x is 3.

2. In the equation $f(x) = 4x + 3$, find $f(3)$.

Statement 2, which uses $f(x)$ notation, is much more concise.

We can think of a function as a machine that takes some input x and turns it into some output $f(x)$, as shown in Figure 2-34(a). The machine shown in Figure 2-34(b) turns the input number 2 into the output value -3 and turns the input number 6 into the output value -11. The set of numbers that we can put into the machine is the domain of the function, and the set of numbers that comes out is the range.

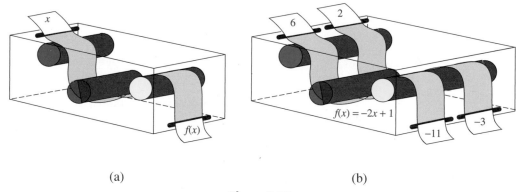

(a)   (b)

Figure 2-34

The letter f used in the notation $y = f(x)$ represents the word *function*. However, other letters can be used to represent functions. The notations $y = g(x)$ and $y = h(x)$ also denote functions involving the independent variable x.

In Example 4, the equation $y = g(x) = x^2 - 2x$ determines a function, because every possible value of x gives a single value of $g(x)$.

EXAMPLE 4   Let $g(x) = x^2 - 2x$. Find   **a.** $g\left(\frac{2}{5}\right)$,   **b.** $g(s)$,   **c.** $g(s^2)$,   and   **d.** $g(-t)$.

Solution **a.** We replace x with $\frac{2}{5}$:

$$g(x) = x^2 - 2x$$

$$g\left(\frac{2}{5}\right) = \left(\frac{2}{5}\right)^2 - 2\left(\frac{2}{5}\right)$$

$$= \frac{4}{25} - \frac{4}{5}$$

$$= -\frac{16}{25}$$

b. We replace x with s:

$$g(x) = x^2 - 2x$$

$$g(s) = s^2 - 2s$$

c. We replace x with s^2:

$$g(x) = x^2 - 2x$$

$$g(s^2) = (s^2)^2 - 2s^2$$

$$= s^4 - 2s^2$$

d. We replace x with $-t$:

$$g(x) = x^2 - 2x$$

$$g(-t) = (-t)^2 - 2(-t)$$

$$= t^2 + 2t$$

Self Check Let $h(x) = -x^2 + 3$. Find **a.** $h(2)$ and **b.** $h(-a)$. ∎

EXAMPLE 5 Let $f(x) = 4x - 1$. Find **a.** $f(3) + f(2)$ and **b.** $f(a) - f(b)$.

Solution **a.** We find $f(3)$ and $f(2)$ separately.

$$f(x) = 4x - 1 \qquad\qquad f(x) = 4x - 1$$

$$f(3) = 4(3) - 1 \qquad\qquad f(2) = 4(2) - 1$$

$$= 12 - 1 \qquad\qquad\qquad = 8 - 1$$

$$= 11 \qquad\qquad\qquad\qquad = 7$$

We then add the results to obtain $f(3) + f(2) = 11 + 7 = 18$.

b. We find $f(a)$ and $f(b)$ separately.

$$f(x) = 4x - 1 \qquad\qquad f(x) = 4x - 1$$

$$f(a) = 4(a) - 1 \qquad\qquad f(b) = 4b - 1$$

We then subtract the results to obtain

$$f(a) - f(b) = (4a - 1) - (4b - 1)$$

$$= 4a - 1 - 4b + 1$$

$$= 4a - 4b$$

Self Check Let $g(x) = -2x + 3$. Find **a.** $g(-2) + g(3)$ and **b.** $g\left(\frac{1}{2}\right) - g(2)$. ∎

Finding Domains and Ranges of Functions

EXAMPLE 6 Find the domains and ranges of the functions defined by **a.** the ordered pairs $(-2, 4), (0, 6), (2, 8)$ and **b.** the equation $y = \frac{1}{x-2}$.

Solution **a.** The ordered pairs can be placed in a table to show a correspondence between x and y, where a single value of y corresponds to each x.

x	y	
-2	4	4 corresponds to -2.
0	6	6 corresponds to 0.
2	8	8 corresponds to 2.

The domain is the set of numbers x: $\{-2, 0, 2\}$. The range is the set of values y: $\{4, 6, 8\}$.

b. The number 2 cannot be substituted for x, because that would make the denominator equal to zero. Since any real number except 2 can be substituted for x in the equation $y = \frac{1}{x-2}$, the domain is the set of all real numbers but 2. This is the interval $(-\infty, 2) \cup (2, \infty)$.

Since a fraction with a numerator of 1 cannot be 0, the range is the set of all real numbers but 0. This is the interval $(-\infty, 0) \cup (0, \infty)$.

Self Check   Find the domains and ranges of the functions defined by   **a.** the set of ordered pairs $\{(-3, 5), (-2, 7), (1, 11)\}$ and   **b.** the equation $y = \frac{2}{x+3}$.   ∎

The **graph of a function** is the graph of the ordered pairs $(x, f(x))$ that define the function. For the graph of the function shown in Figure 2-35, the domain is shown on the x-axis, and the range is shown on the y-axis. For any x in the domain, there corresponds one value $y = f(x)$ in the range.

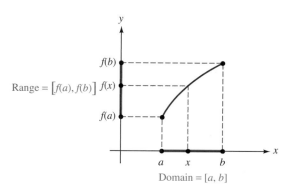

Figure 2-35

EXAMPLE 7   Find the domain and range of the function defined by $y = -2x + 1$.

Solution   We graph the equation as in Figure 2-36. Since every real number x on the x-axis determines a corresponding value of y, the domain is the interval $(-\infty, \infty)$ shown on the x-axis. Since the values of y can be any real number, the range is the interval $(-\infty, \infty)$ shown on the y-axis.

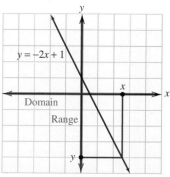

Figure 2-36   ∎

The Vertical Line Test

A **vertical line test** can be used to determine whether the graph of an equation represents a function. If any vertical line intersects a graph more than once, the graph cannot represent a function, because to one number x there would correspond more than one value of y.

The graph in Figure 2-37(a) represents a function, because every vertical line that intersects the graph does so exactly once. The graph in Figure 2-37(b) does not represent a function, because some vertical lines intersect the graph more than once.

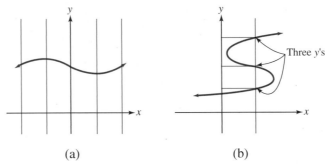

(a)   (b)

Figure 2-37

Linear Functions

In Section 2.1, we graphed equations whose graphs were lines. These equations define basic functions, called **linear functions.**

Linear Functions

A **linear function** is a function defined by an equation that can be written in the form

$$f(x) = mx + b \qquad \text{or} \qquad y = mx + b$$

where m is the slope of the line graph and $(0, b)$ is the y-intercept.

EXAMPLE 8   Solve the equation $3x + 2y = 10$ for y to show that it defines a linear function. Then graph it to find its domain and range.

Solution   We solve the equation for y as follows:

$$3x + 2y = 10$$
$$2y = -3x + 10 \qquad \text{Subtract } 3x \text{ from both sides.}$$
$$y = -\frac{3}{2}x + 5 \qquad \text{Divide both sides by 2.}$$

Because the given equation is written in the form $y = mx + b$, it defines a linear function. The slope of its line graph is $-\frac{3}{2}$, and the y-intercept is $(0, 5)$. The graph appears in Figure 2-38. From the graph, we can see that both the domain and the range are the interval $(-\infty, \infty)$.

A special case of a linear function is the **constant function,** defined by the equation $f(x) = b$, where b is a constant. Its graph, domain, and range are shown in Figure 2-39.

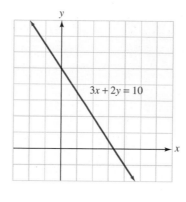

Figure 2-38

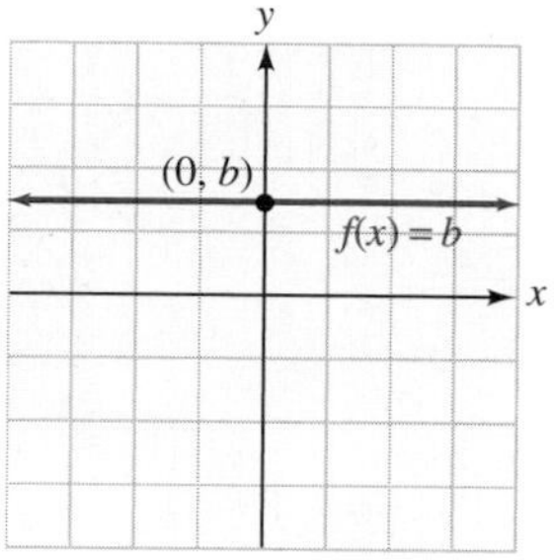

Constant function
Domain: $(-\infty, \infty)$
Range: $\{b\}$

Figure 2-39

Self Check Answers

1. yes **2.** no **3. a.** -5, **b.** 5 **4. a.** -1, **b.** $-a^2 + 3$ **5. a.** 4, **b.** 3
6. a. $\{-3, -2, 1\}$; $\{5, 7, 11\}$, **b.** $(-\infty, -3) \cup (-3, \infty)$, $(-\infty, 0) \cup (0, \infty)$

Orals *Tell whether each equation or inequality determines y to be a function of x.*

1. $y = 2x + 1$ **2.** $y \geq 2x$ **3.** $y^2 = x$

If $f(x) = 2x + 1$, *find*

4. $f(0)$ **5.** $f(1)$ **6.** $f(-2)$

2.4 EXERCISES

REVIEW *Solve each equation.*

1. $\dfrac{y + 2}{2} = 4(y + 2)$

2. $\dfrac{3z - 1}{6} - \dfrac{3z + 4}{3} = \dfrac{z + 3}{2}$

3. $\dfrac{2a}{3} + \dfrac{1}{2} = \dfrac{6a - 1}{6}$

4. $\dfrac{2x + 3}{5} - \dfrac{3x - 1}{3} = \dfrac{x - 1}{15}$

VOCABULARY AND CONCEPTS *Consider the function* $y = f(x) = 5x - 4$. *Fill in the blanks.*

5. Any substitution for x is called an ______ value.

6. The value __ is called an output value.

7. The independent variable is __.

8. The dependent variable is __.

9. A ________ is a correspondence between a set of input values and a set of output values, where each ______ value determines one ______ value.

10. In a function, the set of all inputs is called the ______ of the function.

11. In a function, the set of all output values is called the ______ of the function.

12. The notation $f(3)$ is the value of __ when $x = 3$.

13. The denominator of a fraction can never be __.

14. If a vertical line intersects a graph more than once, the graph ______ represent a function.

15. A linear function is any function that can be written in the form _____.

16. In the function $f(x) = mx + b$, m is the _____ of its graph, and b is the y-coordinate of the _____.

PRACTICE   *Tell whether the equation determines y to be a function of x.*

17. $y = 2x + 3$ **18.** $y = -1$

19. $y = 2x^2$ **20.** $y^2 = x + 1$

21. $y = 3 + 7x^2$ **22.** $y^2 = 3 - 2x$

23. $x = |y|$ **24.** $y = |x|$

Find $f(3)$ and $f(-1)$.

25. $f(x) = 3x$ **26.** $f(x) = -4x$

27. $f(x) = 2x - 3$ **28.** $f(x) = 3x - 5$

29. $f(x) = 7 + 5x$ **30.** $f(x) = 3 + 3x$

31. $f(x) = 9 - 2x$ **32.** $f(x) = 12 + 3x$

Find $f(2)$ and $f(3)$.

33. $f(x) = x^2$ **34.** $f(x) = x^2 - 2$

35. $f(x) = x^3 - 1$ **36.** $f(x) = x^3$

37. $f(x) = (x + 1)^2$ **38.** $f(x) = (x - 3)^2$

39. $f(x) = 2x^2 - x$ **40.** $f(x) = 5x^2 + 2x$

Find $f(2)$ and $f(-2)$.

41. $f(x) = |x| + 2$ **42.** $f(x) = |x| - 5$

43. $f(x) = x^2 - 2$ **44.** $f(x) = x^2 + 3$

45. $f(x) = \dfrac{1}{x + 3}$ **46.** $f(x) = \dfrac{3}{x - 4}$

47. $f(x) = \dfrac{x}{x - 3}$ **48.** $f(x) = \dfrac{x}{x^2 + 2}$

Find $g(w)$ and $g(w + 1)$.

49. $g(x) = 2x$ **50.** $g(x) = -3x$

51. $g(x) = 3x - 5$ **52.** $g(x) = 2x - 7$

Find each value given that $f(x) = 2x + 1$.

53. $f(3) + f(2)$ **54.** $f(1) - f(-1)$

55. $f(b) - f(a)$ **56.** $f(b) + f(a)$

57. $f(b) - 1$ **58.** $f(b) - f(1)$

59. $f(0) + f\left(-\frac{1}{2}\right)$ **60.** $f(a) + f(2a)$

Find the domain and range of each function.

61. $\{(-2, 3), (4, 5), (6, 7)\}$ **62.** $\{(0, 2), (1, 2), (3, 4)\}$

63. $f(x) = \dfrac{1}{x - 4}$ **64.** $f(x) = \dfrac{5}{x + 1}$

Each graph represents a correspondence between x and y. Tell whether the correspondence is a function. If it is, give its domain and range.

65.

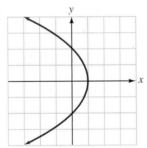

66.

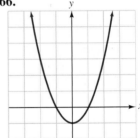

67.

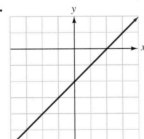

68.

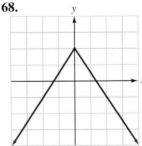

Draw the graph of each linear function and give the domain and range.

69. $f(x) = 2x - 1$

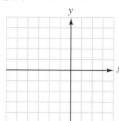

70. $f(x) = -x + 2$

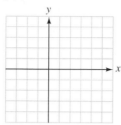

71. $2x - 3y = 6$

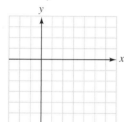

72. $3x + 2y = -6$

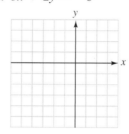

Tell whether each equation defines a linear function.

73. $y = 3x^2 + 2$

74. $y = \dfrac{x - 3}{2}$

75. $x = 3y - 4$

76. $x = \dfrac{8}{y}$

APPLICATIONS

77. Ballistics   A bullet shot straight up is s feet above the ground after t seconds, where $s = f(t) = -16t^2 + 256t$. Find the height of the bullet 3 seconds after it is shot.

78. Artillery fire   A mortar shell is s feet above the ground after t seconds, where $s = f(t) = -16t^2 + 512t + 64$. Find the height of the shell 20 seconds after it is fired.

79. Dolphins   See the illustration. The height h in feet reached by a dolphin t seconds after breaking the surface of the water is given by

$$h = -16t^2 + 32t$$

How long will it take the dolphin to jump out of the water and touch the trainer's hand?

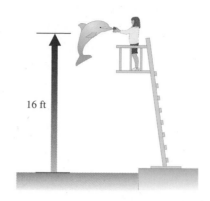

16 ft

80. Forensic medicine   The kinetic energy E of a moving object is given by $E = \frac{1}{2}mv^2$, where m is the mass of the object (in kilograms) and v is the object's velocity (in meters per second). Kinetic energy is measured in joules. Examining the damage done to a victim, a police pathologist determines that the energy of a 3-kilogram mass at impact was 54 joules. Find the velocity at impact.

81. Conversion from degrees Celsius to degrees Fahrenheit   The temperature in degrees Fahrenheit that is equivalent to a temperature in degrees Celsius is given by the function $F(C) = \frac{9}{5}C + 32$. Find the Fahrenheit temperature that is equivalent to 25°C.

82. Conversion from degrees Fahrenheit to degrees Celsius   The temperature in degrees Celsius that is equivalent to a temperature in degrees Fahrenheit is given by the function $C(F) = \frac{5}{9}F - \frac{160}{9}$. Find the Celsius temperature that is equivalent to 14°F.

83. Selling DVD players   An electronics firm manufactures DVD players, receiving \$120 for each unit it makes. If x represents the number of units produced, the income received is determined by the *revenue function* $R(x) = 120x$. The manufacturer has fixed costs of \$12,000 per month and variable costs of \$57.50 for each unit manufactured. Thus, the *cost function* is $C(x) = 57.50x + 12,000$. How many DVD players must the company sell for revenue to equal cost?

84. Selling tires   A tire company manufactures premium tires, receiving \$130 for each tire it makes. If the manufacturer has fixed costs of \$15,512.50 per month and variable costs of \$93.50 for each tire manufactured, how many tires must the company sell for revenue to equal cost?   (*Hint:* See Exercise 83.)

WRITING

85. Explain the concepts of function, domain, and range.

86. Explain why the constant function is a special case of a linear function.

SOMETHING TO THINK ABOUT

Let $f(x) = 2x + 1$ and $g(x) = x^2$. Assume that $f(x) \neq 0$ and $g(x) \neq 0$.

87. Is $f(x) + g(x)$ equal to $g(x) + f(x)$?

88. Is $f(x) - g(x)$ equal to $g(x) - f(x)$?

2.5   Graphs of Other Functions

▌ **Graphs of Nonlinear Functions**   ▌ **Translations of Graphs**
▌ **Reflections of Graphs**

Getting Ready   *Give the slope and the y-intercept of each linear function.*

1. $f(x) = 2x - 3$ 　　　　　　　　　　 **2.** $f(x) = -3x + 4$

Find the value of $f(x)$ when $x = 2$ and $x = -1$.

3. $f(x) = 5x - 4$ 　　　　　　　　　　 **4.** $f(x) = \dfrac{1}{2}x + 3$

In the previous section, we discussed linear functions, functions whose graphs are straight lines. We now extend the discussion to include nonlinear functions, functions whose graphs are not straight lines.

Graphs of Nonlinear Functions

If f is a function whose domain and range are sets of real numbers, its graph is the set of all points $(x, f(x))$ in the xy-plane. In other words, the graph of f is the graph of the equation $y = f(x)$. In this section, we will draw the graphs of many basic functions. The first is $f(x) = x^2$ (or $y = x^2$), often called the **squaring function.**

EXAMPLE 1   Graph the function: $f(x) = x^2$.

Solution   We substitute values for x in the equation and compute the corresponding values of $f(x)$. For example, if $x = -3$, we have

$$f(\boldsymbol{x}) = \boldsymbol{x}^2$$
$$f(\boldsymbol{-3}) = (\boldsymbol{-3})^2 \quad \text{Substitute } -3 \text{ for } x.$$
$$= 9$$

The ordered pair $(-3, 9)$ satisfies the equation and will lie on the graph. We list this pair and the others that satisfy the equation in the table shown in Figure 2-40. We plot the points and draw a smooth curve through them to get the graph, called a **parabola.**

$$f(x) = x^2$$

x	y	$(x, f(x))$
-3	9	$(-3, 9)$
-2	4	$(-2, 4)$
-1	1	$(-1, 1)$
0	0	$(0, 0)$
1	1	$(1, 1)$
2	4	$(2, 4)$
3	9	$(3, 9)$

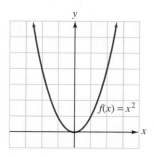

Figure 2-40

From the graph, we see that x can be any real number. This indicates that the domain of the squaring function is the set of real numbers, which is the interval $(-\infty, \infty)$. We can also see that y is always positive or zero. This indicates that the range is the set of nonnegative real numbers, which is the interval $[0, \infty)$.

Self Check   Graph $f(x) = x^2 - 2$ and compare the graph to the graph of $f(x) = x^2$.   ∎

The second basic function is $f(x) = x^3$ (or $y = x^3$), often called the **cubing function.**

EXAMPLE 2   Graph the function: $f(x) = x^3$.

Solution   We substitute values for x in the equation and compute the corresponding values of $f(x)$. For example, if $x = -2$, we have

$$f(x) = x^3$$
$$f(-2) = (-2)^3 \quad \text{Substitute } -2 \text{ for } x.$$
$$= -8$$

The ordered pair $(-2, -8)$ satisfies the equation and will lie on the graph. We list this pair and others that satisfy the equation in the table shown in Figure 2-41. We plot the points and draw a smooth curve through them to get the graph.

$$f(x) = x^3$$

x	y	$(x, f(x))$
-2	-8	$(-2, -8)$
-1	-1	$(-1, -1)$
0	0	$(0, 0)$
1	1	$(1, 1)$
2	8	$(2, 8)$

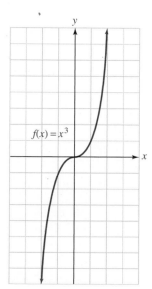

Figure 2-41

From the graph, we can see that x can be any real number. This indicates that the domain of the cubing function is the set of real numbers, which is the interval $(-\infty, \infty)$. We can also see that y can be any real number. This indicates that the range is the set of real numbers, which is the interval $(-\infty, \infty)$.

Self Check Graph $f(x) = x^3 + 1$ and compare the graph to the graph of $f(x) = x^3$. ∎

The third basic function is $f(x) = |x|$ (or $y = |x|$), often called the **absolute value function.**

EXAMPLE 3 Graph the function: $f(x) = |x|$.

Solution We substitute values for x in the equation and compute the corresponding values of $f(x)$. For example, if $x = -3$, we have

$$f(\boldsymbol{x}) = |\boldsymbol{x}|$$
$$f(\boldsymbol{-3}) = |\boldsymbol{-3}| \quad \text{Substitute } -3 \text{ for } x.$$
$$= 3$$

The ordered pair $(-3, 3)$ satisfies the equation and will lie on the graph. We list this pair and others that satisfy the equation in the table shown in Figure 2-42. We plot the points and draw a V-shaped graph through them.

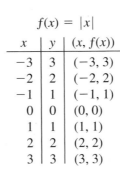

$$f(x) = |x|$$

x	y	$(x, f(x))$
-3	3	$(-3, 3)$
-2	2	$(-2, 2)$
-1	1	$(-1, 1)$
0	0	$(0, 0)$
1	1	$(1, 1)$
2	2	$(2, 2)$
3	3	$(3, 3)$

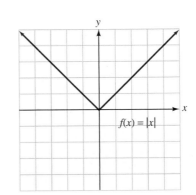

Figure 2-42

From the graph, we see that x can be any real number. This indicates that the domain of the absolute value function is the set of real numbers, which is the interval $(-\infty, \infty)$. We can also see that y is always positive or zero. This indicates that the range is the set of nonnegative real numbers, which is the interval $[0, \infty)$.

Self Check Graph $f(x) = |x - 2|$ and compare the graph to the graph of $f(x) = |x|$. ∎

Accent on Technology

GRAPHING FUNCTIONS

We can graph nonlinear functions with a graphing calculator. For example, to graph $f(x) = x^2$ in a standard window of $[-10, 10]$ for x and $[-10, 10]$ for y, we enter the function by typing $x \wedge 2$ and press the GRAPH key. We will obtain the graph shown in Figure 2-43(a).

To graph $f(x) = x^3$, we enter the function by typing $x \wedge 3$ and press the GRAPH key to obtain the graph in Figure 2-43(b). To graph $f(x) = |x|$, we enter the function by selecting "abs" from the MATH menu, typing x, and pressing the GRAPH key to obtain the graph in Figure 2-43(c).

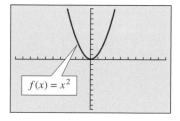

The squaring function

(a)

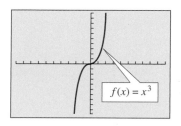

The cubing function

(b)

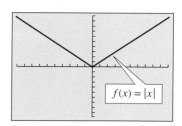

The absolute value function

(c)

Figure 2-43

When using a graphing calculator, we must be sure that the viewing window does not show a misleading graph. For example, if we graph $f(x) = |x|$ in the window $[0, 10]$ for x and $[0, 10]$ for y, we will obtain a misleading graph that looks like a line. (See Figure 2-44.) This is not true. The proper graph is the V-shaped graph shown in Figure 2-43(c).

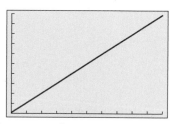

Figure 2-44

Translations of Graphs

Examples 1–3 and their Self Checks suggest that the graphs of different functions may be identical except for their positions in the xy-plane. For example, Figure 2-45 shows the graph of $f(x) = x^2 + k$ for three different values of k. If $k = 0$, we get the graph of $f(x) = x^2$. If $k = 3$, we get the graph of $f(x) = x^2 + 3$, which is identical to the graph of $f(x) = x^2$ except that it is shifted 3 units upward. If $k = -4$, we get the graph of $f(x) = x^2 - 4$, which is identical to the graph of $f(x) = x^2$ except that it is shifted 4 units downward. These shifts are called **vertical translations**.

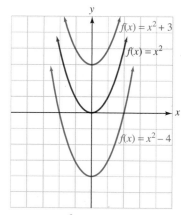

Figure 2-45

In general, we can make these observations.

Vertical Translations

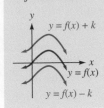

If f is a function and k is a positive number, then

- The graph of $y = f(x) + k$ is identical to the graph of $y = f(x)$ except that it is translated k units upward.
- The graph of $y = f(x) - k$ is identical to the graph of $y = f(x)$ except that it is translated k units downward.

EXAMPLE 4   Graph: $f(x) = |x| + 2$.

Solution   The graph of $f(x) = |x| + 2$ will be the same V-shaped graph as $f(x) = |x|$, except that it is shifted 2 units up. The graph appears in Figure 2-46.

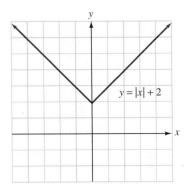

Figure 2-46

Self Check   Graph: $f(x) = |x| - 3$.

Figure 2-47 shows the graph of $f(x) = (x + h)^2$ for three different values of h. If $h = 0$, we get the graph of $f(x) = x^2$. The graph of $f(x) = (x - 3)^2$ is identical to the graph of $f(x) = x^2$, except that it is shifted 3 units to the right. The graph of $f(x) = (x + 2)^2$ is identical to the graph of $f(x) = x^2$, except that it is shifted 2 units to the left. These shifts are called **horizontal translations.**

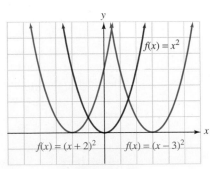

Figure 2-47

In general, we can make these observations.

Horizontal Translations

If f is a function and k is a positive number, then

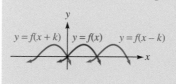

- The graph of $y = f(x - k)$ is identical to the graph of $y = f(x)$ except that it is translated k units to the right.
- The graph of $y = f(x + k)$ is identical to the graph of $y = f(x)$ except that it is translated k units to the left.

EXAMPLE 5   Graph: $f(x) = (x - 2)^2$.

Solution   The graph of $f(x) = (x - 2)^2$ will be the same shape as the graph of $f(x) = x^2$ except that it is shifted 2 units to the right. The graph appears in Figure 2-48.

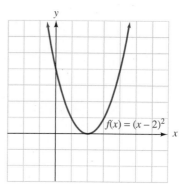

Figure 2-48

Self Check   Graph: $f(x) = (x + 3)^3$.

EXAMPLE 6   Graph: $f(x) = (x - 3)^2 + 2$.

Solution   We can graph this function by translating the graph of $f(x) = x^2$ to the right 3 units and then up 2 units, as shown in Figure 2-49.

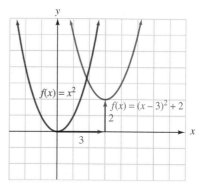

Figure 2-49

Self Check   Graph: $f(x) = |x + 2| - 3$.

Reflections of Graphs

Figure 2-50 shows tables of solutions for $f(x) = x^2$ and for $f(x) = -x^2$. We note that for a given value of x, the corresponding y-values in the tables are *opposites*.

When the values are graphed, we see that the $-$ in $f(x) = -x^2$ has the effect of "flipping" the graph of $f(x) = x^2$ over the x-axis, so that the parabola opens downward. We say that the graph of $f(x) = -x^2$ is a **reflection** of the graph of $f(x) = x^2$ about the x-axis.

$f(x) = x^2$

x	y	$(x, f(x))$
-2	4	$(-2, 4)$
-1	1	$(-1, 1)$
0	0	$(0, 0)$
1	1	$(1, 1)$
2	4	$(2, 4)$

$f(x) = -x^2$

x	y	$(x, f(x))$
-2	-4	$(-2, -4)$
-1	-1	$(-1, -1)$
0	0	$(0, 0)$
1	-1	$(1, -1)$
2	-4	$(2, -4)$

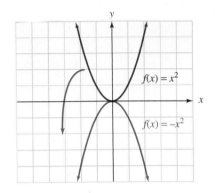

Figure 2-50

EXAMPLE 7   **Reflection of a graph**   Graph: $f(x) = -x^3$.

Solution   To graph $f(x) = -x^3$, we use the graph of $f(x) = x^3$ from Example 2. First, we reflect the portion of the graph of $f(x) = x^3$ in quadrant I to quadrant IV, as shown in Figure 2-51. Then we reflect the portion of the graph of $f(x) = x^3$ that is in quadrant III to quadrant II.

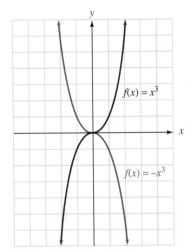

Figure 2-51

Self Check   Graph: $f(x) = -|x|$. ∎

Reflection of a Graph   The graph of $y = -f(x)$ is the graph of $f(x)$ reflected about the x-axis.

PERSPECTIVE

Graphs in Space   In an xy-coordinate system, graphs of equations containing the two variables x and y are lines or curves. Other equations have more than two variables, and graphing them often requires some ingenuity and perhaps the aid of a computer. Graphs of equations with the three variables x, y, and z are viewed in a three-dimensional coordinate system with three axes. The coordinates of points in a three-dimensional coordinate system are ordered triples (x, y, z). For example, the points $P(2, 3, 4)$ and $Q(-1, 2, 3)$ are plotted in Illustration 1.

Graphs of equations in three variables are not lines or curves, but flat planes or curved surfaces. Only the simplest of these equations can be conveniently graphed by hand; a computer provides the best images of others. The graph in Illustration 2 is called a **paraboloid;** it is the three-dimensional version of a parabola. Illustration 3 models a portion of the vibrating surface of a drum head.

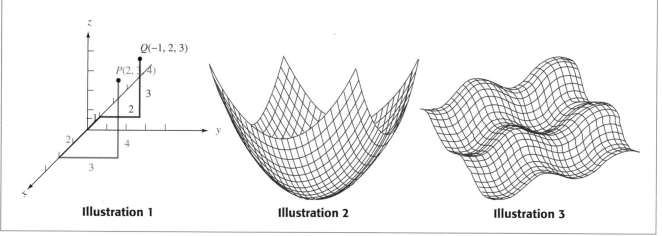

| **Illustration 1** | **Illustration 2** | **Illustration 3** |

Accent on Technology

SOLVING EQUATIONS WITH GRAPHING CALCULATORS

To solve the equation $2(x - 3) + 3 = 7$ with a graphing calculator, we can graph the left-hand side and the right-hand side of the equation in the same window, as shown in Figure 2-52(a). We then trace to find the coordinates of the point where the two graphs intersect, as shown in Figure 2-52(b). We can then zoom and trace again to get Figure 2-52(c). From the figure, we see that $x = 5$.

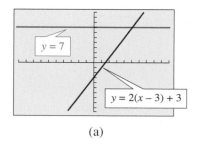

(a)

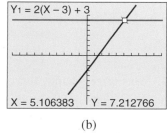

(b)

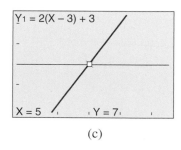

(c)

Figure 2-52

(continued)

We can also solve the equation by using the intersect command found in the CALC menu. To use this method, we first draw the graph shown in Figure 2-52(a) and open the CALC menu and select "5: intersect" to obtain Figure 2-53(a). Then we select a point on the first curve by pressing ENTER, select a point on the second curve by pressing ENTER, and press ENTER again to obtain Figure 2-53(b). From the figure, we see that $x = 5$.

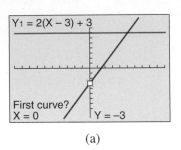

(a)

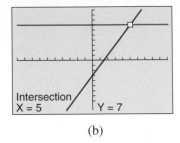

(b)

Figure 2-53

Self Check Answers

1. same shape, but 2 units lower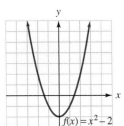

2. same shape, but 1 unit higher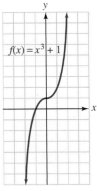

3. same shape, but 2 units to the right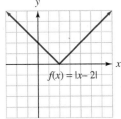

4.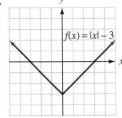

5.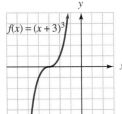

6.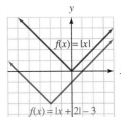

7.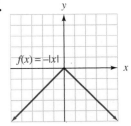

Orals

1. What is the squaring function?

2. What is the cubing function?

3. What is the absolute value function?

4. Describe a parabola.

5. Describe the graph of $f(x) = |x| + 3$.

6. Describe the graph of $f(x) = x^3 - 4$.

7. What is meant by a reflection of a graph?

8. How does the graph of $y = x^2$ compare to the graph of $y = x^2 + 2$?

2.5 EXERCISES

REVIEW

1. List the prime numbers between 40 and 50.
2. State the associative property of addition.

3. State the commutative property of multiplication.

4. What is the additive identity element?
5. What is the multiplicative identity element?
6. Find the multiplicative inverse of $\frac{5}{3}$.

VOCABULARY AND CONCEPTS   *Fill in the blanks.*

7. The function $f(x) = x^2$ is called the _____ function.
8. The function $f(x) = x^3$ is called the _____ function.
9. The function $f(x) = |x|$ is called the _____ function.
10. Shifting the graph of an equation up or down is called a _____ translation.
11. Shifting the graph of an equation to the left or to the right is called a _____ translation.
12. The graph of $f(x) = x^2 + 5$ is the same as the graph of $f(x) = x^2$ except that it is shifted __ units ___.
13. The graph of $f(x) = x^3 - 2$ is the same as the graph of $f(x) = x^3$ except that it is shifted __ units _____.
14. The graph of $f(x) = (x - 5)^3$ is the same as the graph of $f(x) = x^3$ except that it is shifted __ units _____.

15. The graph of $f(x) = (x + 4)^3$ is the same as the graph of $f(x) = x^3$ except that it is shifted __ units _____.

16. To solve an equation with a graphing calculator, graph ___ sides of the equation and find the _____ of the point where the graphs intersect.

PRACTICE   *Graph each function. Check your work with a graphing calculator.*

17. $f(x) = x^2 - 3$

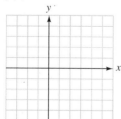

18. $f(x) = x^2 + 2$

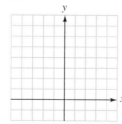

19. $f(x) = (x - 1)^3$

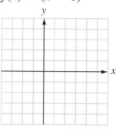

20. $f(x) = (x + 1)^3$

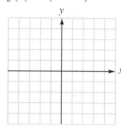

21. $f(x) = |x| - 2$

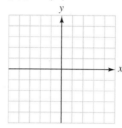

22. $f(x) = |x| + 1$

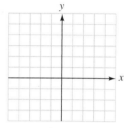

23. $f(x) = |x - 1|$

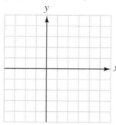

24. $f(x) = |x + 2|$

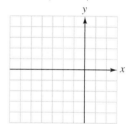

Use a graphing calculator to graph each function, using values of $[-4, 4]$ for x and $[-4, 4]$ for y. The graph is not what it appears to be. Pick a better viewing window and find the true graph.

25. $f(x) = x^2 + 8$

26. $f(x) = x^3 - 8$

27. $f(x) = |x + 5|$

28. $f(x) = |x - 5|$

29. $f(x) = (x - 6)^2$

30. $f(x) = (x + 9)^2$

31. $f(x) = x^3 + 8$

32. $f(x) = x^3 - 12$

Draw each graph using a translation of the graph of
$f(x) = x^2$, $f(x) = x^3$, *or* $f(x) = |x|$.

33. $f(x) = x^2 - 5$

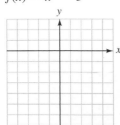

34. $f(x) = x^3 + 4$

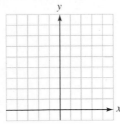

35. $f(x) = (x - 1)^3$

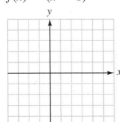

36. $f(x) = (x + 4)^2$

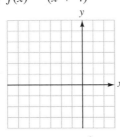

37. $f(x) = |x - 2| - 1$

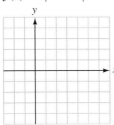

38. $f(x) = (x + 2)^2 - 1$

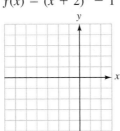

39. $f(x) = (x + 1)^3 - 2$

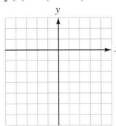

40. $f(x) = |x + 4| + 3$

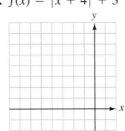

Draw each graph using a translation of the graph of
$f(x) = -x^2$, $f(x) = -x^3$, *or* $f(x) = -|x|$.

41. $f(x) = -|x| + 1$

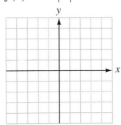

42. $f(x) = -x^3 - 2$

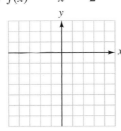

43. $f(x) = -(x - 1)^2$

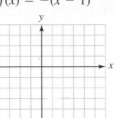

44. $f(x) = -(x + 2)^2$

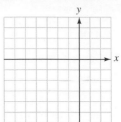

Use a graphing calculator to solve each equation.

45. $3x + 6 = 0$

46. $7x - 21 = 0$

47. $4(x - 1) = 3x$

48. $4(x - 3) - x = x - 6$

49. $11x + 6(3 - x) = 3$

50. $2(x + 2) = 2(1 - x) + 10$

WRITING

51. Explain how to graph an equation by plotting points.

52. Explain why the correct choice of window settings is important when using a graphing calculator.

SOMETHING TO THINK ABOUT *Use a graphing calculator.*

53. Use a graphing calculator with settings of $[-10, 10]$ for x and $[-10, 10]$ for y to graph **a.** $y = -x^2$, **b.** $y = -x^2 + 1$, and **c.** $y = -x^2 + 2$. What do you notice?

54. Use a graphing calculator with settings of $[-10, 10]$ for x and $[-10, 10]$ for y to graph **a.** $y = -|x|$, **b.** $y = -|x| - 1$, and **c.** $y = -|x| - 2$. What do you notice?

55. Graph $y = (x + k)^2 + 1$ for several positive values of k. What do you notice?

56. Graph $y = (x + k)^2 + 1$ for several negative values of k. What do you notice?

57. Graph $y = (kx)^2 + 1$ for several values of k, where $k > 1$. What do you notice?

58. Graph $y = (kx)^2 + 1$ for several values of k, where $0 < k < 1$. What do you notice?

59. Graph $y = (kx)^2 + 1$ for several values of k, where $k < -1$. What do you notice?

60. Graph $y = (kx)^2 + 1$ for several values of k, where $-1 < k < 0$. What do you notice?

Projects

Project 1

The Board of Administrators of Boondocks County has hired your consulting firm to plan a highway. The new highway is to be built in the outback section of the county.

The two main roads in the outback section are Highway N, running in a straight line north and south, and Highway E, running in a straight line east and west. These two highways meet at an intersection that the locals call Four Corners. The only other county road in the area is Slant Road, which runs in a straight line from northwest to southeast, cutting across Highway N north of Four Corners and Highway E east of Four Corners.

The county clerk is unable to find an official map of the area, but there is an old sketch made by the original road designer. It shows that if a rectangular coordinate system is set up using Highways N and E as the axes and Four Corners as the origin, then the equation of the line representing Slant Road is

$$2x + 3y = 12 \quad \text{(where the unit length is 1 mile)}$$

Given this information, the county wants you to do the following:

a. Update the current information by giving the coordinates of the intersections of Slant Road with Highway N and Highway E.
b. Plan a new highway, Country Drive, that will begin 1 mile north of Four Corners and run in a straight line in a generally northeasterly direction, intersecting Slant Road at right angles. The county wants to know the equation of the line representing Country Drive. You should also state the domain on which this equation is valid as a representation of Country Drive.

Project 2

You represent your branch of the Buy-from-Us Corporation. At the company's regional meeting, you must present your revenue and cost reports to the other branch representatives. But disaster strikes! The graphs you had planned to present, containing cost and revenue information for this year and last year, are unlabeled! You cannot immediately recognize which graphs represent costs, which represent revenues, and which repre-

sent which year. Without these graphs, your presentation will not be effective.

The only other information you have with you is in the notes you made for your talk. From these you glean the following financial data about your branch.

1. All cost and revenue figures on the graphs are rounded to the nearest $50,000.
2. Costs for the fourth quarter of last year were $400,000.
3. Revenue was not above $400,000 for any quarter last year.
4. Last year, your branch lost money during the first quarter.
5. This year, your branch made money during three of the four quarters.
6. Profit during the second quarter of this year was $150,000.

And, of course, you know that profit = revenue − cost.

With this information, you must match each of the graphs (Illustrations 1–4) with one of the following titles:

Costs, This Year Costs, Last Year
Revenues, This Year Revenues, Last Year

You should be sure to have sound reasons for your choices—reasons ensuring that no other arrangement of the titles will fit the data. The *last* thing you want to do is present incorrect information.

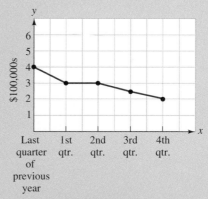

ILLUSTRATION 1

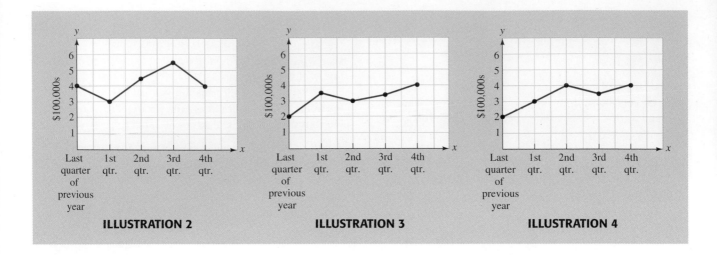

| ILLUSTRATION 2 | ILLUSTRATION 3 | ILLUSTRATION 4 |

Chapter Summary

CONCEPTS	REVIEW EXERCISES
2.1	**Graphing Linear Equations**

1. Use the equation $2x - 3y = 12$ to complete the table.

x	y
-9	
	-8
-3	
	-4
3	
	0
	2

Graph of a vertical line:
   $x = a$
   x-intercept at $(a, 0)$

Graph each equation.

Graph of a horizontal line:
   $y = b$
   y-intercept at $(0, b)$

2. $x + y = 4$

3. $2x - y = 8$

4. $y = 3x + 4$

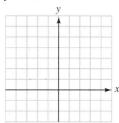

5. $x = 4 - 2y$

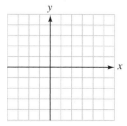

6. $y = 4$

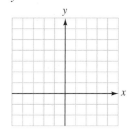

7. $x = -2$

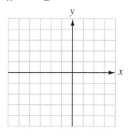

8. $2(x + 3) = x + 2$

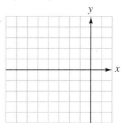

9. $3y = 2(y - 1)$

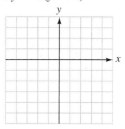

10. Find the midpoint of the line segment joining $P(-3, 5)$ and $Q(6, 11)$.

Midpoint formula:
If $P(x_1, y_1)$ and $Q(x_2, y_2)$, the midpoint of PQ is

$$M\left(\frac{x_1 + x_2}{2}, \frac{y_1 + y_2}{2}\right)$$

| 2.2 | Slope of a Nonvertical Line |

Slope of a nonvertical line:
If $x_2 \neq x_1$,

$$m = \frac{\Delta y}{\Delta x} = \frac{y_2 - y_1}{x_2 - x_1}$$

Horizontal lines have a slope of 0. Vertical lines have no defined slope.

Find the slope of the line passing through points P and Q, if possible.

11. $P(2, 5)$ and $Q(5, 8)$

12. $P(-3, -2)$ and $Q(6, 12)$

13. $P(-3, 4)$ and $Q(-5, -6)$

14. $P(5, -4)$ and $Q(-6, -9)$

15. $P(-2, 4)$ and $Q(8, 4)$

16. $P(-5, -4)$ and $Q(-5, 8)$

Find the slope of the graph of each equation, if one exists.

17. $2x - 3y = 18$

18. $2x + y = 8$

19. $-2(x - 3) = 10$

20. $3y + 1 = 7$

Parallel lines have the same slope. The slopes of two nonvertical perpendicular lines are negative reciprocals.

Tell whether the lines with the given slopes are parallel, perpendicular, or neither.

21. $m_1 = 4, m_2 = -\dfrac{1}{4}$

22. $m_1 = 0.5, m_2 = \dfrac{1}{2}$

23. $m_1 = 0.5, m_2 = -\dfrac{1}{2}$

24. $m_1 = 5, m_2 = -0.2$

25. If the sales of a new business were \$65,000 in its first year and \$130,000 in its fourth year, find the rate of growth in sales per year.

2.3 Writing Equations of Lines

Equations of a line:
Point-slope form:
$$y - y_1 = m(x - x_1)$$
Slope-intercept form:
$$y = mx + b$$
General form:
$$Ax + By = C$$

Write the equation of the line with the given properties. Write the equation in general form.

26. Slope of 3; passing through $P(-8, 5)$

27. Passing through $(-2, 4)$ and $(6, -9)$

28. Passing through $(-3, -5)$; parallel to the graph of $3x - 2y = 7$

29. Passing through $(-3, -5)$; perpendicular to the graph of $3x - 2y = 7$

30. A business purchased a copy machine for \$8,700 and will depreciate it on a straight-line basis over the next 5 years. At the end of its useful life, it will be sold as scrap for \$100. Find its depreciation equation.

2.4 Introduction to Functions

A **function** is a correspondence between a set of input values x and a set of output values y, where exactly one value of y in the range corresponds to each number x in the domain.

$f(k)$ represents the value of $f(x)$ when $x = k$.

Tell whether each equation determines y to be a function of x.

31. $y = 6x - 4$ **32.** $y = 4 - x$

33. $y^2 = x$ **34.** $|y| = x^2$

Assume that $f(x) = 3x + 2$ and $g(x) = x^2 - 4$ and find each value.

35. $f(-3)$ **36.** $g(8)$

37. $g(-2)$ **38.** $f(5)$

The **domain** of a function is the set of input values. The **range** is the set of output values.

Find the domain and range of each function.

39. $f(x) = 4x - 1$

40. $f(x) = 3x - 10$

41. $f(x) = x^2 + 1$

42. $f(x) = \dfrac{4}{2 - x}$

43. $f(x) = \dfrac{7}{x - 3}$

44. $y = 7$

The **vertical line test** can be used to determine whether a graph represents a function.

Use the vertical line test to determine whether each graph represents a function.

45.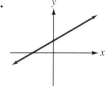

46.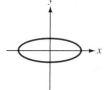

47.

48.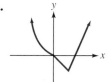

2.5   Graphs of Other Functions

Graphs of nonlinear functions are not lines.

Graph each function.

49. $f(x) = x^2 - 3$

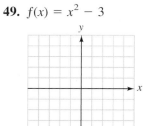

50. $f(x) = |x| - 4$

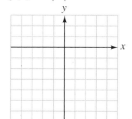

51. $f(x) = (x - 2)^3$

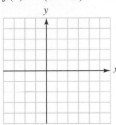

52. $f(x) = (x + 4)^2 - 3$

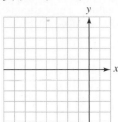

 Use a graphing calculator to graph each function. Compare the results to the answers to Problems 49–52.

53. $f(x) = x^2 - 3$

54. $f(x) = |x| - 4$

55. $f(x) = (x - 2)^3$

56. $f(x) = (x + 4)^2 - 3$

Tell whether each equation defines a linear function.

57. $y = 3x + 2$

58. $y = \dfrac{x + 5}{4}$

59. $4x - 3y = 12$

60. $y = x^2 - 25$

61. Graph: $f(x) = -|x - 3|$.

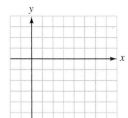

Chapter Test

1. Graph the equation $2x - 5y = 10$.

2. Find the x- and y-intercepts of the graph of $y = \frac{x - 3}{5}$.

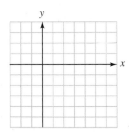

3. Find the midpoint of the line segment.

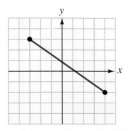

Find the slope of each line, if possible.

4. The line through $(-2, 4)$ and $(6, 8)$

5. The graph of $2x - 3y = 8$

6. The graph of $x = 12$

7. The graph of $y = 12$

8. Write the equation of the line with slope of $\frac{2}{3}$ that passes through $(4, -5)$. Give the answer in slope-intercept form.

9. Write the equation of the line that passes through $(-2, 6)$ and $(-4, -10)$. Give the answer in general form.

10. Find the slope and the y-intercept of the graph of $-2(x - 3) = 3(2y + 5)$.

11. Are the graphs of $4x - y = 12$ and $y = \frac{1}{4}x + 3$ parallel, perpendicular, or neither?

12. Determine whether the graphs of $y = -\frac{2}{3}x + 4$ and $2y = 3x - 3$ are parallel, perpendicular, or neither.

13. Write the equation of the line that passes through the origin and is parallel to the graph of $y = \frac{3}{2}x - 7$.

14. Write the equation of the line that passes through $P(-3, 6)$ and is perpendicular to the graph of $y = -\frac{2}{3}x - 7$.

15. Does $|y| = x$ define y to be a function of x?

16. Find the domain and range of the function $f(x) = |x|$.

17. Find the domain and range of the function $f(x) = x^3$.

Let $f(x) = 3x + 1$ and $g(x) = x^2 - 2$. Find each value.

18. $f(3)$

19. $g(0)$

20. $f(a)$

21. $g(-x)$

Tell whether each graph represents a function.

22.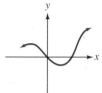

23.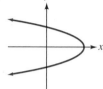

24. Graph: $f(x) = x^2 - 1$.

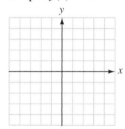

25. Graph: $f(x) = -|x + 2|$.

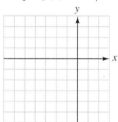

CUMULATIVE REVIEW EXERCISES

Tell which numbers in the set $\left\{-2, 0, 1, 2, \frac{13}{12}, 6, 7, \sqrt{5}, \pi\right\}$ *are in each category.*

1. Natural numbers

2. Whole numbers

3. Rational numbers

4. Irrational numbers

5. Negative numbers

6. Real numbers

7. Prime numbers

8. Composite numbers

9. Even numbers

10. Odd numbers

Graph each set on the number line.

11. $\left\{x \mid -2 < x \leq 5\right\}$

12. $[-5, 0) \cup [3, 6]$

Simplify each expression.

13. $-|5| + |-3|$

14. $\dfrac{|-5| + |-3|}{-|4|}$

Perform the operations.

15. $2 + 4 \cdot 5$

16. $\dfrac{8 - 4}{2 - 4}$

17. $20 \div (-10 \div 2)$

18. $\dfrac{6 + 3(6 + 4)}{2(3 - 9)}$

Evaluate each expression when $x = 2$ and $y = -3$.

19. $-x - 2y$

20. $\dfrac{x^2 - y^2}{2x + y}$

Tell which property of real numbers justifies each statement.

21. $(a + b) + c = a + (b + c)$

22. $3(x + y) = 3x + 3y$

23. $(a + b) + c = c + (a + b)$

24. $(ab)c = a(bc)$

Simplify each expression. Assume that all variables are positive numbers and write all answers without negative exponents.

25. $(x^2 y^3)^4$

26. $\dfrac{c^4 c^8}{(c^5)^2}$

27. $\left(-\dfrac{a^3 b^{-2}}{ab}\right)^{-1}$

28. $\left(\dfrac{-3a^3 b^{-2}}{6a^{-2} b^3}\right)^0$

29. Change 0.00000497 to scientific notation.

30. Change 9.32×10^8 to standard notation.

Solve each equation.

31. $2x - 5 = 11$

32. $\dfrac{2x - 6}{3} = x + 7$

33. $4(y - 3) + 4 = -3(y + 5)$

34. $2x - \dfrac{3(x - 2)}{2} = 7 - \dfrac{x - 3}{3}$

Solve each formula for the indicated variable.

35. $S = \dfrac{n(a + l)}{2}$ for a

36. $A = \dfrac{1}{2}h(b_1 + b_2)$ for h

37. The sum of three consecutive even integers is 90. Find the integers.

38. A rectangle is three times as long as it is wide. If its perimeter is 112 centimeters, find its dimensions.

39. Tell whether the graph of $2x - 3y = 6$ defines a function.

40. Find the slope of the line passing through $P(-2, 5)$ and $Q(8, -9)$.

41. Write the equation of the line passing through $P(-2, 5)$ and $Q(8, -9)$.

42. Write the equation of the line that passes through $P(-2, 3)$ and is parallel to the graph of $3x + y = 8$.

Evaluate each expression, given that $f(x) = 3x^2 + 2$ and $g(x) = 2x - 1$.

43. $f(-1)$

44. $g(0)$

45. $g(t)$

46. $f(-r)$

Graph each equation and tell whether it is a function. If it is a function, give the domain and range.

47. $y = -x^2 + 1$

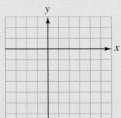

48. $y = \left| \dfrac{1}{2}x - 3 \right|$

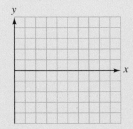

3 Systems of Equations

InfoTrac Project

Do a keyword search on "real estate prices." Write a summary of the article "Cantor Pecorella reports strong demand for Greenwich Village condo tower."

A prospective buyer is trying to decide between buying or renting the cheapest house listed in the article, knowing that, if she buys, upkeep on the condo will cost about 10% of the purchase price each year. Write an equation for the total cost, C for n years, if she buys the condo. If she rents, the rent will be $12,000 a year, with all upkeep expenses paid by the owner. Write an equation for the total cost, C, of renting for n years. Using a system of equations, find the number of years it will take for the cost of renting to equal the cost of purchasing the condo. Find the total cost that will have been put into the condo at this point. What other factors that may help the prospective buyer in her decision has this problem failed to take into consideration?

Complete this project after studying Section 3.2.

© Arvind Garg/CORBIS

Mathematics in Electronics

In a radio, an inductor and a capacitor are used in a resonant circuit to select a wanted radio station at a frequency f and reject all others. The inductance L and the capacitance C determine the inductance reactance X_L and the capacitive reactance X_C of that circuit, where

$$X_L = 2\pi f L \qquad \text{and} \qquad X_C = \frac{1}{2\pi f C}$$

The radio station selected will be at the frequency f, where $X_L = X_C$. Write a formula for f^2 in terms of L and C.

Exercise Set 3.2
Problem 87

151

*We have considered linear equations with the variables
x and y. We found that each equation had infinitely many
solutions (x, y), and that we can graph each equation on
the rectangular coordinate system. In this chapter, we will
discuss many **systems of linear equations** involving two
or three equations.*

| **3.1** | **Solution by Graphing** |

■ **The Graphing Method**   ■ **Consistent Systems**
■ **Inconsistent Systems**   ■ **Dependent Equations**

Getting Ready   *Let $y = -3x + 2$.*

1. Find y when $x = 0$.   **2.** Find y when $x = 3$.

3. Find y when $x = -3$.   **4.** Find y when $x = -\dfrac{1}{3}$.

5. Find five pairs of numbers with a sum of 12.

6. Find five pairs of numbers with a difference of 3.

In the pair of equations

$$\begin{cases} x + 2y = 4 \\ 2x - y = 3 \end{cases} \quad \text{(called a system of equations)}$$

there are infinitely many ordered pairs (x, y) that satisfy the first equation and infinitely many ordered pairs (x, y) that satisfy the second equation. However, there is
only one ordered pair (x, y) that satisfies both equations at the same time. The
process of finding this ordered pair is called *solving the system.*

The Graphing Method

We follow these steps to solve a system of two equations in two variables by graphing.

The Graphing Method

1. On a single set of coordinate axes, graph each equation.

2. Find the coordinates of the point (or points) where the graphs intersect.
These coordinates give the solution of the system.

3. If the graphs have no point in common, the system has no solution.

4. If the graphs of the equations coincide, the system has infinitely many
solutions.

5. Check the solution in both of the original equations.

Consistent Systems

When a system of equations has a solution (as in Example 1), the system is called a **consistent system.**

EXAMPLE 1   Solve the system: $\begin{cases} x + 2y = 4 \\ 2x - y = 3 \end{cases}$.

Solution   We graph both equations on one set of coordinates axes, as shown in Figure 3-1.

$x + 2y = 4$

x	y	(x, y)
4	0	(4, 0)
0	2	(0, 2)
−2	3	(−2, 3)

$2x - y = 3$

x	y	(x, y)
1	−1	(1, −1)
0	−3	(0, −3)
−1	−5	(−1, −5)

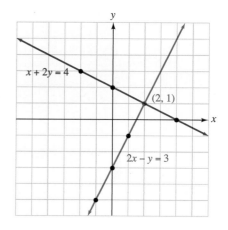

Figure 3-1

Although infinitely many ordered pairs (x, y) satisfy $x + 2y = 4$, and infinitely many ordered pairs (x, y) satisfy $2x - y = 3$, only the coordinates of the point where the graphs intersect satisfy both equations. Since the intersection point has coordinates of (2, 1), the solution is the ordered pair (2, 1), or $x = 2$ and $y = 1$.

When we check the solution, we substitute 2 for x and 1 for y in both equations and verify that (2, 1) satisfies each one.

Self Check   Solve the system: $\begin{cases} 2x + y = 4 \\ x - 3y = -5 \end{cases}$.   ∎

Inconsistent Systems

When a system has no solution (as in Example 2), it is called an **inconsistent system.**

EXAMPLE 2   Solve the system $\begin{cases} 2x + 3y = 6 \\ 4x + 6y = 24 \end{cases}$, if possible.

Solution   We graph both equations on one set of coordinate axes, as shown in Figure 3-2. In this example, the graphs are parallel, because the slopes of the two lines are equal and their y-intercepts are different. We can see that the slope of each line is $-\frac{2}{3}$ by writing each equation in slope-intercept form.

Since the graphs are parallel lines, the lines do not intersect, and the system does not have a solution. It is an inconsistent system.

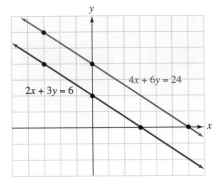

2x + 3y = 6		
x	y	(x, y)
3	0	(3, 0)
0	2	(0, 2)
−3	4	(−3, 4)

4x + 6y = 24		
x	y	(x, y)
6	0	(6, 0)
0	4	(0, 4)
−3	6	(−3, 6)

Figure 3-2

Self Check   Solve the system: $\begin{cases} 2x - 3y = 6 \\ y = \frac{2}{3}x - 3 \end{cases}$.

Dependent Equations

When the equations of a system have different graphs (as in Examples 1 and 2), the equations are called **independent equations.** Two equations with the same graph are called **dependent equations.**

EXAMPLE 3   Solve the system: $\begin{cases} 2y - x = 4 \\ 2x + 8 = 4y \end{cases}$.

Solution   We graph each equation on one set of coordinate axes, as shown in Figure 3-3. Since the graphs coincide, the system has infinitely many solutions. Any ordered pair (x, y) that satisfies one equation satisfies the other also.

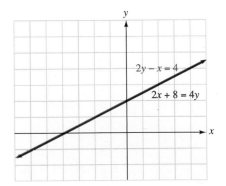

2y − x = 4		
x	y	(x, y)
−4	0	(−4, 0)
0	2	(0, 2)

2x + 8 = 4y		
x	y	(x, y)
−4	0	(−4, 0)
0	2	(0, 2)

Figure 3-3

From the tables in Figure 3-3, we see that $(−4, 0)$ and $(0, 2)$ are solutions. We can find infinitely many solutions by finding more ordered pairs (x, y) that satisfy either equation.

Because the two equations have the same graph, they are dependent equations.

Self Check   Solve the system: $\begin{cases} 2x - y = 4 \\ x = \frac{1}{2}y + 2 \end{cases}$.   ∎

We summarize the possibilities that can occur when two equations, each with two variables, are graphed.

 If the lines are different and intersect, the equations are independent and the system is consistent. **One solution exists.**

 If the lines are different and parallel, the equations are independent and the system is inconsistent. **No solution exists.**

 If the lines coincide, the equations are dependent and the system is consistent. **Infinitely many solutions exist.**

If the equations in two systems are equivalent, the systems are called **equivalent.** In Example 4, we solve a more difficult system by changing it into a simpler equivalent system.

EXAMPLE 4   Solve the system: $\begin{cases} \frac{3}{2}x - y = \frac{5}{2} \\ x + \frac{1}{2}y = 4 \end{cases}$.

Solution   We multiply both sides of $\frac{3}{2}x - y = \frac{5}{2}$ by 2 to eliminate the fractions and obtain the equation $3x - 2y = 5$. We multiply both sides of $x + \frac{1}{2}y = 4$ by 2 to eliminate the fraction and obtain the equation $2x + y = 8$.

The new system

$$\begin{cases} 3x - 2y = 5 \\ 2x + y = 8 \end{cases}$$

is equivalent to the original system and is easier to solve, since it has no fractions. If we graph each equation in the new system, as in Figure 3-4, we see that the coordinates of the point where the two lines intersect are $(3, 2)$. Verify that $x = 3$ and $y = 2$ satisfy each equation in the original system.

$$3x - 2y = 5$$

x	y	(x, y)
0	$-\frac{5}{2}$	$(0, -\frac{5}{2})$
$\frac{5}{3}$	0	$(\frac{5}{3}, 0)$

$$2x + y = 8$$

x	y	(x, y)
4	0	$(4, 0)$
1	6	$(1, 6)$

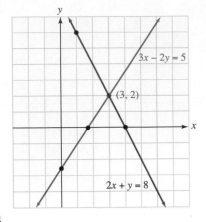

Figure 3-4

Self Check   Solve the system: $\begin{cases} \frac{5}{2}x - y = 2 \\ x + \frac{1}{3}y = 3 \end{cases}$.

Accent on Technology   **SOLVING SYSTEMS BY GRAPHING**

To solve the system

$$\begin{cases} 3x + 2y = 12 \\ 2x - 3y = 12 \end{cases}$$

with a graphing calculator, we first solve each equation for y so we can enter them into a graphing calculator. After solving for y, we obtain the following equivalent system.

$$\begin{cases} y = -\dfrac{3}{2}x + 6 \\ y = \dfrac{2}{3}x - 4 \end{cases}$$

If we use window settings of $[-10, 10]$ for x and $[-10, 10]$ for y, the graphs will look like those in Figure 3-5(a). If we zoom in on the intersection point of the lines and trace, we will get an approximate solution like the one shown in Figure 3-5(b). To get better results, we can do more zooms.

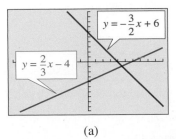

(a)

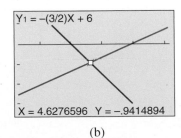

(b)

Figure 3-5

We can also solve this system using the intersect command found in the CALC menu on a TI-83 Plus graphing calculator. To use this method, we first draw the graph of each equation as shown in Figure 3-5(a), then open the CALC menu and select "5: intersect" to obtain Figure 3-6(a). Then we select a point on the first curve by pressing ENTER , select a point on the second curve by pressing ENTER , and press ENTER again to obtain Figure 3-6(b). From the figure, we see that the solution is approximately $(4.6153846, -0.9230769)$.

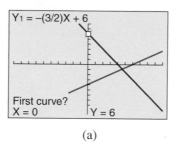
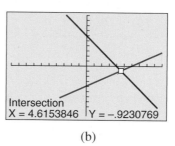

(a)   (b)

Figure 3-6

Verify that the exact solution is $x = \frac{60}{13}$ and $y = -\frac{12}{13}$.

Orals *Tell whether the following systems will have one solution, no solutions, or infinitely many solutions.*

1. $\begin{cases} y = 2x \\ y = 2x + 5 \end{cases}$ **2.** $\begin{cases} y = 2x \\ y = x + x \end{cases}$

3. $\begin{cases} y = 2x \\ y = -2x \end{cases}$ **4.** $\begin{cases} y = 2x + 1 \\ 2x = y \end{cases}$

3.1   EXERCISES

REVIEW   *Write each number in scientific notation.*

1. 93,000,000

2. 0.0000000236

3. 345×10^2

4. 752×10^{-5}

VOCABULARY AND CONCEPTS   *Fill in the blanks.*

5. If two or more equations are considered at the same time, they are called a _____ of equations.

6. When a system of equations has one or more solutions, it is called a _____ system.

7. If a system has no solutions, it is called an _____ system.

8. If two equations have different graphs, they are called _____ equations.

9. Two equations with the same graph are called _____ equations.

10. If the equations in two systems are equivalent, the systems are called _____ systems.

PRACTICE   *Tell whether the ordered pair is a solution of
the system of equations.*

11. $(1, 2)$; $\begin{cases} y = 2x \\ y = \frac{1}{2}x + \frac{3}{2} \end{cases}$

12. $(-1, 2)$; $\begin{cases} y = 3x + 5 \\ y = x + 4 \end{cases}$

13. $(2, -3)$; $\begin{cases} y = \frac{1}{2}x - 2 \\ 3x + 2y = 0 \end{cases}$

14. $(-4, 3)$; $\begin{cases} 4x - y = -19 \\ 3x + 2y = -6 \end{cases}$

Solve each system by graphing, if possible.

15. $\begin{cases} x + y = 6 \\ x - y = 2 \end{cases}$

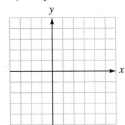

16. $\begin{cases} x - y = 4 \\ 2x + y = 5 \end{cases}$

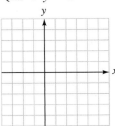

17. $\begin{cases} 2x + y = 1 \\ x - 2y = -7 \end{cases}$

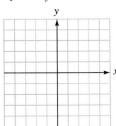

18. $\begin{cases} 3x - y = -3 \\ 2x + y = -7 \end{cases}$

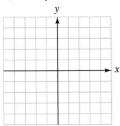

19. $\begin{cases} 2x + 3y = 0 \\ 2x + y = 4 \end{cases}$

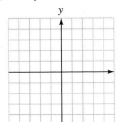

20. $\begin{cases} 3x - 2y = 0 \\ 2x + 3y = 0 \end{cases}$

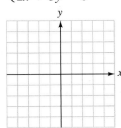

21. $\begin{cases} x = 13 - 4y \\ 3x = 4 + 2y \end{cases}$

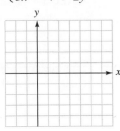

22. $\begin{cases} 3x = 7 - 2y \\ 2x = 2 + 4y \end{cases}$

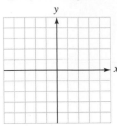

23. $\begin{cases} x = 3 - 2y \\ 2x + 4y = 6 \end{cases}$

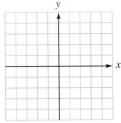

24. $\begin{cases} 3x = 5 - 2y \\ 3x + 2y = 7 \end{cases}$

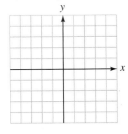

25. $\begin{cases} x = 2 \\ y = \dfrac{4 - x}{2} \end{cases}$

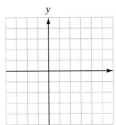

26. $\begin{cases} y = -2 \\ x = \dfrac{4 + 3y}{2} \end{cases}$

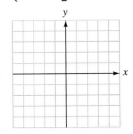

27. $\begin{cases} y = 3 \\ x = 2 \end{cases}$

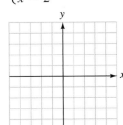

28. $\begin{cases} 2x + 3y = -15 \\ 2x + y = -9 \end{cases}$

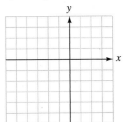

29. $\begin{cases} x = \dfrac{11 - 2y}{3} \\ y = \dfrac{11 - 6x}{4} \end{cases}$

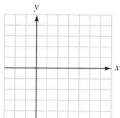

30. $\begin{cases} x = \dfrac{1 - 3y}{4} \\ y = \dfrac{12 + 3x}{2} \end{cases}$

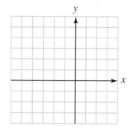

31. $\begin{cases} \dfrac{5}{2}x + y = \dfrac{1}{2} \\ 2x - \dfrac{3}{2}y = 5 \end{cases}$

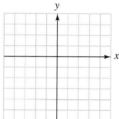

32. $\begin{cases} \dfrac{5}{2}x + 3y = 6 \\ y = \dfrac{24 - 10x}{12} \end{cases}$

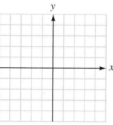

33. $\begin{cases} x = \dfrac{5y - 4}{2} \\ x - \dfrac{5}{3}y + \dfrac{1}{3} = 0 \end{cases}$

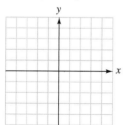

34. $\begin{cases} 2x = 5y - 11 \\ 3x = 2y \end{cases}$

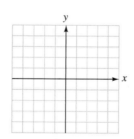

35. $\begin{cases} x = -\dfrac{3}{2}y \\ x = \dfrac{3}{2}y - 2 \end{cases}$

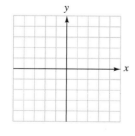

36. $\begin{cases} x = \dfrac{3y - 1}{4} \\ y = \dfrac{4 - 8x}{3} \end{cases}$

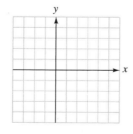

Use a graphing calculator to solve each system. Give all answers to the nearest hundredth.

37. $\begin{cases} y = 3.2x - 1.5 \\ y = -2.7x - 3.7 \end{cases}$

38. $\begin{cases} y = -0.45x + 5 \\ y = 5.55x - 13.7 \end{cases}$

39. $\begin{cases} 1.7x + 2.3y = 3.2 \\ y = 0.25x + 8.95 \end{cases}$

40. $\begin{cases} 2.75x = 12.9y - 3.79 \\ 7.1x - y = 35.76 \end{cases}$

APPLICATIONS

41. Retailing The cost of manufacturing one type of camera and the revenue from the sale of those cameras are shown in the illustration.
  a. From the illustration, find the cost of manufacturing 15,000 cameras.
  b. From the illustration, find the revenue obtained by selling 20,000 cameras.
  c. For what number of cameras will the revenue equal the cost?

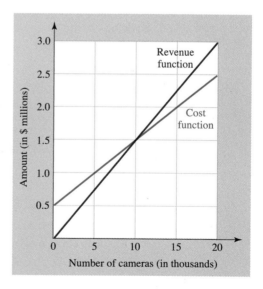

42. Food service
  a. Estimate the point of intersection of the two graphs shown in the illustration on the next page. Express your answer in the form (year, number of meals).

b. What information about dining out does the point of intersection give?

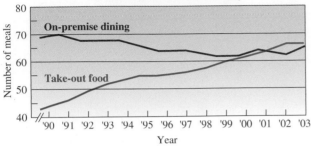

Number of Take-Out and On-Premise Meals Purchased at Commercial Restaurants Per Person Annually

43. Navigation   Two ships are sailing on the same coordinate system. One ship is following a course described by $2x + 3y = 6$, and the other is following a course described by $2x - 3y = 9$.
a. Is there a possibility of a collision?
b. Find the coordinates of the danger point.

c. Is a collision a certainty?

44. Navigation   Two airplanes are flying at the same altitude and in the same coordinate system. One plane is following a course described by $y = \frac{2}{5}x - 2$, and the other is following a course described by $x = \frac{5y + 7}{2}$. Is there a possibility of a collision?

45. Car repairs   Smith Chevrolet charges $50 per hour for labor on car repairs. Lopez Ford charges a diagnosis fee of $30 plus $40 per hour for labor. If the labor on an engine repair costs the same at either shop, how long does the repair take?

46. Landscaping   A landscaper installs some trees and bushes at a bank. He installs 25 plants for a total cost of $1,500. How many trees ($t$) and how many bushes ($b$) did he install if each tree costs $100 and each bush costs $50?

WRITING

47. Explain how to solve a system of two equations in two variables.

48. Can a system of two equations in two variables have exactly two solutions? Why or why not?

SOMETHING TO THINK ABOUT

49. Form an independent system of equations with a solution of $(-5, 2)$.

50. Form a dependent system of equations with a solution of $(-5, 2)$.

3.2   Solution by Elimination

▌ **The Substitution Method**   ▌ **The Addition Method**
▌ **An Inconsistent System**   ▌ **A System with Infinitely Many Solutions**
▌ **Repeating Decimals**   ▌ **Problem Solving**   ▌ **Break-Point Analysis**
▌ **Parallelograms**

Getting Ready   *Remove parentheses.*

1. $3(2x - 7)$ **2.** $-4(3x + 5)$

Substitute $x - 3$ for y and remove parentheses.

3. $3y$ **4.** $-2(y + 2)$

Add the left-hand sides and the right-hand sides of the equations in each system.

5. $\begin{cases} 2x + 5y = 7 \\ 5x - 5y = 8 \end{cases}$ **6.** $\begin{cases} 3a - 4b = 12 \\ -3a - 5b = 15 \end{cases}$

The graphing method provides a way to visualize the process of solving systems of equations. However, it cannot be used to solve systems of higher order, such as three equations, each with three variables. In this section, we will discuss algebraic methods that enable us to solve such systems.

The Substitution Method

To solve a system of two equations (each with two variables) by substitution, we use the following steps.

The Substitution Method

1. If necessary, solve one equation for one of its variables, preferably a variable with a coefficient of 1.
2. Substitute the resulting expression for the variable obtained in Step 1 into the other equation and solve that equation.
3. Find the value of the other variable by substituting the value of the variable found in Step 2 into any equation containing both variables.
4. State the solution.
5. Check the solution in both of the original equations.

EXAMPLE 1   Solve the system: $\begin{cases} 4x + y = 13 \\ -2x + 3y = -17 \end{cases}$.

Solution   **Step 1:**   We solve the first equation for y, because y has a coefficient of 1 and no fractions are introduced.

$$4x + y = 13$$

(1)
$$y = -4x + 13 \qquad \text{Subtract } 4x \text{ from both sides.}$$

Step 2:   We then substitute $-4x + 13$ for y in the second equation of the system and solve for x.

$$-2x + 3y = -17$$
$$-2x + 3(\mathbf{-4x + 13}) = -17 \qquad \text{Substitute } -4x + 13 \text{ for } y.$$
$$-2x - 12x + 39 = -17 \qquad \text{Use the distributive property to remove parentheses.}$$
$$-14x = -56 \qquad \text{Combine like terms and subtract 39 from both sides.}$$
$$x = 4 \qquad \text{Divide both sides by } -14.$$

Step 3:   To find y, we substitute 4 for x in Equation 1 and simplify:

$$y = -4\mathbf{x} + 13$$
$$= -4(\mathbf{4}) + 13 \qquad \text{Substitute 4 for } x.$$
$$= -3$$

Step 4:   The solution is $x = 4$ and $y = -3$, or just $(4, -3)$. The graphs of these two equations would intersect at the point $(4, -3)$.

Step 5:   To verify that this solution satisfies both equations, we substitute $x = 4$ and $y = -3$ into each equation in the system and simplify.

$$4x + y = 13 \qquad\qquad\qquad -2x + 3y = -17$$
$$4(4) + (-3) \stackrel{?}{=} 13 \qquad\qquad -2(4) + 3(-3) \stackrel{?}{=} -17$$
$$16 - 3 \stackrel{?}{=} 13 \qquad\qquad\qquad -8 - 9 \stackrel{?}{=} -17$$
$$13 = 13 \qquad\qquad\qquad\qquad -17 = -17$$

Since the ordered pair $(4, -3)$ satisfies both equations of the system, the solution checks.

Self Check   Solve the system: $\begin{cases} x + 3y = 9 \\ 2x - y = -10 \end{cases}$.   ∎

EXAMPLE 2   Solve the system: $\begin{cases} \frac{4}{3}x + \frac{1}{2}y = -\frac{2}{3} \\ \frac{1}{2}x + \frac{2}{3}y = \frac{5}{3} \end{cases}$.

Solution   First we find an equivalent system without fractions by multiplying both sides of each equation by 6.

$$\begin{aligned}(1)\\(2)\end{aligned} \qquad \begin{cases} 8x + 3y = -4 \\ 3x + 4y = 10 \end{cases}$$

Because no variable in either equation has a coefficient of 1, it is impossible to avoid fractions when solving for a variable. We solve Equation 2 for x.

$$3x + 4y = 10$$
$$3x = -4y + 10 \qquad \text{Subtract } 4y \text{ from both sides.}$$
$$(3) \qquad x = -\frac{4}{3}y + \frac{10}{3} \qquad \text{Divide both sides by 3.}$$

We then substitute $-\frac{4}{3}y + \frac{10}{3}$ for x in Equation 1 and solve for y.

$$8x + 3y = -4$$
$$8\left(-\frac{4}{3}y + \frac{10}{3}\right) + 3y = -4 \qquad \text{Substitute } -\frac{4}{3}y + \frac{10}{3} \text{ for } x.$$
$$-\frac{32}{3}y + \frac{80}{3} + 3y = -4 \qquad \text{Use the distributive property to remove parentheses.}$$
$$-32y + 80 + 9y = -12 \qquad \text{Multiply both sides by 3.}$$
$$-23y = -92 \qquad \text{Combine like terms and subtract 80 from both sides.}$$
$$y = 4 \qquad \text{Divide both sides by } -23.$$

We can find x by substituting 4 for y in Equation 3 and simplifying:

$$x = -\frac{4}{3}y + \frac{10}{3}$$
$$= -\frac{4}{3}(4) + \frac{10}{3} \qquad \text{Substitute 4 for } y.$$
$$= -\frac{6}{3} \qquad\qquad -\frac{16}{3} + \frac{10}{3} = -\frac{6}{3}.$$
$$= -2$$

The solution is the ordered pair $(-2, 4)$. Verify that this solution checks.

Self Check   Solve the system: $\begin{cases} \frac{2}{3}x + \frac{1}{2}y = 1 \\ \frac{1}{3}x - \frac{3}{2}y = 4 \end{cases}$.   ∎

The Addition Method

In the addition method, we combine the equations of the system in a way that will eliminate the terms involving one of the variables.

The Addition Method

1. Write both equations of the system in general form.
2. Multiply the terms of one or both of the equations by constants chosen to make the coefficients of x (or y) differ only in sign.
3. Add the equations and solve the resulting equation, if possible.
4. Substitute the value obtained in Step 3 into either of the original equations and solve for the remaining variable.
5. State the solution obtained in Steps 3 and 4.
6. Check the solution in both of the original equations.

EXAMPLE 3   Solve the system: $\begin{cases} 4x + y = 13 \\ -2x + 3y = -17 \end{cases}$.

Solution   **Step 1:**   This is the system in Example 1. Since both equations are already written in general form, Step 1 is unnecessary.

Step 2:   To solve the system by addition, we multiply the second equation by 2 to make the coefficients of x differ only in sign.

(1) $\begin{cases} 4x + y = 13 \\ -4x + 6y = -34 \end{cases}$

Step 3:   When these equations are added, the terms involving x drop out, and we get

$7y = -21$

$y = -3$ Divide both sides by 7.

Step 4:   To find x, we substitute -3 for y in either of the original equations and solve for x. If we use Equation 1, we have

$4x + y = 13$

$4x + (\mathbf{-3}) = 13$ Substitute -3 for y.

$4x = 16$ Add 3 to both sides.

$x = 4$ Divide both sides by 4.

Step 5:   The solution is $x = 4$ and $y = -3$, or just $(4, -3)$.

Step 6:   The check was completed in Example 1.

Self Check   Solve the system: $\begin{cases} 3x + 2y = 0 \\ 2x - y = -7 \end{cases}$.   ∎

EXAMPLE 4   Solve the system: $\begin{cases} \frac{4}{3}x + \frac{1}{2}y = -\frac{2}{3} \\ \frac{1}{2}x + \frac{2}{3}y = \frac{5}{3} \end{cases}$.

Solution   This is the system in Example 2. To solve it by addition, we find an equivalent system with no fractions by multiplying both sides of each equation by 6 to obtain

(1)
(2) $\begin{cases} 8x + 3y = -4 \\ 3x + 4y = 10 \end{cases}$

To make the y-terms drop out when adding the equations, we multiply both sides of Equation 1 by 4 and both sides of Equation 2 by -3 to get

$\begin{cases} 32x + 12y = -16 \\ -9x - 12y = -30 \end{cases}$

When these equations are added, the y-terms drop out, and we get

$$23x = -46$$
$$x = -2 \qquad \text{Divide both sides by 23.}$$

To find y, we substitute -2 for x in either Equation 1 or Equation 2. If we substitute -2 for x in Equation 2, we get

$$3x + 4y = 10$$
$$3(-2) + 4y = 10 \qquad \text{Substitute } -2 \text{ for } x.$$
$$-6 + 4y = 10 \qquad \text{Simplify.}$$
$$4y = 16 \qquad \text{Add 6 to both sides.}$$
$$y = 4 \qquad \text{Divide both sides by 4.}$$

The solution is $(-2, 4)$.

Self Check   Solve the system: $\begin{cases} \frac{4}{3}x + \frac{1}{2}y = -3 \\ \frac{1}{2}x + \frac{2}{3}y = -\frac{1}{6} \end{cases}$.   ∎

An Inconsistent System

EXAMPLE 5   Solve the system: $\begin{cases} y = 2x + 4 \\ 8x - 4y = 7 \end{cases}$, if possible.

Solution   Because the first equation is already solved for y, we use the substitution method.

$$8x - 4y = 7$$
$$8x - 4(2x + 4) = 7 \qquad \text{Substitute } 2x + 4 \text{ for } y.$$

We then solve this equation for x:

$$8x - 8x - 16 = 7 \qquad \text{Use the distributive property to remove parentheses.}$$
$$-16 = 7 \qquad \text{Combine like terms.}$$

This impossible result shows that the equations in the system are independent and that the system is inconsistent. Since the system has no solution, the graphs of the equations in the system are parallel.

Self Check   Solve the system: $\begin{cases} x = -\frac{5}{2}y + 5 \\ y = -\frac{2}{5}x + 5 \end{cases}$, if possible. ∎

A System with Infinitely Many Solutions

EXAMPLE 6   Solve the system: $\begin{cases} 4x + 6y = 12 \\ -2x - 3y = -6 \end{cases}$.

Solution   Since the equations are written in general form, we use the addition method. To make the *x*-terms drop out when adding the equations, we multiply both sides of the second equation by 2 to get

$$\begin{cases} 4x + 6y = 12 \\ -4x - 6y = -12 \end{cases}$$

After adding the left-hand sides and the right-hand sides, we get

$$0x + 0y = 0$$
$$0 = 0$$

Here, both the *x*- and *y*-terms drop out. The true statement $0 = 0$ shows that the equations in this system are dependent and that the system is consistent.

Note that the equations of the system are equivalent, because when the second equation is multiplied by -2, it becomes the first equation. The line graphs of these equations would coincide. Since any ordered pair that satisfies one of the equations also satisfies the other, there are infinitely many solutions. Verify that three of them are $(0, 2)$, $(3, 0)$, and $(6, -2)$.

Self Check   Solve the system: $\begin{cases} x = -\frac{5}{2}y + 5 \\ y = -\frac{2}{5}x + 2 \end{cases}$. ∎

Repeating Decimals

We have seen how to change fractions into decimal form. By using systems of equations, we can change repeating decimals into fractional form. For example, to write $0.2\overline{54}$ as a fraction, we note that the decimal has a repeating block of two digits and then form an equation by setting *x* equal to the decimal.

(1) $x = 0.2545454 \ldots$

We then form another equation by multiplying both sides of Equation 1 by 10^2.

(2) $100x = 25.4545454 \ldots$ $10^2 = 100$.

We can subtract each side of Equation 1 from the corresponding side of Equation 2 to obtain

$$100x = 25.4\ 54\ 54\ 54\ \ldots$$
$$\underline{x =\ \ 0.2\ 54\ 54\ 54\ \ldots}$$
$$99x = 25.2$$

Finally, we solve $99x = 25.2$ for x and simplify the fraction.

$$x = \frac{25.2}{99} = \frac{25.2 \cdot 10}{99 \cdot 10} = \frac{252}{990} = \frac{\not{18} \cdot 14}{\not{18} \cdot 55} = \frac{14}{55}$$

We can use a calculator to verify that the decimal representation of $\frac{14}{55}$ is $0.2\overline{54}$.

The key step in the solution was multiplying both sides of Equation 1 by 10^2. If there had been n digits in the repeating block of the decimal, we would have multiplied both sides of Equation 1 by 10^n.

Problem Solving

To solve problems using two variables, we follow the same problem-solving strategy discussed in Chapter 1, except that we use two variables and form two equations instead of one.

EXAMPLE 7 **Retail sales** A store advertises two types of cell phones, one selling for $67 and the other for $100. If the receipts from the sale of 36 phones totaled $2,940, how many of each type were sold?

Analyze the problem We can let x represent the number of phones sold for $67 and let y represent the number of phones sold for $100. Then the receipts for the sale of the lower-priced phones are $\$67x$, and the receipts for the sale of the higher-priced phones are $\$100y$.

Form two equations The information of the problem gives the following two equations:

The number of lower-priced phones	plus	the number of higher-priced phones	equals	the total number of phones.
x	$+$	y	$=$	36

The value of the lower-priced phones	plus	the value of the higher-priced phones	equals	the total receipts.
$67x$	$+$	$100y$	$=$	2,940

Solve the system To find out how many of each type of phone were sold, we must solve the system

$$
\begin{aligned}
(1) \quad & \left\{ x + y = 36 \right. \\
(2) \quad & \left. 67x + 100y = 2{,}940 \right.
\end{aligned}
$$

We multiply both sides of Equation 1 by -100, add the resulting equation to Equation 2, and solve for x:

$$-100x - 100y = -3{,}600$$
$$\underline{67x + 100y = 2{,}940}$$
$$-33x = -660$$
$$x = 20 \qquad \text{Divide both sides by } -33.$$

To find y, we substitute 20 for x in Equation 1 and solve for y:

$$x + y = 36$$
$$\mathbf{20} + y = 36 \qquad \text{Substitute 20 for } x.$$
$$y = 16 \qquad \text{Subtract 20 from both sides.}$$

State the conclusion The store sold 20 of the lower-priced phones and 16 of the higher-priced phones.

Check the result If 20 of one type were sold and 16 of the other type were sold, a total of 36 phones were sold.

Since the value of the lower-priced phones is $20(\$67) = \$1,340$ and the value of the higher-priced phones is $16(\$100) = \$1,600$, the total receipts are $\$2,940$. ∎

EXAMPLE 8 **Mixing solutions** How many ounces of a 5% saline solution and how many ounces of a 20% saline solution must be mixed together to obtain 50 ounces of a 15% saline solution?

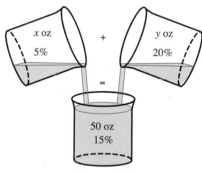

Figure 3-7

Analyze the problem We can let x represent the number of ounces of the 5% solution and let y represent the number of ounces of the 20% solution that are to be mixed. Then the amount of salt in the 5% solution is $0.05x$, and the amount of salt in the 20% solution is $0.20y$. (See Figure 3-7.)

Form two equations The information of the problem gives the following two equations:

The number of ounces of 5% solution	plus	the number of ounces of 20% solution	equals	the total number of ounces in the mixture.
x	+	y	=	50

The salt in the 5% solution	plus	the salt in the 20% solution	equals	the salt in the mixture.
$0.05x$	+	$0.20y$	=	$0.15(50)$

Solve the system To find out how many ounces of each are needed, we solve the following system:

(1) $\begin{cases} x + y = 50 \\ 0.05x + 0.20y = 7.5 \quad 0.15(50) = 7.5 \end{cases}$
(2)

To solve this system by substitution, we can solve Equation 1 for y

$$x + y = 50$$

(3) $\qquad y = 50 - x \quad$ Subtract x from both sides.

and then substitute $50 - x$ for y in Equation 2.

$$0.05x + 0.20y = 7.5$$

$0.05x + 0.20(\mathbf{50 - x}) = 7.5 \qquad$ Substitute $50 - x$ for y.

$5x + 20(50 - x) = 750 \qquad$ Multiply both sides by 100.

$5x + 1,000 - 20x = 750 \qquad$ Use the distributive property to remove parentheses.

$-15x = -250 \qquad$ Combine like terms and subtract 1,000 from both sides.

$$x = \frac{-250}{-15} \qquad$$ Divide both sides by -15.

$$x = \frac{50}{3} \qquad$$ Simplify $\frac{-250}{-15}$.

To find y, we can substitute $\frac{50}{3}$ for x in Equation 3:

$$y = 50 - \mathbf{x}$$

$$= 50 - \frac{\mathbf{50}}{\mathbf{3}} \qquad$$ Substitute $\frac{50}{3}$ for x.

$$= \frac{100}{3}$$

State the conclusion   To obtain 50 ounces of a 15% solution, we must mix $16\frac{2}{3}$ ounces of the 5% solution with $33\frac{1}{3}$ ounces of the 20% solution.

Check the result   We note that $16\frac{2}{3}$ ounces of solution plus $33\frac{1}{3}$ ounces of solution equals the required 50 ounces of solution. We also note that 5% of $16\frac{2}{3} \approx 0.83$, and 20% of $33\frac{1}{3} \approx 6.67$, giving a total of 7.5, which is 15% of 50. ∎

Break-Point Analysis

Running a machine involves both *setup costs* and *unit costs*. Setup costs include the cost of preparing a machine to do a certain job. Unit costs depend on the number of items to be manufactured, including costs of raw materials and labor.

Suppose that a certain machine has a setup cost of $600 and a unit cost of $3 per item. If x items will be manufactured using this machine, the cost will be

$\text{Cost} = 600 + 3x \qquad$ Cost = setup cost + unit cost × the number of items.

Furthermore, suppose that a larger and more efficient machine has a setup cost of $800 and a unit cost of $2 per item. The cost of manufacturing x items using this machine is

$$\text{Cost on larger machine} = 800 + 2x$$

The **break point** is the number of units x that need to be manufactured to make the cost the same using either machine. It can be found by setting the two costs equal to each other and solving for x.

$$600 + 3x = 800 + 2x$$

$$x = 200 \qquad \text{Subtract 600 and } 2x \text{ from both sides.}$$

The break point is 200 units, because the cost using either machine is $1,200 when $x = 200$.

Cost on small machine $= 600 + 3\mathbf{x}$ Cost on large machine $= 800 + 2\mathbf{x}$

$$= 600 + 3(\mathbf{200}) \qquad\qquad\qquad = 800 + 2(\mathbf{200})$$
$$= 600 + 600 \qquad\qquad\qquad\qquad = 800 + 400$$
$$= 1,200 \qquad\qquad\qquad\qquad\quad = 1,200$$

EXAMPLE 9 One machine has a setup cost of $400 and a unit cost of $1.50, and another machine has a setup cost of $500 and a unit cost of $1.25. Find the break point.

Analyze the problem The cost C_1 of manufacturing x units on machine 1 is $1.50x + $400 (the number of units manufactured times $1.50, plus the setup cost of $400). The cost C_2 of manufacturing the same number of units on machine 2 is $1.25x + $500 (the number of units manufactured times $1.25, plus the setup cost of $500). The break point occurs when the costs are equal ($C_1 = C_2$).

Form two equations If x represents the number of items to be manufactured, the cost C_1 using machine 1 is

The cost of using machine 1	equals	the cost of manufacturing x units	plus	the setup cost.
C_1	=	$1.5x$	+	400

The cost C_2 using machine 2 is

The cost of using machine 2	equals	the cost of manufacturing x units	plus	the setup cost.
C_2	=	$1.25x$	+	500

Solve the system To find the break point, we must solve the system $\begin{cases} C_1 = 1.5x + 400 \\ C_2 = 1.25x + 500 \end{cases}$.

Since the break point occurs when $C_1 = C_2$, we can substitute $1.5x + 400$ for C_2 to get

$$1.5x + 400 = 1.25x + 500$$
$$1.5x = 1.25x + 100 \qquad \text{Subtract 400 from both sides.}$$
$$0.25x = 100 \qquad\qquad \text{Subtract } 1.25x \text{ from both sides.}$$
$$x = 400 \qquad\qquad\quad \text{Divide both sides by 0.25.}$$

State the conclusion The break point is 400 units.

Check the result For 400 units, the cost using machine 1 is $400 + 1.5(400) = 400 + 600 = 1,000$. The cost using machine 2 is $500 + 1.25(400) = 500 + 500 = 1,000$. Since the costs are equal, the break point is 400. ∎

Parallelograms

A **parallelogram** is a four-sided figure with its opposite sides parallel. (See Figure 3-8(a).) Here are some important facts about parallelograms.

1. Opposite sides of a parallelogram have the same length.
2. Opposite angles of a parallelogram have the same measure.
3. Consecutive angles of a parallelogram are supplementary.
4. A diagonal of a parallelogram (see Figure 3-8(b)) divides the parallelogram into two *congruent triangles*—triangles with the same shape and same area.
5. In Figure 3-8(b), $\angle 1$ and $\angle 2$, and $\angle 3$ and $\angle 4$, are called pairs of *alternate interior angles*. When a diagonal intersects two parallel sides of a parallelogram, all pairs of alternate interior angles have the same measure.

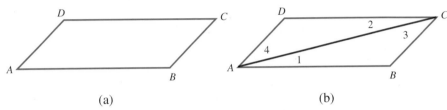

(a)   (b)

Figure 3-8

EXAMPLE 10   Refer to the parallelogram shown in Figure 3-9 and find the values of x and y.

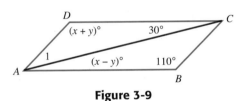

Figure 3-9

Solution   Since diagonal AC intersects two parallel sides, the alternate interior angles that are formed have the same measure. Thus, $(x - y)° = 30°$. Since opposite angles of a parallelogram have the same measure, we know that $(x + y)° = 110°$. We can form the following system of equations and solve it by addition.

(1)
(2)
$$\begin{cases} x - y = 30 \\ x + y = 110 \end{cases}$$

$$2x = 140 \qquad \text{Add Equations 1 and 2.}$$
$$x = 70 \qquad \text{Divide both sides by 2.}$$

We can substitute 70 for x in Equation 2 and solve for y.

$$\boldsymbol{x} + y = 110$$
$$\mathbf{70} + y = 110 \qquad \text{Substitute 70 for } x.$$
$$y = 40 \qquad \text{Subtract 70 from both sides.}$$

Thus, $x = 70$ and $y = 40$.

Self Check   Find the measure of $\angle 1$.   (*Hint:* The sum of the angles of a triangle equals 180°.)

1. $(-3, 4)$ **2.** $(3, -2)$ **3.** $(-2, 3)$ **4.** $(-3, 2)$ **5.** no solution
6. infinitely many solutions; three of them are $(0, 2)$, $(5, 0)$, and $(10, -2)$ **10.** $40°$

Orals   *Solve each system for x.*

1. $\begin{cases} y = 2x \\ x + y = 6 \end{cases}$

2. $\begin{cases} y = -x \\ 2x + y = 4 \end{cases}$

3. $\begin{cases} x - y = 6 \\ x + y = 2 \end{cases}$

4. $\begin{cases} x + y = 4 \\ 2x - y = 5 \end{cases}$

3.2   EXERCISES

REVIEW   *Simplify each expression. Write all answers without using negative exponents.*

1. $(a^2a^3)^2(a^4a^2)^2$

2. $\left(\dfrac{a^2b^3c^4d}{ab^2c^3d^4}\right)^{-3}$

3. $\left(\dfrac{-3x^3y^4}{x^{-5}y^3}\right)^{-4}$

4. $\dfrac{3t^0 - 4t^0 + 5}{5t^0 + 2t^0}$

VOCABULARY AND CONCEPTS   *Fill in the blanks.*

5. Running a machine involves both _____ costs and _____ costs.

6. The _____ point is the number of units that need to be manufactured to make the cost the same on either of two machines.

7. A _____ is a four-sided figure with both pairs of opposite sides parallel.

8. _____ sides of a parallelogram have the same length.

9. _____ angles of a parallelogram have the same measure.

10. _____ angles of a parallelogram are supplementary.

PRACTICE   *Solve each system by substitution, if possible.*

11. $\begin{cases} y = x \\ x + y = 4 \end{cases}$

12. $\begin{cases} y = x + 2 \\ x + 2y = 16 \end{cases}$

13. $\begin{cases} x - y = 2 \\ 2x + y = 13 \end{cases}$

14. $\begin{cases} x - y = -4 \\ 3x - 2y = -5 \end{cases}$

15. $\begin{cases} x + 2y = 6 \\ 3x - y = -10 \end{cases}$

16. $\begin{cases} 2x - y = -21 \\ 4x + 5y = 7 \end{cases}$

17. $\begin{cases} 3x = 2y - 4 \\ 6x - 4y = -4 \end{cases}$

18. $\begin{cases} 8x = 4y + 10 \\ 4x - 2y = 5 \end{cases}$

19. $\begin{cases} 3x - 4y = 9 \\ x + 2y = 8 \end{cases}$

20. $\begin{cases} 3x - 2y = -10 \\ 6x + 5y = 25 \end{cases}$

21. $\begin{cases} 2x + 2y = -1 \\ 3x + 4y = 0 \end{cases}$

22. $\begin{cases} 5x + 3y = -7 \\ 3x - 3y = 7 \end{cases}$

Solve each system by addition, if possible.

23. $\begin{cases} x - y = 3 \\ x + y = 7 \end{cases}$

24. $\begin{cases} x + y = 1 \\ x - y = 7 \end{cases}$

25. $\begin{cases} 2x + y = -10 \\ 2x - y = -6 \end{cases}$

26. $\begin{cases} x + 2y = -9 \\ x - 2y = -1 \end{cases}$

27. $\begin{cases} 2x + 3y = 8 \\ 3x - 2y = -1 \end{cases}$

28. $\begin{cases} 5x - 2y = 19 \\ 3x + 4y = 1 \end{cases}$

29. $\begin{cases} 4x + 9y = 8 \\ 2x - 6y = -3 \end{cases}$

30. $\begin{cases} 4x + 6y = 5 \\ 8x - 9y = 3 \end{cases}$

31. $\begin{cases} 8x - 4y = 16 \\ 2x - 4 = y \end{cases}$

32. $\begin{cases} 2y - 3x = -13 \\ 3x - 17 = 4y \end{cases}$

33. $\begin{cases} x = \dfrac{3}{2}y + 5 \\ 2x - 3y = 8 \end{cases}$

34. $\begin{cases} x = \dfrac{2}{3}y \\ y = 4x + 5 \end{cases}$

Solve each system by any method.

35. $\begin{cases} \dfrac{x}{2} + \dfrac{y}{2} = 6 \\ \dfrac{x}{2} - \dfrac{y}{2} = -2 \end{cases}$

36. $\begin{cases} \dfrac{x}{2} - \dfrac{y}{3} = -4 \\ \dfrac{x}{2} + \dfrac{y}{9} = 0 \end{cases}$

37. $\begin{cases} \dfrac{3}{4}x + \dfrac{2}{3}y = 7 \\ \dfrac{3}{5}x - \dfrac{1}{2}y = 18 \end{cases}$

38. $\begin{cases} \dfrac{2}{3}x - \dfrac{1}{4}y = -8 \\ \dfrac{1}{2}x - \dfrac{3}{8}y = -9 \end{cases}$

39. $\begin{cases} \dfrac{3x}{2} - \dfrac{2y}{3} = 0 \\ \dfrac{3x}{4} + \dfrac{4y}{3} = \dfrac{5}{2} \end{cases}$

40. $\begin{cases} \dfrac{3x}{5} + \dfrac{5y}{3} = 2 \\ \dfrac{6x}{5} - \dfrac{5y}{3} = 1 \end{cases}$

41. $\begin{cases} \dfrac{2}{5}x - \dfrac{1}{6}y = \dfrac{7}{10} \\ \dfrac{3}{4}x - \dfrac{2}{3}y = \dfrac{19}{8} \end{cases}$

42. $\begin{cases} \dfrac{5}{6}x + \dfrac{2}{3}y = \dfrac{7}{6} \\ \dfrac{10}{7}x - \dfrac{4}{9}y = \dfrac{17}{21} \end{cases}$

⊞ *Write each repeating decimal as a fraction. Simplify the answer when possible.*

43. $0.\overline{3}$

44. $0.\overline{29}$

45. $-0.34\overline{89}$

46. $-2.3\overline{47}$

Solve each system for x and y. Solve for $\frac{1}{x}$ and $\frac{1}{y}$ first.

47. $\begin{cases} \dfrac{1}{x} + \dfrac{1}{y} = \dfrac{5}{6} \\ \dfrac{1}{x} - \dfrac{1}{y} = \dfrac{1}{6} \end{cases}$

48. $\begin{cases} \dfrac{1}{x} + \dfrac{1}{y} = \dfrac{9}{20} \\ \dfrac{1}{x} - \dfrac{1}{y} = \dfrac{1}{20} \end{cases}$

49. $\begin{cases} \dfrac{1}{x} + \dfrac{2}{y} = -1 \\ \dfrac{2}{x} - \dfrac{1}{y} = -7 \end{cases}$

50. $\begin{cases} \dfrac{3}{x} - \dfrac{2}{y} = -30 \\ \dfrac{2}{x} - \dfrac{3}{y} = -30 \end{cases}$

APPLICATIONS *Use two variables and two equations to solve each problem.*

51. **Merchandising** A pair of shoes and a sweater cost $98. If the sweater cost $16 more than the shoes, how much did the sweater cost?

52. **Merchandising** A sporting goods salesperson sells 2 fishing reels and 5 rods for $270. The next day, the salesperson sells 4 reels and 2 rods for $220. How much does each cost?

53. **Electronics** Two resistors in the voltage divider circuit in the illustration have a total resistance of 1,375 ohms. To provide the required voltage, R_1 must be 125 ohms greater than R_2. Find both resistances.

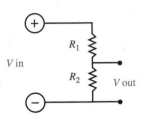

54. **Stowing baggage** A small aircraft can carry 950 pounds of baggage, distributed between two storage compartments. On one flight, the plane is fully loaded, with 150 pounds more baggage in one compartment than the other. How much is stowed in each compartment?

55. **Geometry problem** The rectangular field in the illustration is surrounded by 72 meters of fencing. If the field is partitioned as shown, a total of 88 meters of fencing is required. Find the dimensions of the field.

56. **Geometry** In a right triangle, one acute angle is 15° greater than two times the other acute angle. Find the difference between the angles.

57. Investment income Part of $8,000 was invested at 10% interest and the rest at 12%. If the annual income from these investments was $900, how much was invested at each rate?

58. Investment income Part of $12,000 was invested at 6% interest and the rest at 7.5%. If the annual income from these investments was $810, how much was invested at each rate?

59. Mixing a solution How many ounces of the two alcohol solutions in the illustration must be mixed to obtain 100 ounces of a 12.2% solution?

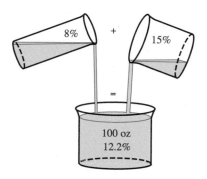

60. Mixing candy How many pounds each of candy shown in the illustration must be mixed to obtain 60 pounds of candy that is worth $3 per pound?

61. Travel A car travels 50 miles in the same time that a plane travels 180 miles. The speed of the plane is 143 mph faster than the speed of the car. Find the speed of the car.

62. Travel A car and a truck leave Rockford at the same time, heading in opposite directions. When they are 350 miles apart, the car has gone 70 miles farther than the truck. How far has the car traveled?

63. Making bicycles A bicycle manufacturer builds racing bikes and mountain bikes, with the per-unit manufacturing costs shown in the illustration. The company has budgeted $15,900 for labor and $13,075 for materials. How many bicycles of each type can be built?

Model	Cost of materials	Cost of labor
Racing	$55	$60
Mountain	$70	$90

64. Farming A farmer keeps some animals on a strict diet. Each animal is to receive 15 grams of protein and 7.5 grams of carbohydrates. The farmer uses two food mixes with nutrients shown in the illustration. How many grams of each mix should be used to provide the correct nutrients for each animal?

Mix	Protein	Carbohydrates
Mix A	12%	9%
Mix B	15%	5%

65. Milling brass plates Two machines can mill a brass plate. One machine has a setup cost of $300 and a cost per plate of $2. The other machine has a setup cost of $500 and a cost per plate of $1. Find the break point.

66. Printing books A printer has two presses. One has a setup cost of $210 and can print the pages of a certain book for $5.98. The other press has a setup cost of $350 and can print the pages of the same book for $5.95. Find the break point.

67. Managing a computer store The manager of a computer store knows that his fixed costs are $8,925 per month and that his unit cost is $850 for every computer sold. If he can sell all the computers he can get for $1,275 each, how many computers must he sell each month to break even?

68. Managing a beauty shop A beauty shop specializing in permanents has fixed costs of $2,101.20 per month. The owner estimates that the cost for each permanent is $23.60. This cost covers labor, chemicals, and electricity. If her shop can give as many permanents as she wants at a price of $44 each, how many must be given each month to break even?

69. Running a small business   A person invests
$18,375 to set up a small business that produces a
piece of computer software that will sell for $29.95.
If each piece can be produced for $5.45, how many
pieces must be sold to break even?

70. Running a record company   Three people invest
$35,000 each to start a record company that will
produce reissues of classic jazz. Each release will be
a set of 3 CDs that will retail for $15 per disc. If each
set can be produced for $18.95, how many sets must
be sold for the investors to make a profit?

*A paint manufacturer can choose between two processes
for manufacturing house paint, with monthly costs shown
in the illustration. Assume that the paint sells for $18 per
gallon.*

Process	Fixed costs	Unit cost (per gallon)
A	$32,500	$13
B	$80,600	$ 5

71. For process A, how many gallons must be sold for the
manufacturer to break even?

72. For process B, how many gallons must be sold for the
manufacturer to break even?

73. If expected sales are 6,000 gallons per month, which
process should the company use?

74. If expected sales are 7,000 gallons per month, which
process should the company use?

*A manufacturer of automobile water pumps is considering
retooling for one of two manufacturing processes, with
monthly fixed costs and unit costs as indicated in the illustration. Each water pump can be sold for $50.*

Process	Fixed costs	Unit cost
A	$12,390	$29
B	$20,460	$17

75. For process A, how many water pumps must be sold
for the manufacturer to break even?

76. For process B, how many water pumps must be sold
for the manufacturer to break even?

77. If expected sales are 550 per month, which process
should be used?

78. If expected sales are 600 per month, which process
should be used?

79. If expected sales are 650 per month, which process
should be used?

80. At what monthly sales level is process B better?

81. Geometry   If two angles are supplementary, their
sum is 180°. If the difference between two supplementary angles is 110°, find the measure of each
angle.

82. Geometry   If two angles are complementary, their
sum is 90°. If one of two complementary angles is
16° greater than the other, find the measure of each
angle.

83. Find x and y in the parallelogram.

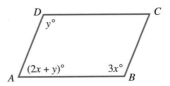

84. Find x and y in the parallelogram.

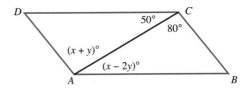

85. Physical therapy   To rehabilitate her knee, an
athlete does leg extensions. Her goal is to regain a full
90° range of motion in this exercise. Use the information in the illustration to determine her current range
of motion in degrees.

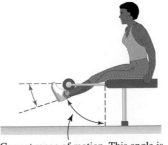

Current range of motion. This angle is
four times larger than the other.

86. Marine Corps The Marine Corps War Memorial in Arlington, Virginia, portrays the raising of the U.S. flag on Iwo Jima during World War II. Find the two angles shown in the illustration, if the measure of one of the angles is 15° less than twice the other.

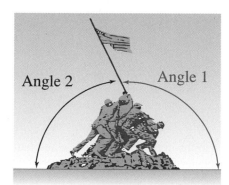

87. Selecting radio frequencies In a radio, an inductor and a capacitor are used in a resonant circuit to select a wanted radio station at a frequency f and reject all others. The inductance L and the capacitance C determine the inductance reactance X_L and the capacitive reactance X_C of that circuit, where

$$X_L = 2\pi f L \quad \text{and} \quad X_C = \frac{1}{2\pi f C}$$

The radio station selected will be at the frequency f where $X_L = X_C$. Write a formula for f^2 in terms of L and C.

88. Choosing salary plans A sales clerk can choose from two salary options:

1. a straight 7% commission
2. $150 + 2\%$ commission

How much would the clerk have to sell for each plan to produce the same monthly paycheck?

WRITING

89. Tell which method you would use to solve the following system. Why?

$$\begin{cases} y = 3x + 1 \\ 3x + 2y = 12 \end{cases}$$

90. Tell which method you would use to solve the following system. Why?

$$\begin{cases} 2x + 4y = 9 \\ 3x - 5y = 20 \end{cases}$$

SOMETHING TO THINK ABOUT

91. Under what conditions will a system of two equations in two variables be inconsistent?

92. Under what conditions will the equations of a system of two equations in two variables be dependent?

3.3	Solutions of Three Equations in Three Variables

▮ Solving Three Equations in Three Variables ▮ A Consistent System
▮ An Inconsistent System ▮ Systems with Dependent Equations
▮ Problem Solving ▮ Curve Fitting

Getting Ready *Tell whether the equation $x + 2y + 3z = 6$ is satisfied by the following values.*

1. $(1, 1, 1)$ 2. $(-2, 1, 2)$
3. $(2, -2, -1)$ 4. $(2, 2, 0)$

We now extend the definition of a linear equation to include equations of the form $ax + by + cz = d$. The solution of a system of three linear equations with three variables is an ordered triple of numbers. For example, the solution of the system

$$\begin{cases} 2x + 3y + 4z = 20 \\ 3x + 4y + 2z = 17 \\ 3x + 2y + 3z = 16 \end{cases}$$

is the triple (1, 2, 3), since each equation is satisfied if $x = 1$, $y = 2$, and $z = 3$.

$$2x + 3y + 4z = 20 \qquad\qquad 3x + 4y + 2z = 17 \qquad\qquad 3x + 2y + 3z = 16$$
$$2(1) + 3(2) + 4(3) \overset{?}{=} 20 \qquad 3(1) + 4(2) + 2(3) \overset{?}{=} 17 \qquad 3(1) + 2(2) + 3(3) \overset{?}{=} 16$$
$$2 + 6 + 12 \overset{?}{=} 20 \qquad\qquad 3 + 8 + 6 \overset{?}{=} 17 \qquad\qquad 3 + 4 + 9 \overset{?}{=} 16$$
$$20 = 20 \qquad\qquad\qquad 17 = 17 \qquad\qquad\qquad 16 = 16$$

The graph of an equation of the form $ax + by + cz = d$ is a flat surface called a **plane.** A system of three linear equations in three variables is consistent or inconsistent, depending on how the three planes corresponding to the three equations intersect. Figure 3-10 illustrates some of the possibilities.

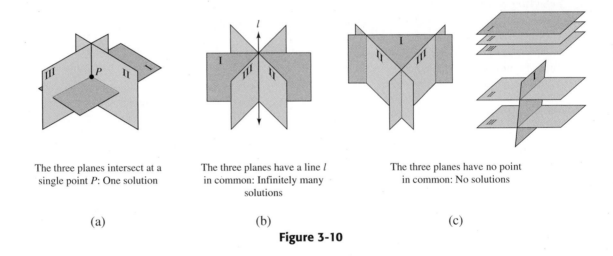

The three planes intersect at a single point P: One solution

(a)

The three planes have a line l in common: Infinitely many solutions

(b)

The three planes have no point in common: No solutions

(c)

Figure 3-10

Solving Three Equations in Three Variables

To solve a system of three linear equations in three variables, we follow these steps.

Solving Three Equations in Three Variables

1. Pick any two equations and eliminate a variable.
2. Pick a different pair of equations and eliminate the same variable.
3. Solve the resulting pair of two equations in two variables.
4. To find the value of the third variable, substitute the values of the two variables found in Step 3 into any equation containing all three variables and solve the equation.
5. Check the solution in all three of the original equations.

A Consistent System

EXAMPLE 1   Solve the system: $\begin{cases} 2x + y + 4z = 12 \\ x + 2y + 2z = 9 \\ 3x - 3y - 2z = 1 \end{cases}$.

Solution   We are given the system

(1)
(2) $\begin{cases} 2x + y + 4z = 12 \\ x + 2y + 2z = 9 \\ 3x - 3y - 2z = 1 \end{cases}$
(3)

If we pick Equations 2 and 3 and add them, the variable z is eliminated:

(2) $x + 2y + 2z = 9$
(3) $\underline{3x - 3y - 2z = 1}$
(4) $4x - y = 10$

We now pick a different pair of equations (Equations 1 and 3) and eliminate z again. If each side of Equation 3 is multiplied by 2 and the resulting equation is added to Equation 1, z is again eliminated:

(1) $2x + y + 4z = 12$
            $\underline{6x - 6y - 4z = 2}$
(5) $8x - 5y = 14$

Equations 4 and 5 form a system of two equations in two variables:

(4) $\begin{cases} 4x - y = 10 \\ 8x - 5y = 14 \end{cases}$
(5)

To solve this system, we multiply Equation 4 by -5 and add the resulting equation to Equation 5 to eliminate y:

            $-20x + 5y = -50$
(5) $\underline{8x - 5y = 14}$
            $-12x = -36$
               $x = 3$ Divide both sides by -12.

To find y, we substitute 3 for x in any equation containing only x and y (such as Equation 5) and solve for y:

(5) $8x - 5y = 14$
            $8(\mathbf{3}) - 5y = 14$ Substitute 3 for x.
            $24 - 5y = 14$ Simplify.
            $-5y = -10$ Subtract 24 from both sides.
            $y = 2$ Divide both sides by -5.

To find z, we substitute 3 for x and 2 for y in an equation containing x, y, and z (such as Equation 1) and solve for z:

(1) $2x + y + 4z = 12$

$2(\mathbf{3}) + \mathbf{2} + 4z = 12$ Substitute 3 for x and 2 for y.

$8 + 4z = 12$ Simplify.

$4z = 4$ Subtract 8 from both sides.

$z = 1$ Divide both sides by 4.

The solution of the system is $(x, y, z) = (3, 2, 1)$. Verify that these values satisfy each equation in the original system.

Self Check   Solve the system: $\begin{cases} 2x + y + 4z = 16 \\ x + 2y + 2z = 11 \\ 3x - 3y - 2z = -9 \end{cases}$.

An Inconsistent System

EXAMPLE 2   Solve the system: $\begin{cases} 2x + y - 3z = -3 \\ 3x - 2y + 4z = 2 \\ 4x + 2y - 6z = -7 \end{cases}$.

Solution   We are given the system of equations

(1)
(2) $\begin{cases} 2x + y - 3z = -3 \\ 3x - 2y + 4z = 2 \\ 4x + 2y - 6z = -7 \end{cases}$
(3)

We can multiply Equation 1 by 2 and add the resulting equation to Equation 2 to eliminate y:

$\qquad 4x + 2y - 6z = -6$

(2) $\underline{3x - 2y + 4z = 2}$

(4) $7x \qquad\quad - 2z = -4$

We now add Equations 2 and 3 to eliminate y again:

(2) $3x - 2y + 4z = 2$

(3) $\underline{4x + 2y - 6z = -7}$

(5) $7x \qquad\quad - 2z = -5$

Equations 4 and 5 form the system

(4)
(5) $\begin{cases} 7x - 2z = -4 \\ 7x - 2z = -5 \end{cases}$

Since $7x - 2z$ cannot equal both -4 and -5, this system is inconsistent; it has no solution.

Self Check   Solve the system: $\begin{cases} 2x + y - 3z = 8 \\ 3x - 2y + 4z = 10 \\ 4x + 2y - 6z = -5 \end{cases}$.

Systems with Dependent Equations

When the equations in a system of two equations in two variables were dependent, the system had infinitely many solutions. This is not always true for systems of three equations in three variables. In fact, a system can have dependent equations and still be inconsistent. Figure 3-11 illustrates the different possibilities.

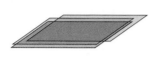
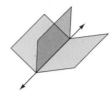
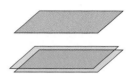

When three planes coincide, the equations are dependent, and there are infinitely many solutions.

When three planes intersect in a common line, the equations are dependent, and there are infinitely many solutions.

When two planes coincide and are parallel to a third plane, the system is inconsistent, and there are no solutions.

(a)

(b)

(c)

Figure 3-11

EXAMPLE 3   Solve the system: $\begin{cases} 3x - 2y + z = -1 \\ 2x + y - z = 5 \\ 5x - y = 4 \end{cases}$.

Solution   We can add the first two equations to get

$$\begin{array}{r} 3x - 2y + z = -1 \\ \underline{2x + y - z = 5} \\ 5x - y = 4 \end{array}$$

(1)

Since Equation 1 is the same as the third equation of the system, the equations of the system are dependent, and there will be infinitely many solutions. From a graphical perspective, the equations represent three planes that intersect in a common line, as shown in Figure 3-11(b).

To write the general solution to this system, we can solve Equation 1 for y to get

$$5x - y = 4$$
$$-y = -5x + 4 \qquad \text{Subtract } 5x \text{ from both sides.}$$
$$y = 5x - 4 \qquad \text{Multiply both sides by } -1.$$

We can then substitute $5x - 4$ for y in the first equation of the system and solve for z to get

$$\begin{array}{ll} 3x - 2\mathbf{y} + z = -1 & \\ 3x - 2(\mathbf{5x - 4}) + z = -1 & \text{Substitute } 5x - 4 \text{ for } y. \\ 3x - 10x + 8 + z = -1 & \text{Use the distributive property to remove parentheses.} \\ -7x + 8 + z = -1 & \text{Combine like terms.} \\ z = 7x - 9 & \text{Add } 7x \text{ and } -8 \text{ to both sides.} \end{array}$$

Since we have found the values of y and z in terms of x, every solution to the system has the form $(x, 5x - 4, 7x - 9)$, where x can be any real number. For example,

If $x = 1$, a solution is $(1, 1, -2)$. $5(1) - 4 = 1$, and $7(1) - 9 = -2$.
If $x = 2$, a solution is $(2, 6, 5)$. $5(2) - 4 = 6$, and $7(2) - 9 = 5$.
If $x = 3$, a solution is $(3, 11, 12)$. $5(3) - 4 = 11$, and $7(3) - 9 = 12$.

Self Check   Solve the system: $\begin{cases} 3x + 2y + z = -1 \\ 2x - y - z = 5 \\ 5x + y = 4 \end{cases}$.   ∎

Problem Solving

EXAMPLE 4   **Manufacturing hammers**   A company makes three types of hammers—good, better, and best. The cost of making each type of hammer is \$4, \$6, and \$7, respectively, and the hammers sell for \$6, \$9, and \$12. Each day, the cost of making 100 hammers is \$520, and the daily revenue from their sale is \$810. How many of each type are manufactured?

Analyze the problem   If we let x represent the number of good hammers, y represent the number of better hammers, and z represent the number of best hammers, we know that

The total number of hammers is $x + y + z$.
The cost of making good hammers is \$4x (\$4 times x hammers).
The cost of making better hammers is \$6y (\$6 times y hammers).
The cost of making best hammers is \$7z (\$7 times z hammers).
The revenue received by selling good hammers is \$6x (\$6 times x hammers).
The revenue received by selling better hammers is \$9y (\$9 times y hammers).
The revenue received by selling best hammers is \$12z (\$12 times z hammers).

Form three equations   Since x represents the number of good hammers made, y represents the number of better hammers made, and z represents the number of best hammers made, we have

The number of good hammers	plus	the number of better hammers	plus	the number of best hammers	equals	the total number of hammers.
x	$+$	y	$+$	z	$=$	100
The cost of good hammers	plus	the cost of better hammers	plus	the cost of best hammers	equals	the total cost.
$4x$	$+$	$6y$	$+$	$7z$	$=$	520
The revenue from good hammers	plus	the revenue from better hammers	plus	the revenue from best hammers	equals	the total revenue.
$6x$	$+$	$9y$	$+$	$12z$	$=$	810

Solve the system   We must now solve the system

$$(1) \quad \begin{cases} x + y + z = 100 \\ 4x + 6y + 7z = 520 \\ 6x + 9y + 12z = 810 \end{cases}$$

(1)
(2)
(3)

If we multiply Equation 1 by -7 and add the result to Equation 2, we get

$$\begin{array}{r} -7x - 7y - 7z = -700 \\ \underline{4x + 6y + 7z = 520} \\ -3x - y = -180 \end{array}$$

(4)

If we multiply Equation 1 by -12 and add the result to Equation 3, we get

$$\begin{array}{r} -12x - 12y - 12z = -1{,}200 \\ \underline{6x + 9y + 12z = 810} \\ -6x - 3y = -390 \end{array}$$

(5)

If we multiply Equation 4 by -3 and add it to Equation 5, we get

$$\begin{array}{r} 9x + 3y = 540 \\ \underline{-6x - 3y = -390} \\ 3x = 150 \\ x = 50 \qquad \text{Divide both sides by 3.} \end{array}$$

To find y, we substitute 50 for x in Equation 4:

$$\begin{aligned} -3\mathbf{x} - y &= -180 \\ -3(\mathbf{50}) - y &= -180 \qquad \text{Substitute 50 for } x. \\ -y &= -30 \qquad \text{Add 150 to both sides.} \\ y &= 30 \qquad \text{Divide both sides by } -1. \end{aligned}$$

To find z, we substitute 50 for x and 30 for y in Equation 1:

$$\begin{aligned} \mathbf{x} + \mathbf{y} + z &= 100 \\ \mathbf{50} + \mathbf{30} + z &= 100 \\ z &= 20 \qquad \text{Subtract 80 from both sides.} \end{aligned}$$

State the conclusion   The company makes 50 good hammers, 30 better hammers, and 20 best hammers each day.

Check the result   Check the solution in each equation in the original system.   ∎

Curve Fitting

EXAMPLE 5   **Curve fitting**   The equation of the parabola shown in Figure 3-12 is of the form $y = ax^2 + bx + c$. Find the equation of the parabola.

Solution   Since the parabola passes through the points shown in the figure, each pair of coordinates satisfies the equation $y = ax^2 + bx + c$. If we substitute the x- and y-values of each point into the equation and simplify, we obtain the following system.

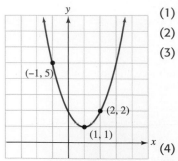

Figure 3-12

(1) $a - b + c = 5$

(2) $a + b + c = 1$

(3) $4a + 2b + c = 2$

If we add Equations 1 and 2, we obtain $2a + 2c = 6$. If we multiply Equation 1 by 2 and add the result to Equation 3, we get $6a + 3c = 12$. We can then divide both sides of $2a + 2c = 6$ by 2 and divide both sides of $6a + 3c = 12$ by 3 to get the system

(4)
(5)
$$\begin{cases} a + c = 3 \\ 2a + c = 4 \end{cases}$$

If we multiply Equation 4 by -1 and add the result to Equation 5, we get $a = 1$. To find c, we can substitute 1 for a in Equation 4 and find that $c = 2$. To find b, we can substitute 1 for a and 2 for c in Equation 2 and find that $b = -2$.

After we substitute these values of a, b, and c into the equation $y = ax^2 + bx + c$, we have the equation of the parabola.

$$y = ax^2 + bx + c$$
$$y = 1x^2 + (-2)x + 2$$
$$y = x^2 - 2x + 2$$ ∎

Self Check Answers

1. $(1, 2, 3)$ **2.** no solution   **3.** infinitely many solutions of the form $(x, 4 - 5x, -9 + 7x)$

Orals   *Is the triple a solution of the system?*

1. $(1, 1, 1)$; $\begin{cases} 2x + y - 3z = 0 \\ 3x - 2y + 4z = 5 \\ 4x + 2y - 6z = 0 \end{cases}$ **2.** $(2, 0, 1)$; $\begin{cases} 3x + 2y - z = 5 \\ 2x - 3y + 2z = 4 \\ 4x - 2y + 3z = 10 \end{cases}$

3.3   EXERCISES 🔊

REVIEW   *Consider the line passing through* $(-2, -4)$ *and* $(3, 5)$.

1. Find the slope of the line.

2. Write the equation of the line in general form.

Find each value if $f(x) = 2x^2 + 1$.

3. $f(0)$

4. $f(-2)$

5. $f(s)$

6. $f(2t)$

VOCABULARY AND CONCEPTS   *Fill in the blanks.*

7. The graph of the equation $2x + 3y + 4z = 5$ is a flat surface called a _____.

8. When three planes coincide, the equations of the system are _____, and there are _____ many solutions.

9. When three planes intersect in a line, the system will have _____ many solutions.

10. When three planes are parallel, the system will have ___ solutions.

PRACTICE   *Tell whether the given triple is a solution of the given system.*

11. $(2, 1, 1)$, $\begin{cases} x - y + z = 2 \\ 2x + y - z = 4 \\ 2x - 3y + z = 2 \end{cases}$

12. $(-3, 2, -1)$, $\begin{cases} 2x + 2y + 3z = -1 \\ 3x + y - z = -6 \\ x + y + 2z = 1 \end{cases}$

Solve each system.

13. $\begin{cases} x + y + z = 4 \\ 2x + y - z = 1 \\ 2x - 3y + z = 1 \end{cases}$ **14.** $\begin{cases} x + y + z = 4 \\ x - y + z = 2 \\ x - y - z = 0 \end{cases}$

15. $\begin{cases} 2x + 2y + 3z = 10 \\ 3x + y - z = 0 \\ x + y + 2z = 6 \end{cases}$ **16.** $\begin{cases} x - y + z = 4 \\ x + 2y - z = -1 \\ x + y - 3z = -2 \end{cases}$

17. $\begin{cases} a + b + 2c = 7 \\ a + 2b + c = 8 \\ 2a + b + c = 9 \end{cases}$ **18.** $\begin{cases} 2a + 3b + c = 2 \\ 4a + 6b + 2c = 5 \\ a - 2b + c = 3 \end{cases}$

19. $\begin{cases} 2x + y - z = 1 \\ x + 2y + 2z = 2 \\ 4x + 5y + 3z = 3 \end{cases}$ **20.** $\begin{cases} 3x - y - 2z = 12 \\ x + y + 6z = 8 \\ 2x - 2y - z = 11 \end{cases}$

21. $\begin{cases} 4x + 3z = 4 \\ 2y - 6z = -1 \\ 8x + 4y + 3z = 9 \end{cases}$ **22.** $\begin{cases} 2x + 3y + 2z = 1 \\ 2x - 3y + 2z = -1 \\ 4x + 3y - 2z = 4 \end{cases}$

23. $\begin{cases} 2x + 3y + 4z = 6 \\ 2x - 3y - 4z = -4 \\ 4x + 6y + 8z = 12 \end{cases}$

24. $\begin{cases} x - 3y + 4z = 2 \\ 2x + y + 2z = 3 \\ 4x - 5y + 10z = 7 \end{cases}$

25. $\begin{cases} x + \frac{1}{3}y + z = 13 \\ \frac{1}{2}x - y + \frac{1}{3}z = -2 \\ x + \frac{1}{2}y - \frac{1}{3}z = 2 \end{cases}$

26. $\begin{cases} x - \frac{1}{5}y - z = 9 \\ \frac{1}{4}x + \frac{1}{5}y - \frac{1}{2}z = 5 \\ 2x + y + \frac{1}{6}z = 12 \end{cases}$

APPLICATIONS

27. Integer problem   The sum of three integers is 18. The third integer is four times the second, and the second integer is 6 more than the first. Find the integers.

28. Integer problem   The sum of three integers is 48. If the first integer is doubled, the sum is 60. If the second integer is doubled, the sum is 63. Find the integers.

29. Geometry   The sum of the angles in any triangle is 180°. In triangle ABC, $\angle A$ is 100° less than the sum of $\angle B$ and $\angle C$, and $\angle C$ is 40° less than twice $\angle B$. Find the measure of each angle.

30. Geometry   The sum of the angles of any four-sided figure is 360°. In the quadrilateral, $\angle A = \angle B$, $\angle C$ is 20° greater than $\angle A$, and $\angle D = 40°$. Find the measure of each angle.

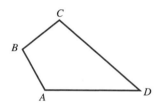

31. Nutritional planning   One unit of each of three foods contains the nutrients shown in the table. How many units of each must be used to provide exactly 11 grams of fat, 6 grams of carbohydrates, and 10 grams of protein?

Food	Fat	Carbohydrates	Protein
A	1	1	2
B	2	1	1
C	2	1	2

32. Nutritional planning One unit of each of three foods contains the nutrients shown in the table. How many units of each must be used to provide exactly 14 grams of fat, 9 grams of carbohydrates, and 9 grams of protein?

Food	Fat	Carbohydrates	Protein
A	2	1	2
B	3	2	1
C	1	1	2

33. Making statues An artist makes three types of ceramic statues at a monthly cost of $650 for 180 statues. The manufacturing costs for the three types are $5, $4, and $3. If the statues sell for $20, $12, and $9, respectively, how many of each type should be made to produce $2,100 in monthly revenue?

34. Manufacturing footballs A factory manufactures three types of footballs at a monthly cost of $2,425 for 1,125 footballs. The manufacturing costs for the three types of footballs are $4, $3, and $2. These footballs sell for $16, $12, and $10, respectively. How many of each type are manufactured if the monthly profit is $9,275? (*Hint:* Profit = income − cost.)

35. Concert tickets Tickets for a concert cost $5, $3, and $2. Twice as many $5 tickets were sold as $2 tickets. The receipts for 750 tickets were $2,625. How many of each price ticket were sold?

36. Mixing nuts The owner of a candy store mixed some peanuts worth $3 per pound, some cashews worth $9 per pound, and some Brazil nuts worth $9 per pound to get 50 pounds of a mixture that would sell for $6 per pound. She used 15 fewer pounds of cashews than peanuts. How many pounds of each did she use?

37. Chainsaw sculpting A north woods sculptor carves three types of statues with a chainsaw. The times required for carving, sanding, and painting a totem pole, a bear, and a deer are shown in the table. How many of each should be produced to use all available labor hours?

	Totem pole	Bear	Deer	Time available
Carving	2 hours	2 hours	1 hour	14 hours
Sanding	1 hour	2 hours	2 hours	15 hours
Painting	3 hours	2 hours	2 hours	21 hours

38. Making clothing A clothing manufacturer makes coats, shirts, and slacks. The times required for cutting, sewing, and packaging each item are shown in the table. How many of each should be made to use all available labor hours?

	Coats	Shirts	Slacks	Time available
Cutting	20 min	15 min	10 min	115 hr
Sewing	60 min	30 min	24 min	280 hr
Packaging	5 min	12 min	6 min	65 hr

39. Curve fitting Find the equation of the parabola shown in the illustration.

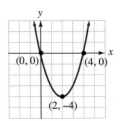

40. Curve fitting Find the equation of the parabola shown in the illustration.

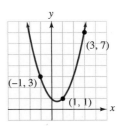

The equation of a circle is of the form $x^2 + y^2 + cx + dy + e = 0.$

41. Curve fitting Find the equation of the circle shown in the illustration.

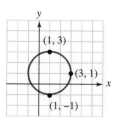

42. Curve fitting Find the equation of the circle shown in the illustration.

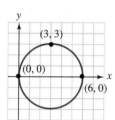

WRITING

43. What makes a system of three equations in three variables inconsistent?

44. What makes the equations of a system of three equations in three variables dependent?

SOMETHING TO THINK ABOUT

45. Solve the system:

$$\begin{cases} x + y + z + w = 3 \\ x - y - z - w = -1 \\ x + y - z - w = 1 \\ x + y - z + w = 3 \end{cases}$$

46. Solve the system:

$$\begin{cases} 2x + y + z + w = 3 \\ x - 2y - z + w = -3 \\ x - y - 2z - w = -3 \\ x + y - z + 2w = 4 \end{cases}$$

3.4 Solution by Matrices

▌ **Matrices** ▌ **Gaussian Elimination**
▌ **Systems with More Equations than Variables**
▌ **Systems with More Variables than Equations**

Getting Ready *Multiply the first row by 2 and add the result to the second row.*

1. 2 3 5
    1 2 3

2. −1 0   4
     2 3 −1

Multiply the first row by −1 and add the result to the second row.

3. 2 3 5
    1 2 3

4. −1 0   4
     2 3 −1

In this section, we will discuss a streamlined method used for solving systems of linear equations. This method involves the use of matrices.

Matrices

Another method of solving systems of equations involves rectangular arrays of numbers called *matrices*.

Matrix   A **matrix** is any rectangular array of numbers.

Some examples of matrices are

$$A = \begin{bmatrix} 1 & 2 & 3 \\ 4 & 5 & 6 \end{bmatrix} \quad B = \begin{bmatrix} 1 & 2 \\ 3 & 4 \\ 5 & 6 \end{bmatrix} \quad C = \begin{bmatrix} 2 & 4 & 6 \\ 8 & 10 & 12 \\ 14 & 16 & 18 \end{bmatrix}$$

The numbers in each matrix are called **elements.** Because matrix A has two rows and three columns, it is called a 2×3 matrix (read "2 by 3" matrix). Matrix B is a 3×2 matrix, because the matrix has three rows and two columns. Matrix C is a 3×3 matrix (three rows and three columns).

Any matrix with the same number of rows and columns, like matrix C, is called a **square matrix.**

To show how to use matrices to solve systems of linear equations, we consider the system

$$\begin{cases} x - 2y - z = 6 \\ 2x + 2y - z = 1 \\ -x - y + 2z = 1 \end{cases}$$

which can be represented by the following matrix, called an **augmented matrix:**

$$\begin{bmatrix} 1 & -2 & -1 & 6 \\ 2 & 2 & -1 & 1 \\ -1 & -1 & 2 & 1 \end{bmatrix}$$

The first three columns of the augmented matrix form a 3×3 matrix called a **coefficient matrix.** It is determined by the coefficients of x, y, and z in the equations of the system. The 3×1 matrix in the last column is determined by the constants in the equations.

Coefficient matrix   *Column of constants*

$$\begin{bmatrix} 1 & -2 & -1 \\ 2 & 2 & -1 \\ -1 & -1 & 2 \end{bmatrix} \qquad \begin{bmatrix} 6 \\ 1 \\ 1 \end{bmatrix}$$

Each row of the augmented matrix represents one equation of the system:

$$\begin{bmatrix} 1 & -2 & -1 & 6 \\ 2 & 2 & -1 & 1 \\ -1 & -1 & 2 & 1 \end{bmatrix} \begin{array}{l} \longleftrightarrow \\ \longleftrightarrow \\ \longleftrightarrow \end{array} \begin{cases} x - 2y - z = 6 \\ 2x + 2y - z = 1 \\ -x - y + 2z = 1 \end{cases}$$

Gaussian Elimination

To solve a 3×3 system of equations by **Gaussian elimination,** we transform an augmented matrix into the following matrix that has all 0's below its main diagonal, which is formed by the elements a, e, and h.

$$\begin{bmatrix} a & b & c & d \\ 0 & e & f & g \\ 0 & 0 & h & i \end{bmatrix} \quad (a, b, c, \dots, i \text{ are real numbers})$$

We can often write a matrix in this form, called **triangular form,** by using the following operations.

Elementary Row Operations

1. Any two rows of a matrix can be interchanged.
2. Any row of a matrix can be multiplied by a nonzero constant.
3. Any row of a matrix can be changed by adding a nonzero constant multiple of another row to it.

- The first row operation corresponds to interchanging two equations of a system.
- The second row operation corresponds to multiplying both sides of an equation by a nonzero constant.
- The third row operation corresponds to adding a nonzero multiple of one equation to another.

None of these operations will change the solution of the given system of equations.

After we have written the matrix in triangular form, we can solve the corresponding system of equations by a back-substitution process, as shown in Example 1.

EXAMPLE 1 Solve the system: $\begin{cases} x - 2y - z = 6 \\ 2x + 2y - z = 1 \\ -x - y + 2z = 1 \end{cases}$.

Solution We can represent the system with the following augmented matrix:

$$\begin{bmatrix} \mathbf{1} & -2 & -1 & 6 \\ 2 & 2 & -1 & 1 \\ -1 & -1 & 2 & 1 \end{bmatrix}$$

To get 0's under the 1 in the first column, we multiply row 1 of the augmented matrix by -2 and add it to row 2 to get a new row 2. We then multiply row 1 by 1 and add it to row 3 to get a new row 3.

$$\begin{bmatrix} 1 & -2 & -1 & 6 \\ 0 & \mathbf{6} & 1 & -11 \\ 0 & -3 & 1 & 7 \end{bmatrix}$$

To get a 0 under the 6 in the second column of the previous matrix, we multiply row 2 by $\frac{1}{2}$ and add it to row 3.

$$\begin{bmatrix} 1 & -2 & -1 & 6 \\ 0 & 6 & 1 & -11 \\ 0 & 0 & \frac{3}{2} & \frac{3}{2} \end{bmatrix}$$

Finally, we multiply row 3 by $\frac{2}{3}$, which is the reciprocal of $\frac{3}{2}$.

$$\begin{bmatrix} 1 & -2 & -1 & 6 \\ 0 & 6 & 1 & -11 \\ 0 & 0 & 1 & 1 \end{bmatrix}$$

The final matrix represents the system of equations

(1)
(2)
(3)
$$\begin{cases} x - 2y - z = 6 \\ 0x + 6y + z = -11 \\ 0x + 0y + z = 1 \end{cases}$$

From Equation 3, we can see that $z = 1$. To find y, we substitute 1 for z in Equation 2 and solve for y:

(2)
$$6y + z = -11$$
$$6y + \mathbf{1} = -11 \qquad \text{Substitute 1 for } z.$$
$$6y = -12 \qquad \text{Subtract 1 from both sides.}$$
$$y = -2 \qquad \text{Divide both sides by 6.}$$

Thus, $y = -2$. To find x, we substitute 1 for z and -2 for y in Equation 1 and solve for x:

(1)
$$x - 2y - z = 6$$
$$x - 2(\mathbf{-2}) - \mathbf{1} = 6 \qquad \text{Substitute 1 for } z \text{ and } -2 \text{ for } y.$$
$$x + 3 = 6 \qquad \text{Simplify.}$$
$$x = 3 \qquad \text{Subtract 3 from both sides.}$$

Thus, $x = 3$. The solution of the given system is $(3, -2, 1)$. Verify that this triple satisfies each equation of the original system.

Self Check   Solve the system: $\begin{cases} x - 2y - z = 2 \\ 2x + 2y - z = -5. \\ -x - y + 2z = 7 \end{cases}$ ∎

Systems with More Equations than Variables

We can use matrices to solve systems that have more equations than variables.

EXAMPLE 2   Solve the system: $\begin{cases} x + y = -1 \\ 2x - y = 7 \\ -x + 2y = -8 \end{cases}$.

Solution   This system can be represented by the following augmented matrix:

$$\begin{bmatrix} \mathbf{1} & 1 & -1 \\ 2 & -1 & 7 \\ -1 & 2 & -8 \end{bmatrix}$$

To get 0's under the 1 in the first column, we multiply row 1 by -2 and add it to row 2. Then we multiply row 1 by 1 and add it to row 3.

$$\begin{bmatrix} 1 & 1 & -1 \\ 0 & \mathbf{-3} & 9 \\ 0 & 3 & -9 \end{bmatrix}$$

To get a 0 under the -3 in the second column, we multiply row 2 by 1 and add it to row 3.

$$\begin{bmatrix} 1 & 1 & -1 \\ 0 & -3 & 9 \\ 0 & 0 & 0 \end{bmatrix}$$

Finally, we multiply row 2 by $-\frac{1}{3}$.

$$\begin{bmatrix} 1 & 1 & -1 \\ 0 & 1 & -3 \\ 0 & 0 & 0 \end{bmatrix}$$

The final matrix represents the system

$$\begin{cases} x + y = -1 \\ 0x + y = -3 \\ 0x + 0y = 0 \end{cases}$$

The third equation can be discarded, because $0x + 0y = 0$ for all x and y. From the second equation, we can read that $y = -3$. To find x, we substitute -3 for y in the first equation and solve for x:

$$x + y = -1$$
$$x + (\mathbf{-3}) = -1 \qquad \text{Substitute } -3 \text{ for } y.$$
$$x = 2 \qquad \text{Add 3 to both sides.}$$

The solution is $(2, -3)$. Verify that this solution satisfies all three equations of the original system.

Self Check Solve the system: $\begin{cases} x + y = 1 \\ 2x - y = 8 \\ -x + 2y = -7 \end{cases}$. ▮

If the last row of the final matrix in Example 2 had been of the form $0x + 0y = k$, where $k \neq 0$, the system would not have a solution. No values of x and y could make the expression $0x + 0y$ equal to a nonzero constant k.

Systems with More Variables than Equations

We can also solve many systems that have more variables than equations.

EXAMPLE 3 Solve the system: $\begin{cases} x + y - 2z = -1 \\ 2x - y + z = -3 \end{cases}$.

Solution This system can be represented by the following augmented matrix.

$$\begin{bmatrix} \mathbf{1} & 1 & -2 & -1 \\ 2 & -1 & 1 & -3 \end{bmatrix}$$

To get a 0 under the 1 in the first column, we multiply row 1 by -2 and add it to row 2.

$$\begin{bmatrix} 1 & 1 & -2 & -1 \\ 0 & -3 & 5 & -1 \end{bmatrix}$$

Then we multiply row 2 by $-\frac{1}{3}$.

$$\begin{bmatrix} 1 & 1 & -2 & -1 \\ 0 & 1 & -\frac{5}{3} & \frac{1}{3} \end{bmatrix}$$

The final matrix represents the system

$$\begin{cases} x + y - 2z = -1 \\ y - \frac{5}{3}z = \frac{1}{3} \end{cases}$$

We add $\frac{5}{3}z$ to both sides of the second equation to obtain

$$y = \frac{1}{3} + \frac{5}{3}z$$

We have not found a specific value for y. However, we have found y in terms of z.

To find a value of x in terms of z, we substitute $\frac{1}{3} + \frac{5}{3}z$ for y in the first equation and simplify to get

$$x + y - 2z = -1$$

$$x + \frac{1}{3} + \frac{5}{3}z - 2z = -1 \qquad \text{Substitute } \tfrac{1}{3} + \tfrac{5}{3}z \text{ for } y.$$

$$x + \frac{1}{3} - \frac{1}{3}z = -1 \qquad \text{Combine like terms.}$$

$$x - \frac{1}{3}z = -\frac{4}{3} \qquad \text{Subtract } \tfrac{1}{3} \text{ from both sides.}$$

$$x = -\frac{4}{3} + \frac{1}{3}z \qquad \text{Add } \tfrac{1}{3}z \text{ to both sides.}$$

A solution of this system must have the form

$$\left(-\frac{4}{3} + \frac{1}{3}z, \;\; \frac{1}{3} + \frac{5}{3}z, \;\; z \right) \qquad \text{This solution is called "a general solution" of the system.}$$

for all values of z. This system has infinitely many solutions, a different one for each value of z. For example,

- If $z = 0$, the corresponding solution is $\left(-\frac{4}{3}, \frac{1}{3}, 0 \right)$.
- If $z = 1$, the corresponding solution is $(-1, 2, 1)$.

Verify that both of these solutions satisfy each equation of the original system.

Self Check   Solve the system: $\begin{cases} x + y - 2z = 11 \\ 2x - y + z = -2 \end{cases}$.

Matrices with the same number of rows and columns can be added. We simply add their corresponding elements. For example,

$$\begin{bmatrix} 2 & 3 & -4 \\ -1 & 2 & 5 \end{bmatrix} + \begin{bmatrix} 3 & -1 & 0 \\ 4 & 3 & 2 \end{bmatrix}$$

$$= \begin{bmatrix} 2+3 & 3+(-1) & -4+0 \\ -1+4 & 2+3 & 5+2 \end{bmatrix}$$

$$= \begin{bmatrix} 5 & 2 & -4 \\ 3 & 5 & 7 \end{bmatrix}$$

To multiply a matrix by a constant, we multiply each element of the matrix by the constant. For example,

$$5 \cdot \begin{bmatrix} 2 & 3 & -4 \\ -1 & 2 & 5 \end{bmatrix}$$

$$= \begin{bmatrix} 5\cdot2 & 5\cdot3 & 5\cdot(-4) \\ 5\cdot(-1) & 5\cdot2 & 5\cdot5 \end{bmatrix}$$

$$= \begin{bmatrix} 10 & 15 & -20 \\ -5 & 10 & 25 \end{bmatrix}$$

Since matrices provide a good way to store information in computers, they are often used in applied problems. For example, suppose there are 66 security officers employed at either the downtown office or the suburban office:

Downtown Office	Male	Female
Day shift	12	18
Night shift	3	0

Suburban Office	Male	Female
Day shift	14	12
Night shift	5	2

The information about the employees is contained in the following matrices.

$$D = \begin{bmatrix} 12 & 18 \\ 3 & 0 \end{bmatrix} \quad \text{and} \quad S = \begin{bmatrix} 14 & 12 \\ 5 & 2 \end{bmatrix}$$

The entry in the first row-first column in matrix D gives the information that 12 males work the day shift at the downtown office. Company management can add the matrices D and S to find corporate-wide totals:

$$D + S = \begin{bmatrix} 12 & 18 \\ 3 & 0 \end{bmatrix} + \begin{bmatrix} 14 & 12 \\ 5 & 2 \end{bmatrix}$$

$$= \begin{bmatrix} 26 & 30 \\ 8 & 2 \end{bmatrix}$$

We interpret the total to mean:

	Male	Female
Day shift	26	30
Night shift	8	2

If one-third of the force in each category at the downtown location retires, the downtown staff would be reduced to $\frac{2}{3}D$ people. We can compute $\frac{2}{3}D$ by multiplying each entry by $\frac{2}{3}$.

$$\frac{2}{3}D = \frac{2}{3}\begin{bmatrix} 12 & 18 \\ 3 & 0 \end{bmatrix}$$

$$= \begin{bmatrix} 8 & 12 \\ 2 & 0 \end{bmatrix}$$

After retirements, downtown staff would be

	Male	Female
Day shift	8	12
Night shift	2	0

Self Check Answers

1. $(1, -2, 3)$ **2.** $(3, -2)$ **3.** infinitely many solutions of the form $\left(3 + \frac{1}{3}z, 8 + \frac{5}{3}z, z\right)$

Orals *Consider the system* $\begin{cases} 3x + 2y = 8 \\ 4x - 3y = 6 \end{cases}$.

1. Find the coefficient matrix. **2.** Find the augmented matrix.

Tell whether each matrix is in triangular form.

3. $\begin{bmatrix} 4 & 1 & 5 \\ 0 & 2 & 7 \\ 0 & 0 & 4 \end{bmatrix}$ **4.** $\begin{bmatrix} 8 & 5 & 2 \\ 0 & 4 & 5 \\ 0 & 7 & 0 \end{bmatrix}$

3.4   EXERCISES

REVIEW   *Write each number in scientific notation.*

1. 93,000,000   **2.** 0.00045
3. 63×10^3 **4.** 0.33×10^3

VOCABULARY AND CONCEPTS   *Fill in the blanks.*

5. A _____ is a rectangular array of numbers.

6. The numbers in a matrix are called its _____.

7. A 3×4 matrix has __ rows and 4 _____.

8. A _____ matrix has the same number of rows as columns.

9. An _____ matrix of a system of equations includes the coefficient matrix and the column of constants.

10. If a matrix has all 0's below its main diagonal, it is written in _____ form.

11. A _____ row operation corresponds to interchanging two equations in a system of equations.

12. A type 2 row operation corresponds to _____ both sides of an equation by a nonzero constant.

13. A type 3 row operation corresponds to adding a _____ multiple of one equation to another.

14. In the Gaussian method of solving systems of equations, we transform the _____ matrix into triangular form and finish the solution by using _____ substitution.

PRACTICE   *Use a row operation to find the missing number in the second matrix.*

15. $\begin{bmatrix} 2 & 1 & 1 \\ 5 & 4 & 1 \end{bmatrix}$ **16.** $\begin{bmatrix} -1 & 3 & 2 \\ 1 & -2 & 3 \end{bmatrix}$
$\begin{bmatrix} 2 & 1 & 1 \\ 3 & 3 & \blacksquare \end{bmatrix}$ $\begin{bmatrix} -1 & 3 & 2 \\ \blacksquare & 1 & 5 \end{bmatrix}$

17. $\begin{bmatrix} 3 & -2 & 1 \\ -1 & 2 & 4 \end{bmatrix}$ **18.** $\begin{bmatrix} 2 & 1 & -3 \\ 2 & 6 & 1 \end{bmatrix}$
$\begin{bmatrix} 3 & -2 & 1 \\ -2 & 4 & \blacksquare \end{bmatrix}$ $\begin{bmatrix} 6 & 3 & \blacksquare \\ 2 & 6 & 1 \end{bmatrix}$

Use matrices to solve each system of equations. If a system has no solution, so indicate.

19. $\begin{cases} x + y = 2 \\ x - y = 0 \end{cases}$ **20.** $\begin{cases} x + y = 3 \\ x - y = -1 \end{cases}$

21. $\begin{cases} x + 2y = -4 \\ 2x + y = 1 \end{cases}$ **22.** $\begin{cases} 2x - 3y = 16 \\ -4x + y = -22 \end{cases}$

23. $\begin{cases} 3x + 4y = -12 \\ 9x - 2y = 6 \end{cases}$ **24.** $\begin{cases} 5x - 4y = 10 \\ x - 7y = 2 \end{cases}$

25. $\begin{cases} 5a = 24 + 2b \\ 5b = 3a + 16 \end{cases}$ **26.** $\begin{cases} 3m = 2n + 16 \\ 2m = -5n - 2 \end{cases}$

27. $\begin{cases} x + y + z = 6 \\ x + 2y + z = 8 \\ x + y + 2z = 9 \end{cases}$ **28.** $\begin{cases} x - y + z = 2 \\ x + 2y - z = 6 \\ 2x - y - z = 3 \end{cases}$

29. $\begin{cases} 2x + y + 3z = 3 \\ -2x - y + z = 5 \\ 4x - 2y + 2z = 2 \end{cases}$

30. $\begin{cases} 3x + 2y + z = 8 \\ 6x - y + 2z = 16 \\ -9x + y - z = -20 \end{cases}$

31. $\begin{cases} 3x - 2y + 4z = 4 \\ x + y + z = 3 \\ 6x - 2y - 3z = 10 \end{cases}$

32. $\begin{cases} 2x + 3y - z = -8 \\ x - y - z = -2 \\ -4x + 3y + z = 6 \end{cases}$

33. $\begin{cases} x + y = 3 \\ 3x - y = 1 \\ 2x + y = 4 \end{cases}$ **34.** $\begin{cases} x - y = -5 \\ 2x + 3y = 5 \\ x + y = 1 \end{cases}$

35. $\begin{cases} 2x - y = 4 \\ x + 3y = 2 \\ -x - 4y = -2 \end{cases}$

36. $\begin{cases} 3x - 2y = 5 \\ x + 2y = 7 \\ -3x - y = -11 \end{cases}$

37. $\begin{cases} 2x + y = 7 \\ x - y = 2 \\ -x + 3y = -2 \end{cases}$

38. $\begin{cases} 3x - y = 2 \\ -6x + 3y = 0 \\ -x + 2y = -4 \end{cases}$

39. $\begin{cases} x + 3y = 7 \\ x + y = 3 \\ 3x + y = 5 \end{cases}$

40. $\begin{cases} x + y = 3 \\ x - 2y = -3 \\ x - y = 1 \end{cases}$

Use matrices to help find a general solution of each system of equations.

41. $\begin{cases} x + 2y + 3z = -2 \\ -x - y - 2z = 4 \end{cases}$

42. $\begin{cases} 2x - 4y + 3z = 6 \\ -4x + 6y + 4z = -6 \end{cases}$

43. $\begin{cases} x - y = 1 \\ y + z = 1 \\ x + z = 2 \end{cases}$

44. $\begin{cases} x + z = 1 \\ x + y = 2 \\ 2x + y + z = 3 \end{cases}$

Remember these facts from geometry.

Two angles whose measures add up to 90° are complementary.

Two angles whose measures add up to 180° are supplementary.

The sum of the measures of the interior angles in a triangle is 180°.

45. Geometry   One angle is 46° larger than its complement. Find the measure of each angle.

46. Geometry   One angle is 28° larger than its supplement. Find the measure of each angle.

47. Geometry   In the illustration, $\angle B$ is 25° more than $\angle A$, and $\angle C$ is 5° less than twice $\angle A$. Find the measure of each angle in the triangle.

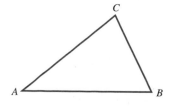

48. Geometry   In the illustration, $\angle A$ is 10° less than $\angle B$, and $\angle B$ is 10° less than $\angle C$. Find the measure of each angle in the triangle.

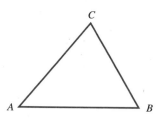

Remember that the equation of a parabola is of the form $y = ax^2 + bx + c$.

49. Curve fitting   Find the equation of the parabola passing through the points $(0, 1)$, $(1, 2)$, and $(-1, 4)$.

50. Curve fitting   Find the equation of the parabola passing through the points $(0, 1)$, $(1, 1)$, and $(-1, -1)$.

WRITING

51. Explain how to check the solution of a system of equations.

52. Explain how to perform a type 3 row operation.

SOMETHING TO THINK ABOUT

53. If the system represented by

$$\begin{bmatrix} 1 & 1 & 0 & 1 \\ 0 & 0 & 1 & 2 \\ 0 & 0 & 0 & k \end{bmatrix}$$

has no solution, what do you know about k?

54. Is it possible for a system with fewer equations than variables to have no solution? Illustrate.

3.5	**Solution by Determinants**
	▌**Determinants** ▌**Cramer's Rule**

Getting Ready   *Find each value.*

1. $3(-4) - 2(5)$ 2. $5(2) - 3(-4)$

3. $2(2 - 5) - 3(5 - 2) + 2(4 - 3)$ 4. $-3(5 - 2) + 2(3 + 1) - 2(5 + 1)$

We now discuss a final method for solving systems of linear equations. This method involves *determinants,* an idea related to the concept of matrices.

Determinants

A **determinant** is a number that is associated with a square matrix. For any square matrix A, the symbol $|A|$ represents the determinant of A.

Value of a 2 × 2 Determinant

If a, b, c, and d are real numbers, the **determinant** of the matrix $\begin{bmatrix} a & b \\ c & d \end{bmatrix}$ is

$$\begin{vmatrix} a & b \\ c & d \end{vmatrix} = ad - bc$$

The determinant of a 2×2 matrix is the number that is equal to the product of the numbers on the major diagonal

$$\begin{vmatrix} a & b \\ c & d \end{vmatrix}$$

minus the product of the numbers on the other diagonal

$$\begin{vmatrix} a & b \\ c & d \end{vmatrix}$$

EXAMPLE 1   Evaluate the determinants: **a.** $\begin{vmatrix} 3 & 2 \\ 6 & 9 \end{vmatrix}$ and **b.** $\begin{vmatrix} -5 & \frac{1}{2} \\ -1 & 0 \end{vmatrix}$.

Solution   **a.** $\begin{vmatrix} 3 & 2 \\ 6 & 9 \end{vmatrix} = 3(9) - 2(6)$ **b.** $\begin{vmatrix} -5 & \frac{1}{2} \\ -1 & 0 \end{vmatrix} = -5(0) - \frac{1}{2}(-1)$

$= 27 - 12$ $= 0 + \frac{1}{2}$

$= 15$ $= \frac{1}{2}$

Self Check   Evaluate the determinant: $\begin{vmatrix} 4 & -3 \\ 2 & 1 \end{vmatrix}$.   ▌

A 3×3 determinant is evaluated by expanding by **minors.**

Value of a 3 × 3 Determinant

$$
\begin{vmatrix} a_1 & b_1 & c_1 \\ a_2 & b_2 & c_2 \\ a_3 & b_3 & c_3 \end{vmatrix} = a_1 \begin{vmatrix} b_2 & c_2 \\ b_3 & c_3 \end{vmatrix} \overset{\substack{\text{Minor} \\ \text{of } a_1 \\ \downarrow}}{} - b_1 \begin{vmatrix} a_2 & c_2 \\ a_3 & c_3 \end{vmatrix} \overset{\substack{\text{Minor} \\ \text{of } b_1 \\ \downarrow}}{} + c_1 \begin{vmatrix} a_2 & b_2 \\ a_3 & b_3 \end{vmatrix} \overset{\substack{\text{Minor} \\ \text{of } c_1 \\ \downarrow}}{}
$$

To find the minor of a_1, we find the determinant formed by crossing out the elements of the matrix that are in the same row and column as a_1:

$$
\begin{vmatrix} a_1 & b_1 & c_1 \\ a_2 & b_2 & c_2 \\ a_3 & b_3 & c_3 \end{vmatrix}
$$

The minor of a_1 is $\begin{vmatrix} b_2 & c_2 \\ b_3 & c_3 \end{vmatrix}$.

To find the minor of b_1, we cross out the elements of the matrix that are in the same row and column as b_1:

$$
\begin{vmatrix} a_1 & b_1 & c_1 \\ a_2 & b_2 & c_2 \\ a_3 & b_3 & c_3 \end{vmatrix}
$$

The minor of b_1 is $\begin{vmatrix} a_2 & c_2 \\ a_3 & c_3 \end{vmatrix}$.

To find the minor of c_1, we cross out the elements of the matrix that are in the same row and column as c_1:

$$
\begin{vmatrix} a_1 & b_1 & c_1 \\ a_2 & b_2 & c_2 \\ a_3 & b_3 & c_3 \end{vmatrix}
$$

The minor of c_1 is $\begin{vmatrix} a_2 & b_2 \\ a_3 & b_3 \end{vmatrix}$.

EXAMPLE 2   Evaluate the determinant: $\begin{vmatrix} 1 & 3 & -2 \\ 2 & 1 & 3 \\ 1 & 2 & 3 \end{vmatrix}$.

Solution

$$
\begin{vmatrix} 1 & 3 & -2 \\ 2 & 1 & 3 \\ 1 & 2 & 3 \end{vmatrix} = 1 \begin{vmatrix} 1 & 3 \\ 2 & 3 \end{vmatrix} \overset{\substack{\text{Minor} \\ \text{of } 1 \\ \downarrow}}{} - 3 \begin{vmatrix} 2 & 3 \\ 1 & 3 \end{vmatrix} \overset{\substack{\text{Minor} \\ \text{of } 3 \\ \downarrow}}{} + (-2) \begin{vmatrix} 2 & 1 \\ 1 & 2 \end{vmatrix} \overset{\substack{\text{Minor} \\ \text{of } -2 \\ \downarrow}}{}
$$

$$
= 1(3 - 6) - 3(6 - 3) - 2(4 - 1)
$$
$$
= -3 - 9 - 6
$$
$$
= -18
$$

Self Check   Evaluate the determinant: $\begin{vmatrix} 2 & -1 & 3 \\ 1 & 2 & -2 \\ 3 & 1 & 1 \end{vmatrix}$.

We can evaluate a 3×3 determinant by expanding it along any row or column. To determine the signs between the terms of the expansion of a 3×3 determinant, we use the following array of signs.

Array of Signs for a
3 × 3 Determinant

$$
\begin{matrix}
+ & - & + \\
- & + & - \\
+ & - & +
\end{matrix}
$$

EXAMPLE 3 Evaluate $\begin{vmatrix} 1 & 3 & -2 \\ 2 & 1 & 3 \\ 1 & 2 & 3 \end{vmatrix}$ by expanding on the middle column.

Solution This is the determinant of Example 2. To expand it along the middle column, we use the signs of the middle column of the array of signs:

$$
\begin{array}{ccc}
\text{Minor} & \text{Minor} & \text{Minor} \\
\text{of 3} & \text{of 1} & \text{of 2} \\
\downarrow & \downarrow & \downarrow
\end{array}
$$

$$
\begin{vmatrix} 1 & \mathbf{3} & -2 \\ 2 & \mathbf{1} & 3 \\ 1 & \mathbf{2} & 3 \end{vmatrix} = -3\begin{vmatrix} 2 & 3 \\ 1 & 3 \end{vmatrix} + 1\begin{vmatrix} 1 & -2 \\ 1 & 3 \end{vmatrix} - 2\begin{vmatrix} 1 & -2 \\ 2 & 3 \end{vmatrix}
$$

$$
= -3(6-3) + 1[3-(-2)] - 2[3-(-4)]
$$
$$
= -3(3) + 1(5) - 2(7)
$$
$$
= -9 + 5 - 14
$$
$$
= -18
$$

As expected, we get the same value as in Example 2.

Self Check Evaluate the determinant: $\begin{vmatrix} 2 & -1 & 3 \\ 1 & 2 & 2 \\ 3 & 1 & 1 \end{vmatrix}$.

Accent on Technology **EVALUATING DETERMINANTS**

To use a graphing calculator to evaluate the determinant in Example 3, we first enter the matrix by pressing the MATRIX key, selecting EDIT, and pressing the ENTER key. We then enter the dimensions and the elements of the matrix to get Figure 3-13(a). We then press 2nd QUIT to clear the screen. We then press MATRIX, select MATH, and press 1 to get Figure 3-13(b). Next, press MATRIX, select NAMES, and press 1 to get Figure 3-13(c). To get the value of the determinant, we now press ENTER to get Figure 3-13(d), which shows that the value of the determinant is −18.

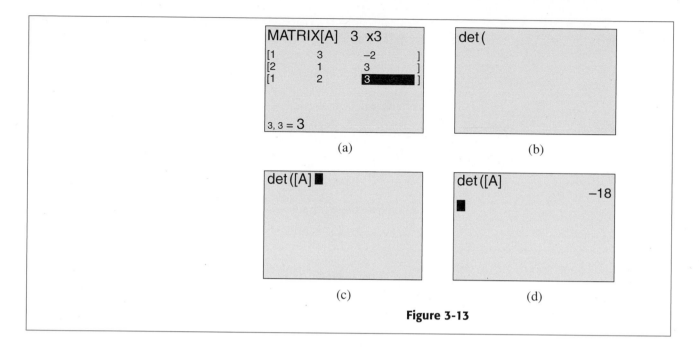

Figure 3-13

Cramer's Rule

The method of using determinants to solve systems of equations is called **Cramer's rule**, named after the 18th-century mathematician Gabriel Cramer. To develop Cramer's rule, we consider the system

$$\begin{cases} ax + by = e \\ cx + dy = f \end{cases}$$

where x and y are variables and a, b, c, d, e, and f are constants.

If we multiply both sides of the first equation by d and multiply both sides of the second equation by $-b$, we can add the equations and eliminate y:

$$\begin{array}{rl} adx + bdy = & ed \\ \underline{-bcx - bdy = -bf} \\ adx - bcx \quad = ed - bf \end{array}$$

To solve for x, we use the distributive property to write $adx - bcx$ as $(ad - bc)x$ on the left-hand side and divide each side by $ad - bc$:

$$(ad - bc)x = ed - bf$$

$$x = \frac{ed - bf}{ad - bc} \quad (ad - bc \neq 0)$$

We can find y in a similar manner. After eliminating the variable x, we get

$$y = \frac{af - ec}{ad - bc} \quad (ad - bc \neq 0)$$

Determinants provide an easy way of remembering these formulas. Note that the denominator for both x and y is

$$\begin{vmatrix} a & b \\ c & d \end{vmatrix} = ad - bc$$

The numerators can be expressed as determinants also:

$$x = \frac{ed - bf}{ad - bc} = \frac{\begin{vmatrix} e & b \\ f & d \end{vmatrix}}{\begin{vmatrix} a & b \\ c & d \end{vmatrix}} \qquad \text{and} \qquad y = \frac{af - ec}{ad - bc} = \frac{\begin{vmatrix} a & e \\ c & f \end{vmatrix}}{\begin{vmatrix} a & b \\ c & d \end{vmatrix}}$$

If we compare these formulas with the original system

$$\begin{cases} ax + by = e \\ cx + dy = f \end{cases}$$

we note that in the expressions for x and y above, the denominator determinant is formed by using the coefficients a, b, c, and d of the variables in the equations. The numerator determinants are the same as the denominator determinant, except that the column of coefficients of the variable for which we are solving is replaced with the column of constants e and f.

Cramer's Rule for Two Equations in Two Variables

The solution of the system $\begin{cases} ax + by = e \\ cx + dy = f \end{cases}$ is given by

$$x = \frac{D_x}{D} = \frac{\begin{vmatrix} e & b \\ f & d \end{vmatrix}}{\begin{vmatrix} a & b \\ c & d \end{vmatrix}} \qquad \text{and} \qquad y = \frac{D_y}{D} = \frac{\begin{vmatrix} a & e \\ c & f \end{vmatrix}}{\begin{vmatrix} a & b \\ c & d \end{vmatrix}}$$

If every determinant is 0, the system is consistent but the equations are dependent.

If $D = 0$ and D_x or D_y is nonzero, the system is inconsistent.

If $D \neq 0$, the system is consistent and the equations are independent.

EXAMPLE 4   Use Cramer's rule to solve $\begin{cases} 4x - 3y = 6 \\ -2x + 5y = 4 \end{cases}$.

Solution   The value of x is the quotient of two determinants. The denominator determinant is made up of the coefficients of x and y:

$$D = \begin{vmatrix} 4 & -3 \\ -2 & 5 \end{vmatrix}$$

To solve for x, we form the numerator determinant from the denominator determinant by replacing its first column (the coefficients of x) with the column of constants (6 and 4).

To solve for y, we form the numerator determinant from the denominator determinant by replacing the second column (the coefficients of y) with the column of constants (6 and 4).

To find the values of x and y, we evaluate each determinant:

$$x = \frac{\begin{vmatrix} 6 & -3 \\ 4 & 5 \end{vmatrix}}{\begin{vmatrix} 4 & -3 \\ -2 & 5 \end{vmatrix}} = \frac{6(5) - (-3)(4)}{4(5) - (-3)(-2)} = \frac{30 + 12}{20 - 6} = \frac{42}{14} = 3$$

$$y = \frac{\begin{vmatrix} 4 & 6 \\ -2 & 4 \end{vmatrix}}{\begin{vmatrix} 4 & -3 \\ -2 & 5 \end{vmatrix}} = \frac{4(4) - 6(-2)}{4(5) - (-3)(-2)} = \frac{16 + 12}{20 - 6} = \frac{28}{14} = 2$$

The solution of this system is (3, 2). Verify that $x = 3$ and $y = 2$ satisfy each equation in the given system.

Self Check   Solve the system: $\begin{cases} 2x - 3y = -16 \\ 3x + 5y = 14 \end{cases}$.

EXAMPLE 5   Use Cramer's rule to solve $\begin{cases} 7x = 8 - 4y \\ 2y = 3 - \frac{7}{2}x \end{cases}$.

Solution   We multiply both sides of the second equation by 2 to eliminate the fraction and write the system in the form

$$\begin{cases} 7x + 4y = 8 \\ 7x + 4y = 6 \end{cases}$$

When we attempt to use Cramer's rule to solve this system for x, we obtain

$$x = \frac{\begin{vmatrix} 8 & 4 \\ 6 & 4 \end{vmatrix}}{\begin{vmatrix} 7 & 4 \\ 7 & 4 \end{vmatrix}} = \frac{8}{0} \quad \text{which is undefined}$$

Since the denominator determinant is 0 and the numerator determinant is not 0, the system is inconsistent. It has no solutions.

We can see directly from the system that it is inconsistent. For any values of x and y, it is impossible that 7 times x plus 4 times y could be both 8 and 6.

Self Check   Solve the system: $\begin{cases} 3x = 8 - 4y \\ y = \dfrac{5}{2} - \dfrac{3}{4}x \end{cases}$.

Cramer's Rule for Three Equations in Three Variables

The solution of the system $\begin{cases} ax + by + cz = j \\ dx + ey + fz = k \\ gx + hy + iz = l \end{cases}$ is given by

$$x = \frac{D_x}{D}, \quad y = \frac{D_y}{D}, \quad \text{and} \quad z = \frac{D_z}{D}$$

where

$$D = \begin{vmatrix} a & b & c \\ d & e & f \\ g & h & i \end{vmatrix} \qquad D_x = \begin{vmatrix} j & b & c \\ k & e & f \\ l & h & i \end{vmatrix}$$

$$D_y = \begin{vmatrix} a & j & c \\ d & k & f \\ g & l & i \end{vmatrix} \qquad D_z = \begin{vmatrix} a & b & j \\ d & e & k \\ g & h & l \end{vmatrix}$$

If every determinant is 0, the system is consistent but the equations are dependent.

If $D = 0$ and D_x or D_y or D_z is nonzero, the system is inconsistent.

If $D \neq 0$, the system is consistent and the equations are independent.

EXAMPLE 6 Use Cramer's rule to solve $\begin{cases} 2x + y + 4z = 12 \\ x + 2y + 2z = 9 \\ 3x - 3y - 2z = 1 \end{cases}$.

Solution The denominator determinant is the determinant formed by the coefficients of the variables. To form the numerator determinants, we substitute the column of constants for the coefficients of the variable being solved for. We form the quotients for x, y, and z and evaluate the determinants:

$$x = \frac{\begin{vmatrix} 12 & 1 & 4 \\ 9 & 2 & 2 \\ 1 & -3 & -2 \end{vmatrix}}{\begin{vmatrix} 2 & 1 & 4 \\ 1 & 2 & 2 \\ 3 & -3 & -2 \end{vmatrix}} = \frac{12\begin{vmatrix} 2 & 2 \\ -3 & -2 \end{vmatrix} - 1\begin{vmatrix} 9 & 2 \\ 1 & -2 \end{vmatrix} + 4\begin{vmatrix} 9 & 2 \\ 1 & -3 \end{vmatrix}}{2\begin{vmatrix} 2 & 2 \\ -3 & -2 \end{vmatrix} - 1\begin{vmatrix} 1 & 2 \\ 3 & -2 \end{vmatrix} + 4\begin{vmatrix} 1 & 2 \\ 3 & -3 \end{vmatrix}} = \frac{12(2) - (-20) + 4(-29)}{2(2) - (-8) + 4(-9)} = \frac{-72}{-24} = 3$$

$$y = \frac{\begin{vmatrix} 2 & 12 & 4 \\ 1 & 9 & 2 \\ 3 & 1 & -2 \end{vmatrix}}{\begin{vmatrix} 2 & 1 & 4 \\ 1 & 2 & 2 \\ 3 & -3 & -2 \end{vmatrix}} = \frac{2\begin{vmatrix} 9 & 2 \\ 1 & -2 \end{vmatrix} - 12\begin{vmatrix} 1 & 2 \\ 3 & -2 \end{vmatrix} + 4\begin{vmatrix} 1 & 9 \\ 3 & 1 \end{vmatrix}}{-24} = \frac{2(-20) - 12(-8) + 4(-26)}{-24} - \frac{-48}{-24} = 2$$

$$z = \frac{\begin{vmatrix} 2 & 1 & 12 \\ 1 & 2 & 9 \\ 3 & -3 & 1 \end{vmatrix}}{\begin{vmatrix} 2 & 1 & 4 \\ 1 & 2 & 2 \\ 3 & -3 & -2 \end{vmatrix}} = \frac{2\begin{vmatrix} 2 & 9 \\ -3 & 1 \end{vmatrix} - 1\begin{vmatrix} 1 & 9 \\ 3 & 1 \end{vmatrix} + 12\begin{vmatrix} 1 & 2 \\ 3 & -3 \end{vmatrix}}{-24} = \frac{2(29) - 1(-26) + 12(-9)}{-24} = \frac{-24}{-24} = 1$$

The solution of this system is (3, 2, 1).

Self Check   Solve the system: $\begin{cases} x + y + 2z = 6 \\ 2x - y + z = 9 \\ x + y - 2z = -6 \end{cases}$.

Self Check Answers

1. 10   **2.** 0   **3.** −20   **4.** (−2, 4)   **5.** no solutions   **6.** (2, −2, 3)

Orals   *Evaluate each determinant.*

1. $\begin{vmatrix} 2 & 1 \\ 1 & 1 \end{vmatrix}$ **2.** $\begin{vmatrix} 0 & 2 \\ 1 & 1 \end{vmatrix}$ **3.** $\begin{vmatrix} 0 & 1 \\ 0 & 1 \end{vmatrix}$

When using Cramer's rule to solve the system $\begin{cases} x + 2y = 5 \\ 2x - y = 4 \end{cases}$,

4. Find the denominator determinant for x.

5. Find the numerator determinant for x.

6. Find the numerator determinant for y.

3.5 EXERCISES

REVIEW   *Solve each equation.*

1. $3(x + 2) - (2 - x) = x - 5$

2. $\frac{3}{7}x = 2(x + 11)$

3. $\frac{5}{3}(5x + 6) - 10 = 0$

4. $5 - 3(2x - 1) = 2(4 + 3x) - 24$

VOCABULARY AND CONCEPTS   *Fill in the blanks.*

5. A determinant is a _____ that is associated with a square matrix.

6. The value of $\begin{vmatrix} a & b \\ c & d \end{vmatrix}$ is _____.

7. The minor of b_1 in $\begin{vmatrix} a_1 & b_1 & c_1 \\ a_2 & b_2 & c_2 \\ a_3 & b_3 & c_3 \end{vmatrix}$ is _____.

8. We can evaluate a determinant by expanding it along any _____ or _____.

9. The denominator determinant for the value of x in the system $\begin{cases} 3x + 4y = 7 \\ 2x - 3y = 5 \end{cases}$ is _____.

10. If the denominator determinant for y in a system of equations is zero, the equations of the system are _____ or the system is _____.

PRACTICE   *Evaluate each determinant.*

11. $\begin{vmatrix} 2 & 3 \\ -2 & 1 \end{vmatrix}$ **12.** $\begin{vmatrix} 3 & -2 \\ -2 & 4 \end{vmatrix}$

13. $\begin{vmatrix} -1 & 2 \\ 3 & -4 \end{vmatrix}$ **14.** $\begin{vmatrix} -1 & -2 \\ -3 & -4 \end{vmatrix}$

15. $\begin{vmatrix} x & y \\ y & x \end{vmatrix}$ **16.** $\begin{vmatrix} x+y & y-x \\ x & y \end{vmatrix}$

17. $\begin{vmatrix} 1 & 0 & 1 \\ 0 & 1 & 0 \\ 1 & 1 & 1 \end{vmatrix}$ **18.** $\begin{vmatrix} 1 & 2 & 0 \\ 0 & 1 & 2 \\ 0 & 0 & 1 \end{vmatrix}$

19. $\begin{vmatrix} -1 & 2 & 1 \\ 2 & 1 & -3 \\ 1 & 1 & 1 \end{vmatrix}$ **20.** $\begin{vmatrix} 1 & 2 & 3 \\ 1 & 2 & 3 \\ 1 & 2 & 3 \end{vmatrix}$

21. $\begin{vmatrix} 1 & -2 & 3 \\ -2 & 1 & 1 \\ -3 & -2 & 1 \end{vmatrix}$ **22.** $\begin{vmatrix} 1 & 1 & 2 \\ 2 & 1 & -2 \\ 3 & 1 & 3 \end{vmatrix}$

23. $\begin{vmatrix} 1 & 2 & 3 \\ 4 & 5 & 6 \\ 7 & 8 & 9 \end{vmatrix}$ **24.** $\begin{vmatrix} 1 & 4 & 7 \\ 2 & 5 & 8 \\ 3 & 6 & 9 \end{vmatrix}$

25. $\begin{vmatrix} a & 2a & -a \\ 2 & -1 & 3 \\ 1 & 2 & -3 \end{vmatrix}$ **26.** $\begin{vmatrix} 1 & 2b & -3 \\ 2 & -b & 2 \\ 1 & 3b & 1 \end{vmatrix}$

27. $\begin{vmatrix} 1 & a & b \\ 1 & 2a & 2b \\ 1 & 3a & 3b \end{vmatrix}$ **28.** $\begin{vmatrix} a & b & c \\ 0 & b & c \\ 0 & 0 & c \end{vmatrix}$

Use Cramer's rule to solve each system, if possible.

29. $\begin{cases} x+y=6 \\ x-y=2 \end{cases}$ **30.** $\begin{cases} x-y=4 \\ 2x+y=5 \end{cases}$

31. $\begin{cases} 2x+y=1 \\ x-2y=-7 \end{cases}$ **32.** $\begin{cases} 3x-y=-3 \\ 2x+y=-7 \end{cases}$

33. $\begin{cases} 2x+3y=0 \\ 4x-6y=-4 \end{cases}$ **34.** $\begin{cases} 4x-3y=-1 \\ 8x+3y=4 \end{cases}$

35. $\begin{cases} y=\dfrac{-2x+1}{3} \\ 3x-2y=8 \end{cases}$ **36.** $\begin{cases} 2x+3y=-1 \\ x=\dfrac{y-9}{4} \end{cases}$

37. $\begin{cases} y=\dfrac{11-3x}{2} \\ x=\dfrac{11-4y}{6} \end{cases}$ **38.** $\begin{cases} x=\dfrac{12-6y}{5} \\ y=\dfrac{24-10x}{12} \end{cases}$

39. $\begin{cases} x=\dfrac{5y-4}{2} \\ y=\dfrac{3x-1}{5} \end{cases}$ **40.** $\begin{cases} y=\dfrac{1-5x}{2} \\ x=\dfrac{3y+10}{4} \end{cases}$

41. $\begin{cases} x+y+z=4 \\ x+y-z=0 \\ x-y+z=2 \end{cases}$ **42.** $\begin{cases} x+y+z=4 \\ x-y+z=2 \\ x-y-z=0 \end{cases}$

43. $\begin{cases} x+y+2z=7 \\ x+2y+z=8 \\ 2x+y+z=9 \end{cases}$ **44.** $\begin{cases} x+2y+2z=10 \\ 2x+y+2z=9 \\ 2x+2y+z=1 \end{cases}$

45. $\begin{cases} 2x+y-z=1 \\ x+2y+2z=2 \\ 4x+5y+3z=3 \end{cases}$ **46.** $\begin{cases} 4x+3z=4 \\ 2y-6z=-1 \\ 8x+4y+3z=9 \end{cases}$

47. $\begin{cases} 2x+y+z=5 \\ x-2y+3z=10 \\ x+y-4z=-3 \end{cases}$ **48.** $\begin{cases} 3x+2y-z=-8 \\ 2x-y+7z=10 \\ 2x+2y-3z=-10 \end{cases}$

49. $\begin{cases} 2x+3y+4z=6 \\ 2x-3y-4z=-4 \\ 4x+6y+8z=12 \end{cases}$

50. $\begin{cases} x-3y+4z-2=0 \\ 2x+y+2z-3=0 \\ 4x-5y+10z-7=0 \end{cases}$

51. $\begin{cases} x+y=1 \\ \dfrac{1}{2}y+z=\dfrac{5}{2} \\ x-z=-3 \end{cases}$ **52.** $\begin{cases} 3x+4y+14z=7 \\ -\dfrac{1}{2}x-y+2z=\dfrac{3}{2} \\ x+\dfrac{3}{2}y+\dfrac{5}{2}z=1 \end{cases}$

53. $\begin{cases} 2x-y+4z+2=0 \\ 5x+8y+7z=-8 \\ x+3y+z+3=0 \end{cases}$

54. $\begin{cases} \dfrac{1}{2}x + y + z + \dfrac{3}{2} = 0 \\ x + \dfrac{1}{2}y + z - \dfrac{1}{2} = 0 \\ x + y + \dfrac{1}{2}z + \dfrac{1}{2} = 0 \end{cases}$

Evaluate each determinant and solve the resulting equation.

55. $\begin{vmatrix} x & 1 \\ 3 & 2 \end{vmatrix} = 1$

56. $\begin{vmatrix} x & -x \\ 2 & -3 \end{vmatrix} = -5$

57. $\begin{vmatrix} x & -2 \\ 3 & 1 \end{vmatrix} = \begin{vmatrix} 4 & 2 \\ x & 3 \end{vmatrix}$

58. $\begin{vmatrix} x & 3 \\ x & 2 \end{vmatrix} = \begin{vmatrix} 3 & 2 \\ 1 & 1 \end{vmatrix}$

APPLICATIONS

59. Making investments A student wants to average a 6.6% return by investing $20,000 in the three stocks listed in the table. Because HiTech is considered to be a high-risk investment, he wants to invest three times as much in SaveTel and HiGas combined as he invests in HiTech. How much should he invest in each stock?

Stock	Rate of return
HiTech	10%
SaveTel	5%
HiGas	6%

60. Making investments See the table. A woman wants to average a $7\frac{1}{3}$% return by investing $30,000 in three certificates of deposit. She wants to invest five times as much in the 8% CD as in the 6% CD. How much should she invest in each CD?

Type of CD	Rate of return
12 month	6%
24 month	7%
36 month	8%

Use a graphing calculator to evaluate each determinant.

61. $\begin{vmatrix} 2 & -3 & 4 \\ -1 & 2 & 4 \\ 3 & -3 & 1 \end{vmatrix}$

62. $\begin{vmatrix} -3 & 2 & -5 \\ 3 & -2 & 6 \\ 1 & -3 & 4 \end{vmatrix}$

63. $\begin{vmatrix} 2 & 1 & -3 \\ -2 & 2 & 4 \\ 1 & -2 & 2 \end{vmatrix}$

64. $\begin{vmatrix} 4 & 2 & -3 \\ 2 & -5 & 6 \\ 2 & 5 & -2 \end{vmatrix}$

WRITING

65. Tell how to find the minor of an element of a determinant.

66. Tell how to find x when solving a system of linear equations by Cramer's rule.

SOMETHING TO THINK ABOUT

67. Show that
$$\begin{vmatrix} x & y & 1 \\ -2 & 3 & 1 \\ 3 & 5 & 1 \end{vmatrix} = 0$$
is the equation of the line passing through $(-2, 3)$ and $(3, 5)$.

68. Show that
$$\frac{1}{2}\begin{vmatrix} 0 & 0 & 1 \\ 3 & 0 & 1 \\ 0 & 4 & 1 \end{vmatrix}$$
is the area of the triangle with vertices at $(0, 0)$, $(3, 0)$, and $(0, 4)$.

Determinants with more than 3 rows and 3 columns can be evaluated by expanding them by minors. The sign array for a 4 × 4 determinant is

$$\begin{matrix} + & - & + & - \\ - & + & - & + \\ + & - & + & - \\ - & + & - & + \end{matrix}$$

Evaluate each determinant.

69. $\begin{vmatrix} 1 & 0 & 2 & 1 \\ 2 & 1 & 1 & 3 \\ 1 & 1 & 1 & 1 \\ 2 & 1 & 1 & 1 \end{vmatrix}$

70. $\begin{vmatrix} 1 & 2 & -1 & 1 \\ -2 & 1 & 3 & -1 \\ 0 & 1 & 1 & 2 \\ 2 & 0 & 3 & 1 \end{vmatrix}$

Projects

Project 1

The number of units of a product that will be produced depends on the unit price of the product. As the unit price gets higher, the product will be produced in greater quantity, because the producer will make more money on each item. The *supply* of the product will grow, and we say that supply *is a function of* (or *depends on*) the unit price. Futhermore, as the price rises, fewer consumers will buy the product, and the *demand* will decrease. The demand for the product is also a function of the unit price.

In this project, we will assume that both supply and demand are *linear* functions of the unit price. Thus, the graph of supply (the *y*-coordinate) versus price (the *x*-coordinate) is a line with positive slope. The graph of the demand function is a line with negative slope. Because these two lines cannot be parallel, they must intersect. The price at which supply equals demand is called the *market price:* At this price, the same number of units of the product will be sold as are manufactured.

You work for Soda Pop Inc. and have the task of analyzing the sales figures for the past year. You have been provided with the following supply and demand functions. (Supply and demand are measured in cases per week; p, the price per case, is measured in dollars.)

The demand for soda is $D(p) = 19{,}000 - 2{,}200p$.

The supply of soda is $S(p) = 3{,}000 + 1{,}080p$.

Both functions are true for values of p from \$3.50 to \$5.75.

Graph both functions on the same set of coordinate axes, being sure to label each graph, and include any other important information. Then write a report for your supervisor that answers the following questions.

a. Explain why producers will be able to sell all of the soda they make when the price is \$3.50 per case. How much money will the producers take in from these sales?

b. How much money will producers take in from sales when the price is \$5.75 per case? How much soda will not be sold?

c. Find the market price for soda (to the nearest cent). How many cases per week will be sold at this price? How much money will the producers take in from sales at the market price?

d. Explain why prices always tend toward the market price. That is, explain why the unit price will rise if the demand is greater than the supply, and why the unit price will fall if supply is greater than demand.

Project 2

Goodstuff Produce Company has two large water canals that feed the irrigation ditches on its fruit farm. One of these canals runs directly north and south, and the other runs directly east and west. The canals cross at the center of the farm property (the origin) and divide the farm into four quadrants. The company is interested in digging some new irrigation ditches in a portion of the northeast quadrant. You have been hired to plan the layout of the new system.

Your design is to make use of ditch Z, which is already present. This ditch runs from a point 300 meters north of the origin to a point 400 meters east of the origin. The owners of Goodstuff want two new ditches dug.

- Ditch A is to begin at a point 100 meters north of the origin and follow a line that travels 3 meters north for every 7 meters it travels east until it intersects ditch Z.

- Ditch B is to run from the origin to ditch Z in such a way that it exactly bisects the area in the northeast quadrant that is south of both ditch Z and ditch A.

You are to provide the equations of the lines that the three ditches follow, as well as the exact location of the gates that will be installed where the ditches intersect one another. Be sure to provide explanations and organized work that will clearly display the desired information and assure the owners of Goodstuff that they will get exactly what they want.

Chapter Summary

CONCEPTS	REVIEW EXERCISES

3.1 Solution by Graphing

If a system of equations has at least one solution, the system is a **consistent system.** Otherwise, the system is an **inconsistent system.**

If the graphs of the equations of a system are distinct, the equations are **independent equations.** Otherwise, the equations are **dependent equations.**

Solve each system by the graphing method.

1. $\begin{cases} 2x + y = 11 \\ -x + 2y = 7 \end{cases}$

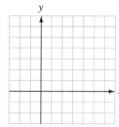

2. $\begin{cases} 3x + 2y = 0 \\ 2x - 3y = -13 \end{cases}$

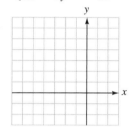

3. $\begin{cases} \dfrac{1}{2}x + \dfrac{1}{3}y = 2 \\ y = 6 - \dfrac{3}{2}x \end{cases}$

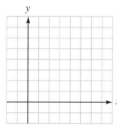

4. $\begin{cases} \dfrac{1}{3}x - \dfrac{1}{2}y = 1 \\ 6x - 9y = 2 \end{cases}$

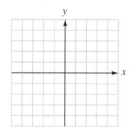

3.2 Solutions by Elimination

To solve a system by *substitution,* solve one equation for a variable. Then substitute the expression found for that variable into the other equation. Then solve for the other variable.

To solve a system by *addition,* combine the equations of the system in a way that will eliminate one of the variables.

Solve each system by substitution.

5. $\begin{cases} y = x + 4 \\ 2x + 3y = 7 \end{cases}$

6. $\begin{cases} y = 2x + 5 \\ 3x - 5y = -4 \end{cases}$

7. $\begin{cases} x + 2y = 11 \\ 2x - y = 2 \end{cases}$

8. $\begin{cases} 2x + 3y = -2 \\ 3x + 5y = -2 \end{cases}$

Solve each system by addition.

9. $\begin{cases} x + y = -2 \\ 2x + 3y = -3 \end{cases}$

10. $\begin{cases} 3x + 2y = 1 \\ 2x - 3y = 5 \end{cases}$

11. $\begin{cases} x + \dfrac{1}{2}y = 7 \\ -2x = 3y - 6 \end{cases}$

12. $\begin{cases} y = \dfrac{x - 3}{2} \\ x = \dfrac{2y + 7}{2} \end{cases}$

| 3.3 | Solutions of Three Equations in Three Variables |

A system of three linear equations in three variables can be solved using a combination of the addition and substitution methods.

Solve each system.

13. $\begin{cases} x + y + z = 6 \\ x - y - z = -4 \\ -x + y - z = -2 \end{cases}$

14. $\begin{cases} 2x + 3y + z = -5 \\ -x + 2y - z = -6 \\ 3x + y + 2z = 4 \end{cases}$

| 3.4 | Solution by Matrices |

A **matrix** is any rectangular array of numbers.

Systems of linear equations can be solved using matrices and the method of **Gaussian elimination.**

Solve each system by using matrices.

15. $\begin{cases} x + 2y = 4 \\ 2x - y = 3 \end{cases}$

16. $\begin{cases} x + y + z = 6 \\ 2x - y + z = 1 \\ 4x + y - z = 5 \end{cases}$

17. $\begin{cases} x + y = 3 \\ x - 2y = -3 \\ 2x + y = 4 \end{cases}$

18. $\begin{cases} x + 2y + z = 2 \\ 2x + 5y + 4z = 5 \end{cases}$

| 3.5 | Solution by Determinants |

A **determinant of a square matrix** is a number.

$\begin{vmatrix} a & b \\ c & d \end{vmatrix} = ad - bc$

$\begin{vmatrix} a_1 & b_1 & c_1 \\ a_2 & b_2 & c_2 \\ a_3 & b_3 & c_3 \end{vmatrix} = a_1 \begin{vmatrix} b_2 & c_2 \\ b_3 & c_3 \end{vmatrix}$

$- b_1 \begin{vmatrix} a_2 & c_2 \\ a_3 & c_3 \end{vmatrix} + c_1 \begin{vmatrix} a_2 & b_2 \\ a_3 & b_3 \end{vmatrix}$

Evaluate each determinant.

19. $\begin{vmatrix} 2 & 3 \\ -4 & 3 \end{vmatrix}$

20. $\begin{vmatrix} -3 & -4 \\ 5 & -6 \end{vmatrix}$

21. $\begin{vmatrix} -1 & 2 & -1 \\ 2 & -1 & 3 \\ 1 & -2 & 2 \end{vmatrix}$

22. $\begin{vmatrix} 3 & -2 & 2 \\ 1 & -2 & -2 \\ 2 & 1 & -1 \end{vmatrix}$

Use Cramer's rule to solve each system.

23. $\begin{cases} 3x + 4y = 10 \\ 2x - 3y = 1 \end{cases}$

24. $\begin{cases} 2x - 5y = -17 \\ 3x + 2y = 3 \end{cases}$

25. $\begin{cases} x + 2y + z = 0 \\ 2x + y + z = 3 \\ x + y + 2z = 5 \end{cases}$

26. $\begin{cases} 2x + 3y + z = 2 \\ x + 3y + 2z = 7 \\ x - y - z = -7 \end{cases}$

Chapter Test

1. Solve $\begin{cases} 2x + y = 5 \\ y = 2x - 3 \end{cases}$
by graphing.

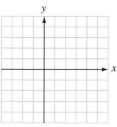

2. Use substitution to solve: $\begin{cases} 2x - 4y = 14 \\ x = -2y + 7 \end{cases}$.

3. Use addition to solve: $\begin{cases} 2x + 3y = -5 \\ 3x - 2y = 12 \end{cases}$.

4. Use any method to solve: $\begin{cases} \dfrac{x}{2} - \dfrac{y}{4} = -4 \\ x + y = -2 \end{cases}$.

Consider the system $\begin{cases} 3(x + y) = x - 3 \\ -y = \dfrac{2x + 3}{3} \end{cases}$.

5. Are the equations of the system dependent or independent?

6. Is the system consistent or inconsistent?

Use an elementary row operation to find the missing number in the second matrix.

7. $\begin{bmatrix} 1 & 2 & -1 \\ 2 & -2 & 3 \end{bmatrix}, \begin{bmatrix} 1 & 2 & -1 \\ -1 & -8 & \end{bmatrix}$

8. $\begin{bmatrix} -1 & 3 & 6 \\ 3 & -2 & 4 \end{bmatrix}, \begin{bmatrix} -1 & 3 & 6 \\ 5 & -8 & \end{bmatrix}$

Consider the system $\begin{cases} x + y + z = 4 \\ x + y - z = 6 \\ 2x - 3y + z = -1 \end{cases}$.

9. Write the augmented matrix that represents the system.

10. Write the coefficient matrix that represents the system.

Use matrices to solve each system.

11. $\begin{cases} x + y = 4 \\ 2x - y = 2 \end{cases}$

12. $\begin{cases} x + y = 2 \\ x - y = -4 \\ 2x + y = 1 \end{cases}$

Evaluate each determinant.

13. $\begin{vmatrix} 2 & -3 \\ 4 & 5 \end{vmatrix}$

14. $\begin{vmatrix} -3 & -4 \\ -2 & 3 \end{vmatrix}$

15. $\begin{vmatrix} 1 & 2 & 0 \\ 2 & 0 & 3 \\ 1 & -2 & 2 \end{vmatrix}$

16. $\begin{vmatrix} 2 & -1 & 1 \\ 3 & 1 & 0 \\ 0 & 1 & 2 \end{vmatrix}$

Consider the system $\begin{cases} x - y = -6 \\ 3x + y = -6 \end{cases}$, *which is to be solved with Cramer's rule.*

17. When solving for x, what is the numerator determinant? (**Don't evaluate it.**)

18. When solving for y, what is the denominator determinant? (**Don't evaluate it.**)

19. Solve the system for x.

20. Solve the system for y.

Consider the system $\begin{cases} x + y + z = 4 \\ x + y - z = 6 \\ 2x - 3y + z = -1 \end{cases}$.

21. Solve for x.

22. Solve for z.

4

Inequalities

🎧 **InfoTrac Project**

Do a subject guide search on "volume." Under "Volume (cubic content)," choose "periodicals." Find the article "Rules of thumb: oil flow, line volume, linefill in barrels, small hole leakage." In the formula for V (speed), "d sup 2" means d^2. If the pipe has an inner diameter of 22 inches, what is the maximum throughput if the speed is no more than 6.3 miles per hour? Now solve the equation for Q to answer the second part of the problem. What is the minimum speed a column of oil must travel through a pipe with an inner diameter of 14 inches to have a throughput of at least 185,000 barrels per day?

Complete this project after studying Section 4.1.

© Royalty-Free/CORBIS

Mathematics in Statistics

**Example 7
Section 4.1**

To estimate the mean property tax paid by homeowners living in Rockford, IL, a researcher selects a random sample of homeowners and computes the mean tax they pay. If the researcher knows that the standard deviation σ of all tax bills is $120, how large must the sample size be before the researcher can be 95% certain that the sample mean will be within $35 of the population mean?

We have previously considered linear equations, statements that two quantities are equal. In this chapter, we will consider mathematical statements that indicate that two quantities are not equal.

4.1	**Linear Inequalities**

▌ **Inequalities**   ▌ **Properties of Inequalities**   ▌ **Linear Inequalities**
▌ **Compound Inequalities Involving "And"**
▌ **Compound Inequalities Involving "Or"**   ▌ **Problem Solving**

Getting Ready   *Graph each set on the number line.*

1. $\{x \mid x > -3\}$ **2.** $\{x \mid x < 4\}$

3. $\{x \mid x \leq 5\}$ **4.** $\{x \mid x \geq -1\}$

In this section, we will review the basic ideas of inequalities, first discussed in Chapter 1. After discussing many properties of inequalities, we will solve linear inequalities and solve problems.

Inequalities

Inequalities are statements indicating that two quantities are unequal. Inequalities can be recognized because they contain one or more of the following symbols.

Inequality Symbols

Inequality	Read as	Example
$a \neq b$	"a is not equal to b."	$5 \neq 9$
$a < b$	"a is less than b."	$2 < 3$
$a > b$	"a is greater than b."	$7 > -5$
$a \leq b$	"a is less than or equal to b."	$-5 \leq 0$ or $-4 \leq -4$
$a \geq b$	"a is greater than or equal to b."	$9 \geq 6$ or $8 \geq 8$
$a \approx b$	"a is approximately equal to b."	$1.14 \approx 1.1$

By definition, $a < b$ means that "a is less than b," but it also means that $b > a$. If a is to the left of b on the number line, then $a < b$.

Properties of Inequalities

There are several basic properties of inequalities.

Trichotomy Property

For any real numbers a and b, exactly one of the following statements is true:

$$a < b, \qquad a = b, \qquad \text{or} \qquad a > b$$

The trichotomy property indicates that exactly one of the following statements is true about any two real numbers. Either

- the first is less than the second,
- the first is equal to the second, or
- the first is greater than the second.

Transitive Property

If a, b, and c are real numbers with $a < b$ and $b < c$, then $a < c$.

If a, b, and c are real numbers with $a > b$ and $b > c$, then $a > c$.

The first part of the transitive property indicates that

- If a first number is less than a second number and
- the second number is less than a third number, then
- the first number is less than the third.

The second part is similar, with the words "is greater than" substituted for "is less than."

Property 1 of Inequalities

Any real number can be added to (or subtracted from) both sides of an inequality to produce another inequality with the same direction.

To illustrate Property 1, we add 4 to both sides of the inequality $3 < 12$ to get

$$3 + 4 < 12 + 4$$
$$7 < 16$$

We note that the $<$ symbol is unchanged (has the same direction).

Subtracting 4 from both sides of $3 < 12$ does not change the direction of the inequality either.

$$3 - 4 < 12 - 4$$
$$-1 < 8$$

Property 2 of Inequalities

If both sides of an inequality are multiplied (or divided) by a positive number, another inequality results with the same direction as the original inequality.

To illustrate Property 2, we multiply both sides of the inequality $-4 < 6$ by 2 to get

$$2(-4) < 2(6)$$
$$-8 < 12$$

The $<$ symbol is unchanged.

Dividing both sides by 2 does not change the direction of the inequality either.

$$\frac{-4}{2} < \frac{6}{2}$$
$$-2 < 3$$

Property 3 of Inequalities

If both sides of an inequality are multiplied (or divided) by a negative number, another inequality results, but with the opposite direction from the original inequality.

To illustrate Property 3, we multiply both sides of the inequality $-4 < 6$ by -2 to get

$$-4 < 6$$
$$-2(-4) > -2(6) \qquad \text{Change} < \text{to} >.$$
$$8 > -12$$

Here, the $<$ symbol changes to a $>$ symbol.

Dividing both sides by -2 also changes the direction of the inequality.

$$-4 < 6$$
$$\frac{-4}{-2} > \frac{6}{-2}$$
$$2 > -3$$

 Comment Remember to change the direction of an inequality symbol every time you multiply or divide both sides by a negative number.

Linear Inequalities

Linear Inequalities

A **linear inequality** in x is any inequality that can be expressed in one of the following forms (with $a \neq 0$).

$$ax + c < 0 \qquad ax + c > 0 \qquad ax + c \leq 0 \qquad \text{or} \qquad ax + c \geq 0$$

We solve linear inequalities just as we solve linear equations, but with one exception. If we multiply or divide both sides by a negative number, we must change the direction of the inequality.

EXAMPLE 1 Solve each linear inequality and graph its solution set: **a.** $3(2x - 9) < 9$ and **b.** $-4(3x + 2) \leq 16$.

Solution   In each part, we use the same steps as for solving equations.

a. $3(2x - 9) < 9$

$\qquad 6x - 27 < 9$ Use the distributive property to remove parentheses.

$\qquad\quad 6x < 36$ Add 27 to both sides.

$\qquad\qquad x < 6$ Divide both sides by 6.

The solution set is $\{x | x < 6\}$, whose graph is shown in Figure 4-1(a). The parenthesis at 6 indicates that 6 is not included in the solution set.

b. $-4(3x + 2) \leq 16$

$\qquad -12x - 8 \leq 16$ Use the distributive property to remove parentheses.

$\qquad\quad -12x \leq 24$ Add 8 to both sides.

$\qquad\qquad x \geq -2$ Divide both sides by -12 and reverse the $\leq$ symbol.

The solution set is $\{x | x \geq -2\}$, whose graph is shown in Figure 4-1(b). The bracket at -2 indicates that -2 is included in the solution set.

$\qquad\qquad\qquad$ (a) $\qquad\qquad\qquad$ (b)

Figure 4-1

Self Check   Solve the linear inequality and graph its solution set: $-3(2x - 4) \leq 24$.   ∎

In Chapter 1, we saw that **interval notation** is another way to express the solution set of an inequality in one variable. Recall that this notation uses parentheses and brackets to indicate whether endpoints of an interval are included or excluded from a solution set.

An interval is called a (an)

- **unbounded interval** if it extends forever in one or more directions. See Figure 4-2(a).

- **open interval** if it is bounded and has no endpoints. See Figure 4-2(b).

- **half-open interval** if it is bounded and has one endpoint. See Figure 4-2(c).

- **closed interval** if it is bounded and has two endpoints. See Figure 4-2(d).

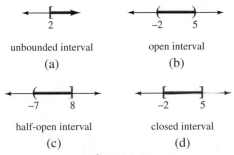

Figure 4-2

Table 4-1 shows the possible intervals that exist when *a* and *b* are real numbers.

Kind of interval	Set notation	Graph	Interval
Unbounded intervals	$\{x\|x > a\}$		(a, ∞)
	$\{x\|x < a\}$		$(-\infty, a)$
	$\{x\|x \geq a\}$		$[a, \infty)$
	$\{x\|x \leq a\}$		$(-\infty, a]$
Open intervals	$\{x\|a < x < b\}$		(a, b)
Half-open intervals	$\{x\|a \leq x < b\}$		$[a, b)$
	$\{x\|a < x \leq b\}$		$(a, b]$
Closed intervals	$\{x\|a \leq x \leq b\}$		$[a, b]$

Table 4-1

EXAMPLE 2 Solve each linear inequality and graph its solution set.

a. $5(x + 2) \geq 3x - 1$ and **b.** $\frac{2}{3}(x + 2) > \frac{4}{5}(x - 3).$

Solution In both parts, we use the same steps as for solving equations.

a. $5(x + 2) \geq 3x - 1$

$5x + 10 \geq 3x - 1$ Use the distributive property to remove parentheses.

$2x + 10 \geq -1$ Subtract $3x$ from both sides.

$2x \geq -11$ Subtract 10 from both sides.

$x \geq -\frac{11}{2}$ Divide both sides by 2.

The solution set is $\{x\|x \geq -\frac{11}{2}\}$, which is the interval $\left[-\frac{11}{2}, \infty\right)$. Its graph is shown in Figure 4-3(a).

b. $\frac{2}{3}(x + 2) > \frac{4}{5}(x - 3)$

$\mathbf{15} \cdot \frac{2}{3}(x + 2) > \mathbf{15} \cdot \frac{4}{5}(x - 3)$ Multiply both sides by 15, the LCD of 3 and 5.

$10(x + 2) > 12(x - 3)$ Simplify.

$10x + 20 > 12x - 36$ Use the distributive property to remove parentheses.

$-2x + 20 > -36$ Subtract $12x$ from both sides.

$-2x > -56$ Subtract 20 from both sides.

$x < 28$ Divide both sides by -2 and reverse the $>$ symbol.

The solution set is $\{x \mid x < 28\}$, which is the interval $(-\infty, 28)$. Its graph is shown in Figure 4-3(b).

(a)   (b)

Figure 4-3

Self Check   Solve the linear inequality and graph its solution set: $\frac{3}{2}(x + 2) < \frac{3}{5}(x - 3)$.

Compound Inequalities Involving "And"

To say that x is between -3 and 8, we write the inequality

$-3 < x < 8$ Read as "-3 is less than x and x is less than 8."

This inequality is called a **compound inequality,** because it is a combination of two inequalities:

$-3 < x$ and $x < 8$

The word *and* indicates that both inequalities are true at the same time.

Compound Inequalities   The inequality $c < x < d$ is equivalent to $c < x$ and $x < d$.

Comment   The inequality $c < x < d$ means $c < x$ and $x < d$. It does not mean $c < x$ or $x < d$.

EXAMPLE 3   Solve the inequality $-3 \le 2x + 5 < 7$ and graph its solution set.

Solution   This inequality means that $2x + 5$ is between -3 and 7. We can solve it by isolating x between the inequality symbols.

$-3 \le 2x + 5 < 7$
$-8 \le 2x < 2$ Subtract 5 from all three parts.
$-4 \le x < 1$ Divide all three parts by 2.

The solution set is $\{x \mid -4 \le x < 1\}$, or in interval notation $[-4, 1)$. The graph is shown in Figure 4-4.

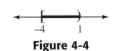

Figure 4-4

Self Check   Solve: $-5 \le 3x - 8 \le 7$. Give the result in interval notation and graph it.   ∎

EXAMPLE 4   Solve: $x + 3 < 2x - 1 < 4x - 3$.

Solution   Since it is impossible to isolate x between the inequality symbols, we solve each of the linear inequalities separately.

$$x + 3 < 2x - 1 \qquad \text{and} \qquad 2x - 1 < 4x - 3$$
$$4 < x \qquad\qquad\qquad 2 < 2x$$
$$x > 4 \qquad\qquad\qquad 1 < x$$
$$\qquad\qquad\qquad\qquad x > 1$$

Figure 4-5

Only those x where $x > 4$ and $x > 1$ are in the solution set. Since all numbers greater than 4 are also greater than 1, the solutions are the numbers x where $x > 4$. Thus, the solution set is the interval $(4, \infty)$, whose graph is shown in Figure 4-5.

Self Check   Solve: $x + 2 < 3x + 1 < 5x + 3$. ∎

Compound Inequalities Involving "Or"

To say that x is less than -5 or greater than 10, we write the inequality

$$x < -5 \qquad \text{or} \qquad x > 10 \qquad \text{Read as "}x\text{ is less than }-5\text{ or }x\text{ is greater than 10."}$$

The word *or* indicates that only one of the inequalities needs to be true to make the entire statement true.

EXAMPLE 5   Solve the compound inequality $x \leq -3$ or $x \geq 8$.

Solution   The word *or* in the statement $x \leq -3$ or $x \geq 8$ indicates that only one of the in-equalities needs to be true to make the statement true. The graph of the inequality is shown in Figure 4-6. The solution set is $(-\infty, -3] \cup [8, \infty)$.

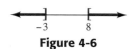

Figure 4-6

Self Check   Solve: $x < -5$ or $x \geq 3$. ∎

 Comment   In the statement $x \leq -3$ or $x \geq 8$, it is incorrect to string the inequali-ties together as $8 \leq x \leq -3$, because that would imply that $8 \leq -3$, which is false.

Problem Solving

EXAMPLE 6   Suppose that a long-distance telephone call costs 36¢ for the first three minutes and 11¢ for each additional minute. For how many minutes can a person talk for less than $2?

Analyze the problem   We can let x represent the total number of minutes that the call can last. Then the cost of the call will be 36¢ for the first three minutes plus 11¢ times the number of addi-tional minutes, where the number of additional minutes is $x - 3$ (the total number of minutes minus 3 minutes). The cost of the call is to be less than $2.

Form an inequality   With this information, we can form the inequality

The cost of the first three minutes	plus	the cost of the additional minutes	is less than	$2.
0.36	+	0.11(x − 3)	<	2

Solve the inequality   We can solve the inequality as follows:

$$0.36 + 0.11(x - 3) < 2$$

$$36 + 11(x - 3) < 200$$ To eliminate the decimal points, multiply both sides by 100.

$$36 + 11x - 33 < 200$$ Use the distributive property to remove parentheses.

$$11x + 3 < 200$$ Combine like terms.

$$11x < 197$$ Subtract 3 from both sides.

$$x < 17.\overline{90}$$ Divide both sides by 11.

State the conclusion   Since the phone company doesn't bill for part of a minute, the longest that the person can talk is 17 minutes.

Check the result   If the call lasts 17 minutes, the customer will be billed $0.36 + $0.11(14) = $1.90. If the call lasts 18 minutes, the customer will be billed $0.36 + $0.11(15) = $2.01.

Accent on Technology   SOLVING LINEAR INEQUALITIES

We can solve linear inequalities with a graphing approach. For example, to solve the inequality $3(2x - 9) < 9$, we can graph $y = 3(2x - 9)$ and $y = 9$ using window settings of $[-10, 10]$ for x and $[-10, 10]$ for y to get Figure 4-7(a). We trace to see that the graph of $y = 3(2x - 9)$ is below the graph of $y = 9$ for x-values in the interval $(-\infty, 6)$. See Figure 4-7(b). This interval is the solution, because in this interval, $3(2x - 9) < 9$.

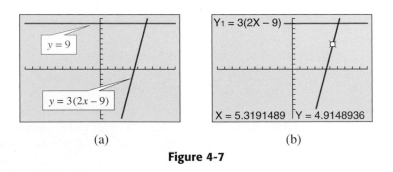

(a)   (b)

Figure 4-7

In statistics, researchers often estimate the mean of a population from the results of a random sample taken from the population.

EXAMPLE 7 A researcher wants to estimate the mean (average) real estate tax paid by home-owners living in Rockford, IL. To do so, he decides to select a *random sample* of homeowners and compute the mean tax paid by the homeowners in that sample. How large must the sample be for the researcher to be 95% certain that his computed sample mean will be within $35 of the true population mean—that is, within $35 of the mean tax paid by all homeowners in the city? Assume that the standard deviation σ of all tax bills in the city is $120.

Solution From elementary statistics, the researcher has the formula

$$\frac{3.84\sigma^2}{N} < E^2$$

where σ^2 is the square of the standard deviation, E is the maximum acceptable error, and N is the sample size. The researcher substitutes 120 for σ and 35 for E in the previous formula and solves for N.

$$\frac{3.84(\mathbf{120})^2}{N} < \mathbf{35}^2$$

$$\frac{55{,}296}{N} < 1{,}225 \qquad \text{Simplify.}$$

$$55{,}296 < 1{,}225N \qquad \text{Multiply both sides by } N.$$

$$45.13959184 < N \qquad \text{Divide both sides by 1,225.}$$

To be 95% certain that the sample mean will be within $35 of the true population mean, the researcher must sample more than 45.13959184 homeowners. Thus, the sample must contain at least 46 homeowners. ∎

Self Check Answers

1. $\{x \mid x \geq -2\}$ ⟵————[————⟶ **2.** $\left(-\infty, -\frac{16}{3}\right)$ ⟵————)————⟶ **3.** $[1, 5]$ ⟵———[———]———⟶
　　　　　　　　　　　　　　　　 -2　　　　　　　　　　　　　　　　　　 $-16/3$　　　　　　　　　　　　　　 1　　5

4. $\left(\frac{1}{2}, \infty\right)$ ⟵————(————⟶ **5.** $(-\infty, -5) \cup [3, \infty)$ ⟵——)——[——⟶
　　　　　　　　　 $1/2$　　　　　　　　　　　　　　　　　　　　　　　　 -5　　3

Orals *Write the solution set of each inequality in both set notation and interval notation.*

1. $2x < 4$ 　　　　　　　　　　　　　　**2.** $3x \geq 9$

3. $-3x > 12$ 　　　　　　　　　　　　 **4.** $-2 < 2x \leq 8$

4.1 EXERCISES 🔊

REVIEW *Simplify each expression.*

1. $\left(\dfrac{t^3 t^5 t^{-6}}{t^2 t^{-4}}\right)^{-3}$ 　　　　　**2.** $\left(\dfrac{a^{-2} b^3 a^5 b^{-2}}{a^6 b^{-5}}\right)^{-4}$

3. A man invests $1,200 in baking equipment to make pies. Each pie requires $3.40 in ingredients. If the man can sell all the pies he can make for $5.95 each, how many pies will he have to make to earn a profit?

4. A woman invested $15,000, part at 7% annual interest and the rest at 8%. If she earned $2,200 in income over a two-year period, how much did she invest at 7%?

VOCABULARY AND CONCEPTS *Fill in the blanks.*

5. The symbol for "is not equal to" is ＿＿.

6. The symbol for "is greater than" is ＿＿.

7. The symbol for "is less than" is ＿＿.

8. The symbol for "is less than or equal to" is ___.
9. The symbol for "is greater than or equal to" is ___.
10. If a and b are two numbers, then $a < b$, _____, or _____.
11. If $a < b$ and $b < c$, then _____.
12. If both sides of an inequality are multiplied by a _____ number, the direction of the inequality remains the same.
13. If both sides of an inequality are divided by a negative number, the direction of the inequality symbol must be _____.
14. $3x + 2 > 7$ is an example of a _____ inequality.
15. The inequality $c < x < d$ is equivalent to _____ and _____.
16. The word "or" between two inequality statements indicates that only ___ of the inequalities needs to be true for the entire statement to be true.
17. The interval $(2, 5)$ is called an _____ interval.
18. The interval $[-5, 3]$ is called a _____ interval.

PRACTICE *Solve each inequality. Give each result in set notation and graph it.*

19. $x + 4 < 5$
20. $x - 5 > 2$

21. $2x + 3 < 9$
22. $5x - 1 \geq 19$

23. $-3x - 1 \leq 5$
24. $-2x + 6 \geq 16$

25. $5x - 3 > 7$
26. $7x - 9 < 5$

27. $-4 < 2x < 8$
28. $-3 \leq 3x < 12$

Solve each inequality. Give each result in interval notation and graph it.

29. $8x + 30 > -2x$
30. $5x - 24 \leq 6$

31. $-3x + 14 \geq 20$
32. $-\frac{1}{2}x + 4 < 32$

33. $4(x + 5) < 12$
34. $-5(x - 2) \geq 15$

35. $3(z - 2) \leq 2(z + 7)$
36. $5(3 + z) > -3(z + 3)$

37. $-11(2 - b) < 4(2b + 2)$

38. $-9(h - 3) + 2h \leq 8(4 - h)$

39. $\frac{1}{2}y + 2 \geq \frac{1}{3}y - 4$
40. $\frac{1}{4}x - \frac{1}{3} \leq x + 2$

41. $\frac{2}{3}x + \frac{3}{2}(x - 5) \leq x$

42. $\frac{5}{9}(x + 3) - \frac{4}{3}(x - 3) \geq x - 1$

43. $-2 < -b + 3 < 5$

44. $2 < -t - 2 < 9$

45. $15 > 2x - 7 > 9$

46. $25 > 3x - 2 > 7$

47. $-6 < -3(x - 4) \leq 24$

48. $-4 \leq -2(x + 8) < 8$

49. $0 \geq \frac{1}{2}x - 4 > 6$

50. $-6 \le \frac{1}{3}a + 1 < 0$

51. $0 \le \frac{4 - x}{3} \le 2$

52. $-2 \le \frac{5 - 3x}{2} \le 2$

53. $x + 3 < 3x - 1 < 2x + 2$

54. $x - 1 \le 2x + 4 \le 3x - 1$

55. $4x \ge -x + 5 \ge 3x - 4$

56. $x + 2 < -\frac{1}{3}x < \frac{1}{2}x$

57. $5(x + 1) \le 4(x + 3) < 3(x - 1)$

58. $-5(2 + x) < 4x + 1 < 3x$

59. $3x + 2 < 8$ or $2x - 3 > 11$

60. $3x + 4 < -2$ or $3x + 4 > 10$

61. $-4(x + 2) \ge 12$ or $3x + 8 < 11$

62. $5(x - 2) \ge 0$ and $-3x < 9$

63. $x < -3$ and $x > 3$

64. $x < 3$ or $x > -3$

APPLICATIONS *Use a calculator to help solve each problem.*

65. Renting a rototiller The cost of renting a rototiller is \$15.50 for the first hour and \$7.95 for each additional hour. How long can a person have the rototiller if the cost is to be less than \$50?

66. Renting a truck How long can a person rent the truck described in the ad if the cost is to be less than \$110?

ACTION
Truck Rental
Big 22 ft Truck
Only \$29.95 for one hour.
\$8.95 for each extra hour.

67. Boating A speedboat is rated to carry 750 lb. If the driver weighs 205 lb and a passenger weighs 175 lb, how many children can safely ride along if they average 90 lb each?

68. Elevators An elevator is rated to carry 900 lb. How many boxes of books can the elevator safely carry if each box weighs 80 lb and the operator weighs 165 lb?

69. Investing money If a woman invests \$10,000 at 8% annual interest, how much more must she invest at 9% so that her annual income will exceed \$1,250?

70. Investing money If a man invests \$8,900 at 5.5% annual interest, how much more must he invest at 8.75% so that his annual income will be more than \$1,500?

71. Buying compact discs A student can afford to spend up to \$330 on a stereo system and some compact discs. If the stereo costs \$175 and the discs are \$8.50 each, find the greatest number of discs he can buy.

72. Buying a computer A student who can afford to spend up to \$2,000 sees the following ad. If she buys a computer, find the greatest number of CD-ROMs that she can buy.

Big Sale!!!!
$1,695.95
All CD-ROMs
$19.95

73. **Averaging grades** A student has scores of 70, 77, and 85 on three exams. What score is needed on a fourth exam to make an average of 80 or better?

74. **Averaging grades** A student has scores of 70, 79, 85, and 88 on four exams. What score does she need on the fifth exam to keep her average above 80?

75. **Planning a work schedule** Nguyen can earn $5 an hour for working at the college library and $9 an hour for construction work. To save time for study, he wants to limit his work to 20 hours a week but still earn more than $125. How many hours can he work at the library?

76. **Scheduling equipment** An excavating company charges $300 an hour for the use of a backhoe and $500 an hour for the use of a bulldozer. (Part of an hour counts as a full hour.) The company employs one operator for 40 hours per week. If the company wants to take in at least $18,500 each week, how many hours per week can it schedule the operator to use a backhoe?

77. **Choosing a medical plan** A college provides its employees with a choice of the two medical plans shown in the table. For what size hospital bills is Plan 2 better than Plan 1? (*Hint:* The cost to the employee includes both the deductible payment and the employee's insurance copayment.)

Plan 1	Plan 2
Employee pays $100	Employee pays $200
Plan pays 70% of the rest	Plan pays 80% of the rest

78. **Choosing a medical plan** To save costs, the college in Exercise 77 raised the employee deductible, as shown in the following table. For what size hospital bills is Plan 2 better than Plan 1? (*Hint:* The cost to the employee includes both the deductible payment and the employee's insurance copayment.)

Plan 1	Plan 2
Employee pays $200	Employee pays $400
Plan pays 70% of the rest	Plan pays 80% of the rest

Use a graphing calculator to solve each inequality.

79. $2x + 3 < 5$

80. $3x - 2 > 4$

81. $5x + 2 \geq -18$

82. $3x - 4 \leq 20$

83. **Choosing sample size** How large would the sample have to be for the researcher in Example 7 to be 95% certain that the true population mean would be within $20 of the sample mean?

84. **Choosing sample size** How large would the sample have to be for the researcher in Example 7 to be 95% certain that the true population mean would be within $10 of the sample mean?

WRITING

85. The techniques for solving linear equations and linear inequalities are similar, yet different. Explain.

86. Explain the concepts of *absolute* inequality and *conditional* inequality.

SOMETHING TO THINK ABOUT

87. If $x > -3$, must it be true that $x^2 > 9$?

88. If $x > 2$, must it be true that $x^2 > 4$?

89. Which of these relations is transitive?
    a. $=$ **b.** $\leq$ **c.** $\not\geq$ **d.** $\neq$

90. The following solution is not correct. Why?

$$\frac{1}{3} > \frac{1}{x}$$

$$3x\left(\frac{1}{3}\right) > 3x\left(\frac{1}{x}\right) \qquad \text{Multiply both sides by } 3x.$$

$$x > 3 \qquad \text{Simplify.}$$

4.2 Equations and Inequalities with Absolute Values

■ **Absolute Value** ■ **Absolute Value Functions**
■ **Equations of the Form $|x| = k$**
■ **Equations with Two Absolute Values**
■ **Inequalities of the Form $|x| < k$**
■ **Inequalities of the Form $|x| > k$**

Getting Ready *Find each value. In Problems 3–4, assume that $x = -5$.*

   1. $-(-5)$ **2.** -0 **3.** $-(x - 5)$ **4.** $-(-x)$

Solve each inequality and give the result in interval notation.

   5. $-4 < x + 1 < 5$ **6.** $x < -3$ or $x > 3$

In this section, we review the definition of absolute value and the graphs of absolute value functions. We will then show how to solve equations and inequalities that contain absolute values.

Absolute Value

Recall the definition of the absolute value of x.

Absolute Value

For any real number x,

    If $x \geq 0$, then $|x| = x$.
    If $x < 0$, then $|x| = -x$.

This definition associates a nonnegative real number with any real number.

- If $x \geq 0$, then x (which is positive or 0) is its own absolute value.
- If $x < 0$, then $-x$ (which is positive) is the absolute value.

Either way, $|x|$ is positive or 0:

    $|x| \geq 0$ **for all real numbers x**

EXAMPLE 1 Find each absolute value: **a.** $|9|$, **b.** $|-5|$, **c.** $|0|$, and **d.** $|2 - \pi|$.

Solution **a.** Since $9 \geq 0$, 9 is its own absolute value: $|\mathbf{9}| = \mathbf{9}$.

b. Since $-5 < 0$, the negative of -5 is the absolute value:

    $|\mathbf{-5}| = -(-5) = \mathbf{5}$

c. Since $0 \geq 0$, 0 is its own absolute value: $|\mathbf{0}| = \mathbf{0}$.

d. Since $\pi \approx 3.14$, it follows that $2 - \pi < 0$. Thus,

$$|2 - \pi| = -(2 - \pi) = \pi - 2$$

Self Check   Find each absolute value: **a.** $|-3|$ and   **b.** $|\pi - 2|$. ∎

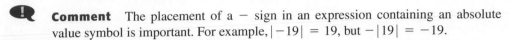

Comment   The placement of a $-$ sign in an expression containing an absolute value symbol is important. For example, $|-19| = 19$, but $-|19| = -19$.

Absolute Value Functions

In Section 2.5, we graphed the absolute value function $y = f(x) = |x|$ and considered many translations of its graph. The work in Example 2 reviews these concepts.

EXAMPLE 2   Graph the function $y = f(x) = |x - 1| + 3$.

Solution   To graph this function, we translate the graph of $y = f(x) = |x|$. We move it 1 unit to the right and 3 units up, as shown in Figure 4-8.

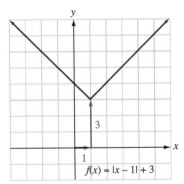

$$f(x) = |x - 1| + 3$$

Figure 4-8

Self Check   Graph: $y = f(x) = |x + 2| - 3$. ∎

We also saw that the graph of $y = f(x) = -|x|$ is the same as the graph of $y = f(x) = |x|$ reflected about the x-axis, as shown in Figure 4-9.

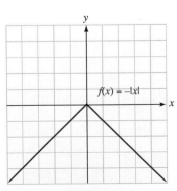

$$f(x) = -|x|$$

Figure 4-9

EXAMPLE 3   Graph the absolute value function $y = f(x) = -|x - 1| + 3$.

Solution   To graph this function, we translate the graph of $y = f(x) = -|x|$. We move it 1 unit to the right and 3 units up, as shown in Figure 4-10.

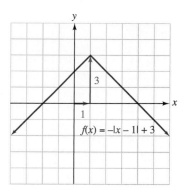

Figure 4-10

Self Check   Graph: $y = f(x) = -|x + 2| - 3$.

Equations of the Form $|x| = k$

In the equation $|x| = 5$, x can be either 5 or -5, because $|5| = 5$ and $|-5| = 5$. In the equation $|x| = 8$, x can be either 8 or -8. In general, the following is true.

Absolute Value Equations

> If $k > 0$, then
>
> $$|x| = k \quad \text{is equivalent to} \quad x = k \quad \text{or} \quad x = -k$$

The absolute value of a number represents the distance on the number line from a point to the origin. The solutions of $|x| = k$ are the coordinates of the two points that lie exactly k units from the origin. See Figure 4-11.

Figure 4-11

The equation $|x - 3| = 7$ indicates that a point on the number line with a coordinate of $x - 3$ is 7 units from the origin. Thus, $x - 3$ can be either 7 or -7.

$$x - 3 = 7 \quad \text{or} \quad x - 3 = -7$$
$$x = 10 \quad | \quad \qquad x = -4$$

Comment   The equation $|x - 3| = 7$ can also be solved using the definition of absolute value.

If $x - 3 \geq 0$, then
$|x - 3| = x - 3$, and
we have

$x - 3 = 7$

$x = 10$

If $x - 3 < 0$, then
$|x - 3| = -(x - 3)$, and
we have

$-(x - 3) = 7$

$-x + 3 = 7$

$-x = 4$

$x = -4$

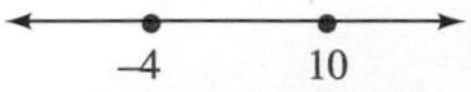

Figure 4-12

The solutions of $|x - 3| = 7$ are 10 and -4, as shown in Figure 4-12.

EXAMPLE 4   Solve the equation: $|3x - 2| = 5$.

Solution   We can write $|3x - 2| = 5$ as

$3x - 2 = 5 \qquad \text{or} \qquad 3x - 2 = -5$

and solve each equation for x:

$3x - 2 = 5 \quad \text{or} \quad 3x - 2 = -5$

$3x = 7 \qquad\qquad 3x = -3$

$x = \dfrac{7}{3} \qquad\qquad x = -1$

Verify that both solutions check.

Self Check   Solve: $|2x - 3| = 7$.

EXAMPLE 5   Solve the equation: $\left| \dfrac{2}{3}x + 3 \right| + 4 = 10$.

Solution   We subtract 4 from both sides to isolate the absolute value on the left-hand side.

$$\left| \frac{2}{3}x + 3 \right| + 4 = 10$$

(1) $$\left| \frac{2}{3}x + 3 \right| = 6 \qquad \text{Subtract 4 from both sides.}$$

We can now write Equation 1 as

$$\frac{2}{3}x + 3 = 6 \qquad \text{or} \qquad \frac{2}{3}x + 3 = -6$$

and solve each equation for x:

$\dfrac{2}{3}x + 3 = 6 \quad \text{or} \quad \dfrac{2}{3}x + 3 = -6$

$\dfrac{2}{3}x = 3 \qquad\qquad \dfrac{2}{3}x = -9$

$2x = 9 \qquad\qquad 2x = -27$

$x = \dfrac{9}{2} \qquad\qquad x = -\dfrac{27}{2}$

Verify that both solutions check.

Self Check   Solve: $\left| \dfrac{3}{5}x - 2 \right| - 3 = 4$.

    Comment   Since the absolute value of a quantity cannot be negative, equations such as $\left| 7x + \frac{1}{2} \right| = -4$ have no solution. Since there are no solutions, their solution sets are empty. Recall that an **empty set** is denoted by the symbol $\varnothing$.

EXAMPLE 6   Solve the equation: $\left| \frac{1}{2}x - 5 \right| - 4 = -4$.

Solution   We first isolate the absolute value on the left-hand side.

$$\left| \frac{1}{2}x - 5 \right| - 4 = -4$$

$$\left| \frac{1}{2}x - 5 \right| = 0 \qquad \text{Add 4 to both sides.}$$

Since 0 is the only number whose absolute value is 0, the binomial $\frac{1}{2}x - 5$ must be 0, and we have

$$\frac{1}{2}x - 5 = 0$$

$$\frac{1}{2}x = 5 \qquad \text{Add 5 to both sides.}$$

$$x = 10 \qquad \text{Multiply both sides by 2.}$$

Verify that 10 satisfies the original equation.

Self Check   Solve: $\left| \frac{2}{3}x + 4 \right| + 4 = 5$.

Equations with Two Absolute Values

The equation $|a| = |b|$ is true when $a = b$ or when $a = -b$. For example,

$$|3| = |3| \quad \text{or} \quad |3| = |-3|$$
$$3 = 3 \quad | \quad 3 = 3$$

In general, the following statement is true.

Equations with Two Absolute Values

If a and b represent algebraic expressions, the equation $|a| = |b|$ is equivalent to the pair of equations

$$a = b \qquad \text{or} \qquad a = -b$$

EXAMPLE 7   Solve the equation: $|5x + 3| = |3x + 25|$.

Solution   This equation is true when $5x + 3 = 3x + 25$, or when $5x + 3 = -(3x + 25)$. We solve each equation for x.

$$5x + 3 = 3x + 25 \quad \text{or} \quad 5x + 3 = -(3x + 25)$$
$$2x = 22 \qquad\qquad 5x + 3 = -3x - 25$$
$$x = 11 \qquad\qquad\qquad 8x = -28$$
$$x = -\frac{28}{8}$$
$$x = -\frac{7}{2}$$

Verify that both solutions check.

Self Check   Solve: $|2x - 3| = |4x + 9|$.   ▮

Inequalities of the Form $|x| < k$

The inequality $|x| < 5$ indicates that a point with a coordinate of x is less than 5 units from the origin. (See Figure 4-13.) Thus, x is between -5 and 5, and

$$|x| < 5 \qquad \text{is equivalent to} \qquad -5 < x < 5$$

The solution set of the inequality $|x| < k \ (k > 0)$ includes the coordinates of the points on the number line that are less than k units from the origin. (See Figure 4-14.)

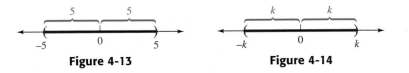

Figure 4-13   **Figure 4-14**

Solving $|x| < k$ and $|x| \leq k$

If $k > 0$, then

$$|x| < k \qquad \text{is equivalent to} \qquad -k < x < k$$
$$|x| \leq k \qquad \text{is equivalent to} \qquad -k \leq x \leq k \quad (k \geq 0)$$

EXAMPLE 8   Solve the inequality: $|2x - 3| < 9$.

Solution   Since $|2x - 3| < 9$ is equivalent to $-9 < 2x - 3 < 9$, we proceed as follows:

$$-9 < 2x - 3 < 9$$
$$-6 < 2x < 12 \qquad \text{Add 3 to all three parts.}$$
$$-3 < x < 6 \qquad \text{Divide all parts by 2.}$$

Any number between -3 and 6, not including either -3 or 6, is in the solution set. This is the interval $(-3, 6)$, whose graph is shown in Figure 4-15.

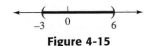

Figure 4-15

Self Check   Solve: $|3x + 2| < 4$.   ▮

Comment   The inequality $|2x - 3| < 9$ can also be solved using the definition of absolute value.

If $2x - 3 \geq 0$, then	If $2x - 3 < 0$, then
$\|2x - 3\| = 2x - 3$,	$\|2x - 3\| = -(2x - 3)$,
and we have	and we have

$$2x - 3 < 9 \qquad\qquad -(2x - 3) < 9$$
$$2x < 12 \qquad\qquad -2x + 3 < 9$$
$$x < 6 \qquad\qquad -2x < 6$$
$$\qquad\qquad\qquad x > -3$$

These results can be written as $-3 < x < 6$.

EXAMPLE 9   Solve the inequality: $|2 - 3x| \leq 5$.

Solution   Since $|2 - 3x| \leq 5$ is equivalent to $-5 \leq 2 - 3x \leq 5$, we proceed as follows:

$$-5 \leq 2 - 3x \leq 5$$

$$-7 \leq -3x \leq 3 \qquad\qquad \text{Subtract 2 from all parts.}$$

$$\frac{7}{3} \geq x \geq -1 \qquad\qquad \text{Divide all three parts by } -3 \text{ and reverse the directions of the inequality symbols.}$$

$$-1 \leq x \leq \frac{7}{3} \qquad\qquad \text{Write an equivalent inequality with the inequality symbols pointing in the opposite direction.}$$

The solution set is the interval $\left[-1, \frac{7}{3}\right]$, whose graph is shown in Figure 4-16.

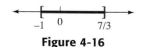

Figure 4-16

Self Check   Solve: $|3 - 2x| \leq 5$.   ∎

Inequalities of the Form $|x| > k$

The inequality $|x| > 5$ indicates that a point with a coordinate of x is more than 5 units from the origin. See Figure 4-17. Thus, $x < -5$ or $x > 5$.

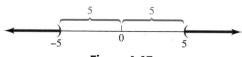

Figure 4-17

In general, the inequality $|x| > k$ can be interpreted to mean that a point with coordinate x is more than k units from the origin. (See Figure 4-18.)

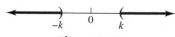

Figure 4-18

Thus,

$$|x| > k \qquad \text{is equivalent to} \qquad x < -k \qquad \text{or} \qquad x > k$$

The *or* indicates an either/or situation. To be in the solution set, x needs to satisfy only one of the two conditions.

Solving $|x| > k$ and $|x| \geq k$

If $k \geq 0$, then

$$|x| > k \qquad \text{is equivalent to} \qquad x < -k \qquad \text{or} \qquad x > k$$
$$|x| \geq k \qquad \text{is equivalent to} \qquad x \leq -k \qquad \text{or} \qquad x \geq k$$

EXAMPLE 10   Solve the inequality: $|5x - 10| > 20$.

Solution   We write the inequality as two separate inequalities and solve each one for x.

$$|5x - 10| > 20 \qquad \text{is equivalent to} \qquad 5x - 10 < -20 \quad \text{or} \quad 5x - 10 > 20$$

$$
\begin{array}{lcl}
5x - 10 < -20 & \text{or} & 5x - 10 > 20 \\
5x < -10 & & 5x > 30 \qquad \text{Add 10 to both sides.} \\
x < -2 & & x > 6 \qquad \text{Divide both sides by 5.}
\end{array}
$$

Thus, x is either less than -2 or greater than 6:

$$x < -2 \qquad \text{or} \qquad x > 6$$

This is the interval $(-\infty, -2) \cup (6, \infty)$, whose graph appears in Figure 4-19.

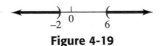

Figure 4-19

Self Check   Solve: $|3x + 4| > 13$.

   Comment   The inequality $|5x - 10| > 20$ can also be solved using the definition of absolute value.

If $5x - 10 \geq 0$, then   If $5x - 10 < 0$, then
$|5x - 10| = 5x - 10$, $|5x - 10| = -(5x - 10)$,
and we have   and we have

$$
\begin{array}{ll}
5x - 10 > 20 & \qquad\qquad -(5x - 10) > 20 \\
5x > 30 & \qquad\qquad -5x + 10 > 20 \\
x > 6 & \qquad\qquad -5x > 10 \\
& \qquad\qquad x < -2
\end{array}
$$

Thus, $x < -2$ or $x > 6$.

EXAMPLE 11   Solve the inequality: $\left| \dfrac{3 - x}{5} \right| \geq 6$.

Solution   We write the inequality as two separate inequalities:

$$\left|\frac{3-x}{5}\right| \geq 6 \quad \text{is equivalent to} \quad \frac{3-x}{5} \leq -6 \quad \text{or} \quad \frac{3-x}{5} \geq 6$$

Then we solve each one for x:

$$\frac{3-x}{5} \leq -6 \quad \text{or} \quad \frac{3-x}{5} \geq 6$$

$3 - x \leq -30$	$3 - x \geq 30$	Multiply both sides by 5.
$-x \leq -33$	$-x \geq 27$	Subtract 3 from both sides.
$x \geq 33$	$x \leq -27$	Divide both sides by -1 and reverse the direction of the inequality symbol.

The solution set is $(-\infty, -27] \cup [33, \infty)$, whose graph appears in Figure 4-20.

Figure 4-20

Self Check   Solve: $\left|\dfrac{3-x}{6}\right| \geq 5$.

EXAMPLE 12   Solve the inequality: $\left|\dfrac{2}{3}x - 2\right| - 3 > 6.$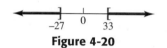

Solution   We begin by adding 3 to both sides to isolate the absolute value on the left-hand side. We then proceed as follows:

$$\left|\frac{2}{3}x - 2\right| - 3 > 6$$

$$\left|\frac{2}{3}x - 2\right| > 9 \qquad \text{Add 3 to both sides to isolate the absolute value.}$$

$$\frac{2}{3}x - 2 < -9 \quad \text{or} \quad \frac{2}{3}x - 2 > 9$$

$\dfrac{2}{3}x < -7$	$\dfrac{2}{3}x > 11$	Add 2 to both sides.
$2x < -21$	$2x > 33$	Multiply both sides by 3.
$x < -\dfrac{21}{2}$	$x > \dfrac{33}{2}$	Divide both sides by 2.

The solution set is $\left(-\infty, -\frac{21}{2}\right) \cup \left(\frac{33}{2}, \infty\right)$, whose graph appears in Figure 4-21.

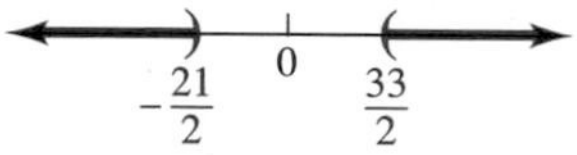

Figure 4-21

Self Check   Solve: $\left|\dfrac{3}{4}x + 2\right| - 1 > 3$.

EXAMPLE 13   Solve the inequality: $|3x - 5| \geq -2$.

Solution   Since the absolute value of any number is nonnegative, and since any nonnegative number is greater than -2, the inequality is true for all x. The solution set is $(-\infty, \infty)$, whose graph appears in Figure 4-22.

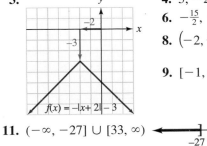

Figure 4-22

Self Check   Solve: $|3x - 5| \leq -2$, if possible.

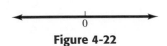

Accent on Technology   SOLVING ABSOLUTE VALUE INEQUALITIES

We can solve many absolute value inequalities by a graphing method. For example, to solve $|2x - 3| < 9$, we graph the equations $y = |2x - 3|$ and $y = 9$ on the same coordinate system. If we use window settings of $[-5, 15]$ for x and $[-5, 15]$ for y, we get the graph shown in Figure 4-23.

The inequality $|2x - 3| < 9$ will be true for all x-coordinates of points that lie on the graph of $y = |2x - 3|$ and below the graph of $y = 9$. By using the trace feature, we can see that these values of x are in the interval $(-3, 6)$.

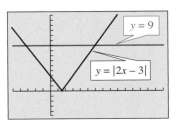

Figure 4-23

Self Check Answers

1. a. 3,   **b.** $\pi - 2$ **2.**

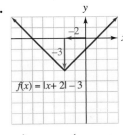

$f(x) = |x + 2| - 3$

3.

$f(x) = -|x + 2| - 3$

4. $5, -2$ **5.** $15, -\frac{25}{3}$

6. $-\frac{15}{2}, -\frac{9}{2}$ **7.** $-1, -6$

8. $\left(-2, \frac{2}{3}\right)$

9. $[-1, 4]$

10. $\left(-\infty, -\frac{17}{3}\right) \cup (3, \infty)$

11. $(-\infty, -27] \cup [33, \infty)$

12. $(-\infty, -8) \cup \left(\frac{8}{3}, \infty\right)$

13. no solution

Orals   *Find each absolute value.*

1. $|-5|$ **2.** $-|5|$ **3.** $-|-6|$ **4.** $-|4|$

Solve each equation or inequality.

5. $|x| = 8$ **6.** $|x| = -5$

7. $|x| < 8$ **8.** $|x| > 8$

9. $|x| \geq 4$ **10.** $|x| \leq 7$

4.2   EXERCISES

REVIEW   *Solve each equation or formula.*

1. $3(2a - 1) = 2a$

2. $\dfrac{t}{6} - \dfrac{t}{3} = -1$

3. $\dfrac{5x}{2} - 1 = \dfrac{x}{3} + 12$

4. $4b - \dfrac{b + 9}{2} = \dfrac{b + 2}{5} - \dfrac{8}{5}$

5. $A = p + prt$ for t

6. $P = 2w + 2l$ for l

VOCABULARY AND CONCEPTS   *Fill in the blanks.*

7. If $x \geq 0, |x| = $ ___.

8. If $x < 0, |x| = $ ___.

9. $|x| \geq $ ___ for all real numbers x.

10. The graph of the function $y = f(x) = |x - 2| - 4$ is the same as the graph of $y = f(x) = |x|$ except that it has been translated 2 units to the _____ and 4 units _____.

11. The graph of $y = f(x) = -|x|$ is the same as the graph of $y = f(x) = |x|$ except that it has been _____ about the x-axis.

12. If $k > 0$, then $|x| = k$ is equivalent to _____.

13. $|a| = |b|$ is equivalent to _____.

14. If $k > 0$, then $|x| < k$ is equivalent to _____.

15. If $k \geq 0$, then $|x| \geq k$ is equivalent to _____.

16. The equation $|x - 4| < -5$ has ___ solutions.

PRACTICE   *Find the value of each expression.*

17. $|8|$ **18.** $|-18|$

19. $-|2|$ **20.** $-|-20|$

21. $-|-30|$ **22.** $-|25|$

23. $|\pi - 4|$ **24.** $|2\pi - 4|$

Graph each absolute value function.

25. $f(x) = |x| - 2$ **26.** $f(x) = -|x| + 1$

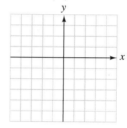 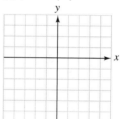

27. $f(x) = -|x + 4|$ **28.** $f(x) = |x - 1| + 2$

 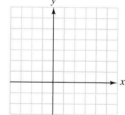

Solve each equation, if possible.

29. $|x| = 4$ **30.** $|x| = 9$

31. $|x - 3| = 6$ **32.** $|x + 4| = 8$

33. $|2x - 3| = 5$ **34.** $|4x - 4| = 20$

35. $|3x + 2| = 16$ **36.** $|5x - 3| = 22$

37. $\left| \dfrac{7}{2}x + 3 \right| = -5$ **38.** $|2x + 10| = 0$

39. $\left| \dfrac{x}{2} - 1 \right| = 3$ **40.** $\left| \dfrac{4x - 64}{4} \right| = 32$

41. $|3 - 4x| = 5$ **42.** $|8 - 5x| = 18$

43. $|3x + 24| = 0$ **44.** $|x - 21| = -8$

45. $\left|\dfrac{3x + 48}{3}\right| = 12$ **46.** $\left|\dfrac{x}{2} + 2\right| = 4$

47. $|x + 3| + 7 = 10$ **48.** $|2 - x| + 3 = 5$

49. $\left|\dfrac{3}{5}x - 4\right| - 2 = -2$ **50.** $\left|\dfrac{3}{4}x + 2\right| + 4 = 4$

51. $|2x + 1| = |3x + 3|$ **52.** $|5x - 7| = |4x + 1|$

53. $|3x - 1| = |x + 5|$ **54.** $|3x + 1| = |x - 5|$

55. $|2 - x| = |3x + 2|$ **56.** $|4x + 3| = |9 - 2x|$

57. $\left|\dfrac{x}{2} + 2\right| = \left|\dfrac{x}{2} - 2\right|$ **58.** $|7x + 12| = |x - 6|$

59. $\left|x + \dfrac{1}{3}\right| = |x - 3|$ **60.** $\left|x - \dfrac{1}{4}\right| = |x + 4|$

61. $|3x + 7| = -|8x - 2|$

62. $-|17x + 13| = |3x - 14|$

Solve each inequality. Write the solution set in interval notation and graph it.

63. $|2x| < 8$ **64.** $|3x| < 27$

65. $|x + 9| \leq 12$ **66.** $|x - 8| \leq 12$

67. $|3x + 2| \leq -3$ **68.** $|3x - 2| < 10$

69. $|4x - 1| \leq 7$ **70.** $|5x - 12| < -5$

71. $|3 - 2x| < 7$ **72.** $|4 - 3x| \leq 13$

73. $|5x| > 5$ **74.** $|7x| > 7$

75. $|x - 12| > 24$ **76.** $|x + 5| \geq 7$

77. $|3x + 2| > 14$ **78.** $|2x - 5| > 25$

79. $|4x + 3| > -5$ **80.** $|4x + 3| > 0$

81. $|2 - 3x| \geq 8$ **82.** $|-1 - 2x| > 5$

83. $-|2x - 3| < -7$ **84.** $-|3x + 1| < -8$

85. $|8x - 3| > 0$ **86.** $|7x + 2| > -8$

87. $\left|\dfrac{x - 2}{3}\right| \leq 4$ **88.** $\left|\dfrac{x - 2}{3}\right| > 4$

89. $|3x + 1| + 2 < 6$ **90.** $|3x - 2| + 2 \geq 0$

91. $3|2x + 5| \geq 9$ **92.** $-2|3x - 4| < 16$

93. $|5x - 1| + 4 \leq 0$ **94.** $-|5x - 1| + 2 < 0$

95. $\left|\dfrac{1}{3}x + 7\right| + 5 > 6$ **96.** $\left|\dfrac{1}{2}x - 3\right| - 4 < 2$

97. $\left| \dfrac{1}{5}x - 5 \right| + 4 > 4$ **98.** $\left| \dfrac{1}{6}x + 6 \right| + 2 < 2$

99. $\left| \dfrac{1}{7}x + 1 \right| \leq 0$ **100.** $|2x + 1| + 2 \leq 2$

101. $\left| \dfrac{x - 5}{10} \right| \leq 0$ **102.** $\left| \dfrac{3}{5}x - 2 \right| + 3 \leq 3$

Write each inequality as an inequality using absolute values.

103. $-4 < x < 4$

104. $x < -4 \text{ or } x > 4$

105. $x + 3 < -6 \text{ or } x + 3 > 6$

106. $-5 \leq x - 3 \leq 5$

APPLICATIONS

107. Springs   The weight on the spring shown in the illustration oscillates up and down according to the formula $|d - 5| \leq 1$, where d is the distance of the weight above the ground. Solve the formula for d and give the range of heights that the weight is above the ground.

5 ft

108. Aviation   An airplane is cruising at 30,000 ft and has been instructed to maintain an altitude that is described by the inequality $|a - 30{,}000| \leq 1{,}500$. Solve the inequality for a and give the range of altitudes at which the plane can fly.

109. Finding temperature ranges   The temperatures on a summer day satisfied the inequality $|t - 78°| \leq 8°$, where t is a temperature in degrees Fahrenheit. Express the range of temperatures as a compound inequality.

110. Finding operating temperatures   A car CD player has an operating temperature of $|t - 40°| < 80°$, where t is a temperature in degrees Fahrenheit. Express this range of temperatures as a compound inequality.

111. Range of camber angles   The specification for a car state that the camber angle c of its wheels should be $0.6° \pm 0.5°$. Express this range with an inequality containing absolute value symbols.

112. Tolerance of sheet steel   A sheet of steel is to be 0.25 inch thick with a tolerance of 0.015 inch. Express this specification with an inequality containing absolute value symbols.

WRITING

113. Explain how to find the absolute value of a given number.

114. Explain why the equation $|x| + 5 = 0$ has no solution.

115. Explain the use of parentheses and brackets when graphing inequalities.

116. If $k > 0$, explain the differences between the solution sets of $|x| < k$ and $|x| > k$.

SOMETHING TO THINK ABOUT

117. For what values of k does $|x| + k = 0$ have exactly two solutions?

118. For what value of k does $|x| + k = 0$ have exactly one solution?

119. Under what conditions is $|x| + |y| > |x + y|$?

120. Under what conditions is $|x| + |y| = |x + y|$?

4.3   Linear Inequalities in Two Variables

- **Graphing Linear Inequalities**   ■ **Graphing Compound Inequalities**
- **Problem Solving**

Getting Ready   *Do the coordinates of the points satisfy the equation $y = 5x + 2$?*

   1. $(0, 2)$ **2.** $\left(-\frac{2}{5}, 0\right)$ **3.** $(3, 18)$ **4.** $(-3, -13)$

In this section, we will show how to solve inequalities that have two variables. We begin by discussing their graphs.

Graphing Linear Inequalities

Inequalities such as

$$2x + 3y < 6 \qquad \text{or} \qquad 3x - 4y \geq 9$$

are called linear inequalities in two varibles. In general, we have the following definition:

Linear Inequalities

A **linear inequality** in x and y is any inequality that can be written in the form

$$Ax + By < C \text{ or } Ax + By > C \text{ or } Ax + By \leq C \text{ or } Ax + By \geq C$$

where A, B, and C are real numbers and A and B are not both 0.

The **graph of a linear inequality** in x and y is the graph of all ordered pairs (x, y) that satisfy the inequality.

The inequality $y > 3x + 2$ is an example of a linear inequality in two variables because it can be written in the form $-3x + y > 2$. To graph it, we first graph the related equation $y = 3x + 2$ as shown in Figure 4-24(a). This boundary line divides the coordinate plane into two half planes, one on either side of the line.

To find which half-plane is the graph of $y > 3x + 2$, we can substitute the coordinates $(0, 0)$ into the inequality and simplify.

$$y > 3x + 2$$
$$\mathbf{0} > 3(\mathbf{0}) + 2 \qquad \text{Substitute 0 for } x \text{ and 0 for } y.$$
$$0 \not> 2$$

Since the coordinates don't satisfy $y > 3x + 2$, the origin is not part of the graph. Thus, the half-plane on the other side of the broken line is the graph, which is shown in Figure 4-24(b).

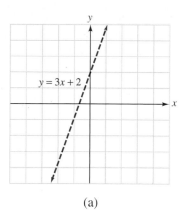 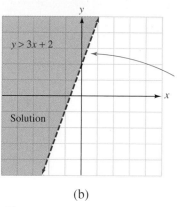

(a)   (b)

Figure 4-24

EXAMPLE 1   Graph the inequality: $2x - 3y \leq 6$.

Solution   This inequality is the combination of the inequality $2x - 3y < 6$ and the equation $2x - 3y = 6$.

We start by graphing $2x - 3y = 6$ to find the boundary line. This time, we draw the solid line shown in Figure 4-25(a), because equality is permitted. To decide which half-plane represents $2x - 3y < 6$, we check to see whether the coordinates of the origin satisfy the inequality.

$$2x - 3y < 6$$
$$2(\mathbf{0}) - 3(\mathbf{0}) < 6 \qquad \text{Substitute 0 for } x \text{ and 0 for } y.$$
$$0 < 6$$

Since the coordinates satisfy the inequality, the origin is in the half-plane that is the graph of $2x - 3y < 6$. The graph is shown in Figure 4-25(b).

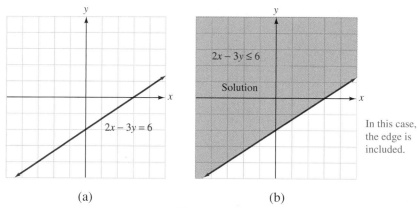

(a)   (b)

Figure 4-25

Self Check   Graph: $3x - 2y \geq 6$.   ∎

EXAMPLE 2   Graph the inequality: $y < 2x$.

Solution   We graph $y = 2x$, as shown in Figure 4-26(a). Because it is not part of the inequality, we draw the edge as a broken line.

To decide which half-plane is the graph of $y < 2x$, we check to see whether the coordinates of some fixed point satisfy the inequality. We cannot use the origin as a test point, because the edge passes through the origin. However, we can choose a different point—say, $(3, 1)$.

$$y < 2x$$
$$\mathbf{1} < 2(\mathbf{3}) \qquad \text{Substitute 1 for } y \text{ and 3 for } x.$$
$$1 < 6$$

Since $1 < 6$ is a true inequality, the point $(3, 1)$ satisfies the inequality and is in the graph, which is shown in Figure 4-26(b).

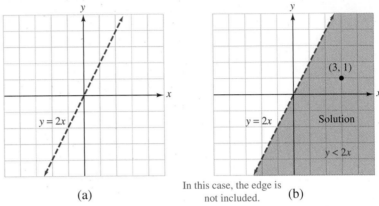

(a)

In this case, the edge is not included.   (b)

Figure 4-26

Self Check   Graph: $y > 2x$.

Graphing Compound Inequalities

EXAMPLE 3   Graph the inequality: $2 < x \leq 5$.

Solution   The inequality $2 < x \leq 5$ is equivalent to the following two inequalities:

$$2 < x \qquad \text{and} \qquad x \leq 5$$

Its graph will contain all points in the plane that satisfy the inequalities $2 < x$ and $x \leq 5$ simultaneously. These points are in the shaded region of Figure 4-27.

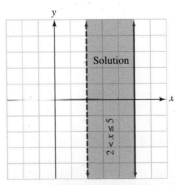

Figure 4-27

Self Check   Graph: $-2 \leq x < 3$.   ∎

Problem Solving

EXAMPLE 4   **Earning money**   Rick has two part-time jobs, one paying $7 per hour and the other paying $5 per hour. He must earn at least $140 per week to pay his expenses while attending college. Write an inequality that shows the various ways he can schedule his time to achieve his goal.

Analyze the problem   If we let x represent the number of hours per week he works on the first job, he will earn $7x$ per week on the first job. If we let y represent the number of hours per week he works on the second job, he will earn $5y$ per week on the second job. To achieve his goal, the sum of these two incomes must be at least $140.

Form an inequality   Since x represents the number of hours per week he works on the first job and y represents the number of hours he works per week on the second job, we have

The hourly rate on the first job	times	the hours worked on the first job	plus	the hourly rate on the second job	times	the hours worked on the second job	is greater than or equal to	$140.
$7	·	x	+	$5	·	y	≥	$140

Solve the inequality   The graph of $7x + 5y \geq 140$ is shown in Figure 4-28. Any point in the shaded region indicates a way that he can schedule his time and earn $140 or more per week. For example, if he works 10 hours on the first job and 15 hours on the second job, he will earn

$$\$7(10) + \$5(15) = \$70 + \$75$$
$$= \$145$$

If he works 5 hours on the first job and 25 hours on the second job, he will earn

$$\$7(5) + \$5(25) = \$35 + \$125$$
$$= \$160$$

Since Rick cannot work a negative number of hours, the graph has no meaning when x or y is negative, so only the first quadrant of the graph is shown.

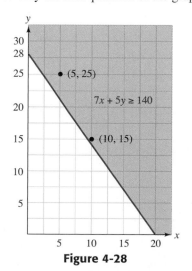

Figure 4-28

∎

Accent on Technology GRAPHING INEQUALITIES

Some calculators (such as the TI-83 Plus) have a graphing-style icon in the $y =$ editor. See Figure 4-29(a). Some of the different graphing styles are as follows.

\	line	A straight line or curved graph is shown.
◥	above	Shading covers the area above a graph.
◣	below	Shading covers the area below a graph.

We can change the icon by placing the cursor on it and pressing ENTER.

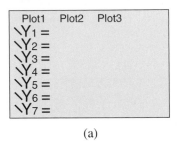

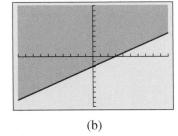

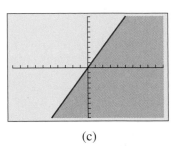

   (a)   (b)   (c)

Figure 4-29

To graph the inequality of Example 1 using window settings of $x = [-10, 10]$ and $y = [-10, 10]$, we change the graphing-style icon to "above" (◥), enter the equation $2x - 3y = 6$ as $y = \frac{2}{3}x - 2$, and press GRAPH to get Figure 4-29(b).

To graph the inequality of Example 2 using window settings of $x = [-10, 10]$ and $y = [-10, 10]$, we change the graphing-style icon to "below" (◣), enter the equation $y = 2x$, and press GRAPH to get Figure 4-29(c).

If your calculator does not have a graphing-style icon, you can graph linear inequalities with a shade feature. To do so, consult your owner's manual.

Note that graphing calculators do not distinguish between solid and dashed lines to show whether or not the edge of a region is included within the graph.

Self Check Answers

1.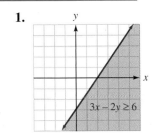

2.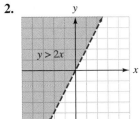

3.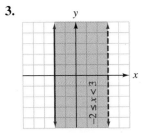

Orals *Do the points satisfy* $2x + 3y < 12$?

1. $(0, 0)$ **2.** $(3, 2)$ **3.** $(2, 3)$ **4.** $(-1, 4)$

Do the points satisfy $3x - 2y \geq 12$?

5. $(0, 0)$ **6.** $(3, 2)$ **7.** $(2, -3)$ **8.** $(5, 1)$

4.3 EXERCISES

REVIEW *Solve each system.*

1. $\begin{cases} x + y = 4 \\ x - y = 2 \end{cases}$ **2.** $\begin{cases} 2x - y = -4 \\ x + 2y = 3 \end{cases}$

3. $\begin{cases} 3x + y = 3 \\ 2x - 3y = 13 \end{cases}$ **4.** $\begin{cases} 2x - 5y = 8 \\ 5x + 2y = -9 \end{cases}$

VOCABULARY AND CONCEPTS *Fill in the blanks.*

5. $3x + 2y < 12$ is an example of a _____ inequality.

6. Graphs of linear inequalities in two variables are _____.

7. The boundary line of a half-plane is called an ____.

8. If $y < \frac{1}{2}x - 2$ and $y > \frac{1}{2}x - 2$ are false, then _____.

PRACTICE *Graph each inequality.*

9. $y > x + 1$

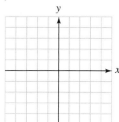

10. $y < 2x - 1$

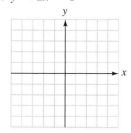

11. $y \geq x$

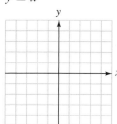

12. $y \leq 2x$

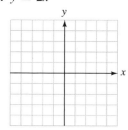

13. $2x + y \leq 6$

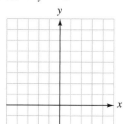

14. $x - 2y \geq 4$

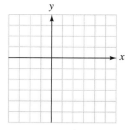

15. $3x \geq -y + 3$

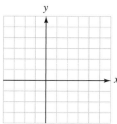

16. $2x \leq -3y - 12$

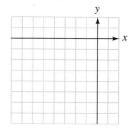

17. $y \geq 1 - \frac{3}{2}x$

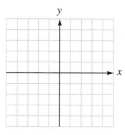

18. $y < \frac{1}{3}x - 1$

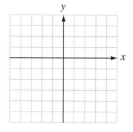

19. $x < 4$

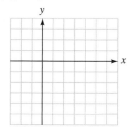

20. $y \geq -2$

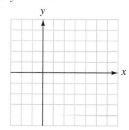

21. $-2 \le x < 0$

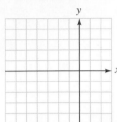

22. $-3 < y \le -1$

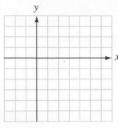

23. $y < -2$ or $y > 3$

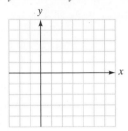

24. $-x \le 1$ or $x \ge 2$

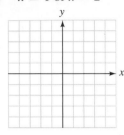

Find the equation of the boundary line or lines. Then give the inequality whose graph is shown.

25.

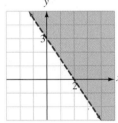

26.

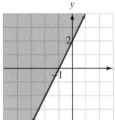

27.

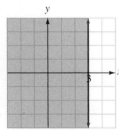

28.

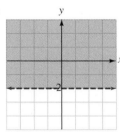

29.

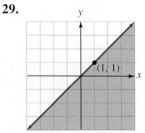

30.

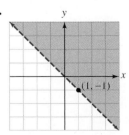

31.

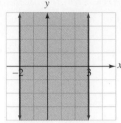

32.

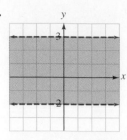

33.

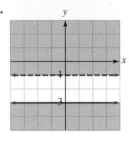

34.

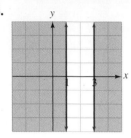

 Use a graphing calculator to graph each inequality.

35. $y < 0.27x - 1$

36. $y > -3.5x + 2.7$

37. $y \ge -2.37x + 1.5$

38. $y \le 3.37x - 1.7$

APPLICATIONS   *Graph each inequality for nonnegative values of x and y. Then give some ordered pairs that satisfy the inequality.*

39. Figuring taxes   On average, it takes an accountant 1 hour to complete a simple tax return and 3 hours to complete a complicated return. If the accountant wants to work no more than 9 hours per day, use the illustration to graph an inequality that shows the possible ways that simple returns (x) and complicated returns (y) can be completed each day.

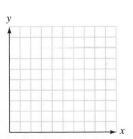

40. Selling trees During a sale, a garden store sold more than $2,000 worth of trees. If a 6-foot maple costs $100 and a 5-foot pine costs $125, use the illustration to graph an inequality that shows the possible ways that maple trees (x) and pine trees (y) were sold.

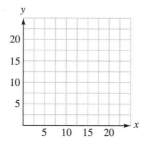

41. Choosing housekeepers One housekeeper charges $6 per hour, and another charges $7 per hour. If Sarah can afford no more than $42 per week to clean her house, use the illustration to graph an inequality that shows the possible ways that she can hire the first housekeeper (x) and the second housekeeper (y).

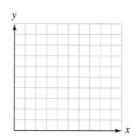

42. Making sporting goods A sporting goods manufacturer allocates at least 1,200 units of time per day to make fishing rods and reels. If it takes 10 units of time to make a rod and 15 units of time to make a reel, use the illustration to graph an inequality that shows the possible ways to schedule the time to make rods (x) and reels (y).

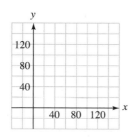

43. Investing A woman has up to $6,000 to invest. If stock in Traffico sells for $50 per share and stock in Cleanco sells for $60 per share, use the illustration to graph an inequality that shows the possible ways that she can buy shares of Traffico (x) and Cleanco (y).

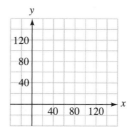

44. Buying concert tickets Tickets to a concert cost $6 for reserved seats and $4 for general admission. If receipts must be at least $10,200 to meet expenses, use the illustration to graph an inequality that shows the possible ways that the box office can sell reserved seats (x) and general admission tickets (y).

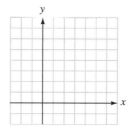

WRITING

45. Explain how to decide where to draw the boundary of the graph of a linear inequality, and whether to draw it as a solid or a broken line.

46. Explain how to decide which side of the boundary of the graph of a linear inequality should be shaded.

SOMETHING TO THINK ABOUT

47. Can an inequality be an identity, one that is satisfied by all (x, y) pairs? Illustrate.

48. Can an inequality have no solutions? Illustrate.

4.4   Systems of Inequalities

❚ Systems of Inequalities   ❚ Problem Solving

Getting Ready   *Do the coordinates satisfy the inequalities $y > 2x - 3$ and $x - y > -1$?*

1. $(0, 0)$ **2.** $(1, 1)$ **3.** $(-2, 0)$ **4.** $(0, 4)$

In the previous section, we learned how to graph linear inequalities. In this section, we will learn how to solve systems of linear inequalities.

Systems of Inequalities

We now consider the graphs of systems of inequalities in the variables x and y. These graphs will usually be the intersection of half-planes.

EXAMPLE 1   Graph the solution set of $\begin{cases} x + y \leq 1 \\ 2x - y > 2 \end{cases}$.

Solution   On one set of coordinate axes, we graph each inequality as shown in Figure 4-30.
    The graph of $x + y \leq 1$ includes the line graph of the equation $x + y = 1$ and all points below it. Since the edge is included, we draw it as a solid line.
    The graph of $2x - y > 2$ contains the points below the graph of the equation $2x - y = 2$. Since the edge is not included, we draw it as a broken line.
    The area where the half-planes intersect represents the solution of the system of inequalities, because any point in that region has coordinates that will satisfy both inequalities.

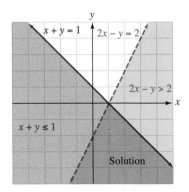

$x + y = 1$		
x	y	(x, y)
0	1	$(0, 1)$
1	0	$(1, 0)$

$2x - y = 2$		
x	y	(x, y)
0	-2	$(0, -2)$
1	0	$(1, 0)$

Figure 4-30

Self Check   Graph the solution set of $\begin{cases} x + y \geq 1 \\ 2x - y < 2 \end{cases}$.

❚

EXAMPLE 2   Graph the solution set of $\begin{cases} y < x^2 \\ y > \dfrac{x^2}{4} - 2 \end{cases}$.

Solution The graph of $y = x^2$ is the red, dashed parabola shown in Figure 4-31. The points with coordinates that satisfy $y < x^2$ are the points that lie below the parabola.

The graph of $y = \frac{x^2}{4} - 2$ is the blue, dashed parabola. This time, the points that lie above the parabola satisfy the inequality. Thus, the solution of the system is the area between the parabolas.

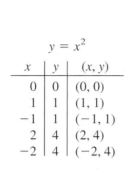

$y = x^2$

x	y	(x, y)
0	0	$(0, 0)$
1	1	$(1, 1)$
-1	1	$(-1, 1)$
2	4	$(2, 4)$
-2	4	$(-2, 4)$

$y = \dfrac{x^2}{4} - 2$

x	y	(x, y)
0	-2	$(0, -2)$
2	-1	$(2, -1)$
-2	-1	$(-2, -1)$
4	2	$(4, 2)$
-4	2	$(-4, 2)$

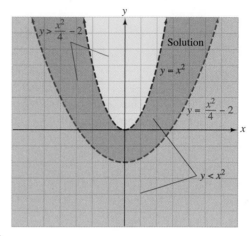

Figure 4-31

Self Check Graph the solution set of $\begin{cases} y \geq x^2 \\ y \leq x + 2 \end{cases}$. ∎

Accent on Technology

SOLVING SYSTEMS OF INEQUALITIES

To solve the system of Example 1, we use window settings of $x = [-10, 10]$ and $y = [-10, 10]$. To graph $x + y \leq 1$, we enter the equation $x + y = 1$ ($y = -x + 1$) and change the graphing-style icon to below (◣). To graph $2x - y > 2$, we enter the equation $2x - y = 2$ ($y = 2x - 2$) and change the graphing-style icon to below (◣). Finally, we press $\boxed{\text{GRAPH}}$ to obtain Figure 4-32(a).

To solve the system of Example 2, we enter the equation $y = x^2$ and change the graphing-style icon to below (◣). We then enter the equation $y = \frac{x^2}{4} - 2$ and change the graphing-style icon to above (◤). Finally, we press $\boxed{\text{GRAPH}}$ to obtain Figure 4-32(b).

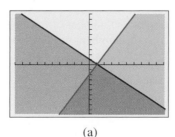

(a)

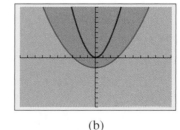

(b)

Figure 4-32

EXAMPLE 3   Graph the solution set of $\begin{cases} x \geq 1 \\ y \geq x \\ 4x + 5y < 20 \end{cases}$.

Solution   The graph of $x \geq 1$ includes the points that lie on the graph of $x = 1$ and to the right, as shown in Figure 4-33(a).

The graph of $y \geq x$ includes the points that lie on the graph of $y = x$ and above it, as shown in Figure 4-33(b).

The graph of $4x + 5y < 20$ includes the points that lie below the graph of $4x + 5y = 20$, as shown in Figure 4-33(c).

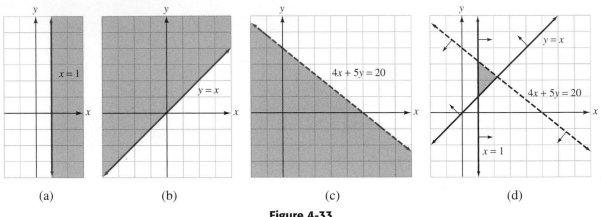

(a)   (b)   (c)   (d)

Figure 4-33

If we merge these graphs onto one set of coordinate axes, we see that the graph of the system includes the points that lie within a shaded triangle, together with the points on two of the three sides of the triangle, as shown in Figure 4-33(d).

Self Check   Graph the solution set of $\begin{cases} x \geq 0 \\ y \leq 0 \\ y \geq -2 \end{cases}$.   ∎

Problem Solving

EXAMPLE 4   **Landscaping**   A homeowner budgets from $300 to $600 for trees and bushes to landscape his yard. After shopping around, he finds that good trees cost $150 and mature bushes cost $75. What combinations of trees and bushes can he afford?

Analyze the problem   If x represents the number of trees purchased, $150x$ will be the cost of the trees. If y represents the number of bushes purchased, $75y$ will be the cost of the bushes. We know that the sum of these costs is to be from $300 to $600.

Form two inequalities   We let x represent the number of trees purchased and y represent the number of bushes purchased. We can then form the following system of inequalities:

The cost of a tree	times	the number of trees purchased	plus	the cost of a bush	times	the number of bushes purchased	is greater than or equal to	$300.
$150	·	x	+	$75	·	y	≥	$300

The cost of a tree	times	the number of trees purchased	plus	the cost of a bush	times	the number of bushes purchased	is less than or equal to	$600.
$150	·	x	+	$75	·	y	≤	$600

Solve the system   We graph the system

$$\begin{cases} 150x + 75y \geq 300 \\ 150x + 75y \leq 600 \end{cases}$$

as in Figure 4-34. The coordinates of each point shown in the graph give a possible combination of trees (x) and bushes (y) that can be purchased.

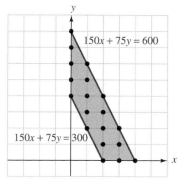

Figure 4-34

State the conclusion   These possibilities are

$$(0, 4), (0, 5), (0, 6), (0, 7), (0, 8)$$
$$(1, 2), (1, 3), (1, 4), (1, 5), (1, 6)$$
$$(2, 0), (2, 1), (2, 2), (2, 3), (2, 4)$$
$$(3, 0), (3, 1), (3, 2), (4, 0)$$

Only these points can be used, because the homeowner cannot buy a portion of a tree.

Check the result   Check some of the ordered pairs to verify that they satisfy both inequalities.   ∎

Self Check Answers

1.

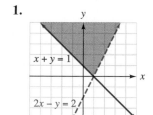

2.

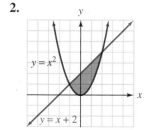

3.

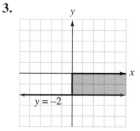

Orals   *Do the coordinates* $(1, 1)$ *satisfy both inequalities?*

1. $\begin{cases} y < x + 1 \\ y > x - 1 \end{cases}$

2. $\begin{cases} y > 2x - 3 \\ y < -x + 3 \end{cases}$

4.4 EXERCISES

REVIEW *Solve each formula for the given variable.*

1. $A = p + prt$ for r

2. $C = \dfrac{5}{9}(F - 32)$ for F

3. $z = \dfrac{x - \mu}{\sigma}$ for x

4. $P = 2l + 2w$ for w

5. $l = a + (n - 1)d$ for d

6. $z = \dfrac{x - \mu}{\sigma}$ for μ

VOCABULARY AND CONCEPTS *Fill in the blanks.*

7. To solve a system of inequalities by graphing, we graph each inequality. The solution is the region where the graphs _____.

8. If an edge is included in the graph of an inequality, we draw it as a _____ graph.

PRACTICE *Graph the solution set of each system.*

9. $\begin{cases} y < 3x + 2 \\ y < -2x + 3 \end{cases}$

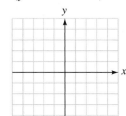

10. $\begin{cases} y \le x - 2 \\ y \ge 2x + 1 \end{cases}$

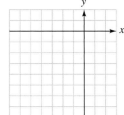

11. $\begin{cases} 3x + 2y > 6 \\ x + 3y \le 2 \end{cases}$

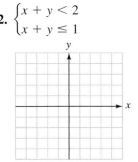

12. $\begin{cases} x + y < 2 \\ x + y \le 1 \end{cases}$

13. $\begin{cases} 3x + y \le 1 \\ -x + 2y \ge 6 \end{cases}$

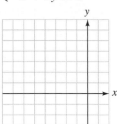

14. $\begin{cases} x + 2y < 3 \\ 2x + 4y < 8 \end{cases}$

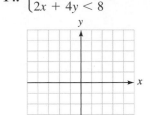

15. $\begin{cases} 2x - y > 4 \\ y < -x^2 + 2 \end{cases}$

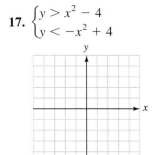

16. $\begin{cases} x \le y^2 \\ y \ge x \end{cases}$

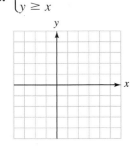

17. $\begin{cases} y > x^2 - 4 \\ y < -x^2 + 4 \end{cases}$

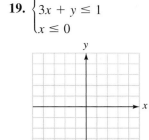

18. $\begin{cases} x \ge y^2 \\ y \ge x^2 \end{cases}$

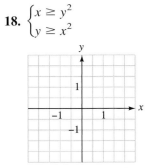

19. $\begin{cases} 2x + 3y \le 6 \\ 3x + y \le 1 \\ x \le 0 \end{cases}$

20. $\begin{cases} 2x + y \le 2 \\ y \ge x \\ x \ge 0 \end{cases}$

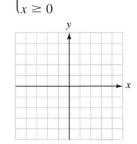

21. $\begin{cases} x - y < 4 \\ y \le 0 \\ x \ge 0 \end{cases}$

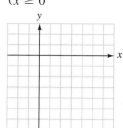

22. $\begin{cases} x + y \le 4 \\ x \ge 0 \\ y \ge 0 \end{cases}$

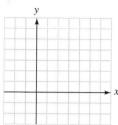

23. $\begin{cases} x \ge 0 \\ y \ge 0 \\ 9x + 3y \le 18 \\ 3x + 6y \le 18 \end{cases}$

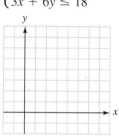

24. $\begin{cases} x + y \ge 1 \\ x - y \le 1 \\ x - y \ge 0 \\ x \le 2 \end{cases}$

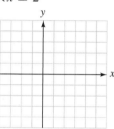

Use a graphing calculator to solve each system.

25. $\begin{cases} y < 3x + 2 \\ y < -2x + 3 \end{cases}$
(See Exercise 9.)

26. $\begin{cases} y > x^2 - 4 \\ y < -x^2 + 4 \end{cases}$
(See Exercise 17.)

APPLICATIONS *Graph a system of inequalities and give two possible solutions to each problem.*

27. Buying compact discs Melodic Music has compact discs on sale for either $10 or $15. If a customer wants to spend at least $30 but no more than $60 on CDs, use the illustration to graph a system of inequalities that will show the possible ways a customer can buy $10 CDs ($x$) and $15 CDs ($y$).

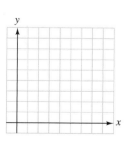

28. Buying boats Dry Boat Works wholesales aluminum boats for $800 and fiberglass boats for $600. Northland Marina wants to order at least $2,400 worth, but no more than $4,800 worth of boats. Use the illustration to graph a system of inequalities that will show the possible combinations of aluminum boats (x) and fiberglass boats (y) that can be ordered.

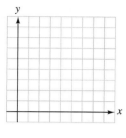

29. Buying furniture A distributor wholesales desk chairs for $150 and side chairs for $100. Best Furniture wants to order no more than $900 worth of chairs, including more side chairs than desk chairs. Use the illustration to graph a system of inequalities that will show the possible combinations of desk chairs (x) and side chairs (y) that can be ordered.

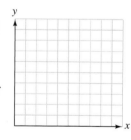

30. Ordering furnace equipment Bolden Heating Company wants to order no more than $2,000 worth of electronic air cleaners and humidifiers from a wholesaler that charges $500 for air cleaners and $200 for humidifiers. If Bolden wants more humidifiers than air cleaners, use the illustration to graph a system of inequalities that will show the possible combinations of air cleaners (x) and humidifiers (y) that can be ordered.

31. When graphing a system of linear inequalities, explain how to decide which region to shade.

32. Explain how a system of two linear inequalities might have no solution.

SOMETHING TO THINK ABOUT

33. The solution of a system of inequalities in two variables is *bounded* if it is possible to draw a circle around it. Can the solution of two linear inequalities be bounded?

34. The solution of $\begin{cases} y \ge |x| \\ y \le k \end{cases}$ has an area of 25. Find k.

4.5   Linear Programming

■ Linear Programming   ■ Applications of Linear Programming

Getting Ready   *Evaluate $2x + 3y$ for each pair of coordinates.*

1. $(0, 0)$ **2.** $(3, 0)$ **3.** $(2, 2)$ **4.** $(0, 4)$

We now use our knowledge of solving systems of inequalities to solve linear programming problems.

Linear Programming

Linear programming is a mathematical technique used to find the optimal allocation of resources in the military, business, telecommunications, and other fields. It got its start during World War II when it became necessary to move huge quantities of people, materials, and supplies as efficiently and economically as possible.

To solve a linear program, we maximize (or minimize) a function (called the **objective function**) subject to given conditions on its variables. These conditions (called **constraints**) are usually given as a system of linear inequalities. For example, suppose that the annual profit (in millions of dollars) earned by a business is given by the equation $P = y + 2x$ and that x and y are subject to the following constraints:

$$\begin{cases} 3x + y \le 120 \\ x + y \le 60 \\ x \ge 0 \\ y \ge 0 \end{cases}$$

To find the maximum profit P that can be earned by the business, we solve the system of inequalities as shown in Figure 4-35(a) and find the coordinates of each corner point of the region R, called a **feasibility region.** We can then write the profit equation

$$P = y + 2x \qquad \text{in the form} \qquad y = -2x + P$$

The equation $y = -2x + P$ is the equation of a set of parallel lines, each with a slope of -2 and a y-intercept of P. To find the line that passes through region R and provides the maximum value of P, we refer to Figure 4-35(b) and locate the line with the greatest y-intercept. Since line l has the greatest y-intercept and intersects region R at the corner point (30, 30), the maximum value of P (subject to the given constraints) is

$$P = y + 2x$$
$$= 30 + 2(30)$$
$$= 90$$

Thus, the maximum profit P that can be earned is \$90 million. This profit occurs when $x = 30$ and $y = 30$.

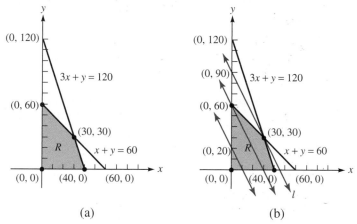

(a)   (b)

Figure 4-35

PERSPECTIVE

How to Solve It

As a young student, George Polya (1888–1985) enjoyed mathematics and understood the solutions presented by his teachers. However, Polya had questions still asked by mathematics students today: "Yes, the solution works, but how is it possible to come up with such a solution? How could I discover such things by myself?" These questions still concerned him years later when, as Professor of Mathematics at Stanford University, he developed an approach to teaching mathematics that was very popular with faculty and students. His book, *How to Solve It,* became a bestseller.

Polya's problem-solving approach involves four steps.

George Polya
(1888–1985)

- *Understand the problem.* What is the unknown? What information is known? What are the conditions?

- *Devise a plan.* Have you seen anything like it before? Do you know any related problems you have solved before? If you can't solve the proposed problem, can you solve a similar but easier problem?

- *Carry out the plan.* Check each step. Can you explain why each step is correct?

- *Look back.* Examine the solution. Can you check the result? Can you use the result, or the method, to solve any other problem?

The preceding discussion illustrates the following important fact.

Maximum or Minimum of an Objective Function

If a linear function, subject to the constraints of a system of linear inequalities in two variables, attains a maximum or a minimum value, that value will occur at a corner point or along an entire edge of the region R that represents the solution of the system.

EXAMPLE 1   If $P = 2x + 3y$, find the maximum value of P subject to the following constraints:

$$\begin{cases} x + y \le 4 \\ 2x + y \le 6 \\ x \ge 0 \\ y \ge 0 \end{cases}$$

Solution   We solve the system of inequalities to find the feasibility region R shown in Figure 4-36. The coordinates of its corner points are $(0, 0)$, $(3, 0)$, $(0, 4)$, and $(2, 2)$.

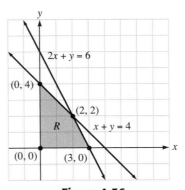

Figure 4-36

Since the maximum value of P will occur at a corner of R, we substitute the coordinates of each corner point into the objective function $P = 2x + 3y$ and find the one that gives the maximum value of P.

Point	$P = 2x + 3y$
$(0, 0)$	$P = 2(\mathbf{0}) + 3(\mathbf{0}) = 0$
$(3, 0)$	$P = 2(\mathbf{3}) + 3(\mathbf{0}) = 6$
$(2, 2)$	$P = 2(\mathbf{2}) + 3(\mathbf{2}) = 10$
$(0, 4)$	$P = 2(\mathbf{0}) + 3(\mathbf{4}) = 12$

The maximum value $P = 12$ occurs when $x = 0$ and $y = 4$.

Self Check   Find the maximum value of $P = 4x + 3y$, subject to the constraints of Example 1.

EXAMPLE 2   If $P = 3x + 2y$, find the minimum value of P subject to the following constraints:

$$\begin{cases} x + y \geq 1 \\ x - y \leq 1 \\ x - y \geq 0 \\ x \leq 2 \end{cases}$$

Solution   We refer to the feasibility region shown in Figure 4-37 with corner points at $\left(\frac{1}{2}, \frac{1}{2}\right)$, $(2, 2)$, $(2, 1)$, and $(1, 0)$.

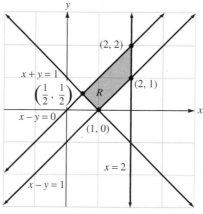

Figure 4-37

Since the minimum value of P occurs at a corner point of region R, we substitute the coordinates of each corner point into the objective function $P = 3x + 2y$ and find the one that gives the minimum value of P.

Point	$P = 3x + 2y$
$\left(\frac{1}{2}, \frac{1}{2}\right)$	$P = 3\left(\frac{1}{2}\right) + 2\left(\frac{1}{2}\right) = \frac{5}{2}$
$(2, 2)$	$P = 3(\mathbf{2}) + 2(\mathbf{2}) = 10$
$(2, 1)$	$P = 3(\mathbf{2}) + 2(\mathbf{1}) = 8$
$(1, 0)$	$P = 3(\mathbf{1}) + 2(\mathbf{0}) = 3$

The minimum value $P = \frac{5}{2}$ occurs when $x = \frac{1}{2}$ and $y = \frac{1}{2}$.

Self Check   Find the minimum value of $P = 2x + y$, subject to the constraints of Example 2.

Applications of Linear Programming

Linear programming problems can be complex and involve hundreds of variables. In this section, we will consider a few simple problems. Since they involve only two variables, we can solve them using graphical methods.

EXAMPLE 3 **Maximizing income** An accountant prepares tax returns for individuals and for small businesses. On average, each individual return requires 3 hours of her time and 1 hour of computer time. Each business return requires 4 hours of her time and 2 hours of computer time. Because of other business considerations, her time is limited to 240 hours, and the computer time is limited to 100 hours. If she earns a profit of $80 on each individual return and a profit of $150 on each business return, how many returns of each type should she prepare to maximize her profit?

Solution First, we organize the given information into a table.

	Individual tax return	Business tax return	Time available
Accountant's time	3	4	240 hours
Computer time	1	2	100 hours
Profit	$80	$150	

Then we solve the problem using the following steps.

Find the objective function Suppose that x represents the number of individual returns to be completed and y represents the number of business returns to be completed. Since each of the x individual returns will earn an $80 profit, and each of the y business returns will earn a $150 profit, the total profit is given by the equation

$$P = 80x + 150y$$

Find the feasibility region Since the number of individual returns and business returns cannot be negative, we know that $x \geq 0$ and $y \geq 0$.

Since each of the x individual returns will take 3 hours of her time, and each of the y business returns will take 4 hours of her time, the total number of hours she will work will be $(3x + 4y)$ hours. This amount must be less than or equal to her available time, which is 240 hours. Thus, the inequality $3x + 4y \leq 240$ is a constraint on the accountant's time.

Since each of the x individual returns will take 1 hour of computer time, and each of the y business returns will take 2 hours of computer time, the total number of hours of computer time will be $(x + 2y)$ hours. This amount must be less than or equal to the available computer time, which is 100 hours. Thus, the inequality $x + 2y \leq 100$ is a constraint on the computer time.

We have the following constraints on the values of x and y.

$$\begin{cases} x \geq 0 & \text{The number of individual returns is nonnegative.} \\ y \geq 0 & \text{The number of business returns is nonnegative.} \\ 3x + 4y \leq 240 & \text{The accountant's time must be less than or equal to 240 hours.} \\ x + 2y \leq 100 & \text{The computer time must be less than or equal to 100 hours.} \end{cases}$$

To find the feasibility region, we graph each of the constraints to find region R, as in Figure 4-38. The four corner points of this region have coordinates of $(0, 0)$, $(80, 0)$, $(40, 30)$, and $(0, 50)$.

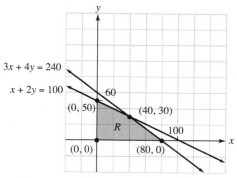

$3x + 4y = 240$

$x + 2y = 100$

Figure 4-38

Find the maximum profit To find the maximum profit, we substitute the coordinates of each corner point into the objective function $P = 80x + 150y$.

Point	$P = 80x + 150y$
$(0, 0)$	$P = 80(0) + 150(0) = 0$
$(80, 0)$	$P = 80(80) + 150(0) = 6{,}400$
$(40, 30)$	$P = 80(40) + 150(30) = 7{,}700$
$(0, 50)$	$P = 80(0) + 150(50) = 7{,}500$

From the table, we can see that the accountant will earn a maximum profit of $7,700 if she prepares 40 individual returns and 30 business returns. ∎

EXAMPLE 4 **Diet problem** Vigortab and Robust are two diet supplements. Each Vigortab tablet costs 50¢ and contains 3 units of calcium, 20 units of Vitamin C, and 40 units of iron. Each Robust tablet costs 60¢ and contains 4 units of calcium, 40 units of Vitamin C, and 30 units of iron. At least 24 units of calcium, 200 units of Vitamin C, and 120 units of iron are required for the daily needs of one patient. How many tablets of each supplement should be taken daily for a minimum cost? Find the daily minimum cost.

Solution First, we organize the given information into a table.

	Vigortab	Robust	Amount required
Calcium	3	4	24
Vitamin C	20	40	200
Iron	40	30	120
Cost	50¢	60¢	

Find the objective function We can let x represent the number of Vigortab tablets to be taken daily and y the corresponding number of Robust tablets. Because each of the x Vigortab tablets will cost 50¢, and each of the *y* Robust tablets will cost 60¢, the total cost will be given by the equation

$$C = 0.50x + 0.60y \qquad 50¢ = \$0.50 \text{ and } 60¢ = \$0.60.$$

Find the feasibility region   Since there are requirements for calcium, Vitamin C, and iron, there is a constraint for each. Note that neither *x* nor *y* can be negative.

$$\begin{cases} 3x + 4y \geq 24 & \text{The amount of calcium must be greater than or equal to 24 units.} \\ 20x + 40y \geq 200 & \text{The amount of Vitamin C must be greater than or equal to 200 units.} \\ 40x + 30y \geq 120 & \text{The amount of iron must be greater than or equal to 120 units.} \\ x \geq 0, y \geq 0 & \text{The number of tablets taken must be greater than or equal to 0.} \end{cases}$$

We graph the inequalities to find the feasibility region and the coordinates of its corner points, as in Figure 4-39.

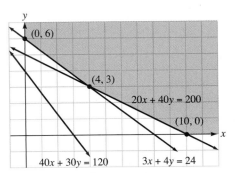

Figure 4-39

Find the minimum cost   In this case, the feasibility region is not bounded on all sides. The coordinates of the corner points are (0, 6), (4, 3), and (10, 0). To find the minimum cost, we substitute each pair of coordinates into the objective function.

Point	$C = 0.50x + 0.60y$
(0, 6)	$C = 0.50(0) + 0.60(6) = 3.60$
(4, 3)	$C = 0.50(4) + 0.60(3) = 3.80$
(10, 0)	$C = 0.50(10) + 0.60(0) = 5.00$

A minimum cost will occur if no Vigortab and 6 Robust tablets are taken daily. The minimum daily cost is $3.60. ∎

EXAMPLE 5   **Production-schedule problem**   A television program director must schedule comedy skits and musical numbers for prime-time variety shows. Each comedy skit requires 2 hours of rehearsal time, costs $3,000, and brings in $20,000 from the show's sponsors. Each musical number requires 1 hour of rehearsal time, costs $6,000, and generates $12,000. If 250 hours are available for rehearsal, and $600,000 is budgeted for comedy and music, how many segments of each type should be produced to maximize income? Find the maximum income.

Solution   First, we organize the given information into a table.

	Comedy	Musical	Available
Rehearsal time (hours)	2	1	250
Cost (in $1,000s)	3	6	600
Generated income (in $1,000s)	20	12	

Find the objective function   We can let x represent the number of comedy skits and y the number of musical numbers to be scheduled. Since each of the x comedy skits generates $20 thousand, the income generated by the comedy skits is $20x$ thousand. The musical numbers produce $12y$ thousand. The objective function to be maximized is

$$V = 20x + 12y$$

Find the feasibility region   Since there are limits on rehearsal time and budget, there is a constraint for each. Note that neither x nor y can be negative.

$$\begin{cases} 2x + y \le 250 & \text{The total rehearsal time must be less than or equal to 250 hours.} \\ 3x + 6y \le 600 & \text{The total cost must be less than or equal to \$600 thousand.} \\ x \ge 0, y \ge 0 & \text{The numbers of skits and musical numbers must be greater than or equal to 0.} \end{cases}$$

We graph the inequalities to find the feasibility region shown in Figure 4-40 and find the coordinates of each corner point.

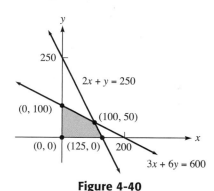

Figure 4-40

Find the maximum income   The coordinates of the corner points of the feasible region are (0, 0), (0, 100), (100, 50), and (125, 0). To find the maximum income, we substitute each pair of coordinates into the objective function.

Corner point	$V = 20x + 12y$
(0, 0)	$V = 20(0) + 12(0) = 0$
(0, 100)	$V = 20(0) + 12(100) = 1{,}200$
(100, 50)	$V = 20(100) + 12(50) = 2{,}600$
(125, 0)	$V = 20(125) + 12(0) = 2{,}500$

Maximum income will occur if 100 comedy skits and 50 musical numbers are scheduled. The maximum income will be 2,600 thousand dollars, or $2,600,000. ∎

1. 14   **2.** $\frac{3}{2}$

                Orals   *Evaluate* $P = 2x + 5y$ *when*

         1. $x = 0, y = 5$ **2.** $x = 2, y = 1$

        Find the corner points of the region determined by

3. $\begin{cases} x \geq 0 \\ y \geq 0 \\ x + y \leq 3 \end{cases}$ **4.** $\begin{cases} x \geq 0 \\ y \geq 0 \\ x + 2y \leq 4 \end{cases}$

4.5 EXERCISES

REVIEW   *Consider the line passing through* $(-2, 4)$ *and* $(5, 7)$.

1. Find the slope of line.

2. Write the equation of line in general form.

3. Write the equation of line in slope-intercept form.

4. Write the equation of the line that passes through the origin and is parallel to line.

VOCABULARY AND CONCEPTS   *Fill in the blanks.*

5. In a linear program, the inequalities are called _____.

6. Ordered pairs that satisfy the constraints of a linear program are called _____ solutions.

7. The function to be maximized (or minimized) in a linear program is called the _____ function.

8. The objective function of a linear program attains a maximum (or minimum), subject to the constraints, at a _____ or along an _____ of the feasibility region.

PRACTICE   *Maximize P subject to the following constraints.*

9. $P = 2x + 3y$

$\begin{cases} x \geq 0 \\ y \geq 0 \\ x + y \leq 4 \end{cases}$

10. $P = 3x + 2y$

$\begin{cases} x \geq 0 \\ y \geq 0 \\ x + y \leq 4 \end{cases}$

11. $P = y + \frac{1}{2}x$

$\begin{cases} x \geq 0 \\ y \geq 0 \\ 2y - x \leq 1 \\ y - 2x \geq -2 \end{cases}$

12. $P = 4y - x$

$\begin{cases} x \leq 2 \\ y \geq 0 \\ x + y \geq 1 \\ 2y - x \leq 1 \end{cases}$

13. $P = 2x + y$

$\begin{cases} y \geq 0 \\ y - x \leq 2 \\ 2x + 3y \leq 6 \\ 3x + y \leq 3 \end{cases}$

14. $P = x - 2y$

$\begin{cases} x + y \leq 5 \\ y \leq 3 \\ x \leq 2 \\ x \geq 0 \\ y \geq 0 \end{cases}$

15. $P = 3x - 2y$

$\begin{cases} x \leq 1 \\ x \geq -1 \\ y - x \leq 1 \\ x - y \leq 1 \end{cases}$

16. $P = x - y$

$\begin{cases} 5x + 4y \leq 20 \\ y \leq 5 \\ x \geq 0 \\ y \geq 0 \end{cases}$

Minimize P subject to the following constraints.

17. $P = 5x + 12y$

$\begin{cases} x \geq 0 \\ y \geq 0 \\ x + y \leq 4 \end{cases}$

18. $P = 3x + 6y$

$\begin{cases} x \geq 0 \\ y \geq 0 \\ x + y \leq 4 \end{cases}$

19. $P = 3y + x$
$$\begin{cases} x \geq 0 \\ y \geq 0 \\ 2y - x \leq 1 \\ y - 2x \geq -2 \end{cases}$$

20. $P = 5y + x$
$$\begin{cases} x \leq 2 \\ y \geq 0 \\ x + y \geq 1 \\ 2y - x \leq 1 \end{cases}$$

21. $P = 6x + 2y$
$$\begin{cases} y \geq 0 \\ y - x \leq 2 \\ 2x + 3y \leq 6 \\ 3x + y \leq 3 \end{cases}$$

22. $P = 2y - x$
$$\begin{cases} x \geq 0 \\ y \geq 0 \\ x + y \leq 5 \\ x + 2y \geq 2 \end{cases}$$

23. $P = 2x - 2y$
$$\begin{cases} x \leq 1 \\ x \geq -1 \\ y - x \leq 1 \\ x - y \leq 1 \end{cases}$$

24. $P = y - 2x$
$$\begin{cases} x + 2y \leq 4 \\ 2x + y \leq 4 \\ x + 2y \geq 2 \\ 2x + y \geq 2 \end{cases}$$

APPLICATIONS   *Write the objective function and the inequalities that describe the constraints in each problem. Graph the feasibility region, showing the corner points. Then find the maximum or minimum value of the objective function.*

25. Making furniture   Two woodworkers, Tom and Carlos, bring in $100 for making a table and $80 for making a chair. On average, Tom must work 3 hours and Carlos 2 hours to make a chair. Tom must work 2 hours and Carlos 6 hours to make a table. If neither wishes to work more than 42 hours per week, how many tables and how many chairs should they make each week to maximize their income? Find the maximum income.

	Table	Chair	Time available
Income ($)	100	80	
Tom's time (hr)	2	3	42
Carlos's time (hr)	6	2	42

26. Making crafts   Two artists, Nina and Rob, make yard ornaments. They bring in $80 for each wooden snowman they make and $64 for each wooden Santa Claus. On average, Nina must work 4 hours and Rob 2 hours to make a snowman. Nina must work 3 hours and Rob 4 hours to make a Santa Claus. If neither wishes to work more than 20 hours per week, how many of each ornament should they make each week to maximize their income? Find the maximum income.

	Snowman	Santa Claus	Time available
Income ($)	80	64	
Nina's time (hr)	4	3	20
Rob's time (hr)	2	4	20

27. Inventories   An electronics store manager stocks from 20 to 30 IBM-compatible computers and from 30 to 50 Macintosh computers. There is room in the store to stock up to 60 computers. The manager receives a commission of $50 on the sale of each IBM-compatible computer and $40 on the sale of each Macintosh computer. If the manager can sell all of the computers, how many should she stock to maximize her commissions? Find the maximum commission.

Inventory	IBM	Macintosh
Minimum	20	30
Maximum	30	50
Commission	$50	$40

28. Diet problem   A diet requires at least 16 units of vitamin C and at least 34 units of Vitamin B complex. Two food supplements are available that provide these nutrients in the amounts and costs shown in the table. How much of each should be used to minimize the cost?

Supplement	Vitamin C	Vitamin B	Cost
A	3 units/g	2 units/g	3¢/g
B	2 units/g	6 units/g	4¢/g

29. **Production**   Manufacturing VCRs and TVs requires the use of the electronics, assembly, and finishing departments of a factory, according to the following schedule:

	Hours for VCR	Hours for TV	Hours available per week
Electronics	3	4	180
Assembly	2	3	120
Finishing	2	1	60

Each VCR has a profit of $40, and each TV has a profit of $32. How many VCRs and TVs should be manufactured weekly to maximize profit? Find the maximum profit.

30. **Production problem**   A company manufactures one type of computer chip that runs at 800 MHz and another that runs at 900 MHz. The company can make a maximum of 50 fast chips per day and a maximum of 100 slow chips per day. It takes 6 hours to make a fast chip and 3 hours to make a slow chip, and the company's employees can provide up to 360 hours of labor per day. If the company makes a profit of $20 on each 900-MHz chip and $27 on each 800-MHz chip, how many of each type should be manufactured to earn the maximum profit?

31. **Financial planning**   A stockbroker has $200,000 to invest in stocks and bonds. She wants to invest at least $100,000 in stocks and at least $50,000 in bonds. If stocks have an annual yield of 9% and bonds have an annual yield of 7%, how much should she invest in each to maximize her income? Find the maximum return.

32. **Production**   A small country exports soybeans and flowers. Soybeans require 8 workers per acre, flowers require 12 workers per acre, and 100,000 workers are available. Government contracts require that there be at least 3 times as many acres of soybeans as flowers planted. It costs $250 per acre to plant soybeans and $300 per acre to plant flowers, and there is a budget of $3 million. If the profit from soybeans is $1,600 per acre and the profit from flowers is $2,000 per acre, how many acres of each crop should be planted to maximize profit? Find the maximum profit.

WRITING

33. What is meant by the constraints of a linear program?

34. What is meant by a feasible solution of a linear program?

SOMETHING TO THINK ABOUT

35. Try to construct a linear programming problem. What difficulties do you encounter?

36. Try to construct a linear programming problem that will have a maximum at every point along an edge of the feasibility region.

Projects

Project 1

A farmer is building a machine shed onto his barn, as shown in the illustration. It is to be 12 feet wide, and h_2 must be no more than 20 feet. In order for all of the shed to be useful for storing machinery, h_1 must be at least 6 feet. For the roof to shed rain and melting snow adequately, the slope of the roof must be at least $\frac{1}{2}$, but to be easily shingled, it must have a slope that is no greater than 1.

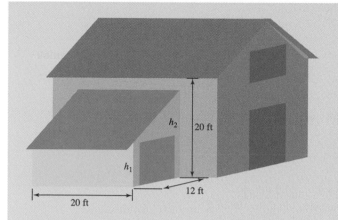

a. Represent on a graph all of the possible values for h_1 and h_2, subject to the constraints listed above.

b. The farmer wishes to minimize the construction costs while still making sure that the shed is large enough for his purposes. He does this by setting a lower bound on the volume of the shed (3,000 cubic feet) and then minimizing the surface area of the walls that must be built. The volume of the shed can be expressed in a formula that contains h_1 and h_2. Derive this formula, and include the volume restriction in your design constraints. Then find the dimen-

sions that will minimize the total area of the two ends of the shed and the outside wall. (The inner wall is already present as a wall of the barn and therefore involves no new cost.)

Project 2

Knowing any three points on the graph of a parabolic function is enough to determine the equation of that parabola. It follows that for any three points that could possibly lie on a parabola, there is exactly one parabola that passes through those points, and this parabola will have the equation $y = ax^2 + bx + c$ for appropriate a, b, and c.

a. In order for a set of three points to lie on the graph of a parabolic function, no two of the points can have the same x-coordinate, and not all three can have the same y-coordinate. Explain why we need these restrictions.

b. Suppose that the points $(1, 3)$ and $(2, 6)$ are on the graph of a parabola. What restrictions would have to be placed on a and b to guarantee that the y-intercept of the parabola has an absolute value of 4 or less? Can $(-1, 8)$ be a third point on such a parabola?

Chapter Summary

CONCEPTS	REVIEW EXERCISES
4.1	**Linear Inequalities**

Trichotomy property:
$a < b$, $a = b$, or $a > b$

Transitive properties:
If $a < b$ and $b < c$, then $a < c$.
If $a > b$ and $b > c$, then $a > c$.

Properties of inequality:
If a and b are real numbers and $a < b$, then
$$a + c < b + c$$
$$a - c < b - c$$
$$ac < bc \quad (c > 0)$$

Graph each inequality and give each solution set in interval notation.

1. $5(x - 2) \le 5$

2. $3x + 4 > 10$

3. $\frac{1}{3}x - 2 \ge \frac{1}{2}x + 2$

4. $\frac{7}{4}(x + 3) < \frac{3}{8}(x - 3)$

5. $3 < 3x + 4 < 10$

6. $4x > 3x + 2 > x - 3$

$ac > bc$ $(c < 0)$

$\dfrac{a}{c} < \dfrac{b}{c}$ $(c > 0)$

$\dfrac{a}{c} > \dfrac{b}{c}$ $(c < 0)$

$c < x < d$ is equivalent to $c < x$ and $x < d$.

7. $-5 \leq 2x - 3 < 5$

8. A woman invests \$10,000 at 6% annual interest. How much more must she invest at 7% so that her annual income is at least \$2,000?

4.2   Equations and Inequalities with Absolute Values

If $x \geq 0, |x| = x$.
If $x < 0, |x| = -x$.

Find each absolute value.

9. $|-7|$

10. $|8|$

11. $-|7|$

12. $-|-12|$

Graph each function.

13. $f(x) = |x + 1| - 3$

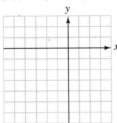

14. $f(x) = |x - 2| + 1$

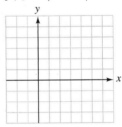

If $k > 0, |x| = k$ is equivalent to $x = k$ or $x = -k$

Solve and check each equation.

15. $|3x + 1| = 10$

16. $\left| \dfrac{3}{2}x - 4 \right| = 9$

$|a| = |b|$ is equivalent to $a = b$ or $a = -b$.

17. $\left| \dfrac{2 - x}{3} \right| = 4$

18. $|3x + 2| = |2x - 3|$

19. $|5x - 4| = |4x - 5|$

20. $\left| \dfrac{3 - 2x}{2} \right| = \left| \dfrac{3x - 2}{3} \right|$

If $k > 0, |x| < k$ is equivalent to $-k < x < k$.

Solve each inequality. Give the solution in interval notation and graph it.

21. $|2x + 7| < 3$

22. $|5 - 3x| \leq 14$

$|x| > k$ is equivalent to $x < -k$ or $x > k$.

23. $\left| \dfrac{2}{3}x + 14 \right| < 0$

24. $\left| \dfrac{1 - 5x}{3} \right| > 7$

25. $|3x - 8| \geq 4$

26. $\left| \dfrac{3}{2}x - 14 \right| \geq 0$

| **4.3** | **Linear Inequalities in Two Variables** |

To graph a linear inequality in x and y, graph the boundary line, and then use a test point to decide which side of the boundary should be shaded.

Graph each inequality on the coordinate plane.

27. $2x + 3y > 6$

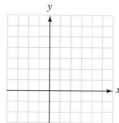

28. $y \leq 4 - x$

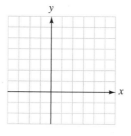

29. $-2 < x < 4$

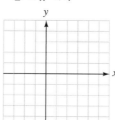

30. $y \leq -2$ or $y > 1$

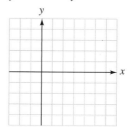

| **4.4** | **Systems of Inequalities** |

Systems of inequalities can be solved by graphing.

Graph the solution set of each system of inequalities.

31. $\begin{cases} y \geq x + 1 \\ 3x + 2y < 6 \end{cases}$

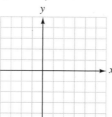

32. $\begin{cases} y \geq x^2 - 4 \\ y < x + 3 \end{cases}$

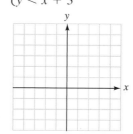

| **4.5** | **Linear Programming** |

If a linear function, subject to the constraints of a system of linear inequalities in two variables, attains a maximum or a minimum value, that value will occur at a corner or along an entire edge of the region R that represents the solution of the system.

33. Maximize $P = 2x + y$ subject to $\begin{cases} x \geq 0 \\ y \geq 0 \\ x + y \leq 3 \end{cases}$

34. A company manufactures fertilizers X and Y. Each 50-pound bag requires three ingredients, which are available in the limited quantities shown below.

Ingredient	Number of pounds in fertilizer X	Number of pounds in fertilizer Y	Total number of pounds available
Nitrogen	6	10	20,000
Phosphorus	8	6	16,400
Potash	6	4	12,000

The profit on each bag of fertilizer X is \$6, and on each bag of Y, \$5. How many bags of each should be produced to maximize profit?

Chapter Test

Graph the solution of each inequality and give the solution in interval notation.

1. $-2(2x + 3) \geq 14$

2. $-2 < \dfrac{x - 4}{3} < 4$

Write each expression without absolute value symbols.

3. $|5 - 8|$

4. $|4\pi - 4|$

Graph each function.

5. $f(x) = |x + 1| - 4$

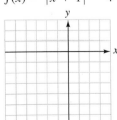

6. $f(x) = |x - 2| + 3$

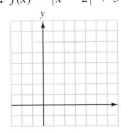

Solve each equation.

7. $|2x + 3| = 11$

8. $|4 - 3x| = 19$

9. $|3x + 4| = |x + 12|$

10. $|3 - 2x| = |2x + 3|$

Graph the solution of each inequality and give the solution in interval notation.

11. $|x + 3| \leq 4$

12. $|2x - 4| > 22$

13. $|4 - 2x| > 2$

14. $|2x - 4| \leq 2$

Graph each inequality.

15. $3x + 2y \geq 6$

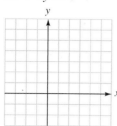

16. $-2 \leq y < 5$

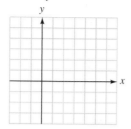

Use graphing to solve each system.

17. $\begin{cases} 2x - 3y \geq 6 \\ y \leq -x + 1 \end{cases}$

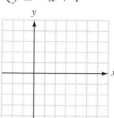

18. $\begin{cases} y \geq x^2 \\ y < x + 3 \end{cases}$

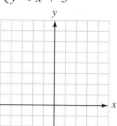

19. Maximize $P = 3x - y$ subject to $\begin{cases} y \geq 1 \\ y \leq 2 \\ y \leq 3x + 1 \\ x \leq 1 \end{cases}$.

CUMULATIVE REVIEW EXERCISES

1. Draw a number line and graph the prime numbers from 50 to 60.

2. Find the additive inverse of -5.

Evaluate each expression when $x = 2$ and $y = -4$.

3. $x - xy$

4. $\dfrac{x^2 - y^2}{3x + y}$

Simplify each expression.

5. $(x^2x^3)^2$

6. $(x^2)^3(x^4)^2$

7. $\left(\dfrac{x^3}{x^5}\right)^{-2}$

8. $\dfrac{a^2b^n}{a^nb^2}$

9. Write 32,600,000 in scientific notation.

10. Write 0.000012 in scientific notation.

Solve each equation, if possible.

11. $3x - 6 = 20$

12. $6(x - 1) = 2(x + 3)$

13. $\dfrac{5b}{2} - 10 = \dfrac{b}{3} + 3$

14. $2a - 5 = -2a + 4(a - 2) + 1$

Tell whether the lines represented by the equations are parallel, perpendicular, or neither.

15. $3x + 2y = 12,\ 2x - 3y = 5$

16. $3x = y + 4,\ y = 3(x - 4) - 1$

17. Write the equation of the line passing through $(-2, 3)$ and perpendicular to the graph of $3x + y = 8$.

18. Solve the formula $A = \frac{1}{2}h(b_1 + b_2)$ for h.

Find each value, given that $f(x) = 3x^2 - x$.

19. $f(2)$

20. $f(-2)$

21. Use graphing to solve $\begin{cases} 2x + y = 5 \\ x - 2y = 0 \end{cases}$.

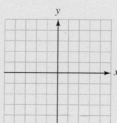

22. Use substitution to solve $\begin{cases} 3x + y = 4 \\ 2x - 3y = -1 \end{cases}$.

23. Use addition to solve: $\begin{cases} x + 2y = -2 \\ 2x - y = 6 \end{cases}$.

24. Solve: $\begin{cases} \dfrac{x}{10} + \dfrac{y}{5} = \dfrac{1}{2} \\ \dfrac{x}{2} - \dfrac{y}{5} = \dfrac{13}{10} \end{cases}$.

25. Solve: $\begin{cases} x + y + z = 1 \\ 2x - y - z = -4 \\ x - 2y + z = 4 \end{cases}$.

26. Solve: $\begin{cases} x + 2y + 3z = 1 \\ 3x + 2y + z = -1 \\ 2x + 3y + z = -2 \end{cases}$.

27. Evaluate: $\begin{vmatrix} 3 & -2 \\ 1 & -1 \end{vmatrix}$.

28. Evaluate: $\begin{vmatrix} 2 & 3 & -1 \\ -1 & -1 & 2 \\ 4 & 1 & -1 \end{vmatrix}$.

Use Cramer's rule to solve each system.

29. $\begin{cases} 4x - 3y = -1 \\ 3x + 4y = -7 \end{cases}$.

30. $\begin{cases} x - 2y - z = -2 \\ 3x + y - z = 6 \\ 2x - y + z = -1 \end{cases}$

Solve each inequality.

31. $-3(x - 4) \geq x - 32$

32. $-8 < -3x + 1 < 10$

Solve each equation.

33. $|4x - 3| = 9$

34. $|2x - 1| = |3x + 4|$

Solve each inequality.

35. $|3x - 2| \leq 4$ **36.** $|2x + 3| - 1 > 4$

Use graphing to solve each inequality.

37. $2x - 3y \leq 12$ **38.** $3 > x \geq -2$

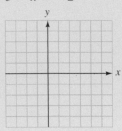   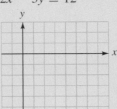

Use graphing to solve each system.

39. $\begin{cases} 3x - 2y < 6 \\ y < -x + 2 \end{cases}$ **40.** $\begin{cases} y < x + 2 \\ 3x + y \leq 6 \end{cases}$

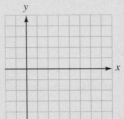   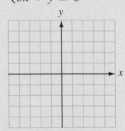

41. To conduct an experiment with mice and rats, a researcher will place the animals into one of two mazes for the number of minutes shown in the table. Find the greatest number of animals that can be used in this experiment.

	Time per mouse	Time per rat	Time available
Maze 1	12 min	8 min	240 min
Maze 2	10 min	15 min	300 min

5

Polynomials and Polynomial Functions

© Royalty-Free/CORBIS

InfoTrac Project

Do a subject guide search on "consumer debt." Under "consumer debt counseling," click on "periodicals." Find the article "Debt counselors rescue bankruptcy chits." Write a summary of the article.

If a consumer charges $1,500 worth of merchandise on his credit card and makes payments of only $25 each month, the balance (without interest) for each month after the first payment can be represented by $B(n) = 1{,}475 - 25n$, where n represents the number of months the $25 payment is paid. The 18% annual interest charged by the bank is equivalent to $1\frac{1}{2}$ percent interest on the balance each month. It can be represented by $I(n) = 0.015(1{,}500 - 25n)$. The total amount owed, including interest, can be represented by the sum of the two functions. How many months will it take to pay off the credit card? Convert your answer into years. Suppose he pays $50 a month. How long will it take to pay off the credit card?

Complete this project after studying Section 5.2.

Mathematics in Computer Science

Computers take more time to do multiplications than additions. To make a computer program run as quickly as possible, computer programmers want to write polynomials such as

$$3x^4 + 2x^3 + 5x^2 + 7x + 1$$

in a form that requires fewer multiplications. Write the polynomial in a form that requires four multiplications.

**Example 10
Section 5.4**

Polynomials can be considered to be the numbers of algebra. In this chapter, we will learn how to add, subtract, and multiply them. Then we will learn how to reverse the operation of multiplication and find the factors of a known product.

5.1	Polynomials and Polynomial Functions

▮ **Polynomials** ▮ **Degree of a Polynomial**
▮ **Evaluating Polynomial Functions**
▮ **Graphing Polynomial Functions**

Getting Ready   *Write each expression using exponents.*

1. $3aabb$ **2.** $-5xxxy$
3. $4pp + 7qq$ **4.** $aaa - bb$

In this section, we discuss important algebraic expressions called *polynomials*. We then examine several polynomial functions.

Polynomials

Algebraic terms are expressions that contain constants and/or variables. Some examples are

$$17, \qquad 9x, \qquad 15y^2, \qquad \text{and} \qquad -24x^4y^5$$

The *numerical coefficient* of 17 is 17. The numerical coefficients of the remaining terms are 9, 15, and -24, respectively.

A **polynomial** is the sum of one or more algebraic terms whose variables have whole-number exponents.

Polynomial in One Variable   A **polynomial in one variable** (say, x) is the sum of one or more terms of the form ax^n, where a is a real number and n is a whole number.

The following expressions are polynomials in x. Note that $17 = 17x^0$.

$$3x^2 + 2x, \qquad \frac{3}{2}x^5 - \frac{7}{3}x^4 - \frac{8}{3}x^3, \quad \text{and} \quad 19x^{20} + \sqrt{3}x^{14} + 4.5x^{11} - 17$$

The following expressions are not polynomials.

$$\frac{2x}{x^2 + 1}, \qquad x^{1/2} - 1, \quad \text{and} \quad x^{-3} + 2x$$

The first expression is the quotient of two polynomials, and the last two have exponents that are not whole numbers.

Polynomial in More than One Variable

A **polynomial in several variables** (say x, y, and z) is the sum of one or more terms of the form $ax^m y^n z^p$; where a is a real number and m, n, and p are whole numbers.

The following expressions are polynomials in more than one variable.

$$3xy, \qquad 5x^2y + 2yz^3 - 3xz, \quad \text{and} \quad u^2v^2w^2 + x^3y^3 + 1$$

A polynomial with one term is called a **monomial,** a polynomial with two terms is called a **binomial,** and a polynomial with three terms is called a **trinomial:**

Monomials (*one term*)	*Binomials* (*two terms*)	*Trinomials* (*three terms*)
$2x^3$	$2x^4 + 5$	$2x^3 + 4x^2 + 3$
a^2b	$-17t^{45} - 3xy$	$3mn^3 - m^2n^3 + 7n$
$3x^3y^5z^2$	$32x^{13}y^5 + 47x^3yz$	$-12x^5y^2 + 13x^4y^3 - 7x^3y^3$

Degree of a Polynomial

Because the variable x occurs three times as a factor in the monomial $2x^3$, the monomial is called a *third-degree monomial* or a *monomial of degree 3*. The monomial $3x^3y^5z^2$ is called a *monomial of degree 10,* because the variables x, y, and z occur as factors a total of ten times. These examples illustrate the following definition.

Degree of a Monomial

If $a \neq 0$, the **degree of ax^n** is n. The degree of a monomial containing several variables is the sum of the exponents on those variables.

EXAMPLE 1   Find the degree of   **a.** $3x^4$,   **b.** $-18x^3y^2z^{12}$,   **c.** $4^7x^2y^3$,   and   **d.** 3.

Solution   **a.** $3x^4$ is a monomial of degree 4, because the variable x occurs as a factor four times.

b. $-18x^3y^2z^{12}$ is a monomial of degree 17, because the sum of the exponents on the variables is 17.

c. $4^7x^2y^3$ is a monomial of degree 5, because the sum of the exponents on the variables is 5.

d. 3 is a monomial of degree 0, because $3 = 3x^0$.

Self Check   Find the degree of   **a.** $-12a^5$ and   **b.** $8a^3b^2$.   ∎

   Comment   Since $a \neq 0$ in the previous definition, 0 has no defined degree.

We define the degree of a polynomial by considering the degrees of each of its terms.

Degree of a Polynomial

The **degree of a polynomial** is the same as the degree of the term in the polynomial with largest degree.

EXAMPLE 2   Find the degree of each polynomial: **a.** $3x^5 + 4x^2 + 7$,   **b.** $7x^2y^8 - 3xy$,
c. $3x + 2y - xy$,   and   **d.** $18x^2y^3 - 12x^7y^2 + 3x^9y^3 - 3$.

Solution   **a.** $3x^5 + 4x^2 + 7$ is a trinomial of degree 5, because the largest degree of the three monomials is 5.

b. $7x^2y^8 - 3xy$ is a binomial of degree 10.

c. $3x + 2y - xy$ is a trinomial of degree 2.

d. $18x^2y^3 - 12x^7y^2 + 3x^9y^3 - 3$ is a polynomial of degree 12.

Self Check   Find the degree of $7a^3b^2 - 14a^2b^4$.   ∎

If the terms of a polynomial in one variable are written so that the exponents decrease as we move from left to right, we say that the terms are written with their exponents in descending order. If the terms are written so that the exponents increase as we move from left to right, we say that the terms are written with their exponents in ascending order.

EXAMPLE 3   Write the exponents of $7x^2 - 5x^4 + 3x + 2x^3 - 1$ in   **a.** descending order and   **b.** ascending order.

Solution   **a.** $-5x^4 + 2x^3 + 7x^2 + 3x - 1$

b. $-1 + 3x + 7x^2 + 2x^3 - 5x^4$

Self Check   Write the exponents of $3x^2 + 2x^4 - 5x - 3$ in descending order.   ∎

In the following polynomial, the exponents on x are in descending order, and the exponents on y are in ascending order.

$$7x^4 - 2x^3y + 4x^2y^2 - 8xy^3 + 12y^4$$

Evaluating Polynomial Functions

A function of the form $y = P(x)$, where $P(x)$ is a polynomial, is called a **polynomial function.** Some examples of polynomial functions are

$$P(x) = -5x - 10 \quad \text{and} \quad Q(t) = 2t^2 + 5t - 2 \qquad \text{Read } P(x) \text{ as "} P \text{ of } x \text{" and } Q(t) \text{ as "} Q \text{ of } t."$$

Polynomial Functions   A **polynomial function in one variable (say, x)** is defined by an equation of the form $y = f(x) = P(x)$, where $P(x)$ is a polynomial in the variable x.

The **degree of the polynomial function** $y = f(x) = P(x)$ is the same as the degree of $P(x)$.

To evaluate a polynomial function at a specific value of its variable, we substitute the value of the variable and simplify. For example, to evaluate $y = P(x)$ at $x = 1$, we substitute 1 for x and simplify.

$$P(x) = x^6 + 4x^5 - 3x^2 + x - 2$$
$$P(1) = (1)^6 + 4(1)^5 - 3(1)^2 + 1 - 2 \quad \text{Substitute 1 for } x.$$
$$= 1 + 4 - 3 + 1 - 2$$
$$= 1$$

Thus, $P(1) = 1$.

EXAMPLE 4   **Height of a rocket**   If a toy rocket is launched straight up with an initial velocity of 128 feet per second, its height h (in feet) above the ground after t seconds is given by the polynomial function

$$P(t) = -16t^2 + 128t \quad \text{The height } h \text{ is the value } P(t).$$

Find the height of the rocket at   **a.** 0 second,   **b.** 3 seconds,   and   **c.** 7.9 seconds.

Solution   **a.** To find the height at 0 second, we substitute 0 for t and simplify.

$$P(t) = -16t^2 + 128t$$
$$P(0) = -16(0)^2 + 128(0)$$
$$= 0$$

At 0 second, the rocket is on the ground waiting to be launched.

b. To find the height at 3 seconds, we substitute 3 for t and simplify.

$$P(3) = -16(3)^2 + 128(3)$$
$$= -16(9) + 384$$
$$= -144 + 384$$
$$= 240$$

At 3 seconds, the height of the rocket is 240 feet.

c. To find the height at 7.9 seconds, we substitute 7.9 for t and simplify.

$$P(7.9) = -16(7.9)^2 + 128(7.9)$$
$$= -16(62.41) + 1{,}011.2$$
$$= -998.56 + 1{,}011.2$$
$$= 12.64$$

At 7.9 seconds, the height is 12.64 feet. It has fallen nearly back to Earth.

Self Check   Find the height of the rocket at 4 seconds.

EXAMPLE 5   For the polynomial function $P(x) = 3x^2 - 2x + 7$, find   **a.** $P(a)$ and   **b.** $P(-2t)$.

Solution   **a.** $P(a) = 3(a)^2 - 2(a) + 7$
$$= 3a^2 - 2a + 7$$

$\qquad$ **b.** $P(-2t) = 3(-2t)^2 - 2(-2t) + 7$

$\qquad\qquad\quad = 12t^2 + 4t + 7$

Self Check   Find $P(2t)$.   ▮

$\qquad$ To evaluate polynomials with more than one variable, we substitute values for the variables into the polynomial and simplify.

EXAMPLE 6   Evaluate $4x^2y - 5xy^3$ at $x = 3$ and $y = -2$.

Solution   We substitute 3 for x and -2 for y and simplify.

$$4x^2y - 5xy^3 = 4(\mathbf{3})^2(-\mathbf{2}) - 5(\mathbf{3})(-\mathbf{2})^3$$
$$= 4(9)(-2) - 5(3)(-8)$$
$$= -72 + 120$$
$$= 48$$

Self Check   Evaluate $3ab^2 - 2a^2b$ at $a = 2$ and $b = -3$.   ▮

Graphing Polynomial Functions

In Sections 2.4 and 2.5, we discussed three basic polynomial functions. The graph of the linear function $y = f(x) = 3x + 1$, the graph of the squaring function $y = f(x) = x^2$, and the graph of the cubing function $y = f(x) = x^3$ are shown in Figure 5-1.

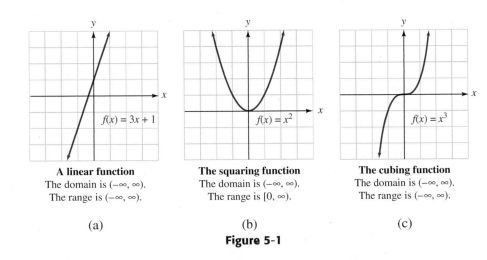

A linear function	The squaring function	The cubing function
The domain is $(-\infty, \infty)$.	The domain is $(-\infty, \infty)$.	The domain is $(-\infty, \infty)$.
The range is $(-\infty, \infty)$.	The range is $[0, \infty)$.	The range is $(-\infty, \infty)$.
(a)	(b)	(c)

Figure 5-1

$\qquad$ We have seen that the graphs of many polynomial functions are translations or reflections of these graphs. For example, the graph of $f(x) = x^2 - 2$ is the graph of $f(x) = x^2$ translated 2 units downward, as shown in Figure 5-2(a). The graph of $f(x) = -x^2$ is the graph of $f(x) = x^2$ reflected about the x-axis, as shown in Figure 5-2(b).

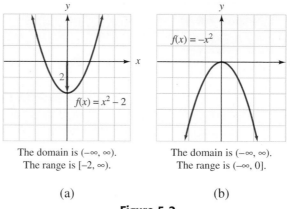

The domain is $(-\infty, \infty)$.
The range is $[-2, \infty)$.

The domain is $(-\infty, \infty)$.
The range is $(-\infty, 0]$.

(a)   (b)

Figure 5-2

In Example 4, we saw that a polynomial function can describe (or model) the height of a rocket. Since the height (h) of the rocket depends on time (t), we say that the height is a function of time, and we can write $h = f(t)$. To graph this function, we can make a table of values, plot the points, and join them with a smooth curve, as shown in Example 7.

EXAMPLE 7   Graph $h = f(t) = -16t^2 + 128t$.

Solution   From Example 4, we saw that

When $t = 3$, then $h = 240$.

When $t = 0$, then $h = 0$.

These ordered pairs and others that satisfy $h = -16t^2 + 128t$ are given in the table shown in Figure 5-3. Here the ordered pairs are pairs (t, h), where t is the time and h is the height.

We plot these pairs on a horizontal t-axis and a vertical h-axis and join the resulting points to get the parabola shown in the figure. From the graph, we can see that 4 seconds into the flight, the rocket attains a maximum height of 256 feet.

$$h = f(t) = -16t^2 + 128t$$

t	$h = f(t)$	$(t, f(t))$
0	0	$(0, 0)$
1	112	$(1, 112)$
2	192	$(2, 192)$
3	240	$(3, 240)$
4	256	$(4, 256)$
5	240	$(5, 240)$
6	192	$(6, 192)$
7	112	$(7, 112)$
8	0	$(8, 0)$

Figure 5-3

Self Check   Use Figure 5-3 to estimate the height of the rocket at 6.5 seconds.

Comment   The parabola shown in Figure 5-3 describes the height of the rocket in relation to time. It does not show the path of the rocket. The rocket goes straight up and then comes straight down.

Accent on Technology **GRAPHING POLYNOMIAL FUNCTIONS**

We can graph polynomial functions with a graphing calculator. For example, to graph $y = P(t) = -16t^2 + 128t$, we can use window settings of $[0, 8]$ for x and $[0, 260]$ for y to get the parabola shown in Figure 5-4(a).

We can trace to estimate the height of the rocket for any number of seconds into the flight. Figure 5-4(b) shows that the height of the rocket 1.6 seconds into the flight is approximately 165 feet.

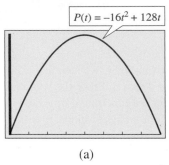

$P(t) = -16t^2 + 128t$

(a)

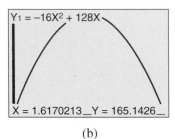

$Y_1 = -16X^2 + 128X$

$X = 1.6170213 \quad Y = 165.1426$

(b)

Figure 5-4

Self Check Answers

1. a. 5, **b.** 5 **2.** 6 **3.** $2x^4 + 3x^2 - 5x - 3$ **4.** 256 ft **5.** $12t^2 - 4t + 7$ **6.** 78 **7.** 150 ft

Orals *Give the degree of each polynomial.*

1. $8x^3$ **2.** $-4x^3y$ **3.** $4x^2y + 2x$ **4.** $3x + 4xyz$

If $P(x) = 2x + 1$, find each value.

5. $P(0)$ **6.** $P(2)$ **7.** $P(-1)$ **8.** $P(-2)$

5.1 **EXERCISES**

REVIEW *Write each expression with a single exponent.*

1. $a^3 a^2$

2. $\dfrac{b^3 b^3}{b^4}$

3. $\dfrac{3(y^3)^{10}}{y^3 y^4}$

4. $\dfrac{4x^{-4} x^5}{2x^{-6}}$

5. The distance from Mars to the sun is about 114,000,000 miles. Express this number in scientific notation.

6. One angstrom is about 0.0000001 millimeter. Express this number in scientific notation.

VOCABULARY AND CONCEPTS *Fill in the blanks.*

7. A polynomial is the ____ of one or more algebraic terms whose variables have _____ number exponents.

8. A monomial is a polynomial with ____ term.

9. A ___ ____ is a polynomial with two terms.

10. A _____ is a polynomial with three terms.

11. The equation $y = P(x)$, where $P(x)$ is a polynomial, defines a function, because each input value x determines ____ output value y.

12. The degree of the polynomial function $y = P(x)$ is the same as the degree of ____.

Classify each polynomial as a monomial, a binomial, a trinomial, or none of these.

13. $3x^2$

14. $2y^3 + 4y^2$

15. $3x^2y - 2x + 3y$

16. $a^2 + b^2$

17. $x^2 - y^2$

18. $\dfrac{17}{2}x^3 + 3x^2 - x - 4$

19. 5

20. $8x^3y^5$

Find the degree of each polynomial.

21. $3x^2 + 2$

22. x^{17}

23. $4x^8 + 3x^2y^4$

24. $19x^2y^4 - y^{10}$

25. $4x^2 - 5y^3z^3t^4$

26. $7x$

27. 121

28. $x^2y^3z^4 + z^{12}$

Write each polynomial with the exponents on x in descending order.

29. $3x - 2x^4 + 7 - 5x^2$

30. $-x^2 + 3x^5 - 7x + 3x^3$

31. $a^2x - ax^3 + 7a^3x^5 - 5a^3x^2$

32. $4x^2y^7 - 3x^5y^2 + 4x^3y^3 - 2x^4y^6 + 5x^6$

Write each polynomial with the exponents on y in ascending order.

33. $4y^2 - 2y^5 + 7y - 5y^3$

34. $y^3 + 3y^2 + 8y^4 - 2$

35. $5x^3y^6 + 2x^4y - 5x^3y^3 + x^5y^7 - 2y^4$

36. $-x^3y^2 + x^2y^3 - 2x^3y + x^7y^6 - 3x^6$

PRACTICE *Consider the polynomial function $P(x) = 2x^2 + x + 2$ and find each value.*

37. $P(0)$

38. $P(1)$

39. $P(-2)$

40. $P(-3)$

The height h, in feet, of a ball shot straight up with an initial velocity of 64 feet per second is given by the polynomial function $h = f(t) = -16t^2 + 64t$. Find the height of the ball after the given number of seconds.

41. 0 second

42. 1 second

43. 2 seconds

44. 4 seconds

Find each value when $x = 2$ and $y = -3$.

45. $x^2 + y^2$

46. $x^3 + y^3$

47. $x^3 - y^3$

48. $x^2 - y^2$

49. $3x^2y + xy^3$

50. $8xy - xy^2$

51. $-2xy^2 + x^2y$

52. $-x^3y - x^2y^2$

Use a calculator to find each value when $x = 3.7$, $y = -2.5$, and $z = 8.9$.

53. x^2y

54. xyz^2

55. $\dfrac{x^2}{z^2}$

56. $\dfrac{z^3}{y^2}$

57. $\dfrac{x + y + z}{xyz}$

58. $\dfrac{x + yz}{xy + z}$

Graph each polynomial function. Check your work with a graphing calculator.

59. $f(x) = x^2 + 2$

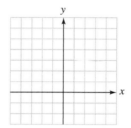

60. $f(x) = x^3 - 2$

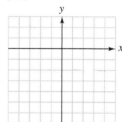
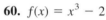

61. $f(x) = -x^3$

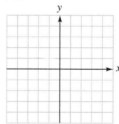

62. $f(x) = -x^2 + 1$

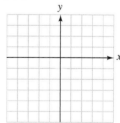
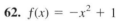

63. $f(x) = -x^3 + x$

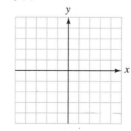

64. $f(x) = x^3 - x$

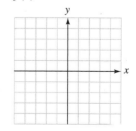

65. $f(x) = x^2 - 2x + 1$

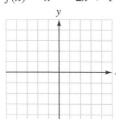

66. $f(x) = x^2 - 2x - 3$

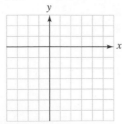

📉 *Use a graphing calculator to graph each polynomial function. Use window settings of $[-4, 6]$ for x and $[-5, 5]$ for y.*

67. $f(x) = 2.75x^2 - 4.7x + 1.5$

68. $f(x) = -2.5x^2 + 1.7x + 3.2$

69. $f(x) = 0.25x^2 - 0.5x - 2.5$

70. $f(x) = 0.37x^2 - 1.4x - 1.5$

🖩 **APPLICATIONS** *The number of feet that a car travels before stopping depends on the driver's reaction time and the braking distance. See the illustration. For one driver, the stopping distance d is given by the polynomial function $d = f(v) = 0.04v^2 + 0.9v$, where v is the speed of the car. Find the stopping distance for each of the following speeds.*

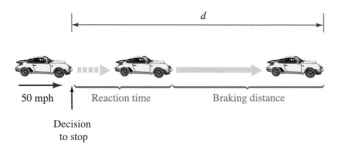

71. 30 mph

73. 60 mph

72. 50 mph

74. 70 mph

A rectangular sheet of metal will be used to make a rain gutter by bending up its sides, as shown in the illustration. Since a cross section is a rectangle, the cross-sectional area is the product of its length and width, and the capacity c of the gutter is a polynomial function of x: $c = f(x) = -2x^2 + 12x$. Find the capacity for each value of x.

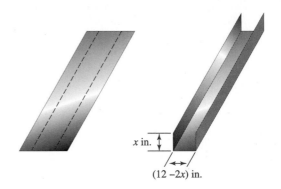

x in.

$(12 - 2x)$ in.

75. 1 inch

77. 3 inches

76. 2 inches

78. 4 inches

79. **Roller coasters** The polynomial function $h(t) = 0.001t^3 - 0.12t^2 + 3.6t + 10$ gives the height h (in feet) during the time t (in seconds) of a portion of the track of a roller coaster.
  a. Find the height of the track when $t = 0$.
  b. Find the height of the track when $t = 10$.
  c. Find the height of the track when $t = 40$.

80. **Customer service** A technical help service of a computer software company has found that on Mondays, the polynomial function $C(t) = -0.0625t^4 + t^3 + 16t$ approximates the number of calls received at any one time. If t represents the time (in hours) since the service opened at 8:00 A.M., how many calls are expected at noon?

WRITING

81. Explain how to find the degree of a polynomial.

82. Explain why $P(x) = x^6 + 4x^5 - 3x^2 + x - 2$ defines a function.

SOMETHING TO THINK ABOUT

83. If $P(x) = x^2 - 5x$, is $P(2) + P(3) = P(2 + 3)$?

84. If $P(x) = x^3 - 3x$, is $P(2) - P(3) = P(2 - 3)$?

85. If $P(x) = x^2 - 2x - 3$, find $P(P(0))$.

86. If $P(x) = 2x^2 - x - 5$, find $P(P(-1))$.

87. Graph $f(x) = x^2$, $f(x) = 2x^2$, and $f(x) = 4x^2$. What do you discover?

88. Graph $f(x) = x^2$, $f(x) = \frac{1}{2}x^2$, and $f(x) = \frac{1}{4}x^2$. What do you discover?

5.2	**Adding and Subtracting Polynomials**

▮ **Combining Like Terms** ▮ **Adding Polynomials**
▮ **Subtracting Polynomials**
▮ **Adding and Subtracting Multiples of Polynomials**

Getting Ready *Use the distributive property to remove parentheses.*

1. $2(x + 3)$ **2.** $-4(2x - 5)$

3. $\frac{5}{2}(2x + 6)$ **4.** $-\frac{2}{3}(3x - 9)$

In this section, we will see how to add and subtract polynomials. To do so, we will first learn how to combine like terms.

Combining Like Terms

Recall that when terms have the same variables with the same exponents, they are called **like** or **similar terms.**

- $3x^2$, $5x^2$, and $7x^2$ are like terms, because they have the same variables with the same exponents.
- $5x^3y^2$, $17x^3y^2$, and $103x^3y^2$ are like terms, because they have the same variables with the same exponents.
- $4x^4y^2$, $12xy^5$, and $98x^7y^9$ are unlike terms. They have the same variables, but with different exponents.
- $3x^4y$ and $5x^4z^2$ are unlike terms. They have different variables.

The distributive property enables us to combine like terms. For example,

$$3x + 7x = (3 + 7)x \qquad \text{Use the distributive property.}$$
$$= 10x \qquad 3 + 7 = 10.$$

$$5x^2y^3 + 22x^2y^3 = (5 + 22)x^2y^3 \qquad \text{Use the distributive property.}$$
$$= 27x^2y^3 \qquad 5 + 22 = 27.$$

$$9xy^4 + 6xy^4 + xy^4 = 9xy^4 + 6xy^4 + 1xy^4 \qquad xy^4 = 1xy^4.$$
$$= (9 + 6 + 1)xy^4 \qquad \text{Use the distributive property.}$$
$$= 16xy^4 \qquad 9 + 6 + 1 = 16.$$

The results of the previous example suggest that to add like terms, *we add their numerical coefficients and keep the same variables with the same exponents.*

 Comment   The terms in the following binomials cannot be combined, because they are not like terms.

$$3x^2 - 5y^2, \qquad -2a^2 + 3a^3, \quad \text{and} \quad 5y^2 + 17xy$$

Adding Polynomials

To add polynomials, we use the distributive property to remove parentheses and combine like terms, whenever possible.

EXAMPLE 1   Add $3x^2 - 2x + 4$ and $2x^2 + 4x - 3$.

Solution

$$(3x^2 - 2x + 4) + (2x^2 + 4x - 3)$$

$= \mathbf{1}(3x^2 - 2x + 4) + \mathbf{1}(2x^2 + 4x - 3)$	Each polynomial has an understood coefficient of 1.
$= 3x^2 - 2x + 4 + 2x^2 + 4x - 3$	Use the distributive property to remove parentheses.
$= 3x^2 + 2x^2 - 2x + 4x + 4 - 3$	Use the commutative property of addition to get the terms involving x^2 together and the terms involving x together.
$= 5x^2 + 2x + 1$	Combine like terms.

Self Check   Add $2a^2 - 3a + 5$ and $5a^2 - 4a - 2$.   ∎

EXAMPLE 2   Add $5x^3y^2 + 4x^2y^3$ and $-2x^3y^2 + 5x^2y^3$.

Solution

$$(5x^3y^2 + 4x^2y^3) + (-2x^3y^2 + 5x^2y^3)$$
$$= \mathbf{1}(5x^3y^2 + 4x^2y^3) + \mathbf{1}(-2x^3y^2 + 5x^2y^3)$$
$$= 5x^3y^2 + 4x^2y^3 - 2x^3y^2 + 5x^2y^3$$
$$= 5x^3y^2 - 2x^3y^2 + 4x^2y^3 + 5x^2y^3$$
$$= 3x^3y^2 + 9x^2y^3$$

Self Check   Add $6a^2b^3 + 5a^3b^2$ and $-3a^2b^3 - 2a^3b^2$.   ∎

The additions in Examples 1 and 2 can be done by aligning the terms vertically.

$$\begin{array}{r} 3x^2 - 2x + 4 \\ \underline{2x^2 + 4x - 3} \\ 5x^2 + 2x + 1 \end{array} \qquad \begin{array}{r} 5x^3y^2 + 4x^2y^3 \\ \underline{-2x^3y^2 + 5x^2y^3} \\ 3x^3y^2 + 9x^2y^3 \end{array}$$

Subtracting Polynomials

To subtract one monomial from another, we add the negative (or opposite) of the monomial that is to be subtracted.

EXAMPLE 3   Perform each subtraction:

a. $8x^2 - 3x^2 = 8x^2 + (-3x^2)$ $\qquad$ b. $3x^2y - 9x^2y = 3x^2y + (-9x^2y)$
$\qquad\qquad\quad = 5x^2$ $\qquad\qquad\qquad\qquad\qquad = -6x^2y$

c. $-5x^5y^3z^2 - 3x^5y^3z^2 = -5x^5y^3z^2 + (-3x^5y^3z^2)$
$\qquad\qquad\qquad\quad = -8x^5y^3z^2$

Self Check   Subtract: a. $-2a^2b^3 - 5a^2b^3$ and   b. $-2a^2b^3 - (-5a^2b^3)$.   ∎

To subtract polynomials, we use the distributive property to remove parentheses and combine like terms, whenever possible.

EXAMPLE 4   Perform each subtraction:

a. $(8x^3y + 2x^2y) - (2x^3y - 3x^2y)$

$\qquad = \mathbf{1}(8x^3y + 2x^2y) - \mathbf{1}(2x^3y - 3x^2y)$ $\qquad$ Insert the understood coefficients of 1.

$\qquad = 8x^3y + 2x^2y - 2x^3y + 3x^2y$ $\qquad$ Use the distributive property to remove parentheses.

$\qquad = 6x^3y + 5x^2y$ $\qquad$ Combine like terms.

b. $(3rt^2 + 4r^2t^2) - (8rt^2 - 4r^2t^2 + r^3t^2)$

$\qquad = \mathbf{1}(3rt^2 + 4r^2t^2) - \mathbf{1}(8rt^2 - 4r^2t^2 + r^3t^2)$ $\qquad$ Insert the understood coefficients of 1.

$\qquad = 3rt^2 + 4r^2t^2 - 8rt^2 + 4r^2t^2 - r^3t^2$ $\qquad$ Use the distributive property to remove parentheses.

$\qquad = -5rt^2 + 8r^2t^2 - r^3t^2$ $\qquad$ Combine like terms.

Self Check   Subtract: $(6a^2b^3 - 2a^2b^2) - (-2a^2b^3 + a^2b^2)$.   ∎

To subtract polynomials in vertical form, we add the negative (or opposite) of the polynomial that is being subtracted.

$$-\dfrac{\begin{array}{r} 8x^3y + 2x^2y \\ 2x^3y - 3x^2y \end{array}}{} \Rightarrow +\dfrac{\begin{array}{r} 8x^3y + 2x^2y \\ -2x^3y + 3x^2y \end{array}}{6x^3y + 5x^2y}$$

Adding and Subtracting Multiples of Polynomials

To add or subtract multiples of one polynomial and another, we use the distributive property to remove parentheses and then combine like terms.

EXAMPLE 5   Simplify: $3(2x^2 + 4x - 7) - 2(3x^2 - 4x - 5)$.

Solution $3(2x^2 + 4x - 7) - 2(3x^2 - 4x - 5)$

$\qquad = 6x^2 + 12x - 21 - 6x^2 + 8x + 10$ $\qquad$ Remove parentheses.

$\qquad = 20x - 11$ $\qquad$ Combine like terms.

Self Check   Simplify: $-2(3a^3 + a^2) - 5(2a^3 - a^2 + a)$.   ∎

1. $7a^2 - 7a + 3$ **2.** $3a^2b^3 + 3a^3b^2$ **3. a.** $-7a^2b^3$, **b.** $3a^2b^3$ **4.** $8a^2b^3 - 3a^2b^2$ **5.** $-16a^3 + 3a^2 - 5a$

Orals   *Combine like terms.*

1. $4x^2 + 5x^2$

2. $3y^2 - 5y^2$

Perform the operations.

3. $(x^2 + 2x + 1) + (2x^2 - 2x + 1)$
4. $(x^2 + 2x + 1) - (2x^2 - 2x + 1)$
5. $(2x^2 - x - 3) + (x^2 - 3x - 1)$
6. $(2x^2 - x - 3) - (x^2 - 3x - 1)$

5.2   EXERCISES

REVIEW   *Solve each inequality. Give the result in interval notation.*

1. $2x + 3 \le 11$

2. $\frac{2}{3}x + 5 > 11$

3. $|x - 4| < 5$

4. $|2x + 1| \ge 7$

VOCABULARY AND CONCEPTS   *Fill in the blanks.*

5. If two algebraic terms have the same variables with the same _____, they are called like terms.
6. $-6a^3b^2$ and $7a^2b^3$ are _____ terms.
7. To add two like monomials, we add their numerical _____ and keep the same variables with the same exponents.
8. To subtract one like monomial from another, we add the _____ of the monomial that is to be subtracted.

Tell whether the terms are like or unlike terms. If they are like terms, combine them.

9. $3x, 7x$

10. $-8x, 3y$

11. $7x, 7y$

12. $3mn, 5mn$

13. $3r^2t^3, -8r^2t^3$

14. $9u^2v, 10u^2v$

15. $9x^2y^3, 3x^2y^2$

16. $27x^6y^4z, 8x^6y^4z^2$

PRACTICE   *Simplify each expression.*

17. $8x + 4x$

18. $-2y + 16y$

19. $5x^3y^2z - 3x^3y^2z$

20. $8wxy - 12wxy$

21. $-2x^2y^3 + 3xy^4 - 5x^2y^3$
22. $3ab^4 - 4a^2b^2 - 2ab^4 + 2a^2b^2$
23. $(3x^2y)^2 + 2x^4y^2 - x^4y^2$
24. $(5x^2y^4)^3 - (5x^3y^6)^2$

Perform each operation.

25. $(3x^2 + 2x + 1) + (-2x^2 - 7x + 5)$
26. $(-2a^2 - 5a - 7) + (-3a^2 + 7a + 1)$

27. $(-a^2 + 2a + 3) - (4a^2 - 2a - 1)$
28. $(x^2 - 3x + 8) - (3x^2 + x + 3)$
29. $(7y^3 + 4y^2 + y + 3) + (-8y^3 - y + 3)$

30. $(6x^3 + 3x - 2) - (2x^3 + 3x^2 + 5)$

31. $(3x^2 + 4x - 3) + (2x^2 - 3x - 1) - (x^2 + x + 7)$

32. $(-2x^2 + 6x + 5) - (-4x^2 - 7x + 2) - (4x^2 + 10x + 5)$
33. $(3x^3 - 2x + 3) + (4x^3 + 3x^2 - 2) + (-4x^3 - 3x^2 + x + 12)$
34. $(x^4 - 3x^2 + 4) + (-2x^4 - x^3 + 3x^2) + (3x^2 + 2x + 1)$

35. $(3y^2 - 2y + 4) + [(2y^2 - 3y + 2) - (y^2 + 4y + 3)]$

36. $(-t^2 - t - 1) - [(t^2 + 3t - 1) - (-2t^2 + 4)]$

Add the polynomials.

37.
$$\begin{array}{r} 3x^3 - 2x^2 + 4x - 3 \\ -2x^3 + 3x^2 + 3x - 2 \\ \underline{5x^3 - 7x^2 + 7x - 12} \end{array}$$

38.
$$\begin{array}{r} 7a^3 \phantom{{}+ 4a^2} + 3a + 7 \\ -2a^3 + 4a^2 \phantom{{}+ 3a} - 13 \\ \underline{3a^3 - 3a^2 + 4a + 5} \end{array}$$

39.
$$\begin{array}{r} -2y^4 - 2y^3 + 4y^2 - 3y + 10 \\ -3y^4 + 7y^3 - y^2 + 14y - 3 \\ -3y^3 - 5y^2 - 5y + 7 \\ \underline{-4y^4 + y^3 - 13y^2 + 14y - 2} \end{array}$$

40.
$$\begin{array}{r} 17t^4 + 3t^3 - 2t^2 - 3t + 4 \\ -12t^4 - 2t^3 + 3t^2 - 5t - 17 \\ -2t^4 - 7t^3 + 4t^2 + 12t - 5 \\ \underline{5t^4 + t^3 + 5t^2 - 13t + 12} \end{array}$$

Subtract the bottom polynomial from the top polynomial.

41.
$$\begin{array}{r} 3x^2 - 4x + 17 \\ \underline{2x^2 + 4x - 5} \end{array}$$

42.
$$\begin{array}{r} -2y^2 - 4y + 3 \\ \underline{3y^2 + 10y - 5} \end{array}$$

43.
$$\begin{array}{r} -5y^3 + 4y^2 - 11y + 3 \\ \underline{-2y^3 - 14y^2 + 17y - 32} \end{array}$$

44.
$$\begin{array}{r} 17x^4 - 3x^2 - 65x - 12 \\ \underline{23x^4 + 14x^2 + 3x - 23} \end{array}$$

Simplify each expression.

45. $3(x + 2) + 2(x - 5)$

46. $-2(x - 4) + 5(x + 1)$

47. $-6(t - 4) - 5(t - 1)$

48. $4(a + 5) - 3(a - 1)$

49. $2(x^3 + x^2) + 3(2x^3 - x^2)$

50. $3(y^2 + 2y) - 4(y^2 - 4)$

51. $-3(2m - n) + 2(m - 3n)$

52. $5(p - 2q) - 4(2p + q)$

53. $-5(2x^3 + 7x^2 + 4x) - 2(3x^3 - 4x^2 - 4x)$

54. $-3(3a^2 + 4b^3 + 7) + 4(5a^2 - 2b^3 + 3)$

55. $4(3z^2 - 4z + 5) + 6(-2z^2 - 3z + 4) - 2(4z^2 + 3z - 5)$

56. $-3(4x^3 - 2x^2 + 4) - 4(3x^3 + 4x^2 + 3x) + 5(3x - 4)$

57. $5(2a^2 + 4a - 2) - 2(-3a^2 - a + 12) - 2(a^2 + 3a - 5)$

58. $-2(2b^2 - 3b + 3) + 3(3b^2 + 2b - 8) - (3b^2 - b + 4)$

APPLICATIONS

59. Write a polynomial that represents the perimeter of the triangle.

60. Write a polynomial that represents the perimeter of the rectangle.

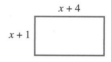

61. **Carpentry**   The wooden plank is $(2x^2 + 5x + 1)$ meters long. If the shorter end is cut off, write a polynomial to express the length of the remaining piece.

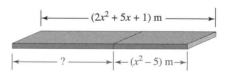

62. **Gardening**   Write a polynomial that expresses how long the hose will be if the shorter pieces are joined together.

Consider the following information. If a house is purchased for $125,000 and is expected to appreciate $1,100 per year, its value y after x years is given by the polynomial function $y = f(x) = 1,100x + 125,000.$

63. Value of a house Find the expected value of the house in 10 years.

64. Value of a house A second house is purchased for $150,000 and is expected to appreciate $1,400 per year.
 a. Find a polynomial function that will give the value y of the house in x years.
 b. Find the value of the second house after 12 years.

65. Value of two houses Find one polynomial function that will give the combined value y of both houses after x years.

66. Value of two houses Find the value of the two houses after 20 years by
 a. substituting 20 into the polynomial functions

$$y = f(x) = 1,100x + 125,000 \text{ and}$$
$$y = f(x) = 1,400x + 150,000$$

 and adding.
 b. substituting 20 into the result of Exercise 65.

Consider the following information. A business bought two cars, one for $16,600 and the other for $19,200. The first car is expected to depreciate $2,100 per year and the second $2,700 per year.

67. Value of a car Find a polynomial function that will give the value of the first car after x years.

68. Value of a car Find a polynomial function that will give the value of the second car after x years.

69. Value of two cars Find one polynomial function that will give the value of both cars after x years.

70. Value of two cars In two ways, find the total value of the cars after 3 years.

WRITING

71. Explain why the terms x^2y and xy^2 are not like terms.

72. Explain how to recognize like terms, and how to add them.

SOMETHING TO THINK ABOUT

73. Find the difference when $3x^2 + 4x - 3$ is subtracted from the sum of $-2x^2 - x + 7$ and $5x^2 + 3x - 1$.

74. Find the difference when $8x^3 + 2x^2 - 1$ is subtracted from the sum of $x^2 + x + 2$ and $2x^3 - x + 9$.

75. Find the sum when $2x^2 - 4x + 3$ minus $8x^2 + 5x - 3$ is added to $-2x^2 + 7x - 4$.

76. Find the sum when $7x^3 - 4x$ minus $x^2 + 2$ is added to $5 + 3x$.

5.3 Multiplying Polynomials

- Multiplying Monomials
- Multiplying a Polynomial by a Monomial
- Multiplying a Polynomial by a Polynomial ▪ The FOIL Method
- Special Products ▪ An Application of Multiplying Polynomials
- Multiplying Expressions That Are Not Polynomials

Getting Ready *Simplify.*

1. $(3a)(4)$
2. $4aaa(a)$
3. $-7a^3 \cdot a$
4. $12a^3a^2$

Use the distributive property to remove parentheses.

5. $2(a + 4)$ **6.** $a(a - 3)$

7. $-4(a - 3)$ **8.** $-2b(b + 2)$

We now discuss how to multiply polynomials.

Multiplying Monomials

In Section 1.3, we saw that to multiply one monomial by another, *we multiply the numerical factors and then multiply the variable factors.*

EXAMPLE 1   Perform each multiplication: **a.** $(3x^2)(6x^3)$,   **b.** $(-8x)(2y)(xy)$,   and   **c.** $(2a^3b)(-7b^2c)(-12ac^4)$.

Solution   We can use the commutative and associative properties of multiplication to rearrange the factors and regroup the numbers.

a. $(3x^2)(6x^3) = 3 \cdot x^2 \cdot 6 \cdot x^3$ **b.** $(-8x)(2y)(xy) = -8 \cdot x \cdot 2 \cdot y \cdot x \cdot y$

$\quad\quad\quad\quad\quad = (3 \cdot 6)(x^2 \cdot x^3)$ $= (-8 \cdot 2) \cdot x \cdot x \cdot y \cdot y$

$\quad\quad\quad\quad\quad = 18x^5$ $= -16x^2y^2$

c. $(2a^3b)(-7b^2c)(-12ac^4) = 2 \cdot a^3 \cdot b \cdot (-7) \cdot b^2 \cdot c \cdot (-12) \cdot a \cdot c^4$

$\quad\quad\quad\quad\quad\quad\quad\quad\quad\quad = 2(-7)(-12) \cdot a^3 \cdot a \cdot b \cdot b^2 \cdot c \cdot c^4$

$\quad\quad\quad\quad\quad\quad\quad\quad\quad\quad = 168a^4b^3c^5$

Self Check   Multiply: **a.** $(-2a^3)(4a^2)$ and   **b.** $(-5b^3)(-3a)(a^2b)$.   ∎

Multiplying a Polynomial by a Monomial

To multiply a polynomial by a monomial, *we use the distributive property and multiply each term of the polynomial by the monomial.*

EXAMPLE 2   Perform each multiplication: **a.** $3x^2(6xy + 3y^2)$,   **b.** $5x^3y^2(xy^3 - 2x^2y)$,   and   **c.** $-2ab^2(3bz - 2az + 4z^3)$

Solution   We can use the distributive property to remove parentheses.

a. $3x^2(6xy + 3y^2) = 3x^2 \cdot 6xy + 3x^2 \cdot 3y^2$

$\quad\quad\quad\quad\quad\quad = 18x^3y + 9x^2y^2$

b. $5x^3y^2(xy^3 - 2x^2y) = 5x^3y^2 \cdot xy^3 - 5x^3y^2 \cdot 2x^2y$

$\quad\quad\quad\quad\quad\quad\quad = 5x^4y^5 - 10x^5y^3$

c. $-2ab^2(3bz - 2az + 4z^3) = -2ab^2 \cdot 3bz - (-2ab^2) \cdot 2az + (-2ab^2) \cdot 4z^3$

$\quad\quad\quad\quad\quad\quad\quad\quad\quad\quad = -6ab^3z + 4a^2b^2z - 8ab^2z^3$

Self Check   Multiply: $-2a^2(a^2 - a + 3)$.   ∎

Multiplying a Polynomial by a Polynomial

To multiply a polynomial by a polynomial, we use the distributive property repeatedly.

 EXAMPLE 3   Perform each multiplication: **a.** $(3x + 2)(4x + 9)$ and
b. $(2a - b)(3a^2 - 4ab + b^2)$.

Solution   We can use the distributive property twice to remove parentheses.

a. $(3x + 2)(4x + 9) = (3x + 2) \cdot 4x + (3x + 2) \cdot 9$

$\qquad\qquad\qquad\quad = 12x^2 + 8x + 27x + 18$

$\qquad\qquad\qquad\quad = 12x^2 + 35x + 18$ Combine like terms.

b. $(2a - b)(3a^2 - 4ab + b^2)$

$\qquad = (2a - b)3a^2 - (2a - b)4ab + (2a - b)b^2$

$\qquad = 6a^3 - 3a^2b - 8a^2b + 4ab^2 + 2ab^2 - b^3$

$\qquad = 6a^3 - 11a^2b + 6ab^2 - b^3$

Self Check   Multiply: $(2a + b)(3a - 2b)$.   ∎

The results of Example 3 suggest that to multiply one polynomial by another, *we multiply each term of one polynomial by each term of the other polynomial and combine like terms, when possible.*

In the next example, we organize the work done in Example 3 vertically.

EXAMPLE 4   Perform each multiplication:

a. $\begin{array}{r} 3x + 2 \\ 4x + 9 \\ \hline 12x^2 + 8x \\ + 27x + 18 \\ \hline 12x^2 + 35x + 18 \end{array}$ $\leftarrow 4x(3x + 2)$
$\leftarrow 9(3x + 2)$

b. $\begin{array}{r} 3a^2 - 4ab + b^2 \\ 2a - b \\ \hline 6a^3 - 8a^2b + 2ab^2 \\ - 3a^2b + 4ab^2 - b^3 \\ \hline 6a^3 - 11a^2b + 6ab^2 - b^3 \end{array}$ $\leftarrow 2a(3a^2 - 4ab + b^2)$
$\leftarrow -b(3a^2 - 4ab + b^2)$

Self Check   Multiply: $\begin{array}{r} 3x^2 + 2x - 5 \\ 2x + 1 \\ \hline \end{array}$ ∎

The FOIL Method

When multiplying two binomials, the distributive property requires that each term of one binomial be multiplied by each term of the other binomial. This fact can be emphasized by drawing lines to show the indicated products. For example, to multiply $3x + 2$ and $x + 4$, we can write

First terms   Last terms

$$(3x + 2)(x + 4) = 3x \cdot x + 3x \cdot 4 + 2 \cdot x + 2 \cdot 4$$

Inner terms $= 3x^2 + 12x + 2x + 8$

Outer terms $= 3x^2 + 14x + 8$ Combine like terms.

We note that

- the product of the **F**irst terms is $3x^2$,
- the product of the **O**uter terms is $12x$,
- the product of the **I**nner terms is $2x$, and
- the product of the **L**ast terms is 8.

This scheme is called the **FOIL** method of multiplying two binomials. FOIL is an acronym for **F**irst terms, **O**uter terms, **I**nner terms, and **L**ast terms. Of course, the resulting terms of the products must be combined, if possible.

To multiply binomials by sight, we find the product of the first terms, then find the products of the outer terms and the inner terms and add them (when possible), and then find the product of the last terms.

EXAMPLE 5   Find each product:

a. $(2x - 3)(3x + 2) = 6x^2 - 5x - 6$

The middle term of $-5x$ in the result comes from combining the outer and inner products of $4x$ and $-9x$:

$$4x + (-9x) = -5x$$

b. $(3x + 1)(3x + 4) = 9x^2 + 15x + 4$

The middle term of $+15x$ in the result comes from combining the products $12x$ and $3x$:

$$12x + 3x = 15x$$

c. $(4x - y)(2x + 3y) = 8x^2 + 10xy - 3y^2$

The middle term of $+10xy$ in the result comes from combining the products $12xy$ and $-2xy$:

$$12xy - 2xy = 10xy$$

Self Check Multiply: $(3a + 4b)(2a - b)$. ∎

Special Products

It is easy to square a binomial using the FOIL method.

EXAMPLE 6 Find each square: **a.** $(x + y)^2$ and **b.** $(x - y)^2$.

Solution We multiply each term of one binomial by each term of the other binomial, and then we combine like terms.

a. $(x + y)^2 = (x + y)(x + y)$

$= x^2 + xy + xy + y^2$ Use the FOIL method.

$= x^2 + 2xy + y^2$ Combine like terms.

We see that the square of the binomial is the square of the first term, plus twice the product of the terms, plus the square of the last term. This product can be illustrated graphically as shown in Figure 5-5.

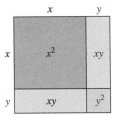

The area of the large square is the product of its length and width: $(x + y)(x + y) = (x + y)^2$.

The area of the large square is also the sum of its four pieces: $x^2 + xy + xy + y^2 = x^2 + 2xy + y^2$.

Thus, $(x + y)^2 = x^2 + 2xy + y^2$.

Figure 5-5

b. $(x - y)^2 = (x - y)(x - y)$

$= x^2 - xy - xy + y^2$ Use the FOIL method.

$= x^2 - 2xy + y^2$ Combine like terms.

We see that the square of the binomial is the square of the first term, minus twice the product of the terms, plus the square of the last term. For a geometric interpretation, see Exercise 115.

Self Check Find the squares: **a.** $(a + 2)^2$ and **b.** $(a - 4)^2$. ∎

EXAMPLE 7 Multiply: $(x + y)(x - y)$.

Solution $(x + y)(x - y) = x^2 - xy + xy - y^2$ Use the FOIL method.

$= x^2 - y^2$ Combine like terms.

From this example, we see that the product of the sum of two quantities and the difference of the same two quantities is the square of the first quantity minus the square of the second quantity. For a geometric interpretation, see Exercise 116.

Self Check Multiply: $(a + 3)(a - 3)$. ▮

The products discussed in Examples 6 and 7 are called **special products.** Since they occur so often, it is useful to learn their forms.

Special Product Formulas

$$(x + y)^2 = (x + y)(x + y) = x^2 + 2xy + y^2$$
$$(x - y)^2 = (x - y)(x - y) = x^2 - 2xy + y^2$$
$$(x + y)(x - y) = x^2 - y^2$$

Because $x^2 + 2xy + y^2 = (x + y)^2$ and $x^2 - 2xy + y^2 = (x - y)^2$, the two trinomials are called **perfect square trinomials.**

 Comment The squares $(x + y)^2$ and $(x - y)^2$ have trinomials for their products. Don't forget to write the middle terms in these products. Remember that

$$(x + y)^2 \neq x^2 + y^2 \quad \text{and that} \quad (x - y)^2 \neq x^2 - y^2$$

Also remember that the product $(x + y)(x - y)$ is the binomial $x^2 - y^2$.

At first, the expression $3[x^2 - 2(x + 3)]$ doesn't look like a polynomial. However, if we remove the parentheses and the brackets, it takes on the form of a polynomial.

$$3[x^2 - 2(x + 3)] = 3[x^2 - 2x - 6]$$
$$= 3x^2 - 6x - 18$$

If an expression has one set of grouping symbols that is enclosed within another set, we always eliminate the inner set first.

EXAMPLE 8 Find the product of $-2[y^3 + 3(y^2 - 2)]$ and $5[y^2 - 2(y + 1)]$.

Solution We change each expression into polynomial form:

$$-2[y^3 + 3(y^2 - 2)] \qquad 5[y^2 - 2(y + 1)]$$
$$-2(y^3 + 3y^2 - 6) \qquad 5(y^2 - 2y - 2)$$
$$-2y^3 - 6y^2 + 12 \qquad 5y^2 - 10y - 10$$

Then we do the multiplication:

$$
\begin{array}{r}
-2y^3 - \;\; 6y^2 + 12 \\
5y^2 - 10y - 10 \\
\hline
-10y^5 - 30y^4 \qquad\quad + \;\; 60y^2 \\
+ 20y^4 + 60y^3 \qquad\quad - 120y \\
+ 20y^3 + \;\; 60y^2 \qquad - 120 \\
\hline
-10y^5 - 10y^4 + 80y^3 + 120y^2 - 120y - 120
\end{array}
$$

Self Check Find the product of $2[a^2 + 3(a - 2)]$ and $3[a^2 + 3(a - 1)]$. ▮

An Application of Multiplying Polynomials

The profit p earned on the sale of one or more items is given by the formula

$$p = r - c$$

where r is the revenue taken in and c is the wholesale cost.

If a salesperson has 12 vacuum cleaners and sells them for \$225 each, the revenue will be $r = \$(12 \cdot 225) = \$2{,}700$. This illustrates the following formula for finding the revenue r:

$$r \quad = \quad \boxed{\begin{array}{c} \text{number of} \\ \text{items sold } (x) \end{array}} \quad \cdot \quad \boxed{\begin{array}{c} \text{selling price of} \\ \text{each item } (p) \end{array}} \quad = xp = px$$

EXAMPLE 9   **Selling vacuum cleaners**   Over the years, a saleswoman has found that the number of vacuum cleaners she can sell depends on price. The lower the price, the more she can sell. She has determined that the number of vacuums (x) that she can sell at a price (p) is related by the equation $x = -\frac{2}{25}p + 28$.

a. Find a formula for the revenue r.

b. How much revenue will be taken in if the vacuums are priced at \$250?

Solution   **a.** To find a formula for revenue, we substitute $-\frac{2}{25}p + 28$ for x in the formula $r = px$ and solve for r.

$$r = p\mathbf{x} \qquad\qquad \text{The formula for revenue.}$$

$$r = p\left(-\frac{2}{25}\mathbf{p} + \mathbf{28}\right) \quad \text{Substitute } -\frac{2}{25}p + 28 \text{ for } x.$$

$$r = -\frac{2}{25}p^2 + 28p \qquad \text{Multiply the polynomials.}$$

b. To find how much revenue will be taken in if the vacuums are priced at \$250, we substitute 250 for p in the formula for revenue.

$$r = -\frac{2}{25}\mathbf{p}^2 + 28\mathbf{p} \qquad\qquad \text{The formula for revenue.}$$

$$r = -\frac{2}{25}(\mathbf{250})^2 + 28(\mathbf{250}) \qquad \text{Substitute 250 for } p.$$

$$= -5{,}000 + 7{,}000$$

$$= 2{,}000$$

The revenue will be \$2,000. ∎

Multiplying Expressions That Are Not Polynomials

The following examples show how to multiply expressions that are not polynomials.

EXAMPLE 10   Find the product of $x^{-2} + y$ and $x^2 - y^{-2}$.

Solution We multiply each term of the second expression by each term of the first expression and then simplify.

$$(x^{-2} + y)(x^2 - y^{-2}) = x^{-2}x^2 - x^{-2}y^{-2} + yx^2 - yy^{-2}$$
$$= x^{-2+2} - x^{-2}y^{-2} + yx^2 - y^{1+(-2)}$$
$$= x^0 - \frac{1}{x^2y^2} + x^2y - y^{-1}$$
$$= 1 - \frac{1}{x^2y^2} + x^2y - \frac{1}{y}$$

Self Check Multiply: $(a^{-3} + b)(a^2 - b^{-1})$.

EXAMPLE 11 Find the product of $x^n + 2x$ and $x^n + 3x^{-n}$.

Solution We multiply each term of the second expression by each term of the first expression and simplify:

$$(x^n + 2x)(x^n + 3x^{-n}) = x^nx^n + x^n(3x^{-n}) + 2x(x^n) + 2x(3x^{-n})$$
$$= x^{n+n} + 3x^{n+(-n)} + 2x^{1+n} + 6xx^{-n}$$
$$= x^{2n} + 3x^0 + 2x^{n+1} + 6x^{1+(-n)}$$
$$= x^{2n} + 3 + 2x^{n+1} + 6x^{1-n}$$

Self Check Multiply: $(a^n + b)(a^n + b^n)$.

Self Check Answers

1. a. $-8a^5$, **b.** $15a^3b^4$ **2.** $-2a^4 + 2a^3 - 6a^2$ **3.** $6a^2 - ab - 2b^2$ **4.** $6x^3 + 7x^2 - 8x - 5$
5. $6a^2 + 5ab - 4b^2$ **6. a.** $a^2 + 4a + 4$, **b.** $a^2 - 8a + 16$ **7.** $a^2 - 9$ **8.** $6a^4 + 36a^3 - 162a + 108$
10. $\frac{1}{a} - \frac{1}{a^3b} + a^2b - 1$ **11.** $a^{2n} + a^nb^n + a^nb + b^{n+1}$

Orals   *Find each product.*

1. $(-2a^2b)(3ab^2)$
2. $(4xy^2)(-2xy)$
3. $3a^2(2a - 1)$
4. $-4n^2(4m - n)$
5. $(x + 1)(2x + 1)$
6. $(3y - 2)(2y + 1)$

5.3   EXERCISES

REVIEW   *Let $a = -2$ and $b = 4$ and find the absolute value of each expression.*

1. $|3a - b|$
2. $|ab - b^2|$
3. $-|a^2b - b^0|$
4. $\left| \dfrac{a^3b^2 + ab}{2(ab)^2 - a^3} \right|$

5. A woman owns 200 shares of ABC Company, valued at \$125 per share, and 350 shares of WD Company, valued at \$75 per share. One day, ABC rose $1\frac{1}{2}$ points and WD fell $1\frac{1}{2}$ points. Find the current value of her portfolio.
6. One light year is approximately 5,870,000,000,000 miles. Write this number in scientific notation.

VOCABULARY AND CONCEPTS   *Fill in the blanks.*

7. To multiply a monomial by a monomial, we multiply the numerical factors and then multiply the _____ factors.
8. To multiply a polynomial by a monomial, we multiply each term of the polynomial by the _____.
9. To multiply a polynomial by a polynomial, we multiply each _____ of one polynomial by each term of the other polynomial.
10. FOIL is an acronym for _____ terms, _____ terms, _____ terms, and _____ terms.
11. $(x + y)^2 = (x + y)(x + y) =$ _____
12. $(x - y)^2 = (x - y)(x - y) =$ _____
13. $(x + y)(x - y) =$ _____
14. $x^2 + 2xy + y^2$ and $x^2 - 2xy + y^2$ are called _____ trinomials.

PRACTICE   *Find each product. Write all answers without using negative exponents.*

15. $(2a^2)(-3ab)$
16. $(-3x^2y)(3xy)$
17. $(-3ab^2c)(5ac^2)$
18. $(-2m^2n)(-4mn^3)$
19. $(4a^2b)(-5a^3b^2)(6a^4)$
20. $(2x^2y^3)(4xy^5)(-5y^6)$
21. $(5x^3y^2)^4\left(\dfrac{1}{5}x^{-2}\right)^2$
22. $(4a^{-2}b^{-1})^2(2a^3b^4)^4$
23. $(-5xx^2)(-3xy)^4$
24. $(-2a^2ab^2)^3(-3ab^2b^2)$
25. $3(x + 2)$
26. $-5(a + b)$
27. $-a(a - b)$
28. $y^2(y - 1)$
29. $3x(x^2 + 3x)$
30. $-2x(3x^2 - 2)$
31. $-2x(3x^2 - 3x + 2)$
32. $3a(4a^2 + 3a - 4)$
33. $5a^2b^3(2a^4b - 5a^0b^3)$
34. $-2a^3b(3a^0b^4 - 2a^2b^3)$
35. $7rst(r^2 + s^2 - t^2)$
36. $3x^2yz(x^2 - 2y + 3z^2)$
37. $4m^2n(-3mn)(m + n)$
38. $-3a^2b^3(2b)(3a + b)$
39. $(x + 2)(x + 3)$
40. $(y - 3)(y + 4)$
41. $(z - 7)(z - 2)$
42. $(x + 3)(x - 5)$
43. $(2a + 1)(a - 2)$
44. $(3b - 1)(2b - 1)$
45. $(3t - 2)(2t + 3)$
46. $(p + 3)(3p - 4)$
47. $(3y - z)(2y - z)$
48. $(2m + n)(3m + n)$
49. $(2x - 3y)(x + 2y)$
50. $(3y + 2z)(y - 3z)$

51. $(3x + y)(3x - 3y)$ **52.** $(2x - y)(3x + 2y)$

53. $(4a - 3b)(2a + 5b)$ **54.** $(3a + 2b)(2a - 7b)$

55. $(x + 2)^2$ **56.** $(x - 3)^2$

57. $(a - 4)^2$ **58.** $(y + 5)^2$

59. $(2a + b)^2$ **60.** $(a - 2b)^2$

61. $(2x - y)^2$ **62.** $(3m + 4n)(3m + 4n)$

63. $(x + 2)(x - 2)$ **64.** $(z + 3)(z - 3)$

65. $(a + b)(a - b)$ **66.** $(p + q)(p - q)$

67. $(2x + 3y)(2x - 3y)$ **68.** $(3a + 4b)(3a - 4b)$

69. $(x - y)(x^2 + xy + y^2)$ **70.** $(x + y)(x^2 - xy + y^2)$

71. $(3y + 1)(2y^2 + 3y + 2)$
72. $(a + 2)(3a^2 + 4a - 2)$
73. $(2a - b)(4a^2 + 2ab + b^2)$
74. $(x - 3y)(x^2 + 3xy + 9y^2)$
75. $(2x - 1)[2x^2 - 3(x + 2)]$
76. $(x + 1)^2[x^2 - 2(x + 2)]$
77. $(a + b)(a - b)(a - 3b)$
78. $(x - y)(x + 2y)(x - 2y)$
79. $x^3(2x^2 + x^{-2})$

80. $x^{-4}(2x^{-3} - 5x^2)$

81. $x^3y^{-6}z^{-2}(3x^{-2}y^2z - x^3y^{-4})$

82. $ab^{-2}c^{-3}(a^{-4}bc^3 + a^{-3}b^4c^3)$

83. $(x^{-1} + y)(x^{-1} - y)$

84. $(x^{-1} - y)(x^{-1} - y)$

85. $(2x^{-3} + y^3)(2x^3 - y^{-3})$

86. $(5x^{-4} - 4y^2)(5x^2 - 4y^{-4})$

87. $x^n(x^{2n} - x^n)$ **88.** $a^{2n}(a^n + a^{2n})$

89. $(x^n + 1)(x^n - 1)$ **90.** $(x^n - a^n)(x^n + a^n)$

91. $(x^n - y^n)(x^n - y^{-n})$ **92.** $(x^n + y^n)(x^n + y^{-n})$

93. $(x^{2n} + y^{2n})(x^{2n} - y^{2n})$ **94.** $(a^{3n} - b^{3n})(a^{3n} + b^{3n})$

95. $(x^n + y^n)(x^n + 1)$ **96.** $(1 - x^n)(x^{-n} - 1)$

Simplify each expression.

97. $3x(2x + 4) - 3x^2$ **98.** $2y - 3y(y^2 + 4)$

99. $3pq - p(p - q)$
100. $-4rs(r - 2) + 4rs$
101. $2m(m - n) - (m + n)(m - 2n)$
102. $-3y(2y + z) + (2y - z)(3y + 2z)$
103. $(x + 3)(x - 3) + (2x - 1)(x + 2)$
104. $(2b + 3)(b - 1) - (b + 2)(3b - 1)$

105. $(3x - 4)^2 - (2x + 3)^2$
106. $(3y + 1)^2 + (2y - 4)^2$
107. $3(x - 3y)^2 + 2(3x + y)^2$
108. $2(x - y^2)^2 - 3(y^2 + 2x)^2$
109. $5(2y - z)^2 + 4(y + 2z)^2$
110. $3(x + 2z)^2 - 2(2x - z)^2$

Use a calculator to find each product.

111. $(3.21x - 7.85)(2.87x + 4.59)$

112. $(7.44y + 56.7)(-2.1y - 67.3)$

113. $(-17.3y + 4.35)^2$

114. $(-0.31x + 29.3)(-81x - 0.2)$

115. Refer to the illustration and answer the following questions.
  a. Find the area of the large square.
  b. Find the area of square I.
  c. Find the area of rectangle II.
  d. Find the area of rectangle III.
  e. Find the area of square IV.

Use the answers to the preceding questions to show that $(x - y)^2 = x^2 - 2xy + y^2$.

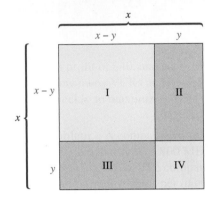

116. Refer to the illustration and answer the following questions.
  a. Find the area of rectangle $ABCD$.
  b. Find the area of rectangle I.
  c. Find the area of rectangle II.

Use the answers to the preceding questions to show that $(x + y)(x - y) = x^2 - y^2$.

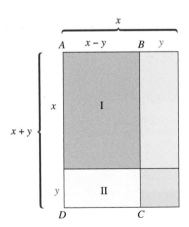

APPLICATIONS   *Assume that the number (x) of televisions a salesman can sell at a certain price (p) is given by the equation* $x = -\frac{1}{5}p + 90$.

117. Find the number of TVs he will sell if the price is $375.

118. If he sold 20 TVs, what was the price?

119. Write a formula for the revenue when x TVs are sold.

120. Find the revenue generated by the sales of the TVs if they are priced at $400 each.

121. Write a polynomial that represents the area of the square garden.

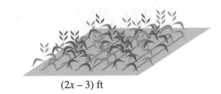

(2x − 3) ft

122. Write a polynomial that represents the area of the rectangular field.

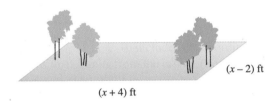

(x − 2) ft

(x + 4) ft

123. Write a polynomial that represents the area of the triangle.

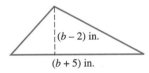

(b − 2) in.

(b + 5) in.

124. Write a polynomial that represents the volume of the cube.

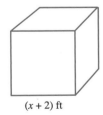

(x + 2) ft

WRITING

125. Explain how to use the FOIL method.

126. Explain how to multiply two trinomials.

SOMETHING TO THINK ABOUT

127. The numbers 0.35×10^7 and 1.96×10^7 both involve the same power of 10. Find their sum.

128. Without converting to standard notation, find the sum: $1.435 \times 10^8 + 2.11 \times 10^7$. (*Hint:* The first number in the previous exercise is not in scientific notation.)

5.4	**The Greatest Common Factor and Factoring by Grouping**

▌ **Prime-Factored Form of a Natural Number**
▌ **Factoring Out the Greatest Common Factor**
▌ **An Application of Factoring** ▌ **Factoring by Grouping** ▌ **Formulas**

Getting Ready *Simplify each expression by removing parentheses.*

1. $4(a + 2)$ **2.** $-5(b - 5)$

3. $a(a + 5)$ **4.** $-b(b - 3)$

5. $6a(a + 2b)$ **6.** $-2p^2(3p + 5)$

In this section, we will reverse the operation of multiplication and show how to find the factors of a known product. The process of finding the individual factors of a product is called **factoring.**

Prime-Factored Form of a Natural Number

If one number a divides a second number b, then a is called a **factor** of b. For example, because 3 divides 24, it is a factor of 24. Each number in the following list is a factor of 24, because each number divides 24.

$$1, \quad 2, \quad 3, \quad 4, \quad 6, \quad 8, \quad 12, \quad \text{and} \quad 24$$

To factor a natural number means to write it as a product of other natural numbers. If each factor is a prime number, the natural number is said to be written in **prime-factored form.** Example 1 shows how to find the prime-factored forms of 60, 84, and 180, respectively.

EXAMPLE 1 Find the prime factorization of each number.

a. $60 = 6 \cdot 10$
$= 2 \cdot 3 \cdot 2 \cdot 5$
$= 2^2 \cdot 3 \cdot 5$

b. $84 = 4 \cdot 21$
$= 2 \cdot 2 \cdot 3 \cdot 7$
$= 2^2 \cdot 3 \cdot 7$

c. $180 = 10 \cdot 18$
$= 2 \cdot 5 \cdot 3 \cdot 6$
$= 2 \cdot 5 \cdot 3 \cdot 3 \cdot 2$
$= 2^2 \cdot 3^2 \cdot 5$

Self Check Find the prime factorization of 120. ▌

The largest natural number that divides 60, 84, and 180 is called the **greatest common factor (GCF)** of the numbers. Because 60, 84, and 180 all have two factors of 2 and one factor of 3, the GCF of these three numbers is $2^2 \cdot 3 = 12$. We note that

$$\frac{60}{12} = 5, \qquad \frac{84}{12} = 7, \qquad \text{and} \qquad \frac{180}{12} = 15$$

There is no natural number greater than 12 that divides 60, 84, and 180.

Algebraic monomials can also have greatest common factors.

EXAMPLE 2   Find the GCF of $6a^2b^3c$, $9a^3b^2c$, and $18a^4c^3$.

Solution   We begin by finding the prime factorization of each monomial:

$$6a^2b^3c = \mathbf{3} \cdot 2 \cdot a \cdot a \cdot b \cdot b \cdot b \cdot c$$
$$9a^3b^2c = \mathbf{3} \cdot 3 \cdot a \cdot a \cdot a \cdot b \cdot b \cdot c$$
$$18a^4c^3 = 2 \cdot \mathbf{3} \cdot 3 \cdot a \cdot a \cdot a \cdot a \cdot c \cdot c \cdot c$$

Since each monomial has one factor of 3, two factors of a, and one factor of c in common, their GCF is

$$3^1 \cdot a^2 \cdot c^1 = 3a^2c$$

Self Check   Find the GCF of $-24x^3y^3$, $3x^3y$, and $18x^2y^2$.   ∎

To find the GCF of several monomials, we follow these steps.

Steps for Finding the GCF

> 1. Find the prime-factored form of each monomial.
> 2. Identify the prime factors and variable factors that are common to each monomial.
> 3. Find the product of the factors found in Step 2 with each factor raised to the smallest power that occurs in any one monomial.

Factoring Out the Greatest Common Factor

We have seen that the distributive property provides a method for multiplying a polynomial by a monomial. For example,

$$2x^3y^3(3x^2 - 4y^3) = 2x^3y^3 \cdot 3x^2 - 2x^3y^3 \cdot 4y^3$$
$$= 6x^5y^3 - 8x^3y^6$$

If the product of a multiplication is $6x^5y^3 - 8x^3y^6$, we can use the distributive property to find the individual factors.

$$6x^5y^3 - 8x^3y^6 = 2x^3y^3 \cdot 3x^2 - 2x^3y^3 \cdot 4y^3$$
$$= 2x^3y^3(3x^2 - 4y^3)$$

Since $2x^3y^3$ is the GCF of the terms of $6x^5y^3 - 8x^3y^6$, this process is called **factoring out the greatest common factor.**

Comment Although it is true that

$$6x^5y^3 - 8x^3y^6 = 2x^3(3x^2y^3 - 4y^6)$$

the polynomial $6x^5y^3 - 8x^3y^6$ is not completely factored, because y^3 can be factored from $3x^2y^3 - 4y^6$. Whenever you factor a polynomial, be sure to factor it completely.

EXAMPLE 3 Factor: $25a^3b + 15ab^3$.

Solution We begin by finding the prime factorization of each monomial:

$$25a^3b = \mathbf{5} \cdot 5 \cdot \mathbf{a} \cdot a \cdot a \cdot \mathbf{b}$$
$$15ab^3 = \mathbf{5} \cdot 3 \cdot \mathbf{a} \cdot \mathbf{b} \cdot b \cdot b$$

Since each term has at least one factor of 5, one factor of a, and one factor of b in common and there are no other common factors, $5ab$ is the GCF of the two terms. We can use the distributive property to factor it out.

$$25a^3b + 15ab^3 = \mathbf{5ab} \cdot 5a^2 + \mathbf{5ab} \cdot 3b^2$$
$$= \mathbf{5ab}(5a^2 + 3b^2)$$

Self Check Factor: $9x^4y^2 - 12x^3y^3$. ∎

EXAMPLE 4 Factor: $3xy^2z^3 + 6xyz^3 - 3xz^2$.

Solution We begin by finding the prime factorization of each monomial:

$$3xy^2z^3 = \mathbf{3} \cdot \mathbf{x} \cdot y \cdot y \cdot \mathbf{z} \cdot \mathbf{z} \cdot z$$
$$6xyz^3 = \mathbf{3} \cdot 2 \cdot \mathbf{x} \cdot y \cdot \mathbf{z} \cdot \mathbf{z} \cdot z$$
$$-3xz^2 = -\mathbf{3} \cdot \mathbf{x} \cdot \mathbf{z} \cdot \mathbf{z}$$

Since each term has one factor of 3, one factor of x, and two factors of z in common, and because there are no other common factors, $3xz^2$ is the GCF of the three terms. We can use the distributive property to factor it out.

$$3xy^2z^3 + 6xyz^3 - 3xz^2 = \mathbf{3xz^2} \cdot y^2z + \mathbf{3xz^2} \cdot 2yz - \mathbf{3xz^2} \cdot 1$$
$$= \mathbf{3xz^2}(y^2z + 2yz - 1)$$

Comment The last term $-3xz^2$ of the given trinomial has an understood coefficient of -1. When the $3xz^2$ is factored out, remember to write the -1.

Self Check Factor: $2a^4b^2 + 6a^3b^2 - 4a^2b$. ∎

A polynomial that cannot be factored is called a **prime polynomial** or an **irreducible polynomial.**

EXAMPLE 5 Factor $3x^2 + 4y + 7$, if possible.

Solution We find the prime factorization of each monomial:

$$3x^2 = 3 \cdot x \cdot x \qquad 4y = 2 \cdot 2 \cdot y \qquad 7 = 7$$

There are no common factors other than 1. This polynomial is an example of a prime polynomial.

Self Check Factor $6a^3 + 7b^2 + 5$, if possible. ∎

EXAMPLE 6 Factor the negative of the GCF from $-6u^2v^3 + 8u^3v^2$.

Solution Because the GCF of the two terms is $2u^2v^2$, the negative of the GCF is $-2u^2v^2$. To factor out $-2u^2v^2$, we proceed as follows:

$$-6u^2v^3 + 8u^3v^2 = -2u^2v^2 \cdot 3v + 2u^2v^2 \cdot 4u$$
$$= -2u^2v^2 \cdot 3v - (-2u^2v^2)4u$$
$$= -2u^2v^2(3v - 4u)$$

Self Check Factor out the negative of the GCF from $-8a^2b^2 - 12ab^3$. ∎

We can also factor out common factors with variable exponents.

EXAMPLE 7 Factor x^{2n} from $x^{4n} + x^{3n} + x^{2n}$.

Solution We write the trinomial as $x^{2n} \cdot x^{2n} + x^{2n} \cdot x^n + x^{2n} \cdot 1$ and factor out x^{2n}.

$$x^{4n} + x^{3n} + x^{2n} = x^{2n} \cdot x^{2n} + x^{2n} \cdot x^n + x^{2n} \cdot 1$$
$$= x^{2n}(x^{2n} + x^n + 1)$$

Self Check Factor a^n from $a^{6n} + a^{3n} - a^n$. ∎

EXAMPLE 8 Factor $a^{-2}b^{-2}$ from $a^{-2}b - a^3b^{-2}$.

Solution We write $a^{-2}b - a^3b^{-2}$ as $a^{-2}b^{-2} \cdot b^3 - a^{-2}b^{-2} \cdot a^5$ and factor out $a^{-2}b^{-2}$.

$$a^{-2}b - a^3b^{-2} = a^{-2}b^{-2} \cdot b^3 - a^{-2}b^{-2} \cdot a^5$$
$$= a^{-2}b^{-2}(b^3 - a^5)$$

Self Check Factor $x^{-3}y^{-2}$ from $x^{-3}y^{-2} - x^6y^{-2}$. ∎

A common factor can have more than one term. For example, in the expression

$$x(a + b) + y(a + b)$$

the binomial $a + b$ is a factor of both terms. We can factor it out to get

$$x(a + b) + y(a + b) = (a + b)x + (a + b)y \qquad \text{Use the commutative property of multiplication.}$$

$$= (a + b)(x + y)$$

EXAMPLE 9 Factor: $a(x - y + z) - b(x - y + z) + 3(x - y + z)$.

Solution We can factor out the GCF of $(x - y + z)$.

$$a(x - y + z) - b(x - y + z) + 3(x - y + z)$$
$$= (x - y + z)a - (x - y + z)b + (x - y + z)3$$
$$= (x - y + z)(a - b + 3)$$

Self Check Factor: $x(a + b - c) - y(a + b - c)$. ∎

An Application of Factoring

EXAMPLE 10 The polynomial $3x^4 + 2x^3 + 5x^2 + 7x + 1$ can be written as

$$3 \cdot x \cdot x \cdot x \cdot x + 2 \cdot x \cdot x \cdot x + 5 \cdot x \cdot x + 7 \cdot x + 1$$

to illustrate that it involves 10 multiplications and 4 additions. Since computers take more time to do multiplications than additions, a programmer wants to write the polynomial in a way that will require fewer multiplications. Write the polynomial so that it contains only four multiplications.

Solution Factor the common x from the first four terms of the polynomial to get

$$3x^4 + 2x^3 + 5x^2 + 7x + 1 = x(3x^3 + 2x^2 + 5x + 7) + 1$$

Now factor the common x from the first three terms of the polynomial in the parentheses to get

$$3x^4 + 2x^3 + 5x^2 + 7x + 1 = x[x(3x^2 + 2x + 5) + 7] + 1$$

Again, factor an x from the first two terms of the polynomial within the set of parentheses to get

(1) $$3x^4 + 2x^3 + 5x^2 + 7x + 1 = x\{x[x(3x + 2) + 5] + 7\} + 1$$

To emphasize that there are now only four multiplications, we rewrite the right-hand side of Equation 1 as

$$x \cdot \{x \cdot [x \cdot (3 \cdot x + 2) + 5] + 7\} + 1$$

The right-hand side of Equation 1 is called the **nested form** of the polynomial. ∎

Factoring by Grouping

Suppose that we wish to factor

$$ac + ad + bc + bd$$

Although no factor is common to all four terms, there is a common factor of a in the first two terms and a common factor of b in the last two terms. We can factor out these common factors to get

$$ac + ad + bc + bd = a(c + d) + b(c + d)$$

We can now factor out the common factor of $c + d$ on the right-hand side:

$$ac + ad + bc + bd = (c + d)(a + b)$$

The grouping in this type of problem is not always unique, but the factorization is. For example, if we write the expression $ac + ad + bc + bd$ in the form

$$ac + bc + ad + bd$$

and factor c from the first two terms and d from the last two terms, we obtain

$$ac + bc + ad + bd = c(a + b) + d(a + b)$$
$$= (a + b)(c + d)$$

The method used in the previous examples is called **factoring by grouping.**

EXAMPLE 11   Factor: $3ax^2 + 3bx^2 + a + 5bx + 5ax + b$.

Solution   Although no factor is common to all six terms, $3x^2$ can be factored out of the first two terms, and $5x$ can be factored out of the fourth and fifth terms to get

$$3ax^2 + 3bx^2 + a + 5bx + 5ax + b = 3x^2(a + b) + a + 5x(b + a) + b$$

This result can be written in the form

$$3ax^2 + 3bx^2 + a + 5bx + 5ax + b = 3x^2(a + b) + 5x(a + b) + (a + b)$$

Since $a + b$ is common to all three terms, it can be factored out to get

$$3ax^2 + 3bx^2 + a + 5bx + 5ax + b = (a + b)(3x^2 + 5x + 1)$$

Self Check   Factor: $4x^3 + 3x^2 + x + 4x^2y + 3xy + y$.   ∎

To factor an expression, it is often necessary to factor more than once, as the following example illustrates.

EXAMPLE 12   Factor: $3x^3y - 4x^2y^2 - 6x^2y + 8xy^2$.

Solution   We begin by factoring out the common factor of xy.

$$3x^3y - 4x^2y^2 - 6x^2y + 8xy^2 = xy(3x^2 - 4xy - 6x + 8y)$$

We can now factor $3x^2 - 4xy - 6x + 8y$ by grouping:

$$3x^3y - 4x^2y^2 - 6x^2y + 8xy^2$$
$$= xy(3x^2 - 4xy - 6x + 8y)$$
$$= xy[x(3x - 4y) - 2(3x - 4y)] \qquad \text{Factor } x \text{ from } 3x^2 - 4xy \text{ and } -2$$
$$\text{from } -6x + 8y.$$
$$= xy(3x - 4y)(x - 2) \qquad\qquad \text{Factor out } 3x - 4y.$$

Because no more factoring can be done, the factorization is complete.

Self Check   Factor: $3a^3b + 3a^2b - 2a^2b^2 - 2ab^2$.   ∎

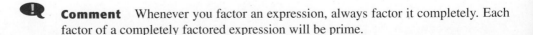

Comment   Whenever you factor an expression, always factor it completely. Each factor of a completely factored expression will be prime.

Formulas

Factoring is often required to solve a literal equation for one of its variables.

EXAMPLE 13 **Electronics** The formula $r_1 r_2 = r r_2 + r r_1$ is used in electronics to relate the combined resistance r of two resistors wired in parallel. The variable r_1 represents the resistance of the first resistor, and the variable r_2 represents the resistance of the second. Solve for r_2.

Solution To isolate r_2 on one side of the equation, we get all terms involving r_2 on the left-hand side and all terms not involving r_2 on the right-hand side. We then proceed as follows:

$$r_1 r_2 = r r_2 + r r_1$$

$$r_1 r_2 - r r_2 = r r_1 \qquad \text{Subtract } r r_2 \text{ from both sides.}$$

$$r_2(r_1 - r) = r r_1 \qquad \text{Factor out } r_2 \text{ on the left-hand side.}$$

$$r_2 = \frac{r r_1}{r_1 - r} \qquad \text{Divide both sides by } r_1 - r.$$

Self Check Solve $A = p + prt$ for p. ∎

Self Check Answers

1. $2^3 \cdot 3 \cdot 5$ **2.** $3x^2 y$ **3.** $3x^3 y^2(3x - 4y)$ **4.** $2a^2 b(a^2 b + 3ab - 2)$ **5.** a prime polynomial
6. $-4ab^2(2a + 3b)$ **7.** $a^n(a^{5n} + a^{2n} - 1)$ **8.** $x^{-3} y^{-2}(1 - x^9)$ **9.** $(a + b - c)(x - y)$
11. $(x + y)(4x^2 + 3x + 1)$ **12.** $ab(3a - 2b)(a + 1)$ **13.** $p = \dfrac{A}{1 + rt}$

Orals *Factor each expression.*

1. $3x^2 - x$ **2.** $7t^3 + 14t^2$
3. $-3a^2 - 6a$ **4.** $-4x^2 + 12x$
5. $3(a + b) + x(a + b)$ **6.** $a(m - n) - b(m - n)$

5.4 EXERCISES 📼

REVIEW *Perform each multiplication.*

1. $(a + 4)(a - 4)$ **2.** $(2b + 3)(2b - 3)$

3. $(4r^2 + 3s)(4r^2 - 3s)$ **4.** $(5a + 2b^3)(5a - 2b^3)$

5. $(m + 4)(m^2 - 4m + 16)$
6. $(p - q)(p^2 + pq + q^2)$

VOCABULARY AND CONCEPTS *Fill in the blanks.*

7. The process of finding the individual factors of a known product is called _____.

8. If a natural number is written as the product of prime factors, we say that it is written in _____ form.

9. The abbreviation GCF stands for

_____.

10. If a polynomial cannot be factored, it is called a _____ polynomial or an _____ polynomial.

PRACTICE *Find the prime-factored form of each number.*

11. 6 **12.** 10

13. 135 **14.** 98

15. 128

16. 357

17. 325

18. 288

Find the GCF of each set of monomials.

19. 36, 48

20. 45, 75

21. 42, 36, 98

22. 16, 40, 60

23. $4a^2b, 8a^3c$

24. $6x^3y^2z, 9xyz^2$

25. $18x^4y^3z^2, -12xy^2z^3$

26. $6x^2y^3, 24xy^3, 40x^2y^2z^3$

Complete each factorization.

27. $3a - 12 = 3(a - \quad)$

28. $5t + 25 = 5(t + \quad)$

29. $8z^2 + 2z = 2z(4z + \quad)$

30. $9t^3 - 3t^2 = 3t^2(3t - \quad)$

Factor each expression, if possible.

31. $2x + 8$

32. $3y - 9$

33. $2x^2 - 6x$

34. $3y^3 + 3y^2$

35. $5xy + 12ab^2$

36. $7x^2 + 14x$

37. $15x^2y - 10x^2y^2$

38. $11m^3n^2 - 12x^2y$

39. $63x^3y^2 + 81x^2y^4$

40. $33a^3b^4c - 16xyz$

41. $14r^2s^3 + 15t^6$

42. $13ab^2c^3 - 26a^3b^2c$

43. $27z^3 + 12z^2 + 3z$

44. $25t^6 - 10t^3 + 5t^2$

45. $24s^3 - 12s^2t + 6st^2$

46. $18y^2z^2 + 12y^2z^3 - 24y^4z^3$

47. $45x^{10}y^3 - 63x^7y^7 + 81x^{10}y^{10}$

48. $48u^6v^6 - 16u^4v^4 - 3u^6v^3$

49. $25x^3 - 14y^3 + 36x^3y^3$

50. $9m^4n^3p^2 + 18m^2n^3p^4 - 27m^3n^4p$

Factor out the negative of the greatest common factor.

51. $-3a - 6$

52. $-6b + 12$

53. $-3x^2 - x$

54. $-4a^3 + a^2$

55. $-6x^2 - 3xy$

56. $-15y^3 + 25y^2$

57. $-18a^2b - 12ab^2$

58. $-21t^5 + 28t^3$

59. $-63u^3v^6z^9 + 28u^2v^7z^2 - 21u^3v^3z^4$

60. $-56x^4y^3z^2 - 72x^3y^4z^5 + 80xy^2z^3$

Factor out the designated factor.

61. x^2 from $x^{n+2} + x^{n+3}$

62. y^3 from $y^{n+3} + y^{n+5}$

63. y^n from $2y^{n+2} - 3y^{n+3}$

64. x^n from $4x^{n+3} - 5x^{n+5}$

65. x^{-2} from $x^4 - 5x^6$

66. y^{-4} from $7y^4 + y$

67. t^{-3} from $t^5 + 4t^{-6}$

68. p^{-5} from $6p^3 - p^{-2}$

69. $4y^{-2n}$ from $8y^{2n} + 12 + 16y^{-2n}$

70. $7x^{-3n}$ from $21x^{6n} + 7x^{3n} + 14$

Factor each expression.

71. $4(x + y) + t(x + y)$

72. $5(a - b) - t(a - b)$

73. $(a - b)r - (a - b)s$

74. $(x + y)u + (x + y)v$

75. $3(m + n + p) + x(m + n + p)$

76. $x(x - y - z) + y(x - y - z)$

77. $(x + y)(x + y) + z(x + y)$

78. $(a - b)^2 + (a - b)$

79. $(u + v)^2 - (u + v)$

80. $a(x - y) - (x - y)^2$

81. $-a(x + y) + b(x + y)$

82. $-bx(a - b) - cx(a - b)$

83. $ax + bx + ay + by$

84. $ar - br + as - bs$

85. $x^2 + yx + 2x + 2y$

86. $2c + 2d - cd - d^2$

87. $3c - cd + 3d - c^2$

88. $x^2 + 4y - xy - 4x$

89. $a^2 - 4b + ab - 4a$

90. $7u + v^2 - 7v - uv$

91. $ax + bx - a - b$

92. $x^2y - ax - xy + a$

93. $x^2 + xy + xz + xy + y^2 + zy$

94. $ab - b^2 - bc + ac - bc - c^2$

95. $mpx + mqx + npx + nqx$

96. $abd - abe + acd - ace$

97. $x^2y + xy^2 + 2xyz + xy^2 + y^3 + 2y^2z$

98. $a^3 - 2a^2b + a^2c - a^2b + 2ab^2 - abc$

99. $2n^4p - 2n^2 - n^3p^2 + np + 2mn^3p - 2mn$

100. $a^2c^3 + ac^2 + a^3c^2 - 2a^2bc^2 - 2bc^2 + c^3$

Solve for the indicated variable.

101. $r_1r_2 = rr_2 + rr_1$ for r_1

102. $r_1r_2 = rr_2 + rr_1$ for r

103. $d_1d_2 = fd_2 + fd_1$ for f

104. $d_1d_2 = fd_2 + fd_1$ for d_1

105. $b^2x^2 + a^2y^2 = a^2b^2$ for a^2

106. $b^2x^2 + a^2y^2 = a^2b^2$ for b^2

107. $S(1 - r) = a - lr$ for r

108. $Sn = (n - 2)180°$ for n

109. $H(a + b) = 2ab$ for a

110. $H(a + b) = 2ab$ for b

111. $3xy - x = 2y + 3$ for y

112. $x(5y + 3) = y - 1$ for y

APPLICATIONS   *Write each polynomial in nested form.*

113. $2x^3 + 5x^2 - 2x + 8$

114. $4x^4 + 5x^2 - 3x + 9$

115. Framing a picture   The dimensions of a family portrait and the frame in which it is mounted are given in the illustration. Write an algebraic expression that describes

a. the area of the picture frame.

b. the area of the portrait.

c. the area of the mat used in the framing. Express the result in factored form.

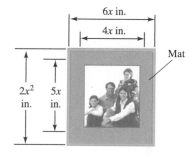

116. Aircraft carriers   The illustration shows the deck of the aircraft carrier *Enterprise.* The rectangular-shaped landing area of $(x^3 + 4x^2 + 5x + 20)$ ft^2 is shaded. If the dimensions can be determined by factoring the expression, what is the width of the landing area?

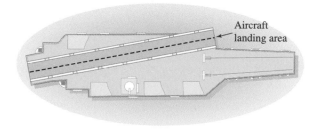

WRITING

117. Explain how to find the greatest common factor of two natural numbers.

118. Explain how to recognize when a number is prime.

SOMETHING TO THINK ABOUT

119. Pick two natural numbers. Divide their product by their greatest common factor. The result is called the **least common multiple** of the two numbers you picked. Why?

120. The number 6 is called a **perfect number,** because the sum of all the divisors of 6 is twice 6: $1 + 2 + 3 + 6 = 12$. Verify that 28 is also a perfect number.

If the greatest common factor of several terms is 1, *the terms are called* **relatively prime.** *Tell whether the terms in each set are relatively prime.*

121. 14, 45
122. 24, 63, 112
123. 60, 28, 36
124. 55, 49, 78
125. $12x^2y, 5ab^3, 35x^2b^3$
126. $18uv, 25rs, 12rsuv$

5.5   The Difference of Two Squares; the Sum and Difference of Two Cubes

▮ Perfect Squares   ▮ The Difference of Two Squares
▮ The Sum and Difference of Two Cubes   ▮ Factoring by Grouping

Getting Ready   *Multiply:*

1. $(a + b)(a - b)$
2. $(5p + q)(5p - q)$
3. $(3m + 2n)(3m - 2n)$
4. $(2a^2 + b^2)(2a^2 - b^2)$
5. $(a - 3)(a^2 + 3a + 9)$
6. $(p + 2)(p^2 - 2p + 4)$

In this section, we will discuss three special types of factoring problems. These types are easy to factor if you learn the appropriate factoring formulas.

Perfect Squares

To factor the difference of two squares, it is helpful to know the first 20 integers that are **perfect squares.**

1, 4, 9, 16, 25, 36, 49, 64, 81, 100, 121, 144, 169, 196, 225, 256, 289, 324, 361, 400

Expressions like $x^6y^4z^2$ are also perfect squares, because they can be written as the square of another quantity:

$$x^6y^4z^2 = (x^3y^2z)^2$$

All exponential expressions that have even-numbered exponents are perfect squares.

The Difference of Two Squares

In section 5.3, we developed the special product formula

(1) $(x + y)(x - y) = x^2 - y^2$

The binomial $x^2 - y^2$ is called the **difference of two squares,** because x^2 represents the square of x, y^2 represents the square of y, and $x^2 - y^2$ represents the difference of these squares.

Equation 1 can be written in reverse order to give a formula for factoring the difference of two squares.

Factoring the Difference of Two Squares

$$x^2 - y^2 = (x + y)(x - y)$$

If we think of the difference of two squares as the square of a **F**irst quantity minus the square of a **L**ast quantity, we have the formula

$$F^2 - L^2 = (F + L)(F - L)$$

and we say: *To factor the square of a* **F**irst *quantity minus the square of a* **L**ast *quantity, we multiply the* **F**irst *plus the* **L**ast *by the* **F**irst *minus the* **L**ast.

EXAMPLE 1   Factor: $49x^2 - 16$.

Solution   We can write $49x^2 - 16$ in the form $(7x)^2 - (4)^2$ and use the formula for factoring the difference of two squares:

$$\mathbf{F}^2 \;-\; \mathbf{L}^2 = (\mathbf{F} \;+\; \mathbf{L})(\mathbf{F} \;-\; \mathbf{L})$$
$$\downarrow \quad\quad \downarrow \quad\quad \downarrow \quad \downarrow \quad \downarrow \quad\quad \downarrow$$
$$(\mathbf{7x})^2 - \mathbf{4}^2 = (\mathbf{7x} + \mathbf{4})(\mathbf{7x} - \mathbf{4}) \quad \text{Substitute } 7x \text{ for F and 4 for L.}$$

We can verify this result by multiplication.

$$(7x + 4)(7x - 4) = 49x^2 - 28x + 28x - 16$$
$$= 49x^2 - 16$$

Self Check   Factor: $81p^2 - 25$.   ∎

Comment   Expressions such as $(7x)^2 + 4^2$, which represent the sum of two squares, cannot be factored in the real-number system. Thus, the binomial $49x^2 + 16$ is a prime binomial.

EXAMPLE 2   Factor: $64a^4 - 25b^2$.

Solution   We can write $64a^4 - 25b^2$ in the form $(8a^2)^2 - (5b)^2$ and use the formula for factoring the difference of two squares.

$$\mathbf{F}^2 \;-\; \mathbf{L}^2 \;=\; (\;\mathbf{F}\; +\; \mathbf{L}\;)(\;\mathbf{F}\; -\; \mathbf{L}\;)$$
$$\downarrow \quad\quad \downarrow \quad\quad\quad \downarrow \quad \downarrow \quad \downarrow \quad\quad \downarrow$$
$$(\mathbf{8a^2})^2 - (\mathbf{5b})^2 = (\mathbf{8a^2} + \mathbf{5b})(\mathbf{8a^2} - \mathbf{5b})$$

Verify by multiplication.

Self Check   Factor: $36r^4 - s^2$.   ∎

EXAMPLE 3   Factor: $x^4 - 1$.

Solution   Because the binomial is the difference of the squares of x^2 and 1, it factors into the sum of x^2 and 1, and the difference of x^2 and 1.

$$x^4 - 1 = (x^2)^2 - (1)^2$$
$$= (x^2 + 1)(x^2 - 1)$$

The factor $x^2 + 1$ is the sum of two squares and is prime. However, the factor $x^2 - 1$ is the difference of two squares and can be factored as $(x + 1)(x - 1)$. Thus,

$$x^4 - 1 = (x^2 + 1)(x^2 - 1)$$
$$= (x^2 + 1)(x + 1)(x - 1)$$

Self Check   Factor: $a^4 - 16$.   ∎

EXAMPLE 4   Factor: $(x + y)^4 - z^4$.

Solution   This expression is the difference of two squares that can be factored:

$$(x + y)^4 - z^4 = [(x + y)^2]^2 - (z^2)^2$$
$$= [(x + y)^2 + z^2][(x + y)^2 - z^2]$$

The factor $(x + y)^2 + z^2$ is the sum of two squares and is prime. However, the factor $(x + y)^2 - z^2$ is the difference of two squares and can be factored as $(x + y + z)(x + y - z)$. Thus,

$$(x + y)^4 - z^4 = [(x + y)^2 + z^2][(x + y)^2 - z^2]$$
$$= [(x + y)^2 + z^2](x + y + z)(x + y - z)$$

Self Check   Factor: $(a - b)^2 - c^2$.   ∎

When possible, we will always factor out a common factor before factoring the difference of two squares. The factoring process is easier when all common factors are factored out first.

EXAMPLE 5   Factor: $2x^4y - 32y$.

Solution

$$\begin{array}{ll} 2x^4y - 32y = 2y(x^4 - 16) & \text{Factor out } 2y. \\ = 2y(x^2 + 4)(x^2 - 4) & \text{Factor } x^4 - 16. \\ = 2y(x^2 + 4)(x + 2)(x - 2) & \text{Factor } x^2 - 4. \end{array}$$

Self Check   Factor: $3a^4 - 3$.   ∎

The Sum and Difference of Two Cubes

The number 64 is called a **perfect cube,** because $4^3 = 64$. To factor the sum or difference of two cubes, it is helpful to know the first ten positive perfect cubes:

1, 8, 27, 64, 125, 216, 343, 512, 729, 1,000

Expressions like $x^9y^6z^3$ are also perfect cubes, because they can be written as the cube of another quantity:

$$x^9y^6z^3 = (x^3y^2z)^3$$

To find formulas for factoring the sum or difference of two cubes, we use the following product formulas:

(2) $\qquad (x + y)(x^2 - xy + y^2) = x^3 + y^3$

(3) $\qquad (x - y)(x^2 + xy + y^2) = x^3 - y^3$

To verify Equation 2, we multiply $x^2 - xy + y^2$ by $x + y$ and see that the product is $x^3 + y^3$.

$$\begin{aligned} (\mathbf{x + y})(x^2 - xy + y^2) &= (\mathbf{x + y})x^2 - (\mathbf{x + y})xy + (\mathbf{x + y})y^2 \\ &= x \cdot x^2 + y \cdot x^2 - x \cdot xy - y \cdot xy \\ &\quad + x \cdot y^2 + y \cdot y^2 \\ &= x^3 + x^2y - x^2y - xy^2 + xy^2 + y^3 \\ &= x^3 + y^3 \quad \text{Combine like terms.} \end{aligned}$$

Equation 3 can also be verified by multiplication.

If we write Equations 2 and 3 in reverse order, we have the formulas for factoring the sum and difference of two cubes.

Sum and Difference of Two Cubes

$$x^3 + y^3 = (x + y)(x^2 - xy + y^2)$$
$$x^3 - y^3 = (x - y)(x^2 + xy + y^2)$$

If we think of the sum of two cubes as the sum of the cube of a **First** quantity plus the cube of a **Last** quantity, we have the formula

$$F^3 + L^3 = (F + L)(F^2 - FL + L^2)$$

To factor the cube of a First quantity plus the cube of a Last quantity, we multiply the sum of the First and Last by

- *the First squared*
- *minus the First times the Last*
- *plus the Last squared.*

The formula for the difference of two cubes is

$$F^3 - L^3 = (F - L)(F^2 + FL + L^2)$$

To factor the cube of a First quantity minus the cube of a Last quantity, we multiply the difference of the First and Last by

- *the First squared*
- *plus the First times the Last*
- *plus the Last squared.*

EXAMPLE 6 Factor: $a^3 + 8$.

Solution Since $a^3 + 8$ can be written as $a^3 + 2^3$, we have the sum of two cubes, which factors as follows:

$$\mathbf{F}^3 + \mathbf{L}^3 = (\mathbf{F} + \mathbf{L})(\mathbf{F}^2 - \mathbf{F}\,\mathbf{L} + \mathbf{L}^2)$$

$$a^3 + 2^3 = (a + 2)(a^2 - a\,2 + 2^2)$$
$$= (a + 2)(a^2 - 2a + 4)$$

Thus, $a^3 + 8 = (a + 2)(a^2 - 2a + 4)$. Check by multiplication.

Self Check   Factor: $p^3 + 27$.   ∎

EXAMPLE 7   Factor: $27a^3 - 64b^3$.

Solution   Since $27a^3 - 64b^3$ can be written as $(3a)^3 - (4b)^3$, we have the difference of two cubes, which factors as follows:

$$\mathbf{F}^3\ -\ \mathbf{L}^3\ = (\mathbf{F} - \mathbf{L})\,(\mathbf{F}^2\ +\ \mathbf{F}\,\mathbf{L}\ +\ \mathbf{L}^2)$$

$$(3a)^3 - (4b)^3 = (3a - 4b)[(3a)^2 + (3a)(4b) + (4b)^2]$$
$$= (3a - 4b)(9a^2 + 12ab + 16b^2)$$

Thus, $27a^3 - 64b^3 = (3a - 4b)(9a^2 + 12ab + 16b^2)$. Check by multiplication.

Self Check   Factor: $8p^3 - 27q^3$.   ∎

EXAMPLE 8   Factor: $a^3 - (c + d)^3$.

Solution $$a^3 - (c + d)^3 = [a - (c + d)][a^2 + a(c + d) + (c + d)^2]$$
$$= (a - c - d)(a^2 + ac + ad + c^2 + 2cd + d^2)$$

Self Check   Factor: $(p + q)^3 - r^3$.   ∎

EXAMPLE 9   Factor: $x^6 - 64$.

Solution   This expression can be considered as the difference of two squares or the difference of two cubes. If we treat it as the difference of two squares, it factors into the product of a sum and a difference.

$$x^6 - 64 = (x^3)^2 - 8^2$$
$$= (x^3 + 8)(x^3 - 8)$$

Each of these factors further, however, for one is the sum of two cubes and the other is the difference of two cubes:

$$x^6 - 64 = (x + 2)(x^2 - 2x + 4)(x - 2)(x^2 + 2x + 4)$$

Self Check   Factor: $x^6 - 1$.   ∎

EXAMPLE 10   Factor: $2a^5 + 128a^2$.

Solution   We first factor out the common factor of $2a^2$ to obtain

$$2a^5 + 128a^2 = 2a^2(a^3 + 64)$$

Then we factor $a^3 + 64$ as the sum of two cubes to obtain

$$2a^5 + 128a^2 = 2a^2(a + 4)(a^2 - 4a + 16)$$

Self Check Factor: $3x^5 + 24x^2$. ▮

EXAMPLE 11 Factor: $16r^{6m} - 54t^{3n}$.

Solution
$$\begin{aligned}
16r^{6m} - 54t^{3n} &= 2(8r^{6m} - 27t^{3n}) && \text{Factor out 2.}\\
&= 2[(2r^{2m})^3 - (3t^n)^3] && \text{Write } 8r^{6m} \text{ as}\\
&&& (2r^{2m})^3 \text{ and } 27t^{3n}\\
&&& \text{as } (3t^n)^3.\\
&= 2[(2r^{2m} - 3t^n)(4r^{4m} + 6r^{2m}t^n + 9t^{2n})] && \text{Factor}\\
&&& (2r^{2m})^3 - (3t^n)^3.
\end{aligned}$$

Self Check Factor: $2a^{3m} - 16b^{3n}$. ▮

Factoring by Grouping

EXAMPLE 12 Factor: $x^2 - y^2 + x - y$.

Solution If we group the first two terms and factor the difference of two squares, we have

$$\begin{aligned}
x^2 - y^2 + x - y &= (x + y)(x - y) + (x - y) && \text{Factor } x^2 - y^2.\\
&= (x - y)(x + y + 1) && \text{Factor out } x - y.
\end{aligned}$$

Self Check Factor: $a^2 - b^2 + a + b$. ▮

Self Check Answers

1. $(9p + 5)(9p - 5)$ **2.** $(6r^2 + s)(6r^2 - s)$ **3.** $(a^2 + 4)(a + 2)(a - 2)$ **4.** $(a - b + c)(a - b - c)$
5. $3(a^2 + 1)(a + 1)(a - 1)$ **6.** $(p + 3)(p^2 - 3p + 9)$ **7.** $(2p - 3q)(4p^2 + 6pq + 9q^2)$
8. $(p + q - r)(p^2 + 2pq + q^2 + pr + qr + r^2)$ **9.** $(x + 1)(x^2 - x + 1)(x - 1)(x^2 + x + 1)$
10. $3x^2(x + 2)(x^2 - 2x + 4)$ **11.** $2(a^m - 2b^n)(a^{2m} + 2a^m b^n + 4b^{2n})$ **12.** $(a + b)(a - b + 1)$

Orals *Factor each expression, if possible.*

1. $x^2 - 1$ **2.** $a^4 - 16$
3. $x^3 + 1$ **4.** $a^3 - 8$
5. $2x^2 - 8$ **6.** $x^4 + 25$

5.5 EXERCISES 🔊

REVIEW *Perform each multiplication.*

1. $(x + 1)(x + 1)$ **5.** $(a + 4)(a + 3)$

2. $(2m - 3)(m - 2)$ **6.** $(3b + 2)(2b - 5)$

3. $(2m + n)(2m + n)$ **7.** $(4r - 3s)(2r - s)$

4. $(3m - 2n)(3m - 2n)$ **8.** $(5a - 2b)(3a + 4b)$

VOCABULARY AND CONCEPTS   *Fill in the blanks.*

9. Write the first ten perfect square natural numbers.
_____

10. $p^2 - q^2 = (p + q)$_____

11. The sum of two squares _____ be factored.

12. Write the first ten perfect cube natural numbers.
_____

13. $p^3 + q^3 = (p + q)$ _____

14. $p^3 - q^3 = (p - q)$ _____

PRACTICE   *Factor each expression, if possible.*

15. $x^2 - 4$

16. $y^2 - 9$

17. $9y^2 - 64$

18. $16x^4 - 81y^2$

19. $x^2 + 25$

20. $144a^2 - b^4$

21. $625a^2 - 169b^4$

22. $4y^2 + 9z^4$

23. $81a^4 - 49b^2$

24. $64r^6 - 121s^2$

25. $36x^4y^2 - 49z^4$

26. $4a^2b^4c^6 - 9d^8$

27. $(x + y)^2 - z^2$

28. $a^2 - (b - c)^2$

29. $(a - b)^2 - c^2$

30. $(m + n)^2 - p^4$

31. $x^4 - y^4$

32. $16a^4 - 81b^4$

33. $256x^4y^4 - z^8$

34. $225a^4 - 16b^8c^{12}$

35. $2x^2 - 288$

36. $8x^2 - 72$

37. $2x^3 - 32x$

38. $3x^3 - 243x$

39. $5x^3 - 125x$

40. $6x^4 - 216x^2$

41. $r^2s^2t^2 - t^2x^4y^2$

42. $16a^4b^3c^4 - 64a^2bc^6$

43. $r^3 + s^3$

44. $t^3 - v^3$

45. $x^3 - 8y^3$

46. $27a^3 + b^3$

47. $64a^3 - 125b^6$

48. $8x^6 + 125y^3$

49. $125x^3y^6 + 216z^9$

50. $1{,}000a^6 - 343b^3c^6$

51. $x^6 + y^6$

52. $x^9 + y^9$

53. $5x^3 + 625$

54. $2x^3 - 128$

55. $4x^5 - 256x^2$

56. $2x^6 + 54x^3$

57. $128u^2v^3 - 2t^3u^2$

58. $56rs^2t^3 + 7rs^2v^6$

59. $(a + b)x^3 + 27(a + b)$

60. $(c - d)r^3 - (c - d)s^3$

Factor each expression. Assume that m and n are natural numbers.

61. $x^{2m} - y^{4n}$

62. $a^{4m} - b^{8n}$

63. $100a^{4m} - 81b^{2n}$

64. $25x^{8m} - 36y^{4n}$

65. $x^{3n} - 8$

66. $a^{3m} + 64$

67. $a^{3m} + b^{3n}$

68. $x^{6m} - y^{3n}$

69. $2x^{6m} + 16y^{3m}$

70. $24 + 3c^{3m}$

Factor each expression by grouping.

71. $a^2 - b^2 + a + b$

72. $x^2 - y^2 - x - y$

73. $a^2 - b^2 + 2a - 2b$

74. $m^2 - n^2 + 3m + 3n$

75. $2x + y + 4x^2 - y^2$

76. $m - 2n + m^2 - 4n^2$

APPLICATIONS

77. Physics   The illustration shows a time-sequence picture of a falling apple. Factor the expression, which gives the difference in the distance fallen by the apple during the time interval from t_1 to t_2 seconds.

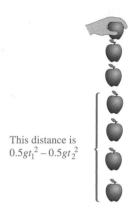

This distance is
$0.5gt_1^2 - 0.5gt_2^2$

78. Darts   A circular dart board has a series of rings around a solid center, called the bullseye. To find the area of the outer white ring, we can use the formula

$$A = \pi R^2 - \pi r^2$$

Factor the expression on the right-hand side of the equation.

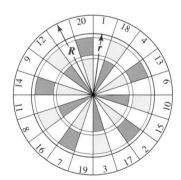

79. Candy   To find the amount of chocolate in the outer coating of a malted-milk ball, we can use the formula for the volume V of the chocolate shell:

$$V = \frac{4}{3}\pi r_1^3 - \frac{4}{3}\pi r_2^3$$

Factor the expression on the right-hand side.

80. Movie stunts   The distance that a stuntman is above the ground t seconds after he jumps from the top of the building shown in the illustration is given by the formula

$$h = 144 - 16t^2$$

Factor the right-hand side of the equation.

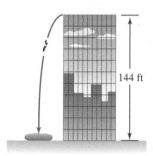

144 ft

WRITING

81. Describe the pattern used to factor the difference of two squares.

82. Describe the patterns used to factor the sum and the difference of two cubes.

SOMETHING TO THINK ABOUT

83. Factor $x^{32} - y^{32}$.

84. Find the error in this proof that $2 = 1$.

$$x = y$$
$$x^2 = xy$$
$$x^2 - y^2 = xy - y^2$$
$$(x + y)(x - y) = y(x - y)$$
$$\frac{(x + y)\cancel{(x - y)}}{\cancel{(x - y)}} = \frac{y\cancel{(x - y)}}{\cancel{x - y}}$$
$$x + y = y$$
$$y + y = y$$
$$2y = y$$
$$\frac{2y}{y} = \frac{y}{y}$$
$$2 = 1$$

5.6   Factoring Trinomials

- ▌ **Special Product Formulas**
- ▌ **Factoring Trinomials with Lead Coefficients of 1**
- ▌ **Factoring Trinomials with Lead Coefficients Other than 1**
- ▌ **Test for Factorability**   ▌ **Using Substitution to Factor Trinomials**
- ▌ **Factoring by Grouping**   ▌ **Using Grouping to Factor Trinomials**

Getting Ready   *Multiply.*

1. $(a + 3)(a - 4)$

2. $(3a + 5)(2a - 1)$

3. $(a + b)(a + 2b)$

4. $(2a + b)(a - b)$

5. $(3a - 2b)(2a - 3b)$

6. $(4a - 3b)^2$

In this section, we will learn how to factor trinomials. We will consider two methods: a guess and check method and a factoring by grouping method.

Special Product Formulas

Many trinomials can be factored by using the following special product formulas.

(1) $$(x + y)(x + y) = x^2 + 2xy + y^2$$

(2) $$(x - y)(x - y) = x^2 - 2xy + y^2$$

To factor the perfect square trinomial $x^2 + 6x + 9$, we note that it can be written in the form $x^2 + 2(3)x + 3^2$. If $y = 3$, this form matches the right-hand side of Equation 1. Thus, $x^2 + 6x + 9$ factors as

$$x^2 + 6x + 9 = x^2 + 2(3)x + 3^2$$
$$= (x + 3)(x + 3)$$

This result can be verified by multiplication:

$$(x + 3)(x + 3) = x^2 + 3x + 3x + 9$$
$$= x^2 + 6x + 9$$

To factor the perfect square trinomial $x^2 - 4xz + 4z^2$, we note that it can be written in the form $x^2 - 2x(2z) + (2z)^2$. If $y = 2z$, this form matches the right-hand side of Equation 2. Thus, $x^2 - 4xz + 4z^2$ factors as

$$x^2 - 4xz + 4z^2 = x^2 - 2x(2z) + (2z)^2$$
$$= (x - 2z)(x - 2z)$$

This result can also be verified by multiplication.

Factoring Trinomials with Lead Coefficients of 1

Since the product of two binomials is often a trinomial, we expect that many trinomials will factor as two binomials. For example, to factor $x^2 + 7x + 12$, we must find two binomials $x + a$ and $x + b$ such that

$$x^2 + 7x + 12 = (x + a)(x + b)$$

where $ab = 12$ and $ax + bx = 7x$.

To find the numbers a and b, we list the possible factorizations of 12 and find the one where the sum of the factors is 7.

The one to choose
↓

$$12(1) \quad 6(2) \quad 4(3) \quad -12(-1) \quad -6(-2) \quad -4(-3)$$

Thus, $a = 4$, $b = 3$, and

$$x^2 + 7x + 12 = (x + a)(x + b)$$

(3) $\qquad x^2 + 7x + 12 = (x + 4)(x + 3)$ Substitute 4 for a and 3 for b.

This factorization can be verified by multiplying $x + 4$ and $x + 3$ and observing that the product is $x^2 + 7x + 12$.

Because of the commutative property of multiplication, the order of the factors in Equation 3 is not important.

To factor trinomials with lead coefficients of 1, we follow these steps.

Factoring Trinomials with Lead Coefficients of 1

1. Write the trinomial in descending powers of one variable.
2. List the factorizations of the third term of the trinomial.
3. Pick the factorization where the sum of the factors is the coefficient of the middle term.

EXAMPLE 1 Factor: $x^2 - 6x + 8$.

Solution Since the trinomial is written in descending powers of x, we can move to Step 2 and list the possible factorizations of the third term, which is 8.

The one to choose
↓

$$8(1) \quad 4(2) \quad -8(-1) \quad -4(-2)$$

In the trinomial, the coefficient of the middle term is -6. The only factorization where the sum of the factors is -6 is $-4(-2)$. Thus, $a = -4$, $b = -2$, and

$$x^2 - 6x + 8 = (x + a)(x + b)$$
$$= (x - 4)(x - 2) \quad \text{Substitute } -4 \text{ for } a \text{ and } -2 \text{ for } b.$$

We can verify this result by multiplication:

$$(x - 4)(x - 2) = x^2 - 2x - 4x + 8$$
$$= x^2 - 6x + 8$$

Self Check Factor: $a^2 - 7a + 12$.

EXAMPLE 2   Factor: $-x + x^2 - 12$.

Solution   We begin by writing the trinomial in descending powers of x:

$$-x + x^2 - 12 = x^2 - x - 12$$

The possible factorizations of the third term are

The one to choose
↓

$$12(-1) \qquad 6(-2) \qquad 4(-3) \qquad 1(-12) \qquad 2(-6) \qquad 3(-4)$$

In the trinomial, the coefficient of the middle term is -1. The only factorization where the sum of the factors is -1 is $3(-4)$. Thus, $a = 3$, $b = -4$, and

$$
\begin{aligned}
-x + x^2 - 12 &= x^2 - x - 12 \\
&= (x + \boldsymbol{a})(x + \boldsymbol{b}) \\
&= (x + \boldsymbol{3})(x - \boldsymbol{4}) \qquad \text{Substitute 3 for } a \text{ and } -4 \text{ for } b.
\end{aligned}
$$

Self Check   Factor: $-3a + a^2 - 10$. ∎

EXAMPLE 3   Factor: $30x - 4xy - 2xy^2$.

Solution   We begin by writing the trinomial in descending powers of y:

$$30x - 4xy - 2xy^2 = -2xy^2 - 4xy + 30x$$

Each term in this trinomial has a common monomial factor of $-2x$, which we will factor out.

$$30x - 4xy - 2xy^2 = -2x(y^2 + 2y - 15)$$

To factor $y^2 + 2y - 15$, we list the factors of -15 and find the pair whose sum is 2.

The one to choose
↓

$$15(-1) \qquad 5(-3) \qquad 1(-15) \qquad 3(-5)$$

The only factorization where the sum of the factors is 2 (the coefficient of the middle term of $y^2 + 2y - 15$) is $5(-3)$. Thus, $a = 5$, $b = -3$, and

$$
\begin{aligned}
30x - 4xy - 2xy^2 &= -2x(y^2 + 2y - 15) \\
&= -2x(y + 5)(y - 3)
\end{aligned}
$$

Self Check   Factor: $18a + 3ab - 3ab^2$. ∎

Comment   In Example 3, be sure to include all factors in the final result. It is a common error to forget to write the $-2x$.

Factoring Trinomials with Lead Coefficients Other than 1

There are more combinations of factors to consider when factoring trinomials with lead coefficients other than 1. To factor $5x^2 + 7x + 2$, for example, we must find two binomials of the form $ax + b$ and $cx + d$ such that

$$5x^2 + 7x + 2 = (ax + b)(cx + d)$$

Since the first term of the trinomial $5x^2 + 7x + 2$ is $5x^2$, the first terms of the binomial factors must be $5x$ and x.

$$5x^2 + 7x + 2 = (5x + b)(x + d)$$

Since the product of the last terms must be 2, and the sum of the products of the outer and inner terms must be $7x$, we must find two numbers whose product is 2 that will give a middle term of $7x$.

$$5x^2 + 7x + 2 = (5x + b)(x + d)$$

$$O + I = 7x$$

Since $2(1)$ and $(-2)(-1)$ give a product of 2, there are four possible combinations to consider:

$(5x + 2)(x + 1)$ $(5x - 2)(x - 1)$

$(5x + 1)(x + 2)$ $(5x - 1)(x - 2)$

Of these possibilities, only the one printed in color gives the correct middle term of $7x$. Thus,

(4) $5x^2 + 7x + 2 = (5x + 2)(x + 1)$

We can verify this result by multiplication:

$$(5x + 2)(x + 1) = 5x^2 + 5x + 2x + 2$$
$$= 5x^2 + 7x + 2$$

Test for Factorability

If a trinomial has the form $ax^2 + bx + c$, with integer coefficients and $a \neq 0$, we can test to see whether it is factorable.

- If the value of $b^2 - 4ac$ is a perfect square, the trinomial can be factored using only integers.
- If the value is not a perfect square, the trinomial cannot be factored using only integers.

For example, $5x^2 + 7x + 2$ is a trinomial in the form $ax^2 + bx + c$ with

$$a = 5, \quad b = 7, \quad \text{and} \quad c = 2$$

For this trinomial, the value of $b^2 - 4ac$ is

$$b^2 - 4ac = 7^2 - 4(5)(2) \quad \text{Substitute 7 for } b, \text{ 5 for } a, \text{ and 2 for } c.$$
$$= 49 - 40$$
$$= 9$$

Since 9 is a perfect square, the trinomial is factorable. Its factorization is shown in Equation 4.

Test for Factorability

> A trinomial of the form $ax^2 + bx + c$, with integer coefficients and $a \neq 0$, will factor into two binomials with integer coefficients if the value of $b^2 - 4ac$ is a perfect square. If $b^2 - 4ac = 0$, the factors will be the same.

EXAMPLE 4   Factor: $3p^2 - 4p - 4$.

Solution   In the trinomial, $a = 3$, $b = -4$, and $c = -4$. To see whether it factors, we evaluate $b^2 - 4ac$.

$$b^2 - 4ac = (-4)^2 - 4(3)(-4) \qquad \text{Substitute } -4 \text{ for } b, 3 \text{ for } a, \text{ and } -4 \text{ for } c.$$
$$= 16 + 48$$
$$= 64$$

Since 64 is a perfect square, it is factorable.

To factor the trinomial, we note that the first terms of the binomial factors must be $3p$ and p to give the first term of $3p^2$.

$$3p^2 - 4p - 4 = (3p + ?)(p + ?)$$

The product of the last terms must be -4, and the sum of the products of the outer terms and the inner terms must be $-4p$.

$$3p^2 - 4p - 4 = (3p + ?)(p + ?)$$
$$\text{O} + \text{I} = -4p$$

Because $1(-4)$, $-1(4)$, and $-2(2)$ all give a product of -4, there are six possible combinations to consider:

$$(3p + 1)(p - 4) \qquad (3p - 4)(p + 1)$$
$$(3p - 1)(p + 4) \qquad (3p + 4)(p - 1)$$
$$(3p - 2)(p + 2) \qquad \mathbf{(3p + 2)(p - 2)}$$

Of these possibilities, only the one printed in color gives the correct middle term of $-4p$. Thus,

$$3p^2 - 4p - 4 = (3p + 2)(p - 2)$$

Self Check   Factor: $4q^2 - 9q - 9$. ∎

EXAMPLE 5   Factor: $4t^2 - 3t - 5$, if possible.

Solution   In the trinomial, $a = 4$, $b = -3$, and $c = -5$. To see whether it is factorable, we evaluate $b^2 - 4ac$ by substituting the values of a, b, and c.

$$b^2 - 4ac = (-3)^2 - 4(4)(-5)$$
$$= 9 + 80$$
$$= 89$$

Since 89 is not a perfect square, the trinomial is not factorable using only integer co-efficients. It is prime.

Self Check   Factor: $5a^2 - 8a + 2$, if possible.   ∎

To factor trinomials, the following hints are helpful.

Factoring a Trinomial

> 1. Write the trinomial in descending powers of one variable.
> 2. Factor out any greatest common factor (including -1 if that is necessary to make the coefficient of the first term positive).
> 3. Test the trinomial for factorability.
> 4. When the sign of the first term of a trinomial is $+$ and the sign of the third term is $+$, the signs between the terms of each binomial factor are the same as the sign of the middle term of the trinomial.
>   When the sign of the first term is $+$ and the sign of the third term is $-$, the signs between the terms of the binomials are opposites.
> 5. Try various combinations of first terms and last terms until you find the one that works.
> 6. Check the factorization by multiplication.

EXAMPLE 6   Factor: $24y + 10xy - 6x^2y$.

Solution   We write the trinomial in descending powers of x and factor out the common factor of $-2y$:

$$24y + 10xy - 6x^2y = -6x^2y + 10xy + 24y$$
$$= -2y(3x^2 - 5x - 12)$$

In the trinomial $3x^2 - 5x - 12$, $a = 3$, $b = -5$, and $c = -12$.

$$b^2 - 4ac = (-5)^2 - 4(3)(-12)$$
$$= 25 + 144$$
$$= 169$$

Since 169 is a perfect square, the trinomial will factor.

Since the sign of the third term of $3x^2 - 5x - 12$ is $-$, the signs between the binomial factors will be opposite. Because the first term is $3x^2$, the first terms of the binomial factors must be $3x$ and x.

$$\overbrace{-2y(3x^2 - 5x - 12) = -2y(3x \qquad)(x \qquad)}^{3x^2}$$

The product of the last terms must be -12, and the sum of the product of the outer terms and the product of the inner terms must be $-5x$.

$$-2y(3x^2 - 5x - 12) = -2y(3x \quad ?)(x \quad ?)$$

$$O + I = -5x$$

Since $1(-12)$, $2(-6)$, $3(-4)$, $12(-1)$, $6(-2)$, and $4(-3)$ all give a product of -12, there are 12 possible combinations to consider.

$(3x + 1)(x - 12)$	$(3x - 12)(x + 1)$
$(3x + 2)(x - 6)$	$(3x - 6)(x + 2)$
$(3x + 3)(x - 4)$	$(3x - 4)(x + 3)$
$(3x + 12)(x - 1)$	$(3x - 1)(x + 12)$
$(3x + 6)(x - 2)$	$(3x - 2)(x + 6)$
The one to choose $\rightarrow$ $(3x + 4)(x - 3)$	$(3x - 3)(x + 4)$

The combinations marked in blue cannot work, because one of the factors has a common factor. This implies that $3x^2 - 5x - 12$ would have a common factor, which it doesn't.

After mentally trying the remaining combinations, we find that only $(3x + 4)(x - 3)$ gives the correct middle term of $-5x$.

$$24y + 10xy - 6x^2y = -2y(3x^2 - 5x - 12)$$
$$= -2y(3x + 4)(x - 3)$$

Verify this result by multiplication.

Self Check Factor: $9b - 6a^2b - 3ab$.

EXAMPLE 7 Factor: $6y + 13x^2y + 6x^4y$.

Solution We write the trinomial in descending powers of x and factor out the common factor of y to obtain

$$6y + 13x^2y + 6x^4y = 6x^4y + 13x^2y + 6y$$
$$= y(6x^4 + 13x^2 + 6)$$

A test of the trinomial $6x^4 + 13x^2 + 6$ will show that it does factor. Since the coefficients of the first and last terms are positive, the signs between the terms in each binomial will be $+$.

Since the first term of the trinomial is $6x^4$, the first terms of the binomial factors must be either $2x^2$ and $3x^2$ or x^2 and $6x^2$.

Since the product of the last terms of the binomial factors must be 6, we must find two numbers whose product is 6 that will lead to a middle term of $13x^2$. After trying some combinations, we find the one that works.

$$6y + 13x^2y + 6x^4y = y(6x^4 + 13x^2 + 6)$$
$$= y(2x^2 + 3)(3x^2 + 2)$$

Verify this result by multiplication.

Self Check Factor: $4b + 11a^2b + 6a^4b$.

EXAMPLE 8 Factor: $x^{2n} + x^n - 2$.

Solution Since the first term is x^{2n}, the first terms of the binomial factors must be x^n and x^n.

$$x^{2n} + x^n - 2 = (x^n \quad \overset{\overset{\displaystyle x^{2n}}{\frown}}{)(x^n} \quad)$$

Since the third term of the trinomial is -2, the last terms of the binomial factors must have opposite signs, have a product of -2, and lead to a middle term of x^n. The only combination that works is

$$x^{2n} + x^n - 2 = (x^n + 2)(x^n - 1)$$

Verify this result by multiplication.

Self Check   Factor: $a^{2n} + a^n - 6$.   ∎

Using Substitution to Factor Trinomials

EXAMPLE 9   Factor: $(x + y)^2 + 7(x + y) + 12$.

Solution   We rewrite the trinomial $(x + y)^2 + 7(x + y) + 12$ as $z^2 + 7z + 12$, where $z = (x + y)$. The trinomial $z^2 + 7z + 12$ factors as $(z + 4)(z + 3)$.

To find the factorization of $(x + y)^2 + 7(x + y) + 12$, we substitute $x + y$ for z in the expression $(z + 4)(z + 3)$ to obtain

$$z^2 + 7z + 12 = (z + 4)(z + 3)$$
$$(x + y)^2 + 7(x + y) + 12 = (x + y + 4)(x + y + 3)$$

Self Check   Factor: $(a + b)^2 - 3(a + b) - 10$.   ∎

Factoring by Grouping

EXAMPLE 10   Factor: $x^2 + 6x + 9 - z^2$.

Solution   We group the first three terms together and factor the trinomial to get

$$x^2 + 6x + 9 - z^2 = (x + 3)(x + 3) - z^2$$
$$= (x + 3)^2 - z^2$$

We can now factor the difference of two squares to get

$$x^2 + 6x + 9 - z^2 = (x + 3 + z)(x + 3 - z)$$

Self Check   Factor: $a^2 + 4a + 4 - b^2$.   ∎

Using Grouping to Factor Trinomials

Factoring by grouping can be used to help factor trinomials of the form $ax^2 + bx + c$. For example, to factor $6x^2 + 7x - 3$, we proceed as follows:

1. We find the product ac: $6(-3) = -18$. This is called the **key number.**

2. We find two factors of the key number -18 whose sum is $b = 7$:

$$9(-2) = -18 \qquad \text{and} \qquad 9 + (-2) = 7$$

3. We use the factors 9 and -2 as coefficients of two terms to be placed between $6x^2$ and -3:

$$6x^2 + 7x - 3 = 6x^2 + 9x - 2x - 3$$

4. We factor by grouping:

$$6x^2 + 9x - 2x - 3 = 3x(2x + 3) - (2x + 3)$$
$$= (2x + 3)(3x - 1) \qquad \text{Factor out } 2x + 3.$$

We can verify this factorization by multiplication.

EXAMPLE 11   Factor: $10x^2 + 13x - 3$.

Solution   Since $a = 10$ and $c = -3$ in the trinomial, $ac = -30$. We must find two factors of -30 whose sum is $+13$. Two such factors are 15 and -2. We use these factors as coefficients of two terms to be placed between $10x^2$ and -3:

$$10x^2 + 15x - 2x - 3$$

Finally, we factor by grouping.

$$10x^2 + 15x - 2x - 3 = 5x(2x + 3) - (2x + 3)$$
$$= (2x + 3)(5x - 1)$$

Thus, $10x^2 + 13x - 3 = (2x + 3)(5x - 1)$.

Self Check   Factor: $15a^2 + 17a - 4$.

Self Check Answers

1. $(a - 4)(a - 3)$ **2.** $(a + 2)(a - 5)$ **3.** $-3a(b + 2)(b - 3)$ **4.** $(4q + 3)(q - 3)$ **5.** a prime polynomial
6. $-3b(2a + 3)(a - 1)$ **7.** $b(2a^2 + 1)(3a^2 + 4)$ **8.** $(a^n + 3)(a^n - 2)$ **9.** $(a + b + 2)(a + b - 5)$
10. $(a + 2 + b)(a + 2 - b)$ **11.** $(3a + 4)(5a - 1)$

Orals   *Factor each expression.*

1. $x^2 + 3x + 2$ **2.** $x^2 + 5x + 4$
3. $x^2 - 5x + 6$ **4.** $x^2 - 3x - 4$
5. $2x^2 + 3x + 1$ **6.** $3x^2 + 4x + 1$

5.6   EXERCISES

REVIEW   *Solve each equation.*

1. $\dfrac{2 + x}{11} = 3$ **2.** $\dfrac{3y - 12}{2} = 9$

3. $\dfrac{2}{3}(5t - 3) = 38$ **4.** $3(p + 2) = 4p$

5. $11r + 6(3 - r) = 3$
6. $2q^2 - 9 = q(q + 3) + q^2$

VOCABULARY AND CONCEPTS   *Fill in the blanks.*

7. $(x + y)(x + y) = x^2 +$ _____
8. $(x - y)(x - y) = x^2 -$ _____
9. $(x + y)(x - y) =$ _____
10. If _____ is a perfect square, the trinomial $ax^2 + bx + c$ will factor using _____ coefficients.

PRACTICE *Complete each factorization.*

11. $x^2 + 5x + 6 = (x + 3)(\underline{\quad})$

12. $x^2 - 6x + 8 = (x - 4)(\underline{\quad})$

13. $x^2 + 2x - 15 = (x + 5)(\underline{\quad})$

14. $x^2 - 3x - 18 = (x - 6)(\underline{\quad})$

15. $2a^2 + 9a + 4 = (\underline{\quad})(a + 4)$

16. $6p^2 - 5p - 4 = (\underline{\quad})(2p + 1)$

17. $4m^2 + 8mn + 3n^2 = (\underline{\quad})(2m + n)$

18. $12r^2 + 5rs - 2s^2 = (\underline{\quad})(3r + 2s)$

Factor each perfect square trinomial.

19. $x^2 + 2x + 1$

20. $y^2 - 2y + 1$

21. $a^2 - 18a + 81$

22. $b^2 + 12b + 36$

23. $4y^2 + 4y + 1$

24. $9x^2 + 6x + 1$

25. $9b^2 - 12b + 4$

26. $4a^2 - 12a + 9$

27. $9z^2 + 24z + 16$

28. $16z^2 - 24z + 9$

Test each trinomial for factorability and factor it, if possible.

29. $x^2 + 9x + 8$

30. $y^2 + 7y + 6$

31. $x^2 - 7x + 10$

32. $c^2 - 7c + 12$

33. $b^2 + 8b + 18$

34. $x^2 - 12x + 35$

35. $x^2 - x - 30$

36. $a^2 + 4a - 45$

37. $a^2 + 5a - 50$

38. $b^2 + 9b - 36$

39. $y^2 - 4y - 21$

40. $x^2 + 4x - 28$

Factor each trinomial. Factor out all common monomials first (including -1 if the lead coefficient is negative). If a trinomial is prime, so indicate.

41. $3x^2 + 12x - 63$

42. $2y^2 + 4y - 48$

43. $a^2b^2 - 13ab^2 + 22b^2$

44. $a^2b^2x^2 - 18a^2b^2x + 81a^2b^2$

45. $b^2x^2 - 12bx^2 + 35x^2$

46. $c^3x^2 + 11c^3x - 42c^3$

47. $-a^2 + 4a + 32$

48. $-x^2 - 2x + 15$

49. $-3x^2 + 15x - 18$

50. $-2y^2 - 16y + 40$

51. $-4x^2 + 4x + 80$

52. $-5a^2 + 40a - 75$

53. $6y^2 + 7y + 2$

54. $6x^2 - 11x + 3$

55. $8a^2 + 6a - 9$

56. $15b^2 + 4b - 4$

57. $6x^2 - 5x - 4$

58. $18y^2 - 3y - 10$

59. $5x^2 + 4x + 1$

60. $6z^2 + 17z + 12$

61. $8x^2 - 10x + 3$

62. $4a^2 + 20a + 3$

63. $a^2 - 3ab - 4b^2$

64. $b^2 + 2bc - 80c^2$

65. $2y^2 + yt - 6t^2$

66. $3x^2 - 10xy - 8y^2$

67. $3x^3 - 10x^2 + 3x$

68. $3t^3 - 3t^2 + t$

69. $-3a^2 + ab + 2b^2$

70. $-2x^2 + 3xy + 5y^2$

71. $-4x^2 - 9 + 12x$

72. $6x + 4 + 9x^2$

73. $5a^2 + 45b^2 - 30ab$

74. $-90x^2 + 2 - 8x$

75. $8x^2z + 6xyz + 9y^2z$

76. $x^3 - 60xy^2 + 7x^2y$

77. $21x^4 - 10x^3 - 16x^2$

78. $16x^3 - 50x^2 + 36x$

79. $x^4 + 8x^2 + 15$

80. $x^4 + 11x^2 + 24$

81. $y^4 - 13y^2 + 30$

82. $y^4 - 13y^2 + 42$

83. $a^4 - 13a^2 + 36$

84. $b^4 - 17b^2 + 16$

85. $z^4 - z^2 - 12$

86. $c^4 - 8c^2 - 9$

87. $4x^3 + x^6 + 3$

88. $a^6 - 2 + a^3$

89. $x^{2n} + 2x^n + 1$

90. $x^{4n} - 2x^{2n} + 1$

91. $2a^{6n} - 3a^{3n} - 2$

92. $b^{2n} - b^n - 6$

93. $x^{4n} + 2x^{2n}y^{2n} + y^{4n}$

94. $y^{6n} + 2y^{3n}z + z^2$

95. $6x^{2n} + 7x^n - 3$

96. $12y^{4n} + 10y^{2n} + 2$

Factor each expression.

97. $(x + 1)^2 + 2(x + 1) + 1$

98. $(a + b)^2 - 2(a + b) + 1$

99. $(a + b)^2 - 2(a + b) - 24$

100. $(x - y)^2 + 3(x - y) - 10$

101. $6(x + y)^2 - 7(x + y) - 20$

102. $2(x - z)^2 + 9(x - z) + 4$

103. $x^2 + 4x + 4 - y^2$

104. $x^2 - 6x + 9 - 4y^2$

105. $x^2 + 2x + 1 - 9z^2$

106. $x^2 + 10x + 25 - 16z^2$

107. $c^2 - 4a^2 + 4ab - b^2$

108. $4c^2 - a^2 - 6ab - 9b^2$

109. $a^2 - b^2 + 8a + 16$

110. $a^2 + 14a - 25b^2 + 49$

111. $4x^2 - z^2 + 4xy + y^2$

112. $x^2 - 4xy - 4z^2 + 4y^2$

Use grouping to help factor each trinomial.

113. $a^2 - 17a + 16$

114. $b^2 - 4b - 21$

115. $2u^2 + 5u + 3$

116. $6y^2 + 5y - 6$

117. $20r^2 - 7rs - 6s^2$

118. $6s^2 + st - 12t^2$

119. $20u^2 + 19uv + 3v^2$

120. $12m^2 + mn - 6n^2$

APPLICATIONS

121. Office furniture The area of the desktop is given by the expression $(4x^2 + 20x - 11)$ in.^{2}. Factor this expression to find the expressions that represent its length and width. Then determine the difference in the length and width of the desktop.

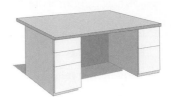

122. Storage The volume of the 8-foot-wide portable storage container is given by the expression $(72x^2 + 120x - 400)$ ft^3. If its dimensions can be determined by factoring the expression, find the height and the length of the container.

WRITING

123. Explain how you would factor -1 from a trinomial.

124. Explain how you would test the polynomial $ax^2 + bx + c$ for factorability.

SOMETHING TO THINK ABOUT

125. Because it is the difference of two squares, $x^2 - q^2$ always factors. Does the test for factorability predict this?

126. The polynomial $ax^2 + ax + a$ factors, because a is a common factor. Does the test for factorability predict this? If not, is there something wrong with the test? Explain.

5.7   Summary of Factoring Techniques

❙ **Factoring Random Polynomials**

Getting Ready   *Factor each polynomial.*

1. $3p^2q - 3pq^2$ **2.** $4p^2 - 9q^2$
3. $p^2 + 5p - 6$ **4.** $6p^2 - 13pq + 6q^2$

In this section, we will discuss ways to approach a randomly chosen factoring problem.

Factoring Random Polynomials

Suppose we wish to factor the trinomial

$$x^2y^2z^3 + 7xy^2z^3 + 6y^2z^3$$

We begin by attempting to identify the problem type. The first type to look for is **factoring out a common monomial.** Because the trinomial has a common monomial factor of y^2z^3, we factor it out:

$$x^2y^2z^3 + 7xy^2z^3 + 6y^2z^3 = y^2z^3(x^2 + 7x + 6)$$

We note that $x^2 + 7x + 6$ is a trinomial that can be factored as $(x + 6)(x + 1)$. Thus,

$$x^2y^2z^3 + 7xy^2z^3 + 6y^2z^3 = y^2z^3(x^2 + 7x + 6)$$
$$= y^2z^3(x + 6)(x + 1)$$

To identify the type of factoring problem, we follow these steps.

Identifying Factoring Problem Types

> **1.** Factor out all common monomial factors.
>
> **2.** If an expression has two terms, check to see if the problem type is
>   **a.** **The difference of two squares:** $x^2 - y^2 = (x + y)(x - y)$
>   **b.** **The sum of two cubes:** $x^3 + y^3 = (x + y)(x^2 - xy + y^2)$
>   **c.** **The difference of two cubes:** $x^3 - y^3 = (x - y)(x^2 + xy + y^2)$
>
> **3.** If an expression has three terms, attempt to factor it as a **trinomial.**
>
> **4.** If an expression has four or more terms, try factoring by **grouping.**
>
> **5.** Continue until each individual factor is prime.
>
> **6.** Check the results by multiplying.

EXAMPLE 1   Factor: $48a^4c^3 - 3b^4c^3$.

Solution   We begin by factoring out the common monomial factor of $3c^3$:

$$48a^4c^3 - 3b^4c^3 = 3c^3(16a^4 - b^4)$$

Since the expression $16a^4 - b^4$ has two terms, we check to see whether it is the difference of two squares, which it is. As the difference of two squares, it factors as $(4a^2 + b^2)(4a^2 - b^2)$.

$$48a^4c^3 - 3b^4c^3 = 3c^3(\mathbf{16a^4 - b^4})$$
$$= 3c^3(\mathbf{4a^2 + b^2})(\mathbf{4a^2 - b^2})$$

The binomial $4a^2 + b^2$ is the sum of two squares and is prime. However, $4a^2 - b^2$ is the difference of two squares and factors as $(2a + b)(2a - b)$.

$$48a^4c^3 - 3b^4c^3 = 3c^3(16a^4 - b^4)$$
$$= 3c^3(4a^2 + b^2)(4a^2 - b^2)$$
$$= 3c^3(4a^2 + b^2)(\mathbf{2a + b})(\mathbf{2a - b})$$

Since each of the individual factors is prime, the factorization is complete.

Self Check   Factor: $3p^4r^3 - 3q^4r^3$.   ∎

EXAMPLE 2   Factor: $x^5y + x^2y^4 - x^3y^3 - y^6$.

Solution   We begin by factoring out the common monomial factor of y:

$$x^5y + x^2y^4 - x^3y^3 - y^6 = y(x^5 + x^2y^3 - x^3y^2 - y^5)$$

Because the expression $x^5 + x^2y^3 - x^3y^2 - y^5$ has four terms, we try factoring by grouping to obtain

$$x^5y + x^2y^4 - x^3y^3 - y^6 = y(x^5 + x^2y^3 - x^3y^2 - y^5) \qquad \text{Factor out } y.$$
$$= y[x^2(\mathbf{x^3 + y^3}) - y^2(\mathbf{x^3 + y^3})] \quad \text{Factor by grouping.}$$
$$= y(\mathbf{x^3 + y^3})(x^2 - y^2) \qquad \text{Factor out } x^3 + y^3.$$

Finally, we factor $x^3 + y^3$ (the sum of two cubes) and $x^2 - y^2$ (the difference of two squares) to obtain

$$x^5y + x^2y^4 - x^3y^3 - y^6 = y(x + y)(x^2 - xy + y^2)(x + y)(x - y)$$

Because each of the individual factors is prime, the factorization is complete.

Self Check   Factor: $a^5p - a^3b^2p + a^2b^3p - b^5p$.   ∎

EXAMPLE 3   Factor: $p^{-4} - p^{-2} - 6$ and write the result without using negative exponents.

Solution   We factor this expression as if it were a trinomial:

$$p^{-4} - p^{-2} - 6 = (p^{-2} - 3)(p^{-2} + 2)$$
$$= \left(\frac{1}{p^2} - 3\right)\left(\frac{1}{p^2} + 2\right)$$

Self Check   Factor: $x^{-4} - x^{-2} - 12$.   ∎

EXAMPLE 4   Factor: $x^3 + 5x^2 + 6x + x^2y + 5xy + 6y$.

Solution   Since there are more than three terms, we try factoring by grouping. We can factor x from the first three terms and y from the last three terms and proceed as follows:

$$x^3 + 5x^2 + 6x + x^2y + 5xy + 6y$$
$$= x(x^2 + 5x + 6) + y(x^2 + 5x + 6)$$
$$= (x^2 + 5x + 6)(x + y) \qquad \text{Factor out } x^2 + 5x + 6.$$
$$= (x + 3)(x + 2)(x + y) \qquad \text{Factor } x^2 + 5x + 6.$$

Self Check   Factor: $a^3 - 5a^2 + 6a + a^2b - 5ab + 6b$.   ∎

EXAMPLE 5   Factor: $x^4 + 2x^3 + x^2 + x + 1$.

Solution   Since there are more than three terms, we try factoring by grouping. We can factor x^2 from the first three terms and proceed as follows:

$$x^4 + 2x^3 + x^2 + x + 1 = x^2(x^2 + 2x + 1) + (x + 1)$$
$$= x^2(x + 1)(x + 1) + (x + 1) \qquad \text{Factor } x^2 + 2x + 1.$$
$$= (x + 1)[x^2(x + 1) + 1] \qquad \text{Factor out } x + 1.$$
$$= (x + 1)(x^3 + x^2 + 1) \qquad \begin{array}{l}\text{Remove inner} \\ \text{parentheses.}\end{array}$$

Self Check   Factor: $a^4 - a^3 - 2a^2 + a - 2$.   ∎

Self Check Answers

1. $3r^3(p^2 + q^2)(p + q)(p - q)$ **2.** $p(a + b)(a^2 - ab + b^2)(a + b)(a - b)$ **3.** $\left(\dfrac{1}{x^2} + 3\right)\left(\dfrac{1}{x^2} - 4\right)$
4. $(a - 2)(a - 3)(a + b)$ **5.** $(a - 2)(a^3 + a^2 + 1)$

Orals   *Factor each expression.*

1. $x^2 - y^2$ **2.** $2x^3 - 4x^4$
3. $x^2 + 4x + 4$ **4.** $x^2 - 5x + 6$
5. $x^3 - 8$ **6.** $x^3 + 8$

5.7   EXERCISES 💾

REVIEW   *Perform the operations.*

1. $(3a^2 + 4a - 2) + (4a^2 - 3a - 5)$

2. $(-4b^2 - 3b - 2) - (3b^2 - 2b + 5)$

3. $5(2y^2 - 3y + 3) - 2(3y^2 - 2y + 6)$

4. $4(3x^2 + 3x + 3) + 3(x^2 - 3x - 4)$

5. $(m + 4)(m - 2)$

6. $(3p + 4q)(2p - 3q)$

VOCABULARY AND CONCEPTS   *Fill in the blanks.*

7. In any factoring problem, always factor out any
_____ first.

8. If an expression has two terms, check to see whether the problem type is the _____ of two squares, the sum of two _____, or the _____ of two cubes.

9. If an expression has three terms, try to factor it as a _____.

10. If an expression has four or more terms, try factoring it by _____.

PRACTICE   *Factor each polynomial, if possible.*

11. $x^2 + 8x + 16$ **12.** $20 + 11x - 3x^2$

13. $8x^3y^3 - 27$

14. $3x^2y + 6xy^2 - 12xy$

15. $xy - ty + xs - ts$

16. $bc + b + cd + d$

17. $25x^2 - 16y^2$

18. $27x^9 - y^3$

19. $12x^2 + 52x + 35$

20. $12x^2 + 14x - 6$

21. $6x^2 - 14x + 8$

22. $12x^2 - 12$

23. $56x^2 - 15x + 1$

24. $7x^2 - 57x + 8$

25. $4x^2y^2 + 4xy^2 + y^2$

26. $100z^2 - 81t^2$

27. $x^3 + (a^2y)^3$

28. $4x^2y^2z^2 - 26x^2y^2z^3$

29. $2x^3 - 54$

30. $4(xy)^3 + 256$

31. $ae + bf + af + be$

32. $a^2x^2 + b^2y^2 + b^2x^2 + a^2y^2$

33. $2(x + y)^2 + (x + y) - 3$

34. $(x - y)^3 + 125$

35. $625x^4 - 256y^4$

36. $2(a - b)^2 + 5(a - b) + 3$

37. $36x^4 - 36$

38. $6x^2 - 63 - 13x$

39. $2x^6 + 2y^6$

40. $x^4 - x^4y^4$

41. $a^4 - 13a^2 + 36$

42. $x^4 - 17x^2 + 16$

43. $x^2 + 6x + 9 - y^2$

44. $x^2 + 10x + 25 - y^8$

45. $4x^2 + 4x + 1 - 4y^2$

46. $9x^2 - 6x + 1 - 25y^2$

47. $x^2 - y^2 - 2y - 1$

48. $a^2 - b^2 + 4b - 4$

49. $x^5 + x^2 - x^3 - 1$

50. $x^5 - x^2 - 4x^3 + 4$

51. $x^5 - 9x^3 + 8x^2 - 72$

52. $x^5 - 4x^3 - 8x^2 + 32$

53. $2x^5z - 2x^2y^3z - 2x^3y^2z + 2y^5z$

54. $x^2y^3 - 4x^2y - 9y^3 + 36y$

55. $x^{2m} - x^m - 6$

56. $a^{2n} - b^{2n}$

57. $a^{3n} - b^{3n}$

58. $x^{3m} + y^{3m}$

59. $x^{-2} + 2x^{-1} + 1$

60. $4a^{-2} - 12a^{-1} + 9$

61. $6x^{-2} - 5x^{-1} - 6$

62. $x^{-4} - y^{-4}$

WRITING

63. What is your strategy for factoring a polynomial?

64. Explain how you can know that your factorization is correct.

SOMETHING TO THINK ABOUT

65. If you have the choice of factoring a polynomial as the difference of two squares or as the difference of two cubes, which do you do first? Why?

66. Can several polynomials have a greatest common factor? Find the GCF of $2x^2 + 7x + 3$ and $x^2 - 2x - 15$.

67. Factor $x^4 + x^2 + 1$.   (*Hint:* Add and subtract x^2.)

68. Factor $x^4 + 7x^2 + 16$.   (*Hint:* Add and subtract x^2.)

| **5.8** | **Solving Equations by Factoring** |

■ **Solving Quadratic Equations**
■ **Solving Higher-Degree Polynomial Equations** ■ **Problem Solving**

Getting Ready *Factor each polynomial.*

1. $2a^2 - 4a$ **2.** $a^2 - 25$
3. $6a^2 + 5a - 6$ **4.** $6a^3 - a^2 - 2a$

In this section, we will learn how to solve quadratic equations by factoring. We will then use this skill to solve problems.

Solving Quadratic Equations

An equation such as $3x^2 + 4x - 7 = 0$ or $-5y^2 + 3y + 8 = 0$ is called a **quadratic** (or **second-degree**) equation.

Quadratic Equations

A **quadratic equation** is any equation that can be written in the form

$$ax^2 + bx + c = 0$$

where a, b, and c are real numbers and $a \neq 0$.

Many quadratic equations can be solved by factoring and then using the **zero-factor property.**

Zero-Factor Property

If a and b are real numbers, then

If $ab = 0$, then $a = 0$ or $b = 0$.

The zero-factor property states that *if the product of two or more numbers is 0, then at least one of the numbers must be 0.*

To solve $x^2 + 5x + 6 = 0$, we factor its left-hand side to obtain

$$(x + 3)(x + 2) = 0$$

Since the product of $x + 3$ and $x + 2$ is 0, then at least one of the factors is 0. Thus, we can set each factor equal to 0 and solve each resulting linear equation for x:

$x + 3 = 0$ or $x + 2 = 0$
$x = -3$ | $x = -2$

To check these solutions, we first substitute -3 and then -2 for x in the equation and verify that each number satisfies the equation.

$$x^2 + 5x + 6 = 0$$
$$(-3)^2 + 5(-3) + 6 \stackrel{?}{=} 0$$
$$9 - 15 + 6 \stackrel{?}{=} 0$$
$$0 = 0$$

$$x^2 + 5x + 6 = 0$$
$$(-2)^2 + 5(-2) + 6 \stackrel{?}{=} 0$$
$$4 - 10 + 6 \stackrel{?}{=} 0$$
$$0 = 0$$

Both -3 and -2 are solutions, because both satisfy the equation.

EXAMPLE 1   Solve: $3x^2 + 6x = 0$.

Solution   We factor the left-hand side, set each factor equal to 0, and solve each resulting equation for x.

$$3x^2 + 6x = 0$$
$$3x(x + 2) = 0 \qquad \text{Factor out the common factor of } 3x.$$
$$3x = 0 \quad \text{or} \quad x + 2 = 0$$
$$x = 0 \quad | \qquad x = -2$$

Verify that both solutions check.

Self Check   Solve: $4p^2 - 12p = 0$.

Comment   In Example 1, do not divide both sides by $3x$, or you will lose the solution $x = 0$.

EXAMPLE 2   Solve: $x^2 - 16 = 0$.

Solution   We factor the difference of two squares on the left-hand side, set each factor equal to 0, and solve each resulting equation.

$$x^2 - 16 = 0$$
$$(x + 4)(x - 4) = 0$$
$$x + 4 = 0 \quad \text{or} \quad x - 4 = 0$$
$$x = -4 \quad | \qquad x = 4$$

Verify that both solutions check.

Self Check   Solve: $a^2 - 81 = 0$.

Many equations that do not appear to be quadratic can be put into quadratic form ($ax^2 + bx + c = 0$) and then solved by factoring.

EXAMPLE 3   Solve: $x = \dfrac{6}{5} - \dfrac{6}{5}x^2$.

Solution   We write the equation in quadratic form and then solve by factoring.

$$x = \frac{6}{5} - \frac{6}{5}x^2$$

$$5x = 6 - 6x^2 \qquad \text{Multiply both sides by 5.}$$

$$6x^2 + 5x - 6 = 0 \qquad \text{Add } 6x^2 \text{ to both sides and subtract 6 from both sides.}$$

$$(3x - 2)(2x + 3) = 0 \qquad \text{Factor the trinomial.}$$

$$3x - 2 = 0 \quad \text{or} \quad 2x + 3 = 0 \qquad \text{Set each factor equal to 0.}$$

$$3x = 2 \qquad\qquad 2x = -3$$

$$x = \frac{2}{3} \qquad\qquad x = -\frac{3}{2}$$

Verify that both solutions check.

Self Check   Solve: $x = \frac{6}{7}x^2 - \frac{3}{7}$.

 Comment   To solve a quadratic equation by factoring, be sure to set the quadratic polynomial equal to 0 before factoring and using the zero-factor property. Do not make the following error:

$$6x^2 + 5x = 6$$

$$x(6x + 5) = 6$$

$$x = 6 \quad \text{or} \quad 6x + 5 = 6 \qquad\qquad \text{If the product of two numbers is 6, neither number need be 6. For example, } 2 \cdot 3 = 6.$$

$$x = \frac{1}{6}$$

Neither solution checks.

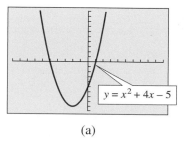 ***Accent on Technology***   **SOLVING QUADRATIC EQUATIONS**

To solve a quadratic equation such as $x^2 + 4x - 5 = 0$ with a graphing calculator, we can use standard window settings of $[-10, 10]$ for x and $[-10, 10]$ for y and graph the quadratic function $y = x^2 + 4x - 5$, as shown in Figure 5-6(a). We can then trace to find the x-coordinates of the x-intercepts of the parabola. See Figure 5-6(b) and 5-6(c). For better results, we can zoom in. Since these are the numbers x that make $y = 0$, they are the solutions of the equation.

We can also find the solutions by using the ZERO command found in the CALC menu.

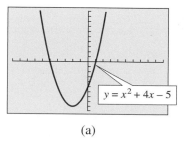

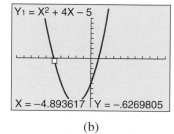

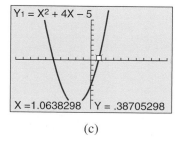

(a)   (b)   (c)

Figure 5-6

Solving Higher-Degree Polynomial Equations

We can solve many polynomial equations with degree greater than 2 by factoring.

EXAMPLE 4 Solve: $6x^3 - x^2 - 2x = 0$

Solution We factor x from the polynomial on the left-hand side and proceed as follows:

$$6x^3 - x^2 - 2x = 0$$
$$x(6x^2 - x - 2) = 0 \qquad \text{Factor out } x.$$
$$x(3x - 2)(2x + 1) = 0 \qquad \text{Factor } 6x^2 - x - 2.$$
$$x = 0 \quad \text{or} \quad 3x - 2 = 0 \quad \text{or} \quad 2x + 1 = 0 \qquad \text{Set each factor equal to 0.}$$
$$x = \frac{2}{3} \qquad\qquad x = -\frac{1}{2}$$

Verify that the solutions check.

Self Check Solve: $5x^3 + 13x^2 - 6x = 0$.

EXAMPLE 5 Solve: $x^4 - 5x^2 + 4 = 0$.

Solution We factor the trinomial on the left-hand side and proceed as follows:

$$x^4 - 5x^2 + 4 = 0$$
$$(x^2 - 1)(x^2 - 4) = 0$$
$$(x + 1)(x - 1)(x + 2)(x - 2) = 0 \qquad \text{Factor } x^2 - 1 \text{ and } x^2 - 4.$$
$$x + 1 = 0 \quad \text{or} \quad x - 1 = 0 \quad \text{or} \quad x + 2 = 0 \quad \text{or} \quad x - 2 = 0$$
$$x = -1 \quad | \quad x = 1 \quad | \quad x = -2 \quad | \quad x = 2$$

Verify that each solution checks.

Self Check Solve: $a^4 - 13a + 36 = 0$.

EXAMPLE 6 Given that $f(x) = 6x^3 + x^2 - 2x$, find all x such that $f(x) = 0$.

Solution We first set $f(x)$ equal to 0.

$$f(x) = 6x^3 + x^2 - 2x$$
$$0 = 6x^3 + x^2 - 2x \qquad \text{Substitute 0 for } f(x).$$

Then we solve for x.

$$6x^3 + x^2 - 2x = 0$$
$$x(6x^2 + x - 2) = 0 \qquad \text{Factor out } x.$$
$$x(2x - 1)(3x + 2) = 0 \qquad \text{Factor } 6x^2 + x - 2.$$
$$x = 0 \quad \text{or} \quad 2x - 1 = 0 \quad \text{or} \quad 3x + 2 = 0 \qquad \text{Set each factor equal to 0.}$$
$$x = \frac{1}{2} \qquad\qquad x = -\frac{2}{3}$$

Verify that $f(0) = 0$, $f\left(\frac{1}{2}\right) = 0$, and $f\left(-\frac{2}{3}\right) = 0$.

Self Check Given that $f(x) = x^4 - 4x^2$, find all x such that $f(x) = 0$.

Accent on Technology   **SOLVING EQUATIONS**

To solve the equation $x^4 - 5x^2 + 4 = 0$ with a graphing calculator, we can use window settings of $[-6, 6]$ for x and $[-5, 10]$ for y and graph the polynomial function $y = x^4 - 5x^2 + 4$ as shown in Figure 5-7. We can then read the values of x that make $y = 0$. They are $x = -2, -1, 1,$ and 2. If the x-coordinates of the x-intercepts were not obvious, we could get their values by using the trace and zoom features. We can also use the ZERO command to find the solutions.

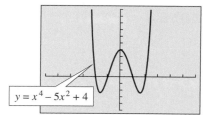

$y = x^4 - 5x^2 + 4$

Figure 5-7

Problem Solving

EXAMPLE 7   **Finding the dimensions of a truss**   The width of the truss shown in Figure 5-8 is 3 times its height. The area of the triangle is 96 square feet. Find its width and height.

Analyze the problem   We can let x be the positive number that represents the height of the truss. Then $3x$ represents its width. We can substitute x for h, $3x$ for b, and 96 for A in the formula for the area of a triangle and solve for x.

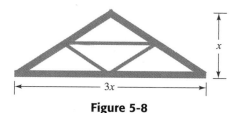

$3x$

x

Figure 5-8

Form and solve an equation

$$A = \frac{1}{2}bh$$

$$96 = \frac{1}{2}(3x)x$$

$$192 = 3x^2 \qquad \text{Multiply both sides by 2.}$$

$$64 = x^2 \qquad \text{Divide both sides by 3.}$$

$$0 = x^2 - 64 \qquad \text{Subtract 64 from both sides.}$$

$$0 = (x + 8)(x - 8) \qquad \text{Factor the difference of two squares.}$$

$$x + 8 = 0 \quad \text{or} \quad x - 8 = 0$$

$$x = -8 \qquad \qquad x = 8$$

State the conclusion   Since the height of a triangle cannot be negative, we must discard the negative solution. Thus, the height of the truss is 8 feet, and its width is 3(8), or 24 feet.

Check the result   The area of a triangle with a width (base) of 24 feet and a height of 8 feet is 96 square feet:

$$A = \frac{1}{2}bh = \frac{1}{2}(\mathbf{24})(\mathbf{8}) = 12(\mathbf{8}) = 96$$ ∎

EXAMPLE 8   **Ballistics**   If the initial velocity of an object thrown straight up into the air is 176 feet per second, when will it hit the ground?

Analyze the problem   The height of an object thrown straight up into the air with an initial velocity of v feet per second is given by the formula

$$h = vt - 16t^2$$

The height h is in feet, and t represents the number of seconds since the object was released.

Form and solve an equation   When the object hits the ground, its height will be 0. Thus, we set h equal to 0, set v equal to 176, and solve for t.

$$\begin{aligned}
\boldsymbol{h} &= \boldsymbol{vt} - \boldsymbol{16t^2} \\
\mathbf{0} &= \mathbf{176}t - 16t^2 \\
0 &= 16t(11 - t) \qquad \text{Factor out } 16t. \\
16t = 0 \quad &\text{or} \quad 11 - t = 0 \qquad \text{Set each factor equal to 0.} \\
t = 0 \quad &| \qquad\quad t = 11
\end{aligned}$$

State the conclusion   When $t = 0$, the object's height above the ground is 0 feet, because it has not been released. When $t = 11$, the height is again 0 feet, because the object has hit the ground. The solution is 11 seconds.

Check the result   Verify that $h = 0$ when $t = 11$.   ∎

1. $0, 3$ **2.** $9, -9$ **3.** $\dfrac{3}{2}, -\dfrac{1}{3}$ **4.** $0, \dfrac{2}{5}, -3$ **5.** $2, -2, 3, -3$ **6.** $0, 0, 2, -2$

Orals   *Solve each equation.*

1. $(x - 2)(x - 3) = 0$ **2.** $(x + 4)(x - 2) = 0$

3. $(x - 2)(x - 3)(x + 1) = 0$ **4.** $(x + 3)(x + 2)(x - 5)(x - 6) = 0$

5.8   **EXERCISES**

1. List the prime numbers less than 10.
2. List the composite numbers between 7 and 17.

3. The formula for the volume of a sphere is $V = \frac{4}{3}\pi r^3$. Find the volume when $r = 21.23$ centimeters. Round the answer to the nearest hundredth.

4. The formula for the volume of a cone is $V = \frac{1}{3}\pi r^2 h$. Find the volume when $r = 12.33$ meters and $h = 14.7$ meters. Round the answer to the nearest hundredth.

VOCABULARY AND CONCEPTS   *Fill in the blanks.*

5. A quadratic equation is any equation that can be written in the form _____ ($a \neq 0$).

6. If a and b are real numbers, then if _____, $a = 0$ or $b = 0$.

PRACTICE   *Solve each equation.*

7. $4x^2 + 8x = 0$

8. $x^2 - 9 = 0$

9. $y^2 - 16 = 0$

10. $5y^2 - 10y = 0$

11. $x^2 + x = 0$

12. $x^2 - 3x = 0$

13. $5y^2 - 25y = 0$

14. $y^2 - 36 = 0$

15. $z^2 + 8z + 15 = 0$

16. $w^2 + 7w + 12 = 0$

17. $y^2 - 7y + 6 = 0$

18. $n^2 - 5n + 6 = 0$

19. $y^2 - 7y + 12 = 0$

20. $x^2 - 3x + 2 = 0$

21. $x^2 + 6x + 8 = 0$

22. $x^2 + 9x + 20 = 0$

23. $3m^2 + 10m + 3 = 0$

24. $2r^2 + 5r + 3 = 0$

25. $2y^2 - 5y + 2 = 0$

26. $2x^2 - 3x + 1 = 0$

27. $2x^2 - x - 1 = 0$

28. $2x^2 - 3x - 5 = 0$

29. $3s^2 - 5s - 2 = 0$

30. $8t^2 + 10t - 3 = 0$

31. $x(x - 6) + 9 = 0$

32. $x^2 + 8(x + 2) = 0$

33. $8a^2 = 3 - 10a$

34. $5z^2 = 6 - 13z$

35. $b(6b - 7) = 10$

36. $2y(4y + 3) = 9$

37. $\dfrac{3a^2}{2} = \dfrac{1}{2} - a$

38. $x^2 = \dfrac{1}{2}(x + 1)$

39. $\dfrac{1}{2}x^2 - \dfrac{5}{4}x = -\dfrac{1}{2}$

40. $\dfrac{1}{4}x^2 + \dfrac{3}{4}x = 1$

41. $x\left(3x + \dfrac{22}{5}\right) = 1$

42. $x\left(\dfrac{x}{11} - \dfrac{1}{7}\right) = \dfrac{6}{77}$

43. $x^3 + x^2 = 0$

44. $2x^4 + 8x^3 = 0$

45. $y^3 - 49y = 0$

46. $2z^3 - 200z = 0$

47. $x^3 - 4x^2 - 21x = 0$

48. $x^3 + 8x^2 - 9x = 0$

49. $z^4 - 13z^2 + 36 = 0$

50. $y^4 - 10y^2 + 9 = 0$

51. $3a(a^2 + 5a) = -18a$

52. $7t^3 = 2t\left(t + \dfrac{5}{2}\right)$

53. $\dfrac{x^2(6x + 37)}{35} = x$

54. $x^2 = -\dfrac{4x^3(3x + 5)}{3}$

Find all x that will make f(x) = 0.

55. $f(x) = x^2 - 49$
56. $f(x) = x^2 + 11x$
57. $f(x) = 2x^2 + 5x - 3$
58. $f(x) = 3x^2 - x - 2$
59. $f(x) = 5x^3 + 3x^2 - 2x$
60. $f(x) = x^4 - 26x^2 + 25$

Use grouping to help solve each equation.

61. $x^3 + 3x^2 - x - 3 = 0$
62. $x^3 - x^2 - 4x + 4 = 0$
63. $2r^3 + 3r^2 - 18r - 27 = 0$
64. $3s^3 - 2s^2 - 3s + 2 = 0$

65. $3y^3 + y^2 = 4(3y + 1)$

66. $w^3 + 16 = w(w + 16)$

67. Integer problem The product of two consecutive even integers is 288. Find the integers.

68. Integer problem The product of two consecutive odd integers is 143. Find the integers.

69. Integer problem The sum of the squares of two consecutive positive integers is 85. Find the integers.

70. Integer problem The sum of the squares of three consecutive positive integers is 77. Find the integers.

APPLICATIONS

71. Geometry Find the perimeter of the rectangle.

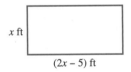

$(w + 4)$ m

w m   96 m²

72. Geometry One side of a rectangle is three times longer than another. If its area is 147 square centimeters, find its dimensions.

73. Geometry Find the dimensions of the rectangle, given that its area is 375 square feet.

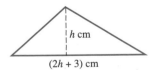

x ft

$(2x - 5)$ ft

74. Geometry Find the height of the triangle, given that its area is 162 square centimeters.

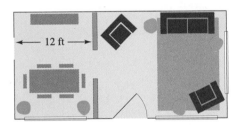

h cm

$(2h + 3)$ cm

75. Fine arts An artist intends to paint a 60-square-foot mural on a large wall as shown in the illustration. Find the dimensions of the mural if the artist leaves a border of uniform width around it.

18 ft

w

11 ft

76. Gardening A woman plans to use one-fourth of her 48-foot-by-100-foot rectangular backyard to plant a garden. Find the perimeter of the garden if the length is to be 40 feet greater than the width.

77. Architecture The rectangular room shown in the illustration is twice as long as it is wide. It is divided into two rectangular parts by a partition, positioned as shown. If the larger part of the room contains 560 square feet, find the dimensions of the entire room.

12 ft

78. Perimeter of a square If the length of one side of a square is increased by 4 inches, the area of the square becomes 9 times greater. Find the perimeter of the original square.

79. Time of flight After how many seconds will an object hit the ground if it is thrown upward with an initial velocity of 160 feet per second?

80. Time of flight After how many seconds will an object hit the ground if it is thrown upward with an initial velocity of 208 feet per second?

81. Ballistics The muzzle velocity of a cannon is 480 feet per second. If a cannonball is fired vertically, at what times will it be at a height of 3,344 feet?

82. Ballistics A slingshot can provide an initial velocity of 128 feet per second. At what times will a stone, shot vertically upward, be 192 feet above the ground?

83. **Winter recreation**   The length of the rectangular ice-skating rink is 20 meters greater than twice its width. Find the width.

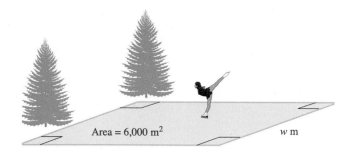

Area = 6,000 m²   *w* m

84. **Carpentry**   A 285-square-foot room is 4 feet longer than it is wide. What length of crown molding is needed to trim the perimeter of the ceiling?

85. **Designing a swimming pool**   Building codes require that the rectangular swimming pool in the illustration be surrounded by a uniform-width walkway of at least 516 square feet. If the length of the pool is 10 feet less than twice the width, how wide should the border be?

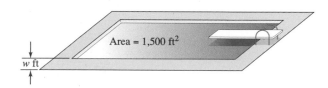

Area = 1,500 ft²

w ft

86. **House construction**   The formula for the area of a trapezoid is $A = \frac{h(B + b)}{2}$. The area of the truss is 44 square feet. Find the height of the truss if the shorter base is the same as the height.

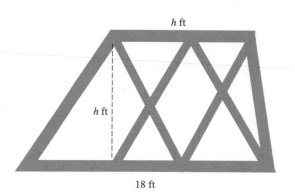

h ft

h ft

18 ft

Use a graphing calculator to find the real solutions of each equation, if any exist. If an answer is not exact, give the answer to the nearest hundredth.

87. $x^2 - 4x + 7 = 0$

88. $2x^2 - 7x + 4 = 0$

89. $-3x^3 - 2x^2 + 5 = 0$

90. $-2x^3 - 3x - 5 = 0$

WRITING

91. Describe the steps for solving an application problem.

92. Explain how to check the solution of an application problem.

SOMETHING TO THINK ABOUT   *Find a quadratic equation with the given roots.*

93. $3, 5$

94. $-2, 6$

95. $0, -5$

96. $\frac{1}{2}, \frac{1}{3}$

Projects

Project 1

Mac operates a bait shop on the Snarly River. It is located between two sharp bends in the river, and the area around Mac's shop has recently become a popular hiking and camping area. Mac has decided to produce some maps of the area for the use of the visitors. Although he knows the region well, he has very little idea of the actual distances from one location to another. What he knows is:

1. Big Falls, a beautiful waterfall on Snarly River, is due east of Mac's.

2. Grandview Heights, a fabulous rock climbing site, is due west of Mac's, right on the river.

3. Foster's General Store, the only sizable camping and climbing outfitter in the area, is located on the river some distance west and north of Mac's.

Mac hires an aerial photographer to take pictures of the area, with some surprising results. If Mac's bait shop is treated as the origin of a coordinate system, with the y-axis running north–south and the x-axis running east–west, then on the domain $-4 \le x \le 4$ (where the units are miles), the river follows the curve

$$P(x) = \frac{1}{4}(x^3 - x^2 - 6x)$$

The aerial photograph is shown in the illustration.

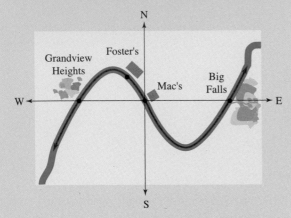

a. Mac would like to include on his maps the exact locations (relative to Mac's) of Big Falls and Grandview Heights. Find these for him, and explain to Mac why your answers must be correct.

b. Mac and Foster have determined that Foster's General Store is 0.7 miles west of the bait shop. Since it is on the river, it is a bit north as well. They decide that to promote business, they will join together to clear a few campsites in the region bordered by the straight-line paths that run between Mac's and Foster's, Mac's and Grandview Heights, and Foster's and Grandview Heights. If they clear 1 campsite for each 40 full acres of area, how many campsites can they put in? (*Hint:* A square mile contains 640 acres.)

c. A path runs in a straight line directly southeast (along the line $y = -x$) from Mac's to the river. How far east and how far south has a hiker on this trail traveled when he or she reaches the river?

Project 2

The rate at which fluid flows through a cylindrical pipe, or any cylinder-shaped tube (an artery for instance), is

$$\text{Velocity of fluid flow} = V = \frac{p}{nL}(R^2 - r^2)$$

where p is the difference in pressure between the two ends of the tube, L is the length of the tube, R is the radius of the tube, and n is the *viscosity constant,* a measure of the thickness of a fluid. Since the variable r represents the distance from the center of the tube, $0 \le r \le R$. In most situations, p, L, R, and n are constants, so V is a function of r.

$$V(r) = \frac{p}{nL}(R^2 - r^2)$$

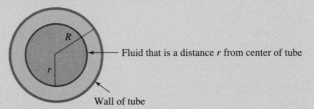

Cross section of tube

Fluid that is a distance r from center of tube

Wall of tube

It can be shown that the velocity of a fluid moving in the tube depends on how far from the center (or how close to the wall of the tube) the fluid is.

a. Consider a pipe with radius 5 centimeters and length 60 centimeters. Suppose that $p = 15$ and $n = 0.001$ (water has a viscosity of approximately 0.001). Find the velocity of the fluid at the center of the pipe. The units of measurement for V will be centimeters per second.

    Find the velocity of the fluid when it is halfway between the center and the wall of the pipe. What percent of the velocity at the center of the pipe does this represent? Where in the pipe is the fluid flowing with a velocity of 4,000 centimeters per second?

b. Suppose that the situation is the same, but the fluid is now machinery oil, with a viscosity of 0.15. Answer the same questions as in part **a,** except find where in the pipe the oil flows with a velocity of 15 cm per second. Note that the oil travels at a much slower speed than water.

c. Medical doctors use various methods to increase the rate of blood flow through an artery. The patient may take a drug that "thins the blood" (lowers its viscosity) or a drug that dilates the artery, or the patient may undergo angioplasty, a surgical procedure that widens the canal through which the blood passes. Explain why each of these increases the velocity V at a given distance r from the center of the artery.

Chapter Summary

CONCEPTS	REVIEW EXERCISES
5.1	**Polynomials and Polynomial Functions**

The **degree of a polynomial** is the degree of the term with highest degree contained within the polynomial.

If $P(x)$ is a polynomial in x then $P(r)$ is the value of the polynomial at $x = r$.

1. Find the degree of $P(x) = 3x^5 + 4x^3 + 2$.

2. Find the degree of $9x^2y + 13x^3y^2 + 8x^4y^4$.

Find each value when $P(x) = -x^2 + 4x + 6$.

3. $P(0)$ **4.** $P(1)$

5. $P(-t)$ **6.** $P(z)$

Graph each function.

7. $f(x) = x^3 - 1$

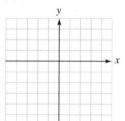

8. $f(x) = x^2 - 2x$

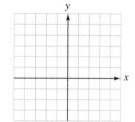

5.2	**Adding and Subtracting Polynomials**

To add polynomials, add their like terms.

Simplify each expression.

9. $(3x^2 + 4x + 9) + (2x^2 - 2x + 7)$

10. $(4x^3 + 4x^2 + 7) - (-2x^3 - x - 2)$

To subtract polynomials, add the negative of the polynomial that is to be subtracted from the other polynomial.

11. $(2x^2 - 5x + 9) - (x^2 - 3) - (-3x^2 + 4x - 7)$

12. $2(7x^3 - 6x^2 + 4x - 3) - 3(7x^3 + 6x^2 + 4x - 3)$

5.3	**Multiplying Polynomials**

To multiply monomials, multiply their numerical factors and multiply their variable factors.

Find each product.

13. $(8a^2b^2)(-2abc)$

14. $(-3xy^2z)(2xz^3)$

To multiply a polynomial by a monomial, multiply each term of the polynomial by the monomial.

Find each product.

15. $2xy^2(x^3y - 4xy^5)$

16. $a^2b(a^2 + 2ab + b^2)$

To multiply polynomials, multiply each term of one polynomial by each term of the other polynomial.

Find each product.

17. $(8x - 5)(2x + 3)$

18. $(3x + 2)(2x - 4)$

19. $(5x - 4)(3x - 2)$

20. $(3x^2 - 2)(x^2 - x + 2)$

5.4   The Greatest Common Factor and Factoring by Grouping

Always factor out common monomial factors as the first step in a factoring problem.

Factor each expression.

21. $4x + 8$

22. $3x^2 - 6x$

23. $5x^2y^3 - 10xy^2$

24. $7a^4b^2 + 49a^3b$

25. $-8x^2y^3z^4 - 12x^4y^3z^2$

Use the distributive property to factor out common monomial factors.

26. $12a^6b^4c^2 + 15a^2b^4c^6$

27. $27x^3y^3z^3 + 81x^4y^5z^2 - 90x^2y^3z^7$

28. $-36a^5b^4c^2 + 60a^7b^5c^3 - 24a^2b^3c^7$

29. Factor x^n from $x^{2n} + x^n$.

30. Factor y^{2n} from $y^{2n} - y^{4n}$.

31. Factor x^{-2} from $x^{-4} - x^{-2}$.

32. Factor a^{-3} from $a^6 + 1$.

33. $5x^2(x + y)^3 - 15x^3(x + y)^4$

34. $-49a^3b^2(a - b)^4 + 63a^2b^4(a - b)^3$

If an expression has four or more terms, try to factor the expression by grouping.

Factor each expression.

35. $xy + 2y + 4x + 8$

36. $ac + bc + 3a + 3b$

37. $x^4 + 4y + 4x^2 + x^2y$

38. $a^5 + b^2c + a^2c + a^3b^2$

Solve for the indicated variable.

39. $S = 2wh + 2wl + 2lh$ for h

40. $S = 2wh + 2wl + 2lh$ for l

5.5 The Difference of Two Squares; the Sum and Difference of Two Cubes

Difference of two squares:
$$x^2 - y^2 = (x + y)(x - y)$$

Factor each expression, if possible.

41. $z^2 - 16$

42. $y^2 - 121$

43. $x^2y^4 - 64z^6$

44. $a^2b^2 + c^2$

45. $(x + z)^2 - t^2$

46. $c^2 - (a + b)^2$

47. $2x^4 - 98$

48. $3x^6 - 300x^2$

Sum of two cubes:
$$x^3 + y^3 = (x + y)(x^2 - xy + y^2)$$

Difference of two cubes:
$$x^3 - y^3 = (x - y)(x^2 + xy + y^2)$$

Factor each expression, if possible.

49. $x^3 + 343$

50. $a^3 - 125$

51. $8y^3 - 512$

52. $4x^3y + 108yz^3$

5.6 Factoring Trinomials

Special product formulas:
$$x^2 + 2xy + y^2 = (x + y)(x + y)$$
$$x^2 - 2xy + y^2 = (x - y)(x - y)$$

Test for factorability:
A trinomial of the form $ax^2 + bx + c \ (a \neq 0)$, will factor with integer coefficients if $b^2 - 4ac$ is a perfect square.

Factor each expression, if possible.

53. $x^2 + 10x + 25$

54. $a^2 - 14a + 49$

Factor each expression, if possible.

55. $y^2 + 21y + 20$

56. $z^2 - 11z + 30$

57. $-x^2 - 3x + 28$

58. $y^2 - 5y - 24$

59. $4a^2 - 5a + 1$

60. $3b^2 + 2b + 1$

61. $7x^2 + x + 2$

62. $-15x^2 + 14x + 8$

63. $y^3 + y^2 - 2y$

64. $2a^4 + 4a^3 - 6a^2$

65. $-3x^2 - 9x - 6$

66. $8x^2 - 4x - 24$

67. $15x^2 - 57xy - 12y^2$

68. $30x^2 + 65xy + 10y^2$

69. $24x^2 - 23xy - 12y^2$

70. $14x^2 + 13xy - 12y^2$

5.7   Summary of Factoring Techniques

Factoring a random expression:

1. Factor out all common monomial factors.

2. If an expression has two terms, check to see if it is

   a. The difference of two squares:

   $x^2 - y^2 = (x + y)(x - y)$

   b. The sum of two cubes:

   $x^3 + y^3 = (x + y)(x^2 - xy + y^2)$

   c. The difference of two cubes:

   $x^3 - y^3 = (x - y)(x^2 + xy + y^2)$

3. If an expression has three terms, attempt to factor it as a **trinomial.**

4. If an expression has four or more terms, try factoring by **grouping.**

5. Continue until each individual factor is prime.

6. Check the results.

Factor each expression, if possible:

71. $x^3 + 5x^2 - 6x$

72. $3x^2y - 12xy - 63y$

73. $z^2 - 4 + zx - 2x$

74. $x^2 + 2x + 1 - p^2$

75. $x^2 + 4x + 4 - 4p^4$

76. $y^2 + 3y + 2 + 2x + xy$

77. $x^{2m} + 2x^m - 3$

78. $x^{-2} - x^{-1} - 2$

| **5.8** | **Solving Equations by Factoring** |

Zero-factor property:
If $xy = 0$, then $x = 0$ or $y = 0$.

Solve each equation.

79. $4x^2 - 3x = 0$ **80.** $x^2 - 36 = 0$

81. $12x^2 + 4x - 5 = 0$ **82.** $7y^2 - 37y + 10 = 0$

83. $t^2(15t - 2) = 8t$ **84.** $3u^3 = u(19u + 14)$

85. Volume   The volume V of the rectangular solid is given by the formula $V = lwh$, where l is its length, w is its width, and h is its height. If the volume is 840 cubic centimeters, the length is 12 centimeters, and the width exceeds the height by 3 centimeters, find the height.

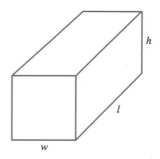

86. Volume of a pyramid   The volume of the pyramid is given by the formula $V = \frac{Bh}{3}$, where B is the area of its base and h is its height. The volume of the pyramid is 1,020 cubic meters. Find the dimensions of its rectangular base if one edge of the base is 3 meters longer than the other and the height of the pyramid is 9 meters.

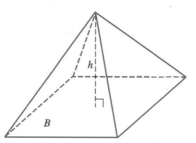

Chapter Test

Find the degree of each polynomial.

1. $3x^3 - 4x^5 - 3x^2 - 5$

2. $3x^5y^3 - x^8y^2 + 2x^9y^4 - 3x^2y^5 + 4$

Let $P(x) = -3x^2 + 2x - 1$ and find each value.

3. $P(2)$

4. $P(-1)$

5. Graph the function $f(x) = x^2 + 2x$.

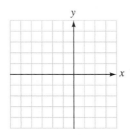

Perform the operations.

6. $(2y^2 + 4y + 3) + (3y^2 - 3y - 4)$

7. $(-3u^2 + 2u - 7) - (u^2 + 7)$

8. $3(2a^2 - 4a + 2) - 4(-a^2 - 3a - 4)$

9. $-2(2x^2 - 2) + 3(x^2 + 5x - 2)$

10. $(3x^3y^2z)(-2xy^{-1}z^3)$

11. $-5a^2b(3ab^3 - 2ab^4)$

12. $(z + 4)(z - 4)$

13. $(3x - 2)(4x + 3)$

Factor each expression.

14. $3xy^2 + 6x^2y$

15. $12a^3b^2c - 3a^2b^2c^2 + 6abc^3$

16. Factor y^n from $x^2y^{n+2} + y^n$.

17. Factor b^n from $a^nb^n - ab^{-n}$.

18. $(u - v)r + (u - v)s$

19. $ax - xy + ay - y^2$

20. $x^2 - 49$

21. $2x^2 - 32$

22. $4y^4 - 64$

23. $b^3 + 125$

24. $b^3 - 27$

25. $3u^3 - 24$

26. $a^2 - 5a - 6$

27. $6b^2 + b - 2$

28. $6u^2 + 9u - 6$

29. $20r^2 - 15r - 5$

30. $x^{2n} + 2x^n + 1$

31. $x^2 + 6x + 9 - y^2$

32. Solve for r: $r_1r_2 - r_2r = r_1r$

33. Solve for x: $x^2 - 5x - 6 = 0$

34. Integer problem   The product of two consecutive positive integers is 156. Find their sum.

35. Preformed concrete   The slab of concrete in the illustration is twice as long as it is wide. The area in which it is placed includes a 1-foot-wide border of 70 square feet. Find the dimensions of the slab.

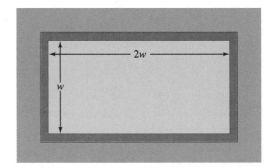

6 Rational Expressions

InfoTrac

Do a keyword search on "oil production." Find the article "Saudi officials fear effects of Iraq's return to oil production," published May 5, 2003. According to the article, what is the major cause of Saudi Arabia's economic instability?

In an oil refinery, two pipes fill a vat. Operating alone, pipe 1 can fill the vat in 5 hours, while pipe 2 can fill the vat in 12 hours.

a. If both pipes are filling the vat at the same time, how soon after opening the valves should the operator return to close the valves to prevent an overflow?

b. If pipe 2 is started at 9 AM and pipe 1 at 11 AM, at what time should the operator return to close the valves?

c. If pipe 1 is started at 9 AM and pipe 2 at 11 AM, when should the operator return?
  Complete this project after studying Section 6.6.

© Royalty-Free/CORBIS

**Exercise Set 6.6
Problem 66**

Mathematics in Engineering

The stiffness of the shaft, shown in the illustration, is given by the formula

$$k = \dfrac{1}{\dfrac{1}{k_1} + \dfrac{1}{k_2}}$$

Section 1   Section 2

where k_1 and k_2 are the individual stiffnesses of each section. If the stiffness k_2 of Section 2 is 4,200,000 in. lb/rad and the design specifications require that the overall stiffness k of the entire shaft be 1,900,000 in. lb/rad, what must the stiffness of Section 1 be?

Rational expressions are the fractions of algebra. In this chapter, we will learn how to simplify, add, subtract, multiply, and divide them.

6.1   Rational Functions and Simplifying Rational Expressions

▋ **Rational Expressions**   ▋ **Rational Functions**
▋ **Finding the Domain and Range of a Rational Function**
▋ **Simplifying Rational Expressions**
▋ **Simplifying Rational Expressions by Factoring Out −1**
▋ **Probability**

Getting Ready   *Simplify each fraction.*

1. $\dfrac{6}{8}$ **2.** $\dfrac{12}{15}$ **3.** $\dfrac{-25}{65}$ **4.** $\dfrac{-49}{-63}$

We begin our discussion of rational expressions by considering rational functions. Then, we will discuss how to simplify rational expressions.

Rational Expressions

Rational expressions are fractions that indicate the quotient of two polynomials. Some examples are

$$\frac{3x}{x-7}, \qquad \frac{5m+n}{8m+16}, \qquad \text{and} \qquad \frac{a^3+2a^2+7}{2a^2-5a+4}$$

Since division by 0 is undefined, the value of a polynomial occurring in the denominator cannot be 0. For example, x cannot be 7 in the rational expression $\frac{3x}{x-7}$, because the denominator would be 0. In $\frac{5m+n}{8m+16}$, m cannot be -2, because that would make the denominator equal to 0.

Rational Functions

Rational expressions often define functions. For example, if the cost of subscribing to an online research service is $6 per month plus $1.50 per hour of access time, the average (mean) hourly cost of the service is the total monthly cost, divided by the number of hours of access time:

$$\overline{c} = \frac{c}{n} = \frac{1.50n+6}{n}$$ $\overline{c}$ is the mean hourly cost, c is the total monthly cost, and n is the number of hours the service is used.

The function

(1) $\qquad \overline{c} = f(n) = \dfrac{1.50n + 6}{n} \quad (n > 0)$

gives the mean hourly cost of using the research service for n hours per month.

Figure 6-1 shows the graph of the rational function $\overline{c} = f(n) = \frac{1.50n + 6}{n}$ $(n > 0)$. Since $n > 0$, the domain of this function is the interval $(0, \infty)$.

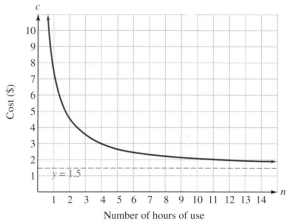

Number of hours of use
Figure 6-1

From the graph, we can see that the mean hourly cost decreases as the number of hours of access time increases. Since the cost of each extra hour of access time is $1.50, the mean hourly cost can approach $1.50 but never drop below it. Thus, the graph of the function approaches the line $y = 1.5$ as n increases without bound. When a graph approaches a line as the dependent variable gets large, we call the line an **asymptote.** The line $y = 1.5$ is a **horizontal asymptote** of the graph.

As n gets smaller and approaches 0, the graph approaches the y-axis but never touches it. The y-axis is a **vertical asymptote** of the graph.

EXAMPLE 1 Find the mean hourly cost when the service described above is used for **a.** 3 hours and **b.** 70.4 hours.

Solution **a.** To find the mean hourly cost for 3 hours of access time, we substitute 3 for n in Equation 1 and simplify:

$$\overline{c} = f(\mathbf{3}) = \frac{1.50(\mathbf{3}) + 6}{\mathbf{3}} = 3.5$$

The mean hourly cost for 3 hours of access time is $3.50.

b. To find the mean hourly cost for 70.4 hours of access time, we substitute 70.4 for n in Equation 1 and simplify:

$$\overline{c} = f(\mathbf{70.4}) = \frac{1.50(\mathbf{70.4}) + 6}{\mathbf{70.4}} = 1.585227273$$

The mean hourly cost for 70.4 hours of access time is approximately $1.59.

Self Check Find the mean hourly cost when the service is used for 5 hours. ∎

Finding the Domain and Range of a Rational Function

Since division by 0 is undefined, any values that make the denominator 0 in a rational function must be excluded from the domain of the function.

EXAMPLE 2   Find the domain of $f(x) = \dfrac{3x + 2}{x^2 + x - 6}$.

Solution   From the set of real numbers, we must exclude any values of x that make the denominator 0. To find these values, we set $x^2 + x - 6$ equal to 0 and solve for x.

$$x^2 + x - 6 = 0$$
$$(x + 3)(x - 2) = 0 \qquad \text{Factor.}$$
$$x + 3 = 0 \quad \text{or} \quad x - 2 = 0 \qquad \text{Set each factor equal to 0.}$$
$$x = -3 \quad | \qquad x = 2 \qquad \text{Solve each linear equation.}$$

Thus, the domain of the function is the set of all real numbers except -3 and 2. In interval notation, the domain is $(-\infty, -3) \cup (-3, 2) \cup (2, \infty)$.

Self Check   Find the domain of $f(x) = \dfrac{x^2 + 1}{x - 2}$.   ∎

Accent on Technology   **FINDING THE DOMAIN AND RANGE OF A FUNCTION**

We can find the domain and range of the function in Example 2 by looking at its graph. If we use window settings of $[-10, 10]$ for x and $[-10, 10]$ for y and graph the function

$$f(x) = \dfrac{3x + 2}{x^2 + x - 6}$$

we will obtain the graph in Figure 6-2(a).

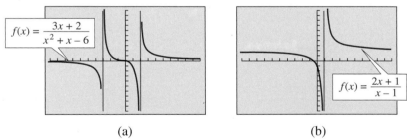

(a)   (b)

Figure 6-2

From the figure, we can see that

- As x approaches -3 from the left, the values of y decrease, and the graph approaches the vertical line $x = -3$.
- As x approaches -3 from the right, the values of y increase, and the graph approaches the vertical line $x = -3$.

We can also see that

- As x approaches 2 from the left, the values of y decrease, and the graph approaches the vertical line $x = 2$.
- As x approaches 2 from the right, the values of y increase, and the graph approaches the vertical line $x = 2$.

The lines $x = -3$ and $x = 2$ are vertical asymptotes. Although the vertical lines in the graph appear to be the graphs of $x = -3$ and $x = 2$, they are not. Graphing calculators draw graphs by connecting dots whose x-coordinates are close together. Often, when two such points straddle a vertical asymptote and their y-coordinates are far apart, the calculator draws a line between them anyway, producing what appears to be a vertical asymptote. If you set your calculator to dot mode instead of connected mode, the vertical lines will not appear.

From Figure 6-2(a), we can also see that

- As x increases to the right of 2, the values of y decrease and approach the value $y = 0$.
- As x decreases to the left of -3, the values of y increase and approach the value $y = 0$.

The line $y = 0$ (the x-axis) is a horizontal asymptote. Graphing calculators do not draw lines that appear to be horizontal asymptotes.

From the graph, we can see that all real numbers x, except -3 and 2, give a value of y. This confirms that the domain of the function is $(-\infty, -3) \cup (-3, 2) \cup (2, \infty)$. We can also see that y can be any value. Thus, the range is $(-\infty, \infty)$.

To find the domain and range of the function $f(x) = \frac{2x + 1}{x - 1}$, we use a calculator to draw the graph shown in Figure 6-2(b). From this graph, we can see that the line $x = 1$ is a vertical asymptote and that the line $y = 2$ is a horizontal asymptote. Since x can be any real number except 1, the domain is the interval $(-\infty, 1) \cup (1, \infty)$. Since y can be any value except 2, the range is $(-\infty, 2) \cup (2, \infty)$.

Simplifying Rational Expressions

Since rational expressions are the fractions of algebra, the familiar rules for arithmetic fractions apply.

Properties of Rational Expressions

If there are no divisions by 0, then

1. $\dfrac{a}{b} = \dfrac{c}{d}$ if and only if $ad = bc$

2. $\dfrac{a}{1} = a$ and $\dfrac{a}{a} = 1$

3. $\dfrac{ak}{bk} = \dfrac{a}{b}$

4. $-\dfrac{a}{b} = \dfrac{-a}{b} = \dfrac{a}{-b}$

To simplify rational expressions, we will use Property 3, which enables us to divide out factors that are common to the numerator and the denominator.

EXAMPLE 3 Simplify: **a.** $\dfrac{10k}{25k^2}$ and **b.** $\dfrac{-8y^3z^5}{6y^4z^3}$.

Solution We find the prime factorization of each numerator and denominator and divide out the common factors:

a. $\dfrac{10k}{25k^2} = \dfrac{5 \cdot 2 \cdot k}{5 \cdot 5 \cdot k \cdot k}$

$= \dfrac{\overset{1}{\cancel{5}} \cdot 2 \cdot \overset{1}{\cancel{k}}}{\cancel{5} \cdot 5 \cdot \cancel{k} \cdot k}$

$= \dfrac{2}{5k}$

b. $\dfrac{-8y^3z^5}{6y^4z^3} = \dfrac{-2 \cdot 4 \cdot y \cdot y \cdot y \cdot z \cdot z \cdot z \cdot z \cdot z}{2 \cdot 3 \cdot y \cdot y \cdot y \cdot y \cdot z \cdot z \cdot z}$

$= \dfrac{-\overset{1}{\cancel{2}} \cdot 4 \cdot \overset{1}{\cancel{y}} \cdot \overset{1}{\cancel{y}} \cdot \overset{1}{\cancel{y}} \cdot \overset{1}{\cancel{z}} \cdot \overset{1}{\cancel{z}} \cdot \overset{1}{\cancel{z}} \cdot z \cdot z}{\cancel{2} \cdot 3 \cdot \cancel{y} \cdot \cancel{y} \cdot \cancel{y} \cdot y \cdot \cancel{z} \cdot \cancel{z} \cdot \cancel{z}}$

$= -\dfrac{4z^2}{3y}$

Self Check Simplify: $\dfrac{-12a^4b^2}{-3ab^4}$.

The rational expressions in Example 3 can also be simplified by using the rules of exponents:

$\dfrac{10k}{25k^2} = \dfrac{5 \cdot 2}{5 \cdot 5} k^{1-2}$

$= \dfrac{2}{5} \cdot k^{-1}$

$= \dfrac{2}{5} \cdot \dfrac{1}{k}$

$= \dfrac{2}{5k}$

$\dfrac{-8y^3z^5}{6y^4z^3} = \dfrac{-2 \cdot 4}{2 \cdot 3} y^{3-4}z^{5-3}$

$= \dfrac{-4}{3} \cdot y^{-1}z^2$

$= -\dfrac{4}{3} \cdot \dfrac{1}{y} \cdot \dfrac{z^2}{1}$

$= -\dfrac{4z^2}{3y}$

EXAMPLE 4 Simplify: $\dfrac{x^2 - 16}{x + 4}$.

Solution We factor $x^2 - 16$ and use the fact that $\dfrac{x+4}{x+4} = 1$.

$\dfrac{x^2 - 16}{x + 4} = \dfrac{\overset{1}{\cancel{(x+4)}}(x - 4)}{1\cancel{(x+4)}}$
$\phantom{\dfrac{x^2 - 16}{x + 4}} \underset{1}{}$

$= \dfrac{x - 4}{1}$

$= x - 4$

Self Check Simplify: $\dfrac{x^2 - 9}{x - 3}$.

Accent on Technology   **CHECKING AN ALGEBRAIC SIMPLIFICATION**

To show that the simplification in Example 4 is correct, we can graph the following functions and show that the graphs are the same.

$$f(x) = \frac{x^2 - 16}{x + 4} \qquad \text{and} \qquad g(x) = x - 4$$

(See Figure 6-3.) Except for the point where $x = -4$, the graphs are the same.

The point where $x = -4$ is excluded from the graph of $f(x) = \frac{x^2 - 16}{x + 4}$, because -4 is not in the domain of f. However, graphing calculators do not show that this point is excluded. The point where $x = -4$ is included in the graph of $g(x) = x - 4$, because -4 is in the domain of g.

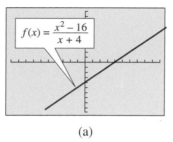   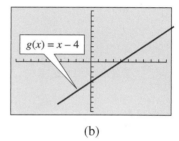

(a)   (b)

Figure 6-3

EXAMPLE 5   Simplify: $\dfrac{2x^2 + 11x + 12}{3x^2 + 11x - 4}$.

Solution   We factor the numerator and denominator and use the fact that $\frac{x + 4}{x + 4} = 1$.

$$\frac{2x^2 + 11x + 12}{3x^2 + 11x - 4} = \frac{(2x + 3)\overset{1}{\cancel{(x + 4)}}}{(3x - 1)\underset{1}{\cancel{(x + 4)}}}$$

$$= \frac{2x + 3}{3x - 1}$$

Comment   Do not divide out the x's in $\frac{2x + 3}{3x - 1}$. The x in the numerator is a factor of the first term only. It is not a factor of the entire numerator. Likewise, the x in the denominator is a factor of the first term only. It is not a factor of the entire denominator.

Self Check   Simplify: $\dfrac{2x^2 + 7x - 15}{2x^2 + 13x + 15}$.   ∎

Simplifying Rational Expressions by Factoring Out −1

To simplify $\frac{b - a}{a - b}$ $(a \neq b)$, we factor -1 from the numerator and divide out any factors common to both the numerator and the denominator:

$$\frac{b-a}{a-b} = \frac{-a+b}{a-b} \quad (a \neq b)$$

$$= \frac{-(a-b)}{(a-b)} \qquad \frac{a-b}{a-b} = 1.$$

$$= \frac{-1}{1}$$

$$= -1$$

In general, we have the following principle.

<table>
<tr><td>**Quotient of a Quantity and Its Opposite**</td><td>The quotient of any nonzero quantity and its negative (or opposite) is -1.</td></tr>
</table>

EXAMPLE 6   Simplify: $\dfrac{3x^2 - 10xy - 8y^2}{4y^2 - xy}$.

Solution   We factor the numerator and denominator and note that because $x - 4y$ and $4y - x$ are negatives, their quotient is -1.

$$\frac{3x^2 - 10xy - 8y^2}{4y^2 - xy} = \frac{(3x + 2y)(x - 4y)}{y(4y - x)} \qquad \frac{x - 4y}{4y - x} = \frac{-1(4y - x)}{4y - x} = -1.$$

$$= \frac{-(3x + 2y)}{y}$$

$$= \frac{-3x - 2y}{y}$$

Self Check   Simplify: $\dfrac{2a^2 - 5ab - 3b^2}{3b^2 - ab}$.

Many rational expressions we shall encounter are already in simplified form. For example, to attempt to simplify

$$\frac{x^2 + xa + 2x + 2a}{x^2 + x - 6}$$

we factor the numerator and denominator and divide out any common factors:

$$\frac{x^2 + xa + 2x + 2a}{x^2 + x - 6} = \frac{x(x + a) + 2(x + a)}{(x - 2)(x + 3)} = \frac{(x + a)(x + 2)}{(x - 2)(x + 3)}$$

Since there are no common factors in the numerator and denominator, the result cannot be simplified.

EXAMPLE 7   Simplify: $\dfrac{(x^2 + 2x)(x^2 + 2x - 3)}{(x^2 + x - 2)(x^2 + 3x)}$.

Solution   We factor the numerator and denominator and divide out all common factors:

$$\frac{(x^2 + 2x)(x^2 + 2x - 3)}{(x^2 + x - 2)(x^2 + 3x)} = \frac{\overset{1}{\cancel{x}}(\overset{1}{\cancel{x+2}})(\overset{1}{\cancel{x+3}})(\overset{1}{\cancel{x-1}})}{(\underset{1}{\cancel{x+2}})(\underset{1}{\cancel{x-1}})\underset{1}{\cancel{x}}(\underset{1}{\cancel{x+3}})} \quad \begin{array}{l} \frac{x}{x} = 1, \frac{x+2}{x+2} = 1, \\[2mm] \frac{x+3}{x+3} = 1, \frac{x-1}{x-1} = 1. \end{array}$$

$$= 1$$

Self Check   Simplify: $\dfrac{(a^2 + 4a)(a^2 - a - 2)}{a(a^2 + 2a - 8)}$.

 Comment   Only factors that are common to the entire numerator and the entire denominator can be divided out. Terms common to both the numerator and denominator cannot be divided out. For example, it is incorrect to divide out the common term of 3 in the following simplification, because doing so gives a wrong answer.

$$\frac{3 + 7}{3} = \frac{\overset{1}{\cancel{3}} + 7}{\underset{1}{\cancel{3}}} = \frac{1 + 7}{1} = 8 \qquad \text{The correct simplification is } \frac{3 + 7}{3} = \frac{10}{3}.$$

The 3's in the fraction $\frac{5 + 3(2)}{3(4)}$ cannot be divided out, because the 3 in the numerator is a factor of the second term only. To be divided out, the 3 must be a factor of the entire numerator.

It is not correct to divide out the y in the rational expression $\dfrac{x^2 y + 6x}{y}$, because y is not a factor of the entire numerator.

Probability

If we toss a coin, it can land as heads or tails. Since the outcomes of heads or tails are equally likely, if we were to toss the same coin many times, we would get heads about half of the time. We say that the **probability** of getting heads on a single toss of a coin is $\frac{1}{2}$.

Activities such as tossing a coin, rolling a die, and drawing a card are called **experiments.** For any experiment, a list of all possible outcomes is called a **sample space.** For example, the sample space S for the experiment of tossing two coins is the set

$$S = \{(H, H), (H, T), (T, H), (T, T)\} \qquad \text{There are 4 possible outcomes.}$$

where the ordered pair (H, T) represents the outcome "heads on the first coin and tails on the second."

An **event** is a subset of the sample space of an experiment. For example, if E is the event "getting at least one heads" in the experiment of tossing two coins, then

$$E = \{(H, H), (H, T), (T, H)\} \qquad \text{There are 3 ways of getting at least one heads.}$$

Because the outcome of getting at least one heads can occur in 3 out of 4 possible ways, we say that the probability of E is $\frac{3}{4}$, and we write

$$P(E) = P(\text{at least one heads}) = \frac{3}{4}$$

Probability of an Event	If a sample space of an experiment has n distinct and equally likely outcomes and E is an event that occurs in s of those ways, then the **probability of E** is

$$P(E) = \frac{s}{n}$$

EXAMPLE 8   Find the probability of the event "tossing a sum of 7 on one toss of two dice."

Solution   We first find the sample space of the experiment "rolling two dice a single time." If we use ordered pair notation and let the first number in each ordered pair be the result on the first die and the second number be the result on the second die, the sample space contains the following elements:

$$(1, 1), (1, 2), (1, 3), (1, 4), (1, 5), \mathbf{(1, 6)}$$
$$(2, 1), (2, 2), (2, 3), (2, 4), \mathbf{(2, 5)}, (2, 6)$$
$$(3, 1), (3, 2), (3, 3), \mathbf{(3, 4)}, (3, 5), (3, 6)$$
$$(4, 1), (4, 2), \mathbf{(4, 3)}, (4, 4), (4, 5), (4, 6)$$
$$(5, 1), \mathbf{(5, 2)}, (5, 3), (5, 4), (5, 5), (5, 6)$$
$$\mathbf{(6, 1)}, (6, 2), (6, 3), (6, 4), (6, 5), (6, 6)$$

Since there are 6 ordered pairs whose numbers give a sum of 7 out of a total of 36 equally likely outcomes, we have

$$P(E) = P(\text{tossing a 7}) = \frac{s}{n} = \frac{6}{36} = \frac{1}{6}$$

Self Check   Find the probability of tossing a sum of 11 on one toss of two dice.   ∎

Self Check Answers

1. \$2.70   **2.** $(-\infty, 2) \cup (2, \infty)$ **3.** $\dfrac{4a^3}{b^2}$ **4.** $x + 3$ **5.** $\dfrac{2x - 3}{2x + 3}$ **6.** $-\dfrac{2a + b}{b}$ **7.** $a + 1$ **8.** $\dfrac{1}{18}$

Orals   Evaluate $f(x) = \dfrac{2x - 3}{x}$ when

1. $x = 1$ **2.** $x = 3$

Give the domain of each function.

3. $f(x) = \dfrac{x + 7}{2x - 4}$ **4.** $f(x) = \dfrac{3x - 4}{x^2 - 9}$

Simplify each expression.

5. $\dfrac{25}{30}$ **6.** $\dfrac{x^2}{xy}$ **7.** $\dfrac{2x - 4}{x - 2}$ **8.** $\dfrac{x - 2}{2 - x}$

6.1 EXERCISES

REVIEW *Factor each expression.*

1. $3x^2 - 9x$

2. $6t^2 - 5t - 6$

3. $27x^6 + 64y^3$

4. $x^2 + ax + 2x + 2a$

VOCABULARY AND CONCEPTS *Fill in the blanks.*

5. If a fraction is the quotient of two polynomials, it is called a _____ expression.

6. The denominator of a fraction can never be __.

7. If a graph approaches a line, the line is called an _____.

8. $\dfrac{a}{b} = \dfrac{c}{d}$ if and only if _____

9. $\dfrac{a}{1} =$ __

10. $\dfrac{a}{a} =$ __, provided that $a \neq$ __

11. $\dfrac{ak}{bk} =$ ___, provided that $b \neq$ __ and $k \neq 0$

12. For an _____, a list of all possible outcomes is called a _____ space.

PRACTICE *The time, t, it takes to travel 600 miles is a function of the mean rate of speed, r: $t = f(r) = \dfrac{600}{r}$. Find t for each value of r.*

13. 30 mph

14. 40 mph

15. 50 mph

16. 60 mph

Suppose the cost (in dollars) of removing $p\%$ of the pollution in a river is given by the function $c = f(p) = \dfrac{50{,}000p}{100 - p}$ $(0 \leq p < 100)$. Find the cost of removing each percent of pollution.

17. 10%

18. 30%

19. 50%

20. 80%

A service club wants to publish a directory of its members. Some investigation shows that the cost of type-setting and photography will be $700 and the cost of printing each directory will be $1.25.

21. Find a function that gives the total cost c of printing x directories.

22. Find a function that gives the mean cost per directory, $\bar{c}$, of printing x directories.

23. Find the total cost of printing 500 directories.

24. Find the mean cost per directory if 500 directories are printed.

25. Find the mean cost per directory if 1,000 directories are printed.

26. Find the mean cost per directory if 2,000 directories are printed.

An electric company charges $7.50 per month plus 9¢ for each kilowatt hour (kwh) of electricity used.

27. Find a function that gives the total cost c of n kwh of electricity.

28. Find a function that gives the mean cost per kwh, $\bar{c}$, when using n kwh.

29. Find the total cost for using 775 kwh.

30. Find the mean cost per kwh when 775 kwh are used.

31. Find the mean cost per kwh when 1,000 kwh are used.

32. Find the mean cost per kwh when 1,200 kwh are used.

The function $f(t) = \dfrac{t^2 + 2t}{2t + 2}$ gives the number of days it would take two construction crews, working together, to frame a house that crew 1 (working alone) could complete in t days and crew 2 (working alone) could complete in $t + 2$ days.

33. If crew 1 can frame a certain house in 15 days, how long would it take both crews working together?

34. If crew 2 can frame a certain house in 20 days, how long would it take both crews working together?

The function $f(t) = \dfrac{t^2 + 3t}{2t + 3}$ gives the number of hours it would take two pipes to fill a pool that the larger pipe (working alone) could fill in t hours and the smaller pipe (working alone) could fill in $t + 3$ hours.

35. If the smaller pipe can fill a pool in 7 hours, how long would it take both pipes to fill the pool?

36. If the larger pipe can fill a pool in 8 hours, how long would it take both pipes to fill the pool?

Use a graphing calculator to graph each function. From the graph, determine its domain.

37. $f(x) = \dfrac{x}{x-2}$

38. $f(x) = \dfrac{x+2}{x}$

39. $f(x) = \dfrac{x+1}{x^2-4}$

40. $f(x) = \dfrac{x-2}{x^2-3x-4}$

Simplify each expression when possible.

41. $\dfrac{12}{18}$

42. $\dfrac{25}{55}$

43. $-\dfrac{112}{36}$

44. $-\dfrac{49}{21}$

45. $\dfrac{288}{312}$

46. $\dfrac{144}{72}$

47. $-\dfrac{244}{74}$

48. $-\dfrac{512}{236}$

49. $\dfrac{12x^3}{3x}$

50. $-\dfrac{15a^2}{25a^3}$

51. $\dfrac{-24x^3y^4}{18x^4y^3}$

52. $\dfrac{15a^5b^4}{21b^3c^2}$

53. $\dfrac{(3x^3)^2}{9x^4}$

54. $\dfrac{8(x^2y^3)^3}{2(xy^2)^2}$

55. $-\dfrac{11x(x-y)}{22(x-y)}$

56. $\dfrac{x(x-2)^2}{(x-2)^3}$

57. $\dfrac{9y^2(y-z)}{21y(y-z)^2}$

58. $\dfrac{-3ab^2(a-b)}{9ab(b-a)}$

59. $\dfrac{(a-b)(c-d)}{(c-d)(a-b)}$

60. $\dfrac{(p+q)(p-r)}{(r-p)(p+q)}$

61. $\dfrac{x+y}{x^2-y^2}$

62. $\dfrac{x-y}{x^2-y^2}$

63. $\dfrac{5x-10}{x^2-4x+4}$

64. $\dfrac{y-xy}{xy-x}$

65. $\dfrac{12-3x^2}{x^2-x-2}$

66. $\dfrac{x^2+2x-15}{x^2-25}$

67. $\dfrac{3x+6y}{x+2y}$

68. $\dfrac{x^2+y^2}{x+y}$

69. $\dfrac{x^3+8}{x^2-2x+4}$

70. $\dfrac{x^2+3x+9}{x^3-27}$

71. $\dfrac{x^2+2x+1}{x^2+4x+3}$

72. $\dfrac{6x^2+x-2}{8x^2+2x-3}$

73. $\dfrac{3m-6n}{3n-6m}$

74. $\dfrac{ax+by+ay+bx}{a^2-b^2}$

75. $\dfrac{4x^2+24x+32}{16x^2+8x-48}$

76. $\dfrac{a^2-4}{a^3-8}$

77. $\dfrac{3x^2-3y^2}{x^2+2y+2x+yx}$

78. $\dfrac{x^2+x-30}{x^2-x-20}$

79. $\dfrac{4x^2+8x+3}{6+x-2x^2}$

80. $\dfrac{6x^2+13x+6}{6-5x-6x^2}$

81. $\dfrac{a^3+27}{4a^2-36}$

82. $\dfrac{a-b}{b^2-a^2}$

83. $\dfrac{2x^2-3x-9}{2x^2+3x-9}$

84. $\dfrac{6x^2-7x-5}{2x^2+5x+2}$

85. $\dfrac{(m+n)^3}{m^2+2mn+n^2}$

86. $\dfrac{x^3-27}{3x^2-8x-3}$

87. $\dfrac{m^3-mn^2}{mn^2+m^2n-2m^3}$

88. $\dfrac{p^3+p^2q-2pq^2}{pq^2+p^2q-2p^3}$

89. $\dfrac{x^4-y^4}{(x^2+2xy+y^2)(x^2+y^2)}$

90. $\dfrac{-4x-4+3x^2}{4x^2-2-7x}$

91. $\dfrac{4a^2 - 9b^2}{2a^2 - ab - 6b^2}$

92. $\dfrac{x^2 + 2xy}{x + 2y + x^2 - 4y^2}$

93. $\dfrac{x - y}{x^3 - y^3 - x + y}$

94. $\dfrac{2x^2 + 2x - 12}{x^3 + 3x^2 - 4x - 12}$

95. $\dfrac{px - py + qx - qy}{px + qx + py + qy}$

96. $\dfrac{6xy - 4x - 9y + 6}{6y^2 - 13y + 6}$

97. $\dfrac{(x^2 - 1)(x + 1)}{(x^2 - 2x + 1)^2}$

98. $\dfrac{(x^2 + 2x + 1)(x^2 - 2x + 1)}{(x^2 - 1)^2}$

99. $\dfrac{(2x^2 + 3xy + y^2)(3a + b)}{(x + y)(2xy + 2bx + y^2 + by)}$

100. $\dfrac{(x - 1)(6ax + 9x + 4a + 6)}{(3x + 2)(2ax - 2a + 3x - 3)}$

Assume that you roll a die one time. Find the probability of each event.

101. Rolling a 2

102. Rolling a 3 or a 4

103. Rolling a 10

104. Rolling an even number

Assume that you draw one card from a standard deck containing 52 cards. Find the probability of each event.

105. Drawing a black card

106. Drawing a jack

107. Drawing an ace

108. Drawing a 4 or a 5

APPLICATIONS

109. Environmental cleanup   The cost (in dollars) of removing $p\%$ of the pollution in a river is approximated by the rational function

$$f(p) = \dfrac{50{,}000p}{100 - p} \quad (0 \le p < 100)$$

Find the cost of removing each percent of pollution.
a. 40%   **b.** 70%

110. Directory costs   The average (mean) cost for a service club to publish a directory of its members is given by the rational function

$$f(x) = \dfrac{1.25x + 800}{x}$$

where x is the number of directories printed. Find the average cost per directory if
a. 700 directories are printed.
b. 2,500 directories are printed.

111. Winning a lottery   Find the probability of winning a lottery with a single ticket if 6,000,000 tickets are sold and each ticket has the same chance of being the winning number.

112. Winning a raffle   Find the probability of winning a raffle if 375 tickets are sold, each with the same chance of being the winning number, and you purchased 3 tickets.

113. Probability of parents having all girls   Find the probability that two parents with three children have all girls.

114. Probability of parents having girls   Find the probability that two parents with three children have exactly two girls.

WRITING

115. Explain how to simplify a rational expression.

116. Explain how to recognize that a rational expression is in lowest terms.

SOMETHING TO THINK ABOUT

117. A student compares an answer of $\frac{a - 3b}{2b - a}$ to an answer of $\frac{3b - a}{a - 2b}$. Are the two answers the same?

118. Is this work correct? Explain.

$$\dfrac{3x^2 + 6}{3y} = \dfrac{\cancel{3}x^2 + 6}{\cancel{3}y} = \dfrac{x^2 + 6}{y}$$

119. In which parts can you divide out the 4's?

  a. $\dfrac{4x}{4y}$ **b.** $\dfrac{4x}{x + 4}$

  c. $\dfrac{4 + x}{4 + y}$ **d.** $\dfrac{4x}{4 + 4y}$

120. In which parts can you divide out the 3's?

  a. $\dfrac{3x + 3y}{3z}$ **b.** $\dfrac{3(x + y)}{3x + y}$

  c. $\dfrac{3x + 3y}{3a - 3b}$ **d.** $\dfrac{x + 3}{3y}$

Interpret each statement.

121. The probability of a man being 20 feet tall is 0.

122. The probability of dying is 1.

123. Explain why all probabilities p are in the interval $0 \le p \le 1$.

6.2 Proportion and Variation

■ **Ratios** ■ **Proportions** ■ **Solving Proportions** ■ **Similar Triangles**
■ **Direct Variation** ■ **Inverse Variation** ■ **Joint Variation**
■ **Combined Variation**

Getting Ready   *Solve each equation.*

1. $30k = 70$ **2.** $\dfrac{k}{4{,}000^2} = 90$

Classify each function as a linear function or a rational function.

3. $f(x) = 3x$ **4.** $f(x) = \dfrac{3}{x}$ $(x > 0)$

In this section, we will discuss ratio and proportion. We will then use our skills to solve variation problems.

Ratios

The quotient of two numbers is often called a **ratio.** For example, the fraction $\frac{2}{3}$ can be read as "the ratio of 2 to 3." Some more examples of ratios are

$$\frac{4x}{7y} \quad \text{(the ratio of } 4x \text{ to } 7y) \qquad \text{and} \qquad \frac{x-2}{3x} \quad \text{(the ratio of } x-2 \text{ to } 3x)$$

Ratios are often used to express **unit costs,** such as the cost per pound of ground round steak.

The cost of a package of ground round. → $\dfrac{\$18.75}{5 \text{ lb}} = \$3.75 \text{ per lb}$ ← The cost per pound.
The weight of the package. →

Ratios are also used to express **rates,** such as an average rate of speed.

A distance traveled → $\dfrac{372 \text{ miles}}{6 \text{ hours}} = 62 \text{ mph}$ ← The average rate of speed.
in a period of time. →

Proportions

An equation indicating that two ratios are equal is called a **proportion.** Two examples of proportions are

$$\frac{1}{4} = \frac{2}{8} \quad \text{and} \quad \frac{4}{7} = \frac{12}{21}$$

In the proportion $\frac{a}{b} = \frac{c}{d}$, a and d are called the **extremes** and b and c are called the **means.**

To develop a fundamental property of proportions, we suppose that

$$\frac{a}{b} = \frac{c}{d}$$

is a proportion and multiply both sides by bd to obtain

$$bd\left(\frac{a}{b}\right) = bd\left(\frac{c}{d}\right)$$

$$\frac{\cancel{b}da}{\cancel{b}} = \frac{b\cancel{d}c}{\cancel{d}}$$

$$ad = bc$$

Thus, if $\frac{a}{b} = \frac{c}{d}$, then $ad = bc$. This shows that in a proportion, the *product of the extremes equals the product of the means.*

Solving Proportions

EXAMPLE 1   Solve: $\dfrac{x + 1}{x} = \dfrac{x}{x + 2}$ $(x \neq 0, -2)$.

Solution

$$\frac{x + 1}{x} = \frac{x}{x + 2}$$

$(x + 1)(x + 2) = x \cdot x$ In a proportion, the product of the extremes equals the product of the means.

$x^2 + 3x + 2 = x^2$ Multiply.

$3x + 2 = 0$ Subtract x^2 from both sides.

$x = -\dfrac{2}{3}$ Subtract 2 from both sides and divide by 3.

Thus, $x = -\frac{2}{3}$.

Self Check   Solve: $\dfrac{x - 1}{x} = \dfrac{x}{x + 3}$ $(x \neq 0, -3)$.   ∎

EXAMPLE 2   Solve: $\dfrac{5a + 2}{2a} = \dfrac{18}{a + 4}$ $(a \neq 0, -4)$.

Solution

$$\frac{5a + 2}{2a} = \frac{18}{a + 4}$$

$(5a + 2)(a + 4) = 2a(18)$ In a proportion, the product of the extremes equals the product of the means.

$5a^2 + 22a + 8 = 36a$ Multiply.

$5a^2 - 14a + 8 = 0$ Subtract $36a$ from both sides.

$(5a - 4)(a - 2) = 0$ Factor.

$$5a - 4 = 0 \quad \text{or} \quad a - 2 = 0 \qquad \text{Set each factor equal to 0.}$$

$$5a = 4 \qquad\qquad\quad a = 2 \qquad \text{Solve each linear equation.}$$

$$a = \frac{4}{5}$$

Thus, $a = \frac{4}{5}$ or $a = 2$.

Self Check Solve: $\dfrac{3x + 1}{12} = \dfrac{x}{x + 2}$ $(x \ne -2)$. ∎

EXAMPLE 3 **Grocery shopping** If 7 pears cost $2.73, how much will 11 pears cost?

Solution We can let c represent the cost of 11 pears. The price per pear of 7 pears is $\frac{\$2.73}{7}$, and the price per pear of 11 pears is $\frac{\$c}{11}$. Since these unit costs are equal, we can set up and solve the following proportion.

$$\frac{2.73}{7} = \frac{c}{11}$$

$$11(2.73) = 7c \qquad \text{In a proportion, the product of the extremes is equal to the product of the means.}$$

$$30.03 = 7c \qquad \text{Multiply.}$$

$$\frac{30.03}{7} = c \qquad \text{Divide both sides by 7.}$$

$$c = 4.29 \qquad \text{Simplify.}$$

Eleven pears will cost $4.29.

Self Check How much will 28 pears cost? ∎

Similar Triangles

If two angles of one triangle have the same measure as two angles of a second triangle, the triangles will have the same shape. In this case, we call the triangles **similar triangles.** Here are some facts about similar triangles.

Similar Triangles

If two triangles are similar, then

1. the three angles of the first triangle have the same measures, respectively, as the three angles of the second triangle.

2. the lengths of all corresponding sides are in proportion.

Since the triangles shown in Figure 6-4 are similar, their corresponding sides are in proportion.

$$\frac{2}{4} = \frac{x}{2x}, \qquad \frac{x}{2x} = \frac{1}{2}, \qquad \frac{1}{2} = \frac{2}{4}$$

The properties of similar triangles often enable us to find the lengths of the sides of a triangle indirectly. For example, on a sunny day we can find the height of a tree and stay safely on the ground.

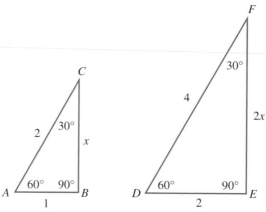

Figure 6-4

EXAMPLE 4   **Height of a tree**   A tree casts a shadow of 29 feet at the same time as a vertical yardstick casts a shadow of 2.5 feet. Find the height of the tree.

Solution   Refer to Figure 6-5, which shows the triangles determined by the tree and its shadow, and the yardstick and its shadow. Because the triangles have the same shape, they are similar, and the measures of their corresponding sides are in proportion. If we let h represent the height of the tree, we can find h by setting up and solving the following proportion.

$$\frac{h}{3} = \frac{29}{2.5}$$

$2.5h = 3(29)$ In a proportion, the product of the extremes is equal to the product of the means.

$2.5h = 87$ Simplify.

$h = 34.8$ Divide both sides by 2.5.

The tree is about 35 feet tall.

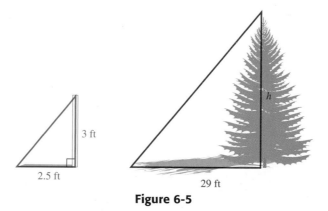

3 ft

2.5 ft

h

29 ft

Figure 6-5   ∎

Direct Variation

To introduce direct variation, we consider the formula

$$C = \pi D$$

for the circumference of a circle, where C is the circumference, D is the diameter, and $\pi \approx 3.14159$. If we double the diameter of a circle, we determine another circle with a larger circumference C_1 such that

$$C_1 = \pi(2D) = 2\pi D = 2C$$

Thus, doubling the diameter results in doubling the circumference. Likewise, if we triple the diameter, we triple the circumference.

In this formula, we say that the variables C and D *vary directly,* or that they are *directly proportional.* This is because as one variable gets larger, so does the other, and in a predictable way. In this example, the constant π is called the *constant of variation* or the *constant of proportionality.*

Direct Variation

> The words "y varies directly with x" or "y is directly proportional to x" mean that $y = kx$ for some nonzero constant k. The constant k is called the **constant of variation** or the **constant of proportionality.**

 Comment   In this section, k will always be a positive number.

Since the formula for direct variation ($y = kx$) defines a linear function, its graph is always a line with a y-intercept at the origin. The graph of $y = kx$ appears in Figure 6-6 for three positive values of k.

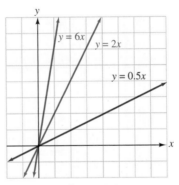

Figure 6-6

One example of direct variation is Hooke's law from physics. Hooke's law states that the distance a spring will stretch varies directly with the force that is applied to it.

If d represents a distance and f represents a force, Hooke's law is expressed mathematically as

$$d = kf$$

where k is the constant of variation. If the spring stretches 10 inches when a weight of 6 pounds is attached, k can be found as follows:

$\boldsymbol{d = kf}$

$\boldsymbol{10} = k(\boldsymbol{6})$ Substitute 10 for d and 6 for f.

$\dfrac{5}{3} = k$ Divide both sides by 6 and simplify.

To find the force required to stretch the spring a distance of 35 inches, we can solve the equation $d = kf$ for f, with $d = 35$ and $k = \frac{5}{3}$.

$$d = kf$$

$$35 = \frac{5}{3}f \quad \text{Substitute 35 for } d \text{ and } \frac{5}{3} \text{ for } k.$$

$$105 = 5f \quad \text{Multiply both sides by 3.}$$

$$21 = f \quad \text{Divide both sides by 5.}$$

The force required to stretch the spring a distance of 35 inches is 21 pounds.

EXAMPLE 5   **Direct variation**   The distance traveled in a given time is directly proportional to the speed. If a car goes 70 miles at 30 mph, how far will it go in the same time at 45 mph?

Solution   The words *distance is directly proportional to speed* can be expressed by the equation

(1) $$d = ks$$

where d is distance, k is the constant of variation, and s is the speed. To find k, we substitute 70 for d and 30 for s, and solve for k.

$$d = ks$$

$$70 = k(30)$$

$$k = \frac{7}{3}$$

To find the distance traveled at 45 mph, we substitute $\frac{7}{3}$ for k and 45 for s in Equation 1 and simplify.

$$d = ks$$

$$d = \frac{7}{3}(45)$$

$$= 105$$

In the time it took to go 70 miles at 30 mph, the car could go 105 miles at 45 mph.

Self Check   How far will the car go in the same time at 60 mph?   ∎

Inverse Variation

In the formula $w = \frac{12}{l}$, w gets smaller as l gets larger, and w gets larger as l gets smaller. Since these variables vary in opposite directions in a predictable way, we say that the variables *vary inversely,* or that they are *inversely proportional.* The constant 12 is the constant of variation.

Inverse Variation   The words "y varies inversely with x" or "y is inversely proportional to x" mean that $y = \frac{k}{x}$ for some nonzero constant k. The constant k is called the **constant of variation.**

The formula for inverse variation $\left(y = \frac{k}{x}\right)$ defines a rational function. The graph of $y = \frac{k}{x}$ appears in Figure 6-7 for three positive values of k.

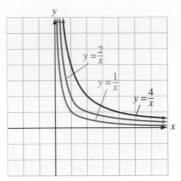

Figure 6-7

Because of gravity, an object in space is attracted to Earth. The force of this attraction varies inversely with the square of the distance from the object to Earth's center.

If f represents the force and d represents the distance, this information can be expressed by the equation

$$f = \frac{k}{d^2}$$

If we know that an object 4,000 miles from Earth's center is attracted to Earth with a force of 90 pounds, we can find k.

$$f = \frac{k}{d^2}$$

$$90 = \frac{k}{4{,}000^2} \qquad \text{Substitute 90 for } f \text{ and 4,000 for } d.$$

$$k = 90(4{,}000^2)$$

$$= 1.44 \times 10^9$$

To find the force of attraction when the object is 5,000 miles from Earth's center, we proceed as follows:

$$f = \frac{k}{d^2}$$

$$f = \frac{1.44 \times 10^9}{5{,}000^2} \qquad \text{Substitute } 1.44 \times 10^9 \text{ for } k \text{ and 5,000 for } d.$$

$$= 57.6$$

The object will be attracted to Earth with a force of 57.6 pounds when it is 5,000 miles from Earth's center.

EXAMPLE 6   **Light intensity**   The intensity I of light received from a light source varies inversely with the square of the distance from the source. If the intensity of a light source 4 feet from an object is 8 candelas, find the intensity at a distance of 2 feet.

Solution   The words *intensity varies inversely with the square of the distance d* can be expressed by the equation

$$I = \frac{k}{d^2}$$

To find k, we substitute 8 for I and 4 for d and solve for k.

$$I = \frac{k}{d^2}$$

$$8 = \frac{k}{4^2}$$

$$128 = k$$

To find the intensity when the object is 2 feet from the light source, we substitute 2 for d and 128 for k and simplify.

$$I = \frac{k}{d^2}$$

$$I = \frac{128}{2^2}$$

$$= 32$$

The intensity at 2 feet is 32 candelas.

Self Check   Find the intensity at a distance of 8 feet.   ∎

Joint Variation

There are times when one variable varies with the product of several variables. For example, the area of a triangle varies directly with the product of its base and height:

$$A = \frac{1}{2}bh$$

Such variation is called *joint variation.*

Joint Variation

> If one variable varies directly with the product of two or more variables, the relationship is called **joint variation.** If y varies jointly with x and z, then $y = kxz$. The nonzero constant k is called the **constant of variation.**

EXAMPLE 7   **Volume of a cone**   The volume V of a cone varies jointly with its height h and the area of its base B. If $V = 6 \text{ cm}^3$ when $h = 3$ cm and $B = 6 \text{ cm}^2$, find V when $h = 2$ cm and $B = 8 \text{ cm}^2$.

Solution   The words V *varies jointly with* h *and* B can be expressed by the equation

$$V = khB \qquad \text{The relationship can also be read as "}V\text{ is directly proportional to the product}$$
$$\text{of } h \text{ and } B.\text{"}$$

We can find k by substituting 6 for V, 3 for h, and 6 for B.

$$V = khB$$

$$6 = k(3)(6)$$

$$6 = k(18)$$

$$\frac{1}{3} = k \qquad \text{Divide both sides by 18; } \frac{6}{18} = \frac{1}{3}.$$

To find V when $h = 2$ and $B = 8$, we substitute these values into the formula $V = \frac{1}{3}hB$.

$$V = \frac{1}{3}hB$$

$$V = \left(\frac{1}{3}\right)(2)(8)$$

$$= \frac{16}{3}$$

The volume is $5\frac{1}{3}$ cm^3.

Combined Variation

Many applied problems involve a combination of direct and inverse variation. Such variation is called **combined variation.**

EXAMPLE 8   **Building highways**   The time it takes to build a highway varies directly with the length of the road, but inversely with the number of workers. If it takes 100 workers 4 weeks to build 2 miles of highway, how long will it take 80 workers to build 10 miles of highway?

Solution   We can let t represent the time in weeks, l represent the length in miles, and w represent the number of workers. The relationship between these variables can be expressed by the equation

$$t = \frac{kl}{w}$$

We substitute 4 for t, 100 for w, and 2 for l to find k:

$$4 = \frac{k(2)}{100}$$

$$400 = 2k \qquad \text{Multiply both sides by 100.}$$

$$200 = k \qquad \text{Divide both sides by 2.}$$

We now substitute 80 for w, 10 for l, and 200 for k in the equation $t = \frac{kl}{w}$ and simplify:

$$t = \frac{kl}{w}$$

$$t = \frac{200(10)}{80}$$

$$= 25$$

It will take 25 weeks for 80 workers to build 10 miles of highway.

Self Check   How long will it take 60 workers to build 6 miles of highway?

Self Check Answers

1. $\frac{3}{2}$ **2.** $\frac{2}{3}$, 1   **3.** \$10.92   **5.** 140 mi   **6.** 2 candelas   **8.** 20 weeks

Orals *Solve each proportion.*

1. $\dfrac{x}{2} = \dfrac{3}{6}$

2. $\dfrac{3}{x} = \dfrac{4}{12}$

3. $\dfrac{5}{7} = \dfrac{2}{x}$

Express each sentence with a formula.

4. a varies directly with b.
5. a varies inversely with b.
6. a varies jointly with b and c.
7. a varies directly with b but inversely with c.

6.2 EXERCISES

REVIEW *Simplify each expression.*

1. $(x^2x^3)^2$

2. $\left(\dfrac{a^3a^5}{a^{-2}}\right)^3$

3. $\dfrac{b^0 - 2b^0}{b^0}$

4. $\left(\dfrac{2r^{-2}r^{-3}}{4r^{-5}}\right)^{-3}$

5. Write 35,000 in scientific notation.
6. Write 0.00035 in scientific notation.
7. Write 2.5×10^{-3} in standard notation.
8. Write 2.5×10^4 in standard notation.

VOCABULARY AND CONCEPTS *Fill in the blanks.*

9. Ratios are used to express _____ and ____.
10. An equation stating that two ratios are equal is called a _____.
11. In a proportion, the product of the _____ is equal to the product of the _____.
12. If two angles of one triangle have the same measure as two angles of a second triangle, the triangles are _____.
13. The equation $y = kx$ indicates _____ variation.
14. The equation $y = \frac{k}{x}$ indicates _____ variation.
15. Inverse variation is represented by a _____ function.
16. Direct variation is represented by a _____ function through the origin.
17. The equation $y = kxz$ indicates _____ variation.
18. The equation $y = \frac{kx}{z}$ indicates _____ variation.

Tell whether the graph represents direct variation, inverse variation, or neither.

19.   20.

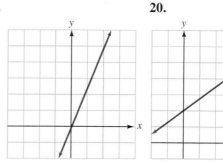

21.   22.

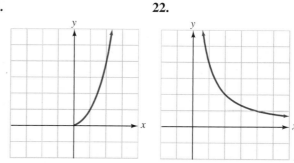

PRACTICE *Solve each proportion for the variable, if possible.*

23. $\dfrac{x}{5} = \dfrac{15}{25}$

24. $\dfrac{4}{y} = \dfrac{6}{27}$

25. $\dfrac{r-2}{3} = \dfrac{r}{5}$

26. $\dfrac{x+1}{x-1} = \dfrac{6}{4}$

27. $\dfrac{3}{n} = \dfrac{2}{n+1}$

28. $\dfrac{4}{x+3} = \dfrac{3}{5}$

29. $\dfrac{5}{5z + 3} = \dfrac{2z}{2z^2 + 6}$ **30.** $\dfrac{9t + 6}{t(t + 3)} = \dfrac{7}{t + 3}$

31. $\dfrac{2}{c} = \dfrac{c - 3}{2}$ **32.** $\dfrac{y}{4} = \dfrac{4}{y}$

33. $\dfrac{2}{3x} = \dfrac{6x}{36}$ **34.** $\dfrac{2}{x + 6} = \dfrac{-2x}{5}$

35. $\dfrac{2(x + 3)}{3} = \dfrac{4(x - 4)}{5}$ **36.** $\dfrac{x + 4}{5} = \dfrac{3(x - 2)}{3}$

37. $\dfrac{1}{x + 3} = \dfrac{-2x}{x + 5}$ **38.** $\dfrac{x - 1}{x + 1} = \dfrac{2}{3x}$

39. $\dfrac{a - 4}{a + 2} = \dfrac{a - 5}{a + 1}$ **40.** $\dfrac{z + 2}{z + 6} = \dfrac{z - 4}{z - 2}$

Express each sentence as a formula.

41. *A* varies directly with the square of *p*.

42. *z* varies inversely with the cube of *t*.

43. *v* varies inversely with the cube of *r*.

44. *r* varies directly with the square of *s*.

45. *B* varies jointly with *m* and *n*.

46. *C* varies jointly with *x*, *y*, and *z*.

47. *P* varies directly with the square of *a*, and inversely with the cube of *j*.

48. *M* varies inversely with the cube of *n*, and jointly with *x* and the square of *z*.

Express each formula in words. In each formula, k is the constant of variation.

49. $L = kmn$

50. $P = \dfrac{km}{n}$

51. $E = kab^2$

52. $U = krs^2t$

53. $X = \dfrac{kx^2}{y^2}$

54. $Z = \dfrac{kw}{xy}$

55. $R = \dfrac{kL}{d^2}$

56. $e = \dfrac{kPL}{A}$

APPLICATIONS   *Set up and solve the required proportion.*

57. Selling shirts   Consider the following ad. How much will 5 shirts cost?

Sale!

Now buy 2

for only $25!

58. Cooking   A recipe requires four 16-ounce bottles of catsup to make two gallons of spaghetti sauce. How many bottles are needed to make 10 gallons of sauce?

59. Gas consumption   A car gets 42 mpg. How much gas will be needed to go 315 miles?

60. Model railroading   An HO-scale model railroad engine is 9 inches long. The HO scale is 87 feet to 1 foot. How long is a real engine?

61. Hobbies   Standard dollhouse scale is 1 inch to 1 foot. Heidi's dollhouse is 32 inches wide. How wide would it be if it were a real house?

62. Staffing   A school board has determined that there should be 3 teachers for every 50 students. How many teachers are needed for an enrollment of 2,700 students?

63. Drafting   In a scale drawing, a 280-foot antenna tower is drawn 7 inches high. The building next to it is drawn 2 inches high. How tall is the actual building?

64. Mixing fuel   The instructions on a can of oil intended to be added to lawnmower gasoline read:

Recommended	Gasoline	Oil
50 to 1	6 gal	16 oz

Are these instructions correct?   (*Hint:* There are 128 ounces in 1 gallon.)

65. Recommended dosage   The recommended child's dose of the sedative hydroxine is 0.006 gram per kilogram of body mass. Find the dosage for a 30-kg child.

66. Body mass   The proper dose of the antibiotic cephalexin in children is 0.025 gram per kilogram of body mass. Find the mass of a child receiving a $1\frac{1}{8}$-gram dose.

Use similar triangles to help solve each problem.

67. Height of a tree   A tree casts a shadow of 28 feet at the same time as a 6-foot man casts a shadow of 4 feet. Find the height of the tree.

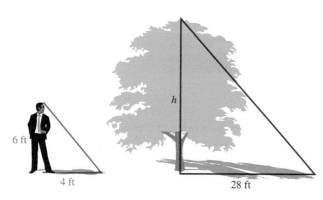

68. Height of a flagpole   A man places a mirror on the ground and sees the reflection of the top of a flagpole. The two triangles in the illustration are similar. Find the height, *h*, of the flagpole.

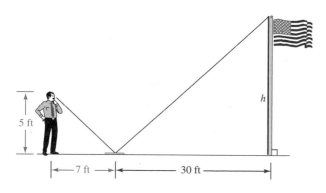

69. Width of a river   Use the dimensions in the illustration to find *w*, the width of the river. The two triangles in the illustration are similar.

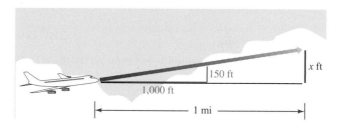

70. Flight path   An airplane ascends 150 feet as it flies a horizontal distance of 1,000 feet. How much altitude will it gain as it flies a horizontal distance of 1 mile?   (*Hint:* 5,280 feet = 1 mile.)

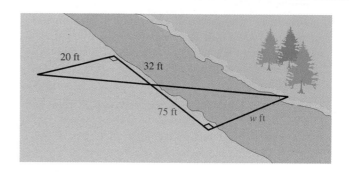

71. Flight path   An airplane descends 1,350 feet as it flies a horizontal distance of 1 mile. How much altitude is lost as it flies a horizontal distance of 5 miles?

72. Ski runs   A ski course with $\frac{1}{2}$ mile of horizontal run falls 100 feet in every 300 feet of run. Find the height of the hill.

Solve each variation problem.

73. Area of a circle   The area of a circle varies directly with the square of its radius. The constant of variation is π. Find the area of a circle with a radius of 6 inches.

74. Falling objects   An object in free fall travels a distance *s* that is directly proportional to the square of the time *t*. If an object falls 1,024 feet in 8 seconds, how far will it fall in 10 seconds?

75. Finding distance   The distance that a car can go is directly proportional to the number of gallons of gasoline it consumes. If a car can go 288 miles on 12 gallons of gasoline, how far can it go on a full tank of 18 gallons?

76. Farming A farmer's harvest in bushels varies directly with the number of acres planted. If 8 acres can produce 144 bushels, how many acres are required to produce 1,152 bushes?

77. Farming The length of time that a given number of bushels of corn will last when feeding cattle varies inversely with the number of animals. If x bushels will feed 25 cows for 10 days, how long will the feed last for 10 cows?

78. Geometry For a fixed area, the length of a rectangle is inversely proportional to its width. A rectangle has a width of 18 feet and a length of 12 feet. If the length is increased to 16 feet, find the width.

79. Gas pressure Under constant temperature, the volume occupied by a gas is inversely proportional to the pressure applied. If the gas occupies a volume of 20 cubic inches under a pressure of 6 pounds per square inch, find the volume when the gas is subjected to a pressure of 10 pounds per square inch.

80. Value of a boat The value of a boat usually varies inversely with its age. If a boat is worth $7,000 when it is 3 years old, how much will it be worth when it is 7 years old?

81. Organ pipes The frequency of vibration of air in an organ pipe is inversely proportional to the length l of the pipe. If a pipe 2 feet long vibrates 256 times per second, how many times per second will a 6-foot pipe vibrate?

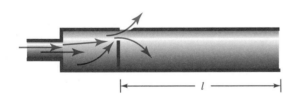

82. Geometry The area of a rectangle varies jointly with its length and width. If both the length and the width are tripled, by what factor is the area multiplied?

83. Geometry The volume of a rectangular solid varies jointly with its length, width, and height. If the length is doubled, the width is tripled, and the height is doubled, by what factor is the volume multiplied?

84. Cost of a trucking company The costs incurred by a trucking company vary jointly with the number of trucks in service and the number of hours they are used. When 4 trucks are used for 6 hours each, the costs are $1,800. Find the costs of using 10 trucks, each for 12 hours.

85. Storing oil The number of gallons of oil that can be stored in a cylindrical tank varies jointly with the height of the tank and the square of the radius of its base. The constant of proportionality is 23.5. Find the number of gallons that can be stored in the cylindrical tank shown in the illustration.

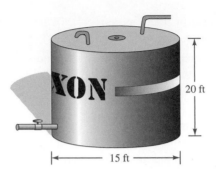

86. Finding the constant of variation A quantity l varies jointly with x and y and inversely with z. If the value of l is 30 when $x = 15$, $y = 5$, and $z = 10$, find k.

87. Electronics The voltage (in volts) measured across a resistor is directly proportional to the current (in amperes) flowing through the resistor. The constant of variation is the **resistance** (in ohms). If 6 volts is measured across a resistor carrying a current of 2 amperes, find the resistance.

88. Electronics The power (in watts) lost in a resistor (in the form of heat) is directly proportional to the square of the current (in amperes) passing through it. The constant of proportionality is the resistance (in ohms). What power is lost in a 5-ohm resistor carrying a 3-ampere current?

89. Building construction The deflection of a beam is inversely proportional to its width and the cube of its depth. If the deflection of a 4-inch-by-4-inch beam is 1.1 inches, find the deflection of a 2-inch-by-8-inch beam positioned as in the illustration.

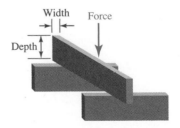

90. Building construction Find the deflection of the
beam in Exercise 89 when the beam is positioned as
in the illustration.

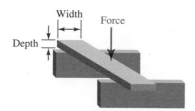

91. Gas pressure The pressure of a certain amount of
gas is directly proportional to the temperature (measured on the Kelvin scale) and inversely proportional
to the volume. A sample of gas at a pressure of 1 at-
mosphere occupies a volume of 1 cubic meter at a
temperature of 273 Kelvin. When heated, the gas
expands to twice its volume, but the pressure remains
constant. To what temperature is it heated?

92. Tension A yo-yo, twirled at the end of a string, is
kept in its circular path by the tension of the string.
The tension T is directly proportional to the square of
the speed s and inversely proportional to the radius r
of the circle. In the illustration, the tension is
32 pounds when the speed is 8 feet/second and the
radius is 6 feet. Find the tension when the speed is
4 feet/second and the radius is 3 feet.

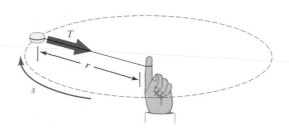

WRITING

93. Explain the terms *means* and *extremes*.

94. Distinguish between a *ratio* and a *proportion*.

95. Explain the term *joint variation*.

96. Explain why the equation $\frac{y}{x} = k$ indicates that y varies
directly with x.

SOMETHING TO THINK ABOUT

97. As temperature increases on the Fahrenheit scale, it
also increases on the Celsius scale. Is this direct variation? Explain.

98. As the cost of a purchase (less than $5) increases, the
amount of change received from a five-dollar bill decreases. Is this inverse variation? Explain.

6.3 Multiplying and Dividing Rational Expressions

■ **Multiplying Rational Expressions** ■ **Dividing Rational Expressions**
■ **Mixed Operations**

Getting Ready *Simplify each fraction.*

1. $\dfrac{45}{30}$ **2.** $\dfrac{72}{180}$ **3.** $\dfrac{600}{450}$ **4.** $\dfrac{210}{45}$

In this section, we will learn how to multiply and divide rational expressions. After
mastering these skills, we will consider mixed operations.

Multiplying Rational Expressions

In Section 6.1, we introduced four basic properties of fractions. We now provide the rule for multiplying fractions.

Multiplying Fractions	If no denominators are 0, then
	$$\frac{a}{b} \cdot \frac{c}{d} = \frac{a \cdot c}{b \cdot d} = \frac{ac}{bd}$$

Thus, *to multiply two fractions, we multiply the numerators and multiply the denominators.*

$$\frac{3}{5} \cdot \frac{2}{7} = \frac{3 \cdot 2}{5 \cdot 7}$$

$$= \frac{6}{35}$$

$$\frac{4}{7} \cdot \frac{5}{8} = \frac{4 \cdot 5}{7 \cdot 8}$$

$$= \frac{\overset{1}{\cancel{2}} \cdot \overset{1}{\cancel{2}} \cdot 5}{7 \cdot \underset{1}{\cancel{2}} \cdot \underset{1}{\cancel{2}} \cdot 2} \qquad \frac{2}{2} = 1.$$

$$= \frac{5}{14}$$

The same rule applies to rational expressions. If $t \neq 0$, then

$$\frac{x^2 y}{t} \cdot \frac{xy^3}{t^3} = \frac{x^2 y \cdot xy^3}{tt^3}$$

$$= \frac{x^2 x \cdot yy^3}{t^4}$$

$$= \frac{x^3 y^4}{t^4}$$

EXAMPLE 1 Find the product of $\dfrac{x^2 - 6x + 9}{x}$ and $\dfrac{x^2}{x - 3}$.

Solution We multiply the numerators and multiply the denominators and then simplify.

$$\frac{x^2 - 6x + 9}{x} \cdot \frac{x^2}{x - 3} = \frac{(x^2 - 6x + 9)x^2}{x(x - 3)} \qquad \text{Multiply the numerators and multiply the denominators.}$$

$$= \frac{(x - 3)(x - 3)xx}{x(x - 3)} \qquad \text{Factor the numerator.}$$

$$= \frac{\overset{1}{\cancel{(x-3)}}(x - 3)\overset{1}{x}x}{\underset{1}{\cancel{x}}\underset{1}{\cancel{(x-3)}}} \qquad \frac{x-3}{x-3} = 1 \text{ and } \frac{x}{x} = 1.$$

$$= x(x - 3)$$

Self Check Multiply: $\dfrac{a^2 + 6a + 9}{a} \cdot \dfrac{a^3}{a + 3}$.

CHECKING AN ALGEBRAIC SIMPLIFICATION

We can check the simplification in Example 1 by graphing the rational functions $f(x) = \left(\frac{x^2 - 6x + 9}{x}\right)\left(\frac{x^2}{x - 3}\right)$ and $g(x) = x(x - 3)$ and observing that the graphs are the same. Note that 0 and 3 are not included in the domain of the first function.

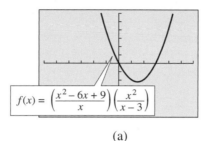

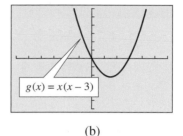

(a)   (b)

Figure 6-8

EXAMPLE 2   Multiply: $\dfrac{x^2 - x - 6}{x^2 - 4} \cdot \dfrac{x^2 + x - 6}{x^2 - 9}$.

Solution
$$\frac{x^2 - x - 6}{x^2 - 4} \cdot \frac{x^2 + x - 6}{x^2 - 9}$$

$$= \frac{(x^2 - x - 6)(x^2 + x - 6)}{(x^2 - 4)(x^2 - 9)}$$ Multiply the numerators and multiply the denominators.

$$= \frac{(x - 3)(x + 2)(x + 3)(x - 2)}{(x + 2)(x - 2)(x + 3)(x - 3)}$$ Factor the polynomials.

$$= \frac{\overset{1}{\cancel{(x - 3)}}\overset{1}{\cancel{(x + 2)}}\overset{1}{\cancel{(x + 3)}}\overset{1}{\cancel{(x - 2)}}}{\underset{1}{\cancel{(x + 2)}}\underset{1}{\cancel{(x - 2)}}\underset{1}{\cancel{(x + 3)}}\underset{1}{\cancel{(x - 3)}}}$$ $\dfrac{x - 3}{x - 3} = 1, \dfrac{x + 2}{x + 2} = 1, \dfrac{x + 3}{x + 3} = 1,$ and $\dfrac{x - 2}{x - 2} = 1.$

$$= 1$$

Self Check   Multiply: $\dfrac{a^2 + a - 6}{a^2 - 9} \cdot \dfrac{a^2 - a - 6}{a^2 - 4}$.

   Comment   Note that when all factors divide out, the result is 1, not 0.

   EXAMPLE 3   Multiply: $\dfrac{6x^2 + 5x - 4}{2x^2 + 5x + 3} \cdot \dfrac{8x^2 + 6x - 9}{12x^2 + 7x - 12}$.

Solution
$$\frac{6x^2 + 5x - 4}{2x^2 + 5x + 3} \cdot \frac{8x^2 + 6x - 9}{12x^2 + 7x - 12}$$

$$= \frac{(6x^2 + 5x - 4)(8x^2 + 6x - 9)}{(2x^2 + 5x + 3)(12x^2 + 7x - 12)}$$ Multiply the numerators and multiply the denominators.

$$= \frac{(3x + 4)(2x - 1)(4x - 3)(2x + 3)}{(2x + 3)(x + 1)(3x + 4)(4x - 3)}$$ Factor the polynomials.

$$= \frac{\overset{1}{\cancel{(3x+4)}}(2x-1)\overset{1}{\cancel{(4x-3)}}\overset{1}{\cancel{(2x+3)}}}{\underset{1}{\cancel{(2x+3)}}(x+1)\underset{1}{\cancel{(3x+4)}}\underset{1}{\cancel{(4x-3)}}} \qquad \frac{3x+4}{3x+4}=1, \frac{4x-3}{4x-3}=1, \text{ and }$$

$$\frac{2x+3}{2x+3}=1.$$

$$= \frac{2x-1}{x+1}.$$

Self Check   Multiply: $\dfrac{2a^2+5a-12}{2a^2+11a+12} \cdot \dfrac{2a^2-3a-9}{2a^2-a-3}$.   ∎

EXAMPLE 4   Multiply: $(2x-x^2) \cdot \dfrac{x}{x^2-5x+6}$.

Solution $(2x-x^2) \cdot \dfrac{x}{x^2-5x+6}$

$$= \frac{2x-x^2}{1} \cdot \frac{x}{x^2-5x+6} \qquad \text{Write } 2x-x^2 \text{ as } \frac{2x-x^2}{1}.$$

$$= \frac{(2x-x^2)x}{1(x^2-5x+6)} \qquad \text{Multiply the fractions.}$$

$$= \frac{x\overset{-1}{\cancel{(2-x)}}x}{\underset{1}{\cancel{(x-2)}}(x-3)} \qquad \begin{array}{l}\text{Factor out } x \text{ and note that the quotient of any} \\ \text{nonzero quantity and its negative is } -1.\end{array}$$

$$= \frac{-x^2}{x-3}$$

Since $\frac{-a}{b}=-\frac{a}{b}$, the $-$ sign can be written in front of the fraction. For this reason, the final result can be written as

$$-\frac{x^2}{x-3}$$

Self Check   Multiply: $\dfrac{x^2+5x+6}{(x^2+4x)(x+2)} \cdot x^3+4x^2$.   ∎

Comment   In Examples 1–4, we can obtain the same answers if we factor first and divide out the common factors before we multiply.

Dividing Rational Expressions

Recall the rule for dividing fractions.

Dividing Fractions
If no denominators are 0, then
$$\frac{a}{b} \div \frac{c}{d} = \frac{a}{b} \cdot \frac{d}{c} = \frac{ad}{bc}$$

We can prove this rule as follows:

$$\frac{a}{b} \div \frac{c}{d} = \frac{\dfrac{a}{b}}{\dfrac{c}{d}} \cdot 1 = \frac{\dfrac{a}{b}}{\dfrac{c}{d}} \cdot \frac{\dfrac{d}{c}}{\dfrac{d}{c}} = \frac{\dfrac{a}{b}\cdot\dfrac{d}{c}}{\dfrac{c}{d}\cdot\dfrac{d}{c}} = \frac{\dfrac{a}{b}\cdot\dfrac{d}{c}}{\dfrac{cd}{cd}} = \frac{\dfrac{a}{b}\cdot\dfrac{d}{c}}{1} = \frac{a}{b}\cdot\frac{d}{c}$$

Thus, *to divide two fractions, we can invert the divisor and multiply.*

$$\frac{3}{5} \div \frac{2}{7} = \frac{3}{5}\cdot\frac{7}{2}$$
$$= \frac{3\cdot7}{5\cdot2}$$
$$= \frac{21}{10}$$

$$\frac{4}{7} \div \frac{2}{21} = \frac{4}{7}\cdot\frac{21}{2}$$
$$= \frac{4\cdot21}{7\cdot2}$$
$$= \frac{\overset{1}{\cancel{2}}\cdot2\cdot3\cdot\overset{1}{\cancel{7}}}{\underset{1}{\cancel{7}}\cdot\underset{1}{\cancel{2}}}$$
$$= 6$$

The same rule applies to rational expressions.

$$\frac{x^2}{y^3z^2} \div \frac{x^2}{yz^3} = \frac{x^2}{y^3z^2}\cdot\frac{yz^3}{x^2}$$ Invert the divisor and multiply.

$$= \frac{x^2yz^3}{x^2y^3z^2}$$ Multiply the numerators and the denominators.

$$= x^{2-2}y^{1-3}z^{3-2}$$ To divide exponential expressions with the same base, keep the base and subtract the exponents.

$$= x^0y^{-2}z^1$$
$$= 1\cdot y^{-2}\cdot z \qquad x^0 = 1.$$
$$= \frac{z}{y^2} \qquad y^{-2} = \frac{1}{y^2}.$$

EXAMPLE 5 Divide: $\dfrac{x^3+8}{x+1} \div \dfrac{x^2-2x+4}{2x^2-2}$.

Solution We invert the divisor and multiply.

$$\frac{x^3+8}{x+1} \div \frac{x^2-2x+4}{2x^2-2}$$
$$= \frac{x^3+8}{x+1}\cdot\frac{2x^2-2}{x^2-2x+4}$$
$$= \frac{(x^3+8)(2x^2-2)}{(x+1)(x^2-2x+4)}$$
$$= \frac{(x+2)(\overset{1}{\cancel{x^2-2x+4}})2(\overset{1}{\cancel{x+1}})(x-1)}{(\underset{1}{\cancel{x+1}})(\underset{1}{\cancel{x^2-2x+4}})} \qquad \frac{x^2-2x+4}{x^2-2x+4}=1, \frac{x+1}{x+1}=1.$$
$$= 2(x+2)(x-1)$$

Self Check Divide: $\dfrac{x^3-8}{x-1} \div \dfrac{x^2+2x+4}{3x^2-3x}$.

EXAMPLE 6   Divide: $\dfrac{x^2 - 4}{x - 1} \div (x - 2)$.

Solution $\dfrac{x^2 - 4}{x - 1} \div (x - 2)$

$= \dfrac{x^2 - 4}{x - 1} \div \dfrac{x - 2}{1}$ Write $x - 2$ as a fraction with a denominator of 1.

$= \dfrac{x^2 - 4}{x - 1} \cdot \dfrac{1}{x - 2}$ Invert the divisor and multiply.

$= \dfrac{x^2 - 4}{(x - 1)(x - 2)}$ Multiply the numerators and the denominators.

$= \dfrac{(x + 2)\overset{1}{\cancel{(x - 2)}}}{(x - 1)\underset{1}{\cancel{(x - 2)}}}$ Factor $x^2 - 4$ and divide out $x - 2$: $\dfrac{x - 2}{x - 2} = 1$.

$= \dfrac{x + 2}{x - 1}$

Self Check   Divide: $\dfrac{a^3 - 9a}{a - 3} \div (a^2 + 3a)$.

Mixed Operations

The following example involves both a division and a multiplication.

EXAMPLE 7   Simplify: $\dfrac{x^2 + 2x - 3}{6x^2 + 5x + 1} \div \dfrac{2x^2 - 2}{2x^2 - 5x - 3} \cdot \dfrac{6x^2 + 4x - 2}{x^2 - 2x - 3}$.

Solution   Since multiplications and divisions are done in order from left to right, we change the division to a multiplication

$$\left(\dfrac{x^2 + 2x - 3}{6x^2 + 5x + 1} \div \dfrac{2x^2 - 2}{2x^2 - 5x - 3} \right) \dfrac{6x^2 + 4x - 2}{x^2 - 2x - 3}$$

$$= \left(\dfrac{x^2 + 2x - 3}{6x^2 + 5x + 1} \cdot \dfrac{2x^2 - 5x - 3}{2x^2 - 2} \right) \dfrac{6x^2 + 4x - 2}{x^2 - 2x - 3}$$

and then multiply the rational expressions and simplify the result.

$$= \dfrac{(x^2 + 2x - 3)(2x^2 - 5x - 3)(6x^2 + 4x - 2)}{(6x^2 + 5x + 1)(2x^2 - 2)(x^2 - 2x - 3)}$$

$$= \dfrac{(x + 3)\overset{1}{\cancel{(x - 1)}}(2x + 1)\overset{1}{\cancel{(x - 3)}}\,2\overset{1}{(3x - 1)}\overset{1}{\cancel{(x + 1)}}}{(3x + 1)\underset{1}{\cancel{(2x + 1)}}\,2\underset{1}{(x + 1)}\underset{1}{\cancel{(x - 1)}}\underset{1}{\cancel{(x - 3)}}\underset{1}{\cancel{(x + 1)}}} \qquad \begin{aligned} &\dfrac{x - 1}{x - 1} = 1, \dfrac{2x + 1}{2x + 1} = 1, \\ &\dfrac{x - 3}{x - 3} = 1, \dfrac{2}{2} = 1, \dfrac{x + 1}{x + 1} = 1. \end{aligned}$$

$$= \dfrac{(x + 3)(3x - 1)}{(3x + 1)(x + 1)}$$

1. $a^2(a + 3)$ **2.** 1   **3.** $\dfrac{a - 3}{a + 1}$ **4.** $x(x + 3)$ **5.** $3x(x - 2)$ **6.** 1

Orals   *Perform the operations and simplify, if possible.*

1. $\dfrac{3}{2} \cdot \dfrac{3}{4}$

2. $\dfrac{3x}{7} \cdot \dfrac{7}{6x}$

3. $\dfrac{x - 2}{y} \cdot \dfrac{y}{x + 2}$

4. $\dfrac{3}{4} \div \dfrac{4}{3}$

5. $\dfrac{5a}{b} \div \dfrac{a}{b}$

6. $\dfrac{x^2 y}{ab} \div \dfrac{2x^2}{ba}$

6.3   EXERCISES

REVIEW   *Perform each operation.*

1. $-2a^2(3a^3 - a^2)$

2. $(2t - 1)^2$

3. $(m^n + 2)(m^n - 2)$

4. $(3b^{-n} + c)(b^{-n} - c)$

VOCABULARY AND CONCEPTS   *Fill in the blanks.*

5. $\dfrac{a}{b} \cdot \dfrac{c}{d} = $ _____ $(b \neq 0, d \neq 0)$

6. $\dfrac{a}{b} \div \dfrac{c}{d} = $ _____ $(b \neq 0, c \neq 0, d \neq 0)$

7. The denominator of a fraction cannot be ___.

8. $\dfrac{a + 1}{a + 1} = $ ___ $(a \neq -1)$

PRACTICE   *Perform the operations and simplify.*

9. $\dfrac{3}{4} \cdot \dfrac{5}{3} \cdot \dfrac{8}{7}$

10. $-\dfrac{5}{6} \cdot \dfrac{3}{7} \cdot \dfrac{14}{25}$

11. $-\dfrac{6}{11} \div \dfrac{36}{55}$

12. $\dfrac{17}{12} \div \dfrac{34}{3}$

13. $\dfrac{x^2 y^2}{cd} \cdot \dfrac{c^{-2} d^2}{x}$

14. $\dfrac{a^{-2} b^2}{x^{-1} y} \cdot \dfrac{a^4 b^4}{x^2 y^3}$

15. $\dfrac{-x^2 y^{-2}}{x^{-1} y^{-3}} \div \dfrac{x^{-3} y^2}{x^4 y^{-1}}$

16. $\dfrac{(a^3)^2}{b^{-1}} \div \dfrac{(a^3)^{-2}}{b^{-1}}$

17. $\dfrac{x^2 + 2x + 1}{x} \cdot \dfrac{x^2 - x}{x^2 - 1}$

18. $\dfrac{a + 6}{a^2 - 16} \cdot \dfrac{3a - 12}{3a + 18}$

19. $\dfrac{2x^2 - x - 3}{x^2 - 1} \cdot \dfrac{x^2 + x - 2}{2x^2 + x - 6}$

20. $\dfrac{9x^2 + 3x - 20}{3x^2 - 7x + 4} \cdot \dfrac{3x^2 - 5x + 2}{9x^2 + 18x + 5}$

21. $\dfrac{x^2 - 16}{x^2 - 25} \div \dfrac{x + 4}{x - 5}$

22. $\dfrac{a^2 - 9}{a^2 - 49} \div \dfrac{a + 3}{a + 7}$

23. $\dfrac{a^2 + 2a - 35}{12x} \div \dfrac{ax - 3x}{a^2 + 4a - 21}$

24. $\dfrac{x^2 - 4}{2b - bx} \div \dfrac{x^2 + 4x + 4}{2b + bx}$

25. $\dfrac{3t^2 - t - 2}{6t^2 - 5t - 6} \cdot \dfrac{4t^2 - 9}{2t^2 + 5t + 3}$

26. $\dfrac{2p^2 - 5p - 3}{p^2 - 9} \cdot \dfrac{2p^2 + 5p - 3}{2p^2 + 5p + 2}$

27. $\dfrac{3n^2 + 5n - 2}{12n^2 - 13n + 3} \div \dfrac{n^2 + 3n + 2}{4n^2 + 5n - 6}$

28. $\dfrac{8y^2 - 14y - 15}{6y^2 - 11y - 10} \div \dfrac{4y^2 - 9y - 9}{3y^2 - 7y - 6}$

29. $(x + 1) \cdot \dfrac{1}{x^2 + 2x + 1}$

30. $\dfrac{x^2 - 4}{x} \div (x + 2)$

31. $(x^2 - x - 2) \cdot \dfrac{x^2 + 3x + 2}{x^2 - 4}$

32. $(2x^2 - 9x - 5) \cdot \dfrac{x}{2x^2 + x}$

33. $(2x^2 - 15x + 25) \div \dfrac{2x^2 - 3x - 5}{x + 1}$

34. $(x^2 - 6x + 9) \div \dfrac{x^2 - 9}{x + 3}$

35. $\dfrac{x^3 + y^3}{x^3 - y^3} \div \dfrac{x^2 - xy + y^2}{x^2 + xy + y^2}$

36. $\dfrac{x^2 - 6x + 9}{4 - x^2} \div \dfrac{x^2 - 9}{x^2 - 8x + 12}$

37. $\dfrac{m^2 - n^2}{2x^2 + 3x - 2} \cdot \dfrac{2x^2 + 5x - 3}{n^2 - m^2}$

38. $\dfrac{x^2 - y^2}{2x^2 + 2xy + x + y} \cdot \dfrac{2x^2 - 5x - 3}{yx - 3y - x^2 + 3x}$

39. $\dfrac{ax + ay + bx + by}{x^3 - 27} \cdot \dfrac{x^2 + 3x + 9}{xc + xd + yc + yd}$

40. $\dfrac{x^2 + 3x + yx + 3y}{x^2 - 9} \cdot \dfrac{x - 3}{x + 3}$

41. $\dfrac{x^2 - x - 6}{x^2 - 4} \cdot \dfrac{x^2 - x - 2}{9 - x^2}$

42. $\dfrac{2x^2 - 7x - 4}{20 - x - x^2} \div \dfrac{2x^2 - 9x - 5}{x^2 - 25}$

43. $\dfrac{2x^2 + 3xy + y^2}{y^2 - x^2} \div \dfrac{6x^2 + 5xy + y^2}{2x^2 - xy - y^2}$

44. $\dfrac{p^3 - q^3}{q^2 - p^2} \cdot \dfrac{q^2 + pq}{p^3 + p^2q + pq^2}$

45. $\dfrac{3x^2y^2}{6x^3y} \cdot \dfrac{-4x^7y^{-2}}{18x^{-2}y} \div \dfrac{36x}{18y^{-2}}$

46. $\dfrac{9ab^3}{7xy} \cdot \dfrac{14xy^2}{27z^3} \div \dfrac{18a^2b^2x}{3z^2}$

47. $(4x + 12) \cdot \dfrac{x^2}{2x - 6} \div \dfrac{2}{x - 3}$

48. $(4x^2 - 9) \div \dfrac{2x^2 + 5x + 3}{x + 2} \div (2x - 3)$

49. $\dfrac{2x^2 - 2x - 4}{x^2 + 2x - 8} \cdot \dfrac{3x^2 + 15x}{x + 1} \div \dfrac{4x^2 - 100}{x^2 - x - 20}$

50. $\dfrac{6a^2 - 7a - 3}{a^2 - 1} \div \dfrac{4a^2 - 12a + 9}{a^2 - 1} \cdot \dfrac{2a^2 - a - 3}{3a^2 - 2a - 1}$

51. $\dfrac{2t^2 + 5t + 2}{t^2 - 4t + 16} \div \dfrac{t + 2}{t^3 + 64} \div \dfrac{2t^3 + 9t^2 + 4t}{t + 1}$

52. $\dfrac{a^6 - b^6}{a^4 - a^3b} \cdot \dfrac{a^3}{a^4 + a^2b^2 + b^4} \div \dfrac{1}{a}$

53. $\dfrac{x^4 - 3x^2 - 4}{x^4 - 1} \cdot \dfrac{x^2 + 3x + 2}{x^2 + 4x + 4}$

54. $\dfrac{x^3 + 2x^2 + 4x + 8}{y^2 - 1} \cdot \dfrac{y^2 + 2y + 1}{x^4 - 16}$

55. $(x^2 - x - 6) \div (x - 3) \div (x - 2)$

56. $(x^2 - x - 6) \div [(x - 3) \div (x - 2)]$

57. $\dfrac{3x^2 - 2x}{3x + 2} \div (3x - 2) \div \dfrac{3x}{3x - 3}$

58. $(2x^2 - 3x - 2) \div \dfrac{2x^2 - x - 1}{x - 2} \div (x - 1)$

59. $\dfrac{2x^2 + 5x - 3}{x^2 + 2x - 3} \div \left(\dfrac{x^2 + 2x - 35}{x^2 - 6x + 5} \div \dfrac{x^2 - 9x + 14}{2x^2 - 5x + 2} \right)$

60. $\dfrac{x^2 - 4}{x^2 - x - 6} \div \left(\dfrac{x^2 - x - 2}{x^2 - 8x + 15} \cdot \dfrac{x^2 - 3x - 10}{x^2 + 3x + 2} \right)$

61. $\dfrac{x^2 - x - 12}{x^2 + x - 2} \div \dfrac{x^2 - 6x + 8}{x^2 - 3x - 10} \cdot \dfrac{x^2 - 3x + 2}{x^2 - 2x - 15}$

62. $\dfrac{4x^2 - 10x + 6}{x^4 - 3x^3} \div \dfrac{2x - 3}{2x^3} \cdot \dfrac{x - 3}{2x - 2}$

APPLICATIONS

63. Area of a triangle Find an expression that represents the area of the triangle.

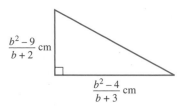

64. Storage capacity Find the capacity of the trunk shown in the illustration.

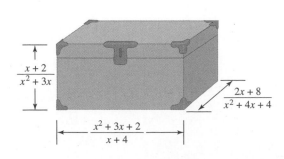

65. Distance, rate, and time Complete the following table.

Rate (mph)	Time (hr)	Distance (m)
$\dfrac{k^2 - k - 6}{k - 4}$	$\dfrac{k^2 - 16}{k^2 - 2k - 3}$	

66. Lab experiments The following table shows data obtained from a physics experiment in which k and k_1 are constants. Complete the table.

Trial	Rate (m/sec)	Time (sec)	Distance (m)
1	$\dfrac{k_1^2 + 3k_1 + 2}{k_1 - 3}$	$\dfrac{k_1^2 - 3k_1}{k_1 + 1}$	
2	$\dfrac{k_2^2 + 6k_2 + 5}{k_2 + 1}$	$k_2 + 6$	$k_2^2 + 11k_2 + 30$

WRITING

67. Explain how to multiply two rational expressions.

68. Explain how to divide one rational expression by another.

SOMETHING TO THINK ABOUT *Insert either a multiplication or a division symbol in each box to make a true statement.*

69. $\dfrac{x^2}{y} \; \blacksquare \; \dfrac{x}{y^2} \; \blacksquare \; \dfrac{x^2}{y^2} = \dfrac{x^3}{y}$ **70.** $\dfrac{x^2}{y} \; \blacksquare \; \dfrac{x}{y^2} \; \blacksquare \; \dfrac{x^2}{y^2} = \dfrac{y^3}{x}$

6.4 Adding and Subtracting Rational Expressions

■ **Adding and Subtracting Rational Expressions with Like Denominators** ■ **Adding and Subtracting Rational Expressions with Unlike Denominators** ■ **Finding the Least Common Denominator** ■ **Mixed Operations**

Getting Ready *Tell whether the fractions are equal.*

1. $\dfrac{3}{5}, \dfrac{27}{45}$ 2. $\dfrac{3}{7}, \dfrac{18}{40}$ 3. $\dfrac{8}{13}, \dfrac{40}{65}$

4. $\dfrac{7}{25}, \dfrac{42}{150}$ 5. $\dfrac{23}{32}, \dfrac{46}{66}$ 6. $\dfrac{8}{9}, \dfrac{64}{72}$

We now discuss how to add and subtract rational expressions. After mastering these skills, we will simplify expressions that involve more than one operation.

Adding and Subtracting Rational Expressions with Like Denominators

Rational expressions with like denominators are added and subtracted according to the following rules.

Adding and Subtracting Rational Expressions

If there are no divisions by 0, then

$$\frac{a}{b} + \frac{c}{b} = \frac{a + c}{b} \qquad \text{and} \qquad \frac{a}{b} - \frac{c}{b} = \frac{a - c}{b}$$

In words, *we add (or subtract) rational expressions with like denominators by adding (or subtracting) the numerators and keeping the common denominator.* Whenever possible, we should simplify the result.

EXAMPLE 1 Simplify: **a.** $\dfrac{17}{22} + \dfrac{13}{22}$, **b.** $\dfrac{3}{2x} + \dfrac{7}{2x}$, and **c.** $\dfrac{4x}{x + 2} - \dfrac{7x}{x + 2}$.

Solution **a.** $\dfrac{17}{22} + \dfrac{13}{22} = \dfrac{17 + 13}{22}$ **b.** $\dfrac{3}{2x} + \dfrac{7}{2x} = \dfrac{3 + 7}{2x}$

$$= \frac{30}{22} \qquad\qquad\qquad\qquad\qquad = \frac{10}{2x}$$

$$= \frac{15 \cdot \cancel{2}}{11 \cdot \cancel{2}} \qquad\qquad\qquad\qquad = \frac{\cancel{2} \cdot 5}{\cancel{2} \cdot x}$$

$$= \frac{15}{11} \qquad\qquad\qquad\qquad\qquad = \frac{5}{x}$$

c. $\dfrac{4x}{x + 2} - \dfrac{7x}{x + 2} = \dfrac{4x - 7x}{x + 2}$

$$= \frac{-3x}{x + 2}$$

Self Check Simplify: **a.** $\dfrac{5}{3a} - \dfrac{2}{3a}$ and **b.** $\dfrac{3a}{a - 2} + \dfrac{2a}{a - 2}$. ∎

Accent on Technology   CHECKING ALGEBRA

We can check the subtraction in part c of Example 1 by graphing the rational functions $f(x) = \frac{4x}{x + 2} - \frac{7x}{x + 2}$, shown in Figure 6-9(a), and $g(x) = \frac{-3x}{x + 2}$, shown in Figure 6-9(b), and observing that the graphs are the same. Note that -2 is not in the domain of either function.

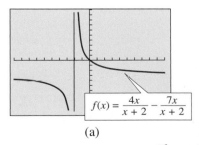

(a)

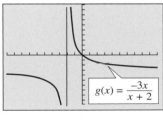

(b)

Figure 6-9

Adding and Subtracting Rational Expressions with Unlike Denominators

To add or subtract rational expressions with unlike denominators, we change them into expressions with a common denominator. When the denominators are negatives (opposites), we can multiply one of them by 1, written in the form $\frac{-1}{-1}$, to get a common denominator.

EXAMPLE 2 Add: $\dfrac{x}{x-y} + \dfrac{y}{y-x}$.

Solution
$$\frac{x}{x-y} + \frac{y}{y-x} = \frac{x}{x-y} + \left(\frac{-1}{-1}\right)\frac{y}{y-x} \qquad \frac{-1}{-1} = 1; \text{ multiplying a fraction by 1 does not change its value.}$$

$$= \frac{x}{x-y} + \frac{-y}{-y+x} \qquad \text{Multiply.}$$

$$= \frac{x}{x-y} + \frac{-y}{x-y} \qquad -y+x = x-y.$$

$$= \frac{x-y}{x-y} \qquad \text{Add the numerators and keep the common denominator.}$$

$$= 1 \qquad \text{Simplify.}$$

Self Check Add: $\dfrac{2a}{a-b} + \dfrac{b}{b-a}$.

When the denominators of two or more rational expressions are different, we often have to multiply one or more of them by 1, written in some appropriate form, to get a common denominator.

EXAMPLE 3 Simplify: **a.** $\dfrac{2}{3} - \dfrac{3}{2}$ and **b.** $\dfrac{3}{x} + \dfrac{4}{y}$.

Solution **a.** $\dfrac{2}{3} - \dfrac{3}{2} = \dfrac{2}{3} \cdot 1 - \dfrac{3}{2} \cdot 1$ \qquad Multiply each fraction by 1.

$$= \frac{2}{3} \cdot \frac{2}{2} - \frac{3}{2} \cdot \frac{3}{3} \qquad \frac{2}{2} = 1 \text{ and } \frac{3}{3} = 1.$$

$$= \frac{2 \cdot 2}{3 \cdot 2} - \frac{3 \cdot 3}{2 \cdot 3} \qquad \text{Multiply the fractions by multiplying their numerators and their denominators.}$$

$$= \frac{4}{6} - \frac{9}{6} \qquad \text{Simplify.}$$

$$= \frac{-5}{6} \qquad \text{Subtract the numerators and keep the common denominator.}$$

$$= -\frac{5}{6}$$

b. $\dfrac{3}{x} + \dfrac{4}{y} = \dfrac{3}{x} \cdot \mathbf{1} + \dfrac{4}{y} \cdot \mathbf{1}$ Multiply each fraction by 1.

$\phantom{\textbf{b.} \dfrac{3}{x} + \dfrac{4}{y}} = \dfrac{3}{x} \cdot \dfrac{\mathbf{y}}{\mathbf{y}} + \dfrac{4}{y} \cdot \dfrac{\mathbf{x}}{\mathbf{x}}$ $\dfrac{y}{y} = 1$ $(y \neq 0)$ and $\dfrac{x}{x} = 1$ $(x \neq 0)$.

$\phantom{\textbf{b.} \dfrac{3}{x} + \dfrac{4}{y}} = \dfrac{3y}{xy} + \dfrac{4x}{xy}$ Multiply the fractions by multiplying their numerators and their denominators.

$\phantom{\textbf{b.} \dfrac{3}{x} + \dfrac{4}{y}} = \dfrac{3y + 4x}{xy}$ Add the numerators and keep the common denominator.

Self Check   Simplify: $\dfrac{5}{a} - \dfrac{7}{b}$.   ∎

EXAMPLE 4   Simplify: **a.** $3 + \dfrac{7}{x-2}$ and   **b.** $\dfrac{4x}{x+2} - \dfrac{7x}{x-2}$.

Solution   **a.** $3 + \dfrac{7}{x-2} = \dfrac{3}{1} + \dfrac{7}{x-2}$ $3 = \dfrac{3}{1}$.

$\phantom{\textbf{a.} 3 + \dfrac{7}{x-2}} = \dfrac{3(x-2)}{1(x-2)} + \dfrac{7}{x-2}$ $\dfrac{x-2}{x-2} = 1$.

$\phantom{\textbf{a.} 3 + \dfrac{7}{x-2}} = \dfrac{3x-6}{x-2} + \dfrac{7}{x-2}$ Remove parentheses.

$\phantom{\textbf{a.} 3 + \dfrac{7}{x-2}} = \dfrac{3x-6+7}{x-2}$ Add the numerators and keep the common denominator.

$\phantom{\textbf{a.} 3 + \dfrac{7}{x-2}} = \dfrac{3x+1}{x-2}$ Simplify.

b. $\dfrac{4x}{x+2} - \dfrac{7x}{x-2}$

$\phantom{\textbf{b.}} = \dfrac{4x(x-2)}{(x+2)(x-2)} - \dfrac{(x+2)7x}{(x+2)(x-2)}$ $\dfrac{x-2}{x-2} = 1$ and $\dfrac{x+2}{x+2} = 1$.

$\phantom{\textbf{b.}} = \dfrac{(4x^2 - 8x) - (7x^2 + 14x)}{(x+2)(x-2)}$ Multiply in the numerators, subtract the numerators, and keep the common denominator.

$\phantom{\textbf{b.}} = \dfrac{4x^2 - 8x - 7x^2 - 14x}{(x+2)(x-2)}$ Use the distributive property to remove parentheses in the numerator.

$\phantom{\textbf{b.}} = \dfrac{-3x^2 - 22x}{(x+2)(x-2)}$ Simplify the numerator.

$\phantom{\textbf{b.}} = \dfrac{-x(3x+22)}{(x+2)(x-2)}$ Factor the numerator to see whether the fraction can be simplified.

Comment   The $-$ sign between the fractions in Step 1 of part b applies to both terms of $7x^2 + 14x$.

Self Check   Simplify: $\dfrac{3a}{a+3} - \dfrac{5a}{a-3}$.   ∎

Finding the Least Common Denominator

When adding fractions with unlike denominators, we change the fractions into fractions having the smallest common denominator possible, called the **least** (or lowest) **common denominator (LCD).**

Suppose we have the fractions $\frac{1}{12}$, $\frac{1}{20}$, and $\frac{1}{35}$. To find the LCD of these fractions, we first find the prime factorizations of each denominator.

$$12 = 4 \cdot 3 = 2^2 \cdot 3$$
$$20 = 4 \cdot 5 = 2^2 \cdot 5$$
$$35 = 5 \cdot 7$$

Since the LCD is the smallest number that can be exactly divided by 12, 20, and 35, it must contain factors of 2^2, 3, 5, and 7.

$$LCD = 2^2 \cdot 3 \cdot 5 \cdot 7 = 420$$

To find the least common denominator of several rational expressions, we follow these steps.

Finding the LCD

1. Prime-factor the denominator of each rational expression.
2. List the different factors of each denominator.
3. Write each factor found in Step 2 to the highest power that occurs in any one factorization.
4. The LCD is the product of the factors to their highest powers found in Step 3.

EXAMPLE 5   Find the LCD of $\dfrac{1}{x^2 + 7x + 6}$, $\dfrac{3}{x^2 - 36}$, and $\dfrac{5}{x^2 + 12x + 36}$.

Solution   We prime-factor each denominator:

$$x^2 + 7x + 6 = (x + 6)(x + 1)$$
$$x^2 - 36 = (x + 6)(x - 6)$$
$$x^2 + 12x + 36 = (x + 6)(x + 6) = (x + 6)^2$$

and list the individual factors:

$$x + 6, \quad x + 1, \quad \text{and} \quad x - 6$$

To find the LCD, we use the highest power of each of these factors:

$$LCD = (x + 6)^2(x + 1)(x - 6)$$

Self Check   Find the LCD of $\dfrac{1}{a^2 - 25}$ and $\dfrac{7}{a^2 + 7a + 10}$.

EXAMPLE 6   Simplify: $\dfrac{x}{x^2 - 2x + 1} + \dfrac{3}{x^2 - 1}$.

Solution   We prime-factor each denominator to find the LCD:

$$x^2 - 2x + 1 = (x - 1)(x - 1) = (x - 1)^2$$
$$x^2 - 1 = (x + 1)(x - 1)$$

The LCD is $(x - 1)^2(x + 1)$.

We now write the rational expressions with their denominators in factored form and change them into expressions with an LCD of $(x - 1)^2(x + 1)$. Then we add.

$$\frac{x}{x^2 - 2x + 1} + \frac{3}{x^2 - 1}$$

$$= \frac{x}{(x - 1)(x - 1)} + \frac{3}{(x + 1)(x - 1)}$$

$$= \frac{x(x + 1)}{(x - 1)(x - 1)(x + 1)} + \frac{3(x - 1)}{(x + 1)(x - 1)(x - 1)} \qquad \frac{x+1}{x+1} = 1, \frac{x-1}{x-1} = 1.$$

$$= \frac{x^2 + x + 3x - 3}{(x - 1)(x - 1)(x + 1)}$$

$$= \frac{x^2 + 4x - 3}{(x - 1)^2(x + 1)} \qquad \text{This result does not simplify.}$$

Self Check   Simplify: $\dfrac{a}{a^2 - 4a + 4} - \dfrac{2}{a^2 - 4}$.   ∎

EXAMPLE 7   Simplify: $\dfrac{3x}{x - 1} - \dfrac{2x^2 + 3x - 2}{(x + 1)(x - 1)}$.

Solution   We write each rational expression in a form having the LCD of $(x + 1)(x - 1)$, remove the resulting parentheses in the first numerator, do the subtraction, and simplify.

(1)
$$\frac{3x}{x - 1} - \frac{2x^2 + 3x - 2}{(x + 1)(x - 1)} = \frac{(x + 1)3x}{(x + 1)(x - 1)} - \frac{2x^2 + 3x - 2}{(x + 1)(x - 1)} \qquad \frac{x+1}{x+1} = 1.$$

$$= \frac{3x^2 + 3x}{(x + 1)(x - 1)} - \frac{2x^2 + 3x - 2}{(x + 1)(x - 1)}$$

$$= \frac{3x^2 + 3x - (2x^2 + 3x - 2)}{(x + 1)(x - 1)}$$

$$= \frac{3x^2 + 3x - 2x^2 - 3x + 2}{(x + 1)(x - 1)}$$

$$= \frac{x^2 + 2}{(x + 1)(x - 1)}$$

Comment   The $-$ sign between the rational expressions in Equation 1 affects every term of the numerator of $2x^2 + 3x - 2$. Whenever we subtract one rational expression from another, we must remember to subtract each term of the numerator in the second expression.

Self Check   Simplify: $\dfrac{2a}{a + 2} - \dfrac{a^2 - 4a + 4}{a^2 + a - 2}$.   ∎

Mixed Operations

EXAMPLE 8   Simplify: $\dfrac{2x}{x^2 - 4} - \dfrac{1}{x^2 - 3x + 2} + \dfrac{x + 1}{x^2 + x - 2}$.

Solution   We factor each denominator to find the LCD:

$$\text{LCD} = (x + 2)(x - 2)(x - 1)$$

We then write each rational expression as one with the LCD as its denominator and do the subtraction and addition.

$$\dfrac{2x}{x^2 - 4} - \dfrac{1}{x^2 - 3x + 2} + \dfrac{x + 1}{x^2 + x - 2}$$

$$= \dfrac{2x}{(x - 2)(x + 2)} - \dfrac{1}{(x - 2)(x - 1)} + \dfrac{x + 1}{(x - 1)(x + 2)}$$

$$= \dfrac{2x(x - 1)}{(x - 2)(x + 2)(x - 1)} - \dfrac{1(x + 2)}{(x - 2)(x - 1)(x + 2)} + \dfrac{(x + 1)(x - 2)}{(x - 1)(x + 2)(x - 2)}$$

$$= \dfrac{2x(x - 1) - 1(x + 2) + (x + 1)(x - 2)}{(x + 2)(x - 2)(x - 1)}$$

$$= \dfrac{2x^2 - 2x - x - 2 + x^2 - x - 2}{(x + 2)(x - 2)(x - 1)}$$

$$= \dfrac{3x^2 - 4x - 4}{(x + 2)(x - 2)(x - 1)}$$

Since the final result simplifies, we have

$$\dfrac{2x}{x^2 - 4} - \dfrac{1}{x^2 - 3x + 2} + \dfrac{x + 1}{x^2 + x - 2}$$

$$= \dfrac{3x^2 - 4x - 4}{(x + 2)(x - 2)(x - 1)}$$

$$= \dfrac{(3x + 2)(x - 2)}{(x + 2)(x - 2)(x - 1)}$$ Factor the numerator and divide out the $x - 2; \dfrac{x - 2}{x - 2} = 1.$

$$= \dfrac{3x + 2}{(x + 2)(x - 1)}$$ ∎

EXAMPLE 9   Simplify: $\left(\dfrac{x^2}{x - 2} + \dfrac{4}{2 - x} \right)^2$.

Solution   We do the addition within the parentheses. Since the denominators are negatives (opposites) of each other, we can write the rational expressions with a common denominator by multiplying both the numerator and denominator of $\frac{4}{2 - x}$ by -1. We can then add, simplify, and square the result.

$$\left(\frac{x^2}{x-2}+\frac{4}{2-x}\right)^2 = \left[\frac{x^2}{x-2}+\frac{(-1)4}{(-1)(2-x)}\right]^2 \quad \frac{-1}{-1}=1.$$

$$= \left[\frac{x^2}{x-2}+\frac{-4}{x-2}\right]^2$$

$$= \left[\frac{x^2-4}{x-2}\right]^2$$

$$= \left[\frac{(x+2)(x-2)}{x-2}\right]^2$$

$$= (x+2)^2$$

$$= x^2+4x+4$$

Self Check Simplify: $\left(\dfrac{a^2}{a-3}-\dfrac{9-6a}{3-a}\right)^2$. ∎

Self Check Answers

1. a. $\dfrac{1}{a}$, **b.** $\dfrac{5a}{a-2}$ **2.** $\dfrac{2a-b}{a-b}$ **3.** $\dfrac{5b-7a}{ab}$ **4.** $\dfrac{-2a(a+12)}{(a+3)(a-3)}$ **5.** $(a+5)(a+2)(a-5)$

6. $\dfrac{a^2+4}{(a-2)^2(a+2)}$ **7.** $\dfrac{a^2+2a-4}{(a+2)(a-1)}$ **9.** a^2-6a+9

Orals *Add or subtract the rational expressions and simplify the result, if possible.*

1. $\dfrac{x}{2}+\dfrac{x}{2}$ **2.** $\dfrac{3a}{4}-\dfrac{a}{4}$ **3.** $\dfrac{x}{x+2}+\dfrac{2}{x+2}$

4. $\dfrac{2a}{a+4}-\dfrac{a-4}{a+4}$ **5.** $\dfrac{2x}{3}+\dfrac{x}{2}$ **6.** $\dfrac{5}{x}-\dfrac{3}{y}$

6.4 EXERCISES 🔊

REVIEW *Graph each interval on a number line.*

1. $(-1, 4]$ **2.** $(-\infty, -5] \cup [4, \infty)$

Solve each formula for the indicated letter.

3. $P = 2l + 2w$; for w

4. $S = \dfrac{a-lr}{1-r}$; for a

VOCABULARY AND CONCEPTS *Fill in the blanks.*

5. $\dfrac{a}{b}+\dfrac{c}{b} = $ _____

6. $\dfrac{a}{b}-\dfrac{c}{b} = $ _____

7. To subtract fractions with like denominators, we _____ the numerators and _____ the common denominator.

8. To add fractions with like denominators, we _____ the numerators and keep the _____ denominator.

9. The abbreviation for the least common denominator is _____.

10. To find the LCD of several fractions, we _____ each denominator and use each factor to the _____ power that it appears in any factorization.

PRACTICE *Perform the operations and simplify the result when possible.*

11. $\dfrac{3}{4}+\dfrac{7}{4}$ **12.** $\dfrac{5}{11}+\dfrac{2}{11}$

13. $\dfrac{10}{33} - \dfrac{21}{33}$

14. $\dfrac{8}{15} - \dfrac{2}{15}$

15. $\dfrac{3}{4y} + \dfrac{8}{4y}$

16. $\dfrac{5}{3z^2} - \dfrac{6}{3z^2}$

17. $\dfrac{3}{a+b} - \dfrac{a}{a+b}$

18. $\dfrac{x}{x+4} + \dfrac{5}{x+4}$

19. $\dfrac{3x}{2x+2} + \dfrac{x+4}{2x+2}$

20. $\dfrac{4y}{y-4} - \dfrac{16}{y-4}$

21. $\dfrac{3x}{x-3} - \dfrac{9}{x-3}$

22. $\dfrac{9x}{x-y} - \dfrac{9y}{x-y}$

23. $\dfrac{5x}{x+1} + \dfrac{3}{x+1} - \dfrac{2x}{x+1}$

24. $\dfrac{4}{a+4} - \dfrac{2a}{a+4} + \dfrac{3a}{a+4}$

25. $\dfrac{3(x^2+x)}{x^2-5x+6} + \dfrac{-3(x^2-x)}{x^2-5x+6}$

26. $\dfrac{2x+4}{x^2+13x+12} - \dfrac{x+3}{x^2+13x+12}$

The denominators of several fractions are given. Find the LCD.

27. 8, 12, 18

28. 10, 15, 28

29. $x^2 + 3x, x^2 - 9$

30. $3y^2 - 6y, 3y(y-4)$

31. $x^3 + 27, x^2 + 6x + 9$

32. $x^3 - 8, x^2 - 4x + 4$

33. $2x^2 + 5x + 3, 4x^2 + 12x + 9, x^2 + 2x + 1$

34. $2x^2 + 5x + 3, 4x^2 + 12x + 9, 4x + 6$

Perform the operations and simplify the result when possible.

35. $\dfrac{1}{2} + \dfrac{1}{3}$

36. $\dfrac{5}{6} + \dfrac{2}{7}$

37. $\dfrac{7}{15} - \dfrac{17}{25}$

38. $\dfrac{8}{9} - \dfrac{5}{12}$

39. $\dfrac{a}{2} + \dfrac{2a}{5}$

40. $\dfrac{b}{6} + \dfrac{3a}{4}$

41. $\dfrac{3a}{2} - \dfrac{4b}{7}$

42. $\dfrac{2m}{3} - \dfrac{4n}{5}$

43. $\dfrac{3}{4x} + \dfrac{2}{3x}$

44. $\dfrac{2}{5a} + \dfrac{3}{2b}$

45. $\dfrac{3a}{2b} - \dfrac{2b}{3a}$

46. $\dfrac{5m}{2n} - \dfrac{3n}{4m}$

47. $\dfrac{a+b}{3} + \dfrac{a-b}{7}$

48. $\dfrac{x-y}{2} + \dfrac{x+y}{3}$

49. $\dfrac{3}{x+2} + \dfrac{5}{x-4}$

50. $\dfrac{2}{a+4} - \dfrac{6}{a+3}$

51. $\dfrac{x+2}{x+5} - \dfrac{x-3}{x+7}$

52. $\dfrac{7}{x+3} + \dfrac{4x}{x+6}$

53. $x + \dfrac{1}{x}$

54. $2 - \dfrac{1}{x+1}$

55. $\dfrac{x+8}{x-3} - \dfrac{x-14}{3-x}$

56. $\dfrac{3-x}{2-x} + \dfrac{x-1}{x-2}$

57. $\dfrac{2a+1}{3a+2} - \dfrac{a-4}{2-3a}$

58. $\dfrac{4}{x-2} + \dfrac{5}{4-x^2}$

59. $\dfrac{x}{x^2+5x+6} + \dfrac{x}{x^2-4}$

60. $\dfrac{x}{3x^2-2x-1} + \dfrac{4}{3x^2+10x+3}$

61. $\dfrac{4}{x^2-2x-3} - \dfrac{x}{3x^2-7x-6}$

62. $\dfrac{2a}{a^2-2a-8} + \dfrac{3}{a^2-5a+4}$

63. $\dfrac{8}{x^2-9} + \dfrac{2}{x-3} - \dfrac{6}{x}$

64. $\dfrac{x}{x^2-4} - \dfrac{x}{x+2} + \dfrac{2}{x}$

65. $\dfrac{x}{x+1} - \dfrac{x}{1-x^2} + \dfrac{1}{x}$

66. $\dfrac{y}{y-2} - \dfrac{2}{y+2} - \dfrac{-8}{4-y^2}$

67. $2x + 3 + \dfrac{1}{x+1}$

68. $x + 1 + \dfrac{1}{x - 1}$

69. $1 + x - \dfrac{x}{x - 5}$

70. $2 - x + \dfrac{3}{x - 9}$

71. $\dfrac{3x}{x - 1} - 2x - x^2$

72. $\dfrac{23}{x - 1} + 4x - 5x^2$

73. $\dfrac{y + 4}{y^2 + 7y + 12} - \dfrac{y - 4}{y + 3} + \dfrac{47}{y + 4}$

74. $\dfrac{x + 3}{2x^2 - 5x + 2} - \dfrac{3x - 1}{x^2 - x - 2}$

75. $\dfrac{3}{x + 1} - \dfrac{2}{x - 1} + \dfrac{x + 3}{x^2 - 1}$

76. $\dfrac{2}{x - 2} + \dfrac{3}{x + 2} - \dfrac{x - 1}{x^2 - 4}$

77. $\dfrac{x - 2}{x^2 - 3x} + \dfrac{2x - 1}{x^2 + 3x} - \dfrac{2}{x^2 - 9}$

78. $\dfrac{2}{x - 1} - \dfrac{2x}{x^2 - 1} - \dfrac{x}{x^2 + 2x + 1}$

79. $\dfrac{5}{x^2 - 25} - \dfrac{3}{2x^2 - 9x - 5} + 1$

80. $\dfrac{3x}{2x - 1} + \dfrac{x + 1}{3x + 2} + \dfrac{2x}{6x^3 + x^2 - 2x}$

81. $\dfrac{3x}{x - 3} + \dfrac{4}{x - 2} - \dfrac{5x}{x^3 - 5x^2 + 6x}$

82. $\dfrac{2x - 1}{x^2 + x - 6} - \dfrac{3x - 5}{x^2 - 2x - 15} + \dfrac{2x - 3}{x^2 - 7x + 10}$

83. $2 + \dfrac{4a}{a^2 - 1} - \dfrac{2}{a + 1}$

84. $\dfrac{a}{a - 1} - \dfrac{a + 1}{2a - 2} + a$

85. $\dfrac{x + 5}{2x^2 - 2} + \dfrac{x}{2x + 2} - \dfrac{3}{x - 1}$

86. $\dfrac{a}{2 - a} + \dfrac{3}{a - 2} - \dfrac{3a - 2}{a^2 - 4}$

87. $\dfrac{a}{a - b} + \dfrac{b}{a + b} + \dfrac{a^2 + b^2}{b^2 - a^2}$

88. $\dfrac{1}{x + y} - \dfrac{1}{x - y} - \dfrac{2y}{y^2 - x^2}$

89. $\dfrac{7n^2}{m - n} + \dfrac{3m}{n - m} - \dfrac{3m^2 - n}{m^2 - 2mn + n^2}$

90. $\dfrac{3b}{2a - b} + \dfrac{2a - 1}{b - 2a} - \dfrac{3a^2 + b}{b^2 - 4ab + 4a^2}$

91. $\dfrac{m + 1}{m^2 + 2m + 1} + \dfrac{m - 1}{m^2 - 2m + 1} + \dfrac{2}{m^2 - 1}$

(*Hint:* Think about this before finding the LCD.)

92. $\dfrac{a + 2}{a^2 + 3a + 2} + \dfrac{a - 1}{a^2 - 1} + \dfrac{3}{a + 1}$

93. $\left(\dfrac{1}{x - 1} + \dfrac{1}{1 - x} \right)^2$

94. $\left(\dfrac{1}{a - 1} - \dfrac{1}{1 - a} \right)^2$

95. $\left(\dfrac{x}{x - 3} + \dfrac{3}{3 - x} \right)^3$

96. $\left(\dfrac{2y}{y + 4} + \dfrac{8}{y + 4} \right)^3$

97. Show that $\dfrac{a}{b} + \dfrac{c}{d} = \dfrac{ad + bc}{bd}$.

98. Show that $\dfrac{a}{b} - \dfrac{c}{d} = \dfrac{ad - bc}{bd}$.

APPLICATIONS

99. Find an expression that represents the total height of the monument.

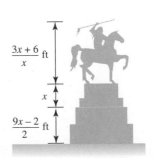

100. Perimeter of a rectangle Write an expression that represents the perimeter of the rectangle.

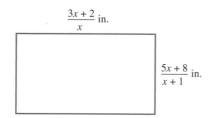

$\dfrac{3x + 2}{x}$ in.

$\dfrac{5x + 8}{x + 1}$ in.

101. Find an expression that represents the width of the ring.

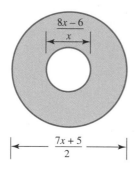

$\dfrac{8x - 6}{x}$

$\dfrac{7x + 5}{2}$

102. Work rates After working together on a common job, one worker completes $\dfrac{x + 2}{4}$ of the work and another completes $\dfrac{x + 3}{6}$ of the work. How much more did the first worker accomplish than the second?

WRITING

103. Explain how to find the least common denominator.

104. Explain how to add two fractions.

SOMETHING TO THINK ABOUT

105. Find the error:

$$\frac{8x + 2}{5} - \frac{3x + 8}{5}$$
$$= \frac{8x + 2 - 3x + 8}{5}$$
$$= \frac{5x + 10}{5}$$
$$= x + 2$$

106. Find the error:

$$\frac{(x + y)^2}{2} + \frac{(x - y)^2}{3}$$
$$= \frac{3 \cdot (x + y)^2}{3 \cdot 2} + \frac{2 \cdot (x - y)^2}{2 \cdot 3}$$
$$= \frac{3x^2 + 3y^2 + 2x^2 - 2y^2}{6}$$
$$= \frac{5x^2 + y^2}{6}$$

6.5	**Complex Fractions**

■ **Simplifying Complex Fractions**

Getting Ready *Use the distributive property to remove parentheses and simplify.*

1. $2\left(1 + \dfrac{1}{2}\right)$ **2.** $12\left(\dfrac{1}{6} - 3\right)$ **3.** $a\left(\dfrac{3}{a} + 2\right)$ **4.** $b\left(\dfrac{5}{b} - 2\right)$

In this section, we will consider **complex fractions,** fractions that have a fraction in their numerator or their denominator. Examples of complex fractions are

$$\frac{\dfrac{3a}{b}}{\dfrac{6ac}{b^2}}, \qquad \frac{\dfrac{2}{x}+1}{3+x}, \qquad \text{and} \qquad \frac{\dfrac{1}{x}+\dfrac{1}{y}}{\dfrac{1}{x}-\dfrac{1}{y}}$$

Simplifying Complex Fractions

We can use two methods to simplify the complex fraction

$$\frac{\dfrac{3a}{b}}{\dfrac{6ac}{b^2}}$$

In one method, we eliminate the rational expressions in the numerator and denominator by writing the complex fraction as a division and using the division rule for fractions:

$$\frac{\dfrac{3a}{b}}{\dfrac{6ac}{b^2}} = \frac{3a}{b} \div \frac{6ac}{b^2}$$

$$= \frac{3a}{b} \cdot \frac{b^2}{6ac} \qquad \text{Invert the divisor and multiply.}$$

$$= \frac{b}{2c} \qquad \text{Multiply and simplify.}$$

In the other method, we eliminate the rational expressions in the numerator and denominator by multiplying the expression by 1, written in the form $\frac{b^2}{b^2}$. We use $\frac{b^2}{b^2}$, because b^2 is the LCD of $\frac{3a}{b}$ and $\frac{6ac}{b^2}$.

$$\frac{\dfrac{3a}{b}}{\dfrac{6ac}{b^2}} = \frac{\boldsymbol{b^2} \cdot \dfrac{3a}{b}}{\boldsymbol{b^2} \cdot \dfrac{6ac}{b^2}} \qquad \frac{b^2}{b^2} = 1.$$

$$= \frac{\dfrac{3ab^2}{b}}{\dfrac{6acb^2}{b^2}}$$

$$= \frac{3ab}{6ac} \qquad \text{Simplify the fractions in the numerator and denominator.}$$

$$= \frac{b}{2c} \qquad \text{Divide out the common factor of } 3a.$$

With either method, the result is the same.

EXAMPLE 1 Simplify: $\dfrac{\frac{2}{x}+1}{3+x}$.

Method 1 We add in the numerator and proceed as follows:

$$\frac{\frac{2}{x}+1}{3+x}=\frac{\frac{2}{x}+\frac{x}{x}}{\frac{3+x}{1}} \qquad \text{Write 1 as }\frac{x}{x}\text{ and }3+x\text{ as }\frac{3+x}{1}.$$

$$=\frac{\frac{2+x}{x}}{\frac{3+x}{1}} \qquad \text{Add }\frac{2}{x}\text{ and }\frac{x}{x}\text{ to get }\frac{2+x}{x}.$$

$$=\frac{2+x}{x}\div\frac{3+x}{1} \qquad \text{Write the complex fraction as a division.}$$

$$=\frac{2+x}{x}\cdot\frac{1}{3+x} \qquad \text{Invert the divisor and multiply.}$$

$$=\frac{2+x}{x^2+3x} \qquad \text{After noting that there are no common factors, multiply the numerators and multiply the denominators.}$$

Method 2 To eliminate the denominator of x, we multiply the numerator and the denominator by x.

$$\frac{\frac{2}{x}+1}{3+x}=\frac{x\left(\frac{2}{x}+1\right)}{x(3+x)} \qquad \frac{x}{x}=1.$$

$$=\frac{x\cdot\frac{2}{x}+x\cdot1}{x\cdot3+x\cdot x} \qquad \text{Use the distributive property to remove parentheses.}$$

$$=\frac{2+x}{x^2+3x} \qquad \text{Simplify.}$$

Self Check Simplify: $\dfrac{\frac{3}{a}+2}{2+a}$.

Accent on Technology   **CHECKING ALGEBRA**

We can check the simplification in Example 1 by graphing the functions

$$f(x)=\frac{\frac{2}{x}+1}{3+x}$$

shown in Figure 6-10(a), and   *(continued)*

$$g(x) = \frac{2 + x}{x^2 + 3x}$$

shown in Figure 6-10(b), and observing that the graphs are the same. Each graph has window settings of $[-5, 3]$ for x and $[-6, 6]$ for y.

$f(x) = \dfrac{\frac{2}{x} + 1}{3 + x}$
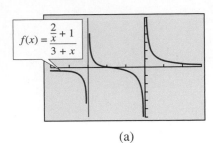

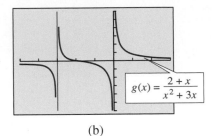
$g(x) = \dfrac{2 + x}{x^2 + 3x}$

(a)   (b)

Figure 6-10

EXAMPLE 2   Simplify: $\dfrac{\dfrac{1}{x} + \dfrac{1}{y}}{\dfrac{1}{x} - \dfrac{1}{y}}$.

Method 1   We add in the numerator and in the denominator and proceed as follows:

$$\frac{\dfrac{1}{x} + \dfrac{1}{y}}{\dfrac{1}{x} - \dfrac{1}{y}} = \frac{\dfrac{1\mathbf{y}}{\mathbf{x}\mathbf{y}} + \dfrac{x1}{xy}}{\dfrac{1\mathbf{y}}{\mathbf{x}\mathbf{y}} - \dfrac{x1}{xy}} \qquad \frac{y}{y} = 1, \frac{x}{x} = 1.$$

$$= \frac{\dfrac{y + x}{xy}}{\dfrac{y - x}{xy}} \qquad \text{Add the fractions in the numerator and subtract the fractions in the denominator.}$$

$$= \frac{y + x}{xy} \div \frac{y - x}{xy} \qquad \text{Write the complex fraction as a division.}$$

$$= \frac{y + x}{xy} \cdot \frac{xy}{y - x} \qquad \text{Invert the divisor and multiply.}$$

$$= \frac{y + x}{y - x} \qquad \text{Multiply the numerators and multiply the denominators and simplify.}$$

Method 2   We multiply the numerator and denominator by xy (the LCD of the rational expressions appearing in the complex fraction) and simplify.

$$\frac{\dfrac{1}{x} + \dfrac{1}{y}}{\dfrac{1}{x} - \dfrac{1}{y}} = \frac{\mathbf{xy}\left(\dfrac{1}{x} + \dfrac{1}{y}\right)}{\mathbf{xy}\left(\dfrac{1}{x} - \dfrac{1}{y}\right)}$$

$$= \frac{xy \cdot \dfrac{1}{x} + xy \cdot \dfrac{1}{y}}{xy \cdot \dfrac{1}{x} - xy \cdot \dfrac{1}{y}} \qquad \text{Use the distributive property to remove parentheses.}$$

$$= \frac{y + x}{y - x} \qquad \text{Simplify.}$$

Self Check   Simplify: $\dfrac{\dfrac{1}{a} - \dfrac{1}{b}}{\dfrac{1}{a} + \dfrac{1}{b}}$.

EXAMPLE 3   Simplify: $\dfrac{x^{-1} + y^{-1}}{x^{-2} - y^{-2}}$.

Method 1   We proceed as follows:

$$\frac{x^{-1} + y^{-1}}{x^{-2} - y^{-2}} = \frac{\dfrac{1}{x} + \dfrac{1}{y}}{\dfrac{1}{x^2} - \dfrac{1}{y^2}} \qquad \text{Write the fraction without using negative exponents.}$$

$$= \frac{\dfrac{y}{xy} + \dfrac{x}{xy}}{\dfrac{y^2}{x^2 y^2} - \dfrac{x^2}{x^2 y^2}} \qquad \text{Get a common denominator in the numerator and denominator.}$$

$$= \frac{\dfrac{y + x}{xy}}{\dfrac{y^2 - x^2}{x^2 y^2}} \qquad \text{Add the fractions in the numerator and denominator.}$$

$$= \frac{y + x}{xy} \div \frac{y^2 - x^2}{x^2 y^2} \qquad \text{Write the fraction as a division.}$$

$$= \frac{y + x}{xy} \cdot \frac{xxyy}{(y - x)(y + x)} \qquad \text{Invert and factor } y^2 - x^2.$$

$$= \frac{(y + x)xxyy}{xy(y - x)(y + x)} \qquad \text{Multiply the numerators and the denominators.}$$

$$= \frac{xy}{y - x} \qquad \text{Divide out the common factors of } x, y, \text{ and } y + x \text{ in the numerator and denominator.}$$

Method 2   We multiply both numerator and denominator by $x^2 y^2$, the LCD of the rational expressions, and proceed as follows:

$$\frac{x^{-1} + y^{-1}}{x^{-2} - y^{-2}} = \frac{\dfrac{1}{x} + \dfrac{1}{y}}{\dfrac{1}{x^2} - \dfrac{1}{y^2}} \qquad \text{Write the fraction without negative exponents.}$$

$$= \frac{x^2 y^2 \left(\dfrac{1}{x} + \dfrac{1}{y} \right)}{x^2 y^2 \left(\dfrac{1}{x^2} - \dfrac{1}{y^2} \right)}$$ Multiply numerator and denominator by $x^2 y^2$.

$$= \frac{x^2 y^2 \cdot \dfrac{1}{x} + x^2 y^2 \cdot \dfrac{1}{y}}{x^2 y^2 \cdot \dfrac{1}{x^2} - x^2 y^2 \cdot \dfrac{1}{y^2}}$$ Use the distributive property to remove parentheses.

$$= \frac{xy^2 + x^2 y}{y^2 - x^2}$$ Simplify.

$$= \frac{xy(y + x)}{(y + x)(y - x)}$$ Factor the numerator and denominator.

$$= \frac{xy}{y - x}$$ Divide out $y + x$, and $\dfrac{y + x}{y + x} = 1$.

Self Check   Simplify: $\dfrac{a^{-2} + b^{-2}}{a^{-1} - b^{-1}}$.

PERSPECTIVE

Each of the complex fractions in the list

$$1 + \frac{1}{2}, \quad 1 + \frac{1}{1 + \dfrac{1}{2}}, \quad 1 + \frac{1}{1 + \dfrac{1}{1 + \dfrac{1}{2}}}, \quad \cdots$$

can be simplified by using the value of the expression preceding it. For example, to simplify the second expression in the list, replace $1 + \frac{1}{2}$ with $\frac{3}{2}$:

$$1 + \frac{1}{1 + \dfrac{1}{2}} = 1 + \frac{1}{\dfrac{3}{2}} = 1 + \frac{2}{3} = \frac{5}{3}$$

To simplify the third expression, we replace $1 + \dfrac{1}{1 + \dfrac{1}{2}}$

with $\dfrac{5}{3}$:

$$1 + \frac{1}{1 + \dfrac{1}{1 + \dfrac{1}{2}}} = 1 + \frac{1}{\dfrac{5}{3}} = 1 + \frac{3}{5} = \frac{8}{5}$$

The complex fractions in the list simplify to the fractions $\frac{3}{2}, \frac{5}{3}, \frac{8}{5}$. The decimal values of these fractions get closer and closer to the irrational number $1.61803398875 \ldots$,

known as the **golden ratio.** This number often appears in the architecture of the ancient Greeks and Egyptians. For example, the width of the stairs in front of the Greek Parthenon (Illustration 1), divided by the building's height, is the golden ratio. The height of the triangular face of the Great Pyramid of Cheops (Illustration 2), divided by the pyramid's width, is also the golden ratio.

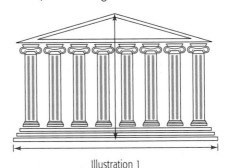

Illustration 1

Illustration 2

Comment $x^{-1} + y^{-1}$ means $\frac{1}{x} + \frac{1}{y}$, and $(x + y)^{-1}$ means $\dfrac{1}{x + y}$. Thus, $x^{-1} + y^{-1} \neq (x + y)^{-1}$.

EXAMPLE 4 Simplify: $\dfrac{\dfrac{2x}{1 - \dfrac{1}{x}} + 3}{3 - \dfrac{2}{x}}$.

Solution We begin by multiplying the numerator and denominator of

$$\dfrac{2x}{1 - \dfrac{1}{x}}$$

by x to eliminate the complex fraction in the numerator.

$$\dfrac{\dfrac{2x}{1 - \dfrac{1}{x}} + 3}{3 - \dfrac{2}{x}} = \dfrac{\dfrac{x2x}{x\left(1 - \dfrac{1}{x}\right)} + 3}{3 - \dfrac{2}{x}} \qquad \dfrac{x}{x} = 1.$$

$$= \dfrac{\dfrac{2x^2}{x \cdot 1 - x \cdot \dfrac{1}{x}} + 3}{3 - \dfrac{2}{x}} \qquad \text{Use the distributive property to remove parentheses.}$$

$$= \dfrac{\dfrac{2x^2}{x - 1} + 3}{3 - \dfrac{2}{x}} \qquad \text{Simplify: } x \cdot 1 = x \text{ and } x \cdot \dfrac{1}{x} = 1.$$

We then multiply the numerator and denominator of the previous fraction by $x(x - 1)$, the LCD of $\frac{2x^2}{x-1}$, 3, and $\frac{2}{x}$, and simplify:

$$\dfrac{\dfrac{2x}{1 - \dfrac{1}{x}} + 3}{3 - \dfrac{2}{x}} = \dfrac{x(x - 1)\left(\dfrac{2x^2}{x - 1} + 3\right)}{x(x - 1)\left(3 - \dfrac{2}{x}\right)} \qquad \dfrac{x}{x} = 1, \dfrac{x-1}{x-1} = 1.$$

$$= \dfrac{x(x - 1) \cdot \dfrac{2x^2}{x - 1} + x(x - 1) \cdot 3}{x(x - 1) \cdot 3 - x(x - 1) \cdot \dfrac{2}{x}} \qquad \text{Use the distributive property.}$$

$$= \dfrac{2x^3 + 3x(x - 1)}{3x(x - 1) - 2(x - 1)} \qquad \text{Simplify.}$$

$$= \dfrac{2x^3 + 3x^2 - 3x}{3x^2 - 5x + 2}$$

This result does not simplify.

Self Check   Simplify: $\dfrac{3 + \dfrac{2}{a}}{\dfrac{2a}{1 + \dfrac{1}{a}} + 3}$.

Self Check Answers

1. $\dfrac{2a + 3}{a^2 + 2a}$ 2. $\dfrac{b - a}{b + a}$ 3. $\dfrac{b^2 + a^2}{ab(b - a)}$ 4. $\dfrac{(3a + 2)(a + 1)}{a(2a^2 + 3a + 3)}$

Orals   *Simplify each complex fraction.*

1. $\dfrac{\frac{3}{4}}{\frac{5}{4}}$ 2. $\dfrac{\frac{a}{b}}{\frac{d}{b}}$ 3. $\dfrac{\frac{3}{4}}{\frac{3}{8}}$

4. $\dfrac{\frac{x + y}{x}}{\frac{x - y}{x}}$ 5. $\dfrac{\frac{x}{y} - 1}{\frac{x}{y}}$ 6. $\dfrac{1 + \frac{a}{b}}{\frac{a}{b}}$

6.5   EXERCISES

REVIEW   *Solve each equation.*

1. $\dfrac{8(a - 5)}{3} = 2(a - 4)$

2. $\dfrac{3t^2}{5} + \dfrac{7t}{10} = \dfrac{3t + 6}{5}$

3. $a^4 - 13a^2 + 36 = 0$

4. $|2x - 1| = 9$

VOCABULARY AND CONCEPTS   *Fill in the blanks.*

5. A _____ fraction is a fraction that has fractions in its numerator or denominator.

6. The fraction $\dfrac{\frac{a}{b}}{\frac{c}{d}}$ is equivalent to $\dfrac{a}{b}$ —— $\dfrac{c}{d}$.

PRACTICE   *Simplify each complex fraction.*

7. $\dfrac{\frac{1}{2}}{\frac{3}{4}}$ 8. $-\dfrac{\frac{3}{4}}{\frac{1}{2}}$

9. $\dfrac{-\frac{2}{3}}{\frac{6}{9}}$ 10. $\dfrac{\frac{11}{18}}{\frac{22}{27}}$

11. $\dfrac{\frac{1}{2} + \frac{1}{3}}{\frac{1}{4}}$ 12. $\dfrac{\frac{1}{4} - \frac{1}{5}}{\frac{1}{3}}$

13. $\dfrac{\frac{1}{2} - \frac{2}{3}}{\frac{2}{3} + \frac{1}{2}}$ 14. $\dfrac{\frac{2}{3} + \frac{4}{5}}{\frac{2}{5} - \frac{1}{3}}$

15. $\dfrac{\dfrac{4x}{y}}{\dfrac{6xz}{y^2}}$

16. $\dfrac{\dfrac{5t^4}{9x}}{\dfrac{2t}{18x}}$

17. $\dfrac{\dfrac{5ab^2}{ab}}{25}$

18. $\dfrac{\dfrac{6a^2b}{4t}}{3a^2b^2}$

19. $\dfrac{\dfrac{x-y}{xy}}{\dfrac{y-x}{x}}$

20. $\dfrac{\dfrac{x^2+5x+6}{3xy}}{\dfrac{x^2-9}{6xy}}$

21. $\dfrac{\dfrac{1}{x}-\dfrac{1}{y}}{xy}$

22. $\dfrac{xy}{\dfrac{1}{x}-\dfrac{1}{y}}$

23. $\dfrac{\dfrac{1}{a}+\dfrac{1}{b}}{\dfrac{1}{a}}$

24. $\dfrac{\dfrac{1}{b}}{\dfrac{1}{a}-\dfrac{1}{b}}$

25. $\dfrac{1+\dfrac{x}{y}}{1-\dfrac{x}{y}}$

26. $\dfrac{\dfrac{x}{y}+1}{1-\dfrac{x}{y}}$

27. $\dfrac{\dfrac{y}{x}-\dfrac{x}{y}}{\dfrac{1}{x}+\dfrac{1}{y}}$

28. $\dfrac{\dfrac{y}{x}-\dfrac{x}{y}}{\dfrac{1}{y}-\dfrac{1}{x}}$

29. $\dfrac{\dfrac{1}{a}-\dfrac{1}{b}}{\dfrac{a}{b}-\dfrac{b}{a}}$

30. $\dfrac{\dfrac{1}{a}+\dfrac{1}{b}}{\dfrac{a}{b}-\dfrac{b}{a}}$

31. $\dfrac{x+1-\dfrac{6}{x}}{\dfrac{1}{x}}$

32. $\dfrac{x-1-\dfrac{2}{x}}{\dfrac{x}{3}}$

33. $\dfrac{5xy}{1+\dfrac{1}{xy}}$

34. $\dfrac{3a}{a+\dfrac{1}{a}}$

35. $\dfrac{1+\dfrac{6}{x}+\dfrac{8}{x^2}}{1+\dfrac{1}{x}-\dfrac{12}{x^2}}$

36. $\dfrac{1-x-\dfrac{2}{x}}{\dfrac{6}{x^2}+\dfrac{1}{x}-1}$

37. $\dfrac{\dfrac{1}{a+1}+1}{\dfrac{3}{a-1}+1}$

38. $\dfrac{2+\dfrac{3}{x+1}}{\dfrac{1}{x}+x+x^2}$

39. $\dfrac{x^{-1}+y^{-1}}{x}$

40. $\dfrac{x^{-1}-y^{-1}}{y}$

41. $\dfrac{y}{x^{-1}-y^{-1}}$

42. $\dfrac{x^{-1}+y^{-1}}{(x+y)^{-1}}$

43. $\dfrac{x^{-1}+y^{-1}}{x^{-1}-y^{-1}}$

44. $\dfrac{(x+y)^{-1}}{x^{-1}+y^{-1}}$

45. $\dfrac{x+y}{x^{-1}+y^{-1}}$

46. $\dfrac{x-y}{x^{-1}-y^{-1}}$

47. $\dfrac{x-y^{-2}}{y-x^{-2}}$

48. $\dfrac{x^{-2}-y^{-2}}{x^{-1}-y^{-1}}$

49. $\dfrac{1+\dfrac{a}{b}}{1-\dfrac{a}{1-\dfrac{a}{b}}}$

50. $\dfrac{1+\dfrac{2}{1+\dfrac{a}{b}}}{1-\dfrac{a}{b}}$

51. $\dfrac{x-\dfrac{1}{x}}{1+\dfrac{1}{\dfrac{1}{x}}}$

52. $\dfrac{\dfrac{a^2+3a+4}{ab}}{2+\dfrac{3+a}{\dfrac{2}{a}}}$

53. $\dfrac{b}{b+\dfrac{2}{2+\dfrac{1}{2}}}$

54. $\dfrac{2y}{y-\dfrac{y}{3-\dfrac{1}{2}}}$

55. $a+\dfrac{a}{1+\dfrac{a}{a+1}}$

56. $b+\dfrac{b}{1-\dfrac{b+1}{b}}$

57. $\dfrac{x - \dfrac{1}{1 - \dfrac{x}{2}}}{\dfrac{3}{x + \dfrac{2}{3}} - x}$

58. $\dfrac{\dfrac{2x}{x - \dfrac{1}{x}} - \dfrac{1}{x}}{2x + \dfrac{2x}{1 - \dfrac{1}{x}}}$

59. $\dfrac{2x + \dfrac{1}{2 - \dfrac{x}{2}}}{\dfrac{4}{\dfrac{x}{2} - 2} - x}$

60. $\dfrac{3x - \dfrac{1}{3 - \dfrac{x}{2}}}{\dfrac{3}{\dfrac{x}{2} - 3} + x}$

Factor each denominator and simplify the complex fraction.

61. $\dfrac{\dfrac{1}{x^2 + 3x + 2} + \dfrac{1}{x^2 + x - 2}}{\dfrac{3x}{x^2 - 1} - \dfrac{x}{x + 2}}$

62. $\dfrac{\dfrac{1}{x^2 - 1} - \dfrac{2}{x^2 + 4x + 3}}{\dfrac{2}{x^2 + 2x - 3} + \dfrac{1}{x + 3}}$

APPLICATIONS

63. Engineering The stiffness k of the shaft is given by the formula

$$k = \dfrac{1}{\dfrac{1}{k_1} + \dfrac{1}{k_2}}$$

Section 1   Section 2

where k_1 and k_2 are the individual stiffnesses of each section. Simplify the complex fraction.

64. Electronics In electronic circuits, resistors oppose the flow of an electric current. To find the total resistance of a parallel combination of the two resistors shown in the illustration, we can use the formula

$$\text{Total resistance} = \dfrac{1}{\dfrac{1}{R_1} + \dfrac{1}{R_2}}$$

where R_1 is the resistance of the first resistor and R_2 is the resistance of the second. Simplify the complex fraction on the right-hand side of the formula.

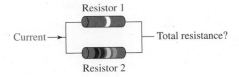

65. Data analysis Use the data in the table to find the average measurement for the three-trial experiment.

	Trial 1	Trial 2	Trial 3
Measurement	$\dfrac{k}{2}$	$\dfrac{k}{3}$	$\dfrac{k}{2}$

66. Transportation If a car travels a distance d_1 at a speed s_1, and then travels a distance d_2 at a speed s_2, the average (mean) speed is given by the formula

$$\bar{s} = \dfrac{d_1 + d_2}{\dfrac{d_1}{s_1} + \dfrac{d_2}{s_2}}$$

Simplify the complex fraction.

WRITING

67. There are two methods used to simplify a complex fraction. Explain one of them.

68. Explain the other method of simplifying a complex fraction.

SOMETHING TO THINK ABOUT

69. Simplify: $(x^{-1}y^{-1})(x^{-1} + y^{-1})^{-1}$.

70. Simplify: $[(x^{-1} + 1)^{-1} + 1]^{-1}$.

6.6   Equations Containing Rational Expressions

▌ Solving Rational Equations   ▌ Formulas   ▌ Problem Solving

Getting Ready   *Solve each proportion.*

1. $\dfrac{3}{x} = \dfrac{6}{9}$
                                        2. $\dfrac{x+1}{6} = \dfrac{2}{x}$

We now show how to solve equations that contain rational expressions. We will then use these skills to solve problems.

Solving Rational Equations

If an equation contains one or more rational expressions, it is called a **rational equation.** Some examples are

$$\frac{3}{5} + \frac{7}{x+2} = 2, \qquad \frac{x+3}{x-3} = \frac{2}{x^2-4}, \qquad \text{and} \qquad \frac{-x^2+10}{x^2-1} + \frac{3x}{x-1} = \frac{2x}{x+1}$$

To solve a rational equation, we can multiply both sides of the equation by the LCD of the rational expressions in the equation to clear it of fractions.

EXAMPLE 1   Solve: $\dfrac{3}{5} + \dfrac{7}{x+2} = 2$.

Solution   We note that x cannot be -2, because this would give a 0 in the denominator of $\frac{7}{x+2}$. If $x \neq -2$, we can multiply both sides of the equation by $5(x+2)$ and simplify to get

$$5(x+2)\left(\frac{3}{5} + \frac{7}{x+2}\right) = 5(x+2)2$$

$$5(x+2)\left(\frac{3}{5}\right) + 5(x+2)\left(\frac{7}{x+2}\right) = 5(x+2)2 \qquad \begin{array}{l}\text{Use the distributive property on}\\ \text{the left-hand side.}\end{array}$$

$$3(x+2) + 5(7) = 10(x+2) \qquad \text{Simplify.}$$

$$3x + 6 + 35 = 10x + 20 \qquad \begin{array}{l}\text{Use the distributive property}\\ \text{and simplify.}\end{array}$$

$$3x + 41 = 10x + 20 \qquad \text{Simplify.}$$

$$-7x = -21 \qquad \begin{array}{l}\text{Add } -10x \text{ and } -41 \text{ to both}\\ \text{sides.}\end{array}$$

$$x = 3 \qquad \text{Divide both sides by } -7.$$

Check: To check, we substitute 3 for x in the original equation and simplify:

$$\frac{3}{5} + \frac{7}{x + 2} = 2$$

$$\frac{3}{5} + \frac{7}{3 + 2} \stackrel{?}{=} 2$$

$$\frac{3}{5} + \frac{7}{5} \stackrel{?}{=} 2$$

$$2 = 2$$

Self Check Solve: $\dfrac{2}{5} + \dfrac{5}{x - 2} = \dfrac{29}{10}$.

Accent on Technology SOLVING EQUATIONS

To use a graphing calculator to approximate the solution of $\frac{3}{5} + \frac{7}{x + 2} = 2$, we graph the functions $f(x) = \frac{3}{5} + \frac{7}{x + 2}$ and $g(x) = 2$. If we use window settings of $[-10, 10]$ for x and $[-10, 10]$ for y, we will obtain the graph shown in Figure 6-11(a).

If we trace and move the cursor close to the intersection point of the two graphs, we will get the approximate value of x shown in Figure 6-11(b). If we zoom twice and trace again, we get the results shown in Figure 6-11(c). Algebra will show that the exact solution is 3, as shown in Example 1.

We can also find the intersection point by using the INTERSECT command found in the CALC menu.

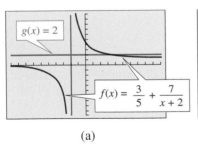
(a)

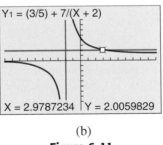
(b)

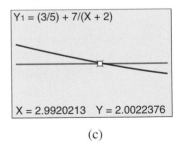
(c)

Figure 6-11

EXAMPLE 2 Solve: $\dfrac{-x^2 + 10}{x^2 - 1} + \dfrac{3x}{x - 1} = \dfrac{2x}{x + 1}$.

Solution We note that x cannot be 1 or -1, because this would give a 0 in a denominator. If $x \neq 1$ and $x \neq -1$, we can clear the equation of denominators by multiplying both sides by the LCD of the three rational expressions and proceed as follows:

$$\frac{-x^2 + 10}{x^2 - 1} + \frac{3x}{x - 1} = \frac{2x}{x + 1}$$

$$\frac{-x^2 + 10}{(x + 1)(x - 1)} + \frac{3x}{x - 1} = \frac{2x}{x + 1} \qquad \text{Factor } x^2 - 1.$$

$$\frac{(x + 1)(x - 1)(-x^2 + 10)}{(x + 1)(x - 1)} + \frac{3x(x + 1)(x - 1)}{x - 1} = \frac{2x(x + 1)(x - 1)}{x + 1} \qquad \begin{array}{l}\text{Multiply both sides by}\\ (x + 1)(x - 1).\end{array}$$

$$-x^2 + 10 + 3x(x + 1) = 2x(x - 1) \qquad \text{Divide out common factors.}$$

$$-x^2 + 10 + 3x^2 + 3x = 2x^2 - 2x \qquad \text{Remove parentheses.}$$

$$2x^2 + 10 + 3x = 2x^2 - 2x \qquad \text{Combine like terms.}$$

$$10 + 3x = -2x \qquad \text{Subtract } 2x^2 \text{ from both sides.}$$

$$10 + 5x = 0 \qquad \text{Add } 2x \text{ to both sides.}$$

$$5x = -10 \qquad \text{Subtract 10 from both sides.}$$

$$x = -2 \qquad \text{Divide both sides by 5.}$$

Verify that -2 is a solution of the original equation. The solution set is $\{-2\}$.
∎

When we multiply both sides of an equation by a quantity that contains a variable, we can get false solutions, called **extraneous solutions.** We must exclude extraneous solutions from the solution set of an equation.

EXAMPLE 3 Solve: $\dfrac{2(x + 1)}{x - 3} = \dfrac{x + 5}{x - 3}.$

Solution We start by noting that x cannot be 3, because this would give a 0 in a denominator. If $x \neq 3$, we can clear the equation of denominators by multiplying both sides by $x - 3$.

$$\frac{2(x + 1)}{x - 3} = \frac{x + 5}{x - 3}$$

$$(x - 3)\frac{2(x + 1)}{x - 3} = (x - 3)\frac{x + 5}{x - 3} \qquad \text{Multiply both sides by } x - 3.$$

$$2(x + 1) = x + 5 \qquad \text{Simplify.}$$

$$2x + 2 = x + 5 \qquad \text{Remove parentheses.}$$

$$x + 2 = 5 \qquad \text{Subtract } x \text{ from both sides.}$$

$$x = 3 \qquad \text{Subtract 2 from both sides.}$$

Since x cannot be 3, the 3 must be discarded. This equation has no solutions. Its solution set is the empty set, $\varnothing$.

Self Check Solve: $\dfrac{6}{x - 6} + 3 = \dfrac{x}{x - 6}.$ ∎

EXAMPLE 4 Solve: $\dfrac{x + 1}{5} - 2 = -\dfrac{4}{x}.$

Solution We note that x cannot be 0, because this would give a 0 in a denominator. If $x \neq 0$, we can clear the equation of denominators by multiplying both sides by $5x$.

$$\frac{x+1}{5} - 2 = -\frac{4}{x}$$

$$5x\left(\frac{x+1}{5} - 2\right) = 5x\left(-\frac{4}{x}\right)$$ Multiply both sides by $5x$.

$$x(x+1) - 10x = -20$$ Remove parentheses and simplify.

$$x^2 + x - 10x = -20$$ Remove parentheses.

$$x^2 - 9x + 20 = 0$$ Combine like terms and add 20 to both sides.

$$(x-5)(x-4) = 0$$ Factor $x^2 - 9x + 20$.

$$x - 5 = 0 \quad \text{or} \quad x - 4 = 0$$ Set each factor equal to 0.

$$x = 5 \quad | \quad x = 4$$

Since 4 and 5 both satisfy the original equation, the solution set is $\{4, 5\}$.

Self Check   Solve: $a + \dfrac{2}{3} = \dfrac{2a - 12}{3(a - 3)}$.

Formulas

Many formulas must be cleared of denominators before we can solve them for a specific variable.

EXAMPLE 5   Solve: $\dfrac{1}{r} = \dfrac{1}{r_1} + \dfrac{1}{r_2}$ for r.

Solution

$$\frac{1}{r} = \frac{1}{r_1} + \frac{1}{r_2}$$

$$\frac{rr_1r_2}{r} = \frac{rr_1r_2}{r_1} + \frac{rr_1r_2}{r_2}$$ Multiply both sides by $r\,r_1\,r_2$.

$$r_1r_2 = rr_2 + rr_1$$ Simplify each fraction.

$$r_1r_2 = r(r_2 + r_1)$$ Factor out r on the right-hand side.

$$\frac{r_1r_2}{r_2 + r_1} = r$$ Divide both sides by $r_2 + r_1$.

$$r = \frac{r_1r_2}{r_2 + r_1}$$

Self Check   Solve: $\dfrac{1}{a} - \dfrac{1}{b} = 1$ for b.

Problem Solving

EXAMPLE 6   **Drywalling a house**   A contractor knows that one crew can drywall a house in 4 days and that another crew can drywall the same house in 5 days. One day must be allowed for the plaster coat to dry. If the contractor uses both crews, can the house be ready for painting in 4 days?

Analyze the problem   Because 1 day is necessary for drying, the drywallers must complete their work in 3 days. Since the first crew can drywall the house in 4 days, it can do $\frac{1}{4}$ of the job in 1 day. Since the second crew can drywall the house in 5 days, it can do $\frac{1}{5}$ of the job in 1 day. If it takes x days for both crews to finish the house, together they can do $\frac{1}{x}$ of the job in 1 day. The amount of work the first crew can do in 1 day plus the amount of work the second crew can do in 1 day equals the amount of work both crews can do in 1 day when working together.

Form an equation   If x represents the number of days it takes for both crews to drywall the house, we can form the equation.

What crew 1 can do in one day	plus	what crew 2 can do in one day	equals	what they can do together in one day.
$\frac{1}{4}$	$+$	$\frac{1}{5}$	$=$	$\frac{1}{x}$

Solve the equation   We can solve this equation as follows.

$$20x\left(\frac{1}{4} + \frac{1}{5}\right) = 20x\left(\frac{1}{x}\right) \qquad \text{Multiply both sides by } 20x.$$

$$5x + 4x = 20 \qquad \text{Remove parentheses and simplify.}$$

$$9x = 20 \qquad \text{Combine like terms.}$$

$$x = \frac{20}{9} \qquad \text{Divide both sides by 9.}$$

State the conclusion   Since it will take $2\frac{2}{9}$ days for both crews to drywall the house and it takes 1 day for drying, it will be ready for painting in $3\frac{2}{9}$ days, which is less than 4 days.
Check the result. ∎

EXAMPLE 7   **Driving to a convention**   A man drove 200 miles to a convention. Because of road construction, his average speed on the return trip was 10 mph less than his average speed going to the convention. If the return trip took 1 hour longer, how fast did he drive in each direction?

Analyze the problem   Because the distance traveled is given by the formula

$$d = rt \qquad (d \text{ is distance, } r \text{ is the rate of speed, and } t \text{ is time})$$

the formula for time is

$$t = \frac{d}{r}$$

We can organize the given information in the chart shown in Figure 6-12.

	Rate	·	Time	=	Distance
Going	r		$\dfrac{200}{r}$		200
Returning	$r - 10$		$\dfrac{200}{r - 10}$		200

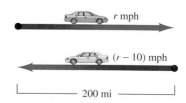

Figure 6-12

Form an equation   Let r represent the average rate of speed going to the meeting. Then $r - 10$ represents the average rate of speed on the return trip. Because the return trip took 1 hour longer, we can form the following equation:

The time it took to travel to the convention	plus 1	equals	the time it took to return.
$\dfrac{200}{r}$	$+\ 1$	$=$	$\dfrac{200}{r - 10}$

Solve the equation   We can solve the equation as follows:

$$r(r - 10)\left(\frac{200}{r} + 1\right) = r(r - 10)\left(\frac{200}{r - 10}\right) \qquad \text{Multiply both sides by } r(r - 10).$$

$$200(r - 10) + r(r - 10) = 200r \qquad \text{Remove parentheses and simplify.}$$

$$200r - 2{,}000 + r^2 - 10r = 200r \qquad \text{Remove parentheses.}$$

$$r^2 - 10r - 2{,}000 = 0 \qquad \text{Subtract } 200r \text{ from both sides.}$$

$$(r - 50)(r + 40) = 0 \qquad \text{Factor } r^2 - 10r - 2{,}000.$$

$$r - 50 = 0 \quad \text{or} \quad r + 40 = 0 \qquad \text{Set each factor equal to 0.}$$

$$r = 50 \quad | \quad \qquad r = -40$$

State the conclusion   We must exclude the solution of -40, because a speed cannot be negative. Thus, the man averaged 50 mph going to the convention, and he averaged $50 - 10$, or 40 mph, returning.

Check the result   At 50 mph, the 200-mile trip took 4 hours. At 40 mph, the return trip took 5 hours, which is 1 hour longer. ∎

EXAMPLE 8   **A river cruise**   The Forest City Queen can make a 9-mile trip down the Rock River and return in a total of 1.6 hours. If the riverboat travels 12 mph in still water, find the speed of the current in the Rock River.

Analyze the problem   We can let c represent the speed of the current. Since the boat travels 12 mph and a current of c mph pushes the boat while it is going downstream, the speed of the boat going downstream is $(12 + c)$ mph. On the return trip, the current pushes against the boat, and its speed is $(12 - c)$ mph. Since $t = \frac{d}{r}$ $\left(\text{time} = \frac{\text{distance}}{\text{rate}}\right)$, the time required for the downstream leg of the trip is $\frac{9}{12 + c}$ hours, and the time required for the upstream leg of the trip is $\frac{9}{12 - c}$ hours.

We can organize this information in the chart shown in Figure 6-13.

Furthermore, we know that the total time required for the round trip is 1.6 hours.

Form an equation   If c represents the speed of the current, then $12 + c$ represents the speed of the boat going downstream, and $12 - c$ represents the speed of the boat going upstream.

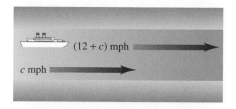

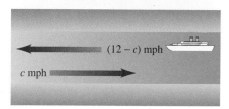

	Rate	·	Time	=	Distance
Going downstream	$12 + c$		$\dfrac{9}{12 + c}$		9
Going upstream	$12 - c$		$\dfrac{9}{12 - c}$		9

Figure 6-13

Since the time going downstream is $\frac{9}{12 + c}$ hours, the time going upstream is $\frac{9}{12 - c}$ hours, and total time is 1.6 hours $\left(\frac{8}{5} \text{ hours}\right)$, we have

The time it takes to travel downstream	plus	the time it takes to travel upstream	equals	the total time for the round trip.
$\dfrac{9}{12 + c}$	$+$	$\dfrac{9}{12 - c}$	$=$	$\dfrac{8}{5}$

Solve the equation   We can multiply both sides of this equation by $5(12 + c)(12 - c)$ to clear it of denominators and proceed as follows:

$$\frac{5(12 + c)(12 - c)9}{12 + c} + \frac{5(12 + c)(12 - c)9}{12 - c} = \frac{5(12 + c)(12 - c)8}{5}$$

$$45(12 - c) + 45(12 + c) = 8(12 + c)(12 - c) \qquad \frac{12 + c}{12 + c} = 1, \frac{12 - c}{12 - c} = 1, \text{ and } \frac{5}{5} = 1.$$

$$540 - 45c + 540 + 45c = 8(144 - c^2) \qquad \text{Multiply.}$$

$$1{,}080 = 1{,}152 - 8c^2 \qquad \text{Combine like terms and multiply.}$$

$$8c^2 - 72 = 0 \qquad \text{Add } 8c^2 \text{ and } -1{,}152 \text{ to both sides.}$$

$$c^2 - 9 = 0 \qquad \text{Divide both sides by 8.}$$

$$(c + 3)(c - 3) = 0 \qquad \text{Factor } c^2 - 9.$$

$$c + 3 = 0 \quad \text{ or } \quad c - 3 = 0 \qquad \text{Set each factor equal to 0.}$$

$$c = -3 \quad | \quad c = 3$$

State the conclusion   Since the current cannot be negative, the apparent solution of -3 must be discarded. Thus, the current in the Rock River is 3 mph.
Check the result.   ∎

1. 4   **3.** 6 is extraneous, no solution   **4.** 1, 2   **5.** $b = \dfrac{a}{1 - a}$

Orals *Solve each equation.*

1. $\dfrac{4}{x} = 2$

2. $\dfrac{9}{y} = 3$

3. $\dfrac{4}{p} + \dfrac{5}{p} = 9$

4. $\dfrac{5}{r} - \dfrac{2}{r} = 1$

5. $\dfrac{4}{y} - \dfrac{1}{y} = 3$

6. $\dfrac{8}{t} - \dfrac{2}{t} = 3$

6.6   EXERCISES

REVIEW *Simplify each expression. Write each answer without using negative exponents.*

1. $(m^2 n^{-3})^{-2}$

2. $\dfrac{a^{-1}}{a^{-1} + 1}$

3. $\dfrac{a^0 + 2a^0 - 3a^0}{(a - b)^0}$

4. $(4x^{-2} + 3)(2x - 4)$

VOCABULARY AND CONCEPTS *Fill in the blanks.*

5. If an equation contains a rational expression, it is called a _____ equation.

6. A false solution to an equation is called an _____ solution.

PRACTICE *Solve each equation. If a solution is extraneous, so indicate.*

7. $\dfrac{1}{4} + \dfrac{9}{x} = 1$

8. $\dfrac{1}{3} - \dfrac{10}{x} = -3$

9. $\dfrac{34}{x} - \dfrac{3}{2} = -\dfrac{13}{20}$

10. $\dfrac{1}{2} + \dfrac{7}{x} = 2 + \dfrac{1}{x}$

11. $\dfrac{3}{y} + \dfrac{7}{2y} = 13$

12. $\dfrac{2}{x} + \dfrac{1}{2} = \dfrac{7}{2x}$

13. $\dfrac{x+1}{x} - \dfrac{x-1}{x} = 0$

14. $\dfrac{2}{x} + \dfrac{1}{2} = \dfrac{9}{4x} - \dfrac{1}{2x}$

15. $\dfrac{7}{5x} - \dfrac{1}{2} = \dfrac{5}{6x} + \dfrac{1}{3}$

16. $\dfrac{x-3}{x-1} - \dfrac{2x-4}{x-1} = 0$

17. $\dfrac{y-3}{y+2} = 3 - \dfrac{1-2y}{y+2}$

18. $\dfrac{3a-5}{a-1} - 2 = \dfrac{2a}{1-a}$

19. $\dfrac{3-5y}{2+y} = \dfrac{3+5y}{2-y}$

20. $\dfrac{x}{x-2} = 1 + \dfrac{1}{x-3}$

21. $\dfrac{a+2}{a+1} = \dfrac{a-4}{a-3}$

22. $\dfrac{z+2}{z+8} - \dfrac{z-3}{z-2} = 0$

23. $\dfrac{x+2}{x+3} - 1 = \dfrac{1}{3 - 2x - x^2}$

24. $\dfrac{x-3}{x-2} - \dfrac{1}{x} = \dfrac{x-3}{x}$

25. $\dfrac{x}{x+2} = 1 - \dfrac{3x+2}{x^2+4x+4}$

26. $\dfrac{3+2a}{a^2+6+5a} + \dfrac{2-5a}{a^2-4} = \dfrac{2-3a}{a^2-6+a}$

27. $\dfrac{2}{x-2} + \dfrac{1}{x+1} = \dfrac{1}{x^2-x-2}$

28. $\dfrac{5}{y-1} + \dfrac{3}{y-3} = \dfrac{8}{y-2}$

29. $\dfrac{a-1}{a+3} - \dfrac{1-2a}{3-a} = \dfrac{2-a}{a-3}$

30. $\dfrac{5}{2z^2+z-3} - \dfrac{2}{2z+3} = \dfrac{z+1}{z-1} - 1$

31. $\dfrac{5}{x+4} + \dfrac{1}{x+4} = x - 1$

32. $\dfrac{2}{x-1} + \dfrac{x-2}{3} = \dfrac{4}{x-1}$

33. $\dfrac{3}{x+1} - \dfrac{x-2}{2} = \dfrac{x-2}{x+1}$

34. $\dfrac{x-4}{x-3} + \dfrac{x-2}{x-3} = x - 3$

35. $\dfrac{2}{x-3} + \dfrac{3}{4} = \dfrac{17}{2x}$

36. $\dfrac{30}{y-2} + \dfrac{24}{y-5} = 13$

37. $\dfrac{x+4}{x+7} - \dfrac{x}{x+3} = \dfrac{3}{8}$

38. $\dfrac{5}{x+4} - \dfrac{1}{3} = \dfrac{x-1}{x}$

Solve each formula for the indicated variable.

39. $\dfrac{1}{p} + \dfrac{1}{q} = \dfrac{1}{f}$ for f

40. $\dfrac{1}{p} + \dfrac{1}{q} = \dfrac{1}{f}$ for p

41. $S = \dfrac{a - lr}{1 - r}$ for r

42. $H = \dfrac{2ab}{a+b}$ for a

43. $\dfrac{1}{R} = \dfrac{1}{r_1} + \dfrac{1}{r_2} + \dfrac{1}{r_3}$ for R

44. $\dfrac{1}{R} = \dfrac{1}{r_1} + \dfrac{1}{r_2} + \dfrac{1}{r_3}$ for r_1

APPLICATIONS

45. Focal length   The design of a camera lens uses the equation

$$\frac{1}{f} = \frac{1}{s_1} + \frac{1}{s_2}$$

which relates the focal length f of a lens to the image distance s_1 and the object distance s_2. Find the focal length of the lens shown in the illustration.   (*Hint:* Convert feet to inches.)

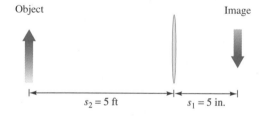

Object   Image

$s_2 = 5$ ft $s_1 = 5$ in.

46. Lensmaker's formula   The focal length f of a lens is given by the lensmaker's formula

$$\frac{1}{f} = 0.6\left(\frac{1}{r_1} + \frac{1}{r_2}\right)$$

where f is the focal length of the lens and r_1 and r_2 are the radii of the two circular surfaces. Find the focal length of the lens shown in the illustration.

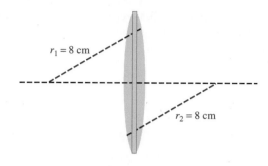

$r_1 = 8$ cm

$r_2 = 8$ cm

47. House painting   If one painter can paint a house in 5 days and another painter can paint the same house in 3 days, how long will it take them to paint the house working together?

48. Reading proof   A proofreader can read 250 pages in 8 hours, and a second proofreader can read 250 pages in 10 hours. If they both work on a 250-page book, can they meet a five-hour deadline?

49. Storing corn   In 10 minutes, a conveyor belt can move 1,000 bushels of corn into the storage bin shown in the illustration. A smaller belt can move 1,000 bushels to the storage bin in 14 minutes. If both belts are used, how long will it take to move 1,000 bushels to the storage bin?

50. Roofing a house   One roofing crew can finish a 2,800-square-foot roof in 12 hours, and another crew can do the job in 10 hours. If they work together, can they finish before a predicted rain in 5 hours?

51. Draining a pool   A drain can empty the swimming pool shown in the illustration in 3 days. A second drain can empty the pool in 2 days. How long will it take to empty the pool if both drains are used?

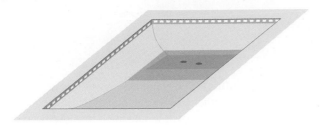

52. Filling a pool   A pipe can fill a pool in 9 hours. If a second pipe is also used, the pool can be filled in 3 hours. How long would it take the second pipe alone to fill the pool?

53. Filling a pond   One pipe can fill a pond in 3 weeks, and a second pipe can fill the pond in 5 weeks. However, evaporation and seepage can empty the pond in 10 weeks. If both pipes are used, how long will it take to fill the pond?

54. Housecleaning   Sally can clean the house in 6 hours, and her father can clean the house in 4 hours. Sally's younger brother, Dennis, can completely mess up the house in 8 hours. If Sally and her father clean and Dennis plays, how long will it take to clean the house?

55. Touring the countryside   A man bicycles 5 mph faster than he can walk. He bicycles 24 miles and walks back along the same route in 11 hours. How fast does he walk?

56. Finding rates   Two trains made the same 315-mile run. Since one train traveled 10 mph faster than the other, it arrived 2 hours earlier. Find the speed of each train.

57. Train travel   A train traveled 120 miles from Freeport to Chicago and returned the same distance in a total time of 5 hours. If the train traveled 20 mph slower on the return trip, how fast did the train travel in each direction?

58. Time on the road   A car traveled from Rockford to Chicago in 3 hours less time than it took a second car to travel from Rockford to St. Louis. If the cars traveled at the same average speed, how long was the first driver on the road?

59. Boating   A man can drive a motorboat 45 miles down the Rock River in the same amount of time that he can drive it 27 miles upstream. Find the speed of the current if the speed of the boat is 12 mph in still water.

60. Rowing a boat   A woman who can row 3 mph in still water rows 10 miles downstream on the Eagle River and returns upstream in a total of 12 hours. Find the speed of the current.

61. Aviation   A plane that can fly 340 mph in still air can fly 200 miles downwind in the same amount of time that it can fly 140 miles upwind. Find the velocity of the wind.

62. Aviation   An airplane can fly 650 miles with the wind in the same amount of time as it can fly 475 miles against the wind. If the wind speed is 40 mph, find the speed of the plane in still air.

63. Discount buying   A repairman purchased some washing-machine motors for a total of $224. When the unit cost decreased by $4, he was able to buy one extra motor for the same total price. How many motors did he buy originally?

64. Price increases   An appliance store manager bought several microwave ovens for a total of $1,800. When her unit cost increased by $25, she was able to buy one less oven for the same total price. How many ovens did she buy originally?

65. Stretching a vacation   A student saved $1,200 for a trip to Europe. By cutting $20 from her daily expenses, she was able to stay three extra days. How long had she originally planned to be gone?

66. Engineering The stiffness of the shaft shown is given by the formula

$$k = \cfrac{1}{\cfrac{1}{k_1} + \cfrac{1}{k_2}}$$

Section 1   Section 2

where k_1 and k_2 are the individual stiffnesses of each section. If the stiffness k_2 of Section 2 is 4,200,000 in. lb/rad, and the design specifications require that the stiffness k of the entire shaft be 1,900,000 in. lb/rad, what must the stiffness k_1 of Section 1 be?

WRITING

67. Why is it necessary to check the solutions of a rational equation?

68. Explain the steps you would use to solve a rational equation.

SOMETHING TO THINK ABOUT

69. Invent a rational equation that has an extraneous solution of 3.

70. Solve: $(x - 1)^{-1} - x^{-1} = 6^{-1}$.

6.7	Dividing Polynomials

- **Dividing a Monomial by a Monomial**
- **Dividing a Polynomial by a Monomial**
- **Dividing a Polynomial by a Polynomial**
- **The Case of the Missing Terms**

Getting Ready   *Simplify the fraction or perform the division.*

1. $\dfrac{16}{24}$ **2.** $-\dfrac{45}{81}$ **3.** $23\overline{)115}$ **4.** $32\overline{)512}$

We have previously seen how to add, subtract, and multiply polynomials. We now consider how to divide them.

Dividing a Monomial by a Monomial

In Example 1, we review how to divide a monomial by a monomial.

EXAMPLE 1   Simplify: $(3a^2b^3) \div (2a^3b)$.

Method 1   We write the expression as a fraction and divide out all common factors:

$$\frac{3a^2b^3}{2a^3b} = \frac{3aabbb}{2aaab}$$

$$= \frac{3\cancel{a}a\cancel{b}bb}{2\cancel{a}aa\cancel{b}}$$

$$= \frac{3b^2}{2a}$$

Method 2   We write the expression as a fraction and use the rules of exponents:

$$\frac{3a^2b^3}{2a^3b} = \frac{3}{2}a^{2-3}b^{3-1}$$

$$= \frac{3}{2}a^{-1}b^2$$

$$= \frac{3}{2}\left(\frac{1}{a}\right)\frac{b^2}{1}$$

$$= \frac{3b^2}{2a}$$

Self Check   Simplify: $\dfrac{6x^3y^2}{8x^2y^3}$. ∎

Dividing a Polynomial by a Monomial

In Example 2, we use the fact that $\frac{a+b}{c} = \frac{a}{c} + \frac{b}{c}$ ($c \neq 0$) to divide a polynomial by a monomial.

EXAMPLE 2   Divide: $4x^3y^2 + 3xy^5 - 12xy$ by $3x^2y^3$.

Solution

$$\frac{4x^3y^2 + 3xy^5 - 12xy}{3x^2y^3} = \frac{4x^3y^2}{3x^2y^3} + \frac{3xy^5}{3x^2y^3} - \frac{12xy}{3x^2y^3} \quad \begin{array}{l}\text{Write the original fraction as}\\ \text{separate fractions.}\end{array}$$

$$= \frac{4x}{3y} + \frac{y^2}{x} - \frac{4}{xy^2} \qquad\qquad \text{Simplify each fraction.}$$

Self Check   Simplify: $\dfrac{8a^3b^4 - 4a^4b^2 + a^2b^2}{4a^2b^2}$. ∎

Dividing a Polynomial by a Polynomial

There is an **algorithm** (a repeating series of steps) to use when the divisor is not a monomial. To use the division algorithm to divide $x^2 + 7x + 12$ by $x + 4$, we write the division in long division form and proceed as follows:

$$x + 4 \overline{)\, x^2 + 7x + 12} \qquad \begin{array}{l}\text{How many times does } x \text{ divide } x^2? \ \dfrac{x^2}{x} = x. \text{ Place the } x \text{ in the}\\ \text{quotient.}\end{array}$$

with x in the quotient.

$$\begin{array}{r} x \\ x + 4 \overline{)\, x^2 + 7x + 12} \\ \underline{x^2 + 4x} \\ 3x + 12 \end{array} \qquad \begin{array}{l}\text{Multiply each term in the divisor by } x \text{ to get } x^2 + 4x,\\ \text{subtract } x^2 + 4x \text{ from } x^2 + 7x, \text{ and bring down the } 12.\end{array}$$

$$\begin{array}{r} x + 3 \\ x + 4 \overline{)\, x^2 + 7x + 12} \\ \underline{x^2 + 4x} \\ 3x + 12 \end{array} \qquad \begin{array}{l}\text{How many times does } x \text{ divide } 3x? \ \dfrac{3x}{x} = +3. \text{ Place the}\\ +3 \text{ in the quotient.}\end{array}$$

$$
\begin{array}{r}
x + 3 \\
x + 4 \overline{)\, x^2 + 7x + 12} \\
\underline{x^2 + 4x } \\
3x + 12 \\
\underline{3x + 12} \\
0
\end{array}
$$

Multiply each term in the divisor by 3 to get $3x + 12$, and subtract $3x + 12$ from $3x + 12$ to get 0.

The division process stops when the result of the subtraction is a constant or a polynomial with degree less than the degree of the divisor. Here, the quotient is $x + 3$ and the remainder is 0.

We can check the quotient by multiplying the divisor by the quotient. The product should be the dividend.

$$
\underbrace{(x + 4)}_{\text{Divisor}} \; \underbrace{(x + 3)}_{\text{quotient}} = \underbrace{x^2 + 7x + 12}_{\text{dividend}} \quad \text{The quotient checks.}
$$

EXAMPLE 3   Divide $2a^3 + 9a^2 + 5a - 6$ by $2a + 3$.

Solution

$$
\begin{array}{r}
a^2 \\
2a + 3 \overline{)\, 2a^3 + 9a^2 + 5a - 6}
\end{array}
$$

How many times does $2a$ divide $2a^3$? $\dfrac{2a^3}{2a} = a^2$. Place a^2 in the quotient.

$$
\begin{array}{r}
a^2 \\
2a + 3 \overline{)\, 2a^3 + 9a^2 + 5a - 6} \\
\underline{2a^3 + 3a^2 } \\
6a^2 + 5a
\end{array}
$$

Multiply each term in the divisor by a^2 to get $2a^3 + 3a^2$, subtract $2a^3 + 3a^2$ from $2a^3 + 9a^2$, and bring down the $5a$.

$$
\begin{array}{r}
a^2 + 3a \\
2a + 3 \overline{)\, 2a^3 + 9a^2 + 5a - 6} \\
\underline{2a^3 + 3a^2 } \\
6a^2 + 5a
\end{array}
$$

How many times does $2a$ divide $6a^2$? $6a^2/2a = 3a$. Place the $+3a$ in the quotient.

$$
\begin{array}{r}
a^2 + 3a \\
2a + 3 \overline{)\, 2a^3 + 9a^2 + 5a - 6} \\
\underline{2a^3 + 3a^2 } \\
6a^2 + 5a \\
\underline{6a^2 + 9a} \\
-4a - 6
\end{array}
$$

Multiply each term in the divisor by $3a$ to get $6a^2 + 9a$, subtract $6a^2 + 9a$ from $6a^2 + 5a$, and bring down -6.

$$
\begin{array}{r}
a^2 + 3a - 2 \\
2a + 3 \overline{)\, 2a^3 + 9a^2 + 5a - 6} \\
\underline{2a^3 + 3a^2 } \\
6a^2 + 5a \\
6a^2 + 9a \\
\underline{-4a - 6}
\end{array}
$$

How many times does $2a$ divide $-4a$? $-4a/2a = -2$. Place the -2 in the quotient.

$$
\begin{array}{r}
a^2 + 3a - 2 \\
2a + 3 \overline{)\, 2a^3 + 9a^2 + 5a - 6} \\
\underline{2a^3 + 3a^2 } \\
6a^2 + 5a \\
\underline{6a^2 + 9a} \\
-4a - 6 \\
\underline{-4a - 6} \\
0
\end{array}
$$

Multiply each term in the divisor by -2 to get $-4a - 6$, and subtract $-4a - 6$ from $-4a - 6$ to get 0.

Since the remainder is 0, the answer is $a^2 + 3a - 2$. We can check the quotient by verifying that

$$\underbrace{(2a + 3)}_{\text{Divisor}} \underbrace{(a^2 + 3a - 2)}_{\text{quotient}} = \underbrace{2a^3 + 9a^2 + 5a - 6}_{\text{dividend}}$$

EXAMPLE 4   Divide $3x^3 + 2x^2 - 3x + 8$ by $x - 2$.

Solution

$$
\require{enclose}
\begin{array}{r}
3x^2 + 8x + 13 \\
x - 2 \enclose{longdiv}{3x^3 + 2x^2 - 3x + 8} \\
\underline{3x^3 - 6x^2} \\
8x^2 - 3x \\
\underline{8x^2 - 16x} \\
13x + 8 \\
\underline{13x - 26} \\
34
\end{array}
$$

This division gives a quotient of $3x^2 + 8x + 13$ and a remainder of 34. It is common to form a fraction with the remainder as the numerator and the divisor as the denominator and to write the result as

$$3x^2 + 8x + 13 + \frac{34}{x - 2}$$

To check, we verify that

$$(x - 2)\left(3x^2 + 8x + 13 + \frac{34}{x - 2}\right) = 3x^3 + 2x^2 - 3x + 8$$

Self Check   Divide: $a - 3\overline{)2a^3 + 3a^2 - a + 2}$.

EXAMPLE 5   Divide $-9x + 8x^3 + 10x^2 - 9$ by $3 + 2x$.

Solution   The division algorithm works best when the polynomials in the dividend and the divisor are written in descending powers of x. We can use the commutative property of addition to rearrange the terms. Then the division is routine:

$$
\require{enclose}
\begin{array}{r}
4x^2 - x - 3 \\
2x + 3 \enclose{longdiv}{8x^3 + 10x^2 - 9x - 9} \\
\underline{8x^3 + 12x^2} \\
- 2x^2 - 9x \\
\underline{- 2x^2 - 3x} \\
- 6x - 9 \\
\underline{- 6x - 9} \\
0
\end{array}
$$

Thus,

$$\frac{-9x + 8x^3 + 10x^2 - 9}{3 + 2x} = 4x^2 - x - 3$$

Self Check   Divide: $2 + 3a\overline{)-4a + 15a^2 + 18a^3 - 4}$.

The Case of the Missing Terms

EXAMPLE 6   Divide $8x^3 + 1$ by $2x + 1$.

Solution   When we write the terms in the dividend in descending powers of x, we see that the terms involving x^2 and x are missing. We must include the terms $0x^2$ and $0x$ in the dividend or leave spaces for them. Then the division is routine.

$$
\begin{array}{r}
4x^2 - 2x + 1 \\
2x + 1 \overline{)\ 8x^3 + 0x^2 + 0x + 1} \\
\underline{8x^3 + 4x^2} \\
-4x^2 + 0x \\
\underline{-4x^2 - 2x} \\
2x + 1 \\
\underline{2x + 1} \\
0
\end{array}
$$

Thus,

$$\frac{8x^3 + 1}{2x + 1} = 4x^2 - 2x + 1$$

Self Check   Divide: $3a - 1 \overline{)\ 27a^3 - 1}$.

EXAMPLE 7   Divide $-17x^2 + 5x + x^4 + 2$ by $x^2 - 1 + 4x$.

Solution   We write the problem with the divisor and the dividend in descending powers of x. After leaving space for the missing term in the dividend, we proceed as follows:

$$
\begin{array}{r}
x^2 - 4x \\
x^2 + 4x - 1 \overline{)\ x^4 \qquad\quad - 17x^2 + 5x + 2} \\
\underline{x^4 + 4x^3 - \quad x^2} \\
-4x^3 - 16x^2 + 5x \\
\underline{-4x^3 - 16x^2 + 4x} \\
x + 2
\end{array}
$$

This division gives a quotient of $x^2 - 4x$ and a remainder of $x + 2$.

$$\frac{-17x^2 + 5x + x^4 + 2}{x^2 - 1 + 4x} = x^2 - 4x + \frac{x + 2}{x^2 + 4x - 1}$$

Self Check   Divide: $\dfrac{2a^2 + 3a^3 + a^4 - 7 + a}{a^2 - 2a + 1}$.

Self Check Answers

1. $\dfrac{3x}{4y}$ **2.** $2ab^2 - a^2 + \dfrac{1}{4}$ **4.** $2a^2 + 9a + 26 + \dfrac{80}{a - 3}$ **5.** $6a^2 + a - 2$ **6.** $9a^2 + 3a + 1$

7. $a^2 + 5a + 11 + \dfrac{18a - 18}{a^2 - 2a + 1}$

Orals   *Divide:*

1. $\dfrac{6x^2y^2}{2xy}$ **2.** $\dfrac{4ab^2 + 8a^2b}{2ab}$ **3.** $\dfrac{x^2 + 2x + 1}{x + 1}$ **4.** $\dfrac{x^2 - 4}{x - 2}$

6.7   EXERCISES

REVIEW   *Remove parentheses and simplify.*

1. $2(x^2 + 4x - 1) + 3(2x^2 - 2x + 2)$

2. $3(2a^2 - 3a + 2) - 4(2a^2 + 4a - 7)$

3. $-2(3y^3 - 2y + 7) - 3(y^2 + 2y - 4)$
    $+ 4(y^3 + 2y - 1)$

4. $3(4y^3 + 3y - 2) + 2(3y^2 - y + 3)$
    $- 5(2y^3 - y^2 - 2)$

VOCABULARY AND CONCEPTS   *Fill in the blanks.*

5. $\dfrac{a}{b} = \dfrac{\quad}{\quad} \cdot a \ (b \neq 0)$

6. The division _____ is a repeating series of steps used to do a long division.

7. divisior $\cdot$ _____ + remainder = dividend

8. If a polynomial is divided by $3a - 2$ and the quotient is $3a^2 + 5$ with a remainder of 6, we usually write the result as $3a^2 + 5 +$ _____ .

PRACTICE   *Perform each division. Write each answer without using negative exponents.*

9. $\dfrac{4x^2y^3}{8x^5y^2}$ **10.** $\dfrac{25x^4y^7}{5xy^9}$

11. $\dfrac{33a^{-2}b^2}{44a^2b^{-2}}$ **12.** $\dfrac{-63a^4b^{-3}}{81a^{-3}b^3}$

13. $\dfrac{45x^{-2}y^{-3}t^0}{-63x^{-1}y^4t^2}$ **14.** $\dfrac{112a^0b^2c^{-3}}{48a^4b^0c^4}$

15. $\dfrac{-65a^{2n}b^nc^{3n}}{-15a^nb^{-n}c}$ **16.** $\dfrac{-32x^{-3n}y^{-2n}z}{40x^{-2}y^{-n}z^{n+1}}$

17. $\dfrac{4x^2 - x^3}{6x}$ **18.** $\dfrac{5y^4 + 45y^3}{15y^2}$

19. $\dfrac{4x^2y^3 + x^3y^2}{6xy}$ **20.** $\dfrac{3a^3y^2 - 18a^4y^3}{27a^2y^2}$

21. $\dfrac{24x^6y^7 - 12x^5y^{12} + 36xy}{48x^2y^3}$

22. $\dfrac{9x^4y^3 + 18x^2y - 27xy^4}{9x^3y^3}$

23. $\dfrac{3a^{-2}b^3 - 6a^2b^{-3} + 9a^{-2}}{12a^{-1}b}$

24. $\dfrac{4x^3y^{-2} + 8x^{-2}y^2 - 12y^4}{12x^{-1}y^{-1}}$

25. $\dfrac{x^ny^n - 3x^{2n}y^{2n} + 6x^{3n}y^{3n}}{x^ny^n}$

26. $\dfrac{2a^n - 3a^nb^{2n} - 6b^{4n}}{a^nb^{n-1}}$

27. $x + 3\overline{\smash{)}x^2 + 5x + 6}$ **28.** $x - 3\overline{\smash{)}x^2 - 5x + 6}$

29. $x + 3\overline{\smash{)}x^2 + 10x + 21}$ **30.** $x + 7\overline{\smash{)}x^2 + 10x + 21}$

31. $\dfrac{6x^2 - x - 12}{2x + 3}$ **32.** $\dfrac{6x^2 - x - 12}{2x - 3}$

33. $\dfrac{3x^3 - 2x^2 + x + 6}{x - 1}$ **34.** $\dfrac{4a^3 + a^2 - 3a + 7}{a + 1}$

35. $\dfrac{6x^3 + 11x^2 - x - 2}{3x - 2}$

36. $\dfrac{6x^3 + 11x^2 - x + 10}{2x + 3}$

37. $\dfrac{6x^3 - x^2 - 6x - 9}{2x - 3}$

38. $\dfrac{16x^3 + 16x^2 - 9x - 5}{4x + 5}$ **39.** $\dfrac{2a + 1 + a^2}{a + 1}$

40. $\dfrac{a - 15 + 6a^2}{2a - 3}$ **41.** $\dfrac{6y - 4 + 10y^2}{5y - 2}$

42. $\dfrac{-10xy + x^2 + 16y^2}{x - 2y}$

43. $\dfrac{-18x + 12 + 6x^2}{x - 1}$

44. $\dfrac{27x + 23x^2 + 6x^3}{2x + 3}$

45. $\dfrac{-9x^2 + 8x + 9x^3 - 4}{3x - 2}$

46. $\dfrac{6x^2 + 8x^3 - 13x + 3}{4x - 3}$

47. $\dfrac{13x + 16x^4 + 3x^2 + 3}{4x + 3}$

48. $\dfrac{3x^2 + 9x^3 + 4x + 4}{3x + 2}$

49. $\dfrac{a^3 + 1}{a - 1}$

50. $\dfrac{27a^3 - 8b^3}{3a - 2b}$

51. $\dfrac{15a^3 - 29a^2 + 16}{3a - 4}$

52. $\dfrac{4x^3 - 12x^2 + 17x - 12}{2x - 3}$

53. $y - 2\overline{)-24y + 24 + 6y^2}$

54. $3 - a\overline{)21a - a^2 - 54}$

55. $2x + y\overline{)32x^5 + y^5}$

56. $3x - y\overline{)81x^4 - y^4}$

57. $x^2 - 2\overline{)x^6 - x^4 + 2x^2 - 8}$

58. $x^2 + 3\overline{)x^6 + 2x^4 - 6x^2 - 9}$

59. $\dfrac{x^4 + 2x^3 + 4x^2 + 3x + 2}{x^2 + x + 2}$

60. $\dfrac{2x^4 + 3x^3 + 3x^2 - 5x - 3}{2x^2 - x - 1}$

61. $\dfrac{x^3 + 3x + 5x^2 + 6 + x^4}{x^2 + 3}$

62. $\dfrac{x^5 + 3x + 2}{x^3 + 1 + 2x}$

Use a calculator to help find each quotient.

63. $x - 2\overline{)9.8x^2 - 3.2x - 69.3}$

64. $2.5x - 3.7\overline{)-22.25x^2 - 38.9x - 16.65}$

APPLICATIONS

65. Find an expression for the length of the longer sides of the rectangle.

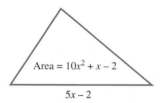

66. Find an expression for the height of the triangle.

Area $= 10x^2 + x - 2$

$5x - 2$

WRITING

67. Explain how to divide a monomial by a monomial.

68. Explain how to check the result of a division problem.

SOMETHING TO THINK ABOUT

69. Since 6 is a factor of 24, 6 divides 24 with no remainder. Decide whether $2x - 3$ is a factor of $10x^2 - x - 21$.

70. Is $x - 1$ a factor of $x^5 - 1$?

Projects

Project 1

Suppose that a motorist usually makes an 80-mile trip at an average speed of 50 mph.

a. How much time does the driver save by increasing her rate by 5 mph? By 10 mph? By 15 mph? Give all answers to the nearest minute.

b. Consider your work in part a, and find a formula that will tell how much time is saved if the motorist travels x mph faster than 50 mph. That is, find an expression involving x that will give the time saved by traveling $(50 + x)$ mph instead of 50 mph.

c. Find a formula that will give the time saved on a trip of d miles by traveling at an average rate that is x mph faster than a usual speed of y mph. When you have this formula, simplify it into a nice, compact form.

d. Test your formula by doing part a over again using the formula. The answers you get should agree with those found earlier.

e. Use your formula to solve the following problem. Every holiday season, Kurt and Ellen travel to visit their relatives, a distance of 980 miles. Under normal circumstances, they can average 60 mph during the trip. However, improved roads will enable them to travel 4 mph faster this year. How much time (to the nearest minute) will they save on this year's trip? How much faster than normal (to the nearest tenth of a mph) would Kurt and Ellen have to travel to save two hours?

Project 2

By cross-fertilization among a number of species of corn, Professor Greenthumb has succeeded in creating a miracle hybrid corn plant. Under ideal conditions, the plant will produce twice as much corn as an ordinary hybrid. However, the new hybrid is very sensitive to the amount of sun and water it receives. If conditions are not ideal, the corn yield diminishes.

To determine what yield the plants will have, daily measurements record the amount of sun the plants receive (the sun index, S) and the amount of water they receive (W, measured in millimeters). Then the daily yield diminution is calculated using the formula

$$\text{Daily yield diminution} = \frac{S^3 + W^3 - (S + W)^2}{(S + W)^3}$$

At the end of a 100-day growing season, the daily yield diminutions are added together to make the total diminution (D). The yield for the year is then determined using the formula

$$\text{Annual yield} = 1 - 0.01D$$

The resulting decimal is the percent of maximum yield that the corn plants will achieve. Remember that maximum yield for these plants is twice the yield of an ordinary corn plant.

As Greenthumb's research assistant, you have been asked to handle a few questions that have come up regarding her research.

a. First, show that if $S = W = 2$, the daily yield diminution is 0. These would be the optimal conditions for Greenthumb's plants.

b. Now suppose that for each day of the 100-day growing season, the sun index is 8, and 6 millimeters of rain fall on the plants. Find the daily yield diminution. To the nearest tenth of a percent, what percentage of the yield of normal plants will the special hybrid plants produce?

c. Show that when the daily yield diminution is 0.5 for each of the 100 days in the growing season, the annual yield will be 0.5 (exactly the yield of ordinary corn plants). Now suppose that through the use of an irrigation system, you arrange for the corn to receive 10 millimeters of rain each day. What would the sun index have to be each day to give an annual yield of 0.5? To answer this question, first simplify the daily yield diminution formula. Do this *before* you substitute any numbers into the formula.

d. Another assistant is also working with Greenthumb's formula, but he is getting different results. Something is wrong; for the situation given in part b, he finds that the daily yield diminution for one day is 34. He simplified Greenthumb's formula first, then substituted in the values for S and W. He shows you the following work. Find all of his mistakes and explain what he did wrong in making each of them.

$$\text{Daily yield diminution} = \frac{S^3 + W^3 - (S + W)^2}{(S + W)^3}$$

$$= \frac{S^3 + W^3 - S^2 + W^2}{(S + W)^3}$$

$$= \frac{(S^3 + W^3) - (S^2 - W^2)}{(S + W)^3}$$

$$= \frac{(S + W) \cdot (S^2 + SW + W^2) - (S + W) \cdot (S - W)}{(S + W)^3}$$

$$= \frac{(S^2 + SW + W^2) - S - W}{(S + W)^2}$$

$$= \frac{S^2 + SW + W^2 - S - W}{S^2 + W^2}$$

$$= SW - S - W$$

So for $S = 8$ and $W = 6$, he gets 34 as the daily yield diminution.

Chapter Summary

CONCEPTS	REVIEW EXERCISES
6.1	**Rational Functions and Simplifying Rational Expressions**

1. Graph the rational function $f(x) = \frac{3x + 2}{x}$ $(x > 0)$. Find the equation of the horizontal and vertical asymptotes.

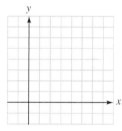

Division by 0 is undefined.

$$\frac{ak}{bk} = \frac{a}{b} \quad (b \neq 0 \text{ and } k \neq 0)$$

To simplify a rational expression, factor the numerator and denominator and divide out all factors common to the numerator and denominator.

Simplify each rational expression.

2. $\dfrac{248x^2y}{576xy^2}$ **3.** $\dfrac{212m^3n}{588m^2n^3}$

4. $\dfrac{x^2 - 49}{x^2 + 14x + 49}$ **5.** $\dfrac{x^2 + 6x + 36}{x^3 - 216}$

6. $\dfrac{x^2 - 2x + 4}{2x^3 + 16}$ **7.** $\dfrac{x - y}{y - x}$

8. $\dfrac{2m - 2n}{n - m}$ **9.** $\dfrac{ac - ad + bc - bd}{d^2 - c^2}$

Find each probability.

10. Rolling an 11 on one roll of two dice

11. Three children, all boys

	6.2

6.2   Proportion and Variation

In a proportion, the product of the extremes is equal to the product of the means.

If two angles of one triangle have the same measure as two angles of a second triangle, the triangles are similar.

Direct variation:

$y = kx$ (*k* is a constant)

Inverse variation:

$y = \dfrac{k}{x}$ (*k* is a constant)

Joint variation:

$y = kxz$ (*k* is a constant)

Combined variation:

$y = \dfrac{kx}{y}$ (*k* is a constant)

Solve each proportion.

12. $\dfrac{x + 1}{8} = \dfrac{4x - 2}{24}$

13. $\dfrac{1}{x + 6} = \dfrac{x + 10}{12}$

14. Find the height of a tree if it casts a 44-foot shadow when a 4-foot tree casts a $2\frac{1}{2}$-foot shadow.

15. Assume that *x* varies directly with *y*. If $x = 12$ when $y = 2$, find *x* when $y = 12$.

16. Assume that *x* varies inversely with *y*. If $x = 24$ when $y = 3$, find *y* when $x = 12$.

17. Assume that *x* varies jointly with *y* and *z*. Find the constant of variation if $x = 24$ when $y = 3$ and $z = 4$.

18. Assume that *x* varies directly with *t* and inversely with *y*. Find the constant of variation if $x = 2$ when $t = 8$ and $y = 64$.

6.3   Multiplying and Dividing Rational Expressions

$\dfrac{a}{b} \cdot \dfrac{c}{d} = \dfrac{ac}{bd}$ $(b, d \neq 0)$

$\dfrac{a}{b} \div \dfrac{c}{d} = \dfrac{a}{b} \cdot \dfrac{d}{c}$ $(b, d, c \neq 0)$

Perform the operations and simplify.

19. $\dfrac{x^2 + 4x + 4}{x^2 - x - 6} \cdot \dfrac{x^2 - 9}{x^2 + 5x + 6}$

20. $\dfrac{x^2 + 3x + 2}{x^2 - x - 6} \cdot \dfrac{3x^2 - 3x}{x^2 - 3x - 4} \div \dfrac{x^2 + 3x + 2}{x^2 - 2x - 8}$

21. $\dfrac{x^2 - x - 6}{x^2 - 3x - 10} \div \dfrac{x^2 - x}{x^2 - 5x} \cdot \dfrac{x^2 - 4x + 3}{x^2 - 6x + 9}$

| **6.4** | **Adding and Subtracting Rational Expressions** |

$$\frac{a}{b} + \frac{c}{b} = \frac{a+c}{b} \quad (b \neq 0)$$

$$\frac{a}{b} - \frac{c}{b} = \frac{a-c}{b} \quad (b \neq 0)$$

To find the LCD of two fractions, factor each denominator and use each factor the greatest number of times that it appears in any one denominator. The product of these factors is the LCD.

Perform the operations and simplify.

22. $\dfrac{5y}{x-y} - \dfrac{3}{x-y}$

23. $\dfrac{3x-1}{x^2+2} + \dfrac{3(x-2)}{x^2+2}$

24. $\dfrac{3}{x+2} + \dfrac{2}{x+3}$

25. $\dfrac{4x}{x-4} - \dfrac{3}{x+3}$

26. $\dfrac{2x}{x+1} + \dfrac{3x}{x+2} + \dfrac{4x}{x^2+3x+2}$

27. $\dfrac{5x}{x-3} + \dfrac{5}{x^2-5x+6} + \dfrac{x+3}{x-2}$

28. $\dfrac{3(x+2)}{x^2-1} - \dfrac{2}{x+1} + \dfrac{4(x+3)}{x^2-2x+1}$

29. $\dfrac{-2(3+x)}{x^2+6x+9} + \dfrac{3(x+2)}{x^2-6x+9} - \dfrac{1}{x^2-9}$

| **6.5** | **Complex Fractions** |

A fraction that has a fraction in its numerator or its denominator is called a **complex fraction.**

Simplify each complex fraction.

30. $\dfrac{\dfrac{3}{x} - \dfrac{2}{y}}{xy}$

31. $\dfrac{\dfrac{1}{x} + \dfrac{2}{y}}{\dfrac{2}{x} - \dfrac{1}{y}}$

32. $\dfrac{2x + 3 + \dfrac{1}{x}}{x + 2 + \dfrac{1}{x}}$

33. $\dfrac{6x + 13 + \dfrac{6}{x}}{6x + 5 - \dfrac{6}{x}}$

34. $\dfrac{1 - \dfrac{1}{x} - \dfrac{2}{x^2}}{1 + \dfrac{4}{x} + \dfrac{3}{x^2}}$

35. $\dfrac{x^{-1} + 1}{x + 1}$

36. $\dfrac{x^{-1} - y^{-1}}{x^{-1} + y^{-1}}$

37. $\dfrac{(x-y)^{-2}}{x^{-2} - y^{-2}}$

6.6 Equations Containing Rational Expressions

Multiplying both sides of an equation by a quantity that contains a variable can lead to false solutions. All possible solutions of a rational equation must be checked.

Solve each equation, if possible.

38. $\dfrac{4}{x} - \dfrac{1}{10} = \dfrac{7}{2x}$

39. $\dfrac{2}{x+5} - \dfrac{1}{6} = \dfrac{1}{x+4}$

40. $\dfrac{2(x-5)}{x-2} = \dfrac{6x+12}{4-x^2}$

41. $\dfrac{7}{x+9} - \dfrac{x+2}{2} = \dfrac{x+4}{x+9}$

Solve each formula for the indicated variable.

42. $\dfrac{x^2}{a^2} - \dfrac{y^2}{b^2} = 1$ for y^2

43. $H = \dfrac{2ab}{a+b}$ for b

44. Trip length Traffic reduced Jim's usual speed by 10 mph, which lengthened his 200-mile trip by 1 hour. Find his usual average speed.

45. Flying speed On a 600-mile trip, a pilot can save 30 minutes by increasing her usual speed by 40 mph. Find her usual speed.

46. Draining a tank If one outlet pipe can drain a tank in 24 hours and another pipe can drain the tank in 36 hours, how long will it take for both pipes to drain the tank?

47. Siding a house Two men have estimated that they can side a house in 8 days. If one of them, who could have sided the house alone in 14 days, gets sick, how long will it take the other man to side the house alone?

6.7 Dividing Polynomials

To find the quotient of two monomials, express the quotient as a fraction and use the rules of exponents to simplify.

Perform each division.

48. $\dfrac{-5x^6y^3}{10x^3y^6}$

49. $\dfrac{30x^3y^2 - 15x^2y - 10xy^2}{-10xy}$

50. $(3x^2 + 13xy - 10y^2) \div (3x - 2y)$

51. $(2x^3 + 7x^2 + 3 + 4x) \div (2x + 3)$

Chapter Test

Simplify each rational expression.

1. $\dfrac{-12x^2y^3z^2}{18x^3y^4z^2}$

2. $\dfrac{2x+4}{x^2-4}$

3. $\dfrac{3y-6z}{2z-y}$

4. $\dfrac{2x^2+7x+3}{4x+12}$

5. Find the probability of tossing a coin three times and getting no heads.

6. Find the height of a tree that casts a shadow of 12 feet when a vertical yardstick casts a shadow of 2 feet.

7. Solve the proportion: $\dfrac{3}{x-2}=\dfrac{x+3}{2x}$.

8. V varies inversely with t. If $V=55$ when $t=20$, find t when $V=75$.

Perform the operations and simplify, if necessary. Write all answers without negative exponents.

9. $\dfrac{x^2y^{-2}}{x^3z^2}\cdot\dfrac{x^2z^4}{y^2z}$

10. $\dfrac{(x+1)(x+2)}{10}\cdot\dfrac{5}{x+2}$

11. $\dfrac{u^2+5u+6}{u^2-4}\cdot\dfrac{u^2-5u+6}{u^2-9}$

12. $\dfrac{x^3+y^3}{4}\div\dfrac{x^2-xy+y^2}{2x+2y}$

13. $\dfrac{xu+2u+3x+6}{u^2-9}\cdot\dfrac{2u-6}{x^2+3x+2}$

14. $\dfrac{a^2+7a+12}{a+3}\div\dfrac{16-a^2}{a-4}$

15. $\dfrac{3t}{t+3}+\dfrac{9}{t+3}$

16. $\dfrac{3w}{w-5}+\dfrac{w+10}{5-w}$

17. $\dfrac{2}{r}+\dfrac{r}{s}$

18. $\dfrac{x+2}{x+1}-\dfrac{x+1}{x+2}$

Simplify each complex fraction.

19. $\dfrac{\dfrac{2u^2w^3}{v^2}}{\dfrac{4uw^4}{uv}}$

20. $\dfrac{\dfrac{x}{y}+\dfrac{1}{2}}{\dfrac{x}{2}-\dfrac{1}{y}}$

Solve each equation.

21. $\dfrac{2}{x-1}+\dfrac{5}{x+2}=\dfrac{11}{x+2}$

22. $\dfrac{u-2}{u-3}+3=u+\dfrac{u-4}{3-u}$

Solve each formula for the indicated variable.

23. $\dfrac{x^2}{a^2}+\dfrac{y^2}{b^2}=1$ for a^2

24. $\dfrac{1}{r}=\dfrac{1}{r_1}+\dfrac{1}{r_2}$ for r_2

25. **Sailing time** A boat sails a distance of 440 nautical miles. If the boat had averaged 11 nautical miles more each day, the trip would have required 2 fewer days. How long did the trip take?

26. **Investing** A student can earn \$300 interest annually by investing in a certificate of deposit at a certain interest rate. If she were to receive an annual interest rate that is 4% higher, she could receive the same annual interest by investing \$2,000 less. How much would she invest at each rate?

27. Divide: $\dfrac{18x^2y^3-12x^3y^2+9xy}{-3xy^4}$.

28. Divide: $(6x^3+5x^2-2)\div(2x-1)$.

CUMULATIVE REVIEW EXERCISES

Simplify each expression.

1. $a^3b^2a^5b^2$

2. $\dfrac{a^3b^6}{a^7b^2}$

3. $\left(\dfrac{2a^2}{3b^4}\right)^{-4}$

4. $\left(\dfrac{x^{-2}y^3}{x^2x^3y^4}\right)^{-3}$

Write each number in standard notation.

5. 4.25×10^4

6. 7.12×10^{-4}

Solve each equation.

7. $\dfrac{a+2}{5} - \dfrac{8}{5} = 4a - \dfrac{a+9}{2}$

8. $\dfrac{3x-4}{6} - \dfrac{x-2}{2} = \dfrac{-2x-3}{3}$

Find the slope of the line with the given properties.

9. Passing through $(-2, 5)$ and $(4, 10)$

10. Has an equation of $3x + 4y = 13$

11. Parallel to a line with equation of $y = 3x + 2$.

12. Perpendicular to a line with equation of $y = 3x + 2$

Let $f(x) = x^2 - 2x$ and find each value.

13. $f(0)$

14. $f(-2)$

15. $f\left(\dfrac{2}{5}\right)$

16. $f(t-1)$

17. Express as a formula: *y varies directly with the product of x and z, but inversely with r.*

18. Does the graph below represent a function?

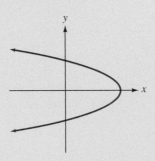

Solve each inequality and graph the solution set.

19. $x - 2 \le 3x + 1 \le 5x - 4$

20. $\left|\dfrac{3a}{5} - 2\right| + 1 \ge \dfrac{6}{5}$

21. Is $3 + x + x^2$ a monomial, a binomial, or a trinomial?

22. Find the degree of $3 + x^2y + 17x^3y^4$.

23. If $f(x) = -3x^3 + x - 4$, find $f(-2)$.

24. Graph: $y = f(x) = 2x^2 - 3$.

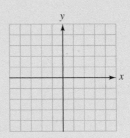

Perform the operations and simplify.

25. $(3x^2 - 2x + 7) + (-2x^2 + 2x + 5) + (3x^2 - 4x + 2)$

26. $(-5x^2 + 3x + 4) - (-2x^2 + 3x + 7)$

27. $(3x + 4)(2x - 5)$

28. $(2x^n - 1)(x^n + 2)$

Factor each expression.

29. $3r^2s^3 - 6rs^4$

30. $5(x - y) - a(x - y)$

31. $xu + yv + xv + yu$

32. $81x^4 - 16y^4$

33. $8x^3 - 27y^6$

34. $6x^2 + 5x - 6$

35. $9x^2 - 30x + 25$

36. $15x^2 - x - 6$

37. $27a^3 + 8b^3$

38. $6x^2 + x - 35$

39. $x^2 + 10x + 25 - y^4$

40. $y^2 - x^2 + 4x - 4$

Solve each equation.

41. $x^3 - 4x = 0$ **42.** $6x^2 + 7 = -23x$

Simplify each expression.

43. $\dfrac{2x^2y + xy - 6y}{3x^2y + 5xy - 2y}$

44. $\dfrac{x^2 - 4}{x^2 + 9x + 20} \div \dfrac{x^2 + 5x + 6}{x^2 + 4x - 5} \cdot \dfrac{x^2 + 3x - 4}{(x - 1)^2}$

45. $\dfrac{2}{x + y} + \dfrac{3}{x - y} - \dfrac{x - 3y}{x^2 - y^2}$

46. $\dfrac{\dfrac{a}{b} + b}{a - \dfrac{b}{a}}$

Solve each equation.

47. $\dfrac{5x - 3}{x + 2} = \dfrac{5x + 3}{x - 2}$

48. $\dfrac{3}{x - 2} + \dfrac{x^2}{(x + 3)(x - 2)} = \dfrac{x + 4}{x + 3}$

Perform the operations.

49. $(x^2 + 9x + 20) \div (x + 5)$

50. $(2x^2 + 4x - x^3 + 3) \div (x - 1)$

7

Radicals and Rational Exponents

InfoTrac

Do a subject guide search on "alternating current." Click on "View periodical references," and find the article "AC fundamentals" from *EC&M Electrical Construction & Maintenance,* Jan 1, 2003, v102. In an alternating current circuit, voltage, current, and impedance are represented by complex numbers. Using the formula found in the article, find the voltage (in volts) of a circuit with a current of $5 - 6.4i$ amperes and an impedance of $3.2 + 4.25i$ ohms. Now find the current of a circuit with voltage of $6.25 - 4.00i$ volts and an impedance of $5.3 + 6.75i$ ohms.

Complete this project after studying Section 7.7.

© Royalty-Free/CORBIS

Mathematics in Photography

Many cameras have an adjustable lens opening, called an **aperture,** that controls the amount of light passing through the circular lens. Various lenses—wide-angle, close-up, and telephoto—are distinguished by their **focal lengths.**

The **f-number** of a lens is its focal length divided by the diameter of its aperture.

**Example 5
Section 7.5**

$$f\text{-number} = \frac{f}{d} \qquad f \text{ is the focal length, and } d \text{ is the diameter of the aperture.}$$

For example, a lens with a focal length of 12 centimeters and an aperture 6 centimeters in diameter has an *f*-number of $\frac{12}{6}$ and is called an *f*/2 lens.

If the area of an aperture is reduced to admit half as much light as an *f*/2 lens, the *f*-number of the lens will change. Find the new *f*-number.

In this chapter, we will reverse the squaring process and learn how to find square roots of numbers. We will also learn how to find other roots of numbers and discuss complex numbers.

7.1 Radical Expressions

- **Square Roots** ▮ **Square Roots of Expressions with Variables**
- **Cube Roots** ▮ ***n*th Roots** ▮ **Square Root and Cube Root Functions**
- **Standard Deviation**

Getting Ready   *Find each power.*

1. 0^2 2. 4^2 3. $(-4)^2$ 4. -4^2

5. $\left(\dfrac{2}{5}\right)^3$ 6. $\left(-\dfrac{3}{4}\right)^4$ 7. $(7xy)^2$ 8. $(7xy)^3$

In this section, we will discuss square roots and other roots of algebraic expressions. We will also consider their related functions.

Square Roots

When solving problems, we must often find what number must be squared to obtain a second number *a*. If such a number can be found, it is called a **square root** of *a*. For example,

- 0 is a square root of 0, because $0^2 = 0$.
- 4 is a square root of 16, because $4^2 = 16$.
- -4 is a square root of 16, because $(-4)^2 = 16$.
- $7xy$ is a square root of $49x^2y^2$, because $(7xy)^2 = 49x^2y^2$.
- $-7xy$ is a square root of $49x^2y^2$, because $(-7xy)^2 = 49x^2y^2$.

All positive numbers have two real-number square roots: one that is positive and one that is negative.

EXAMPLE 1 Find the two square roots of 121.

Solution The two square roots of 121 are 11 and -11, because

$$11^2 = 121 \qquad \text{and} \qquad (-11)^2 = 121$$

Self Check Find the square roots of 144. ∎

To express square roots, we use the symbol $\sqrt{}$, called a **radical sign.** For example,

$$\sqrt{121} = 11 \qquad \text{Read as "The positive square root of 121 is 11."}$$

$$-\sqrt{121} = -11 \qquad \text{Read as "The negative square root of 121 is } -11\text{."}$$

The number under the radical sign is called a **radicand.**

Square Root of a

> If $a > 0$, $\sqrt{a}$ is the positive number whose square is a. In symbols,
>
> $$\left(\sqrt{a}\right)^2 = a$$
>
> The positive number $\sqrt{a}$ is called the **principal square root of a.** The principal square root of 0 is 0: $\sqrt{0} = 0$.

 Comment Remember that the principal square root of a positive number is always positive. Although 5 and -5 are both square roots of 25, only 5 is the principal square root. The radical expression $\sqrt{25}$ represents 5. The radical expression $-\sqrt{25}$ represents -5.

Because of the previous definition, the square root of any number squared is that number. For example,

$$\left(\sqrt{10}\right)^2 = \sqrt{10} \cdot \sqrt{10} = 10 \qquad \left(\sqrt{a}\right)^2 = \sqrt{a} \cdot \sqrt{a} = a.$$

 Comment These examples suggest that if any number a can be factored into two equal factors, either of those factors is a square root of a.

EXAMPLE 2 Simplify each radical:

a. $\sqrt{1} = 1$ 　　　　　　　　　　**b.** $\sqrt{81} = 9$

c. $-\sqrt{81} = -9$ 　　　　　　　**d.** $-\sqrt{225} = -15$

e. $\sqrt{\dfrac{1}{4}} = \dfrac{1}{2}$ 　　　　　　　**f.** $-\sqrt{\dfrac{16}{121}} = -\dfrac{4}{11}$

g. $\sqrt{0.04} = 0.2$ 　　　　　　**h.** $-\sqrt{0.0009} = -0.03$

Self Check Simplify: **a.** $-\sqrt{49}$ and **b.** $\sqrt{\dfrac{25}{49}}$. ∎

Numbers such as 1, 4, 9, 16, 49, and 1,600 are called **integer squares,** because each one is the square of an integer. The square root of every integer square is a rational number.

$$\sqrt{1} = 1, \qquad \sqrt{4} = 2, \qquad \sqrt{9} = 3, \qquad \sqrt{16} = 4, \qquad \sqrt{49} = 7, \qquad \sqrt{1,600} = 40$$

The square roots of many positive integers are not rational numbers. For example, $\sqrt{11}$ is an **irrational number.** To find an approximate value of $\sqrt{11}$ with a calculator, we enter these numbers and press these keys.

11 $\boxed{\text{2nd}}$ $\boxed{\sqrt{}}$ On a scientific calculator.

$\boxed{\text{2nd}}$ $\boxed{\sqrt{}}$ 11 $\boxed{\text{ENTER}}$ On a graphing calculator.

Either way, we will see that

$$\sqrt{11} \approx 3.31662479$$

Square roots of negative numbers are not real numbers. For example, $\sqrt{-9}$ is not a real number, because no real number squared equals -9. Square roots of negative numbers come from a set called **imaginary numbers,** which we will discuss later in this chapter.

Square Roots of Expressions with Variables

If $x \neq 0$, the positive number x^2 has x and $-x$ for its two square roots. To denote the positive square root of $\sqrt{x^2}$, we must know whether x is positive or negative.

If $x > 0$, we can write

$$\sqrt{x^2} = x \qquad \sqrt{x^2} \text{ represents the positive square root of } x^2, \text{ which is } x.$$

If x is negative, then $-x > 0$, and we can write

$$\sqrt{x^2} = -x \qquad \sqrt{x^2} \text{ represents the positive square root of } x^2, \text{ which is } -x.$$

If we don't know whether x is positive or negative, we must use absolute value symbols to guarantee that $\sqrt{x^2}$ is positive.

Definition of $\sqrt{x^2}$

> If x can be any real number then
> $$\sqrt{x^2} = |x|$$

EXAMPLE 3   Simplify each expression. Assume that x can be any real number.

a. $\sqrt{16x^2} = \sqrt{(4x)^2}$ Write $16x^2$ as $(4x)^2$.

$\qquad\qquad = |4x|$ Because $(|4x|)^2 = 16x^2$. Since x could be negative, absolute value symbols are needed.

$\qquad\qquad = 4|x|$ Since 4 is a positive constant in the product $4x$, we can write it outside the absolute value symbols.

b. $\sqrt{x^2 + 2x + 1}$

$\qquad = \sqrt{(x + 1)^2}$ Factor $x^2 + 2x + 1$.

$\qquad = |x + 1|$ Because $(x + 1)^2 = x^2 + 2x + 1$. Since $x + 1$ can be negative (for example, when $x = -5$), absolute value symbols are needed.

c. $\sqrt{x^4} = x^2$ Because $(x^2)^2 = x^4$. Since $x^2 \geq 0$, no absolute value symbols are needed.

Self Check   Simplify: **a.** $\sqrt{25a^2}$ and   **b.** $\sqrt{16a^4}$.   ∎

Cube Roots

The **cube root of x** is any number whose cube is x. For example,

4 is a cube root of 64, because $4^3 = 64$.

$3x^2y$ is a cube root of $27x^6y^3$, because $(3x^2y)^3 = 27x^6y^3$.

$-2y$ is a cube root of $-8y^3$, because $(-2y)^3 = -8y^3$.

Cube Roots
> The **cube root of a** is denoted as $\sqrt[3]{a}$ and is the number whose cube is a. In symbols,
>
> $$\left(\sqrt[3]{a}\right)^3 = a$$

 Comment   The previous definition implies that if a can be factored into three equal factors, any one of those factors is a cube root of a.

We note that 64 has two real-number square roots, 8 and -8. However, 64 has only one real-number cube root, 4, because 4 is the only real number whose cube is 64. Since every real number has exactly one real cube root, it is unnecessary to use absolute value symbols when simplifying cube roots.

Definition of $\sqrt[3]{x^3}$
> If x is any real number, then
>
> $$\sqrt[3]{x^3} = x$$

 EXAMPLE 4   Simplify each radical:

a. $\sqrt[3]{125} = 5$ Because $5^3 = 5 \cdot 5 \cdot 5 = 125$.

b. $\sqrt[3]{\dfrac{1}{8}} = \dfrac{1}{2}$ Because $\left(\dfrac{1}{2}\right)^3 = \dfrac{1}{2} \cdot \dfrac{1}{2} \cdot \dfrac{1}{2} = \dfrac{1}{8}$.

c. $\sqrt[3]{-27x^3} = -3x$ Because $(-3x)^3 = (-3x)(-3x)(-3x) = -27x^3$.

d. $\sqrt[3]{\dfrac{8a^3}{27b^3}} = -\dfrac{2a}{3b}$ Because $\left(-\dfrac{2a}{3b}\right)^3 = \left(-\dfrac{2a}{3b}\right)\left(-\dfrac{2a}{3b}\right)\left(-\dfrac{2a}{3b}\right) = -\dfrac{8a^3}{27b^3}$.

e. $\sqrt[3]{0.216x^3y^6} = 0.6xy^2$ Because $(0.6xy^2)^3 = (0.6xy^2)(0.6xy^2)(0.6xy^2) = 0.216x^3y^6$.

Self Check   Simplify: **a.** $\sqrt[3]{1,000}$,   **b.** $\sqrt[3]{\dfrac{1}{27}}$,   and   **c.** $\sqrt[3]{125a^3}$.   ∎

*n*th Roots

Just as there are square roots and cube roots, there are fourth roots, fifth roots, sixth roots, and so on.

When n is an odd natural number greater than 1, $\sqrt[n]{x}$ represents an **odd root.** Since every real number has only one real nth root when n is odd, we don't need to use absolute value symbols when finding odd roots. For example,

$$\sqrt[5]{243} = 3 \qquad \text{because} \qquad 3^5 = 243$$

$$\sqrt[7]{-128x^7} = -2x \qquad \text{because} \qquad (-2x)^7 = -128x^7$$

When n is an even natural number greater than 1, $\sqrt[n]{x}$ represents an **even root**. In this case, there will be one positive and one negative real nth root. For example, the two real sixth roots of 729 are 3 and -3, because $3^6 = 729$ and $(-3)^6 = 729$. When finding even roots, we use absolute value symbols to guarantee that the principal nth root is positive.

$$\sqrt[4]{(-3)^4} = |-3| = 3$$ Because $3^4 = (-3)^4$. We could also simplify this as follows: $\sqrt[4]{(-3)^4} = \sqrt[4]{81} = 3$.

$$\sqrt[6]{729x^6} = |3x| = 3|x|$$ Because $(3|x|)^6 = 729x^6$. The absolute value symbols guarantee that the sixth root is positive.

In the radical $\sqrt[n]{x}$, n is called the **index** (or **order**) of the radical. When the index is 2, the radical is a square root, and we usually do not write the index.

$$\sqrt[2]{x} = \sqrt{x}$$

 Comment   When n is an even number greater than 1, and $x < 0$, $\sqrt[n]{x}$ is not a real number. For example, $\sqrt[4]{-81}$ is not a real number, because no real number raised to the 4th power is -81.

EXAMPLE 5   Simplify each radical:

a. $\sqrt[4]{625} = 5$, because $5^4 = 625$ Read $\sqrt[4]{625}$ as "the fourth root of 625."

b. $\sqrt[5]{-32} = -2$, because $(-2)^5 = -32$ Read $\sqrt[5]{-32}$ as "the fifth root of -32."

c. $\sqrt[6]{\dfrac{1}{64}} = \dfrac{1}{2}$, because $\left(\dfrac{1}{2}\right)^6 = \dfrac{1}{64}$ Read $\sqrt[6]{\dfrac{1}{64}}$ as "the sixth root of $\dfrac{1}{64}$."

d. $\sqrt[7]{10^7} = 10$, because $10^7 = 10^7$ Read $\sqrt[7]{10^7}$ as "the seventh root of 10^7."

Self Check   Simplify: **a.** $\sqrt[4]{\frac{1}{81}}$ and   **b.** $\sqrt[5]{10^5}$.   ∎

When finding the nth root of an nth power, we can use the following rules.

Definition of $\sqrt[n]{a^n}$

If n is an odd natural number greater than 1, then $\sqrt[n]{a^n} = a$.

If n is an even natural number, then $\sqrt[n]{a^n} = |a|$.

EXAMPLE 6   Simplify each radical. Assume that x can be any real number.

Solution   **a.** $\sqrt[5]{x^5} = x$ Since n is odd, absolute value symbols aren't needed.

b. $\sqrt[4]{16x^4} = |2x| = 2|x|$ Since n is even and x can be negative, absolute value symbols are needed to guarantee that the result is positive.

c. $\sqrt[6]{(x + 4)^6} = |x + 4|$ Absolute value symbols are needed to guarantee that the result is positive.

d. $\sqrt[3]{(x + 1)^3} = x + 1$ Since n is odd, absolute value symbols aren't needed.

e. $\sqrt{(x^2 + 4x + 4)^2} = \sqrt{[(x + 2)^2]^2}$ Factor $x^2 + 4x + 4$.
$$= \sqrt{(x + 2)^4}$$
$$= (x + 2)^2$$ Since $(x + 2)^2$ is always positive, absolute value symbols aren't needed.

Self Check   Simplify: **a.** $\sqrt[4]{16a^4}$ and   **b.** $\sqrt[5]{(a+5)^5}$.   ∎

We summarize the possibilities for $\sqrt[n]{x}$ as follows.

Definition for $\sqrt[n]{x}$

If n is a natural number greater than 1 and x is a real number, then

If $x > 0$, then $\sqrt[n]{x}$ is the positive number such that $\left(\sqrt[n]{x}\right)^n = x$.

If $x = 0$, then $\sqrt[n]{x} = 0$.

If $x < 0$ $\begin{cases} \text{and } n \text{ is odd, then } \sqrt[n]{x} \text{ is the real number such that } \left(\sqrt[n]{x}\right)^n = x. \\ \text{and } n \text{ is even, then } \sqrt[n]{x} \text{ is not a real number.} \end{cases}$

Square Root and Cube Root Functions

Since there is one principal square root for every nonnegative real number x, the equation $f(x) = \sqrt{x}$ determines a function, called the **square root function.**

EXAMPLE 7   Graph $f(x) = \sqrt{x}$ and find its domain and range.

Solution   We can make a table of values and plot points to get the graph shown in Figure 7-1(a), or we can use a graphing calculator with window settings of $[-1, 9]$ for x and $[-2, 5]$ for y to get the graph shown in Figure 7-1(b) Since the equation defines a function, its graph passes the vertical line test.

$f(x) = \sqrt{x}$

x	$f(x)$	$(x, f(x))$
0	0	$(0, 0)$
1	1	$(1, 1)$
4	2	$(4, 2)$
9	3	$(9, 3)$

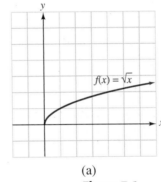

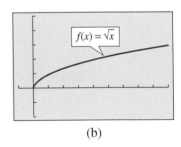

(a)   (b)

Figure 7-1

From either graph, we can see that the domain and the range are the set of non-negative real numbers, which is the interval $[0, \infty)$.

Self Check   Graph $f(x) = \sqrt{x} + 2$ and compare it to the graph of $f(x) = \sqrt{x}$.   ∎

The graphs of many functions are translations or reflections of the square root function. For example, if $k > 0$,

- The graph of $f(x) = \sqrt{x} + k$ is the graph of $f(x) = \sqrt{x}$ translated k units up.
- The graph of $f(x) = \sqrt{x} - k$ is the graph of $f(x) = \sqrt{x}$ translated k units down.
- The graph of $f(x) = \sqrt{x + k}$ is the graph of $f(x) = \sqrt{x}$ translated k units to the left.

- The graph of $f(x) = \sqrt{x - k}$ is the graph of $f(x) = \sqrt{x}$ translated k units to the right.

- The graph of $f(x) = -\sqrt{x}$ is the graph of $f(x) = \sqrt{x}$ reflected about the x-axis.

EXAMPLE 8 Graph $f(x) = -\sqrt{x + 4} - 2$ and find its domain and range.

Solution This graph will be the reflection of $f(x) = \sqrt{x}$ about the x-axis, translated 4 units to the left and 2 units down. See Figure 7-2(a). We can confirm this graph by using a graphing calculator with window settings of $[-5, 6]$ for x and $[-6, 2]$ for y to get the graph shown in Figure 7-2(b).

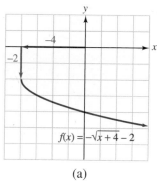

(a)

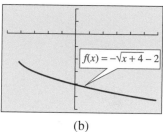

(b)

Figure 7-2

From either graph, we can see that the domain is the interval $[-4, \infty)$ and that the range is the interval $(-\infty, -2]$.

Self Check Graph: $f(x) = \sqrt{x - 2} - 4$. ∎

EXAMPLE 9 **Period of a pendulum** The **period of a pendulum** is the time required for the pendulum to swing back and forth to complete one cycle. (See Figure 7-3.) The period t (in seconds) is a function of the pendulum's length l, which is defined by the formula

$$t = f(l) = 2\pi\sqrt{\frac{l}{32}}$$

Find the period of a pendulum that is 5 feet long.

Solution We substitute 5 for l in the formula and simplify.

$$t = 2\pi\sqrt{\frac{l}{32}}$$

$$t = 2\pi\sqrt{\frac{5}{32}}$$

≈ 2.483647066 Use a calculator.

To the nearest tenth, the period is 2.5 seconds.

Figure 7-3

Self Check To the nearest hundredth, find the period of a pendulum that is 3 feet long. ∎

FINDING THE PERIOD OF A PENDULUM

To solve Example 9 with a graphing calculator with window settings of $[-2, 10]$ for x and $[-2, 10]$ for y, we graph the function $f(x) = 2\pi\sqrt{\dfrac{x}{32}}$, as in Figure 7-4(a). We then trace and move the cursor toward an x value of 5 until we see the coordinates shown in Figure 7-4(b). The period is given by the y-value shown in the screen. By zooming in, we can get better results.

$$f(x) = 2\pi\sqrt{\frac{x}{32}}$$

Y₁ = 2π √(X/32)

X = 5.0212766   Y = 2.4889258

(a)   (b)

Figure 7-4

The equation $f(x) = \sqrt[3]{x}$ defines a **cube root function.** From the graph shown in Figure 7-5(a), we can see that the domain and range of the function $f(x) = \sqrt[3]{x}$ are the set of real numbers. Note that the graph of $f(x) = \sqrt[3]{x}$ passes the vertical line test. Figures 7-5(b) and 7-5(c) show several translations of the cube root function.

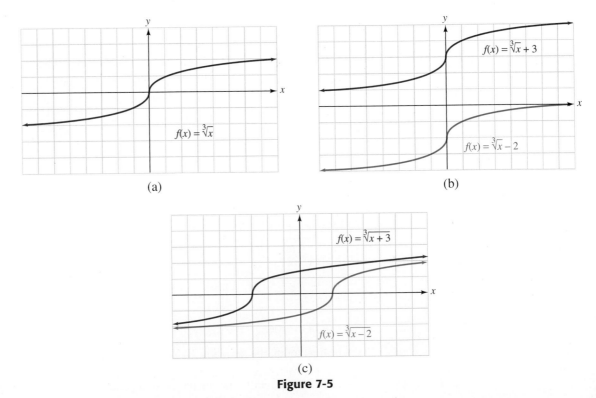

$f(x) = \sqrt[3]{x}$

(a)

$f(x) = \sqrt[3]{x} + 3$

$f(x) = \sqrt[3]{x} - 2$

(b)

$f(x) = \sqrt[3]{x} + 3$

$f(x) = \sqrt[3]{x} - 2$

(c)

Figure 7-5

Standard Deviation

In statistics, the **standard deviation** is used to tell which of a set of distributions is the most variable. To see how to compute the standard deviation of a distribution, we consider the distribution 4, 5, 5, 8, 13 and construct the following table.

Original terms	Mean of the distribution	Differences (original term minus mean)	Squares of the differences from the mean
4	7	−3	9
5	7	−2	4
5	7	−2	4
8	7	1	1
13	7	6	36

The population **standard deviation** of the distribution is the positive square root of the mean of the numbers shown in column 4 of the table.

$$\text{Standard deviation} = \sqrt{\frac{\text{sum of the squares of the differences from the mean}}{\text{number of differences}}}$$

$$= \sqrt{\frac{9 + 4 + 4 + 1 + 36}{5}}$$

$$= \sqrt{\frac{54}{5}}$$

$$\approx 3.286335345 \quad \text{Use a calculator.}$$

To the nearest hundredth, the standard deviation of the given distribution is 3.29.

The symbol for the population standard deviation is σ, the lowercase Greek letter *sigma*.

EXAMPLE 10 Which of the following distributions has the most variability: **a.** 3, 5, 7, 8, 12 or **b.** 1, 4, 6, 11?

Solution We compute the standard deviation of each distribution.

a.

Original terms	Mean of the distribution	Differences (original term minus mean)	Squares of the differences from the mean
3	7	−4	16
5	7	−2	4
7	7	0	0
8	7	1	1
12	7	5	25

$$\sigma = \sqrt{\frac{16 + 4 + 0 + 1 + 25}{5}} = \sqrt{\frac{46}{5}} \approx 3.03$$

b.

Original terms	Mean of the distribution	Differences (original term minus mean)	Squares of the differences from the mean
1	5.5	−4.5	20.25
4	5.5	−1.5	2.25
6	5.5	0.5	0.25
11	5.5	5.5	30.25

$$\sigma = \sqrt{\frac{20.25 + 2.25 + 0.25 + 30.25}{4}} = \sqrt{\frac{53}{4}} \approx 3.64$$

Since the standard deviation for the second distribution is greater than the standard deviation for the first distribution, the second distribution has the greater variability.

Self Check Answers

1. $12, -12$ **2. a.** -7,   **b.** $\dfrac{5}{7}$ **3. a.** $5|a|$,   **b.** $4a^2$ **4. a.** 10,   **b.** $\dfrac{1}{3}$,   **c.** $5a$ **5. a.** $\dfrac{1}{3}$,   **b.** 10

6. a. $2|a|$,   **b.** $a + 5$ **7.** It is 2 units higher.   **8.**   **9.** 1.92 sec

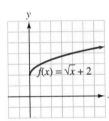

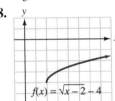

Orals   *Simplify each radical, if possible.*

1. $\sqrt{9}$ **2.** $-\sqrt{16}$ **3.** $\sqrt[3]{-8}$ **4.** $\sqrt[5]{32}$

5. $\sqrt{64x^2}$ **6.** $\sqrt[3]{-27x^3}$

7. $\sqrt{-3}$ **8.** $\sqrt[4]{(x + 1)^8}$

7.1   EXERCISES

REVIEW   *Simplify each rational expression.*

1. $\dfrac{x^2 + 7x + 12}{x^2 - 16}$

2. $\dfrac{a^3 - b^3}{b^2 - a^2}$

Perform the operations.

3. $\dfrac{x^2 - x - 6}{x^2 - 2x - 3} \cdot \dfrac{x^2 - 1}{x^2 + x - 2}$

4. $\dfrac{x^2 - 3x - 4}{x^2 - 5x + 6} \div \dfrac{x^2 - 2x - 3}{x^2 - x - 2}$

5. $\dfrac{3}{m + 1} + \dfrac{3m}{m - 1}$

6. $\dfrac{2x + 3}{3x - 1} - \dfrac{x - 4}{2x + 1}$

VOCABULARY AND CONCEPTS   *Fill in the blanks.*

7. $5x^2$ is the square root of $25x^4$, because _____ $= 25x^4$.

8. 6 is a square root of 36 because _____.

9. The principal square root of x $(x > 0)$ is the _____ square root of x.

10. $\sqrt{x^2} =$ ___.

11. The graph of $f(x) = \sqrt{x} + 3$ is the graph of $f(x) = \sqrt{x}$ translated __ units __.

12. The graph of $f(x) = \sqrt{x + 5}$ is the graph of $f(x) = \sqrt{x}$ translated __ units to the __.

13. $\left(\sqrt[3]{x}\right)^3 =$ __

14. $\sqrt[3]{x^3} =$ __

15. When n is an odd number greater than 1, $\sqrt[n]{x}$ represents an __ root.

16. When n is a positive __ number, $\sqrt[n]{x}$ represents an even root.

17. $\sqrt{0} =$ __

18. The _____ deviation of a set of numbers is the positive square root of the mean of the squares of the differences of the numbers from the mean.

Identify the radicand in each expression.

19. $\sqrt{3x^2}$

20. $5\sqrt{x}$

21. $ab^2\sqrt{a^2 + b^3}$

22. $\dfrac{1}{2}x\sqrt{\dfrac{x}{y}}$

PRACTICE *Find each square root, if possible.*

23. $\sqrt{121}$

24. $\sqrt{144}$

25. $-\sqrt{64}$

26. $-\sqrt{1}$

27. $\sqrt{\dfrac{1}{9}}$

28. $-\sqrt{\dfrac{4}{25}}$

29. $-\sqrt{\dfrac{25}{49}}$

30. $\sqrt{\dfrac{49}{81}}$

31. $\sqrt{-25}$

32. $\sqrt{0.25}$

33. $\sqrt{0.16}$

34. $\sqrt{-49}$

35. $\sqrt{(-4)^2}$

36. $\sqrt{(-9)^2}$

37. $\sqrt{-36}$

38. $-\sqrt{-4}$

Use a calculator to find each square root. Give the answer to four decimal places.

39. $\sqrt{12}$

40. $\sqrt{340}$

41. $\sqrt{679.25}$

42. $\sqrt{0.0063}$

Find each square root. Assume that all variables are unrestricted, and use absolute value symbols when necessary.

43. $\sqrt{4x^2}$

44. $\sqrt{16y^4}$

45. $\sqrt{9a^4}$

46. $\sqrt{16b^2}$

47. $\sqrt{(t + 5)^2}$

48. $\sqrt{(a + 6)^2}$

49. $\sqrt{(-5b)^2}$

50. $\sqrt{(-8c)^2}$

51. $\sqrt{a^2 + 6a + 9}$

52. $\sqrt{x^2 + 10x + 25}$

53. $\sqrt{t^2 + 24t + 144}$

54. $\sqrt{m^2 + 30m + 225}$

Simplify each cube root.

55. $\sqrt[3]{1}$

56. $\sqrt[3]{-8}$

57. $\sqrt[3]{-125}$

58. $\sqrt[3]{512}$

59. $\sqrt[3]{-\dfrac{8}{27}}$

60. $\sqrt[3]{\dfrac{125}{216}}$

61. $\sqrt[3]{0.064}$

62. $\sqrt[3]{0.001}$

63. $\sqrt[3]{8a^3}$

64. $\sqrt[3]{-27x^6}$

65. $\sqrt[3]{-1,000p^3q^3}$

66. $\sqrt[3]{343a^6b^3}$

67. $\sqrt[3]{-\dfrac{1}{8}m^6n^3}$

68. $\sqrt[3]{\dfrac{27}{1,000}a^6b^6}$

69. $\sqrt[3]{0.008z^9}$

70. $\sqrt[3]{0.064s^9t^6}$

Simplify each radical, if possible.

71. $\sqrt[4]{81}$

72. $\sqrt[6]{64}$

73. $-\sqrt[5]{243}$

74. $-\sqrt[4]{625}$

75. $\sqrt[5]{-32}$

76. $\sqrt[6]{729}$

77. $\sqrt[4]{\dfrac{16}{625}}$

78. $\sqrt[5]{-\dfrac{243}{32}}$

79. $-\sqrt[5]{-\dfrac{1}{32}}$

80. $\sqrt[6]{-729}$

81. $\sqrt[4]{-256}$

82. $-\sqrt[4]{\dfrac{81}{256}}$

Simplify each radical. Assume that all variables are unrestricted, and use absolute value symbols where necessary.

83. $\sqrt[4]{16x^4}$

84. $\sqrt[5]{32a^5}$

85. $\sqrt[3]{8a^3}$

86. $\sqrt[6]{64x^6}$

87. $\sqrt[4]{\dfrac{1}{16}x^4}$

88. $\sqrt[4]{\dfrac{1}{81}x^8}$

89. $\sqrt[4]{x^{12}}$

90. $\sqrt[8]{x^{24}}$

91. $\sqrt[5]{-x^5}$

92. $\sqrt[3]{-x^6}$

93. $\sqrt[3]{-27a^6}$

94. $\sqrt[5]{-32x^5}$

Simplify each radical. Assume that all variables are unrestricted, and use absolute value symbols when necessary.

95. $\sqrt[25]{(x + 2)^{25}}$ **96.** $\sqrt[44]{(x + 4)^{44}}$

97. $\sqrt[8]{0.00000001x^{16}y^{8}}$ **98.** $\sqrt[5]{0.00032x^{10}y^{5}}$

Find each value given that $f(x) = \sqrt{x - 4}$.

99. $f(4)$ **100.** $f(8)$

101. $f(20)$ **102.** $f(29)$

Find each value given that $g(x) = \sqrt{x - 8}$.

103. $g(9)$ **104.** $g(17)$

105. $g(8.25)$ **106.** $g(8.64)$

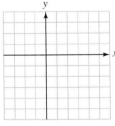 *Find each value given that $f(x) = \sqrt{x^2 + 1}$. Give each answer to four decimal places.*

107. $f(4)$ **108.** $f(6)$

109. $f(2.35)$ **110.** $f(21.57)$

Graph each function and find its domain and range.

111. $f(x) = \sqrt{x + 4}$ **112.** $f(x) = -\sqrt{x - 2}$

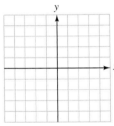 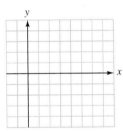

113. $f(x) = -\sqrt{x} - 3$ **114.** $f(x) = \sqrt[3]{x} - 1$

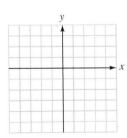

115. Find the standard deviation of the following distribution to the nearest hundredth: 2, 5, 5, 6, 7.

116. Find the standard deviation of the following distribution to the nearest hundredth: 3, 6, 7, 9, 11, 12.

117. **Statistics** In statistics, the formula

$$s_{\bar{x}} = \frac{s}{\sqrt{N}}$$

gives an estimate of the standard error of the mean. Find $s_{\bar{x}}$ to four decimal places when $s = 65$ and $N = 30$.

118. **Statistics** In statistics, the formula

$$\sigma_{\bar{x}} = \frac{\sigma}{\sqrt{N}}$$

gives the standard deviation of means of samples of size N. Find $\sigma_{\bar{x}}$ to four decimal places when $\sigma = 12.7$ and $N = 32$.

APPLICATIONS *Use a calculator to solve each problem.*

119. **Radius of a circle** The radius r of a circle is given by the formula $r = \sqrt{\frac{A}{\pi}}$, where A is its area. Find the radius of a circle whose area is 9π square units.

120. **Diagonal of a baseball diamond** The diagonal d of a square is given by the formula $d = \sqrt{2s^2}$, where s is the length of each side. Find the diagonal of the baseball diamond.

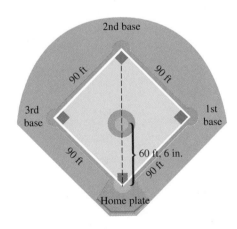

121. **Falling objects** The time t (in seconds) that it will take for an object to fall a distance of s feet is given by the formula

$$t = \frac{\sqrt{s}}{4}$$

If a stone is dropped down a 256-foot well, how long will it take it to hit bottom?

122. Law enforcement Police sometimes use the formula $s = k\sqrt{l}$ to estimate the speed s (in mph) of a car involved in an accident. In this formula, l is the length of the skid in feet, and k is a constant depending on the condition of the pavement. For wet pavement, $k \approx 3.24$. How fast was a car going if its skid was 400 feet on wet pavement?

123. Electronics When the resistance in a circuit is 18 ohms, the current I (measured in amperes) and the power P (measured in watts) are related by the formula

$$I = \sqrt{\frac{P}{18}}$$

Find the current used by an electrical appliance that is rated at 980 watts.

124. Medicine The approximate pulse rate p (in beats per minute) of an adult who is t inches tall is given by the formula

$$p = \frac{590}{\sqrt{t}}$$

Find the approximate pulse rate of an adult who is 71 inches tall.

WRITING

125. If x is any real number, then $\sqrt{x^2} = x$ is not correct. Explain.

126. If x is any real number, then $\sqrt[3]{x^3} = |x|$ is not correct. Explain.

SOMETHING TO THINK ABOUT

127. Is $\sqrt{x^2 - 4x + 4} = x - 2$? What are the exceptions?

128. When is $\sqrt{x^2} \neq x$?

7.2	**Applications of Radicals**

▌ The Pythagorean Theorem ▌ The Distance Formula

Getting Ready *Evaluate each expression.*

1. $3^2 + 4^2$

2. $5^2 + 12^2$

3. $(5 - 2)^2 + (2 + 1)^2$

4. $(111 - 21)^2 + (60 - 4)^2$

In this section, we will discuss the Pythagorean theorem, a theorem that shows the relationship of the sides of a right triangle. We will then use this theorem to develop a formula that gives the distance between two points on the coordinate plane.

The Pythagorean Theorem

If we know the lengths of two legs of a right triangle, we can find the length of the **hypotenuse** (the side opposite the 90° angle) by using the **Pythagorean theorem.**

Pythagorean Theorem

If a and b are the lengths of two legs of a right triangle and c is the length of the hypotenuse, then

$$a^2 + b^2 = c^2$$

In words, the Pythagorean theorem says,

In any right triangle, the square of the hypotenuse is equal to the sum of the squares of the two legs.

Since the lengths of the sides of a triangle are positive numbers, we can use the **square root property of equality** and the Pythagorean theorem to find the length of the third side of any right triangle when the measures of two sides are given.

Square Root Property of Equality

Let a and b represent positive numbers,

$$\text{if } a = b, \text{ then } \sqrt{a} = \sqrt{b}.$$

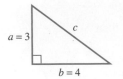

$a = 3$ c

$b = 4$

Figure 7-6

For example, suppose the right triangle shown in Figure 7-6 has legs of length 3 and 4 units. To find the length of the hypotenuse, we use the Pythagorean theorem.

$$a^2 + b^2 = c^2$$
$$3^2 + 4^2 = c^2$$
$$9 + 16 = c^2$$
$$25 = c^2$$
$$\sqrt{25} = \sqrt{c^2} \quad \text{Use the square root property and take the positive square root of both sides.}$$
$$5 = c$$

The length of the hypotenuse is 5 units.

EXAMPLE 1   **Fighting fires**   To fight a forest fire, the forestry department plans to clear a rectangular fire break around the fire, as shown in Figure 7-7. Crews are equipped with mobile communications with a 3,000-yard range. Can crews at points A and B remain in radio contact?

Solution   Points A, B, and C form a right triangle. To find the distance c from point A to point B, we can use the Pythagorean theorem, substituting 2,400 for a and 1,000 for b and solving for c.

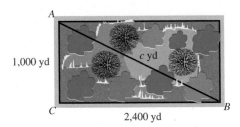

A

1,000 yd c yd

C 2,400 yd B

Figure 7-7

$$a^2 + b^2 = c^2$$
$$2{,}400^2 + 1{,}000^2 = c^2$$
$$5{,}760{,}000 + 1{,}000{,}000 = c^2$$
$$6{,}760{,}000 = c^2$$
$$\sqrt{6{,}760{,}000} = \sqrt{c^2} \quad \text{Take the positive square root of both sides.}$$
$$2{,}600 = c \quad \text{Use a calculator to find the square root.}$$

The two crews are 2,600 yards apart. Because this distance is less than the range of the radios, they can communicate.

Self Check   Can the crews communicate if $b = 1{,}500$ yards?

The Distance Formula

We can use the Pythagorean theorem to develop a formula to find the distance between any two points that are graphed on a rectangular coordinate system.

To find the distance d between points P and Q shown in Figure 7-8, we construct the right triangle PRQ. The distance between P and R is $|x_2 - x_1|$, and the distance between R and Q is $|y_2 - y_1|$. We apply the Pythagorean theorem to the right triangle PRQ to get

$$[d(PQ)]^2 = |x_2 - x_1|^2 + |y_2 - y_1|^2 \qquad \text{Read } d(PQ) \text{ as "the distance between } P \text{ and } Q."$$

$$= (x_2 - x_1)^2 + (y_2 - y_1)^2 \qquad \text{Because } |x_2 - x_1|^2 = (x_2 - x_1)^2 \text{ and } |y_2 - y_1|^2 = (y_2 - y_1)^2.$$

or

(1) $\qquad d(PQ) = \sqrt{(x_2 - x_1)^2 + (y_2 - y_1)^2}$

Equation 1 is called the **distance formula.**

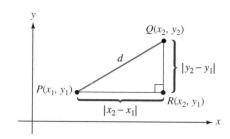

Figure 7-8

Distance Formula	The distance between two points $P(x_1, y_1)$ and $Q(x_2, y_2)$ is given by the formula
	$$d(PQ) = \sqrt{(x_2 - x_1)^2 + (y_2 - y_1)^2}$$

EXAMPLE 2   Find the distance between points $(-2, 3)$ and $(4, -5)$.

Solution   To find the distance, we can use the distance formula by substituting 4 for x_2, -2 for x_1, -5 for y_2, and 3 for y_1.

$$\begin{aligned}
d(PQ) &= \sqrt{(\mathbf{x_2} - \mathbf{x_1})^2 + (y_2 - \mathbf{y_1})^2} \\
&= \sqrt{[\mathbf{4} - (-2)]^2 + (-5 - \mathbf{3})^2} \\
&= \sqrt{(4 + 2)^2 + (-5 - 3)^2} \\
&= \sqrt{6^2 + (-8)^2} \\
&= \sqrt{36 + 64} \\
&= \sqrt{100} \\
&= 10
\end{aligned}$$

The distance between P and Q is 10 units.

Self Check   Find the distance between $P(-2, -2)$ and $Q(3, 10)$.   ∎

EXAMPLE 3   **Building a freeway**   In a city, streets run north and south, and avenues run east and west. Streets and avenues are 850 feet apart. The city plans to construct a straight freeway from the intersection of 25th Street and 8th Avenue to the intersection of 115th Street and 64th Avenue. How long will the freeway be?

Solution   We can represent the roads by the coordinate system in Figure 7-9, where the units on each axis represent 850 feet. We represent the end of the freeway at 25th Street and 8th Avenue by the point $(x_1, y_1) = (25, 8)$. The other end is $(x_2, y_2) = (115, 64)$.

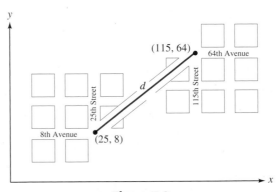

Figure 7-9

We can use the distance formula to find the length of the freeway.

$$d = \sqrt{(x_2 - x_1)^2 + (y_2 - y_1)^2}$$
$$d = \sqrt{(115 - 25)^2 + (64 - 8)^2}$$
$$= \sqrt{90^2 + 56^2}$$
$$= \sqrt{8,100 + 3,136}$$
$$= \sqrt{11,236}$$
$$= 106 \qquad \text{Use a calculator.}$$

Because each unit is 850 feet, the length of the freeway is $106(850) = 90,100$ feet. Since 5,280 feet = 1 mile, we can divide 90,100 by 5,280 to convert 90,100 feet to 17.064394 miles. Thus, the freeway will be about 17 miles long. ∎

EXAMPLE 4   **Bowling**   The velocity, v, of an object after it has fallen d feet is given by the equation $v^2 = 64d$. If an inexperienced bowler lofts the ball 4 feet, with what velocity does it strike the alley?

Solution   We find the velocity by substituting 4 for d in the equation $v^2 = 64d$ and solving for v.

$$v^2 = 64\mathbf{d}$$
$$v^2 = 64(\mathbf{4})$$
$$v^2 = 256$$
$$v = \sqrt{256} \qquad \text{Take the positive square root of both sides.}$$
$$= 16$$

The ball strikes the alley with a velocity of 16 feet per second. ∎

Self Check Answers

1. yes   **2.** 13

Orals   *Evaluate each expression.*

1. $\sqrt{25}$ **2.** $\sqrt{100}$ **3.** $\sqrt{169}$

4. $\sqrt{3^2 + 4^2}$ **5.** $\sqrt{8^2 + 6^2}$ **6.** $\sqrt{5^2 + 12^2}$

7. $\sqrt{5^2 - 3^2}$ **8.** $\sqrt{5^2 - 4^2}$ **9.** $\sqrt{169 - 12^2}$

7.2   EXERCISES

REVIEW   *Find each product.*

1. $(4x + 2)(3x - 5)$

2. $(3y - 5)(2y + 3)$

3. $(5t + 4s)(3t - 2s)$

4. $(4r - 3)(2r^2 + 3r - 4)$

VOCABULARY AND CONCEPTS   *Fill in the blanks.*

5. In a right triangle, the side opposite the 90° angle is called the _____.

6. In a right triangle, the two shorter sides are called ____.

7. If a and b are the lengths of two legs of a right triangle and c is the length of the hypotenuse, then _____.

8. In any right triangle, the square of the hypotenuse is equal to the ____ of the squares of the two ____.

9. With the _____ formula, we can find the distance between two points on a rectangular coordinate system.

10. $d(PQ) = $ _____

PRACTICE   *The lengths of two sides of the right triangle ABC shown in the illustration are given. Find the length of the missing side.*

11. $a = 6$ ft and $b = 8$ ft

12. $a = 10$ cm and $c = 26$ cm

13. $b = 18$ m and $c = 82$ m

14. $b = 7$ ft and $c = 25$ ft

15. $a = 14$ in. and $c = 50$ in.

16. $a = 8$ cm and $b = 15$ cm

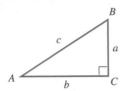

APPLICATIONS   *Give each answer to the nearest tenth.*

17. Geometry   Find the length of the diagonal of one of the faces of the cube.

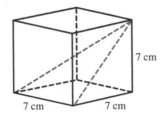

7 cm

7 cm   7 cm

18. Geometry   Find the length of the diagonal of the cube shown in the illustration above.

Find the distance between the given points. If an answer is not exact, use a calculator and give the answer to the nearest tenth.

19. $(0, 0), (3, -4)$

20. $(0, 0), (-6, 8)$

21. $(2, 4), (5, 8)$

22. $(5, 9), (8, 13)$

23. $(-2, -8), (3, 4)$

24. $(-5, -2), (7, 3)$

25. $(6, 8), (12, 16)$

26. $(10, 4), (2, -2)$

27. $(-3, 5), (-5, -5)$

28. $(2, -3), (4, -8)$

29. Geometry   Show that the point $(5, 1)$ is equidistant from points $(7, 0)$ and $(3, 0)$.

30. Geometry   Show that a triangle with vertices at $(2, 3), (-3, 4)$, and $(1, -2)$ is a right triangle. (*Hint:* If the Pythagorean theorem holds, the triangle is a right triangle.)

31. Geometry   Show that a triangle with vertices at $(-2, 4), (2, 8)$, and $(6, 4)$ is isosceles.

32. Geometry   Show that a triangle with vertices at $(-2, 13), (-8, 9)$, and $(-2, 5)$ is isosceles.

APPLICATIONS

33. Sailing   Refer to the sailboat in the illustration. How long must a rope be to fasten the top of the mast to the bow?

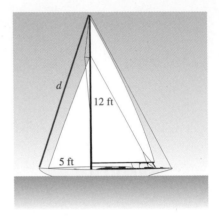

d

12 ft

5 ft

34. Carpentry   The gable end of the roof shown is divided in half by a vertical brace. Find the distance from eaves to peak.

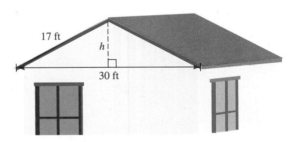

17 ft

h

30 ft

In Exercises 35–38 on the next page, use a calculator. The baseball diamond is a square, 90 feet on a side.

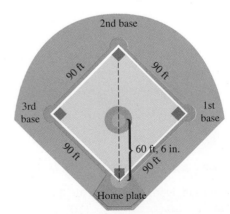

2nd base

90 ft   90 ft

3rd base   1st base

90 ft   90 ft

60 ft, 6 in.

Home plate

35. Baseball How far must a catcher throw the ball to throw out a runner stealing second base?

36. Baseball In baseball, the pitcher's mound is 60 feet, 6 inches from home plate. How far from the mound is second base?

37. Baseball If the third baseman fields a ground ball 10 feet directly behind third base, how far must he throw the ball to throw a runner out at first base?

38. Baseball The shortstop fields a grounder at a point one-third of the way from second base to third base. How far will he have to throw the ball to make an out at first base?

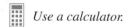 *Use a calculator.*

39. Packing a tennis racket The diagonal d of a rectangular box with dimensions $a \times b \times c$ is given by

$$d = \sqrt{a^2 + b^2 + c^2}$$

Will the racket shown below fit in the shipping carton?

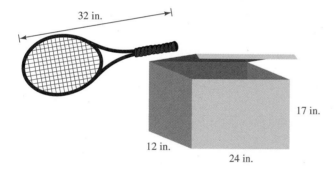

40. Shipping packages A delivery service won't accept a package for shipping if any dimension exceeds 21 inches. An archaeologist wants to ship a 36-inch femur bone. Will it fit in a 3-inch-tall box that has a 21-inch-square base?

41. Shipping packages Can the archaeologist in Exercise 40 ship the femur bone in a cubical box 21 inches on an edge?

42. Reach of a ladder The base of the 37-foot ladder in the illustration is 9 feet from the wall. Will the top reach a window ledge that is 35 feet above the ground?

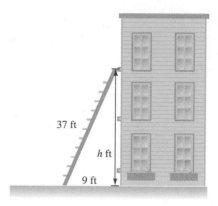

43. Telephone service The telephone cable in the illustration currently runs from A to B to C to D. How much cable is required to run from A to D directly?

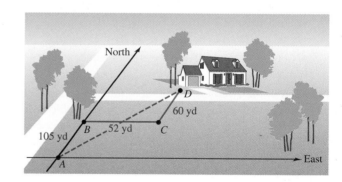

44. Electric service The power company routes its lines as shown in the illustration. How much wire could be saved by going directly from A to E?

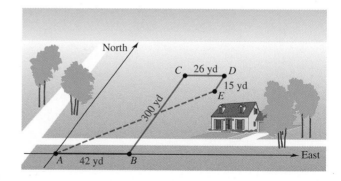

45. Supporting a weight A weight placed on the tight wire pulls the center down 1 foot. By how much is the wire stretched? Round the answer to the nearest hundredth of a foot.

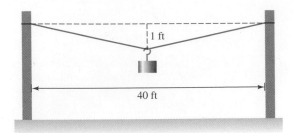

46. Geometry The side, s, of a square with area A square feet is given by the formula $s = \sqrt{A}$. Find the perimeter of a square with an area of 49 square feet.

47. Surface area of a cube The total surface area, A, of a cube is related to its volume, V, by the formula $A = 6\sqrt[3]{V^2}$. Find the surface area of a cube with a volume of 8 cubic centimeters?

48. Area of many cubes A grain of table salt is a cube with a volume of approximately 6×10^{-6} cubic in., and there are about 1.5 million grains of salt in one cup. Find the total surface area of the salt in one cup. (See Exercise 47.)

WRITING

49. State the Pythagorean theorem.

50. Explain the distance formula.

SOMETHING TO THINK ABOUT

51. The formula

$$I = \frac{703w}{h^2}$$

(where w is weight in pounds and h is height in inches) can be used to estimate body mass index, I. The scale shown in the table can be used to judge a person's risk of heart attack. A girl weighing 104 pounds is 54.1 inches tall. Find her estimated body mass index.

20–26	normal
27–29	higher risk
30 and above	very high risk

52. What is the risk of a heart attack for a man who is 6 feet tall and weighs 220 pounds?

7.3 Rational Exponents

■ **Rational Exponents**
■ **Exponential Expressions with Variables in Their Bases**
■ **Fractional Exponents with Numerators Other Than 1**
■ **Negative Fractional Exponents**
■ **Simplifying Radical Expressions**

Getting Ready *Simplify each expression.*

1. x^3x^4

2. $(a^3)^4$

3. $\dfrac{a^8}{a^4}$

4. a^0

5. x^{-4}

6. $(ab^2)^3$

7. $\left(\dfrac{b^2}{c^3}\right)^3$

8. $(a^2a^3)^2$

Rational Exponents

We have seen that positive integer exponents indicate the number of times that a base is to be used as a factor in a product. For example, x^5 means that x is to be used as a factor five times.

$$x^5 = \overbrace{x \cdot x \cdot x \cdot x \cdot x}^{5 \text{ factors of } x}$$

Furthermore, we recall the following properties of exponents.

Rules of Exponents

If there are no divisions by 0, then for all integers m and n,

1. $x^m x^n = x^{m+n}$ **2.** $(x^m)^n = x^{mn}$ **3.** $(xy)^n = x^n y^n$

4. $\left(\dfrac{x}{y}\right)^n = \dfrac{x^n}{y^n}$ **5.** $x^0 = 1 \ (x \neq 0)$ **6.** $x^{-n} = \dfrac{1}{x^n}$

7. $\dfrac{x^m}{x^n} = x^{m-n}$ **8.** $\left(\dfrac{x}{y}\right)^{-n} = \left(\dfrac{y}{x}\right)^n$

To show how to raise bases to fractional powers, we consider the expression $10^{1/2}$. Since fractional exponents must obey the same rules as integer exponents, the square of $10^{1/2}$ is equal to 10.

$$(10^{1/2})^2 = 10^{(1/2)2} \qquad \text{Keep the base and multiply the exponents.}$$
$$= 10^1 \qquad \tfrac{1}{2} \cdot 2 = 1.$$
$$= 10 \qquad 10^1 = 10.$$

However, we have seen that

$$\left(\sqrt{10}\right)^2 = 10$$

Since $(10^{1/2})^2$ and $\left(\sqrt{10}\right)^2$ both equal 10, we define $10^{1/2}$ to be $\sqrt{10}$. Likewise, we define

$$10^{1/3} \text{ to be } \sqrt[3]{10} \qquad \text{and} \qquad 10^{1/4} \text{ to be } \sqrt[4]{10}$$

Rational Exponents

If n is a natural number greater than 1, and $\sqrt[n]{x}$ is a real number, then

$$x^{1/n} = \sqrt[n]{x}$$

EXAMPLE 1   Simplify each expression:

a. $9^{1/2} = \sqrt{9} = 3$ **b.** $-\left(\dfrac{16}{9}\right)^{1/2} = -\sqrt{\dfrac{16}{9}} = -\dfrac{4}{3}$

c. $(-64)^{1/3} = \sqrt[3]{-64} = -4$ **d.** $16^{1/4} = \sqrt[4]{16} = 2$

e. $\left(\dfrac{1}{32}\right)^{1/5} = \sqrt[5]{\dfrac{1}{32}} = \dfrac{1}{2}$ **f.** $0^{1/8} = \sqrt[8]{0} = 0$

g. $-(32x^5)^{1/5} = -\sqrt[5]{32x^5} = -2x$ **h.** $(xyz)^{1/4} = \sqrt[4]{xyz}$

Self Check   Assume that $x > 0$. Simplify: **a.** $16^{1/2}$,   **b.** $\left(\tfrac{27}{8}\right)^{1/3}$,   and   **c.** $-(16x^4)^{1/4}$.   ∎

EXAMPLE 2   Write each radical using a fractional exponent: **a.** $\sqrt[4]{5xyz}$ and   **b.** $\sqrt[5]{\dfrac{xy^2}{15}}$.

Solution   **a.** $\sqrt[4]{5xyz} = (5xyz)^{1/4}$ **b.** $\sqrt[5]{\dfrac{xy^2}{15}} = \left(\dfrac{xy^2}{15}\right)^{1/5}$

Self Check   Write the radical using a fractional exponent: $\sqrt[6]{4ab}$. ∎

Exponential Expressions with Variables in Their Bases

As with radicals, when n is even in the expression $x^{1/n}$ ($n > 1$), there are two real nth roots and we must use absolute value symbols to guarantee that the simplified result is positive.

When n is odd, there is only one real nth root, and we don't need to use absolute value symbols.

When n is even and x is negative, the expression $x^{1/n}$ is not a real number.

EXAMPLE 3   Assume that all variables can be any real number, and simplify each expression.

a. $(-27x^3)^{1/3} = -3x$ Because $(-3x)^3 = -27x^3$. Since n is odd, no absolute value symbols are needed.

b. $(49x^2)^{1/2} = |7x|$ Because $(|7x|)^2 = 49x^2$. Since $7x$ can be negative, absolute value symbols are needed.

   $= 7|x|$

c. $(256a^8)^{1/8} = 2|a|$ Because $(2|a|)^8 = 256a^8$. Since a can be any real number, $2a$ can be negative. Thus, absolute value symbols are needed.

d. $[(y+1)^2]^{1/2} = |y+1|$ Because $|y+1|^2 = (y+1)^2$. Since y can be any real number, $y+1$ can be negative, and the absolute value symbols are needed.

e. $(25b^4)^{1/2} = 5b^2$ Because $(5b^2)^2 = 25b^4$. Since $b^2 \geq 0$, no absolute value symbols are needed.

f. $(-256x^4)^{1/4}$ is not a real number.   Because no real number raised to the 4th power is $-256x^4$.

Self Check   Simplify each expression: **a.** $(625a^4)^{1/4}$ and   **b.** $(b^4)^{1/2}$. ∎

We summarize the cases as follows.

Summary of the Definitions of $x^{1/n}$

If n is a natural number greater than 1 and x is a real number, then

If $x > 0$, then $x^{1/n}$ is the positive number such that $(x^{1/n})^n = x$.
If $x = 0$, then $x^{1/n} = 0$.
If $x < 0$ $\begin{cases} \text{and } n \text{ is odd, then } x^{1/n} \text{ is the real number such that } (x^{1/n})^n = x. \\ \text{and } n \text{ is even, then } x^{1/n} \text{ is not a real number.} \end{cases}$

Fractional Exponents with Numerators Other Than 1

We can extend the definition of $x^{1/n}$ to include fractional exponents with numerators other than 1. For example, since $4^{3/2}$ can be written as $(4^{1/2})^3$, we have

$$4^{3/2} = (4^{1/2})^3 = (\sqrt{4})^3 = 2^3 = 8$$

Thus, we can simplify $4^{3/2}$ by cubing the square root of 4. We can also simplify $4^{3/2}$ by taking the square root of 4 cubed.

$$4^{3/2} = (4^3)^{1/2} = 64^{1/2} = \sqrt{64} = 8$$

In general, we have the following rule.

Changing from Rational Exponents to Radicals

If m and n are positive integers, $x \geq 0$, and $\frac{m}{n}$ is in simplified form, then

$$x^{m/n} = \left(\sqrt[n]{x}\right)^m = \sqrt[n]{x^m}$$

Because of the previous definition, we can interpret $x^{m/n}$ in two ways:

1. $x^{m/n}$ means the mth power of the nth root of x.
2. $x^{m/n}$ means the nth root of the mth power of x.

EXAMPLE 4 Simplify each expression:

a. $\quad 27^{2/3} = \left(\sqrt[3]{27}\right)^2 \qquad$ or $\qquad 27^{2/3} = \sqrt[3]{27^2}$
$\qquad\qquad = 3^2 \qquad\qquad\qquad\qquad\qquad = \sqrt[3]{729}$
$\qquad\qquad = 9 \qquad\qquad\qquad\qquad\qquad\quad = 9$

b. $\quad \left(\dfrac{1}{16}\right)^{3/4} = \left(\sqrt[4]{\dfrac{1}{16}}\right)^3 \qquad$ or $\qquad \left(\dfrac{1}{16}\right)^{3/4} = \sqrt[4]{\left(\dfrac{1}{16}\right)^3}$
$\qquad\qquad\quad = \left(\dfrac{1}{2}\right)^3 \qquad\qquad\qquad\qquad\quad = \sqrt[4]{\dfrac{1}{4{,}096}}$
$\qquad\qquad\quad = \dfrac{1}{8} \qquad\qquad\qquad\qquad\qquad\quad = \dfrac{1}{8}$

c. $\quad (-8x^3)^{4/3} = \left(\sqrt[3]{-8x^3}\right)^4 \qquad$ or $\qquad (-8x^3)^{4/3} = \sqrt[3]{(-8x^3)^4}$
$\qquad\qquad\quad = (-2x)^4 \qquad\qquad\qquad\qquad\qquad = \sqrt[3]{4{,}096x^{12}}$
$\qquad\qquad\quad = 16x^4 \qquad\qquad\qquad\qquad\qquad\quad = 16x^4$

Self Check Simplify: **a.** $16^{3/2}$ and **b.** $(-27x^6)^{2/3}$. ∎

 Comment To avoid large numbers, it is usually better to find the root of the base first, as shown in Example 4.

Negative Fractional Exponents

To be consistent with the definition of negative integer exponents, we define $x^{-m/n}$ as follows.

Definition of $x^{-m/n}$

If m and n are positive integers, $\frac{m}{n}$ is in simplified form, and $x^{1/n}$ is a real number, then

$$x^{-m/n} = \frac{1}{x^{m/n}} \qquad \text{and} \qquad \frac{1}{x^{-m/n}} = x^{m/n} \quad (x \neq 0)$$

EXAMPLE 5 Write each expression without negative exponents, if possible.

a. $64^{-1/2} = \dfrac{1}{64^{1/2}}$

$\phantom{64^{-1/2}} = \dfrac{1}{8}$

b. $16^{-3/2} = \dfrac{1}{16^{3/2}}$

$\phantom{16^{-3/2}} = \dfrac{1}{(16^{1/2})^3}$

$\phantom{16^{-3/2}} = \dfrac{1}{64} \qquad (16^{1/2})^3 = 4^3 = 64.$

c. $(-32x^5)^{-2/5} = \dfrac{1}{(-32x^5)^{2/5}} \quad (x \neq 0)$

$\phantom{(-32x^5)^{-2/5}} = \dfrac{1}{[(-32x^5)^{1/5}]^2}$

$\phantom{(-32x^5)^{-2/5}} = \dfrac{1}{(-2x)^2}$

$\phantom{(-32x^5)^{-2/5}} = \dfrac{1}{4x^2}$

d. $(-16)^{-3/4}$ is not a real number, because $(-16)^{1/4}$ is not a real number.

Self Check Write each expression without negative exponents: **a.** $25^{-3/2}$ and **b.** $(-27a^3)^{-2/3}$.

 Comment By definition, 0^0 is undefined. A base of 0 raised to a negative power is also undefined, because 0^{-2} would equal $\frac{1}{0^2}$, which is undefined since we cannot divide by 0.

We can use the laws of exponents to simplify many expressions with fractional exponents.

EXAMPLE 6 Write all answers without negative exponents. Assume that all variables represent positive numbers. Thus, no absolute value symbols are necessary.

a. $5^{2/7}5^{3/7} = 5^{2/7+3/7} \qquad$ Use the rule $x^m x^n = x^{m+n}$.

$\phantom{5^{2/7}5^{3/7}} = 5^{5/7} \qquad\qquad$ Add: $\frac{2}{7} + \frac{3}{7} = \frac{5}{7}$.

b. $(5^{2/7})^3 = 5^{(2/7)(3)} \qquad$ Use the rule $(x^m)^n = x^{mn}$.

$\phantom{(5^{2/7})^3} = 5^{6/7} \qquad\qquad$ Multiply: $\frac{2}{7}(3) = \frac{2}{7}\left(\frac{3}{1}\right) = \frac{6}{7}$.

c. $(a^{2/3}b^{1/2})^6 = (a^{2/3})^6(b^{1/2})^6 \qquad$ Use the rule $(xy)^n = x^n y^n$.

$\phantom{(a^{2/3}b^{1/2})^6} = a^{12/3}b^{6/2} \qquad\qquad$ Use the rule $(x^m)^n = x^{mn}$ twice.

$\phantom{(a^{2/3}b^{1/2})^6} = a^4 b^3 \qquad\qquad$ Simplify the exponents.

d. $\dfrac{a^{8/3}a^{1/3}}{a^2} = a^{8/3+1/3-2}$ Use the rules $x^m x^n = x^{m+n}$ and $\dfrac{x^m}{x^n} = x^{m-n}$.

$\phantom{\dfrac{a^{8/3}a^{1/3}}{a^2}} = a^{8/3+1/3-6/3}$ $2 = \frac{6}{3}$.

$\phantom{\dfrac{a^{8/3}a^{1/3}}{a^2}} = a^{3/3}$ $\frac{8}{3} + \frac{1}{3} - \frac{6}{3} = \frac{3}{3}$.

$\phantom{\dfrac{a^{8/3}a^{1/3}}{a^2}} = a$ $\frac{3}{3} = 1$.

Self Check   Simplify: **a.** $(x^{1/3}y^{3/2})^6$ and   **b.** $\dfrac{x^{5/3}x^{2/3}}{x^{1/3}}$.   ∎

EXAMPLE 7   Assume that all variables represent positive numbers, and perform the operations. Write all answers without negative exponents.

a. $a^{4/5}(a^{1/5} + a^{3/5}) = a^{4/5}a^{1/5} + a^{4/5}a^{3/5}$ Use the distributive property.

$\phantom{a^{4/5}(a^{1/5} + a^{3/5})} = a^{4/5+1/5} + a^{4/5+3/5}$ Use the rule $x^m x^n = x^{m+n}$.

$\phantom{a^{4/5}(a^{1/5} + a^{3/5})} = a^{5/5} + a^{7/5}$ Simplify the exponents.

$\phantom{a^{4/5}(a^{1/5} + a^{3/5})} = a + a^{7/5}$

 Comment   Note that $a + a^{7/5} \neq a^{1+7/5}$. The expression $a + a^{7/5}$ cannot be simplified, because a and $a^{7/5}$ are not like terms.

b. $x^{1/2}(x^{-1/2} + x^{1/2}) = x^{1/2}x^{-1/2} + x^{1/2}x^{1/2}$ Use the distributive property.

$\phantom{x^{1/2}(x^{-1/2} + x^{1/2})} = x^{1/2-1/2} + x^{1/2+1/2}$ Use the rule $x^m x^n = x^{m+n}$.

$\phantom{x^{1/2}(x^{-1/2} + x^{1/2})} = x^0 + x^1$ Simplify.

$\phantom{x^{1/2}(x^{-1/2} + x^{1/2})} = 1 + x$ $x^0 = 1$.

c. $(x^{2/3} + 1)(x^{2/3} - 1) = x^{4/3} - x^{2/3} + x^{2/3} - 1$ Use the FOIL method.

$\phantom{(x^{2/3} + 1)(x^{2/3} - 1)} = x^{4/3} - 1$ Combine like terms.

d. $(x^{1/2} + y^{1/2})^2 = (x^{1/2} + y^{1/2})(x^{1/2} + y^{1/2})$ Use the FOIL method.

$\phantom{(x^{1/2} + y^{1/2})^2} = x + 2x^{1/2}y^{1/2} + y$ ∎

Simplifying Radical Expressions

We can simplify many radical expressions by using the following steps.

Using Fractional Exponents to Simplify Radicals

1. Change the radical expression into an exponential expression with rational exponents.
2. Simplify the rational exponents.
3. Change the exponential expression back into a radical.

EXAMPLE 8 Simplify: **a.** $\sqrt[4]{3^2}$, **b.** $\sqrt[8]{x^6}$, and **c.** $\sqrt[9]{27x^6y^3}$.

Solution

a. $\sqrt[4]{3^2} = (3^2)^{1/4}$ Change the radical to an exponential expression.

$= 3^{2/4}$ Use the rule $(x^m)^n = x^{mn}$.

$= 3^{1/2}$ $\frac{2}{4} = \frac{1}{2}$.

$= \sqrt{3}$ Change back to radical notation.

b. $\sqrt[8]{x^6} = (x^6)^{1/8}$ Change the radical to an exponential expression.

$= x^{6/8}$ Use the rule $(x^m)^n = x^{mn}$.

$= x^{3/4}$ $\frac{6}{8} = \frac{3}{4}$.

$= (x^3)^{1/4}$ $\frac{3}{4} = 3(\frac{1}{4})$.

$= \sqrt[4]{x^3}$ Change back to radical notation.

c. $\sqrt[9]{27x^6y^3} = (3^3x^6y^3)^{1/9}$ Write 27 as 3^3 and change the radical to an exponential expression.

$= 3^{3/9}x^{6/9}y^{3/9}$ Raise each factor to the $\frac{1}{9}$ power by multiplying the fractional exponents.

$= 3^{1/3}x^{2/3}y^{1/3}$ Simplify each fractional exponent.

$= (3x^2y)^{1/3}$ Use the rule $(xy)^n = x^ny^n$.

$= \sqrt[3]{3x^2y}$ Change back to radical notation.

Self Check Simplify: **a.** $\sqrt[6]{3^3}$ and **b.** $\sqrt[4]{64x^2y^2}$.

Self Check Answers

1. a. 4, **b.** $\frac{3}{2}$, **c.** $-2x$ **2.** $(4ab)^{1/6}$ **3. a.** $5|a|$, **b.** b^2, **4. a.** 64, **b.** $9x^4$ **5. a.** $\frac{1}{125}$, **b.** $\frac{1}{9a^2}$
6. a. x^2y^9 **b.** x^2 **8. a.** $\sqrt{3}$ **b.** $\sqrt{8xy}$

Orals *Simplify each expression.*

1. $4^{1/2}$ **2.** $9^{1/2}$ **3.** $27^{1/3}$ **4.** $1^{1/4}$

5. $4^{3/2}$ **6.** $8^{2/3}$ **7.** $\left(\frac{1}{4}\right)^{1/2}$ **8.** $\left(\frac{1}{4}\right)^{-1/2}$

9. $(8x^3)^{1/3}$ **10.** $(16x^8)^{1/4}$

7.3 EXERCISES

REVIEW *Solve each inequality.*

1. $5x - 4 < 11$

2. $2(3t - 5) \geq 8$

3. $\frac{4}{5}(r - 3) > \frac{2}{3}(r + 2)$

4. $-4 < 2x - 4 \leq 8$

5. How much water must be added to 5 pints of a 20% alcohol solution to dilute it to a 15% solution?

6. A grocer bought some boxes of apples for $70. However, 4 boxes were spoiled. The grocer sold the remaining boxes at a profit of $2 each. How many boxes did the grocer sell if she managed to break even?

VOCABULARY AND CONCEPTS *Fill in the blanks.*

7. $a^4 = $ _____ **8.** $a^ma^n = $ ____

9. $(a^m)^n = $ ____ **10.** $(ab)^n = $ ____

11. $\left(\dfrac{a}{b}\right)^n = $ ___

12. $a^0 = $ ___, provided $a \neq $ ___.

13. $a^{-n} = $ ___ , provided $a \neq $ ___.

14. $\dfrac{a^m}{a^n} = $ _____   **15.** $\left(\dfrac{a}{b}\right)^{-n} = $ _____

16. $x^{1/n} = $ ___

17. $(x^n)^{1/n} = $ ___, provided n is even.

18. $x^{m/n} = \sqrt[n]{x^m} = $ _____

PRACTICE *Change each expression into radical notation.*

19. $7^{1/3}$ **20.** $26^{1/2}$

21. $8^{1/5}$ **22.** $13^{1/7}$

23. $(3x)^{1/4}$ **24.** $(4ab)^{1/6}$

25. $\left(\dfrac{1}{2}x^3 y\right)^{1/4}$ **26.** $\left(\dfrac{3}{4}a^2 b^2\right)^{1/5}$

27. $(4a^2 b^3)^{1/5}$ **28.** $(5pq^2)^{1/3}$

29. $(x^2 + y^2)^{1/2}$ **30.** $(x^3 + y^3)^{1/3}$

Change each radical to an exponential expression.

31. $\sqrt{11}$ **32.** $\sqrt[3]{12}$

33. $\sqrt[4]{3a}$ **34.** $\sqrt[7]{12xy}$

35. $3\sqrt[5]{a}$ **36.** $4\sqrt[3]{p}$

37. $\sqrt[6]{\dfrac{1}{7}abc}$ **38.** $\sqrt[7]{\dfrac{3}{8}p^2 q}$

39. $\sqrt[5]{\dfrac{1}{2}mn}$ **40.** $\sqrt[8]{\dfrac{2}{7}p^2 q}$

41. $\sqrt[3]{a^2 - b^2}$ **42.** $\sqrt{x^2 + y^2}$

Simplify each expression, if possible. Assume that all variables are unrestricted, and use absolute value symbols when necessary.

43. $4^{1/2}$ **44.** $64^{1/2}$

45. $27^{1/3}$ **46.** $125^{1/3}$

47. $16^{1/4}$ **48.** $625^{1/4}$

49. $32^{1/5}$ **50.** $0^{1/5}$

51. $\left(\dfrac{1}{4}\right)^{1/2}$ **52.** $\left(\dfrac{1}{16}\right)^{1/2}$

53. $\left(\dfrac{1}{8}\right)^{1/3}$ **54.** $\left(\dfrac{1}{16}\right)^{1/4}$

55. $-16^{1/4}$ **56.** $-125^{1/3}$

57. $(-27)^{1/3}$ **58.** $(-125)^{1/3}$

59. $(-64)^{1/2}$ **60.** $(-243)^{1/5}$

61. $0^{1/3}$ **62.** $(-216)^{1/2}$

63. $(25y^2)^{1/2}$ **64.** $(-27x^3)^{1/3}$

65. $(16x^4)^{1/4}$ **66.** $(-16x^4)^{1/2}$

67. $(243x^5)^{1/5}$ **68.** $[(x + 1)^4]^{1/4}$

69. $(-64x^8)^{1/4}$ **70.** $[(x + 5)^3]^{1/3}$

71. $36^{3/2}$ **72.** $27^{2/3}$

73. $81^{3/4}$ **74.** $100^{3/2}$

75. $144^{3/2}$ **76.** $1{,}000^{2/3}$

77. $\left(\dfrac{1}{8}\right)^{2/3}$ **78.** $\left(\dfrac{4}{9}\right)^{3/2}$

79. $(25x^4)^{3/2}$ **80.** $(27a^3 b^3)^{2/3}$

81. $\left(\dfrac{8x^3}{27}\right)^{2/3}$ **82.** $\left(\dfrac{27}{64y^6}\right)^{2/3}$

Write each expression without using negative exponents. Assume that all variables represent positive numbers.

83. $4^{-1/2}$ **84.** $8^{-1/3}$

85. $(4)^{-3/2}$ **86.** $25^{-5/2}$

87. $(16x^2)^{-3/2}$ **88.** $(81c^4)^{-3/2}$

89. $(-27y^3)^{-2/3}$ **90.** $(-8z^9)^{-2/3}$

91. $(-32p^5)^{-2/5}$ **92.** $(16q^6)^{-5/2}$

93. $\left(\dfrac{1}{4}\right)^{-3/2}$ **94.** $\left(\dfrac{4}{25}\right)^{-3/2}$

95. $\left(\dfrac{27}{8}\right)^{-4/3}$ **96.** $\left(\dfrac{25}{49}\right)^{-3/2}$

97. $\left(-\dfrac{8x^3}{27}\right)^{-1/3}$ **98.** $\left(\dfrac{16}{81y^4}\right)^{-3/4}$

Perform the operations. Write answers without negative exponents. Assume that all variables represent positive numbers.

99. $5^{4/9}5^{4/9}$ **100.** $4^{2/5}4^{2/5}$

101. $(4^{1/5})^3$ **102.** $(3^{1/3})^5$

103. $\dfrac{9^{4/5}}{9^{3/5}}$

104. $\dfrac{7^{2/3}}{7^{1/2}}$

105. $\dfrac{7^{1/2}}{7^0}$

106. $5^{1/3}5^{-5/3}$

107. $6^{-2/3}6^{-4/3}$

108. $\dfrac{3^{4/3}3^{1/3}}{3^{2/3}}$

109. $\dfrac{2^{5/6}2^{1/3}}{2^{1/2}}$

110. $\dfrac{5^{1/3}5^{1/2}}{5^{1/3}}$

111. $a^{2/3}a^{1/3}$

112. $b^{3/5}b^{1/5}$

113. $(a^{2/3})^{1/3}$

114. $(t^{4/5})^{10}$

115. $(a^{1/2}b^{1/3})^{3/2}$

116. $(a^{3/5}b^{3/2})^{2/3}$

117. $(mn^{-2/3})^{-3/5}$

118. $(r^{-2}s^3)^{1/3}$

119. $\dfrac{(4x^3y)^{1/2}}{(9xy)^{1/2}}$

120. $\dfrac{(27x^3y)^{1/3}}{(8xy^2)^{2/3}}$

121. $(27x^{-3})^{-1/3}$

122. $(16a^{-2})^{-1/2}$

123. $y^{1/3}(y^{2/3} + y^{5/3})$

124. $y^{2/5}(y^{-2/5} + y^{3/5})$

125. $x^{3/5}(x^{7/5} - x^{2/5} + 1)$

126. $x^{4/3}(x^{2/3} + 3x^{5/3} - 4)$

127. $(x^{1/2} + 2)(x^{1/2} - 2)$

128. $(x^{1/2} + y^{1/2})(x^{1/2} - y^{1/2})$

129. $(x^{2/3} - x)(x^{2/3} + x)$

130. $(x^{1/3} + x^2)(x^{1/3} - x^2)$

131. $(x^{2/3} + y^{2/3})^2$

132. $(a^{1/2} - b^{2/3})^2$

133. $(a^{3/2} - b^{3/2})^2$

134. $(x^{-1/2} - x^{1/2})^2$

Use rational exponents to simplify each radical. Assume that all variables represent positive numbers.

135. $\sqrt[6]{p^3}$

136. $\sqrt[8]{q^2}$

137. $\sqrt[4]{25b^2}$

138. $\sqrt[9]{-8x^6}$

WRITING

139. Explain how you would decide whether $a^{1/n}$ is a real number.

140. The expression $(a^{1/2} + b^{1/2})^2$ is not equal to $a + b$. Explain.

SOMETHING TO THINK ABOUT

141. The fraction $\frac{2}{4}$ is equal to $\frac{1}{2}$. Is $16^{2/4}$ equal to $16^{1/2}$? Explain.

142. How would you evaluate an expression with a mixed-number exponent? For example, what is $8^{1\frac{1}{3}}$? What is $25^{2\frac{1}{2}}$? Discuss.

7.4   Simplifying and Combining Radical Expressions

▌ **Properties of Radicals**   ▌ **Simplifying Radical Expressions**
▌ **Adding and Subtracting Radical Expressions**
▌ **Some Special Triangles**

Getting Ready   *Simplify each radical. Assume that all variables represent positive numbers.*

1. $\sqrt{225}$ **2.** $\sqrt{576}$ **3.** $\sqrt[3]{125}$ **4.** $\sqrt[3]{343}$

5. $\sqrt{16x^4}$ **6.** $\sqrt{\dfrac{64}{121}x^6}$ **7.** $\sqrt[3]{27a^3b^9}$ **8.** $\sqrt[3]{-8a^{12}}$

In this section, we will introduce several properties of radicals and use them to simplify radical expressions. Then we will add and subtract radical expressions.

Properties of Radicals

Many properties of exponents have counterparts in radical notation. For example, because $a^{1/n}b^{1/n} = (ab)^{1/n}$, we have

(1) $\sqrt[n]{a}\sqrt[n]{b} = \sqrt[n]{ab}$

For example,

$$\sqrt{5}\sqrt{5} = \sqrt{5 \cdot 5} = \sqrt{5^2} = 5$$

$$\sqrt[3]{7x}\sqrt[3]{49x^2} = \sqrt[3]{7x \cdot 7^2 x^2} = \sqrt[3]{7^3 \cdot x^3} = 7x$$

$$\sqrt[4]{2x^3}\sqrt[4]{8x} = \sqrt[4]{2x^3 \cdot 2^3 x} = \sqrt[4]{2^4 \cdot x^4} = 2x \quad (x > 0)$$

If we rewrite Equation 1, we have the following rule.

Multiplication Property of Radicals

If $\sqrt[n]{a}$ and $\sqrt[n]{b}$ are real numbers, then
$$\sqrt[n]{ab} = \sqrt[n]{a}\sqrt[n]{b}$$

As long as all radicals represent real numbers, *the nth root of the product of two numbers is equal to the product of their nth roots.*

 Comment The multiplication property of radicals applies to the *n*th root of the product of two numbers. There is no such property for sums or differences. For example,

$$\sqrt{9 + 4} \neq \sqrt{9} + \sqrt{4} \qquad\qquad \sqrt{9 - 4} \neq \sqrt{9} - \sqrt{4}$$

$$\sqrt{13} \neq 3 + 2 \qquad\qquad\qquad \sqrt{5} \neq 3 - 2$$

$$\sqrt{13} \neq 5 \qquad\qquad\qquad\qquad \sqrt{5} \neq 1$$

Thus, $\sqrt{a + b} \neq \sqrt{a} + \sqrt{b}$ and $\sqrt{a - b} \neq \sqrt{a} - \sqrt{b}$.

A second property of radicals involves quotients. Because

$$\frac{a^{1/n}}{b^{1/n}} = \left(\frac{a}{b}\right)^{1/n}$$

it follows that

(2) $\dfrac{\sqrt[n]{a}}{\sqrt[n]{b}} = \sqrt[n]{\dfrac{a}{b}} \quad (b \neq 0)$

For example,

$$\frac{\sqrt{8x^3}}{\sqrt{2x}} = \sqrt{\frac{8x^3}{2x}} = \sqrt{4x^2} = 2x \quad (x > 0)$$

$$\frac{\sqrt[3]{54x^5}}{\sqrt[3]{2x^2}} = \sqrt[3]{\frac{54x^5}{2x^2}} = \sqrt[3]{27x^3} = 3x$$

If we rewrite Equation 2, we have the following rule.

Division Property of Radicals

If $\sqrt[n]{a}$ and $\sqrt[n]{b}$ are real numbers, then

$$\sqrt[n]{\frac{a}{b}} = \frac{\sqrt[n]{a}}{\sqrt[n]{b}} \quad (b \neq 0)$$

As long as all radicals represent real numbers, *the nth root of the quotient of two numbers is equal to the quotient of their nth roots.*

Simplifying Radical Expressions

A radical expression is said to be in simplest form when each of the following statements is true.

Simplified Form of a Radical Expression

A radical expression is in simplest form when

1. Each prime and variable factor in the radicand appears to a power that is less than the index of the radical.
2. The radicand contains no fractions or negative numbers.
3. No radicals appear in the denominator of a fraction.

EXAMPLE 1   Simplify: **a.** $\sqrt{12}$,   **b.** $\sqrt{98}$,   and   **c.** $\sqrt[3]{54}$.

Solution   **a.** Recall that squares of integers, such as 1, 4, 9, 16, 25, and 36, are *perfect squares.* To simplify $\sqrt{12}$, we factor 12 so that one factor is the largest perfect square that divides 12. Since 4 is the largest perfect-square factor of 12, we write 12 as $4 \cdot 3$, use the multiplication property of radicals, and simplify.

$$\sqrt{12} = \sqrt{4 \cdot 3} \qquad \text{Write 12 as } 4 \cdot 3.$$
$$= \sqrt{4}\sqrt{3} \qquad \sqrt{4 \cdot 3} = \sqrt{4}\sqrt{3}.$$
$$= 2\sqrt{3} \qquad \sqrt{4} = 2.$$

b. Since the largest perfect-square factor of 98 is 49, we have

$$\sqrt{98} = \sqrt{49 \cdot 2} \qquad \text{Write 98 as } 49 \cdot 2.$$
$$= \sqrt{49}\sqrt{2} \qquad \sqrt{49 \cdot 2} = \sqrt{49}\sqrt{2}.$$
$$= 7\sqrt{2} \qquad \sqrt{49} = 7.$$

c. Numbers that are cubes of integers, such as 1, 8, 27, 64, 125, and 216, are called *perfect cubes.* Since the largest perfect-cube factor of 54 is 27, we have

$$\sqrt[3]{54} = \sqrt[3]{27 \cdot 2} \qquad \text{Write 54 as } 27 \cdot 2.$$
$$= \sqrt[3]{27}\sqrt[3]{2} \qquad \sqrt[3]{27 \cdot 2} = \sqrt[3]{27}\sqrt[3]{2}.$$
$$= 3\sqrt[3]{2} \qquad \sqrt[3]{27} = 3.$$

Self Check   Simplify: **a.** $\sqrt{20}$ and   **b.** $\sqrt[3]{24}$.   ∎

EXAMPLE 2   Simplify: **a.** $\sqrt{\dfrac{15}{49x^2}}$ ($x > 0$)   and   **b.** $\sqrt[3]{\dfrac{10x^2}{27y^6}}$ ($y \neq 0$).

Solution   **a.** We can write the square root of the quotient as the quotient of the square roots and simplify the denominator. Since $x > 0$, we have

$$\sqrt{\frac{15}{49x^2}} = \frac{\sqrt{15}}{\sqrt{49x^2}}$$

$$= \frac{\sqrt{15}}{7x}$$

b. We can write the cube root of the quotient as the quotient of two cube roots. Since $y \neq 0$, we have

$$\sqrt[3]{\frac{10x^2}{27y^6}} = \frac{\sqrt[3]{10x^2}}{\sqrt[3]{27y^6}}$$

$$= \frac{\sqrt[3]{10x^2}}{3y^2}$$

Self Check   Simplify: **a.** $\sqrt{\dfrac{11}{36a^2}}$ ($a > 0$)   and   **b.** $\sqrt[3]{\dfrac{8a^2}{125y^3}}$ ($y \neq 0$).   ∎

EXAMPLE 3   Simplify each expression. Assume that all variables represent positive numbers.

     a. $\sqrt{128a^5}$,   **b.** $\sqrt[3]{24x^5}$,   **c.** $\dfrac{\sqrt{45xy^2}}{\sqrt{5x}}$,   and   **d.** $\dfrac{\sqrt[3]{-432x^5}}{\sqrt[3]{8x}}$.

Solution   **a.** We write $128a^5$ as $64a^4 \cdot 2a$ and use the multiplication property of radicals.

$$\sqrt{128a^5} = \sqrt{64a^4 \cdot 2a} \qquad \text{$64a^4$ is the largest perfect square that divides $128a^5$.}$$

$$= \sqrt{64a^4}\sqrt{2a} \qquad \text{Use the multiplication property of radicals.}$$

$$= 8a^2\sqrt{2a} \qquad \sqrt{64a^4} = 8a^2.$$

b. We write $24x^5$ as $8x^3 \cdot 3x^2$ and use the multiplication property of radicals.

$$\sqrt[3]{24x^5} = \sqrt[3]{8x^3 \cdot 3x^2} \qquad \text{$8x^3$ is the largest perfect cube that divides $24x^5$.}$$

$$= \sqrt[3]{8x^3}\sqrt[3]{3x^2} \qquad \text{Use the multiplication property of radicals.}$$

$$= 2x\sqrt[3]{3x^2} \qquad \sqrt[3]{8x^3} = 2x.$$

c. We can write the quotient of the square roots as the square root of a quotient.

$$\frac{\sqrt{45xy^2}}{\sqrt{5x}} = \sqrt{\frac{45xy^2}{5x}} \qquad \text{Use the quotient property of radicals.}$$

$$= \sqrt{9y^2} \qquad \text{Simplify the fraction.}$$

$$= 3y$$

d. We can write the quotient of the cube roots as the cube root of a quotient.

$$\frac{\sqrt[3]{-432x^5}}{\sqrt[3]{8x}} = \sqrt[3]{\frac{-432x^5}{8x}} \qquad \text{Use the quotient property of radicals.}$$

$$= \sqrt[3]{-54x^4} \qquad \text{Simplify the fraction.}$$

$$= \sqrt[3]{-27x^3 \cdot 2x} \qquad -27x^3 \text{ is the largest perfect cube that divides } -54x^4.$$

$$= \sqrt[3]{-27x^3}\sqrt[3]{2x} \qquad \text{Use the multiplication property of radicals.}$$

$$= -3x\sqrt[3]{2x}$$

Self Check   Simplify: **a.** $\sqrt{98b^3}$,   **b.** $\sqrt[3]{54y^5}$,   and   **c.** $\dfrac{\sqrt{50ab^2}}{\sqrt{2a}}$. Assume that all variables represent positive numbers. ∎

To simplify more complicated radicals, we can use the prime factorization of the radicand to find its perfect-square factors. For example, to simplify $\sqrt{3{,}168x^5y^7}$, we first find the prime factorization of $3{,}168x^5y^7$.

$$3{,}168x^5y^7 = 2^5 \cdot 3^2 \cdot 11 \cdot x^5 \cdot y^7$$

Then we have

$$\sqrt{3{,}168x^5y^7} = \sqrt{2^4 \cdot 3^2 \cdot x^4 \cdot y^6 \cdot 2 \cdot 11 \cdot x \cdot y}$$

$$= \sqrt{2^4 \cdot 3^2 \cdot x^4 \cdot y^6}\sqrt{2 \cdot 11 \cdot x \cdot y} \qquad \text{Write each perfect square under the left radical and each nonperfect square under the right radical.}$$

$$= 2^2 \cdot 3x^2y^3\sqrt{22xy}$$

$$= 12x^2y^3\sqrt{22xy}$$

Adding and Subtracting Radical Expressions

Radical expressions with the same index and the same radicand are called **like** or **similar radicals.** For example, $3\sqrt{2}$ and $2\sqrt{2}$ are like radicals. However,

$3\sqrt{5}$ and $4\sqrt{2}$ are not like radicals, because the radicands are different.

$3\sqrt{5}$ and $2\sqrt[3]{5}$ are not like radicals, because the indexes are different.

We can often combine like terms. For example, to simplify the expression $3\sqrt{2} + 2\sqrt{2}$, we use the distributive property to factor out $\sqrt{2}$ and simplify.

$$\mathbf{3\sqrt{2} + 2\sqrt{2} = (3 + 2)\sqrt{2}}$$

$$= 5\sqrt{2}$$

Radicals with the same index but different radicands can often be written as like radicals. For example, to simplify the expression $\sqrt{27} - \sqrt{12}$, we simplify both radicals and combine the like radicals.

$$\sqrt{27} - \sqrt{12} = \sqrt{9 \cdot 3} - \sqrt{4 \cdot 3}$$

$$= \sqrt{9}\sqrt{3} - \sqrt{4}\sqrt{3} \qquad \sqrt{ab} = \sqrt{a}\sqrt{b}.$$

$$= 3\sqrt{3} - 2\sqrt{3} \qquad \sqrt{9} = 3 \text{ and } \sqrt{4} = 2.$$
$$= (3 - 2)\sqrt{3} \qquad \text{Factor out } \sqrt{3}.$$
$$= \sqrt{3}$$

As the previous examples suggest, we can use the following rule to add or subtract radicals.

Adding and Subtracting Radicals	To add or subtract radicals, simplify each radical and combine all like radicals. To combine like radicals, add the coefficients and keep the common radical.

 EXAMPLE 4   Simplify: $2\sqrt{12} - 3\sqrt{48} + 3\sqrt{3}$.

Solution   We simplify each radical separately and combine like radicals.
$$2\sqrt{12} - 3\sqrt{48} + 3\sqrt{3} = 2\sqrt{4 \cdot 3} - 3\sqrt{16 \cdot 3} + 3\sqrt{3}$$
$$= 2\sqrt{4}\sqrt{3} - 3\sqrt{16}\sqrt{3} + 3\sqrt{3}$$
$$= 2(2)\sqrt{3} - 3(4)\sqrt{3} + 3\sqrt{3}$$
$$= 4\sqrt{3} - 12\sqrt{3} + 3\sqrt{3}$$
$$= (4 - 12 + 3)\sqrt{3}$$
$$= -5\sqrt{3}$$

Self Check   Simplify: $3\sqrt{75} - 2\sqrt{12} + 2\sqrt{48}$.   ∎

EXAMPLE 5   Simplify: $\sqrt[3]{16} - \sqrt[3]{54} + \sqrt[3]{24}$.

Solution   We simplify each radical separately and combine like radicals:
$$\sqrt[3]{16} - \sqrt[3]{54} + \sqrt[3]{24} = \sqrt[3]{8 \cdot 2} - \sqrt[3]{27 \cdot 2} + \sqrt[3]{8 \cdot 3}$$
$$= \sqrt[3]{8}\sqrt[3]{2} - \sqrt[3]{27}\sqrt[3]{2} + \sqrt[3]{8}\sqrt[3]{3}$$
$$= 2\sqrt[3]{2} - 3\sqrt[3]{2} + 2\sqrt[3]{3}$$
$$= -\sqrt[3]{2} + 2\sqrt[3]{3}$$

Comment   We cannot combine $-\sqrt[3]{2}$ and $2\sqrt[3]{3}$, because the radicals have different radicands.

Self Check   Simplify: $\sqrt[3]{24} - \sqrt[3]{16} + \sqrt[3]{54}$.   ∎

EXAMPLE 6   Simplify: $\sqrt[3]{16x^4} + \sqrt[3]{54x^4} - \sqrt[3]{-128x^4}$.

Solution   We simplify each radical separately, factor out $\sqrt[3]{2x}$, and simplify.
$$\sqrt[3]{16x^4} + \sqrt[3]{54x^4} - \sqrt[3]{-128x^4}$$
$$= \sqrt[3]{8x^3 \cdot 2x} + \sqrt[3]{27x^3 \cdot 2x} - \sqrt[3]{-64x^3 \cdot 2x}$$
$$= \sqrt[3]{8x^3}\sqrt[3]{2x} + \sqrt[3]{27x^3}\sqrt[3]{2x} - \sqrt[3]{-64x^3}\sqrt[3]{2x}$$
$$= 2x\sqrt[3]{2x} + 3x\sqrt[3]{2x} + 4x\sqrt[3]{2x}$$

$$= (2x + 3x + 4x)\sqrt[3]{2x}$$

$$= 9x\sqrt[3]{2x}$$

Self Check   Simplify: $\sqrt{32x^3} + \sqrt{50x^3} - \sqrt{18x^3}$ $(x > 0)$.   ∎

Some Special Triangles

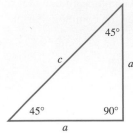

Figure 7-10

An **isosceles right triangle** is a right triangle with two legs of equal length. If we know the length of one leg of an isosceles right triangle, we can use the Pythagorean theorem to find the length of the hypotenuse. Since the triangle shown in Figure 7-10 is a right triangle, we have

$$c^2 = a^2 + a^2 \qquad \text{Use the Pythagorean theorem.}$$

$$c^2 = 2a^2 \qquad \text{Combine like terms.}$$

$$\sqrt{c^2} = \sqrt{2a^2} \qquad \text{Take the positive square root of both sides.}$$

$$c = a\sqrt{2} \qquad \sqrt{2a^2} = \sqrt{2}\sqrt{a^2} = \sqrt{2}a = a\sqrt{2}. \text{ No absolute value symbols are needed, because } a \text{ is positive.}$$

Thus, *in an isosceles right triangle, the length of the hypotenuse is the length of one leg times* $\sqrt{2}$.

EXAMPLE 7   If one leg of the isosceles right triangle shown in Figure 7-10 is 10 feet long, find the length of the hypotenuse.

Solution   Since the length of the hypotenuse is the length of a leg times $\sqrt{2}$, we have

$$c = 10\sqrt{2}$$

The length of the hypotenuse is $10\sqrt{2}$ feet. To two decimal places, the length is 14.14 feet.

Self Check   Find the length of the hypotenuse of an isosceles right triangle if one leg is 12 meters long.   ∎

If the length of the hypotenuse of an isosceles right triangle is known, we can use the Pythagorean theorem to find the length of each leg.

EXAMPLE 8   Find the length of each leg of the isosceles right triangle shown in Figure 7-11.

Solution   We use the Pythagorean theorem.

$$c^2 = a^2 + a^2$$

$$25^2 = 2a^2 \qquad \text{Substitute 25 for } c \text{ and combine like terms.}$$

$$\frac{625}{2} = a^2 \qquad \text{Square 25 and divide both sides by 2.}$$

$$\sqrt{\frac{625}{2}} = a \qquad \text{Take the positive square root of both sides.}$$

$$a = 17.67766953 \qquad \text{Use a calculator.}$$

Figure 7-11

To two decimal places, the length is 17.68 units.   ∎

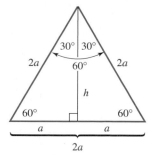

Figure 7-12

From geometry, we know that an **equilateral triangle** is a triangle with three sides of equal length and three 60° angles. If an **altitude** is drawn upon the base of an equilateral triangle, as shown in Figure 7-12, it bisects the base and divides the triangle into two 30°–60°–90° triangles. We can see that the shortest leg of each 30°–60°–90° triangle is *a* units long. Thus,

The shorter leg of a 30°–60°–90° right triangle is half as long as its hypotenuse.

We can find the length of the altitude, *h*, by using the Pythagorean theorem.

$$a^2 + h^2 = (2a)^2$$
$$a^2 + h^2 = 4a^2 \qquad (2a)^2 = (2a)(2a) = 4a^2.$$
$$h^2 = 3a^2 \qquad \text{Subtract } a^2 \text{ from both sides.}$$
$$h = \sqrt{3a^2} \qquad \text{Take the positive square root of both sides.}$$
$$h = a\sqrt{3} \qquad \sqrt{3a^2} = \sqrt{3}\sqrt{a^2} = a\sqrt{3}. \text{ No absolute value symbols are needed, because } a \text{ is positive.}$$

Thus,

The length of the longer leg is the length of the shorter side times $\sqrt{3}$.

EXAMPLE 9   Find the length of the hypotenuse and the longer leg of the right triangle shown in Figure 7-13.

Solution   Since the shorter leg of a 30°–60°–90° right triangle is half as long as its hypotenuse, the hypotenuse is 12 centimeters long.

Since the length of the longer leg is the length of the shorter leg times $\sqrt{3}$, the longer leg is $6\sqrt{3}$ (about 10.39) centimeters long.

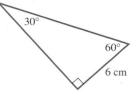

Figure 7-13

Self Check   Find the length of the hypotenuse and the longer leg of a 30°–60°–90° right triangle if the shorter leg is 8 centimeters long.

EXAMPLE 10   Find the length of each leg of the triangle shown in Figure 7-14.

Solution   Since the shorter leg of a 30°–60°–90° right triangle is half as long as its hypotenuse, the shorter leg is $\frac{9}{2}$ centimeters long.

Since the length of the longer leg is the length of the shorter leg times $\sqrt{3}$, the longer leg is $\frac{9}{2}\sqrt{3}$ (or about 7.79) centimeters long.

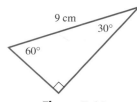

Figure 7-14

Self Check Answers

1. a. $2\sqrt{5}$,   **b.** $2\sqrt[3]{3}$ **2. a.** $\dfrac{\sqrt{11}}{6a}$,   **b.** $\dfrac{2\sqrt[3]{a^2}}{5y}$ **3. a.** $7b\sqrt{2b}$,   **b.** $3y\sqrt[3]{2y^2}$,   **c.** $5b$ **4.** $19\sqrt{3}$

5. $2\sqrt[3]{3} + \sqrt[3]{2}$ **6.** $6x\sqrt{2x}$ **7.** $12\sqrt{2}$ m   **9.** 16 cm, $8\sqrt{3}$ cm

Orals *Simplify:*

1. $\sqrt{7}\sqrt{7}$

2. $\sqrt[3]{4^2}\sqrt[3]{4}$

3. $\dfrac{\sqrt[3]{54}}{\sqrt[3]{2}}$

Simplify each expression. Assume that $b \neq 0$.

4. $\sqrt{18}$

5. $\sqrt[3]{16}$

6. $\sqrt[3]{\dfrac{3x^2}{64b^6}}$

Combine like terms:

7. $3\sqrt{3} + 4\sqrt{3}$

8. $5\sqrt{7} - 2\sqrt{7}$

9. $2\sqrt[3]{9} + 3\sqrt[3]{9}$

10. $10\sqrt[5]{4} - 2\sqrt[5]{4}$

7.4   EXERCISES

REVIEW   *Perform each operation.*

1. $3x^2y^3(-5x^3y^{-4})$

2. $-2a^2b^{-2}(4a^{-2}b^4 - 2a^2b + 3a^3b^2)$

3. $(3t + 2)^2$

4. $(5r - 3s)(5r + 2s)$

5. $2p - 5\overline{)6p^2 - 7p - 25}$

6. $3m + n\overline{)6m^3 - m^2n + 2mn^2 + n^3}$

VOCABULARY AND CONCEPTS   *Fill in the blanks.*

7. $\sqrt[n]{ab} = $ _____

8. $\sqrt[n]{\dfrac{a}{b}} = $ ____

PRACTICE   *Simplify each expression. Assume that all variables represent positive numbers.*

9. $\sqrt{6}\sqrt{6}$

10. $\sqrt{11}\sqrt{11}$

11. $\sqrt{t}\sqrt{t}$

12. $-\sqrt{z}\sqrt{z}$

13. $\sqrt[3]{5x^2}\sqrt[3]{25x}$

14. $\sqrt[4]{25a}\sqrt[4]{25a^3}$

15. $\dfrac{\sqrt{500}}{\sqrt{5}}$

16. $\dfrac{\sqrt{128}}{\sqrt{2}}$

17. $\dfrac{\sqrt{98x^3}}{\sqrt{2x}}$

18. $\dfrac{\sqrt{75y^5}}{\sqrt{3y}}$

19. $\dfrac{\sqrt{180ab^4}}{\sqrt{5ab^2}}$

20. $\dfrac{\sqrt{112ab^3}}{\sqrt{7ab}}$

21. $\dfrac{\sqrt[3]{48}}{\sqrt[3]{6}}$

22. $\dfrac{\sqrt[3]{64}}{\sqrt[3]{8}}$

23. $\dfrac{\sqrt[3]{189a^4}}{\sqrt[3]{7a}}$

24. $\dfrac{\sqrt[3]{243x^7}}{\sqrt[3]{9x}}$

Simplify each radical. Assume that all variables represent positive numbers.

25. $\sqrt{20}$

26. $\sqrt{8}$

27. $-\sqrt{200}$

28. $-\sqrt{250}$

29. $\sqrt[3]{80}$

30. $\sqrt[3]{270}$

31. $\sqrt[3]{-81}$

32. $\sqrt[3]{-72}$

33. $\sqrt[4]{32}$

34. $\sqrt[4]{48}$

35. $\sqrt[5]{96}$

36. $\sqrt[5]{256}$

37. $\sqrt{\dfrac{7}{9}}$

38. $\sqrt{\dfrac{3}{4}}$

39. $\sqrt[3]{\dfrac{7}{64}}$

40. $\sqrt[3]{\dfrac{4}{125}}$

41. $\sqrt[4]{\dfrac{3}{10,000}}$

42. $\sqrt[5]{\dfrac{4}{243}}$

43. $\sqrt[5]{\dfrac{3}{32}}$

44. $\sqrt[6]{\dfrac{5}{64}}$

45. $\sqrt{50x^2}$

46. $\sqrt{75a^2}$

47. $\sqrt{32b}$

48. $\sqrt{80c}$

49. $-\sqrt{112a^3}$

50. $\sqrt{147a^5}$

51. $\sqrt{175a^2b^3}$ **52.** $\sqrt{128a^3b^5}$

53. $-\sqrt{300xy}$ **54.** $\sqrt{200x^2y}$

55. $\sqrt[3]{-54x^6}$ **56.** $-\sqrt[3]{-81a^3}$

57. $\sqrt[3]{16x^{12}y^3}$ **58.** $\sqrt[3]{40a^3b^6}$

59. $\sqrt[4]{32x^{12}y^4}$ **60.** $\sqrt[5]{64x^{10}y^5}$

61. $\sqrt{\dfrac{z^2}{16x^2}}$ **62.** $\sqrt{\dfrac{b^4}{64a^8}}$

63. $\sqrt[4]{\dfrac{5x}{16z^4}}$ **64.** $\sqrt[3]{\dfrac{11a^2}{125b^6}}$

Simplify and combine like radicals. All variables represent positive numbers.

65. $4\sqrt{2x} + 6\sqrt{2x}$ **66.** $6\sqrt[3]{5y} + 3\sqrt[3]{5y}$

67. $8\sqrt[5]{7a^2} - 7\sqrt[5]{7a^2}$ **68.** $10\sqrt[6]{12xyz} - \sqrt[6]{12xyz}$

69. $\sqrt{3} + \sqrt{27}$ **70.** $\sqrt{8} + \sqrt{32}$

71. $\sqrt{2} - \sqrt{8}$ **72.** $\sqrt{20} - \sqrt{125}$

73. $\sqrt{98} - \sqrt{50}$ **74.** $\sqrt{72} - \sqrt{200}$

75. $3\sqrt{24} + \sqrt{54}$ **76.** $\sqrt{18} + 2\sqrt{50}$

77. $\sqrt[3]{24} + \sqrt[3]{3}$ **78.** $\sqrt[3]{16} + \sqrt[3]{128}$

79. $\sqrt[3]{32} - \sqrt[3]{108}$ **80.** $\sqrt[3]{80} - \sqrt[3]{10{,}000}$

81. $2\sqrt[3]{125} - 5\sqrt[3]{64}$ **82.** $3\sqrt[3]{27} + 12\sqrt[3]{216}$

83. $14\sqrt[4]{32} - 15\sqrt[4]{162}$ **84.** $23\sqrt[4]{768} + \sqrt[4]{48}$

85. $3\sqrt[4]{512} + 2\sqrt[4]{32}$ **86.** $4\sqrt[4]{243} - \sqrt[4]{48}$

87. $\sqrt{98} - \sqrt{50} - \sqrt{72}$

88. $\sqrt{20} + \sqrt{125} - \sqrt{80}$

89. $\sqrt{18} + \sqrt{300} - \sqrt{243}$

90. $\sqrt{80} - \sqrt{128} + \sqrt{288}$

91. $2\sqrt[3]{16} - \sqrt[3]{54} - 3\sqrt[3]{128}$

92. $\sqrt[4]{48} - \sqrt[4]{243} - \sqrt[4]{768}$

93. $\sqrt{25y^2z} - \sqrt{16y^2z}$

94. $\sqrt{25yz^2} + \sqrt{9yz^2}$

95. $\sqrt{36xy^2} + \sqrt{49xy^2}$

96. $3\sqrt{2x} - \sqrt{8x}$

97. $2\sqrt[3]{64a} + 2\sqrt[3]{8a}$

98. $3\sqrt[4]{x^4y} - 2\sqrt[4]{x^4y}$

99. $\sqrt{y^5} - \sqrt{9y^5} - \sqrt{25y^5}$

100. $\sqrt{8y^7} + \sqrt{32y^7} - \sqrt{2y^7}$

101. $\sqrt[5]{x^6y^2} + \sqrt[5]{32x^6y^2} + \sqrt[5]{x^6y^2}$

102. $\sqrt[3]{xy^4} + \sqrt[3]{8xy^4} - \sqrt[3]{27xy^4}$

103. $\sqrt{x^2 + 2x + 1} + \sqrt{x^2 + 2x + 1}$

104. $\sqrt{4x^2 + 12x + 9} + \sqrt{9x^2 + 6x + 1}$

Find the missing lengths in each triangle. Give each answer to two decimal places.

105.
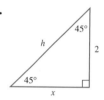

106.
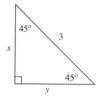

107.
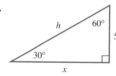

108.
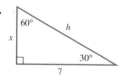

109.
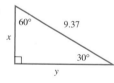

110.
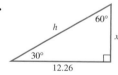

111.
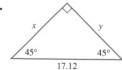

112.
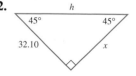

WRITING

113. Explain how to recognize like radicals.

114. Explain how to combine like radicals.

SOMETHING TO THINK ABOUT

115. Can you find any numbers a and b such that
$$\sqrt{a+b} = \sqrt{a} + \sqrt{b}?$$

116. Find the sum:
$$\sqrt{3} + \sqrt{3^2} + \sqrt{3^3} + \sqrt{3^4} + \sqrt{3^5}.$$

7.5   Multiplying and Dividing Radical Expressions

▌ Multiplying a Monomial by a Monomial
▌ Multiplying a Polynomial by a Monomial
▌ Multiplying a Polynomial by a Polynomial ▌ Problem Solving
▌ Rationalizing Denominators ▌ Rationalizing Numerators

Getting Ready   *Perform each operation and simplify, if possible.*

1. a^3a^4 **2.** $\dfrac{b^5}{b^2}$ **3.** $a(a-2)$ **4.** $3b^2(2b+3)$

5. $(a+2)(a-5)$ **6.** $(2a+3b)(2a-3b)$

We now learn how to multiply and divide radical expressions. Then, we will use these skills to solve problems.

Multiplying a Monomial by a Monomial

Radical expressions with the same index can be multiplied and divided.

EXAMPLE 1   Multiply $3\sqrt{6}$ by $2\sqrt{3}$.

Solution   We use the commutative and associative properties of multiplication to multiply the coefficients and the radicals separately. Then we simplify any radicals in the product, if possible.

$$
\begin{aligned}
3\sqrt{6}\cdot 2\sqrt{3} &= 3(2)\sqrt{6}\sqrt{3} && \text{Multiply the coefficients and multiply the radicals.}\\
&= 6\sqrt{18} && 3(2)=6 \text{ and } \sqrt{6}\sqrt{3}=\sqrt{18}.\\
&= 6\sqrt{9}\sqrt{2} && \sqrt{18}=\sqrt{9\cdot 2}=\sqrt{9}\sqrt{2}.\\
&= 6(3)\sqrt{2} && \sqrt{9}=3.\\
&= 18\sqrt{2}
\end{aligned}
$$

Self Check   Multiply $-2\sqrt{7}$ by $5\sqrt{2}$.   ▌

Multiplying a Polynomial by a Monomial

To multiply a polynomial by a monomial, we use the distributive property to remove parentheses and then simplify each resulting term, if possible.

EXAMPLE 2   Multiply: $3\sqrt{3}\left(4\sqrt{8} - 5\sqrt{10}\right)$.

Solution $\mathbf{3\sqrt{3}\left(4\sqrt{8} - 5\sqrt{10}\right)}$

$= \mathbf{3\sqrt{3} \cdot 4\sqrt{8} - 3\sqrt{3} \cdot 5\sqrt{10}}$ Use the distributive property.

$= 12\sqrt{24} - 15\sqrt{30}$ Multiply the coefficients and multiply the radicals.

$= 12\sqrt{4}\sqrt{6} - 15\sqrt{30}$

$= 12(2)\sqrt{6} - 15\sqrt{30}$

$= 24\sqrt{6} - 15\sqrt{30}$

Self Check   Multiply: $4\sqrt{2}\left(3\sqrt{5} - 2\sqrt{8}\right)$.

Multiplying a Polynomial by a Polynomial

To multiply a binomial by a binomial, we use the FOIL method.

EXAMPLE 3   Multiply: $\left(\sqrt{7} + \sqrt{2}\right)\left(\sqrt{7} - 3\sqrt{2}\right)$.

Solution $\left(\sqrt{7} + \sqrt{2}\right)\left(\sqrt{7} - 3\sqrt{2}\right)$

$= \left(\sqrt{7}\right)^2 - 3\sqrt{7}\sqrt{2} + \sqrt{2}\sqrt{7} - 3\sqrt{2}\sqrt{2}$

$= 7 - 3\sqrt{14} + \sqrt{14} - 3(2)$

$= 7 - 2\sqrt{14} - 6$

$= 1 - 2\sqrt{14}$

Self Check   Multiply: $\left(\sqrt{5} + 2\sqrt{3}\right)\left(\sqrt{5} - \sqrt{3}\right)$.

Technically, the expression $\sqrt{3x} - \sqrt{5}$ is not a polynomial, because the variable does not have a whole-number exponent $\left(\sqrt{3x} = 3^{1/2}x^{1/2}\right)$. However, we will multiply such expressions as if they were polynomials.

EXAMPLE 4   Multiply: $\left(\sqrt{3x} - \sqrt{5}\right)\left(\sqrt{2x} + \sqrt{10}\right)$.

Solution $\left(\sqrt{3x} - \sqrt{5}\right)\left(\sqrt{2x} + \sqrt{10}\right)$

$= \sqrt{3x}\sqrt{2x} + \sqrt{3x}\sqrt{10} - \sqrt{5}\sqrt{2x} - \sqrt{5}\sqrt{10}$

$= \sqrt{6x^2} + \sqrt{30x} - \sqrt{10x} - \sqrt{50}$

$= \sqrt{6}\sqrt{x^2} + \sqrt{30x} - \sqrt{10x} - \sqrt{25}\sqrt{2}$

$= \sqrt{6}x + \sqrt{30x} - \sqrt{10x} - 5\sqrt{2}$

Self Check   Multiply: $\left(\sqrt{x} + 1\right)\left(\sqrt{x} - 3\right)$.

Comment   It is important to draw radical signs carefully so that they completely cover the radicand, but no more than the radicand. To avoid confusion, we often write an expression such as $\sqrt{6x}$ in the form $x\sqrt{6}$.

Problem Solving

Figure 7-15

EXAMPLE 5   **Photography**   Many camera lenses (see Figure 7-15) have an adjustable opening called the *aperture,* which controls the amount of light passing through the lens. The *f-number* of a lens is its *focal length* divided by the diameter of its circular aperture:

$$f\text{-number} = \frac{f}{d} \qquad f \text{ is the focal length, and } d \text{ is the diameter of the aperture.}$$

A lens with a focal length of 12 centimeters and an aperture with a diameter of 6 centimeters has an *f*-number of $\frac{12}{6}$ and is an *f*/2 lens. If the area of the aperture is reduced to admit half as much light, the *f*-number of the lens will change. Find the new *f*-number.

Solution   We first find the area of the aperture when its diameter is 6 centimeters.

$$A = \pi r^2 \qquad \text{The formula for the area of a circle.}$$

$$A = \pi(3)^2 \qquad \text{Since a radius is half the diameter, substitute 3 for } r.$$

$$A = 9\pi$$

When the size of the aperture is reduced to admit half as much light, the area of the aperture will be $\frac{9\pi}{2}$ square centimeters. To find the diameter of a circle with this area, we proceed as follows:

$$A = \pi r^2 \qquad \text{The formula for the area of a circle.}$$

$$\frac{9\pi}{2} = \pi\left(\frac{d}{2}\right)^2 \qquad \text{Substitute } \frac{9\pi}{2} \text{ for } A \text{ and } \frac{d}{2} \text{ for } r.$$

$$\frac{9\pi}{2} = \frac{\pi d^2}{4} \qquad \left(\frac{d}{2}\right)^2 = \frac{d^2}{4}.$$

$$18 = d^2 \qquad \text{Multiply both sides by 4, and divide both sides by } \pi.$$

$$d = 3\sqrt{2} \qquad \sqrt{18} = \sqrt{9}\sqrt{2} = 3\sqrt{2}.$$

Since the focal length of the lens is still 12 centimeters and the diameter is now $3\sqrt{2}$ centimeters, the new *f*-number of the lens is

$$f\text{-number} = \frac{f}{d} = \frac{12}{3\sqrt{2}} \qquad \text{Substitute 12 for } f \text{ and } 3\sqrt{2} \text{ for } d.$$

$$\approx 2.828427125 \qquad \text{Use a calculator.}$$

The lens is now an *f*/2.8 lens. ∎

Rationalizing Denominators

To divide radical expressions, we **rationalize the denominator** of a fraction to replace the denominator with a rational number. For example, to divide $\sqrt{70}$ by $\sqrt{3}$, we write the division as the fraction

$$\frac{\sqrt{70}}{\sqrt{3}}$$

To eliminate the radical in the denominator, we multiply the numerator and the denominator by a number that will give a perfect square under the radical in the denominator. Because $3 \cdot 3 = 9$ and 9 is a perfect square, $\sqrt{3}$ is such a number.

$$\frac{\sqrt{70}}{\sqrt{3}} = \frac{\sqrt{70} \cdot \sqrt{3}}{\sqrt{3} \cdot \sqrt{3}} \qquad \text{Multiply numerator and denominator by } \sqrt{3}.$$

$$= \frac{\sqrt{210}}{3} \qquad \text{Multiply the radicals.}$$

Since there is no radical in the denominator and $\sqrt{210}$ cannot be simplified, the expression $\frac{\sqrt{210}}{3}$ is in simplest form, and the division is complete.

EXAMPLE 6   Rationalize the denominator: **a.** $\sqrt{\dfrac{20}{7}}$ and   **b.** $\dfrac{4}{\sqrt[3]{2}}$.

Solution   **a.** We write the square root of the quotient as the quotient of two square roots

$$\sqrt{\frac{20}{7}} = \frac{\sqrt{20}}{\sqrt{7}}$$

and proceed as follows:

$$\frac{\sqrt{20}}{\sqrt{7}} = \frac{\sqrt{20} \cdot \sqrt{7}}{\sqrt{7} \cdot \sqrt{7}} \qquad \text{Multiply numerator and denominator by } \sqrt{7}.$$

$$= \frac{\sqrt{140}}{7} \qquad \text{Multiply the radicals.}$$

$$= \frac{2\sqrt{35}}{7} \qquad \text{Simplify } \sqrt{140}: \sqrt{140} = \sqrt{4 \cdot 35} = \sqrt{4}\sqrt{35} = 2\sqrt{35}.$$

b. Since the denominator is a cube root, we multiply the numerator and the denominator by a number that will give a perfect cube under the radical sign. Since $2 \cdot 4 = 8$ is a perfect cube, $\sqrt[3]{4}$ is such a number.

$$\frac{4}{\sqrt[3]{2}} = \frac{4 \cdot \sqrt[3]{4}}{\sqrt[3]{2} \cdot \sqrt[3]{4}} \qquad \text{Multiply numerator and denominator by } \sqrt[3]{4}.$$

$$= \frac{4\sqrt[3]{4}}{\sqrt[3]{8}} \qquad \text{Multiply the radicals in the denominator.}$$

$$= \frac{4\sqrt[3]{4}}{2} \qquad \sqrt[3]{8} = 2.$$

$$= 2\sqrt[3]{4} \qquad \text{Simplify.}$$

Self Check   Rationalize the denominator: $\dfrac{5}{\sqrt[4]{3}}$.

EXAMPLE 7 Rationalize the denominator: $\dfrac{\sqrt[3]{5}}{\sqrt[3]{18}}$.

Solution We multiply the numerator and the denominator by a number that will result in a perfect cube under the radical sign in the denominator.

Since 216 is the smallest perfect cube that is divisible by 18 ($216 \div 18 = 12$), multiplying the numerator and the denominator by $\sqrt[3]{12}$ will give the smallest possible perfect cube under the radical in the denominator.

$$\dfrac{\sqrt[3]{5}}{\sqrt[3]{18}} = \dfrac{\sqrt[3]{5} \cdot \sqrt[3]{12}}{\sqrt[3]{18} \cdot \sqrt[3]{12}} \quad \text{Multiply numerator and denominator by } \sqrt[3]{12}.$$

$$= \dfrac{\sqrt[3]{60}}{\sqrt[3]{216}} \quad \text{Multiply the radicals.}$$

$$= \dfrac{\sqrt[3]{60}}{6} \quad \sqrt[3]{216} = 6.$$

Self Check Rationalize the denominator: $\dfrac{\sqrt{5}}{\sqrt{18}}$.

EXAMPLE 8 Rationalize the denominator of $\dfrac{\sqrt{5xy^2}}{\sqrt{xy^3}}$ (x and y are positive numbers).

Solution

Method 1

$$\dfrac{\sqrt{5xy^2}}{\sqrt{xy^3}} = \sqrt{\dfrac{5xy^2}{xy^3}}$$

$$= \sqrt{\dfrac{5}{y}}$$

$$= \dfrac{\sqrt{5}}{\sqrt{y}}$$

$$= \dfrac{\sqrt{5}\sqrt{y}}{\sqrt{y}\sqrt{y}}$$

$$= \dfrac{\sqrt{5y}}{y}$$

Method 2

$$\dfrac{\sqrt{5xy^2}}{\sqrt{xy^3}} = \sqrt{\dfrac{5xy^2}{xy^3}}$$

$$= \sqrt{\dfrac{5}{y}}$$

$$= \sqrt{\dfrac{5 \cdot y}{y \cdot y}}$$

$$= \dfrac{\sqrt{5y}}{\sqrt{y^2}}$$

$$= \dfrac{\sqrt{5y}}{y}$$

Self Check Rationalize the denominator: $\dfrac{\sqrt{4ab^3}}{\sqrt{2a^2b^2}}$.

To rationalize the denominator of a fraction with square roots in a binomial denominator, we multiply its numerator and denominator by the *conjugate* of its denominator. Conjugate binomials are binomials with the same terms but with opposite signs between their terms.

Conjugate Binomials The conjugate of $a + b$ is $a - b$, and the conjugate of $a - b$ is $a + b$.

EXAMPLE 9   Rationalize the denominator: $\dfrac{1}{\sqrt{2}+1}$.

Solution   We multiply the numerator and denominator of the fraction by $\sqrt{2}-1$, which is the conjugate of the denominator.

$$\dfrac{1}{\sqrt{2}+1} = \dfrac{1(\sqrt{2}-1)}{(\sqrt{2}+1)(\sqrt{2}-1)} \qquad \dfrac{\sqrt{2}-1}{\sqrt{2}-1} = 1.$$

$$= \dfrac{\sqrt{2}-1}{(\sqrt{2})^2 - 1} \qquad (\sqrt{2}+1)(\sqrt{2}-1) = (\sqrt{2})^2 - 1.$$

$$= \dfrac{\sqrt{2}-1}{2-1} \qquad (\sqrt{2})^2 = 2.$$

$$= \sqrt{2}-1 \qquad \dfrac{\sqrt{2}-1}{2-1} = \dfrac{\sqrt{2}-1}{1} = \sqrt{2}-1.$$

Self Check   Rationalize the denominator: $\dfrac{2}{\sqrt{3}+1}$.

EXAMPLE 10   Rationalize the denominator: $\dfrac{\sqrt{x}+\sqrt{2}}{\sqrt{x}-\sqrt{2}}$ $(x > 0)$.

Solution   We multiply the numerator and denominator by $\sqrt{x}+\sqrt{2}$, which is the conjugate of the denominator, and simplify.

$$\dfrac{\sqrt{x}+\sqrt{2}}{\sqrt{x}-\sqrt{2}} = \dfrac{(\sqrt{x}+\sqrt{2})(\sqrt{x}+\sqrt{2})}{(\sqrt{x}-\sqrt{2})(\sqrt{x}+\sqrt{2})}$$

$$= \dfrac{x + \sqrt{2x} + \sqrt{2x} + 2}{x - 2} \qquad \text{Use the FOIL method.}$$

$$= \dfrac{x + 2\sqrt{2x} + 2}{x - 2}$$

Self Check   Rationalize the denominator: $\dfrac{\sqrt{x}-\sqrt{2}}{\sqrt{x}+\sqrt{2}}$.

Rationalizing Numerators

In calculus, we sometimes have to rationalize a numerator by multiplying the numerator and denominator of the fraction by the conjugate of the numerator.

EXAMPLE 11   Rationalize the numerator: $\dfrac{\sqrt{x}-3}{\sqrt{x}}$ $(x > 0)$.

Solution   We multiply the numerator and denominator by $\sqrt{x}+3$, which is the conjugate of the numerator.

$$\frac{\sqrt{x} - 3}{\sqrt{x}} = \frac{(\sqrt{x} - 3)(\sqrt{x} + 3)}{\sqrt{x}(\sqrt{x} + 3)}$$

$$= \frac{x + 3\sqrt{x} - 3\sqrt{x} - 9}{x + 3\sqrt{x}}$$

$$= \frac{x - 9}{x + 3\sqrt{x}}$$

The final expression is not in simplified form. However, this nonsimplified form is sometimes desirable in calculus.

Self Check   Rationalize the numerator: $\dfrac{\sqrt{x} + 3}{\sqrt{x}}$.

Self Check Answers

1. $-10\sqrt{14}$ **2.** $12\sqrt{10} - 32$ **3.** $-1 + \sqrt{15}$ **4.** $x - 2\sqrt{x} - 3$ **6.** $\dfrac{5\sqrt[4]{27}}{3}$ **7.** $\dfrac{\sqrt{10}}{6}$ **8.** $\dfrac{\sqrt{2ab}}{a}$

9. $\sqrt{3} - 1$ **10.** $\dfrac{x - 2\sqrt{2x} + 2}{x - 2}$ **11.** $\dfrac{x - 9}{x - 3\sqrt{x}}$

Orals   *Simplify:*

1. $\sqrt{3}\sqrt{3}$ **2.** $\sqrt[3]{2}\sqrt[3]{2}\sqrt[3]{2}$ **3.** $\sqrt{3}\sqrt{9}$

4. $\sqrt{a^3 b}\sqrt{ab}$ $(b > 0)$ **5.** $3\sqrt{2}(\sqrt{2} + 1)$ **6.** $(\sqrt{2} + 1)(\sqrt{2} - 1)$

7. $\dfrac{1}{\sqrt{2}}$ **8.** $\dfrac{1}{\sqrt{3} - 1}$

7.5   EXERCISES

REVIEW   *Solve each equation.*

1. $\dfrac{2}{3 - a} = 1$

2. $5(s - 4) = -5(s - 4)$

3. $\dfrac{8}{b - 2} + \dfrac{3}{2 - b} = -\dfrac{1}{b}$

4. $\dfrac{2}{x - 2} + \dfrac{1}{x + 1} = \dfrac{1}{(x + 1)(x - 2)}$

VOCABULARY AND CONCEPTS   *Fill in the blanks.*

5. To multiply $2\sqrt{7}$ by $3\sqrt{5}$, we multiply __ by 3 and then multiply ____ by __.

6. To multiply $2\sqrt{5}(3\sqrt{8} + \sqrt{3})$, we use the _____ property to remove parentheses and simplify each resulting term.

7. To multiply $(\sqrt{3} + \sqrt{2})(\sqrt{3} - 2\sqrt{2})$, we can use the _____ method.

8. The conjugate of $\sqrt{x} + 1$ is _____.

9. To rationalize the denominator of $\dfrac{1}{\sqrt{3} - 1}$, multiply both the numerator and denominator by the _____ of the denominator.

10. To rationalize the numerator of $\dfrac{\sqrt{5} + 2}{\sqrt{5} - 2}$, multiply both the numerator and denominator by _____.

PRACTICE   *Perform each multiplication and simplify, if possible. All variables represent positive numbers.*

11. $\sqrt{2}\sqrt{8}$ **12.** $\sqrt{3}\sqrt{27}$

13. $\sqrt{5}\sqrt{10}$ **14.** $\sqrt{7}\sqrt{35}$

15. $2\sqrt{3}\sqrt{6}$ **16.** $3\sqrt{11}\sqrt{33}$

17. $\sqrt[3]{5}\sqrt[3]{25}$ **18.** $\sqrt[3]{7}\sqrt[3]{49}$

19. $\left(3\sqrt[3]{9}\right)\left(2\sqrt[3]{3}\right)$ **20.** $\left(2\sqrt[3]{16}\right)\left(-\sqrt[3]{4}\right)$

21. $\sqrt[3]{2}\sqrt[3]{12}$ **22.** $\sqrt[3]{3}\sqrt[3]{18}$

23. $\sqrt{ab^3}\sqrt{ab}$ **24.** $\sqrt{8x}\sqrt{2x^3y}$

25. $\sqrt{5ab}\sqrt{5a}$ **26.** $\sqrt{15rs^2}\sqrt{10r}$

27. $\sqrt[3]{5r^2s}\sqrt[3]{2r}$ **28.** $\sqrt[3]{3xy^2}\sqrt[3]{9x^3}$

29. $\sqrt[3]{a^5b}\sqrt[3]{16ab^5}$ **30.** $\sqrt[3]{3x^4y}\sqrt[3]{18x}$

31. $\sqrt{x(x+3)}\sqrt{x^3(x+3)}$

32. $\sqrt{y^2(x+y)}\sqrt{(x+y)^3}$

33. $\sqrt[3]{6x^2(y+z)^2}\sqrt[3]{18x(y+z)}$

34. $\sqrt[3]{9x^2y(z+1)^2}\sqrt[3]{6xy^2(z+1)}$

35. $3\sqrt{5}\left(4-\sqrt{5}\right)$ **36.** $2\sqrt{7}\left(3\sqrt{7}-1\right)$

37. $3\sqrt{2}\left(4\sqrt{3}+2\sqrt{7}\right)$ **38.** $-\sqrt{3}\left(\sqrt{7}-\sqrt{5}\right)$

39. $-2\sqrt{5x}\left(4\sqrt{2x}-3\sqrt{3}\right)$

40. $3\sqrt{7t}\left(2\sqrt{7t}+3\sqrt{3t^2}\right)$

41. $\left(\sqrt{2}+1\right)\left(\sqrt{2}-3\right)$

42. $\left(2\sqrt{3}+1\right)\left(\sqrt{3}-1\right)$

43. $\left(4\sqrt{x}+3\right)\left(2\sqrt{x}-5\right)$

44. $\left(7\sqrt{y}+2\right)\left(3\sqrt{y}-5\right)$

45. $\left(\sqrt{5z}+\sqrt{3}\right)\left(\sqrt{5z}+\sqrt{3}\right)$

46. $\left(\sqrt{3p}-\sqrt{2}\right)\left(\sqrt{3p}+\sqrt{2}\right)$

47. $\left(\sqrt{3x}-\sqrt{2y}\right)\left(\sqrt{3x}+\sqrt{2y}\right)$

48. $\left(\sqrt{3m}+\sqrt{2n}\right)\left(\sqrt{3m}+\sqrt{2n}\right)$

49. $\left(2\sqrt{3a}-\sqrt{b}\right)\left(\sqrt{3a}+3\sqrt{b}\right)$

50. $\left(5\sqrt{p}-\sqrt{3q}\right)\left(\sqrt{p}+2\sqrt{3q}\right)$

51. $\left(3\sqrt{2r}-2\right)^2$ **52.** $\left(2\sqrt{3t}+5\right)^2$

53. $-2\left(\sqrt{3x}+\sqrt{3}\right)^2$ **54.** $3\left(\sqrt{5x}-\sqrt{3}\right)^2$

Rationalize each denominator. All variables represent positive numbers.

55. $\sqrt{\dfrac{1}{7}}$ **56.** $\sqrt{\dfrac{5}{3}}$

57. $\sqrt{\dfrac{2}{3}}$ **58.** $\sqrt{\dfrac{3}{2}}$

59. $\dfrac{\sqrt{5}}{\sqrt{8}}$ **60.** $\dfrac{\sqrt{3}}{\sqrt{50}}$

61. $\dfrac{\sqrt{8}}{\sqrt{2}}$ **62.** $\dfrac{\sqrt{27}}{\sqrt{3}}$

63. $\dfrac{1}{\sqrt[3]{2}}$ **64.** $\dfrac{2}{\sqrt[3]{6}}$

65. $\dfrac{3}{\sqrt[3]{9}}$ **66.** $\dfrac{2}{\sqrt[3]{a}}$

67. $\dfrac{\sqrt[3]{2}}{\sqrt[3]{9}}$ **68.** $\dfrac{\sqrt[3]{9}}{\sqrt[3]{54}}$

69. $\dfrac{\sqrt{8x^2y}}{\sqrt{xy}}$ **70.** $\dfrac{\sqrt{9xy}}{\sqrt{3x^2y}}$

71. $\dfrac{\sqrt{10xy^2}}{\sqrt{2xy^3}}$ **72.** $\dfrac{\sqrt{5ab^2c}}{\sqrt{10abc}}$

73. $\dfrac{\sqrt[3]{4a^2}}{\sqrt[3]{2ab}}$ **74.** $\dfrac{\sqrt[3]{9x}}{\sqrt[3]{3xy}}$

75. $\dfrac{1}{\sqrt[4]{4}}$ **76.** $\dfrac{1}{\sqrt[5]{2}}$

77. $\dfrac{1}{\sqrt[5]{16}}$ **78.** $\dfrac{4}{\sqrt[4]{32}}$

79. $\dfrac{1}{\sqrt{2}-1}$ **80.** $\dfrac{3}{\sqrt{3}-1}$

81. $\dfrac{\sqrt{2}}{\sqrt{5}+3}$ **82.** $\dfrac{\sqrt{3}}{\sqrt{3}-2}$

83. $\dfrac{\sqrt{3}+1}{\sqrt{3}-1}$ **84.** $\dfrac{\sqrt{2}-1}{\sqrt{2}+1}$

85. $\dfrac{\sqrt{7} - \sqrt{2}}{\sqrt{2} + \sqrt{7}}$

86. $\dfrac{\sqrt{3} + \sqrt{2}}{\sqrt{3} - \sqrt{2}}$

87. $\dfrac{2}{\sqrt{x} + 1}$

88. $\dfrac{3}{\sqrt{x} - 2}$

89. $\dfrac{x}{\sqrt{x} - 4}$

90. $\dfrac{2x}{\sqrt{x} + 1}$

91. $\dfrac{2z - 1}{\sqrt{2z} - 1}$

92. $\dfrac{3t - 1}{\sqrt{3t} + 1}$

93. $\dfrac{\sqrt{x} - \sqrt{y}}{\sqrt{x} + \sqrt{y}}$

94. $\dfrac{\sqrt{x} + \sqrt{y}}{\sqrt{x} - \sqrt{y}}$

Rationalize each numerator. All variables represent positive numbers.

95. $\dfrac{\sqrt{3} + 1}{2}$

96. $\dfrac{\sqrt{5} - 1}{2}$

97. $\dfrac{\sqrt{x} + 3}{x}$

98. $\dfrac{2 + \sqrt{x}}{5x}$

99. $\dfrac{\sqrt{x} + \sqrt{y}}{\sqrt{x}}$

100. $\dfrac{\sqrt{x} - \sqrt{y}}{\sqrt{x} + \sqrt{y}}$

APPLICATIONS

101. **Photography**   We have seen that a lens with a focal length of 12 centimeters and an aperture $3\sqrt{2}$ centimeters in diameter is an $f/2.8$ lens. Find the f-number if the area of the aperture is again cut in half.

102. **Photography**   A lens with a focal length of 12 centimeters and an aperture 3 centimeters in diameter is an $f/4$ lens. Find the f-number if the area of the aperture is cut in half.

WRITING

103. Explain how to simplify a fraction with the monomial denominator $\sqrt[3]{3}$.

104. Explain how to simplify a fraction with the monomial denominator $\sqrt[3]{9}$.

SOMETHING TO THINK ABOUT   *Assume that x is a rational number.*

105. Change the numerator of $\dfrac{\sqrt{x} - 3}{4}$ to a rational number.

106. Rationalize the numerator: $\dfrac{2\sqrt{3x} + 4}{\sqrt{3x} - 1}$.

7.6 Radical Equations

- **The Power Rule**   ■ **Equations Containing One Radical**
- **Equations Containing Two Radicals**
- **Equations Containing Three Radicals**

Getting Ready   *Find each power.*

1. $\left(\sqrt{a}\right)^2$

2. $\left(\sqrt{5x}\right)^2$

3. $\left(\sqrt{x + 4}\right)^2$

4. $\left(\sqrt[4]{y - 3}\right)^4$

In this section, we will solve equations that contain radicals. To do so, we will use the **power rule**.

The Power Rule

The Power Rule

If x, y, and n are real numbers and $x = y$, then

$$x^n = y^n$$

If we raise both sides of an equation to the same power, the resulting equation might not be equivalent to the original equation. For example, if we square both sides of the equation

(1) $\qquad x = 3 \qquad$ With a solution set of $\{3\}$.

we obtain the equation

(2) $\qquad x^2 = 9 \qquad$ With a solution set of $\{3, -3\}$.

Equations 1 and 2 are not equivalent, because they have different solution sets, and the solution -3 of Equation 2 does not satisfy Equation 1. Since raising both sides of an equation to the same power can produce an equation with roots that don't satisfy the original equation, we must check each suspected solution in the original equation.

Equations Containing One Radical

EXAMPLE 1 Solve: $\sqrt{x + 3} = 4$.

Solution To eliminate the radical, we apply the power rule by squaring both sides of the equation, and proceed as follows:

$$\sqrt{x + 3} = 4$$
$$\left(\sqrt{x + 3}\right)^2 = (4)^2 \qquad \text{Square both sides.}$$
$$x + 3 = 16$$
$$x = 13 \qquad \text{Subtract 3 from both sides.}$$

We must check the apparent solution of 13 to see whether it satisfies the original equation.

Check: $\sqrt{x + 3} = 4$

$$\sqrt{\mathbf{13} + 3} \stackrel{?}{=} 4 \qquad \text{Substitute 13 for } x.$$
$$\sqrt{16} \stackrel{?}{=} 4$$
$$4 = 4$$

Since 13 satisfies the original equation, it is a solution.

Self Check Solve: $\sqrt{a - 2} = 3$. ∎

To solve an equation with radicals, we follow these steps.

Solving an Equation Containing Radicals	1. Isolate one radical expression on one side of the equation.
	2. Raise both sides of the equation to the power that is the same as the index of the radical.
	3. Solve the resulting equation. If it still contains a radical, go back to Step 1.
	4. Check the possible solutions to eliminate the ones that do not satisfy the original equation.

EXAMPLE 2   **Height of a bridge**   The distance d (in feet) that an object will fall in t seconds is given by the formula

$$t = \sqrt{\dfrac{d}{16}}$$

To find the height of a bridge, a man drops a stone into the water (see Figure 7-16). If it takes the stone 3 seconds to hit the water, how far above the river is the bridge?

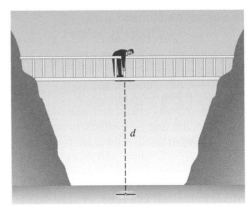

Figure 7-16

Solution   We substitute 3 for t in the formula and solve for d.

$$t = \sqrt{\dfrac{d}{16}}$$

$$3 = \sqrt{\dfrac{d}{16}}$$

$$9 = \dfrac{d}{16} \qquad \text{Square both sides.}$$

$$144 = d \qquad \text{Multiply both sides by 16.}$$

The bridge is 144 feet above the river.

Self Check   How high is the bridge if it takes 4 seconds for the stone to hit the water?

EXAMPLE 3   Solve: $\sqrt{3x + 1} + 1 = x$.

Solution   We first subtract 1 from both sides to isolate the radical. Then, to eliminate the radical, we square both sides of the equation and proceed as follows:

$$\sqrt{3x + 1} + 1 = x$$

$$\sqrt{3x + 1} = x - 1 \qquad \text{Subtract 1 from both sides.}$$

$$\left(\sqrt{3x + 1}\right)^2 = (x - 1)^2 \qquad \text{Square both sides to eliminate the square root.}$$

$$3x + 1 = x^2 - 2x + 1 \qquad \begin{array}{l}(x - 1)^2 \neq x^2 - 1. \text{ Instead, } (x - 1)^2 = \\ (x - 1)(x - 1) = x^2 - x - x + 1 = x^2 - 2x + 1.\end{array}$$

$$0 = x^2 - 5x \qquad \text{Subtract } 3x \text{ and 1 from both sides.}$$

$$0 = x(x - 5) \qquad \text{Factor } x^2 - 5x.$$

$$x = 0 \quad \text{or} \quad x - 5 = 0 \qquad \text{Set each factor equal to 0.}$$

$$x = 0 \quad | \qquad x = 5$$

We must check each apparent solution to see whether it satisfies the original equation.

Check: $\sqrt{3x + 1} + 1 = x$ $\qquad\qquad\qquad$ $\sqrt{3x + 1} = x$

$\quad\sqrt{3(0) + 1} + 1 \stackrel{?}{=} 0$ $\qquad\qquad$ $\sqrt{3(5) + 1} + 1 \stackrel{?}{=} 5$

$\qquad\quad\sqrt{1} + 1 \stackrel{?}{=} 0$ $\qquad\qquad\qquad$ $\sqrt{16} + 1 \stackrel{?}{=} 5$

$\qquad\qquad\quad 2 \neq 0$ $\qquad\qquad\qquad\qquad\qquad 5 = 5$

Since 0 does not check, it must be discarded. The only solution of the original equation is 5.

Self Check   Solve: $\sqrt{4x + 1} + 1 = x$. ∎

Accent on Technology   **SOLVING EQUATIONS CONTAINING RADICALS**

To find approximate solutions for $\sqrt{3x + 1} + 1 = x$ with a graphing calculator, we can use window settings of $[-5, 10]$ for x and $[-2, 8]$ for y and graph the functions $f(x) = \sqrt{3x + 1} + 1$ and $g(x) = x$, as in Figure 7-17(a). We then trace to find the approximate x-coordinate of their intersection point, as in Figure 7-17(b). After repeated zooms, we will see that $x = 5$.

We can also find the x-coordinate of the intersection point by using the INTERSECT command found in the CALC menu.

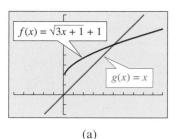

(a)

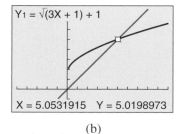

(b)

Figure 7-17

EXAMPLE 4 Solve: $\sqrt[3]{x^3 + 7} = x + 1$.

Solution To eliminate the radical, we cube both sides of the equation and proceed as follows:

$$\sqrt[3]{x^3 + 7} = x + 1$$

$$\left(\sqrt[3]{x^3 + 7}\right)^3 = (x + 1)^3 \qquad \text{Cube both sides to eliminate the cube root.}$$

$$x^3 + 7 = x^3 + 3x^2 + 3x + 1$$

$$0 = 3x^2 + 3x - 6 \qquad \text{Subtract } x^3 \text{ and 7 from both sides.}$$

$$0 = x^2 + x - 2 \qquad \text{Divide both sides by 3.}$$

$$0 = (x + 2)(x - 1)$$

$$x + 2 = 0 \quad \text{or} \quad x - 1 = 0$$

$$x = -2 \quad | \qquad x = 1$$

We check each apparent solution to see whether it satisfies the original equation.

Check: $\qquad \sqrt[3]{x^3 + 7} = x + 1 \qquad\qquad\qquad \sqrt[3]{x^3 + 7} = x + 1$

$$\sqrt[3]{(-2)^3 + 7} \overset{?}{=} -2 + 1 \qquad\qquad \sqrt[3]{1 + 7} \overset{?}{=} 1 + 1$$

$$\sqrt[3]{-8 + 7} \overset{?}{=} -1 \qquad\qquad\qquad \sqrt[3]{8} \overset{?}{=} 2$$

$$\sqrt[3]{-1} \overset{?}{=} -1 \qquad\qquad\qquad\qquad 2 = 2$$

$$-1 = -1$$

Both solutions satisfy the original equation.

Self Check Solve: $\sqrt[3]{x^3 + 8} = x + 2$. ∎

Equations Containing Two Radicals

When more than one radical appears in an equation, it is often necessary to apply the power rule more than once.

EXAMPLE 5 Solve: $\sqrt{x} + \sqrt{x + 2} = 2$.

Solution To remove the radicals, we square both sides of the equation. Since this is easier to do if one radical is on each side of the equation, we subtract $\sqrt{x}$ from both sides to isolate one radical on one side of the equation.

$$\sqrt{x} + \sqrt{x + 2} = 2$$

$$\sqrt{x + 2} = 2 - \sqrt{x} \qquad \text{Subtract } \sqrt{x} \text{ from both sides.}$$

$$\left(\sqrt{x + 2}\right)^2 = \left(2 - \sqrt{x}\right)^2 \qquad \text{Square both sides to eliminate the square root.}$$

$$x + 2 = 4 - 4\sqrt{x} + x \qquad (2 - \sqrt{x})(2 - \sqrt{x}) = 4 - 2\sqrt{x} - 2\sqrt{x} + x = 4 - 4\sqrt{x} + x.$$

$$2 = 4 - 4\sqrt{x} \qquad \text{Subtract } x \text{ from both sides.}$$

$$-2 = -4\sqrt{x} \qquad \text{Subtract 4 from both sides.}$$

$$\frac{1}{2} = \sqrt{x} \qquad \text{Divide both sides by } -4.$$

$$\frac{1}{4} = x \qquad \text{Square both sides.}$$

Check: $\sqrt{x} + \sqrt{x+2} = 2$

$$\sqrt{\frac{1}{4}} + \sqrt{\frac{1}{4} + 2} \overset{?}{=} 2$$

$$\frac{1}{2} + \sqrt{\frac{9}{4}} \overset{?}{=} 2$$

$$\frac{1}{2} + \frac{3}{2} \overset{?}{=} 2$$

$$2 = 2$$

The solution checks.

Self Check   Solve: $\sqrt{a} + \sqrt{a+3} = 3$.

Accent on Technology   SOLVING EQUATIONS CONTAINING RADICALS

To find approximate solutions for $\sqrt{x} + \sqrt{x+2} = 5$ with a graphing calculator, we use window settings of $[-2, 10]$ for x and $[-2, 8]$ for y and graph the functions $f(x) = \sqrt{x} + \sqrt{x+2}$ and $g(x) = 5$, as in Figure 7-18(a). We then trace to find an approximation of the x-coordinate of their intersection point, as in Figure 7-18(b). From the figure, we can see that $x \approx 5.15$. We can zoom to get better results.

We can also find the x-coordinate of the intersection point by using the INTERSECT command found in the CALC menu.

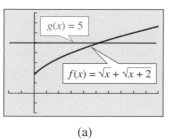

(a)   (b)

Figure 7-18

Equations Containing Three Radicals

EXAMPLE 6   Solve: $\sqrt{x+2} + \sqrt{2x} = \sqrt{18-x}$.

Solution   In this case, it is impossible to isolate one radical on each side of the equation, so we begin by squaring both sides. Then we proceed as follows.

$$\sqrt{x+2} + \sqrt{2x} = \sqrt{18-x}$$

$$\left(\sqrt{x+2} + \sqrt{2x}\right)^2 = \left(\sqrt{18-x}\right)^2 \qquad \text{Square both sides to eliminate one square root.}$$

$$x + 2 + 2\sqrt{x+2}\sqrt{2x} + 2x = 18 - x$$

$$2\sqrt{x+2}\sqrt{2x} = 16 - 4x \qquad \text{Subtract } 3x \text{ and 2 from both sides.}$$

$$\sqrt{x+2}\sqrt{2x} = 8 - 2x \qquad \text{Divide both sides by 2.}$$

$$\left(\sqrt{x+2}\sqrt{2x}\right)^2 = \left(8 - 2x\right)^2 \qquad \text{Square both sides to eliminate the other square roots.}$$

$$(x+2)2x = 64 - 32x + 4x^2$$

$$2x^2 + 4x = 64 - 32x + 4x^2$$

$$0 = 2x^2 - 36x + 64 \qquad \text{Write the equation in quadratic form.}$$

$$0 = x^2 - 18x + 32 \qquad \text{Divide both sides by 2.}$$

$$0 = (x - 16)(x - 2) \qquad \text{Factor the trinomial.}$$

$$x - 16 = 0 \quad \text{or} \quad x - 2 = 0 \qquad \text{Set each factor equal to 0.}$$

$$x = 16 \quad | \quad x = 2$$

Verify that 2 satisfies the equation, but 16 does not. Thus, the only solution is 2.

Self Check   Solve: $\sqrt{3x+4} + \sqrt{x+9} = \sqrt{x+25}$. ∎

Self Check Answers

1. 11   **2.** 256 ft   **3.** 6; 0 is extraneous   **4.** 0, -2 **5.** 1   **6.** 0

Orals   *Solve each equation.*

1. $\sqrt{x+2} = 3$ **2.** $\sqrt{x-2} = 1$

3. $\sqrt[3]{x+1} = 1$ **4.** $\sqrt[3]{x-1} = 2$

5. $\sqrt[4]{x-1} = 2$ **6.** $\sqrt[5]{x+1} = 2$

7.6   EXERCISES

REVIEW   *If $f(x) = 3x^2 - 4x + 2$, find each quantity.*

1. $f(0)$ **2.** $f(-3)$

3. $f(2)$ **4.** $f\left(\dfrac{1}{2}\right)$

VOCABULARY AND CONCEPTS   *Fill in the blanks.*

5. If x, y, and n are real numbers and $x = y$, then _____.

6. When solving equations containing radicals, try to _____ one radical expression on one side of the equation.

7. To solve the equation $\sqrt{x+4} = 5$, we first _____ both sides.

8. To solve the equation $\sqrt[3]{x+4} = 2$, we first _____ both sides.

9. Squaring both sides of an equation can introduce _____ solutions.

10. Always remember to _____ the solutions of an equation containing radicals to eliminate any _____ solutions.

PRACTICE   *Solve each equation. Write all solutions and cross out those that are extraneous.*

11. $\sqrt{5x-6} = 2$

12. $\sqrt{7x-10} = 12$

13. $\sqrt{6x+1} + 2 = 7$

14. $\sqrt{6x+13} - 2 = 5$

15. $2\sqrt{4x+1} = \sqrt{x+4}$

16. $\sqrt{3(x+4)} = \sqrt{5x-12}$

17. $\sqrt[3]{7n-1} = 3$

18. $\sqrt[3]{12m+4} = 4$

19. $\sqrt[4]{10p + 1} = \sqrt[4]{11p - 7}$

20. $\sqrt[4]{10y + 2} = 2\sqrt[4]{2}$

21. $x = \dfrac{\sqrt{12x - 5}}{2}$

22. $x = \dfrac{\sqrt{16x - 12}}{2}$

23. $\sqrt{x + 2} = \sqrt{4 - x}$

24. $\sqrt{6 - x} = \sqrt{2x + 3}$

25. $2\sqrt{x} = \sqrt{5x - 16}$

26. $3\sqrt{x} = \sqrt{3x + 12}$

27. $r - 9 = \sqrt{2r - 3}$

28. $-s - 3 = 2\sqrt{5 - s}$

29. $\sqrt{-5x + 24} = 6 - x$

30. $\sqrt{-x + 2} = x - 2$

31. $\sqrt{y + 2} = 4 - y$

32. $\sqrt{22y + 86} = y + 9$

33. $\sqrt{x}\sqrt{x + 16} = 15$

34. $\sqrt{x}\sqrt{x + 6} = 4$

35. $\sqrt[3]{x^3 - 7} = x - 1$

36. $\sqrt[3]{x^3 + 56} - 2 = x$

37. $\sqrt[4]{x^4 + 4x^2 - 4} = -x$

38. $\sqrt[4]{8x - 8} + 2 = 0$

39. $\sqrt[4]{12t + 4} + 2 = 0$

40. $u = \sqrt[4]{u^4 - 6u^2 + 24}$

41. $\sqrt{2y + 1} = 1 - 2\sqrt{y}$

42. $\sqrt{u} + 3 = \sqrt{u - 3}$

43. $\sqrt{y + 7} + 3 = \sqrt{y + 4}$

44. $1 + \sqrt{z} = \sqrt{z + 3}$

45. $\sqrt{v} + \sqrt{3} = \sqrt{v + 3}$

46. $\sqrt{x} + 2 = \sqrt{x + 4}$

47. $2 + \sqrt{u} = \sqrt{2u + 7}$

48. $5r + 4 = \sqrt{5r + 20} + 4r$

49. $\sqrt{6t + 1} - 3\sqrt{t} = -1$

50. $\sqrt{4s + 1} - \sqrt{6s} = -1$

51. $\sqrt{2x + 5} + \sqrt{x + 2} = 5$

52. $\sqrt{2x + 5} + \sqrt{2x + 1} + 4 = 0$

53. $\sqrt{z - 1} + \sqrt{z + 2} = 3$

54. $\sqrt{16v + 1} + \sqrt{8v + 1} = 12$

55. $\sqrt{x - 5} - \sqrt{x + 3} = 4$

56. $\sqrt{x + 8} - \sqrt{x - 4} = -2$

57. $\sqrt{x + 1} + \sqrt{3x} = \sqrt{5x + 1}$

58. $\sqrt{3x} - \sqrt{x + 1} = \sqrt{x - 2}$

59. $\sqrt{\sqrt{a} + \sqrt{a + 8}} = 2$

60. $\sqrt{\sqrt{2y} - \sqrt{y - 1}} = 1$

61. $\dfrac{6}{\sqrt{x + 5}} = \sqrt{x}$

62. $\dfrac{\sqrt{2x}}{\sqrt{x + 2}} = \sqrt{x - 1}$

63. $\sqrt{x + 2} + \sqrt{2x - 3} = \sqrt{11 - x}$

64. $\sqrt{8 - x} - \sqrt{3x - 8} = \sqrt{x - 4}$

APPLICATIONS

65. Highway design   A curve banked at 8° will accommodate traffic traveling s mph if the radius of the curve is r feet, according to the formula $s = 1.45\sqrt{r}$. If engineers expect 65-mph traffic, what radius should they specify?

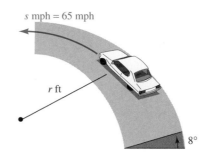

66. Horizon distance   The higher a lookout tower is built, the farther an observer can see. That distance d (called the *horizon distance,* measured in miles) is related to the height h of the observer (measured in feet) by the formula $d = 1.4\sqrt{h}$. How tall must a lookout tower be to see the edge of the forest, 25 miles away?

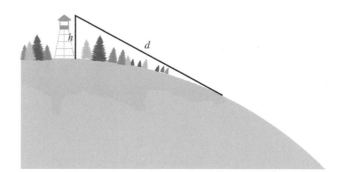

67. Generating power   The power generated by a windmill is related to the velocity of the wind by the formula

$$v = \sqrt[3]{\frac{P}{0.02}}$$

where P is the power (in watts) and v is the velocity of the wind (in mph). Find the speed of the wind when the windmill is generating 500 watts of power.

68. Carpentry   During construction, carpenters often brace walls as shown in the illustration, where the length of the brace is given by the formula

$$l = \sqrt{f^2 + h^2}$$

If a carpenter nails a 10-ft brace to the wall 6 feet above the floor, how far from the base of the wall should he nail the brace to the floor?

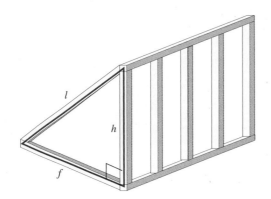

Use a graphing calculator.

69. Depreciation   The formula

$$r = 1 - \sqrt[n]{\frac{T}{C}}$$

gives the annual depreciation rate r of a car that had an original cost of C dollars, a useful life of n years, and a trade-in value of T dollars. Find the annual depreciation rate of a car that cost $22,000 and was sold 5 years later for $9,000. Give the result to the nearest percent.

70. Savings accounts   The interest rate r earned by a savings account after n compoundings is given by the formula

$$\sqrt[n]{\frac{V}{P}} - 1 = r$$

where V is the current value and P is the original principal. What interest rate r was paid on an account in which a deposit of $1,000 grew to $1,338.23 after 5 compoundings?

71. Marketing   The number of wrenches that will be produced at a given price can be predicted by the formula $s = \sqrt{5x}$, where s is the supply (in thousands) and x is the price (in dollars). If the demand, d, for wrenches can be predicted by the formula $d = \sqrt{100 - 3x^2}$, find the equilibrium price.

72. Marketing   The number of footballs that will be produced at a given price can be predicted by the formula $s = \sqrt{23x}$, where s is the supply (in thousands) and x is the price (in dollars). If the demand, d, for footballs can be predicted by the formula $d = \sqrt{312 - 2x^2}$, find the equilibrium price.

73. Medicine   The resistance R to blood flow through an artery can be found using the formula

$$r = \sqrt[4]{\frac{8kl}{\pi R}}$$

where r is the radius of the artery, k is the viscosity of blood, and l is the length of the artery. Solve the formula for R.

74. Generating power   The power P generated by a windmill is given by the formula

$$s = \sqrt[3]{\frac{P}{0.02}}$$

where s is the speed of the wind. Solve the formula for P.

WRITING

75. If both sides of an equation are raised to the same power, the resulting equation might not be equivalent to the original equation. Explain.

76. Explain why you must check each apparent solution of a radical equation.

SOMETHING TO THINK ABOUT

77. Solve: $\sqrt[3]{2x} = \sqrt{x}$.   (*Hint:* Square and then cube both sides.)

78. Solve: $\sqrt[4]{x} = \sqrt{\frac{x}{4}}$.

7.7 Complex Numbers

- **Imaginary Numbers**   ■ **Simplifying Imaginary Numbers**
- **Complex Numbers**   ■ **Arithmetic of Complex Numbers**
- **Rationalizing the Denominator**   ■ **Powers of *i***
- **Absolute Value of a Complex Number**

Getting Ready   *Perform the following operations.*

1. $(3x + 5) + (4x - 5)$ **2.** $(3x + 5) - (4x - 5)$

3. $(3x + 5)(4x - 5)$ **4.** $(3x + 5)(3x - 5)$

We have seen that square roots of negative numbers are not real numbers. However, there is a broader set of numbers, called the **complex numbers,** in which negative numbers do have square roots. In this section, we will discuss this broader set of numbers.

Imaginary Numbers

Consider the number $\sqrt{-3}$. Since no real number squared is -3, $\sqrt{-3}$ is not a real number. For years, people believed that numbers like

$$\sqrt{-1}, \qquad \sqrt{-3}, \qquad \sqrt{-4}, \qquad \text{and} \qquad \sqrt{-9}$$

were nonsense. In the 17th century, René Descartes (1596–1650) called them **imaginary numbers.** Today, imaginary numbers have many important uses, such as describing the behavior of alternating current in electronics.

The imaginary number $\sqrt{-1}$ is often denoted by the letter i:

$$i = \sqrt{-1}$$

Because i represents the square root of -1, it follows that

$$i^2 = -1$$

PERSPECTIVE

The Pythagoreans (ca. 500 B.C.) understood the universe as a harmony of whole numbers. They did not classify fractions as numbers, and were upset that $\sqrt{2}$ was not the ratio of whole numbers. For 2,000 years, little progress was made in the understanding of the various kinds of numbers.

The father of algebra, François Vieta (1540–1603), understood the whole numbers, fractions, and certain irrational numbers. But he was unable to accept negative numbers, and certainly not imaginary numbers.

René Descartes (1596–1650) thought these numbers to be nothing more than figments of his imagination, so he called them *imaginary numbers.* Leonhard Euler (1707–1783) used the letter i for $\sqrt{-1}$; Augustin Cauchy (1789–1857) used the term *conjugate;* and Carl Gauss (1777–1855) first used the word *complex.*

Today, we accept complex numbers without question, but it took many centuries and the work of many mathematicians to make them respectable.

Simplifying Imaginary Numbers

If we assume that multiplication of imaginary numbers is commutative and associative, then

$$(2i)^2 = 2^2 i^2$$
$$= 4(-1) \quad i^2 = -1.$$
$$= -4$$

Since $(2i)^2 = -4$, $2i$ is a square root of -4, and we can write

$$\sqrt{-4} = 2i$$

This result can also be obtained by using the multiplication property of radicals:

$$\sqrt{-4} = \sqrt{4(-1)} = \sqrt{4}\sqrt{-1} = 2i$$

We can use the multiplication property of radicals to simplify any imaginary number. For example,

$$\sqrt{-25} = \sqrt{25(-1)} = \sqrt{25}\sqrt{-1} = 5i$$
$$\sqrt{\frac{-100}{49}} = \sqrt{\frac{100}{49}(-1)} = \frac{\sqrt{100}}{\sqrt{49}}\sqrt{-1} = \frac{10}{7}i$$

These examples illustrate the following rule.

Properties of Radicals

> If at least one of a and b is a nonnegative real number, then
>
> $$\sqrt{ab} = \sqrt{a}\sqrt{b} \qquad \text{and} \qquad \sqrt{\frac{a}{b}} = \frac{\sqrt{a}}{\sqrt{b}} \quad (b \neq 0)$$

 Comment If a and b are negative, then $\sqrt{ab} \neq \sqrt{a}\sqrt{b}$. For example, if $a = -16$ and $b = -4$,

$$\sqrt{(-16)(-4)} = \sqrt{64} = 8 \qquad \text{but} \qquad \sqrt{(-16)}\sqrt{(-4)} = (4i)(2i) =$$
$$8i^2 = 8(-1) = -8$$

The correct solution is -8.

Complex Numbers

The imaginary numbers are a subset of a set of numbers called the **complex numbers.**

Complex Numbers

> A **complex number** is any number that can be written in the form $a + bi$, where a and b are real numbers and $i = \sqrt{-1}$.
>   In the complex number $a + bi$, a is called the **real part,** and b is called the **imaginary part.**

If $b = 0$, the complex number $a + bi$ is a real number. If $b \neq 0$ and $a = 0$, the complex number $0 + bi$ (or just bi) is an imaginary number.

Any imaginary number can be expressed in bi form. For example,

$$\sqrt{-1} = i$$
$$\sqrt{-9} = \sqrt{9(-1)} = \sqrt{9}\sqrt{-1} = 3i$$
$$\sqrt{-3} = \sqrt{3(-1)} = \sqrt{3}\sqrt{-1} = \sqrt{3}i$$

 Comment   The expression $\sqrt{3}i$ is often written as $i\sqrt{3}$ to make it clear that i is not part of the radicand. Don't confuse $\sqrt{3}i$ with $\sqrt{3i}$.

The relationship between the real numbers, the imaginary numbers, and the complex numbers is shown in Figure 7-19.

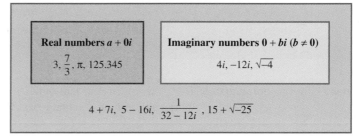

Figure 7-19

Equality of Complex Numbers

The complex numbers $a + bi$ and $c + di$ are equal if and only if

$$a = c \quad \text{and} \quad b = d$$

Because of the previous definition, complex numbers are equal when their real parts are equal and their imaginary parts are equal.

EXAMPLE 1   **a.** $2 + 3i = \sqrt{4} + \dfrac{6}{2}i$ because $2 = \sqrt{4}$ and $3 = \dfrac{6}{2}$.

b. $4 - 5i = \dfrac{12}{3} - \sqrt{25}i$ because $4 = \dfrac{12}{3}$ and $-5 = -\sqrt{25}$.

c. $x + yi = 4 + 7i$ if and only if $x = 4$ and $y = 7$.   ∎

Arithmetic of Complex Numbers

Addition and Subtraction of Complex Numbers

Complex numbers are added and subtracted as if they were binomials:

$$(a + bi) + (c + di) = (a + c) + (b + d)i$$
$$(a + bi) - (c + di) = (a + bi) + (-c - di) = (a - c) + (b - d)i$$

EXAMPLE 2   Perform the operations:

a. $(8 + 4i) + (12 + 8i) = 8 + 4i + 12 + 8i$
$$= 20 + 12i$$

b. $(7 - 4i) + (9 + 2i) = 7 - 4i + 9 + 2i$
$$= 16 - 2i$$

c. $(-6 + i) - (3 - 4i) = -6 + i - 3 + 4i$
$$= -9 + 5i$$

d. $(2 - 4i) - (-4 + 3i) = 2 - 4i + 4 - 3i$
$$= 6 - 7i$$

Self Check   Perform the operations: **a.** $(3 - 5i) + (-2 + 7i)$ and
b. $(3 - 5i) - (-2 + 7i)$.   ∎

To multiply a complex number by an imaginary number, we use the distributive property to remove parentheses and simplify. For example,

$$-5i(4 - 8i) = -5i(4) - (-5i)8i \quad \text{Use the distributive property.}$$
$$= -20i + 40i^2 \quad \text{Simplify.}$$
$$= -20i + 40(-1) \quad \text{Remember that } i^2 = -1.$$
$$= -40 - 20i$$

To multiply two complex numbers, we use the following definition.

Multiplying Complex Numbers

Complex numbers are multiplied as if they were binomials, with $i^2 = -1$:

$$(a + bi)(c + di) = ac + adi + bci + bdi^2$$
$$= ac + adi + bci + bd(-1)$$
$$= (ac - bd) + (ad + bc)i$$

EXAMPLE 3   Multiply the complex numbers:

a. $(2 + 3i)(3 - 2i) = 6 - 4i + 9i - 6i^2$ Use the FOIL method.
$$= 6 + 5i + 6 \quad\quad i^2 = -1, \text{ and combine } -4i \text{ and } 9i.$$
$$= 12 + 5i$$

b. $(3 + i)(1 + 2i) = 3 + 6i + i + 2i^2$ Use the FOIL method.
$$= 3 + 7i - 2 \quad\quad i^2 = -1, \text{ and combine } 6i \text{ and } i.$$
$$= 1 + 7i$$

c. $(-4 + 2i)(2 + i) = -8 - 4i + 4i + 2i^2$ Use the FOIL method.
$$= -8 - 2 \quad\quad i^2 = -1, \text{ and combine } -4i \text{ and } 4i.$$
$$= -10$$

Self Check   Multiply: $(-2 + 3i)(3 - 2i)$.   ∎

The next two examples show how to write complex numbers in $a + bi$ form. It is common to use $a - bi$ as a substitute for $a + (-b)i$.

EXAMPLE 4   Write each number in $a + bi$ form:

a. $7 = 7 + 0i$ **b.** $3i = 0 + 3i$

c. $4 - \sqrt{-16} = 4 - \sqrt{-1(16)}$ **d.** $5 + \sqrt{-11} = 5 + \sqrt{-1(11)}$

$$= 4 - \sqrt{16}\sqrt{-1} \qquad\qquad\qquad = 5 + \sqrt{11}\sqrt{-1}$$

$$= 4 - 4i \qquad\qquad\qquad\qquad\quad\;\; = 5 + \sqrt{11}i$$

Self Check   Write $3 - \sqrt{-25}$ in $a + bi$ form.   ∎

Complex Conjugates   The complex numbers $a + bi$ and $a - bi$ are called **complex conjugates.**

For example,

$3 + 4i$ and $3 - 4i$ are complex conjugates.

$5 - 7i$ and $5 + 7i$ are complex conjugates.

EXAMPLE 5   Find the product of $3 + i$ and its complex conjugate.

Solution   The complex conjugate of $3 + i$ is $3 - i$. We can find the product as follows:

$$(3 + i)(3 - i) = 9 - 3i + 3i - i^2 \qquad \text{Use the FOIL method.}$$

$$= 9 - i^2 \qquad\qquad\qquad \text{Combine like terms.}$$

$$= 9 - (-1) \qquad\qquad\;\; i^2 = -1.$$

$$= 10$$

Self Check   Multiply: $(2 + 3i)(2 - 3i)$.   ∎

The product of the complex number $a + bi$ and its complex conjugate $a - bi$ is the real number $a^2 + b^2$, as the following work shows:

$$(a + bi)(a - bi) = a^2 - abi + abi - b^2i^2 \qquad \text{Use the FOIL method.}$$

$$= a^2 - b^2(-1) \qquad\qquad\qquad i^2 = -1.$$

$$= a^2 + b^2$$

Rationalizing the Denominator

To divide complex numbers, we often have to rationalize a denominator.

EXAMPLE 6   Divide and write the result in $a + bi$ form: $\dfrac{1}{3 + i}$.

Solution   We can rationalize the denominator by multiplying the numerator and the denominator by the complex conjugate of the denominator.

$$\frac{1}{3+i} = \frac{1}{3+i} \cdot \frac{\mathbf{3-i}}{\mathbf{3-i}} \qquad \frac{3-i}{3-i} = 1.$$

$$= \frac{3-i}{9-3i+3i-i^2} \qquad \text{Multiply the numerators and multiply the denominators.}$$

$$= \frac{3-i}{9-(-1)} \qquad i^2 = -1.$$

$$= \frac{3-i}{10}$$

$$= \frac{3}{10} - \frac{1}{10}i$$

Self Check   Rationalize the denominator: $\dfrac{1}{5-i}$. ∎

EXAMPLE 7   Write $\dfrac{3-i}{2+i}$ in $a+bi$ form.

Solution   We multiply the numerator and the denominator of the fraction by the complex conjugate of the denominator.

$$\frac{3-i}{2+i} = \frac{3-i}{2+i} \cdot \frac{\mathbf{2-i}}{\mathbf{2-i}} \qquad \frac{2-i}{2-i} = 1.$$

$$= \frac{6-3i-2i+i^2}{4-2i+2i-i^2} \qquad \text{Multiply the numerators and multiply the denominators.}$$

$$= \frac{5-5i}{4-(-1)} \qquad 6+i^2 = 6-1 = 5.$$

$$= \frac{5(1-i)}{5} \qquad \text{Factor out 5 in the numerator.}$$

$$= 1-i \qquad \text{Simplify.}$$

Self Check   Rationalize the denominator: $\dfrac{2+i}{5-i}$. ∎

EXAMPLE 8   Write $\dfrac{4+\sqrt{-16}}{2+\sqrt{-4}}$ in $a+bi$ form.

Solution $$\frac{4+\sqrt{-16}}{2+\sqrt{-4}} = \frac{4+4i}{2+2i} \qquad \text{Write each number in } a+bi \text{ form.}$$

$$= \frac{\overset{1}{2(\cancel{2+2i})}}{\underset{1}{\cancel{2+2i}}} \qquad \text{Factor out 2 in the numerator and simplify.}$$

$$= 2+0i$$

Self Check   Divide: $\dfrac{3+\sqrt{-25}}{2+\sqrt{-9}}$. ∎

Comment To avoid mistakes, always put complex numbers in $a + bi$ form before doing any complex number arithmetic.

Powers of i

The powers of i produce an interesting pattern:

$$i = \sqrt{-1} = i \qquad\qquad i^5 = i^4 i = 1i = i$$
$$i^2 = \left(\sqrt{-1}\right)^2 = -1 \qquad\qquad i^6 = i^4 i^2 = 1(-1) = -1$$
$$i^3 = i^2 i = -1i = -i \qquad\qquad i^7 = i^4 i^3 = 1(-i) = -i$$
$$i^4 = i^2 i^2 = (-1)(-1) = 1 \qquad\qquad i^8 = i^4 i^4 = (1)(1) = 1$$

The pattern continues: $i, -1, -i, 1, \ldots$ ..

 EXAMPLE 9 Simplify: i^{29}.

Solution We note that 29 divided by 4 gives a quotient of 7 and a remainder of 1. Thus, $29 = 4 \cdot 7 + 1$, and

$$i^{29} = i^{4 \cdot 7 + 1} \qquad 4 \cdot 7 + 1 = 29.$$
$$= \left(i^4\right)^7 \cdot i \qquad i^{4 \cdot 7 + 1} = i^{4 \cdot 7} \cdot i^1 = (i^4)^7 \cdot i.$$
$$= 1^7 \cdot i \qquad i^4 = 1.$$
$$= i$$

Self Check Simplify: i^{31}. ∎

The results of Example 9 illustrate the following fact.

Powers of i

If n is a natural number that has a remainder of r when divided by 4, then

$$i^n = i^r$$

When n is divisible by 4, the remainder r is 0 and $i^0 = 1$.

 EXAMPLE 10 Simplify: i^{55}.

Solution We divide 55 by 4 and get a remainder of 3. Therefore,

$$i^{55} = i^3 = -i$$

Self Check Simplify: i^{62}. ∎

EXAMPLE 11   Simplify each expression:

a. $2i^2 + 4i^3 = 2(-1) + 4(-i)$
$\qquad\qquad = -2 - 4i$

b. $\dfrac{3}{2i} = \dfrac{3}{2i} \cdot \dfrac{i}{i}$ $\dfrac{i}{i} = 1.$

$\qquad = \dfrac{3i}{2i^2}$

$\qquad = \dfrac{3i}{2(-1)}$

$\qquad = \dfrac{3i}{-2}$

$\qquad = 0 - \dfrac{3}{2}i$

c. $-\dfrac{5}{i} = -\dfrac{5}{i} \cdot \dfrac{i^3}{i^3}$ $\dfrac{i^3}{i^3} = 1.$

$\qquad = -\dfrac{5(-i)}{1}$

$\qquad = 5i$

$\qquad = 0 + 5i$

d. $\dfrac{6}{i^3} = \dfrac{6i}{i^3 i}$ $\dfrac{i}{i} = 1.$

$\qquad = \dfrac{6i}{i^4}$

$\qquad = \dfrac{6i}{1}$

$\qquad = 6i$

$\qquad = 0 + 6i$

Self Check   Simplify: **a.** $3i^3 - 2i^2$ and   **b.** $\dfrac{2}{3i}.$ ∎

Absolute Value of a Complex Number

Absolute Value of a Complex Number

The **absolute value** of the complex number $a + bi$ is $\sqrt{a^2 + b^2}$. In symbols,

$$|a + bi| = \sqrt{a^2 + b^2}$$

EXAMPLE 12   Find each absolute value:

a. $|3 + 4i| = \sqrt{3^2 + 4^2}$
$\qquad\quad = \sqrt{9 + 16}$
$\qquad\quad = \sqrt{25}$
$\qquad\quad = 5$

b. $|3 - 4i| = \sqrt{3^2 + (-4)^2}$
$\qquad\quad = \sqrt{9 + 16}$
$\qquad\quad = \sqrt{25}$
$\qquad\quad = 5$

c. $|-5 - 12i| = \sqrt{(-5)^2 + (-12)^2}$
$\qquad\qquad = \sqrt{25 + 144}$
$\qquad\qquad = \sqrt{169}$
$\qquad\qquad = 13$

d. $|a + 0i| = \sqrt{a^2 + 0^2}$
$\qquad\quad = \sqrt{a^2}$
$\qquad\quad = |a|$

Self Check   Evaluate: $|5 + 12i|.$ ∎

!   **Comment**   Note that $|a + bi| = \sqrt{a^2 + b^2}$, not
$\qquad\qquad \cancel{|a + bi| = \sqrt{a^2 + (bi)^2}}$

Self Check Answers

2. a. $1 + 2i$, **b.** $5 - 12i$ **3.** $13i$ **4.** $3 - 5i$ **5.** 13 **6.** $\dfrac{5}{26} + \dfrac{1}{26}i$ **7.** $\dfrac{9}{26} + \dfrac{7}{26}i$ **8.** $\dfrac{21}{13} + \dfrac{1}{13}i$

9. $-i$ **10.** -1 **11. a.** $2 - 3i$, **b.** $0 - \dfrac{2}{3}i$ **12.** 13

Orals *Write each imaginary number in bi form:*

1. $\sqrt{-49}$ **2.** $\sqrt{-64}$ **3.** $\sqrt{-100}$ **4.** $\sqrt{-81}$

Simplify each power of i :

5. i^3 **6.** i^2 **7.** i^4 **8.** i^5

Find each absolute value:

9. $|3 + 4i|$ **10.** $|5 - 12i|$

7.7 EXERCISES

REVIEW *Perform each operation.*

1. $\dfrac{x^2 - x - 6}{9 - x^2} \cdot \dfrac{x^2 + x - 6}{x^2 - 4}$

2. $\dfrac{3x + 4}{x - 2} + \dfrac{x - 4}{x + 2}$

3. Wind speed A plane that can fly 200 mph in still air makes a 330-mile flight with a tail wind and returns, flying into the same wind. Find the speed of the wind if the total flying time is $3\frac{1}{3}$ hours.

4. Finding rates A student drove a distance of 135 miles at an average speed of 50 mph. How much faster would he have to drive on the return trip to save 30 minutes of driving time?

VOCABULARY AND CONCEPTS *Fill in the blanks.*

5. $\sqrt{-1}$, $\sqrt{-3}$, and $\sqrt{-4}$ are examples of _____ numbers.

6. $\sqrt{-1} = \_$ **7.** $i^2 = \_\_\_$

8. $i^3 = \_\_\_$ **9.** $i^4 = \_$

10. $\sqrt{ab} = _____(b \neq 0)$, provided a and b are not both negative.

11. $\sqrt{\dfrac{a}{b}} = \_\_\_\_$, provided a and b are not both negative.

12. $3 + 5i$, $2 - 7i$, and $5 - \dfrac{1}{2}i$ are examples of _____ numbers.

13. The real part of $5 + 7i$ is $\_$. The imaginary part is $\_$.

14. $a + bi = c + di$ if and only if $a = \_$ and $b = \_$.

15. $a + bi$ and $a - bi$ are called complex _____.

16. $|a + bi| = _____$

PRACTICE *Write each imaginary number in bi form.*

17. $\sqrt{-9}$ **18.** $\sqrt{-16}$

19. $\sqrt{-36}$ **20.** $\sqrt{-81}$

21. $\sqrt{-7}$ **22.** $\sqrt{-11}$

Tell whether the complex numbers are equal.

23. $3 + 7i$, $\sqrt{9} + (5 + 2)i$

24. $\sqrt{4} + \sqrt{25}i$, $2 - (-5)i$

25. $8 + 5i$, $2^3 + \sqrt{25}i^3$

26. $4 - 7i$, $-4i^2 + 7i^3$

27. $\sqrt{4} + \sqrt{-4}$, $2 - 2i$

28. $\sqrt{-9} - i$, $4i$

Perform the operations. Write all answers in a + bi form.

29. $(3 + 4i) + (5 - 6i)$

30. $(5 + 3i) - (6 - 9i)$

31. $(7 - 3i) - (4 + 2i)$

32. $(8 + 3i) + (-7 - 2i)$

33. $(8 + 5i) + (7 + 2i)$

34. $(-7 + 9i) - (-2 - 8i)$

35. $(1 + i) - 2i + (5 - 7i)$

36. $(-9 + i) - 5i + (2 + 7i)$

37. $(5 + 3i) - (3 - 5i) + \sqrt{-1}$

38. $(8 + 7i) - (-7 - \sqrt{-64}) + (3 - i)$

39. $(-8 - \sqrt{3}i) - (7 - 3\sqrt{3}i)$

40. $(2 + 2\sqrt{2}i) + (-3 - \sqrt{2}i)$

41. $3i(2 - i)$ **42.** $-4i(3 + 4i)$

43. $-5i(5 - 5i)$ **44.** $2i(7 + 2i)$

45. $(2 + i)(3 - i)$ **46.** $(4 - i)(2 + i)$

47. $(2 - 4i)(3 + 2i)$ **48.** $(3 - 2i)(4 - 3i)$

49. $(2 + \sqrt{2}i)(3 - \sqrt{2}i)$

50. $(5 + \sqrt{3}i)(2 - \sqrt{3}i)$

51. $(8 - \sqrt{-1})(-2 - \sqrt{-16})$

52. $(-1 + \sqrt{-4})(2 + \sqrt{-9})$

53. $(2 + i)^2$ **54.** $(3 - 2i)^2$

55. $(2 + 3i)^2$ **56.** $(1 - 3i)^2$

57. $i(5 + i)(3 - 2i)$ **58.** $i(-3 - 2i)(1 - 2i)$

59. $(2 + i)(2 - i)(1 + i)$

60. $(3 + 2i)(3 - 2i)(i + 1)$

61. $(3 + i)[(3 - 2i) + (2 + i)]$

62. $(2 - 3i)[(5 - 2i) - (2i + 1)]$

Write each expression in $a + bi$ form.

63. $\dfrac{1}{i}$ **64.** $\dfrac{1}{i^3}$

65. $\dfrac{4}{5i^3}$ **66.** $\dfrac{3}{2i}$

67. $\dfrac{3i}{8\sqrt{-9}}$ **68.** $\dfrac{5i^3}{2\sqrt{-4}}$

69. $\dfrac{-3}{5i^5}$ **70.** $\dfrac{-4}{6i^7}$

71. $\dfrac{5}{2 - i}$ **72.** $\dfrac{26}{3 - 2i}$

73. $\dfrac{13i}{5 + i}$ **74.** $\dfrac{2i}{5 + 3i}$

75. $\dfrac{-12}{7 - \sqrt{-1}}$ **76.** $\dfrac{4}{3 + \sqrt{-1}}$

77. $\dfrac{5i}{6 + 2i}$ **78.** $\dfrac{-4i}{2 - 6i}$

79. $\dfrac{3 - 2i}{3 + 2i}$ **80.** $\dfrac{2 + 3i}{2 - 3i}$

81. $\dfrac{3 + 2i}{3 + i}$ **82.** $\dfrac{2 - 5i}{2 + 5i}$

83. $\dfrac{\sqrt{5} - \sqrt{3}i}{\sqrt{5} + \sqrt{3}i}$

84. $\dfrac{\sqrt{3} + \sqrt{2}i}{\sqrt{3} - \sqrt{2}i}$

85. $\left(\dfrac{i}{3 + 2i}\right)^2$

86. $\left(\dfrac{5 + i}{2 + i}\right)^2$ **87.** $\dfrac{i(3 - i)}{3 + i}$

88. $\dfrac{5 + 3i}{i(3 - 5i)}$

89. $\dfrac{(2 - 5i) - (5 - 2i)}{5 - i}$

90. $\dfrac{5i}{(5 + 2i) + (2 + i)}$

Simplify each expression.

91. i^{21} **92.** i^{19}

93. i^{27} **94.** i^{22}

95. i^{100} **96.** i^{42}

97. i^{97} **98.** i^{200}

Find each value.

99. $|6 + 8i|$ **100.** $|12 + 5i|$

101. $|12 - 5i|$ **102.** $|3 - 4i|$

103. $|5 + 7i|$ **104.** $|6 - 5i|$

105. $\left|\dfrac{3}{5} - \dfrac{4}{5}i\right|$ **106.** $\left|\dfrac{5}{13} + \dfrac{12}{13}i\right|$

107. Show that $1 - 5i$ is a solution of $x^2 - 2x + 26 = 0$.

108. Show that $3 - 2i$ is a solution of $x^2 - 6x + 13 = 0$.

109. Show that i is a solution of $x^4 - 3x^2 - 4 = 0$.

110. Show that $2 + i$ is *not* a solution of $x^2 + x + 1 = 0$.

APPLICATIONS *In electronics, the formula $V = IR$ is called **Ohm's law**. It gives the relationship in a circuit between the voltage V (in volts), the current I (in amperes), and the resistance R (in ohms).*

111. Electronics Find V when $I = 2 - 3i$ amperes and $R = 2 + i$ ohms.

112. Electronics Find R when $I = 3 - 2i$ amperes and $V = 18 + i$ volts.

*In electronics, the formula $Z = \frac{V}{I}$ is used to find the **impedance** Z of a circuit, where V is the voltage and I is the current.*

113. Electronics Find the impedance of a circuit when the voltage is $1.7 + 0.5i$ and the current is $0.5i$.

114. Electronics Find the impedance of a circuit when the voltage is $1.6 - 0.4i$ and the current is $-0.2i$.

WRITING

115. Tell how to decide whether two complex numbers are equal.

116. Define the complex conjugate of a complex number.

SOMETHING TO THINK ABOUT

117. Rationalize the numerator: $\dfrac{3 - i}{2}$.

118. Rationalize the numerator: $\dfrac{2 + 3i}{2 - 3i}$

Projects

Project 1

The size of a television screen is measured along the diagonal of its screen, as shown in the illustrations. The screen of a traditional TV has an aspect ratio of $4:3$. This means that the ratio of the width of the screen to its height is $\frac{4}{3}$. The screen of a wide-screen set has an aspect ratio of $16:9$. This means that the ratio of the width of the screen to its is height is $\frac{16}{9}$.

a. Find the width and height of the traditional-screen set shown in the illustration. $\left(Hint: \frac{4}{3} = \frac{4x}{3x}.\right)$
b. Find the viewing area of the traditional-screen set in square inches.
c. Find the width and height of the wide-screen set shown in the illustration.
d. Find the viewing area of the wide-screen set in square inches.
e. Which set has the larger viewing area? Give the answer as a percent.

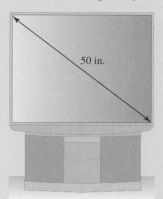

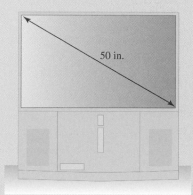

Project 2

Tom and Brian arrange to have a bicycle race. Each leaves his own house at the same time and rides to the other's house, whereupon the winner of the race calls his own house and leaves a message for the loser. A map of the race is shown in the illustration. Brian stays on the highway, averaging 21 mph. Tom knows that he and Brian are evenly matched when biking on the highway, so he cuts across country for the first part of his trip, averaging 15 mph. When Tom reaches the highway at point A, he turns right and follows the highway, averaging 21 mph.

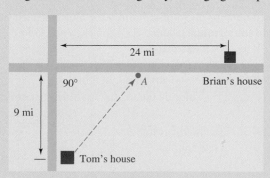

Tom and Brian never meet during the race, and amazingly, the race is a tie. Each of them calls the other at exactly the same moment!

a. How long (to the nearest second) did it take each person to complete the race?

b. How far from the intersection of the two highways is point *A*? (*Hint:* Set the travel times for Brian and Tom equal to each other. You may find two answers, but only one of them matches all of the information.)

c. Show that if Tom had started straight across country for Brian's house (in order to minimize the distance he had to travel), he would have lost the race. By how much time (to the nearest second) would he have lost? Then show that if Tom had biked across country to a point 9 miles from the intersection of the two highways, he would have won the race. By how much time (to the nearest second) would he have won?

Chapter Summary

CONCEPTS	REVIEW EXERCISES
7.1	**Radical Expressions**

<table>
<tr><td>

If n is an even natural number,
$$\sqrt[n]{a^n} = |a|$$

If n is an odd natural number, greater than 1,
$$\sqrt[n]{a^n} = a$$

If n is a natural number greater than 1 and x is a real number, then

   If $x > 0$, then $\sqrt[n]{x}$ is the positive number such that $\left(\sqrt[n]{x}\right)^n = x$.

   If $x = 0$, then $\sqrt[n]{x} = 0$.

   If $x < 0$, and n is odd, $\sqrt[n]{x}$ is the real number such that $\left(\sqrt[n]{x}\right)^n = x$.

   If $x < 0$, and n is even, $\sqrt[n]{x}$ is not a real number.

</td><td>

Simplify each radical. Assume that x can be any number.

1. $\sqrt{49}$ **2.** $-\sqrt{121}$

3. $-\sqrt{36}$ **4.** $\sqrt{225}$

5. $\sqrt[3]{-27}$ **6.** $-\sqrt[3]{216}$

7. $\sqrt[4]{625}$ **8.** $\sqrt[5]{-32}$

9. $\sqrt{25x^2}$ **10.** $\sqrt{x^2 + 4x + 4}$

11. $\sqrt[3]{27a^6b^3}$ **12.** $\sqrt[4]{256x^8y^4}$

Graph each function.

13. $f(x) = \sqrt{x + 2}$ **14.** $f(x) = -\sqrt{x - 1}$

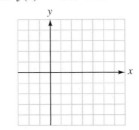   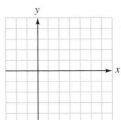

</td></tr>
</table>

15. $f(x) = -\sqrt{x} + 2$

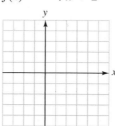

16. $f(x) = -\sqrt[3]{x} + 3$

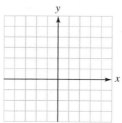

Consider the distribution 4, 8, 12, 16, 20.

17. Find the mean of the distribution.

18. Find the standard deviation.

7.2 Applications of Radicals

The Pythagorean theorem:
If a and b are the lengths of the legs of a right triangle and c is the length of the hypotenuse, then
$$a^2 + b^2 = c^2$$

In Exercises 19–20, the horizon distance d (measured in miles) is related to the height h (measured in feet) of the observer by the formula $d = 1.4\sqrt{h}$.

19. View from a submarine A submarine's periscope extends 4.7 feet above the surface. How far away is the horizon?

20. View from a submarine How far out of the water must a submarine periscope extend to provide a 4-mile horizon?

21. Sailing A technique called *tacking* allows a sailboat to make progress into the wind. A sailboat follows the course in the illustration. Find d, the distance the boat advances into the wind.

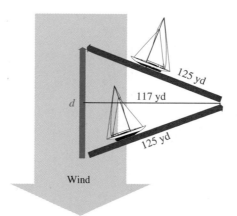

22. Communications   Some campers 3,900 yards from a highway are talking to truckers on a citizen's band radio with an 8,900-yard range. Over what length of highway can these conversations take place?

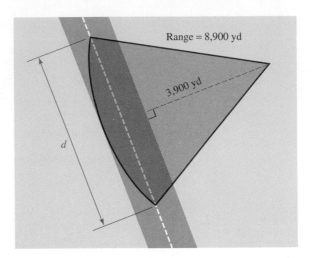

Range = 8,900 yd

3,900 yd

d

The distance formula:

$$d(PQ) = \sqrt{(x_2 - x_1)^2 + (y_2 - y_1)^2}$$

23. Find the distance between points $(0, 0)$ and $(5, -12)$.

24. Find the distance between points $(-4, 6)$ and $(-2, 8)$. Give the result to the nearest hundredth.

7.3	Rational Exponents

If n ($n > 1$) is a natural number and $\sqrt[n]{x}$ is a real number, then $x^{1/n} = \sqrt[n]{x}$.

If n is even, $(x^n)^{1/n} = |x|$.

If n is a natural number greater than 1 and x is a real number, then

   If $x > 0$, then $x^{1/n}$ is the positive number such that $(x^{1/n})^n = x$.

   If $x = 0$, then $x^{1/n} = 0$.

   If $x < 0$, and n is odd, then $x^{1/n}$ is the real number such that $(x^{1/n})^n = x$.

   If $x < 0$ and n is even, then $x^{1/n}$ is not a real number.

Simplify each expression, if possible. Assume that all variables represent positive numbers.

25. $25^{1/2}$

26. $-36^{1/2}$

27. $9^{3/2}$

28. $16^{3/2}$

29. $(-8)^{1/3}$

30. $-8^{2/3}$

31. $8^{-2/3}$

32. $8^{-1/3}$

33. $-49^{5/2}$

34. $\dfrac{1}{25^{5/2}}$

35. $\left(\dfrac{1}{4}\right)^{-3/2}$

36. $\left(\dfrac{4}{9}\right)^{-3/2}$

37. $(27x^3y)^{1/3}$

38. $(81x^4y^2)^{1/4}$

39. $(25x^3y^4)^{3/2}$

40. $(8u^2v^3)^{-2/3}$

If m and n are positive integers and $x > 0$,

$$x^{m/n} = \sqrt[n]{x^m} = \left(\sqrt[n]{x}\right)^m$$

$$x^{-m/n} = \frac{1}{x^{m/n}}$$

$$\frac{1}{x^{-m/n}} = x^{m/n} \quad (x \ne 0)$$

Perform the multiplications. Assume that all variables represent positive numbers and write all answers without negative exponents.

41. $5^{1/4}5^{1/2}$

42. $a^{3/7}a^{2/7}$

43. $u^{1/2}(u^{1/2} - u^{-1/2})$

44. $v^{2/3}(v^{1/3} + v^{4/3})$

45. $(x^{1/2} + y^{1/2})^2$

46. $(a^{2/3} + b^{2/3})(a^{2/3} - b^{2/3})$

Simplify each expression. Assume that all variables are positive.

47. $\sqrt[6]{5^2}$

48. $\sqrt[8]{x^4}$

49. $\sqrt[9]{27a^3b^6}$

50. $\sqrt[4]{25a^2b^2}$

7.4   Simplifying and Combining Radical Expressions

Properties of radicals:

$$\sqrt[n]{ab} = \sqrt[n]{a}\,\sqrt[n]{b}$$

$$\sqrt[n]{\frac{a}{b}} = \frac{\sqrt[n]{a}}{\sqrt[n]{b}} \quad (b \ne 0)$$

Simplify each expression. Assume all variables represent positive numbers.

51. $\sqrt{240}$

52. $\sqrt[3]{54}$

53. $\sqrt[4]{32}$

54. $\sqrt[5]{96}$

55. $\sqrt{8x^3}$

56. $\sqrt{18x^4y^3}$

57. $\sqrt[3]{16x^5y^4}$

58. $\sqrt[3]{54x^7y^3}$

59. $\dfrac{\sqrt{32x^3}}{\sqrt{2x}}$

60. $\dfrac{\sqrt[3]{16x^5}}{\sqrt[3]{2x^2}}$

61. $\sqrt[3]{\dfrac{2a^2b}{27x^3}}$

62. $\sqrt{\dfrac{17xy}{64a^4}}$

Like radicals can be combined by addition and subtraction:

$$3\sqrt{2} + 5\sqrt{2} = 8\sqrt{2}$$

Radicals that are not similar can often be simplified to radicals that are similar and then combined:

$$\sqrt{2} + \sqrt{8} = \sqrt{2} + \sqrt{4}\sqrt{2}$$
$$= \sqrt{2} + 2\sqrt{2}$$
$$= 3\sqrt{2}$$

Simplify and combine like radicals. Assume that all variables represent positive numbers.

63. $\sqrt{2} + \sqrt{8}$

64. $\sqrt{20} - \sqrt{5}$

65. $2\sqrt[3]{3} - \sqrt[3]{24}$

66. $\sqrt[4]{32} + 2\sqrt[4]{162}$

67. $2x\sqrt{8} + 2\sqrt{200x^2} + \sqrt{50x^2}$

68. $3\sqrt{27a^3} - 2a\sqrt{3a} + 5\sqrt{75a^3}$

69. $\sqrt[3]{54} - 3\sqrt[3]{16} + 4\sqrt[3]{128}$

70. $2\sqrt[4]{32x^5} + 4\sqrt[4]{162x^5} - 5x\sqrt[4]{512x}$

In an isosceles right triangle, the length of the hypotenuse is the length of one leg times $\sqrt{2}$.

71. Find the length of the hypotenuse of an isosceles right triangle whose legs measure 7 meters.

72. The hypotenuse of a 30°–60°–90° triangle measures $12\sqrt{3}$ centimeters. Find the length of each leg.

The shorter leg of a 30°–60°–90° triangle is half as long as the hypotenuse. The longer leg is the length of the shorter leg times $\sqrt{3}$.

Find x to two decimal places.

73.

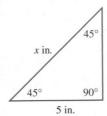

74.

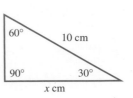

7.5 Multiplying and Dividing Radical Expressions

If two radicals have the same index, they can be multiplied:
$$\sqrt{3x}\sqrt{6x} = \sqrt{18x^2} \ (x>0)$$
$$= 3x\sqrt{2}$$

Simplify each expression. Assume that all variables represent positive numbers.

75. $(2\sqrt{5})(3\sqrt{2})$

76. $2\sqrt{6}\sqrt{216}$

77. $\sqrt{9x}\sqrt{x}$

78. $\sqrt[3]{3}\sqrt[3]{9}$

79. $-\sqrt[3]{2x^2}\sqrt[3]{4x}$

80. $-\sqrt[4]{256x^5y^{11}}\sqrt[4]{625x^9y^3}$

81. $\sqrt{2}(\sqrt{8}-3)$

82. $\sqrt{2}(\sqrt{2}+3)$

83. $\sqrt{5}(\sqrt{2}-1)$

84. $\sqrt{3}(\sqrt{3}+\sqrt{2})$

85. $(\sqrt{2}+1)(\sqrt{2}-1)$

86. $(\sqrt{3}+\sqrt{2})(\sqrt{3}+\sqrt{2})$

87. $(\sqrt{x}+\sqrt{y})(\sqrt{x}-\sqrt{y})$

88. $(2\sqrt{u}+3)(3\sqrt{u}-4)$

To rationalize the binomial denominator of a fraction, multiply the numerator and the denominator by the conjugate of the binomial in the denominator.

Rationalize each denominator

89. $\dfrac{1}{\sqrt{3}}$

90. $\dfrac{\sqrt{3}}{\sqrt{5}}$

91. $\dfrac{x}{\sqrt{xy}}$

92. $\dfrac{\sqrt[3]{uv}}{\sqrt[3]{u^5v^7}}$

93. $\dfrac{2}{\sqrt{2}-1}$

94. $\dfrac{\sqrt{2}}{\sqrt{3}-1}$

95. $\dfrac{2x-32}{\sqrt{x}+4}$

96. $\dfrac{\sqrt{a}+1}{\sqrt{a}-1}$

Rationalize each numerator.

97. $\dfrac{\sqrt{3}}{5}$

98. $\dfrac{\sqrt[3]{9}}{3}$

99. $\dfrac{3-\sqrt{x}}{2}$

100. $\dfrac{\sqrt{a}-\sqrt{b}}{\sqrt{a}}$

7.6 Radical Equations

The power rule:

If $x = y$, then $x^n = y^n$.

Raising both sides of an equation to the same power can lead to extraneous solutions. Be sure to check all suspected solutions.

Solve each equation.

101. $\sqrt{y+3} = \sqrt{2y-19}$

102. $u = \sqrt{25u-144}$

103. $r = \sqrt{12r-27}$

104. $\sqrt{z+1} + \sqrt{z} = 2$

105. $\sqrt{2x+5} - \sqrt{2x} = 1$

106. $\sqrt[3]{x^3+8} = x+2$

7.7 Complex Numbers

Complex numbers: If $a, b, c,$ and d are real numbers and $i^2 = -1$,

$a + bi = c + di$ if and only if $a = c$ and $b = d$

$(a+bi) + (c+di)$
$= (a+c) + (b+d)i$

$(a+bi)(c+di)$
$= (ac-bd) + (ad+bc)i$

$|a+bi| = \sqrt{a^2+b^2}$

Perform the operations and give all answers in $a + bi$ form.

107. $(5+4i) + (7-12i)$

108. $(-6-40i) - (-8+28i)$

109. $\left(-32 + \sqrt{-144}\right) - \left(64 + \sqrt{-81}\right)$

110. $\left(-8 + \sqrt{-8}\right) + \left(6 - \sqrt{-32}\right)$

111. $(2-7i)(-3+4i)$

112. $(-5+6i)(2+i)$

113. $\left(5 - \sqrt{-27}\right)\left(-6 + \sqrt{-12}\right)$

114. $\left(2 + \sqrt{-128}\right)\left(3 - \sqrt{-98}\right)$

115. $\dfrac{3}{4i}$

116. $\dfrac{-2}{5i^3}$

117. $\dfrac{6}{2+i}$

118. $\dfrac{7}{3-i}$

119. $\dfrac{4+i}{4-i}$

120. $\dfrac{3-i}{3+i}$

121. $\dfrac{3}{5 + \sqrt{-4}}$

122. $\dfrac{2}{3 - \sqrt{-9}}$

123. $|9 + 12i|$

124. $|24 - 10i|$

Simplify.

125. i^{12}

126. i^{583}

Chapter Test

Find each root.

1. $\sqrt{49}$

2. $\sqrt[3]{64}$

3. $\sqrt{4x^2}$

4. $\sqrt[3]{8x^3}$

Graph each function and find its domain and range.

5. $f(x) = \sqrt{x - 2}$

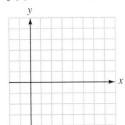

6. $f(x) = \sqrt[3]{x} + 3$

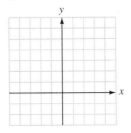

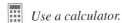 *Use a calculator.*

7. Shipping crates   The diagonal brace on the shipping crate shown in the illustration is 53 inches. Find the height, h, of the crate.

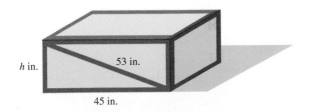

8. Pendulums   The 2-meter pendulum rises 0.1 meter at the extremes of its swing. Find the width w of the swing.

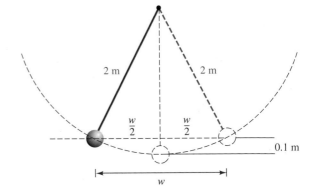

Find the distance between the points.

9. $(6, 8), (0, 0)$

10. $(-2, 5), (22, 12)$

Simplify each expression. Assume that all variables represent positive numbers, and write answers without using negative exponents.

11. $16^{1/4}$

12. $27^{2/3}$

13. $36^{-3/2}$

14. $\left(-\dfrac{8}{27}\right)^{-2/3}$

15. $\dfrac{2^{5/3}2^{1/6}}{2^{1/2}}$

16. $\dfrac{(8x^3y)^{1/2}(8xy^5)^{1/2}}{(x^3y^6)^{1/3}}$

Simplify each expression. Assume that all variables represent positive numbers.

17. $\sqrt{48}$

18. $\sqrt{250x^3y^5}$

19. $\dfrac{\sqrt[3]{24x^{15}y^4}}{\sqrt[3]{y}}$

20. $\sqrt{\dfrac{3a^5}{48a^7}}$

Simplify each expression. Assume that the variables are unrestricted.

21. $\sqrt{12x^2}$

22. $\sqrt{8x^6}$

23. $\sqrt[3]{81x^3}$

24. $\sqrt{18x^4y^9}$

Simplify and combine like radicals. Assume that all variables represent positive numbers.

25. $\sqrt{12} - \sqrt{27}$

26. $2\sqrt[3]{40} - \sqrt[3]{5{,}000} + 4\sqrt[3]{625}$

27. $2\sqrt{48y^5} - 3y\sqrt{12y^3}$

28. $\sqrt[4]{768z^5} + z\sqrt[4]{48z}$

Perform each operation and simplify, if possible. All variables represent positive numbers.

29. $-2\sqrt{xy}\left(3\sqrt{x} + \sqrt{xy^3}\right)$

30. $\left(3\sqrt{2} + \sqrt{3}\right)\left(2\sqrt{2} - 3\sqrt{3}\right)$

Rationalize each denominator.

31. $\dfrac{1}{\sqrt{5}}$

32. $\dfrac{3t - 1}{\sqrt{3t} - 1}$

Rationalize each numerator.

33. $\dfrac{\sqrt{3}}{\sqrt{7}}$

34. $\dfrac{\sqrt{a} + \sqrt{b}}{\sqrt{a} - \sqrt{b}}$

Solve and check each equation.

35. $\sqrt[3]{6n + 4} - 4 = 0$

36. $1 - \sqrt{u} = \sqrt{u - 3}$

Perform the operations. Give all answers in $a + bi$ form.

37. $(2 + 4i) + (-3 + 7i)$

38. $\left(3 - \sqrt{-9}\right) - \left(-1 + \sqrt{-16}\right)$

39. $2i(3 - 4i)$

40. $(3 + 2i)(-4 - i)$

41. $\dfrac{1}{i\sqrt{2}}$

42. $\dfrac{2 + i}{3 - i}$

8 Quadratic Functions and Inequalities

InfoTrac Project

Do a keyword search on 'Pendulum.' Find an article entitled "Huygens's clocks revisited. (Science Observer.)" Write a summary of the article and identify what centuries-old puzzle has been solved and by what university.

Using the formula $l = \dfrac{32t^2}{4\pi^2}$, where l represents the length (in feet) and t the time (in seconds) to swing through one cycle, find the time it takes for the 30-ft pendulum at the University of South Carolina in Aiken to complete one cycle. Give your answer to the nearest hundredth of a second. Calculate the time for a 15-ft pendulum to complete a cycle. How are the answers related? *Hint:* Find the ratio of the two answers and square it

Complete this project after studying Section 8.1.

© Owen Franken/CORBIS

Mathematics in Chemistry

A weak acid (0.1 M concentration) breaks down into free cations (the hydrogen ion, H^+) and anions (A^-). When this acid dissociates, the following equilibrium equation is established:

$$(1) \quad \frac{[H^+][A^-]}{[HA]} = 4 \times 10^{-4}$$

where $[H^+]$, the hydrogen ion concentration, is equal to $[A^-]$, the anion concentration, and $[HA]$ is the concentration of the undissociated acid itself. Find $[H^+]$ at equilibrium.

Exercise Set 8.2 Problem 63

We have discussed how to solve linear equations and certain quadratic equations in which the quadratic expression is factorable. In this chapter, we will discuss more general methods for solving quadratic equations, and we will consider their graphs.

8.1 Solving Quadratic Equations by Completing the Square

- ▮ **Solving Quadratic Equations by Factoring**
- ▮ **The Square Root Property**   ▮ **Completing the Square**
- ▮ **Solving Equations by Completing the Square**   ▮ **Problem Solving**

Getting Ready   *Factor each expression.*

1. $x^2 - 25$ **2.** $b^2 - 81$

3. $6x^2 + x - 2$ **4.** $4x^2 - 4x - 3$

We begin this section by reviewing how to solve quadratic equations by factoring. We will then discuss how to solve these equations by completing the square and use these skills to solve problems.

Solving Quadratic Equations by Factoring

A **quadratic equation** is an equation of the form $ax^2 + bx + c = 0$ $(a \neq 0)$, where a, b, and c are real numbers. We will solve the first two examples by factoring.

EXAMPLE 1   Solve: $x^2 = 9$.

Solution   To solve this quadratic equation by factoring, we proceed as follows:

$$x^2 = 9$$
$$x^2 - 9 = 0 \qquad \text{Subtract 9 from both sides.}$$
$$(x + 3)(x - 3) = 0 \qquad \text{Factor the binomial.}$$
$$x + 3 = 0 \quad \text{or} \quad x - 3 = 0 \qquad \text{Set each factor equal to 0.}$$
$$x = -3 \quad | \quad x = 3 \qquad \text{Solve each linear equation.}$$

Check: *For $x = -3$*   *For $x = 3$*

$$x^2 = 9 \qquad\qquad\qquad x^2 = 9$$
$$(-3)^2 \stackrel{?}{=} 9 \qquad\qquad (3)^2 \stackrel{?}{=} 9$$
$$9 = 9 \qquad\qquad\qquad 9 = 9$$

Self Check   Solve: $p^2 = 64$.   ▌

EXAMPLE 2   Solve: $6x^2 - 7x - 3 = 0$.

Solution
$$6x^2 - 7x - 3 = 0$$
$$(2x - 3)(3x + 1) = 0 \qquad \text{Factor.}$$
$$2x - 3 = 0 \quad \text{or} \quad 3x + 1 = 0 \qquad \text{Set each factor equal to 0.}$$
$$x = \frac{3}{2} \qquad\qquad x = -\frac{1}{3} \qquad \text{Solve each linear equation.}$$

Check: *For $x = \frac{3}{2}$*   *For $x = -\frac{1}{3}$*

$$6x^2 - 7x - 3 = 0 \qquad\qquad\qquad 6x^2 - 7x - 3 = 0$$

$$6\left(\frac{3}{2}\right)^2 - 7\left(\frac{3}{2}\right) - 3 \stackrel{?}{=} 0 \qquad\qquad 6\left(-\frac{1}{3}\right)^2 - 7\left(-\frac{1}{3}\right) - 3 \stackrel{?}{=} 0$$

$$6\left(\frac{9}{4}\right) - 7\left(\frac{3}{2}\right) - 3 \stackrel{?}{=} 0 \qquad\qquad 6\left(\frac{1}{9}\right) - 7\left(-\frac{1}{3}\right) - 3 \stackrel{?}{=} 0$$

$$\frac{27}{2} - \frac{21}{2} - \frac{6}{2} \stackrel{?}{=} 0 \qquad\qquad\qquad \frac{2}{3} + \frac{7}{3} - \frac{9}{3} \stackrel{?}{=} 0$$

$$0 = 0 \qquad\qquad\qquad\qquad\qquad 0 = 0$$

Self Check   Solve: $6m^2 - 5m + 1 = 0$.   ▌

Unfortunately, many quadratic expressions do not factor easily. For example, it would be difficult to solve $2x^2 + 4x + 1 = 0$ by factoring, because $2x^2 + 4x + 1$ cannot be factored by using only integers.

The Square Root Property

To develop general methods for solving all quadratic equations, we first solve $x^2 = c$ by a method similar to the one used in Example 1. If $c > 0$, we can find the real solutions of $x^2 = c$ as follows:

$$x^2 = c$$
$$x^2 - c = 0 \qquad\qquad \text{Subtract } c \text{ from both sides.}$$
$$x^2 - \left(\sqrt{c}\right)^2 = 0 \qquad\qquad c = \left(\sqrt{c}\right)^2.$$
$$\left(x + \sqrt{c}\right)\left(x - \sqrt{c}\right) = 0 \qquad\qquad \text{Factor the difference of two squares.}$$
$$x + \sqrt{c} = 0 \quad \text{or} \quad x - \sqrt{c} = 0 \qquad \text{Set each factor equal to 0.}$$
$$x = -\sqrt{c} \quad | \quad x = \sqrt{c} \qquad \text{Solve each linear equation.}$$

The two solutions of $x^2 = c$ are $x = \sqrt{c}$ and $x = -\sqrt{c}$.

The Square Root Property   If $c > 0$, the equation $x^2 = c$ has two real solutions. They are
$$x = \sqrt{c} \qquad \text{or} \qquad x = -\sqrt{c}$$

We often use the symbol $\pm\sqrt{c}$ to represent the two solutions $\sqrt{c}$ and $-\sqrt{c}$. The symbol $\pm\sqrt{c}$ is read as the positive or negative square root of c.

EXAMPLE 3   Solve: $x^2 - 12 = 0$.

Solution   We can write the equation as $x^2 = 12$ and use the square root property.

$$x^2 - 12 = 0$$
$$x^2 = 12 \qquad \text{Add 12 to both sides.}$$
$$x = \sqrt{12} \quad \text{or} \quad x = -\sqrt{12} \qquad \text{Use the square root property.}$$
$$x = 2\sqrt{3} \quad | \quad x = -2\sqrt{3} \qquad \sqrt{12} = \sqrt{4}\sqrt{3} = 2\sqrt{3}.$$

The solutions can be written as $\pm 2\sqrt{3}$. Verify that each one satisfies the equation.

Self Check   Solve: $x^2 - 18 = 0$.   ∎

EXAMPLE 4   Solve: $(x - 3)^2 = 16$.

Solution
$$(x - 3)^2 = 16$$
$$x - 3 = \sqrt{16} \quad \text{or} \quad x - 3 = -\sqrt{16} \qquad \text{Use the square root property.}$$
$$x - 3 = 4 \quad | \quad x - 3 = -4 \qquad \sqrt{16} = 4 \text{ and } -\sqrt{16} = -4.$$
$$x = 3 + 4 \quad | \quad x = 3 - 4 \qquad \text{Add 3 to both sides.}$$
$$x = 7 \quad | \quad x = -1 \qquad \text{Simplify.}$$

Verify that each solution satisfies the equation.

Self Check   Solve: $(x + 2)^2 = 9$.   ∎

In the following example, the solutions are imaginary numbers.

EXAMPLE 5   Solve: $9x^2 + 25 = 0$.

Solution
$$9x^2 + 25 = 0$$
$$x^2 = -\frac{25}{9} \qquad \begin{array}{l}\text{Subtract 25 from both sides and divide}\\\text{both sides by 9.}\end{array}$$
$$x = \sqrt{-\frac{25}{9}} \quad \text{or} \quad x = -\sqrt{-\frac{25}{9}} \qquad \text{Use the square root property.}$$
$$x = \sqrt{\frac{25}{9}}\sqrt{-1} \quad \Big| \quad x = -\sqrt{\frac{25}{9}}\sqrt{-1} \qquad \sqrt{-\frac{25}{9}} = \sqrt{\frac{25}{9}(-1)} = \sqrt{\frac{25}{9}}\sqrt{-1}.$$
$$x = \frac{5}{3}i \quad \Big| \quad x = -\frac{5}{3}i \qquad \sqrt{\frac{25}{9}} = \frac{5}{3}; \sqrt{-1} = i.$$

Check:
$$9x^2 + 25 = 0 \qquad\qquad\qquad 9x^2 + 25 = 0$$
$$9\left(\frac{5}{3}i\right)^2 + 25 \overset{?}{=} 0 \qquad\qquad 9\left(-\frac{5}{3}i\right)^2 + 25 \overset{?}{=} 0$$
$$9\left(\frac{25}{9}\right)i^2 + 25 \overset{?}{=} 0 \qquad\qquad 9\left(\frac{25}{9}\right)i^2 + 25 \overset{?}{=} 0$$
$$25(-1) + 25 \overset{?}{=} 0 \qquad\qquad 25(-1) + 25 \overset{?}{=} 0$$
$$0 = 0 \qquad\qquad\qquad\qquad 0 = 0$$

Self Check   Solve: $4x^2 + 36 = 0$.   ∎

Completing the Square

All quadratic equations can be solved by a method called **completing the square.** This method is based on the special products

$$x^2 + 2ax + a^2 = (x + a)^2 \qquad \text{and} \qquad x^2 - 2ax + a^2 = (x - a)^2$$

The trinomials $x^2 + 2ax + a^2$ and $x^2 - 2ax + a^2$ are both perfect-square trinomials, because both factor as the square of a binomial. In each case, the coefficient of the first term is 1 and if we take one-half of the coefficient of x in the middle term and square it, we obtain the third term.

$$\left[\frac{1}{2}(2a)\right]^2 = a^2 \qquad\qquad \left[\frac{1}{2}(2a)\right]^2 = (a)^2 = a^2.$$

$$\left[\frac{1}{2}(-2a)\right]^2 = (-a)^2 = a^2 \qquad \left[\frac{1}{2}(-2a)\right]^2 = (-a)^2 = a^2.$$

EXAMPLE 6   Add a number to make each binomial a perfect square trinomial: **a.** $x^2 + 10x$, **b.** $x^2 - 6x$,   and   **c.** $x^2 - 11x$.

Solution   **a.** To make $x^2 + 10x$ a perfect-square trinomial, we find one-half of 10 to get 5, square 5 to get 25, and add 25 to $x^2 + 10x$.

$$x^2 + 10x + \left[\frac{1}{2}(10)\right]^2 = x^2 + 10x + (5)^2$$

$$= x^2 + 10x + 25 \qquad \text{Note that } x^2 + 10x + 25 = (x + 5)^2.$$

b. To make $x^2 - 6x$ a perfect-square trinomial, we find one-half of -6 to get -3, square -3 to get 9, and add 9 to $x^2 - 6x$.

$$x^2 - 6x + \left[\frac{1}{2}(-6)\right]^2 = x^2 - 6x + (-3)^2$$

$$= x^2 - 6x + 9 \qquad \text{Note that } x^2 - 6x + 9 = (x - 3)^2.$$

c. To make $x^2 - 11x$ a perfect-square trinomial, we find one-half of -11 to get $-\frac{11}{2}$, square $-\frac{11}{2}$ to get $\frac{121}{4}$, and add $\frac{121}{4}$ to $x^2 - 11x$.

$$x^2 - 11x + \left[\frac{1}{2}(-11)\right]^2$$

$$= x^2 - 11x + \left(-\frac{11}{2}\right)^2$$

$$= x^2 - 11x + \frac{121}{4} \qquad \text{Note that } x^2 - 11x + \frac{121}{4} = \left(x - \frac{11}{2}\right)^2.$$

Self Check   Add a number to make $a^2 - 5a$ a perfect-square trinomial.   ∎

To see geometrically why completing the square works on $x^2 + 10x$, we refer to Figure 8-1(a), which shows a polygon with an area of $x^2 + 10x$. The only way to turn the polygon into a square is to divide the area of $10x$ into two areas of $5x$ and then reassemble the polygon as shown in Figure 8-1(b). To fill in the missing corner, we must add a square with an area of $5^2 = 25$. Thus, we complete the square.

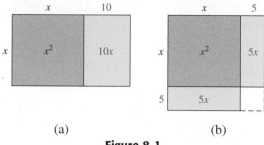

(a)   (b)

Figure 8-1

Solving Equations by Completing the Square

To solve an equation of the form $ax^2 + bx + c = 0$ $(a \neq 0)$ by completing the square, we use the following steps.

Completing the Square

1. Make sure that the coefficient of x^2 is 1. If it isn't, make it 1 by dividing both sides of the equation by the coefficient of x^2.

2. If necessary, add a number to both sides of the equation to place the constant term on the right-hand side of the equal sign.

3. Complete the square:
   a. Find one-half of the coefficient of x and square it.
   b. Add the square to both sides of the equation.

4. Factor the trinomial square and combine like terms.

5. Solve the resulting equation using the square root property.

EXAMPLE 7   Use completing the square to solve $x^2 + 8x + 7 = 0$.

Solution   **Step 1**   In this example, the coefficient of x^2 is already 1.

Step 2   We add -7 to both sides to place the constant on the right-hand side of the equal sign:

$$x^2 + 8x + 7 = 0$$
$$x^2 + 8x = -7$$

Step 3   The coefficient of x is 8, one-half of 8 is 4, and $4^2 = 16$. To complete the square, we add 16 to both sides.

$$x^2 + 8x \mathbf{+ 16} = \mathbf{16} - 7$$

(1) $$x^2 + 8x + 16 = 9 \qquad 16 - 7 = 9.$$

Step 4   Since the left-hand side of Equation 1 is a perfect-square trinomial, we can factor it to get $(x + 4)^2$.

$$x^2 + 8x + 16 = 9$$

(2) $$(x + 4)^2 = 9$$

Step 5   We then solve Equation 2 using the square root property.

$$(x + 4)^2 = 9$$
$$x + 4 = \sqrt{9} \quad \text{or} \quad x + 4 = -\sqrt{9}$$
$$x + 4 = 3 \quad \bigg| \quad x + 4 = -3$$
$$x = -1 \quad \bigg| \quad x = -7$$

Check both solutions. Note that this equation could be solved by factoring.

Self Check   Solve: $a^2 + 5a + 4 = 0$.   ∎

EXAMPLE 8   Solve: $6x^2 + 5x - 6 = 0$.

Solution   **Step 1**   To make the coefficient of x^2 equal to 1, we divide both sides by 6.

$$6x^2 + 5x - 6 = 0$$
$$\frac{6x^2}{6} + \frac{5}{6}x - \frac{6}{6} = \frac{0}{6} \qquad \text{Divide both sides by 6.}$$
$$x^2 + \frac{5}{6}x - 1 = 0 \qquad \text{Simplify.}$$

Step 2   We add 1 to both sides to place the constant on the right-hand side.

$$x^2 + \frac{5}{6}x = 1$$

Step 3   The coefficient of x is $\frac{5}{6}$, one-half of $\frac{5}{6}$ is $\frac{5}{12}$, and $\left(\frac{5}{12}\right)^2 = \frac{25}{144}$. To complete the square, we add $\frac{25}{144}$ to both sides.

$$x^2 + \frac{5}{6}x + \frac{25}{144} = 1 + \frac{25}{144}$$

(3) $\qquad x^2 + \frac{5}{6}x + \frac{25}{144} = \frac{169}{144} \qquad\qquad 1 + \frac{25}{144} = \frac{144}{144} + \frac{25}{144} = \frac{169}{144}.$

Step 4   Since the left-hand side of Equation 3 is a perfect-square trinomial, we can factor it to get $\left(x + \frac{5}{12}\right)^2$.

(4) $\qquad \left(x + \frac{5}{12}\right)^2 = \frac{169}{144}$

Step 5   We can solve Equation 4 by using the square root property.

$$x + \frac{5}{12} = \sqrt{\frac{169}{144}} \qquad \text{or} \quad x + \frac{5}{12} = -\sqrt{\frac{169}{144}}$$
$$x + \frac{5}{12} = \frac{13}{12} \qquad\qquad\qquad x + \frac{5}{12} = -\frac{13}{12}$$
$$x = -\frac{5}{12} + \frac{13}{12} \qquad\qquad x = -\frac{5}{12} - \frac{13}{12}$$
$$x = \frac{8}{12} \qquad\qquad\qquad\qquad x = -\frac{18}{12}$$
$$x = \frac{2}{3} \qquad\qquad\qquad\qquad x = -\frac{3}{2}$$

Check both solutions. Note that this equation could be solved by factoring.

Self Check   Solve: $6p^2 - 5p - 6 = 0$.

EXAMPLE 9   Solve: $2x^2 + 4x + 1 = 0$.

Solution

$$2x^2 + 4x + 1 = 0$$

$$x^2 + 2x + \frac{1}{2} = \frac{0}{2}$$ Divide both sides by 2 to make the coefficient of x^2 equal to 1.

$$x^2 + 2x = -\frac{1}{2}$$ Subtract $\frac{1}{2}$ from both sides.

$$x^2 + 2x \mathbf{+ 1} = \mathbf{1} - \frac{1}{2}$$ Square half the coefficient of x and add it to both sides.

$$(x + 1)^2 = \frac{1}{2}$$ Factor and combine like terms.

$$x + 1 = \sqrt{\frac{1}{2}} \qquad \text{or} \quad x + 1 = -\sqrt{\frac{1}{2}}$$ Use the square root property.

$$x + 1 = \frac{\sqrt{2}}{2} \qquad \qquad x + 1 = -\frac{\sqrt{2}}{2}$$ $\sqrt{\frac{1}{2}} = \frac{1}{\sqrt{2}} = \frac{1 \cdot \sqrt{2}}{\sqrt{2}\sqrt{2}} = \frac{\sqrt{2}}{2}.$

$$x = -1 + \frac{\sqrt{2}}{2} \qquad \qquad x = -1 - \frac{\sqrt{2}}{2}$$

$$x = \frac{-2 + \sqrt{2}}{2} \qquad \qquad x = \frac{-2 - \sqrt{2}}{2}$$

These solutions can be written as $x = \dfrac{-2 \pm \sqrt{2}}{2}$. Check both solutions.

Self Check   Solve: $3x^2 + 6x + 1 = 0$.

In the next example, the solutions are complex numbers.

EXAMPLE 10   Solve: $3x^2 + 2x + 2 = 0$.

Solution

$$3x^2 + 2x + 2 = 0$$

$$x^2 + \frac{2}{3}x + \frac{2}{3} = \frac{0}{3}$$ Divide both sides by 3 to make the coefficient of x^2 equal to 1.

$$x^2 + \frac{2}{3}x = -\frac{2}{3}$$ Subtract $\frac{2}{3}$ from both sides.

$$x^2 + \frac{2}{3}x \mathbf{+ \frac{1}{9}} = \mathbf{\frac{1}{9}} - \frac{2}{3}$$ Square half the coefficient of x and add it to both sides.

$$\left(x + \frac{1}{3}\right)^2 = -\frac{5}{9}$$ Factor and combine terms: $\frac{1}{9} - \frac{2}{3} = \frac{1}{9} - \frac{6}{9} = -\frac{5}{9}.$

$$x + \frac{1}{3} = \sqrt{-\frac{5}{9}} \qquad \text{or} \quad x + \frac{1}{3} = -\sqrt{-\frac{5}{9}} \qquad \text{Use the square root property.}$$

$$x + \frac{1}{3} = \sqrt{\frac{5}{9}}\sqrt{-1} \qquad\qquad x + \frac{1}{3} = -\sqrt{\frac{5}{9}}\sqrt{-1} \qquad \sqrt{-\frac{5}{9}} = \sqrt{\frac{5}{9}(-1)} =$$

$$\sqrt{\frac{5}{9}}\sqrt{-1}.$$

$$x + \frac{1}{3} = \frac{\sqrt{5}}{3}i \qquad\qquad x + \frac{1}{3} = -\frac{\sqrt{5}}{3}i \qquad \sqrt{\frac{5}{9}} = \frac{\sqrt{5}}{\sqrt{9}} = \frac{\sqrt{5}}{3}.$$

$$x = -\frac{1}{3} + \frac{\sqrt{5}}{3}i \qquad\qquad x = -\frac{1}{3} - \frac{\sqrt{5}}{3}i \qquad \text{Subtract } \frac{1}{3} \text{ from both sides.}$$

These solutions can be written as $x = -\dfrac{1}{3} \pm \dfrac{\sqrt{5}}{3}i.$

Self Check   Solve: $x^2 + 4x + 6 = 0$.   ∎

Problem Solving

When you deposit money in a bank account, it earns interest. If you leave the money in the account, the earned interest is deposited back into the account and also earns interest. When this is the case, the account is earning **compound interest.** There is a formula we can use to compute the amount in an account at any time t.

Formula for Compound Interest

> If P dollars is deposited in an account and interest is paid once a year at an annual rate r, the amount A in the account after t years is given by the formula
>
> $A = P(1 + r)^t$

EXAMPLE 11   **Saving money**   A woman invests $10,000 in an account. Find the annual interest rate if the account will be worth $11,025 in 2 years.

Solution   We substitute 11,025 for A, 10,000 for P, and 2 for t in the compound interest formula and solve for r.

$$A = P(1 + r)^t$$

$$11{,}025 = 10{,}000(1 + r)^2 \qquad \text{Substitute.}$$

$$\frac{11{,}025}{10{,}000} = (1 + r)^2 \qquad \text{Divide both sides by 10,000.}$$

$$1.1025 = (1 + r)^2 \qquad \tfrac{11{,}025}{10{,}000} = 1.1025.$$

$$1 + r = 1.05 \quad \text{or} \quad 1 + r = -1.05 \qquad \text{Use the square root property, } \sqrt{1.1025} = 1.05.$$

$$r = 0.05 \quad | \qquad\quad r = -2.05 \qquad \text{Subtract 1 from both sides.}$$

Since an interest rate cannot be negative, we must discard the result of -2.05. Thus, the annual interest rate is 0.05, or 5%.

We can check this result by substituting 0.05 for r, 10,000 for P, and 2 for t in the formula and confirming that the deposit of \$10,000 will grow to \$11,025 in 2 years.

$$A = P(1 + r)^t = \mathbf{10{,}000}(1 + \mathbf{0.05})^2 = 10{,}000(1.1025) = 11{,}025 \quad \blacksquare$$

Self Check Answers

1. $8, -8$ **2.** $\dfrac{1}{3}, \dfrac{1}{2}$ **3.** $\pm 3\sqrt{2}$ **4.** $1, -5$ **5.** $\pm 3i$ **6.** $a^2 - 5a + \dfrac{25}{4}$ **7.** $-1, -4$ **8.** $-\dfrac{2}{3}, \dfrac{3}{2}$

9. $\dfrac{-3 \pm \sqrt{6}}{3}$ **10.** $-2 \pm i\sqrt{2}$

Orals *Solve each equation.*

1. $x^2 = 49$ **2.** $x^2 = 10$

Find the number that must be added to the binomial to make it a perfect square trinomial.

3. $x^2 + 4x$ **4.** $x^2 - 6x$ **5.** $x^2 - 3x$ **6.** $x^2 + 5x$

8.1 EXERCISES

REVIEW *Solve each equation or inequality.*

1. $\dfrac{t+9}{2} + \dfrac{t+2}{5} = \dfrac{8}{5} + 4t$

2. $\dfrac{1-5x}{2x} + 4 = \dfrac{x+3}{x}$

3. $3(t-3) + 3t \le 2(t+1) + t + 1$

4. $-2(y+4) - 3y + 8 \ge 3(2y-3) - y$

VOCABULARY AND CONCEPTS *Fill in the blanks.*

5. If $c > 0$, the solutions of $x^2 = c$ are _____ and _____.

6. To complete the square on x in $x^2 + 6x$, find one-half of __, square it to get __, and add __ to get _____.

7. The symbol $\pm$ is read as _____.

8. The formula for compound interest is _____.

PRACTICE *Use factoring to solve each equation.*

9. $6x^2 + 12x = 0$ **10.** $5x^2 + 11x = 0$

11. $2y^2 - 50 = 0$ **12.** $4y^2 - 64 = 0$

13. $r^2 + 6r + 8 = 0$ **14.** $x^2 + 9x + 20 = 0$

15. $7x - 6 = x^2$ **16.** $5t - 6 = t^2$

17. $2z^2 - 5z + 2 = 0$ **18.** $2x^2 - x - 1 = 0$

19. $6s^2 + 11s - 10 = 0$ **20.** $3x^2 + 10x - 8 = 0$

Use the square root property to solve each equation.

21. $x^2 = 36$ **22.** $x^2 = 144$

23. $z^2 = 5$ **24.** $u^2 = 24$

25. $3x^2 - 16 = 0$ **26.** $5x^2 - 49 = 0$

27. $(y+1)^2 = 1$ **28.** $(y-1)^2 = 4$

29. $(s-7)^2 - 9 = 0$ **30.** $(t+4)^2 = 16$

31. $(x+5)^2 - 3 = 0$ **32.** $(x+3)^2 - 7 = 0$

33. $(x-2)^2 - 5 = 0$ **34.** $(x-5)^2 - 11 = 0$

35. $p^2 + 16 = 0$ **36.** $q^2 + 25 = 0$

37. $4m^2 + 81 = 0$ **38.** $9n^2 + 121 = 0$

Use completing the square to solve each equation.

39. $x^2 + 2x - 8 = 0$ **40.** $x^2 + 6x + 5 = 0$

41. $x^2 - 6x + 8 = 0$ **42.** $x^2 + 8x + 15 = 0$

43. $x^2 + 5x + 4 = 0$ **44.** $x^2 - 11x + 30 = 0$

45. $x + 1 = 2x^2$ **46.** $-2 = 2x^2 - 5x$

47. $6x^2 + 11x + 3 = 0$ **48.** $6x^2 + x - 2 = 0$

49. $9 - 6r = 8r^2$ **50.** $11m - 10 = 3m^2$

51. $\dfrac{7x + 1}{5} = -x^2$ **52.** $\dfrac{3x^2}{8} = \dfrac{1}{8} - x$

53. $p^2 + 2p + 2 = 0$ **54.** $x^2 - 6x + 10 = 0$

55. $y^2 + 8y + 18 = 0$ **56.** $t^2 + t + 3 = 0$

57. $3m^2 - 2m + 3 = 0$ **58.** $4p^2 + 2p + 3 = 0$

Find all x that will make f(x) = 0.

59. $f(x) = 2x^2 + x - 5$ **60.** $f(x) = 3x^2 - 2x - 4$

61. $f(x) = x^2 + x - 3$ **62.** $f(x) = x^2 + 2x - 4$

APPLICATIONS

63. Falling objects The distance s (in feet) that an object will fall in t seconds is given by the formula $s = 16t^2$. How long will it take an object to fall 256 feet?

64. Pendulums The time (in seconds) it takes a pendulum to swing back and forth to complete one cycle is related to its length l (in feet) by the formula:

$$l = \frac{32t^2}{4\pi^2}$$

How long will it take a 5-foot pendulum to swing through one cycle? Give the result to the nearest hundredth.

65. Law enforcement To estimate the speed s (in mph) of a car involved in an accident, police often use the formula $s^2 = 10.5l$, where l is the length of any skid mark. How fast was a car going that was involved in an accident and left skid marks of 500 feet? Give the result to the nearest tenth.

66. Medicine The approximate pulse rate (in beats per minute) of an adult who is t inches tall is given by the formula

$$p^2 = \frac{348{,}100}{t}$$

Find the pulse rate of an adult who is 64 inches tall.

67. Saving money A student invests $8,500 in a savings account drawing interest that is compounded annually. Find the annual rate if the money grows to $9,193.60 in 2 years.

68. Saving money A woman invests $12,500 in a savings account drawing interest that is compounded annually. Find the annual rate if the money grows to $14,045 in 2 years.

WRITING

69. Explain how to complete the square.

70. Tell why a cannot be 0 in the quadratic equation $ax^2 + bx + c = 0$.

SOMETHING TO THINK ABOUT

71. What number must be added to $x^2 + \sqrt{3}x$ to make it a perfect-square trinomial?

72. Solve $x^2 + \sqrt{3}x - \frac{1}{4} = 0$ by completing the square.

<table>
<tr><td>**8.2**</td><td>## Solving Quadratic Equations by the Quadratic Formula</td></tr>
</table>

▌ The Quadratic Formula   ▌ Solving Formulas   ▌ Problem Solving

Getting Ready   *Add a number to each binomial to complete the square. Then write the resulting trinomial as the square of a binomial.*

1. $x^2 + 12x$ **2.** $x^2 - 7x$

Evaluate $\sqrt{b^2 - 4ac}$ for the following values.

3. $a = 6, b = 1, c = -2$ **4.** $a = 4, b = -4, c = -3$

Solving quadratic equations by completing the square is often tedious. Fortunately, there is an easier way. In this section, we will develop a formula, called the **quadratic formula,** that we can use to solve quadratic equations with a minimum of effort. To develop this formula, we will complete the square.

The Quadratic Formula

To develop a formula to solve quadratic equations, we will solve the general quadratic equation $ax^2 + bx + c = 0$ $(a \neq 0)$ by completing the square.

$$ax^2 + bx + c = 0$$

$$\frac{ax^2}{a} + \frac{bx}{a} + \frac{c}{a} = \frac{0}{a} \qquad \text{To make the coefficient of } x^2 \text{ equal to 1, we divide both sides by } a.$$

$$x^2 + \frac{bx}{a} = -\frac{c}{a} \qquad \frac{0}{a} = 0; \text{ subtract } \frac{c}{a} \text{ from both sides.}$$

$$x^2 + \frac{b}{a}x + \left(\frac{b}{2a}\right)^2 = \left(\frac{b}{2a}\right)^2 - \frac{c}{a} \qquad \text{Complete the square on } x \text{ by adding } \left(\frac{b}{2a}\right)^2 \text{ to both sides.}$$

$$x^2 + \frac{b}{a}x + \frac{b^2}{4a^2} = \frac{b^2}{4a^2} - \frac{4ac}{4aa} \qquad \text{Remove parentheses and get a common denominator on the right-hand side.}$$

(1)
$$\left(x + \frac{b}{2a}\right)^2 = \frac{b^2 - 4ac}{4a^2} \qquad \text{Factor the left-hand side and add the fractions on the right-hand side.}$$

We can solve Equation 1 using the square root property.

$$x + \frac{b}{2a} = \sqrt{\frac{b^2 - 4ac}{4a^2}} \qquad \text{or} \quad x + \frac{b}{2a} = -\sqrt{\frac{b^2 - 4ac}{4a^2}}$$

$$x + \frac{b}{2a} = \frac{\sqrt{b^2 - 4ac}}{2a} \qquad\qquad x + \frac{b}{2a} = -\frac{\sqrt{b^2 - 4ac}}{2a}$$

$$x = -\frac{b}{2a} + \frac{\sqrt{b^2 - 4ac}}{2a} \qquad\qquad x = -\frac{b}{2a} - \frac{\sqrt{b^2 - 4ac}}{2a}$$

$$= \frac{-b + \sqrt{b^2 - 4ac}}{2a} \qquad\qquad = \frac{-b - \sqrt{b^2 - 4ac}}{2a}$$

These two solutions give the **quadratic formula.**

The Quadratic Formula

The solutions of $ax^2 + bx + c = 0$ $(a \neq 0)$ are given by the formula

$$x = \frac{-b \pm \sqrt{b^2 - 4ac}}{2a}$$

Comment Be sure to draw the fraction bar under both parts of the numerator, and be sure to draw the radical sign exactly over $b^2 - 4ac$. Don't write the quadratic formula as

$$x = -b \pm \frac{\sqrt{b^2 - 4ac}}{2a} \qquad \text{or as} \qquad x = -b \pm \sqrt{\frac{b^2 - 4ac}{2a}}$$

EXAMPLE 1 Solve: $2x^2 - 3x - 5 = 0$.

Solution In this equation $a = 2$, $b = -3$, and $c = -5$.

$$x = \frac{-b \pm \sqrt{b^2 - 4ac}}{2a}$$

$$= \frac{-(-3) \pm \sqrt{(-3)^2 - 4(2)(-5)}}{2(2)} \qquad \text{Substitute 2 for } a, -3 \text{ for } b, \text{ and } -5 \text{ for } c.$$

$$= \frac{3 \pm \sqrt{9 + 40}}{4}$$

$$= \frac{3 \pm \sqrt{49}}{4}$$

$$= \frac{3 \pm 7}{4}$$

$$x = \frac{3 + 7}{4} \quad \text{or} \quad x = \frac{3 - 7}{4}$$

$$x = \frac{10}{4} \qquad\qquad x = \frac{-4}{4}$$

$$x = \frac{5}{2} \qquad\qquad x = -1$$

Check both solutions. Note that this equation can be solved by factoring.

Self Check Solve: $3x^2 - 5x - 2 = 0$.

EXAMPLE 2 Solve: $2x^2 + 1 = -4x$.

Solution We begin by writing the equation in $ax^2 + bx + c = 0$ form (called **standard form**) before identifying a, b, and c.

$$2x^2 + 4x + 1 = 0$$

In this equation, $a = 2$, $b = 4$, and $c = 1$.

$$x = \frac{-b \pm \sqrt{b^2 - 4ac}}{2a}$$

$$= \frac{-4 \pm \sqrt{4^2 - 4(2)(1)}}{2(2)} \qquad \text{Substitute 2 for } a, 4 \text{ for } b, \text{ and 1 for } c.$$

$$= \frac{-4 \pm \sqrt{16 - 8}}{4}$$

$$= \frac{-4 \pm \sqrt{8}}{4}$$

$$= \frac{-4 \pm 2\sqrt{2}}{4} \qquad\qquad \sqrt{8} = \sqrt{4 \cdot 2} = \sqrt{4}\sqrt{2} = 2\sqrt{2}.$$

$$= \frac{-2 \pm \sqrt{2}}{2} \qquad\qquad \frac{-4 \pm 2\sqrt{2}}{4} = \frac{2(-2 \pm \sqrt{2})}{4} = \frac{-2 \pm \sqrt{2}}{2}.$$

Note that these solutions can be written as $x = -1 \pm \dfrac{\sqrt{2}}{2}$.

Self Check Solve $3x^2 - 2x - 3 = 0$. ∎

The solutions to the next example are complex numbers.

EXAMPLE 3 Solve: $x^2 + x = -1$.

Solution We begin by writing the equation in standard form before identifying a, b, and c.

$$x^2 + x + 1 = 0$$

In this equation, $a = 1$, $b = 1$, and $c = 1$:

$$x = \frac{-b \pm \sqrt{b^2 - 4ac}}{2a}$$

$$= \frac{-1 \pm \sqrt{1^2 - 4(1)(1)}}{2(1)} \qquad \text{Substitute 1 for } a, \text{ 1 for } b, \text{ and 1 for } c.$$

$$= \frac{-1 \pm \sqrt{1 - 4}}{2}$$

$$= \frac{-1 \pm \sqrt{-3}}{2}$$

$$= \frac{-1 \pm \sqrt{3}i}{2}$$

Note that these solutions can be written as $x = -\dfrac{1}{2} \pm \dfrac{\sqrt{3}}{2}i$.

Self Check Solve: $a^2 + 2a + 3 = 0$. ∎

Solving Formulas

EXAMPLE 4 An object thrown straight up with an initial velocity of v_0 feet per second will reach a height of s feet in t seconds according to the formula $s = -16t^2 + v_0t$. Solve the formula for t.

Solution   We begin by writing the equation in standard form:

$$s = -16t^2 + v_0 t$$
$$16t^2 - v_0 t + s = 0$$

Then we can use the quadratic formula to solve for t.

$$t = \frac{-b \pm \sqrt{b^2 - 4ac}}{2a}$$

$$t = \frac{-(-v_0) \pm \sqrt{(-v_0)^2 - 4(16)(s)}}{2(16)}$$ Substitute into the quadratic formula.

$$t = \frac{v_0 \pm \sqrt{v_0^2 - 64s}}{32}$$ Simplify.

Thus, $t = \dfrac{v_0 \pm \sqrt{v_0^2 - 64s}}{32}$.   ∎

Problem Solving

EXAMPLE 5   **Dimensions of a rectangle**   Find the dimensions of the rectangle shown in Figure 8-2, given that its area is 253 cm^2.

Solution   If we let w represent the width of the rectangle, then $w + 12$ represents its length. Since the area of the rectangle is 253 square centimeters, we can form the equation

$$w(w + 12) = 253$$ Area of a rectangle = width · length.

and solve it as follows:

$$w(w + 12) = 253$$
$$w^2 + 12w = 253$$ Use the distributive property to remove parentheses.
$$w^2 + 12w - 253 = 0$$ Subtract 253 from both sides.

Solution by factoring

$$(w - 11)(w + 23) = 0$$
$$w - 11 = 0 \quad \text{or} \quad w + 23 = 0$$
$$w = 11 \quad | \quad w = -23$$

Solution by quadratic formula

$$w = \frac{-12 \pm \sqrt{12^2 - 4(1)(-253)}}{2(1)}$$

$$= \frac{-12 \pm \sqrt{144 + 1{,}012}}{2}$$

$$= \frac{-12 \pm \sqrt{1{,}156}}{2}$$

$$= \frac{-12 \pm 34}{2}$$

$$w = 11 \quad \text{or} \quad w = -23$$

Since the rectangle cannot have a negative width, we discard the solution of -23. Thus, the only solution is $w = 11$. Since the rectangle is 11 centimeters wide and $(11 + 12)$ centimeters long, its dimensions are 11 centimeters by 23 centimeters.

Check: 23 is 12 more than 11, and the area of a rectangle with dimensions of 23 centimeters by 11 centimeters is 253 square centimeters.   ∎

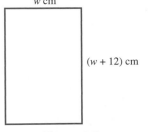

w cm

$(w + 12)$ cm

Figure 8-2

1. $2, -\dfrac{1}{3}$ **2.** $\dfrac{1}{3} \pm \dfrac{\sqrt{10}}{3}$ **3.** $-1 \pm i\sqrt{2}$

Orals   *Identify a, b, and c in each quadratic equation.*

1. $3x^2 - 4x + 7 = 0$ **2.** $-2x^2 + x - 5$

8.2 EXERCISES

REVIEW   *Solve for the indicated variable.*

1. $Ax + By = C$ for y **2.** $R = \dfrac{kL}{d^2}$ for L

Simplify each radical.

3. $\sqrt{24}$ **4.** $\sqrt{288}$

5. $\dfrac{3}{\sqrt{3}}$ **6.** $\dfrac{1}{2 - \sqrt{3}}$

VOCABULARY AND CONCEPTS   *Fill in the blanks.*

7. In the equation $3x^2 - 2x + 6 = 0$, $a = $ __, $b = $ __, and $c = $ __.

8. The solutions of $ax^2 + bx + c = 0$ $(a \neq 0)$ are given by the quadratic formula, which is

$x = $ _____ .

PRACTICE   *Use the quadratic formula to solve each equation.*

9. $x^2 + 3x + 2 = 0$ **10.** $x^2 - 3x + 2 = 0$

11. $x^2 - 2x - 15 = 0$ **12.** $x^2 - 2x - 35 = 0$

13. $x^2 + 12x = -36$ **14.** $y^2 - 18y = -81$

15. $2x^2 - x - 3 = 0$ **16.** $3x^2 - 10x + 8 = 0$

17. $5x^2 + 5x + 1 = 0$ **18.** $4w^2 + 6w + 1 = 0$

19. $8u = -4u^2 - 3$ **20.** $4t + 3 = 4t^2$

21. $16y^2 + 8y - 3 = 0$ **22.** $16x^2 + 16x + 3 = 0$

23. $\dfrac{x^2}{2} + \dfrac{5}{2}x = -1$ **24.** $-3x = \dfrac{x^2}{2} + 2$

25. $x^2 + 2x + 2 = 0$ **26.** $x^2 + 3x + 3 = 0$

27. $2x^2 + x + 1 = 0$ **28.** $3x^2 + 2x + 1 = 0$

29. $3x^2 - 4x = -2$ **30.** $2x^2 + 3x = -3$

31. $3x^2 - 2x = -3$ **32.** $5x^2 = 2x - 1$

Find all x that will make $f(x) = 0$.

33. $f(x) = 4x^2 + 4x - 19$ **34.** $f(x) = 9x^2 + 12x - 8$

35. $f(x) = 3x^2 + 2x + 2$ **36.** $f(x) = 4x^2 + x + 1$

Use the quadratic formula and a scientific calculator to solve each equation. Give all answers to the nearest hundredth.

37. $0.7x^2 - 3.5x - 25 = 0$

38. $-4.5x^2 + 0.2x + 3.75 = 0$

Solve each formula for the indicated variable.

39. $C = \dfrac{N^2 - N}{2}$, for N

(the formula for a selection sort in data processing)

40. $A = 2\pi r^2 + 2\pi hr$, for r

(the formula for the surface area of a right circular-cylinder)

Solve each problem.

41. Integer problem The product of two consecutive even positive integers is 288. Find the integers. (*Hint:* If one integer is x, the next consecutive even integer is $x + 2$.)

42. Integer problem The product of two consecutive odd negative integers is 143. Find the integers. (*Hint:* If one integer is x, the next consecutive odd integer is $x + 2$.)

43. Integer problem The sum of the squares of two consecutive positive integers is 85. Find the integers. (*Hint:* If one integer is x, the next consecutive positive integer is $x + 1$.)

44. Integer problem The sum of the squares of three consecutive positive integers is 77. Find the integers. (*Hint:* If one integer is x, the next consecutive positive integer is $x + 1$, and the third is $x + 2$.)

Note that a and b are solutions to the equation
$(x - a)(x - b) = 0$.

45. Find a quadratic equation that has a solution set of $\{3, 5\}$.

46. Find a quadratic equation that has a solution set of $\{-4, 6\}$.

47. Find a third-degree equation that has a solution set of $\{2, 3, -4\}$.

48. Find a fourth-degree equation that has a solution set of $\{3, -3, 4, -4\}$.

APPLICATIONS

49. Dimensions of a rectangle The rectangle shown below has an area of 96 square feet. Find its dimensions.

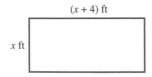

$(x + 4)$ ft

x ft

50. Dimensions of a window The area of the window shown below is 77 square feet. Find its dimensions.

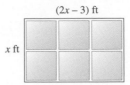

$(2x - 3)$ ft

x ft

51. Side of a square The area of a square is numerically equal to its perimeter. Find the length of each side of the square.

52. Perimeter of a rectangle A rectangle is 2 inches longer than it is wide. Numerically, its area exceeds its perimeter by 11. Find the perimeter.

53. Base of a triangle The height of a triangle is 5 centimeters longer than three times its base. Find the base of the triangle if its area is 6 square centimeters.

54. Height of a triangle The height of a triangle is 4 meters longer than twice its base. Find the height if the area of the triangle is 15 square meters.

55. Finding rates A woman drives her snowmobile 150 miles at the rate of r mph. She could have gone the same distance in 2 hours less time if she had increased her speed by 20 mph. Find r.

56. Finding rates Jeff bicycles 160 miles at the rate of r mph. The same trip would have taken 2 hours longer if he had decreased his speed by 4 mph. Find r.

57. Pricing concert tickets Tickets to a rock concert cost \$4, and the projected attendance is 300 persons. It is further projected that for every 10¢ increase in ticket price, the average attendance will decrease by 5. At what ticket price will the nightly receipts be \$1,248?

58. Setting bus fares A bus company has 3,000 passengers daily, paying a 25¢ fare. For each 5¢ increase in fare, the company estimates that it will lose 80 passengers. What increase in fare will produce a \$994 daily revenue?

59. Computing profit   The *Gazette's* profit is $20 per year for each of its 3,000 subscribers. Management estimates that the profit per subscriber will increase by 1¢ for each additional subscriber over the current 3,000. How many subscribers will bring a total profit of $120,000?

60. Finding interest rates   A woman invests $1,000 in a mutual fund for which interest is compounded annually at a rate r. After one year, she deposits an additional $2,000. After two years, the balance in the account is

$$\$1,000(1 + r)^2 + \$2,000(1 + r)$$

If this amount is $3,368.10, find r.

61. Framing a picture   The frame around the picture in the illustration has a constant width. How wide is the frame if its area equals the area of the picture?

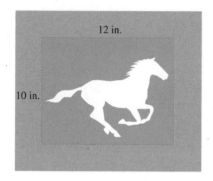

12 in.

10 in.

62. Metal fabrication   A box with no top is to be made by cutting a 2-inch square from each corner of the square sheet of metal shown in the illustration. After bending up the sides, the volume of the box is to be 200 cubic inches. How large should the piece of metal be?

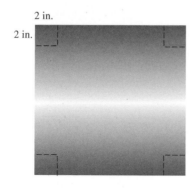

2 in.

2 in.

Use a calculator.

63. Chemistry   A weak acid (0.1 M concentration) breaks down into free cations (the hydrogen ion, H^+) and anions (A^-). When this acid dissociates, the following equilibrium equation is established:

$$\frac{[H^+][A^-]}{[HA]} = 4 \times 10^{-4}$$

where $[H^+]$, the hydrogen ion concentration, is equal to $[A^-]$, the anion concentration. $[HA]$ is the concentration of the undissociated acid itself. Find $[H^+]$ at equilibrium.   (*Hint:* If $[H^+] = x$, then $[HA] = 0.1 - x$.)

64. Chemistry   A saturated solution of hydrogen sulfide (0.1 M concentration) dissociates into cation $[H^+]$ and anion $[HS^-]$, where $[H^+] = [HS^-]$. When this solution dissociates, the following equilibrium equation is established:

$$\frac{[H^+][HS^-]}{[HHS]} = 1.0 \times 10^{-7}$$

Find $[H^+]$.   (*Hint:* If $[H^+] = x$, then $[HHS] = 0.1 - x$.)

WRITING

65. Explain why $x = -b \pm \dfrac{\sqrt{b^2 - 4ac}}{2a}$ is not a correct statement of the quadratic formula.

66. Explain why $x = \dfrac{b \pm \sqrt{b^2 - 4ac}}{a}$ is not a correct statement of the quadratic formula.

SOMETHING TO THINK ABOUT   *All of the equations we have solved so far have had rational-number coefficients. However, the quadratic formula can be used to solve quadratic equations with irrational or even imaginary coefficients. Try solving each of the following equations.*

67. $x^2 + 2\sqrt{2}x - 6 = 0$

68. $\sqrt{2}x^2 + x - \sqrt{2} = 0$

69. $x^2 - 3ix - 2 = 0$

70. $ix^2 + 3x - 2i = 0$

8.3   The Discriminant and Equations
That Can Be Written in Quadratic Form

▌ **The Discriminant**
▌ **Equations That Can Be Written in Quadratic Form**
▌ **Solutions of a Quadratic Equation**

Getting Ready   *Evaluate $b^2 - 4ac$ for the following values.*

1. $a = 2, b = 3,$ and $c = -1$ **2.** $a = -2, b = 4,$ and $c = -3$

We can use part of the quadratic formula to predict the type of solutions, if any, that a quadratic equation will have. We don't even have to solve the equation.

The Discriminant

Suppose that the coefficients a, b, and c in the equation $ax^2 + bx + c = 0\ (a \neq 0)$ are real numbers. Then the solutions of the equation are given by the quadratic formula

$$x = \frac{-b \pm \sqrt{b^2 - 4ac}}{2a} \quad (a \neq 0)$$

If $b^2 - 4ac \geq 0$, the solutions are real numbers. If $b^2 - 4ac < 0$, the solutions are nonreal complex numbers. Thus, the value of $b^2 - 4ac$, called the **discriminant,** determines the type of solutions for a particular quadratic equation.

The Discriminant   If a, b, and c are real numbers and

If $b^2 - 4ac$ is . . .	*then the solutions are . . .*
positive,	real numbers and unequal.
0,	real numbers and equal.
negative,	nonreal complex numbers and complex conjugates.

If a, b, and c are rational numbers and

If $b^2 - 4ac$ is . . .	*then the solutions are . . .*
a perfect square greater than 0,	rational numbers and unequal.
positive and not a perfect square,	irrational numbers and unequal.

EXAMPLE 1   Determine the type of solutions for the equation $x^2 + x + 1 = 0$.

Solution   We calculate the discriminant:

$$b^2 - 4ac = 1^2 - 4(1)(1) \quad a = 1, b = 1, \text{ and } c = 1.$$
$$= -3$$

Since $b^2 - 4ac < 0$, the solutions are nonreal complex conjugates.

Self Check   Determine the type of solutions for $x^2 + x - 1 = 0$.   ∎

EXAMPLE 2   Determine the type of solutions for the equation $3x^2 + 5x + 2 = 0$.

Solution   We calculate the discriminant:

$$b^2 - 4ac = \mathbf{5}^2 - 4(\mathbf{3})(\mathbf{2}) \qquad a = 3, b = 5, \text{ and } c = 2.$$
$$= 25 - 24$$
$$= 1$$

Since $b^2 - 4ac > 0$ and $b^2 - 4ac$ is a perfect square, the solutions are rational and unequal.

Self Check   Determine the type of solutions for $4x^2 - 10x + 25 = 0$.   ∎

EXAMPLE 3   What value of k will make the solutions of the equation $kx^2 - 12x + 9 = 0$ equal?

Solution   We calculate the discriminant:

$$b^2 - 4ac = (\mathbf{-12})^2 - 4(\mathbf{k})(\mathbf{9}) \qquad a = k, b = -12, \text{ and } c = 9.$$
$$= 144 - 36k$$
$$= -36k + 144$$

Since the solutions are to be equal, we let $-36k + 144 = 0$ and solve for k.

$$-36k + 144 = 0$$
$$-36k = -144 \qquad \text{Subtract 144 from both sides.}$$
$$k = 4 \qquad \text{Divide both sides by } -36.$$

If $k = 4$, the solutions will be equal. Verify this by solving $4x^2 - 12x + 9 = 0$ and showing that the solutions are equal.

Self Check   What value of k will make the solutions of $kx^2 - 20x + 25 = 0$ equal?   ∎

Equations That Can Be Written in Quadratic Form

Many equations can be written in quadratic form and then solved with the techniques used for solving quadratic equations. For example, we can solve $x^4 - 5x^2 + 4 = 0$ as follows:

$$x^4 - 5x^2 + 4 = 0$$
$$(\mathbf{x^2})^2 - 5(\mathbf{x^2}) + 4 = 0$$
$$y^2 - 5y + 4 = 0 \qquad \text{Let } y = x^2.$$
$$(y - 4)(y - 1) = 0 \qquad \text{Factor } y^2 - 5y + 4.$$
$$y - 4 = 0 \quad \text{or} \quad y - 1 = 0 \qquad \text{Set each factor equal to 0.}$$
$$y = 4 \quad | \quad \qquad y = 1$$

Since $x^2 = y$, it follows that $x^2 = 4$ or $x^2 = 1$. Thus,

$$x^2 = 4 \qquad \text{or} \qquad x^2 = 1$$
$$x = 2 \quad \text{or} \quad x = -2 \quad | \quad x = 1 \quad \text{or} \quad x = -1$$

This equation has four solutions: 1, -1, 2, and -2. Verify that each one satisfies the original equation. Note that this equation can be solved by factoring.

EXAMPLE 4   Solve: $x - 7x^{1/2} + 12 = 0$.

Solution   If y^2 is substituted for x and y is substituted for $x^{1/2}$, the equation

$$x - 7x^{1/2} + 12 = 0$$

becomes a quadratic equation that can be solved by factoring:

$$y^2 - 7y + 12 = 0 \qquad \text{Substitute } y^2 \text{ for } x \text{ and } y \text{ for } x^{1/2}.$$
$$(y - 3)(y - 4) = 0 \qquad \text{Factor.}$$
$$y - 3 = 0 \quad \text{or} \quad y - 4 = 0 \quad \text{Set each factor equal to 0.}$$
$$y = 3 \quad | \quad y = 4$$

Because $x = y^2$, it follows that

$$x = 3^2 \quad \text{or} \quad x = 4^2$$
$$= 9 \quad | \quad = 16$$

Verify that both solutions satisfy the original equation.

Self Check   Solve: $x + x^{1/2} - 6 = 0$. Be sure to check your solutions.   ∎

EXAMPLE 5   Solve: $\dfrac{24}{x} + \dfrac{12}{x + 1} = 11$.

Solution   Since the denominator cannot be 0, x cannot be 0 or -1. If either 0 or -1 appears as a suspected solution, it is extraneous and must be discarded.

$$\frac{24}{x} + \frac{12}{x + 1} = 11$$

$$x(x + 1)\left(\frac{24}{x} + \frac{12}{x + 1}\right) = x(x + 1)11 \qquad \text{Multiply both sides by } x(x + 1).$$

$$24(x + 1) + 12x = (x^2 + x)11 \qquad \text{Simplify.}$$

$$24x + 24 + 12x = 11x^2 + 11x \qquad \text{Use the distributive property to remove parentheses.}$$

$$36x + 24 = 11x^2 + 11x \qquad \text{Combine like terms.}$$

$$0 = 11x^2 - 25x - 24 \qquad \text{Subtract } 36x \text{ and } 24 \text{ from both sides.}$$

$$0 = (11x + 8)(x - 3) \qquad \text{Factor } 11x^2 - 25x - 24.$$

$$11x + 8 = 0 \qquad \text{or} \quad x - 3 = 0 \qquad \text{Set each factor equal to 0.}$$

$$x = -\frac{8}{11} \quad \bigg| \qquad x = 3$$

Verify that $-\frac{8}{11}$ and 3 satisfy the original equation.

Self Check   Solve: $\dfrac{12}{x} + \dfrac{6}{x+3} = 5$. ∎

EXAMPLE 6   Solve the formula $s = 16t^2 - 32$ for t.

Solution   We proceed as follows:

$$s = 16t^2 - 32$$

$$s + 32 = 16t^2 \qquad \text{Add 32 to both sides.}$$

$$\frac{s + 32}{16} = t^2 \qquad \text{Divide both sides by 16.}$$

$$t^2 = \frac{s + 32}{16} \qquad \text{Write } t^2 \text{ on the left-hand side.}$$

$$t = \pm\sqrt{\frac{s + 32}{16}} \qquad \text{Apply the square root property.}$$

$$t = \pm\frac{\sqrt{s + 32}}{\sqrt{16}} \qquad \sqrt{\frac{a}{b}} = \frac{\sqrt{a}}{\sqrt{b}}.$$

$$t = \pm\frac{\sqrt{s + 32}}{4}$$

Self Check   Solve $a^2 + b^2 = c^2$ for a. ∎

Solutions of a Quadratic Equation

Solutions of a Quadratic Equation

If r_1 and r_2 are the solutions of the quadratic equation $ax^2 + bx + c = 0$, with $a \neq 0$, then

$$r_1 + r_2 = -\frac{b}{a} \qquad \text{and} \qquad r_1 r_2 = \frac{c}{a}$$

Proof   We note that the solutions to the equation are given by the quadratic formula

$$r_1 = \frac{-b + \sqrt{b^2 - 4ac}}{2a} \qquad \text{and} \qquad r_2 = \frac{-b - \sqrt{b^2 - 4ac}}{2a}$$

Thus,

$$r_1 + r_2 = \frac{-b + \sqrt{b^2 - 4ac}}{2a} + \frac{-b - \sqrt{b^2 - 4ac}}{2a}$$

$$= \frac{-b + \sqrt{b^2 - 4ac} - b - \sqrt{b^2 - 4ac}}{2a} \qquad \begin{array}{l}\text{Keep the denominator and}\\\text{add the numerators.}\end{array}$$

$$= -\frac{2b}{2a}$$

$$= -\frac{b}{a}$$

and

$$r_1 r_2 = \frac{-b + \sqrt{b^2 - 4ac}}{2a} \cdot \frac{-b - \sqrt{b^2 - 4ac}}{2a}$$

$$= \frac{b^2 - (b^2 - 4ac)}{4a^2}$$ Multiply the numerators and multiply the denominators.

$$= \frac{b^2 - b^2 + 4ac}{4a^2}$$

$$= \frac{4ac}{4a^2} \qquad\qquad b^2 - b^2 = 0.$$

$$= \frac{c}{a}$$

It can also be shown that if

$$r_1 + r_2 = -\frac{b}{a} \qquad \text{and} \qquad r_1 r_2 = \frac{c}{a}$$

then r_1 and r_2 are solutions of $ax^2 + bx + c = 0$. We can use this fact to check the solutions of quadratic equations.

EXAMPLE 7 Show that $\frac{3}{2}$ and $-\frac{1}{3}$ are solutions of $6x^2 - 7x - 3 = 0$.

Solution Since $a = 6$, $b = -7$, and $c = -3$, we have

$$-\frac{b}{a} = -\frac{-7}{6} = \frac{7}{6} \qquad \text{and} \qquad \frac{c}{a} = \frac{-3}{6} = -\frac{1}{2}$$

Since $\frac{3}{2} + \left(-\frac{1}{3}\right) = \frac{7}{6}$ and $\left(\frac{3}{2}\right)\left(-\frac{1}{3}\right) = -\frac{1}{2}$, these numbers are solutions. Solve the equation to verify that the roots are $\frac{3}{2}$ and $-\frac{1}{3}$.

Self Check Are $-\frac{3}{2}$ and $\frac{1}{3}$ solutions of $6x^2 + 7x - 3 = 0$?

Self Check Answers

1. real numbers that are irrational and unequal   **2.** nonreal numbers that are complex conjugates   **3.** 4   **4.** 4
5. $3, -\dfrac{12}{5}$ **6.** $a = \pm\sqrt{c^2 - b^2}$ **7.** yes

Orals *Find $b^2 - 4ac$ when*

1. $a = 1, b = 1, c = 1$ **2.** $a = 2, b = 1, c = 1$

Determine the type of solutions for

3. $x^2 - 4x + 1 = 0$ **4.** $8x^2 - x + 2 = 0$

Are the following numbers solutions of $x^2 - 7x + 6 = 0$?

5. $1, 5$ **6.** $1, 6$

8.3 EXERCISES

REVIEW *Solve each equation.*

1. $\dfrac{1}{4} + \dfrac{1}{t} = \dfrac{1}{2t}$ **2.** $\dfrac{p-3}{3p} + \dfrac{1}{2p} = \dfrac{1}{4}$

3. Find the slope of the line passing through $(-2, -4)$ and $(3, 5)$.

4. Write the equation of the line passing through $(-2, -4)$ and $(3, 5)$ in general form.

VOCABULARY AND CONCEPTS *Consider the equation $ax^2 + bx + c = 0$ $(a \neq 0)$, and fill in the blanks.*

5. The discriminant is _____.

6. If $b^2 - 4ac < 0$, the solutions of the equation are nonreal complex _____.

7. If $b^2 - 4ac$ is a nonzero perfect square, the solutions are _____ numbers and _____.

8. If r_1 and r_2 are the solutions of the equation, then $r_1 + r_2 =$ ___ and $r_1 r_2 =$ ___.

PRACTICE *Use the discriminant to determine what type of solutions exist for each quadratic equation.* **Do not solve the equation.**

9. $4x^2 - 4x + 1 = 0$ **10.** $6x^2 - 5x - 6 = 0$

11. $5x^2 + x + 2 = 0$ **12.** $3x^2 + 10x - 2 = 0$

13. $2x^2 = 4x - 1$ **14.** $9x^2 = 12x - 4$

15. $x(2x - 3) = 20$ **16.** $x(x - 3) = -10$

Find the values of k that will make the solutions of each given quadratic equation equal.

17. $x^2 + kx + 9 = 0$

18. $kx^2 - 12x + 4 = 0$

19. $9x^2 + 4 = -kx$

20. $9x^2 - kx + 25 = 0$

21. $(k - 1)x^2 + (k - 1)x + 1 = 0$

22. $(k + 3)x^2 + 2kx + 4 = 0$

23. $(k + 4)x^2 + 2kx + 9 = 0$

24. $(k + 15)x^2 + (k - 30)x + 4 = 0$

25. Use the discriminant to determine whether the solutions of $1{,}492x^2 + 1{,}776x - 1{,}984 = 0$ are real numbers.

26. Use the discriminant to determine whether the solutions of $1{,}776x^2 - 1{,}492x + 1{,}984 = 0$ are real numbers.

27. Determine k such that the solutions of $3x^2 + 4x = k$ are nonreal complex numbers.

28. Determine k such that the solutions of $kx^2 - 4x = 7$ are nonreal complex numbers.

Solve each equation.

29. $x^4 - 17x^2 + 16 = 0$ **30.** $x^4 - 10x^2 + 9 = 0$

31. $x^4 - 3x^2 = -2$ **32.** $x^4 - 29x^2 = -100$

33. $x^4 = 6x^2 - 5$ **34.** $x^4 = 8x^2 - 7$

35. $2x^4 - 10x^2 = -8$ **36.** $3x^4 + 12 = 15x^2$

37. $2x^4 + 24 = 26x^2$ **38.** $4x^4 = -9 + 13x^2$

39. $2x + x^{1/2} - 3 = 0$ **40.** $2x - x^{1/2} - 1 = 0$

41. $3x + 5x^{1/2} + 2 = 0$ **42.** $3x - 4x^{1/2} + 1 = 0$

43. $x^{2/3} + 5x^{1/3} + 6 = 0$ **44.** $x^{2/3} - 7x^{1/3} + 12 = 0$

45. $x^{2/3} - 2x^{1/3} - 3 = 0$ **46.** $x^{2/3} + 4x^{1/3} - 5 = 0$

47. $x + 5 + \dfrac{4}{x} = 0$ **48.** $x - 4 + \dfrac{3}{x} = 0$

49. $x + 1 = \dfrac{20}{x}$ **50.** $x + \dfrac{15}{x} = 8$

51. $\dfrac{1}{x-1} + \dfrac{3}{x+1} = 2$ **52.** $\dfrac{6}{x-2} - \dfrac{12}{x-1} = -1$

53. $\dfrac{1}{x+2} + \dfrac{24}{x+3} = 13$ **54.** $\dfrac{3}{x} + \dfrac{4}{x+1} = 2$

55. $x^{-4} - 2x^{-2} + 1 = 0$ **56.** $4x^{-4} + 1 = 5x^{-2}$

57. $x + \dfrac{2}{x-2} = 0$ **58.** $x + \dfrac{x+5}{x-3} = 0$

Solve each equation for the indicated variable.

59. $x^2 + y^2 = r^2$ for x

60. $x^2 + y^2 = r^2$ for y

61. $I = \dfrac{k}{d^2}$ for d

62. $V = \dfrac{1}{3}\pi r^2 h$ for r

63. $xy^2 + 3xy + 7 = 0$ for y

64. $kx = ay - x^2$ for x

65. $\sigma = \sqrt{\dfrac{\Sigma x^2}{N} - \mu^2}$ for μ^2

66. $\sigma = \sqrt{\dfrac{\Sigma x^2}{N} - \mu^2}$ for N

Solve each equation and verify that the sum of the solutions is $-\frac{b}{a}$ and that the product of the solutions is $\frac{c}{a}$.

67. $12x^2 - 5x - 2 = 0$ **68.** $8x^2 - 2x - 3 = 0$

69. $2x^2 + 5x + 1 = 0$ **70.** $3x^2 + 9x + 1 = 0$

71. $3x^2 - 2x + 4 = 0$ **72.** $2x^2 - x + 4 = 0$

73. $x^2 + 2x + 5 = 0$ **74.** $x^2 - 4x + 13 = 0$

WRITING

75. Describe how to predict what type of solutions the equation $3x^2 - 4x + 5 = 0$ will have.

76. How is the discriminant related to the quadratic formula?

SOMETHING TO THINK ABOUT

77. Can a quadratic equation with integer coefficients have one real and one complex solution? Why?

78. Can a quadratic equation with complex coefficients have one real and one complex solution? Why?

8.4   Graphs of Quadratic Functions

▮ Quadratic Functions ▮ Graphs of $f(x) = ax^2$
▮ Graphs of $f(x) = ax^2 + c$ ▮ Graphs of $f(x) = a(x - h)^2$
▮ Graphs of $f(x) = a(x - h)^2 + k$ ▮ Graphs of $f(x) = ax^2 + bx + c$
▮ Problem Solving ▮ The Variance

Getting Ready   If $y = f(x) = 3x^2 + x - 2$, *find each value.*

1. $f(0)$ **2.** $f(1)$ **3.** $f(-1)$ **4.** $f(-2)$

If $x = -\frac{b}{2a}$, *find x when a and b have the following values.*

5. $a = 3$ and $b = -6$ **6.** $a = 5$ and $b = -40$

In this section, we consider graphs of second-degree polynomial functions, called *quadratic functions.*

Quadratic Functions

The graph shown in Figure 8-3 shows the height (in relation to time) of a toy rocket launched straight up into the air.

 Comment Note that the graph describes the height of the rocket, not the path of the rocket. The rocket goes straight up and comes straight down.

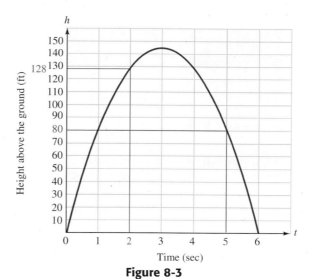

Figure 8-3

From the graph, we can see that the height of the rocket 2 seconds after it was launched is about 128 feet and that the height of the rocket 5 seconds after it was launched is 80 feet.

The parabola shown in Figure 8-3 is the graph of a *quadratic function*.

Quadratic Functions

A **quadratic function** is a second-degree polynomial function of the form

$$y = f(x) = ax^2 + bx + c \quad (a \neq 0)$$

where a, b, and c are real numbers.

We begin the discussion of graphing quadratic functions by considering the graph of $f(x) = ax^2 + bx + c$, where $b = 0$ and $c = 0$.

Graphs of $f(x) = ax^2$

EXAMPLE 1 Graph: **a.** $f(x) = x^2$, **b.** $g(x) = 3x^2$, and **c.** $h(x) = \dfrac{1}{3}x^2$.

Solution We can make a table of ordered pairs that satisfy each equation, plot each point, and join them with a smooth curve, as in Figure 8-4. We note that the graph of $h(x) = \frac{1}{3}x^2$ is wider than the graph of $f(x) = x^2$, and that the graph of $g(x) = 3x^2$ is narrower than the graph of $f(x) = x^2$. In the function $f(x) = ax^2$, the smaller the value of $|a|$, the wider the graph.

$$f(x) = x^2$$

x	$f(x)$	$(x, f(x))$
-2	4	$(-2, 4)$
-1	1	$(-1, 1)$
0	0	$(0, 0)$
1	1	$(1, 1)$
2	4	$(2, 4)$

$$g(x) = 3x^2$$

x	$g(x)$	$(x, g(x))$
-2	12	$(-2, 12)$
-1	3	$(-1, 3)$
0	0	$(0, 0)$
1	3	$(1, 3)$
2	12	$(2, 12)$

$$h(x) = \tfrac{1}{3}x^2$$

x	$h(x)$	$(x, h(x))$
-2	$\frac{4}{3}$	$\left(-2, \frac{4}{3}\right)$
-1	$\frac{1}{3}$	$\left(-1, \frac{1}{3}\right)$
0	0	$(0, 0)$
1	$\frac{1}{3}$	$\left(1, \frac{1}{3}\right)$
2	$\frac{4}{3}$	$\left(2, \frac{4}{3}\right)$

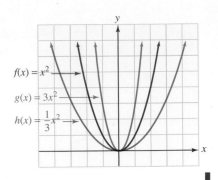

Figure 8-4

If we consider the graph of $f(x) = -3x^2$, we will see that it opens downward and has the same shape as the graph of $g(x) = 3x^2$.

EXAMPLE 2   Graph: $f(x) = -3x^2$.

Solution   We make a table of ordered pairs that satisfy the equation, plot each point, and join them with a smooth curve, as in Figure 8-5.

$$f(x) = -3x^2$$

x	$f(x)$	$(x, f(x))$
-2	-12	$(-2, -12)$
-1	-3	$(-1, -3)$
0	0	$(0, 0)$
1	-3	$(1, -3)$
2	-12	$(2, -12)$

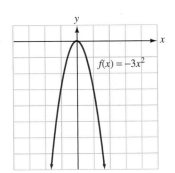

Figure 8-5

Self Check   Graph: $f(x) = -\tfrac{1}{3}x^2$.

The graphs of quadratic functions are called **parabolas.** They open upward when $a > 0$ and downward when $a < 0$. The lowest point of a parabola that opens upward, or the highest point of a parabola that opens downward, is called the **vertex** of the parabola. The vertex of the parabola shown in Figure 8-5 is the point $(0, 0)$.

The vertical line, called an **axis of symmetry,** that passes through the vertex divides the parabola into two congruent halves. The axis of symmetry of the parabola shown in Figure 8-5 is the y-axis.

Graphs of $f(x) = ax^2 + c$

EXAMPLE 3   Graph: **a.** $f(x) = 2x^2$,   **b.** $g(x) = 2x^2 + 3$,   and   **c.** $h(x) = 2x^2 - 3$.

Solution   We make a table of ordered pairs that satisfy each equation, plot each point, and join them with a smooth curve, as in Figure 8-6. We note that the graph of $g(x) = 2x^2 + 3$ is identical to the graph of $f(x) = 2x^2$, except that it has been translated 3 units upward. The graph of $h(x) = 2x^2 - 3$ is identical to the graph of $f(x) = 2x^2$, except that it has been translated 3 units downward.

$f(x) = 2x^2$

x	$f(x)$	$(x, f(x))$
-2	8	$(-2, 8)$
-1	2	$(-1, 2)$
0	0	$(0, 0)$
1	2	$(1, 2)$
2	8	$(2, 8)$

$g(x) = 2x^2 + 3$

x	$g(x)$	$(x, g(x))$
-2	11	$(-2, 11)$
-1	5	$(-1, 5)$
0	3	$(0, 3)$
1	5	$(1, 5)$
2	11	$(2, 11)$

$h(x) = 2x^2 - 3$

x	$h(x)$	$(x, h(x))$
-2	5	$(-2, 5)$
-1	-1	$(-1, -1)$
0	-3	$(0, -3)$
1	-1	$(1, -1)$
2	5	$(2, 5)$

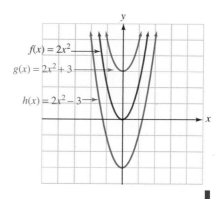

Figure 8-6

The results of Example 3 confirm the following facts, which we have previously discussed.

Vertical Translations of Graphs

If f is a function and k is a positive number, then
- The graph of $y = f(x) + k$ is identical to the graph of $y = f(x)$, except that it is translated k units upward.
- The graph of $y = f(x) - k$ is identical to the graph of $y = f(x)$, except that it is translated k units downward.

Graphs of $f(x) = a(x - h)^2$

EXAMPLE 4   Graph: **a.** $f(x) = 2x^2$,   **b.** $g(x) = 2(x - 3)^2$,   and   **c.** $h(x) = 2(x + 3)^2$.

Solution   We make a table of ordered pairs that satisfy each equation, plot each point, and join them with a smooth curve, as in Figure 8-7. We note that the graph of $g(x) = 2(x - 3)^2$ is identical to the graph of $f(x) = 2x^2$, except that it has been translated 3 units to the right. The graph of $h(x) = 2(x + 3)^2$ is identical to the graph of $f(x) = 2x^2$, except that it has been translated 3 units to the left.

$f(x) = 2x^2$

x	$f(x)$	$(x, f(x))$
-2	8	$(-2, 8)$
-1	2	$(-1, 2)$
0	0	$(0, 0)$
1	2	$(1, 2)$
2	8	$(2, 8)$

$g(x) = 2(x - 3)^2$

x	$g(x)$	$(x, g(x))$
1	8	$(1, 8)$
2	2	$(2, 2)$
3	0	$(3, 0)$
4	2	$(4, 2)$
5	8	$(5, 8)$

$h(x) = 2(x + 3)^2$

x	$h(x)$	$(x, h(x))$
-5	8	$(-5, 8)$
-4	2	$(-4, 2)$
-3	0	$(-3, 0)$
-2	2	$(-2, 2)$
-1	8	$(-1, 8)$

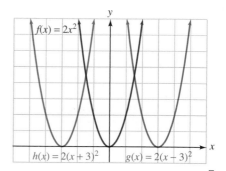

Figure 8-7

The results of Example 4 confirm the following facts.

Horizontal Translations of Graphs

If f is a function and h is a positive number, then
- The graph of $y = f(x - h)$ is identical to the graph of $y = f(x)$, except that it is translated h units to the right.
- The graph of $y = f(x + h)$ is identical to the graph of $y = f(x)$, except that it is translated h units to the left.

Graphs of $f(x) = a(x - h)^2 + k$

EXAMPLE 5   Graph: $f(x) = 2(x - 3)^2 - 4$.

Solution   The graph of $f(x) = 2(x - 3)^2 - 4$ is identical to the graph of $g(x) = 2(x - 3)^2$, except that it has been translated 4 units downward. The graph of $g(x) = 2(x - 3)^2$ is identical to the graph of $h(x) = 2x^2$, except that it has been translated 3 units to the right. Thus, to graph $f(x) = 2(x - 3)^2 - 4$, we can graph $h(x) = 2x^2$ and shift it 3 units to the right and then 4 units downward, as shown in Figure 8-8.

The vertex of the graph is the point $(3, -4)$, and the axis of symmetry is the line $x = 3$.

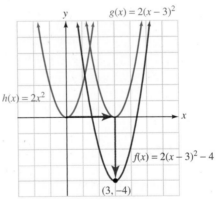

Figure 8-8

Self Check   Graph: $f(x) = 2(x + 3)^2 + 1$.

The results of Example 5 confirm the following facts.

Vertex and Axis of Symmetry of a Parabola

The graph of the function

$$y = f(x) = a(x - h)^2 + k \quad (a \neq 0)$$

is a parabola with vertex at (h, k). (See Figure 8-9.)

The parabola opens upward when $a > 0$ and downward when $a < 0$. The axis of symmetry is the line $x = h$.

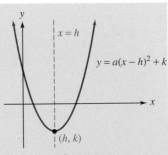

Figure 8-9

Graphs of $f(x) = ax^2 + bx + c$

To graph functions of the form $f(x) = ax^2 + bx + c$, we can complete the square to write the function in the form $f(x) = a(x - h)^2 + k$.

 EXAMPLE 6   Graph: $f(x) = 2x^2 - 4x - 1$.

Solution   We complete the square on x to write the function in the form $f(x) = a(x - h)^2 + k$.

$$f(x) = 2x^2 - 4x - 1$$

$$f(x) = 2(x^2 - 2x) - 1 \qquad \text{Factor 2 from } 2x^2 - 4x.$$

$$f(x) = \mathbf{2}(x^2 - 2x \mathbf{+ 1}) - 1 \mathbf{- 2} \qquad \begin{array}{l}\text{Complete the square on } x. \text{ Since this adds 2 to the} \\ \text{right-hand side, we also subtract 2 from the right-} \\ \text{hand side.}\end{array}$$

(1) $$f(x) = 2(x - 1)^2 - 3 \qquad \text{Factor } x^2 - 2x + 1 \text{ and combine like terms.}$$

From Equation 1, we can see that the vertex will be at the point $(1, -3)$. We can plot the vertex and a few points on either side of the vertex and draw the graph, which appears in Figure 8-10.

$$f(x) = 2x^2 - 4x - 1$$

x	$f(x)$	$(x, f(x))$
-1	5	$(-1, 5)$
0	-1	$(0, -1)$
1	-3	$(1, -3)$
2	-1	$(2, -1)$
3	5	$(3, 5)$

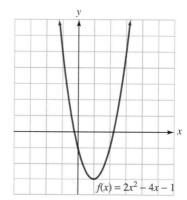

$$f(x) = 2x^2 - 4x - 1$$

Figure 8-10

Self Check   Graph: $f(x) = 2x^2 - 4x + 1$.

Accent on Technology   **GRAPHING QUADRATIC FUNCTIONS**

To use a graphing calculator to graph $f(x) = 0.7x^2 + 2x - 3.5$, we can use window settings of $[-10, 10]$ for x and $[-10, 10]$ for y, enter the function, and press GRAPH to obtain Figure 8-11(a).

To find approximate coordinates of the vertex of the graph, we trace to move the cursor near the lowest point of the graph as shown in Figure 8-11(b). By zooming in twice and tracing as in Figure 8-11(c), we can see that the vertex is a point whose coordinates are approximately $(-1.422872, -4.928549)$.

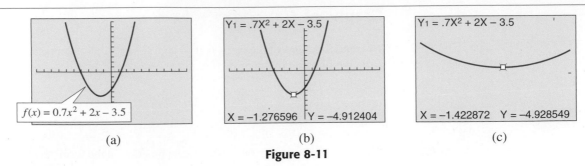

Figure 8-11

The solutions of the quadratic equation $0.7x^2 + 2x - 3.5 = 0$ are the numbers x that will make $f(x) = 0$ in the function $f(x) = 0.7x^2 + 2x - 3.5$. To approximate these numbers, we graph the function as shown in Figure 8-12(a) and find the x-intercepts by tracing to move the cursor near each x-intercept, as in Figures 8-12(b) and 8-12(c). From the graphs, we can read the approximate value of the x-coordinate of each x-intercept. For better results, we can zoom in.

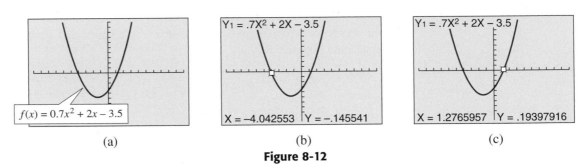

Figure 8-12

We can also solve the equation by using the ZERO command found in the CALC menu. We first graph the function $f(x) = 0.7x^2 + 2x - 3.5$ as in Figure 8-13(a). We then select 2 in the CALC menu to get Figure 8-13(b). We enter -5 for a left guess and press ENTER. We then enter -2 for a right guess and press ENTER. After pressing ENTER again, we will obtain Figure 8-13(c). We can find the second solution in a similar way.

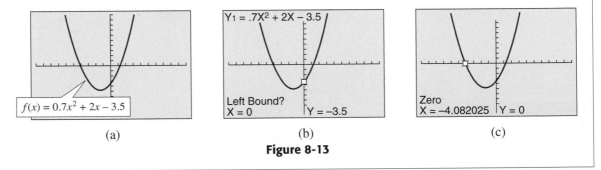

Figure 8-13

Problem Solving

EXAMPLE 7 **Ballistics** The ball shown in Figure 8-14(a) is thrown straight up with a velocity of 128 feet per second. The function $s = h(t) = -16t^2 + 128t$ gives the relation between s (the number of feet the ball is above the ground) and t (the time measured in seconds). How high will the ball go, and when will it hit the ground?

Solution The graph of $s = -16t^2 + 128t$ is a parabola. Since the coefficient of t^2 is negative, it opens downward, and the maximum height of the ball is given by the s-coordinate of the vertex of the parabola. We can find the coordinates of the vertex by completing the square:

$$
\begin{aligned}
s &= -16t^2 + 128t \\
&= -16(t^2 - 8t) && \text{Factor out } -16. \\
&= -16(t^2 - 8t \mathbf{+ 16 - 16}) && \text{Add and subtract 16.} \\
&= -16(t^2 - 8t + 16) + 256 && (-16)(-16) = 256. \\
&= -16(t - \mathbf{4})^2 + \mathbf{256} && \text{Factor } t^2 - 8t + 16.
\end{aligned}
$$

From the result, we can see that the coordinates of the vertex are $(\mathbf{4}, \mathbf{256})$. Since $t = 4$ and $s = 256$ are the coordinates of the vertex, the ball reaches a maximum height of 256 feet in 4 seconds.

From the graph, we can see that the ball will hit the ground in 8 seconds, because the height is 0 when $t = 8$.

To solve this problem with a graphing calculator with window settings of $[0, 10]$ for x and $[0, 300]$ for y, we graph the function $h(t) = -16t^2 + 128t$ to get the graph in Figure 8-14(b). By using trace and zoom, we can determine that the ball reaches a height of 256 feet in 4 seconds and that the ball will hit the ground in 8 seconds.

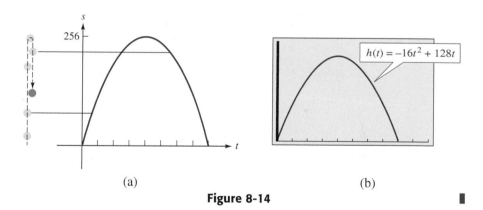

(a)   (b)

Figure 8-14

EXAMPLE 8 **Maximizing area** A man wants to build the rectangular pen shown in Figure 8-15(a) to house his dog. If he uses one side of his barn, find the maximum area that he can enclose with 80 feet of fencing.

Solution If we let the width of the pen be w, the length is represented by $80 - 2w$.
We can find the maximum value of A as follows:

$$
\begin{aligned}
A &= (80 - 2w)w && A = lw. \\
&= 80w - 2w^2 && \text{Remove parentheses.} \\
&= -2(w^2 - 40w) && \text{Factor out } -2. \\
&= -2(w^2 - 40w \mathbf{+ 400 - 400}) && \text{Subtract and add 400.} \\
&= -2(w^2 - 40w + 400) + 800 && -2(-400) = 800. \\
&= -2(w - 20)^2 + 800 && \text{Factor } w^2 - 40w + 400.
\end{aligned}
$$

Thus, the coordinates of the vertex of the graph of the quadratic function are (20, 800), and the maximum area is 800 square feet.

To solve this problem using a graphing calculator with window settings of [0, 50] for x and [0, 1,000] for y, we graph the function $A(w) = -2w^2 + 80w$ to get the graph in Figure 8-15(b). By using trace and zoom, we can determine that the maximum area is 800 square feet when the width is 20 feet.

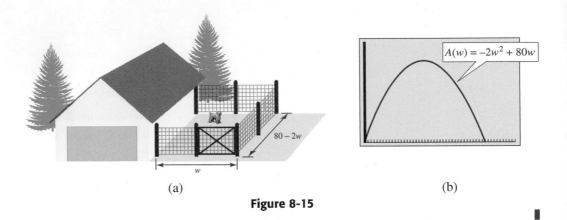

(a)   (b)

Figure 8-15

The Variance

In statistics, the square of the standard deviation is called the **variance.**

EXAMPLE 9   If p is the probability that a person selected at random has AIDS, then $1 - p$ is the probability that the person does not have AIDS. If 100 people in Minneapolis are randomly sampled, we know from statistics that the variance of this type of sample distribution will be $100p(1 - p)$. What value of p will maximize the variance?

Solution   The variance is given by the function

$$v(p) = 100p(1 - p) \qquad \text{or} \qquad v(p) = -100p^2 + 100p$$

Since all probabilities have values between 0 to 1, including 0 and 1, we use window settings of [0, 1] for x when graphing the function $v(p) = -100p^2 + 100p$ on a graphing calculator. If we also use window settings of [0, 30] for y, we will obtain the graph shown in Figure 8-16(a). After using trace and zoom to obtain Figure 8-16(b), we can see that a probability of 0.5 will give the maximum variance.

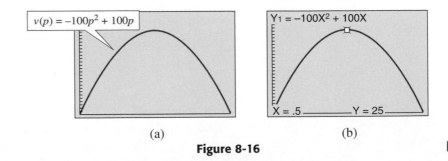

(a)   (b)

Figure 8-16

Self Check Answers

2.

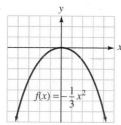

$f(x) = -\frac{1}{3}x^2$

5.

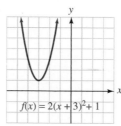

$f(x) = 2(x + 3)^2 + 1$

6.

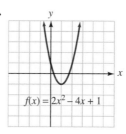

$f(x) = 2x^2 - 4x + 1$

Orals   *Tell whether the graph of each equation opens up or down.*

1. $y = -3x^2 + x - 5$ **2.** $y = 4x^2 + 2x - 3$

3. $y = 2(x - 3)^2 - 1$ **4.** $y = -3(x + 2)^2 + 2$

Find the vertex of the parabola determined by each equation.

5. $y = 2(x - 3)^2 - 1$ **6.** $y = -3(x + 2)^2 + 2$

8.4   EXERCISES

REVIEW   *Find the value of x.*

1.

$(3x + 5)°$ $(5x - 15)°$

2. Lines r and s are parallel.

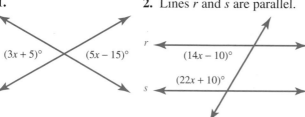

$(14x - 10)°$

$(22x + 10)°$

3. Madison and St. Louis are 385 miles apart. One train leaves Madison and heads toward St. Louis at the rate of 30 mph. Three hours later, a second train leaves Madison, bound for St. Louis. If the second train travels at the rate of 55 mph, in how many hours will the faster train overtake the slower train?

4. A woman invests $25,000, some at 7% annual interest and the rest at 8%. If the annual income from both investments is $1,900, how much is invested at the higher rate?

VOCABULARY AND CONCEPTS   *Fill in the blanks.*

5. A quadratic function is a second-degree polynomial function that can be written in the form _____, where _____.

6. The graphs of quadratic functions are called _____.

7. The highest (or the lowest) point on a parabola is called the _____.

8. A vertical line that divides a parabola into two halves is called an _____ of symmetry.

9. The graph of $y = f(x) + k$ $(k > 0)$ is identical to the graph of $y = f(x)$, except that it is translated k units _____.

10. The graph of $y = f(x) - k$ $(k > 0)$ is identical to the graph of $y = f(x)$, except that it is translated k units _____.

11. The graph of $y = f(x - h)$ $(h > 0)$ is identical to the graph of $y = f(x)$, except that it is translated h units _____.

12. The graph of $y = f(x + h)$ $(h > 0)$ is identical to the graph of $y = f(x)$, except that it is translated h units _____.

13. The graph of $y = f(x) = ax^2 + bx + c$ $(a \neq 0)$ opens _____ when $a > 0$.

14. In statistics, the square of the standard deviation is called the _____.

PRACTICE   *Graph each function.*

15. $f(x) = x^2$

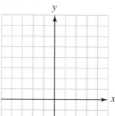

16. $f(x) = -x^2$

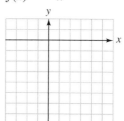

17. $f(x) = x^2 + 2$

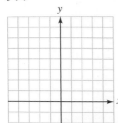

18. $f(x) = x^2 - 3$

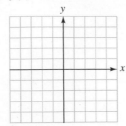

19. $f(x) = -(x - 2)^2$

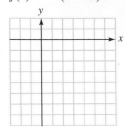

20. $f(x) = (x + 2)^2$

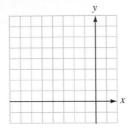

21. $f(x) = (x - 3)^2 + 2$

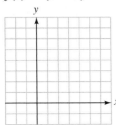

22. $f(x) = (x + 1)^2 - 2$

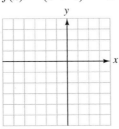

23. $f(x) = x^2 + x - 6$

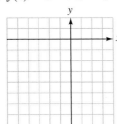

24. $f(x) = x^2 - x - 6$

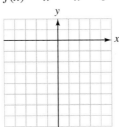

25. $f(x) = -2x^2 + 4x + 1$

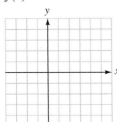

26. $f(x) = -2x^2 + 4x + 3$

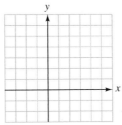

Find the coordinates of the vertex and the axis of symmetry of the graph of each equation. If necessary, complete the square on x to write the equation in the form $y = a(x - h)^2 + k$. **Do not graph the equation.**

27. $y = (x - 1)^2 + 2$

28. $y = 2(x - 2)^2 - 1$

29. $y = 2(x + 3)^2 - 4$

30. $y = -3(x + 1)^2 + 3$

31. $y = -3x^2$

32. $y = 3x^2 - 3$

33. $y = 2x^2 - 4x$

34. $y = 3x^2 + 6x$

35. $y = -4x^2 + 16x + 5$

36. $y = 5x^2 + 20x + 25$

37. $y - 7 = 6x^2 - 5x$

38. $y - 2 = 3x^2 + 4x$

39. The equation $y - 2 = (x - 5)^2$ represents a quadratic function whose graph is a parabola. Find its vertex.

40. Show that $y = ax^2$, where $a \neq 0$, represents a quadratic function whose vertex is at the origin.

Use a graphing calculator to find the coordinates of the vertex of the graph of each quadratic function. Give results to the nearest hundredth.

41. $y = 2x^2 - x + 1$

42. $y = x^2 + 5x - 6$

43. $y = 7 + x - x^2$

44. $y = 2x^2 - 3x + 2$

Use a graphing calculator to solve each equation. If a result is not exact, give the result to the nearest hundredth.

45. $x^2 + x - 6 = 0$

46. $2x^2 - 5x - 3 = 0$

47. $0.5x^2 - 0.7x - 3 = 0$

48. $2x^2 - 0.5x - 2 = 0$

APPLICATIONS

49. Ballistics   If a ball is thrown straight up with an initial velocity of 48 feet per second, its height s after t seconds is given by the equation $s = 48t - 16t^2$. Find the maximum height attained by the ball and the time it takes for the ball to return to Earth.

50. Ballistics From the top of the building, a ball is thrown straight up with an initial velocity of 32 feet per second. The equation $s = -16t^2 + 32t + 48$ gives the height s of the ball t seconds after it is thrown. Find the maximum height reached by the ball and the time it takes for the ball to hit the ground.

51. Maximizing area Find the dimensions of the rectangle of maximum area that can be constructed with 200 feet of fencing. Find the maximum area.

52. Fencing a field A farmer wants to fence in three sides of a rectangular field with 1,000 feet of fencing. The other side of the rectangle will be a river. If the enclosed area is to be maximum, find the dimensions of the field.

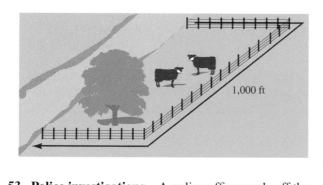

53. Police investigations A police officer seals off the scene of a car collision using a roll of yellow police tape that is 300 feet long. What dimensions should be used to seal off the maximum rectangular area around the collision? What is the maximum area?

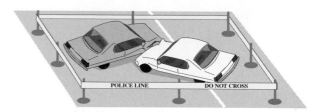

54. Operating costs The cost C in dollars of operating a certain concrete-cutting machine is related to the number of minutes n the machine is run by the function

$$C(n) = 2.2n^2 - 66n + 655$$

For what number of minutes is the cost of running the machine a minimum? What is the minimum cost?

Use a graphing calculator to help solve each problem.

55. Maximizing revenue The revenue R received for selling x stereos is given by the equation

$$R = -\frac{x^2}{1,000} + 10x$$

Find the number of stereos that must be sold to obtain the maximum revenue.

56. Maximizing revenue In Exercise 55, find the maximum revenue.

57. Maximizing revenue The revenue received for selling x radios is given by the formula

$$R = -\frac{x^2}{728} + 9x$$

How many radios must be sold to obtain the maximum revenue? Find the maximum revenue.

58. Maximizing revenue The revenue received for selling x stereos is given by the formula

$$R = -\frac{x^2}{5} + 80x - 1,000$$

How many stereos must be sold to obtain the maximum revenue? Find the maximum revenue.

59. Maximizing revenue When priced at $30 each, a toy has annual sales of 4,000 units. The manufacturer estimates that each $1 increase in cost will decrease sales by 100 units. Find the unit price that will maximize total revenue. (*Hint:* Total revenue = price · the number of units sold.)

60. Maximizing revenue When priced at $57, one type of camera has annual sales of 525 units. For each $1 the camera is reduced in price, management expects to sell an additional 75 cameras. Find the unit price that will maximize total revenue. (*Hint:* Total revenue = price · the number of units sold.)

61. Finding the variance If p is the probability that a person sampled at random has high blood pressure, $1 - p$ is the probability that the person doesn't. If 50 people are sampled at random, the variance of the sample will be $50p(1 - p)$. What two probabilities p will give a variance of 9.375?

62. Finding the variance If p is the probability that a person sampled at random smokes, then $1 - p$ is the probability that the person doesn't. If 75 people are sampled at random, the variance of the sample will be $75p(1 - p)$. What two probabilities p will give a variance of 12?

WRITING

63. The graph of $y = ax^2 + bx + c$ ($a \neq 0$) passes the vertical line test. Explain why this shows that the equation defines a function.

64. The graph of $x = y^2 - 2y$ is a parabola. Explain why its graph does not represent a function.

SOMETHING TO THINK ABOUT

65. Can you use a graphing calculator to find solutions of the equation $x^2 + x + 1 = 0$? What is the problem? How do you interpret the result?

66. Complete the square on x in the equation $y = ax^2 + bx + c$ and show that the vertex of the parabolic graph is the point with coordinates of

$$\left(-\frac{b}{2a}, c - \frac{b^2}{4a} \right)$$

8.5 Quadratic and Other Nonlinear Inequalities

▪ **Solving Quadratic Inequalities** ▪ **Solving Other Inequalities**
▪ **Graphs of Nonlinear Inequalities in Two Variables**

Getting Ready *Factor each trinomial.*

1. $x^2 + 2x - 15$ **2.** $x^2 - 3x + 2$

We have previously solved linear inequalities. We will now discuss how to solve quadratic and other inequalities.

Solving Quadratic Inequalities

Quadratic inequalities in one variable, say x, are inequalities that can be written in one of the following forms, where $a \neq 0$:

$$ax^2 + bx + c < 0 \qquad\qquad ax^2 + bx + c > 0$$
$$ax^2 + bx + c \leq 0 \qquad\qquad ax^2 + bx + c \geq 0$$

To solve one of these inequalities, we must find its solution set. For example, to solve

$$x^2 + x - 6 < 0$$

we must find the values of x that make the inequality true. To find these values, we can factor the trinomial to obtain

$$(x + 3)(x - 2) < 0$$

Since the product of $x + 3$ and $x - 2$ is to be less than 0, their values must be opposite in sign. This will happen when one of the factors is positive and the other is negative.

To keep track of the sign of $x + 3$, we can construct the following graph.

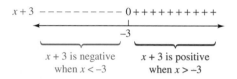

To keep track of the sign of $x - 2$, we can construct the following graph.

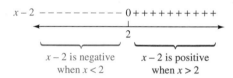

We can merge these graphs as shown in Figure 8-17 and note where the signs of the factors are opposite. This occurs in the interval $(-3, 2)$. Therefore, the product $(x + 3)(x - 2)$ will be less than 0 when

$$-3 < x < 2$$

The graph of the solution set is shown on the number line in the figure.

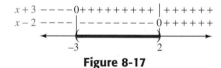

Figure 8-17

EXAMPLE 1   Solve: $x^2 + 2x - 3 \geq 0$.

Solution   We factor the trinomial to get $(x - 1)(x + 3)$ and construct a sign chart, as in Figure 8-18.

Figure 8-18

- $x - 1$ is 0 when $x = 1$, is positive when $x > 1$, and is negative when $x < 1$.
- $x + 3$ is 0 when $x = -3$, is positive when $x > -3$, and is negative when $x < -3$.

The product of $x - 1$ and $x + 3$ will be greater than 0 when the signs of the binomial factors are the same. This occurs in the intervals $(-\infty, -3)$ and $(1, \infty)$. The numbers -3 and 1 are also included, because they make the product equal to 0. Thus, the solution set is

$$(-\infty, -3] \cup [1, \infty) \qquad \text{or} \qquad x \leq -3 \text{ or } x \geq 1$$

The graph of the solution set is shown on the number line in Figure 8-18.

Self Check   Solve $x^2 + 2x - 15 > 0$ and graph the solution set.   ∎

Solving Other Inequalities

Making a sign chart is useful for solving many inequalities that are neither linear nor quadratic.

EXAMPLE 2   Solve: $\dfrac{1}{x} < 6$.

Solution   We subtract 6 from both sides to make the right-hand side equal to 0. We then find a common denominator and add the fractions:

$$\frac{1}{x} < 6$$

$$\frac{1}{x} - 6 < 0 \qquad \text{Subtract 6 from both sides.}$$

$$\frac{1}{x} - \frac{6x}{x} < 0 \qquad \text{Get a common denominator.}$$

$$\frac{1 - 6x}{x} < 0 \qquad \text{Subtract the numerators and keep the common denominator.}$$

We now make a sign chart, as in Figure 8-19.

- The denominator x is 0 when $x = 0$, is positive when $x > 0$, and is negative when $x < 0$.
- The numerator $1 - 6x$ is 0 when $x = \frac{1}{6}$, is positive when $x < \frac{1}{6}$, and is negative when $x > \frac{1}{6}$.

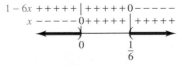

Figure 8-19

The fraction $\frac{1-6x}{x}$ will be less than 0 when the numerator and denominator are opposite in sign. This occurs in the interval

$$(-\infty, 0) \cup \left(\frac{1}{6}, \infty\right) \quad \text{or} \quad x < 0 \text{ or } x > \frac{1}{6}$$

The graph of this interval is shown in Figure 8-19.

Self Check Solve: $\frac{3}{x} > 5$. ∎

 Comment Since we don't know whether x is positive, 0, or negative, multiplying both sides of the inequality $\frac{1}{x} < 6$ by x is a three-case situation:

- If $x > 0$, then $1 < 6x$.
- If $x = 0$, then the fraction $\frac{1}{x}$ is undefined.
- If $x < 0$, then $1 > 6x$.

If you multiply both sides by x and solve $1 < 6x$, you are only considering one case and will get only part of the answer.

 EXAMPLE 3 Solve: $\dfrac{x^2 - 3x + 2}{x - 3} \geq 0$.

Solution We write the fraction with the numerator in factored form.

$$\frac{(x-2)(x-1)}{x-3} \geq 0$$

To keep track of the signs of the binomials, we construct the sign chart shown in Figure 8-20. The fraction will be positive in the intervals where all factors are positive, or where two factors are negative. The numbers 1 and 2 are included, because they make the numerator (and thus the fraction) equal to 0. The number 3 is not included, because it gives a 0 in the denominator.

The solution is the interval $[1, 2] \cup (3, \infty)$. The graph appears in Figure 8-20.

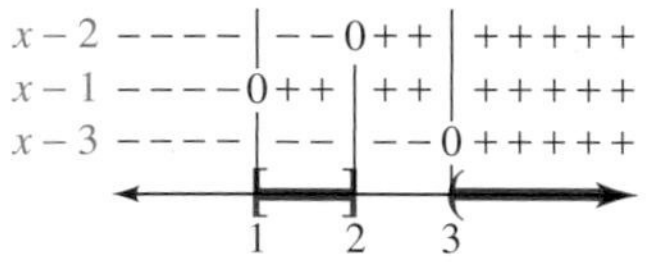

$$
\begin{array}{l}
x-2 \;\; ----\,|--0++|\;+++++ \\
x-1 \;\; ----0++|++\;|\;+++++ \\
x-3 \;\; ----\,|--\;\;|--0+++++ \\
\end{array}
$$

Figure 8-20

Self Check Solve: $\dfrac{x+2}{x^2 - 2x - 3} > 0$ and graph the solution set. ∎

EXAMPLE 4 Solve: $\dfrac{3}{x-1} < \dfrac{2}{x}$.

Solution We subtract $\frac{2}{x}$ from both sides to get 0 on the right-hand side and proceed as follows:

$$\frac{3}{x-1} < \frac{2}{x}$$

$$\frac{3}{x-1} - \frac{2}{x} < 0 \qquad \text{Subtract } \frac{2}{x} \text{ from both sides.}$$

$$\frac{3x}{(x-1)x} - \frac{2(x-1)}{x(x-1)} < 0 \qquad \text{Get a common denominator.}$$

$$\frac{3x - 2x + 2}{x(x-1)} < 0 \qquad \text{Keep the denominator and subtract the numerators.}$$

$$\frac{x+2}{x(x-1)} < 0 \qquad \text{Combine like terms.}$$

We can keep track of the signs of the three factors with the sign chart shown in Figure 8-21. The fraction will be negative in the intervals with either one or three negative factors. The numbers 0 and 1 are not included, because they give a 0 in the denominator, and the number -2 is not included, because it does not satisfy the inequality.

The solution is the interval $(-\infty, -2) \cup (0, 1)$, as shown in Figure 8-21.

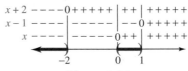

Figure 8-21

Self Check   Solve $\dfrac{2}{x+1} > \dfrac{1}{x}$ and graph the solution set.   ∎

Accent on Technology   **SOLVING INEQUALITIES**

To approximate the solutions of $x^2 + 2x - 3 \geq 0$ (Example 1) by graphing, we can use window settings of $[-10, 10]$ for x and $[-10, 10]$ for y and graph the quadratic function $y = x^2 + 2x - 3$, as in Figure 8-22. The solution of the inequality will be those numbers x for which the graph of $y = x^2 + 2x - 3$ lies above or on the x-axis. We can trace to find that this interval is $(-\infty, -3] \cup [1, \infty)$.

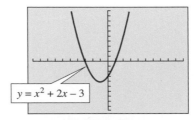

$y = x^2 + 2x - 3$

Figure 8-22

To approximate the solutions of $\frac{3}{x-1} < \frac{2}{x}$ (Example 4), we first write the inequality in the form

$$\frac{3}{x-1} - \frac{2}{x} < 0$$

Then we use window settings of $[-5, 5]$ for x and $[-3, 3]$ for y and graph the function $y = \frac{3}{x-1} - \frac{2}{x}$, as in Figure 8-23(a). The solution of the inequality will be those numbers x for which the graph lies below the x-axis.

We can trace to see that the graph is below the x-axis when x is less than -2. Since we cannot see the graph in the interval $0 < x < 1$, we redraw the graph using window settings of $[-1, 2]$ for x and $[-25, 10]$ for y. See Figure 8-23(b).

We can now see that the graph is below the x-axis in the interval $(0, 1)$. Thus, the solution of the inequality is the union of two intervals:

$$(-\infty, -2) \cup (0, 1)$$

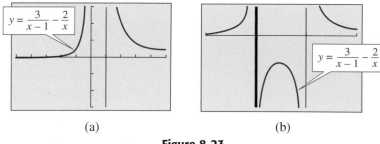

(a)   (b)

Figure 8-23

Graphs of Nonlinear Inequalities in Two Variables

We now consider the graphs of nonlinear inequalities in two variables.

EXAMPLE 5   Graph: $y < -x^2 + 4$.

Solution   The graph of $y = -x^2 + 4$ is the parabolic boundary separating the region representing $y < -x^2 + 4$ and the region representing $y > -x^2 + 4$.

We graph $y = -x^2 + 4$ as a broken parabola, because equality is not permitted. Since the coordinates of the origin satisfy the inequality $y < -x^2 + 4$, the point $(0, 0)$ is in the graph. The complete graph is shown in Figure 8-24.

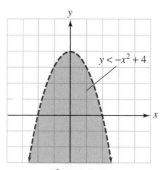

$y < -x^2 + 4$

Figure 8-24

Self Check   Graph: $y \geq -x^2 + 4$.

EXAMPLE 6   Graph: $x \leq |y|$.

Solution   We first graph $x = |y|$ as in Figure 8-25(a), using a solid line because equality is permitted. Since the origin is on the graph, we cannot use it as a test point. However, another point, such as $(1, 0)$, will do. We substitute 1 for x and 0 for y into the inequality to get

$$x \leq |y|$$
$$1 \leq |0|$$
$$1 \leq 0$$

Since $1 \leq 0$ is a false statement, the point $(1, 0)$ does not satisfy the inequality and is not part of the graph. Thus, the graph of $x \leq |y|$ is to the left of the boundary.

The complete graph is shown in Figure 8-25(b).

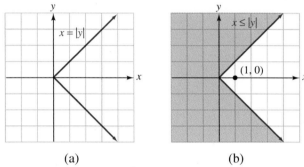

(a)   (b)

Figure 8-25

Self Check   Graph: $x \geq -|y|$.

Self Check Answers

1. $(-\infty, -5) \cup (3, \infty)$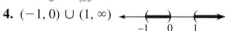

2. $\left(0, \dfrac{3}{5}\right)$

3. $(-2, -1) \cup (3, \infty)$

4. $(-1, 0) \cup (1, \infty)$

5.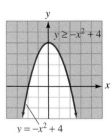

6.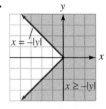

Orals   *Tell where* $x - 2$ *is*

1. 0   **2.** positive   **3.** negative

Tell where $x + 3$ *is*

4. 0   **5.** positive   **6.** negative

Multiply both sides of the equation $\frac{1}{x} < 2$ by x when x is

7. positive   **8.** negative

8.5   EXERCISES

REVIEW   *Write each expression as an equation.*

1. y varies directly with x.

2. y varies inversely with t.

3. t varies jointly with x and y.

4. d varies directly with t but inversely with u^2.

Find the slope of the graph of each equation.

5. $y = 3x - 4$ **6.** $\dfrac{2x - y}{5} = 8$

VOCABULARY AND CONCEPTS   *Fill in the blanks.*

7. When $x > 3$, the binomial $x - 3$ is _____ than zero.

8. When $x < 3$, the binomial $x - 3$ is _____ than zero.

9. If $x = 0$, the fraction $\frac{1}{x}$ is _____.

10. To keep track of the signs of factors in a product or quotient, we can use a _____ chart.

PRACTICE   *Solve each inequality. Give each result in interval notation and graph the solution set.*

11. $x^2 - 5x + 4 < 0$

12. $x^2 - 3x - 4 > 0$

13. $x^2 - 8x + 15 > 0$

14. $x^2 + 2x - 8 < 0$

15. $x^2 + x - 12 \leq 0$

16. $x^2 + 7x + 12 \geq 0$

17. $x^2 + 2x \geq 15$

18. $x^2 - 8x \leq -15$

19. $x^2 + 8x < -16$

20. $x^2 + 6x \geq -9$

21. $x^2 \geq 9$

22. $x^2 \geq 16$

23. $2x^2 - 50 < 0$

24. $3x^2 - 243 < 0$

25. $\dfrac{1}{x} < 2$

26. $\dfrac{1}{x} > 3$

27. $\dfrac{4}{x} \geq 2$

28. $-\dfrac{6}{x} < 12$

29. $-\dfrac{5}{x} < 3$

30. $\dfrac{4}{x} \geq 8$

31. $\dfrac{x^2 - x - 12}{x - 1} < 0$

32. $\dfrac{x^2 + x - 6}{x - 4} \geq 0$

33. $\dfrac{x^2 + x - 20}{x + 2} \geq 0$

34. $\dfrac{x^2 - 10x + 25}{x + 5} < 0$

35. $\dfrac{x^2 - 4x + 4}{x + 4} < 0$

36. $\dfrac{2x^2 - 5x + 2}{x + 2} > 0$

37. $\dfrac{6x^2 - 5x + 1}{2x + 1} > 0$

38. $\dfrac{6x^2 + 11x + 3}{3x - 1} < 0$

39. $\dfrac{3}{x - 2} < \dfrac{4}{x}$

40. $\dfrac{-6}{x + 1} \geq \dfrac{1}{x}$

41. $\dfrac{-5}{x + 2} \geq \dfrac{4}{2 - x}$

42. $\dfrac{-6}{x - 3} < \dfrac{5}{3 - x}$

43. $\dfrac{7}{x - 3} \geq \dfrac{2}{x + 4}$

44. $\dfrac{-5}{x - 4} < \dfrac{3}{x + 1}$

45. $\dfrac{x}{x + 4} \leq \dfrac{1}{x + 1}$

46. $\dfrac{x}{x + 9} \geq \dfrac{1}{x + 1}$

47. $\dfrac{x}{x + 16} > \dfrac{1}{x + 1}$

48. $\dfrac{x}{x + 25} < \dfrac{1}{x + 1}$

49. $(x + 2)^2 > 0$

50. $(x - 3)^2 < 0$

Use a graphing calculator to solve each inequality. Give the answer in interval notation.

51. $x^2 - 2x - 3 < 0$

52. $x^2 + x - 6 > 0$

53. $\dfrac{x + 3}{x - 2} > 0$

54. $\dfrac{3}{x} < 2$

Graph each inequality.

55. $y < x^2 + 1$

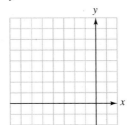

56. $y > x^2 - 3$

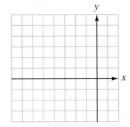

57. $y \leq x^2 + 5x + 6$

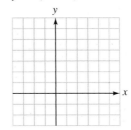

58. $y \geq x^2 + 5x + 4$

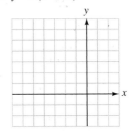

59. $y \geq (x - 1)^2$

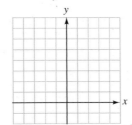

60. $y \leq (x + 2)^2$

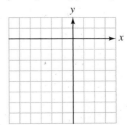

61. $-x^2 - y + 6 > -x$

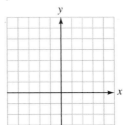

62. $y > (x + 3)(x - 2)$

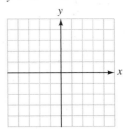

63. $y < |x + 4|$

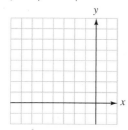

64. $y \geq |x - 3|$

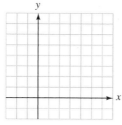

65. $y \leq -|x| + 2$

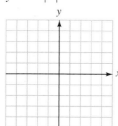

66. $y > |x| - 2$

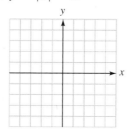

WRITING

67. Explain why $(x - 4)(x + 5)$ will be positive only when the signs of $x - 4$ and $x + 5$ are the same.

68. Tell how to find the graph of $y \geq x^2$.

SOMETHING TO THINK ABOUT

69. Under what conditions will the fraction $\dfrac{(x - 1)(x + 4)}{(x + 2)(x + 1)}$ be positive?

70. Under what conditions will the fraction $\dfrac{(x - 1)(x + 4)}{(x + 2)(x + 1)}$ be negative?

Projects

Project 1

Ballistics is the study of how projectiles fly. The general formula for the height above the ground of an object thrown straight up or down is given by the function

$$h(t) = -16t^2 + v_0t + h_0$$

where h is the object's height (in feet) above the ground t seconds after it is thrown. The initial velocity v_0 is the velocity with which the object is thrown, measured in feet per second. The initial height h_0 is the object's height (in feet) above the ground when it is thrown. (If $v_0 > 0$, the object is thrown upward; if $v_0 < 0$, the object is thrown downward.)

This formula takes into account the force of gravity, but disregards the force of air resistance. It is much more accurate for a smooth, dense ball than for a crumpled piece of paper.

One of the most popular acts of the Bungling Brothers Circus is the Amazing Glendo and his cannon-ball-catching act. A cannon fires a ball vertically into the air; Glendo, standing on a platform above the cannon, uses his catlike reflexes to catch the ball as it passes by on its way toward the roof of the big top. As the balls fly past, they are within Glendo's reach only during a two-foot interval of their upward path.

As an investigator for the company that insures the circus, you have been asked to find answers to the following questions. The answers will determine whether or not Bungling Brothers' insurance policy will be renewed.

a. In the first part of the act, cannonballs are fired from the end of a six-foot cannon with an initial velocity of 80 feet per second. Glendo catches one ball between 40 and 42 feet above the ground. Then he lowers his platform and catches another ball between 25 and 27 feet above the ground.

   i. Show that if Glendo missed a cannonball, it would hit the roof of the 56-foot-tall big top. How long would it take for a ball to hit the big top? To prevent this from happening, a special net near the roof catches and holds any missed cannonballs.

   ii. Find (to the nearest thousandth of a second) how long the cannonballs are within Glendo's reach for each of his catches. Which catch is easier?

Why does your answer make sense? Your company is willing to insure against injuries to Glendo if he has at least 0.025 second to make each catch. Should the insurance be offered?

b. For Glendo's grand finale, the special net at the roof of the big top is removed, making Glendo's catch more significant to the people in the audience, who worry that if Glendo misses, the tent will collapse around them. To make it even more dramatic, Glendo's arms are tied to restrict his reach to a one-foot interval of the ball's flight, and he stands on a platform just under the peak of the big top, so that his catch is made at the very last instant (between 54 and 55 feet above the ground). For this part of the act, however, Glendo has the cannon charged with less gunpowder, so that the muzzle velocity of the cannon is 56 feet per second. Show work to prove that Glendo's big finale is in fact his *easiest* catch, and that even if he misses, the big top is never in any danger of collapsing, so insurance should be offered against injury to the audience.

Project 2

The center of Sterlington is the intersection of Main Street (running east–west) and Due North Road (running north–south). The recreation area for the townspeople is Robin Park, a few blocks from there. The park is bounded on the south by Main Street and on every other side by Parabolic Boulevard, named for its distinctive shape. In fact, if Main Street and Due North Road were used as the axes of a rectangular coordinate system, Parabolic Boulevard would have the equation $y = -(x - 4)^2 + 5$, where each unit on the axes is 100 yards.

The city council has recently begun to consider whether or not to put two walkways through the park. (See Illustration 1.) The walkways would run from two points on Main Street and converge at the northernmost point of the park, dividing the area of the park exactly into thirds.

The city council is pleased with the esthetics of this arrangement but needs to know two important facts.

a. For planning purposes, they need to know exactly where on Main Street the walkways would begin.

b. In order to budget for the construction, they need to know how long the walkways will be.

Provide answers for the city council, along with explanations and work to show that your answers are correct. You will need to use the formula shown in Illustration 2, due to Archimedes (287–212 B.C.), for the area under a parabola but above a line perpendicular to the axis of symmetry of the parabola.

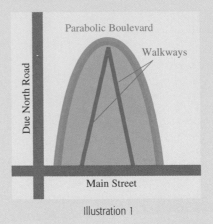

Illustration 1

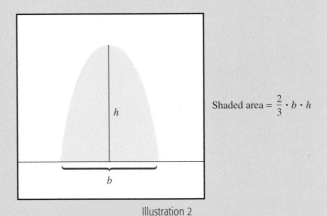

Shaded area $= \frac{2}{3} \cdot b \cdot h$

Illustration 2

Chapter Summary

CONCEPTS	REVIEW EXERCISES
8.1	**Solving Quadratic Equations by Completing the Square**

Square root property:
If $c > 0$, the equation $x^2 = c$ has two real solutions:

$$x = \sqrt{c} \quad \text{and} \quad x = -\sqrt{c}$$

To complete the square, add the square of one-half of the coefficient of x.

Solve each equation by factoring or by using the square root property.

1. $12x^2 + x - 6 = 0$
2. $6x^2 + 17x + 5 = 0$
3. $15x^2 + 2x - 8 = 0$
4. $(x + 2)^2 = 36$

Solve each equation by completing the square.

5. $x^2 + 6x + 8 = 0$
6. $2x^2 - 9x + 7 = 0$
7. $2x^2 - x - 5 = 0$

| **8.2** | **Solving Quadratic Equations by the Quadratic Formula** |

Quadratic formula:

$$x = \frac{-b \pm \sqrt{b^2 - 4ac}}{2a} \quad (a \neq 0)$$

Solve each equation by using the quadratic formula.

8. $x^2 - 8x - 9 = 0$ 9. $x^2 - 10x = 0$

10. $2x^2 + 13x - 7 = 0$ 11. $3x^2 + 20x - 7 = 0$

12. $2x^2 - x - 2 = 0$ 13. $x^2 + x + 2 = 0$

14. **Dimensions of a rectangle**   A rectangle is 2 centimeters longer than it is wide. If both the length and width are doubled, its area is increased by 72 square centimeters. Find the dimensions of the original rectangle.

15. **Dimensions of a rectangle**   A rectangle is 1 foot longer than it is wide. If the length is tripled and the width is doubled, its area is increased by 30 square feet. Find the dimensions of the original rectangle.

16. **Ballistics**   If a rocket is launched straight up into the air with an initial velocity of 112 feet per second, its height after t seconds is given by the formula $h = 112t - 16t^2$, where h represents the height of the rocket in feet. After launch, how long will it be before it hits the ground?

17. **Ballistics**   What is the maximum height of the rocket discussed in Exercise 16?

8.3 The Discriminant and Equations That Can Be Written in Quadratic Form

The discriminant:

If $b^2 - 4ac > 0$, the solutions of $ax^2 + bx + c = 0$ are unequal real numbers.

If $b^2 - 4ac = 0$, the solutions of $ax^2 + bx + c = 0$ are equal real numbers.

If $b^2 - 4ac < 0$, the solutions of $ax^2 + bx + c = 0$ are complex conjugates.

If r_1 and r_2 are solutions of $ax^2 + bx + c = 0$, then

$$r_1 + r_2 = -\frac{b}{a}$$

$$r_1 r_2 = \frac{c}{a}$$

Use the discriminant to determine what types of solutions exist for each equation.

18. $3x^2 + 4x - 3 = 0$

19. $4x^2 - 5x + 7 = 0$

20. Find the values of k that will make the solutions of $(k - 8)x^2 + (k + 16)x = -49$ equal.

21. Find the values of k such that the solutions of $3x^2 + 4x = k + 1$ will be real numbers.

Solve each equation.

22. $x - 13x^{1/2} + 12 = 0$

23. $a^{2/3} + a^{1/3} - 6 = 0$

24. $\dfrac{1}{x + 1} - \dfrac{1}{x} = -\dfrac{1}{x + 1}$

25. $\dfrac{6}{x + 2} + \dfrac{6}{x + 1} = 5$

26. Find the sum of the solutions of the equation $3x^2 - 14x + 3 = 0$.

27. Find the product of the solutions of the equation $3x^2 - 14x + 3 = 0$.

8.4 Graphs of Quadratic Functions

If f is a function and k and h positive numbers, then

The graph of $y = f(x) + k$ is identical to the graph of $y = f(x)$, except that it is translated k units upward.

The graph of $y = f(x) - k$ is identical to the graph of $y = f(x)$, except that it is translated k units downward.

Graph each function and give the coordinates of the vertex of the resulting parabola.

28. $y = 2x^2 - 3$

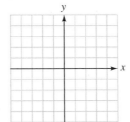

29. $y = -2x^2 - 1$

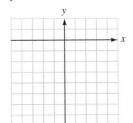

The graph of $y = f(x - h)$ is identical to the graph of $y = f(x)$, except that it is translated h units to the right.

The graph of $y = f(x + h)$ is identical to the graph of $y = f(x)$, except that it is translated h units to the left.

If $a \neq 0$, the graph of $y = a(x - h)^2 + k$ is a parabola with vertex at (h, k). It opens upward when $a > 0$ and downward when $a < 0$.

30. $y = -4(x - 2)^2 + 1$

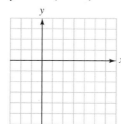

31. $y = 5x^2 + 10x - 1$

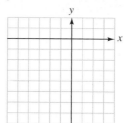

| **8.5** | **Quadratic and Other Nonlinear Inequalities** |

To solve a quadratic inequality in one variable, make a sign chart.

Solve each inequality. Give each result in interval notation and graph the solution set.

32. $x^2 + 2x - 35 > 0$

33. $x^2 + 7x - 18 < 0$

34. $\dfrac{3}{x} \leq 5$

35. $\dfrac{2x^2 - x - 28}{x - 1} > 0$

To solve inequalities with rational expressions, get 0 on the right-hand side, add the fractions, and then factor the numerator and denominator. Then use a sign chart.

Use a graphing calculator to solve each inequality. Compare the results with Review Exercises 32-35.

36. $x^2 + 2x - 35 > 0$

37. $x^2 + 7x - 18 < 0$

38. $\dfrac{3}{x} \leq 5$

39. $\dfrac{2x^2 - x - 28}{x - 1} > 0$

To graph an inequality such as $y > 4x^2 - 3$, first graph the equation $y = 4x^2 - 3$. Then determine which region represents the graph of $y > 4x^2 - 3$.

Graph each inequality.

40. $y < \frac{1}{2}x^2 - 1$

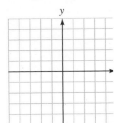

41. $y \geq -|x|$

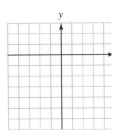

Chapter Test

Solve each equation.

1. $x^2 + 3x - 18 = 0$ **2.** $x(6x + 19) = -15$

Determine what number must be added to each binomial to make it a perfect square.

1. $x^2 + 24x$ **4.** $x^2 - 50x$

5. $x^2 + 5x$ **6.** $x^2 - 7x$

Solve each equation.

7. $x^2 + 4x + 1 = 0$ **8.** $x^2 - 5x - 3 = 0$

9. $x^2 + 12 = 0$ **10.** $x^2 + 2x + 5 = 0$

11. $3x^2 + x - 1 = 0$ **12.** $2x^2 + x = -3$

13. Determine whether the solutions of $3x^2 + 5x + 17 = 0$ are real or nonreal numbers.

14. For what value(s) of k are the solutions of $4x^2 - 2kx + k - 1 = 0$ equal?

15. One leg of a right triangle is 14 inches longer than the other, and the hypotenuse is 26 inches. Find the length of the shorter leg.

16. Solve the equation: $2y - 3y^{1/2} + 1 = 0$.

17. Graph the function $f(x) = \frac{1}{2}x^2 - 4$ and give the coordinates of its vertex.

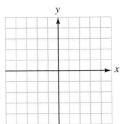

18. Graph the inequality $y \leq -x^2 + 3$.

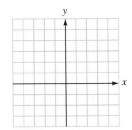

Solve each inequality and graph the solution set.

19. $x^2 - 2x - 8 > 0$

20. $\dfrac{x - 2}{x + 3} \leq 0$

CUMULATIVE REVIEW EXERCISES

Find the domain and range of each function.

1. $f(x) = 2x^2 - 3$

2. $f(x) = -|x - 4|$

Write the equation of the line with the given properties.

3. $m = 3$, passing through $(-2, -4)$

4. Parallel to the graph of $2x + 3y = 6$ and passing through $(0, -2)$

Perform each operation.

5. $(2a^2 + 4a - 7) - 2(3a^2 - 4a)$

6. $(3x + 2)(2x - 3)$

Factor each expression.

7. $x^4 - 16y^4$

8. $15x^2 - 2x - 8$

Solve each equation.

9. $x^2 - 5x - 6 = 0$

10. $6a^3 - 2a = a^2$

Simplify each expression. Assume that all variables represent positive numbers.

11. $\sqrt{25x^4}$

12. $\sqrt{48t^3}$

13. $\sqrt[3]{-27x^3}$

14. $\sqrt[3]{\dfrac{128x^4}{2x}}$

15. $8^{-1/3}$

16. $64^{2/3}$

17. $\dfrac{y^{2/3} y^{5/3}}{y^{1/3}}$

18. $\dfrac{x^{5/3} x^{1/2}}{x^{3/4}}$

Graph each function and give the domain and the range.

19. $f(x) = \sqrt{x - 2}$

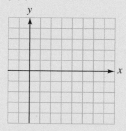

20. $f(x) = -\sqrt{x + 2}$

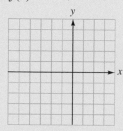

Perform the operations.

21. $(x^{2/3} - x^{1/3})(x^{2/3} + x^{1/3})$ **22.** $(x^{-1/2} + x^{1/2})^2$

Simplify each statement.

23. $\sqrt{50} - \sqrt{8} + \sqrt{32}$

24. $-3\sqrt[4]{32} - 2\sqrt[4]{162} + 5\sqrt[4]{48}$

25. $3\sqrt{2}(2\sqrt{3} - 4\sqrt{12})$ **26.** $\dfrac{5}{\sqrt[3]{x}}$

27. $\dfrac{\sqrt{x} + 2}{\sqrt{x} - 1}$ **28.** $\sqrt[6]{x^3 y^3}$

Solve each equation.

29. $5\sqrt{x + 2} = x + 8$

30. $\sqrt{x} + \sqrt{x + 2} = 2$

31. Find the length of the hypotenuse of the right triangle shown in the illustration.

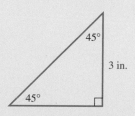

32. Find the length of the hypotenuse of the right triangle shown in the illustration.

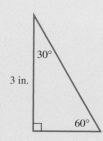

33. Find the distance between $(-2, 6)$ and $(4, 14)$.

34. What number must be added to $x^2 + 6x$ to make a trinomial square?

35. Use the method of completing the square to solve $2x^2 + x - 3 = 0$.

36. Use the quadratic formula to solve
$3x^2 + 4x - 1 = 0$.

37. Graph $f(x) = \frac{1}{2}x^2 + 5$ and find the coordinates of its vertex.

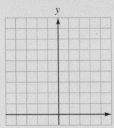

38. Graph $y \leq -x^2 + 3$ and find the coordinates of its vertex.

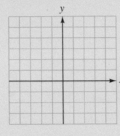

Write each expression as a real number or as a complex number in $a + bi$ form.

39. $(3 + 5i) + (4 - 3i)$

40. $(7 - 4i) - (12 + 3i)$

41. $(2 - 3i)(2 + 3i)$

42. $(3 + i)(3 - 3i)$

43. $(3 - 2i) - (4 + i)^2$

44. $\dfrac{5}{3 - i}$

45. $|3 + 2i|$

46. $|5 - 6i|$

47. For what values of k will the solutions of $2x^2 + 4x = k$ be equal?

48. Solve: $a - 7a^{1/2} + 12 = 0$.

Solve each inequality and graph the solution set on the number line.

49. $x^2 - x - 6 > 0$

50. $x^2 - x - 6 \leq 0$

9 More Functions and Operations on Functions

InfoTrac Project

Do a keyword search on "inverse proportion." Find the article "Upper-extremity motor co-ordination of healthy elderly people." Write a summary of the article.

Suppose the relationship between the number of successful nose touches (y) and age (x) can be represented by the rational function $y = 1{,}710/x$. Make a table of values and let age (x) = 60, 65, 70, 75, and 80 years. How do your answers compare with the results in the article? Would this equation be valid for people younger than 60? Let age (x) = 50, 25, 5, and 0.5 years (6 months). Are your answers realistic?

Complete this project after studying Section 9.3.

© Royalty-Free/CORBIS

Mathematics in Equestrianism

A woman purchases a horse for $2,000. If she plans to keep it at a stable that charges $350 per month for a stall and board, what will be the average cost per month if she keeps the horse for 10 years?

Exercise 62
Exercise Set 9.3

We have seen that the graphs of linear functions are lines and that the graphs of quadratic functions are parabolas. In this chapter, we will discuss symmetries of graphs and consider the graphs of many other types of functions.

9.1 Symmetry and Stretchings of Graphs

- **Graphs of Polynomial Functions**
- **Symmetries of Graphs of Functions** ▪ **Even and Odd Functions**
- **Increasing and Decreasing Functions**
- **Vertical and Horizontal Stretchings of Graphs**

Getting Ready   *Give the degree of each polynomial.*

1. $x^3 + 3x^2 - 2x + 4$ **2.** $x^4 - 5x^2 + 4$

The graphs of higher-degree polynomial functions are curves, often with more than one x-intercept. We now consider methods we can use to graph them.

Graphs of Polynomial Functions

Polynomial Functions

A **polynomial function in one variable** (say x) is defined by an equation of the form $y = f(x) = P(x)$, where $P(x)$ is a polynomial in the variable x.

The **degree of the polynomial function** $f(x) = P(x)$ is the degree of $P(x)$.

EXAMPLE 1   Graph the polynomial function: $f(x) = x^4 - 5x^2 + 4$.

Solution   We plot several points (x, y) whose coordinates satisfy the equation, as in Figure 9-1, and then draw the graph by joining the points with a smooth curve. From the graph, we can see that the domain is $(-\infty, \infty)$, and the range is $[-2.25, \infty)$.

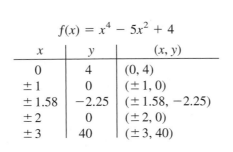

$$f(x) = x^4 - 5x^2 + 4$$

x	y	(x, y)
0	4	$(0, 4)$
± 1	0	$(\pm 1, 0)$
± 1.58	-2.25	$(\pm 1.58, -2.25)$
± 2	0	$(\pm 2, 0)$
± 3	40	$(\pm 3, 40)$

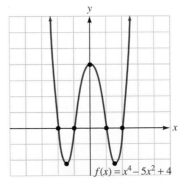

Figure 9-1

Self Check   Graph: $f(x) = x^4 - 10x^2 + 9$.

Accent on Technology   **GRAPHING POLYNOMIAL FUNCTIONS**

To draw the graph of Example 1 with a graphing calculator, we can use window settings of $[-5, 5]$ for x and $[-10, 10]$ for y, enter the equation, and press the $\boxed{\text{GRAPH}}$ key to obtain the graph shown in Figure 9-2.

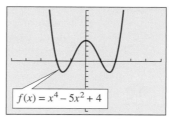

$$f(x) = x^4 - 5x^2 + 4$$

Figure 9-2

The x-intercepts of the graph in Example 1 are the points $(-2, 0)$, $(-1, 0)$, $(1, 0)$, and $(2, 0)$, and the y-intercept is the point $(0, 4)$. Note that if we fold the graph along the y-axis, the left-hand side will exactly match the right-hand side. For this reason, we say that the graph is symmetric about the y-axis.

It is often easier to draw graphs of functions if we know their symmetries.

Symmetries of Graphs of Functions

There are several ways in which a graph can have symmetry.

1. If the point $(-x, y)$ lies on a graph whenever the point (x, y) does, the graph is **symmetric about the y-axis.** See Figure 9-3(a). This implies that a graph is symmetric about the y-axis if we get the same y-coordinate when we evaluate its equation at x or at $-x$.

2. If the point $(-x, -y)$ lies on a graph whenever the point (x, y) does, the graph is **symmetric about the origin.** See Figure 9-3(b). This implies that a graph is symmetric about the origin if we get opposite values of y when we evaluate its equation at x or at $-x$.

3. If the point $(x, -y)$ lies on a graph whenever the point (x, y) does, the graph is **symmetric about the x-axis.** See Figure 9-3(c).

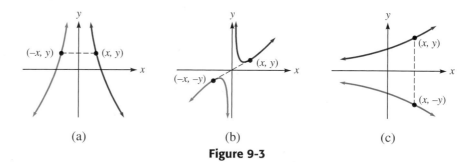

(a)   (b)   (c)

Figure 9-3

To check whether the graph of a function has any symmetries, we can apply the following tests.

Tests for Symmetry of the Graph of a Function	Let x be any number in the domain of f. **1.** To test for *y-axis symmetry,* compute $f(-x)$. If $f(-x) = f(x)$, the graph is symmetric about the y-axis. **2.** To test for *symmetry about the origin,* compute $f(-x)$. If $f(-x) = -f(x)$, the graph is symmetric about the origin. **3.** To test for *x-axis symmetry,* check to see whether $f(x) = -f(x)$. If so, the graph is symmetric about the x-axis.

 Comment   The only function whose graph is symmetric about the x-axis is the function $f(x) = 0$. The graph of any other equation symmetric about the x-axis will not pass the vertical line test and cannot represent a function.

 EXAMPLE 2   Determine the symmetries of the graph of each function.
a. $f(x) = |x|$ and   **b.** $f(x) = x^3 - x$.

Solution   Since neither function is the function $f(x) = 0$, neither one is symmetric about the x-axis.

a. To test $f(x) = |x|$ for y-axis symmetry, we compute $f(-x)$ to see if it is equal to $f(x)$.

$$f(x) = |x| \qquad \text{The original function.}$$
$$f(-x) = |-x| \qquad \text{Substitute } -x \text{ for } x.$$
$$f(-x) = |x| \qquad \text{Simplify: } |-x| = |x|.$$

Since $f(-x) = f(x)$, the graph is symmetric about the y-axis. However, there is no symmetry about the origin because $f(-x) \neq -f(x)$.

b. To test $f(x) = x^3 - x$ for y-axis symmetry, we compute $f(-x)$.

$$f(x) = x^3 - x \qquad \text{The original function.}$$
$$f(-x) = (-x)^3 - (-x) \qquad \text{Substitute } -x \text{ for } x.$$
$$f(-x) = -x^3 + x \qquad \text{Simplify.}$$
$$f(-x) = -(x^3 - x) \qquad \text{Factor out } -1.$$

Since $f(-x) \neq f(x)$, the graph is not symmetric about the y-axis. However, the graph is symmetric about the origin because $f(-x) = -f(x)$.

Self Check   Check $f(x) = x^3 - 9x$ for any symmetries.   ∎

Even and Odd Functions

A function whose graph is symmetric about the y-axis is called an **even function,** as shown in Figure 9-3(a). Since $f(-x) = y$ and $f(x) = y$ in this function, it follows that $f(-x) = f(x)$.

Even Functions

> An **even function** f is a function with the property that $f(-x) = f(x)$ for all x in the domain of f.

In Example 2, $f(x) = |x|$ is an even function because

$$f(-x) = |-x| = |x| = f(x)$$

A function whose graph is symmetric about the origin is called an **odd function,** as shown in Figure 9-3(b). Since $f(-x) = -y$ and $f(x) = y$ in this function, it follows that $f(-x) = -f(x)$.

Odd Functions

> An **odd function** f is a function with the property that $f(-x) = -f(x)$ for all x in the domain of f.

In Example 2, $f(x) = x^3 - x$ is an odd function because

$$f(-x) = (-x)^3 - (-x) = -x^3 + x = -(x^3 - x) = -f(x)$$

EXAMPLE 3 Decide whether each function is even or odd:
a. $f(x) = x^4 + 2$ and   **b.** $f(x) = x^3 + 3x$.

Solution **a.** We find $f(-x)$ and compare it to $f(x)$:

$$f(-x) = (-x)^4 + 2 = x^4 + 2 = f(x)$$

Since $f(-x) = f(x)$, the function is an even function.

b. We find $f(-x)$ and compare it to $f(x)$:

$$f(-x) = (-x)^3 + 3(-x) = -x^3 - 3x = -(x^3 + 3x) = -f(x)$$

Since $f(-x) = -f(x)$, the function is an odd function.

Self Check Decide whether each function is even or odd:
a. $f(x) = x$ and   **b.** $f(x) = x^4 - x^2$ ∎

EXAMPLE 4 Find the intercepts and the symmetries of the graph of $f(x) = x^3 - 4x$. Then graph the function.

Solution To find the *x-intercepts,* we let $y = 0$ and solve for x:

$$y = x^3 - 4x$$
$$0 = x^3 - 4x \qquad\qquad \text{Substitute 0 for } y.$$
$$0 = x(x^2 - 4) \qquad\qquad \text{Factor out } x.$$
$$0 = x(x + 2)(x - 2) \qquad \text{Factor } x^2 - 4.$$

$$x = 0 \quad \text{or} \quad x + 2 = 0 \quad \text{or} \quad x - 2 = 0 \quad \text{Set each factor equal to 0.}$$
$$| \qquad\qquad x = -2 \quad | \qquad x = 2$$

The *x*-intercepts are $(0, 0)$, $(-2, 0)$, and $(2, 0)$.

To find the *y-intercepts,* we let $x = 0$ and solve for y:

$$y = x^3 - 4x$$
$$y = 0^3 - 4(0) \qquad \text{Substitute 0 for } x.$$
$$y = 0$$

The y-intercept is $(0, 0)$.

To test for *y-axis symmetry*, we compute $f(-x)$.

$$f(-x) = (-x)^3 - 4(-x)$$
$$= -x^3 + 4x$$
$$= -(x^3 - 4x) \qquad \text{Factor out } -1.$$
$$= -f(x)$$

Since $f(-x) \neq f(x)$, there is no symmetry about the y-axis. However, because $f(-x) = -f(x)$, the graph is symmetric about the origin. Since $y = x^3 - 4x$ is a function, there is no x-axis symmetry.

To graph the equation, we plot the x-intercepts of $(-2, 0)$, $(0, 0)$, $(2, 0)$, and the y-intercept of $(0, 0)$. We also plot other points for positive values of x and use the symmetry about the origin to draw the rest of the graph, as in Figure 9-4.

From the graph, we can see that the domain is the interval $(-\infty, \infty)$, and the range is the interval $(-\infty, \infty)$.

$f(x) = x^3 - 4x$

x	y	(x, y)
-2	0	$(-2, 0)$
0	0	$(0, 0)$
1	-3	$(1, -3)$
2	0	$(2, 0)$
3	15	$(3, 15)$

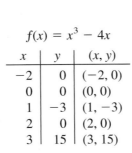

Figure 9-4

Self Check   Graph: $f(x) = x^3 - 9x$.

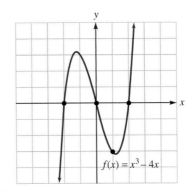

Accent on Technology   **GRAPHING POLYNOMIAL FUNCTIONS**

To draw the graph of Example 4 with a graphing calculator, we can use window settings of $[-5, 5]$ for x and $[-10, 10]$ for y, enter the function, and press the GRAPH key. We will obtain the graph shown in Figure 9-5.

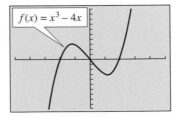

Figure 9-5

Increasing and Decreasing Functions

If the values of a function increase as x increases, we say that the function is *increasing*. (See Figure 9-6(a).) If the values of a function decrease as x increases, we say that the function is *decreasing*. (See Figure 9-6(b).) If the values of a function are constant as x increases, we say that the function is a *constant* function. (See Figure 9-6(c).)

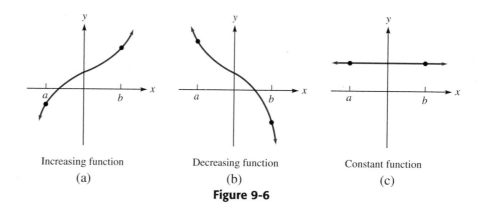

Increasing function	Decreasing function	Constant function
(a)	(b)	(c)

Figure 9-6

We summarize the preceding discussion for functions on the interval (a, b).

Increasing and Decreasing Functions

A function is **increasing on (a, b)** if $f(x_1) < f(x_2)$ whenever $a < x_1 < x_2 < b$.

A function is **decreasing on (a, b)** if $f(x_1) > f(x_2)$ whenever $a < x_1 < x_2 < b$.

A function is **constant on (a, b)** if $f(x_1) = f(x_2)$ for all x_1 and x_2 on (a, b).

EXAMPLE 5 Graph the function $f(x) = |x| - 2$ and tell where it is increasing and where it is decreasing.

Solution The graph of $f(x) = |x| - 2$ is the graph of $f(x) = |x|$, except that it has been translated down 2 units. From the graph, shown in Figure 9-7, we can see that the function is decreasing on the interval $(-\infty, 0)$ and increasing on the interval $(0, \infty)$.

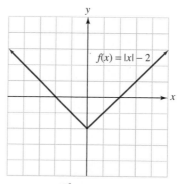

Figure 9-7

Self Check   Graph $f(x) = -|x| + 2$ and tell where it is increasing and where it is decreasing.

Vertical and Horizontal Stretchings of Graphs

Figure 9-8 shows the graphs of $y = f(x) = x^2 - 1$ and $y = 3f(x) = 3(x^2 - 1)$. Because each value of y in $y = 3(x^2 - 1)$ is 3 times greater than the corresponding value in $y = x^2 - 1$, its graph is stretched vertically by a factor of 3. The x-intercepts of both graphs are $(1, 0)$ and $(-1, 0)$.

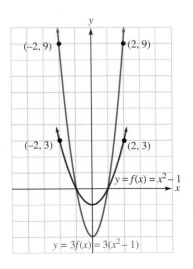

$y = f(x) = x^2 - 1$		
x	y	(x, y)
-2	3	$(-2, 3)$
-1	0	$(-1, 0)$
0	-1	$(0, -1)$
1	0	$(1, 0)$
2	3	$(2, 3)$

$y = 3f(x) = 3(x^2 - 1)$		
x	y	(x, y)
-2	9	$(-2, 9)$
-1	0	$(-1, 0)$
0	-3	$(0, -3)$
1	0	$(1, 0)$
2	9	$(2, 9)$

Figure 9-8

Figure 9-9 shows the graphs of $y = f(x) = x^2 - 1$ and $y = f(3x) = (3x)^2 - 1$. Because each value of x in $y = (3x)^2 - 1$ is $\frac{1}{3}$ of the value of x in $y = x^2 - 1$, the graph of $y = (3x)^2 - 1$ is stretched horizontally by a factor of $\frac{1}{3}$. The y-intercepts of both graphs are $(0, -1)$.

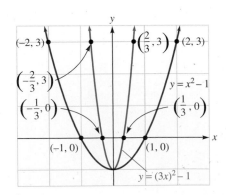

$y = f(x) = x^2 - 1$		
x	y	(x, y)
-2	3	$(-2, 3)$
-1	0	$(-1, 0)$
0	-1	$(0, -1)$
1	0	$(1, 0)$
2	3	$(2, 3)$

$y = f(3x) = (3x)^2 - 1$		
x	y	(x, y)
$-\frac{2}{3}$	3	$\left(-\frac{2}{3}, 3\right)$
$-\frac{1}{3}$	0	$\left(-\frac{1}{3}, 0\right)$
0	-1	$(0, -1)$
$\frac{1}{3}$	0	$\left(\frac{1}{3}, 0\right)$
$\frac{2}{3}$	3	$\left(\frac{2}{3}, 3\right)$

Figure 9-9

In general, we can make the following observations.

Vertical Stretching

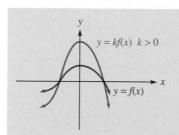

If k is a positive number, then

- The graph of $y = kf(x)$ can be obtained by stretching the graph of $y = f(x)$ vertically by a factor of k.

Horizontal Stretching

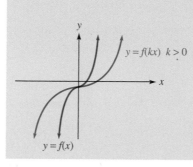

- The graph of $y = f(kx)$ can be obtained by stretching the graph of $y = f(x)$ horizontally by a factor of $\frac{1}{k}$.

Accent on Technology

GRAPHING FUNCTIONS

We can use a graphing calculator to show the effect of a vertical stretching and reflection by graphing $y = x^2 - 4$ and $y = -2(x^2 - 4)$. The graph of $y = x^2 - 4$ is a parabola opening upward with vertex at $(0, -4)$. If that graph is stretched vertically by a factor of 2 and then reflected about the x-axis, the graph of $y = -2(x^2 - 4)$ should be the result. That is what happens, as Figure 9-10 shows. Notice that the x-intercepts of the graphs are the same.

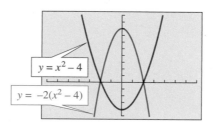

Figure 9-10

Self Check Answers

1.

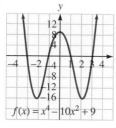

2. symmetric about the origin

3. a. odd,
   b. even

4.

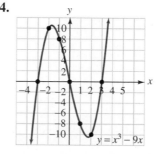

5. increasing on $(-\infty, 0)$,
decreasing on $(0, \infty)$

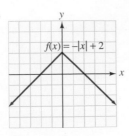

$f(x) = -|x| + 2$

Orals   *Find the symmetries of the graph of each equation.*

1. $y = x^2$ **2.** $y^2 = x$ **3.** $y = x$

4. $y = |x| - 2$

Tell whether each function is even or odd.

5. $y = x^4$ **6.** $y = x^3$

9.1   EXERCISES

REVIEW   *Write each number in scientific notation.*

1. 9,300,000   **2.** 0.00023

Write each number in standard notation.

3. 6.7×10^5 **4.** 6.89×10^{-6}

VOCABULARY AND CONCEPTS   *Fill in the blanks.*

5. A _____ function in x is defined by an equation of the form $y = P(x)$.

6. The degree of the _____ function $y = P(x)$ is the degree of ___.

7. If $(-x, y)$ lies on a graph whenever (x, y) does, the graph is _____ about the _____.

8. If $(-x, -y)$ lies on a graph whenever (x, y) does, the graph is _____ about the _____.

9. If _____ lies on a graph whenever (x, y) does, the graph is _____ about the x-axis.

10. In an ___ function, f, $f(-x) = f(x)$.

11. In an odd function f, $f(-x) = $ _____.

12. A function is _____ on (a, b) if $f(x_1) < f(x_2)$ whenever $a < x_1 < x_2 < b$.

13. A function is _____ on (a, b) if $f(x_1) > f(x_2)$ whenever $a < x_1 < x_2 < b$.

14. A function is _____ on (a, b) if $f(x_1) = f(x_2)$ for all x_1 and x_2 on (a, b).

15. The graph of $y = kf(x)$ can be obtained by stretching the graph of $y = f(x)$ _____ by a factor of k.

16. The graph of $y = f(kx)$ can be obtained by stretching the graph of $y = f(x)$ horizontally by a factor of ___.

PRACTICE   *Find the symmetries of the graph (if any) of each equation. **Do not draw the graph.***

17. $y = x^2 - 1$ **18.** $y = x^3$

19. $y = x^5$ **20.** $y = x^4$

21. $y = -x^2 + 2$ **22.** $y = x^3 + 1$

23. $y = x^2 - x$ **24.** $y^2 = x$

25. $y = -|x + 2|$ **26.** $y = |x| - 3$

27. $|y| = x$ **28.** $y = 2\sqrt{x}$

Tell whether each function is an even function.

29. $y = 3x + 4$ **30.** $y = 3x^2 - 3$

31. $y = x^2$ **32.** $y = -x^6 - 1$

33. $y = x^3 + x$ **34.** $y = 3x^2 - x$

35. $y = |x| + 1$ **36.** $y = \sqrt{x}$

Tell whether each function is an odd function.

37. $y = x$ **38.** $y = -x$

39. $y = x^3 + 1$ **40.** $y = -x^3 + 2$

41. $y = |x| + x$ **42.** $y = \sqrt{x}$

43. $y = x^3 - x$ **44.** $y = 2x^3$

Graph each function and give its domain and range. Check your graph with a graphing calculator.

45. $f(x) = x^4 - 4x^2 + 4$ **46.** $f(x) = x^4 - 2x^2 + 1$

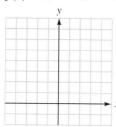   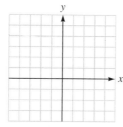

47. $f(x) = x^3 + x$ **48.** $f(x) = x^3 - x$

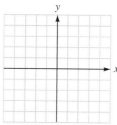   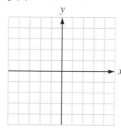

49. $f(x) = x^3 - 3x$ **50.** $f(x) = -x^3 + 2x$

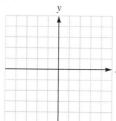   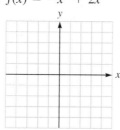

Tell where each function is increasing, decreasing, or constant.

51.   **52.**

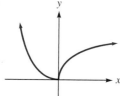   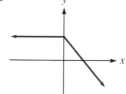

53.   **54.**

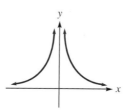   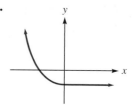

55. $y = f(x) = x^2 - 4x + 4$

56. $y = f(x) = 4 - x^2$

In Exercises 57–60, the graph of each function is a stretching of the graph of $f(x) = x^2$. Graph each function.

57. $f(x) = 2x^2$ **58.** $f(x) = \frac{1}{2}x^2$

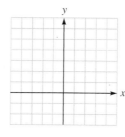   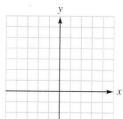

59. $f(x) = -3x^2$ **60.** $f(x) = -\frac{1}{3}x^2$

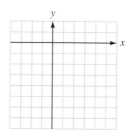   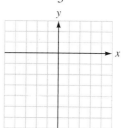

In Exercises 61–64, the graph of each function is a stretching of the graph of $f(x) = x^3$. Graph each function.

61. $f(x) = \left(\frac{1}{2}x\right)^3$ **62.** $f(x) = \frac{1}{8}x^3$

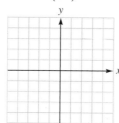   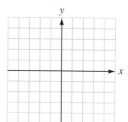

63. $f(x) = -8x^3$

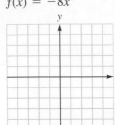

64. $f(x) = (-2x)^3$

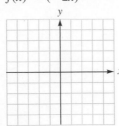

66. Use a graphing calculator to graph $f(x) = kx^2$ for several values of k. Explain what you discover.

SOMETHING TO THINK ABOUT

67. Graph the function $f(x) = x^2 + ax$ for several values of a. How does the graph change?

68. Graph the function $f(x) = x^3 + ax$ for several values of a. How does the graph change?

69. Graph the function $f(x) = (x - a)(x - b)$ for several values of a and b. What is the relationship between the x-intercepts and the equation?

70. Use the insight gained in Exercise 69 to factor $x^3 - 3x^2 - 4x + 12$.

WRITING

65. Use a graphing calculator to graph $f(x) = (kx)^2$ for several values of k. Explain what you discover.

9.2	**Piecewise-Defined Functions and the Greatest Integer Function**

■ Piecewise-Defined Functions   ■ The Greatest Integer Function

Getting Ready

1. Is $f(x) = x^2$ positive or negative when $x > 0$?

2. Is $f(x) = -x^2$ positive or negative when $x < 0$?

3. What is the largest integer that is less than 98.6?

4. What is the largest integer that is less than -2.7?

In this section, we will discuss functions that are defined by using different equations for different parts of their domains. Such functions are called **piecewise-defined functions.**

Piecewise-Defined Functions

A simple piecewise-defined function is $f(x) = |x|$, which can be written in the form

$$f(x) = \begin{cases} x \text{ when } x \geq 0 \\ -x \text{ when } x < 0 \end{cases}$$

When x is in the interval $[0, \infty)$, we use the function $f(x) = x$ to evaluate $|x|$. However, when x is in the interval $(-\infty, 0)$, we use the function $f(x) = -x$ to evaluate $|x|$. The graph of the function is shown in Figure 9-11.

The function, pictured in Figure 9-11, is decreasing on the interval $(-\infty, 0)$ and increasing on the interval $(0, \infty)$.

For $x \geq 0$		
x	y	(x, y)
0	0	$(0, 0)$
1	1	$(1, 1)$
2	2	$(2, 2)$
3	3	$(3, 3)$

For $x < 0$		
x	y	(x, y)
-4	4	$(-4, 4)$
-3	3	$(-3, 3)$
-2	2	$(-2, 2)$
-1	1	$(-1, 1)$

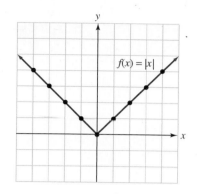

Figure 9-11

EXAMPLE 1   Graph the following piecewise-defined function and tell where it is increasing, decreasing, or constant.

$$f(x) = \begin{cases} x^2 \text{ when } x \leq 0 \\ x \text{ when } 0 < x < 2 \\ -1 \text{ when } x \geq 2 \end{cases}$$

Solution   For each number x, we decide which of these three equations will be used to find the corresponding value of y:

- For numbers $x \leq 0$, $f(x)$ is determined by $f(x) = x^2$, and the graph is the left half of a parabola. (See Figure 9-12.) Since the values of $f(x)$ decrease on this graph as x increases, the function is decreasing on the interval $(-\infty, 0)$.

- For numbers $0 < x < 2$, $f(x)$ is determined by $f(x) = x$, and the graph is part of a line. Since the values of $f(x)$ increase on this graph as x increases, the function is increasing on the interval $(0, 2)$.

- For numbers $x \geq 2$, $f(x)$ is the constant -1, and the graph is part of a horizontal line. Since the values of $f(x)$ remain constant on this line, the function is constant on the interval $(2, \infty)$.

The use of solid and open circles on the graph indicates that $f(x) = -1$ when $x = 2$.
Since every number x determines one value y, the domain of this function is the interval $(-\infty, \infty)$. The range is the interval $[0, \infty) \cup \{-1\}$.

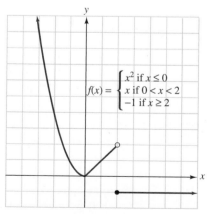

$$f(x) = \begin{cases} x^2 \text{ if } x \leq 0 \\ x \text{ if } 0 < x < 2 \\ -1 \text{ if } x \geq 2 \end{cases}$$

Figure 9-12

Self Check   Graph: $f(x) = \begin{cases} 2x \text{ when } x \le 0 \\ \frac{1}{2}x \text{ when } x > 0 \end{cases}$ ∎

The Greatest Integer Function

The **greatest integer function** is important in computer applications. It is a function determined by the equation

$$y = f(x) = [\![x]\!]$$ Read as "y equals the greatest integer in x."

where the value of y that corresponds to x is the greatest integer that is less than or equal to x. For example,

$$[\![4.7]\!] = 4, \qquad [\![2\tfrac{1}{2}]\!] = 2, \qquad [\![\pi]\!] = 3, \qquad [\![-3.7]\!] = -4, \qquad [\![-5.7]\!] = -6$$

    EXAMPLE 2   Graph: $y = [\![x]\!]$.

Solution   We list several intervals and the corresponding values of the greatest integer function:

$[0, 1)$ $y = [\![x]\!] = 0$ For numbers from 0 to 1, not including 1, the greatest integer in the interval is 0.

$[1, 2)$ $y = [\![x]\!] = 1$ For numbers from 1 to 2, not including 2, the greatest integer in the interval is 1.

$[2, 3)$ $y = [\![x]\!] = 2$ For numbers from 2 to 3, not including 3, the greatest integer in the interval is 2.

In each interval, the values of y are constant, but they jump by 1 at integer values of x. The graph is shown in Figure 9-13. From the graph, we see that the domain is $(-\infty, \infty)$, and the range is the set of integers $\{. \ . \ . \ , -3, -2, -1, 0, 1, 2, 3, . \ . \ .\}$.

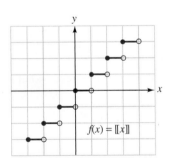

Figure 9-13   ∎

Since the greatest integer function is made up of a series of horizontal line segments, it is an example of a group of functions called **step functions.**

EXAMPLE 3   To print stationery, a printer charges $10 for setup charges, plus $20 for each box. The printer counts any portion of a box as a full box. Graph this step function.

Solution   If we order stationery and cancel before it is printed, the cost will be $10. Thus, the ordered pair (0, 10) will be on the graph.

If we purchase 1 box, the cost will be $10 for setup plus $20 for printing, for a total cost of $30. Thus, the ordered pair (1, 30) will be on the graph.

The cost of $1\frac{1}{2}$ boxes will be the same as the cost of 2 boxes, or $50. Thus, the ordered pairs (1.5, 50) and (2, 50) will be on the graph.

The complete graph is shown in Figure 9-14.

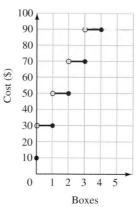

Figure 9-14

Self Check   How much will $3\frac{1}{2}$ boxes cost?   ▮

1. $f(x) = \begin{cases} 2x \text{ when } x \le 0 \\ \frac{1}{2}x \text{ when } x > 0 \end{cases}$

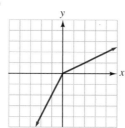

3. $90

Orals   *Consider the function* $f(x) = \begin{cases} x \text{ when } x \le 0 \\ x^2 \text{ when } 0 < x < 4. \\ 5 \text{ when } x > 4 \end{cases}$ *Find each value.*

1. $f(-2)$ **2.** $f(0)$ **3.** $f(2)$ **4.** $f(6)$

9.2   EXERCISES

REVIEW   *Find the value of x. Assume that lines r and s are parallel.*

1.

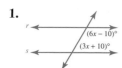

$(6x - 10)°$
$(3x + 10)°$

2.

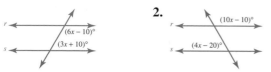

$(10x - 10)°$
$(4x - 20)°$

VOCABULARY AND CONCEPTS   *Fill in the blanks.*

3. Piecewise-defined functions are defined by using different functions for different parts of their _____.

4. When the values of $f(x)$ increase as the values of x increase over an interval, we say that the function is an _____ function over that interval.

5. In a _____ function, the values of _____ are the same.

6. When the values of $f(x)$ decrease as the values of x _____ over an interval, we say that the function is a decreasing function over that interval.

7. When the graph of a function contains a series of horizontal line segments, the function is called a _____ function.

8. The function that gives the largest integer that is less than or equal to a number x is called the _____ function.

9. ___ is the largest integer that is less than 55.7.

10. _____ is the largest integer that is less than -22.3.

PRACTICE *Graph each function and give the intervals on which f is increasing, decreasing, or constant.*

11. $f(x) = \begin{cases} -1 \text{ if } x \le 0 \\ x \text{ if } x > 0 \end{cases}$

12. $f(x) = \begin{cases} -2 \text{ if } x \le 0 \\ x^2 \text{ if } x > 0 \end{cases}$

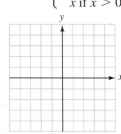
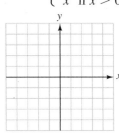

13. $f(x) = \begin{cases} -x \text{ if } x \le 0 \\ x \text{ if } 0 < x < 2 \\ -x \text{ if } x \ge 2 \end{cases}$

14. $f(x) = \begin{cases} -x \text{ if } x < 0 \\ x^2 \text{ if } 0 \le x \le 1 \\ 1 \text{ if } x > 1 \end{cases}$

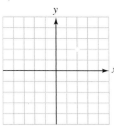
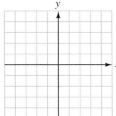

Graph each function.

15. $f(x) = -\lfloor x \rfloor$

16. $f(x) = \lfloor x \rfloor + 2$

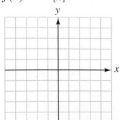
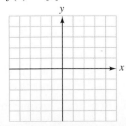

17. $f(x) = 2\lfloor x \rfloor$

18. $f(x) = \left\lfloor \frac{1}{2}x \right\rfloor$

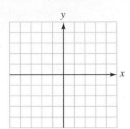
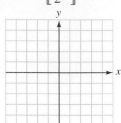

19. **Signum function** Computer programmers use a function, denoted by $f(x) = \text{sgn } x$, that is defined in the following way:

$$f(x) = \begin{cases} -1 \text{ if } x < 0 \\ 0 \text{ if } x = 0 \\ 1 \text{ if } x > 0 \end{cases}$$

Graph this function.

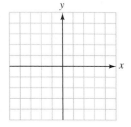

20. **Heaviside unit step function** This function, used in calculus, is defined by

$$f(x) = \begin{cases} 1 \text{ if } x > 0 \\ 0 \text{ if } x < 0 \end{cases}$$

Graph this function.

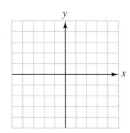

APPLICATIONS

21. **Renting a jet ski** A marina charges \$20 to rent a jet ski for 1 hour, plus \$5 for every extra hour (or portion of an hour). In the illustration, graph the ordered pairs (h, c), where h represents the number of hours and c represents the cost. Find the cost if the ski is used for 2.5 hours.

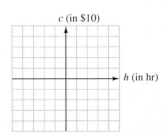

22. **Riding in a taxi**   A cab company charges $3 for a trip up to 1 mile, and $2 for every extra mile (or portion of a mile). In the illustration, graph the ordered pairs (m, c), where m represents the number of miles traveled and c represents the cost. Find the cost to ride $10\frac{1}{4}$ miles.

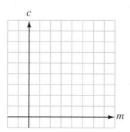

23. **Information access**   Computer access to international data network A costs $10 per day plus $8 per hour or fraction of an hour. Network B charges $15 per day, but only $6 per hour or fraction of an hour. For each network, graph the ordered pairs (t, C), where t represents the connect time and C represents the total cost. (Use the illustration.) Find the minimum daily usage at which it would be more economical to use network B.

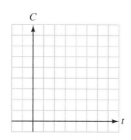

24. **Royalties**   A publisher has agreed to pay the author of a novel a royalty of 7% on sales of the first 50,000 copies and 10% on sales thereafter. If the book sells for $10, express the royalty income I as a function of s, the number of copies sold, and graph the function in the illustration.   (*Hint:* When sales are into the second 50,000 copies, how much was earned on the first 50,000?)

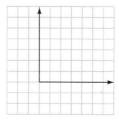

WRITING

25. Tell how to decide whether a function is increasing on the interval (a, b).

26. Describe the greatest integer function.

SOMETHING TO THINK ABOUT

27. Find a piecewise-defined function that is increasing on the interval $(-\infty, -2)$ and decreasing on the interval $(-2, \infty)$.

28. Find a piecewise-defined function that is constant on the interval $(-\infty, 0)$, increasing on the interval $(0, 5)$, and decreasing on the interval $(5, \infty)$.

9.3   More Rational Functions

▮ **Average Monthly Cost**   ▮ **Rational Functions**
▮ **Vertical and Horizontal Asymptotes**   ▮ **Finding Asymptotes**
▮ **Graphing Rational Functions**   ▮ **Graphs with Discontinuities**

Getting Ready   *For what values are the following fractions undefined?*

1. $\dfrac{2}{x - 8}$

2. $\dfrac{-5x}{x + 4}$

3. $\dfrac{x + 2}{(x - 1)(x + 2)}$

4. $\dfrac{3x}{x^2 - 9}$

Average Monthly Cost

In Chapter 6, we saw that rational expressions can define functions. For example, if an equestrian purchases a horse for $2,000 and will pay $350 per month to board it, the average monthly cost of owning the horse is given by the rational function

$$\bar{c} = \frac{C}{n} = \frac{350n + 2,000}{n}$$

$\bar{c}$ is the mean monthly cost, C is the total cost, and n is the number of months the horse is owned.

GRAPHING RATIONAL FUNCTIONS

If we use a graphing calculator with window settings of [0, 60] for x and [0, 2,000] for y to graph the function $f(n) = \frac{350n + 2,000}{n}$, we will obtain the graph shown in Figure 9-15. Note that the graph of the function passes the vertical line test, as expected.

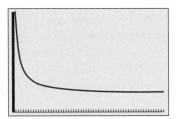

Figure 9-15

From the graph, we can see that the average monthly cost decreases as the number of months of horse ownership increases. Since the cost of boarding the horse is $350 per month, the average monthly cost can approach $350 but never drop below it. The graph of the function approaches the line $y = 350$ as n increases without bound. We have seen that when a graph approaches a line, the line is called an **asymptote.** The line $y = 350$ is a **horizontal asymptote** of the graph.

As n gets smaller and approaches 0, the graph approaches the y-axis but never touches it. The y-axis is called a **vertical asymptote** of the graph.

In this section, we will explore the graphs of rational functions in more detail.

Rational Functions

Rational functions are defined by equations of the form

$$y = \frac{P(x)}{Q(x)}$$

where $P(x)$ and $Q(x)$ are polynomials. Because $Q(x)$ appears in the denominator of a fraction, it cannot equal 0. Thus, the domain of a rational function must exclude all values of x for which $Q(x) = 0$.

Here are some examples of rational functions and their domains:

$$y = f(x) = \frac{3}{x - 2} \qquad\qquad \text{Domain: } (-\infty, 2) \cup (2, \infty)$$

$$y = f(x) = \frac{5x + 2}{x^2 - 4} \qquad\qquad \text{Domain: } (-\infty, -2) \cup (-2, 2) \cup (2, \infty)$$

$$y = f(x) = \frac{3x^2 - 2}{(2x + 1)(x - 3)} \qquad \text{Domain: } \left(-\infty, -\frac{1}{2}\right) \cup \left(-\frac{1}{2}, 3\right) \cup (3, \infty)$$

EXAMPLE 1   Find the domain of $f(x) = \dfrac{3x + 2}{x^2 - 7x + 12}$.

Solution   To find the numbers x that make the denominator 0, we set $x^2 - 7x + 12$ equal to 0 and solve for x.

$$x^2 - 7x + 12 = 0$$
$$(x - 4)(x - 3) = 0 \qquad \text{Factor } x^2 - 7x + 12.$$
$$x - 4 = 0 \quad \text{or} \quad x - 3 = 0 \qquad \text{Set each factor equal to 0.}$$
$$x = 4 \quad | \quad\quad x = 3 \qquad \text{Solve each linear equation.}$$

The domain is the interval $(-\infty, 3) \cup (3, 4) \cup (4, \infty)$.

Self Check   Find the domain of $f(x) = \dfrac{2x - 3}{x^2 - x - 2}$.   ∎

Accent on Technology   FINDING DOMAINS AND RANGES OF RATIONAL FUNCTIONS

We can use graphing calculators to find domains and ranges of rational functions. If we use settings of $[-10, 10]$ for x and $[-10, 10]$ for y and graph $f(x) = \frac{2x + 1}{x - 1}$, we will obtain Figure 9-16.

From the graph, we can see that every real number x except 1 gives a value of y. Thus, the domain of the function is $(-\infty, 1) \cup (1, \infty)$. We can also see that y can be any value except 2. The range of the function is $(-\infty, 2) \cup (2, \infty)$.

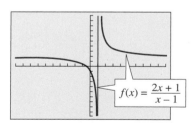

$$f(x) = \frac{2x + 1}{x - 1}$$

Figure 9-16

From Figure 9-16, we can see that

- As x approaches 1 from the left, the values of y decrease, and the graph approaches the vertical line $x = 1$.

- As x approaches 1 from the right, the values of y increase, and the graph approaches the vertical line $x = 1$.

(Continued)

For this reason, the line $x = 1$ is a *vertical asymptote.*

Although the vertical line in Figure 9-16 appears to be the asymptote, it is not. Graphing calculators draw graphs by connecting dots whose x-coordinates are close together. Often when two such points straddle a vertical asymptote and their y-coordinates are far apart, the calculator draws a line between them anyway, producing what appears to be the vertical asymptote shown in the figure. If you set your calculator to "dot" mode instead of "connected" mode, the vertical line will not appear.

From the figure, we can also see that

- As x increases to the right of 1, the values of y decrease and approach the value $y = 2$.
- As x decreases to the left of 1, the values of y increase and approach the value $y = 2$.

The line $y = 2$ is a *horizontal asymptote.* Graphing calculators do not draw lines that appear to be horizontal asymptotes.

Vertical and Horizontal Asymptotes

We have seen that if a graph approaches a line, we call the line an *asymptote.* There are two asymptotes that are apparent in Figure 9-16.

In the following definition, we read $f(x) \rightarrow \infty$ as "$f(x)$ approaches ∞."

Vertical and Horizontal Asymptotes

The line $x = a$ is a **vertical asymptote** for the graph of a function f when

$$f(x) \rightarrow \infty \qquad \text{or} \qquad f(x) \rightarrow -\infty \qquad \text{as} \qquad x \rightarrow a$$

The line $y = b$ is a **horizontal asymptote** for the graph of a function f when

$$f(x) \rightarrow b \qquad \text{as} \qquad x \rightarrow \infty \qquad \text{or as} \qquad x \rightarrow -\infty$$

Figure 9-17 shows several graphs and their asymptotes.

Vertical asymptote at $x = a$

$f(x)$ approaches ∞ as x approaches a from left.

$f(x)$ approaches $-\infty$ as x approaches a from right.

Horizontal asymptote at $y = b$

$f(x)$ approaches b as x approaches ∞ or as x approaches $-\infty$.

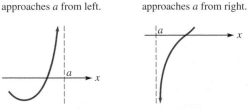

Figure 9-17

Finding Asymptotes

EXAMPLE 2   Find the asymptotes of each function:

$$\textbf{a. } y = \frac{x + 2}{x^2 - 1} \quad \text{and} \quad \textbf{b. } y = \frac{2x^2 + x + 2}{x^2 - 1}.$$

Solution   Each function has a vertical asymptote at $x = 1$ and $x = -1$, because at these values, their denominators are 0 but their numerators are not. We must also find the horizontal asymptotes, if any.

a. Because the degree of the numerator is less than the degree of the denominator, we can find the horizontal asymptote by dividing the numerator and the denominator by x^2, which is the largest power of x in the denominator.

$$y = \frac{x + 2}{x^2 - 1} = \frac{\dfrac{x}{x^2} + \dfrac{2}{x^2}}{\dfrac{x^2}{x^2} - \dfrac{1}{x^2}} = \frac{\dfrac{1}{x} + \dfrac{2}{x^2}}{1 - \dfrac{1}{x^2}}$$

Since $\dfrac{1}{x}, \dfrac{2}{x^2}$, and $\dfrac{1}{x^2}$ all approach 0 as $x \to \infty$, y approaches

$$\frac{0 + 0}{1 - 0} = 0$$

The horizontal asymptote is the line $y = 0$. If the degree of the numerator is less than the degree of the denominator, the horizontal asymptote is always $y = 0$.

b. To find the horizontal asymptote, we do a long division and write the result in the form quotient $+ \frac{\text{remainder}}{\text{divisor}}$.

$$
\begin{array}{r}
2 \\
x^2 - 1 \overline{)\, 2x^2 + x + 2} \\
\underline{2x^2 - 2} \\
x + 4
\end{array}
$$

Thus,

$$y = \frac{2x^2 + x + 2}{x^2 - 1} = 2 + \frac{x + 4}{x^2 - 1}$$

The last fraction approaches 0 as $x \to \infty$, for reasons discussed in part a. So $y \to 2$. The horizontal asymptote is the line $y = 2$. If the degrees of the numerator and denominator are the same, the horizontal asymptote is always $y =$ the lead coefficient of the numerator divided by the lead coefficient of the denominator.

Self Check   Find the asymptotes of $y = \dfrac{x^2}{x^2 - 4}$. ∎

EXAMPLE 3   Find the asymptotes of each function: **a.** $y = \dfrac{3x^3 + 2x^2 + 2}{x^2 - 1}$ and

b. $y = \dfrac{x^4 + x + 2}{x^2 - 1}.$

Solution   Again, the vertical asymptotes are at $x = 1$ and $x = -1$. We must also find the horizontal asymptotes.

a. To look for the horizontal asymptote, we do a long division and write the result in the form quotient $+ \frac{\text{remainder}}{\text{divisor}}$.

$$
\begin{array}{r}
3x + 2 \\
x^2 - 1 \overline{)3x^3 + 2x^2 + 2} \\
\underline{3x^3 - 3x } \\
2x^2 + 3x \\
\underline{2x^2 - 2} \\
3x + 4
\end{array}
$$

Thus,

$$
y = \frac{3x^3 + 2x^2 + 2}{x^2 - 1} = \boldsymbol{3x + 2} + \frac{3x + 4}{x^2 - 1}
$$

The last fraction approaches 0 as $x \to \infty$, and the graph approaches the line $y = 3x + 2$. Since the line is not horizontal, we call the asymptote a **slant asymptote.** If the degree of the numerator is 1 more than the degree of the denominator, there will be a slant asymptote.

b. To look for the horizontal asymptote, we do a long division and write the result in the form quotient $+ \frac{\text{remainder}}{\text{divisor}}$.

$$
\begin{array}{r}
x^2 + 1 \\
x^2 - 1 \overline{)x^4 + x + 2} \\
\underline{x^4 - x^2 } \\
x^2 + x \\
\underline{x^2 - 1} \\
x + 3
\end{array}
$$

Thus,

$$
y = \frac{x^4 + x + 2}{x^2 - 1} = \boldsymbol{x^2 + 1} + \frac{x + 3}{x^2 - 1}
$$

The last fraction approaches 0 as $x \to \infty$, so the curve approaches the parabola $y = x^2 + 1$. Since a parabola is not a line, the graph has no horizontal or slant asymptotes.

Self Check   Find the asymptotes of $y = \dfrac{2x^3 - 3x + 1}{x^2 - 4}$.   ∎

Graphing Rational Functions

We follow these steps to graph the rational function $y = \frac{P(x)}{Q(x)}$, where $P(x)$ and $Q(x)$ are polynomials written in descending powers of x and $\frac{P(x)}{Q(x)}$ is in simplest form.

Strategy for Graphing Rational Functions

$\left(\dfrac{P(x)}{Q(x)} \text{ is in simplified form.}\right)$

Check for symmetry.   If $P(x)$ and $Q(x)$ involve only even powers of x, then $f(x) = f(-x)$, and the graph is symmetric about the y-axis. Otherwise, y-axis symmetry does not exist. Check for symmetry about the origin.

Look for vertical asymptotes.   The real roots of $Q(x) = 0$, if any, determine the vertical asymptotes of the graph.

Look for y-intercepts.   Let $x = 0$. The resulting value of y, if any, is the y-intercept of the graph.

Look for x-intercepts. The real roots of $P(x) = 0$, if any, are the x-intercepts of the graph.

Look for horizontal asymptotes.

- If the degree of $P(x)$ is less than the degree of $Q(x)$, the line $y = 0$ is a horizontal asymptote.
- If the degrees of $P(x)$ and $Q(x)$ are equal, the line $y = \frac{p}{q}$, where p and q are the lead coefficients of $P(x)$ and $Q(x)$, is a horizontal asymptote.
- If the degree of $P(x)$ is greater than the degree of $Q(x)$, there is no horizontal asymptote.

Look for slant asymptotes.   If the degree of $P(x)$ is 1 greater than the degree of $Q(x)$, there is a slant asymptote. To find it, divide $P(x)$ by $Q(x)$ and ignore the remainder.

 Comment   A graph might intersect a horizontal asymptote when $|x|$ is small. However, it will approach but never touch a horizontal asymptote as $x \to \infty$ or $x \to -\infty$.

 EXAMPLE 4   Graph the function $y = f(x) = \dfrac{x^2 - 4}{x^2 - 1}$.

Solution   **Symmetry:**   Because x appears to even powers only, there is symmetry about the y-axis. There is no symmetry about the origin.

Vertical asymptotes: To find the vertical asymptotes, we set the denominator equal to 0 and solve for x.

$$x^2 - 1 = 0$$
$$x + 1 = 0 \quad \text{or} \quad x - 1 = 0$$
$$x = -1 \qquad\qquad x = 1$$

There will be vertical asymptotes at $x = -1$ and $x = 1$.

y-intercept: We can find the y-intercept by setting x equal to 0 and solving for y:

$$y = \frac{x^2 - 4}{x^2 - 1} = \frac{0^2 - 4}{0^2 - 1} = \frac{-4}{-1} = 4$$

The y-intercept is $(0, 4)$.

***x*-intercepts:**   We can find the *x*-intercepts by setting the numerator equal to 0 and solving for *x*:

$$x^2 - 4 = 0$$
$$(x + 2)(x - 2) = 0$$
$$x + 2 = 0 \quad \text{or} \quad x - 2 = 0$$
$$x = -2 \quad | \quad \quad x = 2$$

The *x*-intercepts are $(2, 0)$ and $(-2, 0)$.

Horizontal asymptotes: Since the degrees of the numerator and denominator polynomials are the same, the line

$$y = \frac{1}{1} \quad \begin{array}{l} \text{1 is the lead coefficient of the numerator.} \\ \text{1 is the lead coefficient of the denominator.} \end{array}$$

is a horizontal asymptote. The horizontal asymptote is the line $y = 1$. The graph is shown in Figure 9-18.

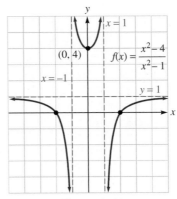

Figure 9-18

Self Check   Use a graphing calculator to graph the function in Example 4.   ∎

EXAMPLE 5   Graph the function $y = f(x) = \dfrac{3x}{x - 2}$.

Solution   **Symmetry:**   Because *x* appears to an odd power, there is no symmetry about the *y*-axis. There is no symmetry about the origin either.

Vertical asymptotes: To find the vertical asymptotes, we set the denominator equal to 0 and solve for *x*. Since the solution is 2, there will be a vertical asymptote at $x = 2$.

***y*-intercept:**   We can find the *y*-intercept by setting *x* equal to 0 and solving for *y*. Since the solution is 0, the *y*-intercept is $(0, 0)$. The graph passes through the origin.

***x*-intercepts:**   We can find the *x*-intercepts by setting the numerator equal to 0 and solving for *x*:

$$3x = 0$$
$$x = 0$$

The only x-intercept is $(0, 0)$.

Horizontal asymptotes: Since the degrees of the numerator and denominator polynomials are the same, the line

$$y = \frac{3}{1} \quad \begin{array}{l} \text{3 is the lead coefficient of the numerator.} \\ \text{1 is the lead coefficient of the denominator.} \end{array}$$

is a horizontal asymptote. The horizontal asymptote is the line $y = 3$.

   To find what happens when x is greater than 2, we pick a value of x greater than 2 and find the corresponding value of y. If $x = 3$, then $y = 9$. After plotting the point $(3, 9)$, we use the intercepts and asymptotes to sketch the graph shown in Figure 9-19.

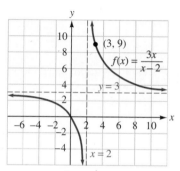

Figure 9-19

Self Check   Use a graphing calculator to graph the function in Example 5.

EXAMPLE 6   Graph the function $y = f(x) = \dfrac{1}{x(x - 1)^2}$.

Solution   **Symmetry:**   The expansion of $x(x - 1)^2$ gives a third-degree polynomial in which x appears to an odd power. So there is no symmetry about the y-axis. Because $y = f(x)$ and $-y = f(-x)$ are not equivalent, there is no symmetry about the origin.

Vertical asymptotes: Since 0 and 1 make the denominator 0, the vertical asymptotes are the lines $x = 0$ ad $x = 1$.

y-intercepts: Since x cannot be 0, the graph has no y-intercept.

x-intercepts: Since y cannot be 0, the graph has no x-intercept.

Horizontal asymptotes: Since the degree of the numerator is less than the degree of the denominator, the line $y = 0$ is a horizontal asymptote.

   The table in Figure 9-20 gives three points lying between the asymptotes. The values of y change sign at $x = 0$: To the left of the y-axis, $y < 0$, and to the right of the y-axis, $y > 0$. This happens because x appears to an odd power as a factor in the denominator. The value of y does not change sign at $x = 1$: To the left and to the right of the asymptote $x = 1$, y is positive. The function behaves this way because $(x - 1)$ appears to an even power in the denominator. The graph of the function appears in Figure 9-20.

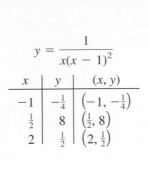

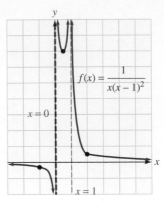

Figure 9-20

Self Check   Using a graphing calculator to graph the function in Example 6.   ∎

EXAMPLE 7   Graph the function $y = f(x) = \dfrac{1}{x^2 + 1}$.

Solution   **Symmetry:**   Because x appears only to an even power, the graph will be symmetric about the y-axis.

Vertical asymptotes: Since no real numbers x make the denominator 0, the graph has no vertical asymptotes.

y-intercepts: Since $f(0) = 1$, the y-intercept of the graph is $(0, 1)$.

x-intercepts: Because the denominator is always positive and the numerator is a positive constant, the graph lies entirely above the x-axis. There are no x-intercepts.

Horizontal asymptotes: Since $f(x) \to 0$ as $x \to \infty$ and $x \to -\infty$, the graph has $y = 0$ as a horizontal asymptote.
    The graph appears in Figure 9-21.

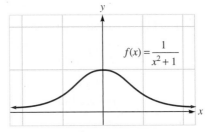

Figure 9-21

Self Check   Use a graphing calculator to graph the function in Example 7.   ∎

EXAMPLE 8   Graph the function $y = f(x) = \dfrac{x^2 + x - 2}{x - 3}$.

Solution   We first factor the numerator of the function

$$y = \frac{(x - 1)(x + 2)}{x - 3}$$

Symmetry: The function is not symmetric about the y-axis or the origin.

Vertical asymptotes: The vertical asymptote is the line $x = 3$.

y-intercepts: The y-intercept is $\left(0, \frac{2}{3}\right)$.

x-intercepts: The x-intercepts are $(1, 0)$ and $(-2, 0)$.

Horizontal asymptotes: We perform a long division and write the expression as

$$y = \frac{x^2 + x - 2}{x - 3} = x + 4 + \frac{10}{x - 3}$$

The fraction $\frac{10}{x-3} \to 0$ as $x \to \infty$. This time the graph does not approach a constant. It approaches the line $y = x + 4$. Because this line is not horizontal, the graph has no horizontal asymptote. However, it does have a slant asymptote: the line $y = x + 4$. The graph appears in Figure 9-22.

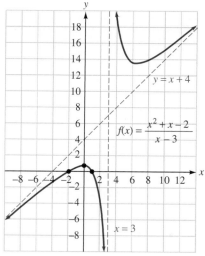

Figure 9-22

Self Check   Use a graphing calculator to graph the function in Example 8.

Graphs with Discontinuities

We have discussed rational functions where the fraction is in simplified form. We now consider a rational function $y = \frac{P(x)}{Q(x)}$ where $P(x)$ and $Q(x)$ have a common factor. Graphs of such functions contain gaps or **discontinuities** that are not the result of vertical asymptotes.

EXAMPLE 9   Find the domain of the function $y = f(x) = \dfrac{x^2 - x - 12}{x - 4}$ and graph it.

Solution   Since a denominator cannot be 0, $x \neq 4$. The domain is the set of all real numbers except 4.

When we factor the numerator of the expression, we see that the numerator and denominator have a common factor of $x - 4$.

$$y = \frac{x^2 - x - 12}{x - 4}$$

$$= \frac{(x + 3)(x - 4)}{x - 4}$$

If $x \neq 4$, the common factor of $x - 4$ can be divided out. The resulting function is equivalent to the original function only when we keep the restriction that $x \neq 4$. Thus,

$$y = \frac{(x + 3)(x - 4)}{x - 4} = x + 3 \quad (x \neq 4)$$

When $x = 4$, the function is not defined. The graph of the function appears in Figure 9-23. It is a line with a point missing—the point with an x-coordinate of 4.

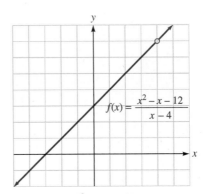

Figure 9-23

Self Check   Use a graphing calculator to graph the function in Example 9.

Self Check Answers

1. $(-\infty, -1) \cup (-1, 2) \cup (2, \infty)$ **4.**
2. vertical asymptote at $x = 2$
and $x = -2$; horizontal
asymptote at $y = 1$
3. vertical asymptotes at $x = 2$
and $x = -2$; slant asymptote
of $y = 2x$

5.

6.

7.    **8.** 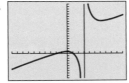   **9.** 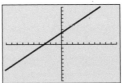   Note that the calculator graph does not indicate the excluded point.

Orals   *Give the domain of each rational function.*

1. $f(x) = \dfrac{7}{x}$

2. $f(x) = \dfrac{x}{x - 5}$

3. $f(x) = \dfrac{4x}{x - 4}$

4. $f(x) = \dfrac{x^2}{(x + 1)(x - 1)}$

9.3 EXERCISES

REVIEW *Perform each operation.*

1. $(2x^2 + 3x) + (x^2 - 2x)$
2. $(3x + 2) - (x^2 + 2)$
3. $(5x + 2)(2x + 5)$
4. $\dfrac{2x^2 + 3x + 1}{x + 1}$
5. If $f(x) = 3x + 2$, find $f(x + 1)$.
6. If $f(x) = x^2 + x$, find $f(2x + 1)$.

VOCABULARY AND CONCEPTS *Fill in the blanks.*

7. When a graph approaches a line, we call the line an _____.

8. A rational function is a function with a polynomial numerator and a _____ polynomial denominator.

9. To find any _____ asymptotes of a rational function, set the denominator polynomial equal to 0 and solve the equation.

10. To find the _____ of a rational function, let $x = 0$ and solve for y.

11. To find the _____ of a rational function, set the numerator equal to 0 and solve the equation.

12. In the function $y = \frac{P(x)}{Q(x)}$, if the degree of $P(x)$ is less than the degree of $Q(x)$, the horizontal asymptote is _____.

13. In the function $y = \frac{P(x)}{Q(x)}$, if the degree of $P(x)$ and $Q(x)$ is _____, the horizontal asymptote is

$$y = \frac{\text{the lead coefficient of the numerator}}{\text{the lead coefficient of the denominator}}$$

14. In a rational function, if the degree of the numerator is 1 greater than the degree of the denominator, the graph will have a _____.

15. A graph can cross a _____ asymptote only when $|x|$ is small.

16. A graph with discontinuities will have _____ points.

Find the domain of each rational function. **Do not graph the function.**

17. $f(x) = \dfrac{x^2}{x - 2}$
18. $f(x) = \dfrac{x^3 - 3x^2 + 1}{x + 3}$

19. $f(x) = \dfrac{2x^2 + 7x - 2}{x^2 - 25}$
20. $f(x) = \dfrac{5x^2 + 1}{x^2 + 5}$

21. $f(x) = \dfrac{x - 1}{x^3 - x}$
22. $f(x) = \dfrac{x + 2}{2x^2 - 9x + 9}$

23. $f(x) = \dfrac{3x^2 + 5}{x^2 + 1}$
24. $f(x) = \dfrac{7x^2 - x + 2}{x^4 + 4}$

Find all vertical, horizontal, and slant asymptotes, x- and y-intercepts, and symmetries, and then graph each function. Check your work with a graphing calculator.

25. $f(x) = \dfrac{1}{x - 2}$
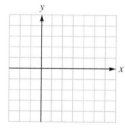

26. $f(x) = \dfrac{3}{x + 3}$
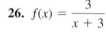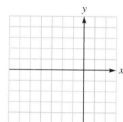

27. $f(x) = \dfrac{x}{x - 1}$
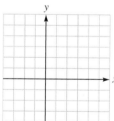

28. $f(x) = \dfrac{x}{x + 2}$
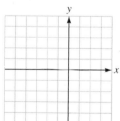

29. $f(x) = \dfrac{x + 1}{x + 2}$
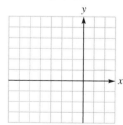

30. $f(x) = \dfrac{x - 1}{x - 2}$
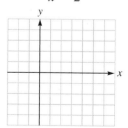

31. $f(x) = \dfrac{2x - 1}{x - 1}$

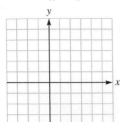

32. $f(x) = \dfrac{3x + 2}{x^2 - 4}$

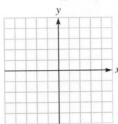

39. $f(x) = \dfrac{x^2 - 9}{x^2}$

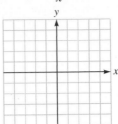

40. $f(x) = \dfrac{3x^2 - 12}{x^2}$

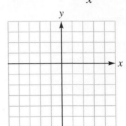

33. $f(x) = \dfrac{x^2 - 9}{x^2 - 4}$

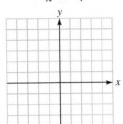

34. $f(x) = \dfrac{x^2 - 4}{x^2 - 9}$

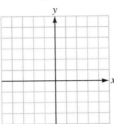

41. $f(x) = \dfrac{x}{(x + 3)^2}$

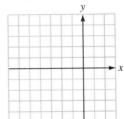

42. $f(x) = \dfrac{x}{(x - 1)^2}$

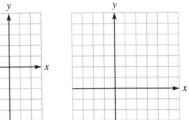

35. $f(x) = \dfrac{x^2 - x - 2}{x^2 - 4x + 3}$

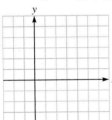

36. $f(x) = \dfrac{x^2 + 7x + 12}{x^2 - 7x + 12}$

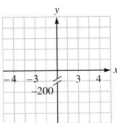

43. $f(x) = \dfrac{x + 1}{x^2(x - 2)}$

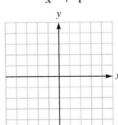

44. $f(x) = \dfrac{x - 1}{x^2(x + 2)^2}$

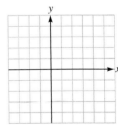

45. $f(x) = \dfrac{x}{x^2 + 1}$

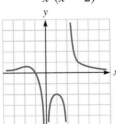

46. $f(x) = \dfrac{x - 1}{x^2 + 2}$

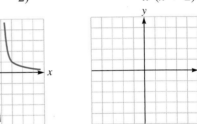

37. $f(x) = \dfrac{x^2 + 2x - 3}{x^3 - 4x}$

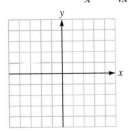

38. $f(x) = \dfrac{3x^2 - 4x + 1}{2x^3 + 3x^2 + x}$

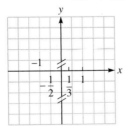

47. $f(x) = \dfrac{3x^2}{x^2 + 1}$

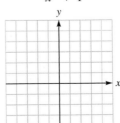

48. $f(x) = \dfrac{x^2 - 9}{2x^2 + 1}$

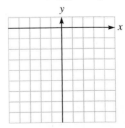

49. $f(x) = \dfrac{x^2 - 2x - 8}{x - 1}$

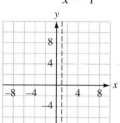

50. $f(x) = \dfrac{x^2 + x - 6}{x + 2}$

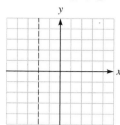

55. $f(x) = \dfrac{x^3 + x}{x}$

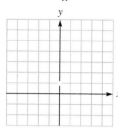

56. $f(x) = \dfrac{x^3 - x^2}{x - 1}$

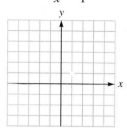

51. $f(x) = \dfrac{x^3 + x^2 + 6x}{x^2 - 1}$

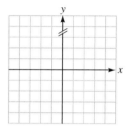

57. $f(x) = \dfrac{x^2 - 2x + 1}{x - 1}$

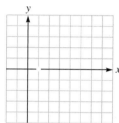

58. $f(x) = \dfrac{2x^2 + 3x - 2}{x + 2}$

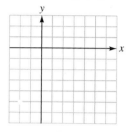

52. $f(x) = \dfrac{x^3 - 2x^2 + x}{x^2 - 4}$

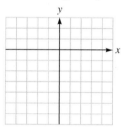

59. $f(x) = \dfrac{x^3 - 1}{x - 1}$

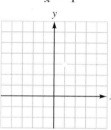

60. $f(x) = \dfrac{x^2 - x}{x^2}$

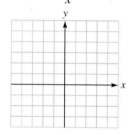

Graph each rational function. Note that the numerator and denominator of each fraction share a common factor.

53. $f(x) = \dfrac{x^2}{x}$

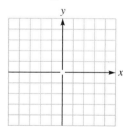

54. $f(x) = \dfrac{x^2 - 1}{x - 1}$

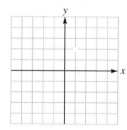

APPLICATIONS   *Refer to the equestrian example that opened this section.*

61. Find the total cost if the owner keeps the horse for 10 years.

62. Find the average cost per month if the owner keeps the horse for 10 years.

63. Find the total cost if the owner keeps the horse for 15 years.

64. Find the average cost per month if the owner keeps the horse for 15 years.

576 Chapter 9 More Functions and Operations on Functions

WRITING

65. Can a rational function have two horizontal asymptotes? Explain.

66. Can a rational function have two slant asymptotes? Explain.

SOMETHING TO THINK ABOUT *In Exercises 67–70, a, b, c, and d are nonzero constants.*

67. Show that the graph of
$$y = \frac{ax + b}{cx^2 + d}$$
has the horizontal asymptote $y = 0$.

68. Show that the graph of
$$y = \frac{ax^2 + b}{cx^2 + d}$$
has the horizontal asymptote $y = \frac{a}{c}$.

69. Show that the graph of
$$y = \frac{ax^3 + b}{cx^2 + d}$$
has the slant asymptote $y = \frac{a}{c}x$.

70. Graph the rational function
$$y = \frac{x^3 + 1}{x}$$
and explain why the curve is said to have a *parabolic asymptote*.

Use a graphing calculator to perform each experiment.

71. Investigate the positioning of the vertical asymptotes of a rational function by graphing $f(x) = \frac{x}{x - k}$ for several values of k. What do you observe?

72. Investigate the positioning of the vertical asymptotes of a rational function by graphing $f(x) = \frac{x}{x^2 - k}$ for $k = 4, 1, -1$, and 0. What do you observe?

73. Find the range of the rational function $f(x) = \frac{kx^2}{x^2 + 1}$ for several values of k. What do you observe?

74. Investigate the positioning of the x-intercepts of a rational function by graphing $f(x) = \frac{x^2 - k}{x}$ for $k = 1, -1$, and 0. What do you observe?

9.4 Partial Fractions

- When the Denominator Has Distinct Linear Factors
- When the Denominator Has Distinct Quadratic Factors
- When the Denominator Has Repeated Linear Factors
- When the Denominator Has Repeated Quadratic Factors
- When the Degree of $P(x)$ is Equal to or Greater Than the Degree of $Q(x)$

Getting Ready *Add the fractions.*

1. $\frac{1}{x} + \frac{2}{x + 1}$

2. $\frac{3}{x + 1} - \frac{2}{x - 1}$

In this section, we discuss how to write complicated fractions as sums of simpler fractions. We begin by reviewing how to add fractions. For example, to find the sum

$$\frac{2}{x} + \frac{6}{x+1} + \frac{-1}{(x+1)^2}$$

we write each fraction with an LCD of $x(x+1)^2$, add the fractions by adding their numerators and keeping the common denominator, and simplify.

$$\frac{2}{x} + \frac{6}{x+1} + \frac{-1}{(x+1)^2} = \frac{2(x+1)^2}{x(x+1)^2} + \frac{6x(x+1)}{(x+1)x(x+1)} + \frac{-1x}{(x+1)^2x}$$

$$= \frac{2x^2 + 4x + 2 + 6x^2 + 6x - x}{x(x+1)^2}$$

$$= \frac{8x^2 + 9x + 2}{x(x+1)^2}$$

To reverse the addition process and write a fraction as the sum of simpler fractions with denominators of smallest possible degree, we must **decompose a fraction into partial fractions.** For example, to decompose the fraction

$$\frac{8x^2 + 9x + 2}{x(x+1)^2}$$

into partial fractions, we will assume that there are constants, A, B, and C such that

$$\frac{8x^2 + 9x + 2}{x(x+1)^2} = \frac{A}{x} + \frac{B}{x+1} + \frac{C}{(x+1)^2}$$

After writing the terms on the right-hand side as fractions with an LCD of $x(x+1)^2$, we add the fractions to get

$$\frac{8x^2 + 9x + 2}{x(x+1)^2} = \frac{A(x+1)^2}{x(x+1)^2} + \frac{Bx(x+1)}{x(x+1)(x+1)} + \frac{Cx}{(x+1)^2x}$$

$$= \frac{Ax^2 + 2Ax + A + Bx^2 + Bx + Cx}{x(x+1)^2}$$

(1) $$\frac{8x^2 + 9x + 2}{x(x+1)^2} = \frac{(A+B)x^2 + (2A+B+C)x + A}{x(x+1)^2}$$
Factor out x^2 from $Ax^2 + Bx^2$. Factor x from $2Ax + Bx + Cx$.

Since the fractions on the left- and right-hand sides of Equation 1 are equal, the coefficients of their polynomial numerators are equal.

$$\begin{cases} A + B & = 8 \quad \text{The coefficients of } x^2. \\ 2A + B + C = 9 \quad \text{The coefficients of } x. \\ A & = 2 \quad \text{The constants.} \end{cases}$$

We can solve this system to find that $A = 2$, $B = 6$, and $C = -1$. Then we know that

$$\frac{8x^2 + 9x + 2}{x(x+1)^2} = \frac{2}{x} + \frac{6}{x+1} + \frac{-1}{(x+1)^2}$$

The method of partial fractions uses the following theorem.

Polynomial Factorization Theorem

> The factorization of any polynomial $Q(x)$ with real coefficients is the product of polynomials of the forms
>
> $$(ax + b)^n \qquad \text{and} \qquad (ax^2 + bx + c)^n$$
>
> where n is a positive integer and $ax^2 + bx + c$ is irreducible over the real numbers.

The example and this theorem illustrate the topic of this section. We begin with an algebraic fraction—like $\frac{P(x)}{Q(x)}$, the quotient of two polynomials—and write it as the sum of two or more fractions with simpler denominators. By the theorem, we know that they will be either first-degree or irreducible second-degree polynomials, or powers of those. We consider each case separately.

We suppose that $P(x)$ and $Q(x)$ have real coefficients, that the degree of $P(x)$ is less than the degree of $Q(x)$, and that the fraction $\frac{P(x)}{Q(x)}$ is in lowest terms.

When the Denominator Has Distinct Linear Factors

EXAMPLE 1 Decompose $\dfrac{9x + 2}{(x + 2)(3x - 2)}$ into partial fractions.

Solution Since each denominator is linear, there are constants A and B such that

$$\frac{9x + 2}{(x + 2)(3x - 2)} = \frac{A}{x + 2} + \frac{B}{3x - 2}$$

After writing the terms on the right-hand side as fractions with an LCD of $(x + 2)(3x - 2)$, we add the fractions to get

$$\frac{9x + 2}{(x + 2)(3x - 2)} = \frac{A(3x - 2)}{(x + 2)(3x - 2)} + \frac{B(x + 2)}{(x + 2)(3x - 2)}$$

$$\frac{9x + 2}{(x + 2)(3x - 2)} = \frac{3Ax - 2A + Bx + 2B}{(x + 2)(3x - 2)}$$

(2) $$\frac{9x + 2}{(x + 2)(3x - 2)} = \frac{(3A + B)x - 2A + 2B}{(x + 2)(3x - 2)} \qquad \text{Factor } x \text{ from } 3Ax + Bx.$$

Since the fractions in Equation 2 are equal, the coefficients of their polynomial numerators are equal, and we have

$$\begin{cases} 9 = 3A + B & \text{The coefficients of } x. \\ 2 = -2A + 2B & \text{The constants.} \end{cases}$$

After solving this system to find that $A = 2$ and $B = 3$, we have

$$\frac{9x + 2}{(x + 2)(3x - 2)} = \frac{2}{x + 2} + \frac{3}{3x - 2}$$

Self Check Decompose $\dfrac{2x + 2}{x(x + 2)}$ into partial fractions. ∎

When the Denominator Has Distinct Quadratic Factors

If $Q(x)$ has a prime quadratic factor of the form $ax^2 + bx + c$, the partial fraction decomposition of $\frac{P(x)}{Q(x)}$ will have a corresponding term of the form

$$\frac{Bx + C}{ax^2 + bx + c}$$

 Comment Be sure that the quadratic factor really is prime—check it with the discriminant. If $b^2 - 4ac$ is negative, the factor is prime.

EXAMPLE 2 Decompose $\dfrac{2x^2 + x + 1}{x^3 + x}$ into partial fractions.

Solution Since the denominator can be written as a product of a linear factor and prime quadratic factor, the partial fractions have the form

$$\frac{2x^2 + x + 1}{x(x^2 + 1)} = \frac{A}{x} + \frac{Bx + C}{x^2 + 1}$$

$$= \frac{A(x^2 + 1)}{x(x^2 + 1)} + \frac{(Bx + C)x}{x(x^2 + 1)}$$

$$= \frac{Ax^2 + A + Bx^2 + Cx}{x(x^2 + 1)}$$

$$= \frac{(A + B)x^2 + Cx + A}{x(x^2 + 1)} \qquad \text{Factor } x^2 \text{ from } Ax^2 + Bx^2.$$

We can equate the corresponding coefficients of the numerators $2x^2 + x + 1$ and $(A + B)x^2 + Cx + A$ to get the system

$$\begin{cases} A + B = 2 & \text{The coefficients of } x^2. \\ C = 1 & \text{The coefficients of } x. \\ A = 1 & \text{To constants.} \end{cases}$$

with solutions $A = 1, B = 1, C = 1$. The partial fraction decomposition is

$$\frac{2x^2 + x + 1}{x(x^2 + 1)} = \frac{A}{x} + \frac{Bx + C}{x^2 + 1}$$

$$= \frac{1}{x} + \frac{1x + 1}{x^2 + 1}$$

$$= \frac{1}{x} + \frac{x + 1}{x^2 + 1}$$

Self Check Decompose $\dfrac{2x^2 + 1}{x^3 + x}$ into partial fractions. ∎

When the Denominator Has Repeated Linear Factors

If $Q(x)$ has n linear factors of $ax + b$, then $(ax + b)^n$ is a factor of $Q(x)$. Each factor of the form $(ax + b)^n$ generates the following sum of n partial fractions:

$$\frac{A}{ax + b} + \frac{B}{(ax + b)^2} + \frac{C}{(ax + b)^3} + \cdots + \frac{D}{(ax + b)^n}$$

EXAMPLE 3   Decompose $\dfrac{3x^2 - x + 1}{x(x - 1)^2}$ into partial fractions.

Solution   In this example, each factor in the denominator is linear. The linear factor x appears once, and the linear factor $x - 1$ appears twice. Thus, there are constants A, B, and C such that

$$\frac{3x^2 - x + 1}{x(x - 1)^2} = \frac{A}{x} + \frac{B}{x - 1} + \frac{C}{(x - 1)^2}$$

After writing the terms on the right-hand side as fractions with an LCD of $x(x - 1)^2$, we combine them to get

$$\frac{3x^2 - x + 1}{x(x - 1)^2} = \frac{A(x - 1)^2}{x(x - 1)^2} + \frac{Bx(x - 1)}{x(x - 1)(x - 1)} + \frac{Cx}{(x - 1)^2 x}$$

$$= \frac{Ax^2 - 2Ax + A + Bx^2 - Bx + Cx}{x(x - 1)^2}$$

$$= \frac{(A + B)x^2 + (-2A - B + C)x + A}{x(x - 1)^2} \qquad \begin{array}{l}\text{Factor } x^2 \text{ from } Ax^2 + Bx^2.\\ \text{Factor } x \text{ from } -2Ax - Bx + Cx.\end{array}$$

Since the fractions are equal, the coefficients of the polynomial numerators are equal, and we have

$$\begin{cases} A + B = 3 & \text{The coefficients of } x^2. \\ -2A - B + C = -1 & \text{The coefficients of } x. \\ A = 1 & \text{The constants.} \end{cases}$$

We can solve this system to find that $A = 1$, $B = 2$, and $C = 3$. So we have

$$\frac{3x^2 - x + 1}{x(x - 1)^2} = \frac{A}{x} + \frac{B}{x - 1} + \frac{C}{(x - 1)^2}$$

$$= \frac{1}{x} + \frac{2}{x - 1} + \frac{3}{(x - 1)^2}$$

Self Check   Decompose $\dfrac{3x^2 + 7x + 1}{x(x + 1)^2}$ into partial fractions.   ∎

When the Denominator Has Repeated Quadratic Factors

If $Q(x)$ has n prime factors of $ax^2 + bx + c$, then $(ax^2 + bx + c)^n$ is a factor. Each factor of the form $(ax^2 + bx + c)^n$ generates a sum of n partial fractions of the form

$$\frac{Ax + B}{ax^2 + bx + c} + \frac{Cx + D}{(ax^2 + bx + c)^2} + \cdots + \frac{Ex + F}{(ax^2 + bx + c)^n}$$

EXAMPLE 4   Decompose $\dfrac{3x^2 + 5x + 5}{(x^2 + 1)^2}$ into partial fractions.

Solution   Since the quadratic factor $x^2 + 1$ is used twice, we must find constants A, B, C, and D such that

$$\frac{3x^2 + 5x + 5}{(x^2 + 1)^2} = \frac{Ax + B}{x^2 + 1} + \frac{Cx + D}{(x^2 + 1)^2}$$

We add the fractions on the right-hand side to get

$$\frac{3x^2 + 5x + 5}{(x^2 + 1)^2} = \frac{(Ax + B)(x^2 + 1)}{(x^2 + 1)(x^2 + 1)} + \frac{Cx + D}{(x^2 + 1)^2}$$

$$= \frac{Ax^3 + Ax + Bx^2 + B + Cx + D}{(x^2 + 1)^2}$$

$$= \frac{Ax^3 + Bx^2 + (A + C)x + B + D}{(x^2 + 1)^2} \qquad \text{Factor } x \text{ from } Ax + Cx.$$

If we add the term $0x^3$ to the numerator on the left-hand side, we can equate the corresponding coefficients in the numerators to get

$$\begin{aligned} A &= 0 & \text{The coefficients of } x^3. \\ B &= 3 & \text{The coefficients of } x^2. \\ A + C &= 5 & \text{The coefficients of } x. \\ B + D &= 5 & \text{The constants.} \end{aligned}$$

with a solution of $A = 0$, $B = 3$, $C = 5$, and $D = 2$. So we have

$$\frac{3x^2 + 5x + 5}{(x^2 + 1)^2} = \frac{Ax + B}{x^2 + 1} + \frac{Cx + D}{(x^2 + 1)^2}$$

$$= \frac{0x + 3}{x^2 + 1} + \frac{5x + 2}{(x^2 + 1)^2}$$

$$= \frac{3}{x^2 + 1} + \frac{5x + 2}{(x^2 + 1)^2}$$

Self Check   Decompose $\dfrac{x^3 + 2x + 3}{(x^2 + 2)^2}$ into partial fractions.

When the Degree of P(x) is Equal to or Greater Than the Degree of Q(x)

When the degree of $P(x)$ is equal to or greater than the degree of $Q(x)$ in the fraction $\frac{P(x)}{Q(x)}$, we perform a long division before decomposing the fraction into partial fractions.

EXAMPLE 5   Decompose $\dfrac{x^2 + 4x + 2}{x^2 + x}$ into partial fractions.

Solution   Because the degree of the numerator and denominator are the same, we must do a long division and express the fraction in quotient $+ \frac{\text{remainder}}{\text{divisor}}$ form:

$$
\begin{array}{r}
1 \\
x^2 + x\,\overline{)\,x^2 + 4x + 2} \\
\underline{x^2 + x} \\
3x + 2
\end{array}
$$

So we can write

(3) $\dfrac{x^2 + 4x + 2}{x^2 + x} = 1 + \dfrac{3x + 2}{x^2 + x}$

Because the degree of the numerator of the fraction on the right-hand side of Equation 3 is less than the degree of the denominator, we can find its partial fraction decomposition:

$$
\begin{aligned}
\frac{3x + 2}{x^2 + x} &= \frac{3x + 2}{x(x + 1)} \\[4pt]
&= \frac{A}{x} + \frac{B}{x + 1} \\[4pt]
&= \frac{A(x + 1) + Bx}{x(x + 1)} \\[4pt]
&= \frac{(A + B)x + A}{x(x + 1)}
\end{aligned}
$$

We equate the corresponding coefficients in the numerator and solve the resulting system of equations to find that the solution is $A = 2, B = 1$. So we have

$$
\frac{x^2 + 4x + 2}{x^2 + x} = 1 + \frac{2}{x} + \frac{1}{x + 1}
$$

Self Check   Decompose $\dfrac{x^2 + x - 1}{x^2 - x}$ into partial fractions.   ∎

Self Check Answers

1. $\dfrac{1}{x} + \dfrac{1}{x + 2}$ **2.** $\dfrac{1}{x} + \dfrac{x}{x^2 + 1}$ **3.** $\dfrac{1}{x} + \dfrac{2}{x + 1} + \dfrac{3}{(x + 1)^2}$ **4.** $\dfrac{x}{x^2 + 2} + \dfrac{3}{(x^2 + 2)^2}$ **5.** $1 + \dfrac{1}{x} + \dfrac{1}{x - 1}$

Orals   *List the denominators that will be used in the decomposition of each fraction.*

1. $\dfrac{3x - 1}{(x + 1)(x - 2)}$ **2.** $\dfrac{x + 5}{x^2 + x}$ **3.** $\dfrac{3x^2 + 2x + 1}{x^3 + x}$ **4.** $\dfrac{2x^2 + x + 4}{(x^2 + 2)^2}$

9.4 EXERCISES 🔲

REVIEW *Simplify each radical expression. Use absolute value symbols if necessary.*

1. $\sqrt{8a^3b}$

2. $\sqrt{x^2 + 6x + 9}$

3. $\sqrt{18x^5} + x^2\sqrt{50x} - 5x^2\sqrt{2x}$

4. $\dfrac{\sqrt{x^2y^3}}{\sqrt{xy^5}}$

Solve each equation.

5. $\sqrt{x - 5} = x - 7$

6. $x - 5 = (x - 7)^2$

VOCABULARY AND CONCEPTS *Fill in the blanks.*

7. A polynomial with real coefficients factors as the product of _____ and _____ factors or powers of those.

8. If the second-degree factors of a polynomial with real coefficients are _____, they don't factor further.

PRACTICE *Decompose each fraction into partial fractions.*

9. $\dfrac{3x - 1}{x(x - 1)}$

10. $\dfrac{4x + 6}{x(x + 2)}$

11. $\dfrac{2x - 15}{x(x - 3)}$

12. $\dfrac{5x + 21}{x(x + 7)}$

13. $\dfrac{3x + 1}{(x + 1)(x - 1)}$

14. $\dfrac{9x - 3}{(x + 1)(x - 2)}$

15. $\dfrac{-2x + 11}{x^2 - x - 6}$

16. $\dfrac{7x + 2}{x^2 + x - 2}$

17. $\dfrac{3x - 23}{x^2 + 2x - 3}$

18. $\dfrac{-x - 17}{x^2 - x - 6}$

19. $\dfrac{9x - 31}{2x^2 - 13x + 15}$

20. $\dfrac{-2x - 6}{3x^2 - 7x + 2}$

21. $\dfrac{4x^2 + 4x - 2}{x(x^2 - 1)}$

22. $\dfrac{x^2 - 6x - 13}{(x + 2)(x^2 - 1)}$

23. $\dfrac{x^2 + x + 3}{x(x^2 + 3)}$

24. $\dfrac{5x^2 + 2x + 2}{x^3 + x}$

25. $\dfrac{3x^2 + 8x + 11}{(x + 1)(x^2 + 2x + 3)}$

26. $\dfrac{-3x^2 + x - 5}{(x + 1)(x^2 + 2)}$

27. $\dfrac{5x^2 + 9x + 3}{x(x + 1)^2}$

28. $\dfrac{2x^2 - 7x + 2}{x(x - 1)^2}$

29. $\dfrac{-2x^2 + x - 2}{x^2(x - 1)}$

30. $\dfrac{x^2 + x + 1}{x^3}$

31. $\dfrac{3x^2 - 13x + 18}{x^3 - 6x^2 + 9x}$

32. $\dfrac{3x^2 + 13x + 20}{x^3 + 4x^2 + 4x}$

33. $\dfrac{x^2 - 2x - 3}{(x - 1)^3}$

34. $\dfrac{x^2 + 8x + 18}{(x + 3)^3}$

35. $\dfrac{x^3 + 4x^2 + 2x + 1}{x^4 + x^3 + x^2}$

36. $\dfrac{3x^3 + 5x^2 + 3x + 1}{x^2(x^2 + x + 1)}$

37. $\dfrac{4x^3 + 5x^2 + 3x + 4}{x^2(x^2 + 1)}$

38. $\dfrac{2x^2 + 1}{x^4 + x^2}$

39. $\dfrac{-x^2 - 3x - 5}{x^3 + x^2 + 2x + 2}$

40. $\dfrac{-2x^3 + 7x^2 + 6}{x^2(x^2 + 2)}$

41. $\dfrac{x^3 + 4x^2 + 3x + 6}{(x^2 + 2)(x^2 + x + 2)}$ **42.** $\dfrac{x^3 + 3x^2 + 2x + 4}{(x^2 + 1)(x^2 + x + 2)}$

51. $\dfrac{2x^4 + 2x^3 + 3x^2 - 1}{(x^2 - x)(x^2 + 1)}$

52. $\dfrac{x^4 - x^3 + 5x^2 + x + 6}{(x^2 + 3)(x^2 + 1)}$

43. $\dfrac{2x^4 + 6x^3 + 20x^2 + 22x + 25}{x(x^2 + 2x + 5)^2}$

WRITING

53. Explain how to decompose $\dfrac{x + 3}{x(x + 1)}$ into partial fractions.

44. $\dfrac{x^3 + 3x^2 + 6x + 6}{(x^2 + x + 5)(x^2 + 1)}$ **45.** $\dfrac{x^3}{x^2 + 3x + 2}$

54. Explain how to decompose $\dfrac{x - 5}{x^2 - 1}$ into partial fractions.

46. $\dfrac{2x^3 + 6x^2 + 3x + 2}{x^3 + x^2}$ **47.** $\dfrac{3x^3 + 3x^2 + 6x + 4}{3x^3 + x^2 + 3x + 1}$

SOMETHING TO THINK ABOUT

55. Is the polynomial $x^3 + 1$ prime?

48. $\dfrac{x^4 + x^3 + 3x^2 + x + 4}{(x^2 + 1)^2}$ **49.** $\dfrac{x^3 + 3x^2 + 2x + 1}{x^3 + x^2 + x}$

56. Decompose $\dfrac{1}{x^3 + 1}$ into partial fractions.

50. $\dfrac{x^4 - x^3 + x^2 - x + 1}{(x^2 + 1)^2}$

9.5   Algebra and Composition of Functions

▮ **Algebra of Functions** ▮ **Composition of Functions**
▮ **The Identity Function** ▮ **The Difference Quotient**
▮ **Problem Solving**

Getting Ready *Assume that $P(x) = 2x + 1$ and $Q(x) = x - 2$. Find each expression.*

1. $P(x) + Q(x)$ **2.** $P(x) - Q(x)$

3. $P(x) \cdot Q(x)$ **4.** $\dfrac{P(x)}{Q(x)}$

Throughout the text, we have talked about functions. In this section, we will show how to add, subtract, multiply, and divide them.

Algebra of Functions

We now consider how functions can be added, subtracted, multiplied, and divided.

Operations on Functions

If the domains and ranges of functions f and g are subsets of the real numbers,

The **sum** of f and g, denoted as $f + g$, is defined by

$$(f + g)(x) = f(x) + g(x)$$

The **difference** of f and g, denoted as $f - g$, is defined by

$$(f - g)(x) = f(x) - g(x)$$

The **product** of f and g, denoted as $f \cdot g$, is defined by

$$(f \cdot g)(x) = f(x)g(x)$$

The **quotient** of f and g, denoted as f/g, is defined by

$$(f/g)(x) = \frac{f(x)}{g(x)} \quad (g(x) \neq 0)$$

The domain of each of these functions is the set of real numbers x that are in the domain of both f and g. In the case of the quotient, there is the further restriction that $g(x) \neq 0$.

EXAMPLE 1 Let $f(x) = 2x^2 + 1$ and $g(x) = 5x - 3$. Find each function and its domain: **a.** $f + g$ and **b.** $f - g$.

Solution **a.** $(f + g)(x) = f(x) + g(x)$
$$= (2x^2 + 1) + (5x - 3)$$
$$= 2x^2 + 5x - 2$$

The domain of $f + g$ is the set of real numbers that are in the domain of both f and g. Since the domain of both f and g is the interval $(-\infty, \infty)$, the domain of $f + g$ is also the interval $(-\infty, \infty)$.

b. $(f - g)(x) = f(x) - g(x)$
$$= (2x^2 + 1) - (5x - 3)$$
$$= 2x^2 + 1 - 5x + 3 \qquad \text{Remove parentheses.}$$
$$= 2x^2 - 5x + 4 \qquad \text{Combine like terms.}$$

Since the domain of both f and g is $(-\infty, \infty)$, the domain of $f - g$ is also the interval $(-\infty, \infty)$.

Self Check Let $f(x) = 3x - 2$ and $g(x) = 2x^2 + 3x$. Find **a.** $f + g$ and **b.** $f - g$. ∎

EXAMPLE 2 Let $f(x) = 2x^2 + 1$ and $g(x) = 5x - 3$. Find each function and its domain: **a.** $f \cdot g$ and **b.** f/g.

Solution **a.** $(f \cdot g)(x) = f(x)g(x)$
$$= (2x^2 + 1)(5x - 3)$$
$$= 10x^3 - 6x^2 + 5x - 3 \quad \text{Multiply.}$$

The domain of $f \cdot g$ is the set of real numbers that are in the domain of both f and g. Since the domain of both f and g is the interval $(-\infty, \infty)$, the domain of $f \cdot g$ is also the interval $(-\infty, \infty)$.

b. $(f/g)(x) = \dfrac{f(x)}{g(x)}$

$\qquad = \dfrac{2x^2 + 1}{5x - 3}$

Since the denominator of the fraction cannot be 0, $x \neq \frac{3}{5}$. The domain of f/g is the interval $\left(-\infty, \frac{3}{5}\right) \cup \left(\frac{3}{5}, \infty\right)$.

Self Check Let $f(x) = 2x^2 - 3$ and $g(x) = x^2 + 1$. Find **a.** $f \cdot g$ and **b.** f/g. ∎

Composition of Functions

We have seen that a function can be represented by a machine: We put in a number from the domain, and a number from the range comes out. For example, if we put the number 2 into the machine shown in Figure 9-24(a), the number $f(2) = 5(2) - 2 = 8$ comes out. In general, if we put x into the machine shown in Figure 9-24(b), the value $f(x)$ comes out.

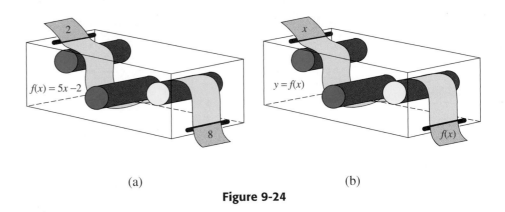

(a)   (b)

Figure 9-24

Often one quantity is a function of a second quantity that depends, in turn, on a third quantity. For example, the cost of a car trip is a function of the gasoline consumed. The amount of gasoline consumed, in turn, is a function of the number of miles driven. Such chains of dependence can be analyzed mathematically as **compositions of functions.**

Suppose that $y = f(x)$ and $y = g(x)$ define two functions. Any number x in the domain of g will produce the corresponding value $g(x)$ in the range of g. If $g(x)$ is in the domain of the function f, then $g(x)$ can be substituted into f, and a corresponding value $f(g(x))$ will be determined. This two-step process defines a new function, called a **composite function,** denoted by $f \circ g$.

The function machines shown in Figure 9-25 illustrate the composition $f \circ g$. When we put a number x into the function g, $g(x)$ comes out. The value $g(x)$ goes into function f, which transforms $g(x)$ into $f(g(x))$. If the function machines for g and f were connected to make a single machine, that machine would be named $f \circ g$.

To be in the domain of the composite function $f \circ g$, a number x has to be in the domain of g. Also, the output of g must be in the domain of f. Thus, the domain of $f \circ g$ consists of those numbers x that are in the domain of g, and for which $g(x)$ is in the domain of f.

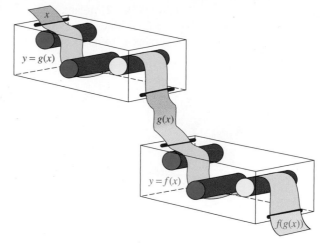

Figure 9-25

Composite Functions

The **composite function** $f \circ g$ is defined by

$$(f \circ g)(x) = f(g(x))$$

The **domain** of $f \circ g$ consists of all those numbers in the domain of g for which $g(x)$ is in the domain of f.

For example, if $f(x) = 4x - 5$ and $g(x) = 3x + 2$, then

$$
\begin{aligned}
(f \circ g)(x) &= f(\mathbf{g(x)}) \\
&= f(\mathbf{3x + 2}) \\
&= 4(\mathbf{3x + 2}) - 5 \\
&= 12x + 8 - 5 \\
&= 12x + 3
\end{aligned}
\qquad
\begin{aligned}
(g \circ f)(x) &= g(\mathbf{f(x)}) \\
&= g(\mathbf{4x - 5}) \\
&= 3(\mathbf{4x - 5}) + 2 \\
&= 12x - 15 + 2 \\
&= 12x - 13
\end{aligned}
$$

Comment   Note that in the previous example, $(f \circ g)(x) \neq (g \circ f)(x)$. This shows that the composition of functions is not commutative.

EXAMPLE 3   Let $f(x) = 2x + 1$ and $g(x) = x - 4$. Find   **a.** $(f \circ g)(9)$,   **b.** $(f \circ g)(x)$,   and **c.** $(g \circ f)(-2)$.

Solution   **a.** $(f \circ g)(9)$ means $f(g(9))$. In Figure 9-26(a), function g receives the number 9, subtracts 4, and releases the number $g(9) = 5$. The 5 then goes into the f function, which doubles 5 and adds 1. The final result, 11, is the output of the composite function $f \circ g$:

$$(f \circ g)(9) = f(\mathbf{g(9)}) = f(\mathbf{5}) = 2(\mathbf{5}) + 1 = 11$$

b. $(f \circ g)(x)$ means $f(g(x))$. In Figure 9-26(a), function g receives the number x, subtracts 4, and releases the number $x - 4$. The $x - 4$ then goes into the f function, which doubles $x - 4$ and adds 1. The final result, $2x - 7$, is the output of the composite function $f \circ g$.

$$(f \circ g)(x) = f(\mathbf{g(x)}) = f(\mathbf{x - 4}) = 2(\mathbf{x - 4}) + 1 = 2x - 7$$

c. $(g \circ f)(-2)$ means $g(f(-2))$. In Figure 9-26(b), function f receives the number -2, doubles it and adds 1, and releases -3 into the g function. Function g subtracts 4 from -3 and releases a final result of -7. Thus,

$$(g \circ f)(-2) = g(f(-2)) = g(-3) = -3 - 4 = -7$$

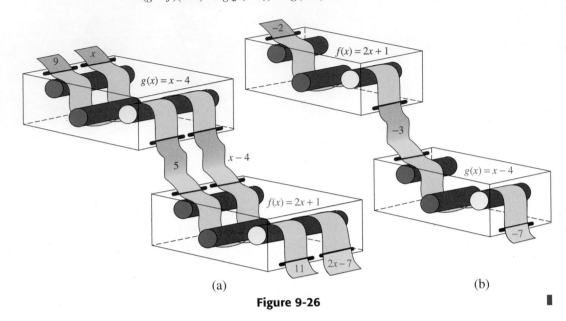

(a)   (b)

Figure 9-26

The Identity Function

The **identity function** is defined by the equation $I(x) = x$. Under this function, the value that corresponds to any real number x is x itself. If f is any function, the composition of f with the identity function is the function f:

$$(f \circ I)(x) = (I \circ f)(x) = f(x)$$

EXAMPLE 4   Let f be any function and I be the identity function, $I(x) = x$. Show that
a. $(f \circ I)(x) = f(x)$ and   **b.** $(I \circ f)(x) = f(x)$.

Solution   **a.** $(f \circ I)(x)$ means $f(I(x))$. Because $I(x) = x$, we have

$$(f \circ I)(x) = f(I(x)) = f(x)$$

b. $(I \circ f)(x)$ means $I(f(x))$. Because I passes any number through unchanged, we have $I(f(x)) = f(x)$ and

$$(I \circ f)(x) = I(f(x)) = f(x)$$

The Difference Quotient

An important function in calculus, called the **difference quotient,** represents the slope of a line that passes through two given points on the graph of a function. The difference quotient is defined as follows:

$$\frac{f(x + h) - f(x)}{h}$$

EXAMPLE 5   If $f(x) = x^2 - 4$, evaluate the difference quotient.

Solution   First, we evaluate $f(x + h)$.

$$f(x) = x^2 - 4$$
$$f(x + h) = (x + h)^2 - 4 \qquad \text{Substitute } x + h \text{ for } x.$$
$$= x^2 + 2xh + h^2 - 4 \qquad (x + h)^2 = x^2 + 2hx + h^2.$$

Then we note that $f(x) = x^2 - 4$. We can now substitute the values of $f(x + h)$ and $f(x)$ into the difference quotient and simplify.

$$\frac{f(x + h) - f(x)}{h} = \frac{(x^2 + 2xh + h^2 - 4) - (x^2 - 4)}{h}$$

$$= \frac{x^2 + 2xh + h^2 - 4 - x^2 + 4}{h} \qquad \text{Remove parentheses.}$$

$$= \frac{2xh + h^2}{h} \qquad \text{Combine like terms.}$$

$$= \frac{h(2x + h)}{h} \qquad \begin{array}{l}\text{Factor out } h \text{ in the} \\ \text{numerator.}\end{array}$$

$$= 2x + h \qquad \text{Divide out } h; \frac{h}{h} = 1.$$

The difference quotient for this function simplifies as $2x + h$.   ∎

Problem Solving

EXAMPLE 6   **Temperature change**   A laboratory sample is removed from a cooler at a temperature of $15°$ Fahrenheit. Technicians are warming the sample at a controlled rate of $3°$ F per hour. Express the sample's Celsius temperature as a function of the time t (in hours) since it was removed from refrigeration.

Solution   The temperature of the sample is $15°$ F when $t = 0$. Because it warms at $3°$ F per hour, it warms $3t°$ after t hours, and the Fahrenheit temperature after t hours is given by the function

$$F(t) = 3t + 15$$

The Celsius temperature is a function of the Fahrenheit temperature, and is given by the formula

$$C(F) = \frac{5}{9}(F - 32)$$

To express the sample's Celsius temperature as a function of time, we find the composition function $C \circ F$.

$$(C \circ F)(t) = C(F(t))$$

$$= \frac{5}{9}(F(t) - 32)$$

$$= \frac{5}{9}[(3t + 15) - 32] \qquad \text{Substitute } 3t + 15 \text{ for } F(t).$$

$$= \frac{5}{9}(3t - 17) \qquad \text{Simplify.}$$

$$= \frac{15}{9}t - \frac{85}{9}$$

$$= \frac{5}{3}t - \frac{85}{9}$$

∎

Self Check Answers

1. a. $2x^2 + 6x - 2$,   **b.** $-2x^2 - 2$ **2. a.** $2x^4 - x^2 - 3$,   **b.** $\dfrac{2x^2 - 3}{x^2 + 1}$

Orals   *If $f(x) = 2x$, $g(x) = 3x$, and $h(x) = 4x$, find*

1. $f + g$	**2.** $h - g$	**3.** $f \cdot h$
4. g/f	**5.** h/f	**6.** $g \cdot h$
7. $(f \circ h)(x)$	**8.** $(f \circ g)(x)$	**9.** $(g \circ h)(x)$

9.5   EXERCISES

REVIEW   *Simplify each expression.*

1. $\dfrac{3x^2 + x - 14}{4 - x^2}$

2. $\dfrac{2x^3 + 14x^2}{3 + 2x - x^2} \cdot \dfrac{x^2 - 3x}{x}$

3. $\dfrac{8 + 2x - x^2}{12 + x - 3x^2} \div \dfrac{3x^2 + 5x - 2}{3x - 1}$

4. $\dfrac{x - 1}{1 + \dfrac{x}{x - 2}}$

VOCABULARY AND CONCEPTS   *Fill in the blanks.*

5. $(f + g)(x) = $ _____

6. $(f - g)(x) = $ _____

7. $(f \cdot g)(x) = $ _____

8. $(f/g)(x) = $ ___ $(g(x) \neq 0)$

9. In Exercises 5–7, the domain of each function is the set of real numbers x that are in the _____ of both f and g.

10. $(f \circ g(x)) = $ _____

11. If I is the identity function, then $(f \circ I)(x) = $ ____.

12. If I is the identity function, then $(I \circ f)(x) = $ ____.

PRACTICE   *Let $f(x) = 3x$ and $g(x) = 4x$. Find each function and its domain.*

13. $f + g$	**14.** $f - g$
15. $f \cdot g$	**16.** f/g
17. $g - f$	**18.** $g + f$
19. g/f	**20.** $g \cdot f$

Let $f(x) = 2x + 1$ and $g(x) = x - 3$. Find each function and its domain.

21. $f + g$	**22.** $f - g$
23. $f \cdot g$	**24.** f/g
25. $g - f$	**26.** $g + f$
27. g/f	**28.** $g \cdot f$

Let $f(x) = 3x - 2$ and $g(x) = 2x^2 + 1$. Find each function and its domain.

29. $f - g$	**30.** $f + g$
31. f/g	**32.** $f \cdot g$

Let $f(x) = x^2 - 1$ and $g(x) = x^2 - 4$. Find each function and its domain.

33. $f - g$ **34.** $f + g$

35. g/f **36.** $g \cdot f$

Let $f(x) = 2x + 1$ and $g(x) = x^2 - 1$. Find each value.

37. $(f \circ g)(2)$ **38.** $(g \circ f)(2)$
39. $(g \circ f)(-3)$ **40.** $(f \circ g)(-3)$
41. $(f \circ g)(0)$ **42.** $(g \circ f)(0)$
43. $(f \circ g)\left(\dfrac{1}{2}\right)$ **44.** $(g \circ f)\left(\dfrac{1}{3}\right)$
45. $(f \circ g)(x)$ **46.** $(g \circ f)(x)$
47. $(g \circ f)(2x)$ **48.** $(f \circ g)(2x)$

Let $f(x) = 3x - 2$ and $g(x) = x^2 + x$. Find each value.

49. $(f \circ g)(4)$ **50.** $(g \circ f)(4)$
51. $(g \circ f)(-3)$ **52.** $(f \circ g)(-3)$
53. $(g \circ f)(0)$ **54.** $(f \circ g)(0)$
55. $(g \circ f)(x)$ **56.** $(f \circ g)(x)$

For each function, find $\dfrac{f(x + h) - f(x)}{h}$.

57. $f(x) = 2x + 3$ **58.** $f(x) = 3x - 5$
59. $f(x) = x^2$ **60.** $f(x) = x^2 - 1$
61. $f(x) = 2x^2 - 1$ **62.** $f(x) = 3x^2$
63. $f(x) = x^2 + x$ **64.** $f(x) = x^2 - x$
65. $f(x) = x^2 + 3x - 4$ **66.** $f(x) = x^2 - 4x + 3$
67. $f(x) = 2x^2 + 3x - 7$ **68.** $f(x) = 3x^2 - 2x + 4$

Find $\dfrac{f(x) - f(a)}{x - a}$.

69. $f(x) = 2x + 3$ **70.** $f(x) = 3x - 5$
71. $f(x) = x^2$ **72.** $f(x) = x^2 - 1$
73. $f(x) = 2x^2 - 1$ **74.** $f(x) = 3x^2$

75. $f(x) = x^2 + x$ **76.** $f(x) = x^2 - x$
77. $f(x) = x^2 + 3x - 4$ **78.** $f(x) = x^2 - 4x + 3$
79. $f(x) = 2x^2 + 3x - 7$ **80.** $f(x) = 3x^2 - 2x + 4$

81. If $f(x) = x + 1$ and $g(x) = 2x - 5$, show that $(f \circ g)(x) \neq (g \circ f)(x)$.
82. If $f(x) = x^2 + 1$ and $g(x) = 3x^2 - 2$, show that $(f \circ g)(x) \neq (g \circ f)(x)$.
83. If $f(x) = x^2 + 2x - 3$, find $f(a)$, $f(h)$, and $f(a + h)$. Then show that $f(a + h) \neq f(a) + f(h)$.
84. If $g(x) = 2x^2 + 10$, find $g(a)$, $g(h)$, and $g(a + h)$. Then show that $g(a + h) \neq g(a) + g(h)$.
85. If $f(x) = x^3 - 1$, find $\dfrac{f(x + h) - f(x)}{h}$.
86. If $f(x) = x^3 + 2$, find $\dfrac{f(x + h) - f(x)}{h}$.

APPLICATIONS

87. Alloys A molten alloy must be cooled slowly to control crystallization. When removed from the furnace, its temperature is 2,700° F, and it will be cooled at 200° per hour. Express the Celsius temperature as a function of the number of hours t since cooling began.

88. Weather forecasting A high pressure area promises increasingly warmer weather for the next 48 hours. The temperature is now 34° Celsius and is expected to rise 1° every 6 hours. Express the Fahrenheit temperature as a function of the number of hours from now. (*Hint: $F = \frac{9}{5}C + 32$.*)

WRITING

89. Explain how to find the domain of f/g.
90. Explain why the difference quotient represents the slope of a line passing through $(x, f(x))$ and $(x + h, f(x + h))$.

SOMETHING TO THINK ABOUT

91. Is composition of functions associative? Choose functions f, g, and h and determine whether $[f \circ (g \circ h)](x) = [(f \circ g) \circ h](x)$.
92. Choose functions f, g, and h and determine whether $f \circ (g + h) = f \circ g + f \circ h$.

9.6   Inverses of Functions

▌ **One-to-One Functions**   ▌ **The Horizontal Line Test**
▌ **Introduction to Inverses of Functions**
▌ **Finding Inverses of Functions**

Getting Ready   *Solve each equation for y.*

1. $x = 3y + 2$ **2.** $x = \dfrac{3}{2}y + 5$

We begin this section by discussing certain functions where each input has a different output. These functions are called **one-to-one functions.**

One-to-One Functions

Recall that for each input into a function, there is a single output. For some functions, different inputs have the same output, as shown in Figure 9-27(a). For other functions, different inputs have different outputs, as shown in Figure 9-27(b).

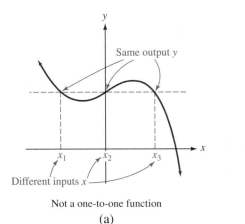

Not a one-to-one function
(a)

A one-to-one function
(b)

Figure 9-27

When every output of a function corresponds to exactly one input, we say that the function is *one-to-one.*

One-to-One Functions
A function is called **one-to-one** if each input value of x in the domain determines a different output value of y in the range.

EXAMPLE 1   Determine whether   **a.** $f(x) = x^2$ and   **b.** $f(x) = x^3$ are one-to-one.

Solution   **a.** The function $f(x) = x^2$ is not one-to-one, because different input values x can determine the same output value y. For example, inputs of 3 and -3 produce the same output value of 9.

$$f(3) = 3^2 = 9 \qquad \text{and} \qquad f(-3) = (-3)^2 = 9$$

b. The function $f(x) = x^3$ is one-to-one, because different input values x determine different output values of y for all x. This is because different numbers have different cubes.

Self Check   Determine whether $f(x) = 2x + 3$ is one-to-one.   ∎

The Horizontal Line Test

A **horizontal line test** can be used to decide whether the graph of a function represents a one-to-one function. If every horizontal line that intersects the graph of a function does so only once, the function is one-to-one. Otherwise, the function is not one-to-one. See Figure 9-28.

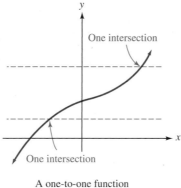

A one-to-one function   Not a one-to-one function

Figure 9-28

EXAMPLE 2   Use the horizontal line test to decide whether the graphs in Figure 9-29 represent one-to-one functions.

Solution   **a.** Because many horizontal lines intersect the graph shown in Figure 9-29(a) twice, the graph does not represent a one-to-one function.

b. Because each horizontal line that intersects the graph in Figure 9-29(b) does so exactly once, the graph does represent a one-to-one function.

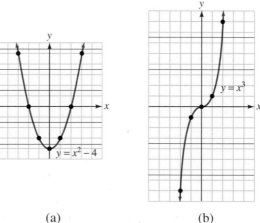

(a)   (b)

Figure 9-29

Self Check Does this graph represent a one-to-one function?

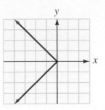

Comment Make sure to use the vertical line test to determine whether the graph represents a function. If it is, use the horizontal line test to determine whether the function is one-to-one.

Introduction to Inverses of Functions

The function defined by $C = \frac{5}{9}(F - 32)$ is the formula that we use to convert degrees Fahrenheit to degrees Celsius. If we substitute a Fahrenheit reading into the formula, a Celsius reading comes out. For example, if we substitute 41° for F, we obtain a Celsius reading of 5°:

$$C = \frac{5}{9}(\mathbf{F} - 32)$$

$$= \frac{5}{9}(\mathbf{41} - 32) \quad \text{Substitute 41 for } F.$$

$$= \frac{5}{9}(9)$$

$$= 5$$

If we want to find a Fahrenheit reading from a Celsius reading, we need a formula into which we can substitute a Celsius reading and have a Fahrenheit reading come out. Such a formula is $F = \frac{9}{5}C + 32$, which takes the Celsius reading of 5° and turns it back into a Fahrenheit reading of 41°.

$$F = \frac{9}{5}C + 32$$

$$= \frac{9}{5}(\mathbf{5}) + 32 \quad \text{Substitute 5 for } C.$$

$$= 41$$

The functions defined by these two formulas do opposite things. The first turns 41° F into 5° Celsius, and the second turns 5° Celsius back into 41° F. For this reason, we say that the functions are *inverses* of each other.

Finding Inverses of Functions

If f is the function determined by the table shown in Figure 9-30(a), it turns the number 1 into 10, 2 into 20, and 3 into 30. Since the inverse of f must turn 10 back into 1, 20 back into 2, and 30 back into 3, it consists of the ordered pairs shown in Figure 9-30(b).

Function f		Inverse of f	
x	y	x	y
1	10	10	1
2	20	20	2
3	30	30	3

Note that the inverse of f is also a function.

↑   ↑   ↑   ↑
Domain   Range   Domain   Range
(a)   (b)

Figure 9-30

We note that the domain of f and the range of its inverse is $\{1, 2, 3\}$. The range of f and the domain of its inverse is $\{10, 20, 30\}$.

This example suggests that to form the inverse of a function f, we simply interchange the coordinates of each ordered pair that determines f. When the inverse of a function is also a function, we call it ***f* inverse** and denote it with the symbol f^{-1}.

Comment The symbol $f^{-1}(x)$ is read as "the inverse of $f(x)$" or just "f inverse." The -1 in the notation $f^{-1}(x)$ is not an exponent. Remember that $f^{-1}(x) \neq \frac{1}{f(x)}$.

Finding the Inverse of a One-to-One Function

If a function is one-to-one, we find its inverse as follows:

1. Replace $f(x)$ with y, if necessary.
2. Interchange the variables x and y.
3. Solve the resulting equation for y.
4. This equation is $y = f^{-1}(x)$.

EXAMPLE 3 If $f(x) = 4x + 2$, find the inverse of f and tell whether it is a function.

Solution To find the inverse, we replace $f(x)$ with y and interchange the positions of x and y.

$$f(x) = 4x + 2$$
$$y = 4x + 2 \quad \text{Replace } f(x) \text{ with } y.$$
$$x = 4y + 2 \quad \text{Interchange the variables } x \text{ and } y.$$

Then we solve the equation for y.

$$x = 4y + 2$$
$$x - 2 = 4y \qquad \text{Subtract 2 from both sides.}$$

(1) $$y = \frac{x - 2}{4} \qquad \text{Divide both sides by 4 and write } y \text{ on the left-hand side.}$$

Because each input x that is substituted into Equation 1 gives one output y, the inverse of f is a function, so we can express it in the form

$$f^{-1}(x) = \frac{x - 2}{4}$$

Self Check If $y = f(x) = -5x - 3$, find the inverse of f and tell whether it is a function. ∎

To emphasize an important relationship between a function and its inverse, we substitute some number x, such as $x = 3$, into the function $f(x) = 4x + 2$ of Example 3. The corresponding value of y produced is

$$y = f(\mathbf{3}) = 4(\mathbf{3}) + 2 = 14$$

If we substitute 14 into the inverse function f^{-1}, the corresponding value of y that is produced is

$$y = f^{-1}(\mathbf{14}) = \frac{\mathbf{14} - 2}{4} = 3$$

Thus, the function f turns 3 into 14, and the inverse function f^{-1} turns 14 back into 3. In general, the composition of a function and its inverse is the identity function.

To prove that $f(x) = 4x + 2$ and $f^{-1}(x) = \frac{x-2}{4}$ are inverse functions, we must show that their composition (in both directions) is the identity function:

$$
\begin{aligned}
(f \circ f^{-1}(x)) &= f(\boldsymbol{f^{-1}(x)}) \\
&= f\left(\frac{\boldsymbol{x-2}}{\boldsymbol{4}}\right) \\
&= 4\left(\frac{\boldsymbol{x-2}}{\boldsymbol{4}}\right) + 2 \\
&= x - 2 + 2 \\
&= x
\end{aligned}
\qquad
\begin{aligned}
(f^{-1} \circ f)(x) &= f^{-1}(\boldsymbol{f(x)}) \\
&= f^{-1}(\boldsymbol{4x + 2}) \\
&= \frac{\boldsymbol{4x + 2} - 2}{4} \\
&= \frac{4x}{4} \\
&= x
\end{aligned}
$$

Thus, $(f \circ f^{-1})(x) = (f^{-1} \circ f)(x) = x$, which is the identity function $I(x)$.

EXAMPLE 4   The set of all pairs (x, y) determined by $3x + 2y = 6$ is a function. Find its inverse function, and graph the function and its inverse on one coordinate system.

Solution   To find the inverse function of $3x + 2y = 6$, we interchange x and y to obtain

$$3y + 2x = 6$$

and then solve the equation for y.

$$
\begin{aligned}
3y + 2x &= 6 \\
3y &= -2x + 6 \qquad \text{Subtract } 2x \text{ from both sides.} \\
y &= -\frac{2}{3}x + 2 \qquad \text{Divide both sides by 3.}
\end{aligned}
$$

Thus, $y = f^{-1}(x) = -\frac{2}{3}x + 2$. The graphs of $3x + 2y = 6$ and $y = f^{-1}(x) = -\frac{2}{3}x + 2$ appear in Figure 9-31.

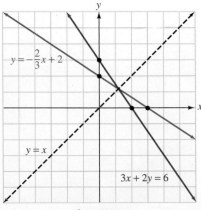

Figure 9-31

Self Check   Find the inverse of the function defined by $2x - 3y = 6$. Graph the function and its inverse on one coordinate system.   ∎

In Example 4, the graphs of $3x + 2y = 6$ and $y = f^{-1}(x) = -\frac{2}{3}x + 2$ are symmetric about the line $y = x$. This is always the case, because when the coordinates (a, b) satisfy an equation, the coordinates (b, a) will satisfy its inverse.

In each example so far, the inverse of a function has been another function. This is not always true, as the following example will show.

EXAMPLE 5   Find the inverse of the function determined by $f(x) = x^2$.

Solution $y = x^2$ Replace $f(x)$ with y.

$x = y^2$ Interchange x and y.

$y = \pm\sqrt{x}$ Use the square root property and write y on the left-hand side.

When the inverse $y = \pm\sqrt{x}$ is graphed as in Figure 9-32, we see that the graph does not pass the vertical line test. Thus, it is not a function.

The graph of $y = x^2$ is also shown in the figure. As expected, the graphs of $y = x^2$ and $y = \pm\sqrt{x}$ are symmetric about the line $y = x$.

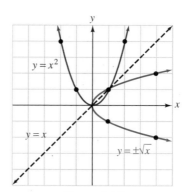

Figure 9-32

Self Check   Find the inverse of the function determined by $f(x) = 4x^2$.   ∎

EXAMPLE 6   Find the inverse of $f(x) = x^3$.

Solution   To find the inverse, we proceed as follows:

$y = x^3$ Replace $f(x)$ with y.

$x = y^3$ Interchange the variables x and y.

$\sqrt[3]{x} = y$ Take the cube root of both sides.

We note that to each number x there corresponds one real cube root. Thus, $y = \sqrt[3]{x}$ represents a function. In $f^{-1}(x)$ notation, we have

$f^{-1}(x) = \sqrt[3]{x}$

Self Check   Find the inverse of $f(x) = x^5$.   ∎

If a function is not one-to-one, we can often make a one-to-one function by restricting its domain.

EXAMPLE 7 Find the inverse of the function defined by $y = x^2$ with $x \geq 0$. Then tell whether the inverse is a function. Graph the function and its inverse.

Solution The inverse of the function $y = x^2$ with $x \geq 0$ is

$x = y^2$ with $y \geq 0$ Interchange the variables x and y.

Before considering the restriction, this equation can be written in the form

$y = \pm \sqrt{x}$ with $y \geq 0$

Since $y \geq 0$, each number x gives only one value of y: $y = \sqrt{x}$. Thus, the inverse is a function.

The graphs of the two functions appear in Figure 9-33. The line $y = x$ is included so that we can see that the graphs are symmetric about the line $y = x$.

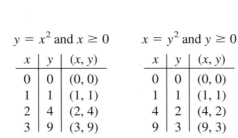

$y = x^2$ and $x \geq 0$				$x = y^2$ and $y \geq 0$		
x	y	(x, y)		x	y	(x, y)
0	0	$(0, 0)$		0	0	$(0, 0)$
1	1	$(1, 1)$		1	1	$(1, 1)$
2	4	$(2, 4)$		4	2	$(4, 2)$
3	9	$(3, 9)$		9	3	$(9, 3)$

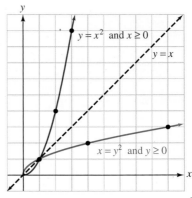

Figure 9-33

1. yes   **2.** no   **3.** $y = -\frac{1}{5}x - \frac{3}{5}$, yes   **4.** $f^{-1}(x) = \frac{3}{2}x + 3$ **5.** $y = \pm\dfrac{\sqrt{x}}{2}$
6. $f^{-1}(x) = \sqrt[5]{x}$

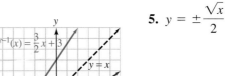

Orals *Find the inverse of each set of ordered pairs.*

1. $\{(1, 2), (2, 3), (5, 10)\}$ **2.** $\{(1, 1), (2, 8), (4, 64)\}$

Find the inverse function of each linear function.

3. $y = \dfrac{1}{2}x$

4. $y = 2x$

Tell whether each function is one-to-one.

5. $y = x^2 - 2$

6. $y = x^3$

9.6   EXERCISES

REVIEW   *Write each complex number in $a + bi$ form or find each value.*

1. $3 - \sqrt{-64}$

2. $(2 - 3i) + (4 + 5i)$

3. $(3 + 4i)(2 - 3i)$

4. $\dfrac{6 + 7i}{3 - 4i}$

5. $|6 - 8i|$

6. $\left|\dfrac{2 + i}{3 - i}\right|$

VOCABULARY AND CONCEPTS   *Fill in the blanks.*

7. A function is called _____ if each input determines a different output.

8. If every _____ line that intersects the graph of a function does so only once, the function is one-to-one.

9. If a one-to-one function turns an input of 2 into an output of 5, the inverse function will turn 5 into __.

10. The symbol $f^{-1}(x)$ is read as _____ or _____.

11. $(f \circ f^{-1})(x) = $ __

12. The graphs of a function and its inverse are symmetrical about the line _____.

PRACTICE   *Determine whether each function is one-to-one.*

13. $f(x) = 2x$

14. $f(x) = |x|$

15. $f(x) = x^4$

16. $f(x) = x^3 + 1$

Each graph represents a function. Use the horizontal line test to decide whether the function is one-to-one.

17.

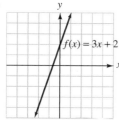

18.

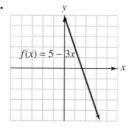

19.

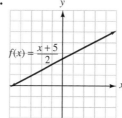

20.

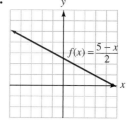

21.

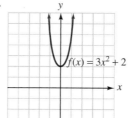

22.

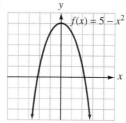

23.

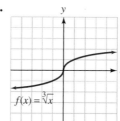

24.

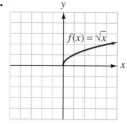

Find the inverse of each set of ordered pairs (x, y) and tell whether the inverse is a function.

25. $\{(3, 2), (2, 1), (1, 0)\}$

26. $\{(4, 1), (5, 1), (6, 1), (7, 1)\}$

27. $\{(1, 2), (2, 3), (1, 3), (1, 5)\}$

28. $\{(-1, -1), (0, 0), (1, 1), (2, 2)\}$

29. $\{(1, 1), (2, 4), (3, 9), (4, 16)\}$

30. $\{(1, 1), (2, 1), (3, 1), (4, 1)\}$

Find the inverse of each function and express it in the form $y = f^{-1}(x)$. Verify each result by showing that $(f \circ f^{-1})(x) = (f^{-1} \circ f)(x) = I(x)$.

31. $f(x) = 3x + 1$

32. $y + 1 = 5x$

33. $x + 4 = 5y$

34. $x = 3y + 1$

35. $f(x) = \dfrac{x - 4}{5}$

36. $f(x) = \dfrac{2x + 6}{3}$

37. $4x - 5y = 20$

38. $3x + 5y = 15$

Find the inverse of each function. Then graph the function and its inverse on one coordinate system. Find the equation of the line of symmetry.

39. $f(x) = 4x + 3$ **40.** $x = 3y - 1$

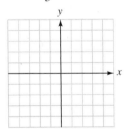   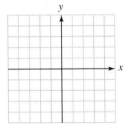

41. $x = \dfrac{y - 2}{3}$ **42.** $f(x) = \dfrac{x + 3}{4}$

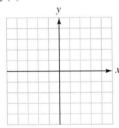   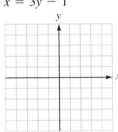

43. $3x - y = 5$ **44.** $2x + 3y = 9$

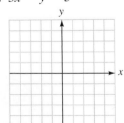   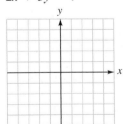

45. $3(x + y) = 2x + 4$ **46.** $-4(y - 1) + x = 2$

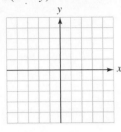   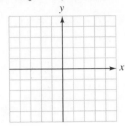

Find the inverse of each function and tell whether it is a function.

47. $y = x^2 + 4$ **48.** $y = x^2 + 5$

49. $y = x^3$ **50.** $xy = 4$

51. $y = |x|$ **52.** $y = \sqrt[3]{x}$

Show that the inverse of the function determined by each equation is also a function. Express it using $f^{-1}(x)$ notation.

53. $f(x) = 2x^3 - 3$

54. $f(x) = \dfrac{3}{x^3} - 1$

Graph each equation and its inverse on one set of coordinate axes. Find the axis of symmetry.

55. $y = x^2 + 1$ **56.** $y = \dfrac{1}{4}x^2 - 3$

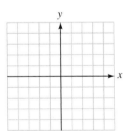   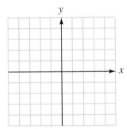

57. $y = \sqrt{x}$ **58.** $y = |x|$

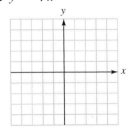   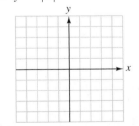

WRITING

59. Explain the purpose of the vertical line test.

60. Explain the purpose of the horizontal line test.

SOMETHING TO THINK ABOUT

61. Find the inverse of $y = \dfrac{x + 1}{x - 1}$.

62. Using the functions of Exercise 61, show that $(f \circ f^{-1})(x) = x$.

Projects

Project 1

Understanding the behavior of a liquid flowing in a pipe has applications in many systems, including home plumbing, oil pipelines, veins and arteries, and hydraulic control lines in aircraft. The *velocity profile* in the illustration shows that fluid next to the pipe wall is moving slowly, but nearer the center of the pipe, flow becomes more rapid.

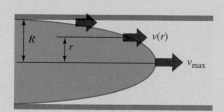

The velocity at a distance r from the pipe's center is given by

$$v(r) = v_{max}\left(1 - \frac{r}{R}\right)^{1/k}$$

where R is the radius of the pipe, v_{max} is the greatest velocity, and k depends on properties of the fluid. For one particular fluid in a 2-inch-diameter pipe, $v_{max} = 8$ and $k = 10$.

a. Graph the function. In the context of this application, what is the domain of the function? (*Hint:* Restrictions on r come from the pipe, not the equation.)

b. Where is the velocity greatest? Does your graph show that this greatest velocity is v_{max}?

c. Graph the function again for $k = 4$. Which graph describes the flow of a thick, syrupy fluid, and which describes a fluid more like water? Why?

Project 2

The illustration shows the geometric basis of the algebraic process of completing the square. Study the figure, where each algebraic expression is related to the area of the corresponding geometric figure. What is the area of the missing piece that will "complete the square"?

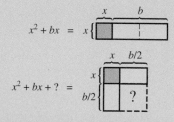

Draw a similar figure that shows geometrically how to complete the square on the expression $x^2 + 8x$.

Project 3

The surface area of a sphere is a function of the radius, given by

(1) $A = 4\pi r^2$

Its volume is also a function of the radius, given by

(2) $V = \dfrac{4}{3}\pi r^3$

a. Express the surface area of a sphere as a function of its volume. (*Hint:* Solve Equation 2 for r and substitute into Equation 1.)

b. Use a graphing calculator to graph the function you found in part a.

c. Insects don't have lungs; they breathe through their skin. Oxygen requirements increase with an insect's volume, but oxygen intake depends on the insect's surface area. Refer to your graph and write a paragraph explaining why insects must be limited in size.

Project 4

Computers can solve very difficult problems quickly. More often, however, computers solve the same simple problem again and again, millions of times. For example, it is not difficult to reconcile one credit card account, but only computers can balance the accounts of millions of credit card holders. Since time is money, it makes sense for computers to perform repeated calculations as quickly as possible. A fraction of a second on each calculation can add up to substantial time savings when the calculation is done a million times.

Computers take more time to perform a multiplication than an addition. Several multiplications are required to evaluate a polynomial. For example, calculating the value of the polynomial $P(x) = 3x^4 + 2x^3 + 5x^2 + 7x + 1$ requires ten multiplications and four additions:

(3) $P(x) = 3 \cdot x \cdot x \cdot x \cdot x + 2 \cdot x \cdot x \cdot x +$
$\qquad 5 \cdot x \cdot x + 7 \cdot x + 1$

By writing the polynomial in a form called **nested form,** we can reduce the number of multiplications and also reduce the time needed for calculation. To do so, we successively factor the variable from those terms where it appears. We can write

$$P(x) = 3x^4 + 2x^3 + 5x^2 + 7x + 1$$

in nested form by factoring x from the first four terms to get

$$P(x) = (3x^3 + 2x^2 + 5x + 7)x + 1$$

We then factor x from $3x^3 + 2x^2 + 5x$ to obtain

$$P(x) = [(3x^2 + 2x + 5)x + 7]x + 1$$

and finally factor x from $3x^2 + 2x$ to get

$$P(x) = \{[(3x + 2)x + 5]x + 7\}x + 1$$

To emphasize the number of multiplications, we write the previous equation as

(4) $P(x) = \{[(3 \cdot x + 2) \cdot x + 5] \cdot x + 7\} \cdot x + 1$

There are still four additions, but now only four multiplications. If this polynomial were to be evaluated for many different values of x, the computer time saved by using nested form could be substantial.

- If a computer takes three times as along to do a multiplication as an addition, how much faster is a calculation of $P(x)$ using nested form instead of standard form?
- Use a calculator and Equation 3 to find $P(-2)$. See if you can store -2 in memory and recall it whenever it is needed.
- Use a calculator and Equation 4 to find $P(-2)$. Which method is easier?
- Write $Q(x) = 5x^3 - 3x^2 + 2x - 4$ in nested form. Find $Q(3)$ in two ways.
- If the coefficients of a polynomial $R(x)$ are $a, b, c, d, \ldots$, describe how to use a calculator to find $R(7)$.

Chapter Summary

CONCEPTS	REVIEW EXERCISES
9.1	**Symmetry and Stretchings of Graphs**
y-axis symmetry: If the point $(-x, y)$ lies on a graph whenever the point (x, y) does, the graph is symmetric about the y-axis.	Find the symmetries of the graph of each equation, if any.
	1. $y = x^2 - 5$ $\qquad$ **2.** $y = x^3 - x$
	3. $x = y^2$ $\qquad$ **4.** $y = x^4 - 2x^2$
Symmetry about the origin: If the point $(-x, -y)$ lies on a graph whenever the point (x, y) does, the graph is symmetric about the origin.	**5.** $y = x^3 + 2$ $\qquad$ **6.** $y = x^4 - x$
	7. $y = -|x| + 5$ $\qquad$ **8.** $y = \sqrt{x} + 1$

***x*-axis symmetry:** If the point $(x, -y)$ lies on a graph whenever the point (x, y) does, the graph is symmetric about the *x*-axis.

A function f with the property that $f(-x) = f(x)$ for all x in the domain of f is called an *even function.*

A function f with the property that $f(-x) = -f(x)$ for all x in the domain of f is called an *odd function.*

A function f is increasing on (a, b) if $f(x_1) < f(x_2)$ whenever $a < x_1 < x_2 < b$.

A function f is decreasing on (a, b) if $f(x_1) > f(x_2)$ whenever $a < x_1 < x_2 < b$.

A function f is constant on (a, b) if $f(x_1) = f(x_2)$ for all x_1 and x_2 on (a, b).

The graph of $y = kf(x)$ is the same as the graph of $y = f(x)$, except that it is stretched vertically by a factor of k.

The graph of $y = f(kx)$ is the same as the graph of $y = f(x)$, except that it is stretched horizontally by a factor of $\frac{1}{k}$.

Tell whether the given function is even or odd.

9. $y = 3x^2$

10. $y = 3x^3$

11. $y = x^3 + x$

12. $y = |x| - 3$

Tell where each graph is increasing, decreasing, or constant.

13.

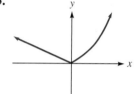

14.

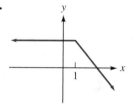

Graph each function along with its basic function.

15. $y = \dfrac{1}{3}x^3$

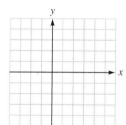

16. $y = (-5x)^3$

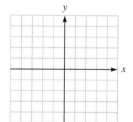

9.2	## Piecewise-Defined Functions and the Greatest Integer Function

Piecewise-defined functions are functions that are defined by using different equations for different parts of their domains.

Graph each function and tell where it is increasing, decreasing, or constant.

17. $y = f(x) = \begin{cases} x & \text{if } x \le 1 \\ -x^2 & \text{if } x > 1 \end{cases}$

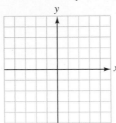

18. $y = f(x) = \begin{cases} x + 3 & \text{if } x \le 0 \\ 3 & \text{if } x > 0 \end{cases}$

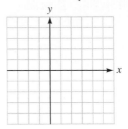

The equation $y = [\![x]\!]$ represents the greatest integer function.

Graph each function.

19. $y = 2[\![x]\!]$

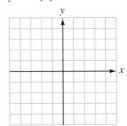

20. $y = [\![x]\!] - 2$

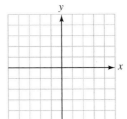

9.3	## More Rational Functions

To graph the rational function $\frac{P(x)}{Q(x)}$,

1. Check for symmetries.
2. Look for vertical asymptotes.
3. Look for the y-intercept.
4. Look for the x-intercepts.
5. Look for horizontal asymptotes.
6. Look for slant asymptotes.

Find all asymptotes of the graph of each rational function.

21. $y = \dfrac{x - 1}{x^2 - 4}$

22. $y = \dfrac{x^2 - 3x - 9}{x - 3}$

Graph each rational function.

23. $y = \dfrac{x}{(x-1)^2}$

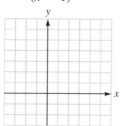

24. $y = \dfrac{(x-1)^2}{x}$

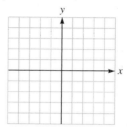

25. $y = \dfrac{x^2 - x - 2}{x^2 + x - 2}$

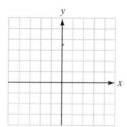

26. $y = \dfrac{x^2 - 4x + 4}{x - 2}$

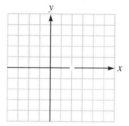

9.4	**Partial Fractions**

The fraction $\frac{P(x)}{Q(x)}$ can often be written as the sum of simpler fractions with denominators determined by the prime factors of $Q(x)$.

Decompose each fraction into partial fractions.

27. $\dfrac{7x + 3}{x^2 + x}$

28. $\dfrac{4x^3 + 3x + x^2 + 2}{x^4 + x^2}$

29. $\dfrac{x^2 + 5}{x^3 + x^2 + 5x}$

30. $\dfrac{x^2 + 1}{(x + 1)^3}$

9.5	**Algebra and Composition of Functions**

Operations with functions:

$(f + g)(x) = f(x) + g(x)$

$(f - g)(x) = f(x) - g(x)$

$(f \cdot g)(x) = f(x)g(x)$

$(f/g)(x) = \dfrac{f(x)}{g(x)} \quad (g(x) \neq 0)$

$(f \circ g)(x) = f(g(x))$

Let $f(x) = 2x$ and $g(x) = x + 1$. Find each function or value.

31. $f + g$

32. $f - g$

33. $f \cdot g$

34. f/g

35. $(f \circ g)(2)$

36. $(g \circ f)(-1)$

37. $(f \circ g)(x)$

38. $(g \circ f)(x)$

| **9.6** | **Inverses of Functions** |

Horizontal line test:
If every horizontal line that intersects the graph of a function does so only once, the function is one-to-one.

To find the inverse of a function, interchange the positions of variables x and y and solve for y.

Graph each function and use the horizontal line test to decide whether the function is one-to-one.

39. $f(x) = 2(x - 3)$

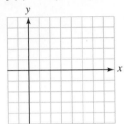

40. $f(x) = x(2x - 3)$

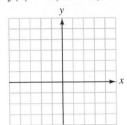

41. $f(x) = -3(x - 2)^2 + 5$

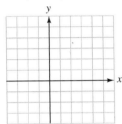

42. $f(x) = |x|$

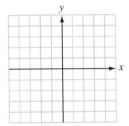

Find the inverse of each function.

43. $f(x) = 6x - 3$

44. $f(x) = 4x + 5$

45. $y = 2x^2 - 1 \ (x \geq 0)$

46. $y = |x|$

Chapter Test

Find the symmetries of the graph of each equation.

1. $y = x^3 - x$

2. $y = x^6 - 5x^4$

Tell whether each function is even or odd.

3. $y = x^2 + |x|$

4. $y = x$

Tell where each function is increasing or decreasing.

5.

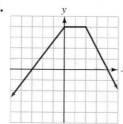

6.

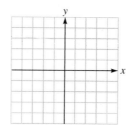

Graph each function.

7. $f(x) = x^3 - 3x$

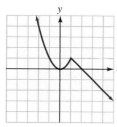

8. $f(x) = x^4 - 6x^2$

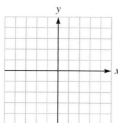

9. $f(x) = \begin{cases} x^2 - 1 \text{ if } x \le 0 \\ -x^2 + 1 \text{ if } x > 0 \end{cases}$ **10.** $f(x) = \llbracket x - 2 \rrbracket$

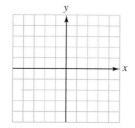

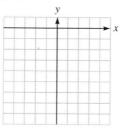

Find all asymptotes of the graph of each rational function. ***Do not graph the functions.***

11. $f(x) = \dfrac{x - 1}{x^2 - 9}$

12. $f(x) = \dfrac{x^2 - 5x - 14}{x - 3}$

Graph each rational function.

13. $f(x) = \dfrac{x^2}{x^2 - 9}$

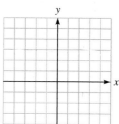

14. $f(x) = \dfrac{3x^2}{x^2 + 1}$

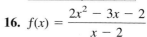

15. $f(x) = \dfrac{x^3 + x}{x^2 - 1}$

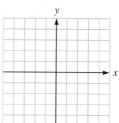

16. $f(x) = \dfrac{2x^2 - 3x - 2}{x - 2}$

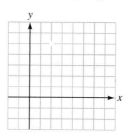

Decompose each fraction into partial fractions.

17. $\dfrac{5x}{(2x - 3)(x + 1)}$

18. $\dfrac{3x^2 + x + 2}{x^3 + 2x}$

Let $f(x) = 4x$ and $g(x) = x - 1$ and find each function.

19. $g + f$

20. $f - g$

21. $g \cdot f$

22. g/f

Let $f(x) = 4x$ and $g(x) = x - 1$ and find each value.

23. $(g \circ f)(1)$

24. $(f \circ g)(0)$

25. $(f \circ g)(-1)$

26. $(g \circ f)(-2)$

Let $f(x) = 4x$ and $g(x) = x - 1$ and find each function.

27. $(f \circ g)(x)$

28. $(g \circ f)(x)$

Find the inverse of each function.

29. $3x + 2y = 12$

30. $y = 3x^2 + 4 \ (x \le 0)$

10 Exponential and Logarithmic Functions

InfoTrac Project

Do a keyword search on "exponential function." Find the article "Mathskit: The exponential function: Fading into the sunset." Read the article through the section titled "Foam." Suppose the equation for the number of bubbles remaining after t seconds can be represented by $N = N_0 e^{-t}$, where N_0 is the number of bubbles originally in the glass. If there were 100,000 bubbles originally, find the number of bubbles remaining after 2 seconds, and after 4 seconds. If there are 15,000 bubbles left in the glass, how much time has lapsed?

Complete this project after studying Section 10.6.

© Getty Images

Mathematics in Medicine

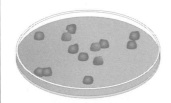

If a medium is inoculated with a bacterial culture that contains 1,000 cells per milliliter, how many generations will pass by the time the culture has grown to a population of 1 million cells per milliliter?

Section 10.6
Example 10

In this chapter, we will discuss two functions that are important in many applications of mathematics. **Exponential functions** *are used to compute compound interest, find radioactive decay, and model population growth.* **Logarithmic functions** *are used to measure acidity of solutions, drug dosage, gain of an amplifier, intensity of earthquakes, and safe noise levels in factories.*

10.1 Exponential Functions

- ∎ **Irrational Exponents** ∎ **Exponential Functions**
- ∎ **Graphing Exponential Functions**
- ∎ **Vertical and Horizontal Translations** ∎ **Compound Interest**

Getting Ready   *Find each value.*

1. 2^3 **2.** $25^{1/2}$ **3.** 5^{-2} **4.** $\left(\dfrac{3}{2}\right)^{-3}$

The graph in Figure 10-1 shows the balance in a bank account in which $10,000 was invested in 2000 at 9% annual interest, compounded monthly. The graph shows that in the year 2010, the value of the account will be approximately $25,000, and in the year 2030, the value will be approximately $147,000. The curve shown in Figure 10-1 is the graph of a function called an **exponential function,** the topic of this section.

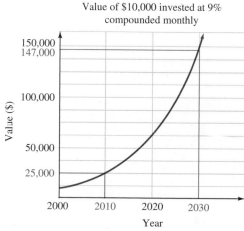

Value of $10,000 invested at 9% compounded monthly

Figure 10-1

Irrational Exponents

We have discussed expressions of the form b^x, where x is a rational number.

$8^{1/2}$ means "the square root of 8."

$5^{1/3}$ means "the cube root of 5."

$3^{-2/5} = \dfrac{1}{3^{2/5}}$ means "the reciprocal of the fifth root of 3^2."

To give meaning to b^x when x is an irrational number, we consider the expression

$5^{\sqrt{2}}$ where $\sqrt{2}$ is the irrational number $1.414213562\ldots$

Each number in the following list is defined, because each exponent is a rational number.

$5^{1.4}, 5^{1.41}, 5^{1.414}, 5^{1.4142}, 5^{1.41421}, \ldots$

Since the exponents are getting closer to $\sqrt{2}$, the numbers in this list are successively better approximations of $5^{\sqrt{2}}$. We can use a calculator to obtain a very good approximation.

Accent on Technology   **EVALUATING EXPONENTIAL EXPRESSIONS**

To find the value of $5^{\sqrt{2}}$ with a scientific calculator, we enter these numbers and press these keys:

5 y^x 2 $\sqrt{\ }$ =

The display will read 9.738517742.

With a graphing calculator, we enter these numbers and press these keys:

5 ^ $\sqrt{\ }$ 2 ENTER

The display will read
```
5^ √(2)
        9.738517742
```

In general, if b is positive and x is a real number, b^x represents a positive number. It can be shown that all of the rules of exponents hold true for irrational exponents.

EXAMPLE 1   Use the rules of exponents to simplify: **a.** $\left(5^{\sqrt{2}}\right)^{\sqrt{2}}$ and   **b.** $b^{\sqrt{3}} \cdot b^{\sqrt{12}}$.

Solution   **a.** $\left(5^{\sqrt{2}}\right)^{\sqrt{2}} = 5^{\sqrt{2}\sqrt{2}}$ Keep the base and multiply the exponents.

$\qquad = 5^2$ $\sqrt{2}\sqrt{2} = \sqrt{4} = 2.$

$\qquad = 25$

b. $b^{\sqrt{3}} \cdot b^{\sqrt{12}} = b^{\sqrt{3}+\sqrt{12}}$ Keep the base and add the exponents.

$\qquad = b^{\sqrt{3}+2\sqrt{3}}$ $\sqrt{12} = \sqrt{4}\sqrt{3} = 2\sqrt{3}.$

$\qquad = b^{3\sqrt{3}}$ $\sqrt{3} + 2\sqrt{3} = 3\sqrt{3}.$

Self Check   Simplify: **a.** $\left(3^{\sqrt{2}}\right)^{\sqrt{8}}$ and   **b.** $b^{\sqrt{2}} \cdot b^{\sqrt{18}}$ ∎

Exponential Functions

If $b > 0$ and $b \neq 1$, the function $y = f(x) = b^x$ is an **exponential function.** Since x can be any real number, its domain is the set of real numbers. This is the interval $(-\infty, \infty)$. Since b is positive, the value of $f(x)$ is positive and the range is the set of positive numbers. This is the interval $(0, \infty)$.

Since $b \neq 1$, an exponential function cannot be the constant function $y = f(x) = 1^x$, in which $f(x) = 1$ for every real number x.

Exponential Functions   An **exponential function with base b** is defined by the equation

$$y = f(x) = b^x \quad (b > 0, b \neq 1, \text{ and } x \text{ is a real number})$$

The **domain of any exponential function** is the interval $(-\infty, \infty)$. The **range** is the interval $(0, \infty)$.

Graphing Exponential Functions

Since the domain and range of $y = f(x) = b^x$ are subsets of real numbers, we can graph exponential functions on a rectangular coordinate system.

EXAMPLE 2   Graph: $f(x) = 2^x$.

Solution   To graph $f(x) = 2^x$, we find several points (x, y) whose coordinates satisfy the equation, plot the points, and join them with a smooth curve, as shown in Figure 10-2.

$f(x) = 2^x$

x	$f(x)$	$(x, f(x))$
-1	$\frac{1}{2}$	$\left(-1, \frac{1}{2}\right)$
0	1	$(0, 1)$
1	2	$(1, 2)$
2	4	$(2, 4)$
3	8	$(3, 8)$

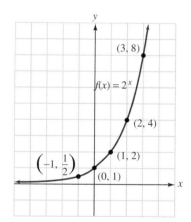

Figure 10-2

By looking at the graph, we can verify that the domain is the interval $(-\infty, \infty)$ and that the range is the interval $(0, \infty)$.

Note that as x decreases, the values of $f(x)$ decrease and approach 0. Thus, the x-axis is the horizontal asymptote of the graph.

Also note that the graph of $f(x) = 2^x$ passes through the points $(0, 1)$ and $(1, 2)$.

Self Check   Graph: $f(x) = 4^x$. ∎

EXAMPLE 3   Graph: $f(x) = \left(\dfrac{1}{2}\right)^x$.

Solution   We find and plot pairs (x, y) that satisfy the equation. The graph of $y = f(x) = \left(\frac{1}{2}\right)^x$ appears in Figure 10-3.

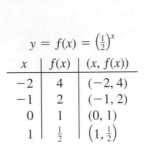

$$y = f(x) = \left(\tfrac{1}{2}\right)^x$$

x	$f(x)$	$(x, f(x))$
-2	4	$(-2, 4)$
-1	2	$(-1, 2)$
0	1	$(0, 1)$
1	$\frac{1}{2}$	$\left(1, \frac{1}{2}\right)$

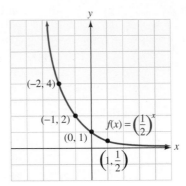

Figure 10-3

By looking at the graph, we can see that the domain is the interval $(-\infty, \infty)$ and that the range is the interval $(0, \infty)$.

In this case, as x increases, the values of $f(x)$ decrease and approach 0. The x-axis is a horizontal asymptote. Note that the graph of $f(x) = \left(\frac{1}{2}\right)^x$ passes through the points $(0, 1)$ and $\left(1, \frac{1}{2}\right)$.

Self Check   Graph: $f(x) = \left(\dfrac{1}{4}\right)^x$.   ∎

Examples 2 and 3 illustrate the following properties of exponential functions.

Properties of Exponential Functions

The **domain** of the exponential function $y = f(x) = b^x$ is the interval $(-\infty, \infty)$.

The **range** is the interval $(0, \infty)$.

The graph has a y-intercept of $(0, 1)$.

The x-axis is an asymptote of the graph.

The graph of $y = f(x) = b^x$ passes through the point $(1, b)$.

EXAMPLE 4   From the graph of $f(x) = b^x$ shown in Figure 10-4, find the value of b.

Solution   We first note that the graph passes through $(0, 1)$. Since the point $(2, 9)$ is on the graph, we substitute 9 for y and 2 for x in the equation $y = b^x$ to get

$$y = b^x$$
$$9 = b^2$$
$$3 = b \qquad \text{Take the positive square root of both sides.}$$

The base b is 3.

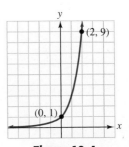

Figure 10-4

Self Check Is this the graph of an exponential function?

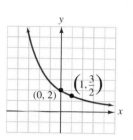

$(0, 2)$ $\left(1, \dfrac{3}{2}\right)$

In Example 2 (where $b = 2$), the values of y increase as the values of x increase. Since the graph rises as we move to the right, we call the function an *increasing function*. When $b > 1$, the larger the value of b, the steeper the curve.

In Example 3 $\left(\text{where } b = \frac{1}{2}\right)$, the values of y decrease as the values of x increase. Since the graph drops as we move to the right, we call the function a *decreasing function*. When $0 < b < 1$, the smaller the value of b, the steeper the curve.

In general, the following is true.

Increasing and Decreasing Functions

If $b > 1$, then $f(x) = b^x$ is an **increasing function.**

If $0 < b < 1$, then $f(x) = b^x$ is a **decreasing function.**

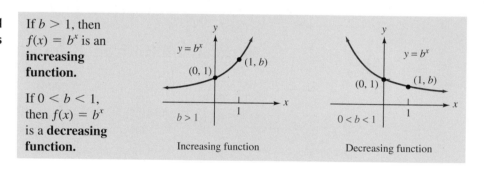

Increasing function

Decreasing function

An exponential function with base b is either increasing (for $b > 1$) or decreasing ($0 < b < 1$). Since different real numbers x determine different values of b^x, exponential functions are one-to-one.

The exponential function defined by

$$y = f(x) = b^x \quad \text{where } b > 0 \text{ and } b \neq 1$$

is one-to-one. Thus,

1. If $b^r = b^s$, then $r = s$.
2. If $r \neq s$, then $b^r \neq b^s$.

Accent on Technology   **GRAPHING EXPONENTIAL FUNCTIONS**

To use a graphing calculator to graph $f(x) = \left(\frac{2}{3}\right)^x$ and $f(x) = \left(\frac{3}{2}\right)^x$, we enter the right-hand sides of the equations. The screen will show the following equations.

$\backslash Y_1 = (2/3) \text{ ^ } X$

$\backslash Y_2 = (3/2) \text{ ^ } X$

If we use window settings of $[-10, 10]$ for x and $[-2, 10]$ for y and press
GRAPH , we will obtain the graph shown in Figure 10-5.

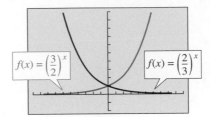

Figure 10-5

The graph of $f(x) = \left(\frac{2}{3}\right)^x$ passes through the points $(0, 1)$ and $\left(1, \frac{2}{3}\right)$.
Since $\frac{2}{3} < 1$, the function is decreasing.
    The graph of $f(x) = \left(\frac{3}{2}\right)^x$ passes through the points $(0, 1)$ and $\left(1, \frac{3}{2}\right)$.
Since $\frac{3}{2} > 1$, the function is increasing.
    Since both graphs pass the horizontal line test, each function is one-to-one.

Vertical and Horizontal Translations

We have seen that when $k > 0$ the graph of

$y = f(x) + k$ is the graph of $y = f(x)$ translated k units upward.

$y = f(x) - k$ is the graph of $y = f(x)$ translated k units downward.

$y = f(x - k)$ is the graph of $y = f(x)$ translated k units to the right.

$y = f(x + k)$ is the graph of $y = f(x)$ translated k units to the left.

EXAMPLE 5   On one set of axes, graph $f(x) = 2^x$ and $f(x) = 2^x + 3$.

Solution   The graph of $f(x) = 2^x + 3$ is identical to the graph of $f(x) = 2^x$, except that it is translated 3 units upward. (See Figure 10-6.)

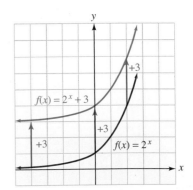

$f(x) = 2^x$

x	y	(x, y)
-4	$\frac{1}{16}$	$\left(-4, \frac{1}{16}\right)$
0	1	$(0, 1)$
2	4	$(2, 4)$

$f(x) = 2^x + 3$

x	y	(x, y)
-4	$3\frac{1}{16}$	$\left(-4, 3\frac{1}{16}\right)$
0	4	$(0, 4)$
2	7	$(2, 7)$

Figure 10-6

Self Check   Graph: $f(x) = 4^x$ and $f(x) = 4^x - 3$.

EXAMPLE 6   On one set of axes, graph $f(x) = 2^x$ and $f(x) = 2^{x+3}$.

Solution   The graph of $f(x) = 2^{x+3}$ is identical to the graph of $f(x) = 2^x$, except that it is translated 3 units to the left. (See Figure 10-7.)

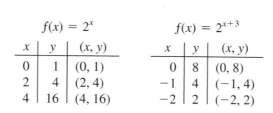

$$f(x) = 2^x$$

x	y	(x, y)
0	1	$(0, 1)$
2	4	$(2, 4)$
4	16	$(4, 16)$

$$f(x) = 2^{x+3}$$

x	y	(x, y)
0	8	$(0, 8)$
-1	4	$(-1, 4)$
-2	2	$(-2, 2)$

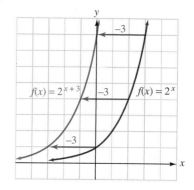

Figure 10-7

Self Check   On one set of axes, graph $f(x) = 4^x$ and $f(x) = 4^{x-3}$. ∎

The graphs of $f(x) = kb^x$ and $f(x) = b^{kx}$ are vertical and horizontal stretchings of the graph of $f(x) = b^x$. To graph these functions, we can plot several points and join them with a smooth curve or use a graphing calculator.

Accent on Technology   **GRAPHING EXPONENTIAL FUNCTIONS**

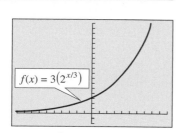

$f(x) = 3\left(2^{x/3}\right)$

Figure 10-8

To use a graphing calculator to graph the exponential function $y = f(x) = 3(2^{x/3})$, we enter the right-hand side of the equation. The display will show the equation

$$\backslash Y_1 = 3(2 \wedge (X/3))$$

If we use window settings of $[-10, 10]$ for x and $[-2, 18]$ for y and press   GRAPH , we will obtain the graph shown in Figure 10-8.

Compound Interest

If we deposit $\$P$ in an account paying an annual simple interest rate r, we can find the amount A in the account at the end of t years by using the formula $A = P + Prt$, or $A = P(1 + rt)$.

Suppose that we deposit $\$500$ in an account that pays interest every six months. Then $P = 500$, and after six months $\left(\frac{1}{2}\text{ year}\right)$, the amount in the account will be

$$A = 500(1 + rt)$$
$$= 500\left(1 + r \cdot \frac{1}{2}\right) \qquad \text{Substitute } \tfrac{1}{2} \text{ for } t.$$
$$= 500\left(1 + \frac{r}{2}\right)$$

The account will begin the second six-month period with a value of $500\left(1 + \frac{r}{2}\right)$. After the second six-month period, the amount will be

$$A = P(1 + rt)$$

$$A = \left[500\left(1 + \frac{r}{2}\right)\right]\left(1 + r \cdot \frac{1}{2}\right) \quad \text{Substitute } 500\left(1 + \frac{r}{2}\right) \text{ for } P \text{ and } \frac{1}{2} \text{ for } t.$$

$$= 500\left(1 + \frac{r}{2}\right)\left(1 + \frac{r}{2}\right)$$

$$= 500\left(1 + \frac{r}{2}\right)^2$$

At the end of a third six-month period, the amount in the account will be

$$A = 500\left(1 + \frac{r}{2}\right)^3$$

In this discussion, the earned interest is deposited back in the account and also earns interest. When this is the case, we say that the account is earning **compound interest.**

Formula for Compound Interest

If P is deposited in an account and interest is paid k times a year at an annual rate r, the amount A in the account after t years is given by

$$A = P\left(1 + \frac{r}{k}\right)^{kt}$$

EXAMPLE 7   **Saving for college**   To save for college, parents invest $12,000 for their newborn child in a mutual fund that should average a 10% annual return. If the quarterly interest is reinvested, how much will be available in 18 years?

Solution   We substitute 12,000 for P, 0.10 for r, and 18 for t into the formula for compound interest and find A. Since interest is paid quarterly, $k = 4$.

$$A = P\left(1 + \frac{r}{k}\right)^{kt}$$

$$A = 12{,}000\left(1 + \frac{0.10}{4}\right)^{4(18)}$$

$$= 12{,}000(1 + 0.025)^{72}$$

$$= 12{,}000(1.025)^{72}$$

$$= 71{,}006.74 \qquad \text{Use a calculator.}$$

In 18 years, the account will be worth $71,006.74.

Self Check   How much would be available if the parents invest $20,000?   ∎

In business applications, the initial amount of money deposited is often called the **present value** (*PV*). The amount to which the money will grow is called the **future value** (*FV*). The interest rate used for each compounding period is the **periodic interest rate** (*i*), and the number of times interest is compounded is the number of **compounding periods** (*n*). Using these definitions, we have an alternate formula for compound interest.

Formula for Compound Interest

$$FV = PV(1 + i)^n$$

This alternate formula appears on business calculators. To use this formula to solve Example 7, we proceed as follows:

$$FV = \boldsymbol{PV}(1 + i)^n$$
$$FV = \boldsymbol{12,000}(1 + 0.025)^{72} \quad i = \tfrac{0.10}{4} = 0.025 \text{ and } n = 4(18) = 72.$$
$$= 71,006.74$$

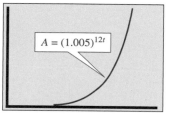

Accent on Technology   **SOLVING INVESTMENT PROBLEMS**

Suppose $1 is deposited in an account earning 6% annual interest, compounded monthly. To use a graphing calculator to estimate how much money will be in the account in 100 years, we can substitute 1 for P, 0.06 for r, and 12 for k into the formula

$$A = P\left(1 + \frac{r}{k}\right)^{kt}$$

$$A = \boldsymbol{1}\left(1 + \frac{0.06}{12}\right)^{12t}$$

and simplify to get

$$A = (1.005)^{12t}$$

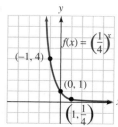

$A = (1.005)^{12t}$

Figure 10-9

We now graph $A = (1.005)^{12t}$ using window settings of $[0, 120]$ for t and $[0, 400]$ for A to obtain the graph shown in Figure 10-9. We can then trace and zoom to estimate that $1 grows to be approximately $397 in 100 years. From the graph, we can see that the money grows slowly in the early years and rapidly in the later years.

Self Check Answers

1. a. 81   **b.** $b^{4\sqrt{2}}$ **2.**

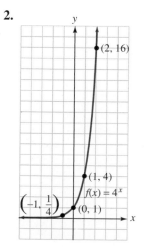

3.

4. No; it doesn't pass through the point $(0, 1)$.

5.

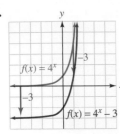

6.

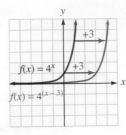

7. $118,344.56

Orals *If x = 2, evaluate each expression.*

1. 2^x **2.** 5^x **3.** $2(3^x)$ **4.** 3^{x-1}

If x = −2, evaluate each expression.

5. 2^x **6.** 5^x **7.** $2(3^x)$ **8.** 3^{x-1}

10.1   EXERCISES

REVIEW   *In the illustration, lines r and s are parallel.*

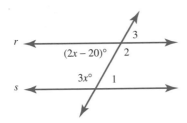

1. Find x.
2. Find the measure of $\angle 1$.
3. Find the measure of $\angle 2$.
4. Find the measure of $\angle 3$.

VOCABULARY AND CONCEPTS   *Fill in the blanks.*

5. If $b > 0$ and $b \neq 1$, $y = f(x) = b^x$ is called an _____ function.

6. The _____ of an exponential function is $(-\infty, \infty)$.

7. The range of an exponential function is the interval _____.

8. The graph of $y = f(x) = 3^x$ passes through the points $(0, \_\_)$ and $(1, \_\_)$.

9. If $b > 1$, then $y = f(x) = b^x$ is an _____ function.

10. If $0 < b < 1$, then $y = f(x) = b^x$ is a _____ function.

11. The formula for compound interest is $A = $ _____.

12. An alternate formula for compound interest is $FV = $ _____.

PRACTICE   *Find each value to four decimal places.*

13. $2^{\sqrt{2}}$ **14.** $7^{\sqrt{2}}$

15. $5^{\sqrt{5}}$ **16.** $6^{\sqrt{3}}$

Simplify each expression.

17. $\left(2^{\sqrt{3}}\right)^{\sqrt{3}}$ **18.** $3^{\sqrt{2}}3^{\sqrt{18}}$

19. $7^{\sqrt{3}}7^{\sqrt{12}}$ **20.** $\left(3^{\sqrt{5}}\right)^{\sqrt{5}}$

Graph each exponential function. Check your work with a graphing calculator.

21. $y = f(x) = 3^x$ **22.** $y = f(x) = 5^x$

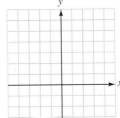 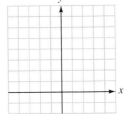

23. $y = f(x) = \left(\dfrac{1}{3}\right)^x$ **24.** $y = f(x) = \left(\dfrac{1}{5}\right)^x$

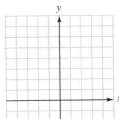 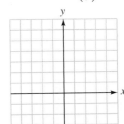

25. $y = f(x) = 3^x - 2$

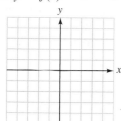

26. $y = f(x) = 2^x + 1$

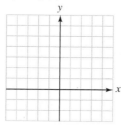

27. $y = f(x) = 3^{x-1}$

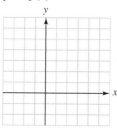

28. $y = f(x) = 2^{x+1}$

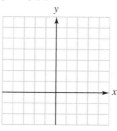

Find the value of b, if any, that would cause the graph of $y = b^x$ to look like the graph indicated.

29.

30.

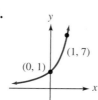

31.

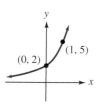

32.

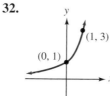

33.

34.

35.

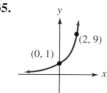

36.

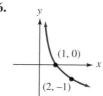

Use a graphing calculator to graph each function. Tell whether the function is an increasing or a decreasing function.

37. $f(x) = \dfrac{1}{2}(3^{x/2})$

38. $f(x) = -3(2^{x/3})$

39. $f(x) = 2(3^{-x/2})$

40. $f(x) = -\dfrac{1}{4}(2^{-x/2})$

APPLICATIONS   *Assume that there are no deposits or withdrawals.*

41. Compound interest   An initial deposit of $10,000 earns 8% interest, compounded quarterly. How much will be in the account after 10 years?

42. Compound interest   An initial deposit of $10,000 earns 8% interest, compounded monthly. How much will be in the account after 10 years?

43. Comparing interest rates   How much more interest could $1,000 earn in 5 years, compounded quarterly, if the annual interest rate were $5\frac{1}{2}\%$ instead of 5%?

44. Comparing savings plans   Which institution in the two ads provides the better investment?

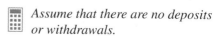

Fidelity Savings & Loan
Earn 5.25%
compounded monthly

Union Trust

Money Market Account
paying 5.35%,
compounded annually

45. Compound interest   If $1 had been invested on July 4, 1776, at 5% interest, compounded annually, what would it be worth on July 4, 2076?

46. Frequency of compounding $10,000 is invested in each of two accounts, both paying 6% annual interest. In the first account, interest compounds quarterly, and in the second account, interest compounds daily. Find the difference between the accounts after 20 years.

47. Radioactive decay   A radioactive material decays according to the formula $A = A_0\left(\frac{2}{3}\right)^t$, where A_0 is the initial amount present and t is measured in years. Find an expression for the amount present in 5 years.

48. Bacteria cultures   A colony of 6 million bacteria is growing in a culture medium. (See the illustration.) The population P after t hours is given by the formula $P = (6 \times 10^6)(2.3)^t$. Find the population after 4 hours.

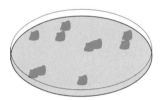

12:00 P.M.   4:00 P.M.

49. Discharging a battery   The charge remaining in a battery decreases as the battery discharges. The charge C (in coulombs) after t days is given by the formula $C = (3 \times 10^{-4})(0.7)^t$. Find the charge after 5 days.

50. Town population   The population of North Rivers is decreasing exponentially according to the formula $P = 3,745(0.93)^t$, where t is measured in years from the present date. Find the population in 6 years, 9 months.

51. Salvage value   A small business purchases a computer for $4,700. It is expected that its value each year will be 75% of its value in the preceding year. If the business disposes of the computer after 5 years, find its salvage value (the value after 5 years).

52. Louisiana Purchase   In 1803, the United States acquired territory from France in the Louisiana Purchase. The country doubled its territory by adding 827,000 square miles of land for $15 million. If the land has appreciated at the rate of 6% each year, what would one square mile of land be worth in 1996?

WRITING

53. If world population is increasing exponentially, why is there cause for concern?

54. How do the graphs of $y = b^x$ differ when $b > 1$ and $0 < b < 1$?

SOMETHING TO THINK ABOUT

55. In the definition of the exponential function, b could not equal 0. Why not?

56. In the definition of the exponential function, b could not be negative. Why not?

10.2   Base-e Exponential Functions

▮ Continuous Compound Interest
▮ Graphing the Exponential Function
▮ Malthusian Population Growth   ▮ The Malthusian Theory

Getting Ready   *Evaluate* $\left(1 + \frac{1}{n}\right)^n$ *for the following values. Round each answer to the nearest hundredth.*

1. $n = 1$ **2.** $n = 2$ **3.** $n = 4$ **4.** $n = 10$

In this section, we will discuss special exponential functions with a base of e, an important number with a value of approximately 2.71.

Continuous Compound Interest

If a bank pays interest twice a year, we say that interest is *compounded semiannually*. If it pays interest four times a year, we say that interest is *compounded quarterly*. If it pays interest continuously (infinitely many times in a year), we say that interest is *compounded continuously*.

To develop the formula for continuous compound interest, we start with the formula

$$A = P\left(1 + \frac{r}{k}\right)^{kt}$$ The formula for compound interest.

and substitute rn for k. Since r and k are positive numbers, so is n.

$$A = P\left(1 + \frac{r}{rn}\right)^{rnt}$$

We can then simplify the fraction $\frac{r}{rn}$ and use the commutative property of multiplication to change the order of the exponents.

$$A = P\left(1 + \frac{1}{n}\right)^{nrt}$$

Finally, we can use a property of exponents to write this formula as

(1) $$A = P\left[\left(1 + \frac{1}{n}\right)^{n}\right]^{rt}$$ Use the property $a^{mn} = (a^m)^n$.

To find the value of $\left(1 + \frac{1}{n}\right)^n$, we use a calculator to evaluate it for several values of n, as shown in Table 10-1.

n	$\left(1 + \frac{1}{n}\right)^n$
1	2
2	2.25
4	2.44140625 . . .
12	2.61303529 . . .
365	2.71456748 . . .
1,000	2.71692393 . . .
100,000	2.71826823 . . .
1,000,000	2.71828046 . . .

Table 10-1

The results suggest that as n gets larger, the value of $\left(1 + \frac{1}{n}\right)^n$ approaches the number 2.71828. . . . This number is called e, which has the following value.

$$e = 2.718281828459 \ldots$$

In continuous compound interest, k (the number of compoundings) is infinitely large. Since k, r, and n are all positive and $k = rn$, as k gets very large (approaches infinity), then so does n. Therefore, we can replace $\left(1 + \frac{1}{n}\right)^n$ in Equation 1 with e to get

$$A = Pe^{rt}$$

Formula for Exponential Growth

If a quantity P increases or decreases at an annual rate r, compounded continuously, then the amount A after t years is given by

$$A = Pe^{rt}$$

If time is measured in years, then r is called the **annual growth rate.** If r is negative, the "growth" represents a decrease.

To compute the amount to which \$12,000 will grow if invested for 18 years at 10% annual interest, compounded continuously, we substitute 12,000 for P, 0.10 for r, and 18 for t in the formula for exponential growth:

$$
\begin{aligned}
A &= \boldsymbol{Pe^{rt}} \\
&= \boldsymbol{12{,}000}e^{0.10(18)} \\
&= 12{,}000e^{1.8} \\
&\approx 72{,}595.76957 \qquad \text{Use a calculator.}
\end{aligned}
$$

After 18 years, the account will contain \$72,595.77. This is \$1,589.03 more than the result in Example 7 in the previous section, where interest was compounded quarterly.

EXAMPLE 1 If \$25,000 accumulates interest at an annual rate of 8%, compounded continuously, find the balance in the account in 50 years.

Solution We substitute 25,000 for P, 0.08 for r, and 50 for t.

$$
\begin{aligned}
A &= \boldsymbol{Pe^{rt}} \\
A &= \boldsymbol{25{,}000}e^{(0.08)(50)} \\
&= 25{,}000e^4 \\
&\approx 1{,}364{,}953.751 \qquad \text{Use a calculator.}
\end{aligned}
$$

In 50 years, the balance will be \$1,364,953.75—over one million dollars.

Self Check Find the balance in 60 years. ∎

The exponential function $y = f(x) = e^x$ is so important that it is often called **the exponential function.**

Graphing the Exponential Function

To graph the exponential function, we plot several points and join them with a smooth curve, as shown in Figure 10-10.

$$y = f(x) = e^x$$

x	$f(x)$	$(x, f(x))$
-2	0.1	$(-2, 0.1)$
-1	0.4	$(-1, 0.4)$
0	1	$(0, 1)$
1	2.7	$(1, 2.7)$
2	7.4	$(2, 7.4)$

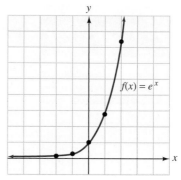

Figure 10-10

Accent on Technology TRANSLATIONS OF THE EXPONENTIAL FUNCTION

Figure 10-11(a) shows the graphs of $f(x) = e^x$, $f(x) = e^x + 5$, and $f(x) = e^x - 3$. To graph these functions with window settings of $[-3, 6]$ for x and $[-5, 15]$ for y, we enter the right-hand sides of the equations after the symbols $Y_1 = $, $Y_2 = $, and $Y_3 = $. The display will show

$$Y_1 = e \wedge (x)$$
$$Y_2 = e \wedge (x) + 5$$
$$Y_3 = e \wedge (x) - 3$$

After graphing these functions, we can see that the graph of $f(x) = e^x + 5$ is 5 units above the graph of $f(x) = e^x$, and that the graph of $f(x) = e^x - 3$ is 3 units below the graph of $f(x) = e^x$.

Figure 10-11(b) shows the calculator graphs of $f(x) = e^x$, $f(x) = e^{x+5}$, and $f(x) = e^{x-3}$. To graph these functions with window settings of $[-7, 10]$ for x and $[-5, 15]$ for y, we enter the right-hand sides of the equations after the symbols $Y_1 = $, $Y_2 = $, $Y_3 = $. The display will show

$$Y_1 = e \wedge (x)$$
$$Y_2 = e \wedge (x + 5)$$
$$Y_3 = e \wedge (x - 3)$$

After graphing these functions, we can see that the graph of $f(x) = e^{x+5}$ is 5 units to the left of the graph of $f(x) = e^x$, and that the graph of $f(x) = e^{x-3}$ is 3 units to the right of the graph of $f(x) = e^x$.

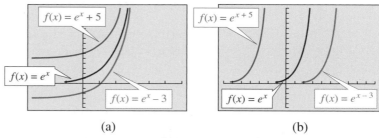

(a)   (b)

Figure 10-11

GRAPHING EXPONENTIAL FUNCTIONS

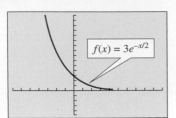

$f(x) = 3e^{-x/2}$

Figure 10-12

Figure 10-12 shows the calculator graph of $f(x) = 3e^{-x/2}$. To graph this function with window settings of $[-7, 10]$ for x and $[-5, 15]$ for y, we enter the right-hand side of the equation after the symbol $Y_1 =$. The display will show the equation

$$Y_1 = 3(e \wedge (-x/2))$$

The calculator graph appears in Figure 10-12. Explain why the graph has a y-intercept of $(0, 3)$.

Malthusian Population Growth

An equation based on the exponential function provides a model for **population growth.** In the **Malthusian model for population growth,** the future population of a colony is related to the present population by the formula $A = Pe^{rt}$.

EXAMPLE 2   **City planning**   The population of a city is currently 15,000, but changing economic conditions are causing the population to decrease by 2% each year. If this trend continues, find the population in 30 years.

Solution   Since the population is decreasing by 2% each year, the annual growth rate is -2%, or -0.02. We can substitute -0.02 for r, 30 for t, and 15,000 for P in the formula for exponential growth and find A.

$$A = Pe^{rt}$$
$$A = 15{,}000e^{-0.02(30)}$$
$$= 15{,}000e^{-0.6}$$
$$= 8{,}232.174541$$

In 30 years, city planners expect a population of approximately 8,232 persons.

Self Check   Find the population in 50 years.   ∎

The Malthusian Theory

The English economist Thomas Robert Malthus (1766–1834) pioneered in population study. He believed that poverty and starvation were unavoidable, because the human population tends to grow exponentially, but the food supply tends to grow linearly.

EXAMPLE 3   Suppose that a country with a population of 1,000 people is growing exponentially according to the formula

$$P = 1{,}000e^{0.02t}$$

where t is in years. Furthermore, assume that the food supply measured in adequate food per day per person is growing linearly according to the formula

$$y = 30.625x + 2,000$$

where x is in years. In how many years will the population outstrip the food supply?

Solution   We can use a graphing calculator, with window settings of [0, 100] for x and [0, 10,000] for y. After graphing the functions, we obtain Figure 10-13(a). If we trace, as in Figure 10-13(b), we can find the point where the two graphs intersect. From the graph, we can see that the food supply will be adequate for about 71 years. At that time, the population of approximately 4,200 people will have problems.

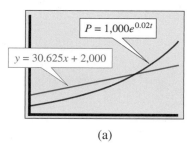
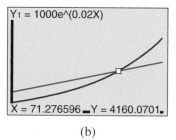

(a)   (b)

Figure 10-13

Self Check   Suppose that the population grows at a 3% rate. For how many years will the food supply be adequate?   ∎

Self Check Answers

1. $3,037,760.44   **2.** 5,518   **3.** about 38 years

Orals   Use a calculator to find each value to the nearest hundredth.

1. e^0 **2.** e^1 **3.** e^2 **4.** e^3

Fill in the blanks.

5. The graph of $f(x) = e^x + 2$ is __ units above the graph of $f(x) = e^x$.

6. The graph of $f(x) = e^{(x-2)}$ is __ units to the right of the graph of $f(x) = e^x$.

10.2   EXERCISES

REVIEW   *Simplify each expression. Assume that all variables represent positive numbers.*

1. $\sqrt{240x^5}$

2. $\sqrt[3]{-125x^5y^4}$

3. $4\sqrt{48y^3} - 3y\sqrt{12y}$

4. $\sqrt[4]{48z^5} + \sqrt[4]{768z^5}$

VOCABULARY AND CONCEPTS   *Fill in the blanks.*

5. To two decimal places, the value of e is ____.

6. The formula for continuous compound interest is $A =$ ____.

7. Since $e > 1$, the base-e exponential function is a(n) _____ function.

8. The graph of the exponential function $y = e^x$ passes through the points (0, 1) and (1, __).

9. The Malthusian population growth formula is _____.

10. The Malthusian prediction is pessimistic, because _____ grows exponentially, but food supplies grow _____.

PRACTICE *Graph each function. Check your work with a graphing calculator. Compare each graph to the graph of $f(x) = e^x$.*

11. $f(x) = e^x + 1$

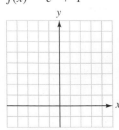

12. $f(x) = e^x - 2$

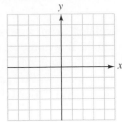

13. $f(x) = e^{(x+3)}$

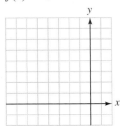

14. $f(x) = e^{(x-5)}$

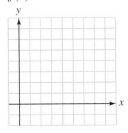

15. $f(x) = -e^x$

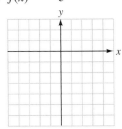

16. $f(x) = -e^x + 1$

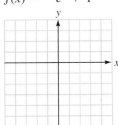

17. $f(x) = 2e^x$

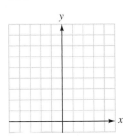

18. $f(x) = \dfrac{1}{2}e^x$

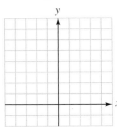

Tell whether the graph of $f(x) = e^x$ could look like the graph shown here.

19.

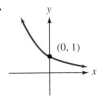

20.

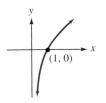

21.

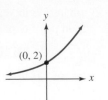

22.

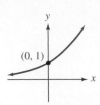

APPLICATIONS *Assume that there are no deposits or withdrawals.*

23. Continuous compound interest An investment of $5,000 earns 6% interest, compounded continuously. What will the investment be worth in 12 years?

24. Continuous compound interest An investment of $6,000 earns 7% interest, compounded continuously. What will the investment be worth in 35 years?

25. Determining the initial deposit An account now contains $12,000 and has been accumulating 7% annual interest, compounded continuously, for 9 years. Find the initial deposit.

26. Determining a previous balance An account now contains $8,000 and has been accumulating 8% annual interest, compounded continuously. How much was in the account 6 years ago?

27. Comparison of compounding methods An initial deposit of $5,000 grows at an annual rate of 8.5% for 5 years. Compare the final balances resulting from continuous compounding and annual compounding.

28. Comparison of compounding methods An initial deposit of $30,000 grows at an annual rate of 8% for 20 years. Compare the final balances resulting from continuous compounding and annual compounding.

Solve each problem.

29. World population growth Earth's population is approximately 6 billion people and is growing at an annual rate of 1.9%. Assuming a Malthusian growth model, find the world population in 30 years.

30. World population growth Earth's population is approximately 6 billion people and is growing at an annual rate of 1.9%. Assuming a Malthusian growth model, find the world population in 40 years.

31. World population growth Assuming a Malthusian growth model and an annual growth rate of 1.9%, by what factor will Earth's current population increase in 50 years? (See Exercise 29.)

32. Population growth The growth of a population is modeled by

$$P = 173e^{0.03t}$$

How large will the population be when $t = 30$?

33. Population decline The decline of a population is modeled by

$$P = 8,000e^{-0.008t}$$

How large will the population be when $t = 20$?

34. Epidemics The spread of foot and mouth disease through a herd of cattle can be modeled by the formula

$$P = P_0e^{0.27t} \quad (t \text{ is in days})$$

If a rancher does not act quickly to treat two cases, how many cattle will have the disease in 10 days?

35. Alcohol absorption In one individual, the blood alcohol level t minutes after drinking two shots of whiskey is given by $P = 0.3(1 - e^{-0.05t})$. Find the blood alcohol level after 15 minutes.

36. Medicine The concentration, x, of a certain drug in an organ after t minutes is given by $x = 0.08(1 - e^{-0.1t})$. Find the concentration of the drug after 30 minutes.

37. Medicine Refer to Exercise 36. Find the initial concentration of the drug.

38. Skydiving Before her parachute opens, a skydiver's velocity v in meters per second is given by $v = 50(1 - e^{-0.2t})$. Find the initial velocity.

39. Skydiving Refer to Exercise 38 and find the velocity after 20 seconds.

40. Free-falling objects After t seconds, a certain falling object has a velocity v given by $v = 50(1 - e^{-0.3t})$. Which is falling faster after 2 seconds, this object or the skydiver in Exercise 38?

41. Depreciation A camping trailer originally purchased for $4,570 is continuously losing value at the rate of 6% per year. Find its value when it is $6\frac{1}{2}$ years old.

42. Depreciation A boat purchased for $7,500 has been continuously decreasing in value at the rate of 2% each year. It is now 8 years, 3 months old. Find its value.

Use a graphing calculator to solve each problem.

43. In Example 3, suppose that better farming methods change the formula for food growth to $y = 31x + 2,000$. How long will the food supply be adequate?

44. In Example 3, suppose that a birth-control program changed the formula for population growth to $P = 1,000e^{0.01t}$. How long will the food supply be adequate?

WRITING

45. Explain why the graph of $f(x) = e^x - 5$ is 5 units below the graph of $f(x) = e^x$.

46. Explain why the graph of $f(x) = e^{(x+5)}$ is 5 units to the left of the graph of $f(x) = e^x$.

SOMETHING TO THINK ABOUT

47. The value of e can be calculated to any degree of accuracy by adding the first several terms of the following list:

$$1, 1, \frac{1}{2}, \frac{1}{2 \cdot 3}, \frac{1}{2 \cdot 3 \cdot 4}, \frac{1}{2 \cdot 3 \cdot 4 \cdot 5}, \cdots$$

The more terms that are added, the closer the sum will be to e. Add the first six numbers in the preceding list. To how many decimal places is the sum accurate?

48. Graph the function defined by the equation

$$y = f(x) = \frac{e^x + e^{-x}}{2}$$

from $x = -2$ to $x = 2$. The graph will look like a parabola, but it is not. The graph, called a **catenary,** is important in the design of power distribution networks, because it represents the shape of a uniform flexible cable whose ends are suspended from the same height. The function is called the **hyperbolic cosine function.**

49. If $e^{t+5} = ke^t$, find k.

50. If $e^{5t} = k^t$, find k.

10.3 Logarithmic Functions

■ Logarithms   ■ Graphs of Logarithmic Functions
■ Vertical and Horizontal Translations   ■ Base-10 Logarithms
■ Electronics   ■ Seismology

Getting Ready *Find each value.*

1. 7^0 **2.** 5^2 **3.** 5^{-2} **4.** $16^{1/2}$

In this section, we consider the inverse function of an exponential function $f(x) = b^x$. This inverse function is called a *logarithmic function.*

Logarithms

Since the exponential function $y = b^x$ is one-to-one, it has an inverse function defined by the equation $x = b^y$. To express this inverse function in the form $y = f^{-1}(x)$, we must solve the equation $x = b^y$ for y. To do this, we need the following definition.

Logarithmic Functions

> If $b > 0$ and $b \neq 1$, the **logarithmic function with base b** is defined by
>
> $$y = \log_b x \quad \text{if and only if} \quad x = b^y$$
>
> The **domain of the logarithmic function** is the interval $(0, \infty)$. The **range** is the interval $(-\infty, \infty)$.

Since the function $y = \log_b x$ is the inverse of the one-to-one exponential function $y = b^x$, the logarithmic function is also one-to-one.

 Comment Since the domain of the logarithmic function is the set of positive numbers, it is impossible to find the logarithm of 0 or the logarithm of a negative number.

The previous definition guarantees that any pair (x, y) that satisfies the equation $y = \log_b x$ also satisfies the equation $x = b^y$.

$\log_4 1 = \mathbf{0}$	because	$1 = 4^{\mathbf{0}}$
$\log_5 25 = \mathbf{2}$	because	$25 = 5^{\mathbf{2}}$
$\log_5 \dfrac{1}{25} = \mathbf{-2}$	because	$\dfrac{1}{25} = 5^{\mathbf{-2}}$
$\log_{16} 4 = \dfrac{\mathbf{1}}{\mathbf{2}}$	because	$4 = 16^{\mathbf{1/2}}$
$\log_2 \dfrac{1}{8} = \mathbf{-3}$	because	$\dfrac{1}{8} = 2^{\mathbf{-3}}$
$\log_b x = \mathbf{y}$	because	$x = b^{\mathbf{y}}$

In each of these examples, the logarithm of a number is an exponent. In fact,

$\log_b x$ is the exponent to which b is raised to get x.

In equation form, we write

$$b^{\log_b x} = x$$

EXAMPLE 1 Find y in each equation: **a.** $\log_6 1 = y$,   **b.** $\log_3 27 = y$,   and   **c.** $\log_5 \frac{1}{5} = y$.

Solution   **a.** We can change the equation $\log_6 1 = y$ into the equivalent exponential equation $6^y = 1$. Since $6^0 = 1$, it follows that $y = 0$. Thus,

$$\log_6 1 = 0$$

b. $\log_3 27 = y$ is equivalent to $3^y = 27$. Since $3^3 = 27$, it follows that $3^y = 3^3$, and $y = 3$. Thus,

$$\log_3 27 = 3$$

c. $\log_5 \frac{1}{5} = y$ is equivalent to $5^y = \frac{1}{5}$. Since $5^{-1} = \frac{1}{5}$, it follows that $5^y = 5^{-1}$, and $y = -1$. Thus,

$$\log_5 \frac{1}{5} = -1$$

Self Check Find y: **a.** $\log_3 9 = y$,   **b.** $\log_2 64 = y$,   and   **c.** $\log_5 \frac{1}{125} = y$.   ∎

EXAMPLE 2 Find x in each equation: **a.** $\log_3 81 = x$,   **b.** $\log_x 125 = 3$,   and   **c.** $\log_4 x = 3$.

Solution   **a.** $\log_3 81 = x$ is equivalent to $3^x = 81$. Because $3^4 = 81$, it follows that $3^x = 3^4$. Thus, $x = 4$.

b. $\log_x 125 = 3$ is equivalent to $x^3 = 125$. Because $5^3 = 125$, it follows that $x^3 = 5^3$. Thus, $x = 5$.

c. $\log_4 x = 3$ is equivalent to $4^3 = x$. Because $4^3 = 64$, it follows that $x = 64$.

Self Check Find x: **a.** $\log_2 32 = x$,   **b.** $\log_x 8 = 3$,   and   **c.** $\log_5 x = 2$.   ∎

EXAMPLE 3 Find x in each equation: **a.** $\log_{1/3} x = 2$,   **b.** $\log_{1/3} x = -2$,   and
c. $\log_{1/3} \frac{1}{27} = x$.

Solution   **a.** $\log_{1/3} x = 2$ is equivalent to $\left(\frac{1}{3}\right)^2 = x$. Thus, $x = \frac{1}{9}$.
b. $\log_{1/3} x = -2$ is equivalent to $\left(\frac{1}{3}\right)^{-2} = x$. Thus,

$$x = \left(\frac{1}{3}\right)^{-2} = 3^2 = 9$$

c. $\log_{1/3} \frac{1}{27} = x$ is equivalent to $\left(\frac{1}{3}\right)^x = \frac{1}{27}$. Because $\left(\frac{1}{3}\right)^3 = \frac{1}{27}$, it follows that $x = 3$.

Self Check Find x: **a.** $\log_{1/4} x = 3$ and   **b.** $\log_{1/4} x = -2$.   ∎

Graphs of Logarithmic Functions

To graph the logarithmic function $y = \log_2 x$, we calculate and plot several points with coordinates (x, y) that satisfy the equation $x = 2^y$. After joining these points with a smooth curve, we have the graph shown in Figure 10-14(a).

To graph $y = \log_{1/2} x$, we calculate and plot several points with coordinates (x, y) that satisfy the equation $x = \left(\frac{1}{2}\right)^y$. After joining these points with a smooth curve, we have the graph shown in Figure 10-14(b).

$y = \log_2 x$

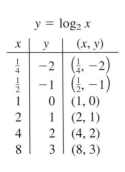

x	y	(x, y)
$\frac{1}{4}$	-2	$\left(\frac{1}{4}, -2\right)$
$\frac{1}{2}$	-1	$\left(\frac{1}{2}, -1\right)$
1	0	$(1, 0)$
2	1	$(2, 1)$
4	2	$(4, 2)$
8	3	$(8, 3)$

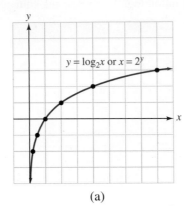

(a)

$y = \log_{1/2} x$

x	y	(x, y)
$\frac{1}{4}$	2	$\left(\frac{1}{4}, 2\right)$
$\frac{1}{2}$	1	$\left(\frac{1}{2}, 1\right)$
1	0	$(1, 0)$
2	-1	$(2, -1)$
4	-2	$(4, -2)$
8	-3	$(8, -3)$

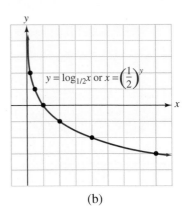

(b)

Figure 10-14

The graphs of all logarithmic functions are similar to those in Figure 10-15. If $b > 1$, the logarithmic function is increasing, as in Figure 10-15(a). If $0 < b < 1$, the logarithmic function is decreasing, as in Figure 10-15(b).

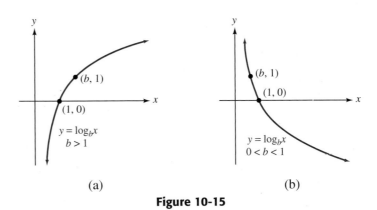

(a)   (b)

Figure 10-15

The graph of $f(x) = \log_b x$ has the following properties.

1. It passes through the point $(1, 0)$.
2. It passes through the point $(b, 1)$.
3. The y-axis is an asymptote.
4. The domain is $(0, \infty)$ and the range is $(-\infty, \infty)$.

The exponential and logarithmic functions are inverses of each other and, therefore, have symmetry about the line $y = x$. The graphs $y = \log_b x$ and $y = b^x$ are shown in Figure 10-16(a) when $b > 1$, and in Figure 10-16(b) when $0 < b < 1$.

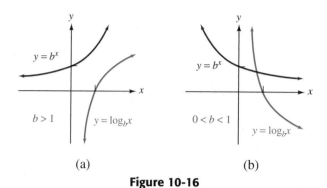

(a)   (b)

Figure 10-16

Vertical and Horizontal Translations

The graphs of many functions involving logarithms are translations of the basic logarithmic graphs.

EXAMPLE 4   Graph the function defined by $y = 3 + \log_2 x$.

Solution   The graph of $y = 3 + \log_2 x$ is identical to the graph of $y = \log_2 x$, except that it is translated 3 units upward. (See Figure 10-17.)

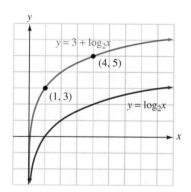

Figure 10-17

Self Check   Graph: $y = \log_3 x - 2$.   ∎

EXAMPLE 5   Graph: $y = \log_{1/2} (x - 1)$.

Solution   The graph of $y = \log_{1/2} (x - 1)$ is identical to the graph of $y = \log_{1/2} x$, except that it is translated 1 unit to the right. (See Figure 10-18.)

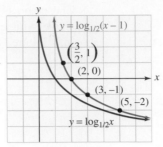

Figure 10-18

Self Check Graph: $y = \log_{1/3}(x + 2)$. ∎

Accent on Technology GRAPHING LOGARITHMIC FUNCTIONS

Graphing calculators can draw graphs of logarithmic functions directly if the base of the logarithmic function is 10 or e. To use a calculator to graph the logarithmic function $f(x) = -2 + \log_{10}\left(\frac{1}{2}x\right)$, we enter the right-hand side of the equation after the symbol $Y_1 = $. The display will show the equation

$$Y_1 = -2 + \log(1/2*x)$$

If we use window settings of $[-1, 5]$ for x and $[-4, 1]$ for y and press GRAPH , we will obtain the graph shown in Figure 10-19.

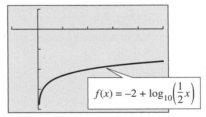

Figure 10-19

Base-10 Logarithms

For computational purposes and in many applications, we will use base-10 logarithms (also called **common logarithms**). When the base b is not indicated in the notation $\log x$, we assume that $b = 10$:

$\log x$ means $\log_{10} x$

Because base-10 logarithms appear so often, it is a good idea to become familiar with the following base-10 logarithms:

$$\log_{10} \frac{1}{100} = -2 \qquad \text{because} \qquad 10^{-2} = \frac{1}{100}$$

$$\log_{10} \frac{1}{10} = -1 \qquad \text{because} \qquad 10^{-1} = \frac{1}{10}$$

$$\log_{10} 1 = 0 \qquad \text{because} \qquad 10^0 = 1$$
$$\log_{10} 10 = 1 \qquad \text{because} \qquad 10^1 = 10$$
$$\log_{10} 100 = 2 \qquad \text{because} \qquad 10^2 = 100$$
$$\log_{10} 1{,}000 = 3 \qquad \text{because} \qquad 10^3 = 1{,}000$$

In general, we have

$$\log_{10} 10^x = x$$

Accent on Technology

FINDING LOGARITHMS

Before calculators, extensive tables provided logarithms of numbers. Today, logarithms are easy to find with a calculator. For example, to find log 32.58 with a scientific calculator, we enter these numbers and press these keys:

32.58 LOG

The display will read **1.51295108**. To four decimal places, log 32.58 = 1.5130.

To use a graphing calculator, we enter these numbers and press these keys:

LOG 32.58 ENTER

The display will read **log (32.58**
1.51295108

EXAMPLE 6 Find x in the equation $\log x = 0.3568$. Round to four decimal places.

Solution The equation $\log x = 0.3568$ is equivalent to $10^{0.3568} = x$. To find x with a scientific calculator, we enter these numbers and press these keys:

10 y^x .3568 =

The display will read **2.274049951**. To four decimal places,

$$x = 2.2740$$

If your calculator has 10^x key, enter .3568 and press it to get the same result.

Self Check Solve: $\log x = 2.7$. Round to four decimal places. ∎

Electronics

Common logarithms are used in electrical engineering to express the voltage gain (or loss) of an electronic device such as an amplifier. The unit of gain (or loss), called the **decibel,** is defined by a logarithmic relation.

Decibel Voltage Gain

If E_O is the output voltage of a device and E_I is the input voltage, the decibel voltage gain is given by

$$\text{dB gain} = 20 \log \frac{E_O}{E_I}$$

EXAMPLE 7   **Finding dB gain**   If the input to an amplifier is 0.4 volt and the output is 50 volts, find the decibel voltage gain of the amplifier.

Solution   We can find the decibel voltage gain by substituting 0.4 for E_I and 50 for E_O into the formula for dB gain:

$$\text{dB gain} = 20 \log \frac{E_O}{E_I}$$

$$\text{dB gain} = 20 \log \frac{50}{0.4}$$

$$= 20 \log 125$$

$$\approx 42 \qquad \text{Use a calculator.}$$

The amplifier provides a 42-decibel voltage gain.   ∎

Seismology

In seismology, common logarithms are used to measure the intensity of earthquakes on the **Richter scale.** The intensity of an earthquake is given by the following logarithmic function.

Richter Scale

If R is the intensity of an earthquake, A is the amplitude (measured in micrometers), and P is the period (the time of one oscillation of Earth's surface, measured in seconds), then

$$R = \log \frac{A}{P}$$

EXAMPLE 8   **Measuring earthquakes**   Find the measure on the Richter scale of an earthquake with an amplitude of 10,000 micrometers (1 centimeter) and a period of 0.1 second.

Solution   We substitute 10,000 for A and 0.1 for P in the Richter scale formula and simplify:

$$R = \log \frac{A}{P}$$

$$R = \log \frac{10,000}{0.1}$$

$$= \log 100,000$$

$$= 5$$

The earthquake measures 5 on the Richter scale.   ∎

Self Check Answers

1. a. 2 b. 6 c. -3 2. a. 5 b. 2 c. 25 3. a. $\dfrac{1}{64}$ b. 16

4. 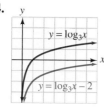   5. 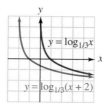   6. 501.1872

Orals *Find the value of x in each equation.*

1. $\log_2 8 = x$ 2. $\log_3 9 = x$ 3. $\log_x 125 = 3$
4. $\log_x 8 = 3$ 5. $\log_4 16 = x$ 6. $\log_x 32 = 5$
7. $\log_{1/2} x = 2$ 8. $\log_9 3 = x$ 9. $\log_x \dfrac{1}{4} = -2$

10.3 EXERCISES

REVIEW *Solve each equation.*

1. $\sqrt[3]{6x + 4} = 4$

2. $\sqrt{3x - 4} = \sqrt{-7x + 2}$

3. $\sqrt{a + 1} - 1 = 3a$ 4. $3 - \sqrt{t - 3} = \sqrt{t}$

VOCABULARY AND CONCEPTS *Fill in the blanks.*

5. The equation $y = \log_b x$ is equivalent to _____.
6. The domain of the logarithmic function is the interval _____.
7. The _____ of the logarithmic function is the interval $(-\infty, \infty)$.
8. $b^{\log_b x} =$ __.
9. Because an exponential function is one-to-one, it has an _____ function.
10. The inverse of an exponential function is called a _____ function.
11. $\log_b x$ is the _____ to which b is raised to get x.
12. The y-axis is an _____ to the graph of $f(x) = \log_b x$.
13. The graph of $y = f(x) = \log_b x$ passes through the points _____ and _____.
14. $\log_{10} 10^x =$ __.   15. dB gain $=$ _____.
16. The intensity of an earthquake is measured by the formula $R =$ _____.

PRACTICE *Write each equation in exponential form.*

17. $\log_3 27 = 3$ 18. $\log_8 8 = 1$

19. $\log_{1/2} \dfrac{1}{4} = 2$ 20. $\log_{1/5} 1 = 0$

21. $\log_4 \dfrac{1}{64} = -3$ 22. $\log_6 \dfrac{1}{36} = -2$

23. $\log_{1/2} \dfrac{1}{8} = 3$ 24. $\log_{1/5} 1 = 0$

Write each equation in logarithmic form.

25. $6^2 = 36$ 26. $10^3 = 1{,}000$

27. $5^{-2} = \dfrac{1}{25}$ 28. $3^{-3} = \dfrac{1}{27}$

29. $\left(\dfrac{1}{2}\right)^{-5} = 32$ 30. $\left(\dfrac{1}{3}\right)^{-3} = 27$

31. $x^y = z$ 32. $m^n = p$

Find each value of x.

33. $\log_2 16 = x$

34. $\log_3 9 = x$

35. $\log_4 16 = x$

36. $\log_6 216 = x$

37. $\log_{1/2} \dfrac{1}{8} = x$

38. $\log_{1/3} \dfrac{1}{81} = x$

39. $\log_9 3 = x$

40. $\log_{125} 5 = x$

41. $\log_{1/2} 8 = x$

42. $\log_{1/2} 16 = x$

43. $\log_7 x = 2$

44. $\log_5 x = 0$

45. $\log_6 x = 1$

46. $\log_2 x = 4$

47. $\log_{25} x = \dfrac{1}{2}$

48. $\log_4 x = \dfrac{1}{2}$

49. $\log_5 x = -2$

50. $\log_3 x = -2$

51. $\log_{36} x = -\dfrac{1}{2}$

52. $\log_{27} x = -\dfrac{1}{3}$

53. $\log_{100} \dfrac{1}{1,000} = x$

54. $\log_{5/2} \dfrac{4}{25} = x$

55. $\log_{27} 9 = x$

56. $\log_{12} x = 0$

57. $\log_x 5^3 = 3$

58. $\log_x 5 = 1$

59. $\log_x \dfrac{9}{4} = 2$

60. $\log_x \dfrac{\sqrt{3}}{3} = \dfrac{1}{2}$

61. $\log_x \dfrac{1}{64} = -3$

62. $\log_x \dfrac{1}{100} = -2$

63. $\log_{2\sqrt{2}} x = 2$

64. $\log_4 8 = x$

65. $2^{\log_2 4} = x$

66. $3^{\log_3 5} = x$

67. $x^{\log_4 6} = 6$

68. $x^{\log_3 8} = 8$

69. $\log 10^3 = x$

70. $\log 10^{-2} = x$

71. $10^{\log x} = 100$

72. $10^{\log x} = \dfrac{1}{10}$

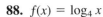 *Use a calculator to find each value, if possible. Give answers to four decimal places.*

73. $\log 8.25$

74. $\log 0.77$

75. $\log 0.00867$

76. $\log 375.876$

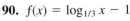 *Use a calculator to find each value of y, if possible. If an answer is not exact, give the answer to two decimal places.*

77. $\log y = 1.4023$

78. $\log y = 2.6490$

79. $\log y = 4.24$

80. $\log y = 0.926$

81. $\log y = -3.71$

82. $\log y = -0.28$

83. $\log y = \log 8$

84. $\log y = \log 7$

Graph each function. Tell whether each function is an increasing or decreasing function.

85. $f(x) = \log_3 x$

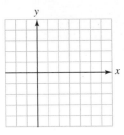

86. $f(x) = \log_{1/3} x$

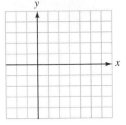

87. $f(x) = \log_{1/2} x$

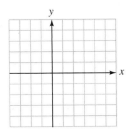

88. $f(x) = \log_4 x$

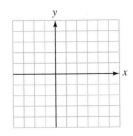

Graph each function.

89. $f(x) = 3 + \log_3 x$

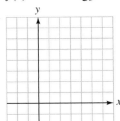

90. $f(x) = \log_{1/3} x - 1$

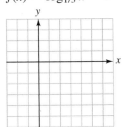

91. $f(x) = \log_{1/2} (x - 2)$

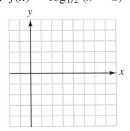

92. $f(x) = \log_4 (x + 2)$

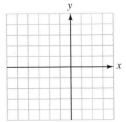

Graph each pair of inverse functions on a single coordinate system.

93. $f(x) = 2^x$

$g(x) = \log_2 x$

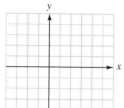

94. $f(x) = \left(\dfrac{1}{2}\right)^x$

$g(x) = \log_{1/2} x$

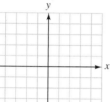

95. $f(x) = \left(\dfrac{1}{4}\right)^x$

$g(x) = \log_{1/4} x$

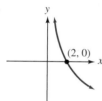

96. $f(x) = 4^x$

$g(x) = \log_4 x$

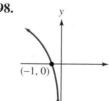

Find the value of b, if any, that would cause the graph of $f(x) = \log_b x$ to look like the graph indicated.

97.

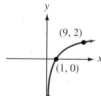

98.

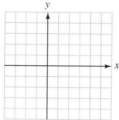

99.

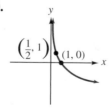

100.

APPLICATIONS *If an answer is not exact, round to the nearest tenth.*

101. Finding the gain of an amplifier Find the dB gain of an amplifier if the input voltage is 0.71 volt when the output voltage is 20 volts.

102. Finding the gain of an amplifier Find the dB gain of an amplifier if the output voltage is 2.8 volts when the input voltage is 0.05 volt.

103. dB gain of an amplifier Find the dB gain of the amplifier.

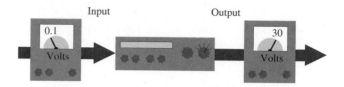

104. dB gain of an amplifier An amplifier produces an output of 80 volts when driven by an input of 0.12 volts. Find the amplifier's dB gain.

105. Earthquakes An earthquake has an amplitude of 5,000 micrometers and a period of 0.2 second. Find its measure on the Richter scale.

106. Earthquakes The period of an earthquake with amplitude of 80,000 micrometers is 0.08 second. Find its measure on the Richter scale.

107. Earthquakes An earthquake has a period of $\frac{1}{4}$ second and an amplitude of 2,500 micrometers. Find its measure on the Richter scale.

108. Earthquakes By what factor must the amplitude of an earthquake change to increase its severity by 1 point on the Richter scale? Assume that the period remains constant.

109. Depreciation Business equipment is often depreciated using the double declining-balance method. In this method, a piece of equipment with a life expectancy of N years, costing $\$C$, will depreciate to a value of $\$V$ in n years, where n is given by the formula

$$n = \frac{\log V - \log C}{\log\left(1 - \dfrac{2}{N}\right)}$$

A computer that cost $\$17,000$ has a life expectancy of 5 years. If it has depreciated to a value of $\$2,000$, how old is it?

110. Depreciation See Exercise 109. A printer worth $\$470$ when new had a life expectancy of 12 years. If it is now worth $\$189$, how old is it?

111. Time for money to grow   If P is invested at the end of each year in an annuity earning annual interest at a rate r, the amount in the account will be A after n years, where

$$n = \frac{\log\left(\dfrac{Ar}{P} + 1\right)}{\log\left(1 + r\right)}$$

If $1,000 is invested each year in an annuity earning 12% annual interest, how long will it take for the account to be worth $20,000?

112. Time for money to grow   If $5,000 is invested each year in an annuity earning 8% annual interest, how long will it take for the account to be worth $50,000? (See Exercise 111.)

WRITING

113. Describe the appearance of the graph of $y = f(x) = \log_b x$ when $0 < b < 1$ and when $b > 1$.

114. Explain why it is impossible to find the logarithm of a negative number.

SOMETHING TO THINK ABOUT

115. Graph $f(x) = -\log_3 x$. How does the graph compare to the graph of $f(x) = \log_3 x$?

116. Find a logarithmic function that passes through the points $(1, 0)$ and $(5, 1)$.

117. Explain why an earthquake measuring 7 on the Richter scale is much worse than an earthquake measuring 6.

10.4   Base-*e* Logarithms

■ **Base-*e* Logarithms**   ■ **Graphing Base-*e* Logarithms**
■ **Doubling Time**

Getting Ready   *Evaluate each expression.*

1. $\log_4 16$ **2.** $\log_2 \dfrac{1}{8}$ **3.** $\log_5 5$ **4.** $\log_7 1$

In this section, we will discuss special logarithmic functions with a base of e.

Base-*e* Logarithms

We have seen the importance of base-e exponential functions in mathematical models of events in nature. Base-e logarithms are just as important. They are called **natural logarithms** or **Napierian logarithms,** after John Napier (1550–1617), and are usually written as $\ln x$, rather than $\log_e x$:

$\ln x$ means $\log_e x$

As with all logarithmic functions, the domain of $y = f(x) = \ln x$ is the interval $(0, \infty)$, and the range is the interval $(-\infty, \infty)$.

To find the base-e logarithms of numbers, we can use a calculator.

Accent on Technology   **EVALUATING LOGARITHMS**

To use a scientific calculator to find the value of ln 9.87, we enter these numbers and press these keys:

9.87 LN

The display will read `2.289499853`. To four decimal places, ln 9.87 = 2.2895.

To use a graphing calculator, we enter these numbers and press these keys:

LN 9.87 ENTER

The display will read `ln (9.87`
`2.289499853`

EXAMPLE 1   Use a calculator to find each value: **a.** ln 17.32   and   **b.** ln (log 0.05).

Solution   **a.** Enter these numbers and press these keys:

Scientific Calculator   *Graphing Calculator*
17.32 LN   LN 17.32 ENTER

Either way, the result is 2.851861903.

b. Enter these numbers and press these keys:

Scientific Calculator   *Graphing Calculator*
0.05 LOG LN   LN (LOG 0.05) ENTER

Either way, we obtain an error, because log 0.05 is a negative number, and we cannot take the logarithm of a negative number.

Self Check   Find each value to four decimal places: **a.** ln π and   **b.** ln $\left(\log \frac{1}{2} \right)$.   ∎

EXAMPLE 2   Solve each equation: **a.** ln $x = 1.335$ and   **b.** ln $x = \log 5.5$. Give each result to four decimal places.

Solution   **a.** The equation ln $x = 1.335$ is equivalent to $e^{1.335} = x$. To use a scientific calculator to find x, enter these numbers and press these keys:

1.335 e^x

The display will read 3.799995946. To four decimal places,

$x = 3.8000$

b. The equation ln $x = \log 5.5$ is equivalent to $e^{\log 5.5} = x$. To use a scientific calculator to find x, press these keys:

5.5 LOG e^x

The display will read 2.096695826. To four decimal places,

$x = 2.0967$

Self Check Solve: **a.** $\ln x = 2.5437$ and **b.** $\log x = \ln 5$.

Graphing Base-e Logarithms

The equation $y = \ln x$ is equivalent to the equation $x = e^y$. To graph $f(x) = \ln x$, we can plot points that satisfy the equation $x = e^y$ and join them with a smooth curve, as shown as Figure 10-20(a). Figure 10-20(b) shows the calculator graph.

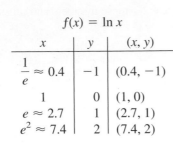

$$f(x) = \ln x$$

x	y	(x, y)
$\dfrac{1}{e} \approx 0.4$	-1	$(0.4, -1)$
1	0	$(1, 0)$
$e \approx 2.7$	1	$(2.7, 1)$
$e^2 \approx 7.4$	2	$(7.4, 2)$

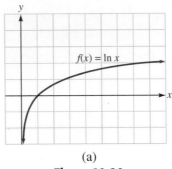

(a)

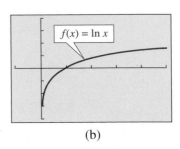

(b)

Figure 10-20

Accent on Technology   GRAPHING LOGARITHMIC FUNCTIONS

Many graphs of logarithmic functions involve translations of the graph of $y = f(x) = \ln x$. For example, Figure 10-21 shows calculator graphs of the functions $f(x) = \ln x$, $f(x) = \ln x + 2$, and $f(x) = \ln x - 3$.

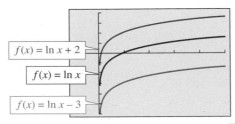

The graph of $f(x) = \ln x + 2$ is 2 units above the graph of $f(x) = \ln x$.

The graph of $f(x) = \ln x - 3$ is 3 units below the graph of $f(x) = \ln x$.

Figure 10-21

Figure 10-22 shows the calculator graphs of the functions $f(x) = \ln x$, $f(x) = \ln (x - 2)$, and $f(x) = \ln (x + 2)$.

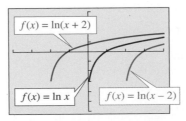

The graph of $f(x) = \ln (x - 2)$ is 2 units to the right of the graph of $f(x) = \ln x$.

The graph of $f(x) = \ln (x + 2)$ is 2 units to the left of the graph of $f(x) = \ln x$.

Figure 10-22

Base-*e* logarithms have many applications.

Doubling Time

If a population grows exponentially at a certain annual rate, the time required for the population to double is called the **doubling time.** In Exercise 51, you will be asked to develop the following formula.

Formula for Doubling Time

If *r* is the annual rate (compounded continuously) and *t* is the time required for a population to double, then

$$t = \frac{\ln 2}{r}$$

EXAMPLE 3   Earth's population is growing at the approximate rate of 2% per year. If this rate continues, in how many years will the population double?

Solution   Because the population is growing at the rate of 2% per year, we substitute 0.02 for *r* into the formula for doubling time and simplify.

$$t = \frac{\ln 2}{r}$$

$$t = \frac{\ln 2}{\mathbf{0.02}}$$

$$\approx 34.65735903$$

The population will double in about 35 years.

Self Check   If the world population's annual growth rate could be reduced to 1.5% per year, what would be the doubling time? ▮

EXAMPLE 4   **Doubling time**   How long will it take $1,000 to double at an annual rate of 8%, compounded continuously?

Solution   We substitute 0.08 for *r* and simplify:

$$t = \frac{\ln 2}{r}$$

$$t = \frac{\ln 2}{\mathbf{0.08}}$$

$$\approx 8.664339757$$

It will take about $8\frac{2}{3}$ years for the money to double.

Self Check   How long will it take at 9%, compounded continuously? ▮

1. **a.** 1.1447   **b.** no value   2. **a.** 12.7267   **b.** 40.6853   3. 46 years   4. about 7.7 years

Orals   1. Write $y = \ln x$ as an exponential equation
2. Write $e^a = b$ as a logarithmic equation
3. Write the formula for doubling time

10.4   EXERCISES

REVIEW   *Write the equation of the required line.*

1. Parallel to $y = 5x + 8$ and passing through the origin

2. Having a slope of 9 and a y-intercept of $(0, 5)$

3. Passing through the point $(3, 2)$ and perpendicular to the line $y = \frac{2}{3}x - 12$

4. Parallel to the line $3x + 2y = 9$ and passing through the point $(-3, 5)$

5. Vertical line through the point $(5, 3)$

6. Horizontal line through the point $(2, 5)$

Simplify each expression.

7. $\dfrac{2x + 3}{4x^2 - 9}$

8. $\dfrac{x + 1}{x} + \dfrac{x - 1}{x + 1}$

9. $\dfrac{x^2 + 3x + 2}{3x + 12} \cdot \dfrac{x + 4}{x^2 - 4}$

10. $\dfrac{1 + \dfrac{y}{x}}{\dfrac{y}{x} - 1}$

VOCABULARY AND CONCEPTS   *Fill in the blanks.*

11. $\ln x$ means ___.

12. The domain of the function $y = f(x) = \ln x$ is the interval ___.

13. The range of the function $y = f(x) = \ln x$ is the interval ___.

14. The graph of $y = f(x) = \ln x$ has the ___ as an asymptote.

15. In the expression $\log x$, the base is understood to be ___.

16. In the expression $\ln x$, the base is understood to be ___.

17. If a population grows exponentially at a rate r, the time it will take the population to double is given by the formula $t =$ ___.

18. The logarithm of a negative number is ___.

PRACTICE   *Use a calculator to find each value, if possible. Express all answers to four decimal places.*

19. $\ln 25.25$ 20. $\ln 0.523$

21. $\ln 9.89$ 22. $\ln 0.00725$

23. $\log (\ln 2)$ 24. $\ln (\log 28.8)$

25. $\ln (\log 0.5)$ 26. $\log (\ln 0.2)$

Use a calculator to find y, if possible. Express all answers to four decimal places.

27. $\ln y = 2.3015$ 28. $\ln y = 1.548$

29. $\ln y = 3.17$ 30. $\ln y = 0.837$

31. $\ln y = -4.72$ 32. $\ln y = -0.48$

33. $\log y = \ln 6$ 34. $\ln y = \log 5$

Tell whether the graph could represent the graph of $y = \ln x$.

35.
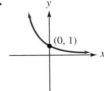

36.
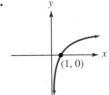

37.

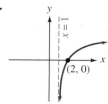

38.

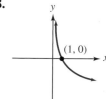

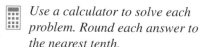 *Use a graphing calculator to graph each function.*

39. $y = -\ln x$ **40.** $y = \ln x^2$

41. $y = \ln(-x)$ **42.** $y = \ln\left(\frac{1}{2}x\right)$

APPLICATIONS *Use a calculator to solve each problem. Round each answer to the nearest tenth.*

43. Population growth   See the ad. How long will it take the population of River City to double?

River City
A growing community

• 6 parks   • 12% annual growth

• 10 churches   • Low crime rate

---

44. Doubling money   How long will it take $1,000 to double if it is invested at an annual rate of 5%, compounded continuously?

45. Population growth   A population growing at an annual rate r will triple in a time t given by the formula

$$t = \frac{\ln 3}{r}$$

How long will it take the population of a town growing at the rate of 12% per year to triple?

46. Tripling money   Find the length of time for $25,000 to triple if invested at 6% annual interest, compounded continuously. (See Exercise 45.)

47. Making Jell-O   After the contents of a package of Jell-O are combined with boiling water, the mixture is placed in a refrigerator whose temperature remains a constant 38° F. Estimate the number of hours t that it will take for the Jell-O to cool to 50° F using the formula

$$t = -\frac{1}{0.9} \ln \frac{50 - T_r}{200 - T_r}$$

where T_r is the temperature of the refrigerator.

48. Forensic medicine   To estimate the number of hours t that a murder victim had been dead, a coroner used the formula

$$t = \frac{1}{0.25} \ln \frac{98.6 - T_s}{82 - T_s}$$

where T_s is the temperature of the surroundings where the body was found. If the crime took place in an apartment where the thermostat was set at 72° F, approximately how long ago did the murder occur?

WRITING

49. The time it takes money to double at an annual rate r, compounded continuously, is given by the formula $t = (\ln 2)/r$. Explain why money doubles more quickly as the rate increases.

50. The time it takes money to triple at an annual rate r, compounded continuously, is given by the formula $t = (\ln 3)/r$. Explain why money triples less quickly as the rate decreases.

SOMETHING TO THINK ABOUT

51. Use the formula $P = P_0 e^{rt}$ to verify that P will be twice P_0 when $t = \frac{\ln 2}{r}$.

52. Use the formula $P = P_0 e^{rt}$ to verify that P will be three times as large as P_0 when $t = \frac{\ln 3}{r}$.

53. Find a formula to find how long it will take money to quadruple.

54. Use a graphing calculator to graph

$$f(x) = \frac{1}{1 + e^{-2x}}$$

and discuss the graph.

<table>
<tr><td>10.5</td><td># Properties of Logarithms</td></tr>
</table>

| **Properties of Logarithms** | **The Change-of-Base Formula** |
| **Chemistry** | **Physiology** |

Getting Ready   *Simplify each expression.*

1. $x^m x^n$ **2.** x^0

3. $(x^m)^n$ **4.** $\dfrac{x^m}{x^n}$

In this section, we will consider many properties of logarithms. We will then use these properties to solve problems.

Properties of Logarithms

Since logarithms are exponents, the properties of exponents have counterparts in the theory of logarithms. We begin with four basic properties.

Properties of Logarithms

> If b is a positive number and $b \neq 1$, then
> **1.** $\log_b 1 = 0$ **2.** $\log_b b = 1$
> **3.** $\log_b b^x = x$ **4.** $b^{\log_b x} = x$ $(x > 0)$

Properties 1 through 4 follow directly from the definition of a logarithm.

1. $\log_b 1 = 0$, because $b^0 = 1$.

2. $\log_b b = 1$, because $b^1 = b$.

3. $\log_b b^x = x$, because $b^x = b^x$.

4. $b^{\log_b x} = x$, because $\log_b x$ is the exponent to which b is raised to get x.

Properties 3 and 4 also indicate that the composition of the exponential and logarithmic functions with the same base (in both directions) is the identity function. This is expected, because the exponential and logarithmic functions with the same base are inverse functions.

EXAMPLE 1   Simplify each expression: **a.** $\log_5 1$,   **b.** $\log_3 3$,   **c.** $\log_7 7^3$,   and   **d.** $b^{\log_b 7}$.

Solution   **a.** By Property 1, $\log_5 1 = 0$, because $5^0 = 1$.

         b. By Property 2, $\log_3 3 = 1$, because $3^1 = 3$.

         c. By Property 3, $\log_7 7^3 = 3$, because $7^3 = 7^3$.

         d. By Property 4, $b^{\log_b 7} = 7$, because $\log_b 7$ is the power to which b is raised to get 7.

Self Check   Simplify: **a.** $\log_4 1$,   **b.** $\log_5 5$,   **c.** $\log_2 2^4$,   and   **d.** $5^{\log_5 2}$.   ∎

The next two properties state that

The logarithm of a product is the sum of the logarithms.

The logarithm of a quotient is the difference of the logarithms.

Properties of Logarithms

If M, N, and b are positive numbers and $b \neq 1$, then

5. $\log_b MN = \log_b M + \log_b N$ **6.** $\log_b \dfrac{M}{N} = \log_b M - \log_b N$

Proof   To prove Property 5, we let $x = \log_b M$ and $y = \log_b N$. We use the definition of logarithms to write each equation in exponential form.

$$M = b^x \qquad \text{and} \qquad N = b^y$$

Then $MN = b^x b^y$, and a property of exponents gives

$$MN = b^{x+y} \qquad b^x b^y = b^{x+y}\text{: Keep the base and add the exponents.}$$

We write this exponential equation in logarithmic form as

$$\log_b MN = x + y$$

Substituting the values of x and y completes the proof.

$$\log_b MN = \log_b M + \log_b N$$

∎

The proof of Property 6 is similar.

Comment   By Property 5 of logarithms, the logarithm of a *product* is equal to the *sum* of the logarithms. The logarithm of a sum or a difference usually does not simplify. In general,

$$\log_b (M + N) \neq \log_b M + \log_b N$$
$$\log_b (M - N) \neq \log_b M - \log_b N$$

By Property 6, the logarithm of a *quotient* is equal to the *difference* of the logarithms. The logarithm of a quotient is not the quotient of the logarithms:

$$\log_b \frac{M}{N} \neq \frac{\log_b M}{\log_b N}$$

Accent on Technology   **VERIFYING PROPERTIES**

We can use a calculator to illustrate Property 5 of logarithms by showing that

$$\ln [(3.7)(15.9)] = \ln 3.7 + \ln 15.9$$

We calculate the left- and right-hand sides of the equation separately and compare the results. To use a calculator to find $\ln [(3.7)(15.9)]$, we enter these numbers and press these keys:

3.7 ☒ 15.9 ═ **LN**   On a scientific calculator.

LN 3.7 ☒ 15.9) **ENTER**   On a graphing calculator.

The display will read $\boxed{4.074651929}$.

To find ln 3.7 + ln 15.9, we enter these numbers and press these keys:

3.7 LN + 15.9 LN =   On a scientific calculator.

LN 3.7) + LN 15.9) ENTER   On a graphing calculator.

The display will read $\boxed{4.074651929}$. Since the left- and right-hand sides are equal, the equation is true.

Two more properties state that

The logarithm of an expression to a power is the power times the logarithm of the expression.

If the logarithms of two numbers are equal, the numbers are equal.

Properties of Logarithms

If M, p, and b are positive numbers and $b \neq 1$, then

7. $\log_b M^p = p \log_b M$

8. If $\log_b x = \log_b y$, then $x = y$.

Proof   To prove Property 7, we let $x = \log_b M$, write the expression in exponential form, and raise both sides to the pth power:

$$M = b^x$$
$$(M)^p = (b^x)^p \quad \text{Raise both sides to the } p\text{th power.}$$
$$M^p = b^{px} \quad \text{Keep the base and multiply the exponents.}$$

Using the definition of logarithms gives

$$\log_b M^p = p\boldsymbol{x}$$

Substituting the value for x completes the proof.

$$\log_b M^p = p \log_b M \qquad \blacksquare$$

Property 8 follows from the fact that the logarithmic function is a one-to-one function. Property 8 will be important in the next section when we solve logarithmic equations.

We can use the properties of logarithms to write a logarithm as the sum or difference of several logarithms.

EXAMPLE 2   Assume that b ($b \neq 1$), x, y, and z are positive numbers. Write each expression in terms of the logarithms of x, y, and z: **a.** $\log_b xyz$ and   **b.** $\log_b \dfrac{xy}{z}$.

Solution   **a.** $\log_b xyz = \log_b (xy)z$

$\qquad\qquad = \log_b (xy) + \log_b z \qquad$ The log of a product is the sum of the logs.

$\qquad\qquad = \log_b x + \log_b y + \log_b z \qquad$ The log of a product is the sum of the logs.

b. $\log_b \dfrac{xy}{z}$

$= \log_b (xy) - \log_b z$ The log of a quotient is the difference of the logs.

$= (\log_b x + \log_b y) - \log_b z$ The log of a product is the sum of the logs.

$= \log_b x + \log_b y - \log_b z$ Remove parentheses.

Self Check   Write $\log_b \dfrac{x}{yz}$ in terms of the logarithms of x, y, and z.   ∎

EXAMPLE 3   Assume that b ($b \neq 1$), x, y, and z are positive numbers. Write each expression in terms of the logarithms of x, y, and z.

     a. $\log_b (x^2 y^3 z)$ and   **b.** $\log_b \dfrac{\sqrt{x}}{y^3 z}$

Solution   **a.** $\log_b (x^2 y^3 z) = \log_b x^2 + \log_b y^3 + \log_b z$ The log of a product is the sum of the logs.

              $= 2 \log_b x + 3 \log_b y + \log_b z$ The log of an expression to a power is the power times the log of the expression.

     b. $\log_b \dfrac{\sqrt{x}}{y^3 z} = \log_b \sqrt{x} - \log_b (y^3 z)$ The log of a quotient is the difference of the logs.

              $= \log_b x^{1/2} - (\log_b y^3 + \log_b z)$ $\sqrt{x} = x^{1/2}$; The log of a product is the sum of the logs.

              $= \dfrac{1}{2} \log_b x - (3 \log_b y + \log_b z)$ The log of a power is the power times the log.

              $= \dfrac{1}{2} \log_b x - 3 \log_b y - \log_b z$ Use the distributive property to remove parentheses.

Self Check   Write $\log_b \sqrt[4]{\dfrac{x^3 y}{z}}$ in terms of the logarithms of x, y, and z.   ∎

     We can use the properties of logarithms to combine several logarithms into one logarithm.

EXAMPLE 4   Assume that b ($b \neq 1$), x, y, and z are positive numbers. Write each expression as one logarithm: **a.** $3 \log_b x + \frac{1}{2} \log_b y$ and   **b.** $\frac{1}{2} \log_b (x - 2) - \log_b y + 3 \log_b z$.

Solution   **a.** $3 \log_b x + \frac{1}{2} \log_b y = \log_b x^3 + \log_b y^{1/2}$ A power times a log is the log of the power.

                        $= \log_b (x^3 y^{1/2})$ The sum of two logs is the log of a product.

     b. $\dfrac{1}{2} \log_b (x - 2) - \log_b y + 3 \log_b z$

         $= \log_b (x - 2)^{1/2} - \log_b y + \log_b z^3$ A power times a log is the log of the power.

         $= \log_b \dfrac{(x - 2)^{1/2}}{y} + \log_b z^3$ The difference of two logs is the log of the quotient.

         $= \log_b \dfrac{z^3 \sqrt{x - 2}}{y}$ The sum of two logs is the log of a product.

Self Check   Write the expression as one logarithm: $2 \log_b x + \frac{1}{2}\log_b y - 2 \log_b (x - y)$.   ∎

We summarize the properties of logarithms as follows.

Properties of Logarithms

If b, M, and N are positive numbers and $b \neq 1$, then

1. $\log_b 1 = 0$

2. $\log_b b = 1$

3. $\log_b b^x = x$

4. $b^{\log_b x} = x$

5. $\log_b MN = \log_b M + \log_b N$

6. $\log_b \dfrac{M}{N} = \log_b M - \log_b N$

7. $\log_b M^p = p \log_b M$

8. If $\log_b x = \log_b y$, then $x = y$.

EXAMPLE 5   Given that $\log 2 \approx 0.3010$ and $\log 3 \approx 0.4771$, find approximations for   **a.** $\log 6$, **b.** $\log 9$,   **c.** $\log 18$,   and   **d.** $\log 2.5$.

Solution   **a.** $\log 6 = \log (2 \cdot 3)$

$= \log 2 + \log 3$ The log of a product is the sum of the logs.

$\approx 0.3010 + 0.4771$ Substitute the value of each logarithm.

≈ 0.7781

b. $\log 9 = \log (3^2)$

$= 2 \log 3$ The log of a power is the power times the log.

$\approx 2(0.4771)$ Substitute the value of $\log 3$.

≈ 0.9542

c. $\log 18 = \log (2 \cdot 3^2)$

$= \log 2 + \log 3^2$ The log of a product is the sum of the logs.

$= \log 2 + 2 \log 3$ The log of a power is the power times the log.

$\approx 0.3010 + 2(0.4771)$

≈ 1.2552

d. $\log 2.5 = \log\left(\dfrac{5}{2}\right)$

$= \log 5 - \log 2$ The log of a quotient is the difference of the logs.

$= \log \dfrac{10}{2} - \log 2$ Write 5 as $\frac{10}{2}$.

$= \log 10 - \log 2 - \log 2$ The log of a quotient is the difference of the logs.

$= 1 - 2 \log 2$ $\log_{10} 10 = 1$.

$\approx 1 - 2(0.3010)$

≈ 0.3980

Self Check   Use the values given in Example 5 and find   **a.** $\log 1.5$ and   **b.** $\log 0.2$.   ∎

The Change-of-Base Formula

If we know the base-a logarithm of a number, we can find its logarithm to some other base b with a formula called the **change-of-base formula.**

Change-of-Base Formula   If a, b, and x are real numbers, then

$$\log_b x = \frac{\log_a x}{\log_a b}$$

Proof   To prove this formula, we begin with the equation $\log_b x = y$.

$$y = \log_b x$$

$$x = b^y \qquad \text{Change the equation from logarithmic to exponential form.}$$

$$\mathbf{\log_a} x = \mathbf{\log_a} b^y \qquad \text{Take the base-}a\text{ logarithm of both sides.}$$

$$\log_a x = y \log_a b \qquad \text{The log of a power is the power times the log.}$$

$$y = \frac{\log_a x}{\log_a b} \qquad \text{Divide both sides by } \log_a b.$$

$$\log_b x = \frac{\log_a x}{\log_a b} \qquad \text{Refer to the first equation and substitute } \log_b x \text{ for } y. \qquad \blacksquare$$

If we know logarithms to base a (for example, $a = 10$), we can find the logarithm of x to a new base b. We simply divide the base-a logarithm of x by the base-a logarithm of b.

 Comment $\dfrac{\log_a x}{\log_a b}$ means that one logarithm is to be divided by the other. They are not to be subtracted.

EXAMPLE 6   Find $\log_4 9$ using base-10 logarithms.

Solution   We can substitute 4 for b, 10 for a, and 9 for x into the change-of-base formula:

$$\log_b \mathbf{x} = \frac{\log_a \mathbf{x}}{\log_a b}$$

$$\log_4 \mathbf{9} = \frac{\log_{10} \mathbf{9}}{\log_{10} 4}$$

$$\approx 1.584962501$$

To four decimal places, $\log_4 9 = 1.5850$.

Self Check   Find $\log_5 3$ using base-10 logarithms. $\blacksquare$

Chemistry

Common logarithms are used to express the acidity of solutions. The more acidic a solution, the greater the concentration of hydrogen ions. This concentration is indicated indirectly by the **pH scale,** or **hydrogen ion index.** The pH of a solution is defined by the following equation.

pH of a Solution   If $[H^+]$ is the hydrogen ion concentration in gram-ions per liter, then

$$pH = -\log [H^+]$$

EXAMPLE 7   **Finding the pH of a solution**   Find the pH of pure water, which has a hydrogen ion concentration of 10^{-7} gram-ions per liter.

Solution   Since pure water has approximately 10^{-7} gram-ions per liter, its pH is

$$pH = -\log [\mathbf{H^+}]$$
$$pH = -\log \mathbf{10^{-7}}$$
$$ = -(-7) \log 10 \qquad \text{The log of a power is the power times the log.}$$
$$ = -(-7) \cdot 1 \qquad \log 10 = 1.$$
$$ = 7 \qquad\qquad\qquad \blacksquare$$

EXAMPLE 8   **Finding hydrogen ion concentration**   Find the hydrogen ion concentration of sea-water if its pH is 8.5.

Solution   To find its hydrogen ion concentration, we substitute 8.5 for the pH and find $[H^+]$.

$$8.5 = -\log [H^+]$$
$$-8.5 = \log [H^+] \qquad \text{Multiply both sides by } -1.$$
$$[H^+] = 10^{-8.5} \qquad \text{Change the equation to exponential form.}$$

We can use a calculator to find that

$$[H^+] \approx 3.2 \times 10^{-9} \text{ gram-ions per liter} \qquad\qquad \blacksquare$$

Physiology

In physiology, experiments suggest that the relationship between the loudness and the intensity of sound is a logarithmic one known as the **Weber–Fechner law.**

Weber–Fechner Law   If L is the apparent loudness of a sound, I is the actual intensity, and k is a constant, then

$$L = k \ln I$$

EXAMPLE 9   **Weber–Fechner law**   Find the increase in intensity that will cause the apparent loudness of a sound to double.

Solution   If the original loudness L_O is caused by an actual intensity I_O, then

(1) $$L_O = k \ln I_O$$

To double the apparent loudness, we multiply both sides of Equation 1 by 2 and use the power rule of logarithms:

$$2L_O = \mathbf{2}k \ln I_O$$
$$ = k \ln (I_O)^2$$

To double the loudness of a sound, the intensity must be squared.

Self Check   What decrease in intensity will cause a sound to be half as loud? $\blacksquare$

Self Check Answers

1. a. 0   **b.** 1   **c.** 4   **d.** 2   **2.** $\log_b x - \log_b y - \log_b z$ **3.** $\frac{1}{4}(3 \log_b x + \log_b y - \log_b z)$

4. $\log_b \dfrac{x^2\sqrt{y}}{(x-y)^2}$ **5. a.** 0.1761   **b.** -0.6990 **6.** 0.6826   **9.** Find the square root of the intensity.

Orals   *Find the value of x in each equation.*

1. $\log_3 9 = x$ **2.** $\log_x 5 = 1$ **3.** $\log_7 x = 3$

4. $\log_2 x = -2$ **5.** $\log_4 x = \dfrac{1}{2}$ **6.** $\log_x 4 = 2$

7. $\log_{1/2} x = 2$ **8.** $\log_9 3 = x$ **9.** $\log_x \dfrac{1}{4} = -2$

10.5   EXERCISES

REVIEW   *Consider the line that passes through* $(-2, 3)$ *and* $(4, -4)$.

1. Find the slope of the line.

2. Find the distance between the points.

3. Find the midpoint of the segment.

4. Write the equation of the line.

VOCABULARY AND CONCEPTS   *Fill in the blanks.*

5. $\log_b 1 = \underline{}$

6. $\log_b b = \underline{}$

7. $\log_b MN = \log_b \underline{} + \log_b \underline{}$

8. $b^{\log_b x} = \underline{}$

9. If $\log_b x = \log_b y$, then $\underline{} = \underline{}$.

10. $\log_b \dfrac{M}{N} = \log_b M \underline{} \log_b N$

11. $\log_b x^p = p \cdot \log_b \underline{}$

12. $\log_b b^x = \underline{}$

13. $\log_b (A + B) \underline{} \log_b A + \log_b B$

14. $\log_b A + \log_b B \underline{} \log_b AB$

15. $\log_4 1 = \underline{}$ **16.** $\log_4 4 = \underline{}$

17. $\log_4 4^7 = \underline{}$ **18.** $4^{\log_4 8} = \underline{}$

19. $5^{\log_5 10} = \underline{}$ **20.** $\log_5 5^2 = \underline{}$

21. $\log_5 5 = \underline{}$ **22.** $\log_5 1 = \underline{}$

23. $\log_7 1 = \underline{}$ **24.** $\log_9 9 = \underline{}$

25. $\log_3 3^7 = \underline{}$ **26.** $5^{\log_5 8} = \underline{}$

27. $8^{\log_8 10} = \underline{}$ **28.** $\log_4 4^2 = \underline{}$

29. $\log_9 9 = \underline{}$ **30.** $\log_3 1 = \underline{}$

PRACTICE   *Use a calculator to verify each equation.*

31. $\log [(2.5)(3.7)] = \log 2.5 + \log 3.7$

32. $\ln \dfrac{11.3}{6.1} = \ln 11.3 - \ln 6.1$

33. $\ln (2.25)^4 = 4 \ln 2.25$

34. $\log 45.37 = \dfrac{\ln 45.37}{\ln 10}$

35. $\log \sqrt{24.3} = \dfrac{1}{2} \log 24.3$

36. $\ln 8.75 = \dfrac{\log 8.75}{\log e}$

Assume that x, y, z, and b ($b \neq 1$) are positive numbers. Use the properties of logarithms to write each expression in terms of the logarithms of x, y, and z.

37. $\log_b xyz$ **38.** $\log_b 4xz$

39. $\log_b \dfrac{2x}{y}$ **40.** $\log_b \dfrac{x}{yz}$

41. $\log_b x^3 y^2$

42. $\log_b xy^2 z^3$

43. $\log_b (xy)^{1/2}$ **44.** $\log_b x^3 y^{1/2}$

45. $\log_b x\sqrt{z}$

46. $\log_b \sqrt{xy}$

47. $\log_b \dfrac{\sqrt[3]{x}}{\sqrt[4]{yz}}$

48. $\log_b \sqrt[4]{\dfrac{x^3 y^2}{z^4}}$

Assume that x, y, z, and b (b ≠ 1) are positive numbers. Use the properties of logarithms to write each expression as the logarithm of a single quantity.

49. $\log_b (x + 1) - \log_b x$

50. $\log_b x + \log_b (x + 2) - \log_b 8$

51. $2 \log_b x + \dfrac{1}{2} \log_b y$

52. $-2 \log_b x - 3 \log_b y + \log_b z$

53. $-3 \log_b x - 2 \log_b y + \dfrac{1}{2} \log_b z$

54. $3 \log_b (x + 1) - 2 \log_b (x + 2) + \log_b x$

55. $\log_b \left(\dfrac{x}{z} + x \right) - \log_b \left(\dfrac{y}{z} + y \right)$

56. $\log_b (xy + y^2) - \log_b (xz + yz) + \log_b z$

Tell whether each statement is true. If a statement is false, explain why.

57. $\log_b 0 = 1$

58. $\log_b (x + y) \neq \log_b x + \log_b y$

59. $\log_b xy = (\log_b x)(\log_b y)$

60. $\log_b ab = \log_b a + 1$

61. $\log_7 7^7 = 7$

62. $7^{\log_7 7} = 7$

63. $\dfrac{\log_b A}{\log_b B} = \log_b A - \log_b B$

64. $\log_b (A - B) = \dfrac{\log_b A}{\log_b B}$

65. $3 \log_b \sqrt[3]{a} = \log_b a$

66. $\dfrac{1}{3} \log_b a^3 = \log_b a$

67. $\log_b \dfrac{1}{a} = -\log_b a$

68. $\log_b 2 = \log_2 b$

*Assume that log 4 = 0.6021, log 7 = 0.8451, and log 9 = 0.9542. Use these values and the properties of logarithms to find each value. **Do not use a calculator.***

69. $\log 28$

70. $\log \dfrac{7}{4}$

71. $\log 2.25$

72. $\log 36$

73. $\log \dfrac{63}{4}$

74. $\log \dfrac{4}{63}$

75. $\log 252$

76. $\log 49$

77. $\log 112$

78. $\log 324$

79. $\log \dfrac{144}{49}$

80. $\log \dfrac{324}{63}$

Use a calculator and the change-of-base formula to find each logarithm to four decimal places.

81. $\log_3 7$

82. $\log_7 3$

83. $\log_{1/3} 3$

84. $\log_{1/2} 6$

85. $\log_3 8$

86. $\log_5 10$

87. $\log_{\sqrt{2}} \sqrt{5}$

88. $\log_\pi e$

APPLICATIONS

89. pH of a solution Find the pH of a solution with a hydrogen ion concentration of 1.7×10^{-5} gram-ions per liter.

90. Hydrogen ion concentration Find the hydrogen ion concentration of a saturated solution of calcium hydroxide whose pH is 13.2.

91. Aquariums To test for safe pH levels in a fresh-water aquarium, a test strip is compared with the scale shown in the illustration. Find the corresponding range in the hydrogen ion concentration.

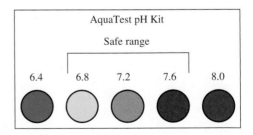

92. pH of pickles The hydrogen ion concentration of sour pickles is 6.31×10^{-4}. Find the pH.

93. Change in loudness   If the intensity of a sound is doubled, find the apparent change in loudness.

94. Change in loudness   If the intensity of a sound is tripled, find the apparent change in loudness.

95. Change in intensity   What change in intensity of sound will cause an apparent tripling of the loudness?

96. Change in intensity   What increase in the intensity of a sound will cause the apparent loudness to be multiplied by 4?

WRITING

97. Explain why $\ln (\log 0.9)$ is undefined.

98. Explain why $\log_b (\ln 1)$ is undefined.

SOMETHING TO THINK ABOUT

99. Show that $\ln (e^x) = x$.

100. If $\log_b 3x = 1 + \log_b x$, find b.

101. Show that $\log_{b^2} x = \dfrac{1}{2} \log_b x$.

102. Show that $e^{x \ln a} = a^x$.

10.6   Exponential and Logarithmic Equations

■ Solving Exponential Equations ■ Solving Logarithmic Equations
■ Radioactive Decay ■ Population Growth

Getting Ready   *Write each expression without using exponents.*

1. $\log x^2$

2. $\log x^{1/2}$

3. $\log x^0$

4. $\log a^b + b \log a$

An **exponential equation** is an equation that contains a variable in one of its exponents. Some examples of exponential equations are

$$3^x = 5, \qquad 6^{x-3} = 2^x, \qquad \text{and} \qquad 3^{2x+1} - 10(3^x) + 3 = 0$$

A **logarithmic equation** is an equation with logarithmic expressions that contain a variable. Some examples of logarithmic equations are

$$\log 2x = 25, \qquad \ln x - \ln (x - 12) = 24, \qquad \text{and} \qquad \log x = \log \frac{1}{x} + 4$$

In this section, we will learn how to solve many of these equations.

Solving Exponential Equations

EXAMPLE 1   Solve: $4^x = 7$.

Solution   Since logarithms of equal numbers are equal, we can take the common logarithm of each side of the equation. The power rule of logarithms then provides a way of moving the variable x from its position as an exponent to a position as a coefficient.

$$4^x = 7$$

$$\log 4^x = \log 7 \qquad \text{Take the common logarithm of each side.}$$

$$x \log 4 = \log 7 \qquad \text{The log of a power is the power times the log.}$$

$$(1) \qquad x = \frac{\log 7}{\log 4} \qquad \text{Divide both sides by log 4.}$$

$$\approx 1.403677461 \qquad \text{Use a calculator.}$$

To four decimal places, $x = 1.4037$.

Self Check Solve: $5^x = 4$. ∎

Comment A careless reading of Equation 1 leads to a common error. The right-hand side of Equation 1 calls for a division, not a subtraction.

$$\frac{\log 7}{\log 4} \quad \text{means} \quad (\log 7) \div (\log 4)$$

It is the expression $\log \frac{7}{4}$ that means $\log 7 - \log 4$.

EXAMPLE 2 Solve: $73 = 1.6(1.03)^t$.

Solution We divide both sides by 1.6 to obtain

$$\frac{73}{1.6} = 1.03^t$$

and solve the equation as in Example 1.

$$1.03^t = \frac{73}{1.6}$$

$$\mathbf{log}\, 1.03^t = \mathbf{log}\, \frac{73}{1.6} \qquad \text{Take the common logarithm of each side.}$$

$$t \log 1.03 = \log \frac{73}{1.6} \qquad \text{The logarithm of a power is the power times the logarithm.}$$

$$t = \frac{\log \frac{73}{1.6}}{\log 1.03} \qquad \text{Divide both sides by log 1.03.}$$

$$t = 129.2493444 \qquad \text{Use a calculator.}$$

To four decimal places, $t = 129.2493$.

Self Check Solve: $47 = 2.5(1.05)^t$. Give the answer to the nearest tenth. ∎

EXAMPLE 3 Solve: $6^{x-3} = 2^x$.

Solution
$$6^{x-3} = 2^x$$

$$\mathbf{log}\, 6^{x-3} = \mathbf{log}\, 2^x \qquad \text{Take the common logarithm of each side.}$$

$$(x - 3) \log 6 = x \log 2 \qquad \text{The log of a power is the power times the log.}$$

$$x \log 6 - 3 \log 6 = x \log 2 \qquad \text{Use the distributive property.}$$

$$x \log 6 - x \log 2 = 3 \log 6 \qquad \text{Add 3 log 6 and subtract } x \log 2 \text{ from both sides.}$$

$$x (\log 6 - \log 2) = 3 \log 6 \qquad \text{Factor out } x \text{ on the left-hand side.}$$

$$x = \frac{3 \log 6}{\log 6 - \log 2} \qquad \text{Divide both sides by log 6 } - \text{ log 2.}$$

$$x \approx 4.892789261 \qquad \text{Use a calculator.}$$

Self Check Solve: $5^{x-2} = 3^x$. ∎

EXAMPLE 4 Solve: $2^{x^2+2x} = \dfrac{1}{2}$.

Solution Since $\frac{1}{2} = 2^{-1}$, we can write the equation in the form

$$2^{x^2+2x} = 2^{-1}$$

Since equal quantities with equal bases have equal exponents, we have

$$x^2 + 2x = -1$$
$$x^2 + 2x + 1 = 0 \qquad \text{Add 1 to both sides.}$$
$$(x + 1)(x + 1) = 0 \qquad \text{Factor the trinomial.}$$
$$x + 1 = 0 \quad \text{or} \quad x + 1 = 0 \qquad \text{Set each factor equal to 0.}$$
$$x = -1 \quad | \quad x = -1$$

Verify that -1 satisfies the equation.

Self Check Solve: $3^{x^2-2x} = \frac{1}{3}$. ∎

Accent on Technology

SOLVING EXPONENTIAL EQUATIONS

To use a graphing calculator to approximate the solutions of $2^{x^2+2x} = \frac{1}{2}$ (see Example 4), we can subtract $\frac{1}{2}$ from both sides of the equation to get

$$2^{x^2+2x} - \frac{1}{2} = 0$$

and graph the corresponding function

$$f(x) = 2^{x^2+2x} - \frac{1}{2}$$

If we use window settings of $[-4, 4]$ for x and $[-2, 6]$ for y, we obtain the graph shown in Figure 10-23(a).

Since the solutions of the equation are its x-intercepts, we can approximate the solutions by zooming in on the values of the x-intercepts, as in Figure 10-23(b). Since $x = -1$ is the only x-intercept, -1 is the only solution. In this case, we have found an exact solution.

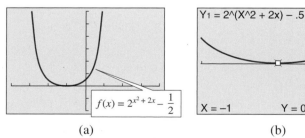

(a)   (b)

Figure 10-23

Solving Logarithmic Equations

In each of the following examples, we use the properties of logarithms to change a logarithmic equation into an algebraic equation.

EXAMPLE 5   Solve: $\log_b (3x + 2) - \log_b (2x - 3) = 0$.

Solution

$$\log_b (3x + 2) - \log_b (2x - 3) = 0$$

$$\log_b (3x + 2) = \log_b (2x - 3) \qquad \text{Add } \log_b (2x - 3) \text{ to both sides.}$$

$$3x + 2 = 2x - 3 \qquad \text{If } \log_b r = \log_b s, \text{ then } r = s.$$

$$x = -5 \qquad \text{Subtract } 2x \text{ and 2 from both sides.}$$

Check: $\log_b (3x + 2) - \log_b (2x - 3) = 0$

$$\log_b [3(-5) + 2] - \log_b [2(-5) - 3] \overset{?}{=} 0$$

$$\log_b (-13) - \log_b (-13) \overset{?}{=} 0$$

Since the logarithm of a negative number does not exist, the apparent solution of -5 must be discarded. This equation has no solutions.

Self Check   Solve: $\log_b (5x + 2) - \log_b (7x - 2) = 0$.   ∎

Comment   Example 5 illustrates that you must check the solutions of a logarithmic equation.

EXAMPLE 6   Solve: $\log x + \log (x - 3) = 1$.

Solution

$$\log x + \log (x - 3) = 1$$

$$\log x (x - 3) = 1 \qquad \text{The sum of two logs is the log of a product.}$$

$$x(x - 3) = 10^1 \qquad \text{Use the definition of logarithms to change the equation to exponential form.}$$

$$x^2 - 3x - 10 = 0 \qquad \text{Remove parentheses and subtract 10 from both sides.}$$

$$(x + 2)(x - 5) = 0 \qquad \text{Factor the trinomial.}$$

$$x + 2 = 0 \quad \text{or} \quad x - 5 = 0 \qquad \text{Set each factor equal to 0.}$$

$$x = -2 \quad | \quad\quad x = 5$$

Check: The number -2 is not a solution, because it does not satisfy the equation (a negative number does not have a logarithm). We will check the remaining number, 5.

$$\log x + \log (x - 3) = 1$$

$$\log 5 + \log (5 - 3) \overset{?}{=} 1 \qquad \text{Substitute 5 for } x.$$

$$\log 5 + \log 2 \overset{?}{=} 1$$

$$\log 10 \overset{?}{=} 1 \qquad \text{The sum of two logs is the log of a product.}$$

$$1 = 1 \qquad \log 10 = 1.$$

Since 5 satisfies the equation, it is a solution.

Self Check   Solve: $\log x + \log (x + 3) = 1$.   ∎

EXAMPLE 7 Solve: $\dfrac{\log (5x - 6)}{\log x} = 2$.

Solution We can multiply both sides of the equation by $\log x$ to get

$$\log (5x - 6) = 2 \log x$$

and apply the power rule of logarithms to get

$$\log (5x - 6) = \log x^2$$

By Property 8 of logarithms, $5x - 6 = x^2$, because they have equal logarithms. Thus,

$$5x - 6 = x^2$$
$$0 = x^2 - 5x + 6$$
$$0 = (x - 3)(x - 2)$$
$$x - 3 = 0 \quad \text{or} \quad x - 2 = 0$$
$$x = 3 \quad | \quad x = 2$$

Verify that both 2 and 3 satisfy the equation.

Self Check Solve: $\dfrac{\log (5x + 6)}{\log x} = 2$.

SOLVING LOGARITHMIC EQUATIONS

To use a graphing calculator to approximate the solutions of $\log x + \log (x - 3) = 1$ (see Example 6), we can subtract 1 from both sides of the equation to get

$$\log x + \log (x - 3) - 1 = 0$$

and graph the corresponding function

$$f(x) = \log x + \log (x - 3) - 1$$

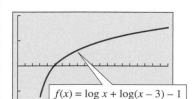

$f(x) = \log x + \log(x - 3) - 1$

Figure 10-24

If we use window settings of $[0, 20]$ for x and $[-2, 2]$ for y, we obtain the graph shown in Figure 10-24. Since the solution of the equation is the x-intercept, we can find the solution by zooming in on the value of the x-intercept or by using the ZERO command. The solution is $x = 5$.

Radioactive Decay

Experiments have determined the time that it takes for half of a sample of a given radioactive material to decompose. This time is a constant, called the material's **half-life.**

When living organisms die, the oxygen/carbon dioxide cycle common to all living things stops, and carbon-14, a radioactive isotope with a half-life of 5,700 years, is no longer absorbed. By measuring the amount of carbon-14 present in an ancient object, archaeologists can estimate the object's age by using the radioactive decay formula.

Radioactive Decay Formula

If A is the amount of radioactive material present at time t, A_0 was the amount present at $t = 0$, and h is the material's half-life, then

$$A = A_0 2^{-t/h}$$

EXAMPLE 8   **Carbon-14 dating**   How old is a wooden statue that retains only one-third of its original carbon-14 content?

Solution   To find the time t when $A = \frac{1}{3}A_0$, we substitute $\frac{A_0}{3}$ for A and 5,700 for h into the radioactive decay formula and solve for t:

$$A = A_0 2^{-t/h}$$

$$\frac{A_0}{3} = A_0 2^{-t/5,700}$$

$$1 = 3(2^{-t/5,700})$$ Divide both sides by A_0 and multiply both sides by 3.

$$\log 1 = \log 3(2^{-t/5,700})$$ Take the common logarithm of each side.

$$0 = \log 3 + \log 2^{-t/5,700}$$ Log 1 = 0, and the log of a product is the sum of the logs.

$$-\log 3 = -\frac{t}{5,700} \log 2$$ Subtract log 3 from both sides and use the power rule of logarithms.

$$5,700\left(\frac{\log 3}{\log 2}\right) = t$$ Multiply both sides by $-\frac{5,700}{\log 2}$.

$$t \approx 9,034.286254$$ Use a calculator.

This statue is approximately 9,000 years old.

Self Check   How old is a statue that retains 25% of its original carbon-14 content?   ∎

Population Growth

Recall that when there is sufficient food and space, populations of living organisms tend to increase exponentially according to the Malthusian growth model.

Malthusian Growth Model

If P is the population at some time t, P_0 is the initial population at $t = 0$, and k depends on the rate of growth, then

$$P = P_0 e^{kt}$$

EXAMPLE 9   **Population growth**   The bacteria in a laboratory culture increased from an initial population of 500 to 1,500 in 3 hours. How long will it take for the population to reach 10,000?

Solution   We substitute 500 for P_0, 1,500 for P, and 3 for t and simplify to find k:

$$P = P_0e^{kt}$$

$$1{,}500 = 500\,(e^{k3}) \qquad \text{Substitute 1,500 for } P, \text{ 500 for } P_0, \text{ and 3 for } t.$$

$$3 = e^{3k} \qquad \text{Divide both sides by 500.}$$

$$3k = \ln 3 \qquad \text{Change the equation from exponential to logarithmic form.}$$

$$k = \frac{\ln 3}{3} \qquad \text{Divide both sides by 3.}$$

To find when the population will reach 10,000, we substitute 10,000 for P, 500 for P_0, and $\frac{\ln 3}{3}$ for k in the equation $P = P_0e^{kt}$ and solve for t:

$$P = P_0e^{kt}$$

$$10{,}000 = 500e^{[(\ln 3)/3]t}$$

$$20 = e^{[(\ln 3)/3]t} \qquad \text{Divide both sides by 500.}$$

$$\left(\frac{\ln 3}{3}\right)t = \ln 20 \qquad \text{Change the equation to logarithmic form.}$$

$$t = \frac{3\ln 20}{\ln 3} \qquad \text{Multiply both sides by } \frac{3}{\ln 3}.$$

$$\approx 8.180499084 \qquad \text{Use a calculator.}$$

The culture will reach 10,000 bacteria in about 8 hours.

Self Check How long will it take to reach 20,000? ∎

EXAMPLE 10 **Generation time** If a medium is inoculated with a bacterial culture that contains 1,000 cells per milliliter, how many generations will pass by the time the culture has grown to a population of 1 million cells per milliliter?

Solution During bacterial reproduction, the time required for a population to double is called the *generation time*. If b bacteria are introduced into a medium, then after the generation time of the organism has elapsed, there are $2b$ cells. After another generation, there are $2(2b)$, or $4b$ cells, and so on. After n generations, the number of cells present will be

(1) $$B = b \cdot 2^n$$

To find the number of generations that have passed while the population grows from b bacteria to B bacteria, we solve Equation 1 for n.

$$\log B = \log\,(b \cdot 2^n) \qquad \text{Take the common logarithm of both sides.}$$

$$\log B = \log b + n\log 2 \qquad \text{Apply the product and power rules of logarithms.}$$

$$\log B - \log b = n\log 2 \qquad \text{Subtract } \log b \text{ from both sides.}$$

$$n = \frac{1}{\log 2}(\log B - \log b) \qquad \text{Multiply both sides by } \frac{1}{\log 2}.$$

(2) $$n = \frac{1}{\log 2}\left(\log \frac{B}{b}\right) \qquad \text{Use the quotient rule of logarithms.}$$

Equation 2 is a formula that gives the number of generations that will pass as the population grows from b bacteria to B bacteria.

To find the number of generations that have passed while a population of 1,000 cells per milliliter has grown to a population of 1 million cells per milliliter, we substitute 1,000 for b and 1,000,000 for B in Equation 2 and solve for n.

$$n = \frac{1}{\log 2} \log \frac{1{,}000{,}000}{1{,}000}$$

$$= \frac{1}{\log 2} \log 1{,}000 \qquad \text{Simplify.}$$

$$= 3.321928095(3) \qquad \frac{1}{\log 2} \approx 3.321928095 \text{ and } \log 1{,}000 = 3.$$

$$= 9.965784285$$

Approximately 10 generations will have passed. ▮

Self Check Answers

1. 0.8614 **2.** 60.1 **3.** $\dfrac{2 \log 5}{\log 5 - \log 3} \approx 6.301320206$ **4.** 1, 1 **5.** 2 **6.** 2 **7.** 6

8. about 11,400 years **9.** about 10 hours

Orals *Solve each equation for x.* **Do not simplify answers.**

1. $3^x = 5$ **2.** $5^x = 3$

3. $2^{-x} = 7$ **4.** $6^{-x} = 1$

5. $\log 2x = \log (x + 2)$ **6.** $\log 2x = 0$

7. $\log x^4 = 4$ **8.** $\log \sqrt{x} = \dfrac{1}{2}$

10.6   EXERCISES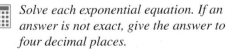

REVIEW *Solve each equation.*

1. $5x^2 - 25x = 0$ **2.** $4y^2 - 25 = 0$
3. $3p^2 + 10p = 8$ **4.** $4t^2 + 1 = -6t$

VOCABULARY AND CONCEPTS *Fill in the blanks.*

5. An equation with a variable in its exponent is called a(n) _____ equation.

6. An equation with a logarithmic expression that contains a variable is a(n) _____ equation.

7. The formula for carbon dating is $A = $ _____.

8. The formula for population growth is $P = $ ____.

PRACTICE *Solve each exponential equation. If an answer is not exact, give the answer to four decimal places.*

9. $4^x = 5$ **10.** $7^x = 12$
11. $e^t = 50$ **12.** $e^{-t} = 0.25$

13. $5 = 2.1(1.04)^t$ **14.** $61 = 1.5(1.02)^t$

15. $13^{x-1} = 2$ **16.** $5^{x+1} = 3$
17. $2^{x+1} = 3^x$ **18.** $5^{x-3} = 3^{2x}$
19. $2^x = 3^x$ **20.** $3^{2x} = 4^x$
21. $7^{x^2} = 10$ **22.** $8^{x^2} = 11$
23. $8^{x^2} = 9^x$ **24.** $5^{x^2} = 2^{5x}$
25. $2^{x^2-2x} = 8$ **26.** $3^{x^2-3x} = 81$
27. $3^{x^2+4x} = \dfrac{1}{81}$ **28.** $7^{x^2+3x} = \dfrac{1}{49}$
29. $4^{x+2} - 4^x = 15$ (*Hint:* $4^{x+2} = 4^x 4^2$.)
30. $3^{x+3} + 3^x = 84$ (*Hint:* $3^{x+3} = 3^x 3^3$.)
31. $2(3^x) = 6^{2x}$
32. $2(3^{x+1}) = 3(2^{x-1})$

Use a calculator to solve each equation, if possible. Give all answers to the nearest tenth.

33. $2^{x+1} = 7$ **34.** $3^{x-1} = 2^x$
35. $2^{x^2-2x} - 8 = 0$ **36.** $3^x - 10 = 3^{-x}$

Solve each logarithmic equation. ***Check all solutions.***

37. $\log 2x = \log 4$ **38.** $\log 3x = \log 9$

39. $\log (3x + 1) = \log (x + 7)$

40. $\log (x^2 + 4x) = \log (x^2 + 16)$

41. $\log (3 - 2x) - \log (x + 24) = 0$

42. $\log (3x + 5) - \log (2x + 6) = 0$

43. $\log \dfrac{4x + 1}{2x + 9} = 0$ **44.** $\log \dfrac{2 - 5x}{2(x + 8)} = 0$

45. $\log x^2 = 2$

46. $\log x^3 = 3$

47. $\log x + \log (x - 48) = 2$

48. $\log x + \log (x + 9) = 1$

49. $\log x + \log (x - 15) = 2$

50. $\log x + \log (x + 21) = 2$

51. $\log (x + 90) = 3 - \log x$

52. $\log (x - 90) = 3 - \log x$

53. $\log (x - 6) - \log (x - 2) = \log \dfrac{5}{x}$

54. $\log (3 - 2x) - \log (x + 9) = 0$

55. $\log x^2 = (\log x)^2$

56. $\log (\log x) = 1$

57. $\dfrac{\log (3x - 4)}{\log x} = 2$

58. $\dfrac{\log (8x - 7)}{\log x} = 2$

59. $\dfrac{\log (5x + 6)}{2} = \log x$

60. $\dfrac{1}{2} \log (4x + 5) = \log x$

61. $\log_3 x = \log_3 \left(\dfrac{1}{x} \right) + 4$

62. $\log_5 (7 + x) + \log_5 (8 - x) - \log_5 2 = 2$

63. $2 \log_2 x = 3 + \log_2 (x - 2)$

64. $2 \log_3 x - \log_3 (x - 4) = 2 + \log_3 2$

65. $\log (7y + 1) = 2 \log (y + 3) - \log 2$

66. $2 \log (y + 2) = \log (y + 2) - \log 12$

Use a graphing calculator to solve each equation. If an answer is not exact, give all answers to the nearest tenth.

67. $\log x + \log (x - 15) = 2$

68. $\log x + \log (x + 3) = 1$

69. $\ln (2x + 5) - \ln 3 = \ln (x - 1)$

70. $2 \log (x^2 + 4x) = 1$

APPLICATIONS

71. Tritium decay The half-life of tritium is 12.4 years. How long will it take for 25% of a sample of tritium to decompose?

72. Radioactive decay In two years, 20% of a radioactive element decays. Find its half-life.

73. Thorium decay An isotope of thorium, ^{227}Th, has a half-life of 18.4 days. How long will it take for 80% of the sample to decompose?

74. Lead decay An isotope of lead, ^{201}Pb, has a half-life of 8.4 hours. How many hours ago was there 30% more of the subtance?

75. Carbon-14 dating The bone fragment shown in the illustration contains 60% of the carbon-14 that it is assumed to have had initially. How old is it?

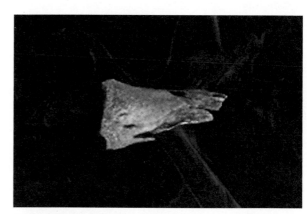

Trudy Haversat and Gary Breschini/Archeological Consulting

76. Carbon-14 dating Only 10% of the carbon-14 in a small wooden bowl remains. How old is the bowl?

77. Compound interest If $500 is deposited in an account paying 8.5% annual interest, compounded semiannually, how long will it take for the account to increase to $800?

78. Continuous compound interest In Exercise 77, how long will it take if the interest is compounded continuously?

79. Compound interest If $1,300 is deposited in a savings account paying 9% interest, compounded quarterly, how long will it take the account to increase to $2,100?

80. Compound interest   A sum of $5,000 deposited in an account grows to $7,000 in 5 years. Assuming annual compounding, what interest rate is being paid?

81. Rule of seventy   A rule of thumb for finding how long it takes an investment to double is called the **rule of seventy.** To apply the rule, divide 70 by the interest rate written as a percent. At 5%, it takes $\frac{70}{5} = 14$ years to double an investment. At 7%, it takes $\frac{70}{7} = 10$ years. Explain why this formula works.

82. Bacterial growth   A bacterial culture grows according to the formula

$$P = P_0 a^t$$

If it takes 5 days for the culture to triple in size, how long will it take to double in size?

83. Rodent control   The rodent population in a city is currently estimated at 30,000. If it is expected to double every 5 years, when will the population reach 1 million?

84. Population growth   The population of a city is expected to triple every 15 years. When can the city planners expect the present population of 140 persons to double?

85. Bacterial culture   A bacterial culture doubles in size every 24 hours. By how much will it have increased in 36 hours?

86. Oceanography   The intensity I of a light a distance x meters beneath the surface of a lake decreases exponentially. From the illustration, find the depth at which the intensity will be 20%.

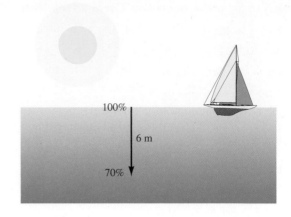

87. Medicine   If a medium is inoculated with a bacterial culture containing 500 cells per milliliter, how many generations will have passed by the time the culture contains 5×10^6 cells per milliliter?

88. Medicine   If a medium is inoculated with a bacterial culture containing 800 cells per milliliter, how many generations will have passed by the time the culture contains 6×10^7 cells per milliliter?

WRITING

89. Explain how to solve the equation $2^x = 7$.

90. Explain how to solve the equation $x^2 = 7$.

SOMETHING TO THINK ABOUT

91. Without solving the following equation, find the values of x that cannot be a solution:

$$\log (x - 3) - \log (x^2 + 2) = 0$$

92. Solve the equation $x^{\log x} = 10,000$.

Projects

Project 1

When an object moves through air, it encounters air resistance. So far, all ballistics problems in this text have ignored air resistance. We now consider the case where an object's fall is affected by air resistance.

At relatively low velocities ($v < 200$ feet per second), the force resisting an object's motion is a constant multiple of the object's velocity:

Resisting force = $f_r = bv$

where b is a constant that depends on the size, shape, and texture of the object, and has units of kilograms per second. This is known as **Stokes' law of resistance.**

In a vacuum, the downward velocity of an object dropped with an initial velocity of 0 feet per second is

$$v(t) = 32t \quad \text{(no air resistance)}$$

t seconds after it is released. However, with air resistance, the velocity is given by the formula

$$v(t) = \frac{32m}{b}(1 - e^{-(b/m)t})$$

where m is the object's mass (in kilograms). There is also a formula for the distance an object falls (in feet) during the first t seconds after release, taking into account air resistance:

$$d(t) = \frac{32m}{b}t - \frac{32m^2}{b^2}(1 - e^{-(b/m)t})$$

Without air resistance, the formula would be

$$d(t) = 16t^2$$

a. Fearless Freda, a renowned skydiving daredevil, performs a practice dive from a hot-air balloon with an altitude of 5,000 feet. With her parachute on, Freda has a mass of 75 kg, so that $b = 15$ kg/sec. How far (to the nearest foot) will Freda fall in 5 seconds? Compare this with the answer you get by disregarding air resistance.

b. What downward velocity (to the nearest ft/sec) does Freda have after she has fallen for 2 seconds? For 5 seconds? Compare these answers with the answers you get by disregarding air resistance.

c. Find Freda's downward velocity after falling for 20, 22, and 25 seconds (without air resistance, Freda would hit the ground in less than 18 seconds). Note that Freda's velocity increases only slightly. This is because for a large enough velocity, the force of air resistance almost counteracts the force of gravity; after Freda has been falling for a few seconds, her velocity becomes nearly constant. The constant velocity that a falling object approaches is called the *terminal velocity*.

$$\text{Terminal velocity} = \frac{32m}{b}$$

Find Freda's terminal velocity for her practice dive.

d. In Freda's show, she dives from a hot-air balloon with an altitude of only 550 feet, and pulls her ripcord when her velocity is 100 feet per second. (She can't tell her speed, but she knows how long it takes to reach that speed.) It takes a fall of 80 more feet for the chute to open fully, but then the chute increases the force of air resistance, making $b = 80$.

After that, Freda's velocity approaches the terminal velocity of an object with this new b-value.

To the nearest hundredth of a second, how long should Freda fall before she pulls the ripcord? To the nearest foot, how close is she to the ground when she pulls the ripcord? How close to the ground is she when the chute takes full effect? At what velocity will Freda hit the ground?

Project 2

If an object at temperature T_0 is surrounded by a constant temperature T_s (for instance, an oven or a large amount of fluid that has a constant temperature), the temperature of the object will change with time t according to the formula

$$T(t) = T_s + (T_0 - T_s)e^{-kt}$$

This is **Newton's law of cooling and warming.** The number k is a constant that depends on how well the object absorbs and dispels heat.

In the course of brewing "yo ho! grog," the dread pirates of Hancock Isle have learned that it is important that their rather disgusting, soupy mash be heated slowly to allow all of the ingredients a chance to add their particular offensiveness to the mixture. However, after the mixture has simmered for several hours, it is equally important that the grog be cooled very quickly, so that it retains its potency. The kegs of grog are then stored in a cool spring.

By trial and error, the pirates have learned that by placing the mash pot into a tub of boiling water (100° C), they can heat the mash in the correct amount of time. They have also learned that they can cool the grog to the temperature of the spring by placing it in ice caves for 1 hour.

With a thermometer, you find that the pirates heat the mash from 20° C to 95° C and then cool the grog from 95° C to 7° C. Calculate how long the pirates cook the mash, and how cold the ice caves are. Assume that $k = 0.5$, and t is measured in hours.

Chapter Summary

CONCEPTS	REVIEW EXERCISES

10.1 · Exponential Functions

CONCEPTS

An exponential function with base b is defined by the equation

$$y = f(x) = b^x$$
$$(b > 0, b \neq 1)$$

REVIEW EXERCISES

Use properties of exponents to simplify.

1. $5^{\sqrt{2}} \cdot 5^{\sqrt{2}}$ **2.** $\left(2^{\sqrt{5}}\right)^{\sqrt{2}}$

Graph the function defined by each equation.

3. $y = 3^x$ **4.** $y = \left(\dfrac{1}{3}\right)^x$

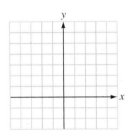   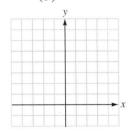

5. The graph of $f(x) = 6^x$ will pass through the points $(0, a)$ and $(1, b)$. Find a and b.

6. Give the domain and range of the function $f(x) = b^x$.

Graph each function by using a translation.

7. $f(x) = \left(\dfrac{1}{2}\right)^x - 2$ **8.** $f(x) = \left(\dfrac{1}{2}\right)^{x+2}$

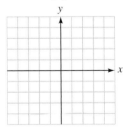   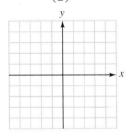

Compound interest

$$A = P\left(1 + \dfrac{r}{k}\right)^{kt}$$

9. ▦ How much will \$10,500 become if it earns 9% per year for 60 years, compounded quarterly?

10.2	**Base-e Exponential Functions**

$e = 2.71828182845904.\ .\ .$

Continuous compound interest
$A = Pe^{rt}$

10. If \$10,500 accumulates interest at an annual rate of 9%, compounded continuously, how much will be in the account in 60 years?

Graph each function.

11. $f(x) = e^x + 1$ **12.** $f(x) = e^{x-3}$

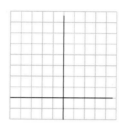   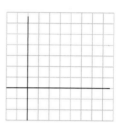

Malthusian population growth
$A = Pe^{rt}$

13. The population of the United States is approximately 275,000,000 people. Find the population in 50 years if $k = 0.015$.

10.3	**Logarithmic Functions**

If $b > 0$ and $b \neq 1$, then
$y = \log_b x$ means $x = b^y$

14. Give the domain and range of the logarithmic function.

Find each value.

15. $\log_3 9$ **16.** $\log_9 \dfrac{1}{3}$

17. $\log_\pi 1$ **18.** $\log_5 0.04$

19. $\log_a \sqrt{a}$ **20.** $\log_a \sqrt[3]{a}$

Find x.

21. $\log_2 x = 5$ **22.** $\log_{\sqrt{3}} x = 4$

23. $\log_{\sqrt{3}} x = 6$ **24.** $\log_{0.1} 10 = x$

25. $\log_x 2 = -\dfrac{1}{3}$ **26.** $\log_x 32 = 5$

27. $\log_{0.25} x = -1$ **28.** $\log_{0.125} x = -\dfrac{1}{3}$

29. $\log_{\sqrt{2}} 32 = x$ **30.** $\log_{\sqrt{5}} x = -4$

31. $\log_{\sqrt{3}} 9\sqrt{3} = x$ **32.** $\log_{\sqrt{5}} 5\sqrt{5} = x$

Graph each function.

33. $f(x) = \log(x - 2)$

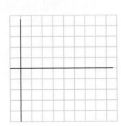

34. $f(x) = 3 + \log x$

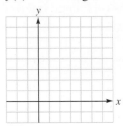

Graph each pair of equations on one set of coordinate axes.

35. $y = 4^x$ and $y = \log_4 x$

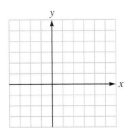

36. $y = \left(\dfrac{1}{3}\right)^x$ and $y = \log_{1/3} x$

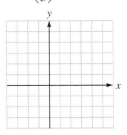

Decibel voltage gain

$$\text{dB gain} = 20 \log \frac{E_O}{E_I}$$

37. An amplifier has an output of 18 volts when the input is 0.04 volt. Find the dB gain.

Richter scale

$$R = \log \frac{A}{p}$$

38. An earthquake had a period of 0.3 second and an amplitude of 7,500 micrometers. Find its measure on the Richter scale.

10.4	**Base-e Logarithms**

$\ln x$ means $\log_e x$.

Use a calculator to find each value to four decimal places.

39. $\ln 452$

40. $\ln(\log 7.85)$

Solve each equation.

41. $\ln x = 2.336$

42. $\ln x = \log 8.8$

Graph each function.

43. $f(x) = 1 + \ln x$

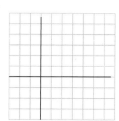

44. $f(x) = \ln (x + 1)$

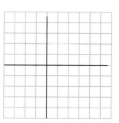

Population doubling time

$$t = \frac{\ln 2}{r}$$

45. How long will it take the population of the United States to double if the growth rate is 3% per year?

10.5   Properties of Logarithms

Properties of logarithms
If b is a positive number and
$b \neq 1$,

1. $\log_b 1 = 0$

2. $\log_b b = 1$

3. $\log_b b^x = x$

4. $b^{\log_b x} = x$

5. $\log_b MN = \log_b M + \log_b N$

6. $\log_b \dfrac{M}{N} = \log_b M - \log_b N$

7. $\log_b M^p = p \log_b M$

8. If $\log_b x = \log_b y$, then
$x = y$.

Simplify each expression.

46. $\log_7 1$ **47.** $\log_7 7$

48. $\log_7 7^3$ **49.** $7^{\log_7 4}$

Simplify each expression.

50. $\ln e^4$ **51.** $\ln 1$

52. $10^{\log_{10} 7}$ **53.** $e^{\ln 3}$

54. $\log_b b^4$ **55.** $\ln e^9$

Write each expression in terms of the logarithms of x, y, and z.

56. $\log_b \dfrac{x^2 y^3}{z^4}$ **57.** $\log_b \sqrt{\dfrac{x}{yz^2}}$

Write each expression as the logarithm of one quantity.

58. $3 \log_b x - 5 \log_b y + 7 \log_b z$

59. $\dfrac{1}{2} \log_b x + 3 \log_b y - 7 \log_b z$

Assume that $\log a = 0.6$, $\log b = 0.36$, and $\log c = 2.4$. Find the value of each expression.

60. $\log abc$ **61.** $\log a^2 b$

62. $\log \dfrac{ac}{b}$ **63.** $\log \dfrac{a^2}{c^3 b^2}$

Change-of-base formula

$$\log_b y = \frac{\log_a y}{\log_a b}$$

pH scale

$$pH = -\log [H^+]$$

Weber–Fechner law

$$L = k \ln I$$

64. 🖩 To four decimal places, find $\log_5 17$.

65. pH of grapefruit   The pH of grapefruit juice is about 3.1. Find its hydrogen ion concentration.

66. Find the decrease in loudness if the intensity is cut in half.

10.6   Exponential and Logarithmic Equations

Solve each equation for x. If an answer is not exact, round to four decimal places.

67. $3^x = 7$ **68.** $5^{x+2} = 625$

69. $25 = 5.5(1.05)^t$ **70.** $4^{2t-1} = 64$

71. $2^x = 3^{x-1}$ **72.** $2^{x^2+4x} = \dfrac{1}{8}$

Solve each equation for x.

73. $\log x + \log (29 - x) = 2$

74. $\log_2 x + \log_2 (x - 2) = 3$

75. $\log_2 (x + 2) + \log_2 (x - 1) = 2$

76. $\dfrac{\log (7x - 12)}{\log x} = 2$

77. $\log x + \log (x - 5) = \log 6$

78. $\log 3 - \log (x - 1) = -1$

79. $e^{x \ln 2} = 9$

80. $\ln x = \ln (x - 1)$

81. $\ln x = \ln (x - 1) + 1$

82. $\ln x = \log_{10} x$ (*Hint:* Use the change-of-base formula.)

Carbon dating

$$A = A_0 2^{-t/h}$$

83. Carbon-14 dating   A wooden statue found in Egypt has a carbon-14 content that is two-thirds of that found in living wood. If the half-life of carbon-14 is 5,700 years, how old is the statue?

Chapter Test

Graph each function.

1. $f(x) = 2^x + 1$

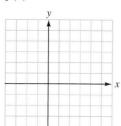

2. $f(x) = 2^{-x}$

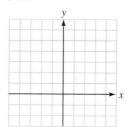

Graph each function.

13. $f(x) = -\log_3 x$

14. $f(x) = \ln x$

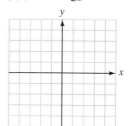
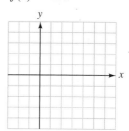

Solve each equation.

3. A radioactive material decays according to the formula $A = A_0(2)^{-t}$. How much of a 3-gram sample will be left in 6 years?

4. An initial deposit of $1,000 earns 6% interest, compounded twice a year. How much will be in the account in one year?

5. Graph the function $f(x) = e^x$.

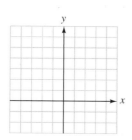

6. An account contains $2,000 and has been earning 8% interest, compounded continuously. How much will be in the account in 10 years?

Find x.

7. $\log_4 16 = x$

8. $\log_x 81 = 4$

9. $\log_3 x = -3$

10. $\log_x 100 = 2$

11. $\log_{3/2} \dfrac{9}{4} = x$

12. $\log_{2/3} x = -3$

Write each expression in terms of the logarithms of a, b, and c.

15. $\log a^2bc^3$

16. $\ln \sqrt{\dfrac{a}{b^2c}}$

Write each expression as a logarithm of a single quantity.

17. $\dfrac{1}{2} \log (a + 2) + \log b - 3 \log c$

18. $\dfrac{1}{3}(\log a - 2 \log b) - \log c$

Assume that log 2 = 0.3010 and log 3 = 0.4771. Find each value. **Do not use a calculator.**

19. $\log 24$

20. $\log \dfrac{8}{3}$

Use the change-of-base formula to find each logarithm. **Do not attempt to simplify the answer.**

21. $\log_7 3$

22. $\log_\pi e$

Tell whether each statement is true. If a statement is not true, explain why.

23. $\log_a ab = 1 + \log_a b$

24. $\dfrac{\log a}{\log b} = \log a - \log b$

25. $\log a^{-3} = \dfrac{1}{3 \log a}$

26. $\ln(-x) = -\ln x$

27. Find the pH of a solution with a hydrogen ion concentration of 3.7×10^{-7}. (*Hint:* $\text{pH} = -\log[\text{H}^+]$.)

28. Find the dB gain of an amplifier when $E_O = 60$ volts and $E_I = 0.3$ volt. (*Hint:* dB gain $= 20 \log(E_O/E_I)$.)

*Solve each equation. **Do not simplify the logarithms.***

29. $5^x = 3$

30. $3^{x-1} = 100^x$

Solve each equation.

31. $\log(5x + 2) = \log(2x + 5)$

32. $\log x + \log(x - 9) = 1$

CUMULATIVE REVIEW EXERCISES

In Exercises 1–16, simplify each expression. Assume that all variables represent positive numbers, and write answers without using negative exponents.

1. $64^{2/3}$

2. $8^{-1/3}$

3. $\dfrac{y^{2/3} y^{5/3}}{y^{1/3}}$

4. $\dfrac{x^{5/3} x^{1/2}}{x^{3/4}}$

5. $(x^{2/3} - x^{1/3})(x^{2/3} + x^{1/3})$

6. $(x^{-1/2} + x^{1/2})^2$

7. $\sqrt[3]{-27x^3}$

8. $\sqrt{48t^3}$

9. $\sqrt[3]{\dfrac{128x^4}{2x}}$

10. $\sqrt{x^2 + 6x + 9}$

11. $\sqrt{50} - \sqrt{8} + \sqrt{32}$

12. $-3\sqrt[4]{32} - 2\sqrt[4]{162} + 5\sqrt[4]{48}$

13. $3\sqrt{2}\left(2\sqrt{3} - 4\sqrt{12}\right)$

14. $\dfrac{5}{\sqrt[3]{x}}$

15. $\dfrac{\sqrt{x} + 2}{\sqrt{x} - 1}$

16. $\sqrt[6]{x^3 y^3}$

Solve each equation.

17. $4\sqrt{a + 5} = 16$

18. $2\sqrt{x} - \sqrt{x + 7} = 2$

19. Use the method of completing the square to solve the equation $2x^2 + x - 3 = 0$.

20. Use the quadratic formula to solve the equation $3x^2 + 4x - 1 = 0$.

Write each complex number in $a + bi$ form.

21. $(3 + 5i) - (4 - 3i)$

22. $(7 - 4i) + (12 + 3i)$

23. $(5 - 2i)(5 + 2i)$

24. $(3 + i)(3 + 3i)$

25. $(4 + i)^2$

26. $\dfrac{2 - 3i}{4 + i}$

Find each value.

27. $|3 - 2i|$

28. $|5 + 6i|$

29. For what values of k will the solutions of $2x^2 + kx = -2$ be equal?

30. Find the coordinates of the vertex of the graph of the equation $y = \frac{1}{2}x^2 - x + 1$.

Let $f(x) = 3x^2 + 2$ and $g(x) = 2x - 1$. Find each value or function.

31. $f(-1)$

32. $(g \circ f)(2)$

33. $(f \circ g)(x)$

34. $(g \circ f)(x)$

Find the inverse of each function.

35. $f(x) = 3x + 2$

36. $f(x) = x^3 + 4$

37. Write $y = \log_2 x$ in exponential form.

38. Write $3^b = a$ in logarithmic notation.

Find x.

39. $\log_x 25 = 2$ **40.** $\log_5 125 = x$

41. $\log_3 x = -3$ **42.** $\log_5 x = 0$

43. Find the inverse of $y = \log_2 x$.

44. If $\log_{10} 10^x = y$, then y equals what quantity?

Assume that $\log 7 = 0.8451$ *and* $\log 14 = 1.1461$. *Evaluate each expression without using a calculator or tables.*

45. $\log 98$ **46.** $\log 2$

11

Solving Polynomial Equations

InfoTrac Project

Do a subject guide search on "cocaine use." Find the article "Chest pain in young may mean cocaine use. (Patients difficult to treat.)" Write a summary of the article.

Suppose the amount of cocaine (in nanograms) remaining in a urine sample after t hours can be represented by the expression $-2t^4 + 5t^3 + 6t + 450$. How much cocaine was in the initial urine sample? Using synthetic division, divide by $t - 4$. Your remainder will be the amount of cocaine in nanograms remaining in a urine sample given 4 hours after ingesting the cocaine. Now let $t = 4$ and substitute 4 for t in $-2t^4 + 5t^3 + 6t + 450$. What do you notice about your answer?

Complete this project after studying Section 11.1.

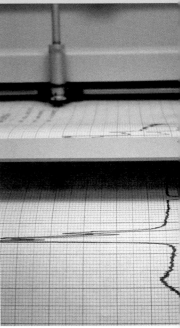

© Photodisc Collection/Getty Images

Mathematics in Farming

To protect cranberry crops from damage during early frosts, growers flood the cranberry bogs.

Three irrigation sources, used together, can flood one bog in a day. If the sources are used one at a time, the second source requires one day longer to flood the bog than the first, and a third requires four days longer than the first. If the bog must be flooded before a freeze that is predicted in three days, can the water in the last two sources be diverted to other bogs?

Example 4
Section 11.3

So far, we have solved linear and quadratic equations. We have also solved some third- and fourth-degree polynomial equations. In this chapter, we will discuss methods to solve higher-degree polynomial equations with real coefficients.

11.1 The Remainder and Factor Theorems; Synthetic Division

■ Zeros of a Polynomial   ■ The Remainder Theorem
■ The Factor Theorem   ■ Synthetic Division
■ Using Synthetic Division to Evaluate Polynomials

Getting Ready   *Evaluate $x^2 - 5x + 6$ for each value of x.*

1. $x = 2$ **2.** $x = 3$ **3.** $x = 4$ **4.** $x = -5$

We have seen that many polynomial equations can be solved by factoring. For example, to solve $x^3 - 3x^2 + 2x = 0$, we proceed as follows:

$$x^3 - 3x^2 + 2x = 0$$
$$x(x^2 - 3x + 2) = 0 \qquad \text{Factor out } x.$$
$$x(x - 1)(x - 2) = 0 \qquad \text{Factor } x^2 - 3x + 2.$$
$$x = 0 \quad \text{or} \quad x - 1 = 0 \quad \text{or} \quad x - 2 = 0 \qquad \text{Set each factor equal to 0.}$$
$$\qquad\qquad x = 1 \qquad\qquad x = 2$$

The solution set of this equation is $\{0, 1, 2\}$.

Zeros of a Polynomial

In general, a **polynomial equation** is an equation that can be written in the form $P(x) = 0$, where

$$P(x) = a_n x^n + a_{n-1} x^{n-1} + a_{n-2} x^{n-2} + \cdots + a_1 x + a_0$$

where n is a natural number and the polynomial is of degree n. A **zero of the polynomial** $P(x)$ is any number r for which $P(r) = 0$. It follows that a zero of $P(x)$ is a root of the polynomial equation $P(x) = 0$. For example, 0, 1, and 2 are zeros of the polynomial

$$P(x) = x^3 - 3x^2 + 2x$$

because

$P(0) = 0^3 - 3(0)^2 + 2(0)$	$P(1) = 1^3 - 3(1)^2 + 2(1)$	$P(2) = 2^3 - 3(2)^2 + 2(2)$
$= 0 - 0 + 0$	$= 1 - 3 + 2$	$= 8 - 12 + 4$
$= 0$	$= 0$	$= 0$

We began the section by showing that the zeros of $x^3 - 3x^2 + 2x$ are the roots of the polynomial equation

$$x^3 - 3x^2 + 2x = 0$$

The Remainder Theorem

There is a relationship between a zero r of a polynomial $P(x)$ and the results of a long division of $P(x)$ by $x - r$. This relationship is best shown with an example.

EXAMPLE 1   Let $P(x) = 3x^3 - 5x^2 + 3x - 10$. Find   **a.** $P(1)$ and   **b.** divide $P(x)$ by $x - 1$.

Solution   **a.** $P(1) = 3(1)^3 - 5(1)^2 + 3(1) - 10$

$$= 3 - 5 + 3 - 10$$

$$= -9$$

b.
$$
\begin{array}{r}
3x^2 - 2x + 1 \\
x - 1 \overline{)3x^3 - 5x^2 + 3x - 10} \\
\underline{3x^3 - 3x^2} \\
-2x^2 + 3x \\
\underline{-2x^2 + 2x} \\
+\;x - 10 \\
\underline{x - 1} \\
-9
\end{array}
$$

Note that the remainder is equal to $P(1)$.

Self Check   Let $P(x) = 2x^2 - 3x + 5$. Find $P(2)$ and divide $P(x)$ by $x - 2$. What do you notice about the results?   ∎

EXAMPLE 2   Let $P(x) = 3x^3 - 5x^2 + 3x - 10$.   **a.** Find $P(-2)$ and   **b.** divide $P(x)$ by $x + 2$.

Solution   **a.** $P(-2) = 3(-2)^3 - 5(-2)^2 + 3(-2) - 10$

$$= 3(-8) - 5(4) + 3(-2) - 10$$

$$= -24 - 20 - 6 - 10$$

$$= -60$$

b.
$$
\begin{array}{r}
3x^2 - 11x + 25 \\
x + 2 \overline{)3x^3 - 5x^2 + 3x - 10} \\
\underline{3x^3 + 6x^2} \\
-11x^2 + 3x \\
\underline{-11x^2 - 22x} \\
25x - 10 \\
\underline{25x + 50} \\
-60
\end{array}
$$

Note that the remainder is equal to $P(-2)$.

Self Check   Let $P(x) = 2x^2 - 3x + 5$. Find $P(-3)$ and divide $P(x)$ by $x + 3$. What do you notice about the results?   ∎

When $P(x)$ was divided by $x - 1$ in Example 1, the remainder was $P(1) = -9$. When $P(x)$ was divided by $x + 2$, or $x - (-2)$, in Example 2, the remainder was $P(-2) = -60$. These results are not coincidental. The **remainder theorem** states that a division of any polynomial $P(x)$ by $x - r$ gives $P(r)$ as the remainder.

The Remainder Theorem   If $P(x)$ is a polynomial, r is any number, and $P(x)$ is divided by $x - r$, the remainder is $P(r)$.

Proof   To divide $P(x)$ by $x - r$, we must find a quotient $Q(x)$ and a remainder $R(x)$ such that

$$\text{Dividend} = \text{divisor} \cdot \text{quotient} + \text{remainder}$$
$$P(x) = (x - r) \cdot Q(x) + R(x)$$

Since the degree of the remainder $R(x)$ must be less than the degree of the divisor $x - r$, and the degree of $x - r$ is 1, $R(x)$ must be a constant R.

In the equation

$$P(x) = (x - r)Q(x) + R$$

the polynomial on the left-hand side is the same as the polynomial on the right-hand side, and the values that they assume for any number x are equal. If we replace x with r, we have

$$P(r) = (r - r)Q(r) + R$$
$$= (0)Q(r) + R$$
$$= R$$

Thus, $P(r) = R$. ∎

EXAMPLE 3   Find the remainder that will occur when $P(x) = 2x^4 - 10x^3 + 17x^2 - 14x - 3$ is divided by $x - 3$.

Solution   By the remainder theorem, the remainder will be $P(3)$.

$$P(x) = 2x^4 - 10x^3 + 17x^2 - 14x - 3$$
$$P(3) = 2(3)^4 - 10(3)^3 + 17(3)^2 - 14(3) - 3 \quad \text{Substitute 3 for } x.$$
$$= 0$$

The remainder will be 0. Although this calculation is tedious, it is easy to do with a calculator.

Self Check   Find the remainder when $P(x)$ is divided by $x - 2$. ∎

PERSPECTIVE

Clay tablets that survive from the early period of the Babylonian civilization, 1800 to 1600 B.C., show that the Babylonians were accomplished mathematicians. Some tablets indicate that the Babylonians knew what we now call the Pythagorean theorem—centuries before the Greeks discovered it. Babylonian merchants used tablets containing multiplication tables and, for division, lists of reciprocals. For other mathematical purposes, the Babylonians had tables of squares, cubes, square roots, and cube roots.

For educating Babylonian mathematics students, other tablets contain long lists of problems and exercises. Some of these problems are practical, but many are puzzle problems to be done for practice or just for fun. Even thousands of years ago, mathematics was not a "spectator sport." Students had to study and do their homework—even word problems. One exercise from a Babylonian tablet asks, "What number added to its reciprocal is 5?" In modern notation, we would solve the equation $x + \frac{1}{x} = 5$.

It is interesting that this feature of mathematics education hasn't changed for thousands of years: We learn mathematics by *doing* mathematics—lots of exercises and lots of practice.

The Factor Theorem

If $R = P(r) = 0$ in the equation $P(x) = (x - r)Q(x) + R$, then $P(x)$ factors as $(x - r)Q(x)$. This fact can help us factor polynomials.

The Factor Theorem

> If $P(x)$ is a polynomial and r is any number, then
>
> If $P(r) = 0$, then $x - r$ is a factor of $P(x)$.
>
> If $x - r$ is a factor of $P(x)$, then $P(r) = 0$.

Proof   **Part 1:**   First, we assume that $P(r) = 0$ and prove that $x - r$ is a factor of $P(x)$. If $P(r) = 0$, then $R = 0$, and the equation $P(x) = (x - r)Q(x) + R$ becomes

$$P(x) = (x - r)Q(x) + 0$$
$$P(x) = (x - r)Q(x)$$

Therefore, $x - r$ divides $P(x)$ evenly, and $x - r$ is a factor of $P(x)$.

Part 2:   Conversely, we assume that $x - r$ is a factor of $P(x)$ and prove that $P(r) = 0$. Because, by assumption, $x - r$ is a factor of $P(x)$, $x - r$ divides $P(x)$ evenly, and the division has a remainder of 0. By the remainder theorem, this remainder is $P(r)$. Hence, $P(r) = 0$. ∎

We can state the factor theorem in a slightly different way:

> *If r is a zero of the polynomial $P(x)$, then $x - r$ is a factor of $P(x)$.*
>
> *If $x - r$ is a factor of $P(x)$, then r is a zero of the polynomial.*

EXAMPLE 4   Determine whether $x + 2$ is a factor of $P(x) = x^4 - 7x^2 - 6x$. If so, find the other factor by long division.

Solution   By the factor theorem, $x + 2$, or $x - (-2)$, is a factor of $P(x)$ if $P(-2) = 0$. So we find the value of $P(-2)$.

$$P(x) = x^4 - 7x^2 - 6x$$
$$P(-2) = (-2)^4 - 7(-2)^2 - 6(-2) \quad \text{Substitute } -2 \text{ for } x.$$
$$= 16 - 28 + 12$$
$$= 0$$

Since $P(-2) = 0$, we know that $x - (-2)$, or $x + 2$, is a factor of $P(x)$.
To find the other factor, we divide $x^4 - 7x^2 - 6x$ by $x + 2$.

$$
\begin{array}{r}
x^3 - 2x^2 - 3x \\
x + 2 \overline{)\, x^4 - 7x^2 - 6x} \\
\underline{x^4 + 2x^3 } \\
-2x^3 - 7x^2 \\
\underline{-2x^3 - 4x^2 } \\
-3x^2 - 6x \\
\underline{-3x^2 - 6x} \\
0
\end{array}
$$

$x^4 - 7x^2 - 6x$ factors as $(x + 2)(x^3 - 2x^2 - 3x)$, although this factorization is not complete.

Self Check   Is $x + 3$ a factor of $P(x)$? ∎

EXAMPLE 5 Let $P(x) = 3x^3 - 5x^2 + 3x - 10$. Show that $P(2) = 0$ and use this fact to help factor $P(x)$.

Solution We find $P(2)$.

$$P(x) = 3x^3 - 5x^2 + 3x - 10$$
$$P(2) = 3(2)^3 - 5(2)^2 + 3(2) - 10 \quad \text{Substitute 2 for } x.$$
$$= 3(8) - 5(4) + 6 - 10$$
$$= 0$$

Since $P(2) = 0$, it follows that $x - 2$ is a factor of $P(x)$. To find the other factor, we divide $P(x)$ by $x - 2$.

$$
\begin{array}{r}
3x^2 + x + 5 \\
x - 2{\overline{\smash{\big)}\,3x^3 - 5x^2 + 3x - 10}} \\
\underline{3x^3 - 6x^2} \\
x^2 + 3x \\
\underline{x^2 - 2x} \\
5x - 10 \\
\underline{5x - 10} \\
0
\end{array}
$$

$3x^3 - 5x^2 + 3x - 10$ factors as $(x - 2)(3x^2 + x + 5)$.

Self Check Show that $P(1) = 0$ and use this fact to help factor $x^3 - 1$. ∎

EXAMPLE 6 Solve: $3x^3 - 5x^2 + 3x - 10 = 0$.

Solution From Example 5, we have

$$3x^3 - 5x^2 + 3x - 10 = 0$$
$$(x - 2)(3x^2 + x + 5) = 0$$

To solve for x, we set each factor equal to 0 and apply the quadratic formula to the equation $3x^2 + x + 5 = 0$.

$$x - 2 = 0 \quad \text{or} \quad 3x^2 + x + 5 = 0$$
$$x = 2 \qquad\qquad x = \frac{-1 \pm \sqrt{1^2 - 4(3)(5)}}{2(3)}$$
$$x = \frac{-1 \pm i\sqrt{59}}{6}$$

The solution set is $\left\{2,\; -\dfrac{1}{6} + \dfrac{\sqrt{59}}{6}i,\; -\dfrac{1}{6} - \dfrac{\sqrt{59}}{6}i\right\}$.

Self Check Solve: $x^3 - 1$. ∎

EXAMPLE 7 Find a polynomial $P(x)$ with zeros of 3, 3, and -5.

Solution We have proved that if r is a zero of a polynomial, then $x - r$ is a factor of the polynomial. In particular, if 3, 3, and -5 are zeros of $P(x)$, then $x - 3, x - 3$, and $x - (-5)$ are factors of $P(x)$.

$$P(x) = (x - 3)(x - 3)(x + 5)$$
$$P(x) = (x^2 - 6x + 9)(x + 5) \quad \text{Multiply } x - 3 \text{ and } x - 3.$$
$$P(x) = x^3 - x^2 - 21x + 45$$

The polynomial $P(x) = x^3 - x^2 - 21x + 45$ has zeros of 3, 3, and -5.

Self Check   Find a polynomial $P(x)$ with zeros of -2, 2, and 3.   ∎

Because 3 occurs twice as a zero in Example 7, we say that 3 is a **zero of multiplicity 2.**

Synthetic Division

Synthetic division is an easy way to divide higher-degree polynomials by binomials of the form $x - r$. To see how it works, we consider the following long division. On the left is a complete division. On the right is a modified version in which the variables have been removed.

$$
x - 3\overline{\smash{\big)}\,
\begin{aligned}
& 2x^2 + 10x + 27 \\
& 2x^3 + 4x^2 - 3x + 10
\end{aligned}}
\qquad\qquad
1 - 3\overline{\smash{\big)}\,
\begin{aligned}
& 2 + 10 + 27 \\
& 2 + 4 - 3 + 10
\end{aligned}}
$$

$$
\begin{array}{r}
2x^3 - 6x^2 \\ \hline
10x^2 - 3x \\
10x^2 - 30x \\ \hline
27x + 10 \\
27x - 81 \\ \hline
\text{(remainder)} \quad 91
\end{array}
\qquad
\begin{array}{r}
2 - 6 \\ \hline
10 - 3 \\
10 - 30 \\ \hline
27 + 10 \\
27 - 81 \\ \hline
\text{(remainder)} \quad 91
\end{array}
$$

We can shorten the work even more by omitting the numbers printed in color.

$$
-3\overline{\smash{\big)}\,
\begin{aligned}
& 2 + 10 + 27 \\
& 2 + 4 - 3 + 10
\end{aligned}}
$$

$$
\begin{array}{r}
- 6 \\ \hline
10 \\
- 30 \\ \hline
27 \\
- 81 \\ \hline
\text{(remainder)} \quad 91
\end{array}
$$

We can then compress the work vertically to get

$$
-3\overline{\smash{\big)}\,
\begin{aligned}
& 2 + 10 + 27 \\
& 2 + 4 - 3 + 10 \\
& - 6 - 30 - 81
\end{aligned}}
$$
$$
\overline{ 10 \quad 27 \quad 91}
$$

If we write the 2 in the quotient on the bottom line, the bottom line gives both the coefficients of the quotient and the remainder. The top line can now be eliminated, and the division appears as

$$
\underline{-3}\,\big|\; 2 + 4 - 3 + 10 \qquad \text{The coefficients of the dividend.}
$$
$$
 - 6 - 30 - 81
$$
$$
\overline{2 \quad 10 \quad 27 \quad 91} \qquad \text{The coefficients of the quotient and the remainder.}
$$

The bottom line was obtained by subtracting the middle line from the top line. If we replace the -3 in the divisor with $+3$, the signs of each number in the middle line will be reversed in the division process. Then the bottom line can be obtained by addition, and we have the final form of the synthetic division.

$$\underline{+3}\begin{array}{rrrr} 2 + & 4 - & 3 + & 10 \\ + & 6 + & 30 + & 81 \\ \hline \mathbf{2} & \mathbf{10} & \mathbf{27} & \mathbf{91} \end{array}$$ The coefficients of the dividend.

The coefficients of the quotient and the remainder.

Thus,

$$\frac{2x^3 + 4x^2 - 3x + 10}{x - 3} = 2x^2 + 10x + 27 + \frac{91}{x - 3}$$

EXAMPLE 8   Use synthetic division to divide $10x + 3x^4 - 8x^3 + 3$ by $x - 2$.

Solution   We first write the terms of $P(x)$ in descending powers of x:

$$3x^4 - 8x^3 + 10x + 3$$

We then write the coefficients of the dividend, with its terms in descending powers of x, and the 2 from the divisor in the following form:

$$\underline{2}\quad 3 \quad -8 \quad \mathbf{0} \quad 10 \quad 3$$ Write 0 for the coefficient of the missing x^2 term.

Then we follow these steps:

$$\underline{2}\begin{array}{rrrrr} 3 & -8 & 0 & 10 & 3 \\ \downarrow & & & & \\ \hline 3 & & & & \end{array}$$ Bring down the 3.

$$\underline{2}\begin{array}{rrrrr} 3 & -8 & 0 & 10 & 3 \\ & 6 & & & \\ \hline 3 & -2 & & & \end{array}$$ Multiply 2 and 3 together to get 6, and add 6 and -8 to get -2.

$$\underline{2}\begin{array}{rrrrr} 3 & -8 & 0 & 10 & 3 \\ & 6 & -4 & & \\ \hline 3 & -2 & -4 & & \end{array}$$ Multiply 2 and -2 together to get -4, and add -4 and 0 to get -4.

$$\underline{2}\begin{array}{rrrrr} 3 & -8 & 0 & 10 & 3 \\ & 6 & -4 & -8 & \\ \hline 3 & -2 & -4 & 2 & \end{array}$$ Multiply 2 and -4 together to get -8, and add -8 and 10 to get 2.

$$\underline{2}\begin{array}{rrrrr} 3 & -8 & 0 & 10 & 3 \\ & 6 & -4 & -8 & 4 \\ \hline 3 & -2 & -4 & 2 & 7 \end{array}$$ Multiply 2 and 2 together to get 4, and add 4 and 3 to get 7.

The coefficients of the quotient and the remainder.

Thus,

$$\frac{10x + 3x^4 - 8x^3 + 3}{x - 2} = 3x^3 - 2x^2 - 4x + 2 + \frac{7}{x - 2}$$

Self Check   Divide: $2x^3 - 4x^2 + 5x - 7$ by $x - 3$.   ∎

Using Synthetic Division to Evaluate Polynomials

EXAMPLE 9   Use synthetic division to find $P(-2)$ when $P(x) = 5x^3 + 3x^2 - 21x - 1$.

Solution   Because of the remainder theorem, $P(-2)$ is the remainder when $P(x)$ is divided by $x - (-2)$. We use synthetic division to find the remainder.

$$
\begin{array}{r|rrrr}
-2 & 5 & 3 & -21 & -1 \\
   &   & -10 &   &   \\
\hline
   & 5 & -7 &   &
\end{array}
\qquad -2(5) = -10;\ 3 + (-10) = -7.
$$

$$
\begin{array}{r|rrrr}
-2 & 5 & 3 & -21 & -1 \\
   &   & -10 & 14 &   \\
\hline
   & 5 & -7 & -7 &
\end{array}
\qquad -2(-7) = 14;\ -21 + 14 = -7.
$$

$$
\begin{array}{r|rrrr}
-2 & 5 & 3 & -21 & -1 \\
   &   & -10 & 14 & 14 \\
\hline
   & 5 & -7 & -7 & 13
\end{array}
\qquad -2(-7) = 14;\ -1 + 14 = 13.
$$

Because the remainder is 13, $P(-2) = 13$.

Self Check   Find $P(3)$.   ∎

EXAMPLE 10   If $P(x) = x^3 - x^2 + x - 1$, find $P(i)$, where $i = \sqrt{-1}$.

Solution   We can use synthetic division.

$$
\begin{array}{r|rrrr}
i & 1 & -1 & +1 & -1 \\
  &   & i & -1-i & +1 \\
\hline
  & 1 & i-1 & -i & 0
\end{array}
$$

Since the remainder is 0, $P(i) = 0$, and i is a zero of $P(x)$.

Self Check   Find $P(-i)$.   ∎

Self Check Answers

1. $P(2)$ is the remainder.   **2.** $P(-3)$ is the remainder.   **3.** -11 **4.** no   **5.** $(x - 1)(x^2 + x + 1)$

6. $1, -\dfrac{1}{2} \pm \dfrac{\sqrt{3}}{2}i$ **7.** $x^3 - 3x^2 - 4x + 12$ **8.** $2x^2 + 2x + 11 + \dfrac{26}{x - 3}$ **9.** 98   **10.** 0

Orals   *Perform each division and find the remainder.*

1. $x - 2\overline{)x^2 + 7x + 6}$ **2.** $x + 3\overline{)x^2 - x + 2}$

11.1   EXERCISES 🔘

REVIEW   *Find the quadrant in which each point lies.*

1. $(3, -2)$ **2.** $(-2, -5)$

3. $(8, \pi)$ **4.** $(-9, 9)$

Find the distance between each pair of points.

5. $(3, -3), (-5, 3)$ **6.** $(-8, 2), (2, -22)$

Find the slope of the line passing through each pair of points.

7. $(3, 5), (-5, -3)$ **8.** $\left(3, \dfrac{3}{5}\right), \left(-\dfrac{3}{5}, 1\right)$

VOCABULARY AND CONCEPTS   *Fill in the blanks.*

9. The variables in a polynomial have _____-number exponents.

10. A zero of $P(x)$ is any number r for which _____.

11. The remainder theorem holds when r is ____ number.

12. If $P(x)$ is a polynomial and $P(x)$ is divided by _____, the remainder will be $P(r)$.

13. If $P(x)$ is a polynomial, then $P(r) = 0$ if and only if $x - r$ is a _____ of $P(x)$.

14. A shortcut method for dividing a polynomial by a binomial of the form $x - r$ is called _____ division.

PRACTICE *Let* $P(x) = 2x^4 - 2x^3 + 5x^2 - 1$. *Find each value by substituting the given value of x into the polynomial and simplifying. Then find the value by doing a long division and finding the remainder.*

15. $P(1)$ **16.** $P(2)$

17. $P(-2)$ **18.** $P(-1)$

Use the remainder theorem to find the remainder when $P(x) = 3x^4 + 5x^3 - 4x^2 - 2x + 1$ *is divided by each binomial.*

19. $x - 12$ **20.** $x + 15$

21. $x + 3.25$ **22.** $x - 7.12$

Use the factor theorem to decide whether each statement is true. If the statement is not true, so indicate.

23. $x - 1$ is a factor of $x^7 - 1$.

24. $x - 2$ is a factor of $x^3 - x^2 + 2x - 8$.

25. $x - 1$ is a factor of $3x^5 + 4x^2 - 7$.

26. $x + 1$ is a factor of $3x^5 + 4x^2 - 7$.

27. $x + 3$ is a factor of $2x^3 - 2x^2 + 1$.

28. $x - 3$ is a factor of $3x^5 - 3x^4 + 5x^2 - 13x - 6$.

29. $x - 1$ is a factor of $x^{1,984} - x^{1,776} + x^{1,492} - x^{1,066}$.

30. $x + 1$ is a factor of $x^{1,984} + x^{1,776} - x^{1,492} - x^{1,066}$.

A partial solution set is given for each equation. Find the complete solution set.

31. $x^3 + 3x^2 - 13x - 15 = 0$; $\{-1\}$

32. $x^3 + 6x^2 + 5x - 12 = 0$; $\{1\}$

33. $x^4 - 2x^3 - 2x^2 + 6x - 3 = 0$; $\{1, 1\}$

34. $x^5 + 4x^4 + 4x^3 - x^2 - 4x - 4 = 0$; $\{1, -2, -2\}$

35. $x^4 - 5x^3 + 7x^2 - 5x + 6 = 0$; $\{2, 3\}$

36. $x^4 + 2x^3 - 3x^2 - 4x + 4 = 0$; $\{1, -2\}$

Find a polynomial with the given zeros.

37. $4, 5$ **38.** $-3, 5$

39. $1, 1, 1$ **40.** $1, 0, -1$

41. $2, 4, 5$ **42.** $7, 6, 3$

43. $1, -1, \sqrt{2}, -\sqrt{2}$

44. $0, 0, 0, \sqrt{3}, -\sqrt{3}$

45. $\sqrt{2}, i, -i$

46. $i, i, 2$

47. $0, 1 + i, 1 - i$

48. $i, 2 + i, 2 - i$

Use synthetic division to express $P(x) = 3x^3 - 2x^2 - 6x - 4$ *in the form* (divisor)(quotient) + remainder *for each divisor.*

49. $x - 1$

50. $x - 2$

51. $x - 3$

52. $x - 4$

53. $x + 1$

54. $x + 2$

55. $x + 3$

56. $x + 4$

Use synthetic division to perform each division.

57. $\dfrac{x^3 + x^2 + x - 3}{x - 1}$

58. $\dfrac{x^3 - x^2 - 5x + 6}{x - 2}$

59. $\dfrac{7x^3 - 3x^2 - 5x + 1}{x + 1}$

60. $\dfrac{2x^3 + 4x^2 - 3x + 8}{x - 3}$

61. $\dfrac{4x^4 - 3x^3 - x + 5}{x - 3}$

62. $\dfrac{x^4 + 5x^3 - 2x^2 + x - 1}{x + 1}$

63. $\dfrac{3x^5 - 768x}{x - 4}$

64. $\dfrac{x^5 - 4x^2 + 4x + 4}{x + 3}$

Let $P(x) = 5x^3 + 2x^2 - x + 1$. *Use synthetic division to find each value.*

65. $P(2)$ **66.** $P(-2)$

67. $P(-5)$ **68.** $P(3)$

Let $P(x) = 2x^4 - x^2 + 2$. Use synthetic division to find each value.

69. $P\left(\dfrac{1}{2}\right)$ **70.** $P\left(\dfrac{1}{3}\right)$

71. $P(i)$ **72.** $P(-i)$

Let $P(x) = x^4 - 8x^3 + 8x + 14x^2 - 15$. Write the terms of $P(x)$ in descending powers of x and use synthetic division to find each value.

73. $P(1)$ **74.** $P(0)$

75. $P(-3)$ **76.** $P(-1)$

Let $P(x) = 8 - 8x^2 + x^5 - x^3$. Write the terms of $P(x)$ in descending powers of x and use synthetic division to find each value.

77. $P(i)$ **78.** $P(-i)$

79. $P(-2i)$ **80.** $P(2i)$

WRITING

81. If $P(2) = 0$ and $P(-2) = 0$, explain why $x^2 - 4$ is a factor of $P(x)$.

82. If $P(x) = x^4 - 3x^3 + kx^2 + 4x - 1$ and $P(2) = 11$, find k.

SOMETHING TO THINK ABOUT

83. Find a_0 if 0 is a zero of
$$P(x) = a_n x^n + a_{n-1}x^{n-1} + \cdots + a_1 x + a_0$$

84. Find a_1 if 0 occurs twice as a zero of
$$P(x) = a_n x^n + a_{n-1}x^{n-1} + \cdots + a_1 x + a_0$$

11.2   Descartes' Rule of Signs and Bounds on Roots

■ **The Fundamental Theorem of Algebra**
■ **The Conjugate Pairs Theorem**   ■ **Descartes' Rule of Signs**
■ **Bounds on Roots**

Getting Ready   *Solve each polynomial equation, give the degree of the polynomial, and tell how many roots the equation has.*

1. $3x - 2 = 7$ **2.** $x^2 - 5x - 6 = 0$

3. $x^3 - 9x = 0$ **4.** $x^4 - 5x^2 + 4 = 0$

The remainder theorem and synthetic division provide a way of verifying that a particular number is a root of a polynomial equation, but they do not provide the roots. We need some guidelines to indicate how many roots to expect, what kind of roots to expect, and where they are located. This section develops several theorems that provide such guidelines.

The Fundamental Theorem of Algebra

Before attempting to find the roots of a polynomial equation, it would be useful to know if any roots exist. This question was answered by Carl Friedrich Gauss (1777–1855) when he proved the **fundamental theorem of algebra.**

The Fundamental Theorem of Algebra

If $P(x)$ is a polynomial with positive degree, then $P(x)$ has at least one zero.

The fundamental theorem of algebra guarantees that polynomials such as

$$2x^3 + 3 \quad \text{and} \quad 32.75x^{1,984} + ix^3 - (2 + i)x - 5$$

all have zeros. Since all polynomials with positive degree have zeros, their corresponding polynomial equations all have roots. They are the zeros of the polynomial.

The next theorem will help us show that every nth-degree polynomial equation has exactly n roots over the complex numbers.

The Polynomial Factorization Theorem

If $n > 0$ and $P(x)$ is an nth-degree polynomial, then $P(x)$ has exactly n linear factors:

$$P(x) = a_n(x - r_1)(x - r_2)(x - r_3) \cdots \cdot (x - r_n)$$

where $r_1, r_2, r_3, \ldots, r_n$ are numbers and a_n is the lead coefficient of $P(x)$.

Proof Let $P(x)$ be a polynomial of degree n ($n > 0$). Because of the fundamental theorem of algebra, we know that $P(x)$ has a zero r_1 and that the equation $P(x) = 0$ has r_1 for a root. By the factor theorem, we know that $x - r_1$ is a factor of $P(x)$. Thus,

$$P(x) = (x - r_1)Q_1(x)$$

If the lead coefficient of the nth-degree polynomial $P(x)$ is a_n, then $Q_1(x)$ is a polynomial of degree $n - 1$ whose lead coefficient is also a_n.

By the fundamental theorem of algebra, we know that $Q_1(x)$ also has a zero, which is r_2. By the factor theorem, $x - r_2$ is a factor of $Q_1(x)$, and

$$P(x) = (x - r_1)(x - r_2)Q_2(x)$$

where $Q_2(x)$ is a polynomial of degree $n - 2$ with lead coefficient a_n.

This process can continue only to n factors of the form $x - r_i$ until the final quotient $Q_n(x)$ is a polynomial of degree $n - n$, or degree 0. Thus, the polynomial $P(x)$ factors completely as

(1) $$P(x) = a_n(x - r_1)(x - r_2)(x - r_3) \cdots \cdot (x - r_n)$$ ∎

If we substitute any one of the numbers $r_1, r_2, r_3, \ldots, r_n$ for x in Equation 1, $P(x)$ will equal 0. Thus, each value of r is a zero of $P(x)$ and a root of the equation $P(x) = 0$. There can be no other roots, because no single factor in Equation 1 is 0 for any value of x not included in the list $r_1, r_2, r_3, \ldots, r_n$.

The values of r in the previous list need not be distinct. Any number r_i that occurs k times as a root of a polynomial equation is called a **root of multiplicity k.**

The following theorem summarizes the previous discussion.

Theorem

If multiple roots are counted individually, the polynomial equation $P(x) = 0$ with degree n ($n > 0$) has exactly n roots among the complex numbers.

The Conjugate Pairs Theorem

Recall that the complex numbers $a + bi$ and $a - bi$ are called **complex conjugates** of each other. The next theorem points out that *complex roots of polynomial equations with real coefficients occur in complex conjugate pairs.*

The Conjugate Pairs Theorem

If a polynomial equation $P(x) = 0$ with real-number coefficients has a complex root $a + bi$ with $b \neq 0$, then its conjugate $a - bi$ is also a root.

EXAMPLE 1   Find a second-degree polynomial equation with real coefficients and a root of $2 + i$.

Solution   Because $2 + i$ is a root, its complex conjugate $2 - i$ is also a root. The equation is

$$[x - (2 + i)][x - (2 - i)] = 0$$
$$(x - 2 - i)(x - 2 + i) = 0$$
$$x^2 - 4x + 5 = 0 \quad \text{Multiply.}$$

The equation $x^2 - 4x + 5 = 0$ will have roots of $2 + i$ and $2 - i$.

Self Check   Find a polynomial equation with real coefficients and a root of $1 - i$.

EXAMPLE 2   Find a fourth-degree polynomial equation with real coefficients and i as a root of multiplicity 2.

Solution   Because i is a root twice and a fourth-degree polynomial equation has four roots, we must find the other two roots. According to the conjugate pairs theorem, the missing roots are the conjugates of the given roots. Thus, the complete solution set is

$$\{i, \quad i, \quad -i, \quad -i\}$$

The equation is

$$(x - i)(x - i)[x - (-i)][x - (-i)] = 0$$
$$(x - i)(x + i)(x - i)(x + i) = 0$$
$$(x^2 + 1)(x^2 + 1) = 0$$
$$x^4 + 2x^2 + 1 = 0$$

Self Check   Find a polynomial equation with real coefficients and $-i$ as a root of multiplicity 2.

EXAMPLE 3   Find a quadratic equation with a double root of i.

Solution   If i is a root twice in a quadratic equation, the equation is

$$(x - i)(x - i) = 0$$
$$x^2 - 2ix - 1 = 0$$

In this equation, the roots are not in complex conjugate pairs. This is not surprising, because the coefficient $2i$ is not a real number.

Self Check   Find a quadratic equation with a double root of $-i$.

Comment   The theorem "Complex roots of real polynomial equations occur in complex conjugate pairs" applies only to polynomial equations with real coefficients.

Descartes' Rule of Signs

René Descartes (1596–1650) is credited with a theorem known as **Descartes' rule of signs,** which enables us to estimate the number of positive, negative, and nonreal roots of a polynomial equation.

   If a polynomial is written in descending powers of x and we scan it from left to right, we say that a variation in sign occurs whenever successive terms have opposite signs. For example, the polynomial

$$P(x) = \overbrace{3x^5 - 2x^4}^{+ \text{ to } -} \overbrace{- 5x^3 + x^2}^{- \text{ to } +} \overbrace{- x}^{+ \text{ to } -} - 9$$

has three variations in sign, and the polynomial

$$P(-x) = 3(-x)^5 - 2(-x)^4 - 5(-x)^3 + (-x)^2 - (-x) - 9$$

$$= \overbrace{-3x^5 - 2x^4 + 5x^3}^{- \text{ to } +} \overbrace{+ x^2 + x - 9}^{+ \text{ to } -}$$

has two variations in sign.

Descartes' Rule of Signs   If $P(x)$ is a polynomial with real coefficients, the number of positive roots of $P(x) = 0$ is either equal to the number of variations in sign of $P(x)$ or less than that by an even number.

The number of negative roots of $P(x) = 0$ is either equal to the number of variations in sign of $P(-x)$ or less than that by an even number.

EXAMPLE 4   Discuss the possibilities for the roots of $P(x) = 3x^3 - 2x^2 + x - 5 = 0$.

Solution   Since there are three variations of sign in $P(x) = 3x^3 - 2x^2 + x - 5 = 0$, there can be either 3 positive roots or only 1 (1 is less than 3 by the even number 2). Because

$$P(-x) = 3(-x)^3 - 2(-x)^2 + (-x) - 5$$
$$= -3x^3 - 2x^2 - x - 5$$

has no variations in sign, there are 0 negative roots. Furthermore, 0 is not a root, because the terms of the polynomial do not have a common factor of x.

   If there are 3 positive roots, then all of the roots are accounted for. If there is 1 positive root, the 2 remaining roots must be nonreal complex numbers. The following chart shows these possibilities.

Number of positive roots	Number of negative roots	Number of nonreal roots
3	0	0
1	0	2

The number of nonreal complex roots is the number needed to bring the total number of roots up to 3.

Self Check   Discuss the possibilities for the roots of $5x^3 + 2x^2 - x + 3 = 0$.   ∎

EXAMPLE 5   Discuss the possibilities for the roots of $P(x) = 5x^5 - 3x^3 - 2x^2 + x - 1 = 0$.

Solution   Since there are three variations of sign in $P(x)$, there are either 3 or 1 positive roots. Because $P(-x) = -5x^5 + 3x^3 - 2x^2 - x - 1$ has two variations in sign, there are 2 or 0 negative roots. The possibilities are shown as follows:

Number of positive roots	Number of negative roots	Number of nonreal roots
1	0	4
3	0	2
1	2	2
3	2	0

In each case, the number of nonreal complex roots is an even number. This is expected, because this polynomial has real coefficients, and its nonreal complex roots will occur in conjugate pairs.

Self Check   Discuss the possibilities for the roots of $5x^5 - 2x^2 - x - 1 = 0$.   ∎

Bounds on Roots

A final theorem provides a way to find **bounds** on the roots of a polynomial equation, enabling us to look for roots where they can be found.

Upper and Lower Bounds on Roots

> Let the lead coefficient of a polynomial $P(x)$ with real coefficients be positive, and do a synthetic division of the coefficients by a positive number c. If each term in the last row of the division is nonnegative, no number greater than c can be a root of $P(x) = 0$. (c is called an **upper bound** of the real roots.)
>
> If $P(x)$ is synthetically divided by a negative number d and the signs in the last row alternate,* no value less than d can be a root of $P(x) = 0$. (d is called a **lower bound** of the real roots.)
>
> *If 0 appears in the third row, that 0 can be assigned either a + or a − sign to help the signs alternate.

EXAMPLE 6   Establish integer bounds for the roots of $2x^3 + 3x^2 - 5x - 7 = 0$.

Solution   We will do several synthetic divisions by positive integers, looking for nonnegative values in the last row. Then we will divide by several negative integers, looking for alternating signs in the last row. Trying 1 first gives

$$\begin{array}{r|rrrr} 1 & 2 & 3 & -5 & -7 \\ & & 2 & 5 & 0 \\ \hline & +2 & +5 & 0 & -7 \end{array}$$

Because one of the signs in the last row is negative, we cannot claim that 1 is an upper bound. We now try 2.

$$\begin{array}{r|rrrr} 2 & 2 & 3 & -5 & -7 \\ & & 4 & 14 & 18 \\ \hline & +2 & +7 & +9 & +11 \end{array}$$

Because the last row is entirely nonnegative, 2 is an upper bound. That is, no number greater than 2 can be a root of the equation.

Now we try some negative divisors, beginning with -3.

$$\begin{array}{r|rrrr} -3 & 2 & 3 & -5 & -7 \\ & & -6 & 9 & -12 \\ \hline & +2 & -3 & +4 & -19 \end{array}$$

Since the signs in the last row alternate, -3 is a lower bound. That is, no number less than -3 can be a root. To see whether there is a greater lower bound, we try -2.

$$\begin{array}{r|rrrr} -2 & 2 & 3 & -5 & -7 \\ & & -4 & 2 & 6 \\ \hline & +2 & -1 & -3 & -1 \end{array}$$

Since the signs in the last row do not alternate, we cannot say that -2 is a lower bound.

All of the real roots of the given equation are in the interval $(-3, 2)$.

Self Check Establish integer bounds for the roots of $2x^3 + 3x^2 - 11x - 7 = 0$. ∎

It is important to understand what the theorem on the bounds of roots says and what it doesn't say. If we divide synthetically by a positive number c and the last row of the synthetic division is entirely nonnegative, the theorem guarantees that c is an upper bound of the roots. However, if the last row contains some negative values, c could still be an upper bound.

If we divide by a negative number d and the signs in the last row alternate, the theorem guarantees that d is a lower bound of the roots. However, if the signs in the last row do not alternate, d could still be a lower bound. This is illustrated in Example 6. It can be shown that the smallest negative root of the equation is approximately -1.81. Thus, -2 is a lower bound for the roots of the equation. However, when we checked -2, the last row of the synthetic division did not have alternating signs. Unfortunately, the theorem does not always determine the best bounds for the roots of the equation.

Self Check Answers

1. $x^2 - 2x + 2 = 0$ **2.** $x^4 + 2x^2 + 1 = 0$ **3.** $x^2 + 2ix - 1 = 0$

4.

Num. of pos. roots	Num. of neg. roots	Num. of nonreal roots
2	1	0
0	1	2

5.

Num. of pos. roots	Num. of neg. roots	Num. of nonreal roots
1	2	2
1	0	4

6. Roots are in $(-4, 3)$.

Orals *Tell how many roots each equation has.*

1. $x^4 - 7x + 12 = 0$

2. $x(x^5 - 3x^2 + 7x) = 0$

Give the number of possible positive roots of each equation.

3. $x^3 + 2x^2 - x - 1 = 0$

4. $3x^3 - 5x^2 + x - 5 = 0$

Give the number of possible negative roots of each equation.

5. $x^4 + 2x^2 + 1 = 0$

6. $3x^6 - 5 = 0$

11.2 EXERCISES

REVIEW *Assume that k represents a positive real number, and complete each sentence.*

1. The graph of $y = f(x - k)$ looks like the graph of $y = f(x)$, except that it has been translated _____.

2. The graph of $y = f(x) - k$ looks like the graph of $y = f(x)$, except that it has been translated _____.

3. The graph of $y = f(-x)$ looks like the graph of $y = f(x)$, except that it has been _____.

4. The graph of $y = -f(x)$ looks like the graph of $y = f(x)$, except that it has been _____.

5. The graph of $y = kf(x)$ looks like the graph of $y = f(x)$, except that it has been _____.

6. The graph of $y = f(kx)$ looks like the graph of $y = f(x)$, except that it has been _____.

VOCABULARY AND CONCEPTS *Fill in the blanks.*

7. If $P(x)$ is a polynomial with positive degree, then $P(x)$ has at least one _____.

8. The statement in Exercise 7 is called the _____.

9. The _____ of $a + bi$ is $a - bi$.

10. The polynomial $6x^4 + 5x^3 - 2x^2 + 3$ has __ variations in sign.

11. The polynomial $(-x)^3 - (-x)^2 - 4$ has __ variations in sign.

12. The equation $7x^4 + 5x^3 - 2x + 1 = 0$ can have at most __ positive roots.

13. The equation $7x^4 + 5x^3 - 2x + 1 = 0$ can have at most __ negative roots.

14. Complex roots occur in complex _____ pairs. (Assume that the equation has real coefficients.)

15. If no number less than d can be a root of $P(x) = 0$, then d is called a(n) _____.

16. If no number greater than c can be a root of $P(x) = 0$, then c is called a(n) _____.

PRACTICE *Tell how many roots each equation has.*

17. $x^{10} = 1$

18. $x^{40} = 1$

19. $3x^4 - 4x^2 - 2x = -7$

20. $-32x^{111} - x^5 = 1$

21. One root of $x(3x^4 - 2) = 12x$ is 0. How many other roots are there?

22. One root of $3x(x^7 - 14x + 3) = 0$ is 0. How many other roots are there?

Write a third-degree polynomial equation with real coefficients and the given roots.

23. $3, -i$

24. $1, i$

25. $2, 2 + i$

26. $-2, 3 - i$

Write a fourth-degree polynomial equation with real coefficients and the given roots.

27. $3, 2, i$

28. $1, 2, 1 + i$

29. $i, 1 - i$

30. $i, 2 - i$

Use Descartes' rule of signs to find the number of possible positive, negative, and nonreal roots of each equation.

31. $3x^3 + 5x^2 - 4x + 3 = 0$

32. $3x^3 - 5x^2 - 4x - 3 = 0$

33. $2x^3 + 7x^2 + 5x + 5 = 0$

34. $-2x^3 - 7x^2 - 5x - 4 = 0$

35. $8x^4 = -5$

36. $-3x^3 = -5$

37. $x^4 + 8x^2 - 5x - 10 = 0$

38. $5x^7 + 3x^6 - 2x^5 + 3x^4 + 9x^3 + x^2 + 1 = 0$

39. $-x^{10} - x^8 - x^6 - x^4 - x^2 - 1 = 0$

40. $x^{10} + x^8 + x^6 + x^4 + x^2 + 1 = 0$

41. $x^9 + x^7 + x^5 + x^3 + x = 0$ (Is 0 a root?)

42. $-x^9 - x^7 - x^5 - x^3 - x = 0$ (Is 0 a root?)

43. $-2x^4 - 3x^2 + 2x + 3 = 0$

44. $-7x^5 - 6x^4 + 3x^3 - 2x^2 + 7x - 4 = 0$

Find integer bounds for the roots of each equation.

45. $x^2 - 2x - 4 = 0$ **46.** $9x^2 - 6x - 1 = 0$

47. $18x^2 - 6x - 1 = 0$ **48.** $2x^2 - 10x - 9 = 0$

49. $6x^3 - 13x^2 - 110x = 0$

50. $12x^3 + 20x^2 - x - 6 = 0$

51. $x^5 + x^4 - 8x^3 - 8x^2 + 15x + 15 = 0$

52. $3x^4 - 5x^3 - 9x^2 + 15x = 0$

53. $3x^5 - 11x^4 - 2x^3 + 38x^2 - 21x - 15 = 0$

54. $3x^6 - 4x^5 - 21x^4 + 4x^3 + 8x^2 + 8x + 32 = 0$

WRITING

55. Explain why the fundamental theorem of algebra guarantees that every polynomial equation of positive degree has at least one root.

56. Explain why the fundamental theorem of algebra and the factor theorem guarantee that an nth-degree polynomial equation has n roots.

SOMETHING TO THINK ABOUT

57. Prove that any odd-degree polynomial equation with real coefficients must have at least one real root.

58. If a, b, c, and d are positive numbers, prove that $ax^4 + bx^2 + cx - d = 0$ has exactly two nonreal roots.

11.3   Rational Roots of Polynomial Equations

▌ **Rational Root Theorem**   ▌ **Finding Possible Rational Roots**
▌ **Finding Rational Roots**   ▌ **An Application**

Getting Ready   *Tell whether the given number could be a root of the given equation.*

1. $x^2 - 4 = 0$;   3   **2.** $x^2 - 5x - 6 = 0$;   4

In this section, we will find rational roots of polynomial equations with integer coefficients. The following theorem enables us to list the possible rational roots of such equations.

Rational Root Theorem

Rational Root Theorem

> Let the polynomial equation
>
> $$P(x) = a_n x^n + a_{n-1} x^{n-1} + a_{n-2} x^{n-2} + \cdots + a_1 x + a_0 = 0$$
>
> have integer coefficients. If the rational number $\frac{p}{q}$ (written in lowest terms) is a root of $P(x) = 0$, then p is a factor of the constant a_0, and q is a factor of the lead coefficient a_n.

Proof   Let $\frac{p}{q}$ (written in lowest terms) be a rational root of $P(x) = 0$. Then the equation is satisfied by $\frac{p}{q}$.

(1) $$a_n \left(\frac{p}{q}\right)^n + a_{n-1}\left(\frac{p}{q}\right)^{n-1} + a_{n-2}\left(\frac{p}{q}\right)^{n-2} + \cdots + a_1\left(\frac{p}{q}\right) + a_0 = 0$$

We can clear Equation 1 of fractions by multiplying both sides by q^n.

(2) $$a_n p^n + a_{n-1} p^{n-1} q + a_{n-2} p^{n-2} q^2 + \cdots + a_1 p q^{n-1} + a_0 q^n = 0$$

We can factor p from all but the last term and subtract $a_0 q^n$ from both sides to get

$$p(a_n p^{n-1} + a_{n-1} p^{n-2} q + a_{n-2} p^{n-3} q^2 + \cdots + a_1 q^{n-1}) = -a_0 q^n$$

Since p is a factor of the left-hand side, it is also a factor of the right-hand side. So p is a factor of $-a_0 q^n$, but because $\frac{p}{q}$ is written in lowest terms, p cannot be a factor of q^n. Therefore, p is a factor of a_0.

We can factor q from all but the first term of Equation 2 and subtract $a_n p^n$ from both sides to get

$$q(a_{n-1} p^{n-1} + a_{n-2} p^{n-2} q + a_{n-3} p^{n-3} q^2 + \cdots + a_0 q^{n-1}) = -a_n p^n$$

Since q is a factor of the left-hand side, it is also a factor of the right-hand side. Because q is not a factor of p^n, it must be a factor of a_n.   ∎

Finding Possible Rational Roots

To illustrate the previous theorem, we consider the equation

$$\frac{1}{2} x^4 + \frac{2}{3} x^3 + 3x^2 - \frac{3}{2} x + 3 = 0$$

Because the theorem requires integer coefficients, we multiply both sides of the equation by 6 to clear it of fractions.

$$3x^4 + 4x^3 + 18x^2 - 9x + 18 = 0$$

By the previous theorem, the only possible numerators for the rational roots of the equation are the factors of the constant term 18:

$$\pm 1, \quad \pm 2, \quad \pm 3, \quad \pm 6, \quad \pm 9, \quad \text{and} \quad \pm 18$$

The only possible denominators are the factors of the lead coefficient 3:

$$\pm 1 \text{ and } \pm 3$$

We can form a list of all possible rational solutions by listing the combinations of possible numerators and denominators:

$$\pm\frac{1}{1}, \quad \pm\frac{2}{1}, \quad \pm\frac{3}{1}, \quad \pm\frac{6}{1}, \quad \pm\frac{9}{1}, \quad \pm\frac{18}{1}, \quad \pm\frac{1}{3}, \quad \pm\frac{2}{3}, \quad \pm\frac{3}{3}, \quad \pm\frac{6}{3}, \quad \pm\frac{9}{3}, \quad \pm\frac{18}{3}$$

Since several of these possibilities are duplicates, we can condense the list to get

Possible rational roots

$$\pm 1, \quad \pm 2, \quad \pm 3, \quad \pm 6, \quad \pm 9, \quad \pm 18, \quad \pm\frac{1}{3}, \quad \pm\frac{2}{3}$$

EXAMPLE 1   Prove that $\sqrt{2}$ is an irrational number.

Solution   We know that $\sqrt{2}$ is a real root of the equation $x^2 - 2 = 0$. Since any rational root of this equation will have the form $\dfrac{\text{factor of the constant } 2}{\text{factor of the lead coefficient } 1}$, the possible rational roots are

$$\pm\frac{1}{1} \quad \text{and} \quad \pm\frac{2}{1}$$

Since none of the numbers $1, -1, 2,$ or -2 satisfies the equation, $\sqrt{2}$ must be irrational.

Self Check   Prove that $\sqrt{3}$ is irrational.   ∎

Finding Rational Roots

EXAMPLE 2   Solve: $P(x) = 2x^3 + 3x^2 - 8x + 3 = 0$.

Solution   Since the equation is of third degree, it has 3 roots. By Descartes' rule of signs, there are two possible combinations of positive, negative, and nonreal roots. They are summarized as follows:

Number of positive roots	Number of negative roots	Number of nonreal roots
2	1	0
0	1	2

Since any rational roots will have the form $\dfrac{\text{factor of the constant } 3}{\text{factor of the lead coefficient } 2}$, the possible rational roots are

$$\pm\frac{3}{1}, \quad \pm\frac{1}{1}, \quad \pm\frac{3}{2}, \quad \pm\frac{1}{2}$$

or, written in order of increasing size,

$$-3, \quad -\frac{3}{2}, \quad -1, \quad -\frac{1}{2}, \quad \frac{1}{2}, \quad 1, \quad \frac{3}{2}, \quad 3$$

We check each possibility to see whether it is a root. We can start with $\frac{3}{2}$.

$$\frac{3}{2} \begin{array}{rrrr} 2 & 3 & -8 & 3 \\ & 3 & 9 & \frac{3}{2} \\ \hline 2 & 6 & 1 & \frac{9}{2} \end{array}$$

Since the remainder is not 0, $\frac{3}{2}$ is not a root and can be crossed off the list. Since every number in the last row of the synthetic division is positive, $\frac{3}{2}$ is an upper bound. So 3 cannot be a root and can be crossed off the list as well.

$$-3, \quad -\frac{3}{2}, \quad -1, \quad -\frac{1}{2}, \quad \frac{1}{2}, \quad 1, \quad \cancel{\frac{3}{2}}, \quad \cancel{3}$$

We now try $\frac{1}{2}$:

$$\frac{1}{2} \begin{array}{rrrr} 2 & 3 & -8 & 3 \\ & 1 & 2 & -3 \\ \hline \mathbf{2} & \mathbf{4} & \mathbf{-6} & 0 \end{array}$$

Since the remainder is 0, $\frac{1}{2}$ is a root. The binomial $x - \frac{1}{2}$ is a factor of $P(x)$, and any remaining roots must be supplied by the remaining factor, which is the quotient $\mathbf{2x^2 + 4x - 6}$. We can find the other roots by solving the equation $2x^2 + 4x - 6 = 0$, called the **depressed equation.**

$$\begin{array}{ll} 2x^2 + 4x - 6 = 0 & \\ x^2 + 2x - 3 = 0 & \text{Divide both sides by 2.} \\ (x - 1)(x + 3) = 0 & \text{Factor } x^2 + 2x - 3. \\ x - 1 = 0 \quad \text{or} \quad x + 3 = 0 & \\ x = 1 \quad | \quad\quad x = -3 & \end{array}$$

The solution set of the equation is $\left\{\frac{1}{2}, 1, -3\right\}$. Note that two positive roots and one negative root is a predicted possibility.

Self Check   Solve: $3x^3 - 10x^2 + 9x - 2 = 0$.   ∎

Accent on Technology   CONFIRMING ROOTS OF AN EQUATION

We can confirm that the roots found in Example 2 are correct by graphing the function $P(x) = 2x^3 + 3x^2 - 8x + 3$ and locating the resulting x-intercepts of the graph. If we use a graphing window of $x = [-4, 4]$ and $y = [-20, 20]$ and graph the function, we will obtain the graph shown in Figure 11-1. From the graph, we can see that the x-intercepts are at $x = -3$, $x = \frac{1}{2}$, and $x = 1$. These are the roots of the equation.

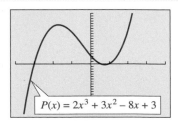

$P(x) = 2x^3 + 3x^2 - 8x + 3$

Figure 11-1

EXAMPLE 3   Solve: $P(x) = x^7 - 2x^6 - 5x^5 + 6x^4 - x^3 + 2x^2 + 5x - 6 = 0$.

Solution   Because the equation is of seventh degree, it has 7 roots. By Descartes' rule of signs, there are six possible combinations of positive, negative, and nonreal roots.

Number of positive roots	Number of negative roots	Number of nonreal roots
5	2	0
3	2	2
1	2	4
5	0	2
3	0	4
1	0	6

The possible rational roots are

$$-6, \quad -3, \quad -2, \quad -1, \quad 1, \quad 2, \quad 3, \quad 6$$

We check each one to eliminate those that don't satisfy the equation, beginning with -3.

$$
\begin{array}{r|rrrrrrr}
-3 & 1 & -2 & -5 & 6 & -1 & 2 & 5 & -6 \\
   &   & -3 & 15 & -30 & 72 & -213 & 633 & -1{,}914 \\
\hline
   & 1 & -5 & 10 & -24 & 71 & -211 & 638 & -1{,}920
\end{array}
$$

Since the last number in the synthetic division is not 0, -3 is not a root and can be crossed off the list. Since the signs in the last row alternate, -3 is a lower bound, and we can cross off -6 as well.

$$-\cancel{6}, \quad -\cancel{3}, \quad -2, \quad -1, \quad 1, \quad 2, \quad 3, \quad 6$$

We now try -2:

$$
\begin{array}{r|rrrrrrr}
-2 & 1 & -2 & -5 & 6 & -1 & 2 & 5 & -6 \\
   &   & -2 & 8 & -6 & 0 & 2 & -8 & 6 \\
\hline
   & 1 & -4 & 3 & 0 & -1 & 4 & -3 & 0
\end{array}
$$

Since the remainder is 0, -2 is a root.

Because this root is negative, we can revise the chart of possibilities to eliminate the possibility that there are 0 negative roots.

Number of positive roots	Number of negative roots	Number of nonreal roots
5	2	0
3	2	2
1	2	4

Because -2 is a root, the factor theorem states that $x - (-2)$, or $x + 2$, is a factor of $P(x)$. Any remaining roots can be found by solving the depressed equation

$$x^6 - 4x^5 + 3x^4 - x^2 + 4x - 3 = 0$$

Because the constant term of this equation is different from the constant term of the original equation, we can cross off other possible rational roots. The number -2 cannot be a root a second time, because it is not a factor of -3. The numbers 2 and 6 are no longer possible roots, because neither is a factor of -3.

The list of possible roots is now

$$-\cancel{6}, \quad -\cancel{3}, \quad -\cancel{2}, \quad -1, \quad 1, \quad \cancel{2}, \quad 3, \quad \cancel{6}$$

Since we know that there is one more negative root, we will synthetically divide the coefficients of the depressed equation by -1.

$$
\begin{array}{r|rrrrrr}
-1 & 1 & -4 & 3 & 0 & -1 & 4 & -3 \\
   &   & -1 & 5 & -8 & 8 & -7 & 3 \\
\hline
   & 1 & -5 & 8 & -8 & 7 & -3 & 0
\end{array}
$$

Since the remainder is 0, -1 is a root, and the solution set so far is $\{-2, -1, \ldots\}$. The number -1 cannot be a root again, because we have found both negative roots. When we cross off -1, we have only two possibilities left.

$$-\cancel{6}, \quad -\cancel{3}, \quad -\cancel{2}, \quad -\cancel{1}, \quad 1, \quad \cancel{2}, \quad 3, \quad \cancel{6}$$

The depressed equation is now $x^5 - 5x^4 + 8x^3 - 8x^2 + 7x - 3 = 0$. We can synthetically divide the coefficients of this equation by 1 to get

$$
\begin{array}{r|rrrrrr}
1 & 1 & -5 & 8 & -8 & 7 & -3 \\
  &   & 1 & -4 & 4 & -4 & 3 \\
\hline
  & 1 & -4 & 4 & -4 & 3 & 0
\end{array}
$$

The depressed equation is now $x^4 - 4x^3 + 4x^2 - 4x + 3 = 0$.

Thus, 1 joins the solution set $\{-2, -1, 1, \ldots\}$. To see whether 1 is a root a second time, we synthetically divide the coefficients of the new depressed equation by 1.

$$
\begin{array}{r|rrrrr}
1 & 1 & -4 & 4 & -4 & 3 \\
  &   & 1 & -3 & 1 & -3 \\
\hline
  & 1 & -3 & 1 & -3 & 0
\end{array}
$$

The depressed equation is now $x^3 - 3x^2 + x - 3 = 0$.

Again, 1 is a root, and the solution set is now $\{-2, -1, 1, 1, \ldots\}$. To see whether 1 is a root a third time, we synthetically divide the coefficients of the new depressed equation by 1.

$$
\begin{array}{r|rrrr}
1 & 1 & -3 & 1 & -3 \\
  &   & 1 & -2 & -1 \\
\hline
  & 1 & -2 & -1 & -4
\end{array}
$$

Since the remainder is not 0, the number 1 is not a root for a third time, and we can cross 1 off the list of possibilities, leaving only 3.

$$-\cancel{6}, \quad -\cancel{3}, \quad -\cancel{2}, \quad -\cancel{1}, \quad \cancel{1}, \quad \cancel{2}, \quad 3, \quad \cancel{6}$$

To see whether 3 is a root, we synthetically divide the coefficients of $x^3 - 3x^2 + x - 3 = 0$ by 3.

$$
\begin{array}{r|rrrr}
3 & 1 & -3 & 1 & -3 \\
  &   & 3 & 0 & 3 \\
\hline
  & 1 & 0 & 1 & 0
\end{array}
$$

Since the remainder is 0, 3 joins the solution set, which is now $\{-2, -1, 1, 1, 3 \ldots\}$.

The depressed equation is now $x^2 + 1 = 0$, which can be solved as a quadratic equation.

$$x^2 + 1 = 0$$
$$x^2 = -1$$
$$x = i \quad \text{or} \quad x = -i$$

The complete solution set is $\{-2, -1, 1, 1, 3, i, -i\}$. The solution set contains 3 positive roots, 2 negative roots, and 2 nonreal roots that are complex conjugates. This combination was one of the predicted possibilities. ∎

CONFIRMING ROOTS OF AN EQUATION

We can confirm that the roots found in Example 3 are correct by graphing the function $P(x) = x^7 - 2x^6 - 5x^5 + 6x^4 - x^3 + 2x^2 + 5x - 6$ and locating the resulting x-intercepts of the graph. If we use a graphing window of $x = [-4, 4]$ and $y = [-130, 50]$ and graph the function, we will obtain the graph shown in Figure 11-2. From the graph, we can see that the x-intercepts are at $x = -2, x = -1, x = 1$, and $x = 3$. We cannot detect the complex roots from the graph.

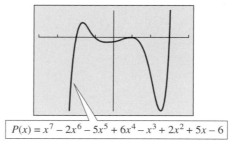

$$P(x) = x^7 - 2x^6 - 5x^5 + 6x^4 - x^3 + 2x^2 + 5x - 6$$

Figure 11-2

An Application

EXAMPLE 4   To protect cranberry crops from the damage of early freezes, growers flood the cranberry bogs. Three irrigation sources, used together, can flood a cranberry bog in one day. If the sources are used one at a time, the second source requires one day longer to flood the bog than the first, and the third requires four days longer than the first. If the bog must be flooded before a freeze that is predicted in three days, can the water in the last two sources be diverted to other bogs?

Solution   Let x represent the number of days it would take the first irrigation source to flood the bog. Then $x + 1$ and $x + 4$ represent the number of days it would take the second and third sources, respectively, to flood the bog.

Because the first source, alone, requires x days to flood the bog, that source could fill $\frac{1}{x}$ of the bog in one day. In one day's time, the remaining sources could flood $\frac{1}{x+1}$ and $\frac{1}{x+4}$ of the bog. This gives the equation

The part of the bog the first source can flood in one day	plus	the part the second source can flood in one day	plus	the part the third source can flood in one day	equals	one bog.
$\dfrac{1}{x}$	$+$	$\dfrac{1}{x+1}$	$+$	$\dfrac{1}{x+4}$	$=$	1

We multiply both sides of the equation by $x(x + 1)(x + 4)$ to clear it of fractions and then simplify to get

$$x(x + 1)(x + 4)\left(\frac{1}{x} + \frac{1}{x + 1} + \frac{1}{x + 4}\right) = 1 \cdot x(x + 1)(x + 4)$$

$$(x + 1)(x + 4) + x(x + 4) + x(x + 1) = x(x + 1)(x + 4)$$

$$x^2 + 5x + 4 + x^2 + 4x + x^2 + x = x^3 + 5x^2 + 4x$$

$$0 = x^3 + 2x^2 - 6x - 4$$

To solve the equation $x^3 + 2x^2 - 6x - 4 = 0$, we first list its possible rational roots, which are the factors of the constant term, -4.

$$-4, \quad -2, \quad -1, \quad 1, \quad 2, \quad \text{and} \quad 4$$

One solution of this equation is $x = 2$, because when we synthetically divide by 2, the remainder is 0:

$$\begin{array}{r|rrrr} 2 & 1 & 2 & -6 & -4 \\ & & 2 & 8 & 4 \\ \hline & 1 & 4 & 2 & 0 \end{array}$$

We find the remaining solutions by using the quadratic formula to solve the depressed equation, which is $x^2 + 4x + 2 = 0$. The two solutions are $-2 + \sqrt{2}$ and $-2 - \sqrt{2}$. Since these numbers are negative and the time it takes to flood the bog cannot be negative, these roots must be discarded. The only meaningful solution is 2.

Since the first source, alone, can flood the bog in two days, and it is three days until the freeze, the other two water sources can be diverted to flood other bogs. ∎

Self Check Answers

1. $\pm\frac{1}{1}$ or $\pm\frac{3}{1}$ do not satisfy $x^2 - 3 = 0$ **2.** $1, 2, \frac{1}{3}$

Orals *Find all rational roots of each equation.*

1. $x^3 + 2x^2 - 2x - 1 = 0$ **2.** $x^3 - 2x^2 + x - 2 = 0$

11.3 EXERCISES 📼

REVIEW *Simplify each radical expression. Assume that all variables represent positive numbers.*

1. $\sqrt{72a^3b^5c}$

2. $\dfrac{5a}{\sqrt{5a}}$

3. $\sqrt{18a^2b} + a\sqrt{50b}$

4. $\left(\sqrt{3} + \sqrt{5b}\right)\left(\sqrt{12} - \sqrt{45b}\right)$

5. $\dfrac{2}{\sqrt{3} - 1}$

6. $\dfrac{\sqrt{11} + \sqrt{x}}{\sqrt{11} - \sqrt{x}}$

VOCABULARY AND CONCEPTS *Fill in the blanks.*

7. The rational roots of the equation $3x^3 + 4x - 7 = 0$ will have the form $\frac{p}{q}$, where p is a factor of ___ and q is a factor of 3.

8. The rational roots of the equation $5x^3 + 3x^2 - 4 = 0$ will have the form $\frac{p}{q}$, where p is a factor of -4 and q is a factor of ___.

9. Consider the synthetic division of $5x^3 - 7x^2 - 3x - 63 = 0$ by 3.

$$\begin{array}{r|rrrr} 3 & 5 & -7 & -3 & -63 \\ & & 15 & 24 & 63 \\ \hline & 5 & 8 & 21 & 0 \end{array}$$

Since the remainder is 0, 3 is a ___ of the equation.

10. In Exercise 9, the depressed equation is

_____.

PRACTICE *Find all rational roots of each equation.*

11. $x^3 - 5x^2 - x + 5 = 0$

12. $x^3 + 7x^2 - x - 7 = 0$

13. $x^3 - 2x^2 - 9x + 18 = 0$

14. $x^3 + 3x^2 - 4x - 12 = 0$

15. $x^3 - 2x^2 - x + 2 = 0$

16. $x^3 + 2x^2 - x - 2 = 0$

17. $x^4 - 10x^3 + 35x^2 - 50x + 24 = 0$

18. $x^4 + 4x^3 + 6x^2 + 4x + 1 = 0$

19. $x^4 + 3x^3 - 13x^2 - 9x + 30 = 0$

20. $x^4 - 8x^3 + 14x^2 + 8x - 15 = 0$

21. $x^5 + 3x^4 - 5x^3 - 15x^2 + 4x + 12 = 0$

22. $x^5 - 3x^4 - 5x^3 + 15x^2 + 4x - 12 = 0$

23. $x^7 - 12x^5 + 48x^3 - 64x = 0$

24. $x^7 + 7x^6 + 21x^5 + 35x^4 + 35x^3 + 21x^2 + 7x + 1 = 0$

25. $3x^3 - 2x^2 + 12x - 8 = 0$

26. $4x^4 - 8x^3 - x^2 + 8x - 3 = 0$

27. $3x^4 - 14x^3 + 11x^2 + 16x - 12 = 0$

28. $2x^4 - x^3 - 2x^2 - 4x - 40 = 0$

29. $12x^4 + 20x^3 - 41x^2 + 20x - 3 = 0$

30. $4x^5 - 12x^4 + 15x^3 - 45x^2 - 4x + 12 = 0$

31. $6x^5 - 7x^4 - 48x^3 + 81x^2 - 4x - 12 = 0$

32. $36x^4 - x^2 + 2x - 1 = 0$

33. $30x^3 - 47x^2 - 9x + 18 = 0$

34. $20x^3 - 53x^2 - 27x + 18 = 0$

35. $15x^3 - 61x^2 - 2x + 24 = 0$

36. $12x^4 + x^3 + 42x^2 + 4x - 24 = 0$

37. $20x^3 - 44x^2 + 9x + 18 = 0$

38. $24x^3 - 82x^2 + 89x - 30 = 0$

Find the other roots given that $1 + i$ is a root of each equation.

39. $x^3 - 5x^2 + 8x - 6 = 0$

40. $x^3 - 2x + 4 = 0$

41. $x^4 - 2x^3 - 7x^2 + 18x - 18 = 0$

42. $x^4 - 2x^3 - 2x^2 + 8x - 8 = 0$

Solve each equation.

43. $x^3 - \dfrac{4}{3}x^2 - \dfrac{13}{3}x - 2 = 0$

44. $x^3 - \dfrac{19}{6}x^2 + \dfrac{1}{6}x + 1 = 0$

45. $x^{-5} - 8x^{-4} + 25x^{-3} - 38x^{-2} + 28x^{-1} - 8 = 0$

46. $1 - x^{-1} - x^{-2} - 2x^{-3} = 0$

APPLICATIONS

47. Parallel resistance If three resistors with resistances of R_1, R_2, and R_3 are wired in parallel, their combined resistance R is given by the formula

$$\frac{1}{R} = \frac{1}{R_1} + \frac{1}{R_2} + \frac{1}{R_3}$$

The design of a voltmeter requires that the resistance R_2 be 10 ohms greater than the resistance R_1, that the resistance R_3 be 50 ohms greater than R_1, and that their combined resistance be 6 ohms. Find the value of each resistance.

48. Fabricating sheet metal The open tray in the illustration is to be manufactured from a 12-by-14-inch rectangular sheet of metal by cutting squares from each corner and folding up the sides. If the volume of the tray is to be 160 cubic inches, what size squares should be cut from each corner?

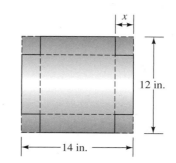

WRITING

49. If n is a positive even integer and c is a positive constant, show that $x^n + c = 0$ has no real roots.

50. If n is a positive even integer and c is a positive constant, show that $x^n - c = 0$ has two real roots.

SOMETHING TO THINK ABOUT

51. Precalculus   A rectangle is inscribed in the parabola $y = 16 - x^2$, as shown in the illustration. Find the point (x, y) if the area of the rectangle is 42 square units.

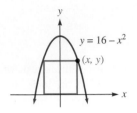

52. Precalculus   One corner of the rectangle in the illustration is at the origin, and the opposite corner (x, y) lies in the first quadrant on the curve $y = x^3 - 2x^2$. Find the point (x, y) if the area of the rectangle is 27 square units.

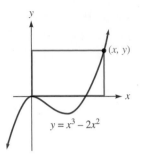

11.4   Irrational Roots of Polynomial Equations

▌ **The Intermediate Value Theorem**   ▌ **The Bisection Method**
▌ **Finding Roots with a Graphing Calculator**

Getting Ready   *Tell whether the value of $x^2 + 5x - 6$ is positive or negative for each value of x.*

1. $x = 2$ **2.** $x = -3$

First-degree equations are easy to solve, and all quadratic equations can be solved with the quadratic formula. There are formulas for solving third- and fourth-degree polynomial equations, although they are complicated. However, there are no formulas for solving polynomial equations of degree 5 or greater. This fact was proved by the Norwegian mathematician Niels Henrik Abel (1802–1829) and (for equations of degree greater than 5) by the French mathematician Evariste Galois (1811–1832).

To solve a higher-degree polynomial equation with integer coefficients, we can use the methods of the previous section to find its rational roots. Once we find them, the final depressed equation would have to be a first- or second-degree equation to enable us to complete the solution. The purpose of this section is to discuss ways of approximating irrational roots of these higher-degree polynomial equations.

The Intermediate Value Theorem

The following theorem, called the **intermediate value theorem,** will lead to a way of locating an interval that contains a root.

The Intermediate Value Theorem

Let $P(x)$ be a polynomial with real coefficients. If $P(a) \neq P(b)$ for $a < b$, then $P(x)$ takes on all values between $P(a)$ and $P(b)$ on the closed interval $[a, b]$.

Justification This theorem becomes clear when we consider the graph of the polynomial $y = P(x)$, shown in Figure 11-3. We have seen that graphs of polynomials are *continuous* curves, a technical term that means, roughly, that they can be drawn without lifting the pencil from the paper. If $P(a) \neq P(b)$, then the continuous curve joining the points $A(a, P(a))$ and $B(b, P(b))$ must take on all values between $P(a)$ and $P(b)$ in the interval $[a, b]$, because the curve has no gaps in it.

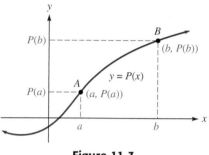

Figure 11-3

The next theorem follows from the intermediate value theorem.

The Location Theorem

Let $P(x)$ be a polynomial with real coefficients. If $P(a)$ and $P(b)$ have opposite signs, there is at least one number r in the interval (a, b) for which $P(r) = 0$.

Proof See Figure 11-4. By the intermediate value theorem, $P(x)$ takes on all values between $P(a)$ and $P(b)$. Since $P(a)$ and $P(b)$ have opposite signs, the number 0 lies between them. Thus, there is a number r between a and b for which $P(r) = 0$. This number r is a zero of $P(x)$, and a root of the equation $P(x) = 0$.

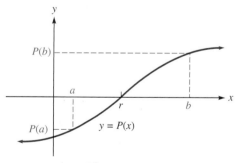

Figure 11-4

The Bisection Method

The location theorem provides a method (called the **bisection method**) for finding the roots of $P(x) = 0$ to any desired degree of accuracy.

Suppose we find, by trial and error, that the numbers x_l and x_r (for left and right) straddle a root—that is, $x_l < x_r$ and $P(x_l)$ and $P(x_r)$ have opposite signs. Further suppose that $P(x_l) < 0$ and $P(x_r) > 0$. We can compute the number c that is halfway between x_l and x_r and then compute $P(c)$. If $P(c) = 0$, we've found a root. However, if $P(c)$ is not 0, we proceed in one of two ways:

1. If $P(c) < 0$, the root r lies between c and x_r, as shown in Figure 11-5(a). In this case, we let c become a new x_l and repeat the procedure.

2. If $P(c) > 0$, the root r lies between x_l and c, as shown in Figure 11-5(b). In this case, we let c become a new x_r and repeat the procedure.

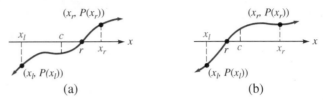

(a)   (b)

Figure 11-5

At any stage in this procedure, the root is contained between the current values of x_l and x_r. If the original bounds were 1 unit apart, after 10 repetitions of this procedure, the root would be between bounds that were 2^{-10} units apart. This means that the zero of $P(x)$ would be within 0.0001 units of either x_l or x_r.

After 20 repetitions, the root would be between bounds that were 2^{-20} units apart. This means that the zero of $P(x)$ would be within 0.000001 of either x_l or x_r.

EXAMPLE 1   Solve $x^2 - 2 = 0$ and express the positive root to the nearest tenth.

Solution   Since $P(1) = -1$ and $P(2) = 2$ have opposite signs, there is a root between 1 and 2. We can let $x_l = 1$ and $x_r = 2$ and compute the midpoint c:

$$c = \frac{x_l + x_r}{2} = \frac{1 + 2}{2} = 1.5$$

Because $P(c) = P(1.5) = 0.25$ is a positive number, we let c become a new x_r and calculate a new midpoint, which we will call c_1.

$$c_1 = \frac{x_l + x_r}{2} = \frac{1 + 1.5}{2} = 1.25$$

Because $P(c_1) = P(1.25) = -0.4375$ is a negative number, we let c_1 become a new x_l and calculate a new midpoint, which we will call c_2.

$$c_2 = \frac{x_l + x_r}{2} = \frac{1.25 + 1.5}{2} = 1.375$$

Because $P(c_2) = P(1.375) = -0.109375$ is a negative number, we let c_2 become a new x_l and calculate a new midpoint, which we will call c_3.

$$c_3 = \frac{x_l + x_r}{2} = \frac{1.375 + 1.5}{2} = 1.4375$$

Because $P(c_3) = P(1.4375) = 0.066406$ is a positive number, we let c_3 become a new x_r and calculate a new midpoint, which we will call c_4.

$$c_4 = \frac{x_l + x_r}{2} = \frac{1.375 + 1.4375}{2} = 1.40625$$

From here on, the first two digits of the midpoints will remain 1.4. The root r of the equation (to the nearest tenth) is 1.4.

Self Check Find the negative root of $x^2 - 2 = 0$ to the nearest tenth. ∎

Finding Roots with a Graphing Calculator

We can approximate the roots of an equation either by using the TRACE and ZOOM capabilities of a graphing calculator or by using the ZERO feature.

Accent on Technology SOLVING EQUATIONS

To use the TRACE and ZOOM capabilities of a graphing calculator to solve the equation $x^4 - 6x^2 + 9 = 0$, we graph the function $y = x^4 - 6x^2 + 9$ and trace and zoom to find the x-intercepts. If we use window settings of $[-6, 6]$ for x and $[-2, 10]$ for y and graph the function, we will obtain Figure 11-6(a). If we trace and move the cursor near the positive x-intercept, we will obtain Figure 11-6(b). If we zoom in and trace, we will obtain Figure 11-6(c). To get better results, we can zoom in and trace again to obtain Figure 11-6(d), which shows that the x-coordinate of the x-intercept is approximately 1.731383. After zooming in and tracing a few more times, we will see that to three decimal places, $x = 1.732$. By symmetry, we know that the negative solution is -1.732.

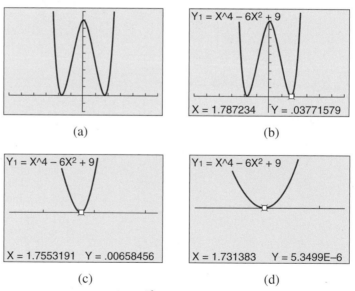

(a)

(b)

(c)

(d)

Figure 11-6

To use the ZERO feature, we first graph the equation to obtain Figure 11-7(a). We then select ZERO under the CALC menu to obtain Figure 11-7(b). To find the positive x-intercept, we guess a left bound of 0 and press ENTER, guess a right bound of 4 and press ENTER, and press ENTER again to obtain Figure 11-7(c), which shows that the x-coordinate of the x-intercept is approximately 1.732051.

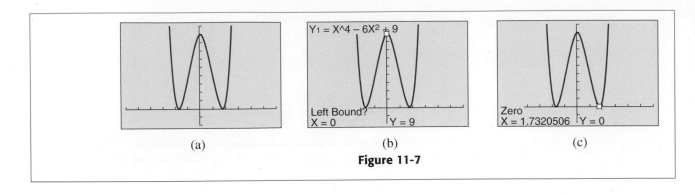

Figure 11-7

The graph shown in Figure 11-7(a) illustrates a situation in which the bisection method does not work. Since the graph doesn't cross the x-axis, the values of the function on opposite sides of the intercepts do not have opposite signs. In this case, the bisection method would be useless.

Self Check Answer

1. -1.4

Orals *Find the mean of each pair of numbers.*

1. $2, 5$ **2.** $2, 3.5$ **3.** $-1, 4$ **4.** $-1, 1.5$

11.4   EXERCISES

REVIEW *Give the equations of all of the asymptotes of each rational function.* ***Don't graph the functions.***

1. $y = \dfrac{5x - 3}{x^2 - 4}$ **2.** $y = \dfrac{5x^2 - 3}{x^2 - 25}$

3. $y = \dfrac{x^2}{x - 2}$ **4.** $y = \dfrac{x + 1}{x^2 + 1}$

VOCABULARY AND CONCEPTS *Fill in the blanks.*

5. If $P(x)$ is a polynomial with real coefficients and $P(a) \neq P(b)$ for $a < b$, then $P(x)$ takes on all values between _____ in the interval $[a, b]$.

6. If $P(x)$ has real coefficients and $P(a)$ and $P(b)$ have opposite signs, there is at least one number r in (a, b) for which _____.

7. In the bisection method, we find two numbers x_l and x_r that straddle a root. We then compute a new guess for the root (a number c) by finding the average of _____.

8. A _____ curve has no gaps in it.

PRACTICE *Show that each equation has at least one real root between the specified numbers.*

9. $2x^2 + x - 3 = 0$; -2 and -1

10. $2x^3 + 17x^2 + 31x - 20 = 0$; -1 and 2

11. $3x^3 - 11x^2 - 14x = 0$; 4 and 5

12. $2x^3 - 3x^2 + 2x - 3 = 0$; 1 and 2

13. $x^4 - 8x^2 + 15 = 0$; 1 and 2

14. $x^4 - 8x^2 + 15 = 0$; 2 and 3

15. $30x^3 + 10 = 61x^2 + 39x$; 2 and 3

16. $30x^3 + 10 = 61x^2 + 39x$; -1 and 0

17. $30x^3 + 10 = 61x^2 + 39x$; 0 and 1

18. $5x^3 - 9x^2 - 4x + 9 = 0$; -1 and 2

Use the bisection method to find the following values to the nearest tenth.

19. The positive root of $x^2 - 3 = 0$.
20. The negative root of $x^2 - 3 = 0$.
21. The negative root of $x^2 - 5 = 0$.
22. The positive root of $x^2 - 5 = 0$.
23. The positive root of $x^3 - x^2 - 2 = 0$.
24. The negative root of $x^3 - x + 2 = 0$.
25. The negative root of $3x^4 + 3x^3 - x^2 - 4x - 4 = 0$.

26. The positive root of
$x^5 + x^4 - 4x^3 - 4x^2 - 5x - 5 = 0$.

Use a graphing calculator to find the distinct real solutions of each equation to the nearest tenth. Which roots, if any, would the bisection method fail to find?

27. $x^2 - 5 = 0$
28. $x^2 - 10x + 25 = 0$

29. $x^3 - 5x^2 + 8x - 4 = 0$

30. $x^3 - 5x^2 - 2x + 10 = 0$

APPLICATIONS

31. Precalculus Use the bisection method or a graphing calculator to find the coordinates of the two points on the graph of $y = x^3$ that lie 1 unit from the origin. (See the illustration.) Give the result to the nearest hundredth.

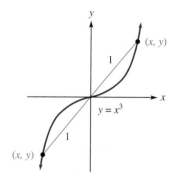

32. Building a crate The width of the shipping crate shown in the illustration is to be 2 feet greater than its height and the length is 5 feet greater than its width, and its volume is to be 170 cubic feet. Use the bisection method or a graphing calculator to find the height of the crate to the nearest tenth of a foot.

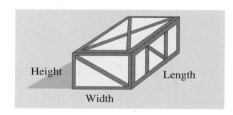

WRITING

33. Explain the intermediate value theorem.

34. Explain the location theorem.

SOMETHING TO THINK ABOUT

35. Explain when the bisection method will not work.

36. Describe how you could tell from a graph that a polynomial equation has no real roots.

Projects

Project 1

Find the dimensions of the right-triangular brace.

Project 2

The bisection method is one of several techniques for finding approximate solutions of equations. Another is **Newton's method,** discovered by Sir Isaac Newton (1642–1727).

The process begins with a guess of a solution and then transforms that guess into a better estimate of the solution. When Newton's method is applied to that estimate, a third solution results—one more accurate than any previous estimate. As the method is applied repeatedly, the results converge on the actual solution.

To solve the quadratic equation $x^2 + x - 3 = 0$, for example, Newton's method uses the fraction

$$\frac{x^2 + 3}{2x + 1}$$

to generate better solutions. To approximate the positive solution, we begin with a guess (3, for example) and substitute 3 for x in the fraction.

$$\frac{x^2 + 3}{2x + 1} = \frac{3^2 + 3}{2(3) + 1} \approx 1.714286$$

The number 1.714286 is a better estimate of the solution than the original guess, 3. We apply Newton's method again and substitute 1.714286 for x.

$$\frac{x^2 + 3}{2x + 1} = \frac{1.714286^2 + 3}{2(1.714286) + 1} \approx 1.341014$$

The estimate 1.341014 is better yet. More passes produce even better accuracy.

- Continue with Newton's method until your results remain unchanged to six decimal places.
- Solve the equation using the quadratic formula. Is Newton's method correct?

To solve the equation $ax^2 + bx + c = 0$, Newton's method uses the fraction

$$\frac{ax^2 - c}{2ax + b}$$

- Solve $3x^2 - 5x - 2 = 0$ using Newton's method. Find two solutions, accurate to six decimal places.

To solve a third-degree equation $ax^3 + bx^2 + cx + d = 0$, Newton's method uses the fraction

$$\frac{2ax^3 + bx^2 - d}{3ax^2 + 2bx + c}$$

- Use this method to find three solutions of $x^3 - x^2 - 5x + 5 = 0$, accurate to six decimal places.

Chapter Summary

CONCEPTS	REVIEW EXERCISES
	11.1 The Remainder and Factor Theorems; Synthetic Division

The remainder theorem:
If $P(x)$ is a polynomial, r is any number, and $P(x)$ is divided by $x - r$, the remainder is $P(r)$.

The factor theorem:
Let $P(x)$ be any polynomial and r any number. Then $P(r) = 0$ if and only if $x - r$ is a factor of $P(x)$.

Let $P(x) = 4x^4 + 2x^3 - 3x^2 - 2$. Find the remainder when $P(x)$ is divided by each binomial.

1. $x - 1$ **2.** $x - 2$

3. $x + 3$ **4.** $x + 2$

Use the factor theorem to decide whether each statement is true.

5. $x - 2$ is a factor of $x^3 + 4x^2 - 2x + 4$.

6. $x + 3$ is a factor of $2x^4 + 10x^3 + 4x^2 + 7x + 21$.

7. $x - 5$ is a factor of $x^5 - 3{,}125$.

8. $x - 6$ is a factor of $x^5 - 6x^4 - 4x + 24$.

Find the polynomial of lowest degree with integer coefficients and the given zeros.

9. $-1, 2$, and $\dfrac{3}{2}$ **10.** $1, -3$, and $\dfrac{1}{2}$

11. $2, -5, i$, and $-i$ **12.** $-3, 2, i$, and $-i$

Use synthetic division to find the quotient and remainder when each polynomial is divided by the given divisor.

13. $3x^4 + 2x^2 + 3x + 7; x - 3$

14. $2x^4 - 3x^2 + 3x - 1; x - 2$

15. $5x^5 - 4x^4 + 3x^3 - 2x^2 + x - 1; x + 2$

16. $4x^5 + 2x^4 - x^3 + 3x^2 + 2x + 1; x + 1$

| 11.2 | Descartes' Rule of Signs and Bounds on Roots |

The fundamental theorem of algebra:

If $P(x)$ is a polynomial with positive degree, then $P(x)$ has at least one zero.

The polynomial equation $P(x) = 0$ with degree n ($n > 0$) has exactly n roots among the complex numbers.

Complex roots of polynomial equations with real coefficients occur in complex conjugate pairs.

Descartes' rule of signs:

If $P(x)$ has real coefficients, the number of positive roots of $P(x) = 0$ is equal to the number of variations in sign of $P(x)$, or is less than that by an even number.

The number of negative roots is equal to the number of variations in sign of $P(-x)$, or is less than that by an even number.

Upper bounds:

Let the lead coefficient of the polynomial $P(x)$ with real coefficients be positive, and do a synthetic division of the coefficients of $P(x)$ by the positive number c. If none of the terms in the bottom row is negative, then c is an upper bound for the real roots of $P(x) = 0$.

Lower bounds:

Let the lead coefficient of the polynomial $P(x)$ with real coefficients be positive, and do a synthetic division of the coefficients of $P(x)$ by the negative number d. If the signs in the bottom row alternate, then d is a lower bound for the real roots of $P(x) = 0$.

How many roots does each equation have?

17. $3x^6 - 4x^5 + 3x + 2 = 0$

18. $2x^6 - 5x^4 + 5x^3 - 4x + x - 12 = 0$

19. $3x^{65} - 4x^{50} + 3x^{17} + 2x = 0$

20. $x^{1,984} - 12 = 0$

Find another root of a polynomial equation with real coefficients if the given quantity is one root.

21. $2 + i$ **22.** $-i$

Find the number of possible positive, negative, and nonreal roots for each equation. **Do not attempt to solve the equation.**

23. $3x^4 + 2x^3 - 4x + 2 = 0$

24. $2x^4 - 3x^3 + 5x^2 + x - 5 = 0$

25. $4x^5 + 3x^4 + 2x^3 + x^2 + x = 7$

26. $3x^7 - 4x^5 + 3x^3 + x - 4 = 0$

27. $x^4 + x^2 + 24{,}567 = 0$

28. $-x^7 - 5 = 0$

Find integer bounds for the roots of each equation.

29. $5x^3 - 4x^2 - 2x + 4 = 0$

30. $x^4 + 3x^3 - 5x^2 - 9x + 1 = 0$

11.3	Rational Roots of Polynomial Equations

If $P(x) = 0$ has integer coefficients and $\frac{p}{q}$ (written in lowest terms) is a root, then p is a factor of the constant, and q is a factor of the lead coefficient.

Find all rational roots of each equation.

31. $2x^3 + 17x^2 + 41x + 30 = 0$

32. $3x^3 + 2x^2 + 2x - 1 = 0$

33. $4x^4 - 25x^2 + 36 = 0$

34. $2x^4 - 11x^3 - 6x^2 + 64x + 32 = 0$

11.4	Irrational Roots of Polynomial Equations

The intermediate value theorem:
Let $P(x)$ be a polynomial with real coefficients. If $P(a) \neq P(b)$ for $a < b$, then $P(x)$ takes on all values between $P(a)$ and $P(b)$ on the closed interval $[a, b]$.

Location theorem:
Let $P(x)$ be a polynomial with real coefficients. If $P(a)$ and $P(b)$ have opposite signs, then there is at least one number r between a and b for which $P(r) = 0$.

Show that each polynomial has a zero between the two given numbers.

35. $5x^3 + 37x^2 + 59x + 18 = 0$; -1 and 0

36. $6x^3 - x^2 - 10x - 3 = 0$; 1 and 2

Use the bisection method to find the positive root of each equation to the nearest tenth.

37. $x^3 - 2x^2 - 9x - 2 = 0$

38. $6x^2 - 13x - 5 = 0$

Use a graphing calculator to find the positive root of each equation to the nearest hundredth.

39. $6x^2 - 7x - 5 = 0$

40. $3x^2 + x - 2 = 0$

41. Designing solar collectors The space available for the installation of three solar collecting panels requires that their lengths differ by the amounts shown, and that the total of their widths be 15 meters. To be equally effective, each panel must measure exactly 60 square meters. Find the dimensions of each panel.

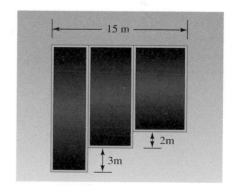

42. Designing a storage tank   The design specifications for a cylindrical storage tank require that its height be 3 feet greater than the radius of its circular base and that the volume of the tank be 19,000 cubic feet. Use the bisection method or a graphing calculator to find the radius of the tank to the nearest hundredth of a foot.

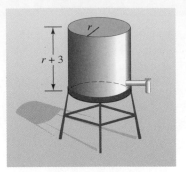

Chapter Test

Use the remainder theorem to find each value.

1. $P(x) = 3x^3 - 9x - 5$; $P(2)$

2. $P(x) = x^5 + 2$; $P(-2)$

Find a polynomial with the given zeros.

3. $5, -1, 0$

4. $i, -i, \sqrt{3}, -\sqrt{3}$

Let $P(x) = 3x^3 - 2x^2 + 4$. Use synthetic division to find each value.

5. $P(1)$

6. $P(-2)$

7. $P\left(-\dfrac{1}{3}\right)$

8. $P(i)$

Use synthetic division to express $P(x) = 2x^3 - 3x^2 - 4x - 1$ in the form (divisor)(quotient) + remainder *for each divisor.*

9. $x - 2$

10. $x + 1$

Use synthetic division to perform each division.

11. $\dfrac{2x^2 - 7x - 15}{x - 5}$

12. $\dfrac{3x^3 + 7x^2 + 2x}{x + 2}$

Write a third-degree polynomial equation with real coefficients and the given roots.

13. $2, i$

14. $1, 2 + i$

Use Descartes' rule of signs to find the number of possible positive, negative, and nonreal roots of each equation.

15. $3x^5 - 2x^4 + 2x^2 - x - 3 = 0$

16. $2x^3 - 5x^2 - 2x - 1 = 0$

Find integer bounds for the roots of each equation.

17. $x^5 - x^4 - 5x^3 + 5x^2 + 4x - 5 = 0$

18. $2x^3 - 11x^2 + 10x + 3 = 0$

Find all roots of the equation.

19. $2x^3 + 3x^2 - 11x - 6 = 0$

Use the bisection method to find the positive root of the equation to the nearest tenth.

20. $x^2 - 11 = 0$

12

Conic Sections and Quadratic Systems

InfoTrac Project

Do a keyword search on "ellipse." Find the article "Steel-drivin' man: Richard Serra's massive new sculptures, as big as houses, create a wholly original spatial drama." Write a summary of the article.

If the center of one of the ellipses is (0, 0), the minor axis is 26.5 feet long, and the major axis is 32.5 feet long, write the equation for the ellipse.

To see a picture of this sculpture visit: http://www.diacenter.org/exhibs_bak/serra/ellipses/index.html

Complete this project after studying Section 12.3.

© Philip Gould/CORBIS

Mathematics in Art

An artist wants to paint a mural on an elliptical background that is 10 feet wide and 6 feet high. To see how to do this construction, see Exercise 53 in Exercise Set 12.3.

Exercise 53
Exercise 12.3

*We have seen that the graphs of linear functions are straight lines, and that the graphs of quadratic functions are parabolas. In this chapter, we will discuss some special curves, called **conic sections.***

12.1 The Circle

▮ **Introduction to the Conic Sections** ▮ **The Circle** ▮ **Problem Solving**

Getting Ready *Square each binomial.*

1. $(x - 2)^2$ **2.** $(x + 4)^2$

What number must be added to each binomial to make it a perfect-square trinomial?

3. $x^2 + 9x$ **4.** $x^2 - 12x$

In this section, we will introduce the conic sections. Then we will discuss the circle in detail.

Introduction to the Conic Sections

The graphs of second-degree equations of the form

$$Ax^2 + Bxy + Cy^2 + Dx + Ey + F = 0$$

were fully investigated in the 17th century by René Descartes (1596–1650) and Blaise Pascal (1623–1662). Descartes discovered that graphs of second-degree equations fall into one of several categories: a point, a pair of lines, a circle, a parabola, an ellipse, a hyperbola, or no graph at all. Because all of these graphs can be formed by the intersection of a plane and a right-circular cone, they are called **conic sections.** See Figure 12-1.

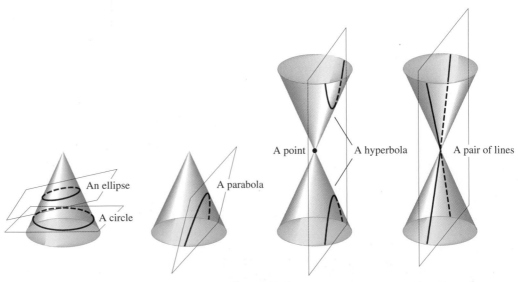

Figure 12-1

Conic sections have many applications. For example, everyone knows the importance of circular wheels and gears, pizza cutters, and ferris wheels.

Parabolas can be rotated to generate dish-shaped surfaces called *paraboloids*. Any light or sound placed at the *focus* of a paraboloid is reflected outward in parallel paths, as shown in Figure 12-2(a). This property makes parabolic surfaces ideal for flashlight and headlight reflectors. It also makes parabolic surfaces good antennas, because signals captured by such antennas are concentrated at the focus. Parabolic mirrors are capable of concentrating the rays of the sun at a single point and thereby generating tremendous heat. This property is used in the design of solar furnaces.

Any object thrown upward and outward travels in a parabolic path, as shown in Figure 12-2(b). In architecture, many arches are parabolic in shape, because this gives strength. Cables that support suspension bridges hang in the form of a parabola. (See Figure 12-2(c).)

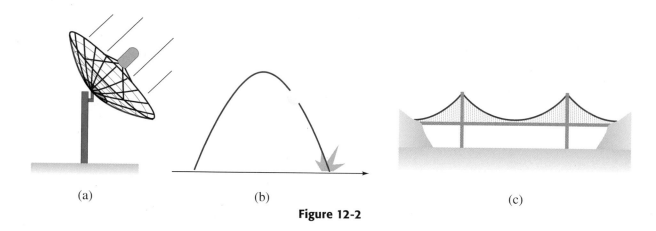

(a)   (b)   (c)

Figure 12-2

Ellipses have optical and acoustical properties that are useful in architecture and engineering. For example, many arches are portions of an ellipse, because the shape is pleasing to the eye. (See Figure 12-3(a).) The planets and many comets have elliptical orbits. (See Figure 12-3(b).) Gears are often cut into elliptical shapes to provide nonuniform motion. (See Figure 12-3(c).)

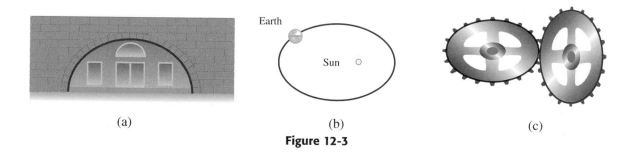

(a)   (b)   (c)

Figure 12-3

Hyperbolas serve as the basis of a navigational system known as LORAN (LOng RAnge Navigation). (See Figure 12-4.) They are also used to find the source of a distress signal, form the basis for the design of hypoid gears, and describe the orbits of some comets.

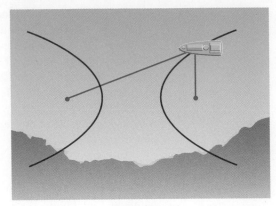

Figure 12-4

The Circle

The Circle

A **circle** is the set of all points in a plane that are a fixed distance from a point called its **center.** The fixed distance is the **radius** of the circle.

To develop the general equation of a circle, we must write the equation of a circle with a radius of r and with center at some point $C(h, k)$, as in Figure 12-5. This task is equivalent to finding all points $P(x, y)$ such that the length of line segment CP is r. We can use the distance formula to find r.

$$r = \sqrt{(x - h)^2 + (y - k)^2}$$

We then square both sides to obtain

(1) $$r^2 = (x - h)^2 + (y - k)^2$$

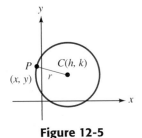

Figure 12-5

Equation 1 is called the **standard form of the equation of a circle** that has a radius of r and center at the point with coordinates (h, k).

Standard Equation of a Circle with Center at (h, k)

Any equation that can be written in the form

$$(x - h)^2 + (y - k)^2 = r^2$$

has a graph that is a circle with radius r and center at point (h, k).

If $r^2 = 0$, the graph reduces to a single point. If $r^2 < 0$, a circle does not exist. If both coordinates of the center are 0, the center of the circle is the origin.

Standard Equation of a Circle with Center at (0, 0)	Any equation that can be written in the form $$x^2 + y^2 = r^2$$ has a graph that is a circle with radius r and center at the origin.

EXAMPLE 1   Find the equation of the circle with radius 5 and center at $C(3, 2)$.

Solution   We substitute 5 for r, 3 for h, and 2 for k in the standard form of a circle and simplify.

$$(x - h)^2 + (y - k)^2 = r^2$$
$$(x - 3)^2 + (y - 2)^2 = 5^2$$
$$x^2 - 6x + 9 + y^2 - 4y + 4 = 25$$
$$x^2 + y^2 - 6x - 4y - 12 = 0$$

The equation is $x^2 + y^2 - 6x - 4y - 12 = 0$.

Self Check   Find the equation of the circle with radius 6 and center at $(2, 3)$.   ∎

EXAMPLE 2   Graph: $x^2 + y^2 = 25$.

Solution   Because this equation can be written in the form $x^2 + y^2 = r^2$, its graph is a circle with center at the origin. Since $r^2 = 25 = 5^2$, the circle has a radius of 5. The graph appears in Figure 12-6.

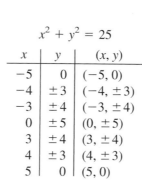

$$x^2 + y^2 = 25$$

x	y	(x, y)
-5	0	$(-5, 0)$
-4	± 3	$(-4, \pm 3)$
-3	± 4	$(-3, \pm 4)$
0	± 5	$(0, \pm 5)$
3	± 4	$(3, \pm 4)$
4	± 3	$(4, \pm 3)$
5	0	$(5, 0)$

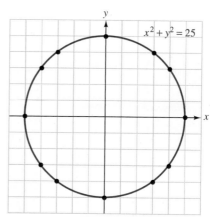

Figure 12-6

Self Check   Graph: $x^2 + y^2 = 4$.   ∎

EXAMPLE 3   Graph: $2x^2 + 2y^2 + 4x + 3y = 3$.

Solution   To identify the curve, we complete the square on x and y and write the equation in standard form.

$$2x^2 + 2y^2 + 4x + 3y = 3$$

$$x^2 + y^2 + 2x + \frac{3}{2}y = \frac{3}{2} \qquad \text{Divide both sides by 2.}$$

$$x^2 + 2x + y^2 + \frac{3}{2}y = \frac{3}{2}$$

To complete the square on x and y, we add 1 and $\frac{9}{16}$ to both sides.

$$x^2 + 2x + \mathbf{1} + y^2 + \frac{3}{2}y + \frac{\mathbf{9}}{\mathbf{16}} = \frac{3}{2} + \mathbf{1} + \frac{\mathbf{9}}{\mathbf{16}}$$

$$(x + 1)^2 + \left(y + \frac{3}{4}\right)^2 = \frac{49}{16}$$

$$[x - (-1)]^2 + \left[y - \left(-\frac{3}{4}\right)\right]^2 = \left(\frac{7}{4}\right)^2$$

We can now see that this result is the equation of a circle that has a radius of $\frac{7}{4}$ and center at $h = -1$ and $k = -\frac{3}{4}$. If we plot the center and draw a circle with a radius of $\frac{7}{4}$ units, we will obtain the circle shown in Figure 12-7.

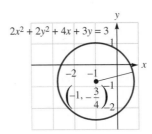

Figure 12-7

Self Check   Write $x^2 + y^2 + 2x - 4y - 11 = 0$ in standard form and graph it.   ∎

Accent on Technology   GRAPHING CIRCLES

Since the graphs of circles fail the vertical line test, their equations do not represent functions. It is somewhat more difficult to use a graphing calculator to graph equations that are not functions. For example, to graph the circle described by $(x - 1)^2 + (y - 2)^2 = 4$, we must split the equation into two functions and graph each one separately. We begin by solving the equation for y.

$$(x - 1)^2 + (y - 2)^2 = 4$$

$$(y - 2)^2 = 4 - (x - 1)^2 \qquad \text{Subtract } (x - 1)^2 \text{ from both sides.}$$

$$y - 2 = \pm\sqrt{4 - (x - 1)^2} \qquad \text{Take the square root of both sides.}$$

$$y = 2 \pm\sqrt{4 - (x - 1)^2} \qquad \text{Add 2 to both sides.}$$

This equation defines two functions. If we use window settings of $[-3, 5]$ for x and $[-3, 5]$ for y and graph the functions

$$y = 2 + \sqrt{4 - (x - 1)^2} \qquad \text{and} \qquad y = 2 - \sqrt{4 - (x - 1)^2}$$

we get the distorted circle shown in Figure 12-8(a). To get a better circle, graphing calculators have a squaring feature that gives an equal unit distance on both the *x*- and *y*-axes. After using this feature, we get the circle shown in Figure 12-8(b).

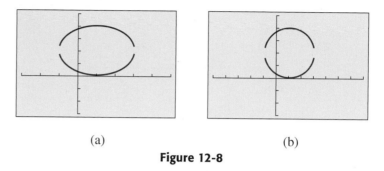

(a)　　　　　　　　(b)

Figure 12-8

Problem Solving

EXAMPLE 4　**Radio translators**　The broadcast area of a television station is bounded by the circle $x^2 + y^2 = 3{,}600$, where x and y are measured in miles. A translator station picks up the signal and retransmits it from the center of a circular area bounded by $(x + 30)^2 + (y - 40)^2 = 1{,}600$. Find the location of the translator and the greatest distance from the main transmitter that the signal can be received.

Solution　The coverage of the television station is bounded by $x^2 + y^2 = 60^2$, a circle centered at the origin with a radius of 60 miles, as shown in Figure 12-9. Because the translator is at the center of the circle $(x + 30)^2 + (y - 40)^2 = 1{,}600$, it is located at $(-30, 40)$, a point 30 miles west and 40 miles north of the television station. The radius of the translator's coverage is $\sqrt{1{,}600}$, or 40 miles.

As shown in Figure 12-9, the greatest distance of reception is the sum of A, the distance from the translator to the television station, and 40 miles, the radius of the translator's coverage.

To find A, we use the distance formula to find the distance between $(x_1, y_1) = (-30, 40)$ and the origin, $(x_2, y_2) = (0, 0)$.

$$A = \sqrt{(x_1 - x_2)^2 + (y_1 - y_2)^2}$$
$$A = \sqrt{(-30 - 0)^2 + (40 - 0)^2}$$
$$= \sqrt{(-30)^2 + 40^2}$$
$$= \sqrt{2{,}500}$$
$$= 50$$

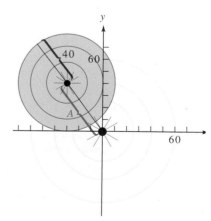

Figure 12-9

The translator is 50 miles from the television station, and it broadcasts the signal an additional 40 miles. The greatest reception distance is 50 + 40, or 90 miles.

Self Check Answers

1. $x^2 + y^2 - 4x - 6y - 23 = 0$ 2.

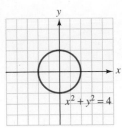

3. $(x + 1)^2 + (y - 2)^2 = 16$

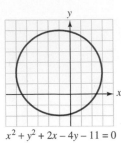

$x^2 + y^2 + 2x - 4y - 11 = 0$

Orals   *Find the center and the radius of each circle.*

1. $x^2 + y^2 = 144$ 2. $x^2 + y^2 = 121$
3. $(x - 2)^2 + y^2 = 16$ 4. $x^2 + (y + 1)^2 = 9$

12.1   EXERCISES

REVIEW   *Insert the number that must be added to make a perfect trinomial square.*

1. $x^2 + 4x +$ ___   2. $y^2 - 12y +$ ___
3. $x^2 - 7x +$ ___   4. $y^2 + 11y +$ ___

Solve each equation by completing the square.

5. $x^2 + 4x = 5$ 6. $y^2 - 12y = 13$

7. $x^2 - 7x - 18 = 0$ 8. $y^2 + 11y = -18$

VOCABULARY AND CONCEPTS   *Fill in the blanks.*

9. A _____ is the set of all points in a _____ that are a fixed distance from a given point.
10. The fixed distance in Exercise 9 is called the _____ of the circle, and the point is called its _____.

Give the coordinates of the circle's center and its radius.

11. $(x - 2)^2 + (y + 5)^2 = 9$ center (_____); radius ___
12. $x^2 + y^2 - 36 = 0$ center (___); radius ___
13. $x^2 + y^2 = 5$ center (___); radius ___
14. $2(x - 9)^2 + 2y^2 = 7$ center (___);

    radius _____

PRACTICE   *Write the equation of each circle.*

15.

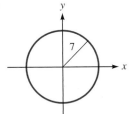

16.

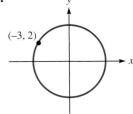

17.

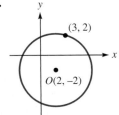

18.

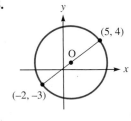

Write the equation of the circle with the following properties.

19. Center at origin; radius 1
20. Center at origin; radius 4
21. Center at (6, 8); radius 5
22. Center at (5, 3); radius 2

23. Center at $(-2, 6)$; radius 12

24. Center at $(5, -4)$; radius 6

25. Center at the origin; diameter of $2\sqrt{2}$

26. Center at the origin; diameter of $8\sqrt{3}$

Graph each equation.

27. $x^2 + y^2 = 9$

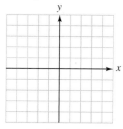

28. $x^2 + y^2 = 16$

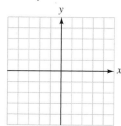

29. $(x - 2)^2 + y^2 = 9$

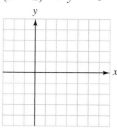

30. $x^2 + (y - 3)^2 = 4$

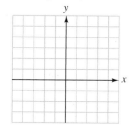

31. $(x - 2)^2 + (y - 4)^2 = 4$

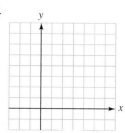

32. $(x - 3)^2 + (y - 2)^2 = 4$

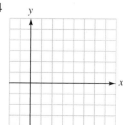

33. $(x + 3)^2 + (y - 1)^2 = 16$

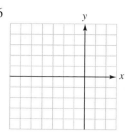

34. $(x - 1)^2 + (y + 4)^2 = 9$

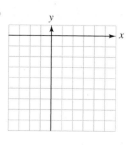

Graph each circle. Give the coordinates of the center.

35. $x^2 + y^2 + 2x - 8 = 0$

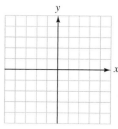

36. $x^2 + y^2 - 4y = 12$

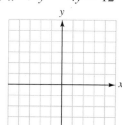

37. $9x^2 + 9y^2 - 12y = 5$

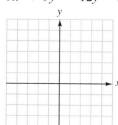

38. $4x^2 + 4y^2 + 4y = 15$

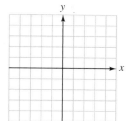

39. $x^2 + y^2 - 2x + 4y = -1$

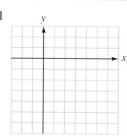

40. $x^2 + y^2 + 4x + 2y = 4$

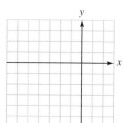

41. $x^2 + y^2 + 6x - 4y = -12$

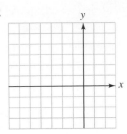

42. $x^2 + y^2 + 8x + 2y = -13$

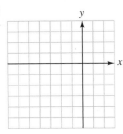

Use a graphing calculator to graph each equation.

43. $3x^2 + 3y^2 = 16$ **44.** $2x^2 + 2y^2 = 9$

45. $(x + 1)^2 + y^2 = 16$ **46.** $x^2 + (y - 2)^2 = 4$

APPLICATIONS

47. Broadcast ranges   A television tower broadcasts a signal with a circular range, as shown. Can a city 50 miles east and 70 miles north of the tower receive the signal?

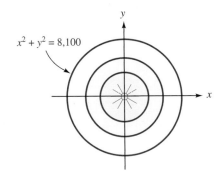

48. Warning sirens   A tornado warning siren can be heard in the circular range as shown. Can a person 4 miles west and 5 miles south of the siren hear its sound?

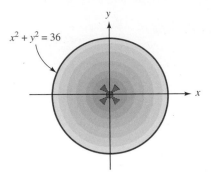

49. Radio translators   Some radio stations extend their broadcast range by installing a translator—a remote device that receives the signal and retransmits it. A station with a broadcast range given by $x^2 + y^2 = 1,600$, where x and y are in miles, installs a translator with a broadcast area bounded by $x^2 + y^2 - 70y + 600 = 0$. Find the greatest distance from the main transmitter that the signal can be received.

50. Ripples in a pond   When a stone is thrown into the center of a pond, the ripples spread out in a circular pattern, moving at a rate of 3 feet per second. If the stone is dropped at the point $(0, 0)$ in the illustration, when will the ripple reach the seagull floating at the point $(15, 36)$?

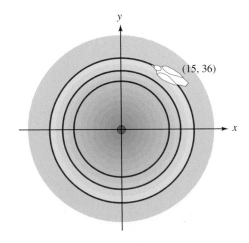

51. Writing equations of circles   An arch is shown in the illustration. Find the equation of the circle of which the outer rim is a part.

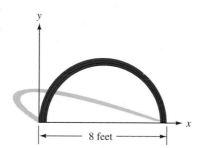

52. Writing equations of circles   The shape of the window is a combination of a rectangle and a semicircle. Find the equation of the circle of which the semicircle is a part.

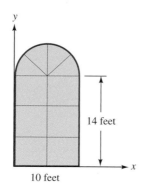

WRITING

53. From the equation of a circle, explain how to determine the radius and the coordinates of its center.

54. Describe three useful applications of the circle.

SOMETHING TO THINK ABOUT   *Write the equation of each circle.*

55. The endpoints of its diameter are $(3, -2)$ and $(3, 8)$.

56. The endpoints of its diameter are $(5, 9)$ and $(-5, -9)$.

57. Radius of 6; center at the intersection of $3x + y = 1$ and $-2x - 3y = 4$

58. Radius of 8; center at the intersection of $x + 2y = 8$ and $2x - 3y = -5$

Find the equation, in the form $(x - h)^2 + (y - k)^2 = r^2$, of the circle passing through the given points.

59. $(0, 8)$, $(5, 3)$, and $(4, 6)$

60. $(-2, 0)$, $(2, 8)$, and $(5, -1)$

12.2	**The Parabola**

■ The Parabola ■ Graphing Equations of Parabolas
■ Problem Solving

Getting Ready   *Square each binomial.*

1. $(p + h)^2$ **2.** $(y - k)^2$

Factor.

3. $4px - 4ph$ **4.** $t^2 - 8t + 16$

We have previously discussed the graphs of polynomial functions that were parabolas opening up or down. We now discuss parabolas that open to the left and to the right and examine the properties of all parabolas in greater detail.

The Parabola

The Parabola

A **parabola** is the set of all points in a plane equidistant from a line l, called the **directrix,** and a fixed point F, called the **focus.**

The point on the parabola that is closest to the directrix is called the **vertex,** and the line passing through the vertex and the focus is called the **axis.** In this section, we consider parabolas that have a vertex at point (h, k) and open to the left, to the right, up, or down.

The parabola shown in Figure 12-10 opens to the right and passes through its vertex $V(h, k)$ and some point $P(x, y)$. Since each point is equidistant from the focus (point F) and the directrix, we can let $d(DV) = d(VF) = p$, where p is a positive constant.

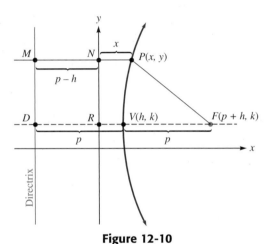

Figure 12-10

Because of the geometry of the figure,

$$d(MP) = p - h + x$$

and by the distance formula,

$$d(PF) = \sqrt{[x - (p + h)]^2 + (y - k)^2}$$

By the definition of the parabola, $d(MP) = d(PF)$. Thus,

$$p - h + x = \sqrt{[x - (p + h)]^2 + (y - k)^2}$$
$$(p - h + x)^2 = [x - (p + h)]^2 + (y - k)^2 \qquad \text{Square both sides.}$$

Finally, we expand the expression on each side of the equation and simplify:

$$p^2 - ph + px - ph + h^2 - hx + px - hx + x^2 = x^2 - 2px - 2hx + p^2 + 2ph + h^2 + (y - k)^2$$
$$-2ph + 2px = -2px + 2ph + (y - k)^2$$
$$4px - 4ph = (y - k)^2$$

(1) $$4p(x - h) = (y - k)^2$$

Equation 1 is one of four **standard equations of a parabola.**

If $p < 0$ in Equation 1, the equation will be a parabola that opens to the left.

Standard Equation of a Parabola with Vertex at (*h*, *k*) that Opens to the Right or Left	The standard equation of a parabola that has vertex at $V(h, k)$ and opening to the right or left is $$(y - k)^2 = 4p(x - h)$$ where p is the distance from the vertex to the focus. If $p > 0$, the parabola opens to the right. If $p < 0$, the parabola opens to the left.

If the parabola has its vertex at the origin, both h and k are 0, and we have the following result.

Standard Equation of a Parabola with Vertex at the Origin that Opens to the Right or Left	The standard equation of a parabola that has vertex at the origin and opening to the right or left is $$y^2 = 4px$$ where p is the distance from the vertex to the focus. If $p > 0$, the parabola opens to the right. If $p < 0$, the parabola opens to the left.

Equations of parabolas that open up or down have the following standard equations.

Standard Equations of the Parabola that Opens Up or Down	*Parabola opening* *Vertex at origin* *Vertex at $V(h, k)$* up or down $x^2 = 4py$ $(x - h)^2 = 4p(y - k)$ If $p > 0$, the parabola opens up. If $p < 0$, the parabola opens down.

EXAMPLE 1 Find the equation of the parabola with vertex at the origin and focus at (3, 0).

Solution A sketch of the parabola is shown in Figure 12-11. Because the focus is to the right of the vertex, the parabola opens to the right, and because the vertex is the origin, the standard equation is $y^2 = 4px$. The distance between the focus and the vertex is $p = 3$. We can substitute 3 for p in the standard equation to get

$$y^2 = 4\mathbf{p}x$$
$$y^2 = 4(\mathbf{3})x$$
$$y^2 = 12x$$

The equation of the parabola is $y^2 = 12x$.

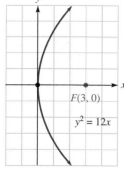

Figure 12-11

Self Check Find the equation of the parabola with vertex at the origin and focus at (−3, 0). ∎

EXAMPLE 2 Find the equation of the parabola that opens upward, has vertex at the point (4, 5), and passes through the point (0, 7).

Solution Because the parabola opens upward, we use the equation

$$(x - h)^2 = 4p(y - k)$$

Since $(h, k) = (4, 5)$ and the point $(0, 7)$ is on the curve, we can substitute 4 for h, 5 for k, 0 for x, and 7 for y in the standard equation and solve for p.

$$(x - h)^2 = 4p(y - k)$$
$$(0 - 4)^2 = 4p(7 - 5)$$
$$16 = 8p$$
$$2 = p$$

To find the equation of the parabola, we substitute 4 for h, 5 for k, and 2 for p in the standard equation and simplify:

$$(x - \mathbf{h})^2 = 4p(y - \mathbf{k})$$
$$(x - \mathbf{4})^2 = 4 \cdot \mathbf{2}(y - \mathbf{5})$$
$$(x - 4)^2 = 8(y - 5)$$

The graph of the equation appears in Figure 12-12.

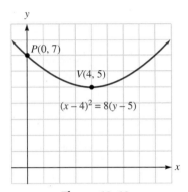

Figure 12-12

Self Check   Find the equation of the parabola that opens upward, has vertex at $(4, 5)$, and passes through $(0, 9)$. ∎

EXAMPLE 3   Find the equations of two parabolas with a vertex at $(2, 4)$ that pass through $(0, 0)$.

Solution   The two parabolas are shown in Figure 12-13.

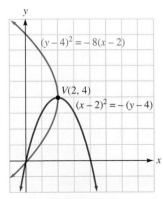

Figure 12-13

Part 1: To find the parabola that opens to the left, we use the equation $(y - k)^2 = 4p(x - h)$. Since the curve passes through the point $(x, y) = (0, 0)$ and the vertex $(h, k) = (2, 4)$, we substitute 0 for x, 0 for y, 2 for h, and 4 for k in the standard equation and solve for p:

$$(\boldsymbol{y} - \boldsymbol{k})^2 = 4p(\boldsymbol{x} - \boldsymbol{h})$$
$$(\boldsymbol{0} - \boldsymbol{4})^2 = 4p(\boldsymbol{0} - \boldsymbol{2})$$
$$16 = -8p$$
$$-2 = p$$

Since $h = 2$, $k = 4$, $p = -2$, and the parabola opens to the left, its equation is

$$(y - \boldsymbol{k})^2 = 4p(x - \boldsymbol{h})$$
$$(y - \boldsymbol{4})^2 = 4(\boldsymbol{-2})(x - \boldsymbol{2})$$
$$(y - 4)^2 = -8(x - 2)$$

Part 2: To find the equation of the parabola that opens down, we use the equation $(x - h)^2 = 4p(y - k)$ and substitute 2 for h, 4 for k, 0 for x, and 0 for y and solve for p:

$$(\boldsymbol{x} - \boldsymbol{h})^2 = 4p(\boldsymbol{y} - \boldsymbol{k})$$
$$(\boldsymbol{0} - \boldsymbol{2})^2 = 4p(\boldsymbol{0} - \boldsymbol{4})$$
$$4 = -16p$$
$$p = -\frac{1}{4}$$

Since $h = 2$, $k = 4$, and $p = -\frac{1}{4}$, the equation is

$$(x - \boldsymbol{h})^2 = 4p(y - \boldsymbol{k})$$
$$(x - \boldsymbol{2})^2 = 4\left(-\frac{1}{4}\right)(y - \boldsymbol{4}) \qquad \text{Substitute 2 for } h, \frac{1}{4} \text{ for } p, \text{ and 4 for } k.$$
$$(x - 2)^2 = -(y - 4)$$
∎

Graphing Equations of Parabolas

EXAMPLE 4 Find the vertex and y-intercepts of the parabola $y^2 + 8x - 4y = 28$ and graph it.

Solution We can complete the square on y to write the equation in standard form:

$$y^2 + 8x - 4y = 28$$
$$y^2 - 4y = -8x + 28 \qquad \text{Subtract } 8x \text{ from both sides.}$$
$$y^2 - 4y + \boldsymbol{4} = -8x + 28 + \boldsymbol{4} \qquad \text{Add 4 to both sides.}$$
$$(2) \qquad (y - 2)^2 = -8(x - 4) \qquad \text{Factor both sides.}$$

Equation 2 represents a parabola opening to the left with vertex at $(4, 2)$. To find the y-intercepts, we substitute 0 for x in Equation 2 and solve for y.

$$(y - 2)^2 = -8(\boldsymbol{x} - 4)$$
$$(y - 2)^2 = -8(\boldsymbol{0} - 4) \qquad \text{Substitute 0 for } x.$$
$$y^2 - 4y + 4 = 32 \qquad \text{Remove parentheses.}$$
$$(3) \qquad y^2 - 4y - 28 = 0$$

We can use the quadratic formula to find that the roots of Equation 3 are $y \approx 7.7$ and $y \approx -3.7$. So the points with coordinates of approximately $(0, 7.7)$ and $(0, -3.7)$ lie on the graph of the parabola. We can use this information and the knowledge that the graph opens to the left and has a vertex at $(4, 2)$ to draw the graph as shown in Figure 12-14.

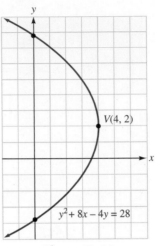

Figure 12-14

Self Check   Find the vertex and y-intercepts of the parabola $y^2 - 8x + 6y = 7$. Then graph it. ∎

Problem Solving

EXAMPLE 5   If a stone is thrown straight up, the equation $s = 128t - 16t^2$ expresses its height in feet t seconds after being thrown. Find the maximum height reached by the stone.

Solution   The graph of $s = 128t - 16t^2$, expressing the height of the stone t seconds after it was thrown, is the parabola shown in Figure 12-15.

To find the maximum height reached by the stone, we find the s-coordinate k of the vertex of the parabola. To find k, we write the equation of the parabola in standard form by completing the square on t.

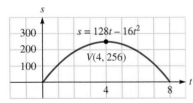

Figure 12-15

$$s = 128t - 16t^2$$

$$16t^2 - 128t = -s \qquad \text{Multiply both sides by } -1.$$

$$t^2 - 8t = \frac{-s}{16} \qquad \text{Divide both sides by 16.}$$

$$t^2 - 8t + 16 = \frac{-s}{16} + 16 \qquad \text{Add 16 to both sides to complete the square.}$$

$$(t - 4)^2 = \frac{-s + 256}{16} \qquad \text{Factor } t^2 - 8t + 16 \text{ and combine like terms.}$$

$$(t - 4)^2 = -\frac{1}{16}(s - 256) \qquad \text{Factor out } -\frac{1}{16}.$$

This equation indicates that the maximum height of 256 feet is reached in 4 seconds.

Comment The parabola shown in Figure 12-15 is not the path of the stone. The stone goes straight up and straight down.

Self Check   At what time will the stone strike the ground?   (*Hint:* Find the *t*-intercept.)   ▮

Self Check Answers

1. $y^2 = -12x$ **2.** $(x - 4)^2 = 4(y - 5)$ **4.** vertex $(-2, -3)$; *y*-intercepts $(0, 1)$, $(0, -7)$
5. 8 seconds

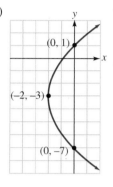

Orals   *Tell whether each parabola opens up or down or left or right.*

1. $y = -3x^2 - 2$ **2.** $y = 7x^2 - 5$
3. $x = -3y^2$ **4.** $x = (y - 3)^2$

12.2   EXERCISES 🖭

REVIEW   *Solve each equation.*

1. $|3x - 4| = 11$ **2.** $\left| \dfrac{4 - 3x}{5} \right| = 12$

3. $|3x + 4| = |5x - 2|$ **4.** $|6 - 4x| = |x + 2|$

VOCABULARY AND CONCEPTS   *Fill in the blanks.*

5. A parabola is the set of all points in a plane equidistant from a line, called the _____, and a fixed point, called the ____.

6. The graph of $x^2 = 4py$ $(p > 0)$ is a _____ with vertex at the ____ and opening _____.

7. The graph of $(y - 2)^2 = 4p(x - 3)$ $(p > 0)$ is a _____ with vertex at ____ and opening to the ____.

8. The graph of $(y - 1)^2 = 4p(x + 3)$ $(p < 0)$ is a _____ with vertex at ____ and opening to the ____.

Tell whether the parabolic graph of the equation opens upward, downward, to the left, or to the right.

9. $y^2 = -4x$ **10.** $y^2 = 10x$
    opens _____   opens _____
11. $x^2 = -8(y - 3)$ **12.** $(x - 2)^2 = (y + 3)$
    opens _____   opens _____

PRACTICE   *Find the equation of each parabola.*

13. Vertex at $(0, 0)$; focus at $(0, 3)$
14. Vertex at $(0, 0)$; focus at $(0, -3)$
15. Vertex at $(0, 0)$; focus at $(-3, 0)$
16. Vertex at $(0, 0)$; focus at $(3, 0)$
17. Vertex at $(3, 5)$; focus at $(3, 2)$

18. Vertex at $(3, 5)$; focus at $(-3, 5)$

19. Vertex at $(3, 5)$; focus at $(3, -2)$

20. Vertex at $(3, 5)$; focus at $(6, 5)$

21. Vertex at $(2, 2)$; passes through $(0, 0)$

22. Vertex at $(-2, -2)$; passes through $(0, 0)$

23. Vertex at $(-4, 6)$; passes through $(0, 3)$

24. Vertex at $(-2, 3)$; passes through $(0, -3)$

25. Vertex at $(6, 8)$; passes through $(5, 10)$ and $(5, 6)$

26. Vertex at $(2, 3)$; passes through $\left(1, \frac{13}{4}\right)$ and $\left(-1, \frac{21}{4}\right)$

27. Vertex at $(3, 1)$; passes through $(4, 3)$ and $(2, 3)$

28. Vertex at $(-4, -2)$; passes through $(-3, 0)$ and $\left(\frac{9}{4}, 3\right)$

Change each equation to standard form, if necessary, and graph it.

29. $y = x^2 + 4x + 5$

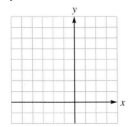

30. $2x^2 - 12x - 7y = 10$

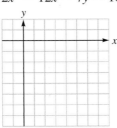

31. $y^2 + 4x - 6y = -1$

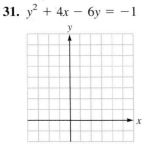

32. $x^2 - 2y - 2x = -7$

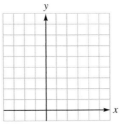

33. $y^2 + 2x - 2y = 5$

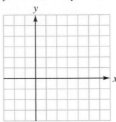

34. $y^2 - 4y = -8x + 20$

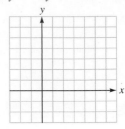

35. $x^2 - 6y + 22 = -4x$

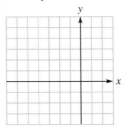

36. $4y^2 - 4y + 16x = 7$

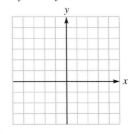

37. $4x^2 - 4x + 32y = 47$

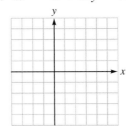

38. $4y^2 - 16x + 17 = 20y$

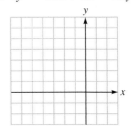

APPLICATIONS

39. Solar furnaces   A parabolic mirror collects rays of
the sun and concentrates them at its focus. In the illustration, how far from the vertex of the parabolic
mirror will it get the hottest? (Assume that all measurements are in feet.)

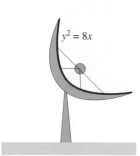

$y^2 = 8x$

40. Searchlight reflectors   A parabolic mirror reflects
light in a beam when the light source is placed at its
focus. In the illustration, how far from the vertex of
the parabolic reflector should the light source be
placed? (Assume that all measurements are in
feet.)

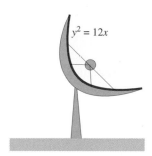

$y^2 = 12x$

41. Design of satellite antennas   The cross section of
the satellite antenna shown in the illustration is a
parabola with the pickup at its focus. Find the distance d from the pickup to the center of the dish.

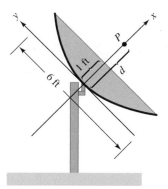

42. Operating a resort   A resort owner plans to
build and rent n cabins for d dollars per week. The
price d that she can charge for each cabin
depends on the number of cabins she builds, where
$d = -45\left(\frac{n}{32} - \frac{1}{2}\right)$. Find the number of cabins that she
should build to maximize her weekly income.

43. Toy rockets   A toy rocket is s feet above the Earth
at the end of t seconds, where $s = -16t^2 + 80\sqrt{3}t$.
Find the maximum height of the rocket.

44. Design of parabolic reflectors   Find the outer diameter (the length $\overline{AB}$) of the parabolic reflector
shown in the illustration.

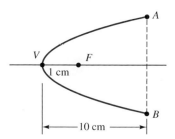

45. Writing equations of parabolas Derive the equation of the parabolic arch shown in the illustration.

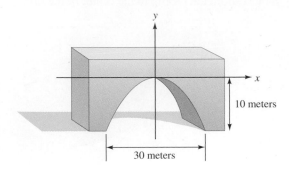

10 meters

30 meters

46. Design of suspension bridges The cable between the towers of the suspension bridge has the shape of a parabola with a vertex 15 feet above the roadway. Find the equation of the parabola.

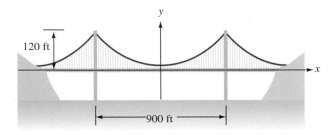

120 ft

900 ft

47. Gateway Arch The Gateway Arch in St. Louis has a shape that approximates a parabola. Find the width w of the arch 200 feet above the ground.

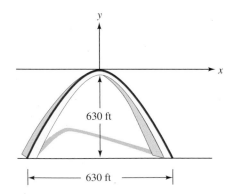

630 ft

630 ft

48. Building tunnels A construction firm plans to build a tunnel whose arch is in the shape of a parabola. The tunnel will span a two-lane highway 8 meters wide. To allow safe passage for most vehicles, the tunnel must be 5 meters high at a distance of 1 meter from the tunnel's edge. Find the maximum height of the tunnel.

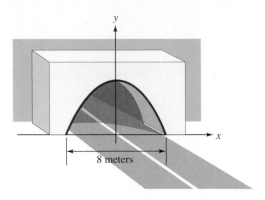

8 meters

WRITING

49. Explain the definition of a parabola, including the concepts of *directrix* and *focus*.

50. Describe three useful applications of parabolas.

SOMETHING TO THINK ABOUT *In Exercises 51–52, find the equation of the parabola passing through the given points. Give the equation in the form $y = ax^2 + bx + c$.*

51. $(1, 8)$, $(-2, -1)$, and $(2, 15)$

52. $(1, -3)$, $(-2, 12)$, and $(-1, 3)$

53. Ballistics A stone tossed upward is s feet above the earth after t seconds, where $s = -16t^2 + 128t$. Show that the stone's height t seconds after it is thrown is equal to its height t seconds before it hits the ground.

54. Ballistics Show that the stone in Exercise 53 reaches its greatest height in one-half of the time it takes until it strikes the ground.

<table>
<tr><td>**12.3**</td><td>**The Ellipse**</td></tr>
</table>

■ **The Ellipse**   ■ **Graphing Equations of Ellipses**

Getting Ready   *Solve each equation for the indicated variable.*

1. $\dfrac{y^2}{b^2} = 1$ for y

2. $\dfrac{x^2}{a^2} = 1$ for x

A third important conic is the ellipse.

The Ellipse

The Ellipse   An ellipse is the set of all points P in the plane such that the sum of the distances from P to two other fixed points F and F' is a positive constant.

In the ellipse shown in Figure 12-16, the fixed points F and F' are called **foci** (each is a **focus**), the midpoint of the chord FF' is called the **center,** and the chord VV' is called the **major axis.** Each of the endpoints V and V' of the major axis is called a **vertex.** The chord BB', perpendicular to the major axis and passing through the center C of the ellipse, is called the **minor axis.**

To derive the equation of the ellipse shown in Figure 12-17, we note that point O is the midpoint of chord FF' and let $d(OF) = d(OF') = c$, where $c > 0$. Then the coordinates of point F are $(c, 0)$, and the coordinates of F' are $(-c, 0)$. We also let $P(x, y)$ be any point on the ellipse.

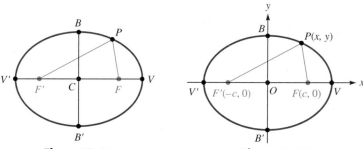

Figure 12-16   **Figure 12-17**

By the definition of an ellipse, $d(F'P) + d(PF)$ must be a positive constant, which we will call $2a$. Thus,

(1) $d(F'P) + d(PF) = 2a$

We can use the distance formula to compute the lengths of $F'P$ and PF,

$$d(F'P) = \sqrt{[x - (-c)]^2 + y^2} \quad \text{and} \quad d(PF) = \sqrt{(x - c)^2 + y^2}$$

and substitute these values into Equation 1 to obtain

$$\sqrt{[x-(-c)]^2+y^2}+\sqrt{(x-c)^2+y^2}=2a$$

or

(2) $\sqrt{[x+c]^2+y^2}=2a-\sqrt{(x-c)^2+y^2}$ Subtract $\sqrt{(x-c)^2+y^2}$ from both sides.

We can square both sides of Equation 2 and simplify to get

$$(x+c)^2+y^2=4a^2-4a\sqrt{(x-c)^2+y^2}+[(x-c)^2+y^2]$$
$$x^2+2cx+c^2+y^2=4a^2-4a\sqrt{(x-c)^2+y^2}+x^2-2cx+c^2+y^2$$
$$4cx=4a^2-4a\sqrt{(x-c)^2+y^2}$$
$$cx=a^2-a\sqrt{(x-c)^2+y^2}$$
$$cx-a^2=-a\sqrt{(x-c)^2+y^2}$$

We square both sides again and simplify to get

$$c^2x^2-2a^2cx+a^4=a^2[(x-c)^2+y^2]$$
$$c^2x^2-2a^2cx+a^4=a^2(x^2-2cx+c^2+y^2)$$
$$c^2x^2-2a^2cx+a^4=a^2x^2-2a^2cx+a^2c^2+a^2y^2$$
$$c^2x^2+a^4=a^2x^2+a^2c^2+a^2y^2$$
$$a^4-a^2c^2=a^2x^2-c^2x^2+a^2y^2$$

(3) $a^2(a^2-c^2)=(a^2-c^2)x^2+a^2y^2$

Because the shortest distance between two points is a straight line segment, $d(F'P)+d(PF)>d(F'F)$. Therefore, $2a>2c$. Thus, $a>c$ and a^2-c^2 is a positive number, which we will call b^2. Letting $b^2=a^2-c^2$ and substituting into Equation 3, we have

(4) $a^2b^2=b^2x^2+a^2y^2$

Dividing both sides of Equation 4 by a^2b^2 gives the equation

$$\frac{x^2}{a^2}+\frac{y^2}{b^2}=1\quad\text{where }a>b>0$$

To find the coordinates of the vertices V and V', we substitute 0 for y and solve for x:

$$\frac{x^2}{a^2}+\frac{y^2}{b^2}=1$$
$$\frac{x^2}{a^2}+\frac{0^2}{b^2}=1$$
$$\frac{x^2}{a^2}=1$$
$$x^2=a^2$$
$$x=a\quad\text{or}\quad x=-a$$

Since the coordinates of V are $(a,0)$ and the coordinates of V' are $(-a,0)$, a is the distance between the center of the ellipse and either of its vertices. Thus, the center of the ellipse is the midpoint of the major axis.

To find the coordinates of B and B', we substitute 0 for x and solve for y:

$$\frac{x^2}{a^2} + \frac{y^2}{b^2} = 1$$

$$\frac{\mathbf{0}^2}{a^2} + \frac{y^2}{b^2} = 1$$

$$y^2 = b^2$$

$$y = b \quad \text{or} \quad y = -b$$

Since the coordinates of B are $(0, b)$ and the coordinates of B' are $(0, -b)$, the distance between the center of the ellipse and either endpoint of the minor axis is b. We have the following results.

The Ellipse: Major Axis on x-Axis, Center at (0, 0)

The standard equation of an ellipse with center at the origin and major axis on the x-axis is

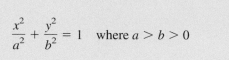

$$\frac{x^2}{a^2} + \frac{y^2}{b^2} = 1 \quad \text{where } a > b > 0$$

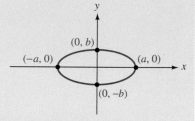

Vertices: $V(a, 0)$ and $V'(-a, 0)$
Ends of the minor axis: $B(0, b)$ and $B'(0, -b)$

The Ellipse: Major Axis on y-Axis, Center at (0, 0)

The standard equation of an ellipse with center at the origin and major axis on the y-axis is

$$\frac{y^2}{a^2} + \frac{x^2}{b^2} = 1 \quad \text{where } a > b > 0$$

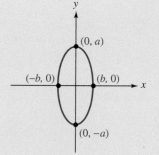

Vertices: $V(0, a)$ and $V'(0, -a)$
Ends of the minor axis: $B(b, 0)$ and $B'(-b, 0)$

To translate the ellipse to a new position centered at the point (h, k) instead of the origin, we replace x and y in the equations with $x - h$ and $y - k$, respectively, to get these results:

The Ellipse: Major Axis Horizontal, Center at (h, k)

The standard equation of an ellipse with center at (h, k) and major axis horizontal is

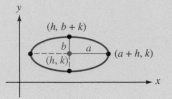

$$\frac{(x-h)^2}{a^2} + \frac{(y-k)^2}{b^2} = 1$$

where $a > b > 0$

Vertices: $V(a+h, k)$ and $V'(-a+h, k)$
Ends of the minor axis: $B(h, b+k)$ and $B'(h, -b+k)$

The Ellipse: Major Axis Vertical, Center at (h, k)

The standard equation of an ellipse with center at (h, k) and major axis vertical is

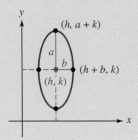

$$\frac{(y-k)^2}{a^2} + \frac{(x-h)^2}{b^2} = 1$$

where $a > b > 0$

Vertices: $V(h, a+k)$ and $V'(h, -a+k)$
Ends of the minor axis: $B(b+h, k)$ and $B'(-b+h, k)$

In all cases, the length of the major axis is $2a$, and the length of the minor axis is $2b$.

EXAMPLE 1   Find the equation of the ellipse with center at the origin, major axis of length 6 units located on the x-axis, and minor axis of length 4 units.

Solution   Since the center is the origin and the length of the major axis is 6, $a = 3$ and the coordinates of the vertices are $(3, 0)$ and $(-3, 0)$, as shown in Figure 12-18.

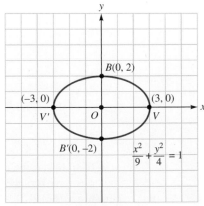

Figure 12-18

Since the length of the minor axis is 4, $b = 2$ and the coordinates of B and B' are $(0, 2)$ and $(0, -2)$. To find the equation, we substitute 3 for a and 2 for b in the standard equation of an ellipse with center at the origin and major axis on the x-axis.

$$\frac{x^2}{a^2} + \frac{y^2}{b^2} = 1$$

$$\frac{x^2}{3^2} + \frac{y^2}{2^2} = 1$$

$$\frac{x^2}{9} + \frac{y^2}{4} = 1$$

Self Check   Find the equation of the ellipse with center at the origin, major axis of length 10 on the x-axis, and minor axis of length 8. ∎

EXAMPLE 2   Find the equation of the ellipse with focus at $(0, 3)$ and vertices V and V' at $(3, 3)$ and $(-5, 3)$.

Solution   Since the midpoint of the major axis is the center of the ellipse, the coordinates of the center are $(-1, 3)$, as in Figure 12-19. Because the major axis is parallel to the x-axis, the standard equation to use is

$$\frac{(x - h)^2}{a^2} + \frac{(y - k)^2}{b^2} = 1$$

where $a > b > 0$

From Figure 12-19, we see that the distance between the center of the ellipse and either vertex is $a = 4$. We also see that the distance between the focus and the center is $c = 1$.

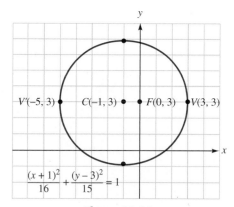

Figure 12-19

Since $b^2 = a^2 - c^2$ in an ellipse, we can substitute 4 for a and 1 for c and find b^2:

$$b^2 = a^2 - c^2$$
$$= 4^2 - 1^2$$
$$= 15$$

To find the equation of the ellipse, we substitute -1 for h, 3 for k, 16 for a^2, and 15 for b^2 in the standard equation and simplify:

$$\frac{(x - h)^2}{a^2} + \frac{(y - k)^2}{b^2} = 1$$

$$\frac{[x - (-1)]^2}{16} + \frac{(y - 3)^2}{15} = 1$$

$$\frac{(x + 1)^2}{16} + \frac{(y - 3)^2}{15} = 1$$

Self Check Find the equation of the ellipse with focus at (3, 1) and vertices at (5, 1) and (−5, 1).

EXAMPLE 3 The orbit of Earth is approximately an ellipse, with the sun at one focus. The ratio of c to a (called the **eccentricity** of the ellipse) is about $\frac{1}{62}$, and the length of the major axis is approximately 186,000,000 miles. How close does Earth get to the sun?

Solution We will assume that the ellipse has its center at the origin and vertices V' and V at (−93,000,000, 0) and (93,000,000, 0), as shown in Figure 12-20. Because the eccentricity $\frac{c}{a}$ is given to be $\frac{1}{62}$ and $a = 93,000,000$, we have

$$\frac{c}{a} = \frac{1}{62}$$

$$c = \frac{1}{62}a$$

$$c = \frac{1}{62}(\mathbf{93{,}000{,}000})$$

$$= 1{,}500{,}000$$

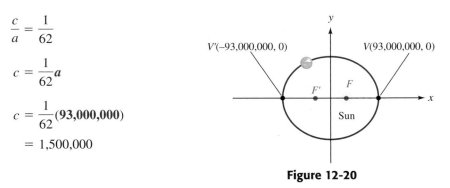

Figure 12-20

The distance $d(FV)$ is the shortest distance between the earth and the sun. (You'll be asked to prove this in the exercises.) Thus,

$$d(FV) = a - c = 93{,}000{,}000 - 1{,}500{,}000 = 91{,}500{,}000 \text{ mi}$$

The Earth's point of closest approach to the sun (called the **perihelion**) is approximately 91.5 million miles.

We can use the eccentricity of an ellipse to judge its shape. If the eccentricity is close to 1, the ellipse is relatively flat, as in Figure 12-21(a). If the eccentricity is close to 0, the ellipse is more circular, as in Figure 12-21(b). Since the eccentricity of the Earth's orbit is $\frac{1}{62}$, the Earth's orbit is almost a circle.

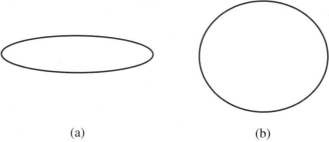

(a)   (b)

Figure 12-21

Graphing Equations of Ellipses

EXAMPLE 4   Graph: $\dfrac{(x+2)^2}{4} + \dfrac{(y-2)^2}{9} = 1$.

Solution   From the equation, we see that the center is at $(-2, 2)$, and the major axis is parallel to the y-axis. Because $a = 3$, the vertices are 3 units above and below the center at points $(-2, 5)$ and $(-2, -1)$. Because $b = 2$, the endpoints of the minor axis are 2 units to the right and left of the center at points $(0, 2)$ and $(-4, 2)$. Using these points, we can sketch the ellipse shown in Figure 12-22.

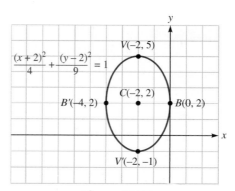

Figure 12-22

Self Check   Graph: $\dfrac{(x-2)^2}{9} + \dfrac{(y+1)^2}{25} = 1$.

EXAMPLE 5   Graph: $4x^2 + 9y^2 - 16x - 18y = 11$.

Solution   We write the equation in standard form by completing the square on x and y:

$$4x^2 + 9y^2 - 16x - 18y = 11$$
$$4x^2 - 16x + 9y^2 - 18y = 11$$
$$4(x^2 - 4x) + 9(y^2 - 2y) = 11$$
$$4(x^2 - 4x + 4) + 9(y^2 - 2y + 1) = 11 + 16 + 9$$
$$4(x - 2)^2 + 9(y - 1)^2 = 36$$
$$\frac{(x-2)^2}{9} + \frac{(y-1)^2}{4} = 1$$

We can now see that the graph is an ellipse with center at $(2, 1)$ and major axis parallel to the x-axis. Because $a = 3$, the vertices are at $(-1, 1)$ and $(5, 1)$. Because $b = 2$, the endpoints of the minor axis are at $(2, -1)$ and $(2, 3)$. Using these points, we can sketch the ellipse shown in Figure 12-23.

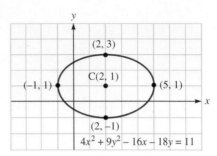

Figure 12-23

Self Check   Graph: $4x^2 + 9y^2 - 8x + 36y = -4$.

Accent on Technology   **GRAPHING ELLIPSES**

To use a graphing calculator to graph

$$\frac{(x + 2)^2}{4} + \frac{(y - 1)^2}{25} = 1$$

we first clear the equation of fractions by multiplying both sides by 100 and solving for y.

$$25(x + 2)^2 + 4(y - 1)^2 = 100 \qquad \text{Multiply both sides by 100.}$$

$$4(y - 1)^2 = 100 - 25(x + 2)^2 \qquad \text{Subtract } 25(x + 2)^2 \text{ from both sides.}$$

$$(y - 1)^2 = \frac{100 - 25(x + 2)^2}{4} \qquad \text{Divide both sides by 4.}$$

$$y - 1 = \pm \frac{\sqrt{100 - 25(x + 2)^2}}{2} \qquad \text{Take the square root of both sides.}$$

$$y = 1 \pm \frac{\sqrt{100 - 25(x + 2)^2}}{2} \qquad \text{Add 1 to both sides.}$$

If we use window settings $[-6, 6]$ for x and $[-6, 6]$ for y and graph the functions

$$y = 1 + \frac{\sqrt{100 - 25(x + 2)^2}}{2} \qquad \text{and} \qquad y = 1 - \frac{\sqrt{100 - 25(x + 2)^2}}{2}$$

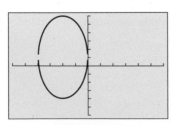

Figure 12-24

we will obtain the ellipse shown in Figure 12-24.

Self Check Answers

1. $\dfrac{x^2}{25} + \dfrac{y^2}{16} = 1$ 2. $\dfrac{x^2}{25} + \dfrac{(y-1)^2}{16} = 1$ 4.

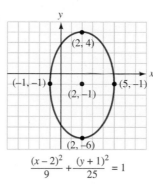

$$\dfrac{(x-2)^2}{9} + \dfrac{(y+1)^2}{25} = 1$$

5.

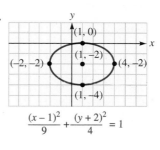

$$\dfrac{(x-1)^2}{9} + \dfrac{(y+2)^2}{4} = 1$$

Orals   *Find the x- and y-intercepts of each ellipse.*

1. $\dfrac{x^2}{9} + \dfrac{y^2}{16} = 1$

2. $\dfrac{x^2}{25} + \dfrac{y^2}{36} = 1$

Find the center of each ellipse.

3. $\dfrac{(x-2)^2}{9} + \dfrac{y^2}{16} = 1$

4. $\dfrac{x^2}{25} + \dfrac{(y+1)^2}{36} = 1$

12.3   EXERCISES

REVIEW   *Perform the operations.*

1. $x(x+4)$

2. $y + (y+3)$

3. $(x+2)^2$

4. $\dfrac{2x-4}{x-2}$

Solve each equation.

5. $x^2 - 4x = 5$

6. $y^2 + 12y = 13$

7. $x^2 - 7x = 0$

8. $y^2 - 81 = 0$

VOCABULARY AND CONCEPTS   *Fill in the blanks.*

9. An ellipse is the set of all points in the plane such that the _____ of the distances from two fixed points is a positive _____.

10. Each of the two fixed points in the definition of an ellipse is called a _____ of the ellipse.

11. The chord that joins the _____ is called the major axis of the ellipse.

12. The chord through the center of an ellipse and perpendicular to the major axis is called the _____ axis.

13. In the ellipse $\dfrac{x^2}{a^2} + \dfrac{y^2}{b^2} = 1$ $(a > b)$, the vertices are $V(\_\_, \_\_)$ and $V'(\_\_, \_\_)$

14. In an ellipse, the relation between a, b, and c is _____.

PRACTICE   *Write the equation of the ellipse that has its center at the origin.*

15. Focus at $(3, 0)$; vertex at $(5, 0)$

16. Focus at $(0, 4)$; vertex at $(0, 7)$

17. Focus at $(0, 1)$; $\frac{4}{3}$ is one-half the length of the minor axis.

18. Focus at $(1, 0)$; $\frac{4}{3}$ is one-half the length of the minor axis.

19. Focus at $(0, 3)$; major axis equal to 8

20. Focus at $(5, 0)$; major axis equal to 12

Write the equation of each ellipse.

21. Center at $(3, 4)$; $a = 3$, $b = 2$; major axis parallel to the y-axis

22. Center at $(3, 4)$; passes through $(3, 10)$ and $(3, -2)$; $b = 2$

23. Center at $(3, 4)$; $a = 3$, $b = 2$; major axis parallel to the x-axis

24. Center at $(3, 4)$; passes through $(8, 4)$ and $(-2, 4)$; $b = 2$

25. Foci at $(-2, 4)$ and $(8, 4)$; $b = 4$

26. Foci at $(8, 5)$ and $(4, 5)$; $b = 3$

27. Vertex at $(6, 4)$; foci at $(-4, 4)$ and $(4, 4)$

28. Center at $(-4, 5)$; $\dfrac{c}{a} = \dfrac{1}{3}$; vertex at $(-4, -1)$

29. Foci at $(6, 0)$ and $(-6, 0)$; $\dfrac{c}{a} = \dfrac{3}{5}$

30. Vertices at $(2, 0)$ and $(-2, 0)$; $\dfrac{2b^2}{a} = 2$

Graph each ellipse.

31. $\dfrac{x^2}{25} + \dfrac{y^2}{49} = 1$ **32.** $4x^2 + y^2 = 4$

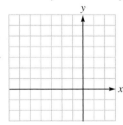

33. $\dfrac{x^2}{16} + \dfrac{(y + 2)^2}{36} = 1$ **34.** $(x - 1)^2 + \dfrac{4y^2}{25} = 4$

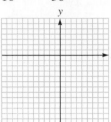   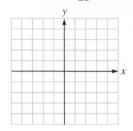

35. $x^2 + 4y^2 - 4x + 8y + 4 = 0$

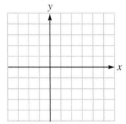

36. $x^2 + 4y^2 - 2x - 16y = -13$

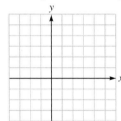

37. $16x^2 + 25y^2 - 160x - 200y + 400 = 0$

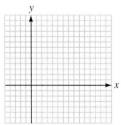

38. $3x^2 + 2y^2 + 7x - 6y = -1$

APPLICATIONS

39. Astronomy   The moon has an orbit that is an ellipse, with Earth at one focus. If the major axis of the orbit is 378,000 miles and the ratio of c to a is approximately $\frac{11}{200}$, how far does the moon get from Earth? (This farthest point in an orbit is called the **apogee.**)

40. Equation of an arch   An arch is a semiellipse 10 meters wide and 3 meters high. Write the equation of the ellipse if the ellipse is centered at the origin.

41. Design of a track   A track is built in the shape of an ellipse with a maximum length of 100 meters and a maximum width of 60 meters. Write the equation of the ellipse and find its **focal width.** That is, find the length of a chord that is perpendicular to the major axis and passes through either focus of the ellipse.

42. Whispering galleries   Any sound from one focus of an ellipse reflects off the ellipse directly back to the other focus. This property explains whispering galleries such as Statuary Hall in Washington, DC. The whispering gallery shown in the illustration has the shape of a semiellipse. Find the distance sound travels as it leaves focus F and returns to focus F'.

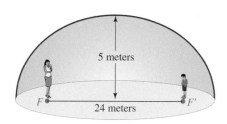

43. Finding the width of a mirror   The oval mirror shown is in the shape of an ellipse. Find the width of the mirror 12 inches above its base.

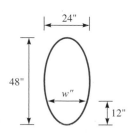

44. Finding the height of a window   The window shown has the shape of an ellipse. Find the height of the window 20 inches from one end.

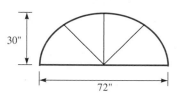

WRITING

45. Explain how to find the x- and y-intercepts of the graph of the ellipse
$$\frac{x^2}{a^2} + \frac{y^2}{b^2} = 1$$

46. Explain the relationship between the center, focus, and vertex of an ellipse.

SOMETHING TO THINK ABOUT

47. What happens to the graph of
$$\frac{x^2}{a^2} + \frac{y^2}{b^2} = 1$$
when $a = b$?

48. Explain why the graph of $x^2 - 2x + y^2 + 4y + 20 = 0$ does not exist.

49. If F is a focus of the ellipse shown in the illustration and B is an endpoint of the minor axis, use the distance formula to prove that the length of segment FB is a.   (*Hint:* In an ellipse, $a^2 - c^2 = b^2$.)

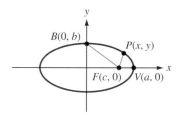

50. If F is a focus of the ellipse shown in the illustration above, and P is any point on the ellipse, use the distance formula to show that the length of FP is $a - \frac{c}{a}x$. (*Hint:* In an ellipse, $a^2 - c^2 = b^2$.)

51. Finding the focal width   In the ellipse shown in the illustration, chord AA' passes through the focus F and is perpendicular to the major axis. Show that the length of AA' (called the **focal width**) is $\frac{2b^2}{a}$.

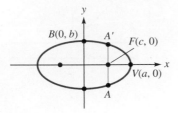

52. Prove that segment FV in Example 3 is the shortest distance between Earth and the sun.   (*Hint:* Refer to Exercise 50.)

53. Constructing an ellipse   We can construct an ellipse by placing two thumbtacks fairly close together, as in the illustration. We then tie each end of a piece of string to a thumbtack, catch the loop with the point of a pencil, and (while keeping the string taut) draw the ellipse.

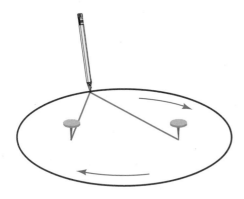

To construct an ellipse that is 10 feet wide and 6 feet high, we must find the length of string to use and the distance between the thumbtacks.

  a. Show that we should use a string that is 10 feet long.

  b. Show that the distance between the thumbtacks should be 8 feet.

Thus, we can construct the ellipse by tying a 10-foot string to thumbtacks that are 8 feet apart.

54. The distance between point $P(x, y)$ and point $(0, 2)$ is $\frac{1}{3}$ of the distance of point P from the line $y = 18$. Find the equation of the curve on which point P lies.

55. Prove that $a > b$ in the development of the standard equation of an ellipse.

56. Show that the expansion of the standard equation of an ellipse is a special case of the general second-degree equation in two variables.

12.4   The Hyperbola

▌ The Hyperbola   ▌ Asymptotes of a Hyperbola
▌ Graphing Equations of Hyperbolas

Getting Ready   *Find the value of y when $\dfrac{x^2}{25} - \dfrac{y^2}{9} = 1$ and x is the given value. Give each result to the nearest tenth.*

  1. $x = 6$ 　　　　　　　　　　　　　　　　**2.** $x = -7$

In this section, we will consider a conic section called a *hyperbola*. Its definition is similar to that of an ellipse, except that we require a constant difference of $2a$ instead of a constant sum.

The Hyperbola

The Hyperbola

A **hyperbola** is the set of all points P in a plane such that the difference of the distances from point P to two other points in the plane is a positive constant.

Points F and F' shown in Figure 12-25 are called the **foci** of the hyperbola, and the midpoint of chord FF' is called the **center.** The points V and V', where the hyperbola intersects FF', are called **vertices.** The segment VV' is called the **transverse axis.**

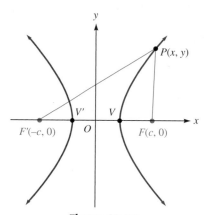

Figure 12-25

To develop the equation of the hyperbola centered at the origin, we note that the origin is the midpoint of chord FF', and we let $d(F'O) = d(OF) = c$, where $c > 0$. Then F is at $(c, 0)$, and F' is at $(-c, 0)$. The definition of a hyperbola requires that $d(F'P) - d(PF) = 2a$, where $2a$ is a positive constant. We use the distance formula to compute the lengths of $F'P$ and PF:

$$d(F'P) = \sqrt{[x - (-c)]^2 + y^2}$$
$$d(PF) = \sqrt{(x - c)^2 + y^2}$$

Substituting these values into the equation $d(F'P) - d(PF) = 2a$ gives

$$\sqrt{(x + c)^2 + y^2} - \sqrt{(x - c)^2 + y^2} = 2a$$

of

$$\sqrt{(x + c)^2 + y^2} = 2a + \sqrt{(x - c)^2 + y^2}$$

After squaring, we have

$$(x + c)^2 + y^2 = 4a^2 + 4a\sqrt{(x - c)^2 + y^2} + (x - c)^2 + y^2$$
$$x^2 + 2cx + c^2 + y^2 = 4a^2 + 4a\sqrt{(x - c)^2 + y^2} + x^2 - 2cx + c^2 + y^2$$
$$4cx = 4a^2 + 4a\sqrt{(x - c)^2 + y^2}$$
$$cx - a^2 = a\sqrt{(x - c)^2 + y^2}$$

Squaring both sides again and simplifying gives

$$c^2x^2 - 2a^2cx + a^4 = a^2(x^2 - 2cx + c^2 + y^2)$$
$$c^2x^2 - 2a^2cx + a^4 = a^2x^2 - 2a^2cx + a^2c^2 + a^2y^2$$
$$c^2x^2 + a^4 = a^2x^2 + a^2c^2 + a^2y^2$$
$$(c^2 - a^2)x^2 - a^2y^2 = a^2(c^2 - a^2) \qquad (1)$$

Because $c > a$ (you will be asked to prove this in the exercises), $c^2 - a^2$ is a positive number. So we can let $b^2 = c^2 - a^2$ and substitute b^2 for $c^2 - a^2$ in Equation 1 to get

$$b^2x^2 - a^2y^2 = a^2b^2$$

We divide both sides of the previous equation by a^2b^2 to get the standard equation for a hyperbola with center at the origin and foci on the x-axis:

$$\frac{x^2}{a^2} - \frac{y^2}{b^2} = 1$$

To find the x-intercepts of the graph, we let $y = 0$ and solve for x. We get

$$\frac{x^2}{a^2} = 1$$
$$x^2 = a^2$$
$$x = a \quad \text{or} \quad x = -a$$

We now know that the x-intercepts are the vertices $V(a, 0)$ and $V'(-a, 0)$. The distance between the center of the hyperbola and either vertex is a, and the center of the hyperbola is the midpoint of the segment $V'V$ as well as of the segment FF'.

We attempt to find the y-intercepts by letting $x = 0$. Then the equation becomes

$$\frac{-y^2}{b^2} = 1 \qquad \text{or} \qquad y^2 = -b^2$$

Because $-b^2$ represents a negative number, and y^2 cannot be negative, the equation has no real solutions. Since there are no y-values corresponding to $x = 0$, the hyperbola does not intersect the y-axis.

This discussion suggests the following results.

Hyperbola: Foci on x-Axis, Center at (0, 0)

The standard equation of a hyperbola with center at the origin and foci on the x-axis is

$$\frac{x^2}{a^2} - \frac{y^2}{b^2} = 1$$

where $a^2 + b^2 = c^2$.

Vertices: $V(a, 0)$ and $V'(-a, 0)$
Foci: $F(c, 0)$ and $F'(-c, 0)$

If the foci are on the y-axis, a similar equation results.

Hyperbola: Foci on *y*-Axis, Center at (0, 0)

The standard equation of a hyperbola with center at the origin and foci on the *y*-axis is

$$\frac{y^2}{a^2} - \frac{x^2}{b^2} = 1$$

where $a^2 + b^2 = c^2$.

Vertices: $V(0, a)$ and $V'(0, -a)$
Foci: $F(0, c)$ and $F'(0, -c)$

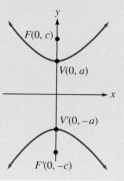

To translate the hyperbola to a new position centered at the point (h, k) instead of the origin, we replace x and y with $x - h$ and $y - k$, respectively. We get the following results.

Hyperbola: Transverse Axis Horizontal, Center at (*h*, *k*)

The standard equation of a hyperbola with center at (h, k) and foci on a line parallel to the *x*-axis is

$$\frac{(x - h)^2}{a^2} - \frac{(y - k)^2}{b^2} = 1$$

where $a^2 + b^2 = c^2$.

Vertices: $V(a + h, k)$ and $V'(-a + h, k)$
Foci: $F(c + h, k)$ and $F'(-c + h, k)$

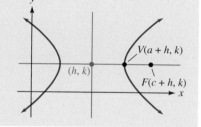

Hyperbola: Transverse Axis Vertical, Center at (*h*, *k*)

The standard equation of a hyperbola with center at (h, k) and foci on a line parallel to the *y*-axis is

$$\frac{(y - k)^2}{a^2} - \frac{(x - h)^2}{b^2} = 1$$

where $a^2 + b^2 = c^2$.

Vertices: $V(h, a + k)$ and $V'(h, -a + k)$
Foci: $F(h, c + k)$ and $F'(h, -c + k)$

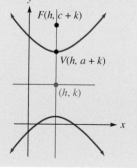

EXAMPLE 1   Write the equation of the hyperbola with vertices $(3, -3)$ and $(3, 3)$ and a focus at $(3, 5)$.

Solution   First, we plot the vertices and focus, as shown in Figure 12-26. Because the foci lie on a vertical line, the equation to use is

$$\frac{(y - k)^2}{a^2} - \frac{(x - h)^2}{b^2} = 1$$

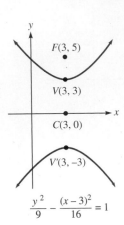

The center of the hyperbola is midway between the vertices V and V'. Thus, the center is point $(3, 0)$; $h = 3$; and $k = 0$. The distance between the vertex and the center is $a = 3$, and the distance between the focus and the center is $c = 5$. We can find b^2 by substituting 3 for a and 5 for c in the following equation to get

$$b^2 = c^2 - a^2 \qquad \text{In a hyperbola, } b^2 = c^2 - a^2.$$
$$b^2 = 5^2 - 3^2$$
$$b^2 = 16$$

Substituting the values of h, k, a^2, and b^2 into the standard equation gives the equation of the hyperbola:

$$\frac{(y - k)^2}{a^2} - \frac{(x - h)^2}{b^2} = 1$$
$$\frac{(y - 0)^2}{9} - \frac{(x - 3)^2}{16} = 1$$
$$\frac{y^2}{9} - \frac{(x - 3)^2}{16} = 1$$

$\dfrac{y^2}{9} - \dfrac{(x-3)^2}{16} = 1$

Figure 12-26

Self Check   Write the equation of the hyperbola with vertices $(3, 1)$ and $(-3, 1)$ and a focus at $(5, 1)$. ∎

Asymptotes of a Hyperbola

The values of a and b are important aids in graphing hyperbolas. To see their value, we consider the hyperbola

$$\frac{x^2}{a^2} - \frac{y^2}{b^2} = 1$$

with center at the origin and vertices at $V(a, 0)$ and $V'(-a, 0)$. We can plot points V, V', $B(0, b)$, and $B'(0, -b)$ and form rectangle $RSQP$, called the **fundamental rectangle,** as shown in Figure 12-27. We can show that the extended diagonals of this rectangle are asymptotes of the hyperbola. In the exercises, you will be asked to show that the equations of these two lines are

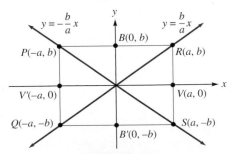

Figure 12-27

$$y = \frac{b}{a}x \quad \text{and} \quad y = -\frac{b}{a}x$$

To show that the extended diagonals are asymptotes of the hyperbola, we solve the equation

$$\frac{x^2}{a^2} - \frac{y^2}{b^2} = 1$$

for y and modify its form as follows:

$$\frac{x^2}{a^2} - \frac{y^2}{b^2} = 1$$

$$b^2x^2 - a^2y^2 = a^2b^2 \qquad \text{Multiply both sides by } a^2b^2.$$

$$y^2 = \frac{b^2x^2 - a^2b^2}{a^2} \qquad \text{Subtract } b^2x^2 \text{ from both sides and divide both sides by } -a^2.$$

$$y^2 = \frac{b^2x^2}{a^2}\left(1 - \frac{a^2}{x^2}\right) \qquad \text{Factor out } b^2x^2 \text{ in the numerator.}$$

(2) $$y = \pm\frac{bx}{a}\sqrt{1 - \frac{a^2}{x^2}} \qquad \text{Take the square root of both sides.}$$

In Equation 2, if a is constant and $|x|$ approaches ∞, then $\frac{a^2}{x^2}$ approaches 0, and $\sqrt{1 - \frac{a^2}{x^2}}$ approaches 1. Thus, the hyperbola approaches the lines

$$y = \frac{b}{a}x \quad \text{and} \quad y = -\frac{b}{a}x$$

Knowing the asymptotes makes it easy to sketch a hyperbola. We simply convert its equation into standard form, find the coordinates of its vertices, and construct the fundamental rectangle with its extended diagonals. Using the vertices and the asymptotes as guides, we can sketch the hyperbola shown in Figure 12-28. The segment BB' is called the **conjugate axis** of the hyperbola.

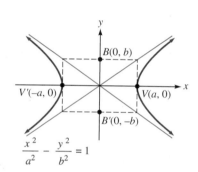

Figure 12-28

Graphing Equations of Hyperbolas

EXAMPLE 2   Graph: $x^2 - y^2 - 2x + 4y = 12$.

Solution   We complete the square on x and y to write the equation into standard form:

$$x^2 - 2x - y^2 + 4y = 12$$
$$x^2 - 2x - (y^2 - 4y) = 12$$
$$x^2 - 2x + \mathbf{1} - (y^2 - 4y + 4) = 12 + \mathbf{1} - 4$$
$$(x - 1)^2 - (y - 2)^2 = 9$$
$$\frac{(x - 1)^2}{9} - \frac{(y - 2)^2}{9} = 1$$

From the standard equation of a hyperbola, we see that the center is $(1, 2)$, that $a = 3$ and $b = 3$, and that the vertices are on a line segment parallel to the x-axis, as shown in Figure 12-29. The vertices V and V' are 3 units to the right and left of the center and have coordinates of $(4, 2)$ and $(-2, 2)$. Points B and B', 3 units above and below the center, have coordinates of $(1, 5)$ and $(1, -1)$. After using points V, V', B, and B' to construct the fundamental rectangle and its extended diagonals, we sketch the graph.

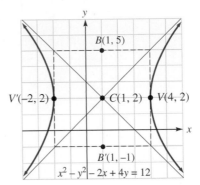

Figure 12-29

Self Check   Graph: $x^2 - y^2 + 6x + 2y = 8$.

Accent on Technology

To graph the equation $x^2 - y^2 - 2x + 4y = 12$ with a graphing calculator, we first solve the equation for y.

$$x^2 - y^2 - 2x + 4y = 12$$
$$-x^2 + y^2 + 2x - 4y = -12$$
$$y^2 - 4y = x^2 - 2x - 12$$
$$y^2 - 4y + \mathbf{4} = x^2 - 2x - 12 + \mathbf{4}$$
$$(y - 2)^2 = x^2 - 2x - 8$$
$$y - 2 = \pm\sqrt{x^2 - 2x - 8}$$
$$y = 2 \pm \sqrt{x^2 - 2x - 8}$$

If we use window settings of $[-4, 6]$ for x and $[-3, 7]$ for y and graph the functions

$$y = 2 + \sqrt{x^2 - 2x - 8} \qquad \text{and} \qquad y = 2 - \sqrt{x^2 - 2x - 8}$$

we will get a graph similar to the one shown in Figure 12-29.

We have considered only those hyperbolas with a major axis that is horizontal or vertical. However, some hyperbolas have nonhorizontal or nonvertical major axes. For example, the graph of the equation $xy = 4$ is a hyperbola with vertices at $(2, 2)$ and $(-2, -2)$, as shown in Figure 12-30.

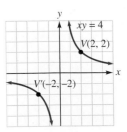

Figure 12-30

Self Check Answers

1. $\dfrac{x^2}{9} - \dfrac{(y-1)^2}{16} = 1$ **2.**

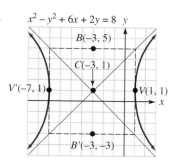

$x^2 - y^2 + 6x + 2y = 8$ y

$B(-3, 5)$

$C(-3, 1)$

$V'(-7, 1)$ $V(1, 1)$

x

$B'(-3, -3)$

Orals   *Find the x- or y-intercepts of each hyperbola.*

1. $\dfrac{x^2}{9} - \dfrac{y^2}{16} = 1$ **2.** $\dfrac{y^2}{25} - \dfrac{x^2}{36} = 1$

12.4   EXERCISES

REVIEW   *Find the inverse of each function. Write the answer in $y = f^{-1}(x)$ form.*

1. $f(x) = 3x - 2$ **2.** $f(x) = \dfrac{x+1}{x}$

3. $f(x) = \dfrac{5x}{x+2}$ **4.** $f(x) = x$

Let $f(x) = x^2 + 1$ and $g(x) = (x+1)^2$. Find each composite function.

5. $f(g(x))$ **6.** $g(f(x))$

7. $f(f(x))$ **8.** $g(g(x))$

VOCABULARY AND CONCEPTS   *Fill in the blanks.*

9. A hyperbola is the set of all points in the plane such that the _____ of the distances from two fixed points is a positive _____.

10. Each of the two fixed points in the definition of a hyperbola is called a _____ of the hyperbola.

11. The vertices of the hyperbola $\dfrac{x^2}{a^2} - \dfrac{y^2}{b^2} = 1$ are (__, __) and (____, __).

12. The vertices of the hyperbola $\dfrac{y^2}{a^2} - \dfrac{x^2}{b^2} = 1$ are (__, __) and (__, ____).

13. The chord that joins the vertices is called the _____ of the hyperbola.

14. In a hyperbola, the relation between a, b, and c is _____.

PRACTICE   *Write the equation of each hyperbola.*

15. Vertices $(5, 0)$ and $(-5, 0)$; focus $(7, 0)$

16. Focus $(3, 0)$; vertex $(2, 0)$; center $(0, 0)$

17. Center $(2, 4)$; $a = 2$, $b = 3$; transverse axis is horizontal

18. Center $(-1, 3)$; vertex $(1, 3)$; focus $(2, 3)$

19. Center $(5, 3)$; vertex $(5, 6)$; passes through $(1, 8)$

20. Foci $(0, 10)$ and $(0, -10)$; $\dfrac{c}{a} = \dfrac{5}{4}$

21. Vertices $(0, 3)$ and $(0, -3)$; $\dfrac{c}{a} = \dfrac{5}{3}$

22. Focus $(4, 0)$; vertex $(2, 0)$; center $(0, 0)$

23. Center $(1, -3)$; $a^2 = 4$; $b^2 = 16$

24. Center $(1, 4)$; focus $(7, 4)$; vertex $(3, 4)$

25. Center at the origin; passes through $(4, 2)$ and $(8, -6)$

26. Center $(3, -1)$; y-intercept -1; x-intercept $3 + \dfrac{3\sqrt{5}}{2}$

Find the area of the fundamental rectangle of each hyperbola.

27. $4(x - 1)^2 - 9(y + 2)^2 = 36$

28. $x^2 - y^2 - 4x - 6y = 6$

29. $x^2 + 6x - y^2 + 2y = -11$

30. $9x^2 - 4y^2 = 18x + 24y + 63$

Write the equation of each hyperbola.

31. Center $(-2, -4)$; $a = 2$; area of fundamental rectangle is 36 square units.

32. Center $(3, -5)$; $b = 6$; area of fundamental rectangle is 24 square units.

33. Vertex $(6, 0)$; one end of conjugate axis at $\left(0, \dfrac{5}{4}\right)$

34. Vertex $(3, 0)$; focus $(-5, 0)$; center $(0, 0)$

Graph each hyperbola.

35. $\dfrac{x^2}{9} - \dfrac{y^2}{4} = 1$

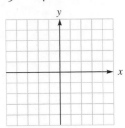

36. $\dfrac{y^2}{4} - \dfrac{x^2}{9} = 1$

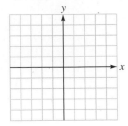

37. $4x^2 - 3y^2 = 36$

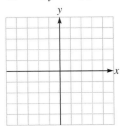

38. $3x^2 - 4y^2 = 36$

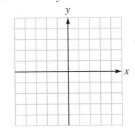

39. $y^2 - x^2 = 1$

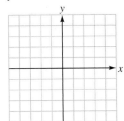

40. $9(y + 2)^2 - 4(x - 1)^2 = 36$

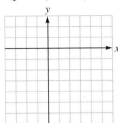

41. $4x^2 - 2y^2 + 8x - 8y = 8$

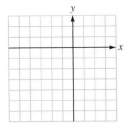

42. $x^2 - y^2 - 4x - 6y = 6$

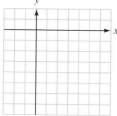

43. $y^2 - 4x^2 + 6y + 32x = 59$

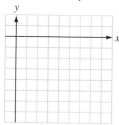

44. $x^2 + 6x - y^2 + 2y = -11$

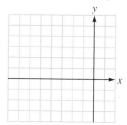

Graph each hyperbola.

45. $-xy = 6$ **46.** $xy = 20$

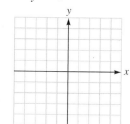   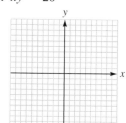

Find the equation of the curve on which point P lies.

47. The difference of the distances between $P(x, y)$ and the points $(-2, 1)$ and $(8, 1)$ is 6.

48. The difference of the distances between $P(x, y)$ and the points $(3, -1)$ and $(3, 5)$ is 5.

49. The distance between point $P(x, y)$ and the point $(0, 3)$ is $\frac{3}{2}$ of the distance between P and the line $y = -2$.

50. The distance between point $P(x, y)$ and the point $(5, 4)$ is $\frac{5}{3}$ of the distance between P and the line $x = -3$.

APPLICATIONS

51. Astronomy   Some comets may have a hyperbolic orbit, with the sun as one focus. When the comet shown in the illustration is far away from Earth, it appears to be approaching Earth along the line $y = 2x$. Find the equation of its orbit if the comet comes within 100 million miles of Earth.

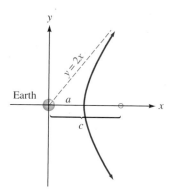

52. Physics   Parallel beams of similarly charged particles are shot from two atomic accelerators 20 meters apart, as shown in the illustration. If the particles were not deflected, the beams would be 2.0×10^{-4} meter apart. However, because the charged particles repel each other, the beams follow the hyperbolic path $y = \frac{k}{x}$, for some k. Find k.

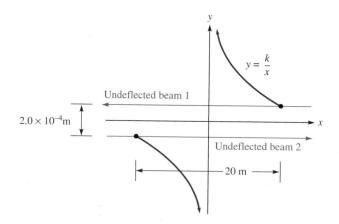

53. Navigation   The LORAN system (LOng RAnge Navigation) uses two radio transmitters 26 miles apart to send simultaneous signals. The navigator on a ship at $P(x, y)$ receives the closer signal first, and determines that the difference of the distances between the ship and each transmitter is 24 miles. That places the ship on a certain curve. Identify the curve and find its equation.

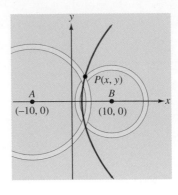

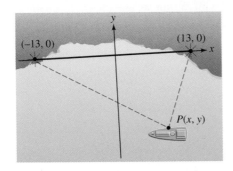

54. Wave propagation   Stones dropped into a calm pond at points A and B create ripples that propagate in widening circles. In the illustration, points A and B are 20 feet apart, and the radii of the circles differ by 12 feet. The point $P(x, y)$ where the circles intersect moves along a curve. Identify the curve and find its equation.

WRITING

55. Explain why the expansion of the standard equation of a hyperbola is a special case of the general equation of second degree with $B = 0$.

56. Describe how you can tell from the equation of a hyperbola whether the transverse axis is vertical or horizontal.

SOMETHING TO THINK ABOUT

57. Prove that $c > a$ for a hyperbola with center at $(0, 0)$ and line segment FF' on the x-axis.

58. Show that the extended diagonals of the fundamental rectangle of the hyperbola

$$\frac{x^2}{a^2} - \frac{y^2}{b^2} = 1 \text{ are } y = \frac{b}{a}x \text{ and } y = -\frac{b}{a}x.$$

12.5   Solving Simultaneous Second-Degree Equations

▌ Solving Systems by Graphing   ▌ Solving Systems Algebraically

Getting Ready   *Add the left-hand sides and the right-hand sides of the following equations.*

1.
$$3x^2 + 2y^2 = 12$$
$$\underline{4x^2 - 3y^2 = 32}$$

2.
$$-7x^2 - 5y^2 = -17$$
$$\underline{12x^2 + 2y^2 = 25}$$

We now discuss techniques for solving systems of two equations in two variables, where at least one of the equations is of second degree.

Solving Systems by Graphing

EXAMPLE 1   Solve $\begin{cases} x^2 + y^2 = 25 \\ 2x + y = 10 \end{cases}$ by graphing.

Solution   The graph of $x^2 + y^2 = 25$ is a circle with center at the origin and radius of 5. The graph of $2x + y = 10$ is a straight line. Depending on whether the line is a secant (intersecting the circle at two points) or a tangent (intersecting the circle at one point) or does not intersect the circle at all, there are two, one, or no solutions to the system, respectively.

After graphing the circle and the line, as shown in Figure 12-31, we see that there are two intersection points $(3, 4)$ and $(5, 0)$. The solutions of the system are

$$\begin{cases} x = 3 \\ y = 4 \end{cases} \quad \text{and} \quad \begin{cases} x = 5 \\ y = 0 \end{cases}$$

Verify that these are exact solutions.

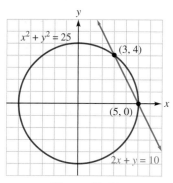

Figure 12-31

Self Check   Solve $\begin{cases} x^2 + y^2 = 25 \\ 2x - y = 5 \end{cases}$ graphically.   ∎

Accent on Technology   **SOLVING SYSTEMS OF EQUATIONS**

To solve the system of equations in Example 1 with a graphing calculator, we must first solve each of the equations for y. From the first equation, $x^2 + y^2 = 25$, we get two equations to graph:

$$y_1 = \sqrt{25 - x^2}$$
$$y_2 = -\sqrt{25 - x^2}$$

From the other equation, $2x + y = 10$, we get a third equation:

$$y_3 = 10 - 2x$$

The graphs of these equations will be similar to those shown in Figure 12-32(a). To find the top solution, we use ZOOM and TRACE to read the coordinates of the point of intersection, as shown in Figure 12-32(b): $x \approx 3$ and

$y \approx 4$. Because the graph is not complete at the other point of intersection shown in Figure 12-32(c), we must use our best judgment. We read the coordinates to be $x \approx 5$ and $y \approx 0$.

We can also find the coordinates of the intersection point by using the INTERSECT feature.

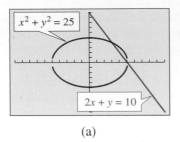

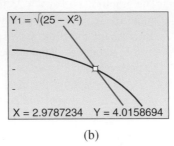

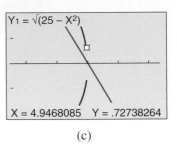

| (a) | (b) | (c) |

Figure 12-32

Solving Systems Algebraically

Algebraic methods can be used to find exact solutions.

EXAMPLE 2   Solve $\begin{cases} x^2 + y^2 = 25 \\ 2x + y = 10 \end{cases}$ algebraically.

Solution   If a system contains one equation of second degree and another of first degree, we can solve it by substitution. Solving the linear equation for y gives

$$2x + y = 10$$
$$y = \mathbf{-2x + 10}$$

We can now substitute $-2x + 10$ for y in the second-degree equation and solve the resulting quadratic equation for x:

$$x^2 + \mathbf{y}^2 = 25$$
$$x^2 + (\mathbf{-2x + 10})^2 = 25$$
$$x^2 + 4x^2 - 40x + 100 = 25 \qquad \text{Remove parentheses.}$$
$$5x^2 - 40x + 75 = 0 \qquad \text{Subtract 25 from both sides and combine like terms.}$$
$$x^2 - 8x + 15 = 0 \qquad \text{Divide both sides by 5.}$$
$$(x - 5)(x - 3) = 0 \qquad \text{Factor } x^2 - 8x + 15.$$
$$x - 5 = 0 \quad \text{or} \quad x - 3 = 0$$
$$x = 5 \quad | \qquad x = 3$$

Because $y = -2x + 10$, if $x = 5$, then $y = 0$; and if $x = 3$, then $y = 4$. The two solutions are

$$\begin{cases} x = 5 \\ y = 0 \end{cases} \quad \text{or} \quad \begin{cases} x = 3 \\ y = 4 \end{cases}$$

Self Check   Solve $\begin{cases} x^2 + y^2 = 25 \\ 2x - y = 5 \end{cases}$ algebraically.   ∎

EXAMPLE 3   Solve $\begin{cases} 4x^2 + 9y^2 = 5 \\ y = x^2 \end{cases}$ algebraically.

Solution   We can solve this system by substitution.

$$4\mathbf{x^2} + 9y^2 = 5$$
$$4\mathbf{y} + 9y^2 = 5 \qquad\qquad \text{Substitute } y \text{ for } x^2.$$
$$9y^2 + 4y - 5 = 0 \qquad\qquad \text{Add } -5 \text{ to both sides.}$$
$$(9y - 5)(y + 1) = 0 \qquad\qquad \text{Factor } 9y^2 + 4y - 5.$$
$$9y - 5 = 0 \quad \text{or} \quad y + 1 = 0$$
$$y = \frac{5}{9} \qquad\qquad\quad y = -1$$

Because $y = x^2$, we can find x by solving the equations

$$x^2 = \frac{5}{9} \qquad \text{and} \qquad x^2 = -\mathbf{1}$$

The solutions of $x^2 = \dfrac{5}{9}$ are

$$x = \frac{\sqrt{5}}{3} \qquad \text{and} \qquad x = -\frac{\sqrt{5}}{3}$$

The equation $x^2 = -1$ has no real solutions. Thus, the solutions of the system are

$$\left(\frac{\sqrt{5}}{3}, \frac{5}{9}\right) \qquad \text{and} \qquad \left(-\frac{\sqrt{5}}{3}, \frac{5}{9}\right)$$

Self Check   Solve: $\begin{cases} x^2 + 3y^2 = 13 \\ x = y^2 - 1 \end{cases}$.   ∎

EXAMPLE 4   Solve $\begin{cases} 3x^2 + 2y^2 = 36 \\ 4x^2 - y^2 = 4 \end{cases}$ algebraically.

Solution   When we have two second-degree equations of the form $ax^2 + by^2 = c$, we can solve the system by eliminating one of the variables by addition. To eliminate the terms involving y^2, we copy the first equation and multiply the second equation by 2 to obtain the following equivalent system.

$$\begin{cases} 3x^2 + 2y^2 = 36 \\ 8x^2 - 2y^2 = 8 \end{cases}$$

We can then add the equations and solve the resulting equation for x:

$$11x^2 = 44$$
$$x^2 = 4$$
$$x = 2 \quad \text{or} \quad x = -2$$

To find y, we substitute 2 for x and then -2 for x in the first equation.

For x = 2	***For x = −2***
$3x^2 + 2y^2 = 36$	$3x^2 + 2y^2 = 36$
$3(2)^2 + 2y^2 = 36$	$3(-2)^2 + 2y^2 = 36$
$12 + 2y^2 = 36$	$12 + 2y^2 = 36$
$2y^2 = 24$	$2y^2 = 24$
$y^2 = 12$	$y^2 = 12$
$y = \sqrt{12} \quad \text{or} \quad y = -\sqrt{12}$	$y = \sqrt{12} \quad \text{or} \quad y = -\sqrt{12}$
$y = 2\sqrt{3} \quad \mid \quad y = -2\sqrt{3}$	$y = 2\sqrt{3} \quad \mid \quad y = -2\sqrt{3}$

The four solutions of this system are

$$\left(2, 2\sqrt{3}\right), \quad \left(2, -2\sqrt{3}\right), \quad \left(-2, 2\sqrt{3}\right), \quad \text{and} \quad \left(-2, -2\sqrt{3}\right)$$

Self Check Solve: $\begin{cases} 2x^2 + y^2 = 23 \\ 3x^2 - 2y^2 = 17 \end{cases}$.

PERSPECTIVE

Satellite dishes and flashlight reflectors are familiar examples of a conic's ability to reflect a beam of light or to concentrate incoming satellite signals at one point. That property is shown in Illustration 1.

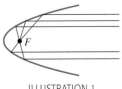

ILLUSTRATION 1

An ellipse has two foci, the points labeled *F* in Illustration 2. Any light or signal that starts at one focus will be reflected to the other. This property is the basis of whispering galleries, where a person standing at one focus can clearly hear another person speaking at the other focus.

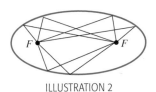

ILLUSTRATION 2

The focal property of the ellipse is also used in **lithotripsy,** a medical procedure for treating kidney stones. The patient is placed in an elliptical tank of water with the kidney stone at one focus. Shock waves from a small controlled explosion at the other focus are concentrated on the stone, pulverizing it.

The hyperbola also has two foci, the two points labeled *F* in Illustration 3. As in the ellipse, light aimed at one focus is reflected toward the other. Hyperbolic mirrors are used in some reflecting telescopes.

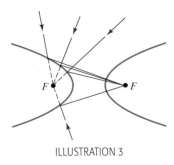

ILLUSTRATION 3

Self Check Answers

1.
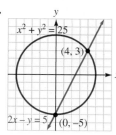

$\begin{cases} x = 4 \\ y = 3 \end{cases}$ and $\begin{cases} x = 0 \\ y = -5 \end{cases}$ **2.** $\begin{cases} x = 4 \\ y = 3 \end{cases}$ and $\begin{cases} x = 0 \\ y = -5 \end{cases}$ **3.** $\left(2, \sqrt{3}\right)$ and $\left(2, -\sqrt{3}\right)$

4. $\left(3, \sqrt{5}\right), \left(3, -\sqrt{5}\right), \left(-3, \sqrt{5}\right), \left(-3, -\sqrt{5}\right)$

Orals   *Give the possible number of solutions of a system when the graphs of the equations are*

1. a line and a parabola
2. a line and a hyperbola
3. a circle and a parabola
4. a circle and a hyperbola

12.5   EXERCISES

REVIEW   *Find the vertical, horizontal, or slant asymptotes of each of the following rational functions.*

1. $y = \dfrac{3x + 1}{x - 1}$

2. $y = \dfrac{x^2 + 3}{x - 1}$

3. $y = \dfrac{3x + 1}{x^2 - 1}$

4. $y = \dfrac{x^2 + 3}{x^2 + 1}$

Tell whether the graph has symmetry about the x-axis, y-axis, or origin, or none of these symmetries.

5. $y = \dfrac{3x^2 + 5}{5x^2 + 3}$

6. $y = \dfrac{3x^3 + 5}{5x^2 + 3}$

7. $y = \dfrac{3x^3}{5x^2 + 3}$

8. $y^2 = \dfrac{3x^3 + 5}{5x^2 + 3}$

VOCABULARY AND CONCEPTS   *Fill in the blanks.*

9. Solutions of systems of second-degree equations are points of intersection of _____ sections.

10. Approximate solutions can be found _____, and exact solutions are found _____.

PRACTICE   *Solve each system of equations by graphing.*

11. $\begin{cases} 8x^2 + 32y^2 = 256 \\ x = 2y \end{cases}$

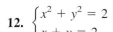

12. $\begin{cases} x^2 + y^2 = 2 \\ x + y = 2 \end{cases}$

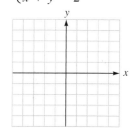

13. $\begin{cases} x^2 + y^2 = 90 \\ y = x^2 \end{cases}$

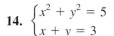

14. $\begin{cases} x^2 + y^2 = 5 \\ x + y = 3 \end{cases}$

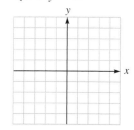

15. $\begin{cases} x^2 + y^2 = 25 \\ 12x^2 + 64y^2 = 768 \end{cases}$ **16.** $\begin{cases} x^2 + y^2 = 13 \\ y = x^2 - 1 \end{cases}$

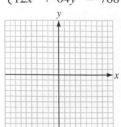   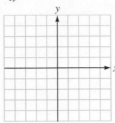

17. $\begin{cases} x^2 - 13 = -y^2 \\ y = 2x - 4 \end{cases}$ **18.** $\begin{cases} x^2 + y^2 = 20 \\ y = x^2 \end{cases}$

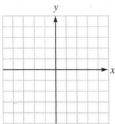   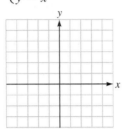

19. $\begin{cases} x^2 - 6x - y = -5 \\ x^2 - 6x + y = -5 \end{cases}$

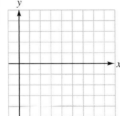

20. $\begin{cases} x^2 - y^2 = -5 \\ 3x^2 + 2y^2 = 30 \end{cases}$

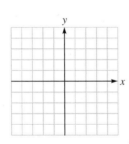

Use a graphing calculator to solve each system of equations.

21. $\begin{cases} y = x + 1 \\ y = x^2 + x \end{cases}$ **22.** $\begin{cases} y = 6 - x^2 \\ y = x^2 - x \end{cases}$

23. $\begin{cases} 6x^2 + 9y^2 = 10 \\ 3y - 2x = 0 \end{cases}$ **24.** $\begin{cases} x^2 + y^2 = 68 \\ y^2 - 3x^2 = 4 \end{cases}$

Solve each system of equations algebraically for real values of x and y.

25. $\begin{cases} 25x^2 + 9y^2 = 225 \\ 5x + 3y = 15 \end{cases}$ **26.** $\begin{cases} x^2 + y^2 = 20 \\ y = x^2 \end{cases}$

27. $\begin{cases} x^2 + y^2 = 2 \\ x + y = 2 \end{cases}$ **28.** $\begin{cases} x^2 + y^2 = 36 \\ 49x^2 + 36y^2 = 1{,}764 \end{cases}$

29. $\begin{cases} x^2 + y^2 = 5 \\ x + y = 3 \end{cases}$ **30.** $\begin{cases} x^2 - x - y = 2 \\ 4x - 3y = 0 \end{cases}$

31. $\begin{cases} x^2 + y^2 = 13 \\ y = x^2 - 1 \end{cases}$ **32.** $\begin{cases} x^2 + y^2 = 25 \\ 2x^2 - 3y^2 = 5 \end{cases}$

33. $\begin{cases} x^2 + y^2 = 30 \\ y = x^2 \end{cases}$ **34.** $\begin{cases} 9x^2 - 7y^2 = 81 \\ x^2 + y^2 = 9 \end{cases}$

35. $\begin{cases} x^2 + y^2 = 13 \\ x^2 - y^2 = 5 \end{cases}$ **36.** $\begin{cases} 2x^2 + y^2 = 6 \\ x^2 - y^2 = 3 \end{cases}$

37. $\begin{cases} x^2 + y^2 = 20 \\ x^2 - y^2 = -12 \end{cases}$ **38.** $\begin{cases} xy = -\dfrac{9}{2} \\ 3x + 2y = 6 \end{cases}$

39. $\begin{cases} y^2 = 40 - x^2 \\ y = x^2 - 10 \end{cases}$ **40.** $\begin{cases} x^2 - 6x - y = -5 \\ x^2 - 6x + y = -5 \end{cases}$

41. $\begin{cases} y = x^2 - 4 \\ x^2 - y^2 = -16 \end{cases}$ **42.** $\begin{cases} 6x^2 + 8y^2 = 182 \\ 8x^2 - 3y^2 = 24 \end{cases}$

43. $\begin{cases} x^2 - y^2 = -5 \\ 3x^2 + 2y^2 = 30 \end{cases}$ **44.** $\begin{cases} \dfrac{1}{x} + \dfrac{1}{y} = 5 \\ \dfrac{1}{x} - \dfrac{1}{y} = -3 \end{cases}$

45. $\begin{cases} \dfrac{1}{x} + \dfrac{2}{y} = 1 \\ \dfrac{2}{x} - \dfrac{1}{y} = \dfrac{1}{3} \end{cases}$ **46.** $\begin{cases} \dfrac{1}{x} + \dfrac{3}{y} = 4 \\ \dfrac{2}{x} - \dfrac{1}{y} = 7 \end{cases}$

47. $\begin{cases} 3y^2 = xy \\ 2x^2 + xy - 84 = 0 \end{cases}$ **48.** $\begin{cases} x^2 + y^2 = 10 \\ 2x^2 - 3y^2 = 5 \end{cases}$

49. $\begin{cases} xy = \dfrac{1}{6} \\ y + x = 5xy \end{cases}$ **50.** $\begin{cases} xy = \dfrac{1}{12} \\ y + x = 7xy \end{cases}$

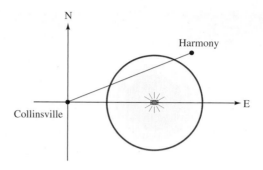

APPLICATIONS

51. Geometry The area of a rectangle is 63 square centimeters, and its perimeter is 32 centimeters. Find the dimensions of the rectangle.

52. Investments Grant receives $225 annual income from one investment. Jeff invested $500 more than Grant, but at an annual rate of 1% less. Jeff's annual income is $240. Find the amount and rate of Grant's investment.

53. Investments Carol receives $67.50 annual income from one investment. John invested $150 more than Carol at an annual rate of $1\frac{1}{2}\%$ more. John's annual income is $94.50. Find the amount and rate of Carol's investment. (*Hint:* There are two answers.)

54. Finding rate and time Jamal drove 306 miles. Jamal's brother made the same trip at a speed 17 miles per hour slower than Jamal did and required an extra $1\frac{1}{2}$ hours. Find Jamal's rate and time.

55. 📟 **Radio reception** A radio station located 120 miles due east of Collinsville has a listening radius of 100 miles. A straight road joins Collinsville with Harmony, a town 200 miles to the east and 100 miles north. How far from Collinsville will a driver first pick up the station?

56. 📟 **Listening range** For how many miles will a driver in Exercise 55 continue to receive the signal?

WRITING

57. Explain how to solve the system $\begin{cases} x^2 + y^2 = 4 \\ x^2 - y^2 = 4 \end{cases}$ graphically.

58. Explain how to solve the system in Exercise 57 algebraically.

SOMETHING TO THINK ABOUT *Consider a second-degree system of equations.*

59. Is it possible for a system of second-degree equations to have no common solution? If so, sketch the graphs of the equations of such a system.

60. Can a system have exactly one solution? If so, sketch the graphs of the equations of such a system.

61. Can a system have exactly two solutions? If so, sketch the possible graphs.

62. Can a system have three solutions? If so, sketch the graphs.

63. Can a system have exactly four solutions? If so, sketch the graphs.

64. Can a system have more than four solutions? If so, sketch the graphs.

Projects

Project 1

To draw a hyperbola, tie a small loop in a long piece of string, large enough to hold the point of a pencil. As in Illustration 1, arrange the string around nails at *A* and *B* and hold a pencil in the loop at *P*. While keeping the string taut, carefully pull both ends together. Explain why the pencil traces a hyperbola. From the measured distance between *A* and *B* and the *difference* of lengths *PA* and *PB*, find the equation of the curve. How would you draw the other branch of the hyperbola?

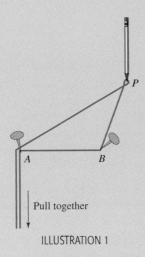

ILLUSTRATION 1

Project 2

Draw a line on a sheet of paper and mark a point *P* about 2 inches away, as in Illustration 2(a). (Use wax paper, and draw the line and point with a fine-tip marker.) Fold the paper over to place the point on the line and make a sharp crease, as in Illustration 2(b). Fold the paper several times, placing *P* at different locations on the line. What curve do these creases seem to describe? What is the relation of the point and the line to that curve? What happens if point *P* is closer to the line? Write a paragraph to describe your findings.

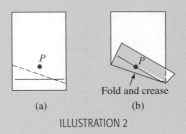

ILLUSTRATION 2

Project 3

Draw a circle on a sheet of paper and mark point *P* inside. Fold and crease several times, placing *P* at several locations on the circle, as in Illustration 3. What curve do the creases describe? What is the significance of point *P*?

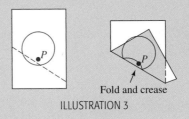

ILLUSTRATION 3

Project 4

This time, draw a circle and mark point *P* on the outside, as in Illustration 4. Fold and crease as before. What curve is described by these creases, and what is the significance of *P*? Do you get two branches of the curve?

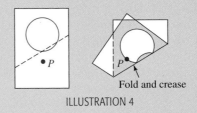

ILLUSTRATION 4

Chapter Summary

CONCEPTS	REVIEW EXERCISES
12.1	**The Circle**

Circle, center at (0, 0), radius r:
$$x^2 + y^2 = r^2$$

Circle, center at (h, k), radius r:
$$(x - h)^2 + (y - k)^2 = r^2$$

General form of a second-degree equation in x and y:
$$Ax^2 + Bxy + Cy^2 + Dx + Ey + F = 0$$

Write the equation of each circle.

1. Center (0, 0); radius 4

2. Center (0, 0); passes through (6, 8)

3. Center (3, −2); radius 5

4. Center (−2, 4); passes through (1, 0)

5. Endpoints of diameter (−2, 4) and (12, 16)

6. Endpoints of diameter (−3, −6) and (7, 10)

Write the equation of each circle in standard form.

7. $x^2 + y^2 - 6x + 4y = 3$

8. $x^2 + 4x + y^2 - 10y = -13$

12.2	**The Parabola**

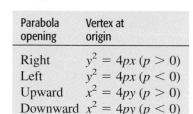

Parabola opening	Vertex at origin
Right	$y^2 = 4px \ (p > 0)$
Left	$y^2 = 4px \ (p < 0)$
Upward	$x^2 = 4py \ (p > 0)$
Downward	$x^2 = 4py \ (p < 0)$

Write the equation of each parabola.

9. Vertex (0, 0); passes through (−8, 4) and (−8, −4)

10. Vertex (0, 0); passes through (−8, 4) and (8, 4)

11. Find the equation of the parabola with vertex at (−2, 3), curve passing through point (−4, −8), and opening downward.

Graph each equation.

12. $x^2 - 4y - 2x + 9 = 0$

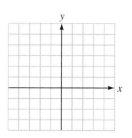

Parabola opening	Vertex at $V(h, k)$
Right	$(y - k)^2 = 4p(x - h)$ $(p > 0)$
Left	$(y - k)^2 = 4p(x - h)$ $(p < 0)$
Upward	$(x - h)^2 = 4p(y - k)$ $(p > 0)$
Downward	$(x - h)^2 = 4p(y - k)$ $(p < 0)$

13. $y^2 - 6y = 4x - 13$

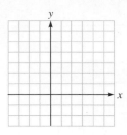

12.3   The Ellipse

Ellipse, center at $(0, 0)$:
Major axis is $2a$; minor axis is $2b$ $(a > b > 0)$.
Center-to-focus distance is c, where $a^2 = b^2 + c^2$.

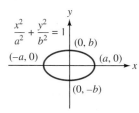

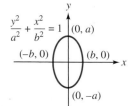

14. Write the equation of the ellipse with center at the origin, major axis that is horizontal and 12 units long, and minor axis 8 units long.

15. Write the equation of the ellipse with center at the origin, major axis that is vertical and 10 units long, and minor axis 4 units long.

Ellipse, center at (h, k):
Major axis is $2a$, minor axis is
$2b$ $(a > b > 0$.

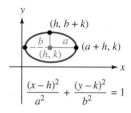

$$\frac{(x-h)^2}{a^2} + \frac{(y-k)^2}{b^2} = 1$$

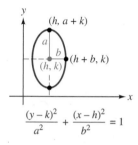

$$\frac{(y-k)^2}{a^2} + \frac{(x-h)^2}{b^2} = 1$$

16. Write the equation of the ellipse with center at point $(-2, 3)$ and curve passing through points $(-2, 0)$ and $(2, 3)$.

17. Write the equation in standard form and graph it.
$4x^2 + y^2 - 16x + 2y = -13$

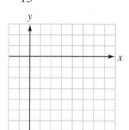

12.4   The Hyperbola

Hyperbola:
Center at $(0, 0)$, center-to-focus
distance is c, where
$a^2 + b^2 = c^2$

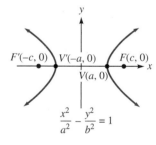

$$\frac{x^2}{a^2} - \frac{y^2}{b^2} = 1$$

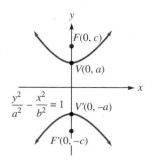

$$\frac{y^2}{a^2} - \frac{x^2}{b^2} = 1$$

18. Write the equation of the hyperbola with center at the origin, passing through points $(-2, 0)$ and $(2, 0)$, and having a focus at $(4, 0)$.

19. Write the equation of the hyperbola with center at the origin, one focus at $(0, 5)$, and one vertex at $(0, 3)$.

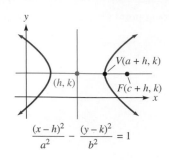

$$\frac{(x-h)^2}{a^2} - \frac{(y-k)^2}{b^2} = 1$$

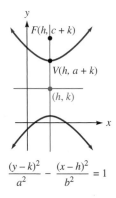

$$\frac{(y-k)^2}{a^2} - \frac{(x-h)^2}{b^2} = 1$$

The extended diagonals of the fundamental rectangle are asymptotes of the graph of a hyperbola.

20. Write the equation of the hyperbola with vertices at point $(-3, 3)$ and $(3, 3)$ and a focus at point $(5, 3)$.

21. Write the equation of the hyperbola with vertices at point $(3, -3)$ and $(3, 3)$ and a focus at point $(3, 5)$.

22. Write the equation of the asymptotes of the hyperbola $\dfrac{x^2}{25} - \dfrac{y^2}{16} = 1$.

23. Write the equation in standard form and graph it.
$$9x^2 - 4y^2 - 16y - 18x = 43$$

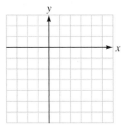

24. Graph: $4xy = 1$.

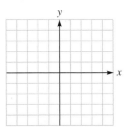

12.5 Solving Simultaneous Second-Degree Equations

Good estimates for solutions of systems of simultaneous second-degree equations can be found by graphing.

25. Solve by graphing: $\begin{cases} 3x^2 + y^2 = 52 \\ x^2 - y^2 = 12 \end{cases}$

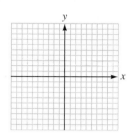

26. Solve by graphing: $\begin{cases} x^2 + y^2 = 16 \\ y = x + 4 \end{cases}$

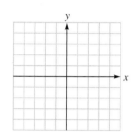

27. Solve by graphing: $\begin{cases} \dfrac{x^2}{16} + \dfrac{y^2}{12} = 1 \\ x^2 - \dfrac{y^2}{3} = 1 \end{cases}$

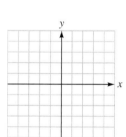

Exact solutions for systems of simultaneous second-degree equations can often be found by algebraic techniques.

28. Solve algebraically: $\begin{cases} 3x^2 + y^2 = 52 \\ x^2 - y^2 = 12 \end{cases}$

29. Solve algebraically: $\begin{cases} x^2 + y^2 = 16 \\ -\sqrt{3}y + 4\sqrt{3} = 3x \end{cases}$

30. Solve algebraically: $\begin{cases} \dfrac{x^2}{16} + \dfrac{y^2}{12} = 1 \\ x^2 - \dfrac{y^2}{2} = 1 \end{cases}$

Chapter Test

In Problems 1–3, write the equation of each circle.

1. Center $(2, 3)$; $r = 3$

2. Ends of diameter at $(-2, -2)$ and $(6, 8)$

3. Center $(2, -5)$, passes through $(7, 7)$

4. Change the equation of the circle
$x^2 + y^2 - 4x + 6y + 4 = 0$ to standard form and graph it.

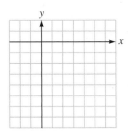

In Problems 5–7, find the equation of each parabola.

5. Vertex $(3, 2)$; focus at $(3, 6)$

6. Vertex $(4, -6)$; passes through $(3, -8)$ and $(3, -4)$

7. Vertex $(2, -3)$; passes through $(0, 0)$

8. Change the equation of the parabola
$x^2 - 6x - 8y = 7$ to standard form and graph it.

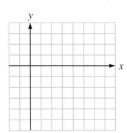

In Problems 9–11, find the equation of each ellipse.

9. Vertex $(10, 0)$, center at the origin, focus at $(6, 0)$

10. Minor axis 24, center at the origin, focus at $(5, 0)$

11. Center $(2, 3)$; passes through $(2, 9)$ and $(0, 3)$

12. Change the equation of the ellipse
$9x^2 + 4y^2 - 18x - 16y - 11 = 0$ to standard form and graph it.

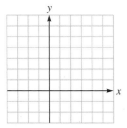

In Problems 13–15, find the equation of each hyperbola.

13. Center at the origin, focus at $(13, 0)$, vertex at $(5, 0)$

14. Vertices $(6, 0)$ and $(-6, 0)$; $\dfrac{c}{a} = \dfrac{13}{12}$

15. Center $(2, -1)$, major axis horizontal and of length 16, distance of 20 between foci

16. Change the equation of the hyperbola
$x^2 - 4y^2 + 16y = 8$ to standard form and graph it.

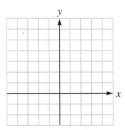

Solve each system algebraically.

17. $\begin{cases} x^2 + y^2 = 23 \\ y = x^2 - 3 \end{cases}$

18. $\begin{cases} 2x^2 - 3y^2 = 9 \\ x^2 + y^2 = 27 \end{cases}$

Complete the square to write each equation in standard form, and identify the curve.

19. $y^2 - 4y - 6x - 14 = 0$

20. $2x^2 + 3y^2 - 4x + 12y + 8 = 0$

CUMULATIVE REVIEW EXERCISES

Graph the function defined by each equation.

1. $f(x) = 3^x - 2$

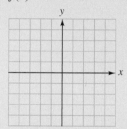

2. $f(x) = 2e^x$

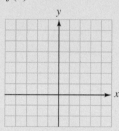

3. $f(x) = \log_3 x$

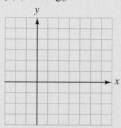

4. $f(x) = \ln(x - 2)$

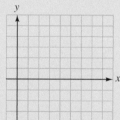

Find each value.

5. $\log_2 64$

6. $\log_{1/2} 8$

7. $\ln e^3$

8. $2^{\log_2 2}$

Write each expression in terms of the logarithms of a, b, and c.

9. $\log abc$

10. $\log \dfrac{a^2 b}{c}$

11. $\log \sqrt{\dfrac{ab}{c^3}}$

12. $\log \dfrac{\sqrt{ab^2}}{c}$

Write each expression as the logarithm of a single quantity.

13. $3 \log a - 3 \log b$

14. $\dfrac{1}{2} \log a + 3 \log b - \dfrac{2}{3} \log c$

Solve each equation.

15. $3^{x+1} = 8$

16. $3^{x-1} = 3^{2x}$

17. $\log x + \log 2 = 3$

18. $\log(x + 1) + \log(x - 1) = 1$

Let $P(x) = 4x^3 + 3x + 2$. Use synthetic division to find each value.

19. $P(1)$ **20.** $P(-2)$

21. $P\left(\dfrac{1}{2}\right)$ **22.** $P(i)$

Tell whether each binomial is a factor of $P(x) = x^3 + 2x^2 - x - 2$. Use synthetic division.

23. $x + 1$ **24.** $x - 2$

25. $x - 1$ **26.** $x + 2$

Tell how many roots each equation has.

27. $x^{12} - 4x^8 + 2x^4 + 12 = 0$

28. $x^{2,000} - 1 = 0$

Tell the number of possible positive, negative, and nonreal roots of each equation.

29. $x^4 + 2x^3 - 3x^2 + x + 2 = 0$

30. $x^4 - 3x^3 - 2x^2 - 3x - 5 = 0$

Solve each equation.

31. $x^3 + x^2 - 9x - 9 = 0$

32. $x^3 - 2x^2 - x + 2 = 0$

Let $f(x) = 3x^2 + 2$ and $g(x) = 2x - 1$. Find each value or composite function.

33. $f(-1)$ **34.** $(g \circ f)(2)$

35. $(f \circ g)(x)$

36. $(g \circ f)(x)$

Find the inverse of each function.

37. $f(x) = 3x - 2$

38. $f(x) = x^3 - 4$

Graph each rational function.

39. $f(x) = \dfrac{x^2}{x^2 - 9}$

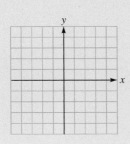

40. $f(x) = \dfrac{3x^2}{x^2 + 1}$

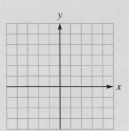

41. $f(x) = \dfrac{x^3 + x}{x^2 - 1}$

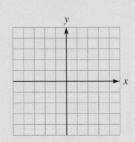

42. $f(x) = \dfrac{2x^2 - 3x - 2}{x - 2}$

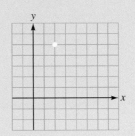

Decompose each fraction into partial fractions.

43. $\dfrac{5x}{(2x - 3)(x + 1)}$

44. $\dfrac{3x^2 + x + 2}{x^3 + 2x}$

13 Natural-Number Functions and Probability

InfoTrac Project

Do a keyword search on "powerball." Find the article "Powerball hits 10-year mark on April 19: The multi-state game has changed lotteries, but not everyone is celebrating." Write a summary of the article.

In Powerball, an official draws five white balls out of a drum that contains 53 white balls and one red ball out of a drum that contains 42 red balls. To win the jackpot, a person must match the numbers on the five white balls and the one red ball. (Order is not important.) What is the probability of winning the jackpot? Explain the statement "The probability of playing and winning the jackpot is the same as the probability of **not** playing and winning the jackpot."

Complete this project after studying Section 13.7.

© Reuters NewMedia, Inc./CORBIS

Mathematics in Finance

An **annuity** is a sequence of equal payments made periodically over a length of time. The sum of the payments and the interest earned during the **term of the annuity** is called the **amount of the annuity.**

After a sales clerk works six months, her employer will begin an annuity for her and will contribute $500 semiannually to a fund that pays 8% annual interest. After she has been employed for two years, what will be the amount of her annuity?

**Example 9
Section 13.4**

767

In this chapter, we introduce several topics that have applications in advanced mathematics and in many occupational areas. The binomial theorem, permutations, and combinations are used in statistics. Arithmetic and geometric sequences are used in the mathematics of finance.

13.1   The Binomial Theorem

■ **Expanding Binomials**   ■ **Pascal's Triangle**   ■ **Factorial Notation**
■ **The Binomial Theorem**
■ **Finding a Particular Term of a Binomial Expansion**

Getting Ready   *Raise each binomial to the indicated power.*

1. $(x + 2)^2$ **2.** $(x - 3)^2$
3. $(x + 1)^3$ **4.** $(x - 2)^3$

In this chapter, we introduce a way to expand powers of binomials of the form $(x + y)^n$. This method leads to ideas needed when working with probability and statistics.

Expanding Binomials

We begin this chapter by introducing a way to expand binomials of the form $(x + y)^n$, where n is a natural number. This method, called the **binomial theorem,** is important when we are working with probability and statistics. Consider the following expansions.

$$(a + b)^0 = 1$$
$$(a + b)^1 = a + b$$
$$(a + b)^2 = a^2 + 2ab + b^2$$
$$(a + b)^3 = a^3 + 3a^2b + 3ab^2 + b^3$$
$$(a + b)^4 = a^4 + 4a^3b + 6a^2b^2 + 4ab^3 + b^4$$
$$(a + b)^5 = a^5 + 5a^4b + 10a^3b^2 + 10a^2b^3 + 5ab^4 + b^5$$
$$(a + b)^6 = a^6 + 6a^5b + 15a^4b^2 + 20a^3b^3 + 15a^2b^4 + 6ab^5 + b^6$$

Four patterns are apparent in the above expansions:

1. Each expansion has one more term than the power of the binomial.
2. The degree of each term in each expansion equals the exponent of the binomial.
3. The first term in each expansion is a raised to the power of the binomial.

13.1 The Binomial Theorem **769**

4. The exponents of a decrease by 1 in each successive term, and the exponents of b, beginning with b^0 in the first term, increase by 1 in each successive term.

Pascal's Triangle

To see another pattern, we write the coefficients in a triangular array:

$$
\begin{array}{c}
(a+b)^0 = \qquad\qquad\qquad 1 \\
(a+b)^1 = \qquad\qquad 1 \quad 1 \\
(a+b)^2 = \qquad\quad 1 \quad 2 \quad 1 \\
(a+b)^3 = \quad\; 1 \quad 3 \quad 3 \quad 1 \\
(a+b)^4 = \quad 1 \quad 4 \quad 6 \quad 4 \quad 1 \\
(a+b)^5 = \; 1 \quad 5 \quad 10 \quad 10 \quad 5 \quad 1 \\
(a+b)^6 = 1 \quad 6 \quad 15 \quad 20 \quad 15 \quad 6 \quad 1
\end{array}
$$

In this array, each entry other than the 1's is the sum of the closest pair of numbers in the line above it. For example, the 20 in the bottom row is the sum of the 10's above it. The first 3 in the fourth row is the sum of the 1 and 2 above it.

This array, called **Pascal's triangle** after Blaise Pascal (1623–1662), continues with the same pattern forever. The next two lines are

$$
\begin{array}{c}
(a+b)^7 = \quad 1 \quad 7 \quad 21 \quad 35 \quad 35 \quad 21 \quad 7 \quad 1 \\
(a+b)^8 = 1 \quad 8 \quad 28 \quad 56 \quad 70 \quad 56 \quad 28 \quad 8 \quad 1
\end{array}
$$

EXAMPLE 1 Expand: $(x+y)^6$.

Solution The first term is x^6, and the exponents of x will decrease by 1 in each successive term. A y will appear in the second term, and the exponents on y will increase by 1 in each successive term, concluding when the term y^6 is reached. The variables in the expansion are

$$x^6 \quad x^5y \quad x^4y^2 \quad x^3y^3 \quad x^2y^4 \quad xy^5 \quad y^6$$

We can use Pascal's triangle to find the coefficients of the variables. Because the binomial is raised to the sixth power, we choose the row in Pascal's triangle where the second entry is 6. The coefficients of the variables are the numbers in that row.

$$1 \quad 6 \quad 15 \quad 20 \quad 15 \quad 6 \quad 1$$

Putting this information together gives the expansion:

$$(x+y)^6 = x^6 + 6x^5y + 15x^4y^2 + 20x^3y^3 + 15x^2y^4 + 6xy^5 + y^6$$

Self Check Expand: $(p+q)^3$. ∎

EXAMPLE 2 Expand: $(x-y)^6$.

Solution We write the binomial as $[x+(-y)]^6$ and substitute $-y$ for y in the result of Example 1.

$$
\begin{aligned}
&[x+(-y)]^6 \\
&= x^6 + 6x^5(-y) + 15x^4(-y)^2 + 20x^3(-y)^3 + 15x^2(-y)^4 + 6x(-y)^5 + (-y)^6 \\
&= x^6 - 6x^5y + 15x^4y^2 - 20x^3y^3 + 15x^2y^4 - 6xy^5 + y^6
\end{aligned}
$$

In general, the signs in the expansion of $(x - y)^n$ alternate. The sign of the first term is $+$, the sign of the second term is $-$, and so on.

Self Check   Expand: $(p - q)^3$. ∎

Factorial Notation

To use the binomial theorem to expand a binomial, we will need **factorial notation.**

Factorial Notation

If n is a natural number, the symbol $n!$ (read as **"n factorial"** or as **"factorial n"**) is defined as

$$n! = n(n - 1)(n - 2)(n - 3) \cdots (3)(2)(1)$$

EXAMPLE 3   Evaluate: **a.** $3!$,   **b.** $6!$,   and   **c.** $10!$.

Solution   **a.** $3! = 3 \cdot 2 \cdot 1 = 6$
  b. $6! = 6 \cdot 5 \cdot 4 \cdot 3 \cdot 2 \cdot 1 = 720$
  c. $10! = 10 \cdot 9 \cdot 8 \cdot 7 \cdot 6 \cdot 5 \cdot 4 \cdot 3 \cdot 2 \cdot 1 = 3{,}628{,}800$

Self Check   Evaluate: **a.** $4!$ and   **b.** $7!$ ∎

There are two fundamental properties of factorials.

Properties of Factorials

1. By definition, $0! = 1$.
2. If n is a natural number, $n(n - 1)! = n!$.

EXAMPLE 4   Show that   **a.** $6 \cdot 5! = 6!$ and   **b.** $8 \cdot 7! = 8!$.

Solution   **a.** $6 \cdot 5! = 6(5 \cdot 4 \cdot 3 \cdot 2 \cdot 1)$ **b.** $8 \cdot 7! = 8(7 \cdot 6 \cdot 5 \cdot 4 \cdot 3 \cdot 2 \cdot 1)$
    $= 6 \cdot 5 \cdot 4 \cdot 3 \cdot 2 \cdot 1$ $= 8 \cdot 7 \cdot 6 \cdot 5 \cdot 4 \cdot 3 \cdot 2 \cdot 1$
    $= 6!$ $= 8!$

Self Check   Show that $4 \cdot 3! = 4!$. ∎

The Binomial Theorem

We can now state the binomial theorem.

The Binomial Theorem

If n is any positive integer, then

$$(a + b)^n = a^n + \frac{n!}{1!(n - 1)!}a^{n-1}b + \frac{n!}{2!(n - 2)!}a^{n-2}b^2$$
$$+ \frac{n!}{3!(n - 3)!}a^{n-3}b^3 + \cdots + b^n$$

In the binomial theorem, the exponents of the variables in each term on the right-hand side follow the familiar patterns:

1. The sum of the exponents of a and b in each term is n.

2. The exponents on a decrease by 1 in each successive term.

3. The exponents on b increase by 1 in each successive term.

However, the method of finding the coefficients is different. Except for the first and last terms, $n!$ is the numerator of each fractional coefficient. When the exponent on b is 2, the factors in the denominator of the fractional coefficient are $2!$ and $(n-2)!$. When the exponent on b is 3, the factors in the denominator are $3!$ and $(n-3)!$, and so on.

EXAMPLE 5 Expand: $(a+b)^5$.

Solution We substitute directly into the binomial theorem.

$$(a+b)^5 = a^5 + \frac{5!}{1!(5-1)!}a^4b + \frac{5!}{2!(5-2)!}a^3b^2 + \frac{5!}{3!(5-3)!}a^2b^3 + \frac{5!}{4!(5-4)!}ab^4 + b^5$$

$$= a^5 + \frac{5 \cdot 4!}{1 \cdot 4!}a^4b + \frac{5 \cdot 4 \cdot 3!}{2 \cdot 1 \cdot 3!}a^3b^2 + \frac{5 \cdot 4 \cdot 3!}{3! \cdot 2 \cdot 1}a^2b^3 + \frac{5 \cdot 4!}{4! \cdot 1}ab^4 + b^5$$

$$= a^5 + 5a^4b + 10a^3b^2 + 10a^2b^3 + 5ab^4 + b^5$$

We note that the coefficients are the same numbers as in the sixth row of Pascal's triangle.

Self Check Expand: $(p+q)^4$. ∎

EXAMPLE 6 Expand: $(2x-3y)^4$.

Solution We first find the expansion of $(a+b)^4$.

$$(a+b)^4 = a^4 + \frac{4!}{1!(4-1)!}a^3b + \frac{4!}{2!(4-2)!}a^2b^2 + \frac{4!}{3!(4-3)!}ab^3 + b^4$$

$$= a^4 + \frac{4 \cdot 3!}{1 \cdot 3!}a^3b + \frac{4 \cdot 3 \cdot 2!}{2 \cdot 1 \cdot 2!}a^2b^2 + \frac{4 \cdot 3!}{3! \cdot 1}ab^3 + b^4$$

$$= a^4 + 4a^3b + 6a^2b^2 + 4ab^3 + b^4$$

We then substitute $2x$ for a and $-3y$ for b in the result.

$$(a+b)^4 = a^4 + 4a^3b + 6a^2b^2 + 4ab^3 + b^4$$
$$[2x+(-3y)]^4 = (2x)^4 + 4(2x)^3(-3y) + 6(2x)^2(-3y)^2 + 4(2x)(-3y)^3 + (-3y)^4$$
$$(2x-3y)^4 = 16x^4 - 96x^3y + 216x^2y^2 - 216xy^3 + 81y^4$$

Self Check Expand: $(3x-2y)^4$. ∎

Finding a Particular Term of a Binomial Expansion

Suppose that we wish to find the fifth term of the expansion of $(a+b)^{11}$. It would be tedious to raise the binomial to the 11th power and then look at the fifth term. The binomial theorem provides an easier way.

EXAMPLE 7 Find the fifth term of the expansion of $(a + b)^{11}$.

Solution In the fifth term, the exponent on b is 4 (the exponent on b is always 1 less than the number of the term). Since the exponent on b added to the exponent on a equals 11, the exponent on a is 7. The variables of the fifth term are a^7b^4.

The number in the numerator of the fractional coefficient is $n!$, which in this case is 11!. The factors in the denominator are 4! and $(11 - 4)!$. The complete fifth term is

$$\frac{11!}{4!(11 - 4)!}a^7b^4 = \frac{11!}{4!7!}a^7b^4$$
$$= \frac{11 \cdot 10 \cdot 9 \cdot 8 \cdot 7!}{4 \cdot 3 \cdot 2 \cdot 1 \cdot 7!}a^7b^4$$
$$= 330a^7b^4$$

Self Check Find the sixth term of the expansion in Example 7. ∎

EXAMPLE 8 Find the sixth term of the expansion of $(a + b)^9$.

Solution In the sixth term, the exponent on b is 5, and the exponent on a is $9 - 5$, or 4. The numerator of the fractional coefficient is 9!, and the factors in the denominator are 5! and $(9 - 5)!$. The sixth term of the expansion is

$$\frac{9!}{5!(9 - 5)!}a^4b^5 = \frac{9 \cdot 8 \cdot 7 \cdot 6 \cdot 5!}{5! \cdot 4!}a^4b^5$$
$$= \frac{9 \cdot 8 \cdot 7 \cdot 6}{4 \cdot 3 \cdot 2 \cdot 1}a^4b^5$$
$$- 126a^4b^5$$

Self Check Find the fifth term of the expansion in Example 8. ∎

EXAMPLE 9 Find the third term of the expansion of $(3x - 2y)^6$.

Solution We begin by finding the third term of the expansion of $(a + b)^6$.

(1) $\qquad \dfrac{6!}{2!(6 - 2)!}a^4b^2 = \dfrac{6 \cdot 5 \cdot 4!}{2 \cdot 1 \cdot 4!}a^4b^2 = 15a^4b^2$

We can then substitute $3x$ for a and $-2y$ for b in Equation 1 to obtain the third term of the expansion of $(3x - 2y)^6$.

$$15a^4b^2 = 15(3x)^4(-2y)^2$$
$$= 15(3)^4(-2)^2x^4y^2$$
$$= 4{,}860x^4y^2$$

Self Check Find the fourth term of the expansion in Example 9. ∎

Self Check Answers

1. $p^3 + 3p^2q + 3pq^2 + q^3$ **2.** $p^3 - 3p^2q + 3pq^2 - q^3$ **3. a.** 24, **b.** 5,040 **4.** $4 \cdot (3 \cdot 2 \cdot 1) = 4!$
5. $p^4 + 4p^3q + 6p^2q^2 + 4pq^3 + q^4$ **6.** $81x^4 - 216x^3y + 216x^2y^2 - 96xy^3 + 16y^4$ **7.** $462a^6b^5$ **8.** $126a^5b^4$
9. $-4{,}320x^3y^3$

Orals   *Find each value.*

1. 1!   **2.** 2!   **3.** 0!   **4.** 5!

Expand each binomial.

5. $(m + n)^2$ **6.** $(m - n)^2$

7. $(p + 2q)^2$ **8.** $(2p - q)^2$

13.1   EXERCISES

REVIEW   *Factor each expression.*

1. $3x^3y^2z^4 - 6xyz^5 + 15x^2yz^2$

2. $3z^2 - 15tz + 12t^2$

3. $a^4 - b^4$

4. $3r^4 - 36r^3 - 135r^2$

Simplify each complex fraction.

5. $\dfrac{\dfrac{1}{x} + \dfrac{1}{3}}{\dfrac{1}{x} - \dfrac{1}{3}}$ **6.** $\dfrac{x - \dfrac{1}{y}}{y - \dfrac{1}{x}}$

VOCABULARY AND CONCEPTS   *Fill in the blanks.*

7. In the expansion of a binomial, there will be one more term than the _____ of the binomial.

8. The _____ of each term in a binomial expansion is the same as the exponent of the binomial.

9. The _____ term in a binomial expansion is the first term raised to the power of the binomial.

10. In the expansion of $(p + q)^n$, the _____ on p decrease by 1 in each successive term.

11. $7! = $ _____ $= 5,040$.

12. $0! = $ __

13. $n \cdot$ _____ $= n!$

14. In the seventh term of $(a + b)^{11}$, the exponent on a is __.

PRACTICE   *Evaluate each expression.*

15. 4!   **16.** $-5!$

17. $3! \cdot 6!$ **18.** $0! \cdot 7!$

19. $6! + 6!$ **20.** $5! - 2!$

21. $\dfrac{9!}{12!}$ **22.** $\dfrac{8!}{5!}$

23. $\dfrac{5! \cdot 7!}{9!}$ **24.** $\dfrac{3! \cdot 5! \cdot 7!}{1! \cdot 8!}$

25. $\dfrac{18!}{6!(18 - 6)!}$ **26.** $\dfrac{15!}{9!(15 - 9)!}$

Use the binomial theorem to expand each binomial.

27. $(a + b)^3$

28. $(a + b)^4$

29. $(a - b)^5$

30. $(x - y)^4$

31. $(2x + y)^3$

32. $(x + 2y)^3$

33. $(x - 2y)^3$

34. $(2x - y)^3$

35. $(2x + 3y)^4$

36. $(2x - 3y)^4$

37. $(x - 2y)^4$

38. $(x + 2y)^4$

39. $(x - 3y)^5$

40. $(3x - y)^5$

41. $\left(\dfrac{x}{2} + y\right)^4$

42. $\left(x + \dfrac{y}{2}\right)^4$

Find the required term in each binomial expansion.

43. $(a + b)^4$; 3rd term   **44.** $(a - b)^4$; 2nd term

45. $(a + b)^7$; 5th term   **46.** $(a + b)^5$; 4th term

47. $(a - b)^5$; 6th term   **48.** $(a - b)^8$; 7th term

49. $(a + b)^{17}$; 5th term   **50.** $(a - b)^{12}$; 3rd term

51. $\left(a - \sqrt{2}\right)^4$; 2nd term   **52.** $\left(a - \sqrt{3}\right)^8$; 3rd term

53. $\left(a + \sqrt{3b}\right)^9$; 5th term   **54.** $\left(\sqrt{2}a - b\right)^7$; 4th term

55. $\left(\dfrac{x}{2} + y\right)^4$; 3rd term   **56.** $\left(m + \dfrac{n}{2}\right)^8$; 3rd term

57. $\left(\dfrac{r}{2} - \dfrac{s}{2}\right)^{11}$; 10th term   **58.** $\left(\dfrac{p}{2} - \dfrac{q}{2}\right)^9$; 6th term

59. $(a + b)^n$; 4th term   **60.** $(a - b)^n$; 5th term

61. $(a + b)^n$; rth term

62. $(a + b)^n$; $(r + 1)$th term

WRITING

63. Define factorial notation and explain how to evaluate 10!.

64. With a calculator, evaluate 69!. Explain why you cannot find 70! with a calculator.

65. Explain how the rth term of a binomial expansion is constructed.

66. Explain the four patterns apparent in a binomial expansion.

SOMETHING TO THINK ABOUT

67. Find the sum of the numbers in each row of the first ten rows of Pascal's triangle. Do you see a pattern?

68. Show that the sum of the coefficients in the binomial expansion of $(x + y)^n$ is 2^n.   (*Hint:* Let $x = y = 1$.)

69. Find the constant term in the expansion of

$$\left(a - \frac{1}{a}\right)^{10}$$

70. Find the coefficient of x^5 in the expansion of

$$\left(x + \frac{1}{x}\right)^9$$

71. If we apply the pattern of coefficients to the coefficient of the first term in the binomial theorem, it would be $\frac{n!}{0!(n - 0)!}$. Show that this expression equals 1.

72. If we apply the pattern of coefficients to the coefficient of the last term in the binomial theorem, it would be $\frac{n!}{n!(n - n)!}$. Show that this expression equals 1.

13.2   Sequences, Series, and Summation Notation

▌ **Sequences** ▌ **Recursive Definition of a Sequence** ▌ **Series**
▌ **Summation Notation**

Getting Ready   *Find each sum.*

1. $1 + 3 + 5 + 7 + 9$ **2.** $5 + 9 + 13 + 17 + 21$

In this section, we will introduce a function whose domain is the set of natural numbers. This function, called a **sequence,** is a list of numbers in a specific order.

Sequences

Sequences   A **sequence** is a function whose domain is the set of natural numbers.

Since a sequence is a function whose domain is the set of natural numbers, we can write its terms as a list of numbers. For example, if n is a natural number, the function defined by $f(n) = 2n - 1$ generates the sequence

$$1, 3, 5, \ldots, 2n - 1, \ldots$$

The number 1 is the first term, 3 is the second term, and $2n - 1$ is the **general,** or **nth term.** If n is a natural number, the function $f(n) = 3n^2 + 1$ generates the sequence

$$4, 13, 28, \ldots, 3n^2 + 1, \ldots$$

The number 4 is the first term, 13 is the second term, 28 is the third term, and $3n^2 + 1$ is the general term.

A constant function such as $g(n) = 1$ is a sequence, because it generates the sequence

$$1, 1, 1, \ldots$$

We seldom use function notation to denote a sequence, because it is often difficult or even impossible to write the general term. In such cases, if there is a pattern that is assumed to be continued, we simply list several terms of the sequence. Some examples of sequences follow:

$$1^2, 2^2, 3^2, \ldots, n^2, \ldots$$
$$3, 9, 19, 33, \ldots, 2n^2 + 1, \ldots$$
$$1, 3, 6, 10, 15, 21, \ldots, \frac{n(n + 1)}{2}, \ldots$$
$$1, 1, 2, 3, 5, 8, 13, 21, \ldots \qquad \textbf{(Fibonacci sequence)}$$
$$2, 3, 5, 7, 11, 13, 17, 19, 23, \ldots \qquad \textbf{(prime numbers)}$$

The **Fibonacci sequence** is named after the 12th-century mathematician Leonardo of Pisa, also known as Fibonacci. After the two 1's in the Fibonacci sequence, each term is the sum of the two terms that immediately precede it. The Fibonacci sequence occurs in many fields, such as the growth patterns of plants, the reproductive habits of bees, and music.

Recursive Definition of a Sequence

A sequence can be defined **recursively** by giving its first term and a rule showing how to obtain the $(n + 1)$th term from the nth term. For example, the information

$$a_1 = 5 \quad \text{(the first term)} \quad \text{and} \quad a_{n+1} = 3a_n - 2 \quad \begin{array}{l}\text{(the rule showing how to}\\\text{get the } (n + 1)\text{th term from}\\\text{the } n\text{th term)}\end{array}$$

defines a sequence recursively. To find the first five terms of this sequence, we proceed as follows:

$$a_1 = 5$$
$$a_2 = 3(\mathbf{a_1}) - 2 = 3(\mathbf{5}) - 2 = 13 \qquad \text{Substitute 5 for } a_1 \text{ and simplify to get } a_2.$$
$$a_3 = 3(\mathbf{a_2}) - 2 = 3(\mathbf{13}) - 2 = 37 \qquad \text{Substitute 13 for } a_2 \text{ and simplify to get } a_3.$$
$$a_4 = 3(\mathbf{a_3}) - 2 = 3(\mathbf{37}) - 2 = 109 \qquad \text{Substitute 37 for } a_3 \text{ and simplify to get } a_4.$$
$$a_5 = 3(\mathbf{a_4}) - 2 = 3(\mathbf{109}) - 2 = 325 \qquad \text{Substitute 109 for } a_4 \text{ and simplify to get } a_5.$$

Series

To add the terms of a sequence, we replace each comma between its terms with a $+$ sign to form a **series.** Because each sequence is infinite, the number of terms in the series associated with it is infinite also. Two examples of infinite series are

$$1^2 + 2^2 + 3^2 + \cdots + n^2 + \cdots$$

and

$$1 + 2 + 3 + 5 + 8 + 13 + 21 + \cdots$$

If the signs between successive terms of an infinite series alternate, the series is called an **alternating infinite series.** Two examples of alternating infinite series are

$$-3 + 6 - 9 + 12 - \cdots + (-1)^n 3n + \cdots$$

and

$$2 - 4 + 8 - 16 + \cdots + (-1)^{n+1} 2^n + \cdots$$

Summation Notation

Summation notation is a shorthand way to indicate the sum of the first n terms, or the **nth partial sum,** of a sequence. For example, the expression

$$\sum_{n=1}^{3} (2n^2 + 1) \qquad \text{The symbol } \Sigma \text{ is the capital letter sigma in the Greek alphabet.}$$

indicates the sum of the three terms obtained if we successively substitute 1, 2, and 3 for n in the expression $2n^2 + 1$.

$$\sum_{n=1}^{3} (2n^2 + 1) = [2(\mathbf{1})^2 + 1] + [2(\mathbf{2})^2 + 1] + [2(\mathbf{3})^2 + 1]$$

$$= 3 + 9 + 19$$

$$= 31$$

EXAMPLE 1   Evaluate: $\displaystyle\sum_{n=1}^{4} (n^2 - 1)$.

Solution   Since n runs from 1 to 4, we substitute 1, 2, 3, and 4 for n in the expression $n^2 - 1$ and find the sum of the resulting terms:

$$\sum_{n=1}^{4} (n^2 - 1) = (\mathbf{1}^2 - 1) + (\mathbf{2}^2 - 1) + (\mathbf{3}^2 - 1) + (\mathbf{4}^2 - 1)$$

$$= 0 + 3 + 8 + 15$$

$$= 26$$

Self Check   Evaluate: $\displaystyle\sum_{n=1}^{5} (n^2 - 1)$. ∎

EXAMPLE 2   Evaluate: $\displaystyle\sum_{n=3}^{5} (3n + 2)$.

Solution   Since n runs from 3 to 5, we substitute 3, 4, and 5 for n in the expression $3n + 2$ and find the sum of the resulting terms:

$$\sum_{n=3}^{5}(3n + 2) = [3(3) + 2] + [3(4) + 2] + [3(5) + 2]$$

$$= 11 + 14 + 17$$

$$= 42$$

Self Check   Evaluate: $\sum_{n=2}^{5}(3n + 2)$.

There are three basic properties of summations. The first states that *the summation of a constant as k runs from 1 to n is n times the constant.*

Summation of a Constant   If c is a constant, then $\sum_{k=1}^{n} c = nc$.

Proof   Because c is a constant, each term is c for each value of k as k runs from 1 to n.

$$\overset{n \text{ number of } c\text{'s}}{\sum_{k=1}^{n} c = \overbrace{c + c + c + c + \cdots + c} = nc}$$

EXAMPLE 3   Evaluate: $\sum_{n=1}^{5} 13$.

Solution $$\sum_{n=1}^{5} 13 = 13 + 13 + 13 + 13 + 13$$

$$= 5(13)$$

$$= 65$$

Self Check   Evaluate: $\sum_{n=1}^{6} 12$.

A second property states that *a constant factor can be brought outside a summation sign.*

Summation of a Product   If c is a constant, then $\sum_{k=1}^{n} cf(k) = c\sum_{k-1}^{n} f(k)$.

Proof $$\sum_{k=1}^{n} cf(k) = cf(1) + cf(2) + cf(3) + \cdots + cf(n)$$

$$= c[f(1) + f(2) + f(3) + \cdots + f(n)] \qquad \text{Factor out } c.$$

$$= c\sum_{k=1}^{n} f(k)$$

EXAMPLE 4   Show that $\displaystyle\sum_{k=1}^{3} 5k^2 = 5\sum_{k=1}^{3} k^2$.

Solution

$$\sum_{k=1}^{3} 5k^2 = \mathbf{5}(1)^2 + \mathbf{5}(2)^2 + \mathbf{5}(3)^2$$
$$= 5 + 20 + 45$$
$$= 70$$

$$5\sum_{k=1}^{3} k^2 = \mathbf{5}[(1)^2 + (2)^2 + (3)^2]$$
$$= 5[1 + 4 + 9]$$
$$= 5(14)$$
$$= 70$$

The quantities are equal.

Self Check   Evaluate: $\displaystyle\sum_{k=1}^{4} 3k$.

The third property states that *the summation of a sum is equal to the sum of the summations.*

Summation of a Sum

$$\sum_{k=1}^{n} [f(k) + g(k)] = \sum_{k=1}^{n} f(k) + \sum_{k=1}^{n} g(k)$$

Proof

$$\sum_{k=1}^{n} [f(k) + g(k)] = [f(1) + g(1)] + [f(2) + g(2)] + [f(3) + g(3)] + \cdots + [f(n) + g(n)]$$
$$= [f(1) + f(2) + f(3) + \cdots + f(n)] + [g(1) + g(2) + g(3) + \cdots + g(n)]$$
$$= \sum_{k=1}^{n} f(k) + \sum_{k=1}^{n} g(k)$$

EXAMPLE 5   Show that $\displaystyle\sum_{k=1}^{3} (k + k^2) = \sum_{k=1}^{3} k + \sum_{k=1}^{3} k^2$.

Solution

$$\sum_{k=1}^{3} (k + k^2) = (1 + 1^2) + (2 + 2^2) + (3 + 3^2)$$
$$= 2 + 6 + 12$$
$$= 20$$

$$\sum_{k=1}^{3} k + \sum_{k=1}^{3} k^2 = (1 + 2 + 3) + (1^2 + 2^2 + 3^2)$$
$$= 6 + 14$$
$$= 20$$

Self Check   Evaluate: $\displaystyle\sum_{k=1}^{3} (k^2 + 2k)$.

EXAMPLE 6 Evaluate $\displaystyle\sum_{k=1}^{5} (2k-1)^2$ directly. Then expand the binomial, apply the previous properties, and evaluate the expression again.

Solution   **Part 1:** $\displaystyle\sum_{k=1}^{5} (2k-1)^2 = 1 + 9 + 25 + 49 + 81 = 165$

Part 2: $\displaystyle\sum_{k=1}^{5} (2k-1)^2 = \sum_{k=1}^{5} (4k^2 - 4k + 1)$

$\displaystyle = \sum_{k=1}^{5} 4k^2 + \sum_{k=1}^{5} (-4k) + \sum_{k=1}^{5} 1$
 The summation of a sum is the sum of the summations.

$\displaystyle = 4\sum_{k=1}^{5} k^2 - 4\sum_{k=1}^{5} k + \sum_{k=1}^{5} 1$
 Bring the constant factors outside the summation signs.

$\displaystyle = 4\sum_{k=1}^{5} k^2 - 4\sum_{k=1}^{5} k + 5$
 The summation of a constant as k runs from 1 to 5 is 5 times that constant.

$= 4(1 + 4 + 9 + 16 + 25)$
$\qquad - 4(1 + 2 + 3 + 4 + 5) + 5$
$= 4(55) - 4(15) + 5$
$= 220 - 60 + 5$
$= 165$

Either way, the sum is 165.

Self Check   Evaluate: $\displaystyle\sum_{k=1}^{4} (2k-1)^2$.

■

1. 50   **2.** 50   **3.** 72   **4.** 30   **5.** 26   **6.** 84

Orals   *Find each sum.*

1. $\displaystyle\sum_{k=1}^{2} k$ **2.** $\displaystyle\sum_{k=2}^{3} k$

13.2   EXERCISES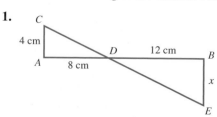

REVIEW   *The triangles are similar. Find the value of x.*

1.
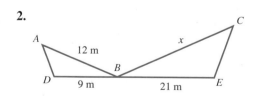

2.

Find the third side of each right triangle.

3.

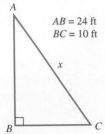

$AB = 24$ ft
$BC = 10$ ft

4.

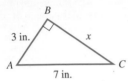

3 in. x

7 in.

VOCABULARY AND CONCEPTS   *Fill in the blanks.*

5. A sequence is a function whose _____ is the set of natural numbers.

6. A _____ is formed when we add the terms of a sequence.

7. _____ is a shorthand way to indicate the sum of one or more terms of a sequence.

8. The symbol $\sum\limits_{k=1}^{5} (k^2 - 3)$ indicates the _____ of the five terms obtained when we successively substitute 1, 2, 3, 4, and 5 for k.

9. $\sum\limits_{k=1}^{5} 6k^2 = \underline{\quad} \sum\limits_{k=1}^{5} k^2$

10. $\sum\limits_{k=1}^{5} (k^2 + 3k) = \sum\limits_{k=1}^{5} k^2 + \underline{\quad}$

11. $\sum\limits_{k=1}^{5} c$, where c is a constant, equals ___.

12. The summation of a sum is equal to the _____ of the summations.

PRACTICE   *Write the first six terms of the sequence defined by each function.*

13. $f(n) = 5n(n-1)$

14. $f(n) = n\left(\dfrac{n-1}{2}\right)\left(\dfrac{n-2}{3}\right)$

Find the next term of each sequence.

15. $1, 6, 11, 16, \ldots$

16. $1, 8, 27, 64, \ldots$

17. $a, a + d, a + 2d, a + 3d, \ldots$

18. $a, ar, ar^2, ar^3, \ldots$

19. $1, 3, 6, 10, \ldots$

20. $20, 17, 13, 8, \ldots$

Find the sum of the first five terms of the sequence with the given general term.

21. n

22. $2k$

23. 3

24. $4k^0$

25. $2\left(\dfrac{1}{3}\right)^n$

26. $(-1)^n$

27. $3n - 2$

28. $2k + 1$

A sequence is defined recursively. Find the first four terms of each sequence.

29. $a_1 = 3$ and $a_{n+1} = 2a_n + 1$

30. $a_1 = -5$ and $a_{n+1} = -a_n - 3$

31. $a_1 = -4$ and $a_{n+1} = \dfrac{a_n}{2}$

32. $a_1 = 0$ and $a_{n+1} = 2a_n^2$

33. $a_1 = k$ and $a_{n+1} = a_n^2$

34. $a_1 = 3$ and $a_{n+1} = ka_n$

35. $a_1 = 8$ and $a_{n+1} = \dfrac{2a_n}{k}$

36. $a_1 = m$ and $a_{n+1} = \dfrac{a_n^2}{m}$

Tell whether each series is an alternating infinite series.

37. $-1 + 2 - 3 + \cdots + (-1)^n n + \cdots$

38. $a + \dfrac{a}{b} + \dfrac{a}{b^2} + \cdots + a\left(\dfrac{1}{b}\right)^{n-1} + \cdots; b = 4$

39. $a + a^2 + a^3 + \cdots + a^n + \cdots; a = 3$

40. $a + a^2 + a^3 + \cdots + a^n + \cdots; a = -2$

Evaluate each sum.

41. $\sum\limits_{k=1}^{5} 2k$

42. $\sum\limits_{k=3}^{6} 3k$

43. $\sum\limits_{k=3}^{4} (-2k^2)$

44. $\sum\limits_{k=1}^{100} 5$

45. $\sum\limits_{k=1}^{5} (3k - 1)$

46. $\sum\limits_{n=2}^{5} (n^2 + 3n)$

47. $\displaystyle\sum_{k=1}^{1,000} \frac{1}{2}$ **48.** $\displaystyle\sum_{x=4}^{5} \frac{2}{x}$

49. $\displaystyle\sum_{x=3}^{4} \frac{1}{x}$

50. $\displaystyle\sum_{x=2}^{6} (3x^2 + 2x) - 3\sum_{x=2}^{6} x^2$

51. $\displaystyle\sum_{x=1}^{4} (4x + 1)^2 - \sum_{x=1}^{4} (4x - 1)^2$

52. $\displaystyle\sum_{x=0}^{10} (2x - 1)^2 + 4\sum_{x=0}^{10} x(1 - x)$

53. $\displaystyle\sum_{x=6}^{8} (5x - 1)^2 + \sum_{x=6}^{8} (10x - 1)$

54. $\displaystyle\sum_{x=2}^{7} (3x + 1)^2 - 3\sum_{x=2}^{7} x(3x + 2)$

WRITING

55. Explain what it means to define something recursively.

56. Explain why the binomial theorem can be stated as

$$\sum_{r=0}^{n} \frac{n!}{r!(n-r)!} a^{n-r} b^r$$

SOMETHING TO THINK ABOUT

57. Find a counterexample to disprove the proposition that the summation of a product is the product of the summations. In other words, prove that

$$\sum_{k=1}^{n} f(k)g(k) \neq \sum_{k=1}^{n} f(k) \sum_{k=1}^{n} g(k)$$

58. Find a counterexample to disprove the proposition that the summation of a quotient is the quotient of the summations. In other words, prove that

$$\sum_{k=1}^{n} \frac{f(k)}{g(k)} \neq \frac{\displaystyle\sum_{k=1}^{n} f(k)}{\displaystyle\sum_{k=1}^{n} g(k)}$$

13.3	**Arithmetic Sequences**

■ **Arithmetic Sequences**   ■ **Arithmetic Means**
■ **The Sum of the First _n_ Terms of an Arithmetic Sequence**
■ **Applications**

Getting Ready   *Complete each table.*

1.

n	$2n + 1$
1	
2	
3	
4	

2.

n	$3n - 5$
3	
4	
5	
6	

The German mathematician Carl Friedrich Gauss (1777–1855) was once a student in the class of a strict teacher. One day, the teacher asked the students to add together all of the natural numbers from 1 through 100. Gauss immediately knew the sum.

He recognized that in the sum

$$1 + 2 + 3 + \cdots + 98 + 99 + 100$$

the first number (1) added to the last number (100) is 101, the second number (2) added to the second from the last number (99) is 101, and the third number (3) added to the third from the last number (98) is 101. Gauss reasoned that there would be fifty pairs of such numbers, and that there would be fifty sums of 101. He multiplied 101 by 50 to get the correct answer of 5,050.

This story illustrates a problem involving the sum of the terms of a sequence, called an **arithmetic sequence,** in which each term except the first is found by adding a constant to the preceding term.

Arithmetic Sequences

Arithmetic Sequences

An **arithmetic sequence** is a sequence of the form

$$a, \quad a + d, \quad a + 2d, \quad a + 3d, \ldots, \quad a + (n - 1)d, \ldots$$

where a is the **first term,** $a + (n - 1)d$ is the **nth term,** and d is the **common difference.**

In this definition, the second term has an addend of d, the third term has an addend of $2d$, the fourth term has an addend of $3d$, and so on. This is why the nth term has an addend of $(n - 1)d$.

EXAMPLE 1   Write the first six terms and the 21st term of an arithmetic sequence with a first term of 7 and a common difference of 5.

Solution   Since the first term a is 7 and the common difference d is 5, the first six terms are

$$7, \quad 7 + 5, \quad 7 + 2(5), \quad 7 + 3(5), \quad 7 + 4(5), \quad 7 + 5(5)$$

or

$$7, \quad 12, \quad 17, \quad 22, \quad 27, \quad 32$$

To find the 21st term, we substitute 21 for n in the formula for the nth term:

$$n\text{th term} = a + (n - 1)d$$
$$21\text{st term} = 7 + (21 - 1)5$$
$$= 7 + (20)5$$
$$= 107$$

The 21st term is 107.

Self Check   Write the first five terms and the 18th term of an arithmetic sequence with a first term of 3 and a common difference of 6.

EXAMPLE 2   Find the 98th term of an arithmetic sequence whose first three terms are 2, 6, and 10.

Solution   Here $a = 2, n = 98$, and $d = 6 - 2 = 10 - 6 = 4$. Because we want to find the 98th term, we substitute these numbers into the formula for the nth term:

$$n\text{th term} = a + (n - 1)d$$
$$98\text{th term} = 2 + (98 - 1)4$$
$$= 2 + (97)4$$
$$= 390$$

Self Check   Write the 50th term of the arithmetic sequence whose first three terms are 3, 8, and 13. ∎

Arithmetic Means

Numbers inserted between a first and last term to form a segment of an arithmetic sequence are called **arithmetic means.** When finding arithmetic means, we consider the last term, l, to be the nth term:

$$l = a + (n - 1)d$$

EXAMPLE 3   Insert three arithmetic means between -3 and 12.

Solution   Since we are inserting three terms between -3 and 12, the total number of terms is five. Thus, $a = -3$, $l = 12$, and $n = 5$. To find the common difference, we substitute -3 for a, 12 for l, and 5 for n in the formula for the last term and solve for d:

$$l = a + (n - 1)d$$
$$12 = -3 + (5 - 1)d$$
$$15 = 4d \qquad \text{Add 3 to both sides and simplify.}$$
$$\frac{15}{4} = d \qquad \text{Divide both sides by 4.}$$

Once we know d, we can find the other terms of the sequence:

$$a + d = -3 + \frac{15}{4} = \frac{3}{4}$$
$$a + 2d = -3 + 2\left(\frac{15}{4}\right) = -3 + \frac{30}{4} = 4\frac{1}{2}$$
$$a + 3d = -3 + 3\left(\frac{15}{4}\right) = -3 + \frac{45}{4} = 8\frac{1}{4}$$

The three arithmetic means are $\frac{3}{4}$, $4\frac{1}{2}$, and $8\frac{1}{4}$.

Self Check   Find three arithmetic means between -5 and 23. ∎

The Sum of the First *n* Terms of an Arithmetic Sequence

To find the sum of the first n terms of an arithmetic sequence, we use the following formula.

Sum of the First *n* Terms of an Arithmetic Sequence

The formula

$$S_n = \frac{n(a + l)}{2}$$

gives the sum of the first *n* terms of an arithmetic sequence. In this formula, *a* is the first term, *l* is the last (or *n*th) term, and *n* is the number of terms.

Proof   We write the first *n* terms of an arithmetic sequence (letting S_n represent their sum). Then we write the same sum in reverse order and add the equations term by term:

$$S_n = a + (a + d) + \cdots + [a + (n - 2)d] + [a + (n - 1)d]$$
$$S_n = [a + (n - 1)d] + [a + (n - 2)d] + \cdots + (a + d) + a$$
$$\overline{2S_n = [2a + (n - 1)d] + [2a + (n - 1)d] + \cdots + [2a + (n - 1)d] + [2a + (n - 1)d]}$$

Because there are *n* equal terms on the right-hand side of the previous equation,

$$2S_n = n[2a + (n - 1)d]$$

or

(1) $$2S_n = n\{a + [a + (n - 1)d]\}$$ Write $2a$ as $a + a$.

We can substitute *l* for $a + (n - 1)d$ in the right-hand side of Equation 1 and divide both sides by 2 to get

$$S_n = \frac{n(a + l)}{2}$$ ∎

EXAMPLE 4   Find the sum of the first 30 terms of the arithmetic sequence $5, 8, 11, \ldots$.

Solution   Here $a = 5, n = 30, d = 3$, and $l = 5 + 29(3) = 92$. Substituting these values into the formula for the sum of an arithmetic sequence gives

$$S_n = \frac{n(a + l)}{2}$$

$$S_{30} = \frac{30(5 + 92)}{2}$$

$$= 15(97)$$

$$= 1,455$$

The sum of the first 30 terms is 1,455.

Self Check   Find the sum of the first 50 terms of the arithmetic sequence $-2, 5, 12, \ldots$.   ∎

Applications

EXAMPLE 5   A student deposits $50 in a non-interest-bearing account and plans to add $7 a week. How much will she have in the account one year after her first deposit?

Solution   Her weekly balances form an arithmetic sequence:

$$50, \quad 57, \quad 64, \quad 71, \quad 78, \ldots$$

with a first term of 50 and a common difference of 7. To find her balance in one year (52 weeks), we substitute 50 for a, 7 for d, and 52 for n in the formula for the last term.

$$l = a + (n - 1)d$$
$$l = \mathbf{50} + (\mathbf{52} - 1)\mathbf{7}$$
$$l = 50 + (51)7$$
$$l = 407$$

After one year, the balance will be $407.

Self Check   How much will the student have in the account after 60 weeks?   ∎

EXAMPLE 6   The equation $s = 16t^2$ represents the distance s (in feet) that an object will fall in t seconds.

- In 1 second, the object will fall 16 feet.
- In 2 seconds, the object will fall 64 feet.
- In 3 seconds, the object will fall 144 feet.

The object fell 16 feet during the first second, 48 feet during the next second, and 80 feet during the third second. Find the distance the object will fall during the 12th second.

Solution   The sequence 16, 48, 80, . . . is an arithmetic sequence with $a = 16$ and $d = 32$. To find the 12th term, we substitute these values into the formula for the last term.

$$l = a + (n - 1)d$$
$$l = \mathbf{16} + (\mathbf{12} - 1)\mathbf{32}$$
$$l = 16 + 11(32)$$
$$l = 368$$

During the 12th second, the object falls 368 feet.

Self Check   How far will the object fall during the 20th second?   ∎

1. 3, 9, 15, 21, 27; 105   **2.** 248   **3.** 2, 9, 16   **4.** 8,475   **5.** $463   **6.** 624 ft

Orals   *Find the next term in each arithmetic sequence.*

1. 2, 6, 10, . . .   **2.** 10, 7, 4, . . .

Find the common difference in each arithmetic sequence.

3. −2, 3, 8, . . .   **4.** 5, −1, −7, . . .

13.3 EXERCISES

REVIEW *Solve each equation.*

1. $x + \sqrt{x+3} = 9$

2. $\dfrac{x+3}{x} + \dfrac{x-3}{5} = 2$

3. $\dfrac{1}{x} + \dfrac{2}{x^2} + \dfrac{1}{x^3} = 0$

4. $x + 3\sqrt{x} - 10 = 0$

5. $x^4 - 1 = 0$

6. $x^4 - 29x^2 + 100 = 0$

VOCABULARY AND CONCEPTS *Fill in the blanks.*

7. An arithmetic sequence is a sequence of the form
$a, a + d, a + 2d, a + 3d, \ldots, a + \underline{\hspace{1cm}} d$

8. In an arithmetic sequence, a is the ____ term, d is the common _____, and n is the _____ of terms.

9. The last term of an arithmetic sequence is given by the formula _____.

10. The formula for the sum of the first n terms of an arithmetic sequence is given by the formula

_____

11. _____ are numbers inserted between a first and last term to form an arithmetic sequence.

12. The formula _____ gives the distance (in feet) that an object will fall in t seconds.

PRACTICE *Write the first six terms of the arithmetic sequences with the given properties.*

13. $a = 1; d = 2$

14. $a = -12; d = -5$

15. $a = 5$; 3rd term is 2

16. $a = 4$; 5th term is 12

17. 7th term is 24; common difference is $\dfrac{5}{2}$

18. 20th term is -49; common difference is -3

Find the missing term in each arithmetic sequence.

19. Find the 40th term of an arithmetic sequence with a first term of 6 and a common difference of 8.

20. Find the 35th term of an arithmetic sequence with a first term of 50 and a common difference of -6.

21. The 6th term of an arithmetic sequence is 28, and the first term is -2. Find the common difference.

22. The 7th term of an arithmetic sequence is -42, and the common difference is -6. Find the first term.

23. Find the 55th term of an arithmetic sequence whose first three terms are $-8, -1$, and 6.

24. Find the 37th term of an arithmetic sequence whose second and third terms are -4 and 6.

25. If the fifth term of an arithmetic sequence is 14 and the second term is 5, find the 15th term.

26. If the fourth term of an arithmetic sequence is 13 and the second term is 3, find the 24th term.

Find the required means.

27. Insert three arithmetic means between 10 and 20.

28. Insert five arithmetic means between 5 and 15.

29. Insert four arithmetic means between -7 and $\dfrac{2}{3}$.

30. Insert three arithmetic means between -11 and -2.

Find the sum of the first n terms of each arithmetic sequence.

31. $5 + 7 + 9 + \cdots$ (to 15 terms)

32. $-3 + (-4) + (-5) + \cdots$ (to 10 terms)

33. $\displaystyle\sum_{n=1}^{20} \left(\frac{3}{2}n + 12 \right)$

34. $\displaystyle\sum_{n=1}^{10} \left(\frac{2}{3}n + \frac{1}{3} \right)$

Solve each problem.

35. Find the sum of the first 30 terms of an arithmetic sequence with 25th term of 10 and a common difference of $\frac{1}{2}$.

36. Find the sum of the first 100 terms of an arithmetic sequence with 15th term of 86 and first term of 2.

37. Find the sum of the first 200 natural numbers.

38. Find the sum of the first 1,000 natural numbers.

APPLICATIONS

39. Borrowing money To pay for college, a student borrows $5,000 interest-free from his father. If he pays his father back at the rate of $200 per month, how much will he still owe after 12 months?

40. Borrowing money If Juanita borrows $5,500 interest-free from her mother to buy a new car and agrees to pay her mother back at the rate of $105 per month, how much will she still owe after four years?

41. Jogging One day, some students jogged $\frac{1}{2}$ mile. Because it was fun, they decided to increase the jogging distance each day by a certain amount. If they jogged $6\frac{3}{4}$ miles on the 51st day, how much was the distance increased each day?

42. Sales The year it incorporated, a company had sales of $237,500. Its sales were expected to increase by $150,000 annually for the next several years. If the forecast was correct, what will sales be in 10 years?

43. Falling objects Find how many feet a brick will travel during the 10th second of its fall.

44. Falling objects If a rock is dropped from the Golden Gate Bridge, how far will it fall in the third second?

45. Pile of logs Several logs are stored in a pile with 20 logs on the bottom layer, 19 on the second layer, 18 on the third layer, and so on. If the top layer has one log, how many logs are in the pile?

46. Theater seating The first row in a movie theater contains 24 seats. As you move toward the back, each row has 1 additional seat. If there are 30 rows, what is the capacity of the theater?

WRITING

47. Can an arithmetic sequence have a first term of 4, a 25th term of 126, and a common difference of $4\frac{1}{4}$? Explain.

48. Can an arithmetic sequence be an alternating sequence? Explain.

SOMETHING TO THINK ABOUT

49. In an arithmetic sequence, can a and d be negative, but l positive?

50. Between 5 and $10\frac{1}{3}$ are three arithmetic means. One of them is 9. Find the other two.

13.4	Geometric Sequences

■ **Geometric Sequences**   ■ **Geometric Means**
■ **Sum of the First _n_ Terms of a Geometric Sequence**
■ **Infinite Geometric Sequences**   ■ **Applications**

Getting Ready   *Complete each table.*

1.

n	$5(2^n)$
1	
2	
3	

2.

n	$6(3^n)$
1	
2	
3	

Another common sequence is the *geometric sequence*.

Geometric Sequences

A **geometric sequence** is a sequence in which each term, except the first, is found by multiplying the preceding term by a constant.

Geometric Sequences

A **geometric sequence** is a sequence of the form

$$a, ar, ar^2, ar^3, \ldots, ar^{n-1}, \ldots$$

where a is the **first term**, ar^{n-1} is the **nth term,** and r is the **common ratio.**

In this definition, the second term of the sequence has a factor of r^1, the third term has a factor of r^2, the fourth term has a factor of r^3, and so on. This explains why the nth term has a factor of r^{n-1}.

EXAMPLE 1 Write the first six terms and the 15th term of the geometric sequence whose first term is 3 and whose common ratio is 2.

Solution We first write the first six terms of the geometric sequence:

$$3, \quad 3(2), \quad 3(2)^2, \quad 3(2)^3, \quad 3(2)^4, \quad 3(2)^5$$

or

$$3, 6, 12, 24, 48, 96$$

To find the 15th term, we substitute 15 for n, 3 for a, and 2 for r in the formula for the nth term:

$$\begin{aligned}
n\text{th term} &= ar^{n-1} \\
15\text{th term} &= 3(2)^{15-1} \\
&= 3(2)^{14} \\
&= 3(16{,}384) \\
&= 49{,}152
\end{aligned}$$

Self Check Write the first five terms of the geometric sequence whose first term is 2 and whose common ratio is 3. Find the 10th term. ∎

EXAMPLE 2 Find the eighth term of a geometric sequence whose first three terms are 9, 3, and 1.

Solution Here $a = 9$, $r = \frac{1}{3}$, and $n = 8$. To find the eighth term, we substitute these values into the formula for the nth term.

$$\begin{aligned}
n\text{th term} &= ar^{n-1} \\
8\text{th term} &= 9\left(\frac{1}{3}\right)^{8-1} \\
&= 9\left(\frac{1}{3}\right)^{7} \\
&= \frac{1}{243}
\end{aligned}$$

Self Check Find the eighth term of a geometric sequence whose first three terms are $\frac{1}{3}$, 1, and 3. ∎

Geometric Means

Numbers inserted between a first and last term to form a segment of a geometric sequence are called **geometric means.** When finding geometric means, we consider the last term, l, to be the nth term.

EXAMPLE 3   Insert two geometric means between 4 and 256.

Solution   The first term is $a = 4$, and because 256 is the fourth term, $n = 4$ and $l = 256$. To find the common ratio, we substitute these values into the formula for the nth term and solve for r:

$$ar^{n-1} = l$$
$$4r^{4-1} = 256$$
$$r^3 = 64$$
$$r = 4$$

The common ratio is 4. The two geometric means are the second and third terms of the geometric sequence:

$$ar = 4 \cdot 4 = 16$$
$$ar^2 = 4 \cdot 4^2 = 4 \cdot 16 = 64$$

The first four terms of the geometric sequence are 4, 16, 64, and 256. The two geometric means between 4 and 256 are 16 and 64.

Self Check   Insert two geometric means between -3 and 192.   ∎

Sum of the First *n* Terms of a Geometric Sequence

To find the sum of the first n terms of a geometric sequence, we use the following formula.

Sum of the First *n* Terms of a Geometric Sequence

The formula

$$S_n = \frac{a - ar^n}{1 - r} \quad (r \neq 1)$$

gives the sum of the first n terms of a geometric sequence. In the formula, S_n is the sum, a is the first term, r is the common ratio, and n is the number of terms.

Proof   We write the sum of the first n terms of the geometric sequence:

(1) $$S_n = a + ar + ar^2 + \cdots + ar^{n-3} + ar^{n-2} + ar^{n-1}$$

Multiplying both sides of this equation by r gives

(2) $$S_n r = ar + ar^2 + \cdots + ar^{n-2} + ar^{n-1} + ar^n$$

We now subtract Equation 2 from Equation 1 and solve for S_n:

$$S_n - S_n r = a - ar^n$$

$$S_n(1 - r) = a - ar^n \quad \text{Factor out } S_n.$$

$$S_n = \frac{a - ar^n}{1 - r} \quad \text{Divide both sides by } 1 - r.$$

EXAMPLE 4 Find the sum of the first six terms of the geometric sequence $8, 4, 2, \ldots$.

Solution Here $a = 8$, $n = 6$, and $r = \frac{1}{2}$. Substituting these values into the formula for the sum of the first n terms of a geometric sequence gives

$$S_n = \frac{a - ar^n}{1 - r}$$

$$S_6 = \frac{8 - 8\left(\frac{1}{2}\right)^6}{1 - \frac{1}{2}}$$

$$= 2\left(\frac{63}{8}\right)$$

$$= \frac{63}{4}$$

The sum of the first six terms is $\dfrac{63}{4}$.

Self Check Find the sum of the first eight terms of the geometric sequence $81, 27, 9, \ldots$.

Infinite Geometric Sequences

Under certain conditions, we can find the sum of all of the terms in an **infinite geometric sequence.** To define this sum, we consider the geometric sequence

$$a, \quad ar, \quad ar^2, \quad \ldots$$

- The first partial sum, S_1, of the sequence is $S_1 = a$.
- The second partial sum, S_2, of the sequence is $S_2 = a + ar$.
- The nth partial sum, S_n, of the sequence is $S_n = a + ar + ar^2 + \cdots + ar^{n-1}$.

If the nth partial sum, S_n, of an infinite geometric sequence approaches some number S as n approaches ∞, then S is called the **sum of the infinite geometric sequence.** The following symbol denotes the sum, S, of an infinite geometric sequence, provided the sum exists.

$$S = \sum_{n=1}^{\infty} ar^{n-1}$$

To develop a formula for finding the sum of all the terms in an infinite geometric sequence, we consider the formula

(3) $$S_n = \frac{a - ar^n}{1 - r} \quad (r \neq 1)$$

If $|r| < 1$ and a is a constant, then as n approaches ∞, ar^n approaches 0, and the term ar^n in Equation 3 can be dropped. This argument gives the following formula.

Sum of the Terms of an Infinite Geometric Sequence

If $|r| < 1$, the sum of the terms of an infinite geometric sequence is given by

$$S = \frac{a}{1 - r}$$

where a is the first term and r is the common ratio.

 Comment If $r \geq 1$, the terms get larger and larger, and the sum does not approach a limit. In this case, the previous formula does not apply.

 EXAMPLE 5 Change $0.\overline{4}$ to a common fraction.

Solution We write the decimal as an infinite geometric series and find its sum:

$$S = \frac{4}{10} + \frac{4}{100} + \frac{4}{1,000} + \frac{4}{10,000} + \cdots$$
$$S = \frac{4}{10} + \frac{4}{10}\left(\frac{1}{10}\right) + \frac{4}{10}\left(\frac{1}{10}\right)^2 + \frac{4}{10}\left(\frac{1}{10}\right)^3 + \cdots$$

Since the common ratio r equals $\frac{1}{10}$, and $\left|\frac{1}{10}\right| < 1$, we can use the formula for the sum of an infinite geometric series:

$$S = \frac{a}{1 - r} = \frac{\frac{4}{10}}{1 - \frac{1}{10}} = \frac{\frac{4}{10}}{\frac{9}{10}} = \frac{4}{9}$$

Long division will verify that $\frac{4}{9} = 0.\overline{4}$.

Self Check Change $0.\overline{7}$ to a common fraction. ∎

Applications

Many of the exponential growth problems discussed in Chapter 10 can be solved using the concepts of geometric sequences.

EXAMPLE 6 A town with a population of 3,500 people has a predicted growth rate of 6% per year for the next 20 years. How many people are expected to live in the town 20 years from now?

Solution Let p_0 be the initial population of the town. After 1 year, the population p_1 will be the initial population (p_0) plus the growth (the product of p_0 and the rate of growth, r).

$$p_1 = p_0 + p_0 r$$
$$= p_0(1 + r) \quad \text{Factor out } p_0.$$

The population p_2 at the end of 2 years will be

$$p_2 = p_1 + p_1 r$$
$$p_2 = \boldsymbol{p_1}(1 + r) \qquad \text{Factor out } p_1.$$
$$p_2 = \boldsymbol{p_0(1 + r)}(1 + r) \qquad \text{Substitute } p_0(1 + r) \text{ for } p_1.$$
$$p_2 = p_0(1 + r)^2$$

The population at the end of the third year will be $p_3 = p_0(1 + r)^3$. Writing the terms in a sequence gives

$$p_0, \quad p_0(1 + r), \quad p_0(1 + r)^2, \quad p_0(1 + r)^3, \quad p_0(1 + r)^4, \ldots$$

This is a geometric sequence with p_0 as the first term and $1 + r$ as the common ratio. In this example, $p_0 = 3{,}500$, $1 + r = 1.06$, and (since the population after 20 years will be the value of the 21st term of the geometric sequence) $n = 21$. We can substitute these values into the formula for the last term of a geometric sequence to get

$$l = \boldsymbol{ar^{n-1}}$$
$$l = \boldsymbol{3{,}500(1.06)^{21-1}}$$
$$l = 3{,}500(1.06)^{20}$$
$$l \approx 11{,}224.97415 \qquad \text{Use a calculator.}$$

The population after 20 years will be approximately 11,225. ∎

EXAMPLE 7 A woman deposits \$2,500 in a bank at 7% annual interest, compounded daily. If the investment is left untouched for 60 years, how much money will be in the account?

Solution We let the initial amount in the account be a_0 and r be the rate. At the end of the first day, the account is worth

$$a_1 = a_0 + a_0\left(\frac{r}{365}\right) = a_0\left(1 + \frac{r}{365}\right)$$

After the second day, the account is worth

$$a_2 = a_1 + a_1\left(\frac{r}{365}\right) = a_1\left(1 + \frac{r}{365}\right) = a_0\left(1 + \frac{r}{365}\right)^2$$

The daily amounts form the following geometric sequence

$$a_0, \quad a_0\left(1 + \frac{r}{365}\right), \quad a_0\left(1 + \frac{r}{365}\right)^2, \quad a_0\left(1 + \frac{r}{365}\right)^3, \ldots$$

where a_0 is the initial deposit and r is the annual rate of interest.

Because interest is compounded daily for 60 years (21,900 days), the amount at the end of 60 years will be the 21,901th term of the sequence.

$$a_{21{,}901} = 2{,}500\left(1 + \frac{0.07}{365}\right)^{21{,}900}$$

We can use a calculator to find that $a_{21{,}901} \approx 166{,}648.71$. ∎

EXAMPLE 8 A pump can remove 20% of the gas in a container with each stroke. Find the percentage of gas that remains in the container after six strokes.

Solution We let V represent the volume of the container. Because each stroke of the pump removes 20% of the gas, 80% remains after each stroke, and we have the geometric sequence

$$V, \quad 0.80V, \quad 0.80(0.80V), \quad 0.80[0.80(0.80V)], \ldots$$

or

$$V, \quad 0.80V, \quad (0.8)^2V, \quad (0.8)^3V, \quad (0.8)^4V, \ldots$$

The amount of gas remaining after six strokes is the seventh term, l, of the sequence:

$$l = \boldsymbol{ar}^{n-1}$$
$$l = \boldsymbol{V(0.8)}^{7-1}$$
$$l = V(0.8)^6$$

We can use a calculator to find that approximately 26% of the gas remains after six strokes. ∎

EXAMPLE 9 **Amount of an annuity**   An **annuity** is a sequence of equal payments made periodically over a length of time. The sum of the payments and the interest earned during the *term* of the annuity is called the *amount* of the annuity.

After a sales clerk works six months, her employer will begin an annuity for her and will contribute $500 every six months to a fund that pays 8% annual interest. After she has been employed for two years, what will be the amount of the annuity?

Solution   Because the payments are to be made semiannually, there will be four payments of $500, each earning a rate of 4% per six-month period. These payments will occur at the end of 6 months, 12 months, 18 months, and 24 months. The first payment, to be made after 6 months, will earn interest for three interest periods. Thus, the amount of the first payment is $500(1.04)^3$. The amounts of each of the four payments after two years are shown in Table 13-1.

The amount of the annuity is the sum of the amounts of the individual payments, a sum of $2,123.23.

Payment (at the end of period)	Amount of payment at the end of 2 years
1	$500(1.04)^3 = $ 562.43
2	$500(1.04)^2 = $ 540.80
3	$500(1.04)^1 = $ 520.00
4	$500 = $ 500.00
	$A_n = $2,123.23

Table 13-1 ∎

1. 2, 6, 18, 54, 162; 39,366   **2.** 729   **3.** 12, −48   **4.** $\frac{3,280}{27}$ **5.** $\frac{7}{9}$

Orals   *Find the next term in each geometric sequence.*

1. 1, 3, 9, . . .   **2.** $1, \dfrac{1}{3}, \dfrac{1}{9}, \ldots$

Find the common ratio in each geometric sequence.

3. $0.2, 0.5, 1.25, \ldots$

4. $\sqrt{3}, 3, 3\sqrt{3}, \ldots$

Find x in each geometric sequence.

5. $2, x, 18, 54, \ldots$

6. $3, x, \dfrac{1}{3}, \dfrac{1}{9}, \ldots$

13.4 EXERCISES

REVIEW *Perform each operation.* $\left(i = \sqrt{-1}.\right)$

1. $(3 + 2i) + (2 - 5i)$

2. $(7 + 8i) - (2 - 5i)$

3. $(3 + i)(3 - i)$

4. $\left(3 + \sqrt{3}i\right)\left(3 - \sqrt{3}i\right)$

5. $\dfrac{2 + 3i}{2 - i}$

6. $(7 + 3i)^2$

7. i^{127}

8. i^{-127}

VOCABULARY AND CONCEPTS *Fill in the blanks.*

9. A geometric sequence is a sequence of the form $a, ar, ar^2, ar^3, \ldots$. The nth term is $a(\underline{\quad})$.

10. In a geometric sequence, a is the _____ term, r is the common _____, and n is the _____ of terms.

11. The last term of a geometric sequence is given by the formula _____.

12. The formula for the sum of the first n terms of a geometric sequence is given by

_____.

13. _____ are numbers inserted between a first and a last term to form a geometric sequence.

14. If $|r| < 1$, the formula _____ gives the sum of the terms of an infinite geometric sequence.

PRACTICE *Write the first four terms of each geometric sequence with the given properties.*

15. $a = 10; r = 2$

16. $a = -3; r = 2$

17. $a = -2$ and $r = 3$

18. $a = 64; r = \dfrac{1}{2}$

19. $a = 3; r = \sqrt{2}$

20. $a = 2; r = \sqrt{3}$

21. $a = 2$; 4th term is 54

22. 3rd term is 4; $r = \dfrac{1}{2}$

Find the requested term of each geometric sequence.

23. Find the sixth term of the geometric sequence whose first three terms are $\frac{1}{4}$, 1, and 4.

24. Find the eighth term of the geometric sequence whose second and fourth terms are 0.2 and 5.

25. Find the fifth term of a geometric sequence whose second term is 6 and whose third term is -18.

26. Find the sixth term of a geometric sequence whose second term is 3 and whose fourth term is $\frac{1}{3}$.

Solve each problem.

27. Insert three positive geometric means between 10 and 20.

28. Insert five geometric means between -5 and 5, if possible.

29. Insert four geometric means between 2 and 2,048.

30. Insert three geometric means between 162 and 2. (There are two possibilities.)

Find the sum of the indicated terms of each geometric sequence.

31. $4, 8, 16, \ldots$ (to 5 terms)

32. $9, 27, 81, \ldots$ (to 6 terms)

33. $2, -6, 18, \ldots$ (to 10 terms)

34. $\dfrac{1}{8}, \dfrac{1}{4}, \dfrac{1}{2}, \ldots$ (to 12 terms)

35. $\displaystyle\sum_{n=1}^{6} 3\left(\dfrac{3}{2}\right)^{n-1}$

36. $\displaystyle\sum_{n=1}^{6} 12\left(-\dfrac{1}{2}\right)^{n-1}$

Find the sum of each infinite geometric sequence.

37. $6 + 4 + \dfrac{8}{3} + \cdots$

38. $8 + 4 + 2 + 1 + \cdots$

39. $\displaystyle\sum_{n=1}^{\infty} 12\left(-\dfrac{1}{2}\right)^{n-1}$ **40.** $\displaystyle\sum_{n=1}^{\infty} \left(\dfrac{1}{3}\right)^{n-1}$

Change each decimal to a common fraction.

41. $0.\overline{5}$ **42.** $0.\overline{6}$

43. $0.\overline{25}$ **44.** $0.\overline{37}$

APPLICATIONS   *Use a calculator to help solve each problem.*

45. Staffing a department   The number of students studying algebra at State College is 623. The Department Chair expects enrollment to increase 10% each year. How many professors will be needed in 8 years to teach algebra if one professor can handle 60 students?

46. Bouncing balls   A Super Ball rebounds to approximately 95% of the height from which it is dropped. If the ball is dropped from a height of 10 meters, how high will it rebound after the 13th bounce?

47. Investing money   If a married couple invests $1,000 in a 1-year certificate of deposit at $6\frac{3}{4}\%$ annual interest, compounded daily, how much interest will be earned during the year?

48. Biology   If a single cell divides into two cells every 30 minutes, how many cells will there be at the end of ten hours?

49. Depreciation   A lawn tractor, costing c dollars when new, depreciates 20% of the previous year's value each year. How much is the lawn tractor worth after 5 years?

50. Financial planning   Maria can invest $1,000 at $7\frac{1}{2}\%$, compounded annually, or at $7\frac{1}{4}\%$, compounded daily. If she invests the money for a year, which is the better investment?

51. Population study   If the population of Earth were to double every 30 years, approximately how many people would there be in the year 3010? (Consider the population in 1990 to be 5 billion and use 1990 as the base year.)

52. Investing money   If Linda deposits $1,300 in a bank at 7% interest, compounded annually, how much will be in the bank 17 years later? (Assume that there are no other transactions on the account.)

53. Real estate appreciation   If a house purchased for $50,000 in 1988 appreciates in value by 6% each year, how much will the house be worth in the year 2010?

54. Compound interest   Find the value of $1,000 left on deposit for 10 years at an annual rate of 7%, compounded annually.

55. Compound interest   Find the value of $1,000 left on deposit for 10 years at an annual rate of 7%, compounded quarterly.

56. Compound interest   Find the value of $1,000 left on deposit for 10 years at an annual rate of 7%, compounded monthly.

57. Compound interest   Find the value of $1,000 left on deposit for 10 years at an annual rate of 7%, compounded daily.

58. Compound interest   Find the value of $1,000 left on deposit for 10 years at an annual rate of 7%, compounded hourly.

59. Saving for retirement   When John was 20 years old, he opened an individual retirement account by investing $2,000 at 11% interest, compounded quarterly. How much will his investment be worth when he is 65 years old?

60. Biology   One bacterium divides into two bacteria every 5 minutes. If two bacteria multiply enough to completely fill a petri dish in 2 hours, how long will it take one bacterium to fill the dish?

61. Mathematical myths   A legend tells of a king who offered to grant the inventor of the game of chess any request. The inventor said, "Simply place one grain of wheat on the first square of a chessboard, two grains on the second, four on the third, and so on, until the board is full. Then give me the wheat." The king agreed. How many grains did the king need to fill the chessboard?

62. Mathematical myths   Estimate the size of the wheat pile in Exercise 61.   (*Hint:* There are about one-half million grains of wheat in a bushel.)

63. Annuities   Find the amount of an annuity if $1,000 is paid semiannually for two years at 6% annual interest. Assume that the first of the four payments is made immediately.

64. Annuities   Note that the amounts shown in Table 13-1 form a geometric sequence. Verify the answer for Example 9 by using the formula for the sum of a geometric sequence.

WRITING

65. Explain how a geometric sequence differs from an arithmetic sequence.

66. Explain why you can sometimes find the sum of the terms of an infinite geometric sequence.

SOMETHING TO THINK ABOUT

67. Does $0.999999 = 1$? Explain.

68. Does $0.999\ldots = 1$? Explain.

13.5   Mathematical Induction

▌The Method of Mathematical Induction

Getting Ready   *Find the value of* $\dfrac{n(1 + n)}{2}$ *for each value of n.*

1. $n = 1$ **2.** $n = 2$ **3.** $n = 3$ **4.** $n = 4$

In Section 13.3, we developed the following formula for the sum of the first n terms of an arithmetic sequence:

$$S_n = \frac{n(a + l)}{2}$$

If we apply this formula to the arithmetic series $1 + 2 + 3 + \cdots + n$, we have

$$S_n = 1 + 2 + 3 + \cdots + n = \frac{n(1 + n)}{2}$$

To see that this formula is true, we can check it for some positive numbers n:

For $n = 1$: $1 = \dfrac{1(1 + 1)}{2}$ is a true statement, because $1 = 1$.

For $n = 2$: $1 + 2 = \dfrac{2(1 + 2)}{2}$ is a true statement, because $3 = 3$.

For $n = 3$: $1 + 2 + 3 = \dfrac{3(1 + 3)}{2}$ is a true statement, because $6 = 6$.

For $n = 6$: $1 + 2 + 3 + 4 + 5 + 6 = \dfrac{6(1 + 6)}{2}$ is a true statement, because $21 = 21$.

However, because the set of positive numbers is infinite, it is impossible to prove the formula by verifying it for all positive numbers. To verify this sequence formula for all positive numbers n, we must have a method of proof called **mathematical induction,** a method first used extensively by Giuseppe Peano (1858–1932).

The Method of Mathematical Induction

Suppose that we are standing in line for a movie and are worrying about whether we will be admitted. Even if the first person gets in, our worries would still be justified. Perhaps there is only room in the theater for a few people.

It would be good news to hear the theater manager say, "If anyone gets in, the next person in line will get in also." However, this promise does not guarantee that anyone will be admitted. Perhaps the theater is already full, and no one will get in.

However, when we see the first person in line walk in, we can conclude that everyone will be admitted, because we know two things:

- The first person was admitted.
- Because of the promise, if the first person is admitted, then so is the second, and when the second person is admitted, then so is the third, and so on until everyone gets in.

This situation is similar to a game played with dominoes. Suppose that some dominoes are placed on end, as in Figure 13-1. When the first domino is knocked over, it knocks over the second. The second domino, in turn, knocks over the third, which knocks over the fourth, and so on until all of the dominoes fall. Two things must happen to guarantee that all of the dominoes fall:

- The first domino must be knocked over.
- Every domino that falls must knock over the next one.

When both conditions are met, it is certain that all of the dominoes will fall.

Figure 13-1

The preceding examples illustrate the basic principle of mathematical induction.

Mathematical Induction

> If a statement involving the natural number n has the following two properties
> **1.** The statement is true for $n = 1$, and
> **2.** If the statement is true for $n = k$, then it is true for $n = k + 1$,
> then the statement is true for all natural numbers.

Mathematical induction provides a way to prove many formulas. Any proof by induction involves two parts. First, we must show that the formula is true for the number 1. Second, we must show that, if the formula is true for any natural number k, then it also is true for the natural number $k + 1$. A proof by induction is complete only when both of these properties are established.

EXAMPLE 1 Use mathematical induction to prove that the following formula is true for every natural number n.

$$1 + 2 + 3 + \cdots + n = \frac{n(n + 1)}{2}$$

Solution **Part 1:** Verify that the formula is true for $n = 1$. When $n = 1$, there is a single term, the number 1, on the left-hand side of the equation. Substituting 1 for n on the right-hand side, we have

$$1 = \frac{n(n + 1)}{2}$$

$$1 = \frac{(1)(1 + 1)}{2}$$

$$1 = 1$$

The formula is true when $n = 1$. Part 1 of the proof is complete.

Part 2:   We assume that the given formula is true when $n = k$. By this assumption, called the **induction hypothesis,** we accept that

(1) $1 + 2 + 3 + \cdots + k = \dfrac{k(k + 1)}{2}$

is a true statement. We must show that the induction hypothesis forces the given formula to be true when $n = k + 1$. We can show this by verifying the statement

(2) $1 + 2 + 3 + \cdots + k + (k + 1) = \dfrac{(k + 1)[(k + 1) + 1]}{2}$

which is obtained from the given formula by replacing n with $k + 1$.

Comparing the left-hand sides of Equations 1 and 2 shows that the left-hand side of Equation 2 contains an extra term of $k + 1$. Thus, we add $k + 1$ to both sides of Equation 1 (which was assumed to be true) to obtain the equation

$$1 + 2 + 3 + \cdots + k + (k + 1) = \frac{k(k + 1)}{2} + (k + 1)$$

Because both terms on the right-hand side of this equation have a common factor of $k + 1$, the right-hand side factors, and the equation can be written as follows:

$$1 + 2 + 3 + \cdots + k + (k + 1) = (k + 1)\left(\frac{k}{2} + 1\right)$$

$$= (k + 1)\left(\frac{k + 2}{2}\right)$$

$$= \frac{(k + 1)(k + 2)}{2}$$

$$= \frac{(k + 1)[(k + 1) + 1]}{2}$$

This final result is Equation 2. Because the truth of Equation 1 implies the truth of Equation 2, Part 2 of the proof is complete. Parts 1 and 2 together establish that the formula is true for any natural number n.   ∎

EXAMPLE 2   Use mathematical induction to prove the following formula for all natural numbers n.

$$1 + 5 + 9 + \cdots + (4n - 3) = n(2n - 1)$$

Solution   **Part 1:**   First we verify the formula for $n = 1$. When $n = 1$, there is a single term, the number 1, on the left-hand side of the equation. Substituting 1 for n on the right-hand side, we have

$$1 = 1[2(1) - 1]$$

$$1 = 1$$

The formula is true for $n = 1$. Part 1 of the proof is complete.

Part 2: We assume that the formula is true for $n = k$. Hence,

(3) $$1 + 5 + 9 + \cdots + (4k - 3) = k(2k - 1)$$

is a true statement. To show that the induction hypothesis guarantees the truth of the formula for $k + 1$ terms, we add the $(k + 1)$th term to both sides of Equation 3. Because the terms on the left-hand side increase by 4, the $(k + 1)$th term is $(4k - 3) + 4$, or $4k + 1$. Adding $4k + 1$ to both sides of Equation 3 gives

$$1 + 5 + 9 + \cdots + (4k - 3) + (4k + 1) = k(2k - 1) + (4k + 1)$$

We can simplify the right-hand side and write the previous equation as follows:

$$1 + 5 + 9 + \cdots + (4k - 3) + [4(k + 1) - 3] = 2k^2 + 3k + 1$$
$$= (k + 1)(2k + 1)$$
$$= (k + 1)[2(k + 1) - 1]$$

Since this result has the same form as the given formula, except that $k + 1$ replaces n, the truth of the formula for $n = k$ implies the truth of the formula for $n = k + 1$. Part 2 of the proof is complete.

Because both of the induction requirements are true, the formula is true for all natural numbers n.   ∎

EXAMPLE 3   Prove: $\dfrac{1}{2} + \dfrac{1}{4} + \dfrac{1}{8} + \cdots + \dfrac{1}{2^n} < 1.$

Solution   **Part 1:** We verify the formula for $n = 1$. When $n = 1$, there is a single term, the fraction $\frac{1}{2}$, on the left-hand side of the equation. Substituting 1 for n on the right-hand side, we have the following true statement:

$$\frac{1}{2} < 1$$

The formula is true for $n = 1$. Part 1 of the proof is complete.

Part 2: We assume that the inequality is true for $n = k$. Thus,

$$\frac{1}{2} + \frac{1}{4} + \frac{1}{8} + \cdots + \frac{1}{2^k} < 1$$

We can multiply both sides of the above inequality by $\frac{1}{2}$ to get

$$\frac{1}{2}\left(\frac{1}{2} + \frac{1}{4} + \frac{1}{8} + \cdots + \frac{1}{2^k} \right) < 1\left(\frac{1}{2} \right)$$

or

$$\frac{1}{4} + \frac{1}{8} + \frac{1}{16} + \cdots + \frac{1}{2^{k+1}} < \frac{1}{2}$$

We now add $\frac{1}{2}$ to both sides of this inequality to get

$$\frac{1}{2} + \frac{1}{4} + \frac{1}{8} + \frac{1}{16} + \cdots + \frac{1}{2^{k+1}} < \frac{1}{2} + \frac{1}{2}$$

or

$$\frac{1}{2} + \frac{1}{4} + \frac{1}{8} + \frac{1}{16} + \cdots + \frac{1}{2^{k+1}} < 1$$

The resulting inequality is the same as the original inequality, except that $k + 1$ appears in place of n. Thus, the truth of the inequality for $n = k$ implies the truth of the inequality for $n = k + 1$. Part 2 of the proof is complete.

Because both of the induction requirements have been verified, this inequality is true for all natural numbers. ∎

Some statements are not true when $n = 1$, but are true for all natural numbers equal to or greater than some given natural number (say, q). In these cases, we verify the given statements for $n = q$ in Part 1 of the induction proof. After establishing Part 2 of the induction proof, the given statement is proved for all natural numbers that are greater than or equal to q.

Orals *Is the equation $3 + 6 + 9 + \cdots + 3n = n(n + 2)$ true for the following values of n?*

1. $n = 1$ **2.** $n = 2$ **3.** $n = 3$

13.5 EXERCISES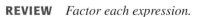

REVIEW *Factor each expression.*

1. $-6a^3b + 9a^2b^2$ **2.** $9a^2 - 4b^2$

3. $10a^2 - 21ab - 10b^2$ **4.** $8a^3 - 27b^3$

VOCABULARY AND CONCEPTS *Fill in the blanks.*

5. Any proof by induction requires ____ parts.

6. Part 1 is to show that the statement is true for _____.

7. Part 2 is to show that the statement is true for _____ whenever it is true for $n = k$.

8. When we assume that a formula is true for $n = k$, we call the assumption the induction _____.

PRACTICE *Verify each formula for $n = 1, 2, 3$, and 4.*

9. $5 + 10 + 15 + \cdots + 5n = \dfrac{5n(n + 1)}{2}$

10. $1^2 + 2^2 + 3^2 + \cdots + n^2 = \dfrac{n(n + 1)(2n + 1)}{6}$

11. $7 + 10 + 13 + \cdots + (3n + 4) = \dfrac{n(3n + 11)}{2}$

12. $1(3) + 2(4) + 3(5) + \cdots + n(n + 2) =$
$\dfrac{n}{6}(n + 1)(2n + 7)$

Prove each formula by mathematical induction, if possible.

13. $2 + 4 + 6 + \cdots + 2n = n(n + 1)$

14. $1 + 3 + 5 + \cdots + (2n - 1) = n^2$

15. $3 + 7 + 11 + \cdots + (4n - 1) = n(2n + 1)$

16. $4 + 8 + 12 + \cdots + 4n = 2n(n + 1)$

17. $10 + 6 + 2 + \cdots + (14 - 4n) = 12n - 2n^2$

18. $8 + 6 + 4 + \cdots + (10 - 2n) = 9n - n^2$

19. $2 + 5 + 8 + \cdots + (3n - 1) = \dfrac{n(3n + 1)}{2}$

20. $3 + 6 + 9 + \cdots + 3n = \dfrac{3n(n + 1)}{2}$

21. $1^2 + 2^2 + 3^2 + \cdots + n^2 = \dfrac{n(n + 1)(2n + 1)}{6}$

22. $1 + 2 + 3 + \cdots + (n - 1) + n + (n - 1) +$
$\cdots + 3 + 2 + 1 = n^2$

23. $\dfrac{1}{3} + 2 + \dfrac{11}{3} + \cdots + \left(\dfrac{5}{3}n - \dfrac{4}{3}\right) = n\left(\dfrac{5}{6}n - \dfrac{1}{2}\right)$

24. $\dfrac{1}{1 \cdot 2} + \dfrac{1}{2 \cdot 3} + \dfrac{1}{3 \cdot 4} + \cdots + \dfrac{1}{n(n + 1)} = \dfrac{n}{n + 1}$

25. $\dfrac{1}{2} + \dfrac{1}{4} + \dfrac{1}{8} + \cdots + \left(\dfrac{1}{2}\right)^n = 1 - \left(\dfrac{1}{2}\right)^n$

26. $\dfrac{1}{3} + \dfrac{2}{9} + \dfrac{4}{27} + \cdots + \dfrac{1}{3}\left(\dfrac{2}{3}\right)^{n-1} = 1 - \left(\dfrac{2}{3}\right)^n$

27. $2^0 + 2^1 + 2^2 + 2^3 + \cdots + 2^{n-1} = 2^n - 1$

28. $1^3 + 2^3 + 3^3 + \cdots + n^3 = \left[\dfrac{n(n + 1)}{2}\right]^2$

29. Prove that $x - y$ is a factor of $x^n - y^n$. (*Hint:* Consider subtracting and adding xy^k to the binomial $x^{k+1} - y^{k+1}$.)

30. Prove that $n < 2^n$.

31. There are $180°$ in the sum of the angles of any triangle. Prove that $(n - 2)180°$ is the sum of the angles of any simple polygon when n is its number of sides. (*Hint:* If a polygon has $k + 1$ sides, it has $k - 2$ sides plus three more sides.)

32. Consider the equation

$$1 + 3 + 5 + \cdots + 2n - 1 = 3n - 2$$

   a. Is the equation true for $n = 1$?
   b. Is the equation true for $n = 2$?
   c. Is the equation true for all natural numbers n?

33. If $1 + 2 + 3 + \cdots + n = \frac{n}{2}(n + 1) + 1$ were true for $n = k$, show that it would be true for $n = k + 1$. Is it true for $n = 1$?

34. Prove that $n + 1 = 1 + n$ for each natural number n.

35. If n is any natural number, prove that $7^n - 1$ is divisible by 6.

36. Prove that $1 + 2n < 3^n$ for $n > 1$.

37. Prove that, if r is a real number where $r \neq 1$, then

$$1 + r + r^2 + \cdots + r^n = \dfrac{1 - r^{n+1}}{1 - r}$$

38. Prove that the formula for the sum of the first n terms of an arithmetic sequence;

$$a + [a + d] + [a + 2d] + [a + (n - 1)d] = \dfrac{n(a + 1)}{2}$$

   where $l = a + (n - 1)d$.

WRITING

39. Explain why we need to verify the formula for $n = 1$.

40. Explain why we need to show that the formula must be true for $n = k + 1$ whenever it is true for $n = k$.

SOMETHING TO THINK ABOUT

41. The expression a^m, where m is a natural number, is defined as the product of m factors of a. An alternative definition of a^m is (Part 1) $a^1 = a$ and (Part 2) $a^{m+1} = a^m \cdot a$. Use induction on n to prove the law of exponents, $a^m a^n = a^{m+n}$.

42. Use induction on n to prove the law of exponents, $(a^m)^n = a^{mn}$. (See Exercise 41.)

13.6   Permutations and Combinations

■ The Multiplication Principle for Events ■ Permutations
■ Formulas for Permutations ■ Combinations
■ Formulas for Combinations ■ The Binomial Theorem
■ Distinguishable Words

Getting Ready   *Evaluate each expression.*

1. $4 \cdot 3 \cdot 2 \cdot 1$

2. $5 \cdot 4 \cdot 3 \cdot 2 \cdot 1$

3. $\dfrac{6 \cdot 5 \cdot 4 \cdot 3 \cdot 2 \cdot 1}{4 \cdot 3 \cdot 2 \cdot 1}$

4. $\dfrac{8 \cdot 7 \cdot 6 \cdot 5 \cdot 4 \cdot 3 \cdot 2 \cdot 1}{2(5 \cdot 4 \cdot 3 \cdot 2 \cdot 1)}$

In this section, we will count the ways items can be arranged and the ways items can be placed in groups. These skills are necessary in the field of statistics.

The Multiplication Principle for Events

Lydia plans to go to dinner and attend a movie. If she has a choice of four restaurants and three movies, in how many ways can she spend her evening? There are four choices of restaurants and, for any one of these choices, there are three choices of movies, as shown in the tree diagram in Figure 13-2.

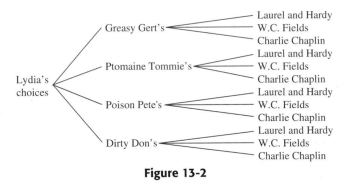

Figure 13-2

The diagram shows that Lydia has 12 ways to spend her evening. One possibility is to eat at Ptomaine Tommie's and watch W. C. Fields. Another is to eat at Dirty Don's and watch Laurel and Hardy.

Any situation that has several outcomes is called an **event.** Lydia's first event (choosing a restaurant) can occur in 4 ways. Her second event (choosing a movie) can occur in 3 ways. Thus, she has $4 \cdot 3$, or 12, ways to spend her evening. This example illustrates the **multiplication principle for events.**

Multiplication Principle for Events

Let E_1 and E_2 be two events. If E_1 can be done in a_1 ways, and if—after E_1 has occurred—E_2 can be done in a_2 ways, then the event "E_1 followed by E_2" can be done in $a_1 \cdot a_2$ ways.

The multiplication principle can be extended to n events.

EXAMPLE 1   If a traveler has four ways to go from New York to Chicago, three ways to go from Chicago to Denver, and six ways to go from Denver to San Francisco, in how many ways can she go from New York to San Francisco?

Solution   We can let E_1 be the event "going from New York to Chicago," E_2 be the event "going from Chicago to Denver," and E_3 be the event "going from Denver to San Francisco." Since there are 4 ways to accomplish E_1, 3 ways to accomplish E_2, and 6 ways to accomplish E_3, the number of routes available is

$$4 \cdot 3 \cdot 6 = 72$$

Self Check   If a person has 4 sweaters and 5 pairs of slacks, how many outfits can he wear?   ∎

Permutations

Suppose we want to arrange 7 books on a shelf. We can fill the first space with any one of the 7 books, the second space with any of the remaining 6 books, the third space with any of the remaining 5 books, and so on, until there is only one space left to fill with the last book. According to the multiplication principle, the number of ways that we can arrange the books is

$$7 \cdot 6 \cdot 5 \cdot 4 \cdot 3 \cdot 2 \cdot 1 = 5{,}040$$

When finding the number of possible arrangements of books on a shelf, we are finding the number of **permutations.** The number of permutations of 7 books, using all the books, is 5,040. The symbol $P(n, r)$ is read as "the number of permutations of n things r at a time." Thus, $P(7, 7) = 5{,}040$.

EXAMPLE 2   Assume that there are 7 signal flags of 7 different colors to hang on a mast. How many different signals can be sent when 3 flags are used?

Solution   We are asked to find $P(7, 3)$, the number of permutations of 7 things using 3 of them. Any one of the 7 flags can hang in the top position on the mast. Any one of the 6 remaining flags can hang in the middle position, and any one of the remaining 5 flags can hang in the bottom position. By the multiplication principle, we have

$$P(7, 3) = 7 \cdot 6 \cdot 5 = 210$$

It is possible to send 210 different signals.

Self Check   How many signals can be sent if two of the seven flags are missing?   ∎

Formulas for Permutations

Although it is correct to write $P(7, 3) = 7 \cdot 6 \cdot 5$, we will change the form of the answer to obtain a convenient formula. To derive this formula, we proceed as follows:

$$P(7, 3) = 7 \cdot 6 \cdot 5$$
$$= \frac{7 \cdot 6 \cdot 5 \cdot 4 \cdot 3 \cdot 2 \cdot 1}{4 \cdot 3 \cdot 2 \cdot 1} \quad \text{Multiply numerator and denominator by } 4 \cdot 3 \cdot 2 \cdot 1.$$
$$= \frac{7!}{4!}$$
$$= \frac{7!}{(7 - 3)!}$$

The generalization of this idea gives the following formula.

Formula for $P(n, r)$ The number of permutations of n things r at a time is given by

$$P(n, r) = \frac{n!}{(n - r)!}$$

EXAMPLE 3   Find   **a.** $P(8, 4)$,   **b.** $P(n, n)$,   and   **c.** $P(n, 0)$.

Solution   **a.** $P(8, 4) = \dfrac{8!}{(8 - 4)!} = \dfrac{8 \cdot 7 \cdot 6 \cdot 5 \cdot 4!}{4!} = 1{,}680$

b. $P(n, n) = \dfrac{n!}{(n - n)!} = \dfrac{n!}{0!} = n!$

c. $P(n, 0) = \dfrac{n!}{(n - 0)!} = \dfrac{n!}{n!} = 1$

Self Check   Find   **a.** $P(7, 5)$ and   **b.** $P(6, 0)$.   ∎

Parts b and c of Example 3 establish the following formulas.

Formulas for $P(n, n)$ and $P(n, 0)$

The number of permutations of n things n at a time and n things 0 at a time are given by the formulas

$$P(n, n) = n! \qquad \text{and} \qquad P(n, 0) = 1$$

EXAMPLE 4   In how many ways can a baseball manager arrange a batting order of 9 players if there are 25 players on the team?

Solution   To find the number of permutations of 25 things 9 at a time, we substitute 25 for n and 9 for r in the formula for finding $P(n, r)$.

$$P(\boldsymbol{n}, r) = \dfrac{\boldsymbol{n}!}{(\boldsymbol{n} - r)!}$$

$$P(\mathbf{25}, \mathbf{9}) = \dfrac{\mathbf{25}!}{(\mathbf{25} - \mathbf{9})!}$$

$$= \dfrac{25!}{16!}$$

$$= \dfrac{25 \cdot 24 \cdot 23 \cdot 22 \cdot 21 \cdot 20 \cdot 19 \cdot 18 \cdot 17 \cdot 16!}{16!}$$

$$\approx 741{,}354{,}768{,}000$$

The number of permutations is approximately 741,354,768,000.

Self Check   In how many ways can the manager arrange a batting order if two players can't play?   ∎

EXAMPLE 5   In how many ways can 5 people stand in a line if 2 people refuse to stand next to each other?

Solution   The total number of ways that 5 people can stand in line is

$$P(5, 5) = 5! = 5 \cdot 4 \cdot 3 \cdot 2 \cdot 1 = 120$$

To find the number of ways that 5 people can stand in line if 2 people insist on standing together, we consider the two people as one person. Then there are 4 people to stand in line, and this can be done in $P(4, 4) = 4! = 24$ ways. However, because either of the two who are paired could be first, there are two arrangements for the pair who insist on standing together. Thus, there are $2 \cdot 4!$, or 48 ways that 5 people can stand in line if 2 people insist on standing together.

The number of ways that 5 people can stand in line if two people refuse to stand together is $5! = 120$ (the total number of ways to line up 5 people) minus $2 \cdot 4! = 48$ (the number of ways to line up the 5 people if 2 do stand together):

$$120 - 48 = 72$$

There are 72 ways to line up the people.

Self Check   In how many ways can 5 people stand in a line if one person demands to be first?

EXAMPLE 6   In how many ways can 5 people be seated at a round table?

Solution   If we were to seat 5 people in a row, there would be $5!$ possible arrangements. However, at a round table, each person has a neighbor to the left and to the right. If each person moves one, two, three, four, or five places to the left, everyone has the same neighbors and the arrangement has not changed. Thus, we must divide $5!$ by 5 to get rid of these duplications. The number of ways that 5 people can be seated at a round table is

$$\frac{5!}{5} = 4! = 4 \cdot 3 \cdot 2 \cdot 1 = 24$$

Self Check   In how many ways can 6 people be seated at a round table?

The results of Example 6 suggest the following fact.

Circular Arrangements   There are $(n - 1)!$ ways to arrange n things in a circle.

Combinations

Suppose that a class of 12 students selects a committee of 3 persons to plan a party. With committees, order is not important. A committee of John, Maria, and Raul is the same as a committee of Maria, Raul, and John. However, if we assume for the moment that order is important, we can find the number of permutations of 12 things 3 at a time.

$$P(12, 3) = \frac{12!}{(12 - 3)!} = \frac{12 \cdot 11 \cdot 10 \cdot 9!}{9!} = 1{,}320$$

However, since we do not care about order, this result of 1,320 ways is too large. Because there are 6 ways ($3! = 6$) of ordering every committee of 3 students, the result of $P(12, 3) = 1{,}320$ is exactly 6 times too big. To get the correct number of committees, we must divide $P(12, 3)$ by 6:

$$\frac{P(12, 3)}{6} = \frac{1{,}320}{6} = 220$$

In cases of selection where order is not important, we are interested in **combinations,** not permutations. The symbols $C(n, r)$ and $\binom{n}{r}$ both mean the number of combinations of n things r at a time.

Formulas for Combinations

If a committee for r people is chosen from a total of n people, the number of possible committees is $C(n, r)$, and there will be $r!$ arrangements of each committee. If we consider the committee as an ordered grouping, the number of orderings is $P(n, r)$. Thus, we have

(1) $r!C(n, r) = P(n, r)$

We can divide both sides of Equation 1 by $r!$ to obtain the formula for finding $C(n, r)$.

$$C(n, r) = \binom{n}{r} = \frac{P(n, r)}{r!} = \frac{n!}{r!(n-r)!}$$

Formula for $C(n, r)$

The number of combinations of n things r at a time is given by

$$C(n, r) = \binom{n}{r} = \frac{n!}{r!(n-r)!}$$

In the exercises, you will be asked to prove the following formulas.

Formulas for $C(n, n)$ and $C(n, 0)$

If n is a whole number, then

$$C(n, n) = 1 \qquad \text{and} \qquad C(n, 0) = 1$$

EXAMPLE 7   If Carla must read 4 books from a reading list of 10 books, how many choices does she have?

Solution   Because the order in which the books are read is not important, we find the number of combinations of 10 things 4 at a time:

$$C(10, 4) = \frac{10!}{4!(10-4)!} = \frac{10 \cdot 9 \cdot 8 \cdot 7 \cdot 6!}{4 \cdot 3 \cdot 2 \cdot 1 \cdot 6!}$$

$$= \frac{10 \cdot 9 \cdot 8 \cdot 7}{4 \cdot 3 \cdot 2}$$

$$= 210$$

Carla has 210 choices.

Self Check   How many choices would Carla have if she had to read 5 books?

EXAMPLE 8 A class consists of 15 men and 8 women. In how many ways can a debate team be chosen with 3 men and 3 women?

Solution There are $C(15, 3)$ ways of choosing 3 men and $C(8, 3)$ ways of choosing 3 women. By the multiplication principle, there are $C(15, 3) \cdot C(8, 3)$ ways of choosing members of the debate team:

$$C(15, 3) \cdot C(8, 3) = \frac{15!}{3!(15 - 3)!} \cdot \frac{8!}{3!(8 - 3)!}$$

$$= \frac{15 \cdot 14 \cdot 13}{6} \cdot \frac{8 \cdot 7 \cdot 6}{6}$$

$$= 25{,}480$$

There are 25,480 ways to choose the debate team.

Self Check In how many ways can the debate team be chosen if it is to have 4 men and 2 women?

The Binomial Theorem

The formula

$$C(n, r) = \frac{n!}{r!(n - r)!}$$

gives the coefficient of the $(r + 1)$th term of the binomial expansion of $(a + b)^n$. This implies that the coefficients of a binomial expansion can be used to solve problems involving combinations. The binomial theorem is restated below—this time listing the $(r + 1)$th term and using combination notation.

Binomial Theorem

If n is any positive integer, then

$$(a + b)^n = \binom{n}{0}a^n + \binom{n}{1}a^{n-1}b + \binom{n}{2}a^{n-2}b^2 + \cdots$$

$$+ \binom{n}{r}a^{n-r}b^r + \cdots + \binom{n}{n}b^n$$

EXAMPLE 9 Use Pascal's triangle to compute $C(7, 5)$.

Solution Consider the eighth row of Pascal's triangle and the corresponding combinations:

$$\begin{array}{cccccccc}
1 & 7 & 21 & 35 & 35 & 21 & 7 & 1 \\
\binom{7}{0} & \binom{7}{1} & \binom{7}{2} & \binom{7}{3} & \binom{7}{4} & \binom{7}{5} & \binom{7}{6} & \binom{7}{7}
\end{array}$$

$C(7, 5) = \binom{7}{5} = 21$.

Self Check Use Pascal's triangle to compute $C(6, 5)$.

Distinguishable Words

A "word" is a distinguishable arrangement of letters. For example, six words can be formed with the letters a, b, and c if each letter is used exactly once. The six words are

abc,   acb,   bac,   bca,   cab,   and   cba

If there are n distinct letters and each letter is used once, the number of distinct words that can be formed is $n! = P(n, n)$. It is more complicated to compute the number of distinguishable words that can be formed with n letters when some of the letters are duplicates.

EXAMPLE 10   Find the number of "words" that can be formed if each of the 6 letters of the word *little* is used once.

Solution   For the moment, assume that the letters of the word *little* are distinguishable: "LitTle." The number of words that can be formed using each letter once is $6! = P(6, 6)$. However, in reality we cannot tell the l's or the t's apart. Therefore, we must divide by a number to get rid of these duplications. Because there are 2! orderings of the two l's and 2! orderings of the two t's, we divide by $2! \cdot 2!$. The number of words that can be formed using each letter of the word *little* is

$$\frac{P(6, 6)}{2! \cdot 2!} = \frac{6!}{2! \cdot 2!} = \frac{6 \cdot 5 \cdot 4 \cdot 3 \cdot 2 \cdot 1}{2 \cdot 1 \cdot 2 \cdot 1} = 180$$

Self Check   How many words can be formed if each letter of the word *balloon* is used once?

PERSPECTIVE

Gambling is an occasional diversion for some and an obsession for others. Whether it is horse racing, slot machines, or state lotteries, the lure of instant riches is hard to resist.

Many states conduct lotteries, and the systems vary. One scheme is typical: For $1, you have two chances to match 6 numbers chosen from 55 numbers and win a grand prize of about $5 million. How likely are you to win? Is it worth $1 to play the game?

To match the six numbers chosen from 55, you must choose the one winning combination out of $C(55, 6)$ possibilities:

$$C(n, r) = \frac{n!}{r!(n - r)!}$$

$$C(55, 6) = \frac{55!}{6!(55 - 6)!}$$

$$= \frac{55 \cdot 54 \cdot 53 \cdot 52 \cdot 51 \cdot 50}{6 \cdot 5 \cdot 4 \cdot 3 \cdot 2}$$

$$= 28,989,675$$

In this game, you have two chances in about 29 million of winning $5 million. Over the long haul, you will win $\frac{2}{29,000,000}$ of the time, so each ticket is worth $\frac{2}{29,000,000}$ of $5,000,000, or about 35¢, if you don't have to share the prize with another winner. For every dollar spent to play the game, you can expect to throw away 65¢. This state lottery is a poor bet. Casinos pay better than 50¢ on the dollar, with some slot machines returning 90¢. "You can't win if you're not in!" is the claim of the lottery promoters. A better claim would be "You won't regret if you don't bet!"

Example 10 illustrates the following general principle.

Distinguishable Words

If a word with n letters has a of one letter, b of another letter, and so on, the number of distinguishable words that can be formed using each letter of the n-letter word exactly once is

$$\frac{n!}{a!b!\cdots}$$

Self Check Answers

1. 20   2. 60   3. **a.** 2,520   **b.** 1   4. approximately 296,541,907,200   5. 24   6. 120   7. 252
8. 38,220   9. 6   10. 1,260

Orals   *Assume that there are 3 books and 5 records.*

1. In how many ways can you pick 1 book and 1 record?

2. In how many ways can 5 soldiers stand in line?

3. Find $P(3, 1)$.

4. Find $P(3, 3)$.

5. Find $C(3, 0)$.

6. $C(3, 3)$.

13.6   EXERCISES

REVIEW   *Find the value of x.*

1. $\log_x 16 = 4$

2. $\log_\pi x = \dfrac{1}{2}$

3. $\log_{\sqrt{7}} 49 = x$

4. $\log_x \dfrac{1}{2} = -\dfrac{1}{3}$

Tell whether the statement is true.

5. $\log_{17} 1 = 0$

6. $\log_5 0 = 1$

7. $\log_b b^b = b$

8. $\dfrac{\log_7 A}{\log_7 B} = \log_7 \dfrac{A}{B}$

VOCABULARY AND CONCEPTS   *Fill in the blanks.*

9. If E_1 and E_2 are two events and E_1 can be done in 4 ways and E_2 can be done in 6 ways, then the event E_1 followed by E_2 can be done in ___ ways.

10. An arrangement of n objects is called a _____.

11. $P(n, r) = $ _____

12. $P(n, n) = $ __

13. $P(n, 0) = $ __

14. There are _____ ways to arrange n things in a circle.

15. Another notation for $C(n, r)$ is ____.

16. $C(n, r) = $ _____

17. $C(n, n) = $ __

18. $C(n, 0) = $ __

19. If a word with n letters has a of one letter, b of another letter, and so on, the number of different words that can be formed is _____

20. Where the order of selection is not important, we are interested in _____, not _____.

PRACTICE   *Evaluate each expression.*

21. $P(7, 4)$

22. $P(8, 3)$

23. $C(7, 4)$

24. $C(8, 3)$

25. $P(5, 5)$

26. $P(5, 0)$

27. $\dbinom{5}{4}$

28. $\dbinom{8}{4}$

29. $\dbinom{5}{0}$

30. $\dbinom{5}{5}$

31. $P(5, 4) \cdot C(5, 3)$

32. $P(3, 2) \cdot C(4, 3)$

33. $\dbinom{5}{3}\dbinom{4}{3}\dbinom{3}{3}$

34. $\dbinom{5}{5}\dbinom{6}{6}\dbinom{7}{7}\dbinom{8}{8}$

35. $\dbinom{68}{66}$

36. $\dbinom{100}{99}$

APPLICATIONS

37. Choosing lunch   A lunchroom has a machine with eight kinds of sandwiches, a machine with four kinds of soda, a machine with both white and chocolate milk, and a machine with three kinds of ice cream. How many different lunches can be chosen? (Consider a lunch to be one sandwich, one drink, and one ice cream.)

38. Manufacturing license plates   How many six-digit license plates can be manufactured if no license plate number begins with 0?

39. Available phone numbers   How many different seven-digit phone numbers can be used in one area code if no phone number begins with 0 or 1?

40. Arranging letters   In how many ways can the letters of the word *number* be arranged?

41. Arranging letters with restrictions   In how many ways can the letters of the word *number* be arranged if the e and r must remain next to each other?

42. Arranging letters with restrictions   In how many ways can the letters of the word *number* be arranged if the e and r cannot be side by side?

43. Arranging letters with repetitions   How many ways can five Scrabble tiles bearing the letters, F, F, F, L, and U be arranged to spell the word *fluff?*

44. Arranging letters with repetitions   How many ways can six Scrabble tiles bearing the letters, B, E, E, E, F, and L be arranged to spell the word *feeble?*

45. Placing people in line   In how many arrangements can 8 women be placed in a line?

46. Placing people in line   In how many arrangements can 5 women and 5 men be placed in a line if the women and men alternate?

47. Placing people in line   In how many arrangements can 5 women and 5 men be placed in a line if all the men line up first?

48. Placing people in line   In how many arrangements can 5 women and 5 men be placed in a line if all the women line up first?

49. Combination locks   How many permutations does a combination lock have if each combination has 3 numbers, no two numbers of the combination are the same, and the lock dial has 30 notches?

50. Combination locks   How many permutations does a combination lock have if each combination has 3 numbers, no two numbers of the combination are the same, and the lock dial has 100 notches?

51. Seating at a table   In how many ways can 8 people be seated at a round table?

52. Seating at a table   In how many ways can 7 people be seated at a round table?

53. Seating at a table   In how many ways can 6 people be seated at a round table if 2 of the people insist on sitting together?

54. Seating arrangements with conditions   In how many ways can 6 people be seated at a round table if 2 of the people refuse to sit together?

55. Arrangements in a circle   In how many ways can 7 children be arranged in a circle if Sally and John want to sit together and Martha and Peter want to sit together?

56. Arrangements in a circle   In how many ways can 8 children be arranged in a circle if Laura, Scott, and Paula want to sit together?

57. Selecting candy bars   In how many ways can 4 candy bars be selected from 10 different candy bars?

58. Selecting birthday cards   In how many ways can 6 birthday cards be selected from 24 different cards?

59. Circuit wiring   A wiring harness containing a red, a green, a white, and a black wire must be attached to a control panel. In how many different orders can the wires be attached?

60. Grading homework   A professor grades homework by randomly checking 7 of the 20 problems assigned. In how many different ways can this be done?

61. Forming words with distinct letters   How many words can be formed from the letters of the word *plastic* if each letter is to be used once?

62. Forming words with repeated letters   How many words can be formed from the letters of the word *banana* if each letter is to be used once?

63. Manufacturing license plates   How many license plates can be made using two different letters followed by four different digits if the first digit cannot be 0 and the letter O is not used?

64. Planning class schedules   If there are seven class periods in a school day, and a typical student takes 5 classes, how many different time patterns are possible for the student?

65. Selecting golf balls   From a bucket containing 6 red and 8 white golf balls, in how many ways can we draw 6 golf balls of which 3 are red and 3 are white?

66. Selecting committees   In how many ways can you select a committee of 3 Republicans and 3 Democrats from a group containing 18 Democrats and 11 Republicans?

67. Selecting committees   In how many ways can you select a committee of 4 Democrats and 3 Republicans from a group containing 12 Democrats and 10 Republicans?

68. Drawing cards   In how many ways can you select a group of 5 red cards and 2 black cards from a deck containing 10 red cards and 8 black cards?

69. Planning dinner   In how many ways can a husband and wife choose 2 different dinners from a menu of 17 dinners?

70. Placing people in line   In how many ways can 7 people stand in a row if 2 of the people refuse to stand together?

71. Geometry   How many lines are determined by 8 points if no 3 points lie on a straight line?

72. Geometry   How many lines are determined by 10 points if no 3 points lie on a straight line?

73. Coaching basketball   How many different teams can a basketball coach start if the entire squad consists of 10 players? (Assume that a starting team has 5 players and each player can play all positions.)

74. Managing baseball   How many different teams can a manager start if the entire squad consists of 25 players? (Assume that a starting team has 9 players and each player can play all positions.)

75. Selecting job applicants   There are 30 qualified applicants for 5 openings in the sales department. In how many different ways can the group of 5 be selected?

76. Sales promotions   If a customer purchases a new stereo system during the spring sale, he may choose any six CDs from 20 classical and 30 jazz selections. In how many ways can the customer choose three of each?

77. Guessing on matching questions   Ten words are to be paired with the correct 10 out of 12 possible definitions. How many ways are there of guessing?

78. Guessing on true-false exams   How many possible ways are there of guessing on a 10-question true-false exam, if it is known that the instructor will have 5 true and 5 false responses?

WRITING

79. Explain how to use Pascal's triangle to find $C(8, 5)$.

80. Explain how to use Pascal's triangle to find $C(10, 8)$.

SOMETHING TO THINK ABOUT

81. Prove that $C(n, n) = 1$.

82. Prove that $C(n, 0) = 1$.

83. Prove that $\binom{n}{r} = \binom{n}{n - r}$.

84. Show that the binomial theorem can be expressed in the form

$$(a + b)^n = \sum_{k=0}^{n} \binom{n}{k} a^{n-k} b^k$$

13.7   Probability

▪ Probability   ▪ Multiplication Property of Probabilities

Getting Ready   *Interpret each statement.*

1. The weather service predicts a 50% chance of snow.

2. The chances of winning the lottery are 2 out of 30 million.

In this section, we will discuss probability, a topic often used in everyday life. To solve probability problems, we will use the material on combinations, studied in the last section.

Probability

The probability that an event will occur is a measure of the likelihood of that event. A tossed coin, for example, can land in two ways, either heads or tails. Because one of these two equally likely outcomes is heads, we expect that out of several tosses, about half will be heads. We say that the probability of obtaining heads in a single toss of the coin is $\frac{1}{2}$.

If records show that out of 100 days with weather conditions like today's, 30 have received rain, the weather service will report, "There is a $\frac{30}{100}$ or 30% probability of rain today."

An **experiment** is a process for which the outcome is uncertain. Tossing a coin, rolling a die, drawing a card, and predicting rain are examples of experiments. For any experiment, the set of all possible outcomes is called a **sample space.**

The sample space, S, for the experiment of tossing two coins is the set

$$S = \{(H, H), (H, T), (T, H), (T, T)\}$$

where the ordered pair (H, T) represents the outcome "heads on the first coin and tails on the second coin." Because there are two possible outcomes for the first coin and two for the second coin, we know (by the multiplication principle for events) that there are $2 \cdot 2 = 4$ possible outcomes. Since there are 4 elements in the sample space S, we write

$$n(S) = 4 \qquad \text{Read as "The number of elements is set } S \text{ is 4."}$$

An **event** associated with an experiment is any subset of the sample space of that experiment. For example, if E is the event "getting at least one heads" in the experiment of tossing two coins, then

$$E = \{(H, H), (H, T), (T, H)\}$$

and $n(E) = 3$. Because the outcome of getting at least one heads can occur in 3 out of 4 possible ways, we say that the **probability** of a favorable outcome is $\frac{3}{4}$.

$$P(E) = P(\text{at least one heads}) = \frac{3}{4}$$

We define the probability of an event as follows.

Probability of an Event   If S is the sample space of an experiment with n distinct and equally likely outcomes, and E is an event that occurs in s of those ways, then the **probability of E** is

$$P(E) = \frac{n(E)}{n(S)} = \frac{s}{n}$$

Because $0 \leq s \leq n$, it follows that $0 \leq \frac{s}{n} \leq 1$. This implies that all probabilities have values from 0 to 1. An event that cannot happen has probability 0. An event that is certain to happen has probability 1.

To say that the probability of tossing heads on one toss of a coin is $\frac{1}{2}$ means that if a fair coin is tossed a large number of times, the ratio of the number of heads to the total number of tosses is nearly $\frac{1}{2}$.

To say that the probability of rolling 5 on one roll of a die is $\frac{1}{6}$ means that as the number of rolls approaches infinity, the ratio of the number of favorable outcomes (rolling a 5) to the total number of outcomes (rolling a 1, 2, 3, 4, 5, or 6) approaches $\frac{1}{6}$.

EXAMPLE 1   Show the sample space of the experiment "rolling two dice a single time."

Solution   We can list ordered pairs and let the first number be the result on the first die and the second number the result on the second die. The sample space, S, is the set with the following elements:

$$(1, 1) \quad (1, 2) \quad (1, 3) \quad (1, 4) \quad (1, 5) \quad (1, 6)$$
$$(2, 1) \quad (2, 2) \quad (2, 3) \quad (2, 4) \quad (2, 5) \quad (2, 6)$$
$$(3, 1) \quad (3, 2) \quad (3, 3) \quad (3, 4) \quad (3, 5) \quad (3, 6)$$
$$(4, 1) \quad (4, 2) \quad (4, 3) \quad (4, 4) \quad (4, 5) \quad (4, 6)$$
$$(5, 1) \quad (5, 2) \quad (5, 3) \quad (5, 4) \quad (5, 5) \quad (5, 6)$$
$$(6, 1) \quad (6, 2) \quad (6, 3) \quad (6, 4) \quad (6, 5) \quad (6, 6)$$

Since there are 6 possible outcomes with the first die and 6 possible outcomes with the second die, we expect $6 \cdot 6 = 36$ equally likely possible outcomes, and we have $n(S) = 36$.

Self Check   How many pairs in the sample space have a sum of 7?   ∎

EXAMPLE 2   Find the probability of the event "rolling a sum of 7 on one roll of two dice."

Solution   The sample space is listed in Example 1. We let E be the set of favorable outcomes, those that give a sum of 7:

$$E = \{(1, 6), (2, 5), (3, 4), (4, 3), (5, 2), (6, 1)\}$$

Since there are 6 favorable outcomes among the 36 equally likely outcomes, $n(E) = 6$, and

$$P(E) = P(\text{rolling a } 7) = \frac{n(E)}{n(S)} = \frac{6}{36} = \frac{1}{6}$$

Self Check   Find the probability of rolling a sum of 4.   ∎

A standard playing deck of 52 cards has two red suits, hearts and diamonds, and two black suits, clubs and spades. Each suit has 13 cards, including an ace, a king, a queen, a jack, and cards numbered from 2 to 10. We will refer to a standard deck of cards in many examples and exercises.

EXAMPLE 3   Find the probability of drawing 5 cards, all hearts, from a standard deck of cards.

Solution   Since the number of ways to draw 5 hearts from the 13 hearts is $C(13, 5)$, we have $n(E) = C(13, 5)$. Since the number of ways to draw 5 cards from the deck is

$C(52, 5)$, we have $n(S) = C(52, 5)$. The probability of drawing 5 hearts is the ratio of the number of favorable outcomes to the number of possible outcomes.

$$P(5 \text{ hearts}) = \frac{C(13, 5)}{C(52, 5)}$$

$$P(5 \text{ hearts}) = \frac{\dfrac{13!}{5!8!}}{\dfrac{52!}{5!47!}}$$

$$= \frac{13!}{5!8!} \cdot \frac{5!47!}{52!}$$

$$= \frac{13 \cdot 12 \cdot 11 \cdot 10 \cdot 9 \cdot 8!}{8!} \cdot \frac{47!}{52 \cdot 51 \cdot 50 \cdot 49 \cdot 48 \cdot 47!}$$

$$= \frac{13 \cdot 12 \cdot 11 \cdot 10 \cdot 9}{52 \cdot 51 \cdot 50 \cdot 49 \cdot 48}$$

$$= \frac{33}{66,640}$$

The probability of drawing 5 hearts is $\dfrac{33}{66,640}$.

Self Check   Find the probability of drawing 6 cards, all diamonds, from the deck.   ∎

Multiplication Property of Probabilities

There is a property of probabilities that is similar to the multiplication principle for events. In the following theorem, we read $P(A \cap B)$ as "the probability of A and B" and $P(B|A)$ as "the probability of B given A." If A and B are events, the set $A \cap B$ contains the outcomes that are in both A and B.

Multiplication Property of Probabilities   If $P(A)$ represents the probability of event A, and $P(B|A)$ represents the probability that event B will occur after event A, then

$$P(A \cap B) = P(A) \cdot P(B|A)$$

EXAMPLE 4   A box contains 40 cubes of the same size. Of these cubes, 17 are red, 13 are blue, and the rest are yellow. If 2 cubes are drawn at random, without replacement, find the probability that 2 yellow cubes will be drawn.

Solution   Of the 40 cubes in the box, 10 are yellow. The probability of getting a yellow cube on the first draw is

$$P(\text{yellow cube on the first draw}) = \frac{10}{40} = \frac{1}{4}$$

Because there is no replacement after the first draw, 39 cubes remain in the box, and 9 of these are yellow. The probability of drawing a yellow cube on the second draw, given that a yellow cube was drawn on the first draw, is

$$P(\text{yellow cube on the second draw}) = \frac{9}{39} = \frac{3}{13}$$

The probability of drawing 2 yellow cubes in succession is the product of the probability of drawing a yellow cube on the first draw and the probability of drawing a yellow cube on the second draw.

$$P(\text{drawing two yellow cubes}) = \frac{1}{4} \cdot \frac{3}{13} = \frac{3}{52}$$

Self Check   Find the probability that 2 blue cubes will be drawn.   ∎

EXAMPLE 5   Repeat Example 3 using the multiplication property of probabilities.

Solution   The probability of drawing a heart on the first draw is $\frac{13}{52}$. The probability of drawing a heart on the second draw *given that we got a heart on the first draw* is $\frac{12}{51}$. The probability is $\frac{11}{50}$ on the third draw, $\frac{10}{49}$ on the fourth draw, and $\frac{9}{48}$ on the fifth draw. By the multiplication property of probabilities,

$$P(\text{5 hearts in a row}) = \frac{13}{52} \cdot \frac{12}{51} \cdot \frac{11}{50} \cdot \frac{10}{49} \cdot \frac{9}{48}$$

$$= \frac{33}{66,640}$$

Self Check   Find the probability of drawing 6 cards, all diamonds, from the deck.   ∎

EXAMPLE 6   In a school, 30% of the students are gifted in mathematics and 10% are gifted in art and mathematics. If a student is gifted in mathematics, find the probability that the student is also gifted in art.

Solution   Let $P(M)$ be the probability that a randomly chosen student is gifted in mathematics, and let $P(M \cap A)$ be the probability that the student is gifted in both art and mathematics. We must find $P(A|M)$, the probability that the student is gifted in art, given that he or she is gifted in mathematics. To do so, we substitute the given values

$$P(M) = 0.3 \quad \text{and} \quad P(M \cap A) = 0.1$$

in the formula for multiplication of probabilities and solve for $P(A|M)$:

$$P(M \cap A) = P(M) \cdot P(A|M)$$

$$0.1 = (0.3)P(A|M)$$

$$P(A|M) = \frac{0.1}{0.3}$$

$$= \frac{1}{3}$$

If a student is gifted in mathematics, there is a probability of $\frac{1}{3}$ that he or she is also gifted in art.

Self Check   If 40% of the students are gifted in art, find the probability that a student gifted in art is also gifted in mathematics.   ∎

1. 6   **2.** $\frac{1}{12}$ **3.** $\frac{33}{391,510}$ **4.** $\frac{1}{10}$ **5.** $\frac{33}{391,510}$ **6.** $\frac{1}{4}$

Orals   *Find each probability.*

1. The probability of getting heads on one toss of a coin.

2. The probability that a day chosen at random will be a Friday.

3. The probability of tossing heads twice on two tosses of a coin.

4. The probability of tossing heads three times on three tosses of a coin.

13.7   EXERCISES

REVIEW   *Solve each equation.*

1. $|x + 3| = 7$

2. $|3x - 2| = |2x - 3|$

Graph each inequality.

3. $|x - 3| < 7$

4. $|3x - 2| \geq |2x - 3|$

VOCABULARY AND CONCEPTS   *Fill in the blanks.*

5. An _____ is any process for which the outcome is uncertain.

6. A list of all possible outcomes for an experiment is called a _____.

7. The probability of an event E is defined as $P(E) =$ $\frac{}{} = \frac{s}{n}$.

8. $P(A \cap B) =$ _____.

PRACTICE   *List the sample space of each experiment.*

9. Rolling one die and tossing one coin

10. Tossing three coins

11. Selecting a letter of the alphabet

12. Picking a one-digit number

An ordinary die is rolled. Find the probability of each event.

13. Rolling a 2

14. Rolling a number greater than 4

15. Rolling a number larger than 1 but less than 6

16. Rolling an odd number

Balls numbered from 1 to 42 are placed in a container and stirred. If one is drawn at random, find the probability of each result.

17. The number is less than 20.

18. The number is less than 50.

19. The number is a prime number.

20. The number is less than 10 or greater than 40.

Refer to the spinner in the illustration. If the spinner is spun, find the probability of each event. Assume that the spinner never stops on a line.

21. The spinner stops on red.

22. The spinner stops on green.

23. The spinner stops on orange.

24. The spinner stops on yellow.

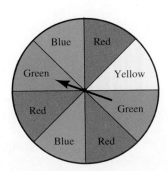

Find the probability of each event.

25. Rolling a sum of 4 on one roll of two dice.

26. Drawing a diamond on one draw from a card deck.

27. Drawing two aces in succession from a card deck if the card is replaced and the deck is shuffled after the first draw.

28. Drawing two aces from a card deck without replacing the card after the first draw.

29. Drawing a red egg from a basket containing 5 red eggs and 7 blue eggs.

30. Getting 2 red eggs in a single scoop from a bucket containing 5 red eggs and 7 yellow eggs.

31. Drawing a bridge hand of 13 cards, all of one suit.

32. Drawing 6 diamonds from a card deck without replacing the cards after each draw.

33. Drawing 5 aces from a card deck without replacing the cards after each draw.

34. Drawing 5 clubs from the black cards in a card deck.

35. Drawing a face card (king, queen, or jack) from a card deck.

36. Drawing 6 face cards in a row from a card deck without replacing the cards after each draw.

37. Drawing 5 orange cubes from a bowl containing 5 orange cubes and 1 beige cube.

38. Rolling a sum of 4 with one roll of three dice.

39. Rolling a sum of 11 with one roll of three dice.

40. Picking, at random, 5 Republicans from a group containing 8 Republicans and 10 Democrats.

41. Tossing 3 heads in 5 tosses of a fair coin.

42. Tossing 5 heads in 5 tosses of a fair coin.

Assume that the probability that an airplane engine will fail during a torture test is $\frac{1}{2}$ and that the aircraft in question has 4 engines.

43. Construct a sample space for the torture test.

44. Find the probability that all engines will survive the test.

45. Find the probability that exactly 1 engine will survive.

46. Find the probability that exactly 2 engines will survive.

47. Find the probability that exactly 3 engines will survive.

48. Find the probability that no engines will survive.

49. Find the sum of the probabilities in Exercises 44 through 48.

Assume that a survey of 282 people is taken to determine the opinions of doctors, teachers, and lawyers on a proposed piece of legislation, with the following results. A person is chosen at random from those surveyed. Find each probability.

	Number that favor	Number that oppose	Number with no opinion	Total
Doctors	70	32	17	119
Teachers	83	24	10	117
Lawyers	23	15	8	46
Total	176	71	35	282

50. The person favors the legislation.

51. A doctor opposes the legislation.

52. A person who opposes the legislation is a lawyer.

53. **Quality control** In a batch of 10 tires, 2 are known to be defective. If 4 tires are chosen at random, find the probability that all 4 tires are good.

54. **Medicine** Out of a group of 9 patients treated with a new drug, 4 suffered a relapse. Find the probability that 3 patients of this group, chosen at random, will remain disease-free.

Use the multiplication property of probabilities.

55. If $P(A) = 0.3$ and $P(B|A) = 0.6$, find $P(A \cap B)$.

56. If $P(A \cap B) = 0.3$ and $P(B|A) = 0.6$, find $P(A)$.

57. **Conditional probability** The probability that a person owns a luxury car is 0.2, and the probability that the owner of such a car also owns a personal computer is 0.7. Find the probability that a person, chosen at random, owns both a luxury car and a computer.

58. Conditional probability   If 40% of the population have completed college, and 85% of college graduates are registered to vote, what percent of the population are college graduates and registered voters?

59. Conditional probability   About 25% of the population watches the evening television news coverage as well as the morning soap operas. If 75% of the population watches the news, what percent of those who watch the news also watch the soaps?

60. Conditional probability   The probability of rain today is 0.40. If it rains, the probability that Bill will forget his raincoat is 0.70. Find the probability that Bill will get wet.

61. Define the concept of sample space.

62. Explain the multiplication property of probabilities.

SOMETHING TO THINK ABOUT

63. If $P(A \cap B) = 0.7$, it is possible that $P(B|A) = 0.6$? Explain.

64. Is it possible that $P(A \cap B) = P(A)$? Explain.

13.8   Computation of Compound Probabilities

❚ **Compound Events**   ❚ **Independent Events**

Getting Ready   *Find each probability.*

1. Drawing an ace on one draw from a standard card deck

2. Drawing a king on one draw from a standard card deck

3. Drawing either an ace or a king on one draw from a standard card deck

4. Drawing either an ace, a king, or a queen on one draw from a standard card deck

Sometimes we must find the probability of one event *or* another, or the probability of one event *and* another. Such events are called **compound events.** As we have seen, if A and B are two events, the probability that A and B will both occur is $P(A \cap B)$. In this section, we also discuss $P(A \cup B)$, the probability that either A or B will occur.

Compound Events

Suppose we want to find the probability of drawing a king or a heart from a standard card deck. If K is the event "drawing a king" and H is the event "drawing a heart," then $P(K) = \frac{4}{52}$, and $P(H) = \frac{13}{52}$. However, the probability of drawing a king *or* a heart is not the sum of these two probabilities. Because the king of hearts was counted twice, once as a king and once as a heart, and because the probability of drawing the king of hearts is $\frac{1}{52}$, we must subtract $\frac{1}{52}$ from the sum of $\frac{4}{52}$ and $\frac{13}{52}$ to get the correct probability.

$$P(\text{king } or \text{ heart}) = P(\text{king}) + P(\text{heart}) - P(\text{king of hearts})$$
$$P(K \cup H) = P(K) + P(H) - P(K \cap H)$$
$$= \frac{4}{52} + \frac{13}{52} - \frac{1}{52}$$
$$= \frac{16}{52}$$
$$= \frac{4}{13}$$

In general, we have the following rule.

$P(A \cup B)$

> If A and B are two events, then
>
> $$P(A \cup B) = P(A) + P(B) - P(A \cap B)$$

If events A and B have no outcomes in common, then $A \cap B = \varnothing$ and $P(A \cap B) = P(\varnothing) = 0$. Such events are **mutually exclusive** (if one event occurs, the other cannot).

$P(A \cup B)$

> If A and B cannot occur simultaneously, then
>
> $$P(A \cup B) = P(A) + P(B)$$

The event $\overline{A}$ (read as "not A") contains all outcomes of the sample space that are not elements of event A. Because the events A and $\overline{A}$ are mutually exclusive,

$$P(A \cup \overline{A}) = P(A) + P(\overline{A})$$

Because either event A or event $\overline{A}$ must happen, $P(A \cup \overline{A}) = 1$. Thus,

$$P(A \cup \overline{A}) = 1$$
$$P(A) + P(\overline{A}) = 1$$
$$P(\overline{A}) = 1 - P(A) \quad \text{Add } -P(A) \text{ to both sides.}$$

This result gives another property of compound probabilities.

$P(\overline{A})$

> If A is any event, then
>
> $$P(\overline{A}) = 1 - P(A)$$

EXAMPLE 1   A counselor tells a student that his probability of earning a grade of D in algebra is $\frac{1}{5}$ and his probability of earning an F is $\frac{1}{25}$. Find the probability that the student earns a C or better.

Solution   Because "earning a D" and "earning an F" are mutually exclusive, the probability of earning a D or F is given by

$$P(D \cup F) = P(D) + P(F) \quad \text{Note that } P(D \text{ and } F) = 0.$$
$$= \frac{1}{5} + \frac{1}{25}$$
$$= \frac{6}{25}$$

The probability that the student will receive a *C* or better is

$$P(C \text{ or better}) = 1 - P(D \cup F)$$

$$= 1 - \frac{6}{25}$$

$$= \frac{19}{25}$$

The probability of earning a *C* or better is $\frac{19}{25}$.

Self Check   Find the probability that the student passes (earns a *D* or better).   ▮

Independent Events

If two events do not influence each other, they are called **independent events.**

Independent Events

The events *A* and *B* are said to be **independent events** if and only if $P(B) = P(B|A)$.

Substituting $P(B)$ for $P(B|A)$ in the multiplication property for probabilities gives a formula for computing probabilities of compound independent events.

Formula for *P(A ∩ B)*

If *A* and *B* are independent events, then

$$P(A \cap B) = P(A) \cdot P(B)$$

The event *A* of "drawing an ace from a standard deck of cards" and the event *B* of "tossing heads" on one toss of a coin are independent events, because neither event influences the other. Consequently,

$$P(A \cap B) = P(A) \cdot P(B)$$

$$P(\text{drawing an ace and tossing heads}) = P(\textbf{drawing an ace}) \cdot P(\textbf{tossing heads})$$

$$= \frac{4}{52} \cdot \frac{1}{2}$$

$$= \frac{1}{26}$$

EXAMPLE 2   The probability that a baseball player can get a hit is $\frac{1}{3}$. Find the probability that she will get three hits in a row.

Solution   Assume that the three times at bat are independent events: One time at bat does not influence her chances of getting a hit on another turn at bat. Because $P(E_1) = \frac{1}{3}$, $P(E_2) = \frac{1}{3}$, and $P(E_3) = \frac{1}{3}$,

$$P(E_1 \cap E_2 \cap E_3) = \frac{1}{3} \cdot \frac{1}{3} \cdot \frac{1}{3} = \frac{1}{27}$$

The probability that she gets three hits in a row is $\frac{1}{27}$.

Self Check The probability that another player can get a hit is $\frac{1}{4}$. Find the probability that she gets four hits in a row. ∎

EXAMPLE 3 A die is rolled three times. Find the probability that the outcome is 6 on the first roll, an even number on the second roll, and an odd prime number on the third roll.

Solution The probability of a 6 on any roll is $P(6) = \frac{1}{6}$. Because there are three even integers represented on a die, the probability of tossing an even number is $P(\text{even number}) = P(E) = \frac{3}{6} = \frac{1}{2}$. Since 3 and 5 are the only odd prime numbers on a die, the probability of rolling an odd prime is $P(\text{odd prime}) = P(O) = \frac{2}{6} = \frac{1}{3}$.

Because these three events are independent, the probability of the events happening in succession is the product of the probabilities:

$$
\begin{aligned}
P(\text{six and even number and odd prime}) &= P(6 \cap E \cap O) \\
&= P(6) \cdot P(E) \cdot P(O) \\
&= \frac{1}{6} \cdot \frac{1}{2} \cdot \frac{1}{3} \\
&= \frac{1}{36}
\end{aligned}
$$

Self Check Find the probability that the outcome is five on the first roll, an odd number on the second roll, and two on the third roll. ∎

EXAMPLE 4 The probability that a drug will cure dandruff is $\frac{1}{8}$. However, if the drug is used, the probability that it will cause side effects is $\frac{1}{6}$. Find the probability that a patient who uses the drug will be cured and will suffer no side effects.

Solution The probability that the drug will cure dandruff is $P(C) = \frac{1}{8}$. The probability of having side effects is $P(E) = \frac{1}{6}$. The probability that the patient will have no side effects is

$$
P(\overline{E}) = 1 - P(E) = 1 - \frac{1}{6} = \frac{5}{6}
$$

Since these events are independent,

$$
\begin{aligned}
P(\text{cure and no side effects}) &= P(C \cap \overline{E}) \\
&= P(C) \cdot P(\overline{E}) \\
&= \frac{1}{8} \cdot \frac{5}{6} \\
&= \frac{5}{48}
\end{aligned}
$$

Self Check Find the probability that the patient will be cured and suffer side effects. ∎

Self Check Answers

1. $\frac{24}{25}$ **2.** $\frac{1}{256}$ **3.** $\frac{1}{72}$ **4.** $\frac{1}{48}$

Orals *Find each probability.*

1. Drawing a red card on one draw from a standard card deck

2. Drawing an ace on one draw from a standard card deck

3. Drawing a red card or an ace on one draw from a standard card deck

4. Drawing a king and tossing heads on one draw from a standard card deck and one toss of a coin

13.8 EXERCISES

REVIEW *Simplify each fraction.*

1. $\dfrac{x^2 - x - 6}{x^2 + 2x}$

2. $\dfrac{x^2 - y^2}{x^2 - xy}$

Perform each operation.

3. $\dfrac{x - 2}{x + 3} \cdot \dfrac{x^2 + 4x + 3}{x^2 - x - 2}$

4. $\dfrac{x - 2}{x + 3} + \dfrac{x - 1}{x - 3}$

VOCABULARY AND CONCEPTS *Fill in the blanks.*

5. A _____ event is one event *or* another or one event *followed* by another.

6. $P(A \cup B) = $ _____

7. If A and B are _____, then $P(A \cup B) = P(A) + P(B)$.

8. The event $\overline{A}$ is read as "_____."

9. $P(\overline{A}) = $ _____

10. Two events, A and B, are called independent events when _____.

11. If A and B are independent events, then $P(A \cap B) = $ _____.

12. If two events do not influence each other, they are called _____ events.

PRACTICE *Assume that you draw one card from a card deck. Find the probability of each event.*

13. Drawing a black card

14. Drawing a jack

15. Drawing a black card or an ace

16. Drawing a red card or a face card

Assume that you draw two cards from a card deck, without replacement. Find the probability of each event.

17. Drawing two aces

18. Drawing three aces

19. Drawing a club and then another black card

20. Drawing a heart and then a spade

Assume that you roll two dice once. Find the probability of each result.

21. Rolling a sum of 7 or 6

22. Rolling a sum of 5 or an even sum

23. Rolling a sum of 10 or an odd sum

24. Rolling a sum of 12 or 1

Assume that you have a bucket containing 7 beige capsules, 3 blue capsules, and 6 green capsules. You make a single draw from the bucket, taking one capsule. Find the probability of each result.

25. Drawing a beige or a blue capsule

26. Drawing a green capsule

27. Not drawing a blue capsule

28. Not drawing either a beige or a blue capsule

In Exercises 29–31, assume that you are using the same bucket of capsules as in Exercises 25–28.

29. On two draws from the bucket, find the probability of drawing a beige capsule followed by a green capsule. (Assume that the capsule is returned to the bucket after the first draw.)

30. On two draws from the bucket, find the probability of drawing one blue capsule and one green capsule. (Assume that the capsule is not returned to the bucket after the first draw.)

31. On three successive draws from the bucket (without replacement), find the probability of failing to draw a beige capsule.

32. Jeff rolls a die and draws one card from a card deck. Find the probability of his rolling a 4 and drawing a four.

33. Birthday problem Three people are in an elevator together. Find the probability that all three were born on the same day of the week.

34. Birthday problem Three people are on a bus together. Find the probability that at least one was born on a different day of the week from the others.

35. Birthday problem Five people are in a room together. Find the probability that all five were born on a different day of the year.

36. Birthday problem Five people are on a bus together. Find the probability that at least two of them were born on the same day of the year.

37. Sharing homework If the probability that Rick will solve a problem is $\frac{1}{4}$ and the probability that Dinah will solve it is $\frac{2}{5}$, find the probability that at least one of them will solve it.

38. Signaling A bugle is used for communication at camp. The call for dinner is based on four pitches and is five notes long. If a child can play these four pitches on a bugle, find the probability that the first five notes that the child plays will call the camp for dinner. (Assume that the child is equally likely to play any of the four pitches each time a note is blown.)

Assume that a woman visits her cabin in Canada. The probability that her lawnmower will start is $\frac{1}{2}$, the probability that her gas-powered saw will start is $\frac{1}{3}$, and the probability that her outboard motor will start is $\frac{3}{4}$. Find each probability.

39. That all three will start

40. That none will start

41. That exactly one will start

42. That exactly two will start

APPLICATIONS

43. Immigration Three children leave Thailand to start a new life in either the United States or France. The probability that May Xao will go to France is $\frac{1}{3}$, that Tou Lia will go to France is $\frac{1}{2}$, and that May Moua will go to France is $\frac{1}{6}$. Find the probability that exactly two of them will end up in the United States.

44. Preparing for the GED The administrators of a program to prepare people for the high school equivalency exam have found that 80% of the students require tutoring in math, 60% need help in English, and 45% need work in both math and English. Find the probability that a student selected at random needs help with either math or English.

45. Insurance losses The insurance underwriters have determined that in any one year, the probability that George will have a car accident is 0.05, and that if he has an accident, the probability that he will be hospitalized is 0.40. Find the probability that George will have an accident but not be hospitalized.

46. Grading homework One instructor grades homework by randomly choosing 3 out of the 15 problems assigned. Bill did only 8 problems. What is the probability that he won't get caught?

WRITING

47. Explain the words *mutually exclusive.*

48. Explain the difference between dependent and independent events and give some examples of each.

SOMETHING TO THINK ABOUT

49. Prove that $P(A\,|\,A) = 1$.

50. Prove that $P(A\,|\,\overline{A}) = 0$.

13.9 Odds and Mathematical Expectation

❚ **Odds** ❚ **Mathematical Expectation**

Getting Ready *Assume that, on average, a horse wins one of every seven races it runs.*

1. In a group of seven races, how many races would we expect the horse to win?

2. In a group of seven races, how many races would we expect the horse to lose?

3. In a race, what is the probability that the horse will win?

4. In a race, what is the probability that the horse will lose?

In this section, we will discuss **mathematical odds,** a concept closely related to probability.

Odds

Odds

> The **odds for an event** is the probability of a favorable outcome divided by the probability of an unfavorable outcome.
>
> The **odds against an event** is the probability of an unfavorable outcome divided by the probability of a favorable outcome.

EXAMPLE 1   The probability that a horse will win a race is $\frac{1}{4}$. Find the odds for and the odds against the horse.

Solution   Because the probability that the horse will win is $\frac{1}{4}$, the probability that the horse will not win is $\frac{3}{4}$. The odds for the horse are

$$\frac{\text{probability of a win}}{\text{probability of a loss}} = \frac{\dfrac{1}{4}}{\dfrac{3}{4}} = \frac{1}{3}$$

or 1 to 3. The odds against the horse are

$$\frac{\text{probability of a loss}}{\text{probability of a win}} = \frac{\dfrac{3}{4}}{\dfrac{1}{4}} = 3$$

or 3 to 1. The odds for an event is the reciprocal of the odds against the event.

Self Check   The probability that a horse will lose a race is $\frac{1}{5}$. Find the odds for and the odds against the horse. ▮

Mathematical Expectation

Suppose we have a chance to play a simple game with the following rules:

1. Roll a single die once.
2. If a six appears, win $3.
3. If a five appears, win $1.
4. If any other number appears, win 50¢.
5. The cost to play (one roll of the die) is $1.

In this game, the probability of any one of the six outcomes—rolling a 6, 5, 4, 3, 2, or 1—is $\frac{1}{6}$, and the winnings are $3, $1, and 50¢. The expected winnings can be found by using the following equation and simplifying the right-hand side:

$$E = \frac{1}{6}(3) + \frac{1}{6}(1) + \frac{1}{6}(0.50) + \frac{1}{6}(0.50) + \frac{1}{6}(0.50) + \frac{1}{6}(0.50)$$

$$= \frac{1}{6}(3 + 1 + 0.50 + 0.50 + 0.50 + 0.50)$$

$$= \frac{1}{6}(6)$$

$$= 1$$

Over the long run, we could expect to win $1 with every play of the game. Since it costs $1 to play the game, the expected gain or loss is 0. Because the expected winnings are equal to the admission price, the game is fair.

Mathematical Expectation

> If a certain event has n different outcomes with probabilities $P_1, P_2, \ldots, P_n$ and the winnings assigned to each outcome are $x_1, x_2, \ldots x_n$, the expected winnings, or **mathematical expectation,** E is given by
>
> $$E = P_1x_1 + P_2x_2 + \cdots + P_nx_n$$

EXAMPLE 2   It costs $1 to play the following game: Roll two dice, collect $5 if we roll a sum of 7, and collect $2 if we roll a sum of 11. All other numbers pay nothing. Is it wise to play the game?

Solution   The probability of rolling a 7 on a single roll of two dice is $\frac{6}{36}$, the probability of rolling an 11 is $\frac{2}{36}$, and the probability of rolling something else is $\frac{28}{36}$. The mathematical expectation is

$$E = \frac{6}{36}(5) + \frac{2}{36}(2) + \frac{28}{36}(0) = \frac{17}{18} \approx \$0.944$$

By playing the game for a long period of time, we can expect to get back about 95¢ for every dollar spent. For the fun of playing, the cost is about 5¢ a game. If the game is enjoyable, it might be worth the expected loss. However, the game is slightly unfair.

Self Check   Is it wise to play the game if you win $4 when you roll a sum of 7 and $6 when you roll a sum of 11?   ∎

Self Check Answers

1. 4 to 1; 1 to 4   **2.** The game is even.

Orals   *Assume one roll of a fair die.*

1. Find the probability of rolling a 3.

2. Find the probability of not rolling a 3.

3. Find the odds for rolling a 3.

4. Find the odds against rolling a 3.

13.9   EXERCISES

REVIEW

1. Find the equation of the line perpendicular to $3x - 2y = 9$ and passing through the point $(1, 1)$.

2. Find the equations of the parabolas with vertex at the point $(3, -2)$ and passing through the origin.

3. Find the equation of the circle with center at $(3, 5)$ and tangent to the x-axis.

4. Write the equation $2x^2 + 4y^2 + 12x - 24y + 46 = 0$ in standard form, and identify the curve.

VOCABULARY AND CONCEPTS   *Fill in the blanks.*

5. The _____ is the probability of a favorable outcome divided by the probability of an unfavorable outcome.

6. The _____ is the probability of an unfavorable outcome divided by the probability of a favorable outcome.

7. If the odds for an event are 1 to 4, the probability of winning is __.

8. Mathematical expectation E is given by $E =$ _____.

PRACTICE   *Assume a single roll of a die.*

9. Find the probability of rolling a 6.
10. Find the odds in favor of rolling a 6.
11. Find the odds against rolling a 6.
12. Find the probability of rolling an even number.
13. Find the odds in favor of rolling an even number.
14. Find the odds against rolling an even number.

Assume a single roll of two dice.

15. Find the probability of rolling a sum of 6.
16. Find the odds in favor of rolling a sum of 6.
17. Find the odds against rolling a sum of 6.
18. Find the probability of rolling an even sum.
19. Find the odds in favor of rolling an even sum.
20. Find the odds against rolling an even sum.

Assume that you are drawing one card from a card deck.

21. Find the odds in favor of drawing a queen.
22. Find the odds against drawing a black card.
23. Find the odds in favor of drawing a face card.
24. Find the odds against drawing a diamond.
25. If the odds in favor of victory are 5 to 2, find the probability of victory.
26. If the odds in favor of victory are 5 to 2, find the odds against victory.
27. If the odds against winning are 90 to 1, find the odds in favor of winning.
28. Find the odds in favor of rolling a 7 on a single toss of two dice.
29. Find the odds against tossing four heads in a row with a fair nickel.
30. Find the odds in favor of a couple having four girl babies in succession. $\left(\text{Assume } P(\text{girl}) = \frac{1}{2}.\right)$
31. The odds against a horse are 8 to 1. Find the probability that the horse will win.
32. The odds against a horse are 1 to 1. Find the probability that the horse will lose.
33. It costs $2 to play the following game:
    a. Draw one card from a card deck.
    b. Collect $5 if an ace is drawn.
    c. Collect $4 if a king is drawn.
    d. Collect nothing for all other cards drawn. Is it wise to play this game? Explain.

34. **Lottery tickets** One thousand tickets are sold for a lottery with two grand prizes of $800. Find a fair price for the tickets.
35. Find the odds against a couple having three baby boys in a row. $\left(\text{Assume } P(\text{boy}) = \frac{1}{2}.\right)$
36. Find the odds in favor of tossing at least three heads in five tosses of a fair coin.
37. Suppose you toss a coin five times and collect $5 if you toss five heads, $4 if you toss four heads, $3 if you toss three heads, and no money for any other combination. How much should you pay to play the game if the game is to be fair?
38. If you toss two dice one time and collect $10 for double 6's and $1 for double 1's, what is a fair price for playing the game?

39. Counting an ace as 1, a face card as 10, and all others at their numerical value, find the expected value if you draw one card from a card deck.

40. Find the expected sum of one roll of two dice.

41. A multiple-choice test of eight questions gives five possible answers for each question. Only one of the answers for each question is right. Find the probability of getting seven right answers by simple guessing.

42. In the situation described in Exercise 41, find the odds in favor of getting seven answers right.

WRITING

43. Define "odds for an event."

44. Explain how odds for an event, odds against the event, and the probability of the event are related.

SOMETHING TO THINK ABOUT

45. If "the odds are against an event," what can be said about its probability?

46. To disguise the unlikely chance of winning, a contest promoter publishes the odds in favor of winning as 0.0000000372 to 1. What are the odds against winning?

Projects

Project 1

By the binomial theorem,

$$(1 + x)^m = 1 + mx + \frac{m(m-1)}{2!}x^2$$

$$+ \frac{m(m-1)(m-2)}{3!}x^3 + \cdots$$

$$+ \frac{m(m-1)\cdots(m-n+1)}{n!}x^n + \cdots$$

- If m is not a positive integer or zero, explain why the coefficients of the terms of the binomial theorem never become zero. In this case, the formula is called the **binomial series.**

- In the binomial series, let $x = -\frac{1}{2}$ and $m = \frac{1}{2}$, so that the left-hand side becomes

$$\left(1 - \frac{1}{2}\right)^{1/2} = \left(\frac{1}{2}\right)^{1/2} = \sqrt{\frac{1}{2}} = \frac{\sqrt{2}}{2}$$

Use a calculator to evaluate this expression.

- On the right-hand side of the equation, also let $x = -\frac{1}{2}$ and $m = \frac{1}{2}$. Calculate the first 4 terms of the series. How close is their sum to the value in the previous equation? Is the sum of the first 5 terms more accurate?

Project 2

If n, a, b, and c are nonnegative integers and $a + b + c = n$, then the **multinomial coefficient** $\binom{n}{a, b, c}$ is defined as $\frac{n!}{a!\, b!\, c!}$.

- Expand $(x + y + z)^3$.

- In this expansion, verify that the coefficient of x^2y is $\binom{3}{2, 1, 0}$ and that the coefficient of xyz is $\binom{3}{1, 1, 1}$. What is the multinomial coefficient of x^3? (*Hint:* $x^3 = x^3y^0z^0$.) What is the multinomial coefficient of xz^2?

- In the expansion of $(w + x + y + z)^7$, what is the coefficient of w^2x^3yz?

- In the expansion of $(x + y)^n$, explain why the multinomial coefficients are just the binomial coefficients.

Project 3

- Find the hockey-stick pattern in the numbers of Pascal's triangle in Illustration 1. What is the missing number in the rightmost hockey stick? Does the pattern work with larger hockey sticks? Experiment and report your conclusions.

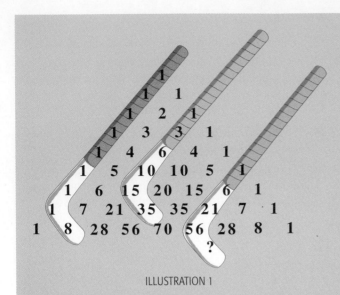

ILLUSTRATION 1

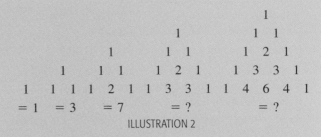

ILLUSTRATION 2

- There are many other patterns hidden in Pascal's triangle. Find some more and share them with your class. Illustration 3 is an idea to get you started

$$
\begin{array}{c}
1 \\
1 \quad 1 \\
1 \quad 2 \quad 1 \\
1 \quad 3 \quad 3 \quad 1 \\
1 \quad 4 \diagup 6 \quad 4 \diagdown 1 \\
1 \quad 5 \langle 10 \quad 10 \quad 5 \rangle 1 \\
1 \quad 6 \quad 15 \diagdown 20 \quad 15 \diagup 6 \quad 1 \\
1 \quad 7 \quad 21 \quad 35 \quad 35 \quad 21 \quad 7 \quad 1 \\
1 \quad 8 \quad 28 \quad 56 \quad 70 \quad 56 \quad 28 \quad 8 \quad 1
\end{array}
$$

ILLUSTRATION 3

- In Illustration 2, find the pattern in the sums of increasingly larger portions of Pascal's triangle. Find the sum of all of the numbers up to and including the row that begins 1   10   45 . . . .

Chapter Summary

CONCEPTS	REVIEW EXERCISES
13.1	**The Binomial Theorem**
$n! = n(n-1)(n-2)\cdots$ $3 \cdot 2 \cdot 1$ $0! = 1 \quad n(n-1)! = n!$	Find each value. **1.** $6!$ **2.** $7! \cdot 0! \cdot 1! \cdot 3!$ **3.** $\dfrac{8!}{7!}$ **4.** $\dfrac{5! \cdot 7! \cdot 8!}{6! \cdot 9!}$
The binomial theorem: If n is any positive integer, then $(a+b)^n = a^n + \dfrac{n!}{1!(n-1)!}a^{n-1}b$ $+ \dfrac{n!}{2!(n-2)!}a^{n-2}b^2 +$ $\dfrac{n!}{3!(n-3)!}a^{n-3}b^3 + \cdots + b^n$	Expand each expression. **5.** $(x+y)^3$ **6.** $(p+q)^4$ **7.** $(a-b)^5$ **8.** $(2a-b)^3$

Find the required term of each expansion.

9. $(a + b)^8$; 4th term

10. $(2x - y)^5$; 3rd term

11. $(x - y)^9$; 7th term

12. $(4x + 7)^6$; 4th term

13.2   Sequences, Series, and Summation Notation

A **sequence** is a function whose domain is the set of natural numbers.

Write the fourth term in each sequence.

13. $0, 7, 26, \ldots, n^3 - 1, \ldots$

14. $\dfrac{3}{2}, 3, \dfrac{11}{2}, \ldots, \dfrac{n^2 + 2}{2}, \ldots$

Find the first four terms of each sequence.

15. $a_1 = 5$ and $a_{n+1} = 3a_n + 2$

16. $a_1 = -2$ and $a_{n+1} = 2a_n^2$

If c is a constant, then

$$\sum_{k=1}^{n} c = nc$$

If c is a constant, then

$$\sum_{k=1}^{n} cf(k) = c \sum_{k=1}^{n} f(k)$$

$$\sum_{k=1}^{n} [f(k) + g(k)] =$$

$$\sum_{k=1}^{n} f(k) + \sum_{k=1}^{n} g(k)$$

Evaluate each expression.

17. $\displaystyle\sum_{k=1}^{4} 3k^2$

18. $\displaystyle\sum_{k=1}^{10} 6$

19. $\displaystyle\sum_{k=5}^{8} (k^3 + 3k^2)$

20. $\displaystyle\sum_{k=1}^{30} \left(\dfrac{3}{2}k - 12\right) - \dfrac{3}{2}\sum_{k=1}^{30} k$

| **13.3** | **Arithmetic Sequences** |

If a is the first term, n is the number of terms, and d is the common difference, then the formula $l = a + (n - 1)d$ gives the last (or nth) term of an arithmetic sequence.

The formula
$$S_n = \frac{n(a + l)}{2}$$
gives the sum of the first n terms of an arithmetic sequence, where S_n is the sum, a is the first term, l is the last (or nth) term, and n is the number of terms.

Find the required term of each arithmetic sequence.

21. 5, 9, 13, . . .; 29th term

22. 8, 15, 22, . . .; 40th term

23. 6, −1, −8, . . .; 15th term

24. $\frac{1}{2}, -\frac{3}{2}, -\frac{7}{2}, \ldots$; 35th term

25. Find three arithmetic means between 2 and 8.

26. Find five arithmetic means between 10 and 100.

Find the sum of the first 40 terms in each sequence.

27. 5, 9, 13, . . .

28. 8, 15, 22, . . .

29. 6, −1, −8, . . .

30. $\frac{1}{2}, -\frac{3}{2}, -\frac{7}{2}, \ldots$

| **13.4** | **Geometric Sequences** |

If a is the first term and r is the common ratio, then the formula $l = ar^{n-1}$ gives the nth (or last) term of a geometric sequence.

The formula
$$S_n = \frac{a - ar^n}{1 - r} \quad (r \neq 1)$$
gives the sum of the first n terms of a geometric sequence, where S_n is the sum, a is the first term, r is the common ratio, and n is the number of terms.

If $|r| < 1$, the formula
$$S = \frac{a}{1 - r}$$
gives the sum of the terms of an infinite geometric sequence, where S is the sum, a is the first term, and r is the common ratio.

Find the required term of each geometric sequence.

31. 81, 27, 9, . . .; 11th term

32. 2, 6, 18, . . .; 9th term

33. $9, \frac{9}{2}, \frac{9}{4}, \ldots$; 15th term

34. $8, -\frac{8}{5}, \frac{8}{25}, \ldots$; 7th term

35. Find three positive geometric means between 2 and 8.

36. Find four geometric means between −2 and 64.

37. Find the positive geometric mean between 4 and 64.

Find the sum of the first 8 terms in each sequence.

38. 81, 27, 9, . . .

39. 2, 6, 18, . . .

40. $9, \frac{9}{2}, \frac{9}{4}, \ldots$

41. $8, -\frac{8}{5}, \frac{8}{25}, \ldots$

42. Find the sum of the first eight terms of the sequence $\frac{1}{3}, 1, 3, \ldots$

43. Find the seventh term of the sequence $2\sqrt{2}, 4, 4\sqrt{2}, \ldots$

Find the sum of each infinite sequence, if possible.

44. $\frac{1}{3}, \frac{1}{6}, \frac{1}{12}, \ldots$

45. $\frac{1}{5}, -\frac{2}{15}, \frac{4}{45}, \ldots$

46. $1, \frac{3}{2}, \frac{9}{4}, \ldots$

47. 0.5, 0.25, 0.125, . . .

Change each decimal into a common fraction.

48. $0.\overline{3}$

49. $0.\overline{9}$

50. $0.\overline{17}$

51. $0.\overline{45}$

52. Investment problems   If Leonard invests \$3,000 in a 6-year certificate of deposit at the annual rate of 7.75%, compounded daily, how much money will be in the account when it matures?

53. College enrollments   The enrollment at Hometown College is growing at the rate of 5% over each previous year's enrollment. If the enrollment is currently 4,000 students, what will it be 10 years from now? What was it 5 years ago?

54. House trailer depreciation   A house trailer that originally cost \$10,000 depreciates in value at the rate of 10% per year. How much will the trailer be worth after 10 years?

13.5	**Mathematical Induction**

Mathematical induction: If a statement involving the natural number n has the two properties that

1. the statement is true for $n = 1$, and

2. if the statement is true for $n = k$, then it is true for $n = k + 1$,

the statement is true for all natural numbers.

55. Verify the following formula for $n = 1$, $n = 2$, $n = 3$, and $n = 4$:

$$1^3 + 2^3 + 3^3 + \cdots + n^3 = \frac{n^2(n + 1)^2}{4}$$

56. Prove the formula given in Exercise 55 by mathematical induction.

13.6	**Permutations and Combinations**

$P(n, r) = \dfrac{n!}{(n - r)!}$

$P(n, n) = n!$

$P(n, 0) = 1$

$C(n, r) = \dbinom{n}{r} = \dfrac{n!}{r!(n - r)!}$

$C(n, n) = 1$

$C(n, 0) = 1$

There are $(n - 1)!$ ways to place n things in a circle.

Evaluate each expression.

57. $P(8, 5)$

58. $C(7, 4)$

59. $0! \cdot 1!$

60. $P(10, 2) \cdot C(10, 2)$

61. $P(8, 6) \cdot C(8, 6)$

62. $C(8, 5) \cdot C(6, 2)$

63. $C(7, 5) \cdot P(4, 0)$

64. $C(12, 10) \cdot C(11, 0)$

65. $\dfrac{P(8, 5)}{C(8, 5)}$

66. $\dfrac{C(8, 5)}{C(13, 5)}$

67. $\dfrac{C(6, 3)}{C(10, 3)}$

68. $\dfrac{C(13, 5)}{C(52, 5)}$

If an *n*-letter word has *a* of one letter, *b* of another letter, and so on, the number of distinguishable words that can be formed using each letter exactly once is

$$\frac{n!}{a!\,b!\cdots}$$

69. In how many ways can 10 teenagers be seated at a round table if 2 girls wish to sit with their boyfriends?

70. How many distinguishable words can be formed from the letters of the word *casserole* if each letter is used exactly once?

13.7	Probability

An event that cannot happen has a probability of 0. An event that is certain to happen has a probability of 1. All other events have probabilities between 0 and 1.

71. Make a tree diagram to illustrate the possible results of tossing a coin four times.

72. In how many ways can you draw a five-card poker hand of 3 aces and 2 kings?

73. Find the probability of drawing the hand described in Exercise 72.

74. Find the probability of not drawing the hand described in Exercise 72.

75. Find the probability of having a 13-card bridge hand consisting of 4 aces, 4 kings, 4 queens, and 1 jack.

76. Find the probability of choosing a committee of 3 men and 2 women from a group of 8 men and 6 women.

77. Find the probability of drawing a club or a spade on one draw from a card deck.

78. Find the probability of drawing a black card or a king on one draw from a card deck.

If *S* is the sample space of an experiment with *n* distinct and equally likely outcomes, and if *E* is an event that occurs in *s* of those ways, then the probability of *E* is

$$P(E) = \frac{n(E)}{n(S)} = \frac{s}{n}$$

$$P(A \cap B) = P(A) \cdot P(B|A)$$

79. Find the probability of getting an ace-high royal flush in hearts (ace, king, queen, jack, and ten of hearts) in poker.

80. Find the probability of being dealt 13 cards of one suit in a bridge hand.

81. Find the probability of getting 3 heads or fewer on 4 tosses of a fair coin.

13.8	**Computation of Compound Probabilities**

If A and B are two events, then
$$P(A \cup B) =$$
$$P(A) + P(B) - P(A \cap B)$$

If A and B cannot occur simultaneously, then
$$P(A \cup B) = P(A) + P(B)$$

If A is any event, then
$$P(\overline{A}) = 1 - P(A)$$

If A and B are independent events, then
$$P(A \cap B) = P(A) \cdot P(B)$$

82. If the probability that a drug cures a certain disease is 0.83 and we give the drug to 800 people with the disease, find the expected number of people who will not be cured.

83. Find the probability of drawing a two or a spade on one draw from a standard card deck.

13.9	**Odds and Mathematical Expectation**

The concepts of probabilities and odds are related.

A game is fair if its cost to play equals the expected winnings.

84. Find the odds against a horse if the probability that the horse will win is $\frac{7}{8}$.

85. Find the odds in favor of a couple having 4 baby girls in a row.

86. If the probability that Joe will marry is $\frac{5}{6}$ and the probability that John will marry is $\frac{3}{4}$, find the odds against either one becoming a husband.

87. If the odds against Priscilla's graduation from college are $\frac{10}{11}$, find the probability that she will graduate.

88. Find the expected earnings if you collect \$1 for every heads you get when you toss a fair coin 4 times.

89. If the total number of subsets that a set with n elements can have is 2^n, explain why
$$\binom{n}{0} + \binom{n}{1} + \binom{n}{2} + \cdots + \binom{n}{n} = 2^n$$

Chapter Test

Find each value.

1. $3! \cdot 0! \cdot 4! \cdot 1!$

2. $\dfrac{2! \cdot 4! \cdot 6! \cdot 8!}{3! \cdot 5! \cdot 7!}$

Find the required term in each expansion.

3. $(x + 2y)^5$; 2nd term

4. $(2a - b)^8$; 7th term

Find each sum.

5. $\displaystyle\sum_{k=1}^{3} (4k + 1)$

6. $\displaystyle\sum_{k=2}^{4} (3k - 21)$

Find the sum of the first ten terms of each sequence.

7. $2, 5, 8, \ldots$

8. $5, 1, -3, \ldots$

9. Find three arithmetic means between 4 and 24.

10. Find two geometric means between -2 and -54.

Find the sum of the first ten terms of each sequence.

11. $\dfrac{1}{4}, \dfrac{1}{2}, 1, \ldots$ **12.** $6, 2, \dfrac{2}{3}, \ldots$

13. A car costing $\$c$ when new depreciates 25% of the previous year's value each year. How much is the car worth after 3 years?

14. A house costing $\$c$ when new appreciates 10% of the previous year's value each year. How much will the house be worth after 4 years?

15. Prove by induction:

$$3 + 4 + 5 + \cdots + (n + 2) = \frac{1}{2}n(n + 5)$$

16. How many six-digit license plates can be made if no plate begins with 0 or 1?

Find each value.

17. $P(7, 2)$ **18.** $P(4, 4)$

19. $C(8, 2)$ **20.** $C(12, 0)$

21. How many ways can 4 men and 4 women stand in line if all the women are first?

22. How many different ways can 6 people be seated at a round table?

23. How many different words can be formed from the letters of the word *bluff* if each letter is used once?

24. Show the sample space of the experiment: toss a fair coin three times.

Find each probability.

25. Rolling a 5 on one roll of a die.

26. Drawing a jack or a queen from a standard card deck.

27. Receiving 5 hearts for a 5-card poker hand.

28. Tossing 2 heads in 5 tosses of a fair coin.

29. If 30% of the population enjoy jazz, and 80% enjoy popular music, and 30% don't like either, what percent of the people like both types of music?

30. If the probability that a person is sick is 0.1, find the probability that the person is well.

Find each probability.

31. Drawing a black card or a face card on one draw from a standard card deck.

32. Rolling a sum of 3 or an even sum with one roll of two dice.

33. If the odds for a horse are 3 to 1, find the probability that the horse will win.

34. Assume a game with the following rules.
  1. Roll a single die once.
  2. If 6 appears, win $4.
  3. If 5 appears, win $2.
  Find the fair price to play the game.

CUMULATIVE REVIEW EXERCISES

Assume that $\log 2 = 0.3010$ *and* $\log 7 = 0.8451$, *and find each value.*

1. $\log 49$

2. $\log \frac{7}{5}$ (*Hint:* $\log 10 = 1$.)

3. Solve: $2 \log 5 + \log x - \log 4 = 2$.

4. Solve: $2^{x+2} = 3^x$.

In Exercises 5–6, use a calculator.

5. Boat depreciation How much will a $9,000 boat be worth after 9 years if it depreciates 12% per year?

6. Find $\log_6 8$.

7. Use graphing to solve $\begin{cases} 2x + y = 5 \\ x - 2y = 0 \end{cases}$.

8. Use substitution to solve $\begin{cases} 3x + y = 4 \\ 2x - 3y = -1 \end{cases}$.

9. Use addition to solve $\begin{cases} x + 2y = -2 \\ 2x - y = 6 \end{cases}$.

10. Use any method to solve $\begin{cases} \dfrac{x}{10} + \dfrac{y}{5} = \dfrac{1}{2} \\ \dfrac{x}{2} - \dfrac{y}{5} = \dfrac{13}{10} \end{cases}$.

11. Evaluate: $\begin{vmatrix} 3 & -2 \\ 1 & -1 \end{vmatrix}$.

12. Use Cramer's rule and solve for y only:
$\begin{cases} 4x - 3y = -1 \\ 3x + 4y = -7 \end{cases}$

13. Solve: $\begin{cases} x + y + z = 1 \\ 2x - y - z = -4 \\ x - 2y + z = 4 \end{cases}$.

14. Solve: $\begin{cases} x + 2y + 3z = 6 \\ 3x + 2y + z = 6 \\ 2x + 3y + z = 6 \end{cases}$.

Let $f(x) = 4x$ and $g(x) = x - 1$. Find each function.

15. $g + f$

16. $f - g$

17. $g \cdot f$

18. g/f

Let $f(x) = 4x$ and $g(x) = x - 1$. Find each value.

19. $(g \circ f)(1)$

20. $(f \circ g)(0)$

21. $(f \circ g)(-1)$

22. $(g \circ f)(-2)$

Let $f(x) = 4x$ and $g(x) = x - 1$. Find each function.

23. $(f \circ g)(x)$

24. $(g \circ f)(x)$

Find the inverse of each function.

25. $3x + 2y = 12$

26. $y = 3x^2 + 4 \ (x \le 0)$

Write the equation of each circle with the given center and radius r.

27. $(0, 0)$; $r = 4$

28. $(2, -3)$; $r = 11$

Complete the square on x and/or y and graph each equation.

29. $x^2 + y^2 - 4y = 12$

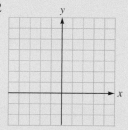

30. $x^2 - 2y - 2x = -7$

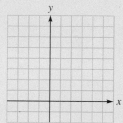

31. $x^2 + 4y^2 + 2x = 3$

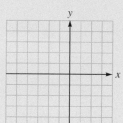

32. $x^2 - 9y^2 - 4x = 5$

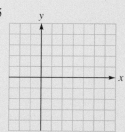

Write the equation of each ellipse.

33. Center $(0, 0)$; horizontal major axis of 12; minor axis of 8

34. Center $(2, 3)$; $a = 5$; $c = 2$; major axis vertical

Write the equation of each hyperbola.

35. Center $(0, 0)$; focus $(3, 0)$; vertex $(2, 0)$

36. Center $(2, 4)$; area of fundamental rectangle is 36 square units; $a = b$; transverse axis parallel to the y-axis

Find the required term of the expansion of $(x + 2y)^8$.

37. 2nd term

38. 6th term

Find each sum.

39. $\displaystyle\sum_{k=1}^{5} 2$

40. $\displaystyle\sum_{k=2}^{6} (3x + 1)$

Find the sum of the first six terms of each sequence.

41. $-2, 1, 4, \ldots$

42. $\dfrac{1}{9}, \dfrac{1}{3}, 1, \ldots$

Find each value.

43. $P(8, 4)$

44. $P(24, 0)$

45. $C(12, 10)$

46. $P(4, 4) \cdot C(6, 6)$

47. In how many ways can 6 men and 4 women be placed in a line if the women line up first?

48. In how many ways can a committee of 4 people be selected from a group of 12 people?

Find each probability.

49. Rolling 11 on one roll of two dice

50. Being dealt an all red 5-card poker hand from a standard deck

51. If the probability that a person is married is 0.6 and the probability that a married person has children is 0.8, find the probability that a randomly chosen person is married with children.

52. The probability that a student earns an A in algebra is 0.3 and the probability that a student earns a B is 0.4. Find the probability that a student earns a C or less.

53. Prove the formula by induction:

$$4 + 7 + 10 + \cdots + (3n + 1) = \frac{n(3n + 5)}{2}$$

APPENDIX I

Matrix Algebra

I.1   Matrix Algebra

In this section, we discuss how to add, subtract, and multiply matrices.

$m \times n$ Matrix

An **$m \times n$ matrix** is a rectangular array of mn numbers arranged in m rows and n columns.

We will use any of the following notations to denote the matrix A:

$$A_{m \times n}, \qquad A, \qquad [a_{ij}], \qquad \begin{bmatrix} a_{11} & a_{12} & a_{13} & \cdots & a_{1n} \\ a_{21} & a_{22} & a_{23} & \cdots & a_{2n} \\ \vdots & & & & \vdots \\ a_{m1} & a_{m2} & a_{m3} & \cdots & a_{mn} \end{bmatrix}$$

The symbol a_{23} represents the entry in row two, column three, of the matrix A. Similarly, the symbol a_{ij} represents in the entry in the ith row and the jth column.

Equal Matrices

If $A = [a_{ij}]$ and $B = [b_{ij}]$ are both $m \times n$ matrices, then

$$A = B \text{ if and only if } a_{ij} = b_{ij}$$

for all i and j, where $i = 1, 2, 3, \ldots, m$ and $j = 1, 2, 3, \ldots, n$.

The previous definition points out that two matrices must be identical to be equal. They must be the same size and have the same corresponding entries.

The Sum of Two Matrices

Let $A = [a_{ij}]$ and $B = [b_{ij}]$ be two $m \times n$ matrices. The sum, $A + B$, is the $m \times n$ matrix C, found by adding the corresponding entries of matrices A and B:

$$A + B = C = [c_{ij}]$$

where $c_{ij} = a_{ij} + b_{ij}$, for $i = 1, 2, 3, \ldots, m$ and $j = 1, 2, 3, \ldots, n$.

EXAMPLE 1   Add the matrices:

$$\begin{bmatrix} 2 & 1 & 3 \\ 1 & -1 & 0 \end{bmatrix} \quad \text{and} \quad \begin{bmatrix} 1 & -1 & 2 \\ -1 & 1 & 5 \end{bmatrix}$$

Solution   Because each matrix is 2×3, their sum is defined and can be calculated by adding their corresponding elements:

$$\begin{bmatrix} 2 & 1 & 3 \\ 1 & -1 & 0 \end{bmatrix} + \begin{bmatrix} 1 & -1 & 2 \\ -1 & 1 & 5 \end{bmatrix} = \begin{bmatrix} 2+1 & 1-1 & 3+2 \\ 1-1 & -1+1 & 0+5 \end{bmatrix}$$

$$= \begin{bmatrix} 3 & 0 & 5 \\ 0 & 0 & 5 \end{bmatrix} \quad\blacksquare$$

EXAMPLE 2   If possible, add the matrices:

$$\begin{bmatrix} 2 & 4 & 3 \\ 1 & 1 & 1 \end{bmatrix} \quad \text{and} \quad \begin{bmatrix} 1 & 2 \\ 2 & 3 \end{bmatrix}$$

Solution   The first matrix is 2×3 and the second is 2×2. Because these matrices are of different sizes, they cannot be added. $\blacksquare$

Several of the properties of real numbers discussed in Chapter 1 apply to matrices also.

Theorem   The addition of two $m \times n$ matrices is commutative.

Proof   Let $A = [a_{ij}]$ and $B = [b_{ij}]$ be $m \times n$ matrices. Then $A + B = C = [c_{ij}]$, where $c_{ij} = a_{ij} + b_{ij}$ for each $i = 1, 2, 3, \ldots, m$ and $j = 1, 2, 3, \ldots, n$. On the other hand, $B + A = D = [d_{ij}]$, where $d_{ij} = b_{ij} + a_{ij}$ for each i and j.

Because each entry in each matrix is a real number, and the addition of real numbers is commutative, it follows that

$$c_{ij} = a_{ij} + b_{ij} = b_{ij} + a_{ij} = d_{ij}$$

for all i and j, $i = 1, 2, 3, \ldots, m$ and $j = 1, 2, 3, \ldots, n$.

By the definition of equality of matrices, $C = D$; therefore, $A + B = B + A$. $\blacksquare$

Theorem   The addition of three $m \times n$ matrices is associative.

The proof of the previous theorem is left as an exercise.

In the collection of all $m \times n$ matrices, there is a matrix called the **zero matrix.**

Zero Matrix   Let A be any $m \times n$ matrix. There is an $m \times n$ matrix **0**, called the **zero matrix** or the **additive identity matrix,** for which

$$A + \mathbf{0} = \mathbf{0} + A = A$$

The matrix **0** consists of m rows and n columns of 0s.

To illustrate the above definition, we note that the matrix

$$\begin{bmatrix} 0 & 0 & 0 \\ 0 & 0 & 0 \\ 0 & 0 & 0 \end{bmatrix}$$

is the 3×3 zero matrix, and that

$$\begin{bmatrix} 0 & 0 & 0 \\ 0 & 0 & 0 \\ 0 & 0 & 0 \end{bmatrix} + \begin{bmatrix} 1 & 2 & 3 \\ 4 & 5 & 6 \\ 7 & 8 & 9 \end{bmatrix} = \begin{bmatrix} 1 & 2 & 3 \\ 4 & 5 & 6 \\ 7 & 8 & 9 \end{bmatrix}$$

Matrices are similar to real numbers in another respect: Every matrix has an additive inverse.

Additive Inverse

Any $m \times n$ matrix A has an **additive inverse,** an $m \times n$ matrix $-A$ with the property that the sum of A and $-A$ is the zero matrix:

$$A + (-A) = (-A) + A = \mathbf{0}$$

The entries of $-A$ are the negatives of the corresponding entries of A.

The additive inverse of the 2×3 matrix $A = \begin{bmatrix} 1 & -3 & 2 \\ 0 & 1 & -5 \end{bmatrix}$ is the matrix

$$-A = \begin{bmatrix} -1 & 3 & -2 \\ 0 & -1 & 5 \end{bmatrix}$$

because their sum is the zero matrix:

$$\begin{aligned} A + (-A) &= \begin{bmatrix} 1 & -3 & 2 \\ 0 & 1 & -5 \end{bmatrix} + \begin{bmatrix} -1 & 3 & -2 \\ 0 & -1 & 5 \end{bmatrix} \\ &= \begin{bmatrix} 1-1 & -3+3 & 2-2 \\ 0+0 & 1-1 & -5+5 \end{bmatrix} \\ &= \begin{bmatrix} 0 & 0 & 0 \\ 0 & 0 & 0 \end{bmatrix} \end{aligned}$$

The Difference of Two Matrices

If A and B are $m \times n$ matrices, their difference, $A - B$, is the sum of A and the additive inverse of B:

$$A - B = A + (-B)$$

EXAMPLE 3

$$\begin{aligned} \begin{bmatrix} 2 & -5 \\ 3 & 1 \end{bmatrix} - \begin{bmatrix} 4 & -5 \\ -3 & 9 \end{bmatrix} &= \begin{bmatrix} 2 & -5 \\ 3 & 1 \end{bmatrix} + \begin{bmatrix} -4 & 5 \\ 3 & -9 \end{bmatrix} \\ &= \begin{bmatrix} -2 & 0 \\ 6 & -8 \end{bmatrix} \end{aligned}$$

∎

We illustrate how to find the product of two matrices by computing the product of a 2 × 3 matrix A and a 3 × 3 matrix B. The result is the 2 × 3 matrix C.

$$AB = \begin{bmatrix} 1 & 2 & 3 \\ 4 & 5 & 6 \end{bmatrix} \cdot \begin{bmatrix} a & b & c \\ d & e & f \\ g & h & i \end{bmatrix} = C$$

Each entry of matrix C is the result of a calculation that involves a row of A and a column of B. For example, the first-row, third-column entry of matrix C is found by keeping a running total of the products of corresponding entries of the first row of A and the third column of B:

$$\begin{bmatrix} \mathbf{1} & \mathbf{2} & \mathbf{3} \\ 4 & 5 & 6 \end{bmatrix} \begin{bmatrix} a & b & \mathbf{c} \\ d & e & \mathbf{f} \\ g & h & \mathbf{i} \end{bmatrix} = \begin{bmatrix} ? & ? & \mathbf{1c + 2f + 3i} \\ ? & ? & ? \end{bmatrix}$$

Similarly, the second-row, second-column entry of matrix C is formed by a calculation involving the second row of A and the second column of B.

$$\begin{bmatrix} 1 & 2 & 3 \\ \mathbf{4} & \mathbf{5} & \mathbf{6} \end{bmatrix} \begin{bmatrix} a & \mathbf{b} & c \\ d & \mathbf{e} & f \\ g & \mathbf{h} & i \end{bmatrix} = \begin{bmatrix} ? & ? & 1c + 2f + 3i \\ ? & \mathbf{4b + 5e + 6h} & ? \end{bmatrix}$$

To calculate the first-row, first-column entry of matrix C, we use the first row of A and the first column of B.

$$\begin{bmatrix} \mathbf{1} & \mathbf{2} & \mathbf{3} \\ 4 & 5 & 6 \end{bmatrix} \cdot \begin{bmatrix} \mathbf{a} & b & c \\ \mathbf{d} & e & f \\ \mathbf{g} & h & i \end{bmatrix} = \begin{bmatrix} \mathbf{1a + 2d + 3g} & ? & 1c + 2f + 3i \\ ? & 4b + 5e + 6h & ? \end{bmatrix}$$

The complete product C is

$$\begin{bmatrix} 1 & 2 & 3 \\ 4 & 5 & 6 \end{bmatrix} \cdot \begin{bmatrix} a & b & c \\ d & e & f \\ g & h & i \end{bmatrix} = \begin{bmatrix} 1a + 2d + 3g & 1b + 2e + 3h & 1c + 2f + 3i \\ 4a + 5d + 6g & 4b + 5e + 6h & 4c + 5f + 6i \end{bmatrix}$$

For the product AB to exist, the number of columns of A must equal the number of rows of B. If the product exists, it will have as many rows as A and as many columns as B:

$$\begin{array}{ccc} A & \cdot & B & = & C \\ m \times n & & n \times p & & m \times p \end{array}$$

These must agree.

The product is $m \times p$.

More formally, we have the following definition.

The Product of Two Matrices

Let $A = [a_{ij}]$ be an $m \times n$ matrix and $B = [b_{ij}]$ be an $n \times p$ matrix. The product, AB, is the $m \times p$ matrix C, found as follows:

$AB = C = [c_{ij}]$, where c_{ij} is the sum of the products of the corresponding entries in the ith row of A and the jth column of B, where $i = 1, 2, 3, \ldots, m$ and $j = 1, 2, 3, \ldots, p$.

EXAMPLE 4   Find C if $AB = \begin{bmatrix} 1 & 2 \\ 3 & 4 \\ 5 & 6 \end{bmatrix} \begin{bmatrix} a & b \\ c & d \end{bmatrix} = C$.

Solution   Because the first matrix is 3×2 and the second matrix is 2×2, the product C exists, and it is a 3×2 matrix. The first-row, first-column entry of C is the total of the products of corresponding entries in the first row of A and the first column of B: $c_{11} = 1a + 2c$. Similarly, c_{12} is computed by using the first row of A and the second column of B: $c_{12} = 1b + 2d$. The entire product is

$$\begin{bmatrix} 1 & 2 \\ 3 & 4 \\ 5 & 6 \end{bmatrix} \begin{bmatrix} a & b \\ c & d \end{bmatrix} = \begin{bmatrix} 1a + 2c & 1b + 2d \\ 3a + 4c & 3b + 4d \\ 5a + 6c & 5b + 6d \end{bmatrix}$$

EXAMPLE 5   Find the product: $\begin{bmatrix} 1 & -1 & 2 \\ 1 & 3 & 0 \\ 0 & 1 & 1 \end{bmatrix} \begin{bmatrix} 2 & 1 \\ 1 & 3 \\ 0 & 1 \end{bmatrix}$.

Solution   Because the matrices are 3×3 and 3×2, the product is a 3×2 matrix.

$$\begin{bmatrix} 1 & -1 & 2 \\ 1 & 3 & 0 \\ 0 & 1 & 1 \end{bmatrix} \begin{bmatrix} 2 & 1 \\ 1 & 3 \\ 0 & 1 \end{bmatrix}$$

$$= \begin{bmatrix} 1 \cdot 2 + (-1) \cdot 1 + 2 \cdot 0 & 1 \cdot 1 + (-1) \cdot 3 + 2 \cdot 1 \\ 1 \cdot 2 + 3 \cdot 1 + 0 \cdot 0 & 1 \cdot 1 + 3 \cdot 3 + 0 \cdot 1 \\ 0 \cdot 2 + 1 \cdot 1 + 1 \cdot 0 & 0 \cdot 1 + 1 \cdot 3 + 1 \cdot 1 \end{bmatrix}$$

$$= \begin{bmatrix} 1 & 0 \\ 5 & 10 \\ 1 & 4 \end{bmatrix}$$

EXAMPLE 6   Find the product: $\begin{bmatrix} 1 & 2 & 3 \end{bmatrix} \begin{bmatrix} 4 \\ 5 \\ 6 \end{bmatrix}$.

Solution   Because the first matrix is 1×3 and the second matrix is 3×1, the product is a 1×1 matrix:

$$\begin{bmatrix} 1 & 2 & 3 \end{bmatrix} \begin{bmatrix} 4 \\ 5 \\ 6 \end{bmatrix} = [1 \cdot 4 + 2 \cdot 5 + 3 \cdot 6] = [32]$$

EXAMPLE 7   If $A = \begin{bmatrix} 1 & 1 \\ 0 & 0 \end{bmatrix}$ and $B = \begin{bmatrix} 0 & 1 \\ 0 & 1 \end{bmatrix}$, calculate AB and BA and thereby show that multiplication of matrices is not commutative.

Solution

$$AB = \begin{bmatrix} 1 & 1 \\ 0 & 0 \end{bmatrix} \begin{bmatrix} 0 & 1 \\ 0 & 1 \end{bmatrix} = \begin{bmatrix} 0 & 2 \\ 0 & 0 \end{bmatrix}$$

$$BA = \begin{bmatrix} 0 & 1 \\ 0 & 1 \end{bmatrix} \begin{bmatrix} 1 & 1 \\ 0 & 0 \end{bmatrix} = \begin{bmatrix} 0 & 0 \\ 0 & 0 \end{bmatrix}$$

Because the products are not equal, matrix multiplication is not commutative. ■

EXAMPLE 8   Find values for x, y, and z such that

$$\begin{bmatrix} 1 & 2 & 3 \\ 2 & -1 & -2 \\ 1 & -3 & -3 \end{bmatrix} \begin{bmatrix} x \\ y \\ z \end{bmatrix} = \begin{bmatrix} 4 \\ 0 \\ -2 \end{bmatrix}$$

Solution   Find the product of the first two matrices and set it equal to the third matrix.

$$\begin{bmatrix} 1 & 2 & 3 \\ 2 & -1 & -2 \\ 1 & -3 & -3 \end{bmatrix} \begin{bmatrix} x \\ y \\ z \end{bmatrix} = \begin{bmatrix} 1x + 2y + 3z \\ 2x - 1y - 2z \\ 1x - 3y - 3z \end{bmatrix} = \begin{bmatrix} 4 \\ 0 \\ -2 \end{bmatrix}$$

The product will be equal to the third matrix if and only if their corresponding components are equal. Set the corresponding components equal to get

$$\begin{cases} x + 2y + 3z = 4 \\ 2x - y - 2z = 0 \\ x - 3y - 3z = -2 \end{cases}$$

Solve this system for x, y, and z to obtain $x = 1$, $y = 0$, and $z = 1$. ■

The number 1 is called *the identity for multiplication* because multiplying a number by 1 does not change that number. There is a **multiplicative identity matrix** with a similar property.

$n \times n$ Identity Matrix

Let A be an $n \times n$ matrix. There is an **$n \times n$ identity matrix I** for which

$$AI = IA = A$$

The matrix I consists of 1's on its diagonal and 0's elsewhere.

$$\begin{bmatrix} 1 & 0 & 0 & \cdots & 0 \\ 0 & 1 & 0 & \cdots & 0 \\ 0 & 0 & 1 & \cdots & 0 \\ \vdots & \vdots & \vdots & & \vdots \\ 0 & 0 & 0 & \cdots & 1 \end{bmatrix}$$

Note that an identity matrix is a **square matrix**—it has the same number of rows and columns.

Example 9 illustrates the previous definition for the 3×3 identity matrix.

EXAMPLE 9   Find   **a.** $\begin{bmatrix} 1 & 0 & 0 \\ 0 & 1 & 0 \\ 0 & 0 & 1 \end{bmatrix} \begin{bmatrix} 1 & 2 & 3 \\ 4 & 5 & 6 \\ 7 & 8 & 9 \end{bmatrix}$ and   **b.** $\begin{bmatrix} 1 & 2 & 3 \\ 4 & 5 & 6 \\ 7 & 8 & 9 \end{bmatrix} \begin{bmatrix} 1 & 0 & 0 \\ 0 & 1 & 0 \\ 0 & 0 & 1 \end{bmatrix}$

Solution   a. $\begin{bmatrix} 1 & 0 & 0 \\ 0 & 1 & 0 \\ 0 & 0 & 1 \end{bmatrix} \begin{bmatrix} 1 & 2 & 3 \\ 4 & 5 & 6 \\ 7 & 8 & 9 \end{bmatrix}$

$$= \begin{bmatrix} 1\cdot1+0\cdot4+0\cdot7 & 1\cdot2+0\cdot5+0\cdot8 & 1\cdot3+0\cdot6+0\cdot9 \\ 0\cdot1+1\cdot4+0\cdot7 & 0\cdot2+1\cdot5+0\cdot8 & 0\cdot3+1\cdot6+0\cdot9 \\ 0\cdot1+0\cdot4+1\cdot7 & 0\cdot2+0\cdot5+1\cdot8 & 0\cdot3+0\cdot6+1\cdot9 \end{bmatrix}$$

$$= \begin{bmatrix} \mathbf{1} & \mathbf{2} & \mathbf{3} \\ \mathbf{4} & \mathbf{5} & \mathbf{6} \\ \mathbf{7} & \mathbf{8} & \mathbf{9} \end{bmatrix}$$

b. $\begin{bmatrix} \mathbf{1} & \mathbf{2} & \mathbf{3} \\ \mathbf{4} & \mathbf{5} & \mathbf{6} \\ \mathbf{7} & \mathbf{8} & \mathbf{9} \end{bmatrix} \begin{bmatrix} 1 & 0 & 0 \\ 0 & 1 & 0 \\ 0 & 0 & 1 \end{bmatrix} = \begin{bmatrix} \mathbf{1} & \mathbf{2} & \mathbf{3} \\ \mathbf{4} & \mathbf{5} & \mathbf{6} \\ \mathbf{7} & \mathbf{8} & \mathbf{9} \end{bmatrix}$ ∎

I.1   EXERCISES

Find values of x and y, if any, that will make the two matrices equal.

1. $\begin{bmatrix} x & y \\ 1 & 3 \end{bmatrix} = \begin{bmatrix} 2 & 5 \\ 1 & 3 \end{bmatrix}$

2. $\begin{bmatrix} x & 5 \\ 3 & y \end{bmatrix} = \begin{bmatrix} 0 & 5 \\ 3 & 2 \end{bmatrix}$

3. $\begin{bmatrix} x & y \\ 1 & 3 \end{bmatrix} = \begin{bmatrix} 2 & 5 \\ 1 & 4 \end{bmatrix}$

4. $\begin{bmatrix} x & y \\ 1 & x+y \end{bmatrix} = \begin{bmatrix} 2 & 1 \\ 1 & 2 \end{bmatrix}$

5. $\begin{bmatrix} x+y & 3+x \\ -2 & 5y \end{bmatrix} = \begin{bmatrix} 3 & 4 \\ -2 & 10 \end{bmatrix}$

6. $\begin{bmatrix} x+y & x-y \\ 2x & 3y \end{bmatrix} = \begin{bmatrix} -x & x-2 \\ -y & 8-y \end{bmatrix}$

7. $\begin{bmatrix} x & 3x \\ y & x+1 \end{bmatrix} = \begin{bmatrix} y & 6 \\ 2 & 3 \end{bmatrix}$

8. $\begin{bmatrix} x \\ y \end{bmatrix} = \begin{bmatrix} 1 & 2 \\ 3 & 4 \end{bmatrix}$

Perform each operation, if possible.

9. $\begin{bmatrix} 2 & 1 & -1 \\ -3 & 2 & 5 \end{bmatrix} + \begin{bmatrix} -3 & 1 & 2 \\ -3 & -2 & -5 \end{bmatrix}$

10. $\begin{bmatrix} 3 & 1 \\ 2 & 2 \end{bmatrix} + \begin{bmatrix} 2 & 1 \\ -1 & 0 \end{bmatrix} + \begin{bmatrix} -5 & -2 \\ -1 & -2 \end{bmatrix}$

11. $\begin{bmatrix} 3 & 2 & 1 \\ -2 & 3 & -3 \\ -4 & -2 & -1 \end{bmatrix} - \begin{bmatrix} -2 & 6 & -2 \\ 5 & 7 & -1 \\ -4 & -6 & 7 \end{bmatrix}$

12. $\begin{bmatrix} -2 & 7 & -3 \\ 3 & 6 & -7 \\ -9 & -2 & -5 \end{bmatrix} + \begin{bmatrix} -5 & -4 & -3 \\ -1 & 2 & 10 \\ -1 & -3 & -4 \end{bmatrix}$

13. $\begin{bmatrix} 1 & 3 & -1 \\ 2 & 1 & 5 \\ 1 & 3 & 0 \end{bmatrix} + \begin{bmatrix} 2 \\ 0 \\ -3 \end{bmatrix}$

14. $\begin{bmatrix} 3 \\ 2 \\ 3 \end{bmatrix} - \begin{bmatrix} -3 & 5 & -6 \\ -3 & -5 & -6 \\ 4 & 6 & -6 \end{bmatrix}$

15. $[1 \quad 2 \quad 3] + [4 \quad 5 \quad 6]$

16. $\begin{bmatrix} 1 \\ 2 \\ 3 \end{bmatrix} + \begin{bmatrix} 4 & -5 & -6 \end{bmatrix}$

17. $\begin{bmatrix} 1 & 3 & -4 \\ 2 & -1 & 3 \\ 1 & 5 & 7 \end{bmatrix} + \begin{bmatrix} 3 & 2 & -8 \\ 9 & 11 & 17 \\ 2 & 1 & 3 \end{bmatrix}$

$- \begin{bmatrix} 1 & 3 & -5 \\ 2 & -9 & 5 \\ 3 & 10 & 11 \end{bmatrix}$

18. $\begin{bmatrix} -3 & -2 & 15 \\ 2 & -5 & -9 \end{bmatrix} - \begin{bmatrix} 3 & 2 & -15 \\ -2 & 5 & -9 \end{bmatrix}$

$+ \begin{bmatrix} 6 & 4 & -30 \\ -3 & 12 & -15 \end{bmatrix}$

Find each product, if possible.

19. $\begin{bmatrix} 2 & 3 \\ 3 & -2 \end{bmatrix} \begin{bmatrix} 1 & 2 \\ 0 & -2 \end{bmatrix}$

20. $\begin{bmatrix} -2 & 3 \\ 3 & -2 \end{bmatrix} \begin{bmatrix} 2 & 4 \\ -5 & 7 \end{bmatrix}$

21. $\begin{bmatrix} -4 & -2 \\ 21 & 0 \end{bmatrix} \begin{bmatrix} -5 & 6 \\ 21 & -1 \end{bmatrix}$

22. $\begin{bmatrix} -5 & 4 \\ 4 & -5 \end{bmatrix} \begin{bmatrix} 6 & -2 \\ 1 & 3 \end{bmatrix}$

23. $\begin{bmatrix} 2 & 1 & 3 \\ 1 & 2 & -1 \\ 0 & 1 & 0 \end{bmatrix} \begin{bmatrix} 1 & 2 & 3 \\ 2 & -2 & 1 \\ 0 & 0 & 1 \end{bmatrix}$

24. $\begin{bmatrix} 2 & 1 & 1 \\ 1 & 1 & 2 \\ 1 & -2 & -1 \end{bmatrix} \begin{bmatrix} 1 & 2 & 3 \\ 1 & 2 & -3 \\ -1 & -1 & 3 \end{bmatrix}$

25. $\begin{bmatrix} 1 & -2 & -3 \end{bmatrix} \begin{bmatrix} 4 \\ -5 \\ -6 \end{bmatrix}$

26. $\begin{bmatrix} 1 \\ -2 \\ -3 \end{bmatrix} \begin{bmatrix} 4 & -5 & -6 \end{bmatrix}$

27. $\begin{bmatrix} 1 & 2 & 3 \end{bmatrix} \begin{bmatrix} 4 & 5 & 6 \end{bmatrix}$

28. $\begin{bmatrix} 2 & 3 & 4 \\ 1 & 2 & 3 \\ -2 & 2 & 2 \end{bmatrix} \begin{bmatrix} -1 \\ 2 \\ 3 \end{bmatrix}$

29. $\begin{bmatrix} 1 & 2 & 3 \end{bmatrix} \begin{bmatrix} 1 & 2 & 3 \\ 4 & 5 & 6 \\ 7 & 8 & 9 \end{bmatrix}$

30. $\begin{bmatrix} 1 & 2 & 3 \\ 1 & 2 & 1 \\ 1 & -1 & -1 \end{bmatrix} \begin{bmatrix} 1 & 2 \\ 2 & 1 \\ 1 & 1 \end{bmatrix}$

Perform the operations.

31. $\begin{bmatrix} 1 & 2 \\ 2 & 3 \end{bmatrix} \left(\begin{bmatrix} 2 & 1 & -5 \\ 1 & 1 & 2 \end{bmatrix} + \begin{bmatrix} -2 & -1 & 6 \\ 0 & -1 & -1 \end{bmatrix} \right)$

32. $\begin{bmatrix} 1 & 2 \\ 2 & 3 \end{bmatrix} \begin{bmatrix} 2 & 1 & -5 \\ 1 & 1 & 2 \end{bmatrix}$

$+ \begin{bmatrix} 1 & 2 \\ 2 & 3 \end{bmatrix} \begin{bmatrix} -2 & -1 & 6 \\ 0 & -1 & -1 \end{bmatrix}$

33. $\begin{bmatrix} 1 & 2 & 3 \\ 2 & 3 & 1 \\ 1 & 2 & 1 \end{bmatrix} \begin{bmatrix} 2 & 1 & 1 \\ 3 & -1 & -1 \\ 2 & -2 & 2 \end{bmatrix}$

$+ \begin{bmatrix} -2 & 3 & 4 \\ 1 & 1 & 1 \\ 0 & 1 & 0 \end{bmatrix}$

34. $\begin{bmatrix} 2 & 1 & 0 \\ 1 & -2 & -1 \\ 1 & 1 & -1 \end{bmatrix} \left(\begin{bmatrix} 1 & 0 & 1 \\ 1 & 1 & 2 \\ 1 & 2 & -1 \end{bmatrix} \right.$

$\left. + \begin{bmatrix} -1 & -1 & 2 \\ 0 & 0 & 1 \\ 1 & 0 & -1 \end{bmatrix} \right)$

35. $\left(\begin{bmatrix} 1 & 2 \\ 2 & 3 \end{bmatrix} \begin{bmatrix} 1 \\ -3 \end{bmatrix} + \right.$

$\left. \begin{bmatrix} -2 \\ 1 \end{bmatrix} \right) \left(\begin{bmatrix} 1 & 2 \end{bmatrix} \begin{bmatrix} 1 \\ -3 \end{bmatrix} + \begin{bmatrix} 4 \end{bmatrix} \right)$

36. $\begin{bmatrix} 1 \\ 2 \end{bmatrix} \begin{bmatrix} -3 & -4 \end{bmatrix} - \begin{bmatrix} 0 & 3 \\ 2 & 1 \end{bmatrix} \begin{bmatrix} 2 & 0 \\ 1 & -1 \end{bmatrix}$

Let $A = \begin{bmatrix} 1 & 3 \\ 2 & 5 \end{bmatrix}$, $B = \begin{bmatrix} -1 \\ 3 \end{bmatrix}$, *and* $C = \begin{bmatrix} 3 & 2 \end{bmatrix}$.
Perform the operations, if possible.

37. $A - BC$

38. $AB + B$

39. $CB - AB$

40. CAB

41. ABC

42. $CA + C$

43. Let $A = \begin{bmatrix} 1 & 1 \\ 1 & 1 \end{bmatrix}$. Find A^7.   (*Hint:* Calculate A^2
and A^3. Is there a pattern?)

44. Let a, b, and c be real numbers. If $ab = ac$ and $a \neq 0$, then $b = c$. Find 2×2 matrices A, B, and C, where $A \neq 0$, to show that such a law does not hold for all matrices.

45. In the real number system, the numbers 0 and 1 are the only numbers that equal their own squares: If $a^2 = a$, then $a = 0$ or $a = 1$. Find a 2×2 matrix A that is neither the zero matrix nor the identity matrix, such that $A^2 = A \cdot A = A$.

46. Another property of the real numbers is that if $ab = 0$, then either $a = 0$ or $b = 0$. To show that this property is not true for matrices, find two nonzero 2×2 matrices, A and B, such that $AB = 0$.

47. Multiplication of three $n \times n$ matrices is associative: $(AB)C = A(BC)$. Verify this by an example chosen from the set of 2×2 matrices.

48. If A is an $m \times n$ matrix and B and C are each $n \times p$ matrices, then $A(B + C) = AB + AC$. Illustrate this with an example chosen from the set of 2×2 matrices, showing that matrix multiplication distributes over matrix addition.

49. Prove that the addition of matrices is associative.

I.2   Matrix Inversion

Two real numbers are called **multiplicative inverses** if their product is the multiplicative identity 1. Some matrices have multiplicative inverses also.

Multiplicative Inverse

If A and B are $n \times n$ matrices, I is the $n \times n$ identity matrix, and

$$AB = BA = I$$

then A and B are called **multiplicative inverses.** Matrix A is the **inverse** of B, and B is the **inverse** of A.

It can be shown that the inverse of a matrix A, if it exists, is unique. The inverse of A is denoted by A^{-1}.

EXAMPLE 1   If $A = \begin{bmatrix} 1 & 1 & 0 \\ 4 & 3 & 0 \\ 2 & 1 & -1 \end{bmatrix}$ and $B = \begin{bmatrix} -3 & 1 & 0 \\ 4 & -1 & 0 \\ -2 & 1 & -1 \end{bmatrix}$, show that A and B are inverses.

Solution   Multiply the matrices in each order to show that the product is the identity matrix.

$$AB = \begin{bmatrix} 1 & 1 & 0 \\ 4 & 3 & 0 \\ 2 & 1 & -1 \end{bmatrix} \begin{bmatrix} -3 & 1 & 0 \\ 4 & -1 & 0 \\ -2 & 1 & -1 \end{bmatrix}$$

$$= \begin{bmatrix} -3+4 & 1-1 & 0 \\ -12+12 & 4-3 & 0 \\ -6+4+2 & 2-1-1 & 1 \end{bmatrix} = \begin{bmatrix} 1 & 0 & 0 \\ 0 & 1 & 0 \\ 0 & 0 & 1 \end{bmatrix}$$

$$BA = \begin{bmatrix} -3 & 1 & 0 \\ 4 & -1 & 0 \\ -2 & 1 & -1 \end{bmatrix} \begin{bmatrix} 1 & 1 & 0 \\ 4 & 3 & 0 \\ 2 & 1 & -1 \end{bmatrix} = \begin{bmatrix} 1 & 0 & 0 \\ 0 & 1 & 0 \\ 0 & 0 & 1 \end{bmatrix}$$ ∎

If a matrix has an inverse, it is called a **nonsingular matrix.** Otherwise, it is called a **singular matrix.** The following theorem, stated without proof, provides a way of calculating the inverse of a nonsingular matrix.

Theorem

> If a sequence of elementary row operations performed on the $n \times n$ matrix A reduces A to the $n \times n$ identity matrix I, then those same row operations, performed in the same order on the identity matrix I, will transform I into A^{-1}. Furthermore, if *no* sequence of row operations will reduce A to I, then A is singular.

To use the previous theorem, we perform elementary row operations on matrix A to change it to the identity matrix, I. At the same time, we perform these elementary row operations on the identity matrix I. This changes I into A^{-1}.

A notation for this process uses an n-row-by-$2n$-column matrix, with matrix A as the left half and matrix I as the right half. If A is nonsingular, the proper row operations performed on $[A \mid I]$ will transform it into $[I \mid A^{-1}]$.

EXAMPLE 2   Find the inverse of matrix A if $A = \begin{bmatrix} 2 & -4 \\ 4 & -7 \end{bmatrix}$.

Solution   Set up a 2×4 matrix with A on the left and I on the right of the broken line:

$$[A \mid I] = \begin{bmatrix} 2 & -4 & \vdots & 1 & 0 \\ 4 & -7 & \vdots & 0 & 1 \end{bmatrix}$$

Perform row operations on the entire matrix to transform the left half into I. Begin by multiplying row 1 by $\frac{1}{2}$ to obtain a new row 1 $\left(\frac{1}{2}R1 \rightarrow R1\right)$ and add -2 times the original row 1 to row 2 to obtain a new row 2 ($-2R1 + R2 \rightarrow R2$).

$$\left(\frac{1}{2}\right)R1 \rightarrow R1$$

$$(-2)R1 + R2 \rightarrow R2$$

$$\begin{bmatrix} 2 & -4 & \vdots & 1 & 0 \\ 4 & -7 & \vdots & 0 & 1 \end{bmatrix} \Leftrightarrow \begin{bmatrix} 1 & -2 & \vdots & \frac{1}{2} & 0 \\ 0 & 1 & \vdots & -2 & 1 \end{bmatrix}$$

$$(2)R2 + R1 \rightarrow R1$$

$$\Leftrightarrow \begin{bmatrix} 1 & 0 & \vdots & -\frac{7}{2} & 2 \\ 0 & 1 & \vdots & -2 & 1 \end{bmatrix}$$

Matrix A has been transformed into I. Thus, the right side of the previous matrix is A^{-1}. Verify this by finding AA^{-1} and $A^{-1}A$ and showing that each product is I:

$$AA^{-1} = \begin{bmatrix} 2 & -4 \\ 4 & -7 \end{bmatrix} \begin{bmatrix} -\frac{7}{2} & 2 \\ -2 & 1 \end{bmatrix} = \begin{bmatrix} 1 & 0 \\ 0 & 1 \end{bmatrix}$$

$$A^{-1}A = \begin{bmatrix} -\frac{7}{2} & 2 \\ -2 & 1 \end{bmatrix} \begin{bmatrix} 2 & -4 \\ 4 & -7 \end{bmatrix} = \begin{bmatrix} 1 & 0 \\ 0 & 1 \end{bmatrix}$$ ∎

EXAMPLE 3   Find the inverse of matrix A if $A = \begin{bmatrix} 1 & 1 & 0 \\ 1 & 2 & 1 \\ 2 & 3 & 2 \end{bmatrix}$.

Solution   Set up a 3×6 matrix with A on the left and I on the right of the broken line:

$$[A \mid I] = \begin{bmatrix} 1 & 1 & 0 & 1 & 0 & 0 \\ 1 & 2 & 1 & 0 & 1 & 0 \\ 2 & 3 & 2 & 0 & 0 & 1 \end{bmatrix}$$

Perform row operations on the entire matrix to transform the left half into I.

$$(-1)R1 + R2 \rightarrow R2$$
$$(-2)R1 + R3 \rightarrow R3$$

$$\begin{bmatrix} 1 & 1 & 0 & 1 & 0 & 0 \\ 1 & 2 & 1 & 0 & 1 & 0 \\ 2 & 3 & 2 & 0 & 0 & 1 \end{bmatrix} \Leftrightarrow \begin{bmatrix} 1 & 1 & 0 & 1 & 0 & 0 \\ 0 & 1 & 1 & -1 & 1 & 0 \\ 0 & 1 & 2 & -2 & 0 & 1 \end{bmatrix}$$

$$(-1)R2 + R1 \rightarrow R1$$
$$(-1)R2 + R3 \rightarrow R3$$

$$\Leftrightarrow \begin{bmatrix} 1 & 0 & -1 & 2 & -1 & 0 \\ 0 & 1 & 1 & -1 & 1 & 0 \\ 0 & 0 & 1 & -1 & -1 & 1 \end{bmatrix}$$

$$R3 + R1 \rightarrow R1$$
$$(-1)R3 + R2 \rightarrow R2$$

$$\Leftrightarrow \begin{bmatrix} 1 & 0 & 0 & 1 & -2 & 1 \\ 0 & 1 & 0 & 0 & 2 & -1 \\ 0 & 0 & 1 & -1 & -1 & 1 \end{bmatrix}$$

The left half has been transformed into the identity matrix, and the right half has become A^{-1}. Thus,

$$A^{-1} = \begin{bmatrix} 1 & -2 & 1 \\ 0 & 2 & -1 \\ -1 & -1 & 1 \end{bmatrix}$$ ∎

EXAMPLE 4   Find the inverse of $A = \begin{bmatrix} 1 & 2 \\ 2 & 4 \end{bmatrix}$, if possible.

Solution   Form the 2×4 matrix

$$[A \mid I] = \begin{bmatrix} 1 & 2 & 1 & 0 \\ 2 & 4 & 0 & 1 \end{bmatrix}$$

and begin to transform the left side of the matrix into the identity matrix I:

$$(-2)R1 + R2 \longrightarrow R2$$

$$\left[\begin{array}{cc:cc} 1 & 2 & 1 & 0 \\ 2 & 4 & 0 & 1 \end{array}\right] \Leftrightarrow \left[\begin{array}{cc:cc} 1 & 2 & 1 & 0 \\ 0 & 0 & -2 & 1 \end{array}\right]$$

In obtaining the second-row, first-column position of A, the entire row of A is "zeroed out." Because it is impossible to transform A to the identity, matrix A is singular and has no inverse. ∎

The next example shows how the inverse of a nonsingular matrix can be used to solve a system of equations.

EXAMPLE 5   Solve the system:

$$\begin{cases} x + y = 3 \\ x + 2y + z = -2 \\ 2x + 3y + 2z = 1 \end{cases}$$

Solution   This system can be written as a single equation involving three matrices.

$$(1) \qquad \begin{bmatrix} 1 & 1 & 0 \\ 1 & 2 & 1 \\ 2 & 3 & 2 \end{bmatrix} \begin{bmatrix} x \\ y \\ z \end{bmatrix} = \begin{bmatrix} 3 \\ -2 \\ 1 \end{bmatrix}$$

The 3×3 matrix on the left is the matrix whose inverse was found in Example 3. Multiply each side of Equation 1 on the left by this inverse to obtain an equivalent system of equations. The solution of this system can be read directly from the matrix to the right of the equals sign:

$$\begin{bmatrix} 1 & -2 & 1 \\ 0 & 2 & -1 \\ -1 & -1 & 1 \end{bmatrix}\begin{bmatrix} 1 & 1 & 0 \\ 1 & 2 & 1 \\ 2 & 3 & 2 \end{bmatrix}\begin{bmatrix} x \\ y \\ z \end{bmatrix} = \begin{bmatrix} 1 & -2 & 1 \\ 0 & 2 & -1 \\ -1 & -1 & 1 \end{bmatrix}\begin{bmatrix} 3 \\ -2 \\ 1 \end{bmatrix}$$

$$\begin{bmatrix} 1 & 0 & 0 \\ 0 & 1 & 0 \\ 0 & 0 & 1 \end{bmatrix}\begin{bmatrix} x \\ y \\ z \end{bmatrix} = \begin{bmatrix} 8 \\ -5 \\ 0 \end{bmatrix}$$

$$\begin{bmatrix} x \\ y \\ z \end{bmatrix} = \begin{bmatrix} 8 \\ -5 \\ 0 \end{bmatrix}$$

The solution of this system of equations is $x = 8$, $y = -5$, $z = 0$. Verify that these results satisfy all three of the original equations. ∎

The equations of Example 5 can be thought of as the matrix equation $AX = B$, where A is the coefficient matrix,

$$A = \begin{bmatrix} 1 & 1 & 0 \\ 1 & 2 & 1 \\ 2 & 3 & 2 \end{bmatrix}$$

X is a column matrix of the variables,

$$X = \begin{bmatrix} x \\ y \\ z \end{bmatrix}$$

and B is a column matrix of the constants from the right sides of the equations,

$$B = \begin{bmatrix} 3 \\ -2 \\ 1 \end{bmatrix}$$

When each side of $AX = B$ is multiplied on the left by A^{-1}, the solution of the system appears as the column of numbers in the matrix that is the product of A^{-1} and B:

$$A^{-1}AX = A^{-1}B$$
$$IX = A^{-1}B$$
$$X = A^{-1}B$$

This method is especially useful for finding solutions of several systems of equations that differ from each other *only* in the column matrix B. If the coefficient matrix A remains unchanged from one system of equations to the next, then A^{-1} needs to be found only once. The solution of each system is found by a single matrix multiplication, $A^{-1}B$.

I.2   EXERCISES

Find the inverse of each matrix, if possible.

1. $\begin{bmatrix} 3 & -4 \\ -2 & 3 \end{bmatrix}$

2. $\begin{bmatrix} 2 & 3 \\ 3 & 5 \end{bmatrix}$

3. $\begin{bmatrix} 3 & 7 \\ 2 & 5 \end{bmatrix}$

4. $\begin{bmatrix} 1 & -2 \\ 2 & -5 \end{bmatrix}$

5. $\begin{bmatrix} 1 & 2 & 3 \\ 2 & 5 & 3 \\ 1 & 0 & 8 \end{bmatrix}$

6. $\begin{bmatrix} 2 & 1 & -1 \\ 2 & 2 & -1 \\ -1 & -1 & 1 \end{bmatrix}$

7. $\begin{bmatrix} 3 & 2 & 1 \\ 1 & 1 & -1 \\ 4 & 3 & 1 \end{bmatrix}$

8. $\begin{bmatrix} -2 & 1 & -3 \\ 2 & 3 & 0 \\ 1 & 0 & 1 \end{bmatrix}$

9. $\begin{bmatrix} 1 & 3 & 5 \\ 0 & 1 & 6 \\ 1 & 4 & 11 \end{bmatrix}$

10. $\begin{bmatrix} 1 & 1 & 1 \\ 2 & 2 & 2 \\ 3 & 3 & 3 \end{bmatrix}$

11. $\begin{bmatrix} 1 & 2 & 3 \\ 0 & 1 & 2 \\ 0 & 0 & 1 \end{bmatrix}$

12. $\begin{bmatrix} 1 & 2 & 3 \\ 0 & 1 & 1 \\ 0 & -1 & 0 \end{bmatrix}$

13. $\begin{bmatrix} 1 & 6 & 4 \\ 1 & -2 & -5 \\ 2 & 4 & -1 \end{bmatrix}$

14. $\begin{bmatrix} 1 & 1 & 1 \\ 1 & 0 & -1 \\ 1 & 2 & 3 \end{bmatrix}$

15. $\begin{bmatrix} 1 & 2 & 3 & 4 \\ 0 & 1 & 2 & 3 \\ 0 & 0 & 1 & 2 \\ 0 & 0 & 0 & 1 \end{bmatrix}$

16. $\begin{bmatrix} 1 & 0 & 0 & 0 \\ 1 & 1 & 0 & 0 \\ 1 & 1 & 1 & 0 \\ 1 & 2 & 2 & 1 \end{bmatrix}$

Use the method of Example 5 to solve each system of equations.

17. $\begin{cases} 3x - 4y = 1 \\ -2x + 3y = 5 \end{cases}$

18. $\begin{cases} 2x + 3y = 7 \\ 3x + 5y = -5 \end{cases}$

19. $\begin{cases} 3x + 7y = 0 \\ 2x + 5y = -10 \end{cases}$

20. $\begin{cases} x - 2y = 12 \\ 2x - 5y = 13 \end{cases}$

21. $\begin{cases} x + 2y + 3z = 1 \\ 2x + 5y + 3z = 3 \\ x \quad\quad + 8z = -2 \end{cases}$

22. $\begin{cases} 2x + y - z = 3 \\ 2x + 2y - z = -1 \\ -x - y + z = 4 \end{cases}$

23. $\begin{cases} 3x + 2y + z = 2 \\ x + y - z = -1 \\ 4x + 3y + z = 0 \end{cases}$

24. $\begin{cases} -2x + y - 3z = 5 \\ 2x + 3y = 1 \\ x \quad\quad + z = -2 \end{cases}$

25. $\begin{cases} x + 2y + 3z = 1 \\ 2x + 2y + 2z = 2x + y \\ z = 3 \end{cases}$

26. $\begin{cases} x + 2y + 3z = 2 \\ x + y + z = x \\ x - y = x \end{cases}$

27. If the $n \times n$ matrix A is nonsingular, and if B and C are $n \times n$ matrices such that $AB = AC$, prove that $B = C$.

28. If B is an $n \times n$ matrix that behaves as an identity ($AB = BA = A$, for any $n \times n$ matrix A), prove that $B = I$.

29. If $A = \begin{bmatrix} 0 & 1 \\ 1 & 0 \end{bmatrix}$, compute $A^2, A^3, A^4, \ldots$. Give a general rule for A^n, where n is a natural number.

30. If A and B are 2×2 matrices, is $(AB)^2 = A^2B^2$? Support your answer.

31. Prove that $\begin{bmatrix} a & b \\ c & d \end{bmatrix}$ has an inverse if and only if $ad - bc \neq 0$.

32. If $A = \begin{bmatrix} 1 & 1 \\ 0 & 1 \end{bmatrix}$, compute A^n for various values of n ($n = 2, 3, 4, \ldots$). What do you notice?

33. If $A = \begin{bmatrix} 1 & 0 \\ 1 & 1 \end{bmatrix}$, compute A^n for various values of n ($n = 2, 3, 4, \ldots$). What do you notice?

34. For what value of x will $\begin{bmatrix} 3 & 8 \\ 6 & x \end{bmatrix}$ not have a multiplicative inverse? (*Hint:* See Exercise 31.)

35. For what values of x will $\begin{bmatrix} x & 8 \\ 2 & x \end{bmatrix}$ not have a multiplicative inverse?

36. Does $(AB)^{-1} = A^{-1}B^{-1}$? Support your answer with an example chosen from 2×2 matrices.

37. Use an example chosen from 2×2 matrices to illustrate that $(AB)^{-1} = B^{-1}A^{-1}$.

38. Let A be any 3×3 matrix. Find a 3×3 matrix E such that the product EA is the result of performing the row operation $R1 \leftrightarrow R2$ on matrix A (the row operation that exchanges rows one and two of matrix A). What is E^{-1}?

39. Let A be any 3×3 matrix. Find another 3×3 matrix E such that the product EA is the result of performing the row operation $(3)R1 + R3 \rightarrow R3$ on matrix A. What is E^{-1}?

Tables

Table A Powers and Roots

n	n^2	$\sqrt{n}$	n^3	$\sqrt[3]{n}$	n	n^2	$\sqrt{n}$	n^3	$\sqrt[3]{n}$
1	1	1.000	1	1.000	51	2,601	7.141	132,651	3.708
2	4	1.414	8	1.260	52	2,704	7.211	140,608	3.733
3	9	1.732	27	1.442	53	2,809	7.280	148,877	3.756
4	16	2.000	64	1.587	54	2,916	7.348	157,464	3.780
5	25	2.236	125	1.710	55	3,025	7.416	166,375	3.803
6	36	2.449	216	1.817	56	3,136	7.483	175,616	3.826
7	49	2.646	343	1.913	57	3,249	7.550	185,193	3.849
8	64	2.828	512	2.000	58	3,364	7.616	195,112	3.871
9	81	3.000	729	2.080	59	3,481	7.681	205,379	3.893
10	100	3.162	1,000	2.154	60	3,600	7.746	216,000	3.915
11	121	3.317	1,331	2.224	61	3,721	7.810	226,981	3.936
12	144	3.464	1,728	2.289	62	3,844	7.874	238,328	3.958
13	169	3.606	2,197	2.351	63	3,969	7.937	250,047	3.979
14	196	3.742	2,744	2.410	64	4,096	8.000	262,144	4.000
15	225	3.873	3,375	2.466	65	4,225	8.062	274,625	4.021
16	256	4.000	4,096	2.520	66	4,356	8.124	287,496	4.041
17	289	4.123	4,913	2.571	67	4,489	8.185	300,763	4.062
18	324	4.243	5,832	2.621	68	4,624	8.246	314,432	4.082
19	361	4.359	6,859	2.668	69	4,761	8.307	328,509	4.102
20	400	4.472	8,000	2.714	70	4,900	8.367	343,000	4.121
21	441	4.583	9,261	2.759	71	5,041	8.426	357,911	4.141
22	484	4.690	10,648	2.802	72	5,184	8.485	373,248	4.160
23	529	4.796	12,167	2.844	73	5,329	8.544	389,017	4.179
24	576	4.899	13,824	2.884	74	5,476	8.602	405,224	4.198
25	625	5.000	15,625	2.924	75	5,625	8.660	421,875	4.217
26	676	5.099	17,576	2.962	76	5,776	8.718	438,976	4.236
27	729	5.196	19,683	3.000	77	5,929	8.775	456,533	4.254
28	784	5.292	21,952	3.037	78	6,084	8.832	474,552	4.273
29	841	5.385	24,389	3.072	79	6,241	8.888	493,039	4.291
30	900	5.477	27,000	3.107	80	6,400	8.944	512,000	4.309
31	961	5.568	29,791	3.141	81	6,561	9.000	531,441	4.327
32	1,024	5.657	32,768	3.175	82	6,724	9.055	551,368	4.344
33	1,089	5.745	35,937	3.208	83	6,889	9.110	571,787	4.362
34	1,156	5.831	39,304	3.240	84	7,056	9.165	592,704	4.380
35	1,225	5.916	42,875	3.271	85	7,225	9.220	614,125	4.397
36	1,296	6.000	46,656	3.302	86	7,396	9.274	636,056	4.414
37	1,369	6.083	50,653	3.332	87	7,569	9.327	658,503	4.431
38	1,444	6.164	54,872	3.362	88	7,744	9.381	681,472	4.448
39	1,521	6.245	59,319	3.391	89	7,921	9.434	704,969	4.465
40	1,600	6.325	64,000	3.420	90	8,100	9.487	729,000	4.481
41	1,681	6.403	68,921	3.448	91	8,281	9.539	753,571	4.498
42	1,764	6.481	74,088	3.476	92	8,464	9.592	778,688	4.514
43	1,849	6.557	79,507	3.503	93	8,649	9.644	804,357	4.531
44	1,936	6.633	85,184	3.530	94	8,836	9.695	830,584	4.547
45	2,025	6.708	91,125	3.557	95	9,025	9.747	857,375	4.563
46	2,116	6.782	97,336	3.583	96	9,216	9.798	884,736	4.579
47	2,209	6.856	103,823	3.609	97	9,409	9.849	912,673	4.595
48	2,304	6.928	110,592	3.634	98	9,604	9.899	941,192	4.610
49	2,401	7.000	117,649	3.659	99	9,801	9.950	970,299	4.626
50	2,500	7.071	125,000	3.684	100	10,000	10.000	1,000,000	4.642

Table B   (continued)

N	0	1	2	3	4	5	6	7	8	9
5.5	.7404	.7412	.7419	.7427	.7435	.7443	.7451	.7459	.7466	.7474
5.6	.7482	.7490	.7497	.7505	.7513	.7520	.7528	.7536	.7543	.7551
5.7	.7559	.7566	.7574	.7582	.7589	.7597	.7604	.7612	.7619	.7627
5.8	.7634	.7642	.7649	.7657	.7664	.7672	.7679	.7686	.7694	.7701
5.9	.7709	.7716	.7723	.7731	.7738	.7745	.7752	.7760	.7767	.7774
6.0	.7782	.7789	.7796	.7803	.7810	.7818	.7825	.7832	.7839	.7846
6.1	.7853	.7860	.7868	.7875	.7882	.7889	.7896	.7903	.7910	.7917
6.2	.7924	.7931	.7938	.7945	.7952	.7959	.7966	.7973	.7980	.7987
6.3	.7993	.8000	.8007	.8014	.8021	.8028	.8035	.8041	.8048	.8055
6.4	.8062	.8069	.8075	.8082	.8089	.8096	.8102	.8109	.8116	.8122
6.5	.8129	.8136	.8142	.8149	.8156	.8162	.8169	.8176	.8182	.8189
6.6	.8195	.8202	.8209	.8215	.8222	.8228	.8235	.8241	.8248	.8254
6.7	.8261	.8267	.8274	.8280	.8287	.8293	.8299	.8306	.8312	.8319
6.8	.8325	.8331	.8338	.8344	.8351	.8357	.8363	.8370	.8376	.8382
6.9	.8388	.8395	.8401	.8407	.8414	.8420	.8426	.8432	.8439	.8445
7.0	.8451	.8457	.8463	.8470	.8476	.8482	.8488	.8494	.8500	.8506
7.1	.8513	.8519	.8525	.8531	.8537	.8543	.8549	.8555	.8561	.8567
7.2	.8573	.8579	.8585	.8591	.8597	.8603	.8609	.8615	.8621	.8627
7.3	.8633	.8639	.8645	.8651	.8657	.8663	.8669	.8675	.8681	.8686
7.4	.8692	.8698	.8704	.8710	.8716	.8722	.8727	.8733	.8739	.8745
7.5	.8751	.8756	.8762	.8768	.8774	.8779	.8785	.8791	.8797	.8802
7.6	.8808	.8814	.8820	.8825	.8831	.8837	.8842	.8848	.8854	.8859
7.7	.8865	.8871	.8876	.8882	.8887	.8893	.8899	.8904	.8910	.8915
7.8	.8921	.8927	.8932	.8938	.8943	.8949	.8954	.8960	.8965	.8971
7.9	.8976	.8982	.8987	.8993	.8998	.9004	.9009	.9015	.9020	.9025
8.0	.9031	.9036	.9042	.9047	.9053	.9058	.9063	.9069	.9074	.9079
8.1	.9085	.9090	.9096	.9101	.9106	.9112	.9117	.9122	.9128	.9133
8.2	.9138	.9143	.9149	.9154	.9159	.9165	.9170	.9175	.9180	.9186
8.3	.9191	.9196	.9201	.9206	.9212	.9217	.9222	.9227	.9232	.9238
8.4	.9243	.9248	.9253	.9258	.9263	.9269	.9274	.9279	.9284	.9289
8.5	.9294	.9299	.9304	.9309	.9315	.9320	.9325	.9330	.9335	.9340
8.6	.9345	.9350	.9355	.9360	.9365	.9370	.9375	.9380	.9385	.9390
8.7	.9395	.9400	.9405	.9410	.9415	.9420	.9425	.9430	.9435	.9440
8.8	.9445	.9450	.9455	.9460	.9465	.9469	.9474	.9479	.9484	.9489
8.9	.9494	.9499	.9504	.9509	.9513	.9518	.9523	.9528	.9533	.9538
9.0	.9542	.9547	.9552	.9557	.9562	.9566	.9571	.9576	.9581	.9586
9.1	.9590	.9595	.9600	.9605	.9609	.9614	.9619	.9624	.9628	.9633
9.2	.9638	.9643	.9647	.9652	.9657	.9661	.9666	.9671	.9675	.9680
9.3	.9685	.9689	.9694	.9699	.9703	.9708	.9713	.9717	.9722	.9727
9.4	.9731	.9736	.9741	.9745	.9750	.9754	.9759	.9763	.9768	.9773
9.5	.9777	.9782	.9786	.9791	.9795	.9800	.9805	.9809	.9814	.9818
9.6	.9823	.9827	.9832	.9836	.9841	.9845	.9850	.9854	.9859	.9863
9.7	.9868	.9872	.9877	.9881	.9886	.9890	.9894	.9899	.9903	.9908
9.8	.9912	.9917	.9921	.9926	.9930	.9934	.9939	.9943	.9948	.9952
9.9	.9956	.9961	.9965	.9969	.9974	.9978	.9983	.9987	.9991	.9996

Table B   Base-10 Logarithms

N	0	1	2	3	4	5	6	7	8	9
1.0	.0000	.0043	.0086	.0128	.0170	.0212	.0253	.0294	.0334	.0374
1.1	.0414	.0453	.0492	.0531	.0569	.0607	.0645	.0682	.0719	.0755
1.2	.0792	.0828	.0864	.0899	.0934	.0969	.1004	.1038	.1072	.1106
1.3	.1139	.1173	.1206	.1239	.1271	.1303	.1335	.1367	.1399	.1430
1.4	.1461	.1492	.1523	.1553	.1584	.1614	.1644	.1673	.1703	.1732
1.5	.1761	.1790	.1818	.1847	.1875	.1903	.1931	.1959	.1987	.2014
1.6	.2041	.2068	.2095	.2122	.2148	.2175	.2201	.2227	.2253	.2279
1.7	.2304	.2330	.2355	.2380	.2405	.2430	.2455	.2480	.2504	.2529
1.8	.2553	.2577	.2601	.2625	.2648	.2672	.2695	.2718	.2742	.2765
1.9	.2788	.2810	.2833	.2856	.2878	.2900	.2923	.2945	.2967	.2989
2.0	.3010	.3032	.3054	.3075	.3096	.3118	.3139	.3160	.3181	.3201
2.1	.3222	.3243	.3263	.3284	.3304	.3324	.3345	.3365	.3385	.3404
2.2	.3424	.3444	.3464	.3483	.3502	.3522	.3541	.3560	.3579	.3598
2.3	.3617	.3636	.3655	.3674	.3692	.3711	.3729	.3747	.3766	.3784
2.4	.3802	.3820	.3838	.3856	.3874	.3892	.3909	.3927	.3945	.3962
2.5	.3979	.3997	.4014	.4031	.4048	.4065	.4082	.4099	.4116	.4133
2.6	.4150	.4166	.4183	.4200	.4216	.4232	.4249	.4265	.4281	.4298
2.7	.4314	.4330	.4346	.4362	.4378	.4393	.4409	.4425	.4440	.4456
2.8	.4472	.4487	.4502	.4518	.4533	.4548	.4564	.4579	.4594	.4609
2.9	.4624	.4639	.4654	.4669	.4683	.4698	.4713	.4728	.4742	.4757
3.0	.4771	.4786	.4800	.4814	.4829	.4843	.4857	.4871	.4886	.4900
3.1	.4914	.4928	.4942	.4955	.4969	.4983	.4997	.5011	.5024	.5038
3.2	.5051	.5065	.5079	.5092	.5105	.5119	.5132	.5145	.5159	.5172
3.3	.5185	.5198	.5211	.5224	.5237	.5250	.5263	.5276	.5289	.5302
3.4	.5315	.5328	.5340	.5353	.5366	.5378	.5391	.5403	.5416	.5428
3.5	.5441	.5453	.5465	.5478	.5490	.5502	.5514	.5527	.5539	.5551
3.6	.5563	.5575	.5587	.5599	.5611	.5623	.5635	.5647	.5658	.5670
3.7	.5682	.5694	.5705	.5717	.5729	.5740	.5752	.5763	.5775	.5786
3.8	.5798	.5809	.5821	.5832	.5843	.5855	.5866	.5877	.5888	.5899
3.9	.5911	.5922	.5933	.5944	.5955	.5966	.5977	.5988	.5999	.6010
4.0	.6021	.6031	.6042	.6053	.6064	.6075	.6085	.6096	.6107	.6117
4.1	.6128	.6138	.6149	.6160	.6170	.6180	.6191	.6201	.6212	.6222
4.2	.6232	.6243	.6253	.6263	.6274	.6284	.6294	.6304	.6314	.6325
4.3	.6335	.6345	.6355	.6365	.6375	.6385	.6395	.6405	.6415	.6425
4.4	.6435	.6444	.6454	.6464	.6474	.6484	.6493	.6503	.6513	.6522
4.5	.6532	.6542	.6551	.6561	.6571	.6580	.6590	.6599	.6609	.6618
4.6	.6628	.6637	.6646	.6656	.6665	.6675	.6684	.6693	.6702	.6712
4.7	.6721	.6730	.6739	.6749	.6758	.6767	.6776	.6785	.6794	.6803
4.8	.6812	.6821	.6830	.6839	.6848	.6857	.6866	.6875	.6884	.6893
4.9	.6902	.6911	.6920	.6928	.6937	.6946	.6955	.6964	.6972	.6981
5.0	.6990	.6998	.7007	.7016	.7024	.7033	.7042	.7050	.7059	.7067
5.1	.7076	.7084	.7093	.7101	.7110	.7118	.7126	.7135	.7143	.7152
5.2	.7160	.7168	.7177	.7185	.7193	.7202	.7210	.7218	.7226	.7235
5.3	.7243	.7251	.7259	.7267	.7275	.7284	.7292	.7300	.7308	.7316
5.4	.7324	.7332	.7340	.7348	.7356	.7364	.7372	.7380	.7388	.7396

Table C (continued)

N	0	1	2	3	4	5	6	7	8	9
5.5	1.7047	.7066	.7084	.7102	.7120	.7138	.7156	.7174	.7192	.7210
5.6	.7228	.7246	.7263	.7281	.7299	.7317	.7334	.7352	.7370	.7387
5.7	.7405	.7422	.7440	.7457	.7475	.7492	.7509	.7527	.7544	.7561
5.8	.7579	.7596	.7613	.7630	.7647	.7664	.7681	.7699	.7716	.7733
5.9	.7750	.7766	.7783	.7800	.7817	.7834	.7851	.7867	.7884	.7901
6.0	1.7918	.7934	.7951	.7967	.7984	.8001	.8017	.8034	.8050	.8066
6.1	.8083	.8099	.8116	.8132	.8148	.8165	.8181	.8197	.8213	.8229
6.2	.8245	.8262	.8278	.8294	.8310	.8326	.8342	.8358	.8374	.8390
6.3	.8405	.8421	.8437	.8453	.8469	.8485	.8500	.8516	.8532	.8547
6.4	.8563	.8579	.8594	.8610	.8625	.8641	.8656	.8672	.8687	.8703
6.5	1.8718	.8733	.8749	.8764	.8779	.8795	.8810	.8825	.8840	.8856
6.6	.8871	.8886	.8901	.8916	.8931	.8946	.8961	.8976	.8991	.9006
6.7	.9021	.9036	.9051	.9066	.9081	.9095	.9110	.9125	.9140	.9155
6.8	.9169	.9184	.9199	.9213	.9228	.9242	.9257	.9272	.9286	.9301
6.9	.9315	.9330	.9344	.9359	.9373	.9387	.9402	.9416	.9430	.9445
7.0	1.9459	.9473	.9488	.9502	.9516	.9530	.9544	.9559	.9573	.9587
7.1	.9601	.9615	.9629	.9643	.9657	.9671	.9685	.9699	.9713	.9727
7.2	.9741	.9755	.9769	.9782	.9796	.9810	.9824	.9838	.9851	.9865
7.3	.9879	.9892	.9906	.9920	.9933	.9947	.9961	.9974	.9988	2.0001
7.4	2.0015	.0028	.0042	.0055	.0069	.0082	.0096	.0109	.0122	.0136
7.5	2.0149	.0162	.0176	.0189	.0202	.0215	.0229	.0242	.0255	.0268
7.6	.0281	.0295	.0308	.0321	.0334	.0347	.0360	.0373	.0386	.0399
7.7	.0412	.0425	.0438	.0451	.0464	.0477	.0490	.0503	.0516	.0528
7.8	.0541	.0554	.0567	.0580	.0592	.0605	.0618	.0631	.0643	.0656
7.9	.0669	.0681	.0694	.0707	.0719	.0732	.0744	.0757	.0769	.0782
8.0	2.0794	.0807	.0819	.0832	.0844	.0857	.0869	.0882	.0894	.0906
8.1	.0919	.0931	.0943	.0956	.0968	.0980	.0992	.1005	.1017	.1029
8.2	.1041	.1054	.1066	.1078	.1090	.1102	.1114	.1126	.1138	.1150
8.3	.1163	.1175	.1187	.1199	.1211	.1223	.1235	.1247	.1258	.1270
8.4	.1282	.1294	.1306	.1318	.1330	.1342	.1353	.1365	.1377	.1389
8.5	2.1401	.1412	.1424	.1436	.1448	.1459	.1471	.1483	.1494	.1506
8.6	.1518	.1529	.1541	.1552	.1564	.1576	.1587	.1599	.1610	.1622
8.7	.1633	.1645	.1656	.1668	.1679	.1691	.1702	.1713	.1725	.1736
8.8	.1748	.1759	.1770	.1782	.1793	.1804	.1815	.1827	.1838	.1849
8.9	.1861	.1872	.1883	.1894	.1905	.1917	.1928	.1939	.1950	.1961
9.0	2.1972	.1983	.1994	.2006	.2017	.2028	.2039	.2050	.2061	.2072
9.1	.2083	.2094	.2105	.2116	.2127	.2138	.2148	.2159	.2170	.2181
9.2	.2192	.2203	.2214	.2225	.2235	.2246	.2257	.2268	.2279	.2289
9.3	.2300	.2311	.2322	.2332	.2343	.2354	.2364	.2375	.2386	.2396
9.4	.2407	.2418	.2428	.2439	.2450	.2460	.2471	.2481	.2492	.2502
9.5	2.2513	.2523	.2534	.2544	.2555	.2565	.2576	.2586	.2597	.2607
9.6	.2618	.2628	.2638	.2649	.2659	.2670	.2680	.2690	.2701	.2711
9.7	.2721	.2732	.2742	.2752	.2762	.2773	.2783	.2793	.2803	.2814
9.8	.2824	.2834	.2844	.2854	.2865	.2875	.2885	.2895	.2905	.2915
9.9	.2925	.2935	.2946	.2956	.2966	.2976	.2986	.2996	.3006	.3016

Table C Base-e Logarithms

N	0	1	2	3	4	5	6	7	8	9
1.0	.0000	.0100	.0198	.0296	.0392	.0488	.0583	.0677	.0770	.0862
1.1	.0953	.1044	.1133	.1222	.1310	.1398	.1484	.1570	.1655	.1740
1.2	.1823	.1906	.1989	.2070	.2151	.2231	.2311	.2390	.2469	.2546
1.3	.2624	.2700	.2776	.2852	.2927	.3001	.3075	.3148	.3221	.3293
1.4	.3365	.3436	.3507	.3577	.3646	.3716	.3784	.3853	.3920	.3988
1.5	.4055	.4121	.4187	.4253	.4318	.4383	.4447	.4511	.4574	.4637
1.6	.4700	.4762	.4824	.4886	.4947	.5008	.5068	.5128	.5188	.5247
1.7	.5306	.5365	.5423	.5481	.5539	.5596	.5653	.5710	.5766	.5822
1.8	.5878	.5933	.5988	.6043	.6098	.6152	.6206	.6259	.6313	.6366
1.9	.6419	.6471	.6523	.6575	.6627	.6678	.6729	.6780	.6831	.6881
2.0	.6931	.6981	.7031	.7080	.7129	.7178	.7227	.7275	.7324	.7372
2.1	.7419	.7467	.7514	.7561	.7608	.7655	.7701	.7747	.7793	.7839
2.2	.7885	.7930	.7975	.8020	.8065	.8109	.8154	.8198	.8242	.8286
2.3	.8329	.8372	.8416	.8459	.8502	.8544	.8587	.8629	.8671	.8713
2.4	.8755	.8796	.8838	.8879	.8920	.8961	.9002	.9042	.9083	.9123
2.5	.9163	.9203	.9243	.9282	.9322	.9361	.9400	.9439	.9478	.9517
2.6	.9555	.9594	.9632	.9670	.9708	.9746	.9783	.9821	.9858	.9895
2.7	.9933	.9969	1.0006	.0043	.0080	.0116	.0152	.0188	.0225	.0260
2.8	1.0296	.0332	.0367	.0403	.0438	.0473	.0508	.0543	.0578	.0613
2.9	.0647	.0682	.0716	.0750	.0784	.0818	.0852	.0886	.0919	.0953
3.0	1.0986	.1019	.1053	.1086	.1119	.1151	.1184	.1217	.1249	.1282
3.1	.1314	.1346	.1378	.1410	.1442	.1474	.1506	.1537	.1569	.1600
3.2	.1632	.1663	.1694	.1725	.1756	.1787	.1817	.1848	.1878	.1909
3.3	.1939	.1969	.2000	.2030	.2060	.2090	.2119	.2149	.2179	.2208
3.4	.2238	.2267	.2296	.2326	.2355	.2384	.2413	.2442	.2470	.2499
3.5	1.2528	.2556	.2585	.2613	.2641	.2669	.2698	.2726	.2754	.2782
3.6	.2809	.2837	.2865	.2892	.2920	.2947	.2975	.3002	.3029	.3056
3.7	.3083	.3110	.3137	.3164	.3191	.3218	.3244	.3271	.3297	.3324
3.8	.3350	.3376	.3403	.3429	.3455	.3481	.3507	.3533	.3558	.3584
3.9	.3610	.3635	.3661	.3686	.3712	.3737	.3762	.3788	.3813	.3838
4.0	1.3863	.3888	.3913	.3938	.3962	.3987	.4012	.4036	.4061	.4085
4.1	.4110	.4134	.4159	.4183	.4207	.4231	.4255	.4279	.4303	.4327
4.2	.4351	.4375	.4398	.4422	.4446	.4469	.4493	.4516	.4540	.4563
4.3	.4586	.4609	.4633	.4656	.4679	.4702	.4725	.4748	.4770	.4793
4.4	.4816	.4839	.4861	.4884	.4907	.4929	.4951	.4974	.4996	.5019
4.5	1.5041	.5063	.5085	.5107	.5129	.5151	.5173	.5195	.5217	.5239
4.6	.5261	.5282	.5304	.5326	.5347	.5369	.5390	.5412	.5433	.5454
4.7	.5476	.5497	.5518	.5539	.5560	.5581	.5602	.5623	.5644	.5665
4.8	.5686	.5707	.5728	.5748	.5769	.5790	.5810	.5831	.5851	.5872
4.9	.5892	.5913	.5933	.5953	.5974	.5994	.6014	.6034	.6054	.6074
5.0	1.6094	.6114	.6134	.6154	.6174	.6194	.6214	.6233	.6253	.6273
5.1	.6292	.6312	.6332	.6351	.6371	.6390	.6409	.6429	.6448	.6467
5.2	.6487	.6506	.6525	.6544	.6563	.6582	.6601	.6620	.6639	.6658
5.3	.6677	.6696	.6715	.6734	.6752	.6771	.6790	.6808	.6827	.6845
5.4	.6864	.6882	.6901	.6919	.6938	.6956	.6974	.6993	.7011	.7029

Use the properties of logarithms and ln 10 = 2.3026 to find logarithms of numbers less than 1 or greater than 10.

Answers to Selected Exercises

Orals (page 11)
5. 2, 3, 5, 7   **6.** 2, 4, 6, 8, 10   **7.** 6   **8.** 10

Exercise 1.1 (page 11)
1. $\frac{3}{4}$ **3.** $\frac{4}{5}$ **5.** $\frac{3}{20}$ **7.** $\frac{14}{9}$ **9.** 1   **11.** $\frac{22}{15}$ **13.** set
15. even   **17.** natural, 1, itself   **19.** 0   **21.** rational
23. <   **25.** ≈   **27.** 1, 2, 9   **29.** −3, 0, 1, 2, 9
31. $\sqrt{3}$ **33.** 2   **35.** 2   **37.** 9
39. ├──┼──●──┼──┼──●──┼──┼──●──┤
          0 1 2 3 4 5 6 7 8
41. ├──┼──●──┼──●──┼──┼──●──┤
         10 11 12 13 14 15 16 17 18
43. 0.875, terminating   **45.** −0.7$\overline{3}$, repeating   **47.** <
49. >   **51.** <   **53.** >   **55.** 12 < 19   **57.** −5 ≥ −6
59. −3 ≤ 5   **61.** 0 > −10
63. ←──(──────→
            3
65. ←──────]──→
                7
67. ←──[──────→
          −5
69. ←──(───)──→
          2   5
71. ←──[───]──→
         −6   9
73. ←──)───(──→
         −3   3
75. ←──]───[──→
         −6   5
77. 20   **79.** −6   **81.** 7

83. 20   **85.** 3 or −3   **87.** $x \ge 0$ **89.**
95. 99

Getting Ready (page 13)
1. 9   **2.** 9   **3.** 12   **4.** 12
5. 5   **6.** 5   **7.** 3   **8.** 6

Orals (page 23)
1. −2   **2.** −7   **3.** −28
4. 28   **5.** −3   **6.** −3

Exercise 1.2 (page 24)
1. ←───(────→   **3.** ←──(────]──→
         4   2   10
5. $45.53   **7.** absolute, common   **9.** change, add
11. negative   **13.** mean, median, mode
15. $(a \cdot b) \cdot c = a \cdot (b \cdot c)$ **17.** $a(b + c) = ab + ac$
19. 1   **21.** −8   **23.** −5   **25.** −7   **27.** 0   **29.** −12
31. 21   **33.** −2   **35.** 4   **37.** $\frac{1}{6}$ **39.** $\frac{11}{10}$ **41.** $-\frac{1}{6}$
43. $-\frac{6}{7}$ **45.** −2   **47.** $\frac{24}{25}$ **49.** 23   **51.** 0   **53.** 2
55. −13   **57.** 1   **59.** 4   **61.** −1   **63.** −4   **65.** −12
67. −20   **69.** 8   **71.** 9   **73.** 12   **75.** −8   **77.** −9
79. $-\frac{5}{4}$ **81.** $-\frac{1}{8}$ **83.** 19,900   **85.** 100.4 ft
87. comm. prop. of add.   **89.** distrib. prop.
91. additive identity prop.   **93.** mult. inverse prop.
95. assoc. prop. of add.   **97.** comm. prop. of mult.
99. assoc. prop. of add.   **101.** distrib. prop.   **103.** $62
105. +4°   **107.** 12°   **109.** 6,900 gal   **111.** +1,325 m
113. $421.88   **115.** $1,211   **117.** 80   **119.** not really
121. 30 cm

Getting Ready (page 27)
1. 4   **2.** 27   **3.** −64   **4.** 81   **5.** $\frac{1}{27}$ **6.** $-\frac{16}{625}$

Orals (page 36)
1. 16   **2.** 27   **3.** x^5 **4.** y^7 **5.** 1   **6.** x^6 **7.** a^6b^3
8. $\dfrac{b^2}{a^4}$ **9.** $\frac{1}{25}$ **10.** x^2 **11.** x^3 **12.** $\dfrac{1}{x^3}$

Exercise 1.3 (page 36)
1. 7   **3.** 1   **5.** base, exponent   **7.** x^{m+n} **9.** $x^n y^n$
11. 1   **13.** x^{m-n} **15.** $A = s^2$ **17.** $A = \frac{1}{2}bh$
19. $A = \pi r^2$ **21.** $V = lwh$ **23.** $V = Bh$ **25.** $V = \frac{1}{3}Bh$
27. base is 5, exponent is 3   **29.** base is x, exponent is 5
31. base is b, exponent is 6   **33.** base is $-mn^2$, exponent is 3
35. 9   **37.** −9   **39.** 9   **41.** $\frac{1}{25}$ **43.** $-\frac{1}{25}$ **45.** $\frac{1}{25}$

47. 1   **49.** 1   **51.** $-32x^5$ **53.** $64x^6$ **55.** x^5 **57.** k^7
59. x^{10} **61.** p^{10} **63.** a^4b^5 **65.** x^5y^4 **67.** x^{28}
69. $\dfrac{1}{b^{72}}$ **71.** $x^{12}y^8$ **73.** $\dfrac{s^3}{r^9}$ **75.** a^{20} **77.** $-\dfrac{1}{d^3}$
79. $27x^9y^{12}$ **81.** $\dfrac{1}{729}m^6n^{12}$ **83.** $\dfrac{a^{15}}{b^{10}}$ **85.** $\dfrac{a^6}{b^4}$ **87.** a^5
89. c^7 **91.** m **93.** a^4 **95.** $3m$ **97.** $\dfrac{64b^{12}}{27a^9}$
99. 1   **101.** $\dfrac{-b^3}{8a^{21}}$ **103.** $\dfrac{27}{8a^{18}}$ **105.** $\dfrac{1}{9x^3}$ **107.** a^{n-1}
109. b^{3n-9} **111.** $\dfrac{1}{a^{n+1}}$ **113.** a^{2-n} **115.** 3.462825992
117. -244.140625 **127.** 108   **129.** $-\frac{1}{216}$ **131.** $\frac{1}{324}$
133. $\frac{27}{8}$ **135.** 15 m²   **137.** 113 cm²   **139.** 45 cm²
141. 300 cm²   **143.** 343 m³   **145.** 360 ft³   **147.** 168 ft³
149. 2,714 m³   **151.** $922,824.13   **157.** $\frac{7}{12}$

Getting Ready (page 39)
1. 10   **2.** 100   **3.** 1,000   **4.** 10,000   **5.** $\frac{1}{100}$
6. $\frac{1}{10,000}$ **7.** 4,000   **8.** $\frac{7}{10,000}$

Orals (page 44)
1. 3.52×10^2 **2.** 5.13×10^3 **3.** 2×10^{-3}
4. 2.5×10^{-4} **5.** 350   **6.** 4,300   **7.** 0.27   **8.** 0.085

Exercise 1.4 (page 44)
1. 0.75   **3.** $1.\overline{4}$ **5.** 89   **7.** 10^n **9.** left
11. 3.9×10^3 **13.** 7.8×10^{-3} **15.** -4.5×10^4
17. -2.1×10^{-4} **19.** 1.76×10^7 **21.** 9.6×10^{-6}
23. 3.23×10^7 **25.** 6.0×10^{-4} **27.** 5.27×10^3
29. 3.17×10^{-4} **31.** 270   **33.** 0.00323   **35.** 796,000
37. 0.00037   **39.** 5.23   **41.** 23,650,000   **43.** 2×10^{11}
45. 1.44×10^7 **47.** 0.04   **49.** 6,000   **51.** 0.64
53. 1.2874×10^{13} **55.** 5.671×10^{10} **57.** 3.6×10^{25}
59. g, x, v, i, r   **61.** 1.19×10^8 cm/hr
63. 1.67248×10^{-18} g   **65.** 1.49×10^{10} in.
67. 5.9×10^4 mi   **69.** 3×10^8 **71.** almost 23 years
73. 2.5×10^{13} mi   **77.** 332

Getting Ready (page 46)
1. 2   **2.** 4   **3.** 3   **4.** 6

Orals (page 54)
1. $9x$ **2.** $2s^2$ **3.** no   **4.** yes   **5.** no   **6.** no   **7.** 3
8. 12   **9.** 5   **10.** 3

Exercise 1.5 (page 54)
1. -64 **3.** 1   **5.** x^8 **7.** $\dfrac{1}{8x^3}$ **9.** equation
11. equivalent   **13.** $c, \frac{b}{c}$ **15.** like   **17.** identity
19. yes   **21.** yes   **23.** 2   **25.** 25   **27.** 3   **29.** 28
31. $\frac{2}{3}$ **33.** 6   **35.** -8 **37.** $\frac{8}{3}$ **39.** yes, $8x$ **41.** no
43. yes, $-2x^2$ **45.** no   **47.** 3   **49.** 13   **51.** 9

53. -4 **55.** -6 **57.** -11 **59.** 13   **61.** -8
63. $-\frac{7}{3}$ **65.** $-\frac{2}{3}$ **67.** -2 **69.** 24   **71.** 6   **73.** 4
75. 3   **77.** 0   **79.** 6   **81.** identity   **83.** -6
85. contradiction   **87.** identity   **89.** $w = \frac{A}{l}$ **91.** $B = \frac{3V}{h}$
93. $t = \frac{I}{pr}$ **95.** $w = \frac{p-2l}{2}$ **97.** $B = \frac{2A}{h} - b$
99. $x = \frac{y-b}{m}$ **101.** $n = \frac{l-a+d}{d}$ **103.** $l = \frac{a-S+Sr}{r}$
105. $l = \frac{2S-na}{n}$ **107.** $m = Fd^2/(GM)$
109. $C = \frac{5}{9}(F - 32)$; $0°, 21.1°, 100°$
111. $n = (C - 6.50)/0.07$; 621, 1,000, 1,692.9 kwh
113. $R = E/I$; $R = 8$ ohms   **115.** $n = 360°/(180° - a)$; 8

Getting Ready (page 57)
1. 20   **2.** 72   **3.** 41   **4.** 7

Orals (page 63)
1. 100   **2.** 200   **3.** $270   **4.** $54x   **5.** 24 m²
6. $l(l-5)$m²

Exercise 1.6 (page 63)
1. $\dfrac{256x^{20}}{81}$ **3.** a^{m+1} **5.** $5x + 4$ **7.** $40x$ **9.** 180°
11. complementary   **13.** right   **15.** vertex   **17.** 32 ft
19. 7 ft, 15 ft   **21.** $355   **23.** 20%   **25.** 233%
27. 300 shares of BB, 200 shares of SS
29. 35 $15 calculators, 50 $67 calculators   **31.** 13
33. 60 mi, 120 mi   **35.** 12 m by 24 m   **37.** 156 ft by 312 ft
39. 10 ft   **41.** 72.5°   **43.** 56°   **45.** 50   **47.** 60°
49. 40°   **51.** 12 in.   **53.** 5 ft   **55.** 90 lb   **57.** 4 ft
59. $-40°$

Getting Ready (page 68)
1. $40   **2.** $0.08x   **3.** $6,000   **4.** $(15,000 - x)$
5. 200 mi   **6.** $20

Orals (page 71)
1. $90   **2.** $0.05x   **3.** $(30,000 - x)$ **4.** $20x
5. 0.08 gal   **6.** $0.04x$ gal

Exercise 1.7 (page 72)
1. 1   **3.** 8   **5.** principal, rate, time
7. value, price, number   **9.** $2,000 at 8%, $10,000 at 9%
11. 10.8%   **13.** $8,250   **15.** 25   **17.** $3\frac{1}{2}$ hr   **19.** $\frac{2}{3}$ hr
21. $\frac{1}{8}$ hr   **23.** $1\frac{1}{2}$ hr   **25.** 4 hr at each rate
27. 20 lb of 95¢ candy; 10 lb of $1.10 candy   **29.** 10 oz
31. 2 gal   **33.** 15   **35.** 30   **37.** $50   **39.** 6 in.

Chapter Summary (page 76)
1. 0, 1, 2, 4   **2.** 1, 2, 4   **3.** $-4, -\frac{2}{3}, 0, 1, 2, 4$
4. $-4, 0, 1, 2, 4$ **5.** π **6.** $-4, -\frac{2}{3}, 0, 1, 2, \pi, 4$
7. $-4, -\frac{2}{3}$ **8.** $1, 2, \pi, 4$ **9.** 2   **10.** 4   **11.** $-4, 0, 2, 4$
12. 1
13.
14.

15.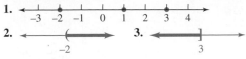

(number line with bracket at −4)

16. (number line with parentheses at −2 and 6)

17. (number line with parentheses at −2 and 3)

18. (number line with brackets at 2 and 6)

19. (number line with parenthesis at 2)

20. (number line with parenthesis at −1)

21. (number line with brackets at 0 and 2)   **22.** 0   **23.** 1

24. 8   **25.** −8   **26.** 8   **27.** −9   **28.** −28   **29.** 57
30. 3   **31.** −9   **32.** 5   **33.** −2   **34.** −13   **35.** 5
36. −13   **37.** −5   **38.** 39   **39.** 53   **40.** 35   **41.** 42
42. 4   **43.** 5   **44.** −12   **45.** −24   **46.** −4   **47.** −2
48. 12   **49.** 27   **50.** 0   **51.** −60   **52.** −1   **53.** 4
54. 15.4   **55.** 15   **56.** 15   **57.** yes   **58.** $-\frac{19}{6}$ **59.** $\frac{2}{15}$
60. $-\frac{4}{3}$ **61.** $\frac{11}{5}$ **62.** distrib. prop.
63. comm. prop. of add.   **64.** assoc. prop. of add.
65. add. identity prop.   **66.** add. inverse prop.
67. comm. prop. of mult.   **68.** assoc. prop. of mult.
69. mult. identity prop.   **70.** mult. inverse prop.
71. double negative rule   **72.** 729   **73.** −64   **74.** −64
75. $-\dfrac{1}{625}$ **76.** $-6x^6$ **77.** $-3x^8$ **78.** $\dfrac{1}{x}$ **79.** x^2
80. $27x^6$ **81.** $256x^{16}$ **82.** $-32x^{10}$ **83.** $243x^{15}$
84. $\dfrac{1}{x^{10}}$ **85.** x^{20} **86.** $\frac{x^6}{9}$ **87.** $\dfrac{16}{x^{16}}$ **88.** x^2
89. x^5 **90.** $\dfrac{1}{a^5}$ **91.** $\dfrac{1}{a^3}$ **92.** $\dfrac{1}{y^7}$ **93.** y^9 **94.** $\dfrac{1}{x}$
95. x^3 **96.** $9x^4y^6$ **97.** $\dfrac{1}{81a^{12}b^8}$ **98.** $\dfrac{64y^9}{27x^6}$ **99.** $\dfrac{64y^3}{125}$
100. 1.93×10^{10} **101.** 2.73×10^{-8} **102.** 72,000,000
103. 0.0000000083   **104.** 4.428×10^{11} **105.** 5
106. −9   **107.** 8   **108.** 7   **109.** 19   **110.** 8   **111.** 12
112. 5   **113.** $r^3 = \dfrac{3V}{4\pi}$ **114.** $h = \dfrac{3V}{\pi r^2}$
115. $x = \dfrac{6v}{ab} - y$ or $x = \dfrac{6v - aby}{ab}$ **116.** $r = \dfrac{V}{\pi h^2} + \dfrac{h}{3}$
117. 5 ft from one end   **118.** 45 m²   **119.** $2\frac{2}{3}$ ft
120. $18,000 at 10%, $7,000 at 9%   **121.** 10 liters
122. $\frac{1}{3}$ hr

Chapter 1 Test (page 81)
1. 1, 2, 5   **2.** $\sqrt{7}$
3. (number line marking −3, −1, 1, 3, 5)
4. (number line marking 2, 3, 5, 7, 9)
5. (number line with parentheses at 4)   **6.** (number line with bracket at −3)
7. (number line brackets at −2 and 4)   **8.** (number line brackets at −1 and 2)
9. −8   **10.** 5   **11.** 2   **12.** 20   **13.** −4   **14.** 1
15. 0.3   **16.** 0.5   **17.** −6   **18.** −10   **19.** 6   **20.** 1
21. comm. prop. of add.   **22.** distrib. prop.   **23.** x^8

24. $8x^6y^9$ **25.** $\dfrac{1}{m^8}$ **26.** $\dfrac{m^4}{n^{10}}$ **27.** 4.7×10^6
28. 2.3×10^{-7} **29.** 653,000   **30.** 0.0245   **31.** −12
32. 6   **33.** $i = \dfrac{f(P-L)}{s}$ **34.** 36 cm²   **35.** $4,000
36. 80 liters

Getting Ready (page 84)
1.  (number line with points at −2, −1, 1, 3)
2. (number line with parentheses at −2)   **3.** (number line with bracket at 3)
4. 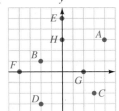 (number line with parentheses at −3 and 2)

Orals (page 95)
1. (3, 0), (0, 3)   **2.** (2, 0), (0, 6)   **3.** (8, 0), (0, 2)
4. (4, 0), (0, −3)   **5.** vertical   **6.** horizontal   **7.** (4, 6)
8. (0, −1)

Exercise 2.1 (page 95)
1. 12   **3.** $\frac{1}{2}$ **5.** 7   **7.** ordered pair   **9.** origin
11. rectangular coordinate   **13.** the y-intercept
15. vertical   **17.** sub 1
19–26.

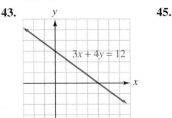

27. (2, 4)   **29.** (−2, −1)   **31.** (4, 0)   **33.** (0, 0)
35. 5, 4, 2   **37.** −5, −3, 3
39.

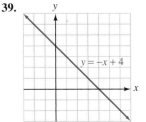

$y = -x + 4$

41.

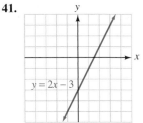

$y = 2x - 3$

43.

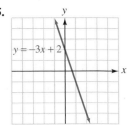

$3x + 4y = 12$

45.

(graph) $y = -3x + 2$

47.

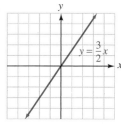

49.

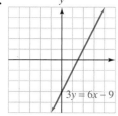

51.

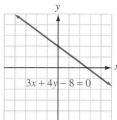

53.

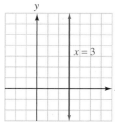

55.

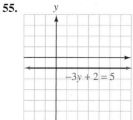

57. $(3, 4)$ **59.** $(9, 12)$
61. $\left(\frac{7}{2}, 6\right)$ **63.** $\left(\frac{1}{2}, -2\right)$
65. $(-4, 0)$ **67.** $(4, 1)$
69. 1.22   **71.** 4.67
73. \$48   **75.** \$3,000
77. \$162,500   **79.** 200
81. 12 mi

83. a. $y = 0.25x + 5$ **b.** 6, 7, 8, 9   **c.** \$10
87. $a = 0, b > 0$

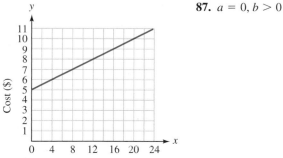

Getting Ready (page 99)
1. 1   **2.** -1 **3.** $\frac{13}{14}$ **4.** $\frac{3}{14}$

Orals (page 106)
1. 3   **2.** 2   **3.** yes   **4.** 5   **5.** yes

Exercise 2.2 (page 106)
1. $x^9 y^6$ **3.** $\frac{x^{12}}{y^8}$ **5.** 1   **7.** y, x **9.** $\frac{y_2 - y_1}{x_2 - x_1}$ **11.** run
13. vertical   **15.** parallel   **17.** 2   **19.** 3   **21.** -1
23. $-\frac{1}{3}$ **25.** 0   **27.** undefined   **29.** -1 **31.** $-\frac{3}{2}$
33. $\frac{3}{4}$ **35.** $\frac{1}{2}$ **37.** 0   **39.** negative   **41.** positive
43. undefined   **45.** perpendicular   **47.** neither
49. parallel   **51.** parallel   **53.** perpendicular   **55.** neither
57. not the same line   **59.** not the same line   **61.** same line

63. $y = 0, m = 0$ **71.** $\frac{1}{165}$ **73.** 0.04, 0.1, 0.16   **75.** $\frac{18}{5}$
77. \$20,000 per yr   **83.** 4

Getting Ready (page 109)
1. 14   **2.** $-\frac{5}{3}$ **3.** $y = 3x - 4$ **4.** $x = \frac{-By - 3}{A}$

Orals (page 118)
1. $y - 3 = 2(x - 2)$ **2.** $y - 8 = 2(x + 3)$
3. $y = -3x + 5$ **4.** $y = -3x - 7$ **5.** parallel
6. perpendicular

Exercise 2.3 (page 119)
1. 6   **3.** -1 **5.** 20 oz   **7.** $y - y_1 = m(x - x_1)$
9. $Ax + By = C$ **11.** perpendicular   **13.** $5x - y = -7$
15. $3x + y = 6$ **17.** $2x - 3y = -11$ **19.** $y = x$
21. $y = \frac{7}{3}x - 3$ **23.** $y = -\frac{9}{5}x + \frac{2}{5}$ **25.** $y = 3x + 17$
27. $y = -7x + 54$ **29.** $y = -4$ **31.** $y = -\frac{1}{2}x + 11$
33. $1, (0, -1)$ **35.** $\frac{2}{3}, (0, 2)$

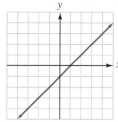

37. $-\frac{2}{3}, (0, 6)$

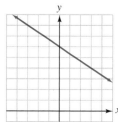

39. $\frac{3}{2}, (0, -4)$ **41.** $-\frac{1}{3}, \left(0, -\frac{5}{6}\right)$
43. $\frac{7}{2}, (0, 2)$ **45.** parallel
47. perpendicular   **49.** parallel
51. perpendicular
53. perpendicular
55. perpendicular   **57.** $y = 4x$
59. $y = 4x - 3$
61. $y = \frac{4}{5}x - \frac{26}{5}$
63. $y = -\frac{1}{4}x$ **65.** $y = -\frac{1}{4}x + \frac{11}{2}$ **67.** $y = -\frac{5}{4}x + 3$
69. perpendicular   **71.** parallel   **73.** $x = -2$ **75.** $x = 5$
77. $y = -\frac{A}{B}x + \frac{C}{B}$ **79.** $y = -2,298x + 19,984$
81. $y = 50,000x + 250,000$ **83.** $y = -\frac{950}{3}x + 1,750$
85. \$490   **87.** \$154,000   **89.** \$180   **99.** $a < 0, b > 0$

Getting Ready (page 122)
1. 1   **2.** 7   **3.** -20 **4.** $-\frac{11}{4}$

Orals (page 129)
1. yes   **2.** no   **3.** no   **4.** 1   **5.** 3   **6.** -3

Exercise 2.4 (page 129)
1. -2 **3.** 2   **5.** input   **7.** x **9.** function, input, output
11. range   **13.** 0   **15.** $y = mx + b$ **17.** yes   **19.** yes
21. yes   **23.** no   **25.** $9, -3$ **27.** $3, -5$ **29.** 22, 2
31. 3, 11   **33.** 4, 9   **35.** 7, 26   **37.** 9, 16   **39.** 6, 15

41. 4, 4 **43.** 2, 2 **45.** $\frac{1}{5}$, 1 **47.** $-2, \frac{2}{5}$
49. $2w, 2w + 2$ **51.** $3w - 5, 3w - 2$ **53.** 12
55. $2b - 2a$ **57.** $2b$ **59.** 1
61. $D = \{-2, 4, 6\}$; $R = \{3, 5, 7\}$
63. $D = (-\infty, 4) \cup (4, \infty)$; $R = (-\infty, 0) \cup (0, \infty)$
65. not a function
67. a function; $D = (-\infty, \infty)$; $R = (-\infty, \infty)$
69. $D = (-\infty, \infty)$; $R = (-\infty, \infty)$

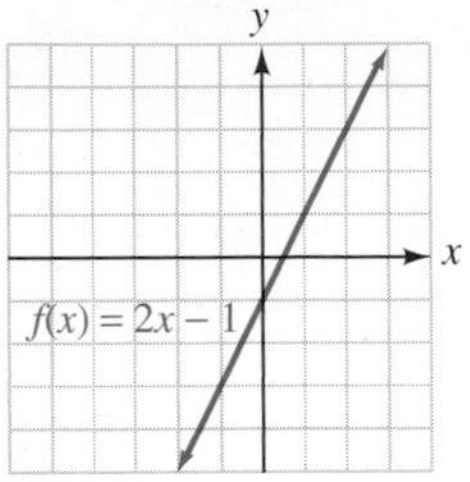

71. $D = (-\infty, \infty)$; $R = (-\infty, \infty)$

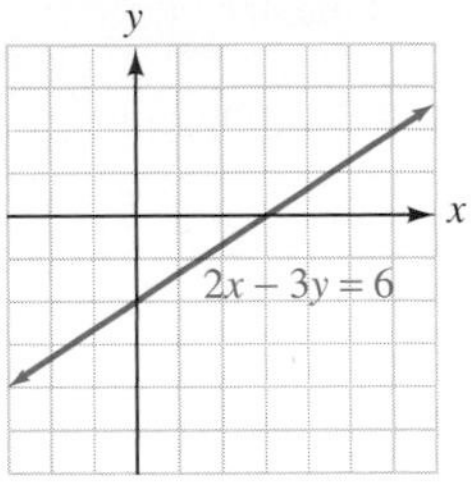

73. no **75.** yes **77.** 624 ft **79.** 1 sec **81.** 77° F
83. 192 **87.** yes

Getting Ready (page 132)

1. 2, (0, −3) **2.** −3, (0, 4) **3.** 6, −9 **4.** 4, $\frac{5}{2}$

Exercise 2.5 (page 141)

1. 41, 43, 47 **3.** $a \cdot b = b \cdot a$ **5.** 1 **7.** squaring
9. absolute value **11.** horizontal **13.** 2, down
15. 4, to the left

17.

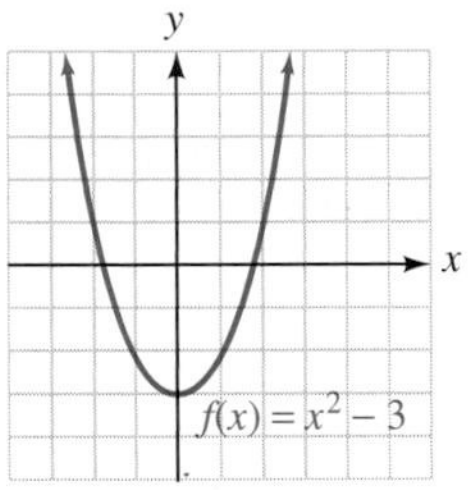

19.

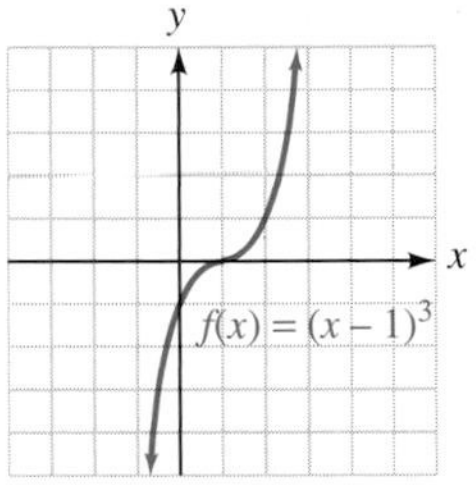

21.

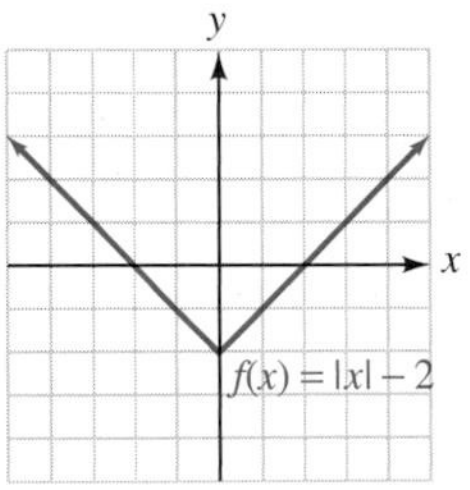

23.

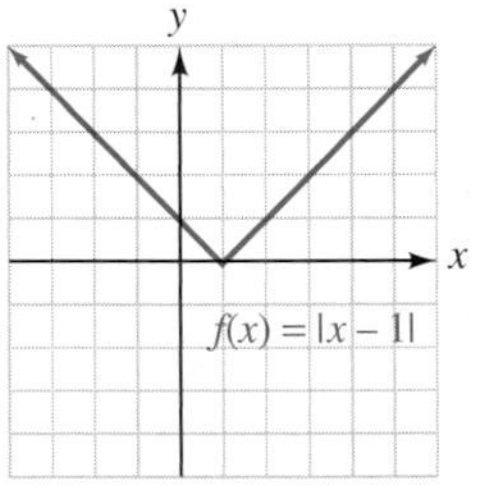

25.

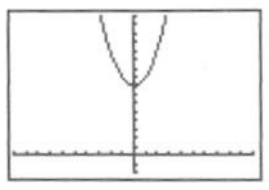

27.

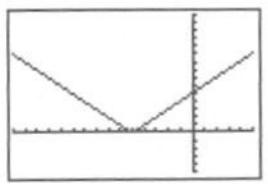

29.

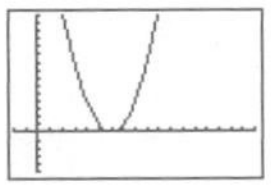

31.

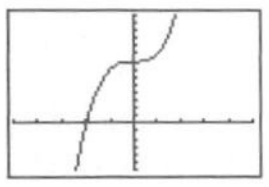

33.

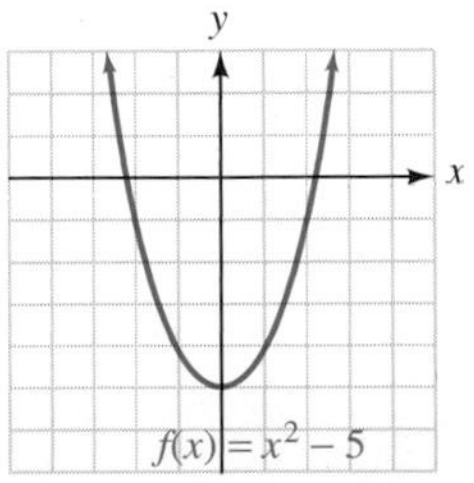

35.

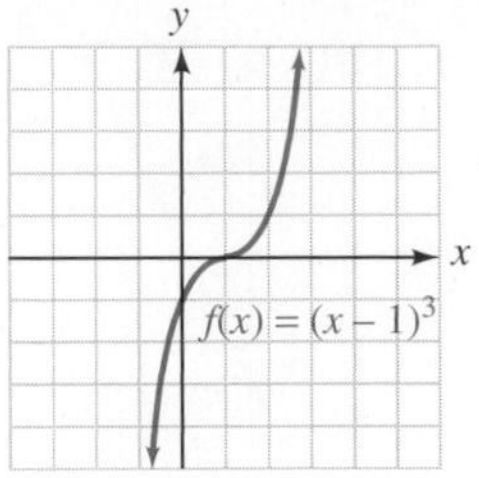

37.

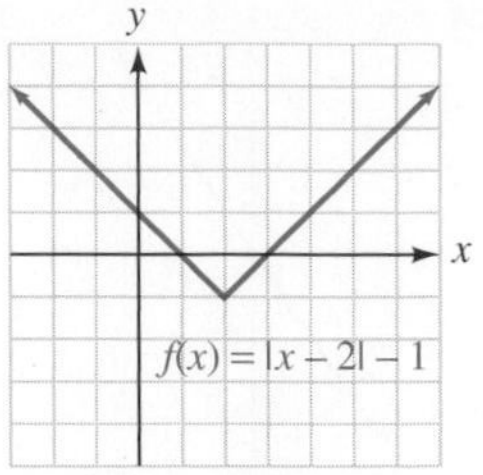

39.

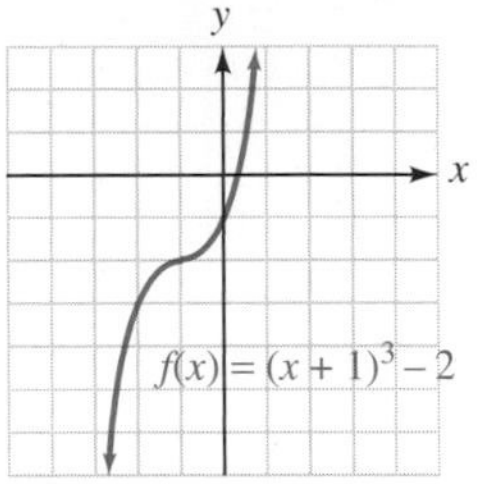

41.

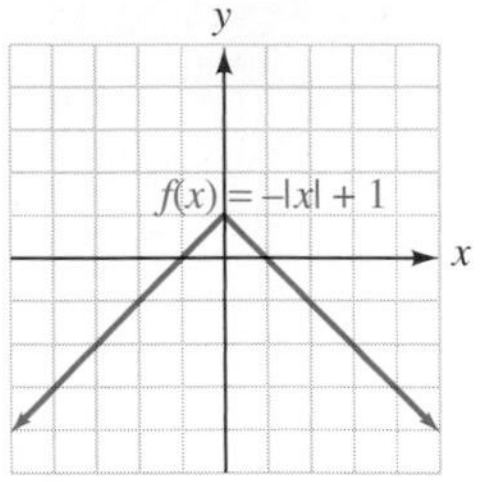

43.

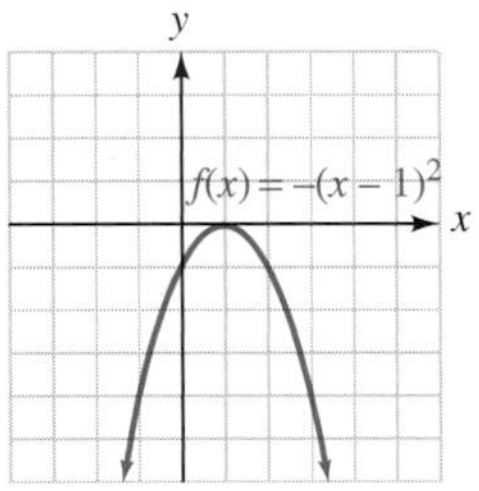

45. −2 **47.** 4 **49.** −3

Chapter Summary (page 144)

1. −10, −6, −6, 0, −2, 6, 9

2.

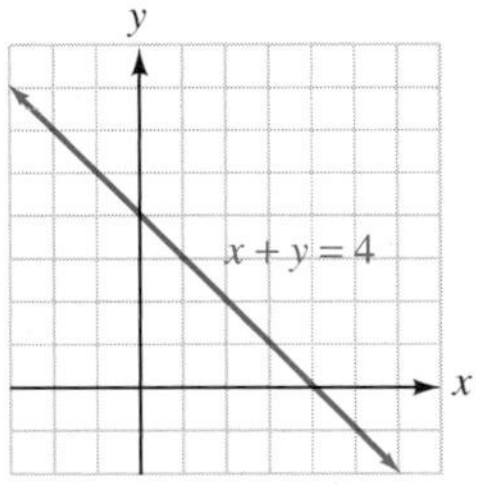

3.

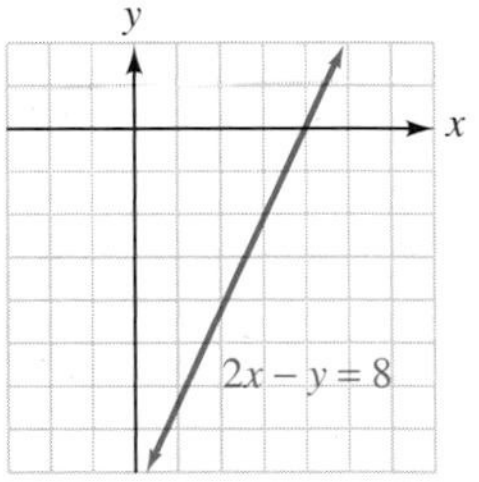

4.

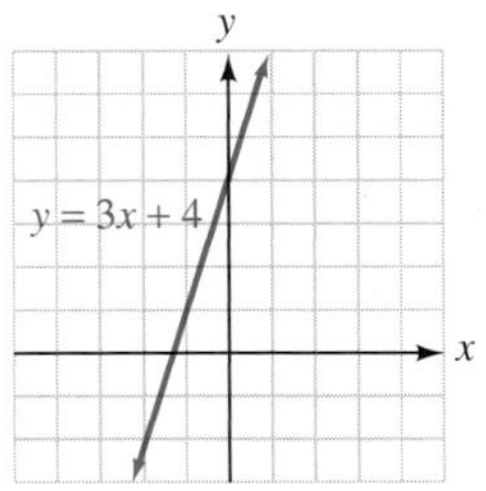

5.

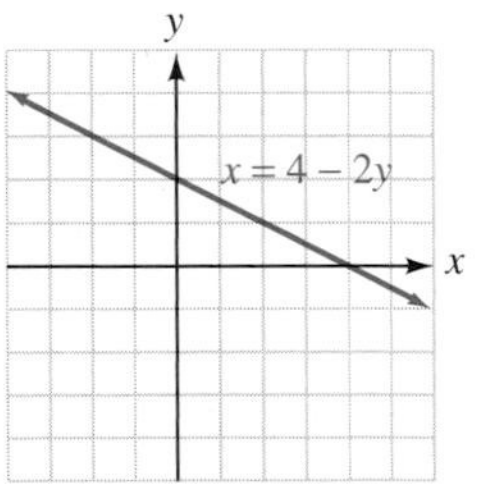

6.

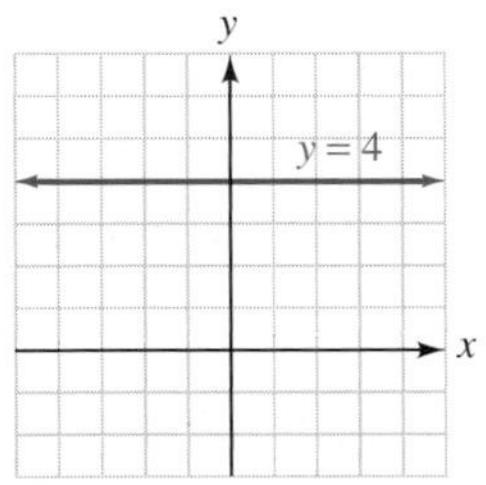

7.

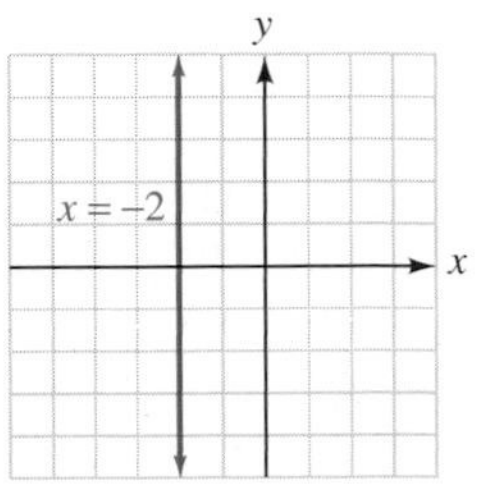

8.

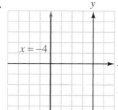

$x = -4$

9.

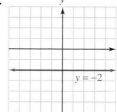

$y = -2$

10. $\left(\frac{3}{2}, 8\right)$ **11.** 1 **12.** $\frac{14}{9}$ **13.** 5 **14.** $\frac{5}{11}$ **15.** 0
16. no defined slope **17.** $\frac{2}{3}$ **18.** -2 **19.** undefined
20. 0 **21.** perpendicular **22.** parallel **23.** neither
24. perpendicular **25.** \$21,666.67 **26.** $3x - y = -29$
27. $13x + 8y = 6$ **28.** $3x - 2y = 1$ **29.** $2x + 3y = -21$
30. $y = -1,720x + 8,700$ **31.** yes **32.** yes **33.** no
34. no **35.** -7 **36.** 60 **37.** 0 **38.** 17
39. D $= (-\infty, \infty)$; R $= (-\infty, \infty)$
40. D $= (-\infty, \infty)$; R $= (-\infty, \infty)$
41. D $= (-\infty, \infty)$; R $= [1, \infty)$
42. D $= (-\infty, 2) \cup (2, \infty)$; R $= (-\infty, 0) \cup (0, \infty)$
43. D $= (-\infty, 3) \cup (3, \infty)$; R $= (-\infty, 0) \cup (0, \infty)$
44. D $= (-\infty, \infty)$; R $= \{7\}$ **45.** a function
46. not a function **47.** not a function **48.** a function
49.

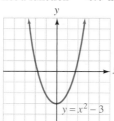

$y = x^2 - 3$

50.

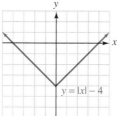

$y = |x| - 4$

51.

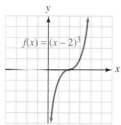

$f(x) = (x - 2)^3$

52.

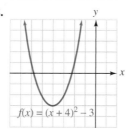

$f(x) = (x + 4)^2 - 3$

53.

54.

55.

56.

57. yes **58.** yes **59.** yes **60.** no

61.

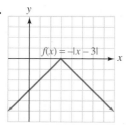

$f(x) = -|x - 3|$

Chapter 2 Test (page 148)

1.

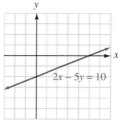

$2x - 5y = 10$

2. x-intercept $(3, 0)$,
y-intercept $\left(0, -\frac{3}{5}\right)$
3. $\left(\frac{1}{2}, \frac{1}{2}\right)$ **4.** $\frac{1}{2}$ **5.** $\frac{2}{3}$
6. undefined **7.** 0
8. $y = \frac{2}{3}x - \frac{23}{3}$
9. $8x - y = -22$
10. $m = -\frac{1}{3}, \left(0, -\frac{3}{2}\right)$
11. neither
12. perpendicular **13.** $y = \frac{3}{2}x$ **14.** $y = \frac{3}{2}x + \frac{21}{2}$
15. no **16.** D $= (-\infty, \infty)$; R $= [0, \infty)$
17. D $= (-\infty, \infty)$; R $= (-\infty, \infty)$ **18.** 10 **19.** -2
20. $3a + 1$ **21.** $x^2 - 2$ **22.** yes **23.** no
24.

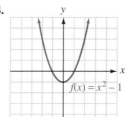

$f(x) = x^2 - 1$

25.

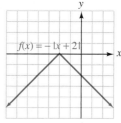

$f(x) = -|x + 2|$

Cumulative Review Exercises (page 149)

1. $1, 2, 6, 7$ **2.** $0, 1, 2, 6, 7$ **3.** $-2, 0, 1, 2, \frac{13}{12}, 6, 7$
4. $\sqrt{5}, \pi$ **5.** -2 **6.** $-2, 0, 1, 2, \frac{13}{12}, 6, 7, \sqrt{5}, \pi$
7. $2, 7$ **8.** 6 **9.** $-2, 0, 2, 6$ **10.** $1, 7$
11.

-2 5

12.

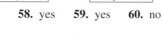

-5 0 3 6

13. -2

14. -2 **15.** 22 **16.** -2 **17.** -4 **18.** -3 **19.** 4
20. -5 **21.** assoc. prop. of add. **22.** distrib. prop.
23. comm. prop. of add. **24.** assoc. prop. of mult.
25. $x^8 y^{12}$ **26.** c^2 **27.** $-\dfrac{b^3}{a^2}$ **28.** 1 **29.** 4.97×10^{-6}
30. $932{,}000{,}000$ **31.** 8 **32.** -27 **33.** -1 **34.** 6
35. $a = \frac{2S}{n} - l$ **36.** $h = \dfrac{2A}{b_1 + b_2}$ **37.** $28, 30, 32$
38. 14 cm by 42 cm **39.** yes **40.** $-\frac{7}{5}$
41. $y = -\frac{7}{5}x + \frac{11}{5}$ **42.** $y = -3x - 3$ **43.** 5
44. -1 **45.** $2t - 1$ **46.** $3r^2 + 2$
47. yes; D $= (-\infty, \infty)$; **48.** yes; D $= (-\infty, \infty)$;
R $= (-\infty, 1]$ R $= [0, \infty)$

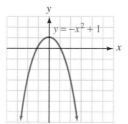

$y = -x^2 + 1$

$y = \left|\frac{1}{2}x - 3\right|$

Getting Ready (page 152)
1. 2   **2.** −7   **3.** 11   **4.** 3

Orals (page 157)
1. no solution   **2.** infinitely many   **3.** one solution
4. no solution

Exercise 3.1 (page 157)
1. 9.3×10^7 **3.** 3.45×10^4 **5.** system   **7.** inconsistent
9. dependent   **11.** yes   **13.** no
15. 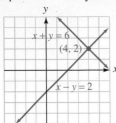   **17.**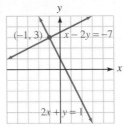

19. 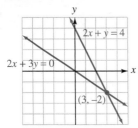   **21.**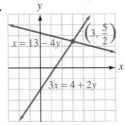

23. 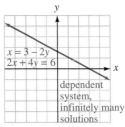   **25.**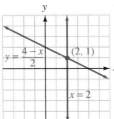

27. 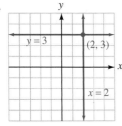   **29.**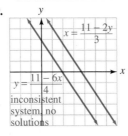

31. 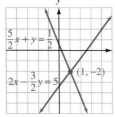   **33.**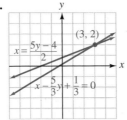

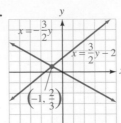
35.

37. $(-0.37, -2.69)$
39. $(-7.64, 7.04)$
41. a. \$2 million
b. \$3 million
c. 10,000 cameras
43. a. yes   **b.** $(3.75, -0.5)$
c. no

45. 3 hr
49. One possible answer is $\begin{cases} x + y = -3 \\ x - y = -7 \end{cases}$.

Getting Ready (page 160)
1. $6x - 21$ **2.** $-12x - 20$ **3.** $3x - 9$ **4.** $-2x + 2$
5. $7x = 15$ **6.** $-9b = 27$

Orals (page 171)
1. 2   **2.** 4   **3.** 4   **4.** 3

Exercise 3.2 (page 171)
1. a^{22} **3.** $\frac{1}{81x^{32}y^4}$ **5.** setup, unit   **7.** parallelogram
9. opposite   **11.** $(2, 2)$ **13.** $(5, 3)$ **15.** $(-2, 4)$
17. no solution   **19.** $\left(5, \frac{3}{2}\right)$ **21.** $\left(-2, \frac{3}{2}\right)$ **23.** $(5, 2)$
25. $(-4, -2)$ **27.** $(1, 2)$ **29.** $\left(\frac{1}{2}, \frac{2}{3}\right)$
31. dependent equations   **33.** no solution   **35.** $(4, 8)$
37. $(20, -12)$ **39.** $\left(\frac{2}{3}, \frac{3}{2}\right)$ **41.** $\left(\frac{1}{2}, -3\right)$ **43.** $\frac{1}{3}$
45. $-\frac{691}{1,980}$ **47.** $(2, 3)$ **49.** $\left(-\frac{1}{3}, 1\right)$ **51.** \$57
53. 625Ω, 750Ω **55.** 16 m by 20 m
57. \$3,000 at 10%, \$5,000 at 12%
59. 40 oz of 8% solution, 60 oz of 15% solution
61. 55 mph   **63.** 85 racing bikes, 120 mountain bikes
65. 200 plates   **67.** 21   **69.** 750
71. 6,500 gal per month   **73.** A (smaller loss)
75. 590 units per month   **77.** A (smaller loss)
79. A   **81.** 35°, 145°   **83.** $x = 22.5, y = 67.5$
85. 72°   **87.** $f^2 = \frac{1}{4\pi^2 LC}$

Getting Ready (page 175)
1. yes   **2.** yes   **3.** no   **4.** yes

Orals (page 182)
1. yes   **2.** no

Exercise 3.3 (page 182)
1. $\frac{9}{5}$ **3.** 1   **5.** $2s^2 + 1$ **7.** plane   **9.** infinitely
11. yes   **13.** $(1, 1, 2)$ **15.** $(0, 2, 2)$ **17.** $(3, 2, 1)$
19. inconsistent system, no solution   **21.** $\left(\frac{3}{4}, \frac{1}{2}, \frac{1}{3}\right)$
23. dependent equations, infinitely many solutions
25. $(2, 6, 9)$ **27.** $-2, 4, 16$
29. $A = 40°, B = 60°, C = 80°$ **31.** 1, 2, 3
33. 30 expensive, 50 middle-priced, 100 inexpensive

35. 250 \$5 tickets, 375 \$3 tickets, 125 \$2 tickets
37. 3 poles, 2 bears, 4 deer   **39.** $y = x^2 - 4x$
41. $x^2 + y^2 - 2x - 2y - 2 = 0$ **45.** $(1, 1, 0, 1)$

Getting Ready (page 185)
1. 5 8 13   **2.** 0 3 7   **3.** $-1 -1 -2$ **4.** 3 3 -5

Orals (page 191)

1. $\begin{bmatrix} 3 & 2 \\ 4 & -3 \end{bmatrix}$ **2.** $\begin{bmatrix} 3 & 2 & 8 \\ 4 & -3 & 6 \end{bmatrix}$ **3.** yes   **4.** no

Exercise 3.4 (page 192)
1. 9.3×10^7 **3.** 6.3×10^4 **5.** matrix   **7.** 3, columns
9. augmented   **11.** type 1   **13.** nonzero   **15.** 0
17. 8   **19.** $(1, 1)$ **21.** $(2, -3)$ **23.** $(0, -3)$ **25.** $(8, 8)$
27. $(1, 2, 3)$ **29.** $(-1, -1, 2)$ **31.** $(2, 1, 0)$ **33.** $(1, 2)$
35. $(2, 0)$ **37.** no solution   **39.** $(1, 2)$
41. $(-6 - z, 2 - z, z)$ **43.** $(2 - z, 1 - z, z)$ **45.** $22°, 68°$
47. $40°, 65°, 75°$ **49.** $y = 2x^2 - x + 1$ **53.** $k \neq 0$

Getting Ready (page 194)
1. -22 **2.** 22   **3.** -13 **4.** -13

Orals (page 201)

1. 1   **2.** -2 **3.** 0   **4.** $\begin{vmatrix} 1 & 2 \\ 2 & -1 \end{vmatrix}$ **5.** $\begin{vmatrix} 5 & 2 \\ 4 & -1 \end{vmatrix}$

6. $\begin{vmatrix} 1 & 5 \\ 2 & 4 \end{vmatrix}$

Exercise 3.5 (page 201)

1. -3 **3.** 0   **5.** number   **7.** $\begin{vmatrix} a_2 & c_2 \\ a_3 & c_3 \end{vmatrix}$ **9.** $\begin{vmatrix} 3 & 4 \\ 2 & -3 \end{vmatrix}$

11. 8   **13.** -2 **15.** $x^2 - y^2$ **17.** 0   **19.** -13
21. 26   **23.** 0   **25.** $10a$ **27.** 0   **29.** $(4, 2)$
31. $(-1, 3)$ **33.** $\left(-\frac{1}{2}, \frac{1}{3}\right)$ **35.** $(2, -1)$ **37.** no solution
39. $\left(5, \frac{14}{5}\right)$ **41.** $(1, 1, 2)$ **43.** $(3, 2, 1)$ **45.** no solution
47. $(3, -2, 1)$ **49.** dependent equations   **51.** $(-2, 3, 1)$
53. no solution   **55.** 2   **57.** 2
59. \$5,000 in HiTech, \$8,000 in SaveTel, \$7,000 in HiGas
61. -23 **63.** 26   **69.** -4

Chapter Summary (page 205)
1. 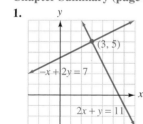   **2.**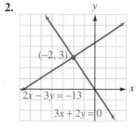

3. 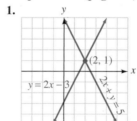   **4.**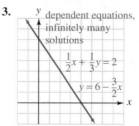

5. $(-1, 3)$ **6.** $(-3, -1)$ **7.** $(3, 4)$ **8.** $(-4, 2)$
9. $(-3, 1)$ **10.** $(1, -1)$ **11.** $(9, -4)$ **12.** $\left(4, \frac{1}{2}\right)$
13. $(1, 2, 3)$ **14.** inconsistent system   **15.** $(2, 1)$
16. $(1, 3, 2)$ **17.** $(1, 2)$ **18.** $(3z, 1 - 2z, z)$ **19.** 18
20. 38   **21.** -3 **22.** 28   **23.** $(2, 1)$ **24.** $(-1, 3)$
25. $(1, -2, 3)$ **26.** $(-3, 2, 2)$

Chapter 3 Test (page 207)
1. 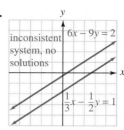   **2.** $(7, 0)$ **3.** $(2, -3)$
4. $(-6, 4)$ **5.** dependent
6. consistent   **7.** 6   **8.** -8

9. $\begin{bmatrix} 1 & 1 & 1 & 4 \\ 1 & 1 & -1 & 6 \\ 2 & -3 & 1 & -1 \end{bmatrix}$ **10.** $\begin{bmatrix} 1 & 1 & 1 \\ 1 & 1 & -1 \\ 2 & -3 & 1 \end{bmatrix}$

11. $(2, 2)$ **12.** $(-1, 3)$ **13.** 22   **14.** -17 **15.** 4
16. 13   **17.** $\begin{vmatrix} -6 & -1 \\ -6 & 1 \end{vmatrix}$ **18.** $\begin{vmatrix} 1 & -1 \\ 3 & 1 \end{vmatrix}$ **19.** -3
20. 3   **21.** 3   **22.** -1

Getting Ready (page 209)
1. 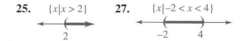   **2.** ⟵ (⟶ 4 **3.** ⟵] ⟶ 5
4. ⟵ [⟶ -1

Orals (page 217)
1. $\{x | x < 2\}, (-\infty, 2)$ **2.** $\{x | x \geq 3\}, [3, \infty)$
3. $\{x | x < -4\}, (-\infty, -4)$ **4.** $\{x | -1 < x \leq 4\}, (-1, 4]$

Exercise 4.1 (page 217)
1. $\dfrac{1}{t^{12}}$ **3.** 471 or more   **5.** $\neq$ **7.** $<$ **9.** $\geq$
11. $a < c$ **13.** reversed   **15.** $c < x, x < d$ **17.** open
19. $\{x | x < 1\}$ **21.** $\{x | x < 3\}$ **23.** $\{x | x \geq -2\}$

25. $\{x | x > 2\}$ **27.** $\{x | -2 < x < 4\}$

29. $(-3, \infty)$

31. $(-\infty, -2]$

33. $(-\infty, -2)$

35. $(-\infty, 20]$

37. $(-\infty, 10)$

39. $[-36, \infty)$

41. $(-\infty, 45/7]$

43. $(-2, 5)$

45. $(8, 11)$

47. $[-4, 6)$

49. no solution

51. $[-2, 4]$

53. $(2, 3)$

55. $[1, 9/4]$

57. $(-\infty, -15)$

59. $(-\infty, 2) \cup (7, \infty)$

61. $(-\infty, 1)$

63. no solution   **65.** 5 hr   **67.** 4

69. more than \$5,000   **71.** 18   **73.** 88 or higher   **75.** 13
77. anything over \$900   **79.** $x < 1$ **81.** $x \geq -4$
83. 139   **87.** no   **89.** a, b, c

Getting Ready (page 221)
1. 5   **2.** 0   **3.** 10   **4.** -5 **5.** $(-5, 4)$
6. $(-\infty, -3) \cup (3, \infty)$

Orals (page 231)
1. 5   **2.** -5 **3.** -6 **4.** -4 **5.** 8 or -8
6. no solution   **7.** $-8 < x < 8$ **8.** $x < -8$ or $x > 8$
9. $x \leq -4$ or $x \geq 4$ **10.** $-7 \leq x \leq 7$

Exercise 4.2 (page 231)
1. $\frac{3}{4}$ **3.** 6   **5.** $t = \frac{A - p}{pr}$ **7.** x **9.** 0   **11.** reflected
13. $a = b$ or $a = -b$ **15.** $x \leq -k$ or $x \geq k$ **17.** 8
19. -2 **21.** -30 **23.** $4 - \pi$
25.

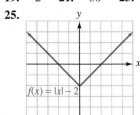

27.

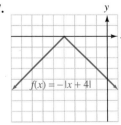

29. 4, -4 **31.** 9, -3 **33.** 4, -1 **35.** $\frac{14}{3}$, -6
37. no solution   **39.** 8, -4 **41.** 2, $-\frac{1}{2}$ **43.** -8
45. -4, -28 **47.** 0, -6 **49.** $\frac{20}{3}$ **51.** -2, $-\frac{4}{5}$
53. 3, -1 **55.** 0, -2 **57.** 0   **59.** $\frac{4}{3}$ **61.** no solution

63. $(-4, 4)$

65. $[-21, 3]$

67. no solution   **69.** $[-3/2, 2]$

71. $(-2, 5)$

73. $(-\infty, -1) \cup (1, \infty)$

75. $(-\infty, -12) \cup (36, \infty)$

77. $(-\infty, -16/3) \cup (4, \infty)$

79. $(-\infty, \infty)$

81. $(-\infty, -2] \cup [10/3, \infty)$

83. $(-\infty, -2) \cup (5, \infty)$

85. $(-\infty, 3/8) \cup (3/8, \infty)$

87. $[-10, 14]$

89. $(-5/3, 1)$

91. $(-\infty, -4] \cup [-1, \infty)$

93. no solution

95. $(-\infty, -24) \cup (-18, \infty)$

97. $(-\infty, 25) \cup (25, \infty)$

99. $[-7, -7]$

101. $[5, 5]$

103. $|x| < 4$ **105.** $|x + 3| > 6$ **107.** 4 ft $\leq d \leq$ 6 ft
109. $70° \leq t \leq 86°$ **111.** $|c - 0.6°| \leq 0.5°$
117. $k < 0$ **119.** x and y must have different signs.

Getting Ready (page 234)
1. yes   **2.** yes   **3.** no   **4.** yes

Orals (page 239)
1. yes   **2.** no   **3.** no   **4.** yes   **5.** no   **6.** no   **7.** yes
8. yes

Exercise 4.3 (page 239)
1. $(3, 1)$ **3.** $(2, -3)$ **5.** linear   **7.** edge
9.

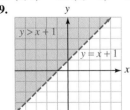

11.

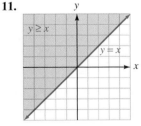

13. 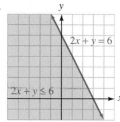   **15.**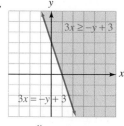

17. 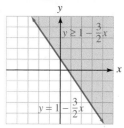   **19.**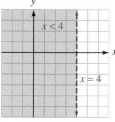

21. 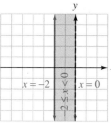   **23.**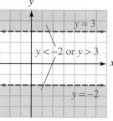

25. $3x + 2y > 6$ **27.** $x \le 3$ **29.** $y \le x$
31. $-2 \le x \le 3$ **33.** $y > -1$ or $y \le -3$
35.   **37.**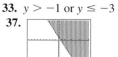

39. $(1, 1), (2, 1), (2, 2)$ **41.** $(2, 2), (3, 3), (5, 1)$

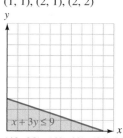   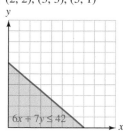

43. $(40, 20), (60, 40), (80, 20)$

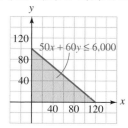

Getting Ready (page 242)
1. yes   **2.** yes   **3.** no   **4.** no

Orals (page 245)
1. yes   **2.** yes

Exercise 4.4 (page 246)
1. $r = \frac{A - p}{pt}$ **3.** $x = z\sigma + \mu$ **5.** $d = \frac{l - a}{n - 1}$ **7.** intersect

9. 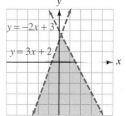   **11.**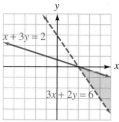

13. 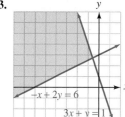   **15.**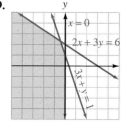

17. 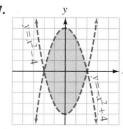   **19.**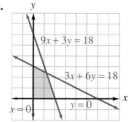

21. 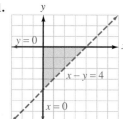   **23.**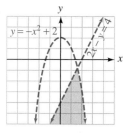

25.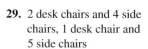

27. 1 \$10 CD and 2 \$15 CDs,   **29.** 2 desk chairs and 4 side
4 \$10 CDs and 1 \$15 CD   chairs, 1 desk chair and
5 side chairs

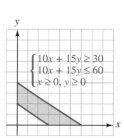   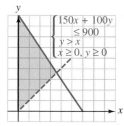

33. no

Getting Ready (page 248)

1. 0   **2.** 6   **3.** 10   **4.** 12

Orals (page 256)

1. 25   **2.** 9   **3.** (0, 0), (0, 3), (3, 0)   **4.** (0, 0), (0, 2), (4, 0)

Exercise 4.5 (page 256)

1. $\frac{3}{7}$ **3.** $y = \frac{3}{7}x + \frac{34}{7}$ **5.** constraints   **7.** objective

9. $P = 12$ at (0, 4)   **11.** $P = \frac{13}{6}$ at $\left(\frac{5}{3}, \frac{4}{3}\right)$

13. $P = \frac{18}{7}$ at $\left(\frac{3}{7}, \frac{12}{7}\right)$ **15.** $P = 3$ at (1, 0)

17. $P = 0$ at (0, 0)   **19.** $P = 0$ at (0, 0)

21. $P = -12$ at (−2, 0)   **23.** $P = -2$ at (1, 2) and (−1, 0)

25. 3 tables, 12 chairs, $1,260

27. 30 IBMs, 30 Macs, $2,700

29. 15 VCRs, 30 TVs, $1,560

31. $150,000 in stocks, $50,000 in bonds; $17,000

Chapter Summary (page 259)

1. (−∞, 3]

2. (2, ∞)

3. (−∞, −24]

4. (−∞, −51/11)

5. (−1/3, 2)

6. (2, ∞)

7. [−1, 4)

8. $20,000 or more   **9.** 7

10. 8   **11.** −7   **12.** −12

13.
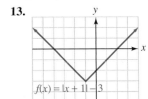
$f(x) = |x + 1| - 3$

14.
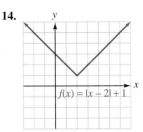
$f(x) = |x - 2| + 1$

15. 3, $-\frac{11}{3}$ **16.** $\frac{26}{3}, -\frac{10}{3}$ **17.** 14, −10   **18.** $\frac{1}{5}, -5$

19. −1, 1   **20.** $\frac{13}{12}$

21. (−5, −2)

22. [−3, 19/3]

23. no solutions   **24.** (−∞, −4) ∪ (22/5, ∞)

25. (−∞, 4/3] ∪ [4, ∞)   **26.** (−∞, ∞)

27.
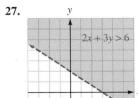
$2x + 3y > 6$
$2x + 3y = 6$

28.
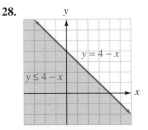
$y = 4 - x$
$y \le 4 - x$

29.
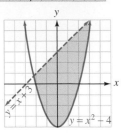
$-2 < x < 4$

30.
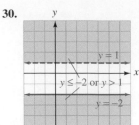
$y = 1$
$y \le -2$ or $y > 1$
$y = -2$

31.
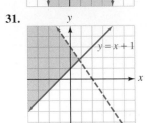
$y = x + 1$
$3x + 2y = 6$

32.
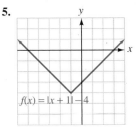
$y = x^2 - 4$

33. max. of 6 at (3, 0)   **34.** 1,000 bags of X, 1,400 bags of Y

Chapter 4 Test (page 262)

1. (−∞, −5]

2. (−2, 16)

3. 3
4. $4\pi - 4$

5.
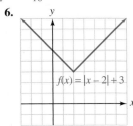
$f(x) = |x + 1| - 4$

6.
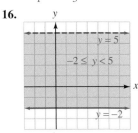
$f(x) = |x - 2| + 3$

7. 4, −7   **8.** −5, $\frac{23}{3}$ **9.** 4, −4   **10.** 0

11. [−7, 1]

12. (−∞, −9) ∪ (13, ∞)

13. (−∞, 1) ∪ (3, ∞)   **14.** [1, 3]

15.
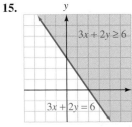
$3x + 2y \ge 6$
$3x + 2y = 6$

16.

$y = 5$
$-2 \le y < 5$
$y = -2$

17.
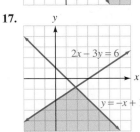
$2x - 3y = 6$
$y = -x + 1$

18.
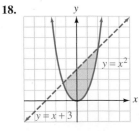
$y = x^2$
$y = x + 3$

19. $P = 2$ at (1, 1)

Cumulative Review Exercises (page 263)

1.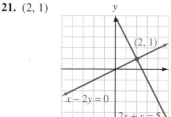
2. 5   **3.** 10   **4.** -6 **5.** x^{10}
6. x^{14} **7.** x^4 **8.** $a^{2-n} b^{n-2}$ **9.** 3.26×10^7
10. 1.2×10^{-5} **11.** $\frac{26}{3}$ **12.** 3   **13.** 6
14. a contradiction   **15.** perpendicular   **16.** parallel
17. $y = \frac{1}{3}x + \frac{11}{3}$ **18.** $h = \dfrac{2A}{b_1 + b_2}$ **19.** 10   **20.** 14
21. $(2, 1)$

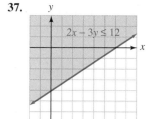

22. $(1, 1)$
23. $(2, -2)$
24. $(3, 1)$
25. $(-1, -1, 3)$
26. $(0, -1, 1)$
27. -1 **28.** 16
29. $(-1, -1)$
30. $(1, 2, -1)$
31. $x \le 11$
32. $-3 < x < 3$ **33.** $3, -\frac{3}{2}$ **34.** $-5, -\frac{3}{5}$
35. $-\frac{2}{3} \le x \le 2$ **36.** $x < -4$ or $x > 1$
37.

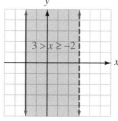

38.

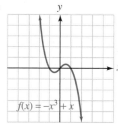

39.

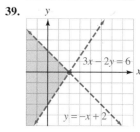

40.

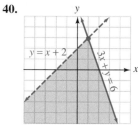

41. 24

Getting Ready (page 266)

1. $3a^2b^2$ **2.** $-5x^3y$ **3.** $4p^2 + 7q^2$ **4.** $a^3 - b^2$

Orals (page 272)

1. 3   **2.** 4   **3.** 3   **4.** 3   **5.** 1   **6.** 5   **7.** -1
8. -3

Exercise 5.1 (page 272)

1. a^5 **3.** $3y^{23}$ **5.** 1.14×10^8 **7.** sum, whole
9. binomial   **11.** one   **13.** monomial   **15.** trinomial
17. binomial   **19.** monomial   **21.** 2   **23.** 8
25. 10   **27.** 0   **29.** $-2x^4 - 5x^2 + 3x + 7$
31. $7a^3x^5 - ax^3 - 5a^3x^2 + a^2x$ **33.** $7y + 4y^2 - 5y^3 - 2y^5$
35. $2x^4y - 5x^3y^3 - 2y^4 + 5x^3y^6 + x^5y^7$ **37.** 2   **39.** 8
41. 0 ft   **43.** 64 ft   **45.** 13   **47.** 35   **49.** -90
51. -48 **53.** -34.225 **55.** 0.17283171
57. -0.12268448

59.

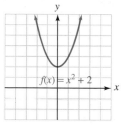

61.

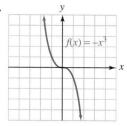

63.

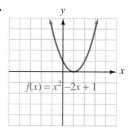

65.

67.   **69.**

71. 63 ft   **73.** 198 ft   **75.** 10 in.^{2} **77.** 18 in.^{2}
79. a. 10 ft   **b.** 35 ft   **c.** 26 ft   **83.** no   **85.** 12

Getting Ready (page 275)

1. $2x + 6$ **2.** $-8x + 20$ **3.** $5x + 15$ **4.** $-2x + 6$

Orals (page 278)

1. $9x^2$ **2.** $-2y^2$ **3.** $3x^2 + 2$ **4.** $-x^2 + 4x$
5. $3x^2 - 4x - 4$ **6.** $x^2 + 2x - 2$

Exercise 5.2 (page 278)

1. $(-\infty, 4]$ **3.** $(-1, 9)$ **5.** exponents
7. coefficients   **9.** like terms, $10x$ **11.** unlike terms
13. like terms, $-5r^2t^3$ **15.** unlike terms   **17.** $12x$
19. $2x^3y^2z$ **21.** $-7x^2y^3 + 3xy^4$ **23.** $10x^4y^2$
25. $x^2 - 5x + 6$ **27.** $-5a^2 + 4a + 4$ **29.** $-y^3 + 4y^2 + 6$
31. $4x^2 - 11$ **33.** $3x^3 - x + 13$ **35.** $4y^2 - 9y + 3$
37. $6x^3 - 6x^2 + 14x - 17$
39. $-9y^4 + 3y^3 - 15y^2 + 20y + 12$ **41.** $x^2 - 8x + 22$
43. $-3y^3 + 18y^2 - 28y + 35$ **45.** $5x - 4$ **47.** $-11t + 29$
49. $8x^3 - x^2$ **51.** $-4m - 3n$ **53.** $-16x^3 - 27x^2 - 12x$
55. $-8z^2 - 40z + 54$ **57.** $14a^2 + 16a - 24$ **59.** $3x + 3$
61. $(x^2 + 5x + 6)$ m   **63.** $136,000
65. $y = 2,500x + 275,000$ **67.** $y = -2,100x + 16,600$
69. $y = -4,800x + 35,800$ **73.** $-2x + 9$
75. $-8x^2 - 2x + 2$

Getting Ready (page 280)

1. $12a$ **2.** $4a^4$ **3.** $-7a^4$ **4.** $12a^5$ **5.** $2a + 8$
6. $a^2 - 3a$ **7.** $-4a + 12$ **8.** $-2b^2 - 4b$

Orals (page 288)

1. $-6a^3b^3$ **2.** $-8x^2y^3$ **3.** $6a^3 - 3a^2$ **4.** $-16mn^2 + 4n^3$
5. $2x^2 + 3x + 1$ **6.** $6y^2 - y - 2$

Exercise 5.3 (page 288)

1. 10 **3.** -15 **5.** \$51,025 **7.** variable **9.** term
11. $x^2 + 2xy + y^2$ **13.** $x^2 - y^2$ **15.** $-6a^3b$
17. $-15a^2b^2c^3$ **19.** $-120a^9b^3$ **21.** $25x^8y^8$ **23.** $-405x^7y^4$
25. $3x + 6$ **27.** $-a^2 + ab$ **29.** $3x^3 + 9x^2$
31. $-6x^3 + 6x^2 - 4x$ **33.** $10a^6b^4 - 25a^2b^6$
35. $7r^3st + 7rs^3t - 7rst^3$ **37.** $-12m^4n^2 - 12m^3n^3$
39. $x^2 + 5x + 6$ **41.** $z^2 - 9z + 14$ **43.** $2a^2 - 3a - 2$
45. $6t^2 + 5t - 6$ **47.** $6y^2 - 5yz + z^2$ **49.** $2x^2 + xy - 6y^2$
51. $9x^2 - 6xy - 3y^2$ **53.** $8a^2 + 14ab - 15b^2$
55. $x^2 + 4x + 4$ **57.** $a^2 - 8a + 16$ **59.** $4a^2 + 4ab + b^2$
61. $4x^2 - 4xy + y^2$ **63.** $x^2 - 4$ **65.** $a^2 - b^2$
67. $4x^2 - 9y^2$ **69.** $x^3 - y^3$ **71.** $6y^3 + 11y^2 + 9y + 2$
73. $8a^3 - b^3$ **75.** $4x^3 - 8x^2 - 9x + 6$

77. $a^3 - 3a^2b - ab^2 + 3b^3$ **79.** $2x^5 + x$ **81.** $\dfrac{3x}{y^4z} - \dfrac{x^6}{y^{10}z^2}$

83. $\dfrac{1}{x^2} - y^2$ **85.** $2x^3y^3 - \dfrac{2}{x^3y^3} + 3$ **87.** $x^{3n} - x^{2n}$

89. $x^{2n} - 1$ **91.** $x^{2n} - \dfrac{x^n}{y^n} - x^ny^n + 1$ **93.** $x^{4n} - y^{4n}$

95. $x^{2n} + x^n + x^ny^n + y^n$ **97.** $3x^2 + 12x$ **99.** $-p^2 + 4pq$
101. $m^2 - mn + 2n^2$ **103.** $3x^2 + 3x - 11$
105. $5x^2 - 36x + 7$ **107.** $21x^2 - 6xy + 29y^2$
109. $24y^2 - 4yz + 21z^2$ **111.** $9.2127x^2 - 7.7956x - 36.0315$
113. $299.29y^2 - 150.51y + 18.9225$ **117.** 15

119. $r = -\dfrac{1}{5}p^2 + 90p$ **121.** $(4x^2 - 12x + 9)$ ft^2

123. $\dfrac{1}{2}(b^2 + 3b - 10)$ in.^{2} **127.** 2.31×10^7

Getting Ready (page 291)

1. $4a + 8$ **2.** $-5b + 25$ **3.** $a^2 + 5a$ **4.** $-b^2 + 3b$
5. $6a^2 + 12ab$ **6.** $-6p^3 - 10p^2$

Orals (page 297)

1. $x(3x - 1)$ **2.** $7t^2(t + 2)$ **3.** $-3a(a + 2)$
4. $-4x(x - 3)$ **5.** $(a + b)(3 + x)$ **6.** $(m - n)(a - b)$

Exercise 5.4 (page 297)

1. $a^2 - 16$ **3.** $16r^4 - 9s^2$ **5.** $m^3 + 64$ **7.** factoring
9. greatest common factor **11.** $2 \cdot 3$ **13.** $3^3 \cdot 5$ **15.** 2^7
17. $5^2 \cdot 13$ **19.** 12 **21.** 2 **23.** $4a^2$ **25.** $6xy^2z^2$
27. 4 **29.** 1 **31.** $2(x + 4)$ **33.** $2x(x - 3)$ **35.** prime
37. $5x^2y(3 - 2y)$ **39.** $9x^2y^2(7x + 9y^2)$ **41.** prime
43. $3z(9z^2 + 4z + 1)$ **45.** $6s(4s^2 - 2st + t^2)$
47. $9x^7y^3(5x^3 - 7y^4 + 9x^3y^7)$ **49.** prime **51.** $-3(a + 2)$
53. $-x(3x + 1)$ **55.** $-3x(2x + y)$ **57.** $-6ab(3a + 2b)$
59. $-7u^2v^3z^2(9uv^3z^7 - 4v^4 + 3uz^2)$ **61.** $x^2(x^n + x^{n+1})$
63. $y^n(2y^2 - 3y^3)$ **65.** $x^{-2}(x^6 - 5x^8)$ **67.** $t^{-3}(t^8 + 4t^{-3})$
69. $4y^{-2n}(2y^{4n} + 3y^{2n} + 4)$ **71.** $(x + y)(4 + t)$
73. $(a - b)(r - s)$ **75.** $(m + n + p)(3 + x)$
77. $(x + y)(x + y + z)$ **79.** $(u + v)(u + v - 1)$
81. $-(x + y)(a - b)$ **83.** $(x + y)(a + b)$
85. $(x + 2)(x + y)$ **87.** $(3 - c)(c + d)$ **89.** $(a + b)(a - 4)$
91. $(a + b)(x - 1)$ **93.** $(x + y)(x + y + z)$
95. $x(m + n)(p + q)$ **97.** $y(x + y)(x + y + 2z)$

99. $n(2n - p + 2m)(n^2p - 1)$ **101.** $r_1 = \dfrac{rr_2}{r_2 - r}$

103. $f = \dfrac{d_1d_2}{d_2 + d_1}$ **105.** $a^2 = \dfrac{b^2x^2}{b^2 - y^2}$ **107.** $r = \dfrac{S - a}{S - l}$

109. $a = \dfrac{Hb}{2b - H}$ **111.** $y = \dfrac{x + 3}{3x - 2}$

113. $x[x(2x + 5) - 2] + 8$ **115. a.** $12x^3$ in.^{2} **b.** $20x^2$ in.^{2}
c. $4x^2(3x - 5)$ in.^{2} **121.** yes **123.** no **125.** yes

Getting Ready (page 300)

1. $a^2 - b^2$ **2.** $25p^2 - q^2$ **3.** $9m^2 - 4n^2$ **4.** $4a^4 - b^4$
5. $a^3 - 27$ **6.** $p^3 + 8$

Orals (page 305)

1. $(x + 1)(x - 1)$ **2.** $(a^2 + 4)(a + 2)(a - 2)$
3. $(x + 1)(x^2 - x + 1)$ **4.** $(a - 2)(a^2 + 2a + 4)$
5. $2(x + 2)(x - 2)$ **6.** prime

Exercise 5.5 (page 305)

1. $x^2 + 2x + 1$ **3.** $4m^2 + 4mn + n^2$ **5.** $a^2 + 7a + 12$
7. $8r^2 - 10rs + 3s^2$ **9.** 1, 4, 9, 16, 25, 36, 49, 64, 81, 100
11. cannot **13.** $(p^2 - pq + q^2)$ **15.** $(x + 2)(x - 2)$
17. $(3y + 8)(3y - 8)$ **19.** prime
21. $(25a + 13b^2)(25a - 13b^2)$ **23.** $(9a^2 + 7b)(9a^2 - 7b)$
25. $(6x^2y + 7z^2)(6x^2y - 7z^2)$ **27.** $(x + y + z)(x + y - z)$
29. $(a - b + c)(a - b - c)$ **31.** $(x^2 + y^2)(x + y)(x - y)$
33. $(16x^2y^2 + z^4)(4xy + z^2)(4xy - z^2)$ **35.** $2(x + 12)(x - 12)$
37. $2x(x + 4)(x - 4)$ **39.** $5x(x + 5)(x - 5)$
41. $t^2(rs + x^2y)(rs - x^2y)$ **43.** $(r + s)(r^2 - rs + s^2)$
45. $(x - 2y)(x^2 + 2xy + 4y^2)$
47. $(4a - 5b^2)(16a^2 + 20ab^2 + 25b^4)$
49. $(5xy^2 + 6z^3)(25x^2y^4 - 30xy^2z^3 + 36z^6)$
51. $(x^2 + y^2)(x^4 - x^2y^2 + y^4)$ **53.** $5(x + 5)(x^2 - 5x + 25)$
55. $4x^2(x - 4)(x^2 + 4x + 16)$
57. $2u^2(4v - t)(16v^2 + 4vt + t^2)$
59. $(a + b)(x + 3)(x^2 - 3x + 9)$ **61.** $(x^m + y^{2n})(x^m - y^{2n})$
63. $(10a^{2m} + 9b^n)(10a^{2m} - 9b^n)$ **65.** $(x^n - 2)(x^{2n} + 2x^n + 4)$
67. $(a^m + b^n)(a^{2m} - a^mb^n + b^{2n})$
69. $2(x^{2m} + 2y^m)(x^{4m} - 2x^{2m}y^m + 4y^{2m})$
71. $(a + b)(a - b + 1)$ **73.** $(a - b)(a + b + 2)$
75. $(2x + y)(1 + 2x - y)$ **77.** $0.5g(t_1 + t_2)(t_1 - t_2)$

79. $\dfrac{4}{3}\pi(r_1 - r_2)(r_1^2 + r_1r_2 + r_2^2)$

83. $(x^{16} + y^{16})(x^8 + y^8)(x^4 + y^4)(x^2 + y^2)(x + y)(x - y)$

Getting Ready (page 308)

1. $a^2 - a - 12$ **2.** $6a^2 + 7a - 5$ **3.** $a^2 + 3ab + 2b^2$
4. $2a^2 - ab - b^2$ **5.** $6a^2 - 13ab + 6b^2$
6. $16a^2 - 24ab + 9b^2$

Orals (page 316)

1. $(x + 2)(x + 1)$ **2.** $(x + 4)(x + 1)$ **3.** $(x - 3)(x - 2)$
4. $(x + 1)(x - 4)$ **5.** $(2x + 1)(x + 1)$ **6.** $(3x + 1)(x + 1)$

Exercise 5.6 (page 316)

1. 31   **3.** 12   **5.** -3 **7.** $2xy + y^2$ **9.** $x^2 - y^2$
11. $x + 2$ **13.** $x - 3$ **15.** $2a + 1$ **17.** $2m + 3n$
19. $(x + 1)^2$ **21.** $(a - 9)^2$ **23.** $(2y + 1)^2$ **25.** $(3b - 2)^2$
27. $(3z + 4)^2$ **29.** $(x + 8)(x + 1)$ **31.** $(x - 2)(x - 5)$
33. prime   **35.** $(x + 5)(x - 6)$ **37.** $(a + 10)(a - 5)$
39. $(y - 7)(y + 3)$ **41.** $3(x + 7)(x - 3)$
43. $b^2(a - 11)(a - 2)$ **45.** $x^2(b - 7)(b - 5)$
47. $-(a - 8)(a + 4)$ **49.** $-3(x - 3)(x - 2)$
51. $-4(x - 5)(x + 4)$ **53.** $(3y + 2)(2y + 1)$
55. $(4a - 3)(2a + 3)$ **57.** $(3x - 4)(2x + 1)$ **59.** prime
61. $(4x - 3)(2x - 1)$ **63.** $(a + b)(a - 4b)$
65. $(2y - 3t)(y + 2t)$ **67.** $x(3x - 1)(x - 3)$
69. $-(3a + 2b)(a - b)$ **71.** $-(2x - 3)^2$ **73.** $5(a - 3b)^2$
75. $z(8x^2 + 6xy + 9y^2)$ **77.** $x^2(7x - 8)(3x + 2)$
79. $(x^2 + 5)(x^2 + 3)$ **81.** $(y^2 - 10)(y^2 - 3)$
83. $(a + 3)(a - 3)(a + 2)(a - 2)$ **85.** $(z^2 + 3)(z + 2)(z - 2)$
87. $(x^3 + 3)(x + 1)(x^2 - x + 1)$ **89.** $(x^n + 1)^2$
91. $(2a^{3n} + 1)(a^{3n} - 2)$ **93.** $(x^{2n} + y^{2n})^2$
95. $(3x^n - 1)(2x^n + 3)$ **97.** $(x + 2)^2$
99. $(a + b + 4)(a + b - 6)$
101. $(3x + 3y + 4)(2x + 2y - 5)$
103. $(x + 2 + y)(x + 2 - y)$ **105.** $(x + 1 + 3z)(x + 1 - 3z)$
107. $(c + 2a - b)(c - 2a + b)$
109. $(a + 4 + b)(a + 4 - b)$ **111.** $(2x + y + z)(2x + y - z)$
113. $(a - 16)(a - 1)$ **115.** $(2u + 3)(u + 1)$
117. $(5r + 2s)(4r - 3s)$ **119.** $(5u + v)(4u + 3v)$
121. $(2x + 11)$ in.; $(2x - 1)$ in.; 12 in.   **125.** yes

Getting Ready (page 319)

1. $3pq(p - q)$ **2.** $(2p + 3q)(2p - 3q)$ **3.** $(p + 6)(p - 1)$
4. $(2p - 3q)(3p - 2q)$

Orals (page 321)

1. $(x + y)(x - y)$ **2.** $2x^3(1 - 2x)$ **3.** $(x + 2)^2$
4. $(x - 3)(x - 2)$ **5.** $(x - 2)(x^2 + 2x + 4)$
6. $(x + 2)(x^2 - 2x + 4)$

Exercise 5.7 (page 321)

1. $7a^2 + a - 7$ **3.** $4y^2 - 11y + 3$ **5.** $m^2 + 2m - 8$
7. common factors   **9.** trinomial   **11.** $(x + 4)^2$
13. $(2xy - 3)(4x^2y^2 + 6xy + 9)$ **15.** $(x - t)(y + s)$
17. $(5x + 4y)(5x - 4y)$ **19.** $(6x + 5)(2x + 7)$
21. $2(3x - 4)(x - 1)$ **23.** $(8x - 1)(7x - 1)$
25. $y^2(2x + 1)(2x + 1)$ **27.** $(x + a^2y)(x^2 - a^2xy + a^4y^2)$
29. $2(x - 3)(x^2 + 3x + 9)$ **31.** $(a + b)(f + e)$
33. $(2x + 2y + 3)(x + y - 1)$
35. $(25x^2 + 16y^2)(5x + 4y)(5x - 4y)$
37. $36(x^2 + 1)(x + 1)(x - 1)$ **39.** $2(x^2 + y^2)(x^4 - x^2y^2 + y^4)$
41. $(a + 3)(a - 3)(a + 2)(a - 2)$
43. $(x + 3 + y)(x + 3 - y)$ **45.** $(2x + 1 + 2y)(2x + 1 - 2y)$
47. $(x + y + 1)(x - y - 1)$
49. $(x + 1)(x^2 - x + 1)(x + 1)(x - 1)$
51. $(x + 3)(x - 3)(x + 2)(x^2 - 2x + 4)$
53. $2z(x + y)(x - y)^2(x^2 + xy + y^2)$ **55.** $(x^m - 3)(x^m + 2)$
57. $(a^n - b^n)(a^{2n} + a^nb^n + b^{2n})$ **59.** $\left(\frac{1}{x} + 1\right)^2$
61. $\left(\frac{3}{x} + 2\right)\left(\frac{2}{x} - 3\right)$ **67.** $(x^2 + x + 1)(x^2 - x + 1)$

Getting Ready (page 323)

1. $2a(a - 2)$ **2.** $(a + 5)(a - 5)$ **3.** $(3a - 2)(2a + 3)$
4. $a(3a - 2)(2a + 1)$

Orals (page 328)

1. 2, 3   **2.** $-4, 2$ **3.** 2, 3, -1 **4.** $-3, -2, 5, 6$

Exercise 5.8 (page 329)

1. 2, 3, 5, 7   **3.** 40,081.00 cm^3 **5.** $ax^2 + bx + c = 0$
7. 0, -2 **9.** 4, -4 **11.** 0, -1 **13.** 0, 5   **15.** $-3, -5$
17. 1, 6   **19.** 3, 4   **21.** $-2, -4$ **23.** $-\frac{1}{3}, -3$ **25.** $\frac{1}{2}, 2$
27. 1, $-\frac{1}{2}$ **29.** 2, $-\frac{1}{3}$ **31.** 3, 3   **33.** $\frac{1}{4}, -\frac{3}{2}$ **35.** 2, $-\frac{5}{6}$
37. $\frac{1}{3}, -1$ **39.** 2, $\frac{1}{2}$ **41.** $\frac{1}{5}, -\frac{5}{3}$ **43.** 0, 0, -1
45. 0, 7, -7 **47.** 0, 7, -3 **49.** 3, -3, 2, -2
51. 0, -2, -3 **53.** 0, $\frac{5}{6}, -7$ **55.** 7, -7 **57.** $\frac{1}{2}, -3$
59. 0, $-1, \frac{2}{5}$ **61.** 1, -1, -3 **63.** 3, -3, $-\frac{3}{2}$
65. 2, -2, $-\frac{1}{3}$ **67.** 16, 18, or -18, -16 **69.** 6, 7
71. 40 m   **73.** 15 ft by 25 ft   **75.** 5 ft by 12 ft
77. 20 ft by 40 ft   **79.** 10 sec   **81.** 11 sec and 19 sec
83. 50 m   **85.** 3 ft   **87.** no solution   **89.** 1
93. $x^2 - 8x + 15 = 0$ **95.** $x^2 + 5x = 0$

Chapter Summary (page 333)

1. 5   **2.** 8   **3.** 6   **4.** 9   **5.** $-t^2 - 4t + 6$
6. $-z^2 + 4z + 6$
7.

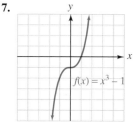

$f(x) = x^3 - 1$

8.

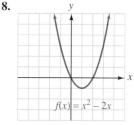

$f(x) = x^2 - 2x$

9. $5x^2 + 2x + 16$ **10.** $6x^3 + 4x^2 + x + 9$
11. $4x^2 - 9x + 19$ **12.** $-7x^3 - 30x^2 - 4x + 3$
13. $-16a^3b^3c$ **14.** $-6x^2y^2z^4$ **15.** $2x^4y^3 - 8x^2y^7$
16. $a^4b + 2a^3b^2 + a^2b^3$ **17.** $16x^2 + 14x - 15$
18. $6x^2 - 8x - 8$ **19.** $15x^2 - 22x + 8$
20. $3x^4 - 3x^3 + 4x^2 + 2x - 4$ **21.** $4(x + 2)$
22. $3x(x - 2)$ **23.** $5xy^2(xy - 2)$ **24.** $7a^3b(ab + 7)$
25. $-4x^3y^3z^2(2z^2 + 3x^2)$ **26.** $3a^2b^4c^2(4a^4 + 5c^4)$
27. $9x^2y^3z^2(3xz + 9x^2y^2 - 10z^5)$
28. $-12a^2b^3c^2(3a^3b - 5a^5b^2c + 2c^5)$ **29.** $x^n(x^n + 1)$
30. $y^{2n}(1 - y^{2n})$ **31.** $x^{-2}(x^{-2} - 1)$ **32.** $a^{-3}(a^9 + a^3)$
33. $5x^2(x + y)^3(1 - 3x^2 - 3xy)$
34. $-7a^2b^2(a - b)^3(7a^2 - 7ab - 9b^2)$ **35.** $(x + 2)(y + 4)$
36. $(a + b)(c + 3)$ **37.** $(x^2 + 4)(x^2 + y)$
38. $(a^3 + c)(a^2 + b^2)$ **39.** $h = \dfrac{S - 2wl}{2w + 2l}$ **40.** $l = \dfrac{S - 2wh}{2w + 2h}$
41. $(z + 4)(z - 4)$ **42.** $(y + 11)(y - 11)$
43. $(xy^2 + 8z^3)(xy^2 - 8z^3)$ **44.** prime
45. $(x + z + t)(x + z - t)$ **46.** $(c + a + b)(c - a - b)$
47. $2(x^2 + 7)(x^2 - 7)$ **48.** $3x^2(x^2 + 10)(x^2 - 10)$
49. $(x + 7)(x^2 - 7x + 49)$ **50.** $(a - 5)(a^2 + 5a + 25)$

51. $8(y - 4)(y^2 + 4y + 16)$ **52.** $4y(x + 3z)(x^2 - 3xz + 9z^2)$
53. $(x + 5)(x + 5)$ **54.** $(a - 7)(a - 7)$
55. $(y + 20)(y + 1)$ **56.** $(z - 5)(z - 6)$
57. $-(x + 7)(x - 4)$ **58.** $(y - 8)(y + 3)$
59. $(4a - 1)(a - 1)$ **60.** prime   **61.** prime
62. $-(5x + 2)(3x - 4)$ **63.** $y(y + 2)(y - 1)$
64. $2a^2(a + 3)(a - 1)$ **65.** $-3(x + 2)(x + 1)$
66. $4(2x + 3)(x - 2)$ **67.** $3(5x + y)(x - 4y)$
68. $5(6x + y)(x + 2y)$ **69.** $(8x + 3y)(3x - 4y)$
70. $(2x + 3y)(7x - 4y)$ **71.** $x(x - 1)(x + 6)$
72. $3y(x + 3)(x - 7)$ **73.** $(z - 2)(z + x + 2)$
74. $(x + 1 + p)(x + 1 - p)$ **75.** $(x + 2 + 2p^2)(x + 2 - 2p^2)$
76. $(y + 2)(y + 1 + x)$ **77.** $(x^m + 3)(x^m - 1)$
78. $\left(\frac{1}{x} - 2\right)\left(\frac{1}{x} + 1\right)$ **79.** $0, \frac{3}{4}$ **80.** $6, -6$ **81.** $\frac{1}{2}, -\frac{5}{6}$
82. $\frac{2}{7}, 5$ **83.** $0, -\frac{2}{3}, \frac{4}{5}$ **84.** $-\frac{2}{3}, 7, 0$ **85.** 7 cm
86. 17 m by 20 m

Chapter 5 Test (page 338)
1. 5   **2.** 13   **3.** -9 **4.** -6
5.

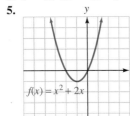

$f(x) = x^2 + 2x$

6. $5y^2 + y - 1$
7. $-4u^2 + 2u - 14$
8. $10a^2 + 22$
9. $-x^2 + 15x - 2$
10. $-6x^4yz^4$
11. $-15a^3b^4 + 10a^3b^5$
12. $z^2 - 16$
13. $12x^2 + x - 6$
14. $3xy(y + 2x)$
15. $3abc(4a^2b - abc + 2c^2)$ **16.** $y^n(x^2y^2 + 1)$
17. $b^n(a^n - ab^{-2n})$ **18.** $(u - v)(r + s)$ **19.** $(a - y)(x + y)$
20. $(x + 7)(x - 7)$ **21.** $2(x + 4)(x \quad 4)$
22. $4(y^2 + 4)(y + 2)(y - 2)$ **23.** $(b + 5)(b^2 - 5b + 25)$
24. $(b - 3)(b^2 + 3b + 9)$ **25.** $3(u - 2)(u^2 + 2u + 4)$
26. $(a - 6)(a + 1)$ **27.** $(3b + 2)(2b - 1)$
28. $3(u + 2)(2u - 1)$ **29.** $5(4r + 1)(r - 1)$ **30.** $(x^n + 1)^2$
31. $(x + 3 + y)(x + 3 - y)$ **32.** $r = \dfrac{r_1 r_2}{r_2 + r_1}$ **33.** $6, -1$
34. 25   **35.** 11 ft by 22 ft

Getting Ready (page 340)
1. $\frac{3}{4}$ **2.** $\frac{4}{5}$ **3.** $-\frac{5}{13}$ **4.** $\frac{7}{9}$

Orals (page 348)
1. -1 **2.** 1   **3.** $(-\infty, 2) \cup (2, \infty)$
4. $(-\infty, -3) \cup (-3, 3) \cup (3, \infty)$ **5.** $\frac{5}{6}$ **6.** $\frac{x}{y}$
7. 2   **8.** -1

Exercise 6.1 (page 349)
1. $3x(x - 3)$ **3.** $(3x^2 + 4y)(9x^4 - 12x^2y + 16y^2)$
5. rational   **7.** asymptote   **9.** a **11.** $\frac{a}{b}, 0$ **13.** 20 hr
15. 12 hr   **17.** $5,555.56   **19.** $50,000
21. $c = f(x) = 1.25x + 700$ **23.** $1,325   **25.** $1.95
27. $c = f(n) = 0.09n + 7.50$ **29.** $77.25   **31.** 9.75¢
33. almost 8 days   **35.** about 2.55 hr

37. $(-\infty, 2) \cup (2, \infty)$ **39.** $(-\infty, -2) \cup (-2, 2) \cup (2, \infty)$

41. $\frac{2}{3}$ **43.** $-\frac{28}{9}$ **45.** $\frac{12}{13}$ **47.** $-\frac{122}{37}$ **49.** $4x^2$
51. $-\dfrac{4y}{3x}$ **53.** x^2 **55.** $-\dfrac{x}{2}$ **57.** $\dfrac{3y}{7(y - z)}$ **59.** 1
61. $\dfrac{1}{x - y}$ **63.** $\dfrac{5}{x - 2}$ **65.** $\dfrac{-3(x + 2)}{x + 1}$ **67.** 3
69. $x + 2$ **71.** $\dfrac{x + 1}{x + 3}$ **73.** $\dfrac{m - 2n}{n - 2m}$ **75.** $\dfrac{x + 4}{2(2x - 3)}$
77. $\dfrac{3(x - y)}{x + 2}$ **79.** $\dfrac{2x + 1}{2 - x}$ **81.** $\dfrac{a^2 - 3a + 9}{4(a - 3)}$
83. in lowest terms   **85.** $m + n$ **87.** $-\dfrac{m + n}{2m + n}$
89. $\dfrac{x - y}{x + y}$ **91.** $\dfrac{2a - 3b}{a - 2b}$ **93.** $\dfrac{1}{x^2 + xy + y^2 - 1}$
95. $\dfrac{x - y}{x + y}$ **97.** $\dfrac{(x + 1)^2}{(x - 1)^3}$ **99.** $\dfrac{3a + b}{y + b}$ **101.** $\dfrac{1}{6}$
103. 0   **105.** $\dfrac{1}{2}$ **107.** $\dfrac{1}{13}$ **109. a.** $33,333.33,
b. $116,666.67   **111.** $\dfrac{1}{6,000,000}$ **113.** $\dfrac{1}{8}$ **117.** yes
119. a, d

Getting Ready (page 352)
1. $\frac{7}{3}$ **2.** 1.44×10^9 **3.** linear function
4. rational function

Orals (page 361)
1. 1   **2.** 9   **3.** $\frac{14}{5}$ **4.** $a = kb$ **5.** $a = \frac{k}{b}$ **6.** $a = kbc$
7. $a = \frac{kb}{c}$

Exercise 6.2 (page 361)
1. x^{10} **3.** -1 **5.** 3.5×10^4 **7.** 0.0025
9. unit costs, rates   **11.** extremes, means   **13.** direct
15. rational   **17.** joint   **19.** direct   **21.** neither   **23.** 3
25. 5   **27.** -3 **29.** 5   **31.** $4, -1$ **33.** $2, -2$ **35.** 39
37. $-\frac{5}{2}, -1$ **39.** no solution   **41.** $A = kp^2$ **43.** $v = k/r^3$
45. $B = kmn$ **47.** $P = ka^2/j^3$
49. L varies jointly with m and n
51. E varies jointly with a and the square of b
53. X varies directly with x^2 and inversely with y^2
55. R varies directly with L and inversely with d^2 **57.** $62.50
59. $7\frac{1}{2}$ gal   **61.** 32 ft   **63.** 80 ft   **65.** 0.18 g   **67.** 42 ft
69. $46\frac{7}{8}$ ft   **71.** 6,750 ft   **73.** 36π in.^{2} **75.** 432 mi
77. 25 days   **79.** 12 in.^{3} **81.** 85.3   **83.** 12
85. 26,437.5 gal   **87.** 3 ohms   **89.** 0.275 in.
91. 546 Kelvin

Getting Ready (page 365)

1. $\frac{3}{2}$ **2.** $\frac{2}{5}$ **3.** $\frac{4}{3}$ **4.** $\frac{14}{3}$

Orals (page 371)

1. $\frac{9}{8}$ **2.** $\frac{1}{2}$ **3.** $\frac{x-2}{x+2}$ **4.** $\frac{9}{16}$ **5.** 5   **6.** $\frac{y}{2}$

Exercise 6.3 (page 371)

1. $-6a^5 + 2a^4$ **3.** $m^{2n} - 4$ **5.** $\frac{ac}{bd}$ **7.** 0   **9.** $\frac{10}{7}$

11. $-\frac{5}{6}$ **13.** $\frac{xy^2 d}{c^3}$ **15.** $-\frac{x^{10}}{y^2}$ **17.** $x+1$ **19.** 1

21. $\frac{x-4}{x+5}$ **23.** $\frac{(a+7)^2(a-5)}{12x^2}$ **25.** $\frac{t-1}{t+1}$ **27.** $\frac{n+2}{n+1}$

29. $\frac{1}{x+1}$ **31.** $(x+1)^2$ **33.** $x-5$ **35.** $\frac{x+y}{x-y}$

37. $-\frac{x+3}{x+2}$ **39.** $\frac{a+b}{(x-3)(c+d)}$ **41.** $-\frac{x+1}{x+3}$

43. $-\frac{2x+y}{3x+y}$ **45.** $-\frac{x^7}{18y^4}$ **47.** $x^2(x+3)$ **49.** $\frac{3x}{2}$

51. $\frac{t+1}{t}$ **53.** $\frac{x-2}{x-1}$ **55.** $\frac{x+2}{x-2}$ **57.** $\frac{x-1}{3x+2}$

59. $\frac{x-7}{x+7}$ **61.** 1   **63.** $\frac{(b-2)(b-3)}{2}$ cm^2

65. $\frac{(k+2)(k+4)}{k+1}$ **69.** $\div$, $\cdot$

Getting Ready (page 373)

1. yes   **2.** no   **3.** yes   **4.** yes   **5.** no   **6.** yes

Orals (page 380)

1. x **2.** $\frac{a}{2}$ **3.** 1   **4.** 1   **5.** $\frac{7x}{6}$ **6.** $\frac{5y-3x}{xy}$

Exercise 6.4 (page 380)

1.

$(-1, 4]$

3. $w = \frac{P-2l}{2}$ **5.** $\frac{a+c}{b}$

7. subtract, keep   **9.** LCD   **11.** $\frac{5}{2}$ **13.** $-\frac{1}{3}$ **15.** $\frac{11}{4y}$

17. $\frac{3-a}{a+b}$ **19.** 2   **21.** 3   **23.** 3   **25.** $\frac{6x}{(x-3)(x-2)}$

27. 72   **29.** $x(x+3)(x-3)$ **31.** $(x+3)^2(x^2-3x+9)$

33. $(2x+3)^2(x+1)^2$ **35.** $\frac{5}{6}$ **37.** $-\frac{16}{75}$ **39.** $\frac{9a}{10}$

41. $\frac{21a-8b}{14}$ **43.** $\frac{17}{12x}$ **45.** $\frac{9a^2-4b^2}{6ab}$ **47.** $\frac{10a+4b}{21}$

49. $\frac{8x-2}{(x+2)(x-4)}$ **51.** $\frac{7x+29}{(x+5)(x+7)}$ **53.** $\frac{x^2+1}{x}$

55. 2   **57.** $\frac{9a^2-11a-10}{(3a+2)(3a-2)}$ **59.** $\frac{2x^2+x}{(x+3)(x+2)(x-2)}$

61. $\frac{-x^2+11x+8}{(3x+2)(x+1)(x-3)}$ **63.** $\frac{-4x^2+14x+54}{x(x+3)(x-3)}$

65. $\frac{x^3+x^2-1}{x(x+1)(x-1)}$ **67.** $\frac{2x^2+5x+4}{x+1}$ **69.** $\frac{x^2-5x-5}{x-5}$

71. $\frac{-x^3-x^2+5x}{x-1}$ **73.** $\frac{-y^2+48y+161}{(y+4)(y+3)}$ **75.** $\frac{2}{x+1}$

77. $\frac{3x+1}{x(x+3)}$ **79.** $\frac{2x^3+x^2-43x-35}{(x+5)(x-5)(2x+1)}$

81. $\frac{3x^2-2x-17}{(x-3)(x-2)}$ **83.** $\frac{2a}{a-1}$ **85.** $\frac{x^2-6x-1}{2(x+1)(x-1)}$

87. $\frac{2b}{a+b}$ **89.** $\frac{7mn^2-7n^3-6m^2+3mn+n}{(m-n)^2}$

91. $\frac{2}{m-1}$ **93.** 0   **95.** 1   **99.** $\frac{11x^2+4x+12}{2x}$ ft

101. $\frac{7x^2-11x+12}{4x}$

Getting Ready (page 383)

1. 3   **2.** -34 **3.** $3+2a$ **4.** $5-2b$

Orals (page 390)

1. $\frac{3}{5}$ **2.** $\frac{a}{d}$ **3.** 2   **4.** $\frac{x+y}{x-y}$ **5.** $\frac{x-y}{x}$ **6.** $\frac{b+a}{a}$

Exercise 6.5 (page 390)

1. 8   **3.** $2, -2, 3, -3$ **5.** complex   **7.** $\frac{2}{3}$ **9.** -1

11. $\frac{10}{3}$ **13.** $-\frac{1}{7}$ **15.** $\frac{2y}{3z}$ **17.** $125b$ **19.** $-\frac{1}{y}$

21. $\frac{y-x}{x^2y^2}$ **23.** $\frac{b+a}{b}$ **25.** $\frac{y+x}{y-x}$ **27.** $y-x$

29. $\frac{-1}{a+b}$ **31.** x^2+x-6 **33.** $\frac{5x^2y^2}{xy+1}$ **35.** $\frac{x+2}{x-3}$

37. $\frac{a-1}{a+1}$ **39.** $\frac{y+x}{x^2y}$ **41.** $\frac{xy^2}{y-x}$ **43.** $\frac{y+x}{y-x}$ **45.** xy

47. $\frac{x^2(xy^2-1)}{y^2(x^2y-1)}$ **49.** $\frac{(b+a)(b-a)}{b(b-a-ab)}$ **51.** $\frac{x-1}{x}$

53. $\frac{5b}{5b+4}$ **55.** $\frac{3a^2+2a}{2a+1}$ **57.** $\frac{(-x^2+2x-2)(3x+2)}{(2-x)(-3x^2-2x+9)}$

59. $\frac{2(x^2-4x-1)}{-x^2+4x+8}$ **61.** $\frac{-2}{x^2-3x-7}$ **63.** $\frac{k_1k_2}{k_2+k_1}$

65. $\frac{4k}{9}$ **69.** $\frac{1}{y+x}$

Getting Ready (page 393)

1. $\frac{9}{2}$ **2.** $3, -4$

Orals (page 400)

1. 2   **2.** 3   **3.** 1   **4.** 3   **5.** 1   **6.** 2

Exercise 6.6 (page 400)

1. $\frac{n^6}{m^4}$ **3.** 0   **5.** rational   **7.** 12   **9.** 40   **11.** $\frac{1}{2}$

13. no solution   **15.** $\frac{17}{25}$ **17.** -2 is extraneous.   **19.** 0

21. 1   **23.** 2   **25.** 2   **27.** $\frac{1}{3}$ **29.** 0   **31.** $2, -5$

33. $-4, 3$ **35.** $6, \frac{17}{3}$ **37.** $1, -11$ **39.** $f = \frac{pq}{q+p}$

41. $r = \frac{S-a}{S-l}$ **43.** $R = \frac{r_1r_2r_3}{r_1r_3+r_1r_2+r_2r_3}$ **45.** $4\frac{8}{13}$ in.

47. $1\frac{7}{8}$ days   **49.** $5\frac{5}{6}$ min   **51.** $1\frac{1}{5}$ days   **53.** $2\frac{4}{13}$ weeks
55. 3 mph   **57.** 60 mph and 40 mph   **59.** 3 mph
61. 60 mph   **63.** 7   **65.** 12 days

Getting Ready (page 403)
1. $\frac{2}{3}$ **2.** $-\frac{5}{9}$ **3.** 5   **4.** 16

Orals (page 408)
1. $3xy$ **2.** $2b + 4a$ **3.** $x + 1$ **4.** $x + 2$

Exercise 6.7 (page 408)

1. $8x^2 + 2x + 4$ **3.** $-2y^3 - 3y^2 + 6y - 6$ **5.** $\dfrac{1}{b}$

7. quotient   **9.** $\dfrac{y}{2x^3}$ **11.** $\dfrac{3b^4}{4a^4}$ **13.** $\dfrac{-5}{7xy^7t^2}$

15. $\dfrac{13a^n b^{2n} c^{3n-1}}{3}$ **17.** $\dfrac{2x}{3} - \dfrac{x^2}{6}$ **19.** $\dfrac{2xy^2}{3} + \dfrac{x^2y}{6}$

21. $\dfrac{x^4 y^4}{2} - \dfrac{x^3 y^9}{4} + \dfrac{3}{4xy^2}$ **23.** $\dfrac{b^2}{4a} - \dfrac{a^3}{2b^4} + \dfrac{3}{4ab}$

25. $1 - 3x^n y^n + 6x^{2n} y^{2n}$ **27.** $x + 2$ **29.** $x + 7$

31. $3x - 5 + \dfrac{3}{2x + 3}$ **33.** $3x^2 + x + 2 + \dfrac{8}{x - 1}$

35. $2x^2 + 5x + 3 + \dfrac{4}{3x - 2}$ **37.** $3x^2 + 4x + 3$

39. $a + 1$ **41.** $2y + 2$ **43.** $6x - 12$ **45.** $3x^2 - x + 2$

47. $4x^3 - 3x^2 + 3x + 1$ **49.** $a^2 + a + 1 + \dfrac{2}{a - 1}$

51. $5a^2 - 3a - 4$ **53.** $6y - 12$
55. $16x^4 - 8x^3 y + 4x^2 y^2 - 2xy^3 + y^4$ **57.** $x^4 + x^2 + 4$
59. $x^2 + x + 1$ **61.** $x^2 + x + 2$

63. $9.8x + 16.4 + \dfrac{-36.5}{x - 2}$ **65.** $3x + 5$ **69.** It is.

Chapter Summary (page 411)
1.

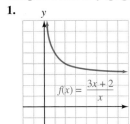

horizontal asymptote: $y = 3$;
vertical asymptote: $x = 0$
2. $\dfrac{31x}{72y}$ **3.** $\dfrac{53m}{147n^2}$ **4.** $\dfrac{x - 7}{x + 7}$

5. $\dfrac{1}{x - 6}$ **6.** $\dfrac{1}{2x + 4}$ **7.** -1

8. -2 **9.** $\dfrac{-a - b}{c + d}$ **10.** $\dfrac{1}{18}$

11. $\dfrac{1}{8}$ **12.** 5   **13.** $-4, -12$

14. 70.4 ft   **15.** 72   **16.** 6   **17.** 2   **18.** 16   **19.** 1

20. $\dfrac{3x(x - 1)}{(x - 3)(x + 1)}$ **21.** 1   **22.** $\dfrac{5y - 3}{x - y}$ **23.** $\dfrac{6x - 7}{x^2 + 2}$

24. $\dfrac{5x + 13}{(x + 2)(x + 3)}$ **25.** $\dfrac{4x^2 + 9x + 12}{(x - 4)(x + 3)}$

26. $\dfrac{5x^2 + 11x}{(x + 1)(x + 2)}$ **27.** $\dfrac{2(3x + 1)}{x - 3}$

28. $\dfrac{5x^2 + 23x + 4}{(x + 1)(x - 1)(x - 1)}$ **29.** $\dfrac{x^2 + 26x + 3}{(x + 3)(x - 3)^2}$

30. $\dfrac{3y - 2x}{x^2 y^2}$ **31.** $\dfrac{y + 2x}{2y - x}$ **32.** $\dfrac{2x + 1}{x + 1}$ **33.** $\dfrac{3x + 2}{3x - 2}$

34. $\dfrac{x - 2}{x + 3}$ **35.** $\dfrac{1}{x}$ **36.** $\dfrac{y - x}{y + x}$ **37.** $\dfrac{x^2 y^2}{(x - y)^2 (y^2 - x^2)}$

38. 5   **39.** $-1, -2$ **40.** 2 and -2 are extraneous.

41. $-1, -12$ **42.** $y^2 = \dfrac{x^2 b^2 - a^2 b^2}{a^2}$ **43.** $b = \dfrac{Ha}{2a - H}$

44. 50 mph   **45.** 200 mph   **46.** $14\frac{2}{5}$ hr   **47.** $18\frac{2}{3}$ days

48. $-\dfrac{x^3}{2y^3}$ **49.** $-3x^2 y + \dfrac{3x}{2} + y$ **50.** $x + 5y$

51. $x^2 + 2x - 1 + \dfrac{6}{2x + 3}$

Chapter 6 Test (page 415)

1. $\dfrac{-2}{3xy}$ **2.** $\dfrac{2}{x - 2}$ **3.** -3 **4.** $\dfrac{2x + 1}{4}$ **5.** $\dfrac{1}{8}$ **6.** 18 ft

7. $6, -1$ **8.** $\dfrac{44}{3}$ **9.** $\dfrac{xz}{y^4}$ **10.** $\dfrac{x + 1}{2}$ **11.** 1

12. $\dfrac{(x + y)^2}{2}$ **13.** $\dfrac{2}{x + 1}$ **14.** -1 **15.** 3   **16.** 2

17. $\dfrac{2s + r^2}{rs}$ **18.** $\dfrac{2x + 3}{(x + 1)(x + 2)}$ **19.** $\dfrac{u^2}{2vw}$ **20.** $\dfrac{2x + y}{xy - 2}$

21. $\dfrac{5}{2}$ **22.** 5; 3 is extraneous   **23.** $a^2 = \dfrac{x^2 b^2}{b^2 - y^2}$

24. $r_2 = \dfrac{rr_1}{r_1 - r}$ **25.** 10 days

26. \$5,000 at 6% and \$3,000 at 10%   **27.** $\dfrac{-6x}{y} + \dfrac{4x^2}{y^2} - \dfrac{3}{y^3}$

28. $3x^2 + 4x + 2$

Cumulative Review Exercises (page 416)
1. $a^8 b^4$ **2.** $\dfrac{b^4}{a^4}$ **3.** $\dfrac{81b^{16}}{16a^8}$ **4.** $x^{21} y^3$ **5.** 42,500
6. 0.000712   **7.** 1   **8.** -2 **9.** $\dfrac{5}{6}$ **10.** $-\dfrac{3}{4}$ **11.** 3
12. $-\dfrac{1}{3}$ **13.** 0   **14.** 8   **15.** $-\dfrac{16}{25}$ **16.** $t^2 - 4t + 3$

17. $y = \dfrac{kxz}{r}$ **18.** no

19. $\left[\dfrac{5}{2}, \infty\right)$ ⟵━━━━[━━➤
5/2

20. $(-\infty, 3] \cup \left[\dfrac{11}{3}, \infty\right)$ ⟵━]━━━━[━➤
3   11/3

21. trinomial   **22.** 7   **23.** 18
24.

y

$y = 2x^2 - 3$

25. $4x^2 - 4x + 14$
26. $-3x^2 - 3$
27. $6x^2 - 7x - 20$
28. $2x^{2n} + 3x^n - 2$
29. $3rs^3(r - 2s)$
30. $(x - y)(5 - a)$
31. $(x + y)(u + v)$

32. $(9x^2 + 4y^2)(3x + 2y)(3x - 2y)$
33. $(2x - 3y^2)(4x^2 + 6xy^2 + 9y^4)$ **34.** $(2x + 3)(3x - 2)$
35. $(3x - 5)^2$ **36.** $(5x + 3)(3x - 2)$
37. $(3a + 2b)(9a^2 - 6ab + 4b^2)$ **38.** $(2x + 5)(3x - 7)$
39. $(x + 5 + y^2)(x + 5 - y^2)$ **40.** $(y + x - 2)(y - x + 2)$
41. $0, 2, -2$ **42.** $-\frac{1}{3}, -\frac{7}{2}$ **43.** $\frac{2x-3}{3x-1}$ **44.** $\frac{x-2}{x+3}$
45. $\frac{4}{x-y}$ **46.** $\frac{a^2 + ab^2}{a^2b - b^2}$ **47.** 0 **48.** -17
49. $x + 4$ **50.** $-x^2 + x + 5 + \frac{8}{x-1}$

Getting Ready (page 419)
1. 0 **2.** 16 **3.** 16 **4.** -16 **5.** $\frac{8}{125}$ **6.** $\frac{81}{256}$
7. $49x^2y^2$ **8.** $343x^3y^3$

Orals (page 428)
1. 3 **2.** -4 **3.** -2 **4.** 2 **5.** $8|x|$ **6.** $-3x$
7. not a real number   **8.** $(x + 1)^2$

Exercise 7.1 (page 428)
1. $\frac{x+3}{x-4}$ **3.** 1 **5.** $\frac{3(m^2 + 2m - 1)}{(m+1)(m-1)}$ **7.** $(5x^2)^2$ **9.** positive
11. 3, up   **13.** x **15.** odd   **17.** 0 **19.** $3x^2$
21. $a^2 + b^3$ **23.** 11 **25.** -8 **27.** $\frac{1}{3}$ **29.** $-\frac{5}{7}$
31. not real   **33.** 0.4 **35.** 4 **37.** not real   **39.** 3.4641
41. 26.0624 **43.** $2|x|$ **45.** $3a^2$ **47.** $|t + 5|$ **49.** $5|b|$
51. $|a + 3|$ **53.** $|t + 12|$ **55.** 1 **57.** -5 **59.** $-\frac{2}{3}$
61. 0.4 **63.** $2a$ **65.** $-10pq$ **67.** $-\frac{1}{2}m^2n$ **69.** $0.2z^3$
71. 3 **73.** -3 **75.** -2 **77.** $\frac{2}{5}$ **79.** $\frac{1}{2}$ **81.** not real
83. $2|x|$ **85.** $2a$ **87.** $\frac{1}{2}|x|$ **89.** $|x^3|$ **91.** $-x$
93. $-3a^2$ **95.** $x + 2$ **97.** $0.1x^2|y|$ **99.** 0 **101.** 4
103. 1 **105.** 0.5 **107.** 4.1231 **109.** 2.5539
111. D: $[-4, \infty)$, R: $[0, \infty)$ **113.** D: $[0, \infty)$, R: $(-\infty, -3]$,

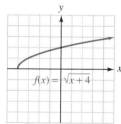

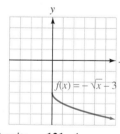

115. 1.67 **117.** 11.8673 **119.** 3 units   **121.** 4 sec
123. about 7.4 amperes

Getting Ready (page 431)
1. 25 **2.** 169 **3.** 18 **4.** $11{,}236$

Orals (page 435)
1. 5 **2.** 10 **3.** 13 **4.** 5 **5.** 10 **6.** 13 **7.** 4
8. 3 **9.** 5

Exercise 7.2 (page 435)
1. $12x^2 - 14x - 10$ **3.** $15t^2 + 2ts - 8s^2$ **5.** hypotenuse
7. $a^2 + b^2 = c^2$ **9.** distance   **11.** 10 ft   **13.** 80 m

15. 48 in.   **17.** 9.9 cm   **19.** 5   **21.** 5   **23.** 13   **25.** 10
27. 10.2   **33.** 13 ft   **35.** about 127 ft   **37.** about 135 ft
39. not quite   **41.** yes   **43.** 173 yd   **45.** 0.05 ft
47. 24 cm²   **51.** about 25

Getting Ready (page 438)
1. x^7 **2.** a^{12} **3.** a^4 **4.** 1 **5.** $\frac{1}{x^4}$ **6.** a^3b^6 **7.** $\frac{b^6}{c^9}$
8. a^{10}

Orals (page 444)
1. 2 **2.** 3 **3.** 3 **4.** 1 **5.** 8 **6.** 4 **7.** $\frac{1}{2}$ **8.** 2
9. $2x$ **10.** $2x^2$

Exercise 7.3 (page 444)
1. $x < 3$ **3.** $r > 28$ **5.** $1\frac{2}{3}$ pints   **7.** $a \cdot a \cdot a \cdot a$
9. a^{mn} **11.** $\frac{a^n}{b^n}$ **13.** $\frac{1}{a^n}, 0$ **15.** $\left(\frac{b}{a}\right)^n$ **17.** $|x|$
19. $\sqrt[3]{7}$ **21.** $\sqrt[5]{8}$ **23.** $\sqrt[4]{3x}$ **25.** $\sqrt[4]{\frac{1}{2}x^3y}$
27. $\sqrt[5]{4a^2b^3}$ **29.** $\sqrt{x^2 + y^2}$ **31.** $11^{1/2}$ **33.** $(3a)^{1/4}$
35. $3a^{1/5}$ **37.** $\left(\frac{1}{7}abc\right)^{1/6}$ **39.** $\left(\frac{1}{2}mn\right)^{1/5}$ **41.** $(a^2 - b^2)^{1/3}$
43. 2 **45.** 3 **47.** 2 **49.** 2 **51.** $\frac{1}{2}$ **53.** $\frac{1}{2}$ **55.** -2
57. -3 **59.** not real   **61.** 0 **63.** $5|y|$ **65.** $2|x|$
67. $3x$ **69.** not real   **71.** 216 **73.** 27 **75.** $1{,}728$
77. $\frac{1}{4}$ **79.** $125x^6$ **81.** $\frac{4x^2}{9}$ **83.** $\frac{1}{2}$ **85.** $\frac{1}{8}$ **87.** $\frac{1}{64x^3}$
89. $\frac{1}{9y^2}$ **91.** $\frac{1}{4p^2}$ **93.** 8 **95.** $\frac{16}{81}$ **97.** $-\frac{3}{2x}$ **99.** $5^{8/9}$
101. $4^{3/5}$ **103.** $9^{1/5}$ **105.** $7^{1/2}$ **107.** $\frac{1}{36}$ **109.** $2^{2/3}$
111. a **113.** $a^{2/9}$ **115.** $a^{3/4}b^{1/2}$ **117.** $\frac{n^{2/5}}{m^{3/5}}$ **119.** $\frac{2x}{3}$
121. $\frac{1}{3}x$ **123.** $y + y^2$ **125.** $x^2 - x + x^{3/5}$ **127.** $x - 4$
129. $x^{4/3} - x^2$ **131.** $x^{4/3} + 2x^{2/3}y^{2/3} + y^{4/3}$
133. $a^3 - 2a^{3/2}b^{3/2} + b^3$ **135.** $\sqrt{x}$ **137.** $\sqrt{5b}$
141. yes

Getting Ready (page 446)
1. 15 **2.** 24 **3.** 5 **4.** 7 **5.** $4x^2$ **6.** $\frac{8}{11}x^3$ **7.** $3ab^3$
8. $-2a^4$

Orals (page 454)
1. 7 **2.** 4 **3.** 3 **4.** $3\sqrt{2}$ **5.** $2\sqrt[3]{2}$ **6.** $\frac{\sqrt[3]{3x^2}}{4b^2}$
7. $7\sqrt{3}$ **8.** $3\sqrt{7}$ **9.** $5\sqrt[3]{9}$ **10.** $8\sqrt[5]{4}$

Exercise 7.4 (page 454)
1. $\frac{-15x^5}{y}$ **3.** $9t^2 + 12t + 4$ **5.** $3p + 4 + \frac{-5}{2p-5}$
7. $\sqrt[n]{a}\sqrt[n]{b}$ **9.** 6 **11.** t **13.** $5x$ **15.** 10 **17.** $7x$
19. $6b$ **21.** 2 **23.** $3a$ **25.** $2\sqrt{5}$ **27.** $-10\sqrt{2}$
29. $2\sqrt[3]{10}$ **31.** $-3\sqrt[3]{3}$ **33.** $2\sqrt[4]{2}$ **35.** $2\sqrt[5]{3}$
37. $\frac{\sqrt{7}}{3}$ **39.** $\frac{\sqrt[3]{7}}{4}$ **41.** $\frac{\sqrt[4]{3}}{10}$ **43.** $\frac{\sqrt[5]{3}}{2}$ **45.** $5x\sqrt{2}$
47. $4\sqrt{2b}$ **49.** $-4a\sqrt{7a}$ **51.** $5ab\sqrt{7b}$
53. $-10\sqrt{3xy}$ **55.** $-3x^2\sqrt[3]{2}$ **57.** $2x^4y\sqrt[3]{2}$

59. $2x^3y\sqrt[4]{2}$ **61.** $\dfrac{z}{4x}$ **63.** $\dfrac{\sqrt[4]{5x}}{2z}$ **65.** $10\sqrt{2x}$
67. $\sqrt[5]{7a^2}$ **69.** $4\sqrt{3}$ **71.** $-\sqrt{2}$ **73.** $2\sqrt{2}$
75. $9\sqrt{6}$ **77.** $3\sqrt[3]{3}$ **79.** $-\sqrt[3]{4}$ **81.** -10
83. $-17\sqrt[4]{2}$ **85.** $16\sqrt[4]{2}$ **87.** $-4\sqrt{2}$ **89.** $3\sqrt{2}+\sqrt{3}$
91. $-11\sqrt[3]{2}$ **93.** $y\sqrt[3]{z}$ **95.** $13y\sqrt{x}$ **97.** $12\sqrt[3]{a}$
99. $-7y^2\sqrt{y}$ **101.** $4x\sqrt[5]{xy^2}$ **103.** $2x+2$
105. $h=2.83, x=2.00$ **107.** $x=8.66, h=10.00$
109. $x=4.69, y=8.11$ **111.** $x=12.11, y=12.11$
115. If $a=0$, then b can be any nonnegative real number. If $b=0$, then a can be any nonnegative real number.

Getting Ready (page 456)
1. a^7 **2.** b^3 **3.** a^2-2a **4.** $6b^3+9b^2$
5. $a^2-3a-10$ **6.** $4a^2-9b^2$

Orals (page 462)
1. 3 **2.** 2 **3.** $3\sqrt{3}$ **4.** a^2b **5.** $6+3\sqrt{2}$ **6.** 1
7. $\dfrac{\sqrt{2}}{2}$ **8.** $\dfrac{\sqrt{3}+1}{2}$

Exercise 7.5 (page 462)
1. 1 **3.** $\frac{1}{3}$ **5.** $2, \sqrt{7}, \sqrt{5}$ **7.** FOIL **9.** conjugate
11. 4 **13.** $5\sqrt{2}$ **15.** $6\sqrt{2}$ **17.** 5 **19.** 18
21. $2\sqrt[3]{3}$ **23.** ab^2 **25.** $5a\sqrt{b}$ **27.** $r\sqrt[3]{10s}$
29. $2a^2b^2\sqrt[3]{2}$ **31.** $x^2(x+3)$ **33.** $3x(y+z)\sqrt[3]{4}$
35. $12\sqrt{5}-15$ **37.** $12\sqrt{6}+6\sqrt{14}$
39. $-8x\sqrt{10}+6\sqrt{15x}$ **41.** $-1-2\sqrt{2}$
43. $8x-14\sqrt{x}-15$ **45.** $5z+2\sqrt{15z}+3$
47. $3x-2y$ **49.** $6a+5\sqrt{3ab}-3b$
51. $18r-12\sqrt{2r}+4$ **53.** $-6x-12\sqrt{x}-6$ **55.** $\dfrac{\sqrt{7}}{7}$
57. $\dfrac{\sqrt{6}}{3}$ **59.** $\dfrac{\sqrt{10}}{4}$ **61.** 2 **63.** $\dfrac{\sqrt[3]{4}}{2}$ **65.** $\sqrt[3]{3}$
67. $\dfrac{\sqrt[3]{6}}{3}$ **69.** $2\sqrt{2x}$ **71.** $\dfrac{\sqrt{5y}}{y}$ **73.** $\dfrac{\sqrt[3]{2ab^2}}{b}$
75. $\dfrac{\sqrt[4]{4}}{2}$ **77.** $\dfrac{\sqrt[5]{2}}{2}$ **79.** $\sqrt{2}+1$ **81.** $\dfrac{3\sqrt{2}-\sqrt{10}}{4}$
83. $2+\sqrt{3}$ **85.** $\dfrac{9-2\sqrt{14}}{5}$ **87.** $\dfrac{2(\sqrt{x}-1)}{x-1}$
89. $\dfrac{x(\sqrt{x}+4)}{x-16}$ **91.** $\sqrt{2z}+1$ **93.** $\dfrac{x-2\sqrt{xy}+y}{x-y}$
95. $\dfrac{1}{\sqrt{3}-1}$ **97.** $\dfrac{x-9}{x(\sqrt{x}-3)}$ **99.** $\dfrac{x-y}{\sqrt{x}(\sqrt{x}-\sqrt{y})}$
101. $f/4$ **105.** $\dfrac{x-9}{4(\sqrt{x}+3)}$

Getting Ready (page 464)
1. a **2.** $5x$ **3.** $x+4$ **4.** $y-3$

Orals (page 470)
1. 7 **2.** 3 **3.** 0 **4.** 9 **5.** 17 **6.** 31

Exercise 7.6 (page 470)
1. 2 **3.** 6 **5.** $x^n=y^n$ **7.** square **9.** extraneous

11. 2 **13.** 4 **15.** 0 **17.** 4 **19.** 8 **21.** $\frac{5}{2}, \frac{1}{2}$ **23.** 1
25. 16 **27.** 14, ~~6~~ **29.** 4, 3 **31.** 2, ~~7~~ **33.** 9, ~~-25~~
35. 2, -1 **37.** -1, ~~1~~ **39.** ~~1~~, no solutions **41.** 0, ~~4~~
43. ~~-3~~, no solutions **45.** 0 **47.** 1, 9 **49.** 4, ~~0~~
51. 2, ~~142~~ **53.** 2 **55.** ~~6~~, no solutions **57.** 0, ~~$\frac{12}{11}$~~
59. 1 **61.** 4, ~~-9~~ **63.** 2, ~~$\frac{21}{2}$~~ **65.** 2,010 ft
67. about 29 mph **69.** 16% **71.** \$5 **73.** $R=\dfrac{8kl}{\pi r^4}$
77. 0, 4

Getting Ready (page 473)
1. $7x$ **2.** $-x+10$ **3.** $12x^2+5x-25$ **4.** $9x^2-25$

Orals (page 481)
1. $7i$ **2.** $8i$ **3.** $10i$ **4.** $9i$ **5.** $-i$ **6.** -1 **7.** 1
8. i **9.** 5 **10.** 13

Exercise 7.7 (page 481)
1. -1 **3.** 20 mph **5.** imaginary **7.** -1 **9.** 1
11. $\dfrac{\sqrt{a}}{\sqrt{b}}$ **13.** 5, 7 **15.** conjugates **17.** $3i$ **19.** $6i$
21. $i\sqrt{7}$ **23.** yes **25.** no **27.** no **29.** $8-2i$
31. $3-5i$ **33.** $15+7i$ **35.** $6-8i$ **37.** $2+9i$
39. $-15+2\sqrt{3}i$ **41.** $3+6i$ **43.** $-25-25i$
45. $7+i$ **47.** $14-8i$ **49.** $8+\sqrt{2}i$ **51.** $-20-30i$
53. $3+4i$ **55.** $-5+12i$ **57.** $7+17i$ **59.** $5+5i$
61. $16+2i$ **63.** $0-i$ **65.** $0+\frac{4}{5}i$ **67.** $\frac{1}{8}-0i$
69. $0+\frac{3}{5}i$ **71.** $2+i$ **73.** $\frac{1}{2}+\frac{5}{2}i$ **75.** $-\frac{42}{25}-\frac{6}{25}i$
77. $\frac{1}{4}+\frac{3}{4}i$ **79.** $\frac{5}{13}-\frac{12}{13}i$ **81.** $\frac{11}{10}+\frac{3}{10}i$ **83.** $\dfrac{1}{4}-\dfrac{\sqrt{15}}{4}i$
85. $-\frac{5}{169}+\frac{12}{169}i$ **87.** $\frac{3}{5}+\frac{4}{5}i$ **89.** $-\frac{6}{13}-\frac{9}{13}i$ **91.** i
93. $-i$ **95.** 1 **97.** i **99.** 10 **101.** 13 **103.** $\sqrt{74}$
105. 1 **111.** $7-4i$ V **113.** $1-3.4i$ **117.** $\frac{5}{3+i}$

Chapter Summary (page 484)
1. 7 **2.** -11 **3.** -6 **4.** 15 **5.** -3 **6.** -6 **7.** 5
8. -2 **9.** $5|x|$ **10.** $|x+2|$ **11.** $3a^2b$ **12.** $4x^2|y|$

13.
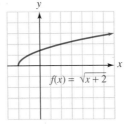
$f(x)=\sqrt{x+2}$

14.
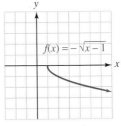
$f(x)=-\sqrt{x-1}$

15.
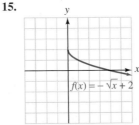
$f(x)=-\sqrt{x}+2$

16.
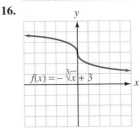
$f(x)=-\sqrt[3]{x}+3$

17. 12   **18.** about 5.7   **19.** 3 mi   **20.** 8.2 ft   **21.** 88 yd
22. 16,000 yd, or about 9 mi   **23.** 13   **24.** 2.83 units
25. 5   **26.** -6 **27.** 27   **28.** 64   **29.** -2 **30.** -4
31. $\frac{1}{4}$ **32.** $\frac{1}{2}$ **33.** $-16{,}807$ **34.** $\frac{1}{3{,}125}$ **35.** 8   **36.** $\frac{27}{8}$
37. $3xy^{1/3}$ **38.** $3xy^{1/2}$ **39.** $125x^{9/2}y^6$ **40.** $\dfrac{1}{4u^{4/3}v^2}$
41. $5^{3/4}$ **42.** $a^{5/7}$ **43.** $u - 1$ **44.** $v + v^2$
45. $x + 2x^{1/2}y^{1/2} + y$ **46.** $a^{4/3} - b^{4/3}$ **47.** $\sqrt[3]{5}$
48. $\sqrt{x}$ **49.** $\sqrt[3]{3ab^2}$ **50.** $\sqrt{5ab}$ **51.** $4\sqrt{15}$
52. $3\sqrt[3]{2}$ **53.** $2\sqrt[4]{2}$ **54.** $2\sqrt[5]{3}$ **55.** $2x\sqrt{2x}$
56. $3x^2y\sqrt{2y}$ **57.** $2xy\sqrt[3]{2x^2y}$ **58.** $3x^2y\sqrt[3]{2x}$
59. $4x$ **60.** $2x$ **61.** $\dfrac{\sqrt[3]{2a^2b}}{3x}$ **62.** $\dfrac{\sqrt{17xy}}{8a^2}$ **63.** $3\sqrt{2}$
64. $\sqrt{5}$ **65.** 0   **66.** $8\sqrt[4]{2}$ **67.** $29x\sqrt{2}$ **68.** $32a\sqrt{3a}$
69. $13\sqrt[3]{2}$ **70.** $-4x\sqrt[4]{2x}$ **71.** $7\sqrt{2}$ m
72. $6\sqrt{3}$ cm, 18 cm   **73.** 7.07 in.   **74.** 8.66 cm
75. $6\sqrt{10}$ **76.** 72   **77.** $3x$ **78.** 3   **79.** $-2x$
80. $-20x^3y^3\sqrt{xy}$ **81.** $4 - 3\sqrt{2}$ **82.** $2 + 3\sqrt{2}$
83. $\sqrt{10} - \sqrt{5}$ **84.** $3 + \sqrt{6}$ **85.** 1   **86.** $5 + 2\sqrt{6}$
87. $x - y$ **88.** $6u + \sqrt{u} - 12$ **89.** $\dfrac{\sqrt{3}}{3}$ **90.** $\dfrac{\sqrt{15}}{5}$
91. $\dfrac{\sqrt{xy}}{y}$ **92.** $\dfrac{\sqrt[3]{u^2}}{u^2v^2}$ **93.** $2\left(\sqrt{2} + 1\right)$ **94.** $\dfrac{\sqrt{6} + \sqrt{2}}{2}$
95. $2\left(\sqrt{x} - 4\right)$ **96.** $\dfrac{a + 2\sqrt{a} + 1}{a - 1}$ **97.** $\dfrac{3}{5\sqrt{3}}$
98. $\dfrac{1}{\sqrt[3]{3}}$ **99.** $\dfrac{9 - x}{2\left(3 + \sqrt{x}\right)}$ **100.** $\dfrac{a - b}{a + \sqrt{ab}}$ **101.** 22
102. 16, 9   **103.** 3, 9   **104.** $\frac{9}{16}$ **105.** 2   **106.** 0, -2
107. $12 - 8i$ **108.** $2 - 68i$ **109.** $-96 + 3i$
110. $-2 - 2\sqrt{2}i$ **111.** $22 + 29i$ **112.** $-16 + 7i$
113. $-12 + 28\sqrt{3}i$ **114.** $118 + 10\sqrt{2}i$ **115.** $0 - \frac{3}{4}i$
116. $0 - \frac{2}{5}i$ **117.** $\frac{12}{5} - \frac{6}{5}i$ **118.** $\frac{21}{10} + \frac{7}{10}i$ **119.** $\frac{15}{17} + \frac{8}{17}i$
120. $\frac{4}{5} - \frac{3}{5}i$ **121.** $\frac{15}{29} - \frac{6}{29}i$ **122.** $\frac{1}{3} + \frac{1}{3}i$ **123.** $15 + 0i$
124. $26 + 0i$ **125.** 1   **126.** $-i$

Chapter 7 Test (page 490)
1. 7   **2.** 4   **3.** $2|x|$ **4.** $2x$
5. $D = [2, \infty), R = [0, \infty)$ **6.** $D = (-\infty, \infty), R = (-\infty, \infty)$

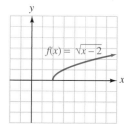
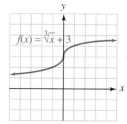

7. 28 in.   **8.** 1.25 m   **9.** 10   **10.** 25   **11.** 2   **12.** 9
13. $\frac{1}{216}$ **14.** $\frac{9}{4}$ **15.** $2^{4/3}$ **16.** $8xy$ **17.** $4\sqrt{3}$
18. $5xy^2\sqrt{10xy}$ **19.** $2x^5y\sqrt[3]{3}$ **20.** $\dfrac{1}{4a}$ **21.** $2|x|\sqrt{3}$
22. $2|x^3|\sqrt{2}$ **23.** $3x\sqrt[3]{3}$ **24.** $3x^2y^4\sqrt{2y}$ **25.** $-\sqrt{3}$

26. $14\sqrt[3]{5}$ **27.** $2y^2\sqrt{3y}$ **28.** $6z\sqrt[4]{3z}$
29. $-6x\sqrt{y} - 2xy^2$ **30.** $3 - 7\sqrt{6}$ **31.** $\frac{\sqrt{5}}{5}$
32. $\sqrt{3t} + 1$ **33.** $\dfrac{3}{\sqrt{21}}$ **34.** $\dfrac{a - b}{a - 2\sqrt{ab} + b}$ **35.** 10
36. $\cancel{4}$, no solutions   **37.** $-1 + 11i$ **38.** $4 - 7i$
39. $8 + 6i$ **40.** $-10 - 11i$ **41.** $0 - \frac{\sqrt{2}}{2}i$ **42.** $\frac{1}{2} + \frac{1}{2}i$

Getting Ready (page 493)
1. $(x + 5)(x - 5)$ **2.** $(b + 9)(b - 9)$ **3.** $(3x + 2)(2x - 1)$
4. $(2x - 3)(2x + 1)$

Orals (page 501)
1. ± 7 **2.** $\pm\sqrt{10}$ **3.** 4   **4.** 9   **5.** $\frac{9}{4}$ **6.** $\frac{25}{4}$

Exercise 8.1 (page 501)
1. 1   **3.** $t \le 4$ **5.** $x = \sqrt{c}, x = -\sqrt{c}$
7. positive or negative   **9.** $0, -2$ **11.** $5, -5$
13. $-2, -4$ **15.** 6, 1   **17.** $2, \frac{1}{2}$ **19.** $\frac{2}{3}, -\frac{5}{2}$ **21.** ± 6
23. $\pm\sqrt{5}$ **25.** $\pm\dfrac{4\sqrt{3}}{3}$ **27.** $0, -2$ **29.** 4, 10
31. $-5 \pm \sqrt{3}$ **33.** $2 \pm \sqrt{5}$ **35.** $\pm 4i$ **37.** $\pm\frac{9}{2}i$
39. $2, -4$ **41.** $2, 4$ **43.** $-1, -4$ **45.** $1, -\frac{1}{2}$
47. $-\frac{1}{3}, -\frac{3}{2}$ **49.** $\frac{3}{4}, -\frac{3}{2}$ **51.** $-\dfrac{7}{10} \pm \dfrac{\sqrt{29}}{10}$ **53.** $-1 \pm i$
55. $-4 \pm i\sqrt{2}$ **57.** $\dfrac{1}{3} \pm \dfrac{2\sqrt{2}}{3}i$ **59.** $-\dfrac{1}{4} \pm \dfrac{\sqrt{41}}{4}$
61. $-\dfrac{1}{2} \pm \dfrac{\sqrt{13}}{2}$ **63.** 4 sec   **65.** 72.5 mph
67. 4%   **71.** $\frac{3}{4}$

Getting Ready (page 503)
1. $x^2 + 12x + 36, (x + 6)^2$ **2.** $x^2 - 7x + \frac{49}{4}, \left(x - \frac{7}{2}\right)^2$
3. 7   **4.** 8

Orals (page 507)
1. $3, -4, 7$ **2.** $-2, 1, -5$

Exercise 8.2 (page 507)
1. $y = \dfrac{-Ax + C}{B}$ **3.** $2\sqrt{6}$ **5.** $\sqrt{3}$ **7.** $3, -2, 6$
9. $-1, -2$ **11.** $-3, 5$ **13.** $-6, -6$ **15.** $-1, \frac{3}{2}$
17. $-\dfrac{1}{2} \pm \dfrac{\sqrt{5}}{10}$ **19.** $-\frac{3}{2}, -\frac{1}{2}$ **21.** $\frac{1}{4}, -\frac{3}{4}$ **23.** $-\dfrac{5}{2} \pm \dfrac{\sqrt{17}}{2}$
25. $-1 \pm i$ **27.** $-\dfrac{1}{4} \pm \dfrac{\sqrt{7}}{4}i$ **29.** $\dfrac{2}{3} \pm \dfrac{\sqrt{2}}{3}i$
31. $\dfrac{1}{3} \pm \dfrac{2\sqrt{2}}{3}i$ **33.** $-\frac{1}{2} \pm \sqrt{5}$ **35.** $-\dfrac{1}{3} \pm \dfrac{\sqrt{5}}{3}i$
37. $8.98, -3.98$ **39.** $N = \dfrac{1 \pm \sqrt{1 + 8C}}{2}$ **41.** 16, 18
43. 6, 7   **45.** $x^2 - 8x + 15 = 0$
47. $x^3 - x^2 - 14x + 24 = 0$ **49.** 8 ft by 12 ft   **51.** 4 units
53. $\frac{4}{3}$ cm   **55.** 30 mph   **57.** \$4.80 or \$5.20   **59.** 4,000

61. 2.26 in.   **63.** about 6.13×10^{-3} M   **67.** $\sqrt{2}, -3\sqrt{2}$
69. $i, 2i$

Getting Ready (page 510)
1. 17   **2.** -8

Orals (page 514)
1. -3 **2.** -7 **3.** irrational and unequal
4. complex conjugates   **5.** no   **6.** yes

Exercise 8.3 (page 515)
1. -2 **3.** $\frac{9}{5}$ **5.** $b^2 - 4ac$ **7.** rational, unequal
9. rational, equal   **11.** complex conjugates   **13.** irrational,
unequal   **15.** rational, unequal   **17.** $6, -6$ **19.** $12, -12$
21. 5   **23.** $12, -3$ **25.** yes   **27.** $k < -\frac{4}{3}$
29. $1, -1, 4, -4$ **31.** $1, -1, \sqrt{2}, -\sqrt{2}$
33. $1, -1, \sqrt{5}, -\sqrt{5}$ **35.** $1, -1, 2, -2$
37. $1, -1, 2\sqrt{3}, -2\sqrt{3}$ **39.** 1   **41.** no solution
43. $-8, -27$ **45.** $-1, 27$ **47.** $-1, -4$ **49.** $4, -5$
51. $0, 2$ **53.** $-1, -\frac{27}{13}$ **55.** $1, 1, -1, -1$ **57.** $1 \pm i$
59. $x = \pm\sqrt{r^2 - y^2}$ **61.** $d = \pm\sqrt{\dfrac{k}{I}} = \pm\dfrac{\sqrt{kI}}{I}$
63. $y = \dfrac{-3x \pm \sqrt{9x^2 - 28x}}{2x}$ **65.** $\mu^2 = \dfrac{\Sigma x^2}{N} - \sigma^2$
67. $\frac{2}{3}, -\frac{1}{4}$ **69.** $-\dfrac{5}{4} \pm \dfrac{\sqrt{17}}{4}$ **71.** $\dfrac{1}{3} \pm \dfrac{\sqrt{11}}{3}i$ **73.** $-1 \pm 2i$
77. no

Getting Ready (page 516)
1. -2 **2.** 2   **3.** 0   **4.** 8   **5.** 1   **6.** 4

Orals (page 525)
1. down   **2.** up   **3.** up   **4.** down   **5.** $(3, -1)$
6. $(-2, 2)$

Exercise 8.4 (page 525)
1. 10   **3.** $3\frac{3}{5}$ hr   **5.** $f(x) = ax^2 + bx + c, a \neq 0$
7. vertex   **9.** upward   **11.** to the right   **13.** upward
15.   **17.**
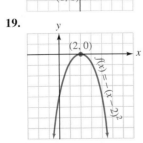

19.   **21.**
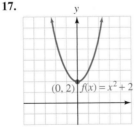
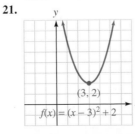

23.   **25.**
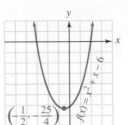
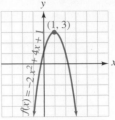

27. $(1, 2), x = 1$ **29.** $(-3, -4), x = -3$ **31.** $(0, 0), x = 0$
33. $(1, -2), x = 1$ **35.** $(2, 21), x = 2$ **37.** $\left(\frac{5}{12}, \frac{143}{24}\right), x = \frac{5}{12}$
39. $(5, 2)$ **41.** $(0.25, 0.88)$ **43.** $(0.5, 7.25)$ **45.** $2, -3$
47. $-1.85, 3.25$ **49.** 36 ft, 3 sec
51. 50 ft by 50 ft, 2,500 ft^2 **53.** 75 ft by 75 ft, 5,625 ft^2
55. 5,000   **57.** 3,276, \$14,742   **59.** \$35
61. 0.25 and 0.75

Getting Ready (page 528)
1. $(x + 5)(x - 3)$ **2.** $(x - 2)(x - 1)$

Orals (page 534)
1. $x = 2$ **2.** $x > 2$ **3.** $x < 2$ **4.** $x = -3$ **5.** $x > -3$
6. $x < -3$ **7.** $1 < 2x$ **8.** $1 > 2x$

Exercise 8.5 (page 535)
1. $y = kx$ **3.** $t = kxy$ **5.** 3   **7.** greater   **9.** undefined
11.   **13.**
$(1, 4)$ $(-\infty, 3) \cup (5, \infty)$
1   4   3   5

15.   **17.**
$[-4, 3]$ $(-\infty, -5] \cup [3, \infty)$
-4 3 -5 3

19. no solutions   **21.** $(-\infty, -3] \cup [3, \infty)$
-3 3

23.   **25.**
$(-5, 5)$ $(-\infty, 0) \cup (1/2, \infty)$
-5 5   0   1/2

27.   **29.**
$(0, 2]$ $(-\infty, -5/3) \cup (0, \infty)$
0   2 $-5/3$ 0

31. $(-\infty, -3) \cup (1, 4)$ **33.** $[-5, -2) \cup [4, \infty)$
-3 1   4 -5 -2 4

35. $(-\infty, -4)$ **37.** $(-1/2, 1/3) \cup (1/2, \infty)$
-4 $-1/2$ 1/3   1/2

39. $(0, 2) \cup (8, \infty)$ **41.** $(-\infty, -2) \cup (2, 18]$
0   2   8 -2 2   18

43. $[-34/5, -4) \cup (3, \infty)$
$-34/5$ -4 3

45. $(-4, -2] \cup (-1, 2]$
-4 -2 -1 2

47. $(-\infty, -16) \cup (-4, -1) \cup (4, \infty)$

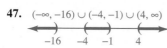

49. $(-\infty, -2) \cup (-2, \infty)$

51. $(-1, 3)$ **53.** $(-\infty, -3) \cup (2, \infty)$

55.

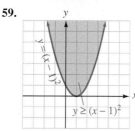

57.

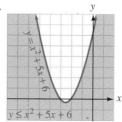

59.

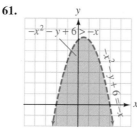

61.

63.

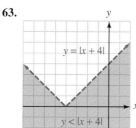

65.

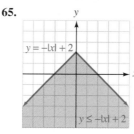

69. when 4 factors are negative, 2 factors are negative, or no factors are negative

Chapter Summary (page 539)

1. $\frac{2}{3}, -\frac{3}{4}$ **2.** $-\frac{1}{3}, -\frac{5}{2}$ **3.** $\frac{2}{3}, -\frac{4}{5}$ **4.** $4, -8$ **5.** $-4, -2$

6. $\frac{7}{2}, 1$ **7.** $\frac{1}{4} \pm \frac{\sqrt{41}}{4}$ **8.** $9, -1$ **9.** $0, 10$ **10.** $\frac{1}{2}, -7$

11. $-7, \frac{1}{3}$ **12.** $\frac{1}{4} \pm \frac{\sqrt{17}}{4}$ **13.** $-\frac{1}{2} \pm \frac{\sqrt{7}}{2}i$

14. 4 cm by 6 cm   **15.** 2 ft by 3 ft   **16.** 7 sec   **17.** 196 ft

18. irrational, unequal   **19.** complex conjugates

20. 12, 152   **21.** $k \geq -\frac{7}{3}$ **22.** 1, 144   **23.** 8, -27

24. 1   **25.** 1, $-\frac{8}{5}$ **26.** $\frac{14}{3}$ **27.** 1

28.

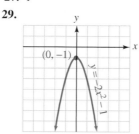

29.

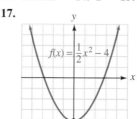

30.

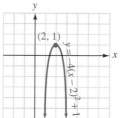

31.

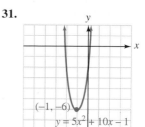

32. $(-\infty, -7) \cup (5, \infty)$

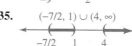

33. $(-9, 2)$

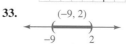

34. $(-\infty, 0) \cup [3/5, \infty)$

35. $(-7/2, 1) \cup (4, \infty)$

36. $x < -7$ or $x > 5$ **37.** $-9 < x < 2$

38. $x < 0$ or $x \geq \frac{3}{5}$ **39.** $-\frac{7}{2} < x < 1$ or $x > 4$

40.

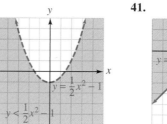

41.

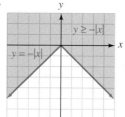

Chapter 8 Test (page 542)

1. $3, -6$ **2.** $-\frac{3}{2}, -\frac{5}{3}$ **3.** 144   **4.** 625   **5.** $\frac{25}{4}$ **6.** $\frac{49}{4}$

7. $-2 \pm \sqrt{3}$ **8.** $\frac{5}{2} \pm \frac{\sqrt{37}}{2}$ **9.** $\pm 2i\sqrt{3}$

10. $-1 \pm 2i$ **11.** $-\frac{1}{6} \pm \frac{\sqrt{13}}{6}$ **12.** $-\frac{1}{4} \pm \frac{\sqrt{23}}{4}i$

13. nonreal   **14.** 2   **15.** 10 in.   **16.** 1, $\frac{1}{4}$

17.

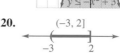

18.

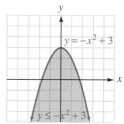

19. $(-\infty, -2) \cup (4, \infty)$ **20.** $(-3, 2]$

Cumulative Review Exercises (page 543)

1. $D = (-\infty, \infty)$; $R = [-3, \infty)$
2. $D = (-\infty, \infty)$; $R = (-\infty, 0]$ **3.** $y = 3x + 2$
4. $y = -\frac{2}{3}x - 2$ **5.** $-4a^2 + 12a - 7$ **6.** $6x^2 - 5x - 6$
7. $(x^2 + 4y^2)(x + 2y)(x - 2y)$ **8.** $(3x + 2)(5x - 4)$
9. $6, -1$ **10.** $0, \frac{2}{3}, -\frac{1}{2}$ **11.** $5x^2$ **12.** $4t\sqrt{3t}$ **13.** $-3x$
14. $4x$ **15.** $\frac{1}{2}$ **16.** 16 **17.** y^2 **18.** $x^{17/12}$
19.
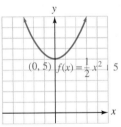
20.
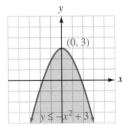

21. $x^{4/3} - x^{2/3}$ **22.** $\frac{1}{x} + 2 + x$ **23.** $7\sqrt{2}$
24. $-12\sqrt[4]{2} + 10\sqrt[4]{3}$ **25.** $-18\sqrt{6}$ **26.** $\dfrac{5\sqrt[3]{x^2}}{x}$
27. $\dfrac{x + 3\sqrt{x} + 2}{x - 1}$ **28.** $\sqrt{xy}$ **29.** $2, 7$ **30.** $\frac{1}{4}$
31. $3\sqrt{2}$ in.   **32.** $2\sqrt{3}$ in.   **33.** 10 **34.** 9
35. $1, -\frac{3}{2}$ **36.** $-\dfrac{2}{3} \pm \dfrac{\sqrt{7}}{3}$

37.
38.
(0, 3)
$f(x) = \frac{1}{2}x^2$ | 5   (0, 5)
$y \le -x^2 + 3$

39. $7 + 2i$ **40.** $-5 - 7i$ **41.** $13 + 0i$ **42.** $12 - 6i$
43. $-12 - 10i$ **44.** $\frac{3}{2} + \frac{1}{2}i$ **45.** $\sqrt{13}$ **46.** $\sqrt{61}$
47. -2 **48.** $9, 16$
49. $(-\infty, -2) \cup (3, \infty)$ **50.** $[-2, 3]$

Getting Ready (page 546)

1. 3 **2.** 4

Orals (page 554)

1. y-axis   **2.** x-axis   **3.** origin   **4.** y-axis   **5.** even
6. odd

Exercise 9.1 (page 554)

1. 9.3×10^6 **3.** $670,000$ **5.** polynomial
7. symmetric, y-axis   **9.** $(x, -y)$, symmetric
11. $-f(x)$ **13.** decreasing   **15.** vertically

17. y-axis   **19.** origin   **21.** y-axis   **23.** none
25. none   **27.** x-axis   **29.** no   **31.** yes   **33.** no
35. yes   **37.** yes   **39.** no   **41.** no   **43.** yes
45.
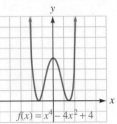
$f(x) = x^4 - 4x^2 + 4$
$D = (-\infty, \infty)$, $R = [0, \infty)$

47.
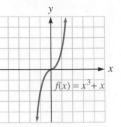
$f(x) = x^3 + x$
$D = (-\infty, \infty)$, $R = (-\infty, \infty)$

49.
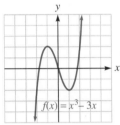
$f(x) = x^3 - 3x$
$D = (-\infty, \infty)$, $R = (-\infty, \infty)$

51. decreasing for $x < 0$, increasing for $x > 0$
53. increasing for $x < 0$, decreasing for $x > 0$
55. decreasing for $x < 2$, increasing for $x > 2$

57.
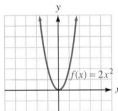
$f(x) = 2x^2$

59.
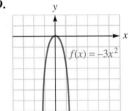
$f(x) = -3x^2$

61.
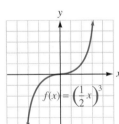
$f(x) = \left(\frac{1}{2}x\right)^3$

63.
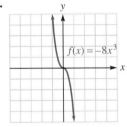
$f(x) = -8x^3$

Getting Ready (page 556)

1. positive   **2.** negative   **3.** 98   **4.** -3

Orals (page 559)

1. -2 **2.** 0 **3.** 4 **4.** 5

Exercise 9.2 (page 559)

1. 20 **3.** domains   **5.** constant, $f(x)$ **7.** step   **9.** 55

11.

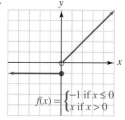

$$f(x) = \begin{cases} -1 \text{ if } x \le 0 \\ x \text{ if } x > 0 \end{cases}$$

constant on $(-\infty, 0)$, increasing on $(0, \infty)$

13.

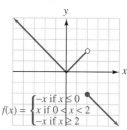

$$f(x) = \begin{cases} -x \text{ if } x \le 0 \\ x \text{ if } 0 < x < 2 \\ -x \text{ if } x \ge 2 \end{cases}$$

decreasing on $(-\infty, 0)$, increasing on $(0, 2)$, decreasing on $(2, \infty)$

15.

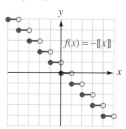

$f(x) = -[\![x]\!]$

17.

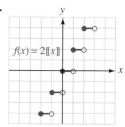

$f(x) = 2[\![x]\!]$

19.

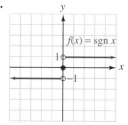

$f(x) = \text{sgn } x$

21. \$30

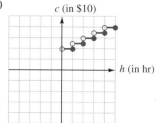

23. After 2 hours, network B is cheaper.

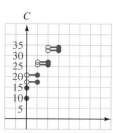

Getting Ready (page 561)
1. 8   **2.** -4 **3.** 1, -2 **4.** 3, -3

Orals (page 572)
1. $(-\infty, 0) \cup (0, \infty)$ **2.** $(-\infty, 5) \cup (5, \infty)$
3. $(-\infty, 4) \cup (4, \infty)$ **4.** $(-\infty, -1) \cup (-1, 1) \cup (1, \infty)$

Exercise 9.3 (page 573)
1. $3x^2 + x$ **3.** $10x^2 + 29x + 10$ **5.** $3x + 5$
7. asymptote   **9.** vertical   **11.** x-intercept
13. the same   **15.** horizontal   **17.** $(-\infty, 2) \cup (2, \infty)$
19. $(-\infty, -5) \cup (-5, 5) \cup (5, \infty)$
21. $(-\infty, -1) \cup (-1, 0) \cup (0, 1) \cup (1, \infty)$ **23.** $(-\infty, \infty)$

25.

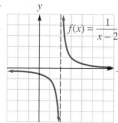

$f(x) = \dfrac{1}{x-2}$

27.

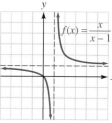

$f(x) = \dfrac{x}{x-1}$

29.

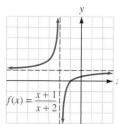

$f(x) = \dfrac{x+1}{x+2}$

31.

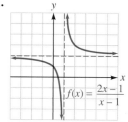

$f(x) = \dfrac{2x-1}{x-1}$

33.

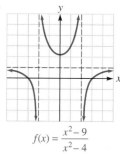

$f(x) = \dfrac{x^2-9}{x^2-4}$

35.

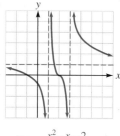

$f(x) = \dfrac{x^2-x-2}{x^2-4x+3}$

37.

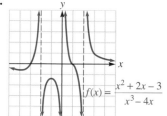

$f(x) = \dfrac{x^2+2x-3}{x^3-4x}$

39.

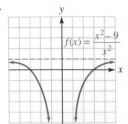

$f(x) = \dfrac{x^2 - 9}{x^2}$

41.

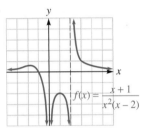

$f(x) = \dfrac{x}{(x+3)^2}$

43.

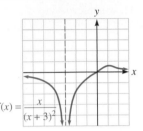

$f(x) = \dfrac{x+1}{x^2(x-2)}$

45.

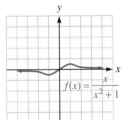

$f(x) = \dfrac{x}{x^2+1}$

47.

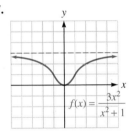

$f(x) = \dfrac{3x^2}{x^2+1}$

49.

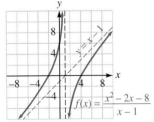

$f(x) = \dfrac{x^2 - 2x - 8}{x-1}$

51.

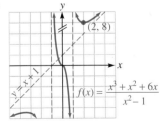

(2, 8)

$f(x) = \dfrac{x^3 + x^2 + 6x}{x^2 - 1}$

53.

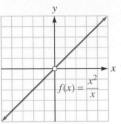

$f(x) = \dfrac{x^2}{x}$

55.

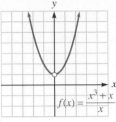

$f(x) = \dfrac{x^3 + x}{x}$

57.

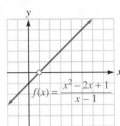

$f(x) = \dfrac{x^2 - 2x + 1}{x - 1}$

59.

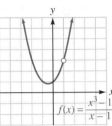

$f(x) = \dfrac{x^3 - 1}{x - 1}$

61. $44,000   **63.** $65,000

Getting Ready (page 576)

1. $\dfrac{3x+1}{x(x+1)}$ **2.** $\dfrac{x-5}{(x+1)(x-1)}$

Orals (page 582)

1. $x+1, x-2$ **2.** $x, x+1$ **3.** x, x^2+1
4. $x^2+2, (x^2+2)^2$

Exercise 9.4 (page 583)

1. $2|a|\sqrt{2ab}$ **3.** $3x^2\sqrt{2x}$ **5.** $x=9$

7. first-degree, second-degree   **9.** $\dfrac{1}{x} + \dfrac{2}{x-1}$

11. $\dfrac{5}{x} - \dfrac{3}{x-3}$ **13.** $\dfrac{1}{x+1} + \dfrac{2}{x-1}$ **15.** $\dfrac{1}{x-3} - \dfrac{3}{x+2}$

17. $\dfrac{8}{x+3} - \dfrac{5}{x-1}$ **19.** $\dfrac{5}{2x-3} + \dfrac{2}{x-5}$

21. $\dfrac{2}{x} + \dfrac{3}{x-1} - \dfrac{1}{x+1}$ **23.** $\dfrac{1}{x} + \dfrac{1}{x^2+3}$

25. $\dfrac{3}{x+1} + \dfrac{2}{x^2+2x+3}$ **27.** $\dfrac{3}{x} + \dfrac{2}{x+1} + \dfrac{1}{(x+1)^2}$

29. $\dfrac{1}{x} + \dfrac{2}{x^2} - \dfrac{3}{x-1}$ **31.** $\dfrac{2}{x} + \dfrac{1}{x-3} + \dfrac{2}{(x-3)^2}$

33. $\dfrac{1}{x-1} - \dfrac{4}{(x-1)^3}$ **35.** $\dfrac{1}{x} + \dfrac{1}{x^2} + \dfrac{2}{x^2+x+1}$

37. $\dfrac{3}{x} + \dfrac{4}{x^2} + \dfrac{x+1}{x^2+1}$ **39.** $-\dfrac{1}{x+1} - \dfrac{3}{x^2+2}$

41. $\dfrac{x+1}{x^2+2} + \dfrac{2}{x^2+x+2}$

43. $\dfrac{1}{x} + \dfrac{x}{x^2+2x+5} + \dfrac{x+2}{(x^2+2x+5)^2}$

45. $x - 3 - \dfrac{1}{x+1} + \dfrac{8}{x+2}$ **47.** $1 + \dfrac{2}{3x+1} + \dfrac{1}{x^2+1}$

49. $1 + \dfrac{1}{x} + \dfrac{x}{x^2+x+1}$ **51.** $2 + \dfrac{1}{x} + \dfrac{3}{x-1} + \dfrac{2}{x^2+1}$

55. No; it's the sum of two cubes.

Getting Ready (page 584)

1. $3x - 1$ **2.** $x + 3$ **3.** $2x^2 - 3x - 2$ **4.** $\dfrac{2x + 1}{x - 2}$

Orals (page 590)

1. $5x$ **2.** x **3.** $8x^2$ **4.** $\dfrac{3}{2}$ **5.** 2 **6.** $12x^2$ **7.** $8x$
8. $6x$ **9.** $12x$

Exercise 9.5 (page 590)

1. $-\dfrac{3x + 7}{x + 2}$ **3.** $\dfrac{x - 4}{3x^2 - x - 12}$ **5.** $f(x) + g(x)$
7. $f(x)g(x)$ **9.** domain   **11.** $f(x)$ **13.** $7x, (-\infty, \infty)$
15. $12x^2, (-\infty, \infty)$ **17.** $x, (-\infty, \infty)$ **19.** $\dfrac{4}{3}, (-\infty, 0) \cup (0, \infty)$
21. $3x - 2, (-\infty, \infty)$ **23.** $2x^2 - 5x - 3, (-\infty, \infty)$
25. $-x - 4, (-\infty, \infty)$ **27.** $\frac{x-3}{2x+1}, \left(-\infty, -\frac{1}{2}\right) \cup \left(-\frac{1}{2}, \infty\right)$
29. $-2x^2 + 3x - 3, (-\infty, \infty)$
31. $(3x - 2)/(2x^2 + 1), (-\infty, \infty)$ **33.** $3, (-\infty, \infty)$
35. $(x^2 - 4)/(x^2 - 1), (-\infty, -1) \cup (-1, 1) \cup (1, \infty)$ **37.** 7
39. 24 **41.** -1 **43.** $-\dfrac{1}{2}$ **45.** $2x^2 - 1$ **47.** $16x^2 + 8x$
49. 58 **51.** 110 **53.** 2 **55.** $9x^2 - 9x + 2$ **57.** 2
59. $2x + h$ **61.** $4x + 2h$ **63.** $2x + h + 1$
65. $2x + h + 3$ **67.** $4x + 2h + 3$ **69.** 2 **71.** $x + a$
73. $2x + 2a$ **75.** $x + a + 1$ **77.** $x + a + 3$
79. $2x + 2a + 3$ **85.** $3x^2 + 3xh + h^2$
87. $C(t) = \frac{5}{9}(2{,}668 - 200t)$

Getting Ready (page 592)

1. $y = \dfrac{x - 2}{3}$ **2.** $y = \dfrac{2x - 10}{3}$

Orals (page 598)

1. $\{(2, 1), (3, 2), (10, 5)\}$ **2.** $\{(1, 1), (8, 2), (64, 4)\}$
3. $f^{-1}(x) = 2x$ **4.** $f^{-1}(x) = \frac{1}{2}x$ **5.** no   **6.** yes

Exercise 9.6 (page 599)

1. $3 - 8i$ **3.** $18 - i$ **5.** 10 **7.** one-to-one
9. 2 **11.** x **13.** yes   **15.** no
17.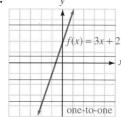
19.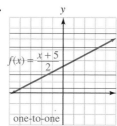
21.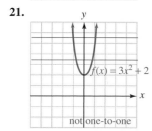
23.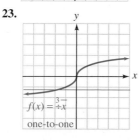

25. $\{(2, 3), (1, 2), (0, 1)\}$; yes
27. $\{(2, 1), (3, 2), (3, 1), (5, 1)\}$; no
29. $\{(1, 1), (4, 2), (9, 3), (16, 4)\}$; yes
31. $f^{-1}(x) = \frac{1}{3}x - \frac{1}{3}$ **33.** $f^{-1}(x) = 5x - 4$
35. $f^{-1}(x) = 5x + 4$ **37.** $f^{-1}(x) = \frac{5}{4}x + 5$
39. 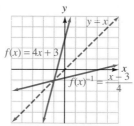   **41.**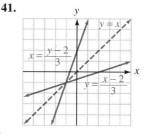

43. 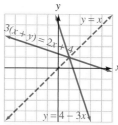   **45.**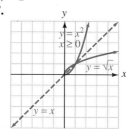

47. $y = \pm\sqrt{x - 4}$, no   **49.** $y = \sqrt[3]{x}$, yes
51. $x = |y|$, no   **53.** $f^{-1}(x) = \sqrt[3]{\dfrac{x + 3}{2}}$
55. 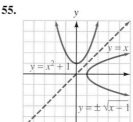   **57.**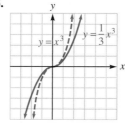

61. $f^{-1}(x) = \dfrac{x + 1}{x - 1}$

Chapter Summary (page 602)

1. y-axis   **2.** origin   **3.** x-axis   **4.** y-axis
5. none   **6.** none   **7.** y-axis   **8.** none
9. even   **10.** odd   **11.** odd   **12.** even
13. decreasing on $(-\infty, 0)$, increasing on $(0, \infty)$
14. constant on $(-\infty, 1)$, decreasing on $(1, \infty)$
15.   **16.**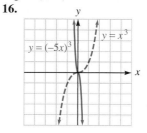

17.

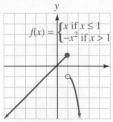

$$f(x) = \begin{cases} x \text{ if } x \le 1 \\ -x^2 \text{ if } x > 1 \end{cases}$$

increasing on $(-\infty, 1)$,
decreasing on $(1, \infty)$

18.

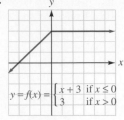

$$y = f(x) = \begin{cases} x + 3 & \text{if } x \le 0 \\ 3 & \text{if } x > 0 \end{cases}$$

increasing on $(-\infty, 0)$,
constant on $(0, \infty)$

19.

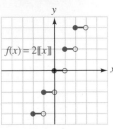

$f(x) = 2[\![x]\!]$

20.

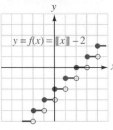

$y = f(x) = [\![x]\!] - 2$

21. vertical asymptotes: $x = 2, x = -2$; horizontal asymptote: $y = 0$ **22.** vertical asymptote: $x = 3$; slant asymptote: $y = x$

23.

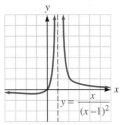

$y = \dfrac{x}{(x-1)^2}$

24.

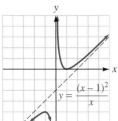

$y = \dfrac{(x-1)^2}{x}$

25.

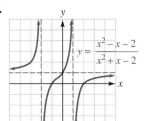

$y = \dfrac{x^2 - x - 2}{x^2 + x - 2}$

26.

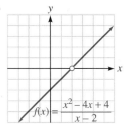

$f(x) = \dfrac{x^2 - 4x + 4}{x - 2}$

27. $\dfrac{3}{x} + \dfrac{4}{x + 1}$ **28.** $\dfrac{3}{x} + \dfrac{2}{x^2} + \dfrac{x - 1}{x^2 + 1}$

29. $\dfrac{1}{x} - \dfrac{1}{x^2 + x + 5}$ **30.** $\dfrac{1}{x + 1} - \dfrac{2}{(x + 1)^2} + \dfrac{2}{(x + 1)^3}$

31. $(f + g)(x) = 3x + 1$ **32.** $(f - g)(x) = x - 1$

33. $(f \cdot g)(x) = 2x^2 + 2x$ **34.** $(f/g)(x) = \dfrac{2x}{x + 1}$ **35.** 6

36. -1 **37.** $2(x + 1)$ **38.** $2x + 1$

39. yes   **40.** no

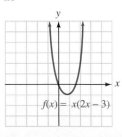

$f(x) = 2(x - 3)$

$f(x) = x(2x - 3)$

41. no   **42.** no

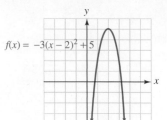

$f(x) = -3(x - 2)^2 + 5$

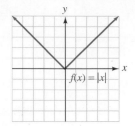

$f(x) = |x|$

43. $f^{-1}(x) = \dfrac{x + 3}{6}$ **44.** $f^{-1}(x) = \dfrac{x - 5}{4}$

45. $f^{-1}(x) = \sqrt{\dfrac{x + 1}{2}}$ **46.** $x = |y|$

Chapter 9 Test (page 606)

1. origin   **2.** y-axis   **3.** even   **4.** odd
5. decreasing on $(-\infty, 0)$, increasing on $(0, 1)$, decreasing on $(1, \infty)$ **6.** increasing on $(-\infty, 0)$, decreasing on $(2, \infty)$

7.

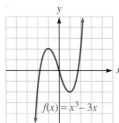

$f(x) = x^3 - 3x$

8.

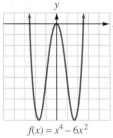

$f(x) = x^4 - 6x^2$

9.

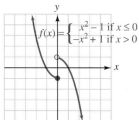

$$f(x) = \begin{cases} x^2 - 1 \text{ if } x \le 0 \\ -x^2 + 1 \text{ if } x > 0 \end{cases}$$

10.

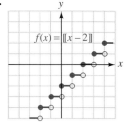

$f(x) = [\![x - 2]\!]$

11. vertical asymptotes: $x = 3, x = -3$; horizontal asymptote: $y = 0$ **12.** vertical asymptote: $x = 3$; slant asymptote: $y = x - 2$

13.

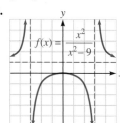

$f(x) = \dfrac{x^2}{x^2 - 9}$

14.

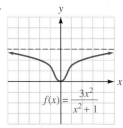

$f(x) = \dfrac{3x^2}{x^2 + 1}$

15.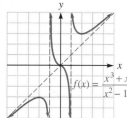

16.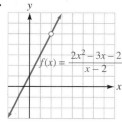

$f(x) = \dfrac{x^3 + x}{x^2 - 1}$

$f(x) = \dfrac{2x^2 - 3x - 2}{x - 2}$

17. $\dfrac{3}{2x - 3} + \dfrac{1}{x + 1}$ **18.** $\dfrac{1}{x} + \dfrac{2x + 1}{x^2 + 2}$

19. $(g + f)(x) = 5x - 1$ **20.** $(f - g)(x) = 3x + 1$

21. $(g \cdot f)(x) = 4x^2 - 4x$ **22.** $(g/f)(x) = \dfrac{x - 1}{4x}$

23. 3   **24.** -4 **25.** -8 **26.** -9 **27.** $4(x - 1)$

28. $4x - 1$ **29.** $y = \dfrac{12 - 2x}{3}$ **30.** $y = -\sqrt{\dfrac{x - 4}{3}}$

Getting Ready (page 609)
1. 8   **2.** 5   **3.** $\frac{1}{25}$ **4.** $\frac{8}{27}$

Orals (page 618)
1. 4   **2.** 25   **3.** 18   **4.** 3   **5.** $\frac{1}{4}$ **6.** $\frac{1}{25}$ **7.** $\frac{2}{9}$ **8.** $\frac{1}{27}$

Exercise 10.1 (page 618)
1. 40   **3.** 120°   **5.** exponential   **7.** $(0, \infty)$
9. increasing   **11.** $P\left(1 + \dfrac{r}{k}\right)^{kt}$ **13.** 2.6651   **15.** 36.5548
17. 8   **19.** $7^{3\sqrt{3}}$

21.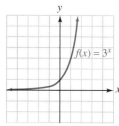

$f(x) = 3^x$

23.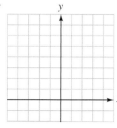

25.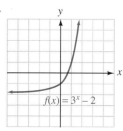

$f(x) = 3^x - 2$

27.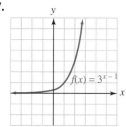

$f(x) = 3^{x-1}$

29. $b = \frac{1}{2}$ **31.** no value   **33.** $b = 2$ **35.** $b = 3$
37. increasing function   **39.** decreasing function

41. $22,080.40   **43.** $32.03   **45.** $2,273,996.13
47. $\frac{32}{243} A_0$ **49.** 5.0421×10^{-5} coulombs   **51.** $1,115.33

Getting Ready (page 620)
1. 2   **2.** 2.25   **3.** 2.44   **4.** 2.59

Orals (page 625)
1. 1   **2.** 2.72   **3.** 7.39   **4.** 20.09   **5.** 2   **6.** 2

Exercise 10.2 (page 625)
1. $4x^2\sqrt{15x}$ **3.** $10y\sqrt{3y}$ **5.** 2.72   **7.** increasing
9. $A = Pe^{rt}$

11.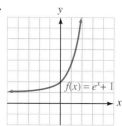

$f(x) = e^x + 1$

13.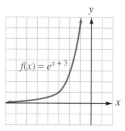

$f(x) = e^{x+3}$

15.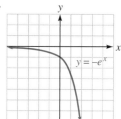

$y = -e^x$

17.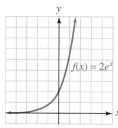

$f(x) = 2e^x$

19. no   **21.** no   **23.** $10,272.17   **25.** $6,391.10
27. $7,518.28 from annual compounding; $7,647.95 from continuous compounding   **29.** 10.6 billion   **31.** 2.6
33. 6,817   **35.** 0.16   **37.** 0   **39.** 49 mps
41. $3,094.15   **43.** 72 yr   **49.** $k = e^5$

Getting Ready (page 628)
1. 1   **2.** 25   **3.** $\frac{1}{25}$ **4.** 4

Orals (page 635)
1. 3   **2.** 2   **3.** 5   **4.** 2   **5.** 2   **6.** 2   **7.** $\frac{1}{4}$ **8.** $\frac{1}{2}$
9. 2

Exercise 10.3 (page 635)
1. 10   **3.** $0; -\frac{5}{9}$ does not check   **5.** $x = b^y$ **7.** range
9. inverse   **11.** exponent   **13.** $(b, 1), (1, 0)$ **15.** $20 \log \frac{E_O}{E_I}$
17. $3^3 = 27$ **19.** $\left(\frac{1}{2}\right)^2 = \frac{1}{4}$ **21.** $4^{-3} = \frac{1}{64}$ **23.** $\left(\frac{1}{2}\right)^3 = \frac{1}{8}$
25. $\log_6 36 = 2$ **27.** $\log_5 \frac{1}{25} = -2$ **29.** $\log_{1/2} 32 = -5$
31. $\log_x z = y$ **33.** 4   **35.** 2   **37.** 3   **39.** $\frac{1}{2}$ **41.** -3
43. 49   **45.** 6   **47.** 5   **49.** $\frac{1}{25}$ **51.** $\frac{1}{6}$ **53.** $-\frac{3}{2}$ **55.** $\frac{2}{3}$
57. 5   **59.** $\frac{3}{2}$ **61.** 4   **63.** 8   **65.** 4   **67.** 4   **69.** 3
71. 100   **73.** 0.9165   **75.** -2.0620 **77.** 25.25
79. 17,378.01   **81.** 0.00   **83.** 8

85. increasing

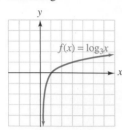

87. decreasing

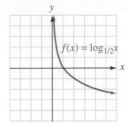

89.

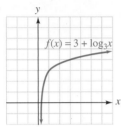

91.

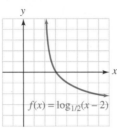

93.

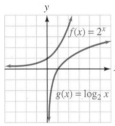

95.

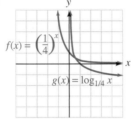

97. no value of b   **99.** 3   **101.** 29.0 dB   **103.** 49.5 dB
105. 4.4   **107.** 4   **109.** 4.2 yr   **111.** 10.8 yr

Getting Ready (page 638)
1. 2   **2.** −3   **3.** 1   **4.** 0

Orals (page 642)
1. $e^y = x$ **2.** $\ln b = a$ **3.** $t = \frac{\ln 2}{r}$

Exercise 10.4 (page 642)

1. $y = 5x$ **3.** $y = -\frac{3}{2}x + \frac{13}{2}$ **5.** $x = 5$ **7.** $\dfrac{1}{2x - 3}$

9. $\dfrac{x + 1}{3(x - 2)}$ **11.** $\log_e x$ **13.** $(-\infty, \infty)$ **15.** 10

17. $\dfrac{\ln 2}{r}$ **19.** 3.2288   **21.** 2.2915   **23.** −0.1592

25. none   **27.** 9.9892   **29.** 23.8075   **31.** 0.0089
33. 61.9098   **35.** no   **37.** no
39.   **41.**   **43.** 5.8 yr
  45. 9.2 yr
  47. about 2.9 hr

Getting Ready (page 644)
1. x^{m+n} **2.** 1   **3.** x^{mn} **4.** x^{m-n}

Orals (page 651)
1. 2   **2.** 5   **3.** 343   **4.** $\frac{1}{4}$ **5.** 2   **6.** 2   **7.** $\frac{1}{4}$ **8.** $\frac{1}{2}$
9. 2

Exercise 10.5 (page 651)
1. $-\frac{7}{6}$ **3.** $\left(1, -\frac{1}{2}\right)$ **5.** 0   **7.** M, N **9.** x, y **11.** x
13. $\neq$ **15.** 0   **17.** 7   **19.** 10   **21.** 1   **23.** 0   **25.** 7
27. 10   **29.** 1   **37.** $\log_b x + \log_b y + \log_b z$
39. $\log_b 2 + \log_b x - \log_b y$ **41.** $3\log_b x + 2\log_b y$
43. $\frac{1}{2}(\log_b x + \log_b y)$ **45.** $\log_b x + \frac{1}{2}\log_b z$

47. $\frac{1}{3}\log_b x - \frac{1}{4}\log_b y - \frac{1}{4}\log_b z$ **49.** $\log_b \dfrac{x + 1}{x}$

51. $\log_b x^2 y^{1/2}$ **53.** $\log_b \dfrac{z^{1/2}}{x^3 y^2}$ **55.** $\log_b \dfrac{\frac{x}{z} + x}{\frac{y}{z} + y} = \log_b \dfrac{x}{y}$

57. false   **59.** false   **61.** true   **63.** false   **65.** true
67. true   **69.** 1.4472   **71.** 0.3521   **73.** 1.1972
75. 2.4014   **77.** 2.0493   **79.** 0.4682   **81.** 1.7712
83. −1.0000   **85.** 1.8928   **87.** 2.3219   **89.** 4.77
91. from 2.5119×10^{-8} to 1.585×10^{-7}
93. It will increase by $k \ln 2$.
95. The intensity must be cubed.

Getting Ready (page 653)
1. $2 \log x$ **2.** $\frac{1}{2} \log x$ **3.** 0   **4.** $2b \log a$

Orals (page 660)
1. $x = \dfrac{\log 5}{\log 3}$ **2.** $x = \dfrac{\log 3}{\log 5}$ **3.** $x = -\dfrac{\log 7}{\log 2}$ **4.** $x = -\dfrac{\log 1}{\log 6}$
5. $x = 2$ **6.** $x = \frac{1}{2}$ **7.** $x = 10$ **8.** $x = 10$

Exercise 10.6 (page 660)
1. 0, 5   **3.** $\frac{2}{3}, -4$ **5.** exponential   **7.** $A_0 2^{-t/h}$
9. 1.1610   **11.** 3.9120   **13.** 22.1184   **15.** 1.2702
17. 1.7095   **19.** 0   **21.** ±1.0878   **23.** 0, 1.0566
25. 3, −1   **27.** −2, −2   **29.** 0   **31.** 0.2789   **33.** 1.8
35. 3, −1   **37.** 2   **39.** 3   **41.** −7   **43.** 4
45. 10, −10   **47.** 50   **49.** 20   **51.** 10   **53.** 10
55. 1, 100   **57.** no solution   **59.** 6   **61.** 9   **63.** 4
65. 1, 7   **67.** 20   **69.** 8   **71.** 5.1 yr   **73.** 42.7 days
75. about 4,200 yr   **77.** 5.6 yr   **79.** 5.4 yr
81. because $\ln 2 \approx 0.7$ **83.** 25.3 yr   **85.** 2.828 times larger
87. 13.3   **91.** $x \leq 3$

Chapter Summary (page 664)
1. $5^{2\sqrt{2}}$ **2.** $2^{\sqrt{10}}$
3.   **4.**

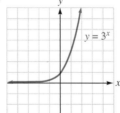

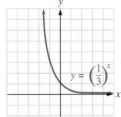

5. $a = 1, b = 6$ **6.** D: $(-\infty, \infty)$, R: $(0, \infty)$

7.
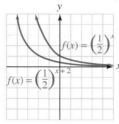

8.
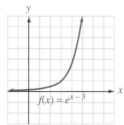

9. $2,189,703.45   **10.** $2,324,767.37

11.
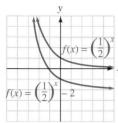

12.
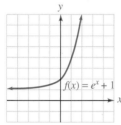

13. about 582,000,000   **14.** D: $(0, \infty)$, R: $(-\infty, \infty)$ **15.** 2
16. $-\frac{1}{2}$ **17.** 0   **18.** -2 **19.** $\frac{1}{2}$ **20.** $\frac{1}{3}$ **21.** 32
22. 9   **23.** 27   **24.** -1 **25.** $\frac{1}{8}$ **26.** 2   **27.** 4   **28.** 2
29. 10   **30.** $\frac{1}{25}$ **31.** 5   **32.** 3

33.
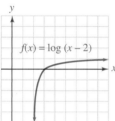

34.
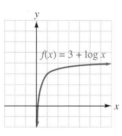

35.
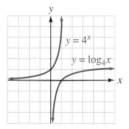

36.
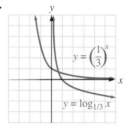

37. 53 dB   **38.** 4.4   **39.** 6.1137   **40.** -0.1111
41. 10.3398   **42.** 2.5715

43.
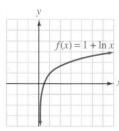

44.
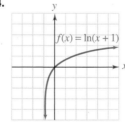

45. 23 yr   **46.** 0   **47.** 1   **48.** 3   **49.** 4
50. 4   **51.** 0   **52.** 7   **53.** 3   **54.** 4
55. 9   **56.** $2\log_b x + 3\log_b y - 4\log_b z$

57. $\frac{1}{2}(\log_b x - \log_b y - 2\log_b z)$ **58.** $\log_b \dfrac{x^3 z^7}{y^5}$

59. $\log_b \dfrac{y^3\sqrt{x}}{z^7}$ **60.** 3.36   **61.** 1.56

62. 2.64   **63.** -6.72 **64.** 1.7604
65. about 7.94×10^{-4} gram-ions per liter   **66.** $k \ln 2$ less
67. $\frac{\log 7}{\log 3} \approx 1.7712$ **68.** 2   **69.** 31.0335   **70.** 2
71. $\frac{\log 3}{\log 3 - \log 2} \approx 2.7095$ **72.** $-1, -3$ **73.** 25, 4   **74.** 4
75. 2   **76.** 4, 3   **77.** 6   **78.** 31   **79.** $\frac{\ln 9}{\ln 2} \approx 3.1699$

80. no solution   **81.** $\dfrac{e}{e-1} \approx 1.5820$ **82.** 1

83. about 3,300 yr

Chapter 10 Test (page 669)

1.
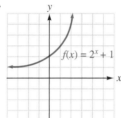

2.
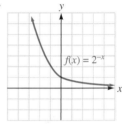

3. $\frac{3}{64}$ g   **4.** $1,060.90   **5.**
6. $4,451.08   **7.** 2
8. 3   **9.** $\frac{1}{27}$ **10.** 10
11. 2   **12.** $\frac{27}{8}$

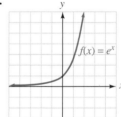

13.
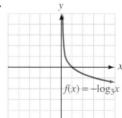

14.
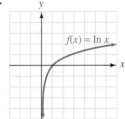

15. $2\log a + \log b + 3\log c$ **16.** $\frac{1}{2}(\ln a - 2\ln b - \ln c)$

17. $\log \dfrac{b\sqrt{a+2}}{c^3}$ **18.** $\log \dfrac{\sqrt[3]{a}}{c\sqrt[3]{b^2}}$ **19.** 1.3801

20. 0.4259   **21.** $\frac{\log 3}{\log 7}$ or $\frac{\ln 3}{\ln 7}$ **22.** $\frac{\log e}{\log \pi}$ or $\frac{\ln e}{\ln \pi}$ **23.** true
24. false   **25.** false   **26.** false   **27.** 6.4   **28.** 46
29. $\frac{\log 3}{\log 5}$ **30.** $\frac{\log 3}{(\log 3) - 2}$ **31.** 1   **32.** 10

Cumulative Review Exercises (page 670)
1. 16   **2.** $\frac{1}{2}$ **3.** y^2 **4.** $x^{17/12}$ **5.** $x^{4/3} - x^{2/3}$
6. $\frac{1}{x} + 2 + x$ **7.** $-3x$ **8.** $4t\sqrt{3t}$ **9.** $4x$ **10.** $x + 3$
11. $7\sqrt{2}$ **12.** $-12\sqrt[4]{2} + 10\sqrt[4]{3}$ **13.** $-18\sqrt{6}$
14. $\dfrac{5\sqrt[3]{x^2}}{x}$ **15.** $\dfrac{x + 3\sqrt{x} + 2}{x - 1}$ **16.** $\sqrt{xy}$ **17.** 11

18. 9 **19.** $1, -\frac{3}{2}$ **20.** $-\frac{2}{3} \pm \frac{\sqrt{7}}{3}$ **21.** $-1 + 8i$

22. $19 - i$ **23.** $29 + 0i$ **24.** $6 + 12i$ **25.** $15 + 8i$
26. $\frac{5}{17} - \frac{14}{17}i$ **27.** $\sqrt{13}$ **28.** $\sqrt{61}$ **29.** 4 **30.** $\left(1, \frac{1}{2}\right)$
31. 5 **32.** 27 **33.** $12x^2 - 12x + 5$ **34.** $6x^2 + 3$

35. $f^{-1}(x) = \dfrac{x - 2}{3}$ **36.** $f^{-1}(x) = \sqrt[3]{x - 4}$ **37.** $2^y = x$

38. $\log_3 a = b$ **39.** 5 **40.** 3 **41.** $\frac{1}{27}$ **42.** 1
43. $y = 2^x$ **44.** x **45.** 1.9912 **46.** 0.301

Getting Ready (page 673)
1. 0 **2.** 0 **3.** 2 **4.** 56

Orals (page 680)
1. 24 **2.** 14

Exercise 11.1 (page 680)
1. QIV **3.** QI **5.** 10 **7.** 1 **9.** whole **11.** any
13. factor **15.** 4 **17.** 67 **19.** 70,249
21. 128.3085938 **23.** true **25.** true **27.** false
29. true **31.** $\{-1, -5, 3\}$ **33.** $\left\{1, 1, \sqrt{3}, -\sqrt{3}\right\}$
35. $\{2, 3, i, -i\}$ **37.** $x^2 - 9x + 20$
39. $x^3 - 3x^2 + 3x - 1$ **41.** $x^3 - 11x^2 + 38x - 40$
43. $x^4 - 3x^2 + 2$ **45.** $x^3 - \sqrt{2}x^2 + x - \sqrt{2}$
47. $x^3 - 2x^2 + 2x$ **49.** $(x - 1)(3x^2 + x - 5) - 9$
51. $(x - 3)(3x^2 + 7x + 15) + 41$
53. $(x + 1)(3x^2 - 5x - 1) - 3$
55. $(x + 3)(3x^2 - 11x + 27) - 85$ **57.** $x^2 + 2x + 3$
59. $7x^2 - 10x + 5 + \frac{-4}{x + 1}$
61. $4x^3 + 9x^2 + 27x + 80 + \frac{245}{x - 3}$
63. $3x^4 + 12x^3 + 48x^2 + 192x$ **65.** 47 **67.** -569
69. $\frac{15}{8}$ **71.** 5 **73.** 0 **75.** 384 **77.** $16 + 2i$
79. $40 - 40i$ **83.** 0

Getting Ready (page 682)
1. 3; first, one **2.** $6, -1$; second, two
3. $0, 3, -3$; third, three **4.** $1, -1, 2, -2$, fourth, four

Orals (page 688)
1. 4 **2.** 6 **3.** one **4.** three or one **5.** none **6.** one

Exercise 11.2 (page 688)
1. k units to the right **3.** reflected about the y-axis
5. stretched vertically by a factor of k **7.** zero
9. conjugate **11.** 0 **13.** 2 **15.** lower bound **17.** 10
19. 4 **21.** 4 **23.** $x^3 - 3x^2 + x - 3 = 0$
25. $x^3 - 6x^2 + 13x - 10 = 0$
27. $x^4 - 5x^3 + 7x^2 - 5x + 6 = 0$
29. $x^4 - 2x^3 + 3x^2 - 2x + 2 = 0$
31. 0 or 2 positive; 1 negative; 0 or 2 nonreal
33. 0 positive; 1 or 3 negative; 0 or 2 nonreal
35. 0 positive; 0 negative; 4 nonreal
37. 1 positive; 1 negative; 2 nonreal
39. 0 positive; 0 negative; 10 nonreal

41. 0 positive; 0 negative; 8 nonreal; yes
43. 1 positive; 1 negative; 2 nonreal **45.** $-2, 4$ **47.** $-1, 1$
49. $-4, 6$ **51.** $-4, 3$ **53.** $-2, 4$

Getting Ready (page 689)
1. no **2.** no

Orals (page 696)
1. 1 **2.** 2

Exercise 11.3 (page 696)
1. $6ab^2\sqrt{2abc}$ **3.** $8a\sqrt{2b}$ **5.** $\sqrt{3} + 1$ **7.** -7
9. root **11.** $1, -1, 5$ **13.** $3, -3, 2$ **15.** $1, -1, 2$
17. $1, 2, 3, 4$ **19.** $2, -5$ **21.** $1, -1, 2, -2, -3$
23. $0, 2, 2, 2, -2, -2, -2$ **25.** $\frac{2}{3}$ **27.** $-1, 2, 3, \frac{2}{3}$
29. $-3, \frac{1}{3}, \frac{1}{2}, \frac{1}{2}$ **31.** $2, 2, -3, -\frac{1}{3}, \frac{1}{2}$ **33.** $\frac{3}{2}, \frac{2}{3}, -\frac{3}{5}$
35. $\frac{2}{3}, -\frac{3}{5}, 4$ **37.** $-\frac{1}{2}, \frac{3}{2}, \frac{6}{5}$ **39.** $3, 1 - i$ **41.** $3, -3, 1 - i$
43. $-\frac{2}{3}, 3, -1$ **45.** $\frac{1}{2}, \frac{1}{2}, \frac{1}{2}, 1, 1$ **47.** 10, 20, 60 ohms
51. $(3, 7)$ or approx. $(1.54, 13.63)$

Getting Ready (page 698)
1. positive **2.** negative

Orals (page 702)
1. 3.5 **2.** 2.75 **3.** 1.5 **4.** 0.25

Exercise 11.4 (page 702)
1. vertical: $x = 2, x = -2$; horizontal: $y = 0$; slant: none **3.**
vertical: $x = 2$; horizontal: none; slant: $y = x + 2$
5. $P(a)$ and $P(b)$ **7.** x_l and x_r
9. $P(-2) - 3$; $P(-1) = -2$ **11.** $P(4) = -40$; $P(5) = 30$
13. $P(1) = 8$; $P(2) = -1$ **15.** $P(2) = -72$; $P(3) = 154$
17. $P(0) = 10$; $P(1) = -60$ **19.** 1.7 **21.** -2.2
23. 1.7 **25.** -1.2 **27.** $-2.2, 2.2$
29. 1, 2; the bisection method fails to find the solution 2
31. $(0.83, 0.56), (-0.83, -0.56)$

Chapter Summary (page 705)
1. 1 **2.** 66 **3.** 241 **4.** 34 **5.** false **6.** false
7. true **8.** true **9.** $2x^3 - 5x^2 - x + 6$
10. $2x^3 + 3x^2 - 8x + 3$ **11.** $x^4 + 3x^3 - 9x^2 + 3x - 10$
12. $x^4 + x^3 - 5x^2 + x - 6$
13. $3x^3 + 9x^2 + 29x + 90$ with remainder of 277
14. $2x^3 + 4x^2 + 5x + 13$ with remainder of 25
15. $5x^4 - 14x^3 + 31x^2 - 64x + 129$ with remainder of -259
16. $4x^4 - 2x^3 + x^2 + 2x$ with remainder of 1 **17.** 6
18. 6 **19.** 65 **20.** 1,984 **21.** $2 - i$ **22.** i
23. 0 or 2 positive; 0 or 2 negative; 0, 2, or 4 nonreal
24. 1 or 3 positive; 1 negative; 0 or 2 nonreal
25. 1 positive; 0, 2, or 4 negative; 0, 2, or 4 nonreal
26. 1 or 3 positive; 0 or 2 negative; 2, 4, or 6 nonreal
27. 0 positive; 0 negative; 4 nonreal
28. 0 positive; 1 negative; 6 nonreal **29.** $-1, 2$ **30.** $-5, 2$
31. $-5, -\frac{3}{2}, -2$ **32.** $\frac{1}{3}$ **33.** $2, -2, \frac{3}{2}, -\frac{3}{2}$
34. $4, 4, -2, -\frac{1}{2}$ **35.** $P(-1) = -9$; $P(0) = 18$

36. $P(1) = -8$; $P(2) = 21$ **37.** 4.2   **38.** 2.5   **39.** 1.67
40. 0.67   **41.** 10 m by 6 m; 12 m by 5 m; 15 m by 4 m
42. 17.27 ft

Chapter Test (page 708)
1. 1   **2.** −30   **3.** $x^3 - 4x^2 - 5x$ **4.** $x^4 - 2x^2 - 3$
5. 5   **6.** −28   **7.** $\frac{11}{3}$ **8.** $6 - 3i$
9. $(x - 2)(2x^2 + x - 2) - 5$
10. $(x + 1)(2x^2 - 5x + 1) - 2$ **11.** $2x + 3$ **12.** $3x^2 + x$
13. $x^3 - 2x^2 + x - 2 = 0$ **14.** $x^3 - 5x^2 + 9x - 5 = 0$
15. 1 or 3 positive; 0 or 2 negative; 0, 2, or 4 nonreal
16. 1 positive; 0 or 2 negative; 0 or 2 nonreal   **17.** −3, 3
18. −1, 6   **19.** 2, −3, $-\frac{1}{2}$ **20.** 3.3

Getting Ready (page 710)
1. $x^2 - 4x + 4$ **2.** $x^2 + 8x + 16$ **3.** $\frac{81}{4}$ **4.** 36

Orals (page 716)
1. (0, 0), 12   **2.** (0, 0), 11   **3.** (2, 0), 4   **4.** (0, −1), 3

Exercise 12.1 (page 716)
1. 4   **3.** $\frac{49}{4}$ **5.** $x = 1, -5$ **7.** $x = -2, 9$
9. circle, plane   **11.** 2, −5; 3   **13.** 0, 0; $\sqrt{5}$
15. $x^2 + y^2 = 49$ **17.** $(x - 2)^2 + (y + 2)^2 = 17$
19. $x^2 + y^2 = 1$ **21.** $(x - 6)^2 + (y - 8)^2 = 25$
23. $(x + 2)^2 + (y - 6)^2 = 144$ **25.** $x^2 + y^2 = 2$
27.
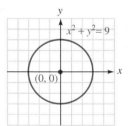
29.
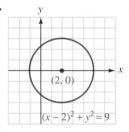
31.
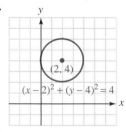
33.
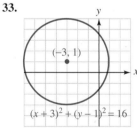
35.
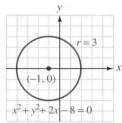
37.
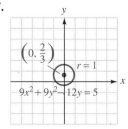

39.
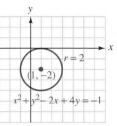
41.
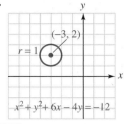

43.

45.

47. yes   **49.** 60 mi   **51.** $(x - 4)^2 + y^2 = 16$
55. $x^2 + y^2 - 6x - 6y - 7 = 0$
57. $(x - 1)^2 + (y + 2)^2 = 36$ **59.** $x^2 + (y - 3)^2 = 25$

Getting Ready (page 719)
1. $p^2 + 2ph + h^2$ **2.** $y^2 - 2ky + k^2$ **3.** $4p(x - h)$
4. $(t - 4)^2$

Orals (page 725)
1. down   **2.** up   **3.** left   **4.** right

Exercise 12.2 (page 725)
1. 5, $-\frac{7}{3}$ **3.** 3, $-\frac{1}{4}$ **5.** directrix, focus
7. parabola, (3, 2), right   **9.** to the left   **11.** downward
13. $x^2 = 12y$ **15.** $y^2 = -12x$
17. $(x - 3)^2 = -12(y - 5)$ **19.** $(x - 3)^2 = -28(y - 5)$
21. $(y - 2)^2 = -2(x - 2)$ or $(x - 2)^2 = -2(y - 2)$
23. $(x + 4)^2 = -\frac{16}{3}(y - 6)$ or $(y - 6)^2 = \frac{9}{4}(x + 4)$
25. $(y - 8)^2 = -4(x - 6)$ **27.** $(x - 3)^2 = \frac{1}{2}(y - 1)$
29.
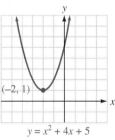
$$y = x^2 + 4x + 5$$
or
$$y - 1 = (x + 2)^2$$
31.
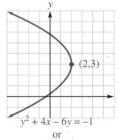
$$y^2 + 4x - 6y = -1$$
or
$$(y - 3)^2 = -4(x - 2)$$
33.
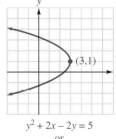
$$y^2 + 2x - 2y = 5$$
or
$$(y - 1)^2 = -2(x - 3)$$
35.
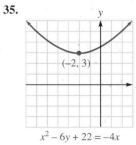
$$x^2 - 6y + 22 = -4x$$
or
$$(x + 2)^2 = 6(y - 3)$$

37.

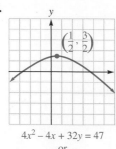

$4x^2 - 4x + 32y = 47$
or
$\left(x - \frac{1}{2}\right)^2 = -8\left(y - \frac{3}{2}\right)$

39. 2 ft **41.** $\frac{9}{4}$ ft
43. 300 ft **45.** $x^2 = \frac{-45}{2}y$
47. about 520 ft
51. $y = x^2 + 4x + 3$

Getting Ready (page 729)
1. $y = \pm b$ **2.** $x = \pm a$

Orals (page 737)
1. $(\pm 3, 0), (0, \pm 4)$ **2.** $(\pm 5, 0), (0, \pm 6)$ **3.** $(2, 0)$
4. $(0, -1)$

Exercise 12.3 (page 737)
1. $x^2 + 4x$ **3.** $x^2 + 4x + 4$ **5.** $x = -1, 5$ **7.** $x = 0, 7$
9. sum, constant **11.** vertices **13.** $a, 0; -a, 0$
15. $\dfrac{x^2}{25} + \dfrac{y^2}{16} = 1$ **17.** $\dfrac{9x^2}{16} + \dfrac{9y^2}{25} = 1$ **19.** $\dfrac{x^2}{7} + \dfrac{y^2}{16} = 1$
21. $\dfrac{(x-3)^2}{4} + \dfrac{(y-4)^2}{9} = 1$ **23.** $\dfrac{(x-3)^2}{9} + \dfrac{(y-4)^2}{4} = 1$
25. $\dfrac{(x-3)^2}{41} + \dfrac{(y-4)^2}{16} = 1$ **27.** $\dfrac{x^2}{36} + \dfrac{(y-4)^2}{20} = 1$
29. $\dfrac{x^2}{100} + \dfrac{y^2}{64} = 1$

31.

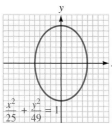

$\dfrac{x^2}{25} + \dfrac{y^2}{49} = 1$

33.

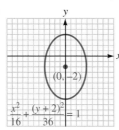

$(0, -2)$

$\dfrac{x^2}{16} + \dfrac{(y+2)^2}{36} = 1$

35.

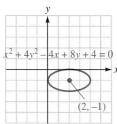

$x^2 + 4y^2 - 4x + 8y + 4 = 0$

$(2, -1)$

37.

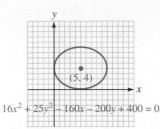

$(5, 4)$

$16x^2 + 25y^2 - 160x - 200y + 400 = 0$

39. 199,395 mi

41. $\dfrac{x^2}{2,500} + \dfrac{y^2}{900} = 1; 36$ m
43. about 20.8 in.

Getting Ready (page 740)
1. $y = \pm 2.0$ **2.** $y = \pm 2.9$

Orals (page 747)
1. $(\pm 3, 0)$ **2.** $(0, \pm 5)$

Exercise 12.4 (page 747)
1. $f^{-1}(x) = \dfrac{x+2}{3}$ **3.** $f^{-1}(x) = \dfrac{2x}{5-x}$
5. $f(g(x)) = (x+1)^4 + 1$ **7.** $f(f(x)) = (x^2+1)^2 + 1$
9. difference, constant **11.** $a, 0; -a, 0$ **13.** transverse axis
15. $\dfrac{x^2}{25} - \dfrac{y^2}{24} = 1$ **17.** $\dfrac{(x-2)^2}{4} - \dfrac{(y-4)^2}{9} = 1$
19. $\dfrac{(y-3)^2}{9} - \dfrac{(x-5)^2}{9} = 1$ **21.** $\dfrac{y^2}{9} - \dfrac{x^2}{16} = 1$
23. $\dfrac{(x-1)^2}{4} - \dfrac{(y+3)^2}{16} = 1$ or $\dfrac{(y+3)^2}{4} - \dfrac{(x-1)^2}{16} = 1$
25. $\dfrac{x^2}{10} - \dfrac{3y^2}{20} - 1$ **27.** 24 sq units **29.** 12 sq units
31. $\dfrac{(x+2)^2}{4} - \dfrac{4(y+4)^2}{81} = 1$ or $\dfrac{(y+4)^2}{4} - \dfrac{4(x+2)^2}{81} = 1$
33. $\dfrac{x^2}{36} - \dfrac{16y^2}{25} = 1$

35.

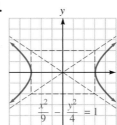

$\dfrac{x^2}{9} - \dfrac{y^2}{4} = 1$

37.

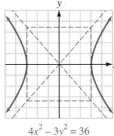

$4x^2 - 3y^2 = 36$

39.

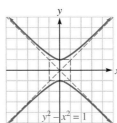

$y^2 - x^2 = 1$

41.

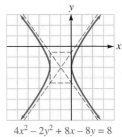

$4x^2 - 2y^2 + 8x - 8y = 8$

43.
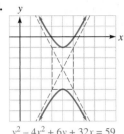
$y^2 - 4x^2 + 6y + 32x = 59$

45.
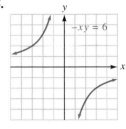
$-xy = 6$

47. $\dfrac{(x-3)^2}{9} - \dfrac{(y-1)^2}{16} = 1$ **49.** $4x^2 - 5y^2 - 60y = 0$

51. $\dfrac{x^2}{100{,}000{,}000^2} - \dfrac{y^2}{200{,}000{,}000^2} = 1$

53. hyperbola: $\dfrac{x^2}{144} - \dfrac{y^2}{25} = 1$

Getting Ready (page 750)
1. $7x^2 - y^2 = 44$ **2.** $5x^2 - 3y^2 = 8$

Orals (page 755)
1. 0, 1, 2   **2.** 0, 1, 2   **3.** 0, 1, 2, 3, 4   **4.** 0, 1, 2, 3, 4

Exercise 12.5 (page 755)
1. vertical: $x = 1$; horizontal: $y = 3$
3. vertical: $x = 1$, $x = -1$; horizontal: $y = 0$ **5.** y-axis
7. origin   **9.** conic

11.
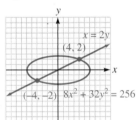
$x = 2y$
$(4, 2)$
$(-4, -2)$ $8x^2 + 32y^2 = 256$

13.
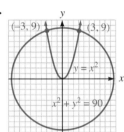
$(-3, 9)$ $(3, 9)$
$y = x^2$
$x^2 + y^2 = 90$

15.
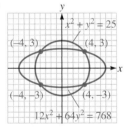
$x^2 + y^2 = 25$
$(-4, 3)$ $(4, 3)$
$(-4, -3)$ $(4, -3)$
$12x^2 + 64y^2 = 768$

17.
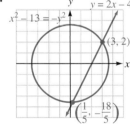
$y = 2x - 4$
$x^2 - 13 = -y^2$
$(3, 2)$
$\left(\dfrac{1}{5}, -\dfrac{18}{5}\right)$

19.
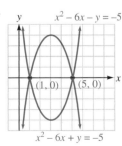
$x^2 - 6x - y = -5$
$(1, 0)$ $(5, 0)$
$x^2 - 6x + y = -5$

21. $(1, 2), (-1, 0)$
23. $(1, 0.67), (-1, -0.67)$
25. $(3, 0), (0, 5)$ **27.** $(1, 1)$
29. $(1, 2), (2, 1)$
31. $(-2, 3), (2, 3)$
33. $\left(\sqrt{5}, 5\right), \left(-\sqrt{5}, 5\right)$

35. $(3, 2), (3, -2), (-3, 2), (-3, -2)$
37. $(2, 4), (2, -4), (-2, 4), (-2, -4)$
39. $\left(-\sqrt{15}, 5\right), \left(\sqrt{15}, 5\right), (-2, -6), (2, -6)$
41. $(0, -4), (-3, 5), (3, 5)$
43. $(-2, 3), (2, 3), (-2, -3), (2, -3)$ **45.** $(3, 3)$
47. $(6, 2), (-6, -2), \left(\sqrt{42}, 0\right), \left(-\sqrt{42}, 0\right)$ **49.** $\left(\dfrac{1}{2}, \dfrac{1}{3}\right), \left(\dfrac{1}{3}, \dfrac{1}{2}\right)$
51. 7 by 9 cm   **53.** either \$750 at 9% or \$900 at 7.5%
55. about 23 mi

Chapter Summary (page 759)
1. $x^2 + y^2 = 16$ **2.** $x^2 + y^2 = 100$
3. $(x-3)^2 + (y+2)^2 = 25$ **4.** $(x+2)^2 + (y-4)^2 = 25$
5. $(x-5)^2 + (y-10)^2 = 85$ **6.** $(x-2)^2 + (y-2)^2 = 89$
7. $(x-3)^2 + (y+2)^2 = 16$ **8.** $(x+2)^2 + (y-5)^2 = 16$
9. $y^2 = -2x$ **10.** $x^2 = 16y$ **11.** $(x+2)^2 = -\dfrac{4}{11}(y-3)$
12.

$(1, 2)$
$x^2 - 4y - 2x + 9 = 0$

13.
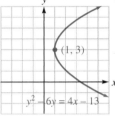
$(1, 3)$
$y^2 - 6y = 4x - 13$

14. $\dfrac{x^2}{36} + \dfrac{y^2}{16} = 1$ **15.** $\dfrac{y^2}{25} + \dfrac{x^2}{4} = 1$

16. $\dfrac{(x+2)^2}{16} + \dfrac{(y-3)^2}{9} = 1$

17. $\dfrac{(x-2)^2}{1} + \dfrac{(y+1)^2}{4} = 1$
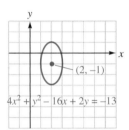
$(2, -1)$
$4x^2 + y^2 - 16x + 2y = -13$

18. $\dfrac{x^2}{4} - \dfrac{y^2}{12} = 1$ **19.** $\dfrac{y^2}{9} - \dfrac{x^2}{16} = 1$
20. $\dfrac{x^2}{9} - \dfrac{(y-3)^2}{16} = 1$ **21.** $\dfrac{y^2}{9} - \dfrac{(x-3)^2}{16} = 1$

22. $y = \pm\dfrac{4}{5}x$

23. $\dfrac{(x-1)^2}{4} - \dfrac{(y+2)^2}{9} = 1$
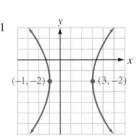
$(-1, -2)$ $(3, -2)$
$9x^2 - 4y^2 - 16y - 18x = 43$

24.

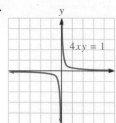

25.

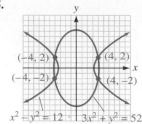

26.

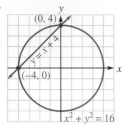

27.

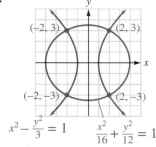

$x^2 - \dfrac{y^2}{3} = 1$ $\dfrac{x^2}{16} + \dfrac{y^2}{12} = 1$

28. $(-4, 2), (-4, -2), (4, 2), (4, -2)$ **29.** $(0, 4), \left(2\sqrt{3}, -2\right)$

30. $(-2, 3), (-2, -3), (2, 3), (2, -3)$

Chapter 12 Test (page 764)

1. $(x - 2)^2 + (y - 3)^2 = 9$ **2.** $(x - 2)^2 + (y - 3)^2 = 41$

3. $(x - 2)^2 + (y + 5)^2 = 169$

4.

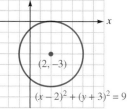

5. $(x - 3)^2 = 16(y - 2)$

6. $(y + 6)^2 = -4(x - 4)$

7. $(x - 2)^2 = \frac{4}{3}(y + 3)$ or $(y + 3)^2 = -\frac{9}{2}(x - 2)$

8.

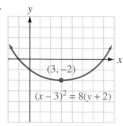

9. $\dfrac{x^2}{100} + \dfrac{y^2}{64} = 1$ **10.** $\dfrac{x^2}{169} + \dfrac{y^2}{144} = 1$ **11.** $\dfrac{x^2}{4} + \dfrac{y^2}{36} = 1$

12.

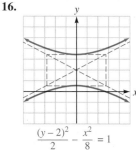

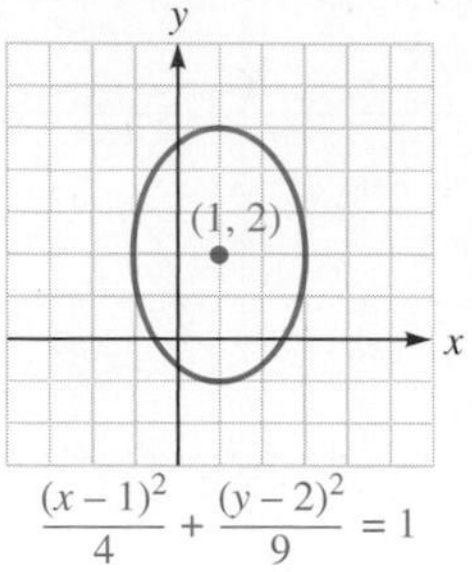

$\dfrac{(x - 1)^2}{4} + \dfrac{(y - 2)^2}{9} = 1$

13. $\dfrac{x^2}{25} - \dfrac{y^2}{144} = 1$

14. $\dfrac{x^2}{36} - \dfrac{4y^2}{25} = 1$

15. $\dfrac{(x - 2)^2}{64} - \dfrac{(y + 1)^2}{36} = 1$

16.

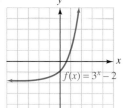

$\dfrac{(y - 2)^2}{2} - \dfrac{x^2}{8} = 1$

17. $\left(\sqrt{7}, 4\right), \left(-\sqrt{7}, 4\right)$

18. $\left(3\sqrt{2}, 3\right), \left(-3\sqrt{2}, 3\right), \left(3\sqrt{2}, -3\right), \left(-3\sqrt{2}, -3\right)$

19. $(y - 2)^2 = 6(x + 3)$; parabola

20. $\dfrac{(x - 1)^2}{3} + \dfrac{(y + 2)^2}{2} = 1$; ellipse

Cumulative Review Exercises (page 765)

1.

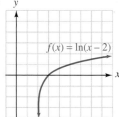

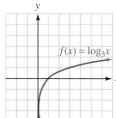

2.

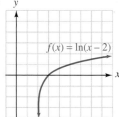

3.

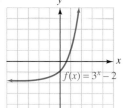

4.

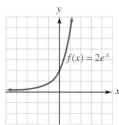

5. 6   **6.** -3 **7.** 3   **8.** 2   **9.** $\log a + \log b + \log c$

10. $2 \log a + \log b - \log c$ **11.** $\frac{1}{2}(\log a + \log b - 3 \log c)$

12. $\frac{1}{2}\log a + \log b - \log c$ **13.** $\log \dfrac{a^3}{b^3}$ **14.** $\log \dfrac{\sqrt{ab^3}}{\sqrt[3]{c^2}}$

15. $x = \dfrac{\log 8}{\log 3} - 1$ **16.** $x = -1$ **17.** $x = 500$

18. $x = \sqrt{11}$ **19.** 9   **20.** -36 **21.** 4   **22.** $2 - i$

23. a factor   **24.** not a factor   **25.** a factor   **26.** a factor
27. 12   **28.** 2,000
29. 2 or 0 positive; 2 or 0 negative; 4, 2, or 0 nonreal
30. 1 positive; 3 or 1 negative; 2 or 0 nonreal   **31.** $-1, -3, 3$
32. $-1, 1, 2$ **33.** 5   **34.** 27   **35.** $12x^2 - 12x + 5$
36. $6x^2 + 3$ **37.** $f^{-1}(x) = \dfrac{x+2}{3}$ **38.** $f^{-1}(x) = \sqrt[3]{x+4}$

39.

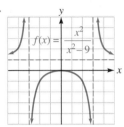

40.

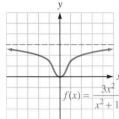

41.

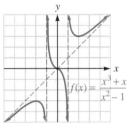

42.

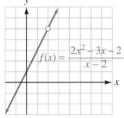

43. $\dfrac{3}{2x-3} + \dfrac{1}{x+1}$ **44.** $\dfrac{1}{x} + \dfrac{2x+1}{x^2+2}$

Getting Ready (page 768)
1. $x^2 + 4x + 4$ **2.** $x^2 - 6x + 9$ **3.** $x^3 + 3x^2 + 3x + 1$
4. $x^3 - 6x^2 + 12x - 8$

Orals (page 773)
1. 1   **2.** 2   **3.** 1   **4.** 120   **5.** $m^2 + 2mn + n^2$
6. $m^2 - 2mn + n^2$ **7.** $p^2 + 4pq + 4q^2$
8. $4p^2 - 4pq + q^2$

Exercise 13.1 (page 773)
1. $3xyz^2(x^2yz^2 - 2z^3 + 5x)$ **3.** $(a^2 + b^2)(a+b)(a-b)$
5. $\dfrac{3+x}{3-x}$ **7.** power   **9.** first   **11.** $7 \cdot 6 \cdot 5 \cdot 4 \cdot 3 \cdot 2 \cdot 1$
13. $(n-1)!$ **15.** 24   **17.** 4,320   **19.** 1,440   **21.** $\frac{1}{1,320}$
23. $\frac{5}{3}$ **25.** 18,564   **27.** $a^3 + 3a^2b + 3ab^2 + b^3$
29. $a^5 - 5a^4b + 10a^3b^2 - 10a^2b^3 + 5ab^4 - b^5$
31. $8x^3 + 12x^2y + 6xy^2 + y^3$
33. $x^3 - 6x^2y + 12xy^2 - 8y^3$
35. $16x^4 + 96x^3y + 216x^2y^2 + 216xy^3 + 81y^4$
37. $x^4 - 8x^3y + 24x^2y^2 - 32xy^3 + 16y^4$
39. $x^5 - 15x^4y + 90x^3y^2 - 270x^2y^3 + 405xy^4 - 243y^5$
41. $\dfrac{x^4}{16} + \dfrac{x^3y}{2} + \dfrac{3x^2y^2}{2} + 2xy^3 + y^4$ **43.** $6a^2b^2$
45. $35a^3b^4$ **47.** $-b^5$ **49.** $2,380a^{13}b^4$ **51.** $-4\sqrt{2}a^3$
53. $1,134a^5b^4$ **55.** $\dfrac{3x^2y^2}{2}$ **57.** $\dfrac{-55r^2s^9}{2,048}$
59. $\dfrac{n!}{3!(n-3)!}a^{n-3}b^3$ **61.** $\dfrac{n!}{(r-1)!(n-r+1)!}a^{n-r+1}b^{r-1}$
69. -252

Getting Ready (page 774)
1. 25   **2.** 65

Orals (page 779)
1. 3   **2.** 5

Exercise 13.2 (page 779)
1. 6 cm   **3.** 26 ft   **5.** domain   **7.** summation notation
9. 6   **11.** $5c$ **13.** 0, 10, 30, 60, 100, 150   **15.** 21
17. $a + 4d$ **19.** 15   **21.** 15   **23.** 15   **25.** $\frac{242}{243}$ **27.** 35
29. 3, 7, 15, 31   **31.** $-4, -2, -1, -\frac{1}{2}$ **33.** k, k^2, k^4, k^8
35. $8, \dfrac{16}{k}, \dfrac{32}{k^2}, \dfrac{64}{k^3}$ **37.** an alternating series
39. not an alternating series   **41.** 30   **43.** -50 **45.** 40
47. 500   **49.** $\frac{7}{12}$ **51.** 160   **53.** 3,725

Getting Ready (page 781)
1. 3, 5, 7, 9   **2.** 4, 7, 10, 13

Orals (page 785)
1. 14   **2.** 1   **3.** 5   **4.** -6

Exercise 13.3 (page 786)
1. 6   **3.** -1 **5.** $1, -1, i, -i$,   **7.** $(n-1)$
9. $l = a + (n-1)d$ **11.** arithmetic means
13. 1, 3, 5, 7, 9, 11   **15.** $5, \frac{7}{2}, 2, \frac{1}{2}, -1, -\frac{5}{2}$
17. $9, \frac{23}{2}, 14, \frac{33}{2}, 19, \frac{43}{2}$ **19.** 318   **21.** 6   **23.** 370
25. 44   **27.** $\frac{25}{2}, 15, \frac{35}{2}$ **29.** $-\frac{82}{15}, -\frac{59}{15}, -\frac{12}{5}, -\frac{13}{15}$ **31.** 285
33. 555   **35.** $157\frac{1}{2}$ **37.** 20,100   **39.** $2,600   **41.** $\frac{1}{8}$ mi
43. 304 ft   **45.** 210

Getting Ready (page 787)
1. 10, 20, 40   **2.** 18, 54, 162

Orals (page 793)
1. 27   **2.** $\frac{1}{27}$ **3.** 2.5   **4.** $\sqrt{3}$ **5.** 6   **6.** 1

Exercise 13.4 (page 794)
1. $5 - 3i$ **3.** 10   **5.** $\frac{1}{5} + \frac{8}{5}i$ **7.** $-i$ **9.** r^{n-1}
11. $l = ar^{n-1}$ **13.** geometric means   **15.** 10, 20, 40, 80
17. $-2, -6, -18, -54$ **19.** $3, 3\sqrt{2}, 6, 6\sqrt{2}$
21. 2, 6, 18, 54   **23.** 256   **25.** -162
27. $10\sqrt[4]{2}, 10\sqrt{2}, 10\sqrt[4]{8}$ **29.** 8, 32, 128, 512   **31.** 124
33. $-29,524$ **35.** $\frac{1,995}{32}$ **37.** 18   **39.** 8   **41.** $\frac{5}{9}$ **43.** $\frac{25}{99}$
45. 23   **47.** $69.82   **49.** about $\frac{1}{3}c$ **51.** about 8.6×10^{19}
53. $180,176.87   **55.** $2,001.60   **57.** $2,013.62
59. $264,094.58   **61.** 1.8447×10^{19} grains   **63.** $4,309.14
67. no

Getting Ready (page 796)
1. 1   **2.** 3   **3.** 6   **4.** 10

Orals (page 800)
1. yes   **2.** no   **3.** no

Exercise 13.5 (page 800)
1. $-3a^2b(2a-3b)$ **3.** $(5a+2b)(2a-5b)$ **5.** two
7. $n=k+1$ **33.** no

Getting Ready (page 801)
1. 24   **2.** 120   **3.** 30   **4.** 168

Orals (page 809)
1. 15   **2.** 120   **3.** 3   **4.** 6   **5.** 1   **6.** 1

Exercise 13.6 (page 809)
1. 2   **3.** 4   **5.** true   **7.** true   **9.** 24   **11.** $\dfrac{n!}{(n-r)!}$
13. 1   **15.** $\dbinom{n}{r}$ **17.** 1   **19.** $\dfrac{n!}{a!\cdot b!\cdots}$ **21.** 840
23. 35   **25.** 120   **27.** 5   **29.** 1   **31.** 1,200   **33.** 40
35. 2,278   **37.** 144   **39.** 8,000,000   **41.** 240   **43.** 6
45. 40,320   **47.** 14,400   **49.** 24,360   **51.** 5,040
53. 48   **55.** 96   **57.** 210   **59.** 24   **61.** 5,040
63. 2,721,600   **65.** 1,120   **67.** 59,400   **69.** 272
71. 28   **73.** 252   **75.** 142,506   **77.** 66

Orals (page 816)
1. $\frac{1}{2}$ **2.** $\frac{1}{7}$ **3.** $\frac{1}{4}$ **4.** $\frac{1}{8}$

Exercise 13.7 (page 816)
1. $-10, 4$ **3.**   **5.** experiment
7. $\dfrac{n(E)}{n(S)}$
9. {(1, H), (2, H), (3, H), (4, H), (5, H), (6, H), (1, T), (2, T), (3, T), (4, T), (5, T), (6, T)}   **11.** {a, b, c, d, e, f, g, h, i, j, k, l, m, n, o, p, q, r, s, t, u, v, w, x, y, z}   **13.** $\frac{1}{6}$ **15.** $\frac{2}{3}$ **17.** $\frac{19}{42}$
19. $\frac{13}{42}$ **21.** $\frac{3}{8}$ **23.** 0   **25.** $\frac{1}{12}$ **27.** $\frac{1}{169}$ **29.** $\frac{5}{12}$
31. about 6.3×10^{-12} **33.** 0   **35.** $\frac{3}{13}$ **37.** $\frac{1}{6}$ **39.** $\frac{1}{8}$
41. $\frac{5}{16}$ **43.** (S = survive, F = fail)   {SSSS, SSSF, SSFS, SFSS, FSSS, SSFF, SFSF, FSSF, SFFS, FSFS, FFSS, SFFF, FSFF, FFSF, FFFS, FFFF}   **45.** $\frac{1}{4}$ **47.** $\frac{1}{4}$ **49.** 1
51. $\frac{32}{119}$ **53.** $\frac{1}{3}$ **55.** 0.18   **57.** 0.14   **59.** about 33%
63. no

Getting Ready (page 818)
1. $\frac{1}{13}$ **2.** $\frac{1}{13}$ **3.** $\frac{2}{13}$ **4.** $\frac{3}{13}$

Orals (page 822)
1. $\frac{1}{2}$ **2.** $\frac{1}{13}$ **3.** $\frac{7}{13}$ **4.** $\frac{1}{26}$

Exercise 13.8 (page 822)
1. $\dfrac{x-3}{x}$ **3.** 1   **5.** compound   **7.** mutually exclusive
9. $1-P(A)$ **11.** $P(A)\cdot P(B)$ **13.** $\frac{1}{2}$ **15.** $\frac{7}{13}$ **17.** $\frac{1}{221}$
19. $\frac{25}{204}$ **21.** $\frac{11}{36}$ **23.** $\frac{7}{12}$ **25.** $\frac{5}{8}$ **27.** $\frac{13}{16}$ **29.** $\frac{21}{128}$

31. $\frac{3}{20}$ **33.** $\frac{1}{49}$ **35.** 0.973   **37.** $\frac{11}{20}$ **39.** $\frac{1}{8}$ **41.** $\frac{3}{8}$
43. $\frac{17}{36}$ **45.** 0.03

Getting Ready (page 823)
1. 1   **2.** 6   **3.** $\frac{1}{7}$ **4.** $\frac{6}{7}$

Orals (page 825)
1. $\frac{1}{6}$ **2.** $\frac{5}{6}$ **3.** 1 to 5   **4.** 5 to 1

Exercise 13.9 (page 826)
1. $2x+3y=5$ **3.** $(x-3)^2+(y-5)^2=25$
5. odds for an event   **7.** $\frac{1}{5}$ **9.** $\frac{1}{6}$ **11.** 5 to 1   **13.** 1 to 1
15. $\frac{5}{36}$ **17.** 31 to 5   **19.** 1 to 1   **21.** 1 to 12
23. 3 to 10   **25.** $\frac{5}{7}$ **27.** 1 to 90   **29.** 15 to 1   **31.** $\frac{1}{9}$
33. No; the expected winnings are $\$\frac{9}{13}$.   **35.** 7 to 1
37. $1.72   **39.** 6.54   **41.** $\frac{32}{390,625}$ **45.** less than $\frac{1}{2}$

Chapter Summary (page 828)
1. 720   **2.** 30,240   **3.** 8   **4.** $\frac{280}{3}$
5. $x^3+3x^2y+3xy^2+y^3$
6. $p^4+4p^3q+6p^2q^2+4pq^3+q^4$
7. $a^5-5a^4b+10a^3b^2-10a^2b^3+5ab^4-b^5$
8. $8a^3-12a^2b+6ab^2-b^3$ **9.** $56a^5b^3$ **10.** $80x^3y^2$
11. $84x^3y^6$ **12.** $439,040x^3$ **13.** 63   **14.** 9
15. 5, 17, 53, 161   **16.** $-2, 8, 128, 32,768$ **17.** 90
18. 60   **19.** 1,718   **20.** -360 **21.** 117   **22.** 281
23. -92 **24.** $-\frac{135}{2}$ **25.** $\frac{7}{2}, 5, \frac{13}{2}$ **26.** 25, 40, 55, 70, 85
27. 3,320   **28.** 5,780   **29.** $-5,220$ **30.** $-1,540$
31. $\frac{1}{729}$ **32.** 13,122   **33.** $\frac{9}{16,384}$ **34.** $\frac{8}{15,625}$
35. $2\sqrt{2}, 4, 4\sqrt{2}$ **36.** $4, -8, 16, -32$ **37.** 16   **38.** $\frac{3,280}{27}$
39. 6,560   **40.** $\frac{2,295}{128}$ **41.** $\frac{520,832}{78,125}$ **42.** $\frac{3,280}{3}$ **43.** $16\sqrt{2}$
44. $\frac{2}{3}$ **45.** $\frac{3}{25}$ **46.** no sum   **47.** 1   **48.** $\frac{1}{3}$ **49.** 1
50. $\frac{17}{99}$ **51.** $\frac{5}{11}$ **52.** $4,775.81   **53.** 6,516; 3,134
54. $3,486.78   **57.** 6,720   **58.** 35   **59.** 1   **60.** 4,050
61. 564,480   **62.** 840   **63.** 21   **64.** 66   **65.** 120
66. $\frac{56}{1,287}$ **67.** $\frac{1}{6}$ **68.** $\frac{33}{66,640}$ **69.** 20,160   **70.** 90,720
72. 24   **73.** $\frac{1}{108,290}$ **74.** $\frac{108,289}{108,290}$ **75.** about 6.3×10^{-12}
76. $\frac{60}{143}$ **77.** $\frac{1}{2}$ **78.** $\frac{7}{13}$ **79.** $\frac{1}{2,598,960}$
80. about $\dfrac{4}{6.350\times10^{11}}$ **81.** $\frac{15}{16}$ **82.** 136
83. $\frac{4}{13}$ **84.** 1 to 7   **85.** 1 to 15   **86.** 1 to 23   **87.** $\frac{11}{21}$
88. $2

Chapter 13 Test (page 833)
1. 144   **2.** 384   **3.** $10x^4y$ **4.** $112a^2b^6$ **5.** 27
6. -36 **7.** 155   **8.** -130 **9.** 9, 14, 19   **10.** $-6, -18$
11. 255.75   **12.** about 9   **13.** about $0.42c$
14. about $1.46c$ **16.** 800,000   **17.** 42   **18.** 24
19. 28   **20.** 1   **21.** 576   **22.** 120   **23.** 60
24. {(H, H, H), (H, H, T), (H, T, H), (H, T, T), (T, H, H), (T, H, T), (T, T, H), (T, T, T)}   **25.** $\frac{1}{6}$ **26.** $\frac{2}{13}$ **27.** $\frac{429}{866,320}$
28. $\frac{5}{16}$ **29.** 40%   **30.** 0.9   **31.** $\frac{3}{13}$ **32.** $\frac{5}{9}$ **33.** $\frac{3}{4}$
34. $1

Cumulative Review Exercises (page 834)

1. 1.6902   **2.** 0.1461   **3.** 16   **4.** $\dfrac{2 \log 2}{\log 3 - \log 2}$

5. \$2,848.31   **6.** 1.16056   **7.** (2, 1)   **8.** (1, 1)

9. (2, −2)   **10.** (3, 1)   **11.** −1   **12.** −1

13. (−1, −1, 3)   **14.** (1, 1, 1)   **15.** $(g + f)(x) = 5x - 1$

16. $(f - g)(x) = 3x + 1$ **17.** $(g \cdot f)(x) = 4x^2 - 4x$

18. $(g/f)(x) = \dfrac{x - 1}{4x}$ **19.** 3   **20.** −4   **21.** −8

22. −9   **23.** $4(x - 1)$ **24.** $4x - 1$ **25.** $y = \dfrac{12 - 2x}{3}$

26. $y = -\sqrt{\dfrac{x - 4}{3}}$ $(y \le 0)$ **27.** $x^2 + y^2 = 16$

28. $(x - 2)^2 + (y + 3)^2 = 121$

29.
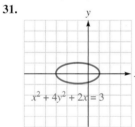
$x^2 + y^2 - 4y = 12$

30.
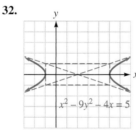
$x^2 - 2y - 2x = -7$

31.
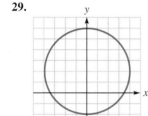
$x^2 + 4y^2 + 2x = 3$

32.
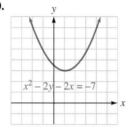
$x^2 - 9y^2 - 4x = 5$

33. $\dfrac{x^2}{36} + \dfrac{y^2}{16} = 1$ **34.** $\dfrac{(x - 2)^2}{21} + \dfrac{(y - 3)^2}{25} = 1$

35. $\dfrac{x^2}{4} - \dfrac{y^2}{5} = 1$ **36.** $\dfrac{(y - 4)^2}{9} - \dfrac{(x - 2)^2}{9} = 1$

37. $16x^7 y$ **38.** $1,792x^3 y^5$ **39.** 10   **40.** 65   **41.** 33

42. $\frac{364}{9}$ **43.** 1,680   **44.** 1   **45.** 66   **46.** 24

47. 17,280   **48.** 495   **49.** $\frac{1}{18}$ **50.** $\frac{506}{19,992} = \frac{253}{9,996}$

51. 0.48   **52.** 0.3

Exercise I.1 (page A-7)

1. 2, 5   **3.** no values   **5.** 1, 2   **7.** 2, 2

9. $\begin{bmatrix} -1 & 2 & 1 \\ -6 & 0 & 0 \end{bmatrix}$ **11.** $\begin{bmatrix} 5 & -4 & 3 \\ -7 & -4 & -2 \\ 0 & 4 & -8 \end{bmatrix}$

13. not possible   **15.** [5 7 9]

17. $\begin{bmatrix} 3 & 2 & -7 \\ 9 & 19 & 15 \\ 0 & -4 & -1 \end{bmatrix}$ **19.** $\begin{bmatrix} 2 & -2 \\ 3 & 10 \end{bmatrix}$

21. $\begin{bmatrix} -22 & -22 \\ -105 & 126 \end{bmatrix}$ **23.** $\begin{bmatrix} 4 & 2 & 10 \\ 5 & -2 & 4 \\ 2 & -2 & 1 \end{bmatrix}$

25. [32]   **27.** not possible   **29.** [30 36 42]

31. $\begin{bmatrix} 2 & 0 & 3 \\ 3 & 0 & 5 \end{bmatrix}$ **33.** $\begin{bmatrix} 12 & -4 & 9 \\ 16 & -2 & 2 \\ 10 & -2 & 1 \end{bmatrix}$ **35.** $\begin{bmatrix} 7 \\ 6 \end{bmatrix}$

37. $\begin{bmatrix} 4 & 5 \\ -7 & -1 \end{bmatrix}$ **39.** not possible   **41.** $\begin{bmatrix} 24 & 16 \\ 39 & 26 \end{bmatrix}$

43. $\begin{bmatrix} 64 & 64 \\ 64 & 64 \end{bmatrix}$ **45.** example: $\begin{bmatrix} 1 & 0 \\ 0 & 0 \end{bmatrix}$

Exercise I.2 (page A-13)

1. $\begin{bmatrix} 3 & 4 \\ 2 & 3 \end{bmatrix}$ **3.** $\begin{bmatrix} 5 & -7 \\ -2 & 3 \end{bmatrix}$ **5.** $\begin{bmatrix} -40 & 16 & 9 \\ 13 & -5 & -3 \\ 5 & -2 & -1 \end{bmatrix}$

7. $\begin{bmatrix} 4 & 1 & -3 \\ -5 & -1 & 4 \\ -1 & -1 & 1 \end{bmatrix}$ **9.** no inverse

11. $\begin{bmatrix} 1 & -2 & 1 \\ 0 & 1 & -2 \\ 0 & 0 & 1 \end{bmatrix}$ **13.** no inverse

15. $\begin{bmatrix} 1 & -2 & 1 & 0 \\ 0 & 1 & -2 & 1 \\ 0 & 0 & 1 & -2 \\ 0 & 0 & 0 & 1 \end{bmatrix}$ **17.** (23, 17)   **19.** (70, −30)

21. (−10, 4, 1)   **23.** (7, −9, −1)   **25.** (4, −6, 3)

29. $A^2 = \begin{bmatrix} 1 & 0 \\ 0 & 1 \end{bmatrix}, A^3 = \begin{bmatrix} 0 & 1 \\ 1 & 0 \end{bmatrix}, A^4 = \begin{bmatrix} 1 & 0 \\ 0 & 1 \end{bmatrix},$

$A^n = \begin{bmatrix} 0 & 1 \\ 1 & 0 \end{bmatrix}$ if n is odd, $A^n = \begin{bmatrix} 1 & 0 \\ 0 & 1 \end{bmatrix}$ if n is even.

33. $A^n = \begin{bmatrix} 1 & 0 \\ n & 1 \end{bmatrix}$ **35.** ±4

39. $E = \begin{bmatrix} 1 & 0 & 0 \\ 0 & 1 & 0 \\ 3 & 0 & 1 \end{bmatrix}, E^{-1} = \begin{bmatrix} 1 & 0 & 0 \\ 0 & 1 & 0 \\ -3 & 0 & 1 \end{bmatrix}$

Index